Mechanics of Laminated Composite Plates

Theory and Analysis

J.N. Reddy
Department of Mechanical Engineering
Texas A&M University
College Station, Texas

CRC

CRC Press
Boca Raton New York London Tokyo

Acquiring Editor: Norm Stanton
Project Editor: Gail Renard
Marketing Manager: Susie Carlisle
Direct Marketing Manager: Becky McEldowney
Cover designer: Dawn Boyd
PrePress: Kevin Luong
Manufacturing: Sheri Schwartz

Library of Congress Cataloging-in-Publication Data

Reddy, J.N. (Junuthula Narasimha), 1945–
Mechanics of laminated composite plates: theory and analysis / by J.N. Reddy.
p. cm.
Includes bibliographical references and index.
ISBN 0-8493-3101-3
1. Plates (Engineering)—Mathematical models. 2. Laminated materials—Mechanical properties—Mathematical models. 3. Composite materials—Mechanical properties—Mathematical models.
I. Title.
TA660.P6R43 1996
624.1′.7765—dc20

96-38476
CIP

International Standard Book Number 0-8493-3101-3
Library of Congress Card Number 96-38476
Printed in the United States of America 1 2 3 4 5 6 7 8 9 0
Printed on acid-free paper

About the Author

J. N. Reddy is the inaugural holder of the *Oscar S. Wyatt Endowed Chair* in the Department of Mechanical Engineering at Texas A&M University, College Station, Texas. Prior to the current position, he was the *Clifton C. Garvin Professor* in the Department of Engineering Science and Mechanics at Virginia Polytechnic Institute and State University, Blacksburg, Virginia. He earned a Ph.D. in Engineering Mechanics, and worked as a postdoctoral fellow at the University of Texas at Austin, as a research scientist for Lockheed Missiles and Space Company during 1974–75, and taught at the University of Oklahoma from 1975 to 1980 and Virginia Polytechnic Institute and State University from 1980 to 1992.

Dr. Reddy has authored numerous papers on the theory and finite element analysis of laminated composite plates and shells, computational fluid mechanics, numerical heat transfer, and applied mathematics. His seminal contributions to mechanics of composite materials and structures are well known through his research on refined structural theories and finite element models of laminated composite plates and shells. Dr. Reddy is the author of nine other books, including: *An Introduction to the Finite Element Method*, McGraw–Hill, 1984 (second edition, 1993); *Energy and Variational Methods in Applied Mechanics*, John Wiley & Sons, 1984; *Applied Functional Analysis and Variational Methods in Engineering*, McGraw–Hill, 1986.

Professor Reddy is the first recipient of the University of Oklahoma College of Engineering's *Award for Outstanding Faculty Achievement in Research*, the *Walter L. Huber Civil Engineering Research Prize* of the American Society of Civil Engineers (ASCE), the *Alumni Research Award* at Virginia Polytechnic Institute, and the *Worcester Reed Warner Medal* and the *Charles Russ Richards Memorial Award* of the American Society of Mechanical Engineers (ASME). Dr. Reddy received a *Technical Achievement Award* from the National Academy of Engineering for "outstanding contributions to engineering education and research." He has also received teaching awards at every academic institution he has served. Dr. Reddy is a *Fellow* of the American Academy of Mechanics (AAM), the American Society of Civil Engineers (ASCE), the American Society of Mechanical Engineers (ASME), and the Aeronautical Society of India (ASI). Dr. Reddy is the editor of the journal *Mechanics of Composite Materials and Structures*, and serves on the editorial boards of over two dozen journals, including *ASME Journal of Applied Mechanics, International Journal for Numerical Methods in Engineering,* and *International Journal for Numerical Methods in Fluids.*

To the Memory of

My parents,
brother,
brother in–law,
father in–law, and
Hans Eggers.

Contents

Preface

The dramatic increase in the use of composite materials in all types of engineering structures (*e.g.*, aerospace, automotive, and underwater structures, as well as in medical prosthetic devices, electronic circuit boards, and sports equipment) and the number of journals and research papers published in the last two decades attest to the fact that there has been a major effort to develop composite material systems, and analyze and design structural components made from composite materials.

The subject of composite materials is truly an interdisciplinary area where chemists, material scientists, chemical engineers, mechanical engineers, and structural engineers contribute to the overall product. The number of students taking courses in composite materials and structures has steadily increased in recent years, and the students are drawn to these courses from a variety of disciplines. The courses offered at universities and the books published on composite materials are of three types: material science, mechanics, and design. The present book belongs to the *mechanics* category.

The motivation for the present book has come from many years of the author's research and teaching in laminated composite structures and from the fact there does not exist a book that contains a detailed coverage of various laminate theories, analytical solutions, and finite element models. The book is largely based on the author's original work on refined theories of laminated composite plates and shells, and analytical and finite element solutions he and his colloborators have developed over the last two decades.

Some mathematical preliminaries, equations of anisotropic elasticity, and virtual work principles and variational methods are reviewed in Chapters 1 through 3. A reader who has had a course in elasticity or energy and variational principles of mechanics may skip these chapters and go directly to Chapter 4, where certain terminology common to composite materials is introduced, followed by a discussion of the constitutive equations of a lamina and transformation of stresses and strains. Readers who have had a basic course in composites may skip Chapter 4 also.

The major journey of the book begins with Chapter 5, where a complete derivation of the equations of motion of the classical and first-order shear deformation laminated plate theories is presented, and laminate stiffness characteristics of selected laminates are discussed. Chapter 6 includes applications of the classical and first-order shear deformation theories to laminated beams and plate strips in cylindrical bending. Here analytical solutions are developed for bending, buckling, natural vibration, and transient response of simple beam and plate structures. Chapter 7 deals with the analysis of specially orthotropic rectangular laminates using the classical laminated plate theory (CLPT). Here, the parametric effects of material anisotropy, lamination scheme, and plate aspect ratio on bending deflections and stresses, buckling loads, vibration frequencies, and transient response are discussed.

Analytical solutions for bending, buckling, natural vibration, and transient response of rectangular laminates based on the Navier and Lévy solution approaches are presented in Chapters 8 and 9 for the classical and first-order shear deformation plate theories (FSDT), respectively. The Rayleigh–Ritz solutions are also discussed for laminates that do not admit the Navier solutions. Chapter 10 deals with finite element analysis of composite laminates. One-dimensional (for beams and plate strips) as well as two-dimensional (plates) finite element models based on CLPT and FSDT are discussed and numerical examples are presented.

Chapters 11 and 12 are devoted to higher-order (third-order) laminate theories and layerwise theories, respectively. Analytical as well as finite element models are discussed. The material included in these chapters is up to date at the time of this writing. Finally, Chapter 13 is concerned about the geometrically nonlinear analysis of composite laminates. Displacement finite element models of laminated plates with the von Kármán nonlinearity are derived, and numerical results are presented for some typical problems.

The book is suitable as a reference for engineers and scientists working in industry and academia, and it can be used as a textbook in a graduate course on theory and/or analysis of composite laminates. It can also be used for a course on stress analysis of laminated composite plates. An introductory course on mechanics of composite materials may prove to be helpful but not necessary because a review of the basics is included in the first four chapters of this book. The first course may cover Chapters 1 through 8 or 9, and a second course may cover Chapters 8 through 13.

The author wishes to thank all his former doctoral students for their research collaboration on the subject. In particular, Chapters 7 through 13 contain results of the research conducted by Drs. Ahmed Khdeir, Stephen Engelstad, Asghar Nosier, and Donald Robbins, Jr. on the development of theories, analytical solutions, and finite element analyses of equivalent single-layer and layerwise theories of composite laminates. The research of the author in composite materials was influenced by many researchers. The author wishes to thank Professor Charles W. Bert of the University of Oklahoma, Professor Robert M. Jones of the Virginia Polytechnic Institute and State University, Professor A. V. Krishna Murty of the Indian Institute of Science, and Dr. Nicholas J. Pagano of Wright–Patterson Air Force Base.

It is also the author's pleasure to acknowledge the help of Mr. Praveen Grama, Mr. Dakshina Moorthy, and Mr. Govind Rengarajan for their help with the proofreading of the manuscript. The author is indebted to Dr. Filis Kokkinos for his dedication and innovative and creative production of the final artwork in this book. Indeed, without his imagination and hundreds of hours of effort the artwork would not have looked as beautiful, professional, and technical as it does.

The author gratefully acknowledges the support of his research in composite materials in the last two decades by the Office of Naval Research (ONR), the Air Force Office of Scientific Research (AFOSR), the U.S. Army Research Office (ARO), the National Aeronautics and Space Administration (NASA Lewis and NASA Langley), the U.S. National Science Foundation (NSF), and the *Oscar S. Wyatt Chair* in the Department of Mechanical Engineering at Texas A&M University. Without this support, it would not have been possible to contribute to the subject of this book. The author is also grateful to Professor G. P. Peterson, a colleague and friend, for his encouragement and support of the author's professional activities at Texas A&M University.

The writing of this book took thousands of hours over the last ten years. Most of these hours came from evenings and holidays that could have been devoted to family matters. While no words of gratitude can replace the time lost with family, it should be recorded that the author is grateful to his wife Aruna for her care, devotion, and love, and to his daughter Anita and son Anil for their understanding and support.

During the long period of writing this book, the author has lost his father, brother, brother in-law, father in-law, and a friend - all suddenly. While death is always imminent, the suddenness makes it more difficult to accept. This book is dedicated to the memory of these individuals.

J. N. Reddy
College Station, Texas

All that not given is lost

Chapter One

Introduction and Preliminaries

1.1 Fiber-Reinforced Composite Materials

Composite materials consist of two or more materials which together produce desirable properties that cannot be achieved with any of the constituents alone. Fiber-reinforced composite materials, for example, consist of high strength and high modulus *fibers* in a *matrix* material. Reinforced steel bars embedded in concrete provide an example of fiber-reinforced composites. In these composites, fibers are the principal load-carrying members, and the matrix material keeps the fibers together, acts as a load-transfer medium between fibers, and protects fibers from being exposed to the environment (*e.g.*, moisture, humidity, etc.).

It is known that fibers are stiffer and stronger than the same material in bulk form, whereas matrix materials have their usual bulk-form properties. Geometrically, fibers have near crystal-sized diameter and a very high length-to-diameter ratio. Short fibers, called *whiskers*, paradoxically exhibit better structural properties than long fibers. To gain a full understanding of the behavior of fibers, matrix materials, and agents that are used to enhance bonding between fibers and matrix and other properties of fiber-reinforced materials, it is necessary to know certain aspects of material science. Since the present study is entirely devoted to mechanics aspects and analysis methods of fiber-reinforced composite materials, no attempt is made here to present basic material science aspects, such as the molecular structure or interatomic forces that hold the matter together. However, an abstract understanding of the material behavior is useful.

Materials are studied at various levels: atomic level, nanolevel, single-crystal level, a group of crystals, and so on. For the purpose of gaining some insight into the material behavior, we consider a basic unit of material as one that has properties, such as the modulus, strength, thermal coefficient of expansion, electrical resistance, etc., whose magnitudes depend on the direction. The directional dependence of properties is a result of the interatomic bonds, which are "stronger" in one direction than in other directions. Materials are "processed" such that the basic units are aligned so that the desired property is maximized in a given direction. Fibers provide an example of such materials. When a property is maximized in one direction, it may be achieved at the expense of the same property in other directions and other properties in the same direction. When materials are processed such that the basic units are randomly oriented, the resulting material tends to have the same value of the property, in an average statistical sense, in all directions. Such materials are called *isotropic* materials. A matrix material, which is made in bulk form, provides an example of isotropic materials. Material scientists are continuously striving to develop better materials for specific applications. The fibers and matrix materials used in composites are either metallic or nonmetallic. The fiber materials in use are common metals like aluminum, copper, iron, nickel, steel, and titanium, and organic materials like glass, carbon, boron, and graphite materials.

Fiber-reinforced composite materials for structural applications are often made in the form of a thin layer, called *lamina*. A lamina is a macro unit of material whose material properties are determined through appropriate laboratory tests. Structural elements, such as bars, beams or plates are then formed by stacking the layers to achieve desired strength and stiffness. Fiber orientation in each lamina and stacking sequence of the layers can be chosen to achieve desired strength and stiffness for a specific application. It is the purpose of the present study to develop equations that describe appropriate kinematics of deformation, govern force equilibrium, and represent the material response of laminated structural elements.

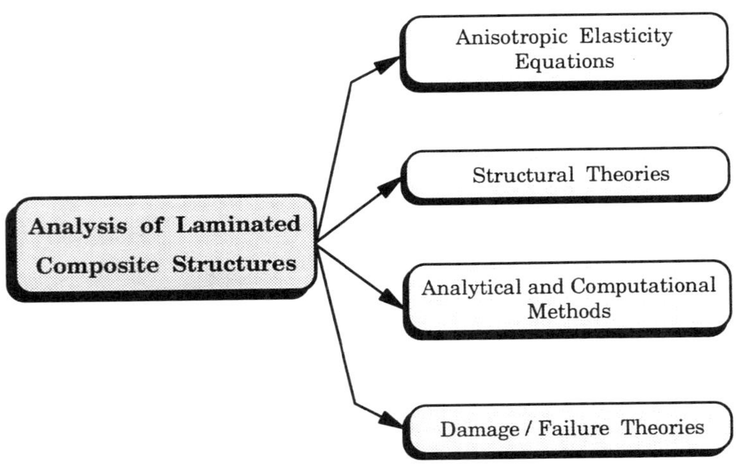

Figure 1.1–1: Basic blocks in the analysis of laminated structures.

Analysis of structural elements made of laminated composite materials involves several steps. As shown in Figure 1.1–1, the analysis requires a knowledge of anisotropic elasticity, structural theories (*i.e.*, kinematics of deformation) of laminates, analytical or computational methods to determine solutions of the governing equations, and failure theories to predict modes of failures and to determine failure loads. A detailed study of the theoretical formulations and solutions of governing equations of laminated composite plate structures constitutes the objective of the present book.

Following this general introduction, a review of vectors and tensors, matrices, coordinate transformations, and integral relations is presented, as they are needed in the sequel. Readers familiar with these topics can skip the remaining portion of this chapter and go directly to Chapter 2 on equations of anisotropic elasticity.

1.2 Vectors and Tensors

1.2–1 Preliminary Comments

The quantities used to express physical laws can be classified into two classes, according to the information needed to specify them completely: scalars and nonscalars. The scalars are given by a single number. Nonscalar quantities require not only a magnitude specified, but also additional information, such as direction. Time, temperature, volume, and mass density provide examples of scalars. Displacement, temperature gradient, force, moment, and acceleration are examples of nonscalars.

The term *vector* is used to imply a nonscalar that has magnitude and "direction", and obeys the parallelogram law of vector addition and rules of scalar multiplication. Vector in modern mathematical analysis is an abstraction of the elementary notion of a *physical vector*, and it is "an element from a linear vector space." While the definition of a vector in abstract analysis does not require the vector to have a magnitude, in nearly all cases of practical interest the vector is endowed with a magnitude. In this book we only need vectors from a special *normed vector space* (see [5,6]), that is, vectors with magnitude. Some nonscalar quantities require the specification of magnitude and two directions. For example, the specification of stress requires not only a force, but also an area upon which the force acts. A stress is a *tensor*. Sometimes a vector is referred to as a tensor of order one, and a tensor of order 2 is called a *dyad*. Stress is a second-order tensor. First- and second-order tensors (*i.e.*, vectors and dyads) will be of primary interest in the present study. We also encounter third-order and fourth-order tensors in the discussion of constitutive equations.

1.2–2 Components of Vectors and Tensors

In the analytical description of a physical phenomenon, a coordinate system in the chosen frame of reference is introduced, and various physical quantities involved in the description are expressed in terms of measurements made in that system. The law thus depends upon the chosen coordinate system and may appear different in another type of coordinate system. The laws of nature, however, should be independent of the choice of a coordinate system, and we may seek to represent the law in a manner independent of a particular coordinate system. A way of doing this is provided by vector and tensor notation. When vector notation is used, a particular coordinate system need not be introduced. Consequently use of vector notation in formulating natural laws leaves them *invariant* to coordinate transformations.

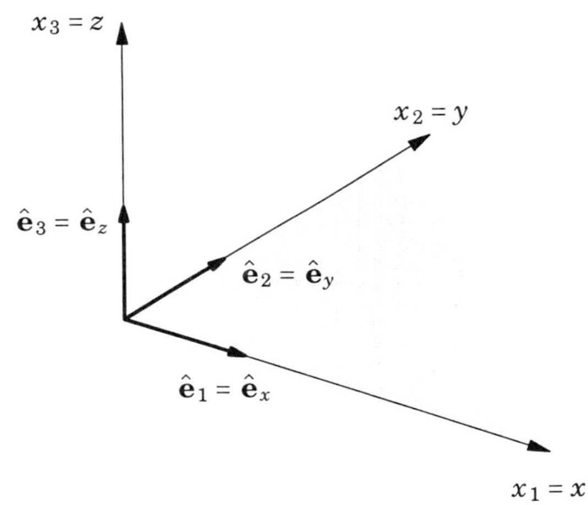

Figure 1.2–1: A rectangular Cartesian coordinate system, $(x_1, x_2, x_3) = (x, y, z)$; $(\hat{\mathbf{e}}_1, \hat{\mathbf{e}}_2, \hat{\mathbf{e}}_3) = (\hat{\mathbf{e}}_x, \hat{\mathbf{e}}_y, \hat{\mathbf{e}}_z)$ are the unit basis vectors.

Often a specific coordinate system is chosen to express governing equations of a problem to facilitate their solution. Then the vector and tensor quantities are expressed in terms of their components in that coordinate system. For example, a vector **A** in a three-dimensional space may be expressed in terms of its components a_1, a_2, and a_3 and *basis vectors* $\mathbf{e}_1$, $\mathbf{e}_2$, and $\mathbf{e}_3$ as

$$\mathbf{A} = a_1\mathbf{e}_1 + a_2\mathbf{e}_2 + a_3\mathbf{e}_3 \qquad (1.2-1)$$

When the basis vectors of a coordinate system are constants, *i.e.*, with fixed lengths and directions, the coordinate system is called a *Cartesian coordinate system.* The general Cartesian system is oblique. When the Cartesian system is orthogonal, it is called *rectangular Cartesian.* When the basis vectors are of unit lengths and mutually orthogonal, they are called *orthonormal.* We denote an orthonormal Cartesian basis by

$$(\hat{\mathbf{e}}_1, \hat{\mathbf{e}}_2, \hat{\mathbf{e}}_3) \quad \text{or} \quad (\hat{\mathbf{e}}_x, \hat{\mathbf{e}}_y, \hat{\mathbf{e}}_z) \tag{1.2-2}$$

The Cartesian coordinates are denoted by

$$(x_1, x_2, x_3) \quad \text{or} \quad (x, y, z) \tag{1.2-3}$$

The familiar rectangular Cartesian coordinate system is shown in Figure 1.2–1. We shall always use a right-hand coordinate system.

A second-order tensor can be expressed in terms of its components as

$$\begin{aligned}\mathbf{\Phi} =& \varphi_{11}\hat{\mathbf{e}}_1\hat{\mathbf{e}}_1 + \varphi_{12}\hat{\mathbf{e}}_1\hat{\mathbf{e}}_2 + \varphi_{13}\hat{\mathbf{e}}_1\hat{\mathbf{e}}_3 \\ &+\varphi_{21}\hat{\mathbf{e}}_2\hat{\mathbf{e}}_1 + \varphi_{22}\hat{\mathbf{e}}_2\hat{\mathbf{e}}_2 + \varphi_{23}\hat{\mathbf{e}}_2\hat{\mathbf{e}}_3 \\ &+\varphi_{31}\hat{\mathbf{e}}_3\hat{\mathbf{e}}_1 + \varphi_{32}\hat{\mathbf{e}}_3\hat{\mathbf{e}}_2 + \varphi_{33}\hat{\mathbf{e}}_3\hat{\mathbf{e}}_3\end{aligned} \tag{1.2-4}$$

Here we have selected a rectangular Cartesian basis to represent the tensor.

The sum of the diagonal terms of a second-order tensor is called the *trace of the tensor*:

$$\text{trace of } \mathbf{\Phi} = \varphi_{11} + \varphi_{22} + \cdots + \varphi_{nn} \tag{1.2-5}$$

The trace of a tensor is an invariant (*i.e.*, its value remains the same in all coordinate systems), called the *first principal invariant* of $\mathbf{\Phi}$, and is denoted by I_1.

1.2–3 Summation Convention

It is convenient to abbreviate a summation of terms by understanding that a repeated index means summation over all values of that index. For example, the component form of vector **A**

$$\mathbf{A} = a^1\mathbf{e}_1 + a^2\mathbf{e}_2 + a^3\mathbf{e}_3$$

where $(\mathbf{e}_1, \mathbf{e}_2, \mathbf{e}_3)$ are basis vectors (not necessarily unit), can be expressed in the form

$$\mathbf{A} = \sum_{j=1}^{3} a^j\mathbf{e}_j = a^j\mathbf{e}_j \tag{1.2-6}$$

The repeated index is a *dummy index* in the sense that any other symbol that is not already used in that expression can be employed:

$$\mathbf{A} = a^j\mathbf{e}_j = a^k\mathbf{e}_k = a^m\mathbf{e}_m \tag{1.2-7}$$

The range of summation is always known in the context of the discussion. For example, in the present context the range of j, k and m is 1 to 3, because we are discussing vectors in a three-dimensional space.

An nth-order tensor can be expressed in a short form using the summation convention:

$$\mathbf{\Phi} = \varphi_{ijk\ell\ldots}\hat{\mathbf{e}}_i\hat{\mathbf{e}}_j\hat{\mathbf{e}}_k\hat{\mathbf{e}}_\ell \cdots \tag{1.2 – 8}$$

Here we have selected a rectangular Cartesian basis to represent the tensor.

A unit second order tensor $\mathbf{I}$ is defined by

$$\mathbf{I} = \delta_{ij}\hat{\mathbf{e}}_i\hat{\mathbf{e}}_j \tag{1.2 – 9}$$

where δ_{ij}, called the *Kronecker delta*, is defined as

$$\delta_{ij} = \begin{cases} 0, & \text{if } i \neq j \\ 1, & \text{if } i = j \end{cases} \tag{1.2 – 10}$$

for any values of i and j.

The *double-dot product* between tensors of second order and higher order is encountered in mechanics. The double-dot product between two second-order tensors $\mathbf{\Phi}$ and $\mathbf{\Psi}$ is defined as

$$\begin{aligned} \mathbf{\Phi} : \mathbf{\Psi} &= (\varphi_{ij}\hat{\mathbf{e}}_i\hat{\mathbf{e}}_j) : (\psi_{mn}\hat{\mathbf{e}}_m\hat{\mathbf{e}}_n) \\ &= \varphi_{ij}\psi_{mn}(\hat{\mathbf{e}}_j \cdot \hat{\mathbf{e}}_m)(\hat{\mathbf{e}}_i \cdot \hat{\mathbf{e}}_n) \\ &= \varphi_{ij}\psi_{mn}\delta_{jm}\delta_{in} \\ &= \varphi_{nm}\psi_{mn} = \varphi_{ij}\psi_{ji} \end{aligned} \tag{1.2 – 11}$$

1.3 Matrices

1.3–1 Definition of a Matrix

In the preceding section we studied components of a second-order tensor. For example, Eq. (1.2–4) contains the components of a second-order tensor $\mathbf{\Phi}$ in a rectangular Cartesian coordinate system. We see that there are nine components of the tensor. The form of the equation suggests writing down the scalars φ_{ij} (jth component in the ith row) in the rectangular array:

$$[\Phi] = \begin{bmatrix} \varphi_{11} & \varphi_{12} & \varphi_{13} \\ \varphi_{21} & \varphi_{22} & \varphi_{23} \\ \varphi_{31} & \varphi_{32} & \varphi_{33} \end{bmatrix}$$

The rectangular array of φ's is called a *matrix*, when it satisfies certain properties.

A *matrix* is a rectangular array of elements. Two matrices are equal if and only if they are identical.

If a matrix has m rows and n columns, we will say that it is m by n ($m \times n$), the number of rows always being listed first. The element in the ith row and jth column of a matrix $[A]$ is generally denoted by a_{ij}, and we will sometimes write $[A] = [a_{ij}]$ to denote this. A square matrix is one that has the same number of rows as columns. An $n \times n$ matrix is said to be of *order n*. The elements of a square matrix for which the row number and the column number are the same (*i.e.*, a_{ij}) are called *diagonal elements* or simply the *diagonal*. A square matrix is said to be a *diagonal matrix* if all of the off-diagonal elements are zero. An *identity matrix*, denoted by $[I]$, is a diagonal matrix whose elements are all 1's. Examples of a diagonal and an identity matrix are given below:

$$\begin{bmatrix} 5 & 0 & 0 & 0 \\ 0 & 2 & 0 & 0 \\ 0 & 0 & 1 & 0 \\ 0 & 0 & 0 & 3 \end{bmatrix}, \qquad [I] = \begin{bmatrix} 1 & 0 & 0 & 0 \\ 0 & 1 & 0 & 0 \\ 0 & 0 & 1 & 0 \\ 0 & 0 & 0 & 1 \end{bmatrix}$$

Note that the components of the identity matrix can be denoted by δ_{ij}. The sum of the diagonal elements is called the *trace* of the matrix.

The *sum* of two matrices of the same size is defined to be a matrix of the same size obtained by simply adding the corresponding elements. If $[A] = [a_{ij}]$ is an $m \times n$ matrix and $[B]$ is an $m \times n$ matrix, their sum is an $m \times n$ matrix, $[C] = [c_{ij}]$, with

$$c_{ij} = a_{ij} + b_{ij} \qquad \text{for all } i, j \tag{1.3 – 1}$$

If $[A]$ is an $m \times n$ matrix, then the $n \times m$ matrix obtained by interchanging its rows and columns is called the *transpose* of $[A]$ and is denoted by $[A]^{\mathrm{T}}$. The following basic properties of a transpose should be noted:

1. $([A]^{\mathrm{T}})^{\mathrm{T}} = [A]$
2. $([A] + [B])^{\mathrm{T}} = [A]^{\mathrm{T}} + [B]^{\mathrm{T}}$

A square matrix $[A]$ of real numbers is said to be *symmetric* if $[A]^{\mathrm{T}} = [A]$. It is said to be *skew symmetric* if $[A]^{\mathrm{T}} = -[A]$. In terms of the elements of $[A]$, these definitions imply that $[A]$ is symmetric if and only if $a_{ij} = a_{ji}$, and it is skew symmetric if and only if $a_{ij} = -a_{ji}$. Note that the diagonal elements of a skew symmetric matrix are always zero since $a_{ij} = -a_{ij}$ implies $a_{ij} = 0$ for $i = j$.

If the matrix has only one row or one column, we will normally use only a single subscript to designate its elements. For example,

$$\{A\} = \begin{Bmatrix} a_1 \\ a_2 \\ a_3 \end{Bmatrix}, \quad \{B\} = \{b_1, b_2, b_3\}$$

denote a column matrix and a row matrix, respectively. Row and column matrices can be used to denote the components of a vector. For example, consider a vector $\mathbf{A} = a_1\hat{\mathbf{e}}_1 + a_2\hat{\mathbf{e}}_2 + a_3\hat{\mathbf{e}}_3$ in a Cartesian system. We can represent $\mathbf{A}$ as a *product* of a row matrix with a column matrix

$$\mathbf{A} = \{a_1 \ a_2 \ a_3\} \begin{Bmatrix} \hat{\mathbf{e}}_1 \\ \hat{\mathbf{e}}_2 \\ \hat{\mathbf{e}}_3 \end{Bmatrix}$$

Note that the vector $\mathbf{A}$ is obtained by multiplying the ith element in the row matrix with the ith element in the column matrix and adding them. This gives us a strong motivation for defining the product of two matrices.

1.3–2 Matrix Multiplication

Let $\{R\} = \{r_1, r_2 \ldots, r_m\}$ be a $1 \times m$ row matrix and let

$$\{C\} = \begin{Bmatrix} c_1 \\ c_2 \\ \vdots \\ c_m \end{Bmatrix}$$

be an $m \times 1$ column matrix. We define the product $\{R\}\{C\}$ to be the scalar

$$\{R\}\{C\} = \{r_1, r_2, \ldots, r_m\} \begin{Bmatrix} c_1 \\ c_2 \\ \vdots \\ c_m \end{Bmatrix} = r_1c_1 + r_2c_2 + \cdots + r_mc_m$$

$$= \sum_{i=1}^{m} r_ic_i \qquad (1.3-2)$$

More generally, let $[A] = [a_{ij}]$ be $m \times n$ and $[B] = [b_{ij}]$ be $n \times p$ matrices. The product $[A][B]$ is defined to be the $m \times p$ matrix $[C] = [c_{ij}]$ with

$$c_{ij} = \{i\,\text{th row of } [A]\} \begin{Bmatrix} j\text{th} \\ \text{column} \\ \text{of } [B] \end{Bmatrix} = \{a_{i1}, a_{i2}, \ldots, a_{in}\} \begin{Bmatrix} b_{1j} \\ b_{2j} \\ \vdots \\ b_{nj} \end{Bmatrix}$$

$$= a_{i1}b_{1j} + a_{i2}b_{2j} + \cdots + a_{in}b_{nj} = \sum_{k=1}^{n} a_{ik}b_{kj} \qquad (1.3-3)$$

The transpose of the product of two matrices is equal to $([A][B])^{\mathrm{T}} = [B]^{\mathrm{T}}[A]^{\mathrm{T}}$ (note the order).

If $[A]$ is an $n \times n$ matrix and $[B]$ is any $n \times n$ matrix such that $[A][B] = [B][A] = [I]$, then $[B]$ is called an *inverse* of $[A]$. If it exists, the inverse of a matrix is unique (a consequence of the associative law). If both $[B]$ and $[C]$ are inverses for $[A]$, then by definition we have

$$[A][B] = [B][A] = [A][C] = [C][A] = [I]$$

Since matrix multiplication is associative, we have

$$\begin{aligned} [B][A][C] &= ([B][A])[C] = [I][C] = [C] \\ &= [B]([A][C]) = [B][I] = [B] \end{aligned}$$

This shows that $[B] = [C]$, and the inverse is unique. The inverse of $[A]$ is denoted by $[A]^{-1}$. A matrix is said to be *singular* if it does not have an inverse.

1.3–3 Static Condensation of Matrix Equations

A *submatrix* of a matrix $[A]$ is a matrix obtained from $[A]$ by deleting certain rows and/or columns of $[A]$. One can also think of *partitioning* $[A]$ into submatrices by certain horizontal and vertical lines. One can multiply partitioned matrices, treating the submatrices like the elements of the matrix. However, we must make sure that all the sums and products of the submatrices are well defined. An example of partitioning of a matrix $[A]$ is shown below:

$$[A] = \begin{bmatrix} a_{11} & a_{12} & a_{13} & a_{14} \\ a_{21} & a_{22} & a_{23} & a_{24} \\ a_{31} & a_{32} & a_{33} & a_{34} \\ a_{41} & a_{42} & a_{43} & a_{44} \end{bmatrix} = \begin{bmatrix} [A^{11}] & [A^{12}] \\ 3 \times 3 & 3 \times 1 \\ [A^{21}] & [A^{22}] \\ 1 \times 3 & 1 \times 1 \end{bmatrix}$$

Consider an $n \times n$ system of matrix equation

$$\begin{bmatrix} a_{11} & a_{12} & a_{13} \cdots a_{1n} \\ a_{21} & a_{22} & a_{23} \cdots a_{2n} \\ a_{31} & a_{32} & a_{33} \cdots a_{3n} \\ \cdots & \cdots & \cdots \\ a_{n1} & a_{n2} & a_{n3} \cdots a_{nn} \end{bmatrix} \begin{Bmatrix} u_1 \\ u_2 \\ u_3 \\ \vdots \\ u_n \end{Bmatrix} = \begin{Bmatrix} f_1 \\ f_2 \\ f_3 \\ \vdots \\ f_n \end{Bmatrix} \qquad (1.3-4)$$

We wish to eliminate a selected set of u's from the system and express the equations in terms of the remaining u's. Such a procedure is known as the *static condensation* of variables, and it is described next.

First, the column of u's is subdivided into two parts, $\{U^1\}$ and $\{U^2\}$, according to what is to be eliminated and what is to be retained. Associated with the selected vectors $\{U^1\}$ and $\{U^2\}$, the matrix equations in Eq. (4) can be represented in submatrix form as

$$\begin{bmatrix} [A^{11}] & [A^{12}] \\ [A^{21}] & [A^{22}] \end{bmatrix} \begin{Bmatrix} \{U^1\} \\ \{U^2\} \end{Bmatrix} = \begin{Bmatrix} \{F^1\} \\ \{F^2\} \end{Bmatrix} \tag{1.3 – 5}$$

Equation (5) represents a pair of two matrix equations:

$$[A^{11}]\{U^1\} + [A^{12}]\{U^2\} = \{F^1\} \tag{1.3 – 6}$$

$$[A^{12}]^{\mathrm{T}}\{U^1\} + [A^{22}]\{U^2\} = \{F^2\} \tag{1.3 – 7}$$

Now suppose that we wish to eliminate $\{U^1\}$. If $[A^{11}]$ is invertible, we can write

$$\{U^1\} = [A^{11}]^{-1}\left(\{F^1\} - [A^{12}]\{U^2\}\right) \tag{1.3 – 8}$$

Then substituting the result into Eq. (7), we obtain

$$\left([A^{22}] - [A^{12}]^{\mathrm{T}}[A^{11}]^{-1}[A^{12}]\right)\{U^2\} = \{F^2\} - [A^{12}]^{\mathrm{T}}[A^{11}]^{-1}\{F^1\}$$

or

$$[\hat{A}^{22}]\{U^2\} = \{\hat{F}^2\} \tag{1.3 – 9}$$

where

$$[\hat{A}^{22}] = [A^{22}] - [A^{12}]^{\mathrm{T}}[A^{11}]^{-1}[A^{12}]\ , \quad \{\hat{F}^2\} = \{F^2\} - [A^{12}]^{\mathrm{T}}[A^{11}]^{-1}\{F^1\} \tag{1.3 – 10}$$

The calculation involves solving for $\{U^2\}$ first, and then, if desired, solving for $\{U^1\}$ next, using Eq. (8).

1.3–4 Determinant of a Matrix

Let $[A] = [a_{ij}]$ be an $n \times n$ matrix. We wish to associate with $[A]$ a scalar that in some sense measures the "size" of $[A]$ and indicates whether or not $[A]$ is nonsingular. The *determinant* of the matrix $[A] = [a_{ij}]$ is defined to be the scalar $|A|$ computed according to the rule

$$|A| \equiv \det[A] = \sum_{i=1}^{n}(-1)^{i+1}a_{i1}|A_{i1}| \tag{1.3 – 11}$$

where $|A_{ij}|$ is the determinant of the $(n-1) \times (n-1)$ matrix that remains on deleting the ith row and the first column of $[A]$. For 1×1 matrices, the determinant is defined according to $|a_{11}| = a_{11}$. For convenience we define the determinant of a zeroth-order matrix to be unity. In the above definition special attention is given to the first column of the matrix $[A]$. We call it the expansion of $|A|$ according to the first column of $[A]$. One can expand $|A|$ according to any column or row:

$$|A| = \sum_{i=1}^{n} (-1)^{i+j} a_{ij} |A_{ij}| \quad \text{for any } j \tag{1.3 – 12}$$

where $|A_{ij}|$ is the determinant of the matrix obtained by deleting the ith row and jth column of matrix $[A]$.

Example 1.3–1:

Let us find the determinant of the matrix

$$[A] = \begin{bmatrix} 2 & 5 & -1 \\ 1 & 4 & 3 \\ 2 & -3 & 5 \end{bmatrix}$$

Using the definition (11), we have

$$\begin{aligned} |A| &= \sum_{i=1}^{3} (-1)^{i+1} a_{i1} |A_{i1}| \\ &= (-1)^2 a_{11} \begin{vmatrix} 4 & 3 \\ -3 & 5 \end{vmatrix} + (-1)^3 a_{21} \begin{vmatrix} 5 & -1 \\ -3 & 5 \end{vmatrix} + (-1)^4 a_{31} \begin{vmatrix} 5 & -1 \\ 4 & 3 \end{vmatrix} \\ &= 2((-1)^2(4)(5) + (-1)^3(-3)3) - 1((-1)^2(5)(5) + (-1)^3(-3)(-1)) \\ &\quad + 2((-1)^2(5)(3) + (-1)^3(4)(-1)) \\ &= 2(20+9) - (25-3) + 2(15+4) = 74 \end{aligned}$$

The cross product of two vectors $\mathbf{A}$ and $\mathbf{B}$ can be expressed as the value of the determinant

$$\mathbf{A} \times \mathbf{B} \equiv \begin{vmatrix} \hat{\mathbf{e}}_1 & \hat{\mathbf{e}}_2 & \hat{\mathbf{e}}_3 \\ A_1 & A_2 & A_3 \\ B_1 & B_2 & B_3 \end{vmatrix} \tag{1.3 – 13a}$$

and the scalar triple product can be expressed as the value of the determinant:

$$\mathbf{A} \cdot (\mathbf{B} \times \mathbf{C}) \equiv \begin{vmatrix} A_1 & A_2 & A_3 \\ B_1 & B_2 & B_3 \\ C_1 & C_2 & C_3 \end{vmatrix} \tag{1.3 – 13b}$$

The verification of these results is left as an exercise for the reader.

1.3–5 Cramer's Rule

For an $n \times n$ matrix $[A]$ the determinant of the $(n-1)\times(n-1)$ submatrix of $[A]$ obtained by deleting row i and column j of $[A]$ is called the *minor* of a_{ij} and is denoted by $M_{ij}(A)$. The quantity $cof_{ij}(A) \equiv (-1)^{i+j} M_{ij}(A)$ is called the *cofactor* of a_{ij}. The determinant of $[A]$ can be cast in terms of the minor and cofactor of a_{ij} as

$$\det[A] = \sum_{i=1}^{n} a_{ij}\ cof_{ij}(A) \qquad (1.3-14)$$

for any value of j.

The *adjunct* (also called *adjoint*) of a matrix $[A]$ is the transpose of the matrix obtained from $[A]$ by replacing each element by its cofactor. The adjunct of $[A]$ is denoted by Adj(A).

Let $[A]$ be an $n \times n$ square matrix with elements a_{ij}. The cofactor of a_{ij}, denoted here by A_{ij}, is related to the minor M_{ij} of a_{ij} by

$$A_{ij} = (-1)^{i+j} M_{ij}$$

By definition, the determinant $|A|$ is given by

$$|A| = \sum_{k=1}^{n} a_{ik}A_{ik} = \sum_{k=1}^{n} a_{kj}A_{kj}\ (\text{no sum on}\, i\, \text{and}\, j)$$

for any value of i and j $\ (i, j \leq n)$. Then we have

$$\sum_{k=1}^{n} a_{rk}A_{ik} = |A|\delta_{ri}, \qquad \sum_{k=1}^{n} a_{ks}A_{kj} = |A|\delta_{sj} \qquad (1.3-15)$$

Using the above result, one can show that the solution to a set of n linear equations in n unknown quantities

$$\sum_{j=1}^{n} a_{ij}u_j = f_i, \qquad i = 1, 2, \ldots, n \qquad (1.3-16)$$

in the case when the determinant of the matrix of coefficients is not zero, $|a_{ij}| \neq 0$, is given by

$$u_j = \frac{1}{|A|}\sum_{i=1}^{n} A_{ij}b_i, \qquad j = 1, 2, \ldots, n \qquad (1.3-17)$$

The result in Eq. (17) is known as *Cramer's rule.*

Example 1.3–2:

Consider the system of equations

$$\begin{bmatrix} 2 & 5 & -1 \\ 1 & 4 & 3 \\ 2 & -3 & 5 \end{bmatrix} \begin{Bmatrix} u_1 \\ u_2 \\ u_3 \end{Bmatrix} = \begin{Bmatrix} f_1 \\ f_2 \\ f_3 \end{Bmatrix} \tag{1.3 – 18}$$

Using Cramer's rule we obtain

$$u_1 = \frac{1}{|A|} \begin{vmatrix} f_1 & 5 & -1 \\ f_2 & 4 & 3 \\ f_3 & -3 & 5 \end{vmatrix}, \; u_2 = \frac{1}{|A|} \begin{vmatrix} 2 & f_1 & -1 \\ 1 & f_2 & 3 \\ 2 & f_3 & 5 \end{vmatrix}, \; u_3 = \frac{1}{|A|} \begin{vmatrix} 2 & 5 & f_1 \\ 1 & 4 & f_2 \\ 2 & -3 & f_3 \end{vmatrix} \tag{1.3 – 19a}$$

where

$$|A| = \begin{vmatrix} 2 & 5 & -1 \\ 1 & 4 & 3 \\ 2 & -3 & 5 \end{vmatrix} = 2(20+9) - 1(25-3) + 2(15+4) = 74 \tag{1.3 – 19b}$$

We obtain

$$\begin{aligned} u_1 &= (29f_1 - 22f_2 + 19f_3)/74 \\ u_2 &= (f_1 + 12f_2 - 7f_3)/74 \\ u_3 &= (-11f_1 + 16f_2 + 3f_3)/74 \end{aligned} \tag{1.3 – 20}$$

1.4 Transformation of Vector and Tensor Components

1.4–1 Vector Transformations

In the analysis of a problem, one must refer all quantities, vectors as well as tensors, to a common coordinate system. Scalars, by definition, are independent of any coordinate system. While vectors and tensors are independent of a particular coordinate system, their components are not. The same vector or tensor can have different components in different coordinate systems. Any two sets of components of a vector can be related by writing one set of components in terms of the other. Such relationships are called transformations.

To establish the rules of the transformation of vector and tensor components, we consider barred $(\bar{x}_1, \bar{x}_2, \bar{x}_3)$ and unbarred (x_1, x_2, x_3) coordinate systems that are related by the equations

$$\begin{aligned} x_1 &= x_1(\bar{x}_1, \bar{x}_2, \bar{x}_3) \\ x_2 &= x_2(\bar{x}_1, \bar{x}_2, \bar{x}_3) \\ x_3 &= x_3(\bar{x}_1, \bar{x}_2, \bar{x}_3) \end{aligned} \tag{1.4 – 1}$$

The inverse transformation is

$$
\begin{aligned}
\bar{x}_1 &= \bar{x}_1(x_1, x_2, x_3) \\
\bar{x}_2 &= \bar{x}_2(x_1, x_2, x_3) \\
\bar{x}_3 &= \bar{x}_3(x_1, x_2, x_3)
\end{aligned}
\qquad (1.4-2)
$$

The bases in the two systems are denoted $(\hat{\mathbf{e}}_1, \hat{\mathbf{e}}_2, \hat{\mathbf{e}}_3)$ and $(\hat{\bar{\mathbf{e}}}_1, \hat{\bar{\mathbf{e}}}_2, \hat{\bar{\mathbf{e}}}_3)$ (see Figure 1.4–1).

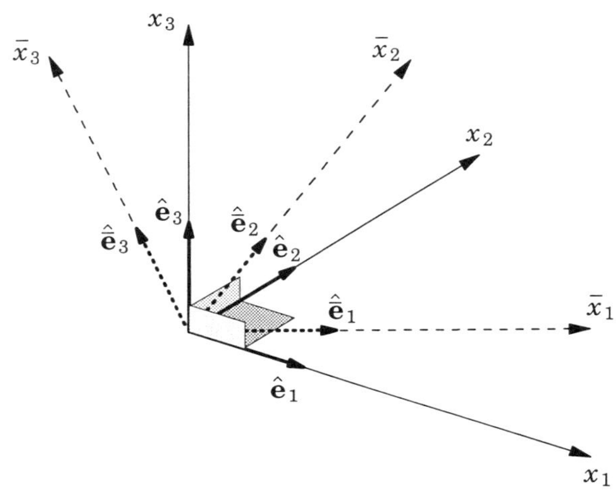

Figure 1.4–1: Unbarred and barred rectangular Cartesian coordinate systems.

The relations between the bases of the barred and unbarred coordinate systems can be written as

$$
\hat{\bar{\mathbf{e}}}_i = (\hat{\bar{\mathbf{e}}}_i \cdot \hat{\mathbf{e}}_k)\ \hat{\mathbf{e}}_k = a_{ik}\hat{\mathbf{e}}_k\ , \quad a_{ik} = \hat{\bar{\mathbf{e}}}_i \cdot \hat{\mathbf{e}}_k \qquad (1.4-3a)
$$

In matrix notation, we can write Eq. (3a) as

$$
\begin{Bmatrix} \hat{\bar{\mathbf{e}}}_1 \\ \hat{\bar{\mathbf{e}}}_2 \\ \hat{\bar{\mathbf{e}}}_3 \end{Bmatrix} = \begin{bmatrix} a_{11} & a_{12} & a_{13} \\ a_{21} & a_{22} & a_{23} \\ a_{31} & a_{32} & a_{33} \end{bmatrix} \begin{Bmatrix} \hat{\mathbf{e}}_1 \\ \hat{\mathbf{e}}_2 \\ \hat{\mathbf{e}}_3 \end{Bmatrix} \qquad (1.4-3b)
$$

where $[A]$ denotes the 3×3 matrix array whose elements are the direction cosines, a_{ij}. We also have the inverse relation

$$
\hat{\mathbf{e}}_j = (\hat{\mathbf{e}}_j \cdot \hat{\bar{\mathbf{e}}}_k)\ \hat{\bar{\mathbf{e}}}_k = a_{kj}\hat{\bar{\mathbf{e}}}_k \qquad (1.4-4a)
$$

or

$$\begin{Bmatrix} \hat{\mathbf{e}}_1 \\ \hat{\mathbf{e}}_2 \\ \hat{\mathbf{e}}_3 \end{Bmatrix} = \begin{bmatrix} a_{11} & a_{21} & a_{31} \\ a_{12} & a_{22} & a_{32} \\ a_{13} & a_{23} & a_{33} \end{bmatrix} \begin{Bmatrix} \hat{\bar{\mathbf{e}}}_1 \\ \hat{\bar{\mathbf{e}}}_2 \\ \hat{\bar{\mathbf{e}}}_3 \end{Bmatrix} \tag{1.4-4b}$$

From Eqs. (3a) and (4a), we note that

$$a_{ik}a_{jk} = \delta_{ij} \tag{1.4-5a}$$

or

$$[A][A]^{\mathrm{T}} = [I] \text{ or } [A]^{-1} = [A]^{\mathrm{T}} \tag{1.4-5b}$$

In other words, $[A]^{\mathrm{T}}$ is equal to its inverse. Such transformations are called orthogonal transformations and $[A]$ is called an orthogonal matrix.

The transformations (3) and (4) between two orthogonal sets of bases also hold for the respective coordinates:

$$\bar{x}_i = a_{ik}x_k \tag{1.4-6a}$$

$$x_j = a_{kj}\bar{x}_k \tag{1.4-6b}$$

Analogous to Eqs. (6a,b), the components of a vector $\mathbf{u}$ in the barred and unbarred coordinate systems are related by the expressions

$$\boxed{\bar{u}_i = a_{ij}u_j \ , \quad u_i = a_{ji}\bar{u}_j} \tag{1.4-7}$$

1.4–2 Tensor Transformations

A second-order tensor can be expressed in two different coordinate systems using the corresponding bases. In a barred system, we have

$$\mathbf{\Phi} = \bar{\varphi}_{mn}\hat{\bar{\mathbf{e}}}_m\hat{\bar{\mathbf{e}}}_n \tag{1.4-8}$$

and in an unbarred system it is expressed as

$$\mathbf{\Phi} = \varphi_{ij}\hat{\mathbf{e}}_i\hat{\mathbf{e}}_j \tag{1.4-9}$$

Using Eq. (4a) for $\hat{\mathbf{e}}_i$ and $\hat{\mathbf{e}}_j$ in Eq. (9) we arrive at the equation

$$\mathbf{\Phi} = \varphi_{ij}a_{mi}a_{nj}\hat{\bar{\mathbf{e}}}_m\hat{\bar{\mathbf{e}}}_n \tag{1.4-10}$$

Comparing Eq. (10) with Eq. (8), we arrive at the relation

$$\boxed{\bar{\varphi}_{mn} = \varphi_{ij} a_{mi} a_{nj}} \tag{1.4-11a}$$

or in matrix form, we have

$$[\bar{\varphi}] = [A][\varphi][A]^T \tag{1.4-11b}$$

The inverse relation can be derived using the orthogonality property of $[A]$: $[A]^{-1} = [A]^T$. Premultiplying both sides of Eq. (11b) with $[A]^{-1} = [A]^T$ and postmultiplying both sides of the resulting equation with $([A]^T)^{-1} = [A]$, we obtain the result

$$[\varphi] = [A]^T[\bar{\varphi}][A] \tag{1.4-12}$$

Next we consider an example of the second-order tensor transformation.

Example 1.4–1:

Suppose that $(\bar{x}_1, \bar{x}_2, \bar{x}_3)$ is obtained from (x_1, x_2, x_3) by rotating the x_1x_2–plane counterclockwise by an angle θ about the x_3–axis (see Figure 1.4–2). Then the two sets of coordinates are related by

$$\begin{aligned}\bar{x}_1 &= x_1 \cos\theta + x_2 \sin\theta \\ \bar{x}_2 &= -x_1 \sin\theta + x_2 \cos\theta \\ \bar{x}_3 &= x_3\end{aligned} \tag{1.4-13a}$$

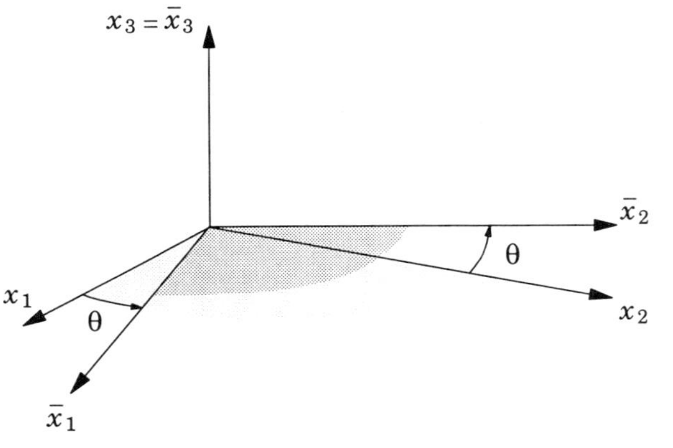

Figure 1.4–2: Transformation of a rectangular Cartesian coordinate system.

or in matrix notation

$$\begin{Bmatrix} \bar{x}_1 \\ \bar{x}_2 \\ \bar{x}_3 \end{Bmatrix} = \begin{bmatrix} \cos\theta & \sin\theta & 0 \\ -\sin\theta & \cos\theta & 0 \\ 0 & 0 & 1 \end{bmatrix} \begin{Bmatrix} x_1 \\ x_2 \\ x_3 \end{Bmatrix} \tag{1.4 – 13b}$$

The inverse relation is given by

$$\begin{Bmatrix} x_1 \\ x_2 \\ x_3 \end{Bmatrix} = \begin{bmatrix} \cos\theta & -\sin\theta & 0 \\ \sin\theta & \cos\theta & 0 \\ 0 & 0 & 1 \end{bmatrix} \begin{Bmatrix} \bar{x}_1 \\ \bar{x}_2 \\ \bar{x}_3 \end{Bmatrix} \tag{1.4 – 13c}$$

Clearly, we have $[A]^{-1} = [A]^T$. The components $\bar{\varphi}_{ij}$ and φ_{ij} of a dyad (or a second-order tensor) $\mathbf{\Phi}$ in the two coordinate systems, $(\bar{x}_1, \bar{x}_2, \bar{x}_3)$ and (x_1, x_2, x_3), respectively, are related by Eq. (11b) and (12). Equation (11b) in the present case gives the relations

$$\begin{aligned}
\bar{\varphi}_{11} &= \varphi_{11}\cos^2\theta + \varphi_{22}\sin^2\theta + (\varphi_{12} + \varphi_{21})\cos\theta\sin\theta \\
\bar{\varphi}_{12} &= (\varphi_{22} - \varphi_{11})\cos\theta\sin\theta + \varphi_{12}\cos^2\theta - \varphi_{21}\sin^2\theta \\
\bar{\varphi}_{13} &= \varphi_{13}\cos\theta + \varphi_{23}\sin\theta \\
\bar{\varphi}_{21} &= (\varphi_{22} - \varphi_{11})\cos\theta\sin\theta + \varphi_{21}\cos^2\theta - \varphi_{12}\sin^2\theta \\
\bar{\varphi}_{22} &= \varphi_{11}\sin^2\theta + \varphi_{22}\cos^2\theta - (\varphi_{12} + \varphi_{21})\cos\theta\sin\theta \\
\bar{\varphi}_{23} &= \varphi_{23}\cos\theta - \varphi_{13}\sin\theta \\
\bar{\varphi}_{31} &= \varphi_{31}\cos\theta + \varphi_{32}\sin\theta \\
\bar{\varphi}_{32} &= \varphi_{32}\cos\theta - \varphi_{31}\sin\theta \\
\bar{\varphi}_{33} &= \varphi_{33}
\end{aligned} \tag{1.4 – 14}$$

The transformation law (11a) is often taken to be the *definition* of a second-order tensor. In other words, $\mathbf{\Phi}$ is a second-order tensor if and only if its components transform according to Eq. (11a). In general, an nth order tensor transforms according to the formula

$$\bar{\varphi}_{mnpq\ldots} = \varphi_{ijk\ell\ldots} a_{mi} a_{nj} a_{pk} a_{q\ell} \cdots \tag{1.4 – 15}$$

1.5 Integral Relations

Relations between volume integrals and surface integrals of the gradient (∇) of a scalar or a vector and divergence $(\nabla\cdot)$ of a vector are needed in the later chapters. We record them here for future reference and use.

Let Ω denote a region in space surrounded by the surface Γ, and let ds be a differential element of the surface whose unit outward normal is denoted by $\hat{\mathbf{n}}$. Let dv be a differential volume element. Let ψ be a scalar function and $\mathbf{A}$ be a vector function defined over the region Ω. Then the following integral identities hold (see Figure 1.5–1):

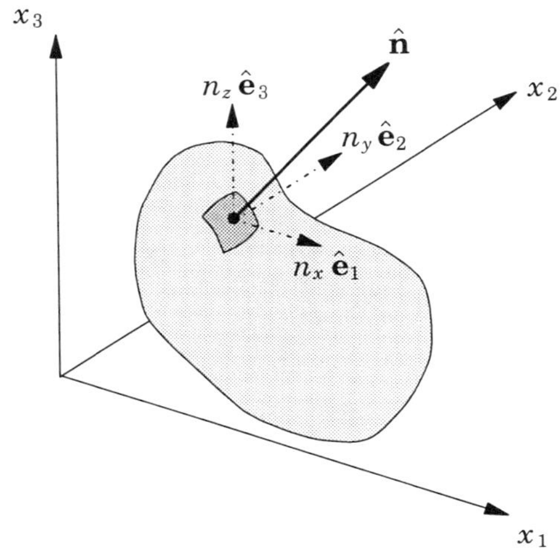

Figure 1.5–1: A solid body with a surface normal vector $\hat{\mathbf{n}}$.

Gradient Theorem

$$\int_\Omega \nabla\psi \, dv = \oint_\Gamma \hat{\mathbf{n}}\psi \, ds \quad \text{(vector form)} \qquad (1.5-1a)$$

$$\int_\Omega \frac{\partial\psi}{\partial x_i} \, dv = \oint_\Gamma n_i\psi \, ds \quad \text{(component form)} \qquad (1.5-1b)$$

Divergence Theorem

$$\int_\Omega (\nabla\cdot\mathbf{A}) \, dv = \oint_\Gamma (\hat{\mathbf{n}}\cdot\mathbf{A}) \, ds \quad \text{(vector form)} \qquad (1.5-2a)$$

$$\int_\Omega \frac{\partial A_i}{\partial x_i} \, dv = \oint_\Gamma n_i A_i \, ds \quad \text{(component form)} \qquad (1.5-2b)$$

In the above integral relations, $\oint_\Gamma$ denotes the integral on the closed boundary Γ of the domain Ω, and the component forms refer to the usual rectangular Cartesian coordinate system. Equations (1) and (2) also hold in two- as well as three-dimensional domains Ω with boundary Γ. In fact, the integral relations (1) and (2) can be expressed concisely in the single statement

$$\int_\Omega (\nabla * F)\, dv = \oint_\Gamma (\hat{\mathbf{n}} * F)\, ds \tag{1.5 – 3}$$

where * denotes an appropriate operation, *i.e.*, gradient, divergence or curl operation, and F is a scalar or vector function.

Some additional integral relations can be derived from Eqs. (1)–(3). Let $\mathbf{A} = \nabla\varphi$ in Eq. (2a) and obtain

$$\int_\Omega \nabla \cdot (\nabla\varphi) dv = \int_\Omega \nabla^2 \varphi\, dv = \oint_\Gamma \hat{\mathbf{n}} \cdot (\nabla\varphi)\, ds \quad \text{(vector form)} \tag{1.5 – 4a}$$

or, in component form

$$\int_\Omega \frac{\partial^2 \varphi}{\partial x_i \partial x_i}\, dv = \oint_\Gamma n_i \frac{\partial \varphi}{\partial x_i}\, ds \tag{1.5 – 4b}$$

The quantity $\hat{\mathbf{n}} \cdot \nabla\varphi$ is called the *normal derivative* of φ on the surface Γ, and is denoted by

$$\begin{aligned}\frac{\partial \varphi}{\partial n} &= \hat{\mathbf{n}} \cdot \nabla\varphi \\ &= n_i \frac{\partial \varphi}{\partial x_i} \\ &= n_x \frac{\partial \varphi}{\partial x} + n_y \frac{\partial \varphi}{\partial y} + n_z \frac{\partial \varphi}{\partial z}\end{aligned} \tag{1.5 – 5}$$

Example 1.5–1: ______________________________

Consider the integral expression

$$I = \int_\Omega \psi \left[\nabla \cdot (f \nabla\varphi)\right]\, dv \tag{1.5 – 6}$$

where f, ψ, and φ are scalar functions of position in Ω. The integral can be expressed in several alternative forms. First, let $f\nabla\varphi \equiv \mathbf{F}$, and note the identity

$$\psi\ \nabla \cdot \mathbf{F} = \nabla \cdot (\psi\mathbf{F}) - \nabla\psi \cdot \mathbf{F} \tag{1.5 – 7}$$

Then

$$\begin{aligned}\int_\Omega \psi(\nabla \cdot \mathbf{F})\, dv &= \int_\Omega [\nabla \cdot (\psi\mathbf{F}) - \nabla\psi \cdot \mathbf{F}]\, dv \\ &= -\int_\Omega \nabla\psi \cdot \mathbf{F} dv + \int_\Omega \nabla \cdot (\psi\mathbf{F})\, ds\end{aligned}$$

The second integral on the right-hand side, by the divergence theorem, is equal to

$$\int_{\Omega} \nabla \cdot (\psi \mathbf{F})\, dv = \oint_{\Gamma} \hat{\mathbf{n}} \cdot (\psi \mathbf{F}) ds \tag{1.5-8}$$

Substituting the definition $\mathbf{F} = f\nabla\varphi$, we obtain

$$\int_{\Omega} \psi[\nabla \cdot (f\nabla\varphi)] dv = -\int_{\Omega} f\nabla\psi \cdot \nabla\varphi\, dv + \oint_{\Gamma} \psi\, f\frac{\partial\varphi}{\partial n}\, ds \tag{1.5-9}$$

Note that the first term on the right-hand side is symmetric in ψ and φ. Therefore, interchanging ψ and φ in Eq. (9), we obtain

$$\int_{\Omega} \varphi[\nabla \cdot (f\nabla\psi)] dv = -\int_{\Omega} f\nabla\varphi \cdot \nabla\psi\, dv + \oint_{\Gamma} \varphi\, f\frac{\partial\psi}{\partial n}\, ds \tag{1.5-10}$$

From Eqs. (9) and (10), we obtain

$$\int_{\Omega} \psi[-\nabla \cdot (f\nabla\varphi)]\, dv = \int_{\Omega} \varphi[-\nabla \cdot (f\nabla\psi)]\, dv + \oint_{\Gamma} f(\varphi\frac{\partial\psi}{\partial n} - \psi\frac{\partial\varphi}{\partial n})\, ds \tag{1.5-11}$$

The integral relations derived in this section are useful in the construction of weak forms as well as in the derivation of the Euler–Lagrange equations associated with a variational principle. Additional integral identities can be derived from those presented above (see Exercises 1.7 and 1.8).

Exercises

1.1 The nine cross-product (or vector product) relations among the basis $(\hat{\mathbf{e}}_1, \hat{\mathbf{e}}_2, \hat{\mathbf{e}}_3)$ can be expressed using the index notation as

$$\hat{\mathbf{e}}_i \times \hat{\mathbf{e}}_j = \epsilon_{ijk}\hat{\mathbf{e}}_k \tag{1}$$

where ϵ_{ijk} is the *permutation symbol*, defined as

$$\epsilon_{ijk} = \begin{cases} 1, & \text{if } i,\ j,\ k \text{ are in cyclic order } (i \neq j \neq k) \\ 0, & \text{if any of } i,\ j,\ k \text{ are repeated} \\ -1, & \text{if } i,\ j,\ k \text{ are not in cyclic order } (i \neq j \neq k) \end{cases} \tag{2}$$

The $\epsilon - \delta$ *identity* relates the Kronecker delta and permutation symbol:

$$\epsilon_{ijk}\epsilon_{mnk} = \delta_{im}\delta_{jn} - \delta_{in}\delta_{jm} \tag{3}$$

Prove the following properties of δ_{ij} and ϵ_{ijk}:

(i) $F_{ij}\delta_{jk} = F_{ik}$

(ii) $\delta_{ii} = 3$, (for i over a range of 1 to 3)

(iii) $\delta_{ij}\delta_{ij} = \delta_{ii}$

(iv) $\epsilon_{ijk}\epsilon_{ijk} = 6$, (for i, j, k over a range of 1 to 3)

(v) $\epsilon_{ijk}A_iA_j = 0$

(vi) $\epsilon_{ijk} = \epsilon_{kij} = \epsilon_{jki} = -\epsilon_{jik} = -\epsilon_{ikj} = -\epsilon_{kji}$ (4)

1.2 Prove the following vector identities using the summation convention and the $\epsilon - \delta$ identity (2) of Problem 1.1. In the first three identities **A, B, C** and **D** denote vectors:

(i) $(\mathbf{A} \times \mathbf{B} \times \mathbf{C} \times \mathbf{D}) = [\mathbf{A} \cdot (\mathbf{C} \times \mathbf{D})]\mathbf{B} - [\mathbf{B} \cdot (\mathbf{C} \times \mathbf{D})]\mathbf{A}$

(ii) $(\mathbf{A} \times \mathbf{B}) \cdot (\mathbf{C} \times \mathbf{D}) = (\mathbf{A} \cdot \mathbf{C})(\mathbf{B} \cdot \mathbf{D}) - (\mathbf{A} \cdot \mathbf{D})(\mathbf{B} \cdot \mathbf{C})$

(iii) $(\mathbf{A} \times \mathbf{B}) \cdot [(\mathbf{B} \times \mathbf{C}) \times (\mathbf{C} \cdot \mathbf{A})] = [\mathbf{A} \cdot (\mathbf{B} \times \mathbf{C})]^2$

(iv) $(\mathbf{AB})^T = (\mathbf{B})^T(\mathbf{A})^T$, where **A** and **B** are dyads

1.3 Find the transformation matrix for the following transformation: rotate 90° about the $\hat{\mathbf{e}}_3$–axis and then rotate 45° about the position assumed by the $\hat{\mathbf{e}}_1$–axis after the first rotation.

1.4 If $[A]$ is a symmetric $n \times n$ matrix, and $[B]$ is any $n \times n$ matrix, show that $[B]^{\mathrm{T}}[A][B]$ is symmetric.

1.5 If $[A]$ and $[B]$ are nonsingular $n \times n$ matrices, prove that:

(i) $[A][B]$ is nonsingular

(ii) $([A][B])^{-1} = [B]^{-1}[A]^{-1}$

1.6 Using Cramer's rule, determine the solution to the following two sets of equations:

$$\begin{aligned} 2x_1 - x_2 - x_3 &= 2 \\ x_1 + 2x_2 + x_3 &= 2 \\ 4x_1 - 8x_2 - 5x_3 &= 2 \end{aligned}$$

$$\begin{aligned} 2x_1 - x_2 &= 1 \\ -x_1 + 4x_2 - x_3 &= 2 \\ -x_2 + 2x_3 &= 1 \end{aligned}$$

1.7 Derive the following integral identity

$$\begin{aligned} -\int_\Omega w_i \left[\frac{\partial}{\partial x_j}\left(\frac{\partial u_i}{\partial x_j} + \frac{\partial u_j}{\partial x_i}\right)\right] dv \\ = \int_\Omega \frac{\partial w_i}{\partial x_j}\left(\frac{\partial u_i}{\partial x_j} + \frac{\partial u_j}{\partial x_i}\right) dv - \oint_\Gamma w_i n_j \left(\frac{\partial u_i}{\partial x_j} + \frac{\partial u_j}{\partial x_i}\right) ds \end{aligned}$$

where w_i and u_i, are functions of position in Ω. The summation convention on repeated subscripts is used.

1.8 Derive the following integral identity

$$\int_{\Omega}(\varphi\nabla^4\psi - \nabla^2\varphi\nabla^2\psi)\, dv = \oint_{\Gamma}\left[\varphi\frac{\partial}{\partial n}(\nabla^2\psi) - \nabla^2\psi\frac{\partial\varphi}{\partial n}\right] ds$$

1.9 If A is an arbitrary vector and $\mathbf{\Phi}$ is an arbitrary second-order tensor, show that

(i) $(\mathbf{I} \times \mathbf{A}) \cdot \mathbf{\Phi} = \mathbf{A} \times \mathbf{\Phi}$, $\mathbf{I}$ = unit tensor

(ii) $(\mathbf{\Phi} \times \mathbf{A})^{\mathrm{T}} = -\mathbf{A} \times \mathbf{\Phi}^{\mathrm{T}}$

1.10 (*Eigenvalues of a Matrix*) The *eigenvalue problem* associated with a square matrix $[A]$ consists of finding the scalars λ, called the *eigenvalues* of $[A]$, such that there exists a nonzero vector $\{X\}$, called the *eigenvector*, satisfying $[A]\{X\} = \lambda\{X\}$, equivalently, $[A - \lambda I]\{X\} = \{0\}$. Show that this homogeneous equation has a nontrivial solution only if $\det[A - \lambda I]$ is zero.

References for Additional Reading

1. Aris, R., *Vectors, Tensors, and the Basic Equations of Fluid Mechanics,* Prentice–Hall, Englewood Cliffs, New Jersey, 1962.
2. Bowen, R. M. and Wang, C. C., *Introduction to Vectors and Tensors*, Vols. I and II, Plenum, New York, 1976.
3. Eisele, J. A., *Applied Matrix and Tensor Analysis,* Wiley–Interscience, New York, 1970.
4. Gantmacher, F. R., *The Theory of Matrices,* Chelsea, New York, 1959.
5. Hildebrand, F. B., *Methods of Applied Mathematics,* Second Edition, Prentice–Hall, Englewood Cliffs, New Jersey, 1965.
6. Jeffreys, H., *Cartesian Tensors*, Cambridge University Press, London, 1965.
7. Kreyszig, E., *Advanced Engineering Mathematics*, 6th Edition, John Wiley, New York, 1988.
8. Reddy, J. N. and Rasmussen, M. L., *Advanced Engineering Analysis*, John Wiley, New York, 1982; reprinted by Krieger, Melbourne, Florida, 1990.
9. Reddy, J. N., *Applied Functional Analysis and Variational Methods in Engineering*, McGraw–Hill, New York, 1986; reprinted by Krieger, Melbourne, Florida, 1992.
10. Young, E. C., *Vector and Tensor Analysis*, Marcel Dekker, New York, 1978.

Chapter Two

Equations of Anisotropic Elasticity

2.1 Classification of Equations

The objective of this chapter is to review the governing equations of an anisotropic elastic body. The equations governing the motion of a solid body can be classified into four basic categories:

(1) Kinematics (strain–displacement equations)

(2) Kinetics (conservation of momenta)

(3) Thermodynamics (first and second laws of thermodynamics)

(4) Constitutive equations (stress–strain relations)

Kinematics is a study of the geometric changes or deformation in a body, without the consideration of forces causing the deformation. *Kinetics* is the study of the static or dynamic equilibrium of forces acting on a body. The thermodynamic principles are concerned with the conservation of energy and relations among heat, mechanical work, and thermodynamic properties of the body. The constitutive equations describe the constitutive behavior of the body, and relate the dependent variables introduced in the kinetic description to those in the kinematic and thermodynamic descriptions. These equations are supplemented by appropriate boundary and initial conditions of the problem.

In the following sections, an overview of the governing equations of an anisotropic elastic body is presented. The strain–displacement relations, equations of motion, and the constitutive equations for an isothermal state (*i.e.*, no change in the temperature of the body) are presented first. Subsequently, the thermodynamic principles are considered only to determine the temperature distribution in a solid body and to account for the effect of nonuniform temperature distribution on the strains.

2.2 Kinematics

2.2–1 Descriptions of Motion

A solid body $\mathcal{B}$ is a set of material particles which can be identified as having one-to-one correspondence with the points of a region Ω of Euclidean point space R^3. The particles of $\mathcal{B}$ are identified by their time-dependent positions relative to the selected frame of reference. The simultaneous position of all material points of $\mathcal{B}$ at a fixed time is called a *configuration* of the structure. A mathematical description of the configuration at any time requires a coordinate system that is fixed in space.

Consider a deformable body (or a structural system) $\mathcal{B}$ of known geometry, constitution, and loading. Under given geometric restrictions and loading, the body will undergo motion and/or deformation (*i.e.*, geometric changes within the body). If the applied loads are time dependent, the deformation of the body will be a function of time, *i.e.*, the geometry of the body will change continuously with time. If the loads are applied slowly so that the deformation is only dependent on the loads, the body will take a definitive shape at the end of each load application. Whether the deformation is time dependent or not, the forces acting on the body will be in equilibrium at all times.

Suppose that the body $\mathcal{B}$ under consideration at time $t = 0$ (or initially) occupies a configuration $\mathcal{C}^0$, in which a particle X of the body $\mathcal{B}$ occupies a position $\mathbf{X}$. Note that X is the name of the particle that occupies the location $\mathbf{X}$ in the reference configuration. At time $t > 0$, the body assumes a new configuration $\mathcal{C}$ and the particle X occupies the position $\mathbf{x}$.

There are two commonly used descriptions of motion and deformation in continuum mechanics. In the *referential* or *Lagrangian description*, the motion of a body $\mathcal{B}$ is referred to a reference configuration $\mathcal{C}^R$. Thus, in the Lagrangian description the current coordinates (x_1, x_2, x_3) are expressed in terms of the reference coordinates (X_1, X_2, X_3) and time t as

$$\mathbf{x} = \mathbf{x}(X_1, X_2, X_3, t) \qquad (2.2-1)$$

Often, the reference configuration $\mathcal{C}^R$ is chosen to be the unstressed state of the body, *i.e.*, $\mathcal{C}^R \equiv \mathcal{C}^0$. The coordinates (X_1, X_2, X_3) are called the *material coordinates.*

In the *spatial* or *Eulerian description* of a body $\mathcal{B}$, the motion is referred to the current configuration $\mathcal{C}$ occupied by the body $\mathcal{B}$. The spatial description focuses attention on a given region of space instead of on a given body of matter, and is the description most used in fluid mechanics, whereas in the Lagrangian description the coordinate system $\mathbf{X}$ is fixed on a given body of matter in its undeformed configuration, and its position $\mathbf{x}$ at any time is referred to the material coordinates X_i. Thus, during a motion of a body $\mathcal{B}$, a representative particle X occupies a succession of points which together form a curve in Euclidean space. This curve is called the path of X and is given parametrically by Eq. (1).

2.2–2 Strains and Strain–Displacement Relations

The phrase *deformation of a body* refers to relative displacements and changes in the geometry experienced by the body. Referred to a rectangular Cartesian frame of reference (X_1, X_2, X_3), every particle X in the body corresponds to a set of coordinates $\mathbf{X} = (X_1,\ X_2,\ X_3)$. When the body is deformed under the action of external forces, the particle X moves to a new position $\mathbf{x} = (x_1, x_2, x_3)$. The displacement of the particle X is given by

$$\mathbf{u} = \mathbf{x} - \mathbf{X} \quad \text{or} \quad u_i = x_i - X_i \tag{2.2 – 2}$$

If the displacement of every particle in the body is known, we can construct the current (deformed) configuration $\mathcal{C}$ from the reference (or undeformed) configuration $\mathcal{C}^0$. In the Lagrangian description, the displacements are expressed in terms of the material coordinates X_i, and we have

$$u_i(X_1, X_2, X_3, t) = x_i(X_1, X_2, X_3, t) - X_i \tag{2.2 – 3}$$

A rigid–body motion is one in which all material particles of the body undergo the same linear and angular displacements. Thus a rigid-body motion does not alter the shape but only the location of a body. A deformable body is one in which the material particles can move relative to each other. The deformation (*i.e.*, relative motion of material particles) of a deformable body can be determined only by considering the change of distance between any two arbitrary but infinitesimally close points of the body.

Consider two neighboring particles P and Q which occupy the positions $P : (X_1, X_2, X_3)$ and $Q : (X_1 + dX_1,\ X_2 + dX_2,\ X_3 + dX_3)$, respectively, in the reference configuration of the body $\mathcal{B}$. The particles are separated by the infinitesimal distance $dS = \sqrt{dX_i dX_i}$ (sum on i) in the undeformed configuration $\mathcal{C}^0$, and $d\mathbf{X}$ is the vector connecting the position of P to the position of Q. These two particles move to new places $\bar{P}$ and $\bar{Q}$, respectively, in the deformed body (see Figure 2.2–1). Suppose that the positions of $\bar{P}$ and $\bar{Q}$ are (x_1, x_2, x_3) and $(x_1 + dx_1,\ x_2 + dx_2,\ x_3 + dx_3)$, respectively. The two particles are now separated by the distance $ds = \sqrt{dx_i dx_i}$ in the deformed configuration $\mathcal{C}$, and $d\mathbf{x}$ is the vector connecting $\bar{P}$ to $\bar{Q}$. The vector $d\mathbf{x}$ can be interpreted as the position occupied by the deformed material vector $d\mathbf{X}$. When the material vector $d\mathbf{X}$ is small but finite, the line vector $d\mathbf{x}$ in general does not coincide exactly with the deformed position of $d\mathbf{X}$, which lies along a curve in the deformed body.

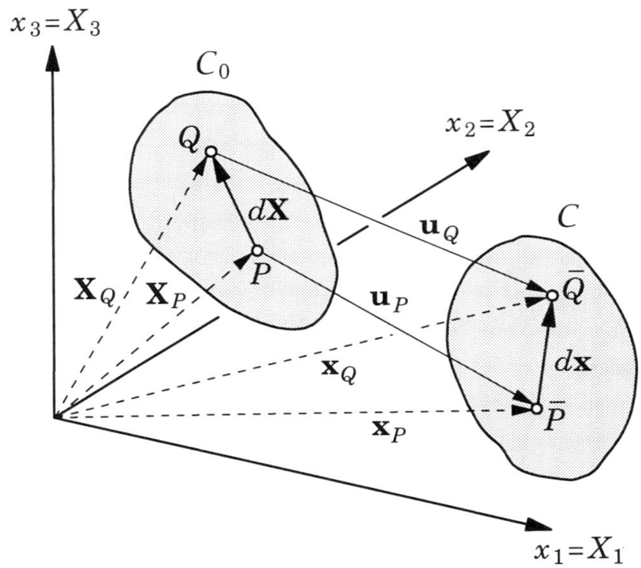

Figure 2.2–1: Kinematics of deformation of a continuous medium.

The deformation of a body can be measured in terms of the strain tensor $\mathbf{E}$, which is defined such that it gives the change in the square of the length of the material vector $d\mathbf{X}$

$$2d\mathbf{X} \cdot \mathbf{E} \cdot d\mathbf{X} \equiv (ds)^2 - (dS)^2 = d\mathbf{x} \cdot d\mathbf{x} - d\mathbf{X} \cdot d\mathbf{X} \tag{2.2 - 4a}$$

and in rectangular Cartesian component form we have

$$2dX_i E_{ij} dX_j \equiv (ds)^2 - (dS)^2 = dx_i dx_i - dX_i dX_i \tag{2.2 - 4b}$$

In Eq. (4b) and in the equations that follow, the *summation convention* on repeated indices is used, and the range of summation is 1 to 3. This convention will be used throughout this chapter and the component forms are for rectangular Cartesian systems unless stated otherwise. The tensor $\mathbf{E}$ is known as the *Green–Lagrange strain tensor.*

In order to express the strains in terms of the displacements, we use Eq. (2) and write

$$\mathbf{x} = \mathbf{X} + \mathbf{u}(X_1, X_2, X_3, t) \tag{2.2 - 5}$$

Since $\mathbf{x}$ is a function of $\mathbf{X}$, its total differential is given by [using the chain rule of differentiation and Eq. (5)]

$$d\mathbf{x} \equiv d\mathbf{X} \cdot \nabla\mathbf{x} = d\mathbf{X} \cdot (\mathbf{I} + \nabla\mathbf{u}) \tag{2.2 - 6}$$

where ∇ denotes the gradient operator with respect to the material coordinates, $\mathbf{X}$. Now the strain tensor or its components from Eqs. (4a,b) can be expressed in terms of the displacement vector or its components with the help of Eq. (6):

$$
\begin{aligned}
2d\mathbf{X} \cdot \mathbf{E} \cdot d\mathbf{X} &= d\mathbf{x} \cdot d\mathbf{x} - d\mathbf{X} \cdot d\mathbf{X} \\
&= d\mathbf{X} \cdot (\mathbf{I} + \nabla\mathbf{u}) \cdot [d\mathbf{X} \cdot (\mathbf{I} + \nabla\mathbf{u})] - d\mathbf{X} \cdot d\mathbf{X} \\
&= d\mathbf{X} \cdot (\mathbf{I} + \nabla\mathbf{u}) \cdot (\mathbf{I} + \nabla\mathbf{u})^T \cdot d\mathbf{X} - d\mathbf{X} \cdot d\mathbf{X} \\
&= d\mathbf{X} \cdot \left[(\mathbf{I} + \nabla\mathbf{u}) \cdot (\mathbf{I} + \nabla\mathbf{u})^T - \mathbf{I}\right] \cdot d\mathbf{X} \qquad (2.2-7)
\end{aligned}
$$

Thus the Green (or Green–Lagrange) strain tensor $\mathbf{E}$ is given in terms of the displacement gradients as

$$
\begin{aligned}
\mathbf{E} &= \frac{1}{2}\left[(\mathbf{I} + \nabla\mathbf{u}) \cdot (\mathbf{I} + \nabla\mathbf{u})^T - \mathbf{I}\right] \\
&= \frac{1}{2}\left[\nabla\mathbf{u} + (\nabla\mathbf{u})^T + \nabla\mathbf{u} \cdot (\nabla\mathbf{u})^T\right] \qquad (2.2-8a)
\end{aligned}
$$

The rectangular Cartesian component form is given by

$$
\begin{aligned}
E_{jk} &= \frac{1}{2}\left[\left(\delta_{ij} + \frac{\partial u_i}{\partial X_j}\right)\left(\delta_{ik} + \frac{\partial u_i}{\partial X_k}\right) - \delta_{jk}\right] \\
&= \frac{1}{2}\left(\frac{\partial u_j}{\partial X_k} + \frac{\partial u_k}{\partial X_j} + \frac{\partial u_m}{\partial X_j}\frac{\partial u_m}{\partial X_k}\right) \qquad (2.2-8b)
\end{aligned}
$$

Note that the Green–Lagrange strain tensor is symmetric, $\mathbf{E} = \mathbf{E}^T (E_{ij} = E_{ji})$. The strain components defined in Eq. (8) are called finite strain components because no assumption concerning the smallness (compared to unity) of the strains is made. It should be noted that if the reference coordinate system is curvilinear (*i.e.*, X_i are curvilinear, instead of rectangular Cartesian coordinates), Eq. (8) takes the form

$$
E_{jk} = \frac{1}{2}(u_{j;k} + u_{k;j} + u_{m;j}u^m_{;k}) \qquad (2.2-9)
$$

where $u_{j;k}$ denotes the covariant derivative (see [1–4]) of the covariant components of the displacement vector, and $u^m_{;k}$ is the covariant derivative of the contravariant components of the displacement vector.

We note that $(ds)^2 - (dS)^2 = 0$ if and only if $E_{jk} = 0$. Thus a deformation in which the length of every line element remains unchanged is a rigid-body motion. Therefore, a necessary and sufficient condition for a motion of a body to be a rigid-body motion is that all components of the strain tensor E_{ij} be zero throughout the body.

The six Cartesian components of strain are given by

$$
\begin{aligned}
E_{11} &= \frac{\partial u_1}{\partial X_1} + \frac{1}{2}\left[\left(\frac{\partial u_1}{\partial X_1}\right)^2 + \left(\frac{\partial u_2}{\partial X_1}\right)^2 + \left(\frac{\partial u_3}{\partial X_1}\right)^2\right] \\
E_{22} &= \frac{\partial u_2}{\partial X_2} + \frac{1}{2}\left[\left(\frac{\partial u_1}{\partial X_2}\right)^2 + \left(\frac{\partial u_2}{\partial X_2}\right)^2 + \left(\frac{\partial u_3}{\partial X_2}\right)^2\right] \\
E_{33} &= \frac{\partial u_3}{\partial X_3} + \frac{1}{2}\left[\left(\frac{\partial u_1}{\partial X_3}\right)^2 + \left(\frac{\partial u_2}{\partial X_3}\right)^2 + \left(\frac{\partial u_3}{\partial X_3}\right)^2\right] \\
E_{12} &= \frac{1}{2}\left(\frac{\partial u_1}{\partial X_2} + \frac{\partial u_2}{\partial X_1} + \frac{\partial u_1}{\partial X_1}\frac{\partial u_1}{\partial X_2} + \frac{\partial u_2}{\partial X_1}\frac{\partial u_2}{\partial X_2} + \frac{\partial u_3}{\partial X_1}\frac{\partial u_3}{\partial X_2}\right) \\
E_{13} &= \frac{1}{2}\left(\frac{\partial u_1}{\partial X_3} + \frac{\partial u_3}{\partial X_1} + \frac{\partial u_1}{\partial X_1}\frac{\partial u_1}{\partial X_3} + \frac{\partial u_2}{\partial X_1}\frac{\partial u_2}{\partial X_3} + \frac{\partial u_3}{\partial X_1}\frac{\partial u_3}{\partial X_3}\right) \\
E_{23} &= \frac{1}{2}\left(\frac{\partial u_2}{\partial X_3} + \frac{\partial u_3}{\partial X_2} + \frac{\partial u_1}{\partial X_2}\frac{\partial u_1}{\partial X_3} + \frac{\partial u_2}{\partial X_2}\frac{\partial u_2}{\partial X_3} + \frac{\partial u_3}{\partial X_2}\frac{\partial u_3}{\partial X_3}\right)
\end{aligned} \tag{2.2-10}
$$

If the displacement gradients are so small, $|u_{i,j}| << 1$, that their squares and products are negligible compared to $|u_{i,j}|$, then the Green–Lagrange strain tensor reduces to the *infinitesimal strain tensor*:

$$\varepsilon = \frac{1}{2}\left[\nabla \mathbf{u} + (\nabla \mathbf{u})^T\right] \tag{2.2-11}$$

or

$$\varepsilon_{ij} = \frac{1}{2}\left(\frac{\partial u_i}{\partial X_j} + \frac{\partial u_j}{\partial X_i}\right) \tag{2.2-12}$$

and the difference between $\frac{\partial u_i}{\partial x_j}$ and $\frac{\partial u_i}{\partial X_j}$ vanishes. The explicit form of the infinitesimal strain components (11) is given by

$$\varepsilon_{11} = \frac{\partial u_1}{\partial x_1}\,, \quad \varepsilon_{22} = \frac{\partial u_2}{\partial x_2}\,, \quad \varepsilon_{33} = \frac{\partial u_3}{\partial x_3}\,, \quad \gamma_{12} \equiv 2\varepsilon_{12} = \frac{\partial u_1}{\partial x_2} + \frac{\partial u_2}{\partial x_1}$$

$$\gamma_{13} \equiv 2\varepsilon_{13} = \frac{\partial u_1}{\partial x_3} + \frac{\partial u_3}{\partial x_1}\,, \quad \gamma_{23} \equiv 2\varepsilon_{23} = \frac{\partial u_2}{\partial x_3} + \frac{\partial u_3}{\partial x_2} \tag{2.2-13}$$

where γ_{ij} denote the engineering shear strains.

Example 2.2–1:

(a) A square block is deformed as shown by dotted lines in Figure 2.2–2a. Assuming that the body is very thin and the strains (due to the Poisson effect) associated with the thickness direction are negligible, we wish to determine the two-dimensional strains.

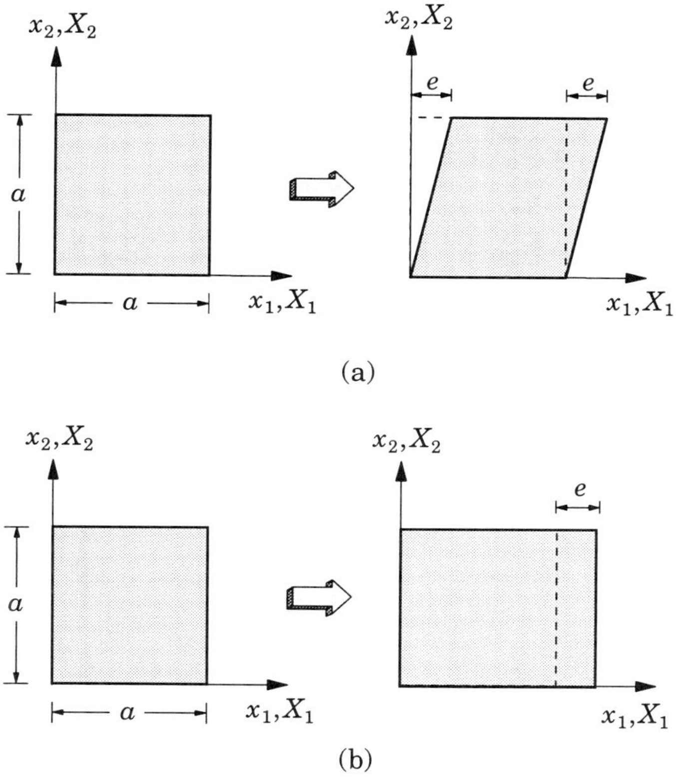

Figure 2.2–2: Reference (undeformed) and deformed configurations of a square block. (a) Pure shear deformation. (b) Pure extensional deformation.

A material particle which occupied position (X_1, X_2, X_3) in the undeformed body takes the position (x_1, x_2, x_3) in the deformed body. The current coordinates of the material particle can be expressed in terms of its original position as

$$x_1 = X_1 + \frac{e}{a}X_2, \quad x_2 = X_2, \quad x_3 = X_3 \tag{2.2 - 14}$$

The displacements are

$$u_1 \equiv x_1 - X_1 = \frac{e}{a}X_2, \quad u_2 \equiv x_2 - X_2 = 0, \quad u_3 \equiv x_3 - X_3 = 0 \tag{2.2 - 15}$$

Then the Lagrangian strains can be computed using Eq. (10). The only nonzero strain component is ($e = 0.2$cm and $a = 10$cm)

$$E_{12} = \frac{e}{2a} = 0.01 \ \ \text{cm/cm} \tag{2.2 - 16}$$

Note that the strain field is linear.

(b) Consider a square block, deformed as shown by dotted lines in Figure 2.2–2b. The current coordinates of the material particle occupying position (X_1, X_2, X_3) in the undeformed body can be expressed as

$$x_1 = X_1 + \frac{e}{a}X_1, \quad x_2 = X_2, \quad x_3 = X_3 \tag{2.2-17}$$

The displacements are

$$u_1 \equiv x_1 - X_1 = \frac{e}{a}X_1, \quad u_2 \equiv x_2 - X_2 = 0, \quad u_3 \equiv x_3 - X_3 = 0 \tag{2.2-18}$$

The only nonzero Lagrangian strain is

$$E_{11} = \frac{e}{a} + \frac{1}{2}\left(\frac{e}{a}\right)^2 = (0.02 + 0.0002) \text{ cm/cm} \tag{2.2-19}$$

The strain is nonlinear. The nonlinear part of the strain is less than 0.02 percent.

2.2–3 Compatibility Equations

By definition, the components of the strain tensor can be computed from a differentiable displacement field using Eq. (12). However, if the *six* components of strain tensor are given and if we are required to find the *three* displacement components, the strains given should be such that a unique solution to the six differential equations relating the strains and displacements exists. The existence of a unique solution is guaranteed if the infinitesimal strain components satisfy the following six compatibility conditions:

$$\frac{\partial^2 \varepsilon_{ij}}{\partial x_m \partial x_n} + \frac{\partial^2 \varepsilon_{mn}}{\partial x_i \partial x_j} - \frac{\partial^2 \varepsilon_{im}}{\partial x_j \partial x_n} - \frac{\partial^2 \varepsilon_{jn}}{\partial x_i \partial x_m} = 0 \tag{2.2-20}$$

for any $i, j, m, n = 1, 2, 3$. For the two-dimensional case, Eq. (20) reduces to the following single compatibility equation

$$\frac{\partial^2 \varepsilon_{11}}{\partial x_2^2} + \frac{\partial^2 \varepsilon_{22}}{\partial x_1^2} - 2\frac{\partial^2 \varepsilon_{12}}{\partial x_1 \partial x_2} = 0 \tag{2.2-21}$$

It should be noted that the strain compatibility equations are satisfied automatically when the strains are computed from a displacement field. Thus, one needs to verify the compatibility conditions only when the strains are computed from stresses that are in equilibrium.

2.3 Kinetics

2.3–1 Stress and Its Measures

Stress at a point is a measure of force per unit area. The force per unit area acting on an elemental area ds of the deformed body is called the *stress vector* acting on the element. The concept also applies to the surface created by slicing the deformed body with a plane. The stress vector on the area element Δs is the vector $\Delta\mathbf{f}/\Delta s$, which has the same direction as $\Delta\mathbf{f}$ but with magnitude equal to $|\Delta\mathbf{f}|/\Delta s$. The stress vector at a point P on Δs is defined by

$$\mathbf{T} \equiv \lim_{\Delta s \to 0} \frac{\Delta\mathbf{f}}{\Delta s} \tag{2.3 – 1}$$

Equation (1) indicates that the stress vector at a point depends on the force vector (its direction and magnitude) and the surface area. The surface area in turn depends on the orientation of the plane used to slice the body. For example, consider a constant cross-section member. If it is sliced with a plane perpendicular to its axis, the resulting surface area A_0 is the same as the area of cross section of the member. However, if it is sliced, at the same point, with a plane oriented at an angle $0° < \theta < 90°$ to the vertical axis, the surface area A obtained will be different: $A = A_0/\cos\theta$. Thus the stress vector at a point depends on the force vector as well as the orientation of the surface on which it is acting. For this reason, the stress vector acting on a plane with normal $\hat{\mathbf{n}}$ is denoted by $\mathbf{T}^{(\hat{\mathbf{n}})}$.

The state of stress at a point inside a body can be expressed in terms of stress vectors on three mutually perpendicular planes, say planes perpendicular to the rectangular coordinate axes. Let $\mathbf{T}^{(i)}$ denote the stress vector at a point P on a plane perpendicular to the x_i-axis (see Figure 2.3–1a). Each vector $\mathbf{T}^{(i)}$, $(i = 1, 2, 3)$ can be resolved into components along the coordinate lines

$$\mathbf{T}^{(i)} = \sigma_{ij}\hat{\mathbf{e}}_j \quad (i \;=\; 1, 2, 3) \tag{2.3 – 2}$$

where the symbol σ_{ij} denotes the component of the stress vector $\mathbf{T}^{(i)}$ at the point P along the x_j–direction, and $\hat{\mathbf{e}}_j$ is the unit base vector along the x_j–direction. There are a total of nine stress components σ_{ij}, $(i, j = 1, 2, 3)$ at the point P. It can be shown that the quantities σ_{ij} transform like the components of a second–order tensor, called the stress tensor, σ. The stress components σ_{ij} are shown on three perpendicular planes in Figure 2.3–1b. Note that the cube that is used to indicate the stress components has no dimensions; it is a point cube so that all nine components are acting at the point P.

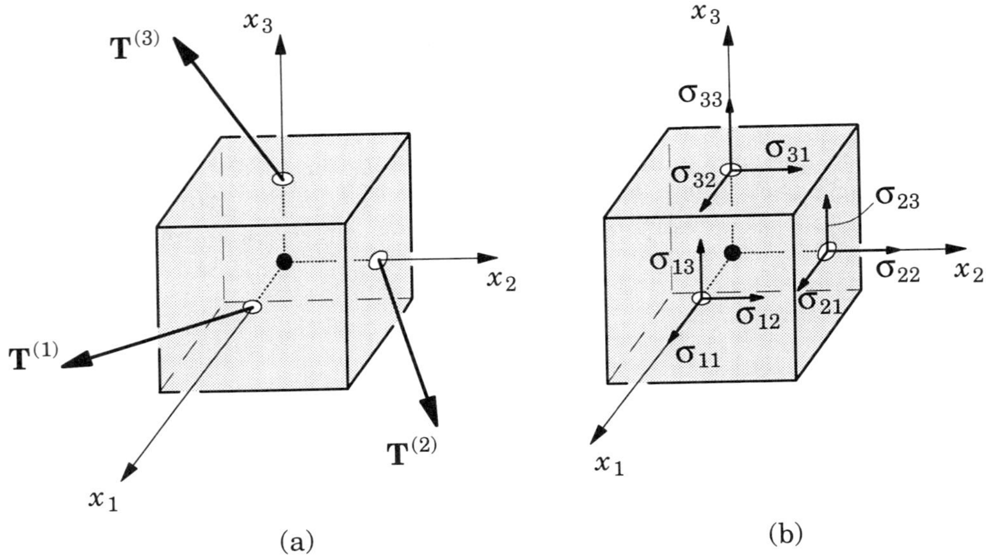

Figure 2.3–1: Stress vectors and components of stress tensor on coordinate planes. (a) Stress vectors. (b) Stress components.

It can be shown, using the linear momentum principle or Newton's second law of motion, that the stress vector $\mathbf{T}^{(\hat{\mathbf{n}})}$ on a plane with unit normal $\hat{\mathbf{n}}$ is related to the stress tensor σ by the relation

$$\mathbf{T}^{(\hat{\mathbf{n}})} = \hat{\mathbf{n}} \cdot \sigma = \sigma^{\mathrm{T}} \cdot \hat{\mathbf{n}} \quad \text{or} \quad \{T^{(\hat{\mathbf{n}})}\} = [\sigma]^{\mathrm{T}}\{n\} \tag{2.3 – 3}$$

Equation (3) is called the *Cauchy stress formula* and σ is known as the *Cauchy stress tensor.*

In the above discussion, stress vector $\mathbf{T}^{(\hat{\mathbf{n}})}$ at a point in a deformed body is measured as the force per unit area in the deformed body. The area element Δs in the deformed body corresponds to an area element ΔS in the reference configuration, in as much the same way $\mathbf{x}$ is the position of a material particle X in the deformed body whose position in the reference configuration was $\mathbf{X}$. Stress vectors can be *referred* to either the reference coordinate system (X_1, X_2, X_3) or the spatial coordinate system (x_1, x_2, x_3). The stress that is measured in the deformed configuration $\mathcal{C}$ and referred to the coordinate system in the deformed body is called the *Cauchy stress* or *true stress*, σ. The components σ_{ij} can be interpreted as the jth component of the force per unit area in the current configuration $\mathcal{C}$ acting on a surface segment whose outward normal at $\mathbf{x}$ is in the ith direction.

A suitable strain measure to use with the Cauchy stress tensor σ would be the infinitesimal strain tensor ε of Eq. (2.2–12). Since we defined strain tensor $\mathbf{E}$ as a function of the material point $\mathbf{X}$ in a reference state, the stress must also be expressed as a function of the material point. The stress measure that is used in nonlinear analysis of solid bodies is the second Piola–Kirchhoff stress tensor, which is measured in the deformed body but referred to the material coordinates, X_i. The components of the second Piola–Kirchhoff stress tensor are denoted by P_{ij}. The second Piola–Kirchhoff stress tensor is symmetric whenever the Cauchy stress tensor is symmetric, *i.e.*, when no distributed moments are present in the body. For small deformation problems, the difference between the two measures of stress disappears.

2.3–2 Equations of Motion

Consider a given mass of a material body $\mathcal{B}$, at any instant occupying a volume Ω bounded by a surface Γ. Suppose that the body is acted upon by external forces $\mathbf{T}$ (per unit surface area) and $\mathbf{f}$ (per unit volume). The principle of conservation of linear momentum states that the rate of change of the total linear momentum of a given continuous medium equals the vector sum of all the external forces acting on the body $\mathcal{B}$, which initially occupied a configuration $\mathcal{C}^0$, provided Newton's third law of action and reaction governs the internal forces. The principle leads to the following result:

$$\nabla \cdot \sigma + \mathbf{f} = \rho \frac{d^2\mathbf{x}}{dt^2} \qquad (2.3-4)$$

where ∇ is the gradient operator with respect to the coordinates $\mathbf{x}$ and ρ is the material density in the deformed body. In writing Eq. (4), all variables are assumed to be functions of the current coordinates $\mathbf{x}$ and time t. Note that the total time derivative $d\mathbf{x}/dt$ is equal to the velocity vector $\mathbf{v}$, and $d^2\mathbf{x}/dt^2$ is total time derivative of $\mathbf{v}$, which is equal to

$$\frac{d\mathbf{v}}{dt} = \frac{\partial \mathbf{v}}{\partial t} + \mathbf{v} \cdot \nabla \mathbf{v} \qquad (2.3-5a)$$

and in component form,

$$\frac{dv_i}{dt} = \frac{\partial v_i}{\partial t} + v_j \frac{\partial v_i}{\partial x_j} \qquad (2.3-5b)$$

Thus the total time derivative of a function in a spatial coordinate system consists of two parts: *instantaneous* (or partial) time derivative and *convective* time derivative. The partial time derivative is a measure of the change in the quantity with time, whereas the convective time derivative denotes the change at the spatial location brought by a particle that instantaneously occupies the position. In a material description the total time derivative is the same as the partial time derivative because the particle under consideration remains unchanged: $d\mathbf{x}/dt = \partial \mathbf{X}/\partial t + \partial \mathbf{u}/\partial t = 0 + \partial \mathbf{u}/\partial t$.

The equations of motion (4) can be expressed in terms of the second Piola–Kirchhoff stress tensor $\mathbf{P}$ by referring all variables to the reference coordinates $\mathbf{X}$

$$\nabla \cdot (\mathbf{P} \cdot \nabla \mathbf{x}) + \mathbf{f} = \rho_0 \frac{\partial^2 \mathbf{u}}{\partial t^2} \qquad (2.3-6a)$$

or

$$\nabla \cdot [\mathbf{P} \cdot (\mathbf{I} + \nabla \mathbf{u})] + \mathbf{f} = \rho_0 \frac{\partial^2 \mathbf{u}}{\partial t^2} \qquad (2.3-6b)$$

where ρ_0 is the density in the reference configuration. In component form, we have

$$\frac{\partial}{\partial X_j}\left(P_{jm}\frac{\partial x_i}{\partial X_m}\right) + f_i = \rho_0 \frac{\partial^2 u_i}{\partial t^2} \qquad (2.3-7a)$$

or

$$\frac{\partial}{\partial X_j}\left[P_{jm}\left(\delta_{im} + \frac{\partial u_i}{\partial X_m}\right)\right] + f_i = \rho_0 \frac{\partial^2 u_i}{\partial t^2} \qquad (2.3-7b)$$

Note that the equations of motion are nonlinear in both descriptions of motion. They are nonlinear in the spatial description because the material time derivative of velocity contains a *convective* part [*i.e.*, the second term in Eq. (5)] that depends on the velocity vector $\mathbf{v}$. For very slow motions, the convective term is negligible and the equations of motion become linear, unless the constitutive behavior is nonlinear. Equations of motion in the material description are nonlinear due to the large strains (*i.e.*, due to $\partial x_i/\partial X_m$). When the strains are small, *i.e.*, $\partial x_i/\partial X_m = \delta_{im}$, the equations of motion in terms of stresses become linear in the material description. Even when the stress equations of motion are linear, nonlinearity can come from the nonlinear strain–displacement relations (2.2–10) and/or nonlinear stress–strain equations.

When the displacements and displacement gradients are small, then the second Piola–Kirchhoff stress tensor $\mathbf{P}$ is approximately equal to the Cauchy stress tensor σ:

$$\frac{\rho_0}{\rho} \approx 1, \quad \mathbf{P} \approx \sigma$$

The equations of motion under the small–displacement assumption become

$$\nabla \cdot \sigma + \mathbf{f} = \rho_0 \frac{\partial^2 \mathbf{u}}{\partial t^2} \quad \left(\frac{\partial \sigma_{ji}}{\partial X_j} + f_i = \rho_0 \frac{\partial^2 u_i}{\partial t^2}\right) \qquad (2.3-8)$$

As stated above, Eq. (8) becomes nonlinear when expressed in terms of displacement components if the strain–displacement relations or stress–strain relations are nonlinear.

For coordinate systems other than the rectangular Cartesian systems, the strain–displacement relations and the equations of motion can be derived from the vector forms (2.2–8a), (2.2–11), (6b) and (8) by expressing σ or $\mathbf{P}$, $\mathbf{f}$, $\mathbf{u}$, and ∇ in the chosen coordinate system (see Exercises 7 and 9). The vector forms of the governing equations are invariant, *i.e.*, independent of the choice of the coordinate system.

The principle of conservation of moment of momentum, in the absence of any distributed moments, leads to the symmetry of the stress tensor:

$$\sigma = \sigma^{\mathrm{T}} \quad (\sigma_{ij} = \sigma_{ji}) \ , \quad \mathbf{P} = \mathbf{P}^{\mathrm{T}} \quad (P_{ij} = P_{ji}) \tag{2.3 – 9}$$

Thus there are only six independent components of the Cauchy and second Piola–Kirchhoff stress tensors. Since the Cauchy and second Piola–Kirchhoff stress tensors are second-order and symmetric, we may write them (especially when writing on a board or a paper) with a "double arrow" notation as

$$\overset{\leftrightarrow}{\sigma} \ , \quad \overset{\leftrightarrow}{P} \tag{2.3 – 10a}$$

This notation is meaningful and descriptive of the nature of these tensors; the notation indicates that the quantities are dyads (*i.e.*, having two base vectors) and they are symmetric:

$$\overset{\leftrightarrow}{\sigma} = \hat{\mathbf{e}}_i \sigma_{ij} \hat{\mathbf{e}}_j \ , \quad \overset{\leftrightarrow}{P} = \hat{\mathbf{e}}_i P_{ij} \hat{\mathbf{e}}_j \tag{2.3 – 10b}$$

Note that the equations of motion or equilibrium contain three equations relating six stress components and therefore cannot be solved for all six components uniquely. Additional equations are required. These include the strain–displacement relations discussed in Section 2.2 and constitutive relations or stress–strain relations to be discussed in the next section.

Example 2.3–1: ______________________________

Consider the following stress field in a body that is in equilibrium:

$$\sigma_{11} = c_1 X_1 + c_2 X_2 + c_3 X_1 X_2$$

$$\sigma_{12} = -\frac{c_3}{2}(X_2)^2 - c_1 X_2, \quad \sigma_{22} = c_4 X_1 + c_1 X_2$$

and all other components of stress are zero. We wish to determine if the stress field satisfies the equations of equilibrium in the presence of body forces, $f_1 = 0, f_2 = -c_1$, and $f_3 = 0$. We assume that the body experienced only a small deformation. We have

$$
\begin{aligned}
0 &= \frac{\partial \sigma_{11}}{\partial X_1} + \frac{\partial \sigma_{12}}{\partial X_2} + \frac{\partial \sigma_{13}}{\partial X_3} + f_1 \\
&= (c_1 + c_3 X_2) + (-c_1 - c_3 X_2) + 0 + 0
\end{aligned}
$$

$$
\begin{aligned}
0 &= \frac{\partial \sigma_{12}}{\partial X_1} + \frac{\partial \sigma_{22}}{\partial X_2} + \frac{\partial \sigma_{23}}{\partial X_3} + f_2 \\
&= 0 + c_1 + 0 + 0 + (-c_1)
\end{aligned}
$$

Thus, the first two equations of equilibrium are identically satisfied for any choice of constants, c_1, c_2, c_3, and c_4. The third equation of equilibrium is trivially satisfied.

Example 2.3–2:

Consider the cantilevered beam under an end load (see Figure 2.3–2). The bending moment at any distance X_1 is given by $M_{11} = P(L - X_1)$. Then the stress component σ_{11} can be calculated using the flexure stress formula from elementary strength of materials:

$$
\sigma_{11} = -\frac{M_{11} X_3}{I_{22}} = -\frac{P}{I_{22}}(L - X_1) X_3 \equiv c_1 (L - X_1) X_3 \tag{2.3 - 11}
$$

where I_{22} is area moment of inertia about the X_2–axis. Assuming a two-dimensional state of stress (with respect to the X_1 and X_3 coordinates) in the beam, we wish to determine the stress components σ_{13} and σ_{33} in the absence of body forces. Since the stress components σ_{12}, σ_{22}, and σ_{23} are assumed to be zero, the first equation of equilibrium yields

$$
\frac{\partial \sigma_{13}}{\partial X_3} = -\frac{\partial \sigma_{11}}{\partial X_1} = c_1 X_3
$$

Integration with respect to X_3 gives

$$
\sigma_{13} = \frac{c_1}{2}(X_3)^2 + f \tag{2.3 - 12}
$$

where f is a function of X_1 only. The second equation of equilibrium is trivially satisfied. The third equation of equilibrium gives

$$
\frac{\partial \sigma_{33}}{\partial X_3} = -\frac{\partial \sigma_{13}}{\partial X_1} = -\frac{df}{dX_1}
$$

Integration with respect to X_3 yields

$$
\sigma_{33} = -\frac{df}{dX_1} X_3 + g \tag{2.3 - 13}
$$

where g is a function of X_1 only.

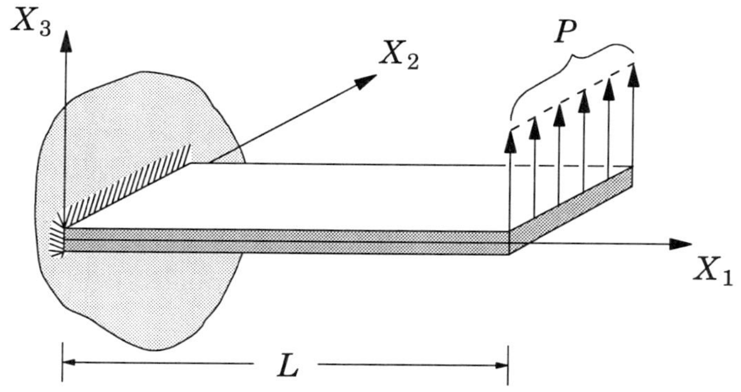

Figure 2.3–2: A cantilevered beam (*i.e.*, fixed at one end and no support at the other end) under an end load.

The functions f and g can be determined using the boundary conditions of the beam. Note that σ_{13} and σ_{33} must be zero on the top and bottom surfaces of the beam (*i.e.*, at $X_3 = \pm h/2$). Vanishing of σ_{33} at $X_3 = \pm h/2$ gives

$$-\frac{df}{dX_1}\frac{h}{2} + g = 0, \quad -\frac{df}{dX_1}(-\frac{h}{2}) + g = 0$$

which imply that

$$-\frac{df}{dX_1} = 0, \quad g = 0, \quad \text{or} \quad f = c_2 \quad \text{and} \quad g = 0$$

Vanishing of σ_{13} at $X_3 = \pm h/2$ gives

$$c_2 = -\frac{c_1 h^2}{8}$$

Thus the two–dimensional state of stress is given by

$$\sigma_{11} = -\frac{P}{I_{22}}(L - X_1)X_3, \; \sigma_{13} = \frac{Ph^2}{8I_{22}}\left[1 - \left(\frac{2X_3}{h}\right)^2\right], \; \sigma_{33} = 0 \qquad (2.3-14)$$

Since the stress field is derived from stress equilibrium equations, it is necessary to see if the strain compatibility condition (21) is satisfied. Suppose that the strains $\varepsilon_{11}, \varepsilon_{13}$, and ε_{33} are related to the stress components σ_{11}, σ_{13}, and σ_{33} by the relations (see the next section for details)

$$\begin{aligned}
\varepsilon_{11} &= C_{11}\sigma_{11} + C_{13}\sigma_{33} + C_{15}\sigma_{13} \\
\varepsilon_{33} &= C_{13}\sigma_{11} + C_{33}\sigma_{33} + C_{35}\sigma_{13} \\
\varepsilon_{13} &= C_{15}\sigma_{11} + C_{35}\sigma_{33} + C_{55}\sigma_{13}
\end{aligned}$$

Then

$$\varepsilon_{11} = -C_{11}\frac{P}{I_{22}}(L - X_1)X_3 + C_{15}\frac{Ph^2}{8I_{22}}\left[1 - \left(\frac{2X_3}{h}\right)^2\right]$$
$$\varepsilon_{33} = -C_{13}\frac{P}{I_{22}}(L - X_1)X_3 + C_{35}\frac{Ph^2}{8I_{22}}\left[1 - \left(\frac{2X_3}{h}\right)^2\right]$$
$$\varepsilon_{13} = -C_{15}\frac{P}{I_{22}}(L - X_1)X_3 + C_{55}\frac{Ph^2}{8I_{22}}\left[1 - \left(\frac{2X_3}{h}\right)^2\right] \qquad (2.3-15)$$

Substituting these strain components into the compatibility equation [see Eq. (2.2–21)]

$$\frac{\partial^2 \varepsilon_{11}}{\partial X_3^2} + \frac{\partial^2 \varepsilon_{33}}{\partial X_1^2} - 2\frac{\partial^2 \varepsilon_{13}}{\partial X_1 \partial X_3} = 0 \qquad (2.3-16)$$

we obtain

$$0 + 0 + 2C_{15}\frac{P}{I_{22}} = 0$$

Thus the strains are compatible only if $C_{15} = 0$.

In the next section we consider constitutive equations of a linear elastic body. These include isothermal elasticity, thermoelasticity, and electroelasticity. The constitutive equations presented in the next section are of considerable interest in the study of laminated composite plates.

2.4 Constitutive Equations

2.4–1 Generalized Hooke's Law

The kinematic relations and the mechanical and thermodynamic principles are applicable to any continuum irrespective of its physical constitution. Here we consider equations characterizing the individual material and its reaction to applied loads. These equations are called the *constitutive equations.* The formulation of the constitutive equations for a given material is guided by certain rules (*i.e.*, constitutive axioms). We will not discuss them here but we will review the linear constitutive relations for solids undergoing small deformations (see References 1–4 for details).

A material body is said to be *homogeneous* if the material properties are the same throughout the body. In a *heterogeneous* body, the material properties are a function of position. An *anisotropic* body is one which has different values of a material property in different directions at a point, *i.e.*, material properties are directionally dependent. An *isotropic* body is one for which every material property in all directions at a point is the same. An isotropic or anisotropic material can be nonhomogeneous or homogeneous.

A material body is said to be *ideally elastic* when the body recovers, under isothermal conditions, its original form completely upon removal of the forces causing deformation, and there is a one-to-one relationship between the state of stress and the state of strain. The constitutive equations described here do not include creep at constant stress and stress relaxation at constant strain. Thus, the material coefficients that specify the constitutive relationship between the stress and strain components are assumed to be constant during the deformation. This does not automatically imply that we neglect temperature effects on deformation. We account for the thermal expansion of the material which can produce strains or stresses as large as those produced by the applied mechanical forces. The dependence of the constitutive properties on temperature and strains can be accounted for when required. Here we review the basic constitutive equations of linear elasticity (*i.e.*, generalized Hooke's law) for small displacements. A discussion of the thermoelastic constitutive relations is also presented in the sequel.

The generalized Hooke's law relates the nine components of stress to the nine components of strain as

$$\sigma_{ij} = C_{ijk\ell}\varepsilon_{k\ell} \tag{2.4-1}$$

where $\varepsilon_{k\ell}$ are the infinitesimal strain components, σ_{ij} are the Cauchy stress components, and $C_{ijk\ell}$ are the material coefficients. The nine equations in Eq. (1) contain 81 material coefficients. However, due to the symmetry of both σ_{ij} and $\varepsilon_{k\ell}$, it follows that

$$C_{ijk\ell} = C_{jik\ell}, \quad C_{ijk\ell} = C_{ij\ell k} \tag{2.4-2}$$

Hence, there are only 36 independent elastic coefficients. The number of independent constants for a general anisotropic material can be reduced to 21, as described next.

Equation (1) can be expressed in the following convenient form using the single-subscript notation for stress and strain components and double-subscript notation for the elastic constants:

$$\begin{Bmatrix} \sigma_1 \\ \sigma_2 \\ \sigma_3 \\ \sigma_4 \\ \sigma_5 \\ \sigma_6 \end{Bmatrix} = \begin{bmatrix} C_{11} & C_{12} & C_{13} & C_{14} & C_{15} & C_{16} \\ C_{21} & C_{22} & C_{23} & C_{24} & C_{25} & C_{26} \\ C_{31} & C_{32} & C_{33} & C_{34} & C_{35} & C_{36} \\ C_{41} & C_{42} & C_{43} & C_{44} & C_{45} & C_{46} \\ C_{51} & C_{52} & C_{53} & C_{54} & C_{55} & C_{56} \\ C_{61} & C_{62} & C_{63} & C_{64} & C_{65} & C_{66} \end{bmatrix} \begin{Bmatrix} \varepsilon_1 \\ \varepsilon_2 \\ \varepsilon_3 \\ \varepsilon_4 \\ \varepsilon_5 \\ \varepsilon_6 \end{Bmatrix} \tag{2.4-3a}$$

or

$$\sigma_i = C_{ij}\varepsilon_j \tag{2.4-3b}$$

where C_{ij} are the elastic coefficients, and

$$\sigma_1 = \sigma_{11}, \ \sigma_2 = \sigma_{22}, \ \sigma_3 = \sigma_{33}, \ \sigma_4 = \sigma_{23}, \ \sigma_5 = \sigma_{13}, \ \sigma_6 = \sigma_{12}$$

$$\varepsilon_1 = \varepsilon_{11}, \ \varepsilon_2 = \varepsilon_{22}, \ \varepsilon_3 = \varepsilon_{33}, \ \varepsilon_4 = 2\varepsilon_{23}, \ \varepsilon_5 = 2\varepsilon_{13}, \ \varepsilon_6 = 2\varepsilon_{12} \qquad (2.4-4)$$

The two-subscript components C_{ij} are obtained from $C_{ijk\ell}$ by the following change of subscripts:

$$11 \to 1 \quad 22 \to 2 \quad 33 \to 3 \quad 23 \to 4 \quad 13 \to 5 \quad 12 \to 6$$

The resulting C_{ij} are also symmetric, $C_{ij} = C_{ji}$. Thus, there are only 21 (*i.e.*, sum of all upper or lower diagonal terms: 6+5+4+3+2+1=21) independent coefficients of the matrix $[C]$. The single-subscript notation (4) used for stresses and strains is called the *contracted notation* or the Voigt–Kelvin notation.

When elastic coefficients $C_{ijk\ell}$ at a point remain unchanged for every pair of coordinate systems which are mirror images of each other in a plane, that plane is called a *plane of material symmetry* at that point. If the elastic coefficients are invariant after a transformation $[T]$, then (*i.e.*, $C_{ijk\ell}$ are components of a fourth-order tensor)

$$C_{ijk\ell} = T_{im}T_{jn}T_{kp}T_{\ell q}C_{mnpq} \qquad (2.4-5)$$

The above equation provides relations between components of $\mathbf{C}$ and thus reduces the number of independent coefficients. We consider a few such transformations next.

In the following discussion, the coordinate systems used are solely for the discussion of material symmetries, and they differ in general from those used for the problem description, such as those discussed in Sections 2.2 and 2.3. In later chapters, we use the rectangular Cartesian coordinates (x, y, z) formulate the governing equations, and use (1,2,3) or (x_1, x_2, x_3) for the material coordinates.

2.4–2 Monoclinic Materials

When the elastic coefficients at a point have the same value for every pair of coordinate systems which are the mirror images of each other with respect to a certain plane, the material is called a *monoclinic material.* If the plane of symmetry is $x_3 = z = 0$, then the transformation matrix (which converts the old coordinate system to the new one) is given by

$$[T] = \begin{bmatrix} 1 & 0 & 0 \\ 0 & 1 & 0 \\ 0 & 0 & -1 \end{bmatrix} \quad (\text{or } T_{11} = T_{22} = 1, \ T_{33} = -1, \ T_{ij} = 0 \ for \ i \neq j) \quad (2.4-6)$$

Using this transformation in Eq. (5), we obtain

$$[C] = \begin{bmatrix} C_{11} & C_{12} & C_{13} & 0 & 0 & C_{16} \\ C_{12} & C_{22} & C_{23} & 0 & 0 & C_{26} \\ C_{13} & C_{23} & C_{33} & 0 & 0 & C_{36} \\ 0 & 0 & 0 & C_{44} & C_{45} & 0 \\ 0 & 0 & 0 & C_{45} & C_{55} & 0 \\ C_{16} & C_{26} & C_{36} & 0 & 0 & C_{66} \end{bmatrix} \qquad (2.4-7)$$

Clearly there are only 13 independent coefficients.

2.4–3 Orthotropic Materials

When three mutually orthogonal planes of material symmetry exist, the number of elastic coefficients is reduced to 9, and such materials are called *orthotropic.* The stress–strain relations for an orthotropic material take the form

$$\begin{Bmatrix} \sigma_1 \\ \sigma_2 \\ \sigma_3 \\ \sigma_4 \\ \sigma_5 \\ \sigma_6 \end{Bmatrix} = \begin{bmatrix} C_{11} & C_{12} & C_{13} & 0 & 0 & 0 \\ C_{12} & C_{22} & C_{23} & 0 & 0 & 0 \\ C_{13} & C_{23} & C_{33} & 0 & 0 & 0 \\ 0 & 0 & 0 & C_{44} & 0 & 0 \\ 0 & 0 & 0 & 0 & C_{55} & 0 \\ 0 & 0 & 0 & 0 & 0 & C_{66} \end{bmatrix} \begin{Bmatrix} \varepsilon_1 \\ \varepsilon_2 \\ \varepsilon_3 \\ \varepsilon_4 \\ \varepsilon_5 \\ \varepsilon_6 \end{Bmatrix} \qquad (2.4-8)$$

The elastic coefficients C_{ij} in Eq. (8) are related to the engineering constants, E_i, ν_{ij} and G_{ij} by

$$C_{11} = E_1 \cdot \frac{1-\nu_{23}\nu_{32}}{\Delta}, \quad C_{12} = E_1 \cdot \frac{\nu_{21}+\nu_{31}\nu_{23}}{\Delta} = E_2 \cdot \frac{\nu_{12}+\nu_{32}\nu_{13}}{\Delta}$$

$$C_{13} = E_1 \cdot \frac{\nu_{31}+\nu_{21}\nu_{32}}{\Delta} = E_3 \cdot \frac{\nu_{13}+\nu_{12}\nu_{23}}{\Delta}, \quad C_{22} = E_2 \cdot \frac{1-\nu_{13}\nu_{31}}{\Delta}$$

$$C_{23} = E_2 \cdot \frac{\nu_{32}+\nu_{12}\nu_{31}}{\Delta} = E_3 \cdot \frac{\nu_{23}+\nu_{21}\nu_{13}}{\Delta}, \quad C_{33} = E_3 \cdot \frac{1-\nu_{12}\nu_{21}}{\Delta}$$

$$C_{44} = G_{23}, \quad C_{55} = G_{13}, \quad C_{66} = G_{12}$$

$$\Delta = 1 - \nu_{12}\nu_{21} - \nu_{23}\nu_{32} - \nu_{31}\nu_{13} - 2\nu_{21}\nu_{32}\nu_{13} \qquad (2.4-9)$$

where

E_1, E_2, E_3 = Young's moduli in 1, 2, and 3 directions, respectively,

ν_{ij} = Poisson's ratio – defined as ratio of transverse strain in the jth direction to the axial strain in the ith direction, when stressed in the i-direction $-\varepsilon_j/\varepsilon_i$ (for $\sigma_i = \sigma$ and all other stresses are zero) $(i, j = 1, 2, 3)$

G_{23}, G_{13}, G_{12} = shear moduli in the 2-3, 1-3, and 1-2 planes, respectively.

In addition, the following *reciprocal relationship* holds among the principal moduli E_i and Poisson's ratios ν_{ij}:

$$\frac{\nu_{ij}}{E_i} = \frac{\nu_{ji}}{E_j} \quad (i, j = 1, 2, 3) \qquad (2.4-10)$$

Example 2.4–1:

The material properties of graphite fabric–carbon matrix layers, which are characterized as orthotropic, are:

$$E_1 = 25.1 \times 10^6 \text{ psi}, \quad E_2 = 4.8 \times 10^6 \text{ psi}, \quad E_3 = 0.75 \times 10^6 \text{ psi}$$

$$G_{12} = 1.36 \times 10^6 \text{ psi}, \quad G_{13} = 1.2 \times 10^6 \text{ psi}, \quad G_{23} = 0.47 \times 10^6 \text{ psi}$$

$$\nu_{12} = 0.036, \ \nu_{13} = 0.25, \ \nu_{23} = 0.171 \qquad (2.4-11a)$$

The matrix of elastic coefficients for the material can be calculated using Eq. (9) as

$$[C] = \begin{bmatrix} 25.16 & 0.2063 & 0.1934 & 0 & 0 & 0 \\ 0.2063 & 4.824 & 0.1304 & 0 & 0 & 0 \\ 0.1934 & 0.1304 & 4.832 & 0 & 0 & 0 \\ 0 & 0 & 0 & 0.47 & 0 & 0 \\ 0 & 0 & 0 & 0 & 1.2 & 0 \\ 0 & 0 & 0 & 0 & 0 & 1.36 \end{bmatrix} \text{msi} \qquad (2.4-11b)$$

2.4–4 Transversely Isotropic Materials

Transversely isotropic materials are of special importance in the study of fiber-reinforced composite materials. A material is said to be *transversely isotropic* if it is invariant with respect to an arbitrary rotation about a given axis. If the axis of rotation is x_3, then $C_{ijk\ell}$ must be invariant under the transformation. For this case, the transformation matrix is given by

$$[T] = \begin{bmatrix} \cos\theta & \sin\theta & 0 \\ -\sin\theta & \cos\theta & 0 \\ 0 & 0 & 1 \end{bmatrix} \qquad (2.4-12)$$

where θ is an arbitrary angle of rotation, measured in the counterclockwise direction, of the x_1x_2–plane about the x_3–axis. Clearly, a transversely isotropic material is a special case of an orthotropic material and whose material coefficients are the same with respect to axes 1 and 2. The material coefficient matrix takes the form

$$[C] = \begin{bmatrix} C_{11} & C_{12} & C_{13} & 0 & 0 & 0 \\ C_{12} & C_{11} & C_{13} & 0 & 0 & 0 \\ C_{13} & C_{13} & C_{33} & 0 & 0 & 0 \\ 0 & 0 & 0 & C_{44} & 0 & 0 \\ 0 & 0 & 0 & 0 & C_{44} & 0 \\ 0 & 0 & 0 & 0 & 0 & C_{66} \end{bmatrix}, \quad C_{66} = \frac{1}{2}(C_{11} - C_{12}) \qquad (2.4-13)$$

Thus, there are five independent constants that characterize transverse isotropy.

2.4–5 Isotropic Materials

Finally, when there exist no preferred directions in the material (*i.e.*, the material has infinite number of planes of material symmetry), the number of independent elastic coefficients reduces to 2. Such materials are called *isotropic*. In this special case, Eq. (1) can be expressed in the form

$$\sigma_{ij} = [\lambda \delta_{ij}\delta_{k\ell} + \mu(\delta_{ik}\delta_{j\ell} + \delta_{i\ell}\delta_{jk})]\varepsilon_{k\ell} \qquad (2.4-14)$$

where λ and μ are the Lamé elastic constants and δ_{ij} is the Kronecker delta. The Lamé constants are related to the shear modulus G, Young's modulus E, and Poisson's ratio ν as follows:

$$\mu \equiv G = \frac{E}{2(1+\nu)}, \quad \lambda = \frac{\nu E}{(1+\nu)(1-2\nu)} \qquad (2.4-15)$$

The coefficients C_{ij} for isotropic materials have the following meaning:

$$C_{11} = C_{22} = C_{33} = \lambda + 2\mu, \quad C_{12} = C_{13} = C_{23} = \lambda$$

$$C_{44} = C_{55} = C_{66} = \mu \qquad (2.4-16)$$

and all other C_{ij} are zero.

This completes the discussion of the constitutive equations for linear elastic materials in the absence of thermal effects. We will return to these relations in Chapter 4. In the next section we present constitutive equations of thermoelasticity and electroelasticty, which couple thermal and electrical effects to mechanical effects. The thermoelasticity and electroelasticity constitutive relations are of interest, for example, in the study of smart materials and structures. While a limited use of these relations is made in the present book, they are presented here for the sake of completeness.

2.5 Equations of Thermoelasticity and Electroelasticity

2.5–1 Thermodynamic Principles

Of the four principles of thermodynamics, the first law of thermodynamics and the second law of thermodynamics are important in the study of deformable solids. The first law of thermodynamics, also known as the *principle of conservation of energy*, states that the time rate of change of the total energy is equal to the sum of the rate of work done by applied forces and the change of heat content per unit time. The second law of thermodynamics places restrictions on the interconvertibility of heat and work done. For irreversible processes, the second law states that the entropy production is positive.

The thermodynamic principles can be expressed, in the Lagrangian description of deformation of solid bodies, as

$$\rho c_v \frac{\partial T}{\partial t} = -\nabla \cdot \mathbf{q} + Q + \sigma : \dot{\varepsilon} \tag{2.5-1}$$

where T is the temperature, $\mathbf{q}$ is the heat flux vector, Q is the internal heat generation (measured per unit volume), ρ is the density, c_v is the specific heat at constant volume or constant strain, σ is the stress tensor, and $\dot{\varepsilon}$ is the *strain rate tensor* (or time rate of the strain tensor).

Equation (1), termed the generalized *heat conduction equation*, is used to determine the temperature distribution in the body. The viscous dissipation couples the thermal problem to the stress problem. Even when the viscous dissipation is neglected, the thermal problem is coupled to the stress problem through constitutive relations, as explained in the next section.

The thermal problem for the solid requires the temperature or the heat flux to be specified on all parts of the boundary enclosing the heat transfer region as

$$T = \hat{T}(s,t) \quad \text{on} \quad \Gamma_T \tag{2.5-2}$$

$$\mathbf{n} \cdot \mathbf{q} + h_c(T - T_c) = \hat{q}_n(s,t) \quad \text{on} \quad \Gamma_q \tag{2.5-3}$$

where Γ is the total boundary enclosing the heat transfer region, $\Gamma = \Gamma_T \cup \Gamma_q$, $\Gamma_T \cap \Gamma_q =$ empty, h_c is the convective heat transfer coefficient, T_c is a reference (or sink) temperature for convective transfer, $\hat{q}_n$ is the specified boundary flux, and s denotes the position of a point on the boundary.

2.5–2 Thermoelasticity

The thermoelastic problem is governed by the strain–displacement equations of Section 2.2, equations of motion of Section 2.3, thermodynamic equations of Section 2.5–1, and the constitutive equations to be given in this section. The constitutive equation of the thermal problem is the well known Fourier's heat conduction law, which states that heat flux is proportional to the gradient of temperature:

$$\mathbf{q} = -\mathbf{k} \cdot \nabla T \quad \text{or} \quad q_i = -k_{ij} \frac{\partial T}{\partial x_j} \tag{2.5 – 4}$$

where $\mathbf{k}$ denotes the *thermal conductivity tensor* of order two. The negative sign in Eq. (4) indicates that heat flows from higher temperatures to lower temperatures.

The constitutive equation of thermoelasticity is derived by assuming the existence of the *Helmholtz free-energy function* $\Psi_0 = \Psi_0(\varepsilon_{ij}, T)$ (see [7,8])

$$\begin{aligned}\Psi_0(\varepsilon_{ij}, T) &= U_0 - \eta T \\ &= \frac{1}{2} C_{ijk\ell}\, \varepsilon_{ij}\, \varepsilon_{k\ell} - \beta_{ij}\, \varepsilon_{ij} \theta + \frac{1}{2T_0} \rho c_v \theta^2\end{aligned} \tag{2.5 – 5}$$

such that

$$\sigma_{ij} = \frac{\partial \Psi_0}{\partial \varepsilon_{ij}} = C_{ijk\ell}\, \varepsilon_{k\ell} - \beta_{ij}\, \theta \tag{2.5 – 6}$$

where $\theta = T - T_0$, T_0 is the reference temperature, η is the entropy density, and β_{ij} are material coefficients. It is assumed that η and σ_{ij} are initially zero. Equation (6) is known as the *Duhamel–Neumann law* for an anisotropic body. Inverting relations (6), we obtain

$$\varepsilon_{ij} = S_{ijk\ell}\, \sigma_{k\ell} + \alpha_{ij}\, \theta \tag{2.5 – 7}$$

where $S_{ijk\ell}$ are the elastic compliances, and α_{ij} are the thermal coefficients of expansion and related to β_{ij} by $\beta_{ij} = C_{ijk\ell}\, \alpha_{k\ell}$.

2.5–3 Hygrothermal Elasticity

Temperature and moisture concentration in fiber-reinforced composites cause reductions of both strength and stiffness [9–11]. Therefore, it is important to determine the temperature and moisture concentration in composite laminates under given initial and boundary conditions. As described in the previous section, the heat conduction problem described by equations (1)–(4) can be used to determine the temperature field.

The moisture concentration problem is mathematically similar to the heat transfer problem. The moisture concentration c in a solid is described by Fick's second law:

$$\frac{\partial c}{\partial t} = -\nabla \cdot \mathbf{q}_f + \phi_f \tag{2.5 – 8a}$$

$$\mathbf{q}_f = -\mathbf{D} \cdot \nabla c \qquad (2.5-8b)$$

where $\mathbf{D}$ denotes the *mass diffusivity tensor* of order two, $\mathbf{q}_f$ is the flux vector, and ϕ_f is the moisture source in the domain. The negative sign in Eq. (8b) indicates that moisture seeps from higher concentration to lower concentration. The boundary conditions involve specifying the moisture concentration or the flux normal to the boundary:

$$c = \hat{c}(s,t) \quad \text{on} \quad \Gamma_1 \qquad (2.5-9a)$$

$$\mathbf{n} \cdot \mathbf{q}_f = \hat{q}_f(s,t) \quad \text{on} \quad \Gamma_2 \qquad (2.5-9b)$$

where $\Gamma = \Gamma_1 \cup \Gamma_2, \Gamma_1 \cap \Gamma_2 =$ empty and quantities with a hat are specified functions on the respective boundaries.

The moisture–induced strains $\{\varepsilon\}^M$ are given by

$$\{\varepsilon\}^M = \{\alpha_M\}c \qquad (2.5-10)$$

where $\{\alpha_M\}$ is the vector of *coefficients of hygroscopic expansion*. Thus the hygrothermal strains have the same form as the thermal strains [see Eq. (7)]. The total strains are given by

$$\{\varepsilon\} = [S]\{\sigma\} + \{\alpha_T\}(T - T_0) + \{\alpha_M\}(c - c_0) \qquad (2.5-11)$$

where T_0 and c_0 are reference values from which the strains and stresses are measured. In view of the similarity between the thermal and moisture strains, we will use only thermal strains to show their contribution to governing equations in Chapter 5.

2.5–4 Electroelasticity

Electroelasticity deals with the phenomena caused by interactions between electric and mechanical fields. The *piezoelectric effect* is one such phenomenon, and it is concerned with the effect of the electric charge on the deformation [12]. A laminated structure with piezoelectric laminae receives actuation through an applied electric field, and the piezoelectric laminae send electric signals that are used to measure the motion or deformation of the laminate. In these problems, the electric charge that is applied to actuate a structure provides an additional body force to the stress analysis problem, much the same way a temperature field induces a body force through thermal strains.

The piezoelectric effect is described by the *polarization vector* $\mathbf{P}$, which represents the electric moment per unit volume or polarization charge per unit area. It is related to the stress tensor by the relation (see [12–14])

$$\mathbf{P} = \mathbf{d} \cdot \sigma \quad \text{or} \quad P_i = d_{ijk}\sigma_{jk} \tag{2.5 – 12}$$

where $\mathbf{d}$ is the third-order tensor of piezoelectric moduli. The inverse effect relates the electric field vector $\mathcal{E}$ to the linear strain tensor ε by

$$\varepsilon = \mathcal{E} \cdot \mathbf{d} \quad \text{or} \quad \varepsilon_{ij} = d_{kij}\mathcal{E}_k \tag{2.5 – 13}$$

Note that d_{kij} is symmetric with respect to indices i and j because of the symmetry of ε_{ij} (note that $i, j, k = 1, 2, 3$).

The *pyroelectic effect* is another phenomenon that relates temperature changes to polarization of a material. For a small temperature change ΔT, the change in polarization vector $\Delta\mathbf{P}$ is given by

$$\Delta\mathbf{P} = \mathbf{p}\Delta T \tag{2.5 – 14}$$

where $\mathbf{p}$ is the vector of pyroelectric coefficients.

The coupling between the mechanical, thermal, and electrical fields can be established using thermodynamical principles and Maxwell's relations. Analogous to the strain energy function U_0 for elasticity and the Helmholtz free-energy function Ψ_0 for thermoelasticity, we assume the existence of a function Φ_0

$$\begin{aligned}\Phi_0(\varepsilon_{ij}, \mathcal{E}_i, T) &= U_0 - \mathcal{E} \cdot \mathbf{D} - \eta T \\ &= \frac{1}{2}C_{ijk\ell}\, \varepsilon_{ij}\, \varepsilon_{k\ell} - e_{ijk}\, \varepsilon_{ij}\mathcal{E}_k - \beta_{ij}\, \varepsilon_{ij}\theta \\ &\quad - \frac{1}{2}\epsilon_{k\ell}\, \mathcal{E}_k\mathcal{E}_\ell - p_k\mathcal{E}_k\theta - \frac{\rho c_v}{2T_0}\theta^2\end{aligned} \tag{2.5 – 15a}$$

which is called the electric *Gibbs free-energy function* or *enthalpy function*, such that

$$\sigma_{ij} = \frac{\partial\Phi_0}{\partial\varepsilon_{ij}}, \quad D_i = -\frac{\partial\Phi_0}{\partial\mathcal{E}_i}, \quad \eta = -\frac{\partial\Phi_0}{\partial T} = -\frac{\partial\Phi_0}{\partial\theta} \tag{2.5 – 15b}$$

where σ_{ij} are the components of the stress tensor, D_i are the components of the electric displacement vector, and η is the entropy. Use of Eq. (15a) in Eq. (15b) gives the constitutive equations of a deformable piezoelectric medium:

$$\sigma_{ij} = C_{ijk\ell}\, \varepsilon_{k\ell} - e_{ijk}\mathcal{E}_k - \beta_{ij}\theta \tag{2.5 – 16a}$$

$$D_k = e_{ijk}\, \varepsilon_{ij} + \epsilon_{k\ell}\mathcal{E}_\ell + p_k\theta \tag{2.5 – 16b}$$

$$\eta = \beta_{ij}\, \varepsilon_{ij} + p_k\mathcal{E}_k + \frac{\rho c_v}{T_0}\theta \tag{2.5 – 16c}$$

where $C_{ijk\ell}$ are the elastic moduli, e_{ijk} are the piezoelectric moduli, ϵ_{ij} are the dielectric constants, p_k are the pyroelectric constants, β_{ij} are the temperature expansion coefficients, c_v is the specific heat per unit mass, and T_0 is the reference temperature. In single-subscript notation, Eq. (16) can be expressed as

$$\sigma_i = C_{ij}\varepsilon_j - e_{ik}\mathcal{E}_k - \beta_i\theta \quad (2.5-17a)$$

$$D_k = e_{kj}\varepsilon_j + \epsilon_{k\ell}\mathcal{E}_\ell + p_k\theta \quad (2.5-17b)$$

$$\eta = \beta_i\varepsilon_i + p_k\mathcal{E}_k + \frac{\rho c_v}{T_0}\theta \quad (2.5-17c)$$

Note that the range of summation in (17) is different for different terms: $i,j = 1,2,\cdots,6; k,\ell = 1,2,3$. For the general anisotropic material, there are 21 independent elastic constants, 18 piezoelectric constants, 6 dielectric constants, 3 pyroelectric constants, and 6 thermal expansion coefficients.

Maxwell's equation governing the electric displacement vector is given by

$$\nabla \cdot \mathbf{D} = 0 \quad (2.5-18)$$

It is often assumed that the electric field $\mathcal{E}$ is derivable from an electric scalar potential function ψ:

$$\mathcal{E} = -\nabla\psi \quad (2.5-19)$$

This assumption allows us to write Eq. (18), in view of Eq. (17b), as

$$\frac{\partial}{\partial x_1}\left(\epsilon_{11}\frac{\partial\psi}{\partial x_1}\right) + \frac{\partial}{\partial x_2}\left(\epsilon_{22}\frac{\partial\psi}{\partial x_2}\right) + \frac{\partial}{\partial x_3}\left(\epsilon_{33}\frac{\partial\psi}{\partial x_3}\right) + f_e = 0 \quad (2.5-20a)$$

where

$$f_e = -\frac{\partial}{\partial x_k}\left(e_{k\ell}\varepsilon_\ell + p_k\theta\right) \quad (2.5-20b)$$

2.6 Summary

In this chapter a review of the linear and nonlinear strain–displacement relations, equations of motion in terms of stresses and displacements, compatibility conditions on strains, and linear constitutive equations of elasticity, thermoelasticity and electroelasticity is presented. For the small displacements case, the strain–displacement relations, the equations of motion, the heat conduction equation, the moisture diffusion equation, the electric potential equation, and the constitutive relations for linear thermo–electro–elasticity in the rectangular Cartesian coordinates (x_1, x_2, x_3) are summarized here. From Chapter 4 onwards, we use (x_1, x_2, x_3) for the principal material coordinates, and (x, y, z) for the problem description. We use the notation (u, v, w) to denote the displacements, σ_{xx}, σ_{xy} and so on for stresses, and ε_{xx}, ε_{xy} and so on for strains in the (x, y, z) coordinate system.

The constitutive equations are written in terms of variables and coefficients in the principal materials coordinates (x_1, x_2, x_3).

Since the heat conduction, moisture diffusion, and electric potential are governed by the same form of differential equation, all three are stated here in the form of a single transport equation. In summarizing the equations, it is tacitly assumed that the temperature, moisture diffusion, and electric scalar potential are determined by solving the corresponding governing equations (*i.e.*, Fourier's heat conduction equation, Fick's second law, and Maxwell's equation). Of course, in general, these equations are coupled to each other through constitutive equations.

Strain–Displacement Equations:

$$\varepsilon_{xx} = \frac{\partial u}{\partial x}\,, \quad \varepsilon_{yy} = \frac{\partial v}{\partial y}\,, \quad \varepsilon_{zz} = \frac{\partial w}{\partial z}$$

$$\gamma_{yz} = \frac{\partial v}{\partial z} + \frac{\partial w}{\partial y}\,, \quad \gamma_{xz} = \frac{\partial u}{\partial z} + \frac{\partial w}{\partial x}\,, \quad \gamma_{xy} = \frac{\partial u}{\partial y} + \frac{\partial v}{\partial x} \tag{2.6 – 1}$$

Equations of Motion ($\sigma_{ij} = \sigma_{ji}$):

$$\frac{\partial \sigma_{xx}}{\partial x} + \frac{\partial \sigma_{xy}}{\partial y} + \frac{\partial \sigma_{xz}}{\partial z} + f_x = \rho_0 \frac{\partial^2 u}{\partial t^2}$$

$$\frac{\partial \sigma_{xy}}{\partial x} + \frac{\partial \sigma_{yy}}{\partial y} + \frac{\partial \sigma_{yz}}{\partial z} + f_y = \rho_0 \frac{\partial^2 v}{\partial t^2}$$

$$\frac{\partial \sigma_{xz}}{\partial x} + \frac{\partial \sigma_{yz}}{\partial y} + \frac{\partial \sigma_{zz}}{\partial z} + f_z = \rho_0 \frac{\partial^2 w}{\partial t^2} \tag{2.6 – 2}$$

Transport Equation ($\varphi = T, c,$ or ψ):

$$\frac{\partial}{\partial x}\left(a_{xx}\frac{\partial \varphi}{\partial x} + a_{xy}\frac{\partial \varphi}{\partial y}\right) + \frac{\partial}{\partial y}\left(a_{xy}\frac{\partial \varphi}{\partial x} + a_{yy}\frac{\partial \varphi}{\partial y}\right) + \frac{\partial}{\partial z}\left(a_{zz}\frac{\partial \varphi}{\partial z}\right) + f = a_0 \frac{\partial \varphi}{\partial t} \tag{2.6 – 3}$$

Constitutive Relations (Monoclinic Piezoelectric Materials)

$$\begin{Bmatrix} \sigma_1 \\ \sigma_2 \\ \sigma_3 \\ \sigma_4 \\ \sigma_5 \\ \sigma_6 \end{Bmatrix} = \begin{bmatrix} C_{11} & C_{12} & C_{13} & 0 & 0 & C_{16} \\ C_{12} & C_{22} & C_{23} & 0 & 0 & C_{26} \\ C_{13} & C_{23} & C_{33} & 0 & 0 & C_{36} \\ 0 & 0 & 0 & C_{44} & C_{45} & 0 \\ 0 & 0 & 0 & C_{45} & C_{55} & 0 \\ C_{16} & C_{26} & C_{36} & 0 & 0 & C_{66} \end{bmatrix} \begin{Bmatrix} \varepsilon_1 - \alpha_{11}\Delta T \\ \varepsilon_2 - \alpha_{22}\Delta T \\ \varepsilon_3 - \alpha_{33}\Delta T \\ \varepsilon_4 \\ \varepsilon_5 \\ \varepsilon_6 - 2\alpha_{12}\Delta T \end{Bmatrix}$$

$$+\begin{bmatrix} 0 & 0 & e_{31} \\ 0 & 0 & e_{32} \\ 0 & 0 & e_{33} \\ e_{14} & e_{24} & 0 \\ e_{15} & e_{25} & 0 \\ 0 & 0 & e_{36} \end{bmatrix} \begin{Bmatrix} \frac{\partial\psi}{\partial x} \\ \frac{\partial\psi}{\partial y} \\ \frac{\partial\psi}{\partial z} \end{Bmatrix} \qquad (2.6-4)$$

$$\begin{Bmatrix} D_1 \\ D_2 \\ D_3 \end{Bmatrix} = \begin{bmatrix} 0 & 0 & 0 & e_{14} & e_{15} & 0 \\ 0 & 0 & 0 & e_{24} & e_{25} & 0 \\ e_{31} & e_{32} & e_{33} & 0 & 0 & e_{36} \end{bmatrix} \begin{Bmatrix} \varepsilon_1 \\ \varepsilon_2 \\ \varepsilon_3 \\ \varepsilon_4 \\ \varepsilon_5 \\ \varepsilon_6 \end{Bmatrix}$$

$$-\begin{bmatrix} \epsilon_{11} & \epsilon_{12} & 0 \\ \epsilon_{12} & \epsilon_{22} & 0 \\ 0 & 0 & \epsilon_{33} \end{bmatrix}^{(k)} \begin{Bmatrix} \frac{\partial\psi}{\partial x} \\ \frac{\partial\psi}{\partial y} \\ \frac{\partial\psi}{\partial z} \end{Bmatrix} \qquad (2.6-5)$$

Equations (1)–(5) are to be solved for a given geometry, loading, and initial and boundary conditions.

Exercises

2.1 Write the position of an arbitrary point (x_1, x_2, x_3) in the deformed body (solid lines) in terms of its coordinates in the undeformed body (broken lines) and compute the nonlinear Lagrangian strains for the body shown in Figure P2.1.

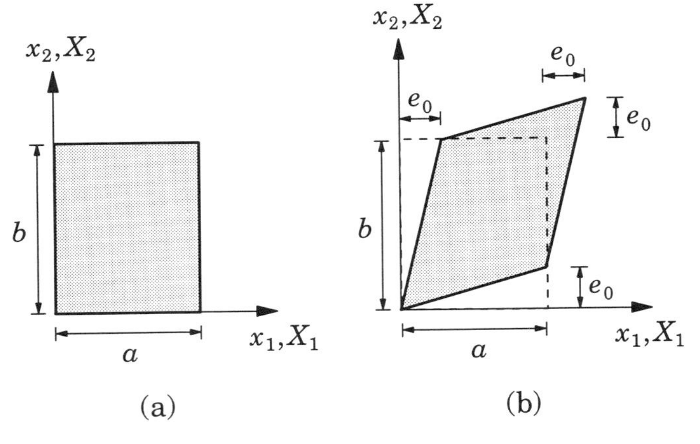

Figure P2.1

2.2 Write the position of an arbitrary point (x_1, x_2, x_3) in the deformed body (solid lines) in terms of its coordinates in the undeformed body (broken lines) and compute the nonlinear Lagrangian strains for the body shown in Figure P2.2.

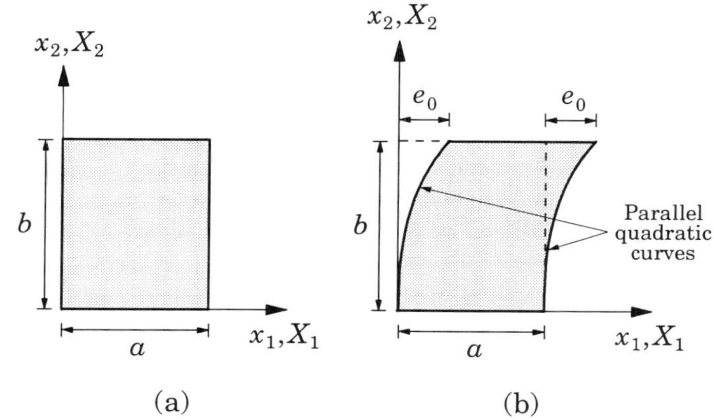

Figure P2.2

2.3 Compute the axial strain in the line element AB and the shear strain at point O of the rectangular block shown in Figure P2.3 using the engineering definitions.

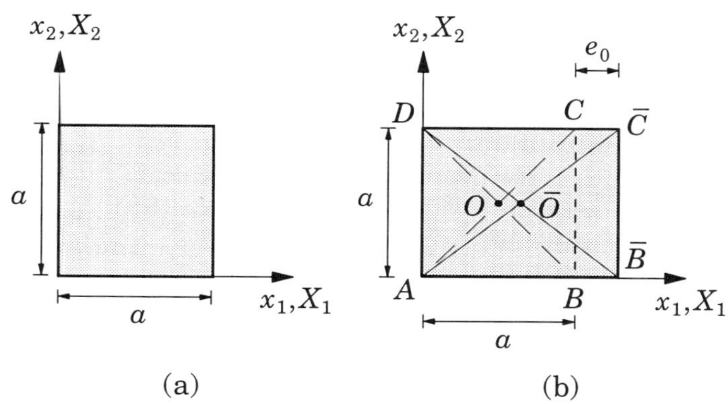

Figure P2.3

2.4 Compute the nonlinear strain components E_{ij} associated with the displacement field

$$u_1 = \frac{e_o}{b} X_2 \ , \ u_2 = \frac{e_o}{a} X_1 \ , \ u_3 = 0$$

where e_o, a, and b are constants.

2.5 Find the linear strains associated with the 2–D displacement field

$$
\begin{aligned}
u_1 &= -\frac{Px_1^2x_2}{2EI} + \frac{PL^2}{2EL}x_2 + \frac{(2+\nu)P}{6EI}x_2^3 \\
u_2 &= \frac{Ph^2(1+\nu)}{EI}(L-x_1) + \frac{\nu P}{2EI}x_1x_2^2 + \frac{P}{6EI}x_1^3 - \frac{PL^2}{2EI}x_1 + \frac{PL^3}{3EI}
\end{aligned}
$$

where P, h, ν, and EI are constants.

2.6 Find the linear strains associated with the 2–D displacement field

$$
\begin{aligned}
u_1 &= -c_0x_1x_2 + c_1x_2 + c_2x_3 + c_4 \\
u_2 &= \frac{1}{2}c_0\left[\nu\left(x_2^2 - x_1^2\right) + x_1^2\right] + c_4x_3 + c_5x_1 + c_6 \\
u_3 &= 0
\end{aligned}
$$

where $c_0, c_1, \cdots, c_6$ are constants.

2.7 Use the definition (2.2–11) and the vector form of the displacement field and the del operator (∇) in the cylindrical coordinate system

$$
\mathbf{u} = u_r\hat{\mathbf{e}}_r + u_\theta\hat{\mathbf{e}}_\theta + u_z\hat{\mathbf{e}}_z \quad \text{and} \quad \nabla = \hat{\mathbf{e}}_r\frac{\partial}{\partial r} + \hat{\mathbf{e}}_\theta\frac{1}{r}\frac{\partial}{\partial \theta} + \hat{\mathbf{e}}_z\frac{\partial}{\partial z}
$$

to compute the linear strain–displacement relations in the cylindrical coordinate system:

$$
\begin{aligned}
\varepsilon_{rr} &= \frac{\partial u_r}{\partial r}, \quad \gamma_{r\theta} = \frac{1}{r}\frac{\partial u_r}{\partial \theta} + \frac{\partial u_\theta}{\partial r} - \frac{u_\theta}{r} \\
\varepsilon_{\theta\theta} &= \frac{1}{r}\frac{\partial u_\theta}{\partial \theta} + \frac{u_r}{r}, \quad \gamma_{\theta z} = \frac{1}{r}\frac{\partial u_z}{\partial \theta} + \frac{\partial u_\theta}{\partial z} \\
\varepsilon_{zz} &= \frac{\partial u_z}{\partial z}, \quad \gamma_{rz} = \frac{\partial u_r}{\partial z} + \frac{\partial u_z}{\partial r}
\end{aligned}
$$

Note that the derivatives of $\hat{\mathbf{e}}_r$ and $\hat{\mathbf{e}}_\theta$ with respect to θ are nonzero:

$$
\frac{\partial \hat{\mathbf{e}}_r}{\partial \theta} = \hat{\mathbf{e}}_\theta, \quad \frac{\partial \hat{\mathbf{e}}_\theta}{\partial \theta} = -\hat{\mathbf{e}}_r
$$

2.8 Show that in order to have a valid displacement field corresponding to a given infinitesimal strain tensor ε, it must satisfy the compatibility relation

$$
\nabla \times (\nabla \times \varepsilon)^T = \mathbf{0} \quad \text{or} \quad \varepsilon_{imp}\varepsilon_{jnq}\varepsilon_{ij,mn} = 0
$$

2.9 Use the del operator (∇) and the dyadic form of σ in the cylindrical coordinate system (r, θ, z) to express the equations of motion (2.3–8) in the cylindrical coordinate system:

$$
\frac{\partial \sigma_{rr}}{\partial r} + \frac{1}{r}\frac{\partial \sigma_{r\theta}}{\partial \theta} + \frac{\partial \sigma_{rz}}{\partial z} + \frac{\sigma_{rr} - \sigma_{\theta\theta}}{r} + f_r = \rho_0\frac{\partial^2 u_r}{\partial t^2}
$$

$$\frac{\partial \sigma_{r\theta}}{\partial r} + \frac{1}{r}\frac{\partial \sigma_{\theta\theta}}{\partial \theta} + \frac{\partial \sigma_{\theta z}}{\partial z} + 2\frac{\sigma_{r\theta}}{r} + f_\theta = \rho_0 \frac{\partial^2 u_\theta}{\partial t^2}$$

$$\frac{\partial \sigma_{rz}}{\partial r} + \frac{1}{r}\frac{\partial \sigma_{\theta z}}{\partial \theta} + \frac{\partial \sigma_{zz}}{\partial z} + \frac{\sigma_{rz}}{r} + f_z = \rho_0 \frac{\partial^2 u_z}{\partial t^2}$$

2.10 Under the assumption that the inplane strains are small, *i.e.*, neglecting the products of stress components with the gradients of the displacements u_1 and u_2, show that the equations of motion (2.3–7b) take the form

$$\frac{\partial P_{11}}{\partial X_1} + \frac{\partial P_{21}}{\partial X_2} + \frac{\partial P_{31}}{\partial X_3} + f_1 = \rho_0 \frac{\partial^2 u_1}{\partial t^2}$$

$$\frac{\partial P_{12}}{\partial X_1} + \frac{\partial P_{22}}{\partial X_2} + \frac{\partial P_{32}}{\partial X_3} + f_2 = \rho_0 \frac{\partial^2 u_2}{\partial t^2}$$

$$\frac{\partial}{\partial X_1}\left(P_{11}\frac{\partial u_3}{\partial X_1} + P_{12}\frac{\partial u_3}{\partial X_2} + P_{13}\right) + \frac{\partial}{\partial X_2}\left(P_{21}\frac{\partial u_3}{\partial X_1} + P_{22}\frac{\partial u_3}{\partial X_2} + P_{23}\right)$$
$$+ \frac{\partial}{\partial X_3}\left(P_{31}\frac{\partial u_3}{\partial X_1} + P_{32}\frac{\partial u_3}{\partial X_2} + P_{33}\right) + f_3 = \rho_0 \frac{\partial^2 u_3}{\partial t^2}$$

References for Additional Reading

1. Eringen, A. C., *Mechanics of Continua*, John Wiley, New York, 1967.
2. Fung, Y. C., *Foundations of Solid Mechanics*, Prentice–Hall, Englewood Cliffs, New Jersey, 1965.
3. Malvern, L. E., *Introduction to the Mechanics of a Continuous Medium*, Prentice–Hall, Englewood Cliffs, New Jersey, 1969.
4. Reddy, J. N. and Rasmussen, M. L., *Advanced Engineering Analysis*, John Wiley, New York, 1982; reprinted by Krieger, Melbourne, Florida, 1990.
5. Reddy, J. N., *Energy and Variational Methods in Applied Mechanics*, John Wiley, New York, 1984.
6. Lekhnitskii, S. G., *Theory of Elasticity of an Anisotropic Elastic Body*, Holden–Day, San Francisco, 1963.
7. Nowinski, J. L., *Theory of Thermoelasticity with Applications*, Sijthoff & Noordhoff, Alphen aan den Rijn, Netherlands, 1978.
8. Carslaw, H. S. and Jaeger, J. C., *Conduction of Heat in Solids*, Second Edition, Oxford University Press, London, U. K., 1959.
9. Browning, C. E., Husman, G. E., and Whitney, J. M., "Moisture Effects in Epoxy Matrix Composites," *Composite Materials: Testing and Design, Fourth Conference,* ASTM STP **617**, 481–496, American Society for Testing and Materials, Philadelphia, PA, 1977.

10. Shen, C. H. and Springer, G. S., "Moisture Absorption and Desorption of Composite Materials," *Journal of Composite Materials,* **10**, 2–20 (1976).

11. Jost, W., *Diffusion in Solids, Liquids, and Gases,* Academic Press, New York, 1952.

12. Tiersten, H. F., *Linear Piezoelectric Plate Vibrations,* Plenum, New York, 1969.

13. Penfield, P., Jr. and Hermann, A. H., *Electrodynamics of Moving Media,* Research Monograph No. 40, The M. I. T. Press, Cambridge, Massachusetts, 1967.

14. Parton, V. Z. and Kudryavtsev, B. A., *Engineering Mechanics of Composite Structures,* CRC Press, Boca Raton, Florida, 1993.

15. Reddy, J. N. (ed.), *Mechanics of Composite Materials. Selected Works of Nicholas J. Pagano*, Kluwer Academic Publishers, Netherlands, 1994.

Chapter Three

Virtual Work Principles and Variational Methods

3.1 Virtual Work

3.1–1 Introduction

In the analytical description of a physical process, often assumptions are made concerning how the process works. The governing equations are then derived using the laws (or principles) of physics. The set of equations thus obtained is called a *mathematical model* of the process. For solid mechanics problems some of the laws of physics take several alternative forms. For example, the principle of conservation of linear momentum, which requires that the vector sum of all applied forces acting on a body be equal to the total time rate of momentum of the body (see Section 2.3), is known in mechanics as Newton's second law and it is also derivable from the *principle of virtual displacements*. In the former case, the vector sum of all forces and moments acting on a body is set to zero to obtain the equations of motion. In the latter, the work done by actual forces in moving through virtual displacements that are consistent with the geometric constraints of a body is set to zero to obtain the equations of motion. The latter is known as an energy principle. It is useful in (a) deriving governing equations and boundary conditions, and (b) obtaining approximate solutions by variational methods. The phrase "variational formulations" will be used here to include variational principles as well as integral statements that are equivalent to governing equations.

The use of Newton's laws to determine the governing equations of a structural problem requires the isolation of a typical volume element of the structure with all its applied and reactive forces (*i.e.*, the free-body diagram of the element). For simple mechanical systems for which the free-body diagram can be set up, the vector approach provides an easy and direct way of deriving the governing equations. However, for complicated systems the procedure becomes more cumbersome and intractable. In addition, the type of boundary conditions to be used in conjunction with the derived equations is not always clear. In such cases, energy principles provide alternative means to obtain the governing equations and their solution.

In the context of the present study, the principles of virtual work will be used to derive the equations of motion of laminated plates. The Rayleigh–Ritz and Galerkin methods will be used to determine solutions to certain plate problems. The virtual work principles and variational methods also have direct relation to the finite element method, which is recognized as the most powerful numerical analysis procedure in determining approximate solutions to all practical structural problems. We begin with the concepts of virtual displacements and forces and work and energy.

3.1–2 Virtual Displacements and Forces

From purely geometrical considerations, a given mechanical system can take many possible configurations consistent with the geometric constraints of the system. Of all the possible configurations, only one corresponds to the actual configuration; it is this configuration that satisfies Newton's second law (*i.e.*, equations of equilibrium or motion of the system). The set of configurations that satisfy the geometric constraints but not necessarily Newton's second law is called the *set of admissible configurations*. These configurations are restricted to a neighborhood of the true configuration so that they are obtained from infinitesimal variations of the true configuration. During such variations, the geometric constraints of the system are not violated and all the forces are fixed at their actual values. When a mechanical system experiences such variations in its configuration, it is said to undergo *virtual displacements* from its true or actual configuration. These displacements need not have any relationship to the actual displacements that might occur due to a change in the applied loads. The displacements are called virtual because they are *imagined* to take place (*i.e.*, hypothetical) while the actual loads acting at their fixed values. The virtual displacements at the boundary points at which the geometric conditions (or displacements) are specified are necessarily zero.

Analogous to the concept of virtual displacements, one can think of virtual forces on a system. The virtual forces must be a set of forces that are in equilibrium among themselves. These forces can be either internal or external, and need not have any relation to the actual forces of the system. For example, a bar fixed at one end and subjected to an axial load at the other end can be imagined to have a variety of virtual forces. One such set is a force δP applied at the right and left ends in opposite directions. For a cantilever beam of length L, a virtual force system consists of applying a virtual point load δF acting upward at the free end and downward at the fixed end, and a moment of $\delta M = \delta F \cdot L$ acting clockwise at the fixed end. The set $(\delta F,\ -\delta F,\ \delta M)$ of virtual forces are in self equilibrium as can be verified by summing virtual forces and moments.

3.1–3 External and Internal Virtual Work

The work done by the actual forces through virtual displacements or the work done by virtual forces through actual displacements is called *virtual work.* The virtual work done by actual forces $\mathbf{F}$ in a body Ω_0 in moving through the virtual displacements $\delta \mathbf{u}$ is given by

$$\delta W = \int_{\Omega_0} \mathbf{F} \cdot \delta \mathbf{u} \, dv \qquad (3.1-1a)$$

where dv denotes the volume element in the material body Ω_0. Here all variables are referred to the reference configuration $\mathcal{C}^0$. Similarly, the virtual work done by virtual forces $\delta \mathbf{F}$ in moving through the actual displacement $\mathbf{u}$ is

$$\delta W^* = \int_{\Omega_0} \delta \mathbf{F} \cdot \mathbf{u} \, dv \qquad (3.1-1b)$$

External Virtual Work

The external virtual work done due to virtual displacements $\delta \mathbf{u}$ in a solid body Ω_0 subjected to body forces $\mathbf{f}$ per unit volume and surface tractions $\mathbf{T}$ per unit area of the boundary Γ_σ is given by

$$\delta V = -\left(\int_{\Omega_0} \mathbf{f} \cdot \delta \mathbf{u} \, dv + \int_{\Gamma_\sigma} \mathbf{T} \cdot \delta \mathbf{u} \, ds \right) \qquad (3.1-2a)$$

where ds denotes a surface element and Γ_σ denotes the portion of the boundary on which stresses are specified. The negative sign in Eq. (2) indicates that the work is performed on the body. It is understood that the displacements are specified on the remaining portion $\Gamma_u = \Gamma - \Gamma_\sigma$ of the boundary Γ. Therefore, the virtual displacements are zero on Γ_u, irrespective of whether $\mathbf{u}$ is specified to be zero or not. For example, a bar fixed at one end ($x = 0$) and subjected to an axial load at the other end ($x = L$) can be imagined to have a virtual displacement $\delta u(x)$, provided $\delta u(0) = 0$, because the actual displacement is specified at $x = 0$.

The external virtual work done due to virtual body forces $\delta \mathbf{f}$ and surface tractions $\delta \mathbf{T}$ on Γ_u of a solid body Ω_0 in moving through the actual displacements $\mathbf{u}$ is given by

$$\delta V^* = -\left(\int_{\Omega_0} \delta \mathbf{f} \cdot \mathbf{u} \, dv + \int_{\Gamma_u} \delta \mathbf{T} \cdot \mathbf{u} \, ds \right) \qquad (3.1-2b)$$

Internal Virtual Work

The forces applied on a body cause it to deform. The relative movement of the material particles in the body can be measured in terms of strains and the body experiences internal stresses (see Sections 2.2 and 2.3). The forces associated with the stress field move the material particles through displacements corresponding to the strain field in the body, and hence work is done. The work done by these internal forces in moving through displacements of the material particles is called *internal work.* Note that the work done on the body is responsible for the internal work.

The internal virtual work due to the virtual displacement $\delta \mathbf{u}$ can be computed as follows. Suppose that an infinitesimal material element of volume $d\Omega \equiv dV = dX_1 dX_2 dX_3$ of the body experiences virtual strains δE_{ij} due to the virtual displacements δu_i, where [see Eq. (2.2–8b)]

$$\delta E_{ij} = \frac{1}{2}(\delta u_{i,j} + \delta u_{j,i} + u_{m,j}\ \delta u_{m,i} + u_{m,i}\ \delta u_{m,j}), \quad u_{i,j} \equiv \frac{\partial u_i}{\partial X_j} \qquad (3.1-3)$$

The work done by the force due to actual stress P_{11}, for example, in moving through the virtual displacement $\delta u_1 = \delta E_{11} dX_1$ is

$$P_{11} dX_2 dX_3 \cdot \delta E_{11} dX_1 = P_{11} \delta E_{11} dv$$

Here P_{ij} denote the components of the second Piola–Kirchhoff stress tensor and E_{ij} denote the components of the Green–Lagrange strain tensor, both are measured in the deformed body but referred to the undeformed configuration. Similarly, the work done by the force due to stress P_{12} in shearing the body is

$$P_{12} dX_2 dX_3 \cdot 2\delta E_{12} dX_1 = P_{12} 2\delta E_{12} dv$$

Thus, the total virtual work done by forces due to all the stresses in a volume element (that originally occupied the material element dv) in moving through their respective displacements is

$$(P_{11} \cdot \delta E_{11} + P_{22} \cdot \delta E_{22} + \cdots + P_{12} \cdot 2\delta E_{12}) dv = P_{ij} \cdot \delta E_{ij} dv \qquad (3.1-4)$$

The total internal virtual work done is obtained by integrating the above expression over the entire volume of the body

$$\delta U = \int_{\Omega_0} P_{ij}\ \delta E_{ij}\ dv \qquad (3.1-5)$$

Equation (5) is valid for any material body irrespective of its constitutive behavior. The expression in Eq. (5) is called the *virtual strain energy* of a deformable body.

The internal virtual work done by virtual stresses δP_{ij} in moving through the actual strains E_{ij} is

$$\delta U^* = \int_{\Omega_0} E_{ij}\ \delta P_{ij}\ dv \qquad (3.1-6)$$

The expression in Eq. (6) is also known as the *virtual complementary strain energy*. The virtual forces $(\delta f_i, \delta T_i)$ and virtual stresses (δP_{ij}) should be such that the stress equilibrium equations (2.3–7b) and stress boundary conditions (2.3–3) are satisfied:

$$[\delta P_{jm}(\delta_{im} + u_{i,m})]_{,j} + \delta f_i = 0 \quad \text{in} \quad \Omega_0$$

$$[\delta P_{jm}(\delta_{im} + u_{i,m})]n_j = \delta T_i \quad \text{on} \quad \Gamma_\sigma \tag{3.1 - 7}$$

3.2 The Variational Operator and Functionals

3.2–1 Properties of the Variational Operator

The delta symbol δ used in conjunction with virtual displacements and forces can be interpreted as an operator, called the *variational operator*. It is used to denote a variation (or change) in a given quantity, *i.e.*, δu denotes a variation in u. Thus δ is an operator that produces virtual change or variation δu in a dependent variable u, in much the same way as dx denotes a change in x. The operator proves to be very useful in constructing virtual work statements and deriving governing equations from virtual work principles, as will be shown shortly.

Let $\mathbf{u}$ be the true displacement vector that defines an actual configuration of a given material body, and suppose that $\mathbf{u} = \hat{\mathbf{u}}$ on portion Γ_u of the total boundary Γ. Then an admissible configuration is of the form

$$\bar{\mathbf{u}} = \mathbf{u} + \alpha \mathbf{v}, \quad \alpha \text{ is a small real number} \tag{3.2 - 1}$$

where $\mathbf{v}$ is an arbitrary vector function. Since $\bar{\mathbf{u}}$ also satisfies the geometric condition $\bar{\mathbf{u}} = \hat{\mathbf{u}}$ on Γ_u, it follows that

$$\mathbf{v} = \mathbf{0} \quad \text{on} \quad \Gamma_u \tag{3.2 - 2}$$

Since α is small, $\alpha \mathbf{v}$ is an infinitesimal variation of the displacement vector $\mathbf{u}$; the variation is consistent with the geometric constraints of the system in the sense that $\mathbf{v}$ *satisfies the homogeneous form of the specified displacement boundary conditions.* The set of all such functions $\mathbf{v}$ is called the *set of admissible variations.*

Equation (1) defines a set of varied configurations; an infinite number of configurations $\bar{\mathbf{u}}$ can be generated for a fixed $\mathbf{v}$ by assigning different values to α. All these paths satisfy the specified geometric boundary conditions on boundary Γ_u. For example, consider the axial deformation of a bar, and let $u(x)$ be the true configuration with $u(0) = \hat{u}$. Then all functions $\bar{u}$ such that $\bar{u}(0) = \hat{u}$ are admissible configurations (see Figure 3.1–1), and all functions

v such that $v(0) = 0$ are admissible variations. Clearly, $ax + bx^2 + \cdots$ for any arbitrary constants $a, b, \cdots$ is an admissible variation. For a cantilever beam, *i.e.*, a beam fixed at one end ($x = 0$) and free at the other end ($x = L$), an admissible variation is of the form $v = ax^2 + bx^3 + \cdots$ (see Figure 3.1–2).

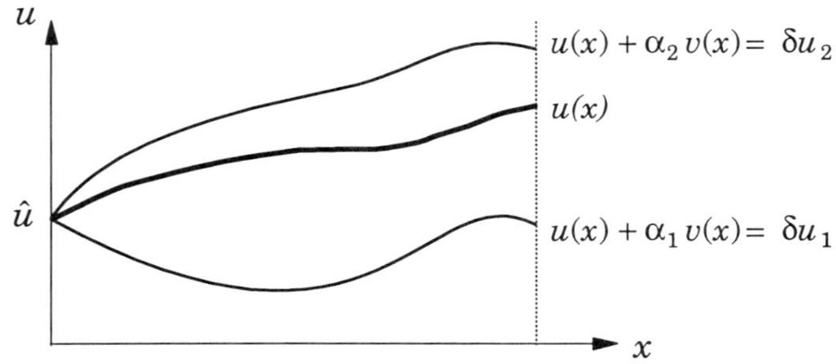

Figure 3.1–1: Set of admissible configurations of a deformable bar.

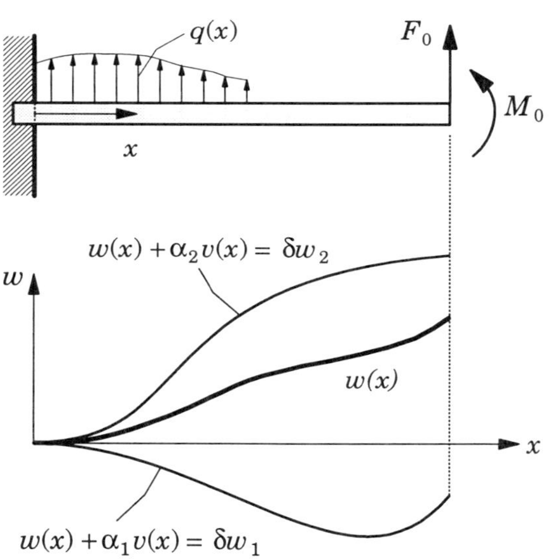

Figure 3.1–2: A cantilever beam and admissible configurations.

For any $\mathbf{v}$ all configurations $\bar{\mathbf{u}}$ of the form in Eq. (1) reduce to the actual configuration ($\mathbf{u}$) when α is set to zero. Therefore for any *fixed* position $\mathbf{x}$, $\alpha\mathbf{v}$ can be viewed as a change or *variation* in the actual configuration $\mathbf{u}$. This variation is denoted by $\delta\mathbf{u}$:

$$\delta \mathbf{u} \equiv \alpha \mathbf{v} \tag{3.2 – 3}$$

and $\delta \mathbf{u}$ is called the *first variation* of $\mathbf{u}$. We have, by definition (3), the identity

$$\delta(\nabla \mathbf{u}) = \alpha(\nabla \mathbf{v}) = \nabla(\alpha \mathbf{v}) = \nabla(\delta \mathbf{u}) \tag{3.2 – 4a}$$

or

$$\delta(u_{i,j}) = \alpha v_{i,j} = (\alpha v_i)_{,j} = (\delta u_i)_{,j} \tag{3.2 – 4b}$$

where, for example, $u_{i,j} = \partial u_i / \partial x_j$. Thus the differential and the variational operators commute: $(\delta)(\nabla)(\cdot) = (\nabla)(\delta)(\cdot)$.

There is an analogy between the variational operator δ and the total differential operator d. To see this, consider a function F of the dependent variable u and its derivative $u' = du/dx$ in one dimension. The total differential of F, for fixed x, is

$$dF = \frac{\partial F}{\partial x} dx + \frac{\partial F}{\partial u} du + \frac{\partial F}{\partial u'} du' \tag{3.2 – 5}$$

The variation of F is

$$\delta F = \frac{\partial F}{\partial u} \cdot \alpha v + \frac{\partial F}{\partial u'} \cdot \alpha v' + \frac{1}{2} \frac{\partial^2 F}{\partial u^2} (\alpha v)^2 + \frac{1}{2} \left(\frac{\partial^2 F}{\partial u'^2} \right) (\alpha v')^2 + \cdots \tag{3.2 – 6a}$$

Since α is very small, terms involving squares and higher powers of α can be neglected. We have

$$\delta F = \frac{\partial F}{\partial u} \delta u + \frac{\partial F}{\partial u'} \delta u' \tag{3.2 – 6b}$$

Since x is fixed during the variation of u to $u + \delta u$, we have $dx = 0$ in Eq. (5) and the analogy between δF in Eq. (6b) and dF in Eq. (5) becomes apparent: the variational operator, δ, *is a differential operator with respect to the dependent variable*, u. Indeed, the laws of variation of sums, products, ratios, powers, and so forth, are completely analogous to the corresponding laws of differentiation. For example, if $F_1 = F_1(u)$ and $F_2 = F_2(u)$, we have

(a) $\delta(F_1 \pm F_2) = \delta F_1 \pm \delta F_2$

(b) $\delta(F_1 \; F_2) = \delta F_1 \; F_2 + F_1 \; \delta F_2$

(c) $\delta(\frac{F_1}{F_2}) = (\delta F_1 \; F_2 - F_1 \; \delta F_2)/F_2^2$

(d) $\delta(F_1)^n = n(F_1)^{n-1} \delta F_1$ (3.2–7a)

If $G = G(u, v, w)$ is a function of several dependent variables (and possibly their derivatives), the total variation is the sum of partial variations:

$$\delta G = \delta_u G + \delta_v G + \delta_w G \tag{3.2 - 7b}$$

where, for example, δ_u denotes the partial variation of G with respect to u.

The variational operator can be interchanged with differential and integral operators (commutativity):

$$\text{(a)} \quad \delta\left(\frac{du}{dx}\right) = \alpha\frac{dv}{dx} = \frac{d}{dx}(\alpha v) = \frac{d}{dx}(\delta u)$$
$$\text{(b)} \quad \delta\left(\int_\Omega u \; d\Omega\right) = \alpha\int_\Omega v \; d\Omega = \int_\Omega \alpha v \; d\Omega = \int_\Omega \delta u \; d\Omega \tag{3.2 - 8}$$

3.2–2 Functionals

Integral expressions whose integrands are functions of dependent variables and their derivatives are called functionals. Mathematically, a *functional* is a real number (or scalar) obtained by operating on functions (dependent variables) from a given set (or vector space). Thus, a functional $I(\cdot)$ is an operator which maps functions u of a vector space H into a real number $I(u)$ in the set of real numbers, R:

$$I : H \to R \tag{3.2 - 9}$$

For example, the integral expression

$$I(u) = \int_0^L \left[au(x) + bu'(x) + cu''(x)\right] dx \tag{3.2 - 10}$$

qualifies as a functional for all integrable and square-integrable functions $u(x)$. Note that $I(u)$ is a number whose value depends on the choice of u.

A functional is said to be *linear* if

$$I(\alpha u + \beta v) = \alpha I(u) + \beta I(v) \tag{3.2 - 11}$$

for all constants α and β and dependent variables u and v. A *quadratic functional* is one which satisfies the relation

$$I(\alpha u) = \alpha^2 I(u) \tag{3.2 - 12}$$

for all constants α and dependent variable u.

The first variation of a functional of u (and its derivatives) can be calculated using the definition in Eq. (6b). Let $I(u)$ denote the integral defined in the interval (a, b)

$$I(u) = \int_a^b F(x, u, u') \; dx \tag{3.2 - 13}$$

where F is a function, in general, of x, u and $du/dx \equiv u'$. The first variation of the functional I is

$$\delta I = \delta \int_a^b F\ dx = \int_a^b \delta F\ dx = \int_a^b \left(\frac{\partial F}{\partial u} \delta u + \frac{\partial F}{\partial u'} \delta u' \right) dx \qquad (3.2-14)$$

where Eq. (6b) is used in arriving at the last step. Thus, the variation of a functional can be readily calculated.

3.3 Extrema of Functionals

3.3–1 Fundamental Lemma of Variational Calculus

The *fundamental lemma of calculus of variations* is useful in obtaining differential equations from integral statements. The lemma can be stated as follows: for any integrable function G, if the statement

$$\int_a^b G \cdot \eta\ dx = 0 \qquad (3.3-1)$$

holds for any arbitrary continuous function $\eta(x)$, for all x in (a, b), then it follows that $G = 0$ in (a, b). A mathematical proof of the lemma can be found in most books on variational calculus. A simple proof of the lemma follows. Since η is arbitrary, it can be replaced by G. We have

$$\int_a^b G^2\ dx = 0$$

Since an integral of a positive function is positive, the above statement implies that $G = 0$.

A more general statement of the fundamental lemma is as follows: If η is arbitrary in $a < x < b$ and $\eta(a)$ is arbitrary, then the statement

$$\int_a^b G\eta\ dx + B(a)\eta(a) = 0 \qquad (3.3-2)$$

implies that $G = 0$ in $a < x < b$ and $B(a) = 0$.

3.3–2 Euler–Lagrange Equations

Consider the question of finding the extremum (*i.e.*, minimum or maximum) of the functional

$$I(u) = \int_a^b F(x, u, u')\ dx, \quad u(a) = u_a, \quad u(b) = u_b \qquad (3.3-3)$$

The necessary condition for the functional to have a minimum or maximum is (analogous to minima or maxima of functions) that its first variation be zero:

$$\delta I = 0 \tag{3.3-4}$$

In most of our study in this book we shall be interested in the use of Eq. (4), not in finding the extremum of $I(u)$, because condition (4) provides the means to the determination of the governing equations and their solution by the variational methods. We will not be interested in determining the maximum or minimum value of the functional itself.

Returning to the condition $\delta I = 0$, we have

$$0 = \int_a^b \left(\frac{\partial F}{\partial u} \delta u + \frac{\partial F}{\partial u'} \delta u' \right) \, dx$$

Note that $\delta u' = \delta(du/dx) = d(\delta u)/dx$. We cannot use the fundamental lemma in the above equation because it is not in the form of Eq. (1) or Eq. (2). To recast the above equation in the form of Eq. (1), we integrate the second term by parts and obtain

$$\begin{aligned} 0 &= \int_a^b \left(\frac{\partial F}{\partial u} \delta u + \frac{\partial F}{\partial u'} \delta u' \right) \, dx \\ &= \int_a^b \left(\frac{\partial F}{\partial u} \delta u + \frac{\partial F}{\partial u'} \frac{d\delta u}{dx} \right) \, dx \\ &= \int_a^b \left[\frac{\partial F}{\partial u} - \frac{d}{dx} \left(\frac{\partial F}{\partial u'} \right) \right] \delta u \; dx + \left[\frac{\partial F}{\partial u'} \delta u \right]_a^b \end{aligned} \tag{3.3-5}$$

Let us first examine the boundary expression:

$$\left[\frac{\partial F}{\partial u'} \right] \cdot \delta u$$

There are two parts to this expression: a varied quantity and its coefficient. The variable u that is subjected to variation is called the *primary variable*. The coefficient of the varied quantity, *i.e.*, the expression next to δu in the boundary term is called a *secondary variable*. The product of the primary variable (or its variation) with the secondary variable often represents the work done (or virtual work done). The specification of the primary variable at a boundary point is termed the *essential boundary condition*, and the specification of the secondary variable ($\partial F / \partial u'$) is called the *natural boundary condition*. In solid mechanics, these are known as the *geometric* and *force* boundary conditions, respectively. All admissible variations must satisfy the homogeneous form of the essential (or geometric) boundary conditions: $\delta u(a) = 0$ and $\delta u(b) = 0$. Elsewhere, $a < x < b$, δu is arbitrary.

Returning to Eq. (5), we note that the boundary terms drop out because of the conditions on δu. We have

$$0 = \int_a^b \left[\frac{\partial F}{\partial u} - \frac{d}{dx}\left(\frac{\partial F}{\partial u'}\right)\right] \delta u \, dx$$

which must hold for any δu in (a,b). In view of the fundamental lemma of calculus of variations ($\eta = \delta u$), it follows that

$$G \equiv \frac{\partial F}{\partial u} - \frac{d}{dx}\left(\frac{\partial F}{\partial u'}\right) = 0 \ \text{ in } \ a < x < b \qquad (3.3-6)$$

Thus the necessary condition for $I(u)$ to be an extremum at $u = u(x)$ is that $u(x)$ be the solution of Eq. (6). Equation (6) is known as the *Euler–Lagrange equation* associated with the functional $I(u)$, *i.e.*, $\delta I = 0$ is equivalent to Eq. (6).

If $u(a) = u_a$ and $\delta u(b)$ is arbitrary (*i.e.*, $u(a)$ is specified but u is not specified at $x = b$), then $\delta u(a) = 0$ and we have from Eq. (5) the result

$$0 = \int_a^b \left[\frac{\partial F}{\partial u} - \frac{d}{dx}\left(\frac{\partial F}{\partial u'}\right)\right] \delta u \, dx + \left(\frac{\partial F}{\partial u'}\right)_{x=b} \delta u(b) \qquad (3.3-7)$$

Since δu is arbitrary in (a,b) and $\delta u(b)$ is arbitrary, the above equation implies, in view of Eq. (2), that both the integral expression and the boundary term be zero separately:

$$\frac{\partial F}{\partial u} - \frac{d}{dx}\left(\frac{\partial F}{\partial u'}\right) = 0, \quad a < x < b \qquad (3.3-8a)$$

$$\left(\frac{\partial F}{\partial u'}\right) = 0 \ \text{ at } x = b \qquad (3.3-8b)$$

Both Eq. (8a) and Eq. (8b) are called the Euler–Lagrange equations. Note that the boundary conditions that are a part of the Euler–Lagrange equations always belong to the class of natural boundary conditions.

3.4 Virtual Work Principles

3.4–1 Introduction

Recall that the virtual work due to virtual displacements is the work done by actual forces in displacing the body through virtual displacements that are consistent with the geometric constraints. All applied forces are kept constant during the virtual displacements. Consider a rigid body acted upon by a set of applied forces $\mathbf{F}_1$, $\mathbf{F}_2$, ...$\mathbf{F}_n$, and suppose that the points of application of these forces are subjected to the virtual displacements $\delta\mathbf{u}_1$, $\delta\mathbf{u}_2$, ..., $\delta\mathbf{u}_n$, respectively. The virtual displacement $\delta\mathbf{u}_i$ has no relation to $\delta\mathbf{u}_j$ for $i \neq j$. The external virtual work done by the virtual displacements is

$$\delta V = -[\mathbf{F}_1 \cdot \delta\mathbf{u}_1 + \mathbf{F}_2 \cdot \delta\mathbf{u}_2 + \cdots + \mathbf{F}_n \cdot \delta\mathbf{u}_n] = -\sum_{i=1}^{n} \mathbf{F}_i \cdot \delta\mathbf{u}_i \qquad (3.4-1)$$

The internal virtual work done δU is zero because a rigid body does not undergo any strains (hence virtual strains are zero). In addition, the virtual displacements $\delta\mathbf{u}_1$, $\delta\mathbf{u}_2$, $\cdots$, $\delta\mathbf{u}_n$ should all be the same, say $\delta\mathbf{u}$, for a rigid body. Thus, we have

$$\delta V = -\sum_{i=1}^{n} \mathbf{F}_i \cdot \delta\mathbf{u}_i = -\left(\sum_{i=1}^{n} \mathbf{F}_i\right) \cdot \delta\mathbf{u} \quad \text{and} \quad \delta U = 0 \qquad (3.4-2)$$

But by Newton's second law, the vector sum of the forces acting on a body in static equilibrium is zero. This implies that the total virtual work, $\delta U + \delta V$, is equal to zero. Thus, for a body in equilibrium the total virtual work done due to virtual displacements is zero. This statement is known as *the principle of virtual displacements.* The principle also holds for continuous, deformable bodies, for which δU is not zero. In this section the principle of virtual displacements and its special case are described, since they play an important role in the formulation of theories (*e.g.*, plate theories) and their analysis by variational methods of approximation.

3.4–2 The Principle of Virtual Displacements

Consider a continuous body $\mathcal{B}$ in equilibrium under the action of body forces $\mathbf{f}$ and surface tractions $\mathbf{T}$. Let the reference configuration be the initial configuration $\mathcal{C}^0$, whose volume is denoted as Ω_0. Suppose that over portion Γ_u of the total boundary Γ of the region Ω_0 the displacements are specified to be $\hat{\mathbf{u}}$, and on portion Γ_σ the tractions are specified to be $\hat{\mathbf{T}}$. The boundary portions Γ_u and Γ_σ are disjoint (*i.e.*, do not overlap), and their sum is the total boundary Γ. Let $\mathbf{u}$ be the displacement vector corresponding to the equilibrium configuration of the body, and let $\mathbf{P}$ and $\mathbf{E}$ be the associated second Piola–Kirchhoff stress and Green–Lagrange strain tensors, respectively. In the case of small strains and small displacements, we use σ and ε in place of $\mathbf{P}$ and $\mathbf{E}$, respectively. The set of admissible configurations are defined by sufficiently differentiable functions that satisfy the geometric boundary conditions: $\mathbf{u} = \hat{\mathbf{u}}$ on Γ_u.

If the body is in equilibrium, then of all admissible configurations, the actual one corresponding to the equilibrium configuration makes the total virtual work done zero. In order to determine the equations governing the equilibrium configuration $\mathcal{C}$, we let the body experience a virtual displacement $\delta\mathbf{u}$ from the true configuration $\mathcal{C}$. The virtual displacements are arbitrary, continuous functions except that they satisfy the homogeneous form of geometric boundary conditions, *i.e.*, they must belong to the set of admissible variations.

The principle of virtual displacements can be stated as: *if a continuous body is in equilibrium, the virtual work of all actual forces in moving through a virtual displacement is zero*:

$$\delta U + \delta V \equiv \delta W = 0 \tag{3.4-3}$$

Just as we derived the Euler–Lagrange equations associated with the statement $\delta I = 0$, we can derive them for the statement in Eq. (3). However, first we must identify δU and δV for a given problem. The principle of virtual work is independent of any constitutive law and applies to both elastic (linear and nonlinear) and inelastic continua.

For a solid body, the external and internal virtual work expressions are given in Eqs. (3.1-2) and (3.1-4), respectively. The principle can be expressed as

$$\int_{\Omega_0} \mathbf{P} : \delta\mathbf{E}\; dv - \int_{\Omega_0} \mathbf{f}\cdot\delta\mathbf{u}\; dv - \int_{\Gamma_\sigma} \mathbf{T}\cdot\delta\mathbf{u}\; ds = 0 \tag{3.4-4}$$

where $\mathbf{P} : \delta\mathbf{E}$ denotes the "double dot product," Ω_0 is the volume of the undeformed body, and dv and ds denote the volume and surface elements of Ω_0. Writing in terms of the Cartesian rectangular components, Eq. (4) takes the form

$$\boxed{\int_{\Omega_0} (P_{ij}\delta E_{ij} - f_i\delta u_i)\; dv - \int_{\Gamma_\sigma} T_i\delta u_i\; ds = 0} \tag{3.4-5}$$

where the summation on repeated subscripts is implied.

The Euler–Lagrange equations associated with the statement (5) of the principle of virtual displacements are nothing but the equilibrium equations of the 3-D elasticity theory. Recall the strain–displacement equations from Eq. (2.2–8). The virtual strains δE_{ij} are related to the virtual displacements δu_i by

$$\delta E_{ij} = \mathcal{E}\left[\frac{1}{2}(u_{i,j} + u_{j,i} + u_{m,i}u_{m,j})\right], \quad u_{i,j} \equiv \frac{\partial u_i}{\partial X_j} \tag{3.4-6}$$

Substituting δE_{ij} from the above equation into Eq. (5), and using the divergence theorem, Eq. (1.5–2), to transfer differentiation from δu_i to its coefficient, one obtains

$$0 = \int_{\Omega_0} \left[\frac{1}{2} P_{ij} (\delta u_{i,j} + \delta u_{j,i} + u_{m,i} \delta u_{m,j} + \delta u_{m,i} u_{m,j}) - f_i \delta u_i \right] dv - \int_{\Gamma_\sigma} T_i \delta u_i \, ds$$
$$= \int_{\Omega_0} [P_{ij} (\delta_{mi} + u_{m,i}) \delta u_{m,j} - f_i \delta u_i] \, dv - \int_{\Gamma_\sigma} T_i \delta u_i \, ds$$
$$= -\int_{\Omega_0} \{[P_{ij} (\delta_{mi} + u_{m,i})]_{,j} + f_m\} \delta u_m dv - \int_{\Gamma_\sigma} T_m \delta u_m \, ds$$
$$+ \oint_{\Gamma} P_{ij} (\delta_{mi} + u_{m,i}) n_j \delta u_m \, ds \qquad (3.4-7)$$

Since $\delta u_m = 0$ on Γ_u, we have

$$0 = -\int_{\Omega_0} \{[P_{ij} (\delta_{mi} + u_{m,i})]_{,j} + f_m\} \delta u_m \, dv + \int_{\Gamma_\sigma} [P_{ij} (\delta_{mi} + u_{m,i}) n_j - T_m] \delta u_m \, ds \qquad (3.4-8)$$

Because the virtual displacements are arbitrary in Ω_0 and on Γ_σ, Eq. (8) yields the following equations [*cf.*, Eq. (2.3–7b)]

$$[P_{ij} (\delta_{mi} + u_{m,i})]_{,j} + f_m = 0 \text{ in } \Omega_0 \qquad (3.4-9)$$

$$[P_{ij} (\delta_{mi} + u_{m,i})] n_j - T_m = 0 \text{ on } \Gamma_\sigma \qquad (3.4-10)$$

Equations (9) and (10) are the Euler–Lagrange equations associated with the principle of virtual displacements for a body undergoing large deformations. The stress boundary conditions in Eq. (10) are the natural boundary conditions. The principle of virtual displacements is applicable to any continuous body with arbitrary constitutive behavior (*i.e.*, elastic or inelastic).

For small strains and displacements case, Eqs. (9) and (10) reduce to

$$\sigma_{mj,j} + f_m = 0 \text{ in } \Omega_0 \qquad (3.4-11)$$

$$\sigma_{mj} n_j - T_m = 0 \text{ on } \Gamma_\sigma \qquad (3.4-12)$$

where σ_{mj} denote the stress components and no distinction is made between the second Piola–Kirchhoff stress components and Cauchy stress components.

Example 3.4–1:

Consider the bending of a cantilevered beam of length L, modulus E and moment of inertia I, and subjected to a distributed transverse load q, an end moment M_0 (counterclockwise), and a transverse load F_0 (see Figure 3.4–1). Under the assumption of small strains and displacements ($x \approx X$), we derive the governing differential equation and the natural boundary conditions of the problem using the principle of virtual displacements.

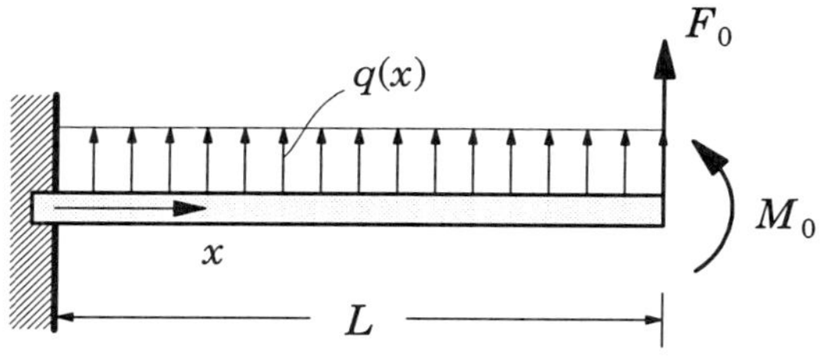

Figure 3.4–1: A cantilever beam with distributed transverse load q, and concentrated load F_0 and moment M_0 at the right end.

The geometric constraints of the beam require the transverse deflection w_0 and slope at the fixed end to be zero:

$$w_0(0) = \frac{dw_0}{dx}(0) = 0 \qquad (3.4-13)$$

It will be shown that w_0 and dw_0/dx are the primary variables and their specification constitutes the essential boundary conditions. The set of admissible configurations consists of functions that satisfy Eq. (13); *i.e.*, any admissible variation $\delta w_0(x)$ of the beam should satisfy the conditions (see Figure 3.1–2)

$$\delta w_0(0) = 0, \quad \frac{d\delta w_0}{dx}(0) = 0 \qquad (3.4-14)$$

In addition to the conditions (14), an admissible variation $\delta w_0(x)$ must be twice differentiable as required by the expression for internal virtual work done. Thus, for the problem at hand, the set of admissible variations consists of functions that satisfy Eq. (14).

The internal virtual work done per unit length of the beam by the actual internal force $\sigma_{xx}\ dA$ in moving through the virtual displacements $\delta\varepsilon_{xx}\ dx$ is given by

$$\sigma_{xx} dA \cdot \delta\varepsilon_{xx} dx$$

The total internal virtual work done is

$$\delta U = \int_0^L \int_A \sigma_{xx}\ \delta\varepsilon_{xx}\ dAdx \qquad (3.4-15)$$

where all other stresses are assumed to be zero, *i.e.*, the Euler–Bernoulli assumptions are invoked. The actual strain in the beam is given by (from the kinematics of the Euler–Bernoulli beam theory)

$$\varepsilon_{xx} = -\frac{z}{R} \approx -z\frac{d^2 w_0}{dx^2} \qquad (3.4-16)$$

where R denotes the radius of curvature of the beam, which is approximated as

$$\frac{1}{R} = \frac{d^2 w_0}{dx^2}\left[1 + \left(\frac{dw_0}{dx}\right)^2\right]^{-\frac{3}{2}} \approx \frac{d^2 w_0}{dx^2}$$

for the small strain case. The virtual strain $\delta\varepsilon_{xx}$ is related to the virtual displacement δw_0 by $\delta\varepsilon_{xx} = -z(d^2\delta w_0/dx^2)$. Substituting this expression into the internal virtual work, we obtain

$$\begin{aligned}\delta U &= -\int_0^L \left(\int_A \sigma_{xx} z \, dA\right) \frac{d^2\delta w_0}{dx^2} \, dx \\ &= -\int_0^L M \frac{d^2 \delta w_0}{dx^2} \, dx \end{aligned} \qquad (3.4-17)$$

where M is the bending moment

$$M(x) = \int_A \sigma_{xx} \; z \, dA \qquad (3.4-18)$$

The virtual work done by the external distributed force $q(x)$ in moving through the displacement δw_0 is

$$-\int_0^L q(x) \cdot \delta w_0 \, dx$$

The virtual work done by the moment M_0 at $x = L$ in rotating through the virtual rotation $\frac{d\delta w_0}{dx}(L)$ is

$$-M_0 \frac{d\delta w_0}{dx}(L)$$

and the virtual work done by the load F_0 in moving through the virtual displacement $\delta w_0(L)$ is

$$-F_0 \delta w_0(L)$$

Total external virtual work done is

$$\delta V = -\left\{\int_0^L q(x)\delta w_0 \, dx + M_0 \frac{d\delta w_0}{dx}(L) + F_0 \delta w_0(L)\right\} \qquad (3.4-19)$$

The principle of virtual displacements states that if the beam is in equilibrium we must have

$$\delta U + \delta V = 0$$

or

$$\int_0^L \left(-M \frac{d^2\delta w_0}{dx^2} - q\delta w_0\right) dx - M_0 \frac{d\delta w_0}{dx}(L) - F_0 \delta w_0(L) = 0 \qquad (3.4-20)$$

To obtain the Euler–Lagrange equations associated with the virtual work statement (20), integrate the first term by parts twice

$$\int_0^L \left(-\frac{d^2M}{dx^2} - q\right)\delta w_0 dx + \left[-M\frac{d\delta w_0}{dx} + \frac{dM}{dx}\delta w_0\right]_0^L - M_0\frac{d\delta w_0}{dx}(L) - F_0\delta w_0(L) = 0$$

Note from the boundary terms that w_0 and dw_0/dx are primary variables and dM/dx and M are the secondary variables of the problem. We have

$$\int_0^L \left(-\frac{d^2M}{dx^2} - q\right)\delta w_0\, dx - [M(L) + M_0]\frac{d\delta w_0}{dx}(L) + M(0)\frac{d\delta w_0}{dx}(0)$$
$$+ \left[\frac{dM}{dx}(L) - F_0\right]\delta w_0(L) - \frac{dM}{dx}(0)\delta w_0(0) = 0$$

Because of conditions (14) on δw_0 and $d\delta w_0/dx$, the second and fourth boundary expressions vanish. We have

$$0 = \int_0^L \left(-\frac{d^2M}{dx^2} - q\right)\delta w_0\, dx - [M(L) + M_0]\frac{d\delta w_0}{dx}(L) + \left[\frac{dM}{dx}(L) - F_0\right]\delta w_0(L) \quad (3.4-21)$$

Since δw_0 is arbitrary in $(0 < x < L)$ and is arbitrary at $x = L$, we obtain, using the fundamental lemma, the Euler–Lagrange equations

$$-\frac{d^2M}{dx^2} - q = 0\ ,\ 0 < x < L \quad (3.4-22a)$$

$$M(L) + M_0 = 0 \quad \frac{dM}{dx}(L) - F_0 = 0 \quad (3.4-22b)$$

Note that the natural boundary conditions involve specifying generalized forces. In the present problem the moment at $x = L$ is specified to be M_0 and the shear force, dM/dx, is specified to be F_0 at $x = L$.

The bending moment $M(x)$ can be expressed in terms of the transverse deflection $w(x)$ using Eqs. (18) and (16) and a stress–strain relation. For linear elastic material, we assume $\sigma_{xx} = E\varepsilon_{xx}$ and obtain

$$M(x) = -EI\frac{d^2w_0}{dx^2} \quad (3.4-23)$$

We note, once again, that the principle of virtual displacements is valid for any material body. When the body is assumed to obey Hooke's law, the principle of virtual displacements can be expressed as one of minimizing a quadratic functional, namely, the total potential energy.

3.4–3 The Principle of Minimum Total Potential Energy

A special case of the principle of virtual displacements that deals with linear as well as nonlinear elastic bodies is known as the *principle of minimum total potential energy.* For elastic bodies (in the absence of temperature variations) there exists a *strain energy density* function U_0 such that

$$\mathbf{P} = \frac{\partial U_0}{\partial \mathbf{E}} \quad \text{or} \quad P_{ij} = \frac{\partial U_0}{\partial E_{ij}} \qquad (3.4-24)$$

Equation (24) represents the constitutive equation of an hyperelastic material. The strain energy density U_0 is a single-valued function of strains at a point, and is assumed to be positive definite. The statement of the principle of virtual displacements, Eq. (5), can be expressed in terms of the strain energy density U_0:

$$\int_{\Omega_0} \frac{\partial U_0}{\partial \mathbf{E}} : \delta \mathbf{E}\, dv - \left[\int_{\Omega_0} \mathbf{f}\cdot\delta\mathbf{u}\, dv + \int_{\Gamma_\sigma} \mathbf{T}\cdot\delta\mathbf{u}\, ds\right] = 0 \qquad (3.4-25a)$$

or, in component form,

$$\int_{\Omega_0} \frac{\partial U_0}{\partial E_{ij}}\delta E_{ij}\, dv - \left[\int_{\Omega_0} f_i\delta u_i\, dv + \int_{\Gamma_\sigma} T_i\delta u_i\, ds\right] = 0 \qquad (3.4-25b)$$

The first integral is equal to

$$\int_{\Omega_0} \delta U_0\, dv = \delta U$$

where U is the internal strain energy functional

$$U = \int_{\Omega_0} U_0\, dv \qquad (3.4-26a)$$

Suppose that there exists a potential V whose first variation is

$$\begin{aligned}\delta V &= -\left[\int_{\Omega_0} \mathbf{f}\cdot\delta\mathbf{u}\, dv + \int_{\Gamma_\sigma} \mathbf{T}\cdot\delta\mathbf{u}\, ds\right]\\ &= -\left[\int_{\Omega_0} f_i\delta u_i\, dv + \int_{\Gamma_\sigma} T_i\delta u_i\, ds\right]\end{aligned} \qquad (3.4-26b)$$

Then the principle of virtual work takes the form

$$\delta U + \delta V = \delta(U+V) \equiv \delta\Pi = 0 \qquad (3.4-27)$$

The sum $U + V = \Pi$ is called the *total potential energy* of the elastic body. The statement in Eq. (27) is known as the *principle of minimum total potential energy.* It states that *of all admissible displacements, those which satisfy the equilibrium equations make the total potential energy a minimum:*

$$\Pi(\mathbf{u}) \leq \Pi(\bar{\mathbf{u}}) \tag{3.4-28}$$

where $\mathbf{u}$ is the true solution and $\bar{\mathbf{u}}$ is any admissible displacement field. The equality holds only if $\mathbf{u} = \bar{\mathbf{u}}$.

Example 3.4–2:

We consider the cantilever beam problem of Example 3.4–1 (see Figure 3.4–1). The minimum total potential energy principle requires us to construct the total potential energy (*i.e.*, sum of the strain energy and potential energy due to applied loads) of the beam and set its first variation to zero to obtain the Euler–Lagrange equations of the functional.

The total strain energy stored in the beam is

$$\begin{aligned} U &= \frac{1}{2}\int_0^L \int_A (\sigma_{xx} dA)(\varepsilon_{xx} dx) \\ &= \frac{1}{2}\int_0^L \int_A \sigma_{xx}\varepsilon_{xx}\, dA dx \\ &= -\frac{1}{2}\int_0^L M\frac{d^2w_0}{dx^2}\, dx \\ &= \frac{1}{2}\int_0^L EI\left(\frac{d^2w_0}{dx^2}\right)^2 dx \end{aligned} \tag{3.4-29}$$

where the bending moment–deflection relationship (23) is used to write the last expression for U. The work done by external applied loads q, M_0 and F_0 is

$$V = -\left[\int_0^L q w_0\, dx + M_0\frac{dw_0}{dx}(L) + F_0 w_0(L)\right] \tag{3.4-30}$$

The total potential energy of the beam is given by

$$\Pi = U + V = \int_0^L \left[\frac{EI}{2}\left(\frac{d^2w_0}{dx^2}\right)^2 - q w_0\right] dx - M_0\frac{dw_0}{dx}(L) - F_0 w_0(L) \tag{3.4-31}$$

The total potential energy principle requires that $\delta(U+V) = 0$:

$$\begin{aligned} 0 &= \delta\left[\int_0^L \frac{EI}{2}\left(\frac{d^2w_0}{dx^2}\right)^2 dx - \int_0^L q w_0 dx - M_0\frac{dw_0}{dx}(L) - F_0 w_0(L)\right] \\ &= \int_0^L \left(EI\frac{d^2w_0}{dx^2}\frac{d^2\delta w_0}{dx^2} - q\delta w_0\right) dx - M_0\frac{d\delta w_0}{dx}(L) - F_0\delta w_0(L) \end{aligned}$$

where the properties (b) and (d) of Eq. (3.2–7a) and (a) and (b) of Eq. (3.2–8) of the variational operator δ are used to write the last equation. Integration by parts of the first term, and use of Eq. (14) and the property that δw_0 is arbitrary in $(0, L)$ and at $x = L$, yields the Euler–Lagrange equations

$$\frac{d^2}{dx^2}\left(EI\frac{d^2w_0}{dx^2}\right) - q = 0 \ , \ 0 < x < L \tag{3.4 – 32}$$

$$\left(-EI\frac{d^2w_0}{dx^2}\right)|_{x=L} + M_0 = 0 \tag{3.4 – 33a}$$

$$\frac{d}{dx}\left(-EI\frac{d^2w_0}{dx^2}\right)|_{x=L} - F_0 = 0 \tag{3.4 – 33b}$$

Equations (22a,b), and (32) and (33a,b) are the same when M is replaced by $-EI(d^2w_0/dx^2)$, *i.e.*, when the beam constitutive equation is used.

The minimum property of the total potential energy can be established by considering an arbitrary admissible displacement field, $\bar{w}_0$. For the example problem we have

$$\Pi(\bar{w}) = \int_0^L \left[\frac{EI}{2}\left(\frac{d^2\bar{w}}{dx^2}\right) - q\bar{w}\right] dx - M_0\frac{d\bar{w}}{dx}(L) - F_0\bar{w}(L) \tag{3.4 – 34}$$

where $\bar{w}$ is an admissible transverse deflection, which is of the form

$$\bar{w} = w_0 + \alpha v, \quad \alpha \text{ small}$$

$$v(0) = 0, \quad \frac{dv}{dx}(0) = 0 \tag{3.4 – 35}$$

Then $\Pi(\bar{w}) = \Pi(w_0 + \alpha v)$

$$\begin{aligned}
\Pi(w_0 + \alpha v) &= \int_0^L \left\{\frac{EI}{2}\left[\left(\frac{d^2w_0}{dx^2}\right)^2 + \alpha^2\left(\frac{d^2v}{dx^2}\right)^2 + 2\alpha\frac{d^2w_0}{dx^2}\frac{d^2v}{dx^2}\right] - q(w_0 + \alpha v)\right\}dx \\
&- M_0\left[\frac{dw_0}{dx}(L) + \alpha\frac{dv}{dx}(L)\right] - F_0[w_0(L) + \alpha v(L)] \\
&= \Pi(w_0) + \int_0^L \alpha^2\frac{EI}{2}\left(\frac{d^2v}{dx^2}\right)^2 dx + \alpha\int_0^L \left[EI\frac{d^2w_0}{dx^2}\frac{d^2v}{dx^2} - qv\right] dx \\
&- \alpha M_0\frac{dv}{dx}(L) - \alpha F_0 v(L)
\end{aligned} \tag{a}$$

Now consider the second integral and the boundary term on the right–hand side

$$
\begin{aligned}
&\alpha\left\{\int_0^L \left(EI\frac{d^2w_0}{dx^2}\frac{d^2v}{dx^2} - qv\right)dx - M_0\frac{dv}{dx}(L) - F_0v(L)\right\} \\
&= \alpha\left\{\int_0^L \left[\frac{d^2}{dx^2}\left(EI\frac{d^2w_0}{dx^2}\right) - q\right]vdx + \left[EI\frac{d^2w_0}{dx^2}\frac{dv}{dx} - \frac{d}{dx}\left(EI\frac{dw_0}{dx}\right)v\right]_0^L \right. \\
&\quad \left. - M_0\frac{dv}{dx}(L) - F_0v(L)\right\} \qquad (b)
\end{aligned}
$$

The boundary terms in the square brackets are zero for the lower limit because $v(0) = \frac{dv}{dx}(0) = 0$. For the upper limit we have

$$
\begin{aligned}
EI\frac{d^2w_0}{dx^2}(L)\frac{dv}{dx}(L) - \left[\frac{d}{dx}\left(EI\frac{d^2w_0}{dx^2}\right)\right]_{x=L} v(L) \\
= M_0\frac{dv}{dx}(L) + F_0v(L) \qquad (c)
\end{aligned}
$$

which cancel the remaining boundary terms in Eq. (b). The integral over $(0, L)$ is identically zero because w_0 is the true solution. Thus all terms linear in α in Eq. (a) vanish, giving the result

$$
\begin{aligned}
\Pi(\bar{w}) &= \Pi(w_0) + \int_0^L \alpha^2\frac{EI}{2}\left(\frac{d^2v}{dx^2}\right)^2 \\
&\geq \Pi(w_0)
\end{aligned}
$$

because the integral is positive for all values of $v \neq 0$. When $v = 0$, $\Pi(\bar{w}) = \Pi(w_0)$ and $\bar{w} = w_0$. Thus $\Pi(\bar{w})$ is greater than $\Pi(w_0)$ when $\bar{w} \neq w_0$.

3.5 Variational Methods

3.5–1 Introduction

In Section 3.4 we saw how virtual work and variational principles can be used to obtain governing differential equations and associated boundary conditions. Here we study the direct use of the variational principles in the solution of the underlying equations. The methods to be described here are known as the *classical* variational methods. In these methods, we seek an approximate solution to the problem in terms of adjustable parameters that are determined by substituting the assumed solution into a variational statement equivalent to the governing equations of the problem. Such solution methods are called *direct methods* because the approximate solutions are obtained directly by applying the same variational principle that was used to derive the governing (*i.e.*, Euler–Lagrange) equations.

The assumed solutions in the variational methods are in the form of a finite linear combination of *undetermined parameters* with appropriately chosen functions. This amounts to representing a continuous function by a finite set of functions. Since the solution of a continuum problem in general cannot be represented by a finite set of functions, error is introduced into the solution. Therefore, the solution obtained is an approximation to the true solution of the equations describing a physical problem. As the number of linearly independent terms in the assumed solution is increased, the error in the approximation will be reduced, and the assumed solution converges to the exact solution.

It should be understood that the equations governing a physical problem are themselves approximate. The approximations are introduced by several sources, including the geometry, representation of specified loads and boundary conditions, and material behavior. Therefore, when one thinks of permissible error in an approximate solution, it is understood to be relative to exact solutions of the governing equations that inherently contain the approximations. The variational methods of approximation to be described here are limited to the Rayleigh–Ritz method and the weighted–residual methods (*e.g.*, the least–squares method, collocation method, and so on). The weighted–residual methods will be visited only briefly. Interested readers may consult the references at the end of the chapter for additional details.

3.5–2 The Rayleigh–Ritz Method

As noted in Section 3.4 the principle of virtual displacements gives the equilibrium equations as the Euler–Lagrange equations. These governing equations are in the form of differential equations that are not always tractable by exact methods of solution. There exist a number of approximate methods that can be used to solve differential equations (*e.g.*, finite–difference methods, perturbation methods, etc.). The most direct methods are those which bypass the derivation of the Euler–Lagrange equations, and go directly from a variational statement of the problem to the solution of the Euler–Lagrange equations. One such direct method was proposed by Lord Rayleigh, and a generalization of the method was proposed independently by Ritz. Thus, the Rayleigh–Ritz method is necessarily based on variational statements, such as those provided by the principles of virtual displacements or the minimum total potential energy, which are equivalent to the governing differential equations as well as the natural boundary conditions, and they are also known as the *weak forms*.

The basic idea of the Rayleigh–Ritz method is described here using the principle of virtual displacements or the minimum total potential energy principle. In the Rayleigh–Ritz method we approximate a dependent unknown (*e.g.*, the displacement) u by a finite linear combination of the form

$$u \approx U_N = \sum_{j=1}^{N} c_j \varphi_j + \varphi_0 \qquad (3.5-1)$$

and then determine the parameters c_j by requiring that the principle of virtual displacements hold for the approximate problem. In Eq. (1) c_j denote undetermined parameters, and φ_0 and φ_j are called *approximation functions*, which are appropriately selected functions of position x. Equation (1) can be viewed as a representation of u in a component form; the parameters c_j are the *components* (or coordinates) and φ_j are the *coordinate functions.* Another interpretation of Eq. (1) is provided by the finite Fourier series, where c_j are known as the *Fourier coefficients.*

Properties of Approximation Functions

In order to ensure that the algebraic equations resulting from the Rayleigh–Ritz approximation have a solution, and the approximate solution converges to the true solution of the problem as the number of parameters N is increased, we must choose φ_j $(j = 1, 2, 3, ...N)$ and φ_0 such that they meet the following requirements:

1. φ_0 has the principal purpose of satisfying the specified *essential* (or geometric) boundary conditions associated with the variational formulation; φ_0 plays the role of *particular solution.* It should be the lowest order possible for completeness.

2. φ_j $(j = 1, 2, ..., N)$ should satisfy the following three conditions:

   (a) be continuous as required in the variational statement (*i.e.*, φ_j should be such that it has a nonzero contribution to the virtual work statement);

   (b) satisfy the homogeneous form of the specified essential boundary conditions;

   (c) the set $\{\varphi_j\}$ is linearly independent and complete. $\qquad (3.5-2)$

The completeness property is defined mathematically as follows. Given a function u and a real number $\varepsilon > 0$, the sequence $\{\varphi_j\}$ is said to be *complete* if there exists an integer N (which depends on ε) and scalars $c_1, c_2, \cdots, c_N$ such that

$$\|u - \sum_{j=1}^{N} c_j \varphi_j\| < \varepsilon \qquad (3.5-3)$$

where $\|\cdot\|$ denotes a norm in the vector space of functions u. The set $\{\varphi_j\}$ is called the *spanning set*. A sequence of algebraic polynomials, for example, is complete if it contains terms of all degrees up to the highest degree (N).

Linear independence of a set of functions refers to the property that there exists no nontrivial relation among them, *i.e.*, the relation

$$\alpha_1\varphi_1 + \alpha_2\varphi_2 + \cdots + \alpha_N\varphi_N = 0$$

holds only for all $\alpha_j = 0$. Thus no function is derivable from a linear combination of others in the set.

For polynomial approximations functions, the linear independence and completeness properties require φ_j to be increasingly higher-order polynomials. For example, if φ_1 is a linear polynomial, φ_2 should be a quadratic polynomial, φ_3 should be a cubic polynomial, and so on (but each φ_j need not be complete by itself):

$$\varphi_1 = a_1 + b_1 x,\ \varphi_2 = a_2 + b_2 x + c_2 x^2,\ a_2 + c_2 x^2,\ \text{ or }\ b_2 x + c_2 x^2,\ \cdots$$

The completeness property is essential for the convergence of the Rayleigh–Ritz approximation (see Reddy [11], p. 262).

Since the natural boundary conditions of the problem are included in the variational statements, we require the Rayleigh–Ritz approximation U_N to satisfy only the specified essential boundary conditions of the problem. This is done by selecting φ_i to satisfy the homogeneous form and φ_0 to satisfy the actual form of the essential boundary conditions. For instance, if u is specified to be $\hat{u}$ on the boundary $x = L$, we require

$$\varphi_0 = \hat{u} \ \text{ at }\ x = L \ \text{ and }\ \varphi_i = 0 \ \text{ at }\ x = L \ \text{ for }\ i = 1, 2, ..., N$$

The requirement on φ_i to satisfy the homogeneous form of the specified essential boundary conditions follows from the approximation adopted in Eq. (1). Since $U_N = \hat{u}$ and $\varphi_0 = \hat{u}$ at $x = L$, we have

$$U_N(L) = \sum_{j=1}^{N} c_j\varphi_j(L) + \varphi_0(L)$$

$$\hat{u} = \sum_{j=1}^{N} c_j\varphi_j(L) + \hat{u}$$

and, therefore, it follows that $\sum_{j=1}^{N} c_j\varphi_j(L) = 0$. Since this condition must hold for any set of parameters c_j, it follows that

$$\varphi_j(L) = 0 \ \text{ for }\ j = 1, 2, \cdots, N$$

Note that when the specified values are zero, *i.e.*, $\hat{u} = 0$, there is no need to include φ_0 (or equivalently, $\varphi_0 = 0$); however, φ_j are still required to satisfy the specified (homogeneous) essential boundary conditions.

The conditions in Eq. (2) provide guidelines for selecting the coordinate functions; they do not give any formula for generating the functions. As a general rule, coordinate functions should be selected from the admissible set, from the lowest order to a desirable order without missing any intermediate admissible terms in the representation of U_N(*i.e.*, satisfy the completeness property). The function φ_0 has no other role to play than to satisfy specified (nonhomogeneous) essential boundary conditions; there are no continuity conditions on φ_0. Therefore, one should select the lowest order φ_o that satisfies the essential boundary conditions.

Ritz Equations for the Parameters

Once the functions φ_0 and φ_i are selected, the parameters c_j in Eq. (1) are determined by requiring U_N to minimize the total potential energy functional Π (or satisfy the principle of virtual work) of the problem: $\delta\Pi(U_N) = 0$. Note that $\Pi(U_N)$ is now a real-valued function of variables, $c_1, c_2, \cdots, c_N$. Hence minimization of the functional $\Pi(U_N)$ is reduced to the minimization of a function of several variables:

$$0 = \delta\Pi(U_N) = \delta\Pi(c_i) = \sum_{i=1}^{N} \frac{\partial \Pi}{\partial c_i}\delta c_i \quad \text{or} \quad \frac{\partial \Pi}{\partial c_i} = 0 \qquad (3.5-4)$$

This gives N algebraic equations in the N coefficients $(c_1, c_2, ..., c_N)$

$$0 = \frac{\partial \Pi}{\partial c_i} = \sum_{j=1}^{N} R_{ij}c_j - F_i \quad \text{or} \quad [R]\{c\} = \{F\} \qquad (3.5-5)$$

where R_{ij} and F_i are known coefficients that depend on the problem parameters (*e.g.*, geometry, material coefficients, and loads) and the approximation functions. These coefficients will be defined for each problem discussed in the sequel. Equations (5) are then solved for $\{c\}$ and substituted back into Eq. (1) to obtain the N-parameter Rayleigh–Ritz solution.

Some general features of the Rayleigh–Ritz method based on the principle of virtual displacements are listed below:

1. If the coordinate functions φ_i are selected to satisfy the conditions (2), the assumed approximation for the displacements converges to the true solution with an increase in the number of parameters (*i.e.*, as $N \to \infty$). A mathematical proof of such an assertion can be found in the textbook by Reddy [11].

2. For increasing values of N, the previously computed coefficients of the algebraic equations (5) remain unchanged, provided the previously selected coordinate functions are not changed. One must add only the newly computed coefficients to the system of equations.

3. If the resulting algebraic equations are symmetric, one needs to compute only upper or lower diagonal elements in the coefficient matrix, $[R]$. The symmetry of the coefficient matrix depends on the variational statement of the problem.

4. If the variational (or virtual work) statement is nonlinear in u, then the resulting algebraic equations will also be nonlinear in the parameters c_i. To solve such nonlinear equations, a variety of numerical methods are available (*e.g.*, Newton's method, the Newton–Raphson method, the Picard method), which will be discussed later in this book (see Chapter 13).

5. Since the strains are computed from an approximate displacement field, the strains and stresses are generally less accurate than the displacement.

6. The equilibrium equations of the problem are satisfied only in the energy sense, not in the differential equation sense. Therefore the displacements obtained from the Rayleigh–Ritz approximation, in general do not satisfy the equations of equilibrium pointwise, unless the solution converged to the exact solution.

7. Since a continuous system is approximated by a finite number of coordinates (or degrees of freedom), the approximate system is less flexible than the actual system. Consequently, the displacements obtained from the total potential energy by the Rayleigh–Ritz method converge to the exact displacements from below:

$$U_1 < U_2 < ... < U_N < U_M ... < u(\text{exact}), \quad \text{for } M > N$$

   where U_N denotes the N-parameter Rayleigh–Ritz approximation of u obtained from the principle of virtual displacements or the minimum total potential energy principle. It should be noted that the displacements obtained from the Rayleigh–Ritz method based on the total complementary energy principle provide the upper bound.

8. The Rayleigh–Ritz method can be applied, in principle, to any physical problem that can be cast in a *weak form* – a form that contains the natural boundary conditions of the problem. In particular, the Rayleigh–Ritz method can be applied to all structural problems since a virtual work statement exists.

Example 3.5–1:

Consider the cantilever beam shown in Figure 3.4–1. We set up the coordinate system such that the origin is at the fixed end. For this case the geometric (or essential) boundary conditions are

$$w_0(0) = 0, \quad \frac{dw_0}{dx}(0) = 0$$

The force (or natural) boundary conditions can be arbitrary. For example, the beam can be subjected to uniformly distributed transverse load $q(x) = q_0$, concentrated point load F_0, and moment M_0, as in Figure 3.4–1. The applied loads will have no bearing on the selection of φ_0 and φ_i. The applied loads will enter the analysis through the expression for the external work done [see Eq. (3.4–19)], which will alter the expression for the coefficients F_i of Eq. (5).

An N-parameter Rayleigh–Ritz approximation of the transverse deflection $w_0(x)$ is chosen in the form

$$w_0(x) \approx W_N = \sum_{j=1}^{N} c_j\varphi_j + \varphi_0 \tag{3.5 – 6}$$

Since the specified essential boundary conditions are homogeneous, $\varphi_0 = 0$. Next, we must select φ_i to satisfy the homogeneous form of the specified essential boundary conditions

$$\varphi_i(0) = 0 \text{ and } \frac{d\varphi_i}{dx}(0) = 0 \tag{3.5 – 7}$$

and φ_i must be differentiable as required by the total potential energy functional in Eq. (3.4–31) of Example 3.4–2. Since there are two conditions to satisfy, we begin with $\varphi_1 = a + bx + cx^2$ and determine two of the three constants using Eq. (7). The third constant will remain arbitrary. Conditions (7) give $a = 0$ and $b = 0$. Hence, $\varphi_1(x) = cx^2$. We can take $c = 1$ (or it will be absorbed into c_1). Since $\varphi_1, \varphi_2, \cdots, \varphi_N$ are linearly independent and complete, we choose

$$\varphi_2 = x^3, \ \varphi_3 = x^4, \ \cdots, \ \varphi_N = x^{N+1}$$

The Rayleigh–Ritz approximation becomes

$$W_N = c_1x^2 + c_2x^3 + \cdots + c_Nx^{N+1} \tag{3.5 – 8}$$

Substituting expression (8) into Eq. (3.4–31) we obtain Π as a function of the coefficients $c_1, c_2, ..., c_N$:

$$\begin{aligned}\Pi(c_1, c_2, \cdots, c_N) = \int_0^L \Big\{ &\frac{EI}{2}[2c_1 + 6c_2x + \cdots + N(N+1)c_Nx^{N-1}]^2 \\ &- q(c_1x^2 + c_2x^3 + \cdots + c_Nx^{N+1})\Big\} dx \\ &- F_0[c_1x^2 + c_2x^3 + \cdots + c_Nx^{N+1}]_{x=L} \\ &- M_0[2c_1x + 3c_2x^2 + \cdots + (N+1)c_Nx^N]_{x=L}\end{aligned} \tag{3.5 – 9}$$

Using the total potential energy principle, $\delta\Pi = 0$, which requires that Π be a minimum with respect to each of c_1, $c_2, \cdots, c_N$,

$$\delta\Pi = \frac{\partial\Pi}{\partial c_1}\delta c_1 + \frac{\partial\Pi}{\partial c_2}\delta c_2 + \cdots + \frac{\partial\Pi}{\partial c_N}\delta c_N$$

we arrive at the conditions

$$\frac{\partial\Pi}{\partial c_1} = 0, \; \frac{\partial\Pi}{\partial c_2} = 0, \; \cdots, \; \frac{\partial\Pi}{\partial c_N} = 0 \tag{3.5 − 10}$$

The ith equation in (10) has the form

$$\begin{aligned} 0 = \frac{\partial\Pi}{\partial c_i} &= \int_0^L \{EI[2c_1 + 6c_2 x + \cdots + N(N+1)c_N x^{N-1}]i(i+1)x^{i-1} \\ &\quad - q\, x^{i+1}\}dx - F_0 L^{i+1} - M_0(i+1)x^i]_{x=L} \\ &= c_1\left[\int_0^L 2EI \cdot i(i+1)x^{i-1}dx\right] + c_2\left[\int_0^L 6EIx \cdot i(i+1)x^{i-1}dx\right] \\ &\quad + \cdots + c_N\left[\int_0^L EIN(N+1)x^{N-1}i(i+1)x^{i-1}dx\right] = 0 \\ &\quad - \int_0^L q(x)x^{i+1}dx - F_0 L^{i+1} - M_0(i+1)L^i \\ &= c_1 R_{i1} + c_2 R_{i2} + \cdots + c_N R_{iN} - F_i \\ &= \sum_{j=1}^{N} R_{ij}c_j - F_i\,, \quad (i = 1, 2, ..., N) \end{aligned} \tag{3.5 − 11a}$$

where

$$\begin{aligned} R_{ij} &= EI\int_0^L j(j+1)x^{j-1} \cdot i(i+1)x^{i-1}dx \\ F_i &= \int_0^L q(x)x^{i+1}dx + F_0 L^{i+1} + M_0(i+1)L^i \end{aligned} \tag{3.5 − 11b}$$

Equation (11a) can be written in matrix form as

$$[R]\{c\} = \{F\}$$

This completes the formulation of the $N-$parameter Rayleigh–Ritz approximation of a cantilever beam problem.

As an example, we consider a two-parameter approximation. Suppose that the beam is loaded with a uniformly distributed transverse force, $q = q_0$. For this case, we have

$$\begin{aligned} R_{11} &= 4EIL, \quad R_{12} = 6EIL^2 = R_{21}, \quad R_{22} = 12EIL^3 \\ F_1 &= \frac{q_0 L^3}{3} + F_0 L^2 + 2M_0 L, \quad F_2 = \frac{q_0 L^4}{4} + F_0 L^3 + 3M_0 L^2 \end{aligned}$$

and

$$EIL\begin{bmatrix} 4 & 6L \\ 6L & 12L^2 \end{bmatrix}\begin{Bmatrix} c_1 \\ c_2 \end{Bmatrix} = \frac{q_0L^3}{12}\begin{Bmatrix} 4 \\ 3L \end{Bmatrix} + F_0L^2\begin{Bmatrix} 1 \\ L \end{Bmatrix} + M_0L\begin{Bmatrix} 2 \\ 3L \end{Bmatrix}$$

The solution of these equations is

$$c_1 = \frac{1}{EI}\left(\frac{5}{24}q_0L^2 + \frac{1}{2}F_0L + \frac{1}{2}M_0\right), \quad c_2 = -\frac{1}{12EI}\left(q_0L + 2F_0L\right)$$

and the two-parameter Rayleigh–Ritz solution becomes

$$W_2 = \frac{1}{EI}\left(\frac{5}{24}q_0L^2 + \frac{1}{2}F_0L + \frac{1}{2}M_0\right)x^2 - \left(\frac{q_0L}{12EI} + \frac{F_0L}{6EI}\right)x^3 \qquad (3.5-12)$$

The exact solution is

$$w_0(x) = \frac{q_0x^2}{24EI}(6L^2 - 4Lx + x^2) + \frac{F_0}{6EI}(3Lx^2 - x^3) - \frac{M_0}{2EI}x^2 \qquad (3.5-13)$$

The two-parameter solution is exact for the case in which $q_0 = 0$. For $q_0 \neq 0$, the solution is not exact for every x, but the maximum deflection, $w_0(L)$, predicted by the two-parameter Rayleigh–Ritz solution coincides with the exact. The three-parameter solution, with $\phi_3 = x^4$, would be exact for this problem because the exact solution is a fourth-order polynomial.

If we were to choose trigonometric functions for φ_i, we may select the functions

$$\varphi_i = 1 - \cos\frac{(2i-1)\pi x}{2L}$$

This particular choice would not give the exact solution for a finite value of N, because the applied load F_0, when expanded in terms of φ_i, would involve infinite number of terms. Thus, a proper choice of the coordinate functions is important in realizing the exact solution. Of course, both choices (algebraic and trigonometric functions) would yield acceptable results with finite number of terms.

3.5–3 Weighted–Residual Methods

Consider an operator equation in the form

$$A(u) = f \ in \ \Omega$$

$$B_1(u) = \hat{u} \text{ on } \Gamma_1, \ B_2(u) = \hat{g} \text{ on } \Gamma_2 \qquad (3.5-14)$$

where A is a linear or nonlinear differential operator, u is the dependent variable, f is a given force term in the domain Ω, B_1 and B_2 are boundary operators associated with essential and natural boundary conditions of the operator A, and $\hat{u}$ and $\hat{g}$ are specified values on the portions Γ_1 and Γ_2 of the boundary Γ of the domain. An example of Eq. (14) is given by

$$A(u) = -\frac{d}{dx}\left(a\frac{du}{dx}\right), \; B_1(u) = u, \; B_2(u) = a\frac{du}{dx}$$

$$\Gamma_1 \text{ is the point } x = 0, \; \Gamma_2 \text{ is the point } x = L$$

We seek a solution in the form

$$U_N = \sum_{j=1}^{N} c_j\varphi_j + \varphi_0 \tag{3.5 – 15}$$

where the parameters c_j are determined by requiring the residual of the approximation

$$R_N = A\left(\sum_{j=1}^{N} c_j\varphi_j + \varphi_0\right) - f \neq 0 \tag{3.5 – 16a}$$

be orthogonal to N linearly independent set of *weight functions* ψ_i:

$$\int_\Omega \psi_i R_N(c_i, \varphi_i, f)\; d\mathbf{x} = 0, \; i = 1, 2, ..., N \tag{3.5 – 16b}$$

The method based on this procedure is called, for obvious reason, a *weighted–residual method.*

The coordinate function φ_o and φ_i in a weighted–residual method should satisfy the properties in Eq. (2), except that they should satisfy *all* specified boundary conditions:

- φ_o should satisfy *all* specified boundary conditions.
- φ_i should satisfy the homogeneous form of *all* specified boundary conditions. (3.5–17)

The variational statement referred to in Property 2a of (2) is now given in Eq. (16b). Properties in (17) are required because the boundary conditions, both essential and natural, are not included in Eq. (16b). Both properties now require φ_i to be of higher order than those used in the Rayleigh–Ritz method. On the other hand, ψ_i can be any linearly independent set, such as $\{1, x, ...\}$, and no continuity requirements are placed on ψ_i.

Various special cases of the weighted–residual method differ from each other due to the choice of the weight function ψ_i. The most commonly used weight functions are

Galerkin's method:	$\psi_i = \varphi_i$
Least–squares method:	$\psi_i = A(\varphi_i)$
Collocation method:	$\psi_i = \delta(\mathbf{x} - \mathbf{x}_i)$

Here $\delta(\cdot)$ denotes the Dirac delta function. The weighted–residual method in the general form (16b) (with $\psi_i \neq \varphi_i$) is known as the *Petrov–Galerkin method.* Equation (16b) provides N linearly independent equations for the determination of the parameters c_i. If A is a nonlinear operator, the resulting algebraic equations will be nonlinear. Whenever A is linear, we have

$$A\left(\sum_{j=1}^{N} c_j\varphi_j + \varphi_0\right) = \sum_{j=1}^{N} c_j A(\varphi_j) + A(\varphi_0) \qquad (3.5-18)$$

and Eq. (16b) becomes

$$\sum_{j=1}^{N}\left[\int_\Omega \psi_i A(\varphi_j) d\mathbf{x}\right] c_j - \int_\Omega \psi_i \left[f - A(\varphi_0)\right] d\mathbf{x} = 0$$

$$\sum_{j=1}^{N} G_{ij}c_j - q_i = 0, \ i = 1, 2, ..., N \qquad (3.5-19a)$$

where

$$G_{ij} = \int_\Omega \psi_i A(\varphi_j) d\mathbf{x}, \ q_i = \int_\Omega \psi_i \left[f - A(\varphi_0)\right] \ d\mathbf{x} \qquad (3.5-19b)$$

Note that G_{ij} is not symmetric in general, even when $\psi_i = \varphi_i$ (Galerkin's method). It is symmetric when A is a linear operator and $\psi_i = A(\varphi_i)$ (the least–squares method).

It should be noted that in most problems of interest in solid mechanics, the operator A is of the form that permits the use of integration by parts (or gradient/divergence theorems) to transfer part of the differentiation to the weight functions ψ_i, provided ψ_i are sufficiently differentiable. For example, if A is a $2m$th-order linear differential operator, then the product $\psi_i A(\varphi_j)$ can be integrated m times by parts to obtain an expression that involves mth order derivatives of both ψ_i and φ_j and includes the natural boundary conditions of the problem. For problems for which there exists a quadratic form or a virtual work statement, the weighted–residual methods are not recommended as they require higher-order functions. When trigonometric approximation functions are used, this is not a concern.

The Galerkin Method

The Galerkin method is a special case of the Petrov–Galerkin method in which the coordinate functions and the weighted functions are the same ($\varphi_i = \psi_i$). It constitutes a generalization of the Rayleigh–Ritz method. When the governing equation has even order highest derivative, it is possible to construct a weak form of the equation, and use the Rayleigh–Ritz method. If the Galerkin method is used in such cases, it would involve the use of higher-order coordinate functions and the solution of unsymmetric equations.

The Rayleigh–Ritz and Galerkin methods yield the same set of algebraic equations for the following two cases:

1. The specified boundary conditions of the problem are all essential type, and therefore the requirements on φ_i in both methods are the same.
2. The problem has both essential and natural boundary conditions, but the coordinate functions used in the Galerkin method are also used in the Rayleigh–Ritz method.

Least–Squares Method

The least–squares method is a variational method in which the integral of the square of the residual in the approximation of a given differential equation is minimized with respect to the parameters in the approximation:

$$\min I(c_i) \equiv \int_\Omega |R_N(c_j, \varphi_j, f|^2 \, d\mathbf{x} \tag{3.5 - 20a}$$

or

$$\int_\Omega 2R_N(c_j, \varphi_j, f) \frac{\partial R_N}{\partial c_i} \, d\mathbf{x} = 0, \;\; i = 1, 2, ..., N \tag{3.5 - 20b}$$

where R_N is the residual defined in Eq. (16a). Equation (20) provides N algebraic equations for the constants c_i.

First we note that the least–squares method is a special case of the weighted–residual method for the weight function, $\psi_i = 2(\partial R_N/\partial c_i)$ [compare Eqs. (16b) and (20b)]. Therefore, the coordinate functions φ_i should satisfy the same conditions as in the case of the weighted–residual method. Next, if the operator A in the governing equation is linear, the weight function ψ_i becomes

$$\psi_i = \frac{\partial R_N}{\partial c_i} = A(\varphi_i) \tag{3.5 - 21}$$

Then from Eq. (20) we have

$$\sum_{j=1}^{N} \left[\int_{\Omega} A(\varphi_i) A(\varphi_j) d\mathbf{x} \right] c_j - \int_{\Omega} [A(\varphi_i) f - A(\varphi_i) A(\varphi_0)] \; d\mathbf{x} = 0$$

or

$$\sum_{j=1}^{N} L_{ij} c_j - h_i = 0, \; i = 1, 2, ..., N \tag{3.5 - 22a}$$

where

$$L_{ij} = \int_{\Omega} A(\varphi_i) A(\varphi_j) \; d\mathbf{x}, \; h_i = \int_{\Omega} A(\varphi_i) \left[f - A(\varphi_0) \right] \; d\mathbf{x} \tag{3.5 - 22b}$$

Note that the coefficient matrix is symmetric. The least-squares method requires higher-order coordinate functions than the Rayleigh–Ritz method because the coefficient matrix L_{ij} involves the same operator as in the original differential equation and no trading of differentiation can be achieved. For first-order differential equations the least-squares method yields a symmetric coefficient matrix, whereas the Rayleigh–Ritz and Galerkin methods yield unsymmetric coefficient matrices.

3.6 Summary

In this chapter an introduction to the principles of virtual displacements and its special case, the principles of minimum total potential energy, is presented. The virtual work principles provide a means for the derivation of the governing equations of structural systems, provided one can write the internal and external virtual work expressions for the system. They also yield the natural boundary conditions and give the form of the essential and natural boundary conditions. The last feature proves to be very helpful in the derivation of new higher-order plate theories.

A brief but complete introduction to the Rayleigh–Ritz method and weighted–residual methods is also included in this chapter. The Rayleigh–Ritz method makes direct use of the virtual work principles in determining the approximate solutions. The weighted–residual methods, on the other hand, do not require any variational statements. They are based on a weighted–integral statement of the governing equations.

The single most difficult step in all classical variational methods presented in this chapter is the selection of the coordinate functions. The conditions in (3.5–2) and (3.5–17) on the coordinate functions merely provide the guidelines for their selection, but there exists no systematic procedure for constructing the functions. The selection of coordinate functions becomes more difficult for problems with irregular domains (*i.e.*, noncircular and nonrectangular) or discontinuous data (*i.e.*, loading or geometry). Further, the generation of coefficient matrices for the resulting algebraic equations cannot be automated for a *class* of problems that differ from each other only in the geometry of the domain, boundary conditions, or loading. These limitations of the classical variational methods can be overcome by representing a given domain as a collection of geometrically simple subdomains, called *finite elements*, for which we can systematically generate the coordinate functions. One such technique is the finite element method, which is discussed in Chapter 10 of this book.

The principle of virtual displacements will be used in this book to derive governing equations of plates according to various theories, and the Rayleigh–Ritz and Galerkin methods will be used to determine solutions of beam and plate problems. These ideas are also useful in the study of the finite element method. In the finite element method, the principle of virtual displacements or its equivalent is used over the domain of each finite element and the Rayleigh–Ritz method with algebraic approximation functions is used to derive the *finite element model*, which represents the force–displacement relationships of the element.

Exercises

3.1–3.3 Write expressions for the total virtual work done, $\delta W = \delta U + \delta V$, for each of the beam structures shown in Figs. P3.1 through P3.3.

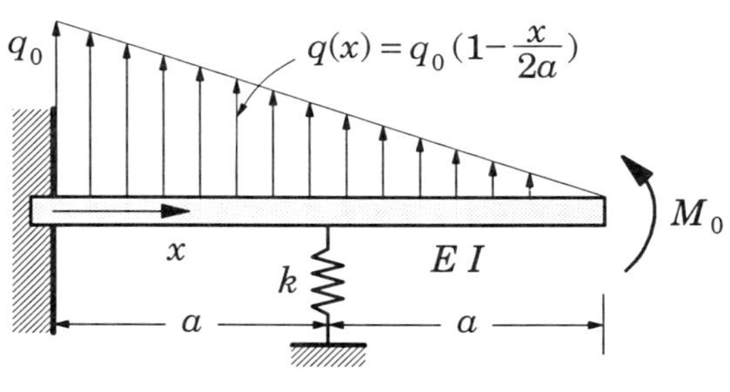

Figure P3.1

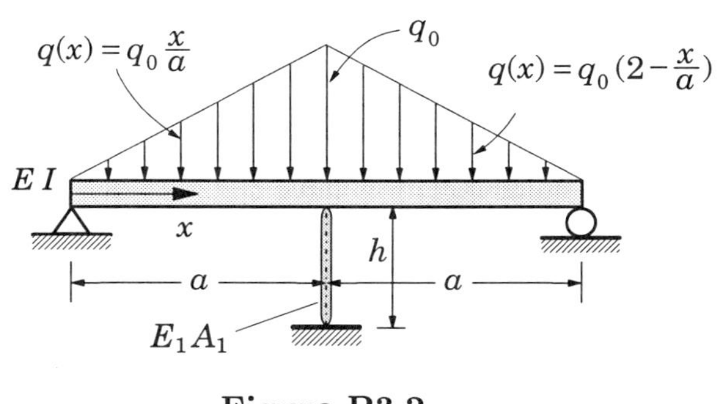

Figure P3.2

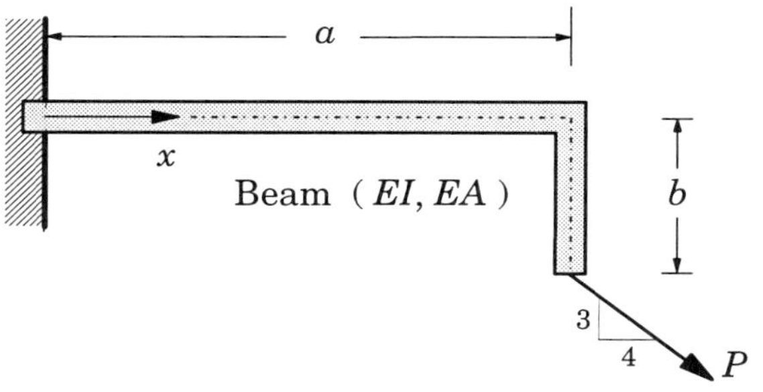

Figure P3.3

3.4–3.6 Write expressions for the total complementary virtual work done, $\delta W^* = \delta U^* + \delta V^*$, for each of the structures shown in Figs. P3.1 through P3.3.

3.7 Compute the strain energy density of a solid body whose material behaves according to the relation

$$\sigma_{ij} = K(\varepsilon_{ij})^n$$

where K is a constant and $n \geq \frac{1}{2}$. Compute the first variation and first partial variations of the following functions of dependent variables. Assume that c_{ij} are not functions of the dependent variables.

Find the Euler–Lagrange equations and the natural boundary conditions associated with each of the following functionals. The dependent variables are listed as the arguments of the functional. All other variables are not functions of the dependent variables.

3.8

$$\Pi(u_0) = \int_0^L \left[\frac{EA}{2}\left(\frac{du_0}{dx}\right)^2 - fu_0\right] dx + \frac{k}{2}[u_0(L)]^2 - Pu_0(L), \quad u_0(0) = 0$$

3.9

$$\Pi(w_0) = \int_0^L \left[\frac{EI}{2}\left(\frac{d^2w_0}{dx^2}\right)^2 + \frac{k}{2}w_0^2 - qw_0\right] dx, \quad w_0(0) = 0, \quad w_0(L) = 0$$

3.10

$$\Pi(u_0, w_0) = \int_0^L \left\{\frac{EA}{2}\left[\frac{du_0}{dx} + \frac{1}{2}\left(\frac{dw_0}{dx}\right)^2\right]^2 + \frac{EI}{2}\left(\frac{d^2w_0}{dx^2}\right)^2\right\} dx - F_0 w_0(L) - Pu_0(L)$$

$$u_0(0) = 0, \; w_0(0) = 0, \; \frac{dw_0}{dx}(0) = 0$$

3.11

$$I(u) = \int_\Omega \left[\frac{a_{11}}{2}\left(\frac{\partial u}{\partial x}\right)^2 + \frac{a_{22}}{2}\left(\frac{\partial u}{\partial y}\right)^2 - fu\right] dxdy - \int_{\Gamma_\sigma} tu \; ds$$

3.12

$$\Pi(w_0) = \int_\Omega \left[\frac{D_{11}}{2}\left(\frac{\partial^2 w_0}{\partial x^2}\right)^2 + D_{22}\left(\frac{\partial^2 w_0}{\partial y^2}\right)^2 + 2D_{12}\left(\frac{\partial^2 w_0}{\partial x \partial y}\right)^2 - qw_0\right] dxdy$$

$$w_0 = 0, \; \frac{\partial w_0}{\partial n} = 0 \;\; \text{on the boundary} \;\; \Gamma$$

3.13

$$\begin{aligned} I(u,v) = \int_\Omega \Bigg\{ \frac{1}{2} \Bigg[& \left(c_{11}\frac{\partial u}{\partial x} + c_{12}\frac{\partial v}{\partial y}\right)^2 + \left(c_{12}\frac{\partial u}{\partial x} + c_{22}\frac{\partial v}{\partial y}\right)^2 \\ & + c_{33}\left(\frac{\partial u}{\partial y} + \frac{\partial v}{\partial x}\right)^2 \Bigg] - f_1 u - f_2 v \Bigg\} dxdy \\ & - \int_{\Gamma_2} (t_1 u + t_2 v) ds \end{aligned}$$

$$u = \hat{u} \text{ and } v = \hat{v} \;\; \text{on} \;\; \Gamma_1, \;\; \Gamma_1 + \Gamma_2 = \Gamma$$

3.14 (*Torsion of Prismatic Members*) Consider torsion of a constant-section cylindrical member made of homogeneous, isotropic and linear elastic material. Assuming that the torque applied, M_z, about the axis of the cylindrical member is small so that the rotation of the cross sections along the length is constant, the displacements of a point (x, y) in the cross section can be expressed in the form

$$u = -y\theta z, \;\; v = x\theta z, \;\; w = w(x,y)$$

where θ is the angle of twist, which is considered as a dependent variable. The work done by the applied torque M_z is

$$V = -M_z \theta L$$

Use the principle of total potential energy to derive the differential equation and natural boundary conditions governing the warping function w.

3.15 (*The Timoshenko Beam Theory*) Starting with the displacement field

$$\begin{aligned} u_1(x,z,t) &= u_0(x,t) + z\phi(x,t) \\ u_2(x,z,t) &= 0 \\ u_3(x,z,t) &= w_0(x,t) \end{aligned}$$

derive the governing differential equations from the principle of virtual displacements for the dynamic case. Assume linear strain–displacement equations.

Ans:

$$\frac{\partial N_{xx}}{\partial x} = I_0 \frac{\partial^2 u_0}{\partial t^2} \tag{1}$$

$$\frac{\partial Q_x}{\partial x} + q = I_0 \frac{\partial^2 w_0}{\partial t^2} \tag{2}$$

$$\frac{\partial M_{xx}}{\partial x} - Q_x = I_2 \frac{\partial^2 \phi}{\partial t^2} \tag{3}$$

where

$$\left\{ \begin{array}{c} N_{xx} \\ M_{xx} \end{array} \right\} = \int_{-h/2}^{h/2} \left\{ \begin{array}{c} 1 \\ z \end{array} \right\} \sigma_{xx}\, dz, \quad Q_x = \int_{-h/2}^{h/2} \sigma_{xz}\, dz \tag{4}$$

$$I_i = \int_{-h/2}^{h/2} \rho(z)^i\, dz \tag{5}$$

3.16 Consider the displacement field

$$\begin{aligned} u_1(x,y,z) &= u_0(x) + y\psi(x) + z\phi(x) \\ u_2(x,y,z) &= v_0(x) \\ u_3(x,y,z) &= w_0(x) \end{aligned}$$

for bending of beams about both the $y-$ and $z-$axes. Assuming that distributed loads of $p(x)$, $q(x)$ and $f(x)$ per unit length are applied along the x, y and z axes, respectively, derive the differential equations of equilibrium in terms of appropriate stress resultants for the small displacement case. Write the equilibrium equations in terms of the displacement (u_0, v_0, w_0, ψ, ϕ) and thermal stress resultants by assuming the stress–strain relations

$$\sigma_{xx} = E(\varepsilon_x - \alpha T), \ \sigma_{xy} = 2G\varepsilon_{xy}, \ \sigma_{xz} = 2G\varepsilon_{xz}$$

3.17 (*The Reddy Third-Order Beam Theory*) Consider the displacement field

$$\begin{aligned} u(x,z,t) &= u_0(x,t) + z\phi(x,t) - \alpha z^3 \left(\phi + \frac{\partial w_0}{\partial x}\right) \\ w(x,z,t) &= w_0(x,t) \end{aligned} \tag{1}$$

where $\alpha = 4/3h^3$. The displacement field accommodates quadratic variation of transverse shear strains and stresses and vanishing of transverse shear shear stress on the top and bottom planes of a beam.

(a) Show that the nonzero linear strains associated with the displacement field (1) are

$$\varepsilon_{xx} = \varepsilon_{xx}^{(0)} + z\varepsilon_{xx}^{(1)} + z^3\varepsilon_{xx}^{(3)}$$
$$\gamma_{xz} = \gamma_{xz}^{(0)} + z^2\gamma_{xz}^{(2)} \tag{2a}$$

where

$$\varepsilon_{xx}^{(0)} = \frac{\partial u_0}{\partial x}\,,\quad \varepsilon_{xx}^{(1)} = \frac{\partial \phi}{\partial x}\,,\quad \varepsilon_{xx}^{(3)} = -\alpha\left(\frac{\partial \phi}{\partial x} + \frac{\partial^2 w_0}{\partial x^2}\right)$$
$$\gamma_{xz}^{(0)} = \phi + \frac{\partial w_0}{\partial x}\,,\quad \gamma_{xz}^{(2)} = -\beta\left(\phi + \frac{\partial w_0}{\partial x}\right) \tag{2b}$$

where $\beta = 3\alpha = 4/h^2$.

(b) Using the dynamic version of the principle of virtual displacements, show that the equations of motion of the theory are given by

$$\frac{\partial N_{xx}}{\partial x} = I_0\frac{\partial^2 u_0}{\partial t^2} \tag{3}$$

$$\frac{\partial \bar{Q}_x}{\partial x} + \frac{\partial}{\partial x}\left(N_{xx}\frac{\partial w_0}{\partial x}\right) + \alpha\frac{\partial^2 P_{xx}}{\partial x^2} + q$$
$$= I_0\frac{\partial^2 w_0}{\partial t^2} + \alpha\left(J_4\frac{\partial^3 \phi}{\partial x \partial t^2} - \alpha I_6\frac{\partial^4 w_0}{\partial x^2 \partial t^2}\right) \tag{4}$$

$$\frac{\partial \bar{M}_{xx}}{\partial x} - \bar{Q}_x = K_2\frac{\partial^2 \phi}{\partial t^2} - \alpha J_4\frac{\partial^3 w_0}{\partial x \partial t^2} \tag{5}$$

where

$$\begin{Bmatrix} N_{xx} \\ M_{xx} \\ P_{xx} \end{Bmatrix} = \int_{-h/2}^{h/2} \begin{Bmatrix} 1 \\ z \\ z^3 \end{Bmatrix} \sigma_{xx}\, dz,\quad \begin{Bmatrix} Q_x \\ R_x \end{Bmatrix} = \int_{-h/2}^{h/2} \begin{Bmatrix} 1 \\ z^2 \end{Bmatrix} \sigma_{xz}\, dz \tag{6}$$

$$\bar{M}_{xx} = M_{xx} - \alpha P_{xx},\quad \bar{Q}_x = Q_x - \beta R_x \tag{7}$$

$$J_i = I_i - \alpha I_{i+2},\quad K_2 = I_2 - 2\alpha I_4 + \alpha^2 I_6,\quad I_i = \int_{-h/2}^{h/2} \rho(z)^i\, dz \tag{8}$$

and (P_{xx}, R_x) denote the higher-order stress resultants. Note that the primary and secondary variables of the theory are

$$\text{Primary Variables}: \quad u_0,\ w_0,\ \frac{\partial w_0}{\partial x},\ \phi \tag{9a}$$
$$\text{Secondary Variables}: \quad N_{xx},\ V_x,\ P_{xx},\ \bar{M}_{xx} \tag{9b}$$

where

$$V_x \equiv \bar{Q}_x + N_{xx}\frac{\partial w_0}{\partial x} + \alpha\left[\frac{\partial P_{xx}}{\partial x} - \left(I_3\ddot{u}_0 + J_4\ddot{\phi}_x - \alpha I_6\frac{\partial \ddot{w}_0}{\partial x}\right)\right] \tag{10}$$

3.18 Suppose that the total displacements (u, v, w) along the three coordinate axes (x, y, z) in a laminated beam can be expressed as

$$\begin{aligned} u(x,z) &= u_0(x) + z\phi_x(x) + z^2\psi_x(x) + z^3\theta_x(x) \\ v(x,z) &= 0 \\ w(x,z) &= w_0(x) + z\phi_z(x) + z^2\psi_z(x) + z^3\theta_z(x) \end{aligned} \tag{1}$$

where (u_0, w_0) denote the displacements of a point $(x, y, 0)$ along the x and z directions, respectively, ϕ_x denotes the rotation of a transverse normal about the $y-$axis, and $\psi_x, \theta_x, \phi_z, \psi_z$, and θ_z are functions of x. Construct the total potential energy functional for the theory. Assume that the beam is subjected to a distributed load $q(x)$ at the top surface of the beam.

3.19 Give the approximation functions φ_1 and φ_0 required in the (i) Rayleigh–Ritz and (ii) weighted–residual methods to solve the following problems:

(a) A bar fixed at the left end and connected to an axial elastic spring (spring constant, k) at the right end.

(b) A beam clamped at the left end and simply supported at the right end.

3.20 Consider a uniform beam fixed at one end and supported by an elastic spring (spring constant k) in the vertical direction. Assume that the beam is loaded by uniformly distributed load q_0. Determine a one-parameter Ritz solution using algebraic functions.

3.21 Use the total potential energy functional in Eq. (3.4–31) to determine a two-parameter Ritz solution of a simply supported beam subjected a transverse point load F_0 at the center. You may use the symmetry about the center $(x = L/2)$ of the beam.

3.22 Consider the buckling of a uniform beam according to the Timoshenko beam theory. The total potential energy functional for the problem can be written as

$$\Pi(w_0, \phi_x) = \frac{1}{2}\int_0^L \left[D\left(\frac{d\phi_x}{dx}\right)^2 + S\left(\frac{dw_0}{dx} - \phi_x\right)^2 - N\left(\frac{dw_0}{dx}\right)^2 \right] dx$$

where $w_0(x)$ is the transverse deflection, ϕ_x is the rotation, D is the flexural stiffness, S is the shear stiffness, and N is the axial compressive load. We wish to determine the critical buckling load P of a simply supported beam using the Rayleigh–Ritz method. Assume one-parameter approximation of the form

$$w_0(x) = c_1 \sin\frac{\pi x}{L}\,, \qquad \phi_x(x) = c_2 \cos\frac{\pi x}{L}$$

and determine the critical buckling load.

3.23 Determine a two-parameter Galerkin solution of the cantilever beam problem in Example 3.5–1.

3.24 Determine a two-parameter collocation solution of the cantilever beam problem in Example 3.5–1. Use collocation points $x = L/2$ and $x = L$.

References for Additional Reading

1. Galerkin, B. G., "Rods and Plates. Series Occurring in Various Questions Concerning the Elastic Equilibrium of Rods and Plates," (in Russian), *Vestnik Inghenerov*, **19**, 897–908 (1915).
2. Lanczos, C., *The Variational Principles of Mechanics*, The University of Toronto Press, Toronto, 1964.
3. Langhaar, H. L., *Energy Methods in Applied Mechanics*, Wiley, New York, 1962.
4. Mikhlin, S. G., *Variational Methods in Mathematical Physics*, (translated from the 1957 Russian edition by T. Boddington) The MacMillan Company, New York, 1964.
5. Mikhlin, S. G., *The Problem of the Minimum of a Quadratic Functional* (translated from the 1952 Russian edition by A. Feinstein), Holden–Day, San Francisco, 1965.
6. Mikhlin, S. G., *An Advanced Course of Mathematical Physics*, American Elsevier, New York, 1970.
7. Oden, J. T. and Reddy, J. N., *Variational Methods in Theoretical Mechanics*, 2nd Edition, Springer–Verlag, Berlin, 1982.
8. Oden, J. T. and Ripperger, E. A., *Mechanics of Elastic Structures*, Second Edition., Hemisphere, New York, 1981.
9. Reddy, J. N. and Rasmussen, M. L., *Advanced Engineering Analysis*, John Wiley, New York, 1982; reprinted by Krieger, Melbourne, Florida, 1991.
10. Reddy, J. N., *Energy and Variational Methods in Applied Mechanics*, John Wiley, New York, 1984.
11. Reddy, J. N., *Applied Functional Analysis and Variational Methods in Engineering*, McGraw–Hill, New York, 1986; reprinted by Krieger, Melbourne, Florida, 1992.
12. Ritz, W., "Über Eine Methode zur *Lösung* Gewisser Varationsprobleme der Mathematischen Physik," *Journal Reine Angew. Math.*, **135**(1), p. 1 (1909).
13. Washizu, K., *Variational Methods in Elasticity and Plasticity*, 3rd Edition, Pergamon Press, New York, 1982.
14. Timoshenko, S. P., "On the Correction for Shear of the Differential Equation for Transverse Vibrations of Prismatic Bars," *Philosophical Magazine*, **41**, 744–746 (1921).
15. Timoshenko, S. P., "On the Transverse Vibrations of Bars of Uniform Cross Section," *Philosophical Magazine*, **43**, 125–131 (1922).
16. Timoshenko, S. P. and Woinowsky–Krieger, S., *Theory of Plates and Shells*, McGraw–Hill, Singapore, 1970.

Chapter Four

Introduction to Composite Materials

4.1 Basic Concepts and Terminology

Composite materials are those formed by combining two or more materials on a macroscopic scale such that they have better engineering properties than the conventional materials, for example, metals. Some of the properties that can be improved by forming a composite material are stiffness, strength, weight reduction, corrosion resistance, thermal properties, fatigue life, and wear resistance. Most man-made composite materials are made from two materials: a reinforcement material called *fiber* and a base material, called *matrix* material.

Composite materials are commonly formed in three different types: (1) *fibrous composites*, which consist of fibers of one material in a matrix material of another; (2) *particulate composites*, which are composed of macrosize particles of one material in a matrix of another; and (3) *laminated composites*, which are made of layers of different materials, including composites of the first two types. The particles and matrix in particulate composites can be either metallic or nonmetallic. Thus there exist four possible combinations: metallic in nonmetallic, nonmetallic in metallic, nonmetallic in nonmetallic, and metallic in metallic.

The stiffness and strength of fibrous composites come from fibers which are stiffer and stronger than the same material in bulk form. Shorter fibers, called whiskers, exhibit better strength and stiffness properties than long fibers. Whiskers are about 1 to 10 microns (*i.e.*, micro inches or μ in.) in diameter and 10 to 100 times as long. Fibers may be 5 microns to 0.005 inches. Some forms of graphite fibers are 5 to 10 microns in diameter, and they are handled as a bundle of several thousand fibers. The matrix material keeps the fibers together, acts as a load-transfer medium between fibers, and protects fibers from being exposed to the environment. Matrix materials have their usual bulk-form properties whereas fibers have directionally dependent properties.

The basic mechanism of load transfer between the matrix and a fiber can be explained by considering a bar of single fiber in a matrix material (see Figure 4.1–1a). The load transfer between the matrix material and fiber takes place through shear stress. When the applied load is tensile, shear stress τ develops on the outer surface of the fiber, and its magnitude decreases from a high value at the end of the fiber to zero at a distance from the end. The tensile stress σ in the fiber cross section has the opposite trend, starting from zero value at the end of the fiber to its maximum at a distance from the end. The two stresses together balance the applied load, P. The distance from the free end to the point at which the normal stress attains its maximum and shear stress becomes zero is known as the *characteristic distance.* The pure tensile state continues along the rest of the fiber.

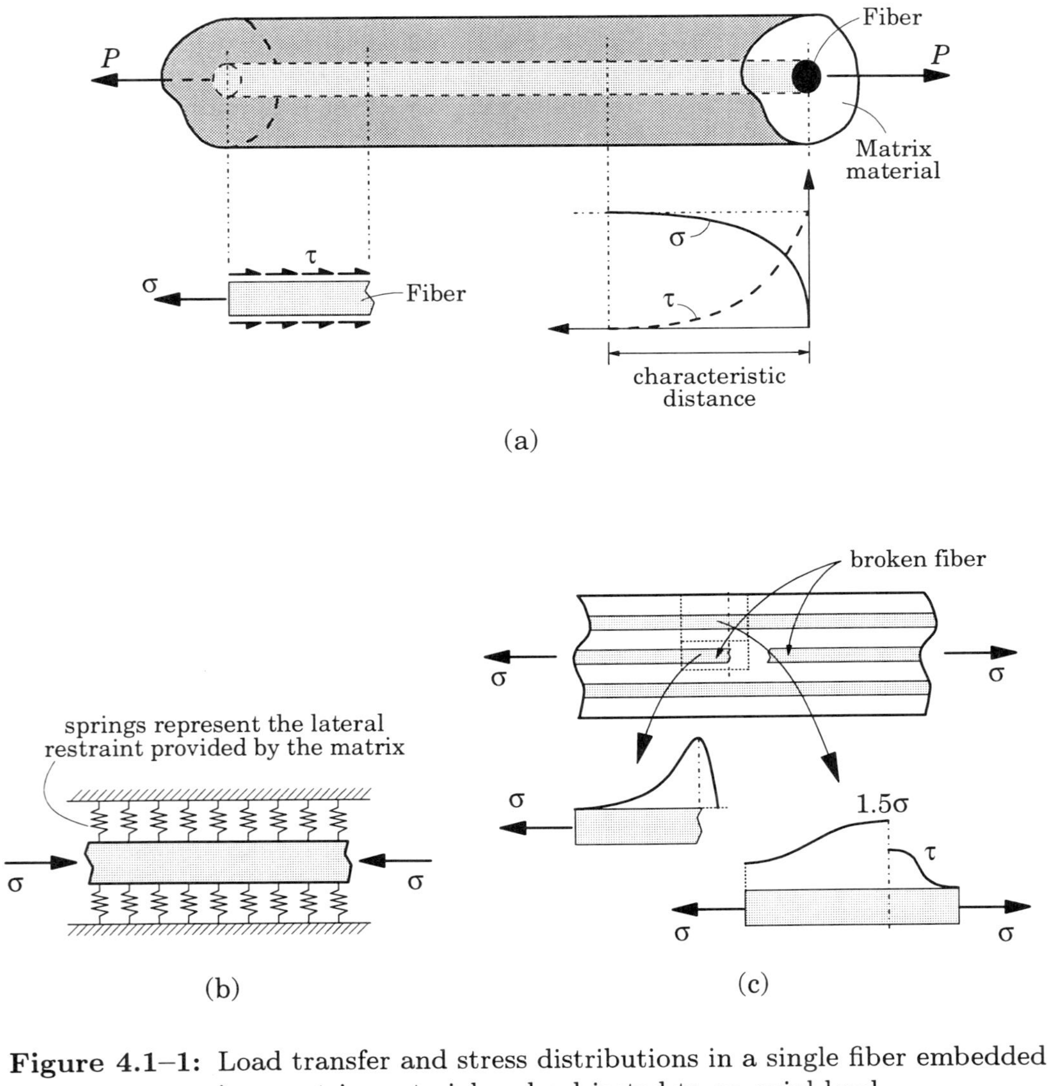

Figure 4.1–1: Load transfer and stress distributions in a single fiber embedded in a matrix material and subjected to an axial load.

When a compressive load is applied, the stresses in the region of characteristic length are reversed in sign; in the compressive region, *i.e.*, rest of the fiber length, the fiber tends to buckle, much like a wire subjected to compressive load. At this stage, the matrix provides a lateral support to reduce the tendency of the fiber to buckle (Figure 4.1–1b). When a fiber is broken, the load carried by the fiber is transferred through shear stress to the neighboring two fibers (see Figure 4.1–1c), elevating the fiber axial stress level to a value of 1.5σ.

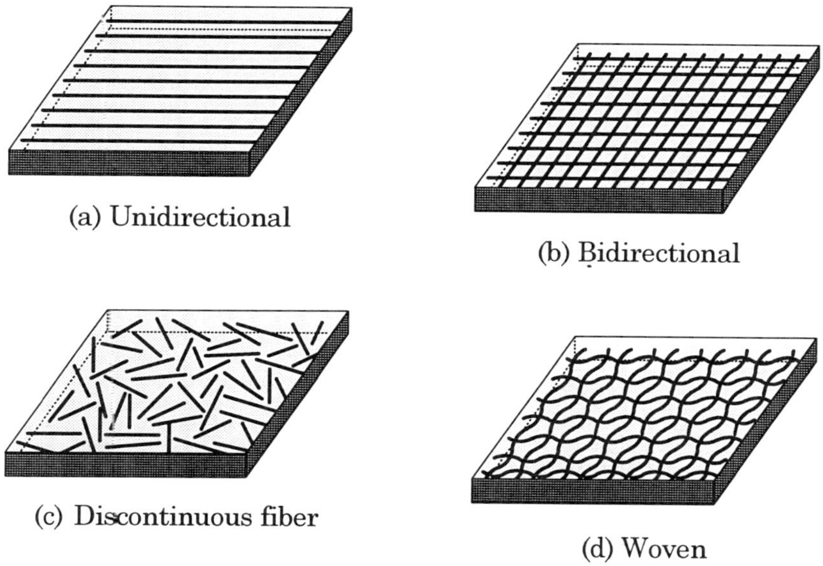

Figure 4.1–2: Various types of fiber-reinforced composite laminae.

A *lamina* or *ply* is a typical sheet of composite material. It represents a fundamental building block. A fiber-reinforced lamina consists of many fibers embedded in a matrix material, which can be a metal like aluminum, or a nonmetal like a thermoset or thermoplastic polymer. Often, coupling (chemical) agents and fillers are added to improve the bonding between fibers and matrix material and increase toughness. The fibers can be continuous or discontinuous, woven, unidirectional, bidirectional, or randomly distributed (see Figure 4.1–2). Unidirectional fiber-reinforced laminae exhibit the highest strength and modulus in the direction of the fibers, but they have very low strength and modulus in the direction transverse to the fibers. A poor bonding between a fiber and matrix results in poor transverse properties and failures in the form of fiber pull out, fiber breakage, and fiber buckling. Discontinuous fiber-reinforced composites have lower strength and modulus than continuous fiber-reinforced composites.

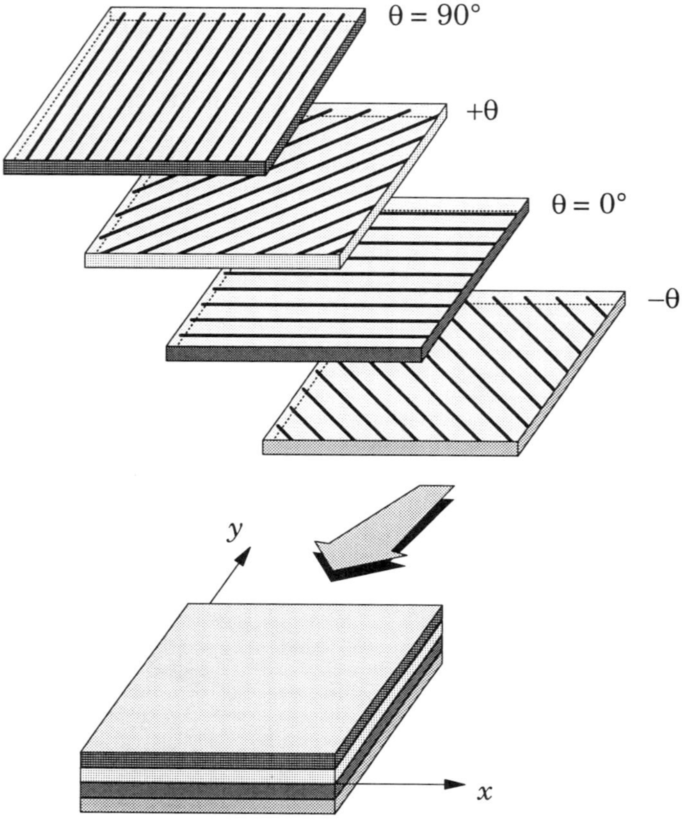

Figure 4.1–3: A laminate made up of laminae with different fiber orientations.

A *laminate* is a collection of laminae stacked to achieve the desired stiffness and thickness. For example, unidirectional fiber-reinforced laminae can be stacked so that the fibers in each lamina are oriented in the same or different directions (see Figure 4.1–3). The sequence of various orientations of a fiber-reinforced composite layer in a laminate is termed the *lamination scheme* or *stacking sequence*. The layers are usually bonded together with the same matrix material as that in a lamina. A unidirectional laminate (*i.e.*, all laminae have the same fiber orientation) will be very strong along the fiber direction and weak in the transverse direction. The laminate will be weak in shear also. If a laminate has layers with fibers oriented at 30° or 45°, it can take shear loads. The lamination scheme and material properties of individual lamina provide an added flexibility to designers to tailor the stiffness and strength of the laminate to match the structural requirements. Laminated composite structures also have disadvantages. Because of the mismatch of material properties between layers, the shear stresses produced between the layers, especially at the edges of a laminate, may cause delamination. Also, during manufacturing of laminates, material defects such as interlaminar voids, delamination, incorrect orientation, damaged fibers, and variation in thickness may be introduced. It is impossible to eliminate manufacturing defects altogether. Therefore, analysis and design procedures should account for the defects.

This book is devoted to the study of laminated structures. Determination of static, transient, vibration, and buckling characteristics of laminates with different lamination schemes, thicknesses, loads, and boundary conditions constitutes the major objective of the study. The information gained from such a study can help designers to select materials, number of layers, and orientations for the best performance in a particular application. The theoretical formulations presented and analysis tools developed in this book should prove to be invaluable to engineers as well as researchers in the field of composite materials.

In the remaining portion of this chapter, we study the mechanical behavior of a single lamina, treating it as a linear, orthotropic, elastic continuum. The generalized Hooke's law is revisited for an orthotropic material, the elastic coefficients of an orthotropic material are expressed in terms of engineering constants of a lamina, and the fiber-matrix interactions in a unidirectional lamina are discussed. Transformation of stresses, strains, and elasticity coefficients from the lamina material coordinates to the problem coordinates are also presented.

4.2 Constitutive Equations of a Lamina

4.2–1 Generalized Hooke's Law: Isothermal Case

In this section we study the mechanical behavior of a typical lamina, which is the basic building block of a composite laminate. In formulating the constitutive equations of a lamina we assume that:

(i) a lamina is a continuum, *i.e.*, no gaps or empty spaces exist.

(ii) a lamina behaves as a linear elastic material.

The first assumption amounts to considering the *macromechanical behavior* of a lamina. If fiber–matrix debonding and fiber breakage, for example, are to be included in the formulation of the constitutive equations of a lamina, then we must consider the *micromechanics* approach, which treats the constituent materials as continua and accounts for the mechanical behavior of the constituents and possibly their interactions. The second assumption implies that the generalized Hooke's law is valid. It should be noted that both assumptions can be removed if we were to develop micromechanical constitutive models for inelastic (*e.g.*, plastic, viscoelastic, etc.) behavior of a lamina.

Composite materials are inherently heterogeneous from the microscopic point of view. From the macroscopic point of view, wherein the material properties of a composite are derived from a weighted average of the constituent materials, fiber and matrix, composite materials are assumed to be homogeneous. The following discussion of constitutive equations is independent of whether the material is homogeneous or not, because the stress–strain relations hold for a typical point in the body.

Table 4.2–1: Correspondence between the tensor and contracted components of stress and strain.

Stresses		Strains	
Tensor Notation	Contracted Notation	Tensor Notation	Contracted Notation
σ_{11}	σ_1	ε_{11}	ε_1
σ_{22}	σ_2	ε_{22}	ε_2
σ_{33}	σ_3	ε_{33}	ε_3
σ_{23}	σ_4	$2\varepsilon_{23}$	ε_4
σ_{13}	σ_5	$2\varepsilon_{13}$	ε_5
σ_{12}	σ_6	$2\varepsilon_{12}$	ε_6

The generalized Hooke's law for an anisotropic material under isothermal conditions is (see Section 2.4) given by

$$\sigma_{ij} = C_{ijk\ell}\varepsilon_{k\ell} \tag{4.2 – 1}$$

or in contracted notation (see Table 4.2–1)

$$\sigma_i = C_{ij}\varepsilon_j \tag{4.2 – 2}$$

where σ_{ij} (σ_i) are the stress components, ε_{ij} (ε_i) are the strain components, and C_{ij} are the material coefficients, all referred to an orthogonal Cartesian coordinate system (x_1, x_2, x_3) fixed in the body.

When materials possess one or more planes of material symmetry, the number of independent elastic coefficients can be reduced, as discussed in detail in Section 2.4. Often fiber-reinforced lamina are characterized as orthotropic, *i.e.*, the material has three orthogonal planes of material symmetry. If the coordinate planes are chosen parallel to the three orthogonal planes of material symmetry, the stress–strain relations and strain–stress relations can be written as

$$\begin{Bmatrix} \sigma_1 \\ \sigma_2 \\ \sigma_3 \\ \sigma_4 \\ \sigma_5 \\ \sigma_6 \end{Bmatrix} = \begin{bmatrix} C_{11} & C_{12} & C_{13} & 0 & 0 & 0 \\ C_{12} & C_{22} & C_{23} & 0 & 0 & 0 \\ C_{13} & C_{23} & C_{33} & 0 & 0 & 0 \\ 0 & 0 & 0 & C_{44} & 0 & 0 \\ 0 & 0 & 0 & 0 & C_{55} & 0 \\ 0 & 0 & 0 & 0 & 0 & C_{66} \end{bmatrix} \begin{Bmatrix} \varepsilon_1 \\ \varepsilon_2 \\ \varepsilon_3 \\ \varepsilon_4 \\ \varepsilon_5 \\ \varepsilon_6 \end{Bmatrix} \tag{4.2 – 3}$$

$$\begin{Bmatrix} \varepsilon_1 \\ \varepsilon_2 \\ \varepsilon_3 \\ \varepsilon_4 \\ \varepsilon_5 \\ \varepsilon_6 \end{Bmatrix} = \begin{bmatrix} S_{11} & S_{12} & S_{13} & 0 & 0 & 0 \\ S_{12} & S_{22} & S_{23} & 0 & 0 & 0 \\ S_{13} & S_{23} & S_{33} & 0 & 0 & 0 \\ 0 & 0 & 0 & S_{44} & 0 & 0 \\ 0 & 0 & 0 & 0 & S_{55} & 0 \\ 0 & 0 & 0 & 0 & 0 & S_{66} \end{bmatrix} \begin{Bmatrix} \sigma_1 \\ \sigma_2 \\ \sigma_3 \\ \sigma_4 \\ \sigma_5 \\ \sigma_6 \end{Bmatrix} \tag{4.2 – 4}$$

where S_{ij} denote the compliance coefficients, $[S] = [C]^{-1}$.

4.2–2 Generalized Hooke's Law: Nonisothermal Case

In much of the present study we shall assume that the material coefficients of a lamina remain unchanged during deformation. When damage in composites is studied, the material coefficients are adjusted/reduced to account for various types of microdamage (*e.g.*, fiber breaks, matrix cracks). When temperature changes occur in the elastic body, we account for the thermal expansion of the material, even though the variation of elastic constants with temperature is neglected. When the strains, geometric changes, and temperature variations are sufficiently small, all governing equations are linear and superposition of mechanical and thermal effects is possible.

The linear thermoelastic constitutive equations can be derived using arguments presented in Section 2.5. For a linear thermoelastic equation, we assume that there exists a function Ψ, called the Helmholtz *free-energy function*, of the form

$$\Psi = -\beta_i(T - T_0)\varepsilon_i + \hat{C}_{ij}\varepsilon_i\varepsilon_j + C_7(T - T_0) + C_8(T - T_0)^2 \tag{4.2 – 5}$$

where C_7 and C_8 are material coefficients, $T - T_0$ denotes the temperature change from reference temperature T_0, and summation on repeated subscripts is taken in the range of 1 to 6. The stresses are then derivable from the free-energy function

$$\sigma_j = \frac{\partial \Psi}{\partial \varepsilon_j} = -\beta_j(T - T_0) + C_{ji}\varepsilon_i \tag{4.2 – 6}$$

Equation (6) is known as the *generalized Duhamel–Neumann form* of Hooke's law. In writing Eq. (6) we have assumed that β_i and C_{ij} are independent of strain and temperature. The parameters β_i are known in terms of the linear coefficients of thermal expansion α_i as

$$\beta_j = C_{ji}\alpha_i \tag{4.2-7}$$

Therefore, the linear thermoelastic constitutive equations take the form

$$\sigma_j = C_{ji}[-\alpha_i(T - T_0) + \varepsilon_i] \tag{4.2-8}$$

$$\varepsilon_j = S_{ji}\sigma_i + \alpha_i(T - T_0) \tag{4.2-9}$$

For orthotropic materials α_4, α_5, and α_6 are zero, and for isotropic behavior we have $\alpha_1 = \alpha_2 = \alpha_3$.

4.2-3 Engineering Constants of Orthotropic Materials

The constitutive equations for a material which has three orthogonal planes of material property symmetry (*i.e.*, orthotropic materials), referred to the principal material directions, are given by Eqs. (3) and (4). Fiber-reinforced materials are often characterized as orthotropic with respect to a rectangular coordinate system in which the fiber direction is parallel to the x_1–axis. In this case, the material stiffness coefficients C_{ij} can be expressed in terms of engineering constants, *i.e.*, Young's moduli, Poisson's ratios, and shear moduli [see Eq. (2.4–9)]. Most often, the material properties are determined in a laboratory in terms of the engineering constants. These constants are measured using simple tests like uniaxial tension or pure shear. Because of their direct and obvious physical meaning, engineering constants are used in place of the more abstract stiffness and compliance coefficients of Eqs. (3) and (4).

For linear elastic behavior, the strains due to the application of individual uniaxial and shear stress components can be superposed. For example, the extensional strain ε_{11} in the principal material (or fiber) direction 1 due to the stress σ_{11} in the same direction is σ_{11}/E_1, where E_1 denotes Young's modulus of the smeared fiber-reinforced material in the fiber direction. The extensional strain ε_{11} due to the stress σ_{22} in direction 2 is $-\nu_{21}\sigma_{22}/E_2$, where ν_{21} is a Poisson's ratio

$$\nu_{21} = -\frac{\varepsilon_{11}}{\varepsilon_{22}}$$

and E_2 is Young's modulus of the smeared material in material direction 2. Similarly, σ_{33} contributes to ε_{11} by the amount $-\nu_{31}\sigma_{33}/E_3$. Hence, the total normal strain ε_{11} due to the simultaneous application of all three normal stress components is

$$\varepsilon_{11} = \frac{\sigma_{11}}{E_1} - \frac{\sigma_{22}\nu_{21}}{E_2} - \frac{\sigma_{33}\nu_{31}}{E_3}$$

Similarly, we can write

$$\varepsilon_{22} = -\frac{\sigma_{11}\nu_{12}}{E_1} + \frac{\sigma_{22}}{E_2} - \frac{\sigma_{33}\nu_{32}}{E_3}$$

$$\varepsilon_{33} = -\frac{\sigma_{11}\nu_{13}}{E_1} - \frac{\sigma_{22}\nu_{23}}{E_2} + \frac{\sigma_{33}}{E_3}$$

The simple shear tests with an orthotropic material give the results

$$2\varepsilon_{12} = \frac{\sigma_{12}}{G_{12}}\ , \quad 2\varepsilon_{13} = \frac{\sigma_{13}}{G_{13}}\ , \quad 2\varepsilon_{23} = \frac{\sigma_{23}}{G_{23}}$$

where $2\varepsilon_{ij}$ $(i \neq j)$ denotes the change in the right angle between two lines parallel to the x_1 and x_2 directions at a point, σ_{ij} $(i \neq j)$ denotes the corresponding shear stress in the $x_i - x_j$ plane, and G_{ij} $(i \neq j)$ are the shear moduli in the $x_i - x_j$ plane.

The strain–stress relations can be expressed in matrix form, using the single–subscript notation (*i.e.*, contracted notation), as

$$\begin{Bmatrix} \varepsilon_1 \\ \varepsilon_2 \\ \varepsilon_3 \\ \varepsilon_4 \\ \varepsilon_5 \\ \varepsilon_6 \end{Bmatrix} = \begin{bmatrix} \frac{1}{E_1} & -\frac{\nu_{21}}{E_2} & -\frac{\nu_{31}}{E_3} & 0 & 0 & 0 \\ -\frac{\nu_{12}}{E_1} & \frac{1}{E_2} & -\frac{\nu_{32}}{E_3} & 0 & 0 & 0 \\ -\frac{\nu_{13}}{E_1} & -\frac{\nu_{23}}{E_2} & \frac{1}{E_3} & 0 & 0 & 0 \\ 0 & 0 & 0 & \frac{1}{G_{23}} & 0 & 0 \\ 0 & 0 & 0 & 0 & \frac{1}{G_{13}} & 0 \\ 0 & 0 & 0 & 0 & 0 & \frac{1}{G_{12}} \end{bmatrix} \begin{Bmatrix} \sigma_1 \\ \sigma_2 \\ \sigma_3 \\ \sigma_4 \\ \sigma_5 \\ \sigma_6 \end{Bmatrix} \tag{4.2 – 10}$$

Comparing Eq. (10) with Eq. (4), we conclude that the 6×6 matrix in Eq. (10) is the compliance matrix $[S]$. Since the compliance matrix $[S]$ is the inverse of the stiffness matrix $[C]$ in Eq. (3), and because the inverse of a symmetric matrix is symmetric, it follows that the compliance matrix $[S]$ is also a symmetric matrix. This in turn implies, in view of Eq. (10), that the following reciprocal relations hold:

$$\frac{\nu_{21}}{E_2} = \frac{\nu_{12}}{E_1}; \quad \frac{\nu_{31}}{E_3} = \frac{\nu_{13}}{E_1}; \quad \frac{\nu_{32}}{E_3} = \frac{\nu_{23}}{E_2} \tag{4.2 – 11a}$$

or, in short [see Eq. (2.4–10)]

$$\frac{\nu_{ij}}{E_i} = \frac{\nu_{ji}}{E_j} \quad \text{(no sum on } i,j) \tag{4.2 – 11b}$$

for $i, j = 1, 2, 3$. Thus, there are only nine independent material coefficients

$$E_1, \ E_2, \ E_3, \ G_{23}, \ G_{13}, \ G_{12}, \ \nu_{12}, \ \nu_{13}, \ \nu_{23} \qquad (4.2-12)$$

for an orthotropic material, since ν_{21}, ν_{31}, and ν_{32} can be expressed in terms of the three principal moduli and Poisson's ratios ν_{12}, ν_{13}, and ν_{23}. Physical arguments can be used to restrict Young's moduli and shear moduli to be positive:

$$E_1, \ E_2, \ E_3, \ G_{12}, \ G_{23}, \ G_{13} > 0 \qquad (4.2-13)$$

For isotropic materials we have $E_1 = E_2 = E_3 = E$, $G_{12} = G_{13} = G_{23} \equiv G$, and $\nu_{12} = \nu_{23} = \nu_{13} \equiv \nu$. Of the three constants (E, ν, G), only two are independent and the third one is related to the other two by the relation

$$G = \frac{E}{2(1+\nu)} \qquad (4.2-14)$$

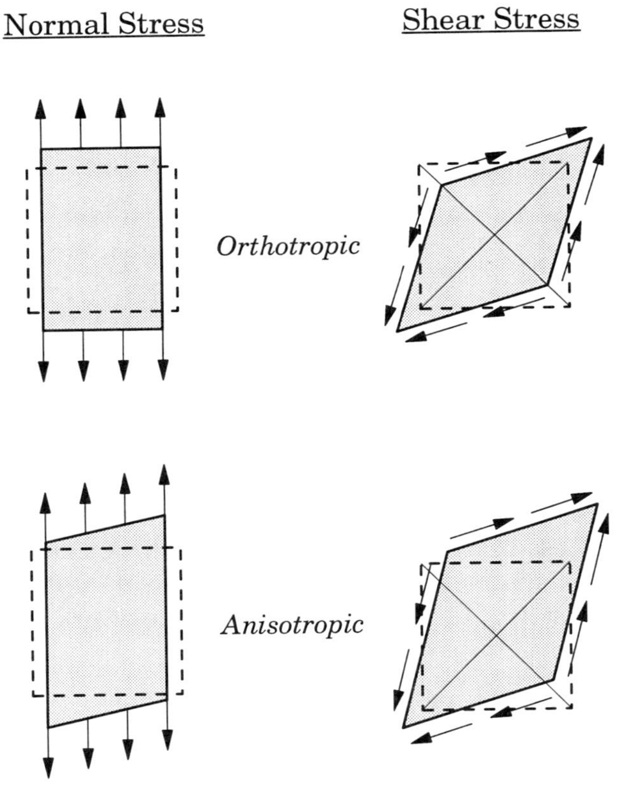

Figure 4.2–1: Deformation of orthotropic and anisotropic rectangular block under uniaxial tension.

A qualitative understanding of the anisotropic behavior of a material can be obtained by simple tension and shear tests. Application of a normal stress to a rectangular block of isotropic or orthotropic material leads to only extension in the direction of the applied stress and contraction perpendicular to it, whereas an anisotropic material experiences extension in the direction of the applied normal stress, contraction perpendicular to it, as well as shearing strain (see Figure 4.2–1). Conversely, the application of a shearing stress to an anisotropic material causes shearing strain as well as normal strains. Normal stress applied to an orthotropic material at an angle to its principal material directions causes it to behave like an anisotropic material. The coupling between the two loading modes and the two deformation modes plays a significant role in the testing, analysis, and design of composite materials.

4.2–4 Characterization of a Unidirectional Lamina

A unidirectional fiber-reinforced lamina is treated as an orthotropic material whose material symmetry planes are parallel and transverse to the fiber direction. The material coordinate axis x_1 is taken to be parallel to the fiber, the x_2–axis transverse to the fiber direction in the plane of the lamina, and the x_3–axis is perpendicular to the plane of the lamina (see Figure 4.2–2). The orthotropic material properties of a lamina are obtained either by the theoretical approach or by laboratory testing.

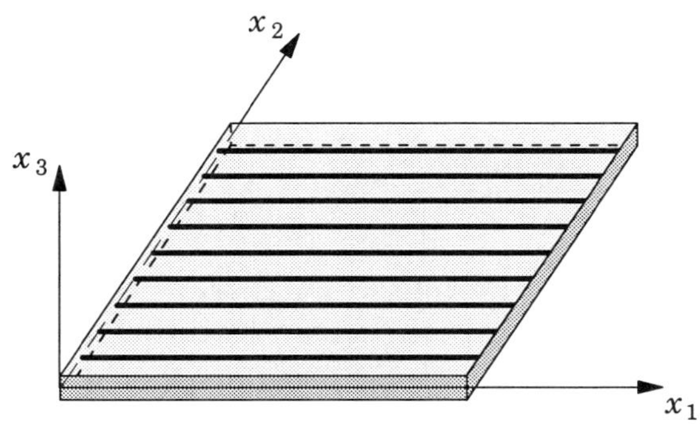

Figure 4.2–2: A unidirectional fiber-reinforced lamina with the material coordinate system (x_1, x_2, x_3).

The theoretical approach, called a *micromechanics approach*, used to determine the engineering constants of a continuous fiber-reinforced composite material is based on the following assumptions:

1. Perfect bonding exists between fibers and matrix.
2. Fibers are parallel, and uniformly distributed throughout.
3. The matrix is free of voids or microcracks and initially in a stress-free state.
4. Both fibers and matrix are isotropic and obey Hooke's law.
5. The applied loads are either parallel or perpendicular to the fiber direction.

The moduli and Poisson's ratio of a fiber-reinforced material can be expressed in terms of the moduli, Poisson's ratios, and volume fractions of the constituents. To this end, let

E_f = modulus of the fiber

E_m = modulus of the matrix

ν_f = Poisson's ratio of the fiber

ν_m = Poisson's ratio of the matrix

V_f = fiber volume fraction

V_m = matrix volume fraction

Then it can be shown that the lamina engineering constants are given by

$$E_1 = E_f V_f + E_m V_m, \quad \nu_{12} = \nu_f V_f + \nu_m V_m$$

$$E_2 = \frac{E_f E_m}{E_f V_m + E_m V_f}, \quad G_{12} = \frac{G_f G_m}{G_f V_m + G_m V_f} \tag{4.2-15}$$

where E_1 is the longitudinal modulus, E_2 is transverse modulus, ν_{12} is the major Poisson's ratio, and G_{12} is the shear modulus, and

$$G_f = \frac{E_f}{2(1+\nu_f)}, \quad G_m = \frac{E_m}{2(1+\nu_m)} \tag{4.2-16}$$

The engineering parameters E_1, E_2, E_3, G_{12}, G_{13}, G_{23}, ν_{12}, ν_{13}, and ν_{23} of an orthotropic material can be determined experimentally using an appropriate test specimen made of the material. For example, E_1 and ν_{12} of a fiber-reinforced material are measured using a uniaxial test specimen. The specimen consists of several layers of the material with fibers in each layer being aligned with the longitudinal direction. The specimen is then loaded along the longitudinal direction and strains along and perpendicular to the fiber directions are measured using strain gauges (see Figure 4.2–3). If the load applied is P, the stress along the fiber direction is $\sigma_{11} = P/A$, where A denotes the cross–sectional area of the specimen. For different values of the load, σ_{11} can be computed and longitudinal and lateral strains ε_{11} and ε_{22} can be recorded. Then E_1 denotes the slope of the experimental relation between σ_{11} and ε_{11}. The slope of the experimental relation between the lateral contraction ε_{22} and longitudinal extension ε_{11} is Poisson's ratio ν_{12}. Similarly, other constants can be determined from appropriate tests. The values of the engineering constants for several materials are presented in Tables 4.2–2 and 4.2–3.

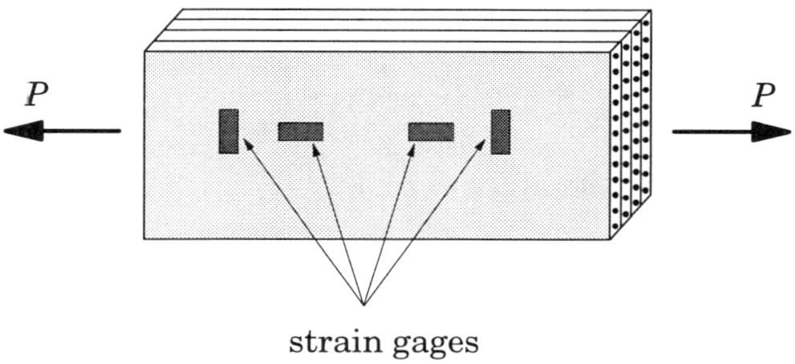

Figure 4.2–3: Mechanical characterization of a laminated specimen.

Table 4.2–2: Values of the engineering constants for several materials* (the properties of polymeric materials vary with processing and therefore the values given here may differ from those found elsewhere).

Material	E_1	E_2	G_{12}	G_{13}	G_{23}	ν_{12}
Aluminum	10.6	10.6	3.38	3.38	3.38	0.33
Copper	18.0	18.0	6.39	6.39	6.39	0.33
Steel	30.0	30.0	11.24	11.24	11.24	0.29
Gr.–Ep (AS)	20.0	1.3	1.03	1.03	0.90	0.30
Gr.–Ep (T)	19.0	1.5	1.00	0.90	0.90	0.22
Gl.–Ep (1)	7.8	2.6	1.30	1.30	0.50	0.25
Gl.–Ep (2)	5.6	1.2	0.60	0.60	0.50	0.26
Br.–Ep	30.0	3.0	1.00	1.00	0.60	0.30

*Moduli are in msi = million psi; 1 psi = 6.895 kN/m^2.

Gr.–Ep (AS) = graphite–epoxy (AS/3501); Gr.–Ep (T) = graphite–epoxy (T300/934); Gl.–Ep = glass–epoxy; Br.–Ep = boron–epoxy.

Table 4.2–3: Values of additional engineering constants for the materials listed in Table 4.2–2.*

Material	E_3	ν_{13}	ν_{23}	α_1	α_2
Aluminum	10.6	0.33	0.33	13.1	13.1
Copper	18.0	0.33	0.33	18.0	18.0
Steel	30.0	0.29	0.29	10.0	10.0
Gr.–Ep (AS)	1.3	0.30	0.49	1.0	30.0
Gr.–Ep (T)	1.5	0.22	0.49	−0.167	15.6
Gl.–Ep (1)	2.6	0.25	0.34	3.5	11.4
Gl.–Ep (2)	1.3	0.26	0.34	4.8	12.3
Br.–Ep	3.0	0.25	0.25	2.5	8.0

* Units of E_3 are msi, and the units of α_1 and α_2 are 10^{-6} in./in./°F.

4.3 Transformation of Stresses and Strains

4.3–1 Coordinate Transformations

In the preceding sections we discussed constitutive equations for a linear orthotropic material. The constitutive relations for orthotropic materials were written in terms of the stress and strain components that are referred to the coordinate system which coincides with the principal material coordinate system. The coordinate system used in the solution of a problem, in general, does not coincide with the material coordinate system. Further, composite laminates have several layers, each with different orientation of their material coordinates with respect to the laminate coordinates. Thus, there is a need to establish transformation relations among stresses and strains in one coordinate system to the corresponding quantities in another coordinate system. These relations can be used to transform constitutive equations from the material coordinates to the coordinates used in the solution of a problem.

In forming flat laminates, fiber-reinforced lamina are stacked with their x_1x_2–planes parallel but each having its own fiber direction. If the z–coordinate of the problem is oriented normal to the plane of the laminate, the x_3–coordinate of each lamina we will always coincide with the z–coordinate of the problem. Thus we have a special type of coordinate transformation between the material coordinates and the coordinates used in the problem description (see Example 1.4–1).

Let (x, y, z) denote the coordinates used in the solution of a problem, and let (x_1, x_2, x_3) be the principal material coordinates such that (x_1, x_2, x_3)

is obtained from (x, y, z) by rotating the xy–plane counterclockwise (when looking down on the lamina) by an angle θ about the z–axis (see Figure 4.3-1). Conversely, (x, y, z) is obtained from (x_1, x_2, x_3) by rotating the x_1x_2–plane clockwise by an angle θ. The coordinates of a material point in the coordinate systems are related by $(z = x_3)$ (see Figure 4.3–1):

$$\begin{Bmatrix} x_1 \\ x_2 \\ x_3 \end{Bmatrix} = \begin{bmatrix} \cos\theta & \sin\theta & 0 \\ -\sin\theta & \cos\theta & 0 \\ 0 & 0 & 1 \end{bmatrix} \begin{Bmatrix} x \\ y \\ z \end{Bmatrix} = [A] \begin{Bmatrix} x \\ y \\ z \end{Bmatrix} \tag{4.3-1}$$

The inverse of Eq. (1) is

$$\begin{Bmatrix} x \\ y \\ z \end{Bmatrix} = \begin{bmatrix} \cos\theta & -\sin\theta & 0 \\ \sin\theta & \cos\theta & 0 \\ 0 & 0 & 1 \end{bmatrix} \begin{Bmatrix} x_1 \\ x_2 \\ x_3 \end{Bmatrix} = [A]^{\mathrm{T}} \begin{Bmatrix} x_1 \\ x_2 \\ x_3 \end{Bmatrix} \tag{4.3-2}$$

Note that the inverse of $[A]$ is equal to its transpose: $[A]^{-1} = [A]^{\mathrm{T}}$.

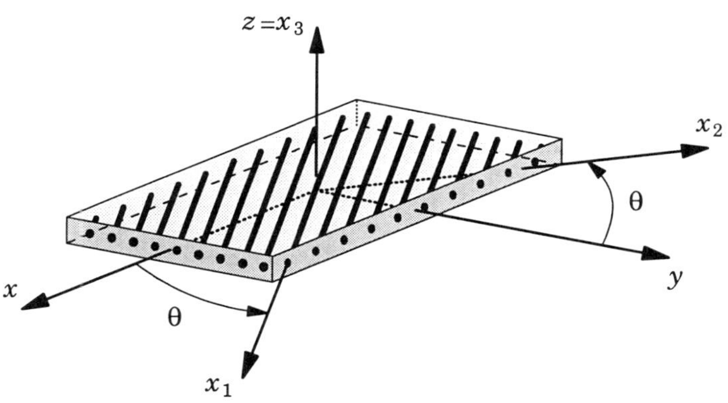

Figure 4.3–1: A fiber-reinforced lamina with global and material coordinate systems.

The transformation relations (1) and (2) are also valid for the unit vectors associated with the two coordinate systems [see Eqs. (1.4–3b) and (1.4–4b); here (x_1, x_2, x_3) is the barred coordinate system and (x, y, z) is the unbarred coordinate system]:

$$\begin{Bmatrix} \hat{\mathbf{e}}_1 \\ \hat{\mathbf{e}}_2 \\ \hat{\mathbf{e}}_3 \end{Bmatrix} = [A] \begin{Bmatrix} \hat{\mathbf{e}}_x \\ \hat{\mathbf{e}}_y \\ \hat{\mathbf{e}}_z \end{Bmatrix}, \quad \begin{Bmatrix} \hat{\mathbf{e}}_x \\ \hat{\mathbf{e}}_y \\ \hat{\mathbf{e}}_z \end{Bmatrix} = [A]^{\mathrm{T}} \begin{Bmatrix} \hat{\mathbf{e}}_1 \\ \hat{\mathbf{e}}_2 \\ \hat{\mathbf{e}}_3 \end{Bmatrix} \tag{4.3-3}$$

4.3–2 Transformation of Stress Components

Next we consider the relationship between the components of stress in the two coordinate systems. Let σ denote the stress tensor, which has components $\sigma_{11}, \sigma_{12}, \cdots, \sigma_{33}$ in the material coordinates (x_1, x_2, x_3) and components $\sigma_{xx}, \sigma_{xy}, \cdots, \sigma_{zz}$ in the problem (or laminate) coordinates (x, y, z). In the notation of Chapter 2, (x_1, x_2, x_3) is the system with bars and (x, y, z) is the system without bars. Since stress is a second-order tensor, we have from Eq. (1.4–11a) the following relation among the two sets of components:

$$(\sigma_{k\ell})_m = a_{ki}a_{\ell j}(\sigma_{ij})_p \tag{4.3-4a}$$

$$(\sigma_{k\ell})_p = a_{ik}a_{j\ell}(\sigma_{ij})_m \tag{4.3-4b}$$

where $(\sigma_{ij})_m$ are the components of the stress tensor in the material coordinates, whereas $(\sigma_{ij})_p$ are the components of the same stress tensor in the problem coordinates, and a_{ij} are the direction cosines

$$a_{ij} = (\hat{\mathbf{e}}_i)_m \cdot (\hat{\mathbf{e}}_j)_p \tag{4.3-4c}$$

and $(\hat{\mathbf{e}}_i)_m$ and $(\hat{\mathbf{e}}_i)_p$ are the orthonormal basis vectors in the material and problem coordinate systems, respectively. Equations (4a,b) can be expressed in convenient matrix forms. Toward this objective, we introduce the 3×3 arrays of the stress components

$$[\sigma]_p = \begin{bmatrix} \sigma_{xx} & \sigma_{xy} & \sigma_{xz} \\ \sigma_{xy} & \sigma_{yy} & \sigma_{yz} \\ \sigma_{xz} & \sigma_{yz} & \sigma_{zz} \end{bmatrix}, \quad [\sigma]_m = \begin{bmatrix} \sigma_{11} & \sigma_{12} & \sigma_{13} \\ \sigma_{12} & \sigma_{22} & \sigma_{23} \\ \sigma_{13} & \sigma_{23} & \sigma_{33} \end{bmatrix} \tag{4.3-5}$$

Now Eqs. (4a,b) can be expressed in matrix form as [see Eqs. (1.4–11b) and (1.4–12)]

$$[\sigma]_m = [A][\sigma]_p[A]^T \tag{4.3-6a}$$

$$[\sigma]_p = [A]^T[\sigma]_m[A] \tag{4.3-6b}$$

where $[A]$ is the 3×3 matrix of direction cosines in Eq. (1).

Equation (6a) provides a means to convert stress components referred to the problem (laminate) coordinate system to those referred to the material (lamina) coordinate system, while Eq. (6b) allows computation of stress components referred to the problem coordinates in terms of stress components referred to the material coordinates. Here we are interested in determining the transformation relations for the coordinate transformation of the form described by Eq. (2).

Carrying out the matrix multiplications in Eq. (6b), we obtain

$$\begin{Bmatrix} \sigma_{xx} \\ \sigma_{yy} \\ \sigma_{zz} \\ \sigma_{yz} \\ \sigma_{xz} \\ \sigma_{xy} \end{Bmatrix} = \begin{bmatrix} \cos^2\theta & \sin^2\theta & 0 & 0 & 0 & -\sin 2\theta \\ \sin^2\theta & \cos^2\theta & 0 & 0 & 0 & \sin 2\theta \\ 0 & 0 & 1 & 0 & 0 & 0 \\ 0 & 0 & 0 & \cos\theta & \sin\theta & 0 \\ 0 & 0 & 0 & -\sin\theta & \cos\theta & 0 \\ \sin\theta\cos\theta & -\sin\theta\cos\theta & 0 & 0 & 0 & \cos^2\theta - \sin^2\theta \end{bmatrix} \begin{Bmatrix} \sigma_1 \\ \sigma_2 \\ \sigma_3 \\ \sigma_4 \\ \sigma_5 \\ \sigma_6 \end{Bmatrix} \tag{4.3-7}$$

or

$$\{\sigma\}_p = [T]\{\sigma\}_m \tag{4.3-8}$$

The inverse relationship between $\{\sigma\}_m$ and $\{\sigma\}_p$, Eq. (6a), is given by

$$\begin{Bmatrix} \sigma_1 \\ \sigma_2 \\ \sigma_3 \\ \sigma_4 \\ \sigma_5 \\ \sigma_6 \end{Bmatrix} = \begin{bmatrix} \cos^2\theta & \sin^2\theta & 0 & 0 & 0 & \sin 2\theta \\ \sin^2\theta & \cos^2\theta & 0 & 0 & 0 & -\sin 2\theta \\ 0 & 0 & 1 & 0 & 0 & 0 \\ 0 & 0 & 0 & \cos\theta & -\sin\theta & 0 \\ 0 & 0 & 0 & \sin\theta & \cos\theta & 0 \\ -\sin\theta\cos\theta & \sin\theta\cos\theta & 0 & 0 & 0 & \cos^2\theta - \sin^2\theta \end{bmatrix} \begin{Bmatrix} \sigma_{xx} \\ \sigma_{yy} \\ \sigma_{zz} \\ \sigma_{yz} \\ \sigma_{xz} \\ \sigma_{xy} \end{Bmatrix} \tag{4.3-9}$$

or

$$\{\sigma\}_m = [R]\{\sigma\}_p \tag{4.3-10}$$

The result in Eq. (9) can also be obtained from Eq. (7) by replacing θ with $-\theta$.

Example 4.3–1:

The stress transformation equations (9) can be derived alternatively by considering the equilibrium of an element of the lamina (see Figure 4.3–2). Consider a wedge element whose slant face is parallel to the fibers. Suppose that the thickness of the lamina is h, and the length of the slant face is ΔS. Then the horizontal and vertical sides of the wedges are of lengths $\Delta S\cos\theta$ and $\Delta S\sin\theta$, respectively. The forces acting on any face of the wedge are obtained by multiplying the stresses acting on the face with the area of the surface.

Suppose that we wish to determine σ_{22} in terms of $(\sigma_{xx}, \sigma_{yy}, \sigma_{xy})$. Then by summing all forces acting on the wedge along coordinate x_2 (*i.e.*, equilibrium of forces along x_2) we obtain

$$\begin{aligned} \sigma_{22}\Delta S\, h - (\sigma_{xx}\Delta S\sin\theta\, h)\sin\theta + (\sigma_{xy}\Delta S\sin\theta\, h)\cos\theta - (\sigma_{yy}\Delta S\cos\theta\, h)\cos\theta \\ + (\sigma_{xy}\Delta S\cos\theta\, h)\sin\theta = 0 \end{aligned}$$

or

$$\sigma_{22} = \sigma_{xx} \sin^2\theta + \sigma_{yy} \cos^2\theta - 2\sigma_{xy} \cos\theta \sin\theta$$

Similarly, summing the forces along x_1 coordinate, we obtain

$$\sigma_{12}\Delta S\ h + (\sigma_{xx}\Delta S \sin\theta\ h) \cos\theta + (\sigma_{xy}\Delta S \sin\theta\ h) \sin\theta - (\sigma_{yy}\Delta S \cos\theta\ h) \sin\theta \\ - (\sigma_{xy}\Delta S \cos\theta\ h) \cos\theta = 0$$

or

$$\sigma_{12} = -\sigma_{xx} \sin\theta \cos\theta + \sigma_{yy} \cos\theta \sin\theta + \sigma_{xy}(\cos^2\theta - \sin^2\theta)$$

Clearly, the expressions for σ_{22} and σ_{12} derived here are the same as those for σ_1 and σ_6, respectively, in Eq. (9). The stress component σ_{11} can be determined in terms of $(\sigma_{xx}, \sigma_{yy}, \sigma_{xy})$ by considering a wedge element whose slant face is perpendicular to the fibers (see Figure 4.3–2). By summing forces along the $x-$ and $y-$coordinates we can obtain stresses σ_{xx} and σ_{xy} in terms of $(\sigma_{11}, \sigma_{22}, \sigma_{12})$.

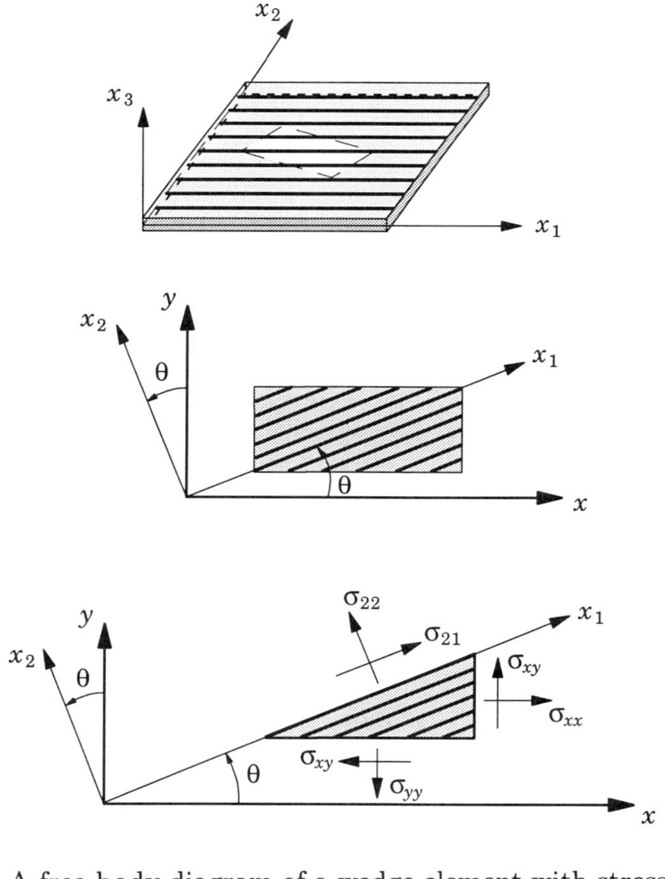

Figure 4.3–2: A free-body diagram of a wedge element with stress components.

Example 4.3–2:

Consider a thin (*i.e.*, the thickness is about one-tenth of the radius), filament wound, closed cylindrical pressure vessel (see Figure 4.3–3). The vessel is of 63.5 cm (25 in.) internal diameter and pressurized to 1.379 MPa (200 psi). We wish to determine the shear and normal forces per unit length of filament winding. Assume a filament winding angle of $\theta = 53.125°$ from the longitudinal axis of the pressure vessel, and use the following material properties, typical of graphite–epoxy material: $E_1 = 140$ MPa (20.3 Msi) , $E_2 = 10$ MPa (1.45 Msi), $G_{12} = 7$ MPa (1.02 Msi) , and $\nu_{12} = 0.3$.

The equations of equilibrium of forces in a structure do not depend on the material properties. Hence, equations derived for the longitudinal (σ_{xx}) and circumferential (σ_{yy}) stresses in a thin-walled cylindrical pressure vessel are valid here:

$$\sigma_{xx} = \frac{pD_i}{4h} \ , \quad \sigma_{yy} = \frac{pD_i}{2h}$$

where p is internal pressure, D_i is internal diameter, and h is thickness of the pressure vessel. We obtain

$$\sigma_{xx} = \frac{1.379 \times 0.635}{4h} = \frac{0.2189}{h} \text{ MPa} \ , \quad \sigma_{yy} = \frac{1.379 \times 0.635}{2h} = \frac{0.4378}{h} \text{ MPa}$$

The shear stress σ_{xy} is zero.

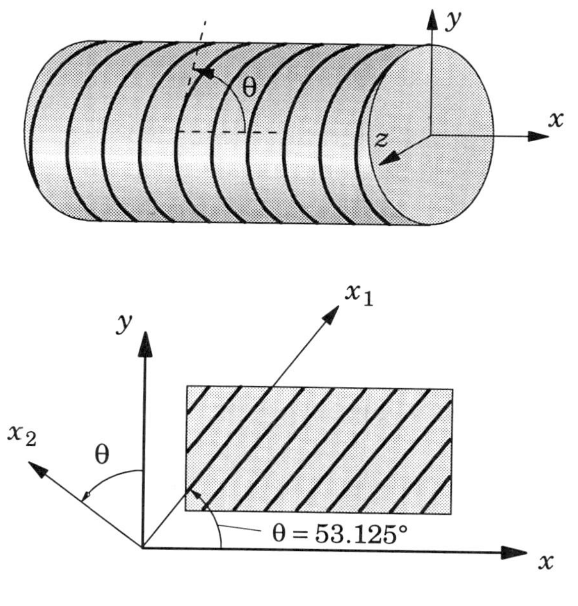

Figure 4.3–3: A filament wound cylindrical pressure vessel.

Next we determine the shear stress along the fiber and the normal stress in the fiber using the transformation equations (9) or from the equations derived in Example 4.3–1. We obtain

$$\sigma_{11} = \frac{0.2189}{h}(0.6)^2 + \frac{0.4378}{h}(0.8)^2 = \frac{0.3590}{h} \text{ MPa}$$

$$\sigma_{22} = \frac{0.2189}{h}(0.8)^2 + \frac{0.4378}{h}(0.6)^2 = \frac{0.2977}{h} \text{ MPa}$$

$$\sigma_{12} = \left(\frac{0.4378}{h} - \frac{0.2189}{h}\right) \times 0.6 \times 0.8 = \frac{0.1051}{h} \text{ MPa}$$

Hence the normal and shear forces per unit length along the fiber-matrix interface are $F_{22} = 0.2977$ MN and $F_{12} = 0.1051$ MN, whereas the force per unit length in the fiber direction is $F_{11} = 0.359$ MN.

4.3–3 Transformation of Strain Components

Since strains are also tensor quantities, transformation equations derived for stresses, Eqs. (6a,b), are also valid for *tensor* components of strains:

$$[\varepsilon]_m = [A][\varepsilon]_p[A]^T \tag{4.3-11a}$$

$$[\varepsilon]_p = [A]^T[\varepsilon]_m[A] \tag{4.3-11b}$$

Therefore, Eqs. (7) and (9) are valid for strains when the stress components are replaced with tensor components of strains from the two coordinate systems. However, the single-column formats in Eqs. (7) and (9) for stresses are not valid for single-column formats of strains because of the definition (see Table 4.2–1):

$$2\varepsilon_{12} = \varepsilon_6,\ 2\varepsilon_{13} = \varepsilon_5,\ 2\varepsilon_{23} = \varepsilon_4 \tag{4.3-12}$$

Slight modification of the results in Eqs. (7) and (9) will yield the proper relations for the engineering components of strains. We have

$$\begin{Bmatrix} \varepsilon_{xx} \\ \varepsilon_{yy} \\ \varepsilon_{zz} \\ 2\varepsilon_{yz} \\ 2\varepsilon_{xz} \\ 2\varepsilon_{xy} \end{Bmatrix} = \begin{bmatrix} \cos^2\theta & \sin^2\theta & 0 & 0 & 0 & -\sin\theta\cos\theta \\ \sin^2\theta & \cos^2\theta & 0 & 0 & 0 & \sin\theta\cos\theta \\ 0 & 0 & 1 & 0 & 0 & 0 \\ 0 & 0 & 0 & \cos\theta & \sin\theta & 0 \\ 0 & 0 & 0 & -\sin\theta & \cos\theta & 0 \\ \sin 2\theta & -\sin 2\theta & 0 & 0 & 0 & \cos^2\theta - \sin^2\theta \end{bmatrix} \begin{Bmatrix} \varepsilon_1 \\ \varepsilon_2 \\ \varepsilon_3 \\ \varepsilon_4 \\ \varepsilon_5 \\ \varepsilon_6 \end{Bmatrix} \tag{4.3-13}$$

The inverse relation is given by

$$\begin{Bmatrix} \varepsilon_1 \\ \varepsilon_2 \\ \varepsilon_3 \\ \varepsilon_4 \\ \varepsilon_5 \\ \varepsilon_6 \end{Bmatrix} = \begin{bmatrix} \cos^2\theta & \sin^2\theta & 0 & 0 & 0 & \sin\theta\cos\theta \\ \sin^2\theta & \cos^2\theta & 0 & 0 & 0 & -\sin\theta\cos\theta \\ 0 & 0 & 1 & 0 & 0 & 0 \\ 0 & 0 & 0 & \cos\theta & -\sin\theta & 0 \\ 0 & 0 & 0 & \sin\theta & \cos\theta & 0 \\ -\sin 2\theta & \sin 2\theta & 0 & 0 & 0 & \cos^2\theta - \sin^2\theta \end{bmatrix} \begin{Bmatrix} \varepsilon_{xx} \\ \varepsilon_{yy} \\ \varepsilon_{zz} \\ 2\varepsilon_{yz} \\ 2\varepsilon_{xz} \\ 2\varepsilon_{xy} \end{Bmatrix} \tag{4.3-14}$$

We note that the transformation matrix $[T]$ in Eq. (8) is the transpose of the square matrix in Eq. (14). Similarly, the transformation matrix in Eq. (13) is the transpose of the matrix $[R]$ in Eq. (10):

$$\{\varepsilon\}_p = [R]^{\mathrm{T}}\{\varepsilon\}_m \tag{4.3-15a}$$

$$\{\varepsilon\}_m = [T]^{\mathrm{T}}\{\varepsilon\}_p \tag{4.3-15b}$$

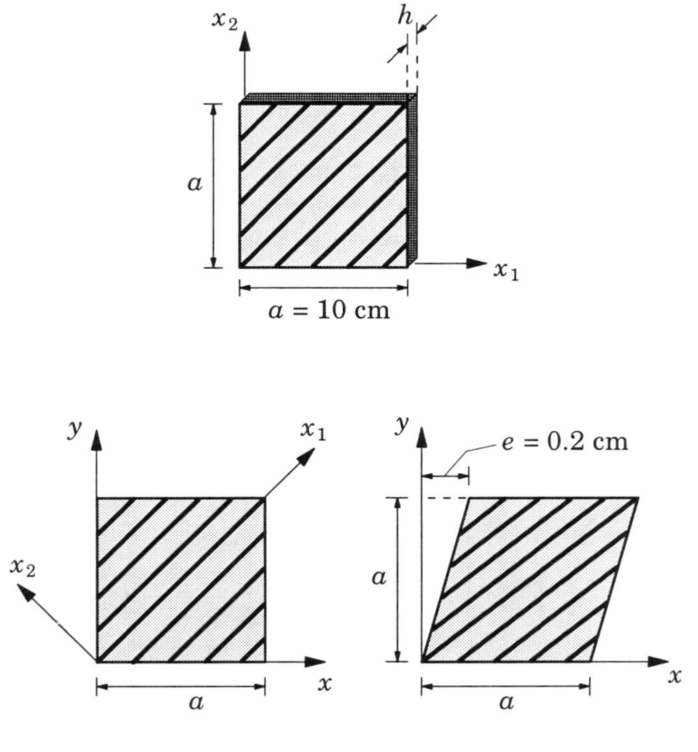

Figure 4.3–4: Deformation of a fiber-reinforced lamina.

Example 4.3–3:

A square lamina of thickness h and planar dimension a is made of glass–epoxy material ($E_1 = 40 \times 10^3$ MPa, $E_2 = 10 \times 10^3$ MPa, $G_{12} = 3.5 \times 10^3$ MPa, and $\nu_{12} = 0.25$). When the lamina is deformed as shown in Figure 4.3–4 (see Example 2.2–1, page 29), we wish to determine the longitudinal strain in the fiber and shear strain at the center of the lamina. The fibers are oriented at 45° to the horizontal.

From Eq. (2.2–16), the only nonzero strain is $\varepsilon_{xy} = 0.01$. Hence, longitudinal strain in the fiber is

$$\varepsilon_{11} = 0 + 0 + 2\varepsilon_{xy}\frac{1}{\sqrt{2}}\frac{1}{\sqrt{2}} = 0.01 \text{ cm/cm}$$

and the shear strain is given by

$$\varepsilon_{12} = 0 + 0 + 2\varepsilon_{xy}\left(\frac{1}{2} - \frac{1}{2}\right) = 0.0$$

Example 4.3–4:

Suppose that the thickness of the cylindrical pressure vessel of Example 4.3–2 is $h = 2$ cm. Then the stress field in the material coordinates becomes

$$\sigma_{11} = 17.95 \text{ MPa}, \quad \sigma_{22} = 14.885 \text{ MPa}, \quad \sigma_{12} = 5.255 \text{ MPa}$$

The strains in the material coordinates can be calculated using the strain–stress relations (4.2–10). We have ($\nu_{21}/E_2 = \nu_{12}/E_1$)

$$\varepsilon_{11} = \frac{\sigma_{11}}{E_1} - \frac{\sigma_{11}\nu_{12}}{E_1} = \frac{17.95}{140} - \frac{14.885 \times 0.3}{140} = 0.0963 \text{ m/m}$$

$$\varepsilon_{22} = -\frac{\sigma_{11}\nu_{12}}{E_2} + \frac{\sigma_{22}}{E_2} = -\frac{17.95 \times 0.3}{10} + \frac{14.885}{10} = 0.95 \text{ m/m}$$

$$\varepsilon_{12} = \frac{\sigma_{12}}{2G_{12}} = \frac{5.255}{2 \times 7} = 0.3757$$

The strains in the pressure vessel with respect to the (x, y) coordinates can be computed using Eq. (13) or (15a):

$$\begin{aligned}
\varepsilon_{xx} &= 0.0963 \times (0.6)^2 + 0.95 \times (0.8)^2 - 0.3757 \times 0.6 \times 0.8 = 0.4623 \text{ m/m} \\
\varepsilon_{yy} &= 0.0963 \times (0.8)^2 + 0.95 \times (0.6)^2 + 0.3757 \times 0.6 \times 0.8 = 0.5840 \text{ m/m} \\
\varepsilon_{xy} &= 2(0.0963 - 0.95) \times (0.6) \times 0.8 + 0.3757[(0.6)^2 - (0.8)^2] = -0.9247
\end{aligned}$$

4.3–4 Transformation of Material Coefficients

In formulating the problem of a laminated structure, we must write the governing equations, with all its variables and coefficients, in the problem coordinates. In the previous section we discussed transformation of coordinates (which are also valid for displacements and forces), stresses, and strains. The only remaining quantities that need to be transformed from the material coordinate system to the problem coordinates are the material stiffnesses C_{ij} and thermal coefficients of expansion α_{ij}.

The material stiffnesses C_{ij} in their original form [see Eq. (4.2–1a)] are the components of a fourth-order tensor. Hence, the tensor transformation law holds. The fourth-order elasticity tensor components $\bar{C}_{ijk\ell}$ in the problem coordinates can be related to the components C_{mnpq} in the material coordinates by the tensor transformation law

$$\bar{C}_{ijk\ell} = a_{im}a_{jn}a_{kp}a_{\ell q}C_{mnpq}$$

However, the above equation involves five matrix multiplications with four-subscript material coefficients. Alternatively, the same result can be obtained by using the stress–strain and strain–stress relations (4.2–3) and (4.2–4), and the stress and strain transformation equations in (8), (10), (13), and (14):

$$\{\sigma\}_p = [T]\{\sigma\}_m = [T][C]_m\{\varepsilon\}_m = [T][C]_m[T]^T\{\varepsilon\}_p \tag{4.3 – 16a}$$

where $[C]_m$ is the 6×6 material stiffness matrix [see Eq. (2.4–3a)] in the material coordinates and $[T]$ is the transformation matrix defined in Eq. (8). Using Hooke's law in the problem coordinate system, we can write

$$\{\sigma\}_p = [C]_p\{\varepsilon\}_p \tag{4.3 – 16b}$$

Hence, it follows from Eqs. (16a) and (16b) that ($[C]_p \equiv [\bar{C}]$ and $[C]_m \equiv [C]$)

$$[\bar{C}] = [T][C][T]^T \tag{4.3 – 17}$$

Equation (17) is valid for general constitutive matrix $[C]$ (*i.e.*, for orthotropic as well as anisotropic). Of course, $[T]$ is the matrix based on the particular transformation (1) (rotation about a transverse normal to the lamina).

Carrying out the matrix multiplications in (17) for the anisotropic case, we obtain

$$\begin{aligned}
\bar{C}_{11} &= C_{11}\cos^4\theta - 4C_{16}\cos^3\theta\sin\theta + 2(C_{12}+2C_{66})\cos^2\theta\sin^2\theta \\
&\quad - 4C_{26}\cos\theta\sin^3\theta + C_{22}\sin^4\theta \\
\bar{C}_{12} &= C_{12}\cos^4\theta + 2(C_{16}-C_{26})\cos^3\theta\sin\theta + (C_{11}+C_{22}-4C_{66})\cos^2\theta\sin^2\theta
\end{aligned}$$

$$
\begin{aligned}
&\quad + 2(C_{26} - C_{16}) \cos\theta \sin^3\theta + C_{12} \sin^4\theta \\
\bar{C}_{13} &= C_{13} \cos^2\theta - 2C_{36} \cos\theta \sin\theta + C_{23} \sin^2\theta \\
\bar{C}_{16} &= C_{16} \cos^4\theta + (C_{11} - C_{12} - 2C_{66}) \cos^3\theta \sin\theta + 3(C_{26} - C_{16}) \cos^2\theta \sin^2\theta \\
&\quad + (2C_{66} + C_{12} - C_{22}) \cos\theta \sin^3\theta - C_{26} \sin^4\theta \\
\bar{C}_{22} &= C_{22} \cos^4\theta + 4C_{26} \cos^3\theta \sin\theta + 2(C_{12} + 2C_{66}) \cos^2\theta \sin^2\theta \\
&\quad + 4C_{16} \cos\theta \sin^3\theta + C_{11} \sin^4\theta \\
\bar{C}_{23} &= C_{23} \cos^2\theta + 2C_{36} \cos\theta \sin\theta + C_{13} \sin^2\theta \\
\bar{C}_{26} &= C_{26} \cos^4\theta + (C_{12} - C_{22} + 2C_{66}) \cos^3\theta \sin\theta + 3(C_{16} - C_{26}) \cos^2\theta \sin^2\theta \\
&\quad + (C_{11} - C_{12} - 2C_{66}) \cos\theta \sin^3\theta - C_{16} \sin^4\theta \\
\bar{C}_{33} &= C_{33} \\
\bar{C}_{36} &= (C_{13} - C_{23}) \cos\theta \sin\theta + C_{36}(\cos^2\theta - \sin^2\theta) \\
\bar{C}_{66} &= 2(C_{16} - C_{26}) \cos^3\theta \sin\theta + (C_{11} + C_{22} - 2C_{12} - 2C_{66}) \cos^2\theta \sin^2\theta \\
&\quad + 2(C_{26} - C_{16}) \cos\theta \sin^3\theta + C_{66}(\cos^4\theta + \sin^4\theta) \\
\bar{C}_{44} &= C_{44} \cos^2\theta + C_{55} \sin^2\theta + 2C_{45} \cos\theta \sin\theta \\
\bar{C}_{45} &= C_{45}(\cos^2\theta - \sin^2\theta) + (C_{55} - C_{44}) \cos\theta \sin\theta \\
\bar{C}_{55} &= C_{55} \cos^2\theta + C_{44} \sin^2\theta - 2C_{45} \cos\theta \sin\theta \\
\bar{C}_{14} &= C_{14} \cos^3\theta + (C_{15} - 2C_{46}) \cos^2\theta \sin\theta + (C_{24} - 2C_{56}) \cos\theta \sin^2\theta + C_{25} \sin^3\theta \\
\bar{C}_{15} &= C_{15} \cos^3\theta - (C_{14} + 2C_{56}) \cos^2\theta \sin\theta + (C_{25} + 2C_{46}) \cos\theta \sin^2\theta - C_{24} \sin^3\theta \\
\bar{C}_{24} &= C_{24} \cos^3\theta + (C_{25} + 2C_{46}) \cos^2\theta \sin\theta + (C_{14} + 2C_{56}) \cos\theta \sin^2\theta + C_{15} \sin^3\theta \\
\bar{C}_{25} &= C_{25} \cos^3\theta + (2C_{56} - C_{24}) \cos^2\theta \sin\theta + (C_{15} - 2C_{46}) \cos\theta \sin^2\theta - C_{14} \sin^3\theta \\
\bar{C}_{34} &= C_{34} \cos\theta + C_{35} \sin\theta \\
\bar{C}_{35} &= C_{35} \cos\theta - C_{34} \sin\theta \\
\bar{C}_{46} &= C_{46} \cos^3\theta + (C_{56} + C_{14} - C_{24}) \cos^2\theta \sin\theta + (C_{15} - C_{25} - C_{46}) \cos\theta \sin^2\theta \\
&\quad - C_{56} \sin^3\theta \\
\bar{C}_{56} &= C_{56} \cos^3\theta + (C_{15} - C_{25} - C_{46}) \cos^2\theta \sin\theta + (C_{24} - C_{14} - C_{56}) \cos\theta \sin^2\theta \\
&\quad + C_{46} \sin^3\theta
\end{aligned}
\tag{4.3-18}
$$

When $[C]$ is the matrix corresponding to an orthotropic material, it has the form shown in Eq. (4.2–3) or (2.4–8); then Eq. (16b) has the explicit form [*cf*. Eq. (2.4–7) for monoclinic materials]

$$
\begin{Bmatrix} \sigma_{xx} \\ \sigma_{yy} \\ \sigma_{zz} \\ \sigma_{yz} \\ \sigma_{xz} \\ \sigma_{xy} \end{Bmatrix} = \begin{bmatrix} \bar{C}_{11} & \bar{C}_{12} & \bar{C}_{13} & 0 & 0 & \bar{C}_{16} \\ \bar{C}_{21} & \bar{C}_{22} & \bar{C}_{23} & 0 & 0 & \bar{C}_{26} \\ \bar{C}_{31} & \bar{C}_{32} & \bar{C}_{33} & 0 & 0 & \bar{C}_{36} \\ 0 & 0 & 0 & \bar{C}_{44} & \bar{C}_{45} & 0 \\ 0 & 0 & 0 & \bar{C}_{45} & \bar{C}_{55} & 0 \\ \bar{C}_{16} & \bar{C}_{26} & \bar{C}_{36} & 0 & 0 & \bar{C}_{66} \end{bmatrix} \begin{Bmatrix} \varepsilon_{xx} \\ \varepsilon_{yy} \\ \varepsilon_{zz} \\ 2\varepsilon_{yz} \\ 2\varepsilon_{xz} \\ 2\varepsilon_{xy} \end{Bmatrix} \tag{4.3-19}
$$

where the $\bar{C}_{ij}$ are the transformed elastic coefficients referred to the (x, y, z) coordinate system, which are related to the elastic coefficients in the material coordinates C_{ij} by Eq. (18). Note that C_{14}, C_{15}, C_{16}, C_{24}, C_{25}, C_{26}, C_{34}, C_{35}, C_{36}, C_{45}, C_{46}, and C_{56} are zero for an orthotropic material.

In order to relate compliance coefficients in the two coordinate systems, consider the strain transformation in Eq. (15a):

$$\begin{aligned}\{\varepsilon\}_p &= [R]^T\{\varepsilon\}_m = [R]^T\left([S]_m\{\sigma\}_m\right) = [R]^T[S]_m\left([R]\{\sigma\}_p\right)\\ &\equiv [S]_p\{\sigma\}_p \end{aligned} \tag{4.3 - 20a}$$

where $[S]_p \equiv [\bar{S}]$ and $[S]_m \equiv [S]$ and

$$[\bar{S}] = [R]^T[S][R] \tag{4.3 - 20b}$$

The transformed compliance coefficients $\bar{S}_{ij}$, referred to the (x, y, z) system, are related to the compliance coefficients S_{ij} in the material coordinates by Eq. (20b) as follows:

$$\begin{aligned}
\bar{S}_{11} &= S_{11}\cos^4\theta - 2S_{16}\cos^3\theta\sin\theta + (2S_{12} + S_{66})\cos^2\theta\sin^2\theta\\
&\quad - 2S_{26}\cos\theta\sin^3\theta + S_{22}\sin^4\theta\\
\bar{S}_{12} &= S_{12}\cos^4\theta + (S_{16} - S_{26})\cos^3\theta\sin\theta + (S_{11} + S_{22} - S_{66})\cos^2\theta\sin^2\theta\\
&\quad + (S_{26} - S_{16})\cos\theta\sin^3\theta + S_{12}\sin^4\theta\\
\bar{S}_{13} &= S_{13}\cos^2\theta - S_{36}\cos\theta\sin\theta + S_{23}\sin^2\theta\\
\bar{S}_{16} &= S_{16}\cos^4\theta + (2S_{11} - 2S_{12} - S_{66})\cos^3\theta\sin\theta + 3(S_{26} - S_{16})\cos^2\theta\sin^2\theta\\
&\quad + (S_{66} + 2S_{12} - 2S_{22})\cos\theta\sin^3\theta - S_{26}\sin^4\theta\\
\bar{S}_{22} &= S_{22}\cos^4\theta + 2S_{26}\cos^3\theta\sin\theta + (2S_{12} + S_{66})\cos^2\theta\sin^2\theta\\
&\quad + 2S_{16}\cos\theta\sin^3\theta + S_{11}\sin^4\theta\\
\bar{S}_{23} &= S_{23}\cos^2\theta + S_{36}\cos\theta\sin\theta + S_{13}\sin^2\theta\\
\bar{S}_{26} &= S_{26}\cos^4\theta + (2S_{12} - 2S_{22} + S_{66})\cos^3\theta\sin\theta + 3(S_{16} - S_{26})\cos^2\theta\sin^2\theta\\
&\quad + (2S_{11} - 2S_{12} - S_{66})\cos\theta\sin^3\theta - S_{16}\sin^4\theta\\
\bar{S}_{33} &= S_{33}\\
\bar{S}_{36} &= 2(S_{13} - S_{23})\cos\theta\sin\theta + S_{36}(\cos^2\theta - \sin^2\theta)\\
\bar{S}_{66} &= S_{66}(\cos^2\theta - \sin^2\theta)^2 + 4(S_{16} - S_{26})(\cos^2\theta - \sin^2\theta)\cos\theta\sin\theta\\
&\quad + 4(S_{11} + S_{22} - 2S_{12})\cos^2\theta\sin^2\theta\\
\bar{S}_{44} &= S_{44}\cos^2\theta + 2S_{45}\cos\theta\sin\theta + S_{55}\sin^2\theta\\
\bar{S}_{45} &= S_{45}(\cos^2\theta - \sin^2\theta) + (S_{55} - S_{44})\cos\theta\sin\theta\\
\bar{S}_{55} &= S_{55}\cos^2\theta + S_{44}\sin^2\theta - 2S_{45}\cos\theta\sin\theta\\
\bar{S}_{14} &= S_{14}\cos^3\theta + (S_{15} - S_{46})\cos^2\theta\sin\theta + (S_{24} - S_{56})\cos\theta\sin^2\theta + S_{25}\sin^3\theta\\
\bar{S}_{15} &= S_{15}\cos^3\theta - (S_{14} + S_{56})\cos^2\theta\sin\theta + (S_{25} + S_{46})\cos\theta\sin^2\theta - S_{24}\sin^3\theta\\
\bar{S}_{24} &= S_{24}\cos^3\theta + (S_{25} + S_{46})\cos^2\theta\sin\theta + (S_{14} + S_{56})\cos\theta\sin^2\theta + S_{15}\sin^3\theta\\
\bar{S}_{25} &= S_{25}\cos^3\theta + (-S_{24} + S_{56})\cos^2\theta\sin\theta + (S_{15} - S_{46})\cos\theta\sin^2\theta - S_{14}\sin^3\theta
\end{aligned}$$

$$
\begin{aligned}
\bar{S}_{34} &= S_{34}\cos\theta + S_{35}\sin\theta \\
\bar{S}_{35} &= S_{35}\cos\theta - S_{34}\sin\theta \\
\bar{S}_{46} &= (2S_{14} - 2S_{24} + S_{56})\cos^2\theta\sin\theta + (2S_{15} - 2S_{25} - S_{46})\cos\theta\sin^2\theta \\
&\quad + S_{46}\cos^3\theta - S_{56}\sin^3\theta \\
\bar{S}_{56} &= (2S_{15} - 2S_{25} - S_{46})\cos^2\theta\sin\theta + (2S_{24} - 2S_{14} - S_{56})\cos\theta\sin^2\theta \\
&\quad + S_{56}\cos^3\theta + S_{46}\sin^3\theta \qquad (4.3-21)
\end{aligned}
$$

For an orthotropic material, the matrix $[S]$ has the form shown in Eq. (4.2–4), and the strain–stress relations in the problem coordinates take the form

$$
\begin{Bmatrix} \varepsilon_{xx} \\ \varepsilon_{yy} \\ \varepsilon_{zz} \\ 2\varepsilon_{yz} \\ 2\varepsilon_{xz} \\ 2\varepsilon_{xy} \end{Bmatrix} = \begin{bmatrix} \bar{S}_{11} & \bar{S}_{12} & \bar{S}_{13} & 0 & 0 & \bar{S}_{16} \\ \bar{S}_{21} & \bar{S}_{22} & \bar{S}_{13} & 0 & 0 & \bar{S}_{26} \\ \bar{S}_{31} & \bar{S}_{32} & \bar{S}_{33} & 0 & 0 & \bar{S}_{36} \\ 0 & 0 & 0 & \bar{S}_{44} & \bar{S}_{45} & 0 \\ 0 & 0 & 0 & \bar{S}_{45} & \bar{S}_{55} & 0 \\ \bar{S}_{16} & \bar{S}_{26} & \bar{S}_{36} & 0 & 0 & \bar{S}_{66} \end{bmatrix} \begin{Bmatrix} \sigma_{xx} \\ \sigma_{yy} \\ \sigma_{zz} \\ \sigma_{yz} \\ \sigma_{xz} \\ \sigma_{xy} \end{Bmatrix} \qquad (4.3-22)
$$

Note that Eq. (22) relates strains to stresses in the problem coordinates while Eq. (4.2–10) relates the strains to stresses in the material coordinates.

The thermal coefficients α_{ij} are the components of a second-order tensor, and therefore they transform like the strain components (because $\alpha_6 = 2\alpha_{12}$, and so on). In the context of the present study, only nonzero components of thermal expansion tensor are $\alpha_{11} \equiv \alpha_1$, $\alpha_{22} \equiv \alpha_2$, and $\alpha_{33} \equiv \alpha_3$. All other components are zero. Hence, following Eq. (7), we can write the transformation relations ($\alpha_{12} = \alpha_{13} = \alpha_{23} = 0$)

$$
\begin{aligned}
\alpha_{xx} &= \alpha_{11}\cos^2\theta + \alpha_{22}\sin^2\theta \\
\alpha_{yy} &= \alpha_{11}\sin^2\theta + \alpha_{22}\cos^2\theta \\
2\alpha_{xy} &= 2(\alpha_{11} - \alpha_{22})\sin\theta\cos\theta \\
2\alpha_{xz} &= 0, \quad 2\alpha_{yz} = 0, \quad \alpha_{zz} = \alpha_{33} \qquad (4.3-23)
\end{aligned}
$$

The transformation relations (18), (21), and (23) are valid for a rectangular coordinate system (x_1, x_2, x_3) which is oriented at an angle θ (in the xy–plane) from the (x, y, z) coordinate system (see Figure 4.3–1). The orientation angle θ is measured counterclockwise from the x–axis to the x_1–axis. The same transformations hold for the coefficients of hygroscopic expansion (see Section 2.5–3).

In summary, Eq. (4.2–3) represents the stress–strain relations in the principal material coordinates (x_1, x_2, x_3) and Eq. (19) represents the stress–strain relations in the (x, y, z) coordinate system. The material coefficients of the lamina in the (x, y, z) coordinate system are related to material coefficients in the material coordinates by Eq. (18). In general, for the kth layer of a laminate, the stress–strain relations in the laminate coordinate system can be written as

$$\{\sigma\}_p^{(k)} = [\bar{C}]^{(k)}\left(\{\varepsilon\}_p^{(k)} - \{\alpha_T\}_p^{(k)}(T - T_0) - \{\alpha_M\}_p^{(k)}(c - c_0)\right)$$
$$\{\varepsilon\}_p^{(k)} = [\bar{S}]^{(k)}\{\sigma\}_p^{(k)} + \{\alpha_T\}_p^{(k)}(T - T_0) + \{\alpha_M\}_p^{(k)}(c - c_0) \quad (4.3-24)$$

where all quantities are referred to the (x, y, z) coordinate system, and $\{\alpha_T\}$ and $\{\alpha_M\}$ are vectors of thermal and hygroscopic coefficients of expansion, respectively.

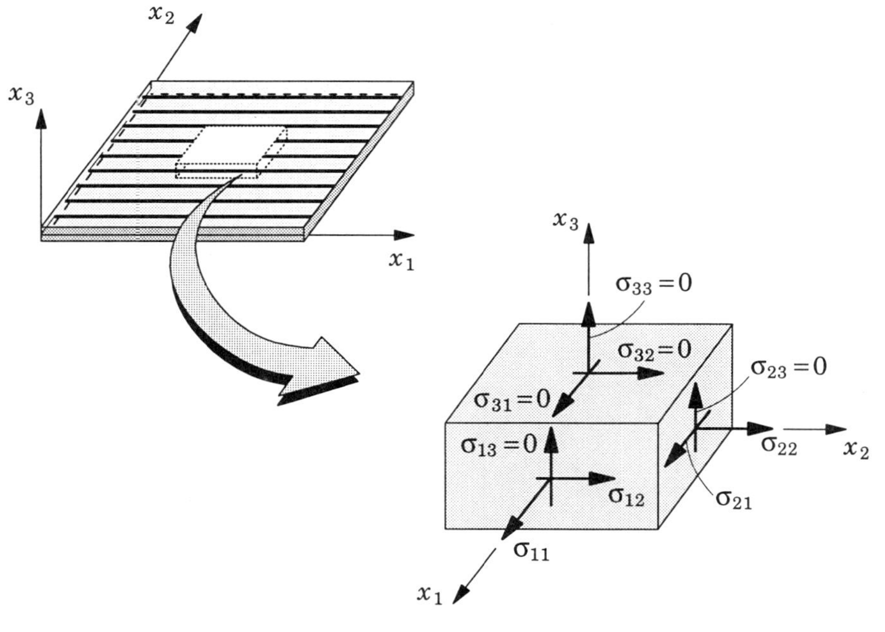

Figure 4.4–1: A lamina in a plane state of stress.

4.4 Plane Stress Constitutive Relations

A plane stress state is defined to be one in which all transverse stresses are negligible. Most laminates are typically thin and experience a plane state of stress. For a lamina in the x_1x_2–plane, the transverse stress components are σ_{33}, σ_{13}, and σ_{23} (see Figure 4.4–1). Although these stress components are small in comparison to σ_{11}, σ_{22}, and σ_{12}, they can induce failures because fiber-reinforced composite laminates are weak in the transverse direction (because the strength providing fibers are in the x_1x_2–plane). Thus, the transverse stress components are not always neglected in the laminate analyses (see Chapter 5). However, whenever they are neglected in a laminate theory, the constitutive equations must be modified to account for the fact that

$$\sigma_3 = 0,\ \sigma_4 = 0,\ \sigma_5 = 0 \quad (\sigma_{33} = 0,\ \sigma_{32} = 0,\ \sigma_{31} = 0) \tag{4.4-1}$$

For an orthotropic lamina, the condition $\sigma_{33} = 0$ results in the following transverse normal strain [from Eq. (4.2–4)]

$$\varepsilon_3 = S_{13}\sigma_1 + S_{23}\sigma_2$$

The zero transverse shear stresses result in zero transverse shear strains:

$$\varepsilon_4 = 0,\ \varepsilon_5 = 0$$

The strain–stress relations for the in–plane stress components $(\sigma_1, \sigma_2, \sigma_6)$ reduce to

$$\begin{Bmatrix} \varepsilon_1 \\ \varepsilon_2 \\ \varepsilon_6 \end{Bmatrix} = \begin{bmatrix} S_{11} & S_{12} & 0 \\ S_{12} & S_{22} & 0 \\ 0 & 0 & S_{66} \end{bmatrix} \begin{Bmatrix} \sigma_1 \\ \sigma_2 \\ \sigma_6 \end{Bmatrix}$$

$$= \begin{bmatrix} \frac{1}{E_1} & -\frac{\nu_{12}}{E_1} & 0 \\ -\frac{\nu_{12}}{E_1} & \frac{1}{E_2} & \\ 0 & 0 & \frac{1}{G_{12}} \end{bmatrix} \begin{Bmatrix} \sigma_1 \\ \sigma_2 \\ \sigma_6 \end{Bmatrix} \tag{4.4-2a}$$

where

$$\varepsilon_3 = S_{13}\sigma_1 + S_{23}\sigma_2 = -\frac{\nu_{13}}{E_1}\sigma_1 - \frac{\nu_{23}}{E_2}\sigma_2 \tag{4.4-2b}$$

The strain–stress relations (2) can be inverted to stress–strain relations

$$\begin{Bmatrix} \sigma_1 \\ \sigma_2 \\ \sigma_6 \end{Bmatrix} = \begin{bmatrix} Q_{11} & Q_{12} & 0 \\ Q_{12} & Q_{22} & 0 \\ 0 & 0 & Q_{66} \end{bmatrix} \begin{Bmatrix} \varepsilon_1 \\ \varepsilon_2 \\ \varepsilon_6 \end{Bmatrix} \tag{4.4-3a}$$

where the Q_{ij}, called the *plane stress-reduced stiffnesses*, are given by

$$\begin{aligned} Q_{11} &= \frac{S_{22}}{S_{11}S_{22} - S_{12}^2} = \frac{E_1}{1 - \nu_{12}\nu_{21}} \\ Q_{12} &= \frac{S_{12}}{S_{11}S_{22} - S_{12}^2} = \frac{\nu_{12}E_2}{1 - \nu_{12}\nu_{21}} \\ Q_{22} &= \frac{S_{11}}{S_{11}S_{22} - S_{12}^2} = \frac{E_2}{1 - \nu_{12}\nu_{21}} \\ Q_{66} &= \frac{1}{S_{66}} = G_{12} \end{aligned} \tag{4.4-3b}$$

Note that the reduced stiffnesses involve four independent material constants: E_1, E_2, ν_{12}, and G_{12}.

When the transverse normal stress is neglected but the transverse shear stresses are included, Eq. (3a) should be appended with the constitutive relations

$$\begin{Bmatrix} \sigma_4 \\ \sigma_5 \end{Bmatrix} = \begin{bmatrix} Q_{44} & 0 \\ 0 & Q_{55} \end{bmatrix} \begin{Bmatrix} \varepsilon_4 \\ \varepsilon_5 \end{Bmatrix} \tag{4.4-3c}$$

where $Q_{44} = C_{44} = G_{23}$ and $Q_{55} = C_{55} = G_{13}$.

For the plane stress case, the thermoelastic constitutive equations for an orthotropic lamina with piezoelectric effect may be expressed as

$$\begin{Bmatrix} \sigma_1 \\ \sigma_2 \\ \sigma_6 \end{Bmatrix} = \begin{bmatrix} Q_{11} & Q_{12} & 0 \\ Q_{12} & Q_{22} & 0 \\ 0 & 0 & Q_{66} \end{bmatrix} \begin{Bmatrix} \varepsilon_1 - \alpha_1\,\Delta T \\ \varepsilon_2 - \alpha_2\,\Delta T \\ \varepsilon_6 \end{Bmatrix} - \begin{bmatrix} 0 & 0 & e_{31} \\ 0 & 0 & e_{32} \\ 0 & 0 & 0 \end{bmatrix} \begin{Bmatrix} \mathcal{E}_1 \\ \mathcal{E}_2 \\ \mathcal{E}_3 \end{Bmatrix} \tag{4.4-4}$$

$$\begin{Bmatrix} \sigma_4 \\ \sigma_5 \end{Bmatrix} = \begin{bmatrix} Q_{44} & 0 \\ 0 & Q_{55} \end{bmatrix} \begin{Bmatrix} \varepsilon_4 \\ \varepsilon_5 \end{Bmatrix} - \begin{bmatrix} 0 & e_{24} & 0 \\ e_{15} & 0 & 0 \end{bmatrix} \begin{Bmatrix} \mathcal{E}_1 \\ \mathcal{E}_2 \\ \mathcal{E}_3 \end{Bmatrix} \tag{4.4-5}$$

$$\begin{Bmatrix} D_1 \\ D_2 \\ D_3 \end{Bmatrix} = \begin{bmatrix} 0 & 0 & 0 & e_{15} & 0 \\ 0 & 0 & e_{24} & 0 & 0 \\ e_{31} & e_{32} & 0 & 0 & 0 \end{bmatrix} \begin{Bmatrix} \varepsilon_1 \\ \varepsilon_2 \\ \varepsilon_4 \\ \varepsilon_5 \\ \varepsilon_6 \end{Bmatrix} + \begin{bmatrix} \epsilon_{11} & 0 & 0 \\ 0 & \epsilon_{22} & 0 \\ 0 & 0 & \epsilon_{33} \end{bmatrix} \begin{Bmatrix} \mathcal{E}_1 \\ \mathcal{E}_2 \\ \mathcal{E}_3 \end{Bmatrix} \tag{4.4-6}$$

where $Q_{ij}^{(k)}$ are the plane stress-reduced stiffnesses and $e_{ij}^{(k)}$ are the piezoelectric moduli of the kth lamina in its material coordinate system, $(\sigma_i, \varepsilon_i, \mathcal{E}_i, D_i)$ are the stress, strain, electric field, and electric displacement components, respectively, referred to the material coordinate system (x_1, x_2, x_3), α_1 and α_2 are the coefficients of thermal expansion along the x_1 and x_2 directions, respectively, and ΔT is the temperature increment from a reference state, $\Delta T = T - T_0$.

The transformed stress–strain relations of an orthotropic lamina in a plane state of stress are

$$\begin{aligned} \begin{Bmatrix} \sigma_{xx} \\ \sigma_{yy} \\ \sigma_{xy} \end{Bmatrix} &= \begin{bmatrix} \bar{Q}_{11} & \bar{Q}_{12} & \bar{Q}_{16} \\ \bar{Q}_{12} & \bar{Q}_{22} & \bar{Q}_{26} \\ \bar{Q}_{16} & \bar{Q}_{26} & \bar{Q}_{66} \end{bmatrix} \left(\begin{Bmatrix} \varepsilon_{xx} \\ \varepsilon_{yy} \\ \gamma_{xy} \end{Bmatrix} - \begin{Bmatrix} \alpha_{xx} \\ \alpha_{yy} \\ 2\alpha_{xy} \end{Bmatrix} \Delta T \right) \\ &\quad + \begin{bmatrix} 0 & 0 & \bar{e}_{31} \\ 0 & 0 & \bar{e}_{32} \\ 0 & 0 & \bar{e}_{36} \end{bmatrix} \begin{Bmatrix} \frac{\partial\psi}{\partial x} \\ \frac{\partial\psi}{\partial y} \\ \frac{\partial\psi}{\partial z} \end{Bmatrix} \end{aligned} \tag{4.4-7}$$

$$\left\{\begin{matrix}\sigma_{yz}\\ \sigma_{xz}\end{matrix}\right\} = \begin{bmatrix}\bar{Q}_{44} & \bar{Q}_{45}\\ \bar{Q}_{45} & \bar{Q}_{55}\end{bmatrix}\left\{\begin{matrix}\gamma_{yz}\\ \gamma_{xz}\end{matrix}\right\} + \begin{bmatrix}\bar{e}_{14} & \bar{e}_{24} & 0\\ \bar{e}_{15} & \bar{e}_{25} & 0\end{bmatrix}\left\{\begin{matrix}\frac{\partial\psi}{\partial x}\\ \frac{\partial\psi}{\partial y}\\ \frac{\partial\psi}{\partial z}\end{matrix}\right\} \tag{4.4-8}$$

$$\left\{\begin{matrix}D_x\\ D_y\\ D_z\end{matrix}\right\} = \begin{bmatrix}0 & 0 & \bar{e}_{14} & \bar{e}_{15} & 0\\ 0 & 0 & \bar{e}_{24} & \bar{e}_{25} & 0\\ \bar{e}_{31} & \bar{e}_{32} & 0 & 0 & \bar{e}_{36}\end{bmatrix}\left\{\begin{matrix}\varepsilon_{xx}\\ \varepsilon_{yy}\\ \gamma_{yz}\\ \gamma_{xz}\\ \gamma_{xy}\end{matrix}\right\} - \begin{bmatrix}\epsilon_{xx} & \epsilon_{xy} & 0\\ \epsilon_{xy} & \epsilon_{yy} & 0\\ 0 & 0 & \epsilon_{zz}\end{bmatrix}\left\{\begin{matrix}\frac{\partial\psi}{\partial x}\\ \frac{\partial\psi}{\partial y}\\ \frac{\partial\psi}{\partial z}\end{matrix}\right\} \tag{4.4-9}$$

where ψ denotes the scalar electric potential [see Eq. (2.5–19)] and

$$\begin{aligned}
\bar{Q}_{11} &= Q_{11}\cos^4\theta + 2(Q_{12} + 2Q_{66})\sin^2\theta\ \cos^2\theta + Q_{22}\sin^4\theta\\
\bar{Q}_{12} &= (Q_{11} + Q_{22} - 4Q_{66})\sin^2\theta\ \cos^2\theta + Q_{12}(\sin^4\theta + \cos^4\theta)\\
\bar{Q}_{22} &= Q_{11}\sin^4\theta + 2(Q_{12} + 2Q_{66})\sin^2\theta\ \cos^2\theta + Q_{22}\cos^4\theta\\
\bar{Q}_{16} &= (Q_{11} - Q_{12} - 2Q_{66})\sin\theta\ \cos^3\theta + (Q_{12} - Q_{22} + 2Q_{66})\sin^3\theta\ cos\theta\\
\bar{Q}_{26} &= (Q_{11} - Q_{12} - 2Q_{66})\sin^3\theta\cos\theta + (Q_{12} - Q_{22} + 2Q_{66})\sin\theta\ \cos^3\theta\\
\bar{Q}_{66} &= (Q_{11} + Q_{22} - 2Q_{12} - 2Q_{66})\sin^2\theta\ \cos^2\theta + Q_{66}(\sin^4\theta + \cos^4\theta)\\
\bar{Q}_{44} &= Q_{44}\cos^2\theta + Q_{55}\sin^2\theta\\
\bar{Q}_{45} &= (Q_{55} - Q_{44})\cos\theta\sin\theta\\
\bar{Q}_{55} &= Q_{55}\cos^2\theta + Q_{44}\sin^2\theta
\end{aligned} \tag{4.4-10}$$

α_{xx}, α_{yy}, and α_{xy} are the transformed thermal coefficients of expansion

$$\begin{aligned}
\alpha_{xx} &= \alpha_1\cos^2\theta + \alpha_2\sin^2\theta\\
\alpha_{yy} &= \alpha_1\sin^2\theta + \alpha_2\cos^2\theta\\
2\alpha_{xy} &= 2(\alpha_1 - \alpha_2)\sin\theta\cos\theta
\end{aligned} \tag{4.4-11}$$

and $\bar{e}_{ij}$ are the transformed piezoelectric moduli, and $\epsilon_{xx}, \epsilon_{xy}$, and ϵ_{xy} are transformed dielectric coefficients

$$\begin{aligned}
&\bar{e}_{31} = e_{31}\cos^2\theta + e_{32}\sin^2\theta,\ \bar{e}_{32} = e_{31}\sin^2\theta + e_{32}\cos^2\theta,\ \bar{e}_{33} = e_{33}\\
&\bar{e}_{36} = (e_{31} - e_{32})\sin\theta\cos\theta,\ \bar{e}_{14} = (e_{15} - e_{24})\sin\theta\cos\theta\\
&\bar{e}_{24} = e_{24}\cos^2\theta + e_{15}\sin^2\theta,\ \bar{e}_{15} = e_{15}\cos^2\theta + e_{24}\sin^2\theta\\
&\bar{e}_{25} = (e_{15} - e_{24})\sin\theta\cos\theta,\ \epsilon_{xx} = \epsilon_{11}\cos^2\theta + \epsilon_{22}\sin^2\theta\\
&\epsilon_{yy} = \epsilon_{11}\sin^2\theta + \epsilon_{22}\cos^2\theta,\ \epsilon_{xy} = (\epsilon_{22} - \epsilon_{11})\sin\theta\cos\theta
\end{aligned} \tag{4.4-12}$$

This completes the development of constitutive relations for an orthotropic lamina in a plane state of stress. It should be noted that the plane stress assumption (1) *does not* imply that ε_3 is zero [see Eq. (2b)] but it does imply ε_4 and ε_5 are zero.

Example 4.4–1:

The material properties of graphite fabric–carbon matrix layers are (see Example 2.4–1):

$$E_1 = 25.1 \times 10^6 \text{ psi}, \ E_2 = 4.8 \times 10^6 \text{ psi}, \ E_3 = 0.75 \times 10^6 \text{ psi}$$

$$G_{12} = 1.36 \times 10^6 \text{ psi}, \ G_{13} = 1.2 \times 10^6 \text{ psi}, \ G_{23} = 0.47 \times 10^6 \text{ psi}$$

$$\nu_{12} = 0.036, \ \nu_{13} = 0.25, \ \nu_{23} = 0.171$$

The matrix of plane stress-reduced elastic coefficients for the material can be calculated using Eq. (3b) as [*cf.* Eq. (2.4–15b)]

$$[Q] = \begin{bmatrix} 25.11 & 0.1728 & 0 & 0 & 0 \\ 0.1728 & 4.8010 & 0 & 0 & 0 \\ 0 & 0 & 0.47 & 0 & 0 \\ 0 & 0 & 0 & 1.20 & 0 \\ 0 & 0 & 0 & 0 & 1.36 \end{bmatrix} \text{ msi} \tag{4.4 – 13}$$

The transformed coefficients for various angles of orientation are given below:

$$[\bar{Q}]_{90} = \begin{bmatrix} 4.8010 & 0.1728 & 0 & 0 & 0 \\ 0.1728 & 25.11 & 0 & 0 & 0 \\ 0 & 0 & 0.47 & 0 & 0 \\ 0 & 0 & 0 & 1.20 & 0 \\ 0 & 0 & 0 & 0 & 1.36 \end{bmatrix} \text{ msi} \tag{4.4 – 14}$$

$$[\bar{Q}]_{45} = \begin{bmatrix} 8.923 & 6.203 & 0 & 0 & 5.076 \\ 6.203 & 8.923 & 0 & 0 & 5.076 \\ 0 & 0 & 0.835 & 0.365 & 0 \\ 0 & 0 & 0.365 & 0.835 & 0 \\ 5.076 & 5.076 & 0 & 0 & 7.390 \end{bmatrix} \text{ msi} \tag{4.4 – 15a}$$

$$[\bar{Q}]_{-45} = \begin{bmatrix} 8.923 & 6.203 & 0 & 0 & -5.076 \\ 6.203 & 8.923 & 0 & 0 & -5.076 \\ 0 & 0 & 0.835 & -0.365 & 0 \\ 0 & 0 & -0.365 & 0.835 & 0 \\ -5.076 & -5.076 & 0 & 0 & 7.390 \end{bmatrix} \text{ msi} \tag{4.4 – 15b}$$

$$[\bar{Q}]_{30} = \begin{bmatrix} 15.51 & 4.696 & 0 & 0 & 7.007 \\ 4.696 & 5.355 & 0 & 0 & 1.785 \\ 0 & 0 & 0.6525 & 0.3161 & 0 \\ 0 & 0 & 0.3161 & 0.6525 & 0 \\ 7.007 & 1.785 & 0 & 0 & 5.883 \end{bmatrix} \text{ msi} \tag{4.4 – 16}$$

Exercises

4.1 (*Apparent Moduli of an Orthotropic Material*) Note that the transformed material compliance matrix $[\bar{S}]$ is relatively full and is in the same form as that for a monoclinic material [see Eq. (2.4–11)]. For an orthotropic material, we have

$$\left\{\begin{array}{c}\varepsilon_{xx}\\ \varepsilon_{yy}\\ 2\varepsilon_{xy}\end{array}\right\}=\begin{bmatrix}\bar{S}_{11} & \bar{S}_{12} & \bar{S}_{16}\\ \bar{S}_{21} & \bar{S}_{22} & \bar{S}_{26}\\ \bar{S}_{16} & \bar{S}_{26} & \bar{S}_{66}\end{bmatrix}\left\{\begin{array}{c}\sigma_{xx}\\ \sigma_{yy}\\ \sigma_{xy}\end{array}\right\} \tag{1}$$

where $\bar{S}_{ij}$ are the transformed compliances defined in Eq. (4.3–21). Guided by the form of the strain–stress relations (4.2–10) in the material coordinates, we can write strain–stress relations in the problem coordinates as

$$\left\{\begin{array}{c}\varepsilon_{xx}\\ \varepsilon_{yy}\\ 2\varepsilon_{xy}\end{array}\right\}=\begin{bmatrix}\frac{1}{E_x} & -\frac{\nu_{yx}}{E_y} & \frac{\eta_{xy,x}}{E_x}\\ -\frac{\nu_{xy}}{E_x} & \frac{1}{E_y} & \frac{\eta_{xy,y}}{E_y}\\ \frac{\eta_{xy,x}}{E_x} & \frac{\eta_{xy,y}}{E_y} & \frac{1}{G_{xy}}\end{bmatrix}\left\{\begin{array}{c}\sigma_{xx}\\ \sigma_{yy}\\ \sigma_{xy}\end{array}\right\} \tag{2}$$

Comparing Eq. (2) with Eq. (1), we note that

$$\frac{1}{E_x}=\bar{S}_{11},\quad -\frac{\nu_{yx}}{E_y}=\bar{S}_{12},\quad \frac{\eta_{xy,y}}{E_y}=\bar{S}_{16},\quad \frac{\eta_{xy,y}}{E_y}=\bar{S}_{26} \tag{3}$$

and so on. Thus, the equivalent modulus of elasticity E_x in the problem coordinates, for example, can be evaluated using the engineering constants in the material coordinate system:

$$\frac{1}{E_x}\equiv\bar{S}_{11}=S_{11}\cos^4\theta+(2S_{12}+S_{66})\sin^2\theta\,\cos^2\theta+S_{22}\sin^4\theta \tag{4a}$$

where

$$S_{11}=\frac{1}{E_1}\,,\quad S_{12}=-\frac{\nu_{21}}{E_2}\,,\quad S_{22}=\frac{1}{E_2}\,,\quad S_{66}=\frac{1}{G_{12}} \tag{4b}$$

Thus, the apparent compliance $\bar{S}_{11}$ in the (x,y,z) coordinate system is contributed by the compliances S_{11}, S_{12}, S_{22}, and S_{66} and the lamination angle θ:

$$\frac{1}{E_x}=\frac{1}{E_1}\cos^4\theta+\left(-\frac{2\nu_{21}}{E_2}+\frac{1}{G_{12}}\right)\sin^2\theta\,\cos^2\theta+\frac{1}{E_2}\sin^4\theta \tag{5}$$

We note that the compliance $\bar{S}_{16}$, which was zero in the material coordinates, is contributed by S_{11}, S_{12}, S_{22}, and S_{66}:

$$\bar{S}_{16}=\left(\frac{2}{E_1}+\frac{2\nu_{21}}{E_2}-\frac{1}{G_{12}}\right)\sin\theta\,\cos^3\theta-\left(\frac{2}{E_2}+\frac{2\nu_{21}}{E_2}-\frac{1}{G_{12}}\right)\sin^3\theta\cos\theta \tag{6}$$

Physically, $\bar{S}_{16}$ represents the normal strain in the x–direction caused by the shear stress in the xy–plane, when all other stresses are zero. Since $\bar{S}_{16} = \bar{S}_{61}$, it also represents the shear strain in the xy–plane caused by the normal stress along the x–direction, when all other stresses are zero. Guided by these observations, Lekhnitskii [4] introduced the following engineering constants, called *coefficients of mutual influence*:

$\eta_{ij,i}$ =characterizes shearing in the $x_i x_j$–plane caused by a normal stress in the x_i–direction $(i \neq j)$

$$= \frac{2\varepsilon_{ij}}{\varepsilon_{ii}}, \quad \text{for } \sigma_{ii} \neq 0 \text{ and all other stresses being zero} \tag{7}$$

The compliance $\bar{S}_{16}$ and $\bar{S}_{26}$ are related, by definition, to the coefficients $\eta_{xy,x}$ and $\eta_{xy,y}$ by

$$\bar{S}_{16} = \eta_{xy,x}/E_x, \quad \bar{S}_{26} = \eta_{xy,y}/E_y \tag{8}$$

Show that

$$\frac{\eta_{xy,x}}{E_x} = \left(\frac{2}{E_1} + \frac{2\nu_{21}}{E_2} - \frac{1}{G_{12}}\right) \sin\theta \, \cos^3\theta - \left(\frac{2}{E_1} + \frac{2\nu_{21}}{E_2} - \frac{1}{G_{12}}\right) \sin^3\theta \cos\theta \tag{9}$$

$$\frac{\eta_{xy,y}}{E_y} = \left(\frac{2}{E_1} + \frac{2\nu_{21}}{E_2} - \frac{1}{G_{12}}\right) \sin^3\theta \, \cos\theta - \left(\frac{2}{E_2} + \frac{2\nu_{21}}{E_2} - \frac{1}{G_{12}}\right) \sin\theta \, \cos^3\theta \tag{10}$$

4.2 (*Continuation of Exercise 1*) Derive an expression for G_{xy} in terms of $E_1, E_2, \nu_{12}, G_{12}$, and θ.

4.3 (*Continuation of Exercise 1*) Show that G_{xy} is a maximum for $\theta = 45°$. Make use of the following trigonometric identities:

$$\cos^4\theta = \frac{1}{8}(3 + 4\cos 2\theta + \cos 4\theta)$$

$$\sin^4\theta = \frac{1}{8}(3 - 4\cos 2\theta + \cos 4\theta)$$

$$\cos^2\theta \, \sin^2\theta = \frac{1}{4}(1 - \cos 4\theta)$$

4.4 (*Continuation of Exercise 1*) Show that the coefficient of mutual influence is zero at $\theta = 0°$ and $\theta = 90°$.

4.5 (*Continuation of Exercise 1*) Show that the moduli E_x (and E_y) varies between E_1 and E_2, but it can either exceed or get smaller than both E_1 and E_2.

4.6 (*Continuation of Exercise 1*) Derive the expression for E_x in terms of $E_1, E_2, \nu_{12}, G_{12}, \alpha_1, \alpha_2$, and θ for the nonisothermal case.

4.7 (*Continuation of Exercise 1*) Derive the expression for G_{xy} in terms of $E_1, E_2, \nu_{12}, G_{12}, \alpha_1, \alpha_2$, and θ for the nonisothermal case.

4.8 Show that the following combinations of stiffness coefficients are invariant:

$$
\begin{aligned}
U_1 &= \frac{1}{8}\left(3Q_{11} + 3Q_{22} + 2Q_{12} + 4Q_{66}\right)\\
U_2 &= \frac{1}{2}\left(Q_{11} - Q_{22}\right)\\
U_3 &= \frac{1}{8}\left(Q_{11} + Q_{22} - 2Q_{12} - 4Q_{66}\right)\\
U_4 &= \frac{1}{8}\left(Q_{11} + Q_{22} + 6Q_{12} - 4Q_{66}\right)
\end{aligned}
$$

4.9 Show that the following combinations of compliance coefficients are invariant:

$$
\begin{aligned}
V_1 &= \frac{1}{8}\left(3S_{11} + 3S_{22} + 2S_{12} + S_{66}\right)\\
V_2 &= \frac{1}{2}\left(S_{11} - S_{22}\right)\\
V_3 &= \frac{1}{8}\left(S_{11} + S_{22} - 2S_{12} - S_{66}\right)\\
V_4 &= \frac{1}{8}\left(S_{11} + S_{22} + 6S_{12} - S_{66}\right)
\end{aligned}
$$

4.10 Rewrite the transformation equations (4.4–10) in terms of the invariants of Exercise 8. *Ans:*

$$
\begin{aligned}
\bar{Q}_{11} &= U_1 + U_2\cos 2\theta + U_3\cos 4\theta\\
\bar{Q}_{12} &= U_4 - U_3\cos 4\theta\\
\bar{Q}_{22} &= U_1 - U_2\cos 2\theta + U_3\cos 4\theta\\
\bar{Q}_{16} &= \frac{1}{2}U_2\sin 2\theta + U_3\sin 4\theta\\
\bar{Q}_{26} &= \frac{1}{2}U_2\sin 2\theta - U_3\sin 4\theta\\
\bar{Q}_{66} &= \frac{1}{2}\left(U_1 - U_4\right) - U_3\cos 4\theta
\end{aligned}
$$

4.11 Verify the transformation relations for the piezoelectric moduli given in Eq. (4.4–12).

4.12 Consider a square, graphite–epoxy lamina of length 8 in., width 2 in., and thickness 0.005 in., and subjected to an axial load of 1000 lbs. Determine the transverse normal strain ε_3. Assume that the load is applied parallel to the fibers, and use $E_1 = 20$ msi, $E_2 = 1.3$ msi, $G_{12} = G_{13} = 1.03$ msi, $G_{23} = 0.9$ msi, $\nu_{12} = \nu_{13} = 0.3$, and $\nu_{23} = 0.49$.

4.13 Compute the numerical values of the reduced stiffnesses Q_{ij} for the graphite–epoxy material of Exercise 12. *Ans:*

$$[Q] = \begin{bmatrix} 20.118 & 0.392 & 0 & 0 & 0 \\ 0.392 & 1.308 & 0 & 0 & 0 \\ 0 & 0 & 1.030 & 0 & 0 \\ 0 & 0 & 0 & 0.900 & 0 \\ 0 & 0 & 0 & 0 & 1.030 \end{bmatrix} \text{msi}$$

4.14 The material properties of AS/3501 graphite–epoxy material layers are

$$E_1 = 140 \times 10^3 \text{ MPa}, \; E_2 = 10 \times 10^3 \text{ MPa}, \; G_{12} = 7 \times 10^3 \text{ MPa}$$

$$G_{13} = 7 \times 10^3 \text{ MPa}, \; G_{23} = 7 \times 10^3 \text{ MPa}, \; \nu_{12} = 0.3$$

$$\alpha_1 = 1.0 \times 10^{-6} \text{ m/m/°K}, \; \alpha_2 = 30 \times 10^{-6} \text{ m/m/°K}$$

Show that

$$[Q] = \begin{bmatrix} 140.9 & 3.019 & 0 & 0 & 0 \\ 3.019 & 10.06 & 0 & 0 & 0 \\ 0 & 0 & 7.00 & 0 & 0 \\ 0 & 0 & 0 & 7.00 & 0 \\ 0 & 0 & 0 & 0 & 7.00 \end{bmatrix} \text{GPa}$$

The transformed coefficients for various angles of orientation are given below:

$$[\bar{Q}]_{90} = \begin{bmatrix} 10.06 & 3.019 & 0 & 0 & 0 \\ 3.019 & 140.9 & 0 & 0 & 0 \\ 0 & 0 & 7.0 & 0 & 0 \\ 0 & 0 & 0 & 7.0 & 0 \\ 0 & 0 & 0 & 0 & 7.0 \end{bmatrix} \text{GPa}$$

$$[\bar{Q}]_{45} = \begin{bmatrix} 46.25 & 32.25 & 0 & 0 & 32.71 \\ 32.25 & 46.25 & 0 & 0 & 32.71 \\ 0 & 0 & 7.0 & 0 & 0 \\ 0 & 0 & 0 & 7.0 & 0 \\ 32.71 & 32.71 & 0 & 0 & 36.23 \end{bmatrix} \text{GPa}$$

Also, compute the transformed thermal coefficients of expansion for $\theta = 45°$.

References for Additional Reading

1. Ambartsumyan, S. A., *Theory of Anisotropic Shells*, NASA Report TT F–118, 1964.
2. Ambartsumyan, S. A., *Theory of Anisotropic Plates*, Izdat. Nauka, Moskva (1967), English translation by Technomic, Stamford, CT, 1969.
3. Jones, R. M., *Mechanics of Composite Materials*, Scripta Book Co., Washington, D.C., 1975.
4. Lekhnitskii, S. G., *Theory of Elasticity of an Anisotropic Body*, Mir Publishers, Moscow (Translated from the revised 1977 Russian Edition), 1982.
5. Christensen, R. M., *Mechanics of Composite Materials*, John Wiley & Sons, New York, 1979.
6. Tsai, S. W. and Hahn, H. T., *Introduction to Composite Materials*, Technomic, Lancaster, Pennsylvania, 1980.
7. Agarwal, B. D. and Broutman, L. J., *Analysis and Performance of Fiber Composites*, John Wiley & Sons, New York, 1980.
8. Reddy, J. N., *Energy and Variational Methods in Applied Mechanics,* John Wiley and Sons, New York, 1984.
9. Vinson, J. R. and Sierakowski, R. L., *The Behavior of Structures Composed of Composite Materials,* Kluwer, Netherlands, 1986.
10. Mallick, P. K., *Fiber-Reinforced Composites. Materials, Manufacturing, and Design*, Marcel Dekker, New York, 1988.
11. Gibson, R. F., *Principles of Composite Material Mechanics,* McGraw–Hill, New York, 1994.
12. Ochoa, O. O. and Reddy, J. N., *Finite Element Analysis of Composite Laminates*, Kluwer, Netherlands, 1992.
13. Parton, V. Z. and Kudryavtsev, B. A., *Engineering Mechanics of Composite Structures,* CRC Press, Boca Raton, Florida, 1993.
14. Reddy, J. N. (ed.), *Mechanics of Composite Materials. Selected Works of Nicholas J. Pagano*, Kluwer, Netherlands, 1994.
15. Reddy, J. N. and Miravete, A., *Practical Analysis of Composite Laminates,* CRC Press, Boca Raton, Florida, 1995.
16. Adams, D. F. and Doner, D. R., "Longitudinal Shear Loading of a Unidirectional Composite," *Journal of Composite Materials,* **1**, 4–17 (1967).
17. Adams, D. F. and Doner, D. R., "Transverse Normal Loading of a Unidirectional Composite," *Journal of Composite Materials,* **1**, 152–164 (1967).
18. Ishikawa, T., Koyama, K., and Kobayashi, S., "Thermal Expansion Coefficients of Unidirectional Composites," *Journal of Composite Materials,* **12**, 153–168 (1978).
19. Halpin, J. C. and Tsai, S. W., "Effects of Environmental Factors on Composite Materials," AFML–TR–67–423, Air Force Flight Mechanics Laboratory, Dayton, Ohio, 1969.

Chapter Five

Classical and First-Order Theories of Laminated Composite Plates

5.1 Introduction

5.1–1 Preliminary Comments

Composite laminates are formed by stacking layers of different composite materials and/or fiber orientation. By construction, composite laminates have their planar dimensions one to two orders of magnitude larger than their thickness. Often laminates are used in applications that require axial and bending strengths. Therefore, composite laminates are treated as plate elements.

The objective of this chapter is to develop two commonly used laminate plate theories, namely the classical and first-order shear deformation theories. To provide a background for the theories discussed in this chapter, an overview of the literature on laminate plate theories, existing at the time of this writing, is included here.

5.1–2 Classification of Structural Theories

The analyses of composite plates in the past have been based on one of the following approaches:

(1) Equivalent single-layer theories (2-D)

    (a) Classical laminate theory

    (b) Shear deformation laminate theories

(2) Three-dimensional elasticity theory (3-D)

    (a) Traditional 3-D elasticity formulations

    (b) Layerwise theories

(3) Multiple model methods (2-D and 3-D)

The equivalent single-layer (ESL) theories are derived from the 3-D elasticity theory by making suitable assumptions concerning the kinematics of deformation or the stress state through the thickness of the laminate. These assumptions allow the reduction of a 3-D problem to a 2-D problem. In the three-dimensional elasticity theory or in a layerwise theory, each layer is modeled as a 3-D solid. Here we present a review of the equivalent single-layer theories. Literature reviews and development of the governing equations of third-order shear theories, layerwise theories, and multiple models will be presented in later chapters (see Chapters 11 and 12).

5.2 An Overview of ESL Laminate Theories

The equivalent single-layer laminate theories are those in which a heterogeneous laminated plate is treated as a statically equivalent, single layer having a complex constitutive behavior, reducing the 3-D continuum problem to a 2-D problem. The ESL theories are developed by assuming the form of the displacement field or stress field as a linear combination of unknown functions and the thickness coordinate [1–13]:

$$\varphi_i(x,y,z,t) = \sum_{j=0}^{N} (z)^j \varphi_i^j(x,y,t) \tag{5.2-1}$$

where φ_i is the ith component of displacement or stress, (x,y) are the in-plane coordinates, z is the thickness coordinate, t denotes the time, and φ_i^j are functions to be determined.

When φ_i are displacements, then the equations governing φ_i^j are determined by the principle of virtual displacements (or its dynamic version when time dependency is to be included):

$$0 = \int_0^T (\delta U + \delta V - \delta K)\, dt \tag{5.2-2}$$

where $\delta U, \delta V$, and δK denote the virtual strain energy, virtual work done by external applied forces, and the virtual kinetic energy, respectively. These quantities are determined in terms of actual stresses and virtual strains, which depend on the assumed displacement functions, φ_i and their variations. For plate structures, laminated or not, the integration over the domain of the plate is represented as the (tensor) product of integration over the plane of the plate and integration over the thickness of the plate, because of the explicit nature of the assumed displacement field in the thickness coordinate:

$$\int_{Vol.} (\cdot)\, dV = \int_{-\frac{h}{2}}^{\frac{h}{2}} \int_{\Omega_0} (\cdot)\, d\Omega\, dz \tag{5.2-3}$$

where h denotes the total thickness of the plate and Ω_0 denotes the undeformed midplane of the plate, which is chosen as the reference plane. Since all functions are explicit in the thickness coordinate, the integration over plate thickness is carried out explicitly, reducing the problem to a two dimensional one. Consequently, the Euler–Lagrange equations of Eq. (2) consist of differential equations involving the dependent variables $\varphi_i^j(x, y, t)$ and thickness-averaged stress resultants, $R_{ij}^{(m)}$:

$$R_{ij}^{(m)} = \int_{-\frac{h}{2}}^{\frac{h}{2}} (z)^m \sigma_{ij} dz \tag{5.2-4}$$

The resultants can be written in terms of φ_i with the help of the assumed constitutive equations (stress–strain relations) and strain–displacement relations. More complete development of this procedure is forthcoming in this chapter.

The same approach is used when φ_i denote stress components, except that the basis of the derivation of the governing equations is the principle of virtual forces. In the present book, the stress-based theories will not be developed. Readers interested in stress-based theories may consult the book by Panc [14].

The simplest ESL laminate theory is the *classical laminated plate theory* (or CLPT) [15–20], which is an extension of the Kirchhoff (classical) plate theory to laminated composite plates. It is based on the displacement field

$$\begin{aligned}
u(x, y, z, t) &= u_0(x, y, t) - z\frac{\partial w_0}{\partial x} \\
v(x, y, z, t) &= v_0(x, y, t) - z\frac{\partial w_0}{\partial y} \\
w(x, y, z, t) &= w_0(x, y, t)
\end{aligned} \tag{5.2-5}$$

where (u_0, v_0, w_0) are the displacement components along the (x, y, z) coordinate directions, respectively, of a point on the midplane ($i.e., z = 0$). The displacement field (5) implies that straight lines normal to the xy–plane before deformation remain straight and normal to the midsurface after deformation. The Kirchhoff assumption amounts to neglecting both transverse shear and transverse normal effects, *i.e.*, deformation is due entirely to bending and in-plane stretching.

The next theory in the hierarchy of ESL laminate theories is the *first-order shear deformation theory* (or FSDT) [21–27], which is based on the displacement field

$$\begin{aligned}
u(x, y, z, t) &= u_0(x, y, t) + z\phi_x(x, y, t) \\
v(x, y, z, t) &= v_0(x, y, t) + z\phi_y(x, y, t) \\
w(x, y, z, t) &= w_0(x, y, t)
\end{aligned} \tag{5.2-6}$$

where ϕ_x and $-\phi_y$ denote rotations about the y and x axes, respectively. The FSDT extends the kinematics of the CLPT by including a gross transverse shear deformation in its kinematic assumptions, *i.e.*, the transverse shear strain is assumed to be constant with respect to the thickness coordinate. Inclusion of this rudimentary form of shear deformation allows the normality restriction of the classical laminate theory to be relaxed. The first-order shear deformation theory requires shear correction factors (see [28–32]), which are difficult to determine for arbitrarily laminated composite plate structures. The shear correction factors depend not only on the lamination and geometric parameters, but also on the loading and boundary conditions.

In both CLPT and FSDT, the plane-stress state assumption is used and plane-stress reduced form of the constitutive law of Section 4.4 is used. In both theories the inextensibility and/or straightness of transverse normals can be removed. Such extensions lead to second- and higher-order theories of plates.

Second- and higher-order ESL laminated plate theories use higher-order polynomials [*i.e.*, $N > 1$ in Eq. (1)] in the expansion of the displacement components through the thickness of the laminate (see [33–38], among many others). The higher-order theories introduce additional unknowns that are often difficult to interpret in physical terms. The second-order theory with transverse inextensibility is based on the displacement field

$$\begin{aligned} u(x,y,z,t) &= u_0(x,y,t) + z\phi_x(x,y,t) + z^2\psi_x(x,y,t) \\ v(x,y,z,t) &= v_0(x,y,t) + z\phi_y(x,y,t) + z^2\psi_y(x,y,t) \\ w(x,y,z,t) &= w_0(x,y,t) \end{aligned} \tag{5.2 – 7}$$

The *third-order laminate theory of Reddy* [38] with transverse inextensibility is based on the displacement field

$$\begin{aligned} u(x,y,z,t) &= u_0(x,y,t) + z\phi_x(x,y,t) + z^3\left(-\frac{4}{3h^2}\right)\left(\phi_x + \frac{\partial w_0}{\partial x}\right) \\ v(x,y,z,t) &= v_0(x,y,t) + z\phi_y(x,y,t) + z^3\left(-\frac{4}{3h^2}\right)\left(\phi_y + \frac{\partial w_0}{\partial y}\right) \\ w(x,y,z,t) &= w_0(x,y,t) \end{aligned} \tag{5.2 – 8}$$

The displacement field accommodates quadratic variation of transverse shear strains (and hence stresses) and vanishing of transverse shear stresses on the top and bottom of a general laminate composed of monoclinic layers. Thus there is no need to use shear correction factors in a third-order theory. The theory was generalized in Reference 39. The third-order theories provide a slight increase in accuracy relative to the FSDT solution, at the expense of a significant increase in computational effort. Further, finite element models of third-order theories that satisfy the vanishing of transverse shear stresses on the bounding planes require continuity of the transverse deflection and its derivatives between elements. Complete derivations of the governing equations of the third-order theories and their solutions are presented in Chapter 11.

In addition to their inherent simplicity and low computational cost, the ESL models often provide sufficiently accurate description of global response for thin to moderately thick laminates, *e.g.*, gross deflections, critical buckling loads, and fundamental vibration frequencies and associated mode shapes. Of the ESL theories, the FSDT with transverse extensibility appears to provide the best compromise of solution accuracy, economy, and simplicity. However, the ESL models have limitations that prevent them from being used to solve the whole spectrum of composite laminate problems. First, the accuracy of the global response predicted by the ESL models deteriorates as the laminate becomes thicker. Second, the ESL models are often incapable of accurately describing the state of stress and strain at the ply level near geometric and material discontinuities or near regions of intense loading – the areas where accurate stresses are needed most. In such cases, 3-D theories or multiple model approaches are required.

This completes an overview of various ESL theories. In the remaining sections of this chapter, we study the classical plate theory and first-order shear deformation plate theory for laminated plates. The third-order laminate theory and the layerwise theory developed and advanced by the author are presented in Chapters 11 and 12, respectively.

5.3 The Classical Laminated Plate Theory

5.3–1 Assumptions

The classical laminated plate theory is an extension of the classical plate theory to composite laminates. In the *classical laminated plate theory* (CLPT) it is assumed† that the *Kirchhoff hypothesis* holds:

(1) Straight lines perpendicular to the midsurface (*i.e.*, transverse normals) before deformation remain straight after deformation.

(2) The transverse normals do not experience elongation (*i.e.*, they are inextensible).

(3) The transverse normals rotate such that they remain perpendicular to the midsurface after deformation.

---

† An *assumption* is that which is necessary for the development of the mathematical model, whereas a *restriction* is not a necessary condition for the development of the theory.

The first two assumptions imply that the transverse displacement is independent of the transverse (or thickness) coordinate and the transverse normal strain ε_{zz} is zero. The third assumption results in zero transverse shear strains, $\varepsilon_{xz} = 0, \varepsilon_{yz} = 0$.

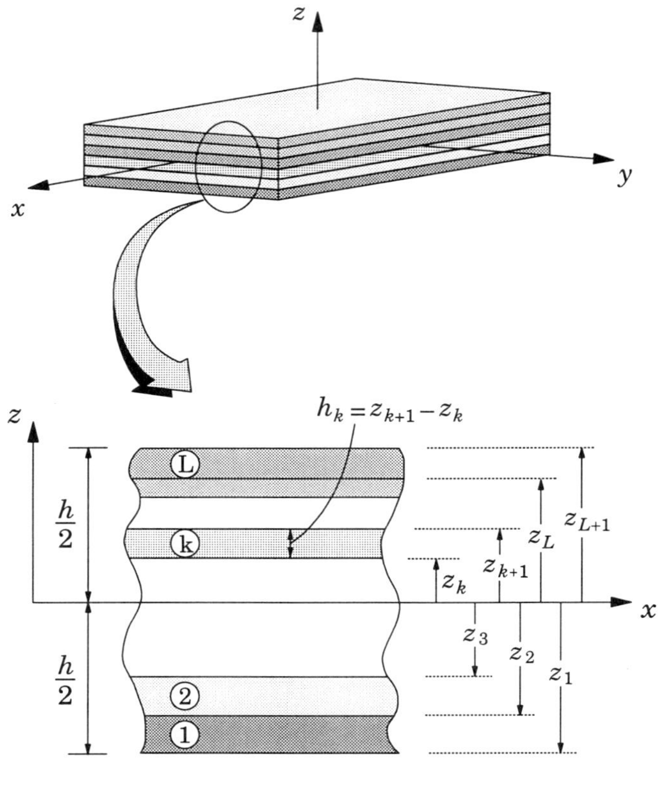

Figure 5.3–1: Coordinate system and layer numbering used for a typical laminated plate.

5.3–2 Displacements and Strains

Consider a plate of total thickness h composed of N orthotropic layers with the principal material coordinates (x_1^k, x_2^k, x_3^k) of the kth lamina oriented at an angle θ_k to the laminate coordinate, x. Although not necessary, it is convenient to take the xy–plane of the problem in the undeformed midplane Ω_0 of the laminate (see Figure 5.3–1). The z–axis is taken positive upward from the midplane. The kth layer is located between the points $z = z_k$ and $z = z_{k+1}$ in the thickness direction. The total domain $\bar{\Omega}_0$ of the laminate is the tensor product of $\Omega_0 \times (-h/2, h/2)$. The boundary of $\bar{\Omega}_0$ consists of top surface $S_t(z = h/2)$ and bottom surfaces $S_b(z = -h/2)$, and the edge $\bar{\Gamma} \equiv \Gamma \times (-h/2, h/2)$ of the laminate. In general, Γ is a curved surface, with outward normal $\hat{\mathbf{n}} = n_x\hat{\mathbf{e}}_x + n_y\hat{\mathbf{e}}_y$. Different parts of the boundary $\bar{\Gamma}$ are subjected to, in general, a combination of generalized forces and generalized displacements. A discussion of the boundary conditions is presented in the sequel.

In formulating the theory, we make the certain assumptions or place restrictions, as stated here:

- The layers are perfectly bonded together (assumption).
- The material of each layer is linearly elastic and has two planes of material symmetry (*i.e.*, orthotropic) (restriction).
- Each layer is of uniform thickness (restriction).
- The strains and displacements are small (restriction).
- The transverse shear stresses on the top and bottom surfaces of the laminate are zero (restriction).

By the Kirchhoff assumptions, a material point occupying the position (x, y, z) in the undeformed laminate moves to the position $(x+u, y+v, z+w)$ in the deformed laminate, where (u, v, w) are the components of the total displacement vector $\mathbf{u}$ along the (x, y, z) coordinates. We have

$$\mathbf{u} = u\hat{\mathbf{e}}_x + v\hat{\mathbf{e}}_y + w\hat{\mathbf{e}}_z \tag{5.3-1}$$

where $(\hat{\mathbf{e}}_x, \hat{\mathbf{e}}_y, \hat{\mathbf{e}}_z)$ are unit vectors along the (x, y, z) coordinates. Due to small strain and small displacement assumption, no distinction is made between the material coordinates and spatial coordinates, between the finite Green strain tensor and infinitesimal strain tensor, and between the second Piola–Kirchhoff stress tensor and the Cauchy stress tensor (see Chapter 3). The Kirchhoff hypothesis requires the displacements (u, v, w) to be such that (see Figure 5.3–2)

$$\begin{aligned} u(x, y, z, t) &= u_0(x, y, t) - z\frac{\partial w_0}{\partial x} \\ v(x, y, z, t) &= v_0(x, y, t) - z\frac{\partial w_0}{\partial y} \\ w(x, y, z, t) &= w_0(x, y, t) \end{aligned} \tag{5.3-2}$$

where (u_0, v_0, w_0) are the displacements along the coordinate lines of a material point on the xy–plane. Note that the form of the displacement field (1) allows reduction of the 3-D problem to one of studying the deformation of the reference plane $z = 0$ (or midplane). Once the midplane displacements (u_0, v_0, w_0) are known, the displacements of any arbitrary point (x, y, z) in the 3-D continuum can be determined using Eq. (1).

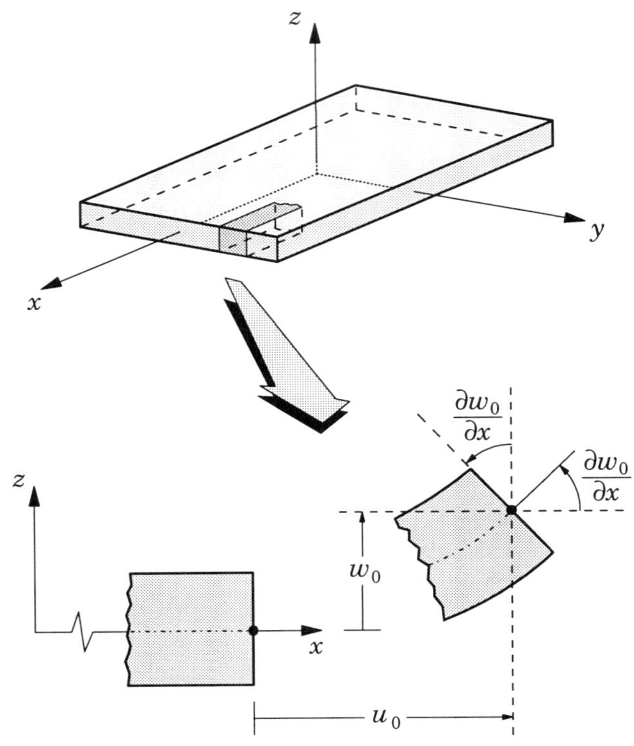

Figure 5.3–2: Undeformed and deformed geometries of an edge of a plate under the Kirchhoff assumptions.

The strains associated with the displacement field (2) can be computed using either the nonlinear strain–displacement relations (2.2–10) or the linear strain–displacement relations (2.2–13). The nonlinear strains are given by

$$E_{xx} = \frac{\partial u}{\partial x} + \frac{1}{2}\left[\left(\frac{\partial u}{\partial x}\right)^2 + \left(\frac{\partial v}{\partial x}\right)^2 + \left(\frac{\partial w}{\partial x}\right)^2\right]$$

$$E_{yy} = \frac{\partial v}{\partial y} + \frac{1}{2}\left[\left(\frac{\partial u}{\partial y}\right)^2 + \left(\frac{\partial v}{\partial y}\right)^2 + \left(\frac{\partial w}{\partial y}\right)^2\right]$$

$$E_{zz} = \frac{\partial w}{\partial z} + \frac{1}{2}\left[\left(\frac{\partial u}{\partial z}\right)^2 + \left(\frac{\partial v}{\partial z}\right)^2 + \left(\frac{\partial w}{\partial z}\right)^2\right]$$

$$E_{xy} = \frac{1}{2}\left(\frac{\partial u}{\partial y} + \frac{\partial v}{\partial x} + \frac{\partial u}{\partial x}\frac{\partial u}{\partial y} + \frac{\partial v}{\partial x}\frac{\partial v}{\partial y} + \frac{\partial w}{\partial x}\frac{\partial w}{\partial y}\right)$$

$$E_{xz} = \frac{1}{2}\left(\frac{\partial u}{\partial z} + \frac{\partial w}{\partial x} + \frac{\partial u}{\partial x}\frac{\partial u}{\partial z} + \frac{\partial v}{\partial x}\frac{\partial v}{\partial z} + \frac{\partial w}{\partial x}\frac{\partial w}{\partial z}\right)$$
$$E_{yz} = \frac{1}{2}\left(\frac{\partial v}{\partial z} + \frac{\partial w}{\partial y} + \frac{\partial u}{\partial y}\frac{\partial u}{\partial z} + \frac{\partial v}{\partial y}\frac{\partial v}{\partial z} + \frac{\partial w}{\partial y}\frac{\partial w}{\partial z}\right) \tag{5.3-3}$$

If the components of the displacement gradients are of the order ϵ, *i.e.*,

$$\frac{\partial u}{\partial x}, \frac{\partial u}{\partial y}, \frac{\partial v}{\partial x}, \frac{\partial v}{\partial y}, \frac{\partial w}{\partial z} = \mathrm{O}(\epsilon) \tag{5.3-4}$$

then the small strain assumption implies that terms of the order ϵ^2 are negligible in the strains. Terms of order ϵ^2 are

$$\left(\frac{\partial u}{\partial x}\right)^2, \left(\frac{\partial u}{\partial y}\right)^2, \left(\frac{\partial u}{\partial z}\right)^2, \left(\frac{\partial u}{\partial x}\right)\left(\frac{\partial u}{\partial y}\right), \left(\frac{\partial u}{\partial x}\right)\left(\frac{\partial u}{\partial z}\right), \left(\frac{\partial u}{\partial y}\right)\left(\frac{\partial u}{\partial z}\right)$$

$$\left(\frac{\partial v}{\partial x}\right)^2, \left(\frac{\partial v}{\partial y}\right)^2, \left(\frac{\partial v}{\partial z}\right)^2, \left(\frac{\partial v}{\partial x}\right)\left(\frac{\partial v}{\partial y}\right), \left(\frac{\partial v}{\partial x}\right)\left(\frac{\partial v}{\partial z}\right), \left(\frac{\partial v}{\partial y}\right)\left(\frac{\partial v}{\partial z}\right)$$

$$\left(\frac{\partial w}{\partial x}\right)\left(\frac{\partial w}{\partial z}\right), \left(\frac{\partial w}{\partial y}\right)\left(\frac{\partial w}{\partial z}\right), \left(\frac{\partial w}{\partial z}\right)^2 \tag{5.3-5}$$

If the rotations $\partial w_0/\partial x$ and $\partial w_0/\partial y$ of transverse normals are moderate (say 10°–15°), then the following terms are small but *not* negligible compared to ϵ:

$$\left(\frac{\partial w}{\partial x}\right)^2, \left(\frac{\partial w}{\partial y}\right)^2, \frac{\partial w}{\partial x}\frac{\partial w}{\partial y} \tag{5.3-6}$$

and they should be included in the strain–displacement relations. Thus for small strains and moderate rotations case the strain–displacement relations (3) take the form

$$\varepsilon_{xx} = \frac{\partial u}{\partial x} + \frac{1}{2}\left(\frac{\partial w}{\partial x}\right)^2, \quad \varepsilon_{xy} = \frac{1}{2}\left(\frac{\partial u}{\partial y} + \frac{\partial v}{\partial x} + \frac{\partial w}{\partial x}\frac{\partial w}{\partial y}\right)$$

$$\varepsilon_{xz} = \frac{1}{2}\left(\frac{\partial u}{\partial z} + \frac{\partial w}{\partial x}\right), \quad \varepsilon_{yy} = \frac{\partial v}{\partial y} + \frac{1}{2}\left(\frac{\partial w}{\partial y}\right)^2$$

$$\varepsilon_{yz} = \frac{1}{2}\left(\frac{\partial v}{\partial z} + \frac{\partial w}{\partial y}\right), \quad \varepsilon_{zz} = \frac{\partial w}{\partial z} \tag{5.3-7}$$

where, for this special case of geometric nonlinearity (*i.e.*, small strains but moderate rotations), the notation ε_{ij} is used in place of E_{ij}. The corresponding second Piola–Kirchhoff stresses will be denoted σ_{ij}.

For the assumed displacement field in Eq. (2), $\partial w/\partial z = 0$. In view of the assumptions in Eqs. (4)–(6), the strains in Eq. (7) reduce to

$$
\boxed{
\begin{aligned}
\varepsilon_{xx} &= \frac{\partial u_0}{\partial x} + \frac{1}{2}\left(\frac{\partial w_0}{\partial x}\right)^2 - z\frac{\partial^2 w_0}{\partial x^2}\\
\varepsilon_{xy} &= \frac{1}{2}\left(\frac{\partial u_0}{\partial y} + \frac{\partial v_0}{\partial x} + \frac{\partial w_0}{\partial x}\frac{\partial w_0}{\partial y}\right) - 2z\frac{\partial^2 w_0}{\partial x \partial y}\\
\varepsilon_{yy} &= \frac{\partial v_0}{\partial y} + \frac{1}{2}\left(\frac{\partial w_0}{\partial y}\right)^2 - z\frac{\partial^2 w_0}{\partial y^2}\\
\varepsilon_{xz} &= \frac{1}{2}\left(-\frac{\partial w_0}{\partial x} + \frac{\partial w_0}{\partial x}\right) = 0\\
\varepsilon_{yz} &= \frac{1}{2}\left(-\frac{\partial w_0}{\partial y} + \frac{\partial w_0}{\partial y}\right) = 0\\
\varepsilon_{zz} &= 0
\end{aligned}
}
\qquad (5.3-8)
$$

The strains in Eqs. (8) are called the *von Kármán strains*, and the associated plate theory is termed the *von Kármán plate theory.* Note that the transverse strains $(\varepsilon_{xz}, \varepsilon_{yz}, \varepsilon_{zz})$ are identically zero in the classical plate theory.

The first three strains in Eq. (8) have the form

$$
\left\{\begin{matrix}\varepsilon_{xx}\\ \varepsilon_{yy}\\ \gamma_{xy}\end{matrix}\right\} = \left\{\begin{matrix}\varepsilon_{xx}^{(0)}\\ \varepsilon_{yy}^{(0)}\\ \gamma_{xy}^{(0)}\end{matrix}\right\} + z\left\{\begin{matrix}\varepsilon_{xx}^{(1)}\\ \varepsilon_{yy}^{(1)}\\ \gamma_{xy}^{(1)}\end{matrix}\right\} \qquad (5.3-9)
$$

$$
\{\varepsilon^0\} = \left\{\begin{matrix}\varepsilon_{xx}^{(0)}\\ \varepsilon_{yy}^{(0)}\\ \gamma_{xy}^{(0)}\end{matrix}\right\} = \left\{\begin{matrix}\frac{\partial u_0}{\partial x} + \frac{1}{2}\left(\frac{\partial w_0}{\partial x}\right)^2\\ \frac{\partial v_0}{\partial y} + \frac{1}{2}\left(\frac{\partial w_0}{\partial y}\right)^2\\ \frac{\partial u_0}{\partial y} + \frac{\partial v_0}{\partial x} + \frac{\partial w_0}{\partial x}\frac{\partial w_0}{\partial y}\end{matrix}\right\}, \quad \{\varepsilon^1\} = \left\{\begin{matrix}\varepsilon_{xx}^{(1)}\\ \varepsilon_{yy}^{(1)}\\ \gamma_{xy}^{(1)}\end{matrix}\right\} = \left\{\begin{matrix}-\frac{\partial^2 w_0}{\partial x^2}\\ -\frac{\partial^2 w_0}{\partial y^2}\\ -2\frac{\partial^2 w_0}{\partial x \partial y}\end{matrix}\right\} \qquad (5.3-10)
$$

where $(\varepsilon_{xx}^{(0)}, \varepsilon_{yy}^{(0)}, \gamma_{xy}^{(0)})$ are the *membrane strains*, and $(\varepsilon_{xx}^{(1)}, \varepsilon_{yy}^{(1)}, \gamma_{xy}^{(1)})$ are the flexural (bending) strains, known as the *curvatures.*

Once the displacements (u_0, v_0, w_0) of the midplane are known, strains at any point (x, y, z) in the plate can be computed using Eqs. (9) and (10). Note from Eq. (9) that all strain components vary linearly through the laminate thickness, and they are independent of the material variations through the laminate thickness (see Figure 5.3–3a). For a fixed value of z, the strains are, in general, nonlinear functions of x and y, and they depend on time t for dynamic problems.

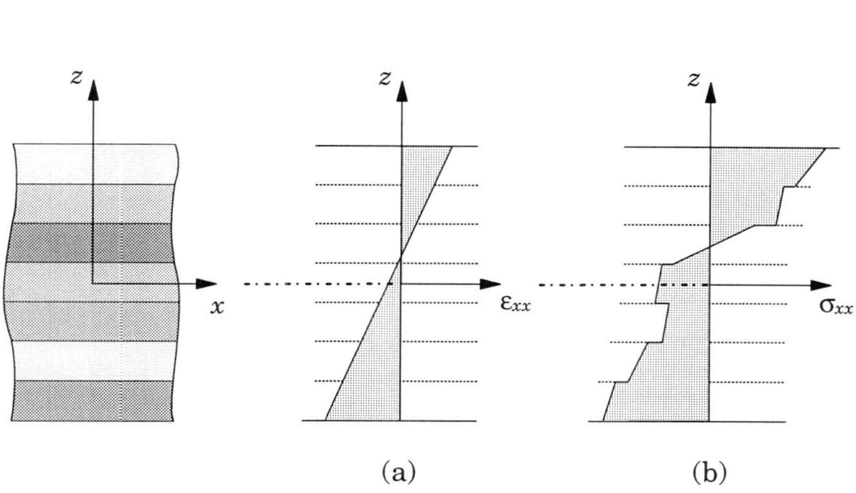

Figure 5.3–3: Variations of strains and stresses through layer and laminate thicknesses. (a) Variation of a typical in-plane strain. (b) Variation of corresponding stress.

5.3–3 Lamina Constitutive Relations

In the classical laminate theory, all three transverse strain components $(\varepsilon_{zz}, \varepsilon_{xz}, \varepsilon_{yz})$ are zero by definition. For a laminate composed of orthotropic layers with their x_1x_2–plane oriented arbitrarily with respect to the xy–plane, the transverse shear stresses $(\sigma_{xz}, \sigma_{yz})$ are also zero. Since $\varepsilon_{zz} = 0$, the transverse normal stress σ_{zz}, although not zero identically, does not appear in the virtual work statement and hence in the equations of motion. Consequently, it amounts to neglecting the transverse normal stress. Thus we have, in theory, a case of both plane strain and plane stress. However, from practical considerations, a thin or moderately thick plate is in a state of plane stress because of thickness being small compared to the in-plane dimensions. Hence, the plane-stress reduced constitutive relations of Section 4.4 may be used.

The linear constitutive relations for the kth orthotropic (piezoelectric) lamina in the principal material coordinates of a lamina are

$$\begin{Bmatrix} \sigma_1 \\ \sigma_2 \\ \sigma_6 \end{Bmatrix}^{(k)} = \begin{bmatrix} Q_{11} & Q_{12} & 0 \\ Q_{12} & Q_{22} & 0 \\ 0 & 0 & Q_{66} \end{bmatrix}^{(k)} \begin{Bmatrix} \varepsilon_1 - \alpha_1\,\Delta T \\ \varepsilon_2 - \alpha_2\,\Delta T \\ \varepsilon_6 \end{Bmatrix} - \begin{bmatrix} 0 & 0 & e_{31} \\ 0 & 0 & e_{32} \\ 0 & 0 & 0 \end{bmatrix}^{(k)} \begin{Bmatrix} \mathcal{E}_1 \\ \mathcal{E}_2 \\ \mathcal{E}_3 \end{Bmatrix}^{(k)} \qquad (5.3-11a)$$

where $Q_{ij}^{(k)}$ are the plane stress-reduced stiffnesses and $e_{ij}^{(k)}$ are the piezoelectric moduli of the kth lamina [*cf.* Eq. (4.4–4)], $(\sigma_i, \varepsilon_i, \mathcal{E}_i)$ are the stress, strain, and electric field components, respectively, referred to the material coordinate system $(x_1,\ x_2,\ x_3)$, α_1 and α_2 are the coefficients of thermal expansion along the x_1 and x_2 directions, respectively, and ΔT is the temperature increment from a reference state, $\Delta T = T - T_0$. When piezoelectric effects are not present, the part containing the piezoelectric moduli $e_{ij}^{(k)}$ should be omitted. The coefficients $Q_{ij}^{(k)}$ are known in terms of the engineering constants of the kth layer:

$$Q_{11} = \frac{E_1}{1-\nu_{12}\nu_{21}}\ , \quad Q_{12} = \frac{\nu_{12}E_2}{1-\nu_{12}\nu_{21}} = \frac{\nu_{21}E_1}{1-\nu_{12}\nu_{21}}$$

$$Q_{22} = \frac{E_2}{1-\nu_{12}\nu_{21}}\ , \quad Q_{66} = G_{12} \tag{5.3-11b}$$

Since the laminate is made of several orthotropic layers, with their material axes oriented arbitrarily with respect to the laminate coordinates, the constitutive equations of each layer must be transformed to the laminate coordinates (x, y, z), as explained in Section 4.3. The stress–strain relations (11a) when transformed to the laminate coordinates (x, y, z) relate the stresses $(\sigma_{xx}, \sigma_{yy}, \sigma_{xy})$ to the strains $(\varepsilon_{xx}, \varepsilon_{yy}, \gamma_{xy})$ and components of the electric displacement vector $(\mathcal{E}_x, \mathcal{E}_y, \mathcal{E}_z)$ in the laminate coordinates [see Eq. (4.4–7)]

$$\begin{Bmatrix} \sigma_{xx} \\ \sigma_{yy} \\ \sigma_{xy} \end{Bmatrix}^{(k)} = \begin{bmatrix} \bar{Q}_{11} & \bar{Q}_{12} & \bar{Q}_{16} \\ \bar{Q}_{12} & \bar{Q}_{22} & \bar{Q}_{26} \\ \bar{Q}_{16} & \bar{Q}_{26} & \bar{Q}_{66} \end{bmatrix}^{(k)} \left(\begin{Bmatrix} \varepsilon_{xx} \\ \varepsilon_{yy} \\ \gamma_{xy} \end{Bmatrix} - \begin{Bmatrix} \alpha_{xx} \\ \alpha_{yy} \\ 2\alpha_{xy} \end{Bmatrix} \Delta T \right)$$
$$- \begin{bmatrix} 0 & 0 & \bar{e}_{31} \\ 0 & 0 & \bar{e}_{32} \\ 0 & 0 & \bar{e}_{36} \end{bmatrix}^{(k)} \begin{Bmatrix} \mathcal{E}_x \\ \mathcal{E}_y \\ \mathcal{E}_z \end{Bmatrix}^{(k)} \tag{5.3-12a}$$

where

$$\begin{aligned}
\bar{Q}_{11} &= Q_{11}\cos^4\theta + 2(Q_{12} + 2Q_{66})\sin^2\theta\cos^2\theta + Q_{22}\sin^4\theta \\
\bar{Q}_{12} &= (Q_{11} + Q_{22} - 4Q_{66})\sin^2\theta\cos^2\theta + Q_{12}(\sin^4\theta + \cos^4\theta) \\
\bar{Q}_{22} &= Q_{11}\sin^4\theta + 2(Q_{12} + 2Q_{66})\sin^2\theta\cos^2\theta + Q_{22}\cos^4\theta \\
\bar{Q}_{16} &= (Q_{11} - Q_{12} - 2Q_{66})\sin\theta\cos^3\theta + (Q_{12} - Q_{22} + 2Q_{66})\sin^3\theta\cos\theta \\
\bar{Q}_{26} &= (Q_{11} - Q_{12} - 2Q_{66})\sin^3\theta\cos\theta + (Q_{12} - Q_{22} + 2Q_{66})\sin\theta\cos^3\theta \\
\bar{Q}_{66} &= (Q_{11} + Q_{22} - 2Q_{12} - 2Q_{66})\sin^2\theta\cos^2\theta + Q_{66}(\sin^4\theta + \cos^4\theta)
\end{aligned} \tag{5.3-12b}$$

and α_{xx}, α_{yy}, and α_{xy} are the transformed thermal coefficients of expansion [see Eq. (4.4–11)]

$$\begin{aligned}\alpha_{xx} &= \alpha_1 \cos^2\theta + \alpha_2 \sin^2\theta \\ \alpha_{yy} &= \alpha_1 \sin^2\theta + \alpha_2 \cos^2\theta \\ 2\alpha_{xy} &= 2(\alpha_1 - \alpha_2)\sin\theta\cos\theta\end{aligned} \qquad (5.3-12c)$$

and $\bar{e}_{ij}$ are the transformed piezoelectric moduli

$$\begin{aligned}\bar{e}_{31} &= e_{31}\cos^2\theta + e_{32}\sin^2\theta \\ \bar{e}_{32} &= e_{31}\sin^2\theta + e_{32}\cos^2\theta \\ \bar{e}_{36} &= (e_{31} - e_{32})\sin\theta\cos\theta\end{aligned} \qquad (5.3-12d)$$

Here θ is the angle measured counterclockwise from the x–coordinate to the x_1–coordinate. Note that stresses are also linear through the thickness of each layer; however, they will have different linear variation in different material layers when $\bar{Q}_{ij}^{(k)}$ change from layer to layer (see Fig. 5.3–3b). If we assume that the temperature increment varies linearly, consistent with the mechanical strains, we can write

$$\Delta T = T^0(x,y,t) + zT^1(x,y,t) \qquad (5.3-13)$$

and the total strains are of the form in Eq. (10a) with

$$\{\varepsilon^0\} = \begin{Bmatrix} \varepsilon_{xx}^{(0)} - \alpha_{xx}T^0 \\ \varepsilon_{yy}^{(0)} - \alpha_{yy}T^0 \\ \gamma_{xy}^{(0)} - 2\alpha_{xy}T^0 \end{Bmatrix}, \quad \{\varepsilon^1\} = \begin{Bmatrix} \varepsilon_{xx}^{(1)} - \alpha_{xx}T^1 \\ \varepsilon_{yy}^{(1)} - \alpha_{yy}T^1 \\ \gamma_{xy}^{(1)} - 2\alpha_{xy}T^1 \end{Bmatrix} \qquad (5.3-14)$$

5.3–4 Equations of Motion

As noted earlier, the transverse strains $(\gamma_{xz}, \gamma_{yz}, \varepsilon_{zz})$ are identically zero in the classical plate theory. Consequently, the transverse shear stresses $(\sigma_{xz}, \sigma_{yz})$ are zero for a laminate made of orthotropic layers if they are computed from the constitutive relations. The transverse normal stress σ_{zz} is not zero by the constitutive relation because of the Poisson effect. However, all three stress components do not enter the formulation because the virtual strain energy of these stresses is zero due to the fact that kinematically consistent virtual strains must be zero [see Eq. (8b)]:

$$\delta\varepsilon_{xz} = 0, \quad \delta\varepsilon_{yz} = 0, \quad \delta\varepsilon_{zz} = 0$$

Whether the transverse stresses are accounted for or not in a theory, they are present in reality to keep the plate in equilibrium. In addition, these stress components may be specified on the boundary. Thus, the transverse stresses do not enter the virtual strain energy expression, but they must be accounted for in the boundary conditions and equilibrium of forces.

Here, the governing equations are derived using the principle of virtual displacements. In the derivations, we account for thermal (and hence, moisture) and piezoelectric effects only with the understanding that the material properties are independent of temperature and electric fields, and that the temperature T and electric displacement vector $\mathcal{E}$ are known functions of position (hence, $\delta T = 0$ and $\delta \mathcal{E} = \mathbf{0}$; see Section 2.5–2). Thus temperature and electric fields enter the formulation only through constitutive equations [see Eq. (2.6–4a)].

The dynamic version of the principle of virtual work [see Eq. (3.4–3)] is

$$0 = \int_0^T (\delta U + \delta V - \delta K)\, dt \tag{5.3 - 15}$$

where the virtual strain energy δU (volume integral of δU_0), virtual work done by applied forces δV, and the virtual kinetic energy δK are given by

$$\begin{aligned}
\delta U &= \int_{\Omega_0} \int_{-\frac{h}{2}}^{\frac{h}{2}} \left(\sigma_{xx}\delta\varepsilon_{xx} + \sigma_{yy}\delta\varepsilon_{yy} + 2\sigma_{xy}\delta\varepsilon_{xy}\right)\, dzdxdy \\
&= \int_{\Omega_0} \left\{ \int_{-\frac{h}{2}}^{\frac{h}{2}} \left[\sigma_{xx}\left(\delta\varepsilon_{xx}^{(0)} + z\delta\varepsilon_{xx}^{(1)}\right) + \sigma_{yy}\left(\delta\varepsilon_{yy}^{(0)} + z\delta\varepsilon_{yy}^{(1)}\right) \right.\right. \\
&\quad \left.\left. + \sigma_{xy}\left(\delta\gamma_{xy}^{(0)} + z\delta\gamma_{xy}^{(1)}\right)\right] dz \right\} dxdy
\end{aligned} \tag{5.3 - 16}$$

$$\begin{aligned}
\delta V &= -\int_{\Omega_0} \left[q_b(x,y)\delta w(x,y,-\frac{h}{2}) + q_t(x,y)\delta w(x,y,\frac{h}{2}) \right] dxdy \\
&\quad - \int_{\Gamma_\sigma} \int_{-\frac{h}{2}}^{\frac{h}{2}} \left[\hat{\sigma}_{nn}\delta u_n + \hat{\sigma}_{ns}\delta u_s + \hat{\sigma}_{nz}\delta w \right] dzds \\
&= -\int_{\Omega_0} \left[\left(q_b(x,y) + q_t(x,y)\right)\delta w_0(x,y) \right] dxdy \\
&\quad - \int_{\Gamma_\sigma} \int_{-\frac{h}{2}}^{\frac{h}{2}} \left[\hat{\sigma}_{nn}\left(\delta u_{0n} - z\frac{\partial \delta w_0}{\partial n}\right) + \hat{\sigma}_{ns}\left(\delta u_{0s} - z\frac{\partial \delta w_0}{\partial s}\right) \right. \\
&\quad \left. + \hat{\sigma}_{nz}\delta w_0 \right] dzds
\end{aligned} \tag{5.3 - 17}$$

$$\begin{aligned}
\delta K &= \int_{\Omega_0} \int_{-\frac{h}{2}}^{\frac{h}{2}} \rho_0 \left[\left(\dot{u}_0 - z\frac{\partial \dot{w}_0}{\partial x}\right)\left(\delta\dot{u}_0 - z\frac{\partial \delta\dot{w}_0}{\partial x}\right) \right. \\
&\quad \left. + \left(\dot{v}_0 - z\frac{\partial \dot{w}_0}{\partial y}\right)\left(\delta\dot{v}_0 - z\frac{\partial \delta\dot{w}_0}{\partial y}\right) + \dot{w}_0\delta\dot{w}_0 \right] dz\; dxdy
\end{aligned} \tag{5.3 - 18}$$

where q_b is the distributed force at the bottom ($z = -h/2$) of the laminate, q_t is the distributed force at the top ($z = h/2$) of the laminate, $(\hat{\sigma}_{nn}, \hat{\sigma}_{ns}, \hat{\sigma}_{nz})$

are the specified stress components on the portion Γ_σ of the boundary Γ, $(\delta u_{0n}, \delta u_{0s})$ are the virtual displacements along the normal and tangential directions, respectively, on the boundary Γ (see Figure 5.3–4), ρ_0 is the density of the plate material, and a superposed dot on a variable indicates time derivative, $\dot{u}_0 = \partial u_0/\partial t$. Details of how (u_{0n}, u_{0s}) and $(\sigma_{nn}, \sigma_{ns})$ are related to (u_0, v_0) and $(\sigma_{xx}, \sigma_{yy}, \sigma_{xy})$, respectively, will be presented shortly.

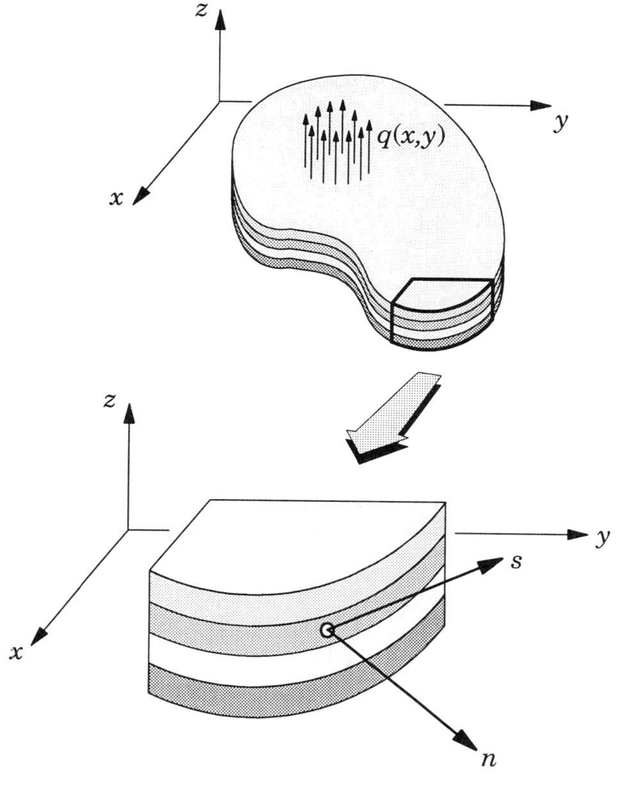

Figure 5.3–4: Geometry of a laminated plate with curved boundary.

The virtual displacements are zero on the portion of the boundary where the corresponding actual displacements are specified. For time-dependent problems, the admissible virtual displacements must also vanish at time $t = 0$ and $t = T$. Since we are interested in the governing differential equations and the form of the natural boundary conditions of the theory, we can assume that the stresses are specified on either a part or whole of the boundary. If a stress component is specified only on a part of the boundary, on the remaining part of the boundary the corresponding virtual displacement must be zero, contributing nothing to the virtual work done.

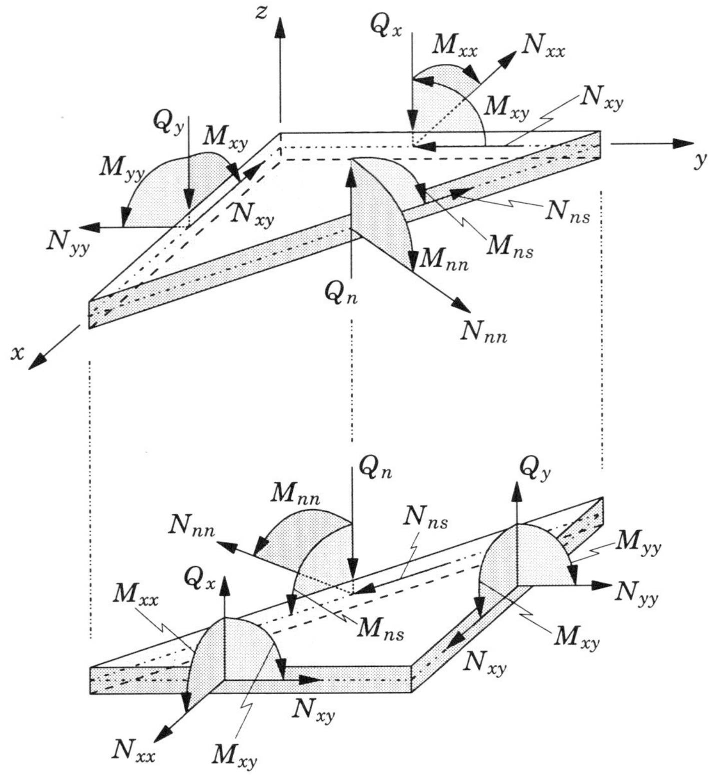

Figure 5.3–5: Force and moment resultants on a plate element.

Substituting for $\delta U, \delta V$, and δK from Eqs. (16)–(18) into the virtual work statement in Eq. (15) and integrating through the thickness of the laminate, we obtain

$$
\begin{aligned}
0 = \int_0^T \Bigg\{ \int_{\Omega_0} \Bigg[& N_{xx}\delta\varepsilon_{xx}^{(0)} + M_{xx}\delta\varepsilon_{xx}^{(1)} + N_{yy}\delta\varepsilon_{yy}^{(0)} + M_{yy}\delta\varepsilon_{yy}^{(1)} + N_{xy}\delta\gamma_{xy}^{(0)} \\
& + M_{xy}\delta\gamma_{xy}^{(1)} - q\delta w_0 - I_0\left(\dot{u}_0\delta\dot{u}_0 + \dot{v}_0\delta\dot{v}_0 + \dot{w}_0\delta\dot{w}_0\right) \\
& + I_1\left(\frac{\partial\delta\dot{w}_0}{\partial x}\dot{u}_0 + \frac{\partial\dot{w}_0}{\partial x}\delta\dot{u}_0 + \frac{\partial\delta\dot{w}_0}{\partial y}\dot{v}_0 + \frac{\partial\dot{w}_0}{\partial y}\delta\dot{v}_0\right) \\
& - I_2\left(\frac{\partial\dot{w}_0}{\partial x}\frac{\partial\delta\dot{w}_0}{\partial x} + \frac{\partial\dot{w}_0}{\partial y}\frac{\partial\delta\dot{w}_0}{\partial y}\right)\Bigg] dxdy \\
- \int_{\Gamma_\sigma} \Bigg(& \hat{N}_{nn}\delta u_{0n} + \hat{N}_{ns}\delta u_{0s} - \hat{M}_{nn}\frac{\partial\delta w_0}{\partial n} \\
& -\hat{M}_{ns}\frac{\partial\delta w_0}{\partial s} + \hat{Q}_n\delta w_0\Bigg) ds \Bigg\} dt
\end{aligned}
\qquad (5.3-19)
$$

where $q = q_b + q_t$ and

$$\left\{\begin{array}{c} N_{xx} \\ N_{yy} \\ N_{xy} \end{array}\right\} = \int_{-\frac{h}{2}}^{\frac{h}{2}} \left\{\begin{array}{c} \sigma_{xx} \\ \sigma_{yy} \\ \sigma_{xy} \end{array}\right\} dz, \quad \left\{\begin{array}{c} M_{xx} \\ M_{yy} \\ M_{xy} \end{array}\right\} = \int_{-\frac{h}{2}}^{\frac{h}{2}} \left\{\begin{array}{c} \sigma_{xx} \\ \sigma_{yy} \\ \sigma_{xy} \end{array}\right\} z\, dz \tag{5.3-20a}$$

$$\left\{\begin{array}{c} \hat{N}_{nn} \\ \hat{N}_{ns} \end{array}\right\} = \int_{-\frac{h}{2}}^{\frac{h}{2}} \left\{\begin{array}{c} \hat{\sigma}_{nn} \\ \hat{\sigma}_{ns} \end{array}\right\} dz, \quad \left\{\begin{array}{c} \hat{M}_{nn} \\ \hat{M}_{ns} \end{array}\right\} = \int_{-\frac{h}{2}}^{\frac{h}{2}} \left\{\begin{array}{c} \hat{\sigma}_{nn} \\ \hat{\sigma}_{ns} \end{array}\right\} z\, dz \tag{5.3-20b}$$

$$\left\{\begin{array}{c} I_0 \\ I_1 \\ I_2 \end{array}\right\} = \int_{-\frac{h}{2}}^{\frac{h}{2}} \left\{\begin{array}{c} 1 \\ z \\ z^2 \end{array}\right\} \rho_0\, dz, \quad \hat{Q}_n = \int_{-\frac{h}{2}}^{\frac{h}{2}} \hat{\sigma}_{nz}\, dz \tag{5.3-20c}$$

The quantities (N_{xx}, N_{yy}, N_{xy}) are called the in-plane *force resultants*, and (M_{xx}, M_{yy}, M_{xy}) are called the *moment resultants* (see Figure 5.3–5); Q_n denotes the transverse force resultant, and (I_0, I_1, I_2) are the mass moments of inertia.

The virtual strains are known in terms of the virtual displacements in the same way as the true strains in terms of the true displacements [see Eq. (8)]:

$$\begin{aligned}
\delta\varepsilon_{xx}^{(0)} &= \frac{\partial \delta u_0}{\partial x} + \frac{\partial w_0}{\partial x}\frac{\partial \delta w_0}{\partial x}, \quad \delta\varepsilon_{xx}^{(1)} = -\frac{\partial^2 \delta w_0}{\partial x^2} \\
\delta\varepsilon_{yy}^{(0)} &= \frac{\partial \delta v_0}{\partial y} + \frac{\partial w_0}{\partial y}\frac{\partial \delta w_0}{\partial y}, \quad \delta\varepsilon_{yy}^{(1)} = -\frac{\partial^2 \delta w_0}{\partial y^2} \\
\delta\gamma_{xy}^{(0)} &= \frac{\partial \delta u_0}{\partial y} + \frac{\partial \delta v_0}{\partial x} + \frac{\partial \delta w_0}{\partial x}\frac{\partial w_0}{\partial y} + \frac{\partial w_0}{\partial x}\frac{\partial \delta w_0}{\partial y} \\
\delta\gamma_{xy}^{(1)} &= -2\frac{\partial^2 \delta w_0}{\partial x \partial y}
\end{aligned} \tag{5.3-21}$$

Substituting for the virtual strains from Eq. (21) into Eq. (19) and integrating by parts to relieve the virtual displacements $(\delta u_0, \delta v_0, \delta w_0)$ in Ω_0 of any differentiation, so that we can use the fundamental lemma of variational calculus, we obtain

$$\begin{aligned}
0 = \int_0^T \Bigg\{ \int_{\Omega_0} \Bigg[&- N_{xx,x}\delta u_0 - \left(N_{xx}\frac{\partial w_0}{\partial x}\right)_{,x} \delta w_0 - M_{xx,xx}\delta w_0 - N_{yy,y}\delta v_0 \\
&- \left(N_{yy}\frac{\partial w_0}{\partial y}\right)_{,y} \delta w_0 - M_{yy,yy}\delta w_0 - N_{xy,y}\delta u_0 - N_{xy,x}\delta v_0 \\
&- \left(N_{xy}\frac{\partial w_0}{\partial y}\right)_{,x} \delta w_0 - \left(N_{xy}\frac{\partial w_0}{\partial x}\right)_{,y} \delta w_0 - 2M_{xy,xy}\delta w_0 - q\delta w_0 \\
&+ I_0\left(\ddot{u}_0\delta u_0 + \ddot{v}_0\delta v_0 + \ddot{w}_0\delta w_0\right) - I_2\left(\frac{\partial^2 \ddot{w}_0}{\partial x^2} + \frac{\partial^2 \ddot{w}_0}{\partial y^2}\right)\delta w_0
\end{aligned}$$

$$
+ I_1\left(\frac{\partial \ddot{u}_0}{\partial x}\delta w_0 - \frac{\partial \ddot{w}_0}{\partial x}\delta u_0 + \frac{\partial \ddot{v}_0}{\partial y}\delta w_0 - \frac{\partial \ddot{w}_0}{\partial y}\delta v_0\right)\Bigg] dxdy
$$

$$
\begin{aligned}
&+ \oint_\Gamma \Bigg[N_{xx}n_x\delta u_0 + \left(N_{xx}\frac{\partial w_0}{\partial x}\right)n_x\delta w_0 - M_{xx}n_x\frac{\partial \delta w_0}{\partial x} + M_{xx,x}n_x\delta w_0 \\
&\quad + N_{yy}n_y\delta v_0 + \left(N_{yy}\frac{\partial w_0}{\partial y}\right)n_y\delta w_0 - M_{yy}n_y\frac{\partial \delta w_0}{\partial y} + M_{yy,y}n_y\delta w_0 \\
&\quad - M_{xy}n_x\frac{\partial \delta w_0}{\partial y} + M_{xy,x}n_y\delta w_0 - M_{xy}n_y\frac{\partial \delta w_0}{\partial x} + M_{xy,y}n_x\delta w_0 \\
&\quad + N_{xy}n_y\delta u_0 + N_{xy}n_x\delta v_0 + N_{xy}\frac{\partial w_0}{\partial y}n_x\delta w_0 + N_{xy}\frac{\partial w_0}{\partial x}n_y\delta w_0\Bigg] ds \\
&- \int_{\Gamma_\sigma}\left(\hat{N}_{nn}\delta u_{0n} + \hat{N}_{ns}\delta u_{0s} - \hat{M}_{nn}\frac{\partial \delta w_0}{\partial n} - \hat{M}_{ns}\frac{\partial \delta w_0}{\partial s} + \hat{Q}_n\delta w_0\right) ds \\
&+ \oint_\Gamma \left[-I_1\left(\ddot{u}_0 n_x + \ddot{v}_0 n_y\right) + I_2\left(\frac{\partial \ddot{w}_0}{\partial x}n_x + \frac{\partial \ddot{w}_0}{\partial y}n_y\right)\right]\delta w_0 \; ds\Bigg\} dt
\end{aligned}
\tag{5.3-22}
$$

where a comma followed by subscripts denotes differentiation with respect to the subscripts: $N_{xx,x} = \partial N_{xx}/\partial x$, and so on. Note that both spatial and time integration–by–parts were used in arriving at the last expression. The terms obtained in Ω_0 but evaluated at $t = 0$ and $t = T$ were set to zero because the virtual displacements are zero there.

Collecting the coefficients of each of the virtual displacements $(\delta u_0, \delta v_0, \delta w_0)$ together and noting that the virtual displacements are zero on Γ_u, we obtain

$$
\begin{aligned}
0 = \int_0^T \Bigg\{ \int_{\Omega_0} \Bigg[&- \left(N_{xx,x} + N_{xy,y} - I_0\ddot{u}_0 + I_1\frac{\partial \ddot{w}_0}{\partial x}\right)\delta u_0 \\
&- \left(N_{xy,x} + N_{yy,y} - I_0\ddot{v}_0 + I_1\frac{\partial \ddot{w}_0}{\partial y}\right)\delta v_0 \\
&- \Bigg(M_{xx,xx} + 2M_{xy,xy} + M_{yy,yy} + \mathcal{N}(w_0) + q \\
&- I_0\ddot{w}_0 - I_1\frac{\partial \ddot{u}_0}{\partial x} - I_1\frac{\partial \ddot{v}_0}{\partial y} + I_2\frac{\partial^2 \ddot{w}_0}{\partial x^2} + I_2\frac{\partial^2 \ddot{w}_0}{\partial y^2}\Bigg)\delta w_0\Bigg] dxdy \\
+ \int_{\Gamma_\sigma} \Bigg[&\left(N_{xx}n_x + N_{xy}n_y\right)\delta u_0 + \left(N_{xy}n_x + N_{yy}n_y\right)\delta v_0 \\
&+ \Bigg(M_{xx,x}n_x + M_{xy,y}n_x + M_{yy,y}n_y + M_{xy,x}n_y + \mathcal{P}(w_0) \\
&-I_1\ddot{u}_0 n_x - I_1\ddot{v}_0 n_y + I_2\frac{\partial \ddot{w}_0}{\partial x}n_x + I_2\frac{\partial \ddot{w}_0}{\partial y}n_y\Bigg)\delta w_0
\end{aligned}
$$

$$- \left(M_{xx}n_x + M_{xy}n_y\right)\frac{\partial \delta w_0}{\partial x} - \left(M_{xy}n_x + M_{yy}n_y\right)\frac{\partial \delta w_0}{\partial y}\bigg] ds$$

$$- \int_{\Gamma_\sigma}\left(\hat{N}_{nn}\delta u_{0n} + \hat{N}_{ns}\delta u_{0s} - \hat{M}_{nn}\frac{\partial \delta w_0}{\partial n} - \hat{M}_{ns}\frac{\partial \delta w_0}{\partial s} + \hat{Q}_n \delta w_0\right) ds\bigg\} dt \tag{5.3-23}$$

where

$$\mathcal{N}(w_0) = \frac{\partial}{\partial x}\left(N_{xx}\frac{\partial w_0}{\partial x} + N_{xy}\frac{\partial w_0}{\partial y}\right) + \frac{\partial}{\partial y}\left(N_{xy}\frac{\partial w_0}{\partial x} + N_{yy}\frac{\partial w_0}{\partial y}\right) \tag{5.3-24a}$$

$$\mathcal{P}(w_0) = \left(N_{xx}\frac{\partial w_0}{\partial x} + N_{xy}\frac{\partial w_0}{\partial y}\right) n_x + \left(N_{xy}\frac{\partial w_0}{\partial x} + N_{yy}\frac{\partial w_0}{\partial y}\right) n_y \tag{5.3-24b}$$

The Euler–Lagrange equations of the theory are obtained by setting the coefficients of δu_0, δv_0, and δw_0 over Ω_0 of Eq. (23) to zero separately:

$$\begin{aligned}
\delta u_0: &\quad \frac{\partial N_{xx}}{\partial x} + \frac{\partial N_{xy}}{\partial y} = I_0 \frac{\partial^2 u_0}{\partial t^2} - I_1 \frac{\partial^2}{\partial t^2}\left(\frac{\partial w_0}{\partial x}\right)\\
\delta v_0: &\quad \frac{\partial N_{xy}}{\partial x} + \frac{\partial N_{yy}}{\partial y} = I_0 \frac{\partial^2 v_0}{\partial t^2} - I_1 \frac{\partial^2}{\partial t^2}\left(\frac{\partial w_0}{\partial y}\right)\\
\delta w_0: &\quad \frac{\partial^2 M_{xx}}{\partial x^2} + 2\frac{\partial^2 M_{xy}}{\partial y \partial x} + \frac{\partial^2 M_{yy}}{\partial y^2} + \mathcal{N}(w_0) + q = I_0 \frac{\partial^2 w_0}{\partial t^2}\\
&\quad - I_2 \frac{\partial^2}{\partial t^2}\left(\frac{\partial^2 w_0}{\partial x^2} + \frac{\partial^2 w_0}{\partial y^2}\right) + I_1 \frac{\partial^2}{\partial t^2}\left(\frac{\partial u_0}{\partial x} + \frac{\partial v_0}{\partial y}\right)
\end{aligned} \tag{5.3-25}$$

The terms involving I_2 are called *rotary* (or *rotatory*) inertia terms, and are often neglected in most books. The term can contribute to higher-order vibration or frequency modes.

Next we obtain the boundary conditions of the theory from Eq. (23). In order to collect the coefficients of the virtual displacements and their derivatives on the boundary, we should express $(\delta u_0, \delta v_0)$ in terms of $(\delta u_{0n}, \delta u_{0s})$. If the unit outward normal vector $\hat{\mathbf{n}}$ is oriented at an angle θ from the x–axis, then its direction cosines are $n_x = \cos\theta$ and $n_y = \sin\theta$. Hence, the transformation between the coordinate system (n, s, r) and (x, y, z) is given by

$$\begin{aligned}
\hat{\mathbf{e}}_x &= \cos\theta\ \hat{\mathbf{e}}_n - \sin\theta\ \hat{\mathbf{e}}_s\\
\hat{\mathbf{e}}_y &= \sin\theta\ \hat{\mathbf{e}}_n + \cos\theta\ \hat{\mathbf{e}}_s\\
\hat{\mathbf{e}}_z &= \hat{\mathbf{e}}_r
\end{aligned} \tag{5.3-26}$$

Therefore, the displacements (u_{0n}, u_{0s}) are related to (u_0, v_0) by

$$u_0 = n_x u_{0n} - n_y u_{0s}, \quad v_0 = n_y u_{0n} + n_x u_{0s} \tag{5.3-27a}$$

Similarly, the normal and tangential derivatives $(w_{0,n}, w_{0,s})$ are related to the derivatives $(w_{0,x}, w_{0,y})$ by

$$\frac{\partial w_0}{\partial x} = n_x \frac{\partial w_0}{\partial n} - n_y \frac{\partial w_0}{\partial s}, \quad \frac{\partial w_0}{\partial y} = n_y \frac{\partial w_0}{\partial n} + n_x \frac{\partial w_0}{\partial s} \tag{5.3-27b}$$

Now we can rewrite the boundary expressions in terms of (u_{0n}, u_{0s}) and $(w_{0,n}, w_{0,s})$. We have

$$\begin{aligned}
&(N_{xx}n_x + N_{xy}n_y)\,\delta u_0 + (N_{xy}n_x + N_{yy}n_y)\,\delta v_0 \\
&= (N_{xx}n_x + N_{xy}n_y)(n_x\delta u_n - n_y\delta u_s) + (N_{xy}n_x + N_{yy}n_y)(n_y\delta u_n + n_x\delta u_s) \\
&= \left(N_{xx}n_x^2 + 2N_{xy}n_xn_y + N_{yy}n_y^2\right)\delta u_n + \left[(N_{yy} - N_{xx})\,n_xn_y + N_{xy}\left(n_x^2 - n_y^2\right)\right]\delta u_s
\end{aligned} \tag{5.3-28a}$$

We recognize that the coefficients of δu_{0n} and δu_{0s} in the right-hand side of the above equation are equal to N_{nn} and N_{ns}, respectively. This follows from the fact that the stresses $(\sigma_{nn}, \sigma_{ns})$ are related to $(\sigma_{xx}, \sigma_{yy}, \sigma_{xy})$ by the transformation in Eq. (4.3–10a):

$$\begin{Bmatrix} \sigma_{nn} \\ \sigma_{ns} \end{Bmatrix} = \begin{bmatrix} n_x^2 & n_y^2 & 2n_xn_y \\ -n_xn_y & n_xn_y & n_x^2 - n_y^2 \end{bmatrix} \begin{Bmatrix} \sigma_{xx} \\ \sigma_{yy} \\ \sigma_{xy} \end{Bmatrix} \tag{5.3-28b}$$

Hence we have

$$\begin{Bmatrix} N_{nn} \\ N_{ns} \end{Bmatrix} = \begin{bmatrix} n_x^2 & n_y^2 & 2n_xn_y \\ -n_xn_y & n_xn_y & n_x^2 - n_y^2 \end{bmatrix} \begin{Bmatrix} N_{xx} \\ N_{yy} \\ N_{xy} \end{Bmatrix} \tag{5.3-29a}$$

$$\begin{Bmatrix} M_{nn} \\ M_{ns} \end{Bmatrix} = \begin{bmatrix} n_x^2 & n_y^2 & 2n_xn_y \\ -n_xn_y & n_xn_y & n_x^2 - n_y^2 \end{bmatrix} \begin{Bmatrix} M_{xx} \\ M_{yy} \\ M_{xy} \end{Bmatrix} \tag{5.3-29b}$$

In view of the above relations, the boundary integrals in Eq. (23) can be written as

$$\begin{aligned}
0 = \int_0^T \int_{\Gamma_\sigma} \Bigg[& \left(N_{nn} - \hat{N}_{nn}\right)\delta u_{0n} + \left(N_{ns} - \hat{N}_{ns}\right)\delta u_{0s} \\
& + \Big(M_{xx,x}n_x + M_{xy,y}n_x + M_{yy,y}n_y + M_{xy,x}n_y + \mathcal{P}(w_0) \\
& - I_1\ddot{u}_0 n_x - I_1\ddot{v}_0 n_y + I_2\frac{\partial \ddot{w}_0}{\partial x}n_x + I_2\frac{\partial \ddot{w}_0}{\partial y}n_y - \hat{Q}_n\Big)\delta w_0 \\
& - \left(M_{nn} - \hat{M}_{nn}\right)\frac{\partial \delta w_0}{\partial n} - \left(M_{ns} - \hat{M}_{ns}\right)\frac{\partial \delta w_0}{\partial s}\Bigg] ds\,dt
\end{aligned} \tag{5.3-30}$$

The natural boundary conditions are then given by

$$N_{nn} - \hat{N}_{nn} = 0 \ , \quad N_{ns} - \hat{N}_{ns} = 0 \ , \quad Q_n - \hat{Q}_n = 0$$

$$M_{nn} - \hat{M}_{nn} = 0 \ , \quad M_{ns} - \hat{M}_{ns} = 0 \qquad (5.3-31a)$$

on Γ_σ, where

$$Q_n \equiv \left(M_{xx,x} + M_{xy,y} - I_1\ddot{u}_0 + I_2\frac{\partial \ddot{w}_0}{\partial x}\right) n_x + \left(M_{yy,y} + M_{xy,x} - I_1\ddot{v}_0 + I_2\frac{\partial \ddot{w}_0}{\partial y}\right) n_y + \mathcal{P}(w_0) \qquad (5.3-31b)$$

Thus the primary variables (*i.e.*, generalized displacements) and secondary variables (*i.e.*, generalized forces) of the theory are

$$\text{primary variables:} \quad u_n,\ u_s,\ w_0,\ \frac{\partial w_0}{\partial n},\ \frac{\partial w_0}{\partial s}$$

$$\text{second variables:} \quad N_{nn},\ N_{ns},\ Q_n,\ M_{nn},\ M_{ns} \qquad (5.3-32)$$

The generalized displacements are specified on Γ_u, which constitutes the essential (or geometric) boundary conditions.

We note that the equations in Eq. (25) have the total spatial differential order of eight. In other words, if the equations are expressed in terms of the displacements (u_0, v_0, w_0), they would contain second-order spatial derivatives of u_0 and v_0 and fourth-order spatial derivatives of w_0. Hence, the classical laminated plate theory is said to be an eighth-order theory. This implies that there should be only eight boundary conditions, whereas Eq. (32) shows five essential and five natural boundary conditions, giving a total of ten boundary conditions. To eliminate this discrepancy, one integrates the tangential derivative term by parts to obtain the boundary term

$$-\oint_\Gamma M_{ns}\frac{\partial \delta w_0}{\partial s}\, ds = \oint_\Gamma \frac{\partial M_{ns}}{\partial s}\delta w_0 \, ds - [M_{ns}\delta w_0]_\Gamma \qquad (5.3-33a)$$

The term in the square bracket is zero since the end points of a closed curve coincide. This term now must be added to Q_n (because it is a coefficient of δw_0):

$$V_n \equiv Q_n + \frac{\partial M_{ns}}{\partial s} \qquad (5.3-33b)$$

which should be balanced by the applied force $\hat{Q}_n$. This boundary condition, $V_n = \hat{Q}_n$, is known as the *Kirchhoff free-edge condition.* The boundary conditions of the classical laminated plate theory are

$$\boxed{\begin{array}{c} u_n,\ u_s,\ w_0,\ \dfrac{\partial w_0}{\partial n}\ \ \text{(essential)} \\[2mm] N_{nn},\ N_{ns},\ V_n,\ M_{nn}\ \ \text{(natural)} \end{array}} \tag{5.3-34}$$

The initial conditions of the theory involve specifying the values of the displacements and their first derivatives with respect to time at $t = 0$:

$$u_n = u_n^0,\ u_s = u_s^0,\ w_0 = w_0^0$$

$$\dot{u}_n = \dot{u}_n^0,\ \dot{u}_s = \dot{u}_s^0,\ \dot{w}_0 = \dot{w}_0^0 \tag{5.3-35}$$

where variables with superscript '0' denotes values at time $t = 0$. We note that both the displacement and velocities must be specified.

This completes the basic development of the classical laminated plate theory for nonlinear and dynamic analyses. As a special case, one can obtain the equations of equilibrium from (25) by setting all terms involving time derivatives to zero. For linear analysis, we set $\mathcal{N}(w_0)$ and $\mathcal{P}(w_0)$ to zero, in addition to setting the nonlinear terms in the strain–displacement equations to zero. Equations (25) are applicable to linear and nonlinear elastic bodies, since the constitutive equations were not utilized in deriving the governing equations of motion.

5.3–5 Laminate Constitutive Equations

Here we derive the constitutive equations that relate the force and moment resultants in Eq. (20a) to the strains of a laminate. To this end, we assume that each layer is orthotropic with respect to its material symmetry lines and obeys Hooke's law, *i.e.*, Eq. (12a) holds for the kth lamina in the problem coordinates. For the moment we consider the case in which the temperature and piezoelectric effects are not included. Although the strains are continuous through the thickness, stresses are not, due to the change in material coefficients through the thickness (*i.e.*, each lamina). Hence, the integration of stresses through the laminate thickness requires laminawise integration. The force resultants are given by

$$\begin{aligned} \left\{\begin{array}{c} N_{xx} \\ N_{yy} \\ N_{xy} \end{array}\right\} &= \sum_{k=1}^{N} \int_{z_k}^{z_{k+1}} \left\{\begin{array}{c} \sigma_{xx} \\ \sigma_{yy} \\ \sigma_{xy} \end{array}\right\} dz \\ &= \sum_{k=1}^{N} \int_{z_k}^{z_{k+1}} \begin{bmatrix} \bar{Q}_{11} & \bar{Q}_{12} & \bar{Q}_{16} \\ \bar{Q}_{12} & \bar{Q}_{22} & \bar{Q}_{26} \\ \bar{Q}_{16} & \bar{Q}_{26} & \bar{Q}_{66} \end{bmatrix}^{(k)} \left\{\begin{array}{c} \varepsilon_{xx}^{(0)} + z\varepsilon_{xx}^{(1)} \\ \varepsilon_{yy}^{(0)} + z\varepsilon_{yy}^{(1)} \\ \gamma_{xy}^{(0)} + z\gamma_{xy}^{(1)} \end{array}\right\} dz \end{aligned}$$

or

$$\begin{Bmatrix} N_{xx} \\ N_{yy} \\ N_{xy} \end{Bmatrix} = \begin{bmatrix} A_{11} & A_{12} & A_{16} \\ A_{12} & A_{22} & A_{26} \\ A_{16} & A_{26} & A_{66} \end{bmatrix} \begin{Bmatrix} \varepsilon_{xx}^{(0)} \\ \varepsilon_{yy}^{(0)} \\ \gamma_{xy}^{(0)} \end{Bmatrix} + \begin{bmatrix} B_{11} & B_{12} & B_{16} \\ B_{12} & B_{22} & B_{26} \\ B_{16} & B_{26} & B_{66} \end{bmatrix} \begin{Bmatrix} \varepsilon_{xx}^{(1)} \\ \varepsilon_{yy}^{(1)} \\ \gamma_{xy}^{(1)} \end{Bmatrix} \tag{5.3 – 36}$$

$$\begin{aligned} \begin{Bmatrix} M_{xx} \\ M_{yy} \\ M_{xy} \end{Bmatrix} &= \sum_{k=1}^{N} \int_{z_k}^{z_{k+1}} \begin{Bmatrix} \sigma_{xx} \\ \sigma_{yy} \\ \sigma_{xy} \end{Bmatrix} z\, dz \\ &= \sum_{k=1}^{N} \int_{z_k}^{z_{k+1}} \begin{bmatrix} \bar{Q}_{11} & \bar{Q}_{12} & \bar{Q}_{16} \\ \bar{Q}_{12} & \bar{Q}_{22} & \bar{Q}_{26} \\ \bar{Q}_{16} & \bar{Q}_{26} & \bar{Q}_{66} \end{bmatrix}^{(k)} \begin{Bmatrix} \varepsilon_{xx}^{(0)} + z\varepsilon_{xx}^{(1)} \\ \varepsilon_{yy}^{(0)} + z\varepsilon_{yy}^{(1)} \\ \gamma_{xy}^{(0)} + z\gamma_{xy}^{(1)} \end{Bmatrix} z\, dz \end{aligned}$$

$$\begin{Bmatrix} M_{xx} \\ M_{yy} \\ M_{xy} \end{Bmatrix} = \begin{bmatrix} B_{11} & B_{12} & B_{16} \\ B_{12} & B_{22} & B_{26} \\ B_{16} & B_{26} & B_{66} \end{bmatrix} \begin{Bmatrix} \varepsilon_{xx}^{(0)} \\ \varepsilon_{yy}^{(0)} \\ \gamma_{xy}^{(0)} \end{Bmatrix} + \begin{bmatrix} D_{11} & D_{12} & D_{16} \\ D_{12} & D_{22} & D_{26} \\ D_{16} & D_{26} & D_{66} \end{bmatrix} \begin{Bmatrix} \varepsilon_{xx}^{(1)} \\ \varepsilon_{yy}^{(1)} \\ \gamma_{xy}^{(1)} \end{Bmatrix} \tag{5.3 – 37}$$

where A_{ij} are called *extensional stiffnesses*, D_{ij} the *bending stiffnesses*, and B_{ij} the *bending–extensional coupling stiffnesses*, which are defined in terms of the lamina stiffnesses $\bar{Q}_{ij}^{(k)}$ as

$$(A_{ij}, B_{ij}, D_{ij}) = \int_{-\frac{h}{2}}^{\frac{h}{2}} \bar{Q}_{ij}(1, z, z^2)dz = \sum_{k=1}^{N} \int_{z_k}^{z_{k+1}} \bar{Q}_{ij}^{(k)}(1, z, z^2)dz \tag{5.3 – 38a}$$

or

$$A_{ij} = \sum_{k=1}^{N} \bar{Q}_{ij}^{(k)}(z_{k+1} - z_k)\ , \quad B_{ij} = \frac{1}{2}\sum_{k=1}^{N} \bar{Q}_{ij}^{(k)}(z_{k+1}^2 - z_k^2)$$

$$D_{ij} = \frac{1}{3}\sum_{k=1}^{N} \bar{Q}_{ij}^{(k)}(z_{k+1}^3 - z_k^3) \tag{5.3 – 38b}$$

Note that $\bar{Q}$'s, and therefore A's, B's, and D's, are, in general, functions of position (x, y). Equations (36) and (37) can be written in a compact form as

$$\begin{Bmatrix} \{N\} \\ \{M\} \end{Bmatrix} = \begin{bmatrix} [A] & [B] \\ [B] & [D] \end{bmatrix} \begin{Bmatrix} \{\varepsilon^0\} \\ \{\varepsilon^1\} \end{Bmatrix} \tag{5.3 – 39}$$

where $\{\varepsilon^0\}$ and $\{\varepsilon^1\}$ are vectors of the membrane and bending strains defined in Eq. (10), and $[A]$, $[B]$, and $[D]$ are the 3×3 symmetric matrices of laminate coefficients defined in Eqs. (38a,b). Values of the laminate stiffnesses for various stacking sequences will be presented in Section 5.5.

For the nonisothermal case, the strains are given by Eq. (14) and the laminate constitutive equations (39) become

$$\left\{\begin{matrix}\{N\}\\ \{M\}\end{matrix}\right\} = \begin{bmatrix}[A] & [B]\\ [B] & [D]\end{bmatrix}\left\{\begin{matrix}\{\varepsilon^0\}\\ \{\varepsilon^1\}\end{matrix}\right\} - \left\{\begin{matrix}\{N^T\}\\ \{M^T\}\end{matrix}\right\} - \left\{\begin{matrix}\{N^P\}\\ \{M^P\}\end{matrix}\right\} \tag{5.3 – 40}$$

where $\{N^T\}$ and $\{M^T\}$ are thermal force resultants

$$\{N^T\} = \sum_{k=1}^{N}\int_{z_k}^{z_{k+1}} [\bar{Q}]^{(k)}\{\bar{\alpha}\}^{(k)}\Delta T\ dz \tag{5.3 – 41a}$$

$$\{M^T\} = \sum_{k=1}^{N}\int_{z_k}^{z_{k+1}} [\bar{Q}]^{(k)}\{\bar{\alpha}\}^{(k)}\Delta T\ z\ dz \tag{5.3 – 41b}$$

and $\{N^P\}$ and $\{M^P\}$ are the piezoelectric resultants

$$\{N^P\} = \sum_{k=1}^{N}\int_{z_k}^{z_{k+1}} [\bar{e}]^{(k)}\{E\}^{(k)}\ dz \tag{5.3 – 42a}$$

$$\{M^P\} = \sum_{k=1}^{N}\int_{z_k}^{z_{k+1}} [\bar{e}]^{(k)}\{E\}^{(k)}\ z\ dz \tag{5.3 – 42b}$$

Relations similar to (41) can be written for hygroscopic effects.

5.3–6 Equations of Motion in Terms of Displacements

The stress resultants (N's and M's) are related to the displacement gradients, temperature increment, and electric field. In the absence of the temperature and electric effects, the force and moment resultants can be expressed in terms of the displacements (u_0, v_0, w_0) by the relations

$$\begin{aligned}\left\{\begin{matrix}N_{xx}\\ N_{yy}\\ N_{xy}\end{matrix}\right\} &= \begin{bmatrix}A_{11} & A_{12} & A_{16}\\ A_{12} & A_{22} & A_{26}\\ A_{16} & A_{26} & A_{66}\end{bmatrix}\left\{\begin{matrix}\frac{\partial u_0}{\partial x} + \frac{1}{2}(\frac{\partial w_0}{\partial x})^2\\ \frac{\partial v_0}{\partial y} + \frac{1}{2}(\frac{\partial w_0}{\partial y})^2\\ \frac{\partial u_0}{\partial y} + \frac{\partial v_0}{\partial x} + \frac{\partial w_0}{\partial x}\frac{\partial w_0}{\partial y}\end{matrix}\right\}\\ &\quad - \begin{bmatrix}B_{11} & B_{12} & B_{16}\\ B_{12} & B_{22} & B_{26}\\ B_{16} & B_{26} & B_{66}\end{bmatrix}\left\{\begin{matrix}\frac{\partial^2 w_0}{\partial x^2}\\ \frac{\partial^2 w_0}{\partial y^2}\\ 2\frac{\partial^2 w_0}{\partial x\partial y}\end{matrix}\right\}\end{aligned} \tag{5.3 – 43}$$

$$\begin{Bmatrix} M_{xx} \\ M_{yy} \\ M_{xy} \end{Bmatrix} = \begin{bmatrix} B_{11} & B_{12} & B_{16} \\ B_{12} & B_{22} & B_{26} \\ B_{16} & B_{26} & B_{66} \end{bmatrix} \begin{Bmatrix} \frac{\partial u_0}{\partial x} + \frac{1}{2}(\frac{\partial w_0}{\partial x})^2 \\ \frac{\partial v_0}{\partial y} + \frac{1}{2}(\frac{\partial w_0}{\partial y})^2 \\ \frac{\partial u_0}{\partial y} + \frac{\partial v_0}{\partial x} + \frac{\partial w_0}{\partial x}\frac{\partial w_0}{\partial y} \end{Bmatrix}$$
$$- \begin{bmatrix} D_{11} & D_{12} & D_{16} \\ D_{12} & D_{22} & D_{26} \\ D_{16} & D_{26} & D_{66} \end{bmatrix} \begin{Bmatrix} \frac{\partial^2 w_0}{\partial x^2} \\ \frac{\partial^2 w_0}{\partial y^2} \\ 2\frac{\partial^2 w_0}{\partial x \partial y} \end{Bmatrix} \qquad (5.3-44)$$

The equations of motion (25) can be expressed in terms of displacements (u_0, v_0, w_0) by substituting for the force and moment resultants from Eqs. (43) and (44). In general, the laminate stiffnesses can be functions of position (x, y) (*i.e.*, nonhomogeneous plates). For homogeneous laminates (*i.e.*, for laminates with constant A's, B's, and D's), the equations of motion (25) take the form

$$A_{11}\left(\frac{\partial^2 u_0}{\partial x^2} + \frac{\partial w_0}{\partial x}\frac{\partial^2 w_0}{\partial x^2}\right) + A_{12}\left(\frac{\partial^2 v_0}{\partial x \partial y} + \frac{\partial w_0}{\partial y}\frac{\partial^2 w_0}{\partial x \partial y}\right)$$
$$+ A_{16}\left(\frac{\partial^2 u_0}{\partial x \partial y} + \frac{\partial^2 v_0}{\partial x^2} + \frac{\partial^2 w_0}{\partial x^2}\frac{\partial w_0}{\partial y} + \frac{\partial w_0}{\partial x}\frac{\partial^2 w_0}{\partial x \partial y}\right)$$
$$- B_{11}\frac{\partial^3 w_0}{\partial x^3} - B_{12}\frac{\partial^3 w_0}{\partial x \partial y^2} - 2B_{16}\frac{\partial^3 w_0}{\partial x^2 \partial y}$$
$$+ A_{16}\left(\frac{\partial^2 u_0}{\partial x \partial y} + \frac{\partial w_0}{\partial x}\frac{\partial^2 w_0}{\partial x \partial y}\right) + A_{26}\left(\frac{\partial^2 v_0}{\partial y^2} + \frac{\partial w_0}{\partial y}\frac{\partial^2 w_0}{\partial y^2}\right)$$
$$+ A_{66}\left(\frac{\partial^2 u_0}{\partial y^2} + \frac{\partial^2 v_0}{\partial x \partial y} + \frac{\partial^2 w_0}{\partial x \partial y}\frac{\partial w_0}{\partial y} + \frac{\partial w_0}{\partial x}\frac{\partial^2 w_0}{\partial y^2}\right)$$
$$- B_{16}\frac{\partial^3 w_0}{\partial x^2 \partial y} - B_{26}\frac{\partial^3 w_0}{\partial y^3} - 2B_{66}\frac{\partial^3 w_0}{\partial x \partial y^2}$$
$$- \left(\frac{\partial N_{xx}^T}{\partial x} + \frac{\partial N_{xy}^T}{\partial y}\right) = I_0\frac{\partial^2 u_0}{\partial t^2} - I_1\frac{\partial^3 w_0}{\partial x \partial t^2} \qquad (5.3-45)$$

$$A_{16}\left(\frac{\partial^2 u_0}{\partial x^2} + \frac{\partial w_0}{\partial x}\frac{\partial^2 w_0}{\partial x^2}\right) + A_{26}\left(\frac{\partial^2 v_0}{\partial x \partial y} + \frac{\partial w_0}{\partial y}\frac{\partial^2 w_0}{\partial x \partial y}\right)$$
$$+ A_{66}\left(\frac{\partial^2 u_0}{\partial x \partial y} + \frac{\partial^2 v_0}{\partial x^2} + \frac{\partial^2 w_0}{\partial x^2}\frac{\partial w_0}{\partial y} + \frac{\partial w_0}{\partial x}\frac{\partial^2 w_0}{\partial x \partial y}\right)$$
$$- B_{16}\frac{\partial^3 w_0}{\partial x^3} - B_{26}\frac{\partial^3 w_0}{\partial x \partial y^2} - 2B_{66}\frac{\partial^3 w_0}{\partial x^2 \partial y}$$
$$+ A_{12}\left(\frac{\partial^2 u_0}{\partial x \partial y} + \frac{\partial w_0}{\partial x}\frac{\partial^2 w_0}{\partial x \partial y}\right) + A_{22}\left(\frac{\partial^2 v_0}{\partial y^2} + \frac{\partial w_0}{\partial y}\frac{\partial^2 w_0}{\partial y^2}\right)$$

$$
\begin{aligned}
&+ A_{26}\left(\frac{\partial^2 u_0}{\partial y^2} + \frac{\partial^2 v_0}{\partial x \partial y} + \frac{\partial^2 w_0}{\partial x \partial y}\frac{\partial w_0}{\partial y} + \frac{\partial w_0}{\partial x}\frac{\partial^2 w_0}{\partial y^2}\right) \\
&- B_{12}\frac{\partial^3 w_0}{\partial x^2 \partial y} - B_{22}\frac{\partial^3 w_0}{\partial y^3} - 2B_{26}\frac{\partial^3 w_0}{\partial x \partial y^2} \\
&- \left(\frac{\partial N_{xy}^T}{\partial x} + \frac{\partial N_{yy}^T}{\partial y}\right) = I_0\frac{\partial^2 v_0}{\partial t^2} - I_1\frac{\partial^3 w_0}{\partial y \partial t^2}
\end{aligned} \qquad (5.3-46)
$$

$$
\begin{aligned}
&B_{11}\left(\frac{\partial^3 u_0}{\partial x^3} + \frac{\partial^2 w_0}{\partial x^2}\frac{\partial^2 w_0}{\partial x^2} + \frac{\partial w_0}{\partial x}\frac{\partial^3 w_0}{\partial x^3}\right) + B_{12}\left(\frac{\partial^3 v_0}{\partial x^2 \partial y} + \frac{\partial^2 w_0}{\partial x \partial y}\frac{\partial^2 w_0}{\partial x \partial y}\right. \\
&\left.+\frac{\partial w_0}{\partial y}\frac{\partial^3 w_0}{\partial x^2 \partial y}\right) + B_{16}\left(\frac{\partial^3 u_0}{\partial x^2 \partial y} + \frac{\partial^3 v_0}{\partial x^3} + \frac{\partial^3 w_0}{\partial x^3}\frac{\partial w_0}{\partial y} + 2\frac{\partial^2 w_0}{\partial x^2}\frac{\partial^2 w_0}{\partial x \partial y}\right. \\
&\left.+\frac{\partial w_0}{\partial x}\frac{\partial^3 w_0}{\partial x^2 \partial y}\right) - D_{11}\frac{\partial^4 w_0}{\partial x^4} - D_{12}\frac{\partial^4 w_0}{\partial x^2 \partial y^2} - 2D_{16}\frac{\partial^4 w_0}{\partial x^3 \partial y} \\
&+ 2B_{16}\left(\frac{\partial^3 u_0}{\partial x^2 \partial y} + \frac{\partial^2 w_0}{\partial x^2}\frac{\partial^2 w_0}{\partial x \partial y} + \frac{\partial w_0}{\partial x}\frac{\partial^3 w_0}{\partial x^2 \partial y}\right) + 2B_{26}\left(\frac{\partial^3 v_0}{\partial x \partial y^2} + \frac{\partial^2 w_0}{\partial x \partial y}\frac{\partial^2 w_0}{\partial y^2}\right. \\
&\left.+\frac{\partial w_0}{\partial y}\frac{\partial^3 w_0}{\partial x \partial y^2}\right) + 2B_{66}\left(\frac{\partial^3 u_0}{\partial x \partial y^2} + \frac{\partial^3 v_0}{\partial x^2 \partial y} + \frac{\partial^3 w_0}{\partial x^2 \partial y}\frac{\partial w_0}{\partial y} + \frac{\partial^2 w_0}{\partial x \partial y}\frac{\partial^2 w_0}{\partial x \partial y}\right. \\
&\left.+ \frac{\partial^2 w_0}{\partial x^2}\frac{\partial^2 w_0}{\partial y^2} + \frac{\partial w_0}{\partial x}\frac{\partial^3 w_0}{\partial x \partial y^2}\right) - 2D_{16}\frac{\partial^4 w_0}{\partial x^3 \partial y} - 2D_{26}\frac{\partial^4 w_0}{\partial x \partial y^3} \\
&- 4D_{66}\frac{\partial^4 w_0}{\partial x^2 \partial y^2} + B_{12}\left(\frac{\partial^3 u_0}{\partial x \partial y^2} + \frac{\partial^2 w_0}{\partial x \partial y}\frac{\partial^2 w_0}{\partial x \partial y} + \frac{\partial w_0}{\partial x}\frac{\partial^3 w_0}{\partial x \partial y^2}\right) \\
&+ B_{22}\left(\frac{\partial^3 v_0}{\partial y^3} + \frac{\partial^2 w_0}{\partial y^2}\frac{\partial^2 w_0}{\partial y^2} + \frac{\partial w_0}{\partial y}\frac{\partial^3 w_0}{\partial y^3}\right) + B_{26}\left(\frac{\partial^3 u_0}{\partial y^3} + \frac{\partial^3 v_0}{\partial x \partial y^2}\right. \\
&\left.+ \frac{\partial^3 w_0}{\partial x \partial y^2}\frac{\partial w_0}{\partial y} + 2\frac{\partial^2 w_0}{\partial x \partial y}\frac{\partial^2 w_0}{\partial y^2} + \frac{\partial w_0}{\partial x}\frac{\partial^3 w_0}{\partial y^3}\right) \\
&- D_{12}\frac{\partial^4 w_0}{\partial x^2 \partial y^2} - D_{22}\frac{\partial^4 w_0}{\partial y^4} - 2D_{26}\frac{\partial^4 w_0}{\partial x \partial y^3} + \mathcal{N}(w_0) + q \\
&- \left(\frac{\partial^2 M_{xx}^T}{\partial x^2} + 2\frac{\partial^2 M_{xy}^T}{\partial y \partial x} + \frac{\partial^2 M_{yy}^T}{\partial y^2}\right) \\
&= I_0\frac{\partial^2 w_0}{\partial t^2} - I_2\frac{\partial^2}{\partial t^2}\left(\frac{\partial^2 w_0}{\partial x^2} + \frac{\partial^2 w_0}{\partial y^2}\right) + I_1\frac{\partial^2}{\partial t^2}\left(\frac{\partial u_0}{\partial x} + \frac{\partial v_0}{\partial y}\right)
\end{aligned} \qquad (5.3-47)
$$

where $\mathcal{N}(w_0)$ was defined in Eq. (24a).

The nonlinear partial differential Eqs. (45)–(47) can be simplified for linear analyses, static analyses, and lamination schemes for which some of the stiffnesses (A_{ij}, B_{ij}, D_{ij}) are zero. These cases will be considered in the sequel. Once the displacements are determined by solving Eqs. (45)–(47), analytically or numerically for a given problem, the strains and stresses in each lamina can be computed using Eqs. (10) and (12), respectively.

Example 5.3–1: (*Cylindrical Bending*)

If a plate is infinitely long in one direction, the plate becomes a *plate strip*. Consider a plate strip that has a finite dimension along the x–axis and subjected to a transverse load $q(x)$ that is uniform at any section parallel to the x–axis. In such a case, the deflection w_0 and displacements (u_0, v_0) of the plate are functions of only x. Therefore, all derivatives with respect to y are zero. In such cases, the deflected surface of the plate strip is cylindrical, and it is referred to as the *cylindrical bending*. For this case, the governing equations (45)–(47) reduce to

$$A_{11}\left(\frac{\partial^2 u_0}{\partial x^2}+\frac{\partial w_0}{\partial x}\frac{\partial^2 w_0}{\partial x^2}\right)+A_{16}\frac{\partial^2 v_0}{\partial x^2}-B_{11}\frac{\partial^3 w_0}{\partial x^3}-\frac{\partial N_{xx}^T}{\partial x}$$
$$=I_0\frac{\partial^2 u_0}{\partial t^2}-I_1\frac{\partial^3 w_0}{\partial x\partial t^2} \qquad (5.3-48a)$$

$$A_{16}\left(\frac{\partial^2 u_0}{\partial x^2}+\frac{\partial w_0}{\partial x}\frac{\partial^2 w_0}{\partial x^2}\right)+A_{66}\frac{\partial^2 v_0}{\partial x^2}-B_{16}\frac{\partial^3 w_0}{\partial x^3}-\frac{\partial N_{xy}^T}{\partial x}=I_0\frac{\partial^2 v_0}{\partial t^2} \qquad (5.3-48b)$$

$$B_{11}\left(\frac{\partial^3 u_0}{\partial x^3}+\frac{\partial^2 w_0}{\partial x^2}\frac{\partial^2 w_0}{\partial x^2}+\frac{\partial w_0}{\partial x}\frac{\partial^3 w_0}{\partial x^3}\right)+B_{16}\frac{\partial^3 v_0}{\partial x^3}-D_{11}\frac{\partial^4 w_0}{\partial x^4}-\frac{\partial^2 M_{xx}^T}{\partial x^2}$$
$$+\frac{\partial}{\partial x}\left(N_{xx}\frac{\partial w_0}{\partial x}\right)+q=I_0\frac{\partial^2 w_0}{\partial t^2}-I_2\frac{\partial^4 w_0}{\partial x^2\partial t^2}+I_1\frac{\partial^3 u_0}{\partial x\partial t^2} \qquad (5.3-48c)$$

Example 5.3–2:

Suppose that a six-layer $(\pm 60/0)_s$ symmetric laminate is subjected to loads such that the only nonzero strains at a point (x, y) are $\varepsilon_{xx}^{(0)} = \varepsilon_0$ in./in. and $\varepsilon_{xx}^{(1)} = \kappa_0$/in. Assume that layers are of thickness 0.005 in. with material properties $E_1 = 7.8$ psi, $E_2 = 2.6$ psi, $G_{12} = G_{13} = 1.3$ psi, $G_{23} = 0.5$ psi, and $\nu_{12} = 0.25$. We wish to determine the state of stress $(\sigma_{xx}, \sigma_{yy}, \sigma_{xy})$ and force resultants in the laminate.

The only nonzero strain is $\varepsilon_{xx} = \varepsilon_0 + z\kappa_0$. Hence, the stresses in kth lamina are given by

$$\sigma_{xx}^{(k)}=\bar{Q}_{11}^{(k)}\left(\varepsilon_0+z\kappa_0\right), \quad \sigma_{yy}^{(k)}=\bar{Q}_{12}^{(k)}\left(\varepsilon_0+z\kappa_0\right), \quad \sigma_{xy}^{(k)}=\bar{Q}_{16}^{(k)}\left(\varepsilon_0+z\kappa_0\right)$$

where

$$[\bar{Q}]_{\pm 60^\circ}=\begin{bmatrix}3.215 & 1.431 & \pm 0.707\\ 1.431 & 5.871 & \pm 1.593\\ \pm 0.707 & \pm 1.593 & 2.068\end{bmatrix}\text{ msi}, \quad [\bar{Q}]_{0^\circ}=\begin{bmatrix}7.966 & 0.664 & 0\\ 0.664 & 2.655 & 0\\ 0 & 0 & 1.3\end{bmatrix}\text{ msi}$$

The stress resultants are given by

$$\begin{Bmatrix} N_{xx} \\ N_{yy} \\ N_{xy} \end{Bmatrix} = \begin{Bmatrix} 0.1440 \\ 0.0353 \\ 0 \end{Bmatrix} \varepsilon_0 \times 10^6 \text{ lbs/in.}, \quad \begin{Bmatrix} M_{xx} \\ M_{yy} \\ M_{xy} \end{Bmatrix} = \begin{Bmatrix} 7.6306 \\ 3.1566 \\ 0.7066 \end{Bmatrix} \kappa_0 \text{ lbs-in.}$$

If $\varepsilon_0 = 1000 \times 10^6$ in./in. and $\kappa_0 = 0$, we have

$$\begin{Bmatrix} N_{xx} \\ N_{yy} \\ N_{xy} \end{Bmatrix} = \begin{Bmatrix} 144 \\ 35.3 \\ 0 \end{Bmatrix} \text{ lbs/in.}, \quad \begin{Bmatrix} M_{xx} \\ M_{yy} \\ M_{xy} \end{Bmatrix} = \begin{Bmatrix} 0 \\ 0 \\ 0 \end{Bmatrix} \text{ lbs-in.}$$

If $\varepsilon_0 = 0$ in./in. and $\kappa_0 = 1.0$ /in., we have

$$\begin{Bmatrix} N_{xx} \\ N_{yy} \\ N_{xy} \end{Bmatrix} = \begin{Bmatrix} 0 \\ 0 \\ 0 \end{Bmatrix} \text{ lbs/in.}, \quad \begin{Bmatrix} M_{xx} \\ M_{yy} \\ M_{xy} \end{Bmatrix} = \begin{Bmatrix} 7.6306 \\ 3.1566 \\ 0.7066 \end{Bmatrix} \text{ lbs-in.}$$

5.4 The First-Order Laminated Plate Theory

5.4–1 Displacements and Strains

In the *first-order shear deformation laminated plate theory* (FSDT), the *Kirchhoff hypothesis* is relaxed by removing the third part; *i.e.*, the transverse normals *do not* remain perpendicular to the midsurface after deformation (see Figure 5.4–1). This amounts to including transverse shear strains in the theory. The inextensibility of transverse normals requires that w not be a function of the thickness coordinate, z.

Under the same assumptions and restrictions as in the classical laminate theory, the displacement field of the first-order theory is of the form

$$\begin{aligned} u(x,y,z,t) &= u_0(x,y,t) + z\phi_x(x,y,t) \\ v(x,y,z,t) &= v_0(x,y,t) + z\phi_y(x,y,t) \\ w(x,y,z,t) &= w_0(x,y,t) \end{aligned} \qquad (5.4-1)$$

where $(u_0, v_0, w_0, \phi_x, \phi_y)$ are unknown functions to be determined. As before, (u_0, v_0, w_0) denote the displacements of a point on the plane $z = 0$. Note that

$$\frac{\partial u}{\partial z} = \phi_x, \quad \frac{\partial v}{\partial z} = \phi_y \qquad (5.4-2a)$$

which indicate that ϕ_x and ϕ_y are the rotations of a transverse normal about the $y-$ and $x-$axes, respectively (see Figure 5.4–1). The notation that ϕ_x

denotes the rotation of a transverse normal about the $y-$axis and ϕ_y denotes the rotation about the $x-$axis may be confusing to some, and they do not follow the right-hand rule. However, the notation has been used extensively in the literature, and we will not depart from it. If (β_x, β_y) denote the rotations about the x and y axes, respectively, that follow the right-hand rule, then

$$\beta_x = \phi_y \ , \quad \beta_y = -\phi_x \tag{5.4 – 2b}$$

The quantities $(u_0, v_0, w_0, \phi_x, \phi_y)$ will be called the *generalized displacements.* For thin plates, *i.e.*, when the plate in-plane characteristic dimension to thickness ratio is on the order 50 or greater, the rotation functions ϕ_x and ϕ_y should approach the respective slopes of the transverse deflection:

$$\phi_x = -\frac{\partial w_0}{\partial x} \ , \quad \phi_y = -\frac{\partial w_0}{\partial y}$$

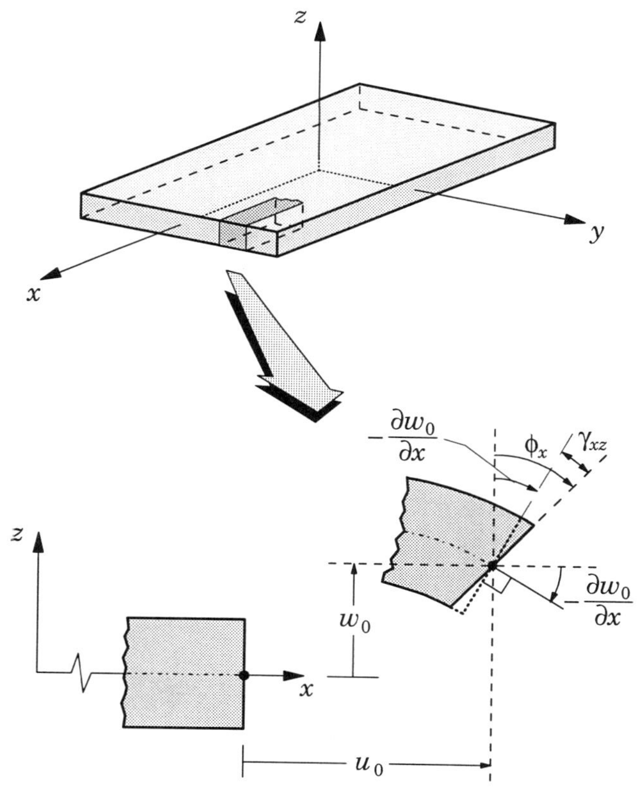

Figure 5.4–1: Undeformed and deformed geometries of an edge of a plate under the assumptions of the first-order plate theory.

The nonlinear strains associated with the displacement field (1) are obtained by using Eq. (1) in Eq. (5.3–7):

$$
\begin{aligned}
\varepsilon_{xx} &= \frac{\partial u_0}{\partial x} + \frac{1}{2}\left(\frac{\partial w_0}{\partial x}\right)^2 + z\frac{\partial \phi_x}{\partial x}\\
\gamma_{xy} &= \left(\frac{\partial u_0}{\partial y} + \frac{\partial v_0}{\partial x} + \frac{\partial w_0}{\partial x}\frac{\partial w_0}{\partial y}\right) + z\left(\frac{\partial \phi_x}{\partial y} + \frac{\partial \phi_y}{\partial x}\right)\\
\varepsilon_{yy} &= \frac{\partial v_0}{\partial y} + \frac{1}{2}\left(\frac{\partial w_0}{\partial y}\right)^2 + z\frac{\partial \phi_y}{\partial y}\\
\gamma_{xz} &= \frac{\partial w_0}{\partial x} + \phi_x, \quad \gamma_{yz} = \frac{\partial w_0}{\partial y} + \phi_y, \quad \varepsilon_{zz} = 0
\end{aligned}
\qquad (5.4-3)
$$

Note that the strains $(\varepsilon_{xx}, \varepsilon_{yy}, \gamma_{xy})$ are linear through the laminate thickness, while the transverse shear strains $(\gamma_{xz}, \gamma_{yz})$ are constant in the first-order laminated plate theory.

The strains in Eq. (3) have the form

$$
\begin{Bmatrix} \varepsilon_{xx}\\ \varepsilon_{yy}\\ \gamma_{yz}\\ \gamma_{xz}\\ \gamma_{xy} \end{Bmatrix} = \begin{Bmatrix} \varepsilon_{xx}^{(0)}\\ \varepsilon_{yy}^{(0)}\\ \gamma_{yz}^{(0)}\\ \gamma_{xz}^{(0)}\\ \gamma_{xy}^{(0)} \end{Bmatrix} + z\begin{Bmatrix} \varepsilon_{xx}^{(1)}\\ \varepsilon_{yy}^{(1)}\\ \gamma_{yz}^{(1)}\\ \gamma_{xz}^{(1)}\\ \gamma_{xy}^{(1)} \end{Bmatrix} = \begin{Bmatrix} \frac{\partial u_0}{\partial x} + \frac{1}{2}\left(\frac{\partial w_0}{\partial x}\right)^2\\ \frac{\partial v_0}{\partial y} + \frac{1}{2}\left(\frac{\partial w_0}{\partial y}\right)^2\\ \frac{\partial w_0}{\partial y} + \phi_y\\ \frac{\partial w_0}{\partial x} + \phi_x\\ \frac{\partial u_0}{\partial y} + \frac{\partial v_0}{\partial x} + \frac{\partial w_0}{\partial x}\frac{\partial w_0}{\partial y} \end{Bmatrix} + z\begin{Bmatrix} \frac{\partial \phi_x}{\partial x}\\ \frac{\partial \phi_y}{\partial y}\\ 0\\ 0\\ \frac{\partial \phi_x}{\partial y} + \frac{\partial \phi_y}{\partial x} \end{Bmatrix} \qquad (5.4-4)
$$

5.4–2 Equations of Motion

The governing equations of the first-order theory will be derived using the dynamic version of the principle of virtual displacements:

$$
0 = \int_0^T (\delta U + \delta V - \delta K)\, dt \qquad (5.4-5)
$$

where the virtual strain energy δU, virtual work done by applied forces δV, and the virtual kinetic energy δK are given by

$$
\begin{aligned}
\delta U = \int_{\Omega_0} \Bigg\{ \int_{-\frac{h}{2}}^{\frac{h}{2}} \Big[&\sigma_{xx}\left(\delta\varepsilon_{xx}^{(0)} + z\delta\varepsilon_{xx}^{(1)}\right) + \sigma_{yy}\left(\delta\varepsilon_{yy}^{(0)} + z\delta\varepsilon_{yy}^{(1)}\right)\\
&+ \sigma_{xy}\left(\delta\gamma_{xy}^{(0)} + z\delta\gamma_{xy}^{(1)}\right) + \sigma_{xz}\delta\gamma_{xz}^{(0)} + \sigma_{yz}\delta\gamma_{yz}^{(0)}\Big]\, dz \Bigg\} dxdy
\end{aligned}
\qquad (5.4-6)
$$

$$\delta V = -\int_{\Omega_0} \left[(q_b + q_t)\,\delta w_0\right] dxdy - \int_{\Gamma_\sigma}\int_{-\frac{h}{2}}^{\frac{h}{2}} \left[\hat{\sigma}_{nn}\left(\delta u_n + z\delta\phi_n\right) + \hat{\sigma}_{ns}\left(\delta u_s + z\delta\phi_s\right) + \hat{\sigma}_{nz}\delta w_0\,\right] dzds \tag{5.4-7}$$

$$\delta K = \int_{\Omega_0}\int_{-\frac{h}{2}}^{\frac{h}{2}} \rho_0 \Big[\left(\dot{u}_0 + z\dot{\phi}_x\right)\left(\delta\dot{u}_0 + z\delta\dot{\phi}_x\right) + \left(\dot{v}_0 + z\dot{\phi}_y\right)\left(\delta\dot{v}_0 + z\delta\dot{\phi}_y\right) + \dot{w}_0\delta\dot{w}_0\Big] dz\; dxdy \tag{5.4-8}$$

where all variables were previously introduced [see Eqs. (5.3–16)–(5.3–18) and the paragraph following the equations].

Substituting for $\delta U, \delta V$, and δK from Eqs. (6)–(8) into the virtual work statement in Eq. (5) and integrating through the thickness of the laminate, we obtain

$$\begin{aligned} 0 = \int_0^T \Bigg\{ \int_{\Omega_0} \Big[& N_{xx}\delta\varepsilon_{xx}^{(0)} + M_{xx}\delta\varepsilon_{xx}^{(1)} + N_{yy}\delta\varepsilon_{yy}^{(0)} + M_{yy}\delta\varepsilon_{yy}^{(1)} + N_{xy}\delta\gamma_{xy}^{(0)} + M_{xy}\delta\gamma_{xy}^{(1)} \\ & + Q_x\delta\gamma_{xz}^{(0)} + Q_y\delta\gamma_{yz}^{(0)} - q\delta w_0 - I_0\left(\dot{u}_0\delta\dot{u}_0 + \dot{v}_0\delta\dot{v}_0 + \dot{w}_0\delta\dot{w}_0\right) \\ & - I_1\left(\dot{\phi}_x\delta\dot{u}_0 + \dot{\phi}_y\delta\dot{v}_0 + \delta\dot{\phi}_x\dot{u}_0 + \delta\dot{\phi}_y\dot{v}_0\right) - I_2\left(\dot{\phi}_x\delta\dot{\phi}_x + \dot{\phi}_y\delta\dot{\phi}_y\right)\Big] dxdy \\ & - \int_{\Gamma_\sigma}\left(\hat{N}_{nn}\delta u_n + \hat{N}_{ns}\delta u_s + \hat{M}_{nn}\delta\phi_n + \hat{M}_{ns}\delta\phi_s + \hat{Q}_n\delta w_0\right) ds \Bigg\} dt \end{aligned} \tag{5.4-9}$$

where $q = q_b + q_t$, the stress resultants $(N_{xx}, N_{yy}, N_{xy}, M_{xx}, M_{yy}, M_{xy})$ and the inertias (I_0, I_1, I_2) are as defined in Eq. (5.3–20), $(N_{nn}, N_{ns}, M_{nn}, M_{ns})$ are as defined in Eq. (29a,b), and

$$\begin{Bmatrix} Q_x \\ Q_y \end{Bmatrix} = \int_{-\frac{h}{2}}^{\frac{h}{2}} \begin{Bmatrix} \sigma_{xz} \\ \sigma_{yz} \end{Bmatrix} dz \tag{5.4-10a}$$

The quantities (Q_x, Q_y) are called the *transverse force resultants.*

Shear Correction Factors

Since the transverse shear strains are represented as constant through the laminate thickness, it follows that the transverse shear stresses will also be constant. It is well known from elementary theory of homogeneous beams that the transverse shear stress varies parabolically through the beam thickness. In composite laminated beams and plates, the transverse shear stresses vary at least quadratically through layer thickness. This discrepancy between the actual stress state and the constant stress state predicted by the first-order theory is often corrected in computing the transverse shear force resultants (Q_x, Q_y) by multiplying the integrals in Eq. (10a) with a parameter K, called *shear correction coefficient:*

$$\begin{Bmatrix} Q_x \\ Q_y \end{Bmatrix} = K \int_{-\frac{h}{2}}^{\frac{h}{2}} \begin{Bmatrix} \sigma_{xz} \\ \sigma_{yz} \end{Bmatrix} dz \qquad (5.4-10b)$$

This amounts to modifying the plate transverse shear stiffnesses. The factor K is computed such that the strain energy due to transverse shear stresses in Eq. (10b) equals the strain energy due to the true transverse stresses predicted by the three-dimensional elasticity theory.

For example, consider a homogeneous beam with rectangular cross section, with width b and height h. The actual shear stress distribution through the thickness of the beam, from a course on mechanics of materials, is given by

$$\sigma_{xz}^c = \frac{3Q}{2bh}\left[1 - \left(\frac{2z}{h}\right)^2\right], \quad -\frac{h}{2} \le z \le \frac{h}{2}$$

where Q is the transverse shear force. The transverse shear stress in the first-order theory is a constant, $\sigma_{xz}^f = Q/bh$. The strain energies due to transverse shear stresses in the two theories are

$$\begin{aligned} U_s^c &= \frac{1}{2G_{13}} \int_A (\sigma_{xz}^c)^2 \, dA = \frac{3Q^2}{5G_{13}bh} \\ U_s^f &= \frac{1}{2G_{13}} \int_A \left(\sigma_{xz}^f\right)^2 dA = \frac{Q^2}{2G_{13}bh} \end{aligned}$$

The shear correction factor is the ratio of U_s^f to U_s^c, which gives $K = 5/6$. The shear correction factor for a general laminate depends on lamina properties and lamination scheme.

Returning to the virtual work statement in Eq. (9), we substitute for the virtual strains into Eq. (9) and integrate by parts to relieve the virtual generalized displacements $(\delta u_0, \delta v_0, \delta w_0, \delta\phi_x, \delta\phi_y)$ in Ω_0 of any differentiation, so that we can use the fundamental lemma of variational calculus; we obtain

$$\begin{aligned} 0 = \int_0^T \int_{\Omega_0} \Big[&- \left(N_{xx,x} + N_{xy,y} - I_0\ddot{u}_0 - I_1\ddot{\phi}_x\right) \delta u_0 \\ &- \left(N_{xy,x} + N_{yy,y} - I_0\ddot{v}_0 - I_1\ddot{\phi}_y\right) \delta v_0 \\ &- \left(M_{xx,x} + M_{xy,y} - Q_x - I_2\ddot{\phi}_x - I_1\ddot{u}_0\right) \delta\phi_x \\ &- \left(M_{xy,x} + M_{yy,y} - Q_y - I_2\ddot{\phi}_y - I_1\ddot{v}_0\right) \delta\phi_y \\ &- \left(Q_{x,x} + Q_{y,y} + \mathcal{N}(w_0) + q - I_0\ddot{w}_0\right) \delta w_0 \Big] dxdy \end{aligned}$$

$$+\int_0^T \int_\Gamma \left[\left(N_{nn} - \hat{N}_{nn}\right) \delta u_n + \left(N_{ns} - \hat{N}_{ns}\right) \delta u_s + \left(Q_n - \hat{Q}_n\right) \delta w_0 \right.$$
$$\left. + \left(M_{nn} - \hat{M}_{nn}\right) \delta\phi_n + \left(M_{ns} - \hat{M}_{ns}\right) \delta\phi_s \right] dsdt \qquad (5.4-11)$$

where $\mathcal{N}(w_0)$ and $\mathcal{P}(w_0)$ were defined in Eq. (5.3–24), and the boundary expressions were arrived by expressing ϕ_x and ϕ_y in terms of the normal and tangential rotations, (ϕ_n, ϕ_s):

$$\phi_x = n_x\phi_n - n_y\phi_s \ , \quad \phi_y = n_y\delta\phi_n + n_x\delta\phi_s \qquad (5.4-12)$$

The Euler–Lagrange equations are obtained by setting the coefficients of δu_0, δv_0, δw_0, $\delta\phi_x$, and $\delta\phi_y$ in Ω_0 to zero separately:

$$\begin{aligned}
\delta u_0 &: \quad \frac{\partial N_{xx}}{\partial x} + \frac{\partial N_{xy}}{\partial y} = I_0\frac{\partial^2 u_0}{\partial t^2} + I_1\frac{\partial^2 \phi_x}{\partial t^2}\\
\delta v_0 &: \quad \frac{\partial N_{xy}}{\partial x} + \frac{\partial N_{yy}}{\partial y} = I_0\frac{\partial^2 v_0}{\partial t^2} + I_1\frac{\partial^2 \phi_y}{\partial t^2}\\
\delta w_0 &: \quad \frac{\partial Q_x}{\partial x} + \frac{\partial Q_y}{\partial y} + \mathcal{N}(w_0) + q = I_0\frac{\partial^2 w_0}{\partial t^2}\\
\delta \phi_x &: \quad \frac{\partial M_{xx}}{\partial x} + \frac{\partial M_{xy}}{\partial y} - Q_x = I_2\frac{\partial^2 \phi_x}{\partial t^2} + I_1\frac{\partial^2 u_0}{\partial t^2}\\
\delta \phi_y &: \quad \frac{\partial M_{xy}}{\partial x} + \frac{\partial M_{yy}}{\partial y} - Q_y = I_2\frac{\partial^2 \phi_y}{\partial t^2} + I_1\frac{\partial^2 v_0}{\partial t^2}
\end{aligned} \qquad (5.4-13)$$

The natural boundary conditions are obtained by setting the coefficients of δu_n, δu_s, δw_0, $\delta\phi_n$, and $\delta\phi_s$ on Γ to zero separately:

$$N_{nn} - \hat{N}_{nn} = 0 \ , \quad N_{ns} - \hat{N}_{ns} = 0 \ , \quad Q_n - \hat{Q}_n = 0$$
$$M_{nn} - \hat{M}_{nn} = 0 \ , \quad M_{ns} - \hat{M}_{ns} = 0 \qquad (5.4-14a)$$

where

$$Q_n \equiv Q_x n_x + Q_y n_y + \mathcal{P}(w_0) \qquad (5.4-14b)$$

Thus the primary and secondary variables of the theory are

primary variables:	$u_n,\ u_s,\ w_0,\ \phi_n,\ \phi_s$
secondary variables:	$N_{nn},\ N_{ns},\ Q_n,\ M_{nn},\ M_{ns}$

(5.4 – 15)

Note that Q_n defined in Eq. (14b) is the same as that defined in Eq. (5.3–31b). This follows from the last two equations of (13).

The initial conditions of the theory involve specifying the values of the displacements and their first derivatives with respect to time at $t = 0$:

$$\begin{aligned} u_n &= u_n^0,\ u_s = u_s^0,\ w_0 = w_0^0,\ \phi_n = \phi_n^0,\ \phi_s = \phi_s^0 \\ \dot{u}_n &= \dot{u}_n^0,\ \dot{u}_s = \dot{u}_s^0,\ \dot{w}_0 = \dot{w}_0^0,\ \dot{\phi}_n = \dot{\phi}_n^0,\ \dot{\phi}_s = \dot{\phi}_s^0 \end{aligned} \tag{5.4-16}$$

for all points in Ω_0.

5.4–3 Laminate Constitutive Equations

The laminate constitutive equations for the first-order theory are obtained using the lamina constitutive equations (5.3–12a) and the following relations:

$$\begin{Bmatrix} \sigma_{yz} \\ \sigma_{xz} \end{Bmatrix}^{(k)} = \begin{bmatrix} \bar{Q}_{44} & \bar{Q}_{45} \\ \bar{Q}_{45} & \bar{Q}_{55} \end{bmatrix}^{(k)} \begin{Bmatrix} \gamma_{yz}^{(0)} \\ \gamma_{xz}^{(0)} \end{Bmatrix} - \begin{bmatrix} \bar{e}_{14} & \bar{e}_{24} & 0 \\ \bar{e}_{15} & \bar{e}_{25} & 0 \end{bmatrix}^{(k)} \begin{Bmatrix} \mathcal{E}_x \\ \mathcal{E}_y \\ \mathcal{E}_z \end{Bmatrix}^{(k)} \tag{5.4-17a}$$

where [see Eq. (4.4–10)]

$$\begin{aligned} \bar{Q}_{44} &= Q_{44}\cos^2\theta + Q_{55}\sin^2\theta \\ \bar{Q}_{45} &= (Q_{55} - Q_{44})\cos\theta\sin\theta \\ \bar{Q}_{55} &= Q_{44}\sin^2\theta + Q_{55}\cos^2\theta \end{aligned} \tag{5.4-17b}$$

$$\begin{aligned} \bar{e}_{14} &= (e_{15} - e_{24})\sin\theta\cos\theta,\ \bar{e}_{24} = e_{24}\cos^2\theta + e_{15}\sin^2\theta \\ \bar{e}_{15} &= e_{15}\cos^2\theta + e_{24}\sin^2\theta,\ \bar{e}_{25} = (e_{15} - e_{24})\sin\theta\cos\theta \end{aligned} \tag{5.4-17c}$$

The laminate constitutive equations in Eqs. (5.3–36) and (5.3–37) are valid also for the first-order laminate theory. In addition, we have the following laminate constitutive equations:

$$\begin{Bmatrix} Q_y \\ Q_x \end{Bmatrix} = K \sum_{k=1}^{N} \int_{z_k}^{z_{k+1}} \begin{Bmatrix} \sigma_{yz} \\ \sigma_{xz} \end{Bmatrix} dz$$

or

$$\boxed{\begin{Bmatrix} Q_y \\ Q_x \end{Bmatrix} = K \begin{bmatrix} A_{44} & A_{45} \\ A_{45} & A_{55} \end{bmatrix} \begin{Bmatrix} \gamma_{yz}^{(0)} \\ \gamma_{xz}^{(0)} \end{Bmatrix} - \begin{Bmatrix} Q_y^P \\ Q_x^P \end{Bmatrix}} \tag{5.4-18}$$

where the extensional stiffnesses A_{44}, A_{45}, and A_{55} are defined by

$$
\begin{aligned}
(A_{44}, A_{45}, A_{55}) &= \int_{-\frac{h}{2}}^{\frac{h}{2}} (\bar{Q}_{44}, \bar{Q}_{45}, \bar{Q}_{55})\ dz \\
&= \sum_{k=1}^{N} \int_{z_k}^{z_{k+1}} (\bar{Q}_{44}^{(k)}, \bar{Q}_{45}^{(k)}, \bar{Q}_{55}^{(k)})\ dz \\
&= \sum_{k=1}^{N} (\bar{Q}_{44}^{(k)}, \bar{Q}_{45}^{(k)}, \bar{Q}_{55}^{(k)})(z_{k+1} - z_k) \qquad (5.4-19a)
\end{aligned}
$$

and the piezoelectric forces Q_x^P and Q_y^P are defined by

$$
\begin{Bmatrix} Q_x^P \\ Q_y^P \end{Bmatrix} = \sum_{k=1}^{N} \int_{z_k}^{z_{k+1}} \begin{bmatrix} \bar{e}_{14} & \bar{e}_{24} & 0 \\ \bar{e}_{15} & \bar{e}_{25} & 0 \end{bmatrix}^{(k)} \begin{Bmatrix} \mathcal{E}_x \\ \mathcal{E}_y \\ \mathcal{E}_z \end{Bmatrix}^{(k)} dz \qquad (5.4-19b)
$$

When thermal and piezoelectric effects are not present, the stress resultants (N's and M's) are related to the generalized displacements $(u_0, v_0, w_0, \phi_x, \phi_y)$ by the relations

$$
\begin{Bmatrix} N_{xx} \\ N_{yy} \\ N_{xy} \end{Bmatrix} = \begin{bmatrix} A_{11} & A_{12} & A_{16} \\ A_{12} & A_{22} & A_{26} \\ A_{16} & A_{26} & A_{66} \end{bmatrix} \begin{Bmatrix} \frac{\partial u_0}{\partial x} + \frac{1}{2}(\frac{\partial w_0}{\partial x})^2 \\ \frac{\partial v_0}{\partial y} + \frac{1}{2}(\frac{\partial w_0}{\partial y})^2 \\ \frac{\partial u_0}{\partial y} + \frac{\partial v_0}{\partial x} + \frac{\partial w_0}{\partial x}\frac{\partial w_0}{\partial y} \end{Bmatrix} + \begin{bmatrix} B_{11} & B_{12} & B_{16} \\ B_{12} & B_{22} & B_{26} \\ B_{16} & B_{26} & B_{66} \end{bmatrix} \begin{Bmatrix} \frac{\partial \phi_x}{\partial x} \\ \frac{\partial \phi_y}{\partial y} \\ \frac{\partial \phi_x}{\partial y} + \frac{\partial \phi_y}{\partial x} \end{Bmatrix} \qquad (5.4-20)
$$

$$
\begin{Bmatrix} M_{xx} \\ M_{yy} \\ M_{xy} \end{Bmatrix} = \begin{bmatrix} B_{11} & B_{12} & B_{16} \\ B_{12} & B_{22} & B_{26} \\ B_{16} & B_{26} & B_{66} \end{bmatrix} \begin{Bmatrix} \frac{\partial u_0}{\partial x} + \frac{1}{2}(\frac{\partial w_0}{\partial x})^2 \\ \frac{\partial v_0}{\partial y} + \frac{1}{2}(\frac{\partial w_0}{\partial y})^2 \\ \frac{\partial u_0}{\partial y} + \frac{\partial v_0}{\partial x} + \frac{\partial w_0}{\partial x}\frac{\partial w_0}{\partial y} \end{Bmatrix} + \begin{bmatrix} D_{11} & D_{12} & D_{16} \\ D_{12} & D_{22} & D_{26} \\ D_{16} & D_{26} & D_{66} \end{bmatrix} \begin{Bmatrix} \frac{\partial \phi_x}{\partial x} \\ \frac{\partial \phi_y}{\partial y} \\ \frac{\partial \phi_x}{\partial y} + \frac{\partial \phi_y}{\partial x} \end{Bmatrix} \qquad (5.4-21)
$$

$$
\begin{Bmatrix} Q_y \\ Q_x \end{Bmatrix} = K \begin{bmatrix} A_{44} & A_{45} \\ A_{45} & A_{55} \end{bmatrix} \begin{Bmatrix} \frac{\partial w_0}{\partial y} + \phi_y \\ \frac{\partial w_0}{\partial x} + \phi_x \end{Bmatrix} \qquad (5.4-22)
$$

When thermal and piezoelectric effects are present, Eqs. (20) and (21) take the same form as Eq. (5.3–40), and Eq. (22) will contain the column of piezoelectric forces given in Eq. (18).

5.4–4 Equations of Motion in Terms of Displacements

The equations of motion (13) can be expressed in terms of displacements $(u_0, v_0, w_0, \phi_x, \phi_y)$ by substituting for the force and moment resultants from Eqs. (20)–(22). For homogeneous laminates, the equations of motion (13) take the form

$$
\begin{aligned}
&A_{11}\left(\frac{\partial^2 u_0}{\partial x^2}+\frac{\partial w_0}{\partial x}\frac{\partial^2 w_0}{\partial x^2}\right)+A_{12}\left(\frac{\partial^2 v_0}{\partial y\partial x}+\frac{\partial w_0}{\partial y}\frac{\partial^2 w_0}{\partial y\partial x}\right)+\\
&A_{16}\left(\frac{\partial^2 u_0}{\partial y\partial x}+\frac{\partial^2 v_0}{\partial x^2}+\frac{\partial^2 w_0}{\partial x^2}\frac{\partial w_0}{\partial y}+\frac{\partial w_0}{\partial x}\frac{\partial^2 w_0}{\partial y\partial x}\right)+\\
&B_{11}\frac{\partial^2 \phi_x}{\partial x^2}+B_{12}\frac{\partial^2 \phi_y}{\partial y\partial x}+B_{16}\left(\frac{\partial^2 \phi_x}{\partial x\partial y}+\frac{\partial^2 \phi_y}{\partial x^2}\right)+\\
&A_{16}\left(\frac{\partial^2 u_0}{\partial x\partial y}+\frac{\partial w_0}{\partial x}\frac{\partial^2 w_0}{\partial x\partial y}\right)+A_{26}\left(\frac{\partial^2 v_0}{\partial y^2}+\frac{\partial w_0}{\partial y}\frac{\partial^2 w_0}{\partial y^2}\right)+\\
&A_{66}\left(\frac{\partial^2 u_0}{\partial y^2}+\frac{\partial^2 v_0}{\partial x\partial y}+\frac{\partial^2 w_0}{\partial x\partial y}\frac{\partial w_0}{\partial y}+\frac{\partial w_0}{\partial x}\frac{\partial^2 w_0}{\partial y^2}\right)+\\
&B_{16}\frac{\partial^2 \phi_x}{\partial x\partial y}+B_{26}\frac{\partial^2 \phi_y}{\partial y^2}+B_{66}\left(\frac{\partial^2 \phi_x}{\partial y^2}+\frac{\partial^2 \phi_y}{\partial y\partial x}\right)-\\
&\left(\frac{\partial N_{xx}^T}{\partial x}+\frac{\partial N_{xy}^T}{\partial y}\right)-\left(\frac{\partial N_{xx}^P}{\partial x}+\frac{\partial N_{xy}^P}{\partial y}\right)=I_0\frac{\partial^2 u_0}{\partial t^2}+I_1\frac{\partial^2 \phi_x}{\partial t^2}
\end{aligned}
\tag{5.4 – 23}
$$

$$
\begin{aligned}
&A_{16}\left(\frac{\partial^2 u_0}{\partial x^2}+\frac{\partial w_0}{\partial x}\frac{\partial^2 w_0}{\partial x^2}\right)+A_{26}\left(\frac{\partial^2 v_0}{\partial y\partial x}+\frac{\partial w_0}{\partial y}\frac{\partial^2 w_0}{\partial y\partial x}\right)+\\
&A_{66}\left(\frac{\partial^2 u_0}{\partial y\partial x}+\frac{\partial^2 v_0}{\partial x^2}+\frac{\partial^2 w_0}{\partial x^2}\frac{\partial w_0}{\partial y}+\frac{\partial w_0}{\partial x}\frac{\partial^2 w_0}{\partial y\partial x}\right)+\\
&B_{16}\frac{\partial^2 \phi_x}{\partial x^2}+B_{26}\frac{\partial^2 \phi_y}{\partial y\partial x}+B_{66}\left(\frac{\partial^2 \phi_x}{\partial x\partial y}+\frac{\partial^2 \phi_y}{\partial x^2}\right)+\\
&A_{12}\left(\frac{\partial^2 u_0}{\partial x\partial y}+\frac{\partial w_0}{\partial x}\frac{\partial^2 w_0}{\partial x\partial y}\right)+A_{22}\left(\frac{\partial^2 v_0}{\partial y^2}+\frac{\partial w_0}{\partial y}\frac{\partial^2 w_0}{\partial y^2}\right)+\\
&A_{26}\left(\frac{\partial^2 u_0}{\partial y^2}+\frac{\partial^2 v_0}{\partial x\partial y}+\frac{\partial^2 w_0}{\partial x\partial y}\frac{\partial w_0}{\partial y}+\frac{\partial w_0}{\partial x}\frac{\partial^2 w_0}{\partial y^2}\right)\\
&B_{12}\frac{\partial^2 \phi_x}{\partial x\partial y}+B_{22}\frac{\partial^2 \phi_y}{\partial y^2}+B_{26}\left(\frac{\partial^2 \phi_x}{\partial y^2}+\frac{\partial^2 \phi_y}{\partial x\partial y}\right)-\\
&\left(\frac{\partial N_{xy}^T}{\partial x}+\frac{\partial N_{yy}^T}{\partial y}\right)-\left(\frac{\partial N_{xy}^P}{\partial x}+\frac{\partial N_{yy}^P}{\partial y}\right)=I_0\frac{\partial^2 v_0}{\partial t^2}+I_1\frac{\partial^2 \phi_y}{\partial t^2}
\end{aligned}
\tag{5.4 – 24}
$$

$$KA_{55}\left(\frac{\partial^2 w_0}{\partial x^2}+\frac{\partial \phi_x}{\partial x}\right)+KA_{45}\left(\frac{\partial^2 w_0}{\partial y\partial x}+\frac{\partial \phi_y}{\partial x}\right)+$$
$$KA_{45}\left(\frac{\partial^2 w_0}{\partial x\partial y}+\frac{\partial \phi_x}{\partial y}\right)+KA_{44}\left(\frac{\partial^2 w_0}{\partial y^2}+\frac{\partial \phi_y}{\partial y}\right)+$$
$$\mathcal{N}(w)+q-\left(\frac{\partial Q_x^P}{\partial x}+\frac{\partial Q_y^P}{\partial y}\right)=I_0\frac{\partial^2 w_0}{\partial t^2} \quad (5.4-25)$$

$$B_{11}\left(\frac{\partial^2 u_0}{\partial x^2}+\frac{\partial w_0}{\partial x}\frac{\partial^2 w_0}{\partial x^2}\right)+B_{12}\left(\frac{\partial^2 v_0}{\partial y\partial x}+\frac{\partial w_0}{\partial y}\frac{\partial^2 w_0}{\partial y\partial x}\right)+$$
$$B_{16}\left(\frac{\partial^2 u_0}{\partial y\partial x}+\frac{\partial^2 v_0}{\partial x^2}+\frac{\partial^2 w_0}{\partial x^2}\frac{\partial w_0}{\partial y}+\frac{\partial w_0}{\partial x}\frac{\partial^2 w_0}{\partial y\partial x}\right)+$$
$$D_{11}\frac{\partial^2 \phi_x}{\partial x^2}+D_{12}\frac{\partial^2 \phi_y}{\partial y\partial x}+D_{16}\left(\frac{\partial^2 \phi_x}{\partial x\partial y}+\frac{\partial^2 \phi_y}{\partial x^2}\right)+$$
$$B_{16}\left(\frac{\partial^2 u_0}{\partial x\partial y}+\frac{\partial w_0}{\partial x}\frac{\partial^2 w_0}{\partial x\partial y}\right)+B_{26}\left(\frac{\partial^2 v_0}{\partial y^2}+\frac{\partial w_0}{\partial y}\frac{\partial^2 w_0}{\partial y^2}\right)+$$
$$B_{66}\left(\frac{\partial^2 u_0}{\partial y^2}+\frac{\partial^2 v_0}{\partial x\partial y}+\frac{\partial^2 w_0}{\partial x\partial y}\frac{\partial w_0}{\partial y}+\frac{\partial w_0}{\partial x}\frac{\partial^2 w_0}{\partial y^2}\right)+$$
$$D_{16}\frac{\partial^2 \phi_x}{\partial x\partial y}+D_{26}\frac{\partial^2 \phi_y}{\partial y^2}+D_{66}\left(\frac{\partial^2 \phi_x}{\partial y^2}+\frac{\partial^2 \phi_y}{\partial y\partial x}\right)-$$
$$KA_{55}\left(\frac{\partial w_0}{\partial x}+\phi_x\right)-KA_{45}\left(\frac{\partial w_0}{\partial y}+\phi_y\right)-$$
$$\left(\frac{\partial M_{xx}^T}{\partial x}+\frac{\partial M_{xy}^T}{\partial y}\right)-\left(\frac{\partial M_{xx}^P}{\partial x}+\frac{\partial M_{xy}^P}{\partial y}-Q_x^P\right)$$
$$=I_2\frac{\partial^2 \phi_x}{\partial t^2}+I_1\frac{\partial^2 u_0}{\partial t^2} \quad (5.4-26)$$

$$B_{16}\left(\frac{\partial^2 u_0}{\partial x^2}+\frac{\partial w_0}{\partial x}\frac{\partial^2 w_0}{\partial x^2}\right)+B_{26}\left(\frac{\partial^2 v_0}{\partial y\partial x}+\frac{\partial w_0}{\partial y}\frac{\partial^2 w_0}{\partial y\partial x}\right)+$$
$$B_{66}\left(\frac{\partial^2 u_0}{\partial y\partial x}+\frac{\partial^2 v_0}{\partial x^2}+\frac{\partial^2 w_0}{\partial x^2}\frac{\partial w_0}{\partial y}+\frac{\partial w_0}{\partial x}\frac{\partial^2 w_0}{\partial y\partial x}\right)+$$
$$D_{16}\frac{\partial^2 \phi_x}{\partial x^2}+D_{26}\frac{\partial^2 \phi_y}{\partial y\partial x}+D_{66}\left(\frac{\partial^2 \phi_x}{\partial x\partial y}+\frac{\partial^2 \phi_y}{\partial x^2}\right)+$$
$$B_{12}\left(\frac{\partial^2 u_0}{\partial x\partial y}+\frac{\partial w_0}{\partial x}\frac{\partial^2 w_0}{\partial x\partial y}\right)+B_{22}\left(\frac{\partial^2 v_0}{\partial y^2}+\frac{\partial w_0}{\partial y}\frac{\partial^2 w_0}{\partial y^2}\right)+$$
$$B_{26}\left(\frac{\partial^2 u_0}{\partial y^2}+\frac{\partial^2 v_0}{\partial x\partial y}+\frac{\partial^2 w_0}{\partial x\partial y}\frac{\partial w_0}{\partial y}+\frac{\partial w_0}{\partial x}\frac{\partial^2 w_0}{\partial y^2}\right)+$$

$$D_{12}\frac{\partial^2 \phi_x}{\partial x \partial y} + D_{22}\frac{\partial^2 \phi_y}{\partial y^2} + D_{26}\left(\frac{\partial^2 \phi_x}{\partial y^2} + \frac{\partial^2 \phi_y}{\partial x \partial y}\right) -$$
$$KA_{45}\left(\frac{\partial w_0}{\partial x} + \phi_x\right) - KA_{44}\left(\frac{\partial w_0}{\partial y} + \phi_y\right) -$$
$$\left(\frac{\partial M_{xy}^T}{\partial x} + \frac{\partial M_{yy}^T}{\partial y}\right) - \left(\frac{\partial M_{xy}^P}{\partial x} + \frac{\partial M_{yy}^P}{\partial y} - Q_y^P\right)$$
$$= I_2\frac{\partial^2 \phi_y}{\partial t^2} + I_1\frac{\partial^2 v_0}{\partial t^2} \qquad (5.4-27)$$

Equations (22)–(27) describe five second-order, nonlinear, partial differential equations in terms of the five generalized displacements. Hence, the first-order laminated plate theory is a tenth-order theory and there are ten boundary conditions, as stated earlier in Eqs. (14) and (15). Note that the displacement field of the classical plate theory can be obtained from that of the first-order theory by setting

$$\phi_x = -\frac{\partial w_0}{\partial x} \quad \text{and} \quad \phi_y = -\frac{\partial w_0}{\partial y} \qquad (5.4-28)$$

Conversely, the relations in Eq. (28) can be used to derive the first-order theory from the classical plate theory via the penalty function method (see Chapter 10).

Example 5.4–1:

The linearized equations of motion for cylindrical bending according to the first-order shear deformation theory are given by setting all derivatives with respect to y in Eqs. (23)–(27):

$$A_{11}\frac{\partial^2 u_0}{\partial x^2} + A_{16}\frac{\partial^2 v_0}{\partial x^2} + B_{11}\frac{\partial^2 \phi_x}{\partial x^2} + B_{16}\frac{\partial^2 \phi_y}{\partial x^2} - \frac{\partial N_{xx}^T}{\partial x} - \frac{\partial N_{xx}^P}{\partial x}$$
$$= I_0\frac{\partial^2 u_0}{\partial t^2} + I_1\frac{\partial^2 \phi_x}{\partial t^2} \qquad (5.4-29)$$

$$A_{16}\frac{\partial^2 u_0}{\partial x^2} + A_{66}\frac{\partial^2 v_0}{\partial x^2} + B_{16}\frac{\partial^2 \phi_x}{\partial x^2} + B_{66}\frac{\partial^2 \phi_y}{\partial x^2} - \frac{\partial N_{xy}^T}{\partial x} - \frac{\partial N_{xy}^P}{\partial x}$$
$$= I_0\frac{\partial^2 v_0}{\partial t^2} + I_1\frac{\partial^2 \phi_y}{\partial t^2} \qquad (5.4-30)$$

$$B_{11}\frac{\partial^2 u_0}{\partial x^2} + B_{16}\frac{\partial^2 v_0}{\partial x^2} + D_{11}\frac{\partial^2 \phi_x}{\partial x^2} + D_{16}\frac{\partial^2 \phi_y}{\partial x^2} - KA_{55}\left(\frac{\partial w_0}{\partial x} + \phi_x\right)$$
$$- KA_{45}\phi_y - \frac{\partial M_{xx}^T}{\partial x} - \frac{\partial M_{xx}^P}{\partial x} + Q_x^P = I_2\frac{\partial^2 \phi_x}{\partial t^2} + I_1\frac{\partial^2 u_0}{\partial t^2} \qquad (5.4-31)$$

$$
\begin{aligned}
& B_{16}\frac{\partial^2 u_0}{\partial x^2} + B_{66}\frac{\partial^2 v_0}{\partial x^2} + D_{16}\frac{\partial^2 \phi_x}{\partial x^2} + D_{66}\frac{\partial^2 \phi_y}{\partial x^2} - KA_{44}\phi_y \\
& - KA_{45}\left(\frac{\partial w_0}{\partial x} + \phi_x\right) - \frac{\partial M_{xy}^T}{\partial x} - \frac{\partial M_{xy}^P}{\partial x} + Q_y^P \\
& = I_2\frac{\partial^2 \phi_y}{\partial t^2} + I_1\frac{\partial^2 v_0}{\partial t^2}
\end{aligned} \tag{5.4 – 32}
$$

$$
\begin{aligned}
& KA_{55}\left(\frac{\partial^2 w_0}{\partial x^2} + \frac{\partial \phi_x}{\partial x}\right) + KA_{45}\frac{\partial \phi_y}{\partial x} + \frac{\partial}{\partial x}\left(N_{xx}\frac{\partial w_0}{\partial x}\right) \\
& + q - \frac{\partial Q_x^P}{\partial x} = I_0\frac{\partial^2 w_0}{\partial t^2}
\end{aligned} \tag{5.4 – 33}
$$

5.5 Laminate Stiffnesses for Selected Laminates

5.5–1 General Discussion

A close examination of the laminate stiffnesses defined in Eqs. (5.3–38) and (5.4–19) show that their values depend on the material stiffnesses, layer thicknesses, and the lamination scheme. Symmetry or antisymmetry of the lamination scheme and material properties about the midplane of the laminate reduce some of the laminate stiffnesses to zero. The book by Jones [40] has an excellent discussion of the laminate stiffnesses for various types of laminated plates. In this section we review selective lamination schemes for their laminate stiffness characteristics.

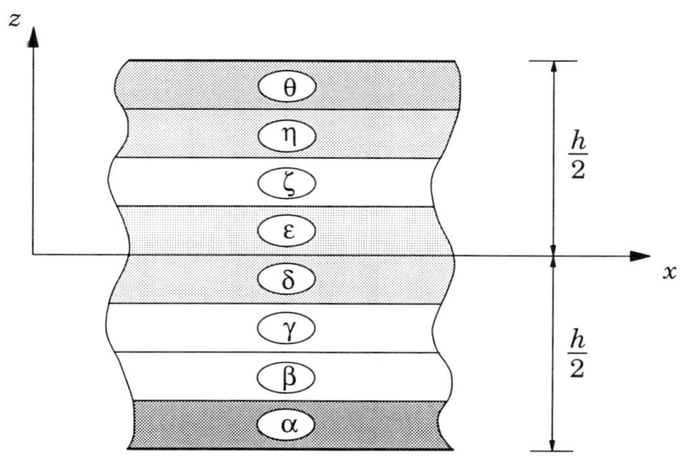

Figure 5.5–1: A laminate with general stacking sequence.

Before we embark on the discussion of laminate stiffnesses, it is useful to introduce the terminology and notation associated with special lamination schemes. The lamination scheme of a laminate will be denoted by $[\alpha/\beta/\gamma/\delta/\varepsilon/\cdots]$, where α is the orientation of the first ply, β is the orientation of the second ply, and so on (see Figure 5.5–1). The plies are counted in the positive z direction (see Figure 5.3–1). Unless stated otherwise, this notation also implies that all layers are of the same thickness and made of the same material.

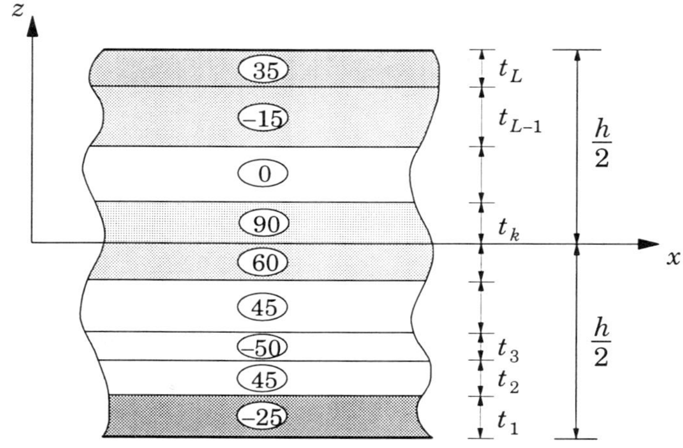

Figure 5.5–2: A general angle-ply laminate.

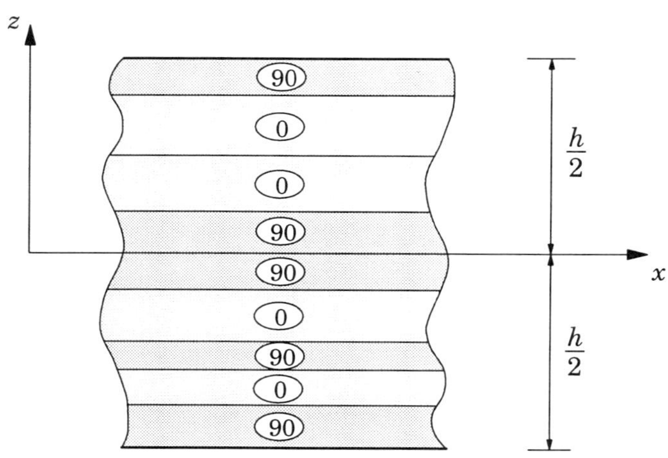

Figure 5.5–3: A cross-ply laminated plate with the 0° and 90° layers.

A general laminate has layers of different orientations θ where $-90^\circ \leq \theta \leq 90^\circ$. For example, (0/15/–35/45/90/–45) is a six-ply laminate. General *angle-ply* laminates (see Figure 5.5–2) have ply orientations of θ and $-\theta$ where $0^\circ \leq \theta \leq 90^\circ$, and with at least one layer having an orientation other than 0° or 90°. An example of angle-ply laminates is provided by (15/–30/0/90/45/–45). *Cross-ply* laminates are those which have ply orientations of 0° or 90° (see Figure 5.5–3). An example of a cross-ply laminate is (0/90/90/0/0/90). For layers with 0° or 90° orientations, the layer stiffnesses $\bar{Q}_{16}$, $\bar{Q}_{26}$, $\bar{Q}_{45}$ are zero. Hence, $A_{16} = A_{26} = A_{45} = D_{16} = D_{26} = 0$.

When ply stacking sequence, material, and geometry (*i.e.*, ply thicknesses) are symmetric about the midplane of the laminate, the laminate is called a *symmetric laminate* (see Figure 5.5–4). For a symmetric laminate, the upper half through the laminate thickness is a mirror image of the lower half. The laminates $(-45/45/45/-45) = (-45/45)_s$ and $(45/-45/-45/45) = (45/-45)_s$, with all layers having the same thickness and material, are examples of a symmetric angle-ply laminate, $(0/90/90/0) = (0/90)_s$ is a symmetric cross-ply laminate, and $(30/-45/0/90/90/0/-45/30)=(30/-45/0/90)_s$ is a general symmetric laminate. Note that symmetric laminates are also denoted by displaying only the lamination scheme of the upper half. The symmetric laminate (30/–45/0/90/90/0/–45/30) is denoted as $(30/-45/0/90)_s$.

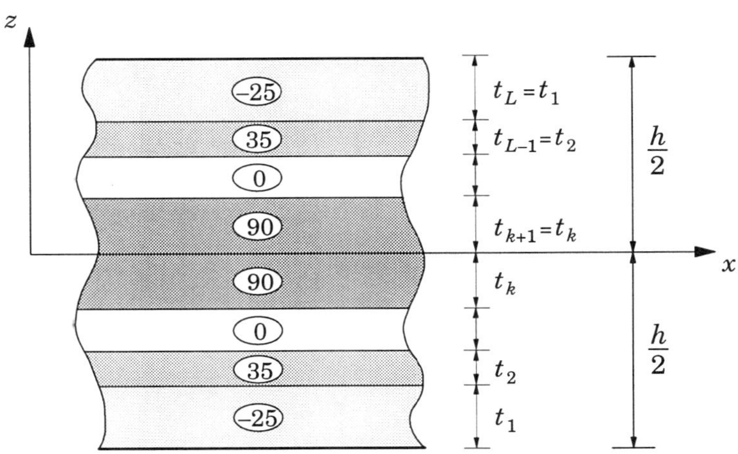

Figure 5.5–4: A symmetric laminate.

An unsymmetric or *asymmetric laminate* is a laminate that is not symmetric. An *antisymmetric* laminate is one whose lamination scheme is antisymmetric and material and thicknesses are symmetric about the midplane. Examples of antisymmetric angle-ply and cross-ply laminates are provided, respectively, by (–30/30/–30/30/–30/30) $\equiv$ $(-30/30)_3$ and (0/90/0/90/0/90) $\equiv$ $(0/90)_3$.

In the following paragraphs we present laminate stiffnesses A_{ij}, B_{ij}, and D_{ij} for various lamination schemes. Laminate stiffnesses A_{ij} depend only the thicknesses and stiffnesses of the layers but not on their placement through the thickness. On the other hand, laminate stiffnesses D_{ij} depend not only on the layer thickness and stiffnesses but also on their location relative to the midplane. For example, both $(0/90)_s$ and $(90/0)_s$ laminates will have the same in-plane stiffnesses A_{ij}. However, $(0/90)_s$ laminate will have larger bending stiffnesses D_{ij} about an axis perpendicular to the fiber direction than the $(90/0)_s$ laminate because the 0° layers are located farther from the midplane in the $(0/90)_s$ laminate. Both A_{ij} and D_{ij} are always positive. Laminate stiffnesses B_{ij} can be negative, depending on the lamination scheme and on the number of layers.

The laminate stiffnesses A_{ij}, B_{ij}, and D_{ij} for any arbitrary laminate can be computed using the definitions in Eq. (5.3–38). Computer programs are available to evaluate the stiffnesses for arbitrary stacking sequence and material properties.

5.5–2 Single-Layer Plates

Here we discuss some special cases of single-layered configurations and their stiffnesses. The special single-layered configurations discussed here include: isotropic, specially orthotropic (*i.e.*, the principal material coordinates coincide with those of the plate), generally orthotropic (*i.e.*, the principal material coordinates *do not* coincide with those of the plate), and anisotropic. The bending–stretching coupling coefficients B_{ij} and the shear stiffnesses A_{16}, A_{26}, D_{16}, and D_{26} can shown to be zero for all single-layer plates except for generally orthotropic and anisotropic single-layer plates.

Single Isotropic Layer

For a single isotropic layer with material constants E and ν $[G = \frac{E}{2(1+\nu)}]$ and thickness h, the nonzero laminate stiffnesses of Eqs. (5.3–38) and (5.4–19) become

$$A_{11} = \frac{Eh}{1-\nu^2}\,, \quad A_{12} = \nu A_{11},\ A_{22} = A_{11}$$

$$A_{66} = \frac{1-\nu}{2}A_{11},\ A_{44} = A_{55} = \frac{1-\nu}{2}KA_{11}$$

$$D_{11} = \frac{Eh^3}{12(1-\nu^2)}\,, \quad D_{12} = \nu D_{11},\ D_{22} = D_{11},\ D_{66} = \frac{1-\nu}{2}D_{11} \qquad (5.5-1)$$

The plate constitutive equations for the classical and first-order theories become

$$\begin{Bmatrix} N_{xx} \\ N_{yy} \\ N_{xy} \end{Bmatrix} = \begin{bmatrix} A_{11} & \nu A_{11} & 0 \\ \nu A_{11} & A_{11} & 0 \\ 0 & 0 & \frac{1-\nu}{2} A_{11} \end{bmatrix} \begin{Bmatrix} \varepsilon_{xx}^{(0)} \\ \varepsilon_{yy}^{(0)} \\ \gamma_{xy}^{(0)} \end{Bmatrix} \tag{5.5-2}$$

$$\begin{Bmatrix} M_{xx} \\ M_{yy} \\ M_{xy} \end{Bmatrix} = \begin{bmatrix} D_{11} & \nu D_{11} & 0 \\ \nu D_{11} & D_{11} & 0 \\ 0 & 0 & \frac{1-\nu}{2} D_{11} \end{bmatrix} \begin{Bmatrix} \varepsilon_{xx}^{(1)} \\ \varepsilon_{yy}^{(1)} \\ \gamma_{xy}^{(1)} \end{Bmatrix} \tag{5.5-3}$$

$$\begin{Bmatrix} Q_y \\ Q_x \end{Bmatrix} = K \frac{1-\nu}{2} \begin{bmatrix} A_{11} & 0 \\ 0 & A_{11} \end{bmatrix} \begin{Bmatrix} \gamma_{yz}^{(0)} \\ \gamma_{xz}^{(0)} \end{Bmatrix} \tag{5.5-4}$$

The nonzero thermal stress resultants $\{N^T\}$ and $\{M^T\}$ are given by

$$N_{xx}^T = N_{yy}^T = \frac{E\alpha}{(1-\nu)} \int_{-\frac{h}{2}}^{\frac{h}{2}} \Delta T \, dz, \quad M_{xx}^T = M_{yy}^T = \frac{E\alpha}{(1-\nu)} \int_{-\frac{h}{2}}^{\frac{h}{2}} \Delta T z \, dz \tag{5.5-5}$$

Single Specially Orthotropic Layer

For a single specially orthotropic layer, the stiffnesses can be expressed in terms of the Q_{ij} and thickness h. The nonzero stiffnesses of Eqs. (5.3–38) and (5.4–19) become

$$A_{11} = Q_{11}h, \quad A_{12} = Q_{12}h, \quad A_{22} = Q_{22}h$$

$$A_{66} = Q_{66}h, \quad A_{44} = KQ_{44}h, \quad A_{55} = KQ_{55}h$$

$$D_{11} = \frac{Q_{11}h^3}{12}, \quad D_{12} = \frac{Q_{12}h^3}{12}, \quad D_{22} = \frac{Q_{22}h^3}{12}, \quad D_{66} = \frac{Q_{66}h^3}{12} \tag{5.5-6}$$

where Q_{ij} are the plane-stress-reduced stiffnesses, and they are given in terms of the engineering constants [see Eq. (4.4–3b)] as

$$Q_{11} = \frac{E_1}{1-\nu_{12}\nu_{21}}, \quad Q_{12} = \frac{\nu_{12}E_2}{1-\nu_{12}\nu_{21}}, \quad Q_{22} = \frac{E_2}{1-\nu_{12}\nu_{21}}$$

$$Q_{66} = G_{12}, \quad Q_{44} = G_{23}, \quad Q_{55} = G_{13} \tag{5.5-7}$$

The plate constitutive equations for the classical and first-order theories become

$$\begin{Bmatrix} N_{xx} \\ N_{yy} \\ N_{xy} \end{Bmatrix} = h \begin{bmatrix} Q_{11} & Q_{12} & 0 \\ Q_{12} & Q_{22} & 0 \\ 0 & 0 & Q_{66} \end{bmatrix} \begin{Bmatrix} \varepsilon_{xx}^{(0)} \\ \varepsilon_{yy}^{(0)} \\ \gamma_{xy}^{(0)} \end{Bmatrix} \tag{5.5-8}$$

$$\left\{\begin{array}{c} M_{xx} \\ M_{yy} \\ M_{xy} \end{array}\right\} = \frac{h^3}{12}\begin{bmatrix} Q_{11} & Q_{12} & 0 \\ Q_{12} & Q_{22} & 0 \\ 0 & 0 & Q_{66} \end{bmatrix}\left\{\begin{array}{c} \varepsilon_{xx}^{(1)} \\ \varepsilon_{yy}^{(1)} \\ \gamma_{xy}^{(1)} \end{array}\right\} \tag{5.5-9}$$

$$\left\{\begin{array}{c} Q_y \\ Q_x \end{array}\right\} = Kh\begin{bmatrix} Q_{44} & 0 \\ 0 & Q_{55} \end{bmatrix}\left\{\begin{array}{c} \gamma_{yz}^{(0)} \\ \gamma_{xz}^{(0)} \end{array}\right\} \tag{5.5-10}$$

The nonzero thermal stress resultants are given by

$$\left\{\begin{array}{c} N_{xx}^T \\ N_{yy}^T \end{array}\right\} = \begin{bmatrix} Q_{11} & Q_{12} \\ Q_{12} & Q_{22} \end{bmatrix}\left\{\begin{array}{c} \alpha_1 \\ \alpha_2 \end{array}\right\}\int_{-\frac{h}{2}}^{\frac{h}{2}} \Delta T\, dz \tag{5.5-11a}$$

$$\left\{\begin{array}{c} M_{xx}^T \\ M_{yy}^T \end{array}\right\} = \begin{bmatrix} Q_{11} & Q_{12} \\ Q_{12} & Q_{22} \end{bmatrix}\left\{\begin{array}{c} \alpha_1 \\ \alpha_2 \end{array}\right\}\int_{-\frac{h}{2}}^{\frac{h}{2}} \Delta T z\, dz \tag{5.5-11b}$$

Single Generally Orthotropic Layer

For a single generally orthotropic layer (*i.e.*, the principal material coordinates do not coincide with those of the plate), the stiffnesses can be expressed in terms of the transformed coefficients $\bar{Q}_{ij}$ and thickness h. The nonzero stiffnesses are $(B_{ij} = 0)$

$$A_{ij} = \bar{Q}_{ij}h, \quad D_{ij} = \frac{\bar{Q}_{ij}h^3}{12}, \quad A_{44} = Kh\bar{Q}_{44}, \; A_{55} = Kh\bar{Q}_{55} \tag{5.5-12}$$

The plate constitutive equations are

$$\left\{\begin{array}{c} N_{xx} \\ N_{yy} \\ N_{xy} \end{array}\right\} = \begin{bmatrix} A_{11} & A_{12} & A_{16} \\ A_{12} & A_{22} & A_{26} \\ A_{16} & A_{26} & A_{66} \end{bmatrix}\left\{\begin{array}{c} \varepsilon_{xx}^{(0)} \\ \varepsilon_{yy}^{(0)} \\ \gamma_{xy}^{(0)} \end{array}\right\} \tag{5.5-13}$$

$$\left\{\begin{array}{c} M_{xx} \\ M_{yy} \\ M_{xy} \end{array}\right\} = \begin{bmatrix} D_{11} & D_{12} & D_{16} \\ D_{12} & D_{22} & D_{26} \\ D_{16} & D_{26} & D_{66} \end{bmatrix}\left\{\begin{array}{c} \varepsilon_{xx}^{(1)} \\ \varepsilon_{yy}^{(1)} \\ \gamma_{xy}^{(1)} \end{array}\right\} \tag{5.5-14}$$

$$\left\{\begin{array}{c} Q_y \\ Q_x \end{array}\right\} = K\begin{bmatrix} A_{44} & A_{45} \\ A_{45} & A_{55} \end{bmatrix}\left\{\begin{array}{c} \gamma_{yz}^{(0)} \\ \gamma_{xz}^{(0)} \end{array}\right\} \tag{5.5-15}$$

The thermal stress resultants for this case are given by

$$\left\{\begin{array}{c} N_{xx}^T \\ N_{yy}^T \\ N_{xy}^T \end{array}\right\} = \begin{bmatrix} \bar{Q}_{11} & \bar{Q}_{12} & \bar{Q}_{16} \\ \bar{Q}_{12} & \bar{Q}_{22} & \bar{Q}_{26} \\ \bar{Q}_{16} & \bar{Q}_{26} & \bar{Q}_{66} \end{bmatrix}\left\{\begin{array}{c} \alpha_{xx} \\ \alpha_{yy} \\ 2\alpha_{xy} \end{array}\right\}\int_{-\frac{h}{2}}^{\frac{h}{2}} \Delta T dz \tag{5.5-16}$$

Similar expression holds for $\{M^T\}$.

If the temperature increment is linear through the layer thickness, $\Delta T = T_0 + zT_1$, the thermal stress resultants have the form

$$\begin{Bmatrix} N_{xx}^T \\ N_{yy}^T \\ N_{xy}^T \end{Bmatrix} = \begin{bmatrix} \bar{Q}_{11} & \bar{Q}_{12} & \bar{Q}_{16} \\ \bar{Q}_{12} & \bar{Q}_{22} & \bar{Q}_{26} \\ \bar{Q}_{16} & \bar{Q}_{26} & \bar{Q}_{66} \end{bmatrix} \begin{Bmatrix} \alpha_{xx} \\ \alpha_{yy} \\ 2\alpha_{xy} \end{Bmatrix} T_0 h$$

$$\begin{Bmatrix} M_{xx}^T \\ M_{yy}^T \\ M_{xy}^T \end{Bmatrix} = \begin{bmatrix} \bar{Q}_{11} & \bar{Q}_{12} & \bar{Q}_{16} \\ \bar{Q}_{12} & \bar{Q}_{22} & \bar{Q}_{26} \\ \bar{Q}_{16} & \bar{Q}_{26} & \bar{Q}_{66} \end{bmatrix} \begin{Bmatrix} \alpha_{xx} \\ \alpha_{yy} \\ 2\alpha_{xy} \end{Bmatrix} \frac{T_1 h^3}{12} \qquad (5.5-17)$$

Single Anisotropic Layer

For a single anisotropic layer, the stiffnesses are expressed in terms of the coefficients C_{ij} and thickness h. The nonzero stiffnesses are ($B_{ij} = 0$)

$$A_{ij} = C_{ij}h, \quad D_{ij} = \frac{C_{ij}h^3}{12} \qquad (5.5-18)$$

for $i, j = 1, 2, 3, 4, 5$ and 6 [see Eq. (2.4–3a)]. The plate constitutive equations are the same as in Eqs. (13)–(16) with the plate stiffnesses given by Eq. (18).

Example 5.5–1:

The material properties of boron–epoxy material layers are

$$E_1 = 30 \times 10^6 \text{ psi}, \; E_2 = E_3 = 3 \times 10^6 \text{ psi}, \; G_{12} = G_{13} = 1.5 \times 10^6 \text{ psi}$$

$$G_{23} = 0.6 \times 10^6 \text{ psi}, \; \nu_{12} = 0.25, \; \nu_{13} = 0.25, \; \nu_{23} = 0.25 \qquad (5.5-19)$$

The matrix of elastic coefficients for the material is

$$[C] = \begin{bmatrix} 30.508 & 1.017 & 1.017 & 0 & 0 & 0 \\ 1.017 & 3.234 & 0.834 & 0 & 0 & 0 \\ 1.017 & 0.834 & 3.234 & 0 & 0 & 0 \\ 0 & 0 & 0 & 0.6 & 0 & 0 \\ 0 & 0 & 0 & 0 & 1.5 & 0 \\ 0 & 0 & 0 & 0 & 0 & 1.5 \end{bmatrix} \text{msi}$$

The plane stress-reduced elastic coefficient matrix in the material coordinates is

$$[Q] = \begin{bmatrix} 30.189 & 0.755 & 0 & 0 & 0 \\ 0.755 & 3.019 & 0 & 0 & 0 \\ 0 & 0 & 0.6 & 0 & 0 \\ 0 & 0 & 0 & 1.5 & 0 \\ 0 & 0 & 0 & 0 & 1.5 \end{bmatrix} \text{msi}$$

The transformed stiffness matrix $[\bar{Q}]$ for $\theta = 60°$ is given by

$$[\bar{Q}]_{60} = \begin{bmatrix} 4.993 & 5.573 & 0 & 0 & 3.101 \\ 5.573 & 18.578 & 0 & 0 & 8.664 \\ 0 & 0 & 1.275 & 0.390 & 0 \\ 0 & 0 & 0.390 & 0.825 & 0 \\ 3.101 & 8.664 & 0 & 0 & 6.318 \end{bmatrix} \text{ msi}$$

The laminate stiffnesses A_{ij} and D_{ij} for $i, j = 1, 2, 6$ may be computed using Eq. (12). The transverse shear stiffnesses A_{44}, A_{45}, and A_{55} are given by $A_{ij} = \bar{Q}_{ij} hK$ for $i, j = 4, 5$, where K is the shear correction coefficient.

Suppose that the thermal coefficients of expansion of the material are

$$\alpha_1 = 2.5 \times 10^{-6} \text{ in./in./°F} \; , \; \alpha_2 = 8.0 \times 10^{-6} \text{ in./in./°F} \qquad (5.5-20)$$

The transformed coefficients are

$$\begin{Bmatrix} \alpha_{xx} \\ \alpha_{yy} \\ 2\alpha_{xy} \end{Bmatrix}_{60°} = \begin{Bmatrix} 6.625 \\ 3.875 \\ -4.763 \end{Bmatrix} \times 10^{-6} \text{ in./in./°F}$$

5.5–3 Symmetric Laminates

When the material properties, locations, and lamination scheme are symmetric about the midplane, the laminate is called a *symmetric laminate.* If a laminate is not symmetric, it is said to be an *unsymmetric laminate.* Due to the symmetry of the layer material coefficients $\bar{Q}_{ij}^{(k)}$, distances z_k, and thicknesses h_k about the midplane of the laminate for *every layer*, the coupling stiffnesses B_{ij} are zero for symmetric laminates (see Figure 5.5–5). The elimination of the coupling between bending and extension simplifies the governing equations. When the strain–displacement equations are linear, the equations governing the in-plane deformation can be uncoupled from those governing bending of symmetric laminates. Further, if there are no applied in-plane forces or displacements, the in-plane deformation (*i.e.*, strains) will be zero, and only the bending equations must be analyzed. From production point of view, symmetric laminates do not have the tendency to twist from the thermally induced contractions that occur during cooling following the curing process.

The force and moment resultants for a symmetric laminate, in general, have the same form as the generally orthotropic single-layer plates [see Eqs. (13)–15)]. For certain special cases of symmetric laminates, the relations between strains and resultants can be further simplified, as explained next.

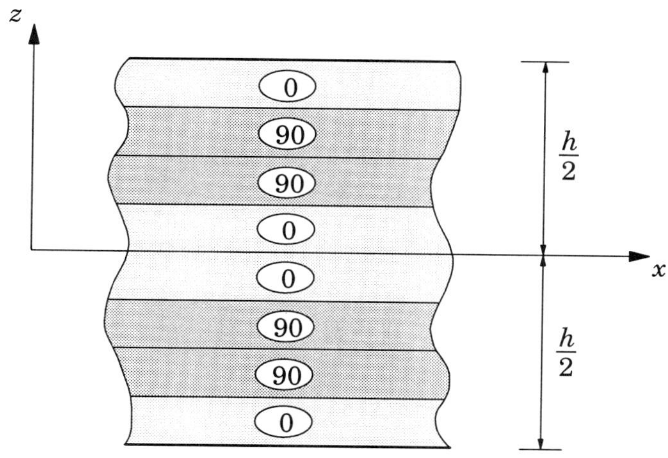

Figure 5.5–5: A symmetric cross-ply laminate.

Symmetric Laminates with Multiple Isotropic Layers

When isotropic layers of possibly different material properties and thicknesses are arranged symmetrically from both a geometric and a material property standpoint, the resulting laminate will have the following laminate constitutive equations for the classical or first-order theories:

$$\begin{Bmatrix} N_{xx} \\ N_{yy} \\ N_{xy} \end{Bmatrix} = \begin{bmatrix} A_{11} & A_{12} & 0 \\ A_{12} & A_{11} & 0 \\ 0 & 0 & A_{66} \end{bmatrix} \begin{Bmatrix} \varepsilon_{xx}^{(0)} \\ \varepsilon_{yy}^{(0)} \\ \gamma_{xy}^{(0)} \end{Bmatrix} \tag{5.5 - 21a}$$

$$\begin{Bmatrix} M_{xx} \\ M_{yy} \\ M_{xy} \end{Bmatrix} = \begin{bmatrix} D_{11} & D_{12} & 0 \\ D_{12} & D_{11} & 0 \\ 0 & 0 & D_{66} \end{bmatrix} \begin{Bmatrix} \varepsilon_{xx}^{(1)} \\ \varepsilon_{yy}^{(1)} \\ \gamma_{xy}^{(1)} \end{Bmatrix} \tag{5.5 - 21b}$$

$$\begin{Bmatrix} Q_y \\ Q_x \end{Bmatrix} = K \begin{bmatrix} A_{44} & 0 \\ 0 & A_{55} \end{bmatrix} \begin{Bmatrix} \gamma_{yz}^{(0)} \\ \gamma_{xz}^{(0)} \end{Bmatrix} \tag{5.5 - 21c}$$

where the laminate stiffnesses A_{ij} and D_{ij} are defined by Eqs. (5.3–38) and (5.4–19) with

$$\bar{Q}_{11}^{(k)} = \bar{Q}_{22}^{(k)} = \frac{E^k}{1-\nu_k^2}\,, \quad \bar{Q}_{16}^{(k)} = \bar{Q}_{26}^{(k)} = 0$$

$$\bar{Q}_{12}^{(k)} = \frac{\nu_k E^k}{1-\nu_k^2}\,, \quad \bar{Q}_{44}^{(k)} = \bar{Q}_{55}^{(k)} = \bar{Q}_{66}^{(k)} = \frac{E^k}{2(1+\nu_k)} \tag{5.5 - 22}$$

The thermal stress resultants for this case are given by

$$\left\{\begin{matrix} N_{xx}^T \\ N_{yy}^T \end{matrix}\right\} = \sum_{k=1}^{L} \int_{z_k}^{z_{k+1}} \begin{bmatrix} \bar{Q}_{11} & \bar{Q}_{12} \\ \bar{Q}_{12} & \bar{Q}_{22} \end{bmatrix}^{(k)} \left\{\begin{matrix} \alpha_{xx} \\ \alpha_{yy} \end{matrix}\right\} \Delta T \, dz \tag{5.5 – 23}$$

and similar expression holds for $\{M^T\}$.

If $\Delta T = T_0 + zT_1$, then Eq. (23) can be written as

$$\left\{\begin{matrix} N_{xx}^T \\ N_{yy}^T \end{matrix}\right\} = T_0 \begin{bmatrix} A_{11} & A_{12} \\ A_{12} & A_{22} \end{bmatrix} \left\{\begin{matrix} \alpha_{xx} \\ \alpha_{yy} \end{matrix}\right\}, \quad \left\{\begin{matrix} M_{xx}^T \\ M_{yy}^T \end{matrix}\right\} = T_1 \begin{bmatrix} D_{11} & D_{12} \\ D_{12} & D_{22} \end{bmatrix} \left\{\begin{matrix} \alpha_{xx} \\ \alpha_{yy} \end{matrix}\right\} \tag{5.5 – 24}$$

Symmetric Laminates with Multiple Specially Orthotropic Layers

A laminate composed of multiple specially orthotropic layers that are symmetrically disposed, both from a material and geometric properties standpoint, about the midplane of the laminate do not exhibit coupling between bending and extension *i.e.*, $B_{ij} = 0$. The laminate constitutive equations are again given by Eqs. (21a,b,c), where the laminate stiffnesses A_{ij} and D_{ij} are defined by Eqs. (5.3–38) and (5.4–19) with

$$\bar{Q}_{11}^{(k)} = \frac{E_1^k}{1 - \nu_{12}^k \nu_{21}^k}, \quad \bar{Q}_{12}^{(k)} = \frac{\nu_{21}^k E_1^k}{1 - \nu_{12}^k \nu_{21}^k}, \quad \bar{Q}_{22}^{(k)} = \frac{E_2^k}{1 - \nu_{12}^k \nu_{21}^k}$$

$$\bar{Q}_{16}^{(k)} = 0, \ \bar{Q}_{26}^{(k)} = 0, \ \bar{Q}_{66}^{(k)} = G_{12}^k, \ \bar{Q}_{44}^{(k)} = G_{23}^k, \ \bar{Q}_{55}^{(k)} = G_{13}^k \tag{5.5 – 25}$$

Such laminates are also called *specially orthotropic laminates.* The thermal stress resultants have the same form as those given in Eq. (23).

A common example of specially orthotropic laminates is provided by the *regular symmetric cross-ply laminates,* which consist of laminae of the same thickness and material properties but have their major principal material coordinates (*i.e.*, x_1 and x_2) alternating at 0° and 90° to the laminate axes x and y: $(0/90/0/90/\cdots)$. The regular symmetric cross-ply laminates necessarily contain an odd number of layers; otherwise, they are not symmetric. Of course, a general symmetric cross-ply laminate can have either an even or odd number of layers: (0/90/0/90/90/0/90/0) or (0/90/90/0/0/90/90/0) (see Figure 5.5–5).

Symmetric Laminates with Multiple Generally Orthotropic Layers

Laminates can be composed of generally orthotropic layers whose major principal material directions are aligned with the laminate axes at an angle θ degrees. If the thicknesses, locations, and material properties of the layers are symmetric about the midplane of the laminate, the coupling between bending and extension is zero, $B_{ij} = 0$, and the laminate constitutive equations are given by Eqs. (13)–(15). Note that the coupling between normal forces and shearing strain, shearing force and normal strains, normal moments and twist, and twisting moment and normal curvatures is not zero for these laminates (*i.e.*, A_{16}, A_{26}, D_{16}, and D_{26} are not zero). An example of a general symmetric laminate with generally orthotropic laminae is provided by $(30/{-60}_3/15_5/{-60}_3/30)$, where the subscript denotes the number of layers of the same orientation and thickness.

Regular symmetric angle-ply laminates are those which have an *odd* number of orthotropic laminae of equal thicknesses and alternating orientations: $(\alpha/{-\alpha}/\alpha/{-\alpha}/\alpha/\cdots)$, $0° < \alpha < 90°$ (see Figure 5.5–6). A general symmetric angle-ply laminate has the form $(\theta/\beta/\gamma/\cdots)_s$, where θ, β, and γ can take any values between $-90°$ and $90°$, and each layer can have any thickness, but they should be symmetrically placed about the midplane. It can be shown that the stiffnesses A_{16}, A_{26}, D_{16}, and D_{26} of a regular symmetric angle-ply laminate are the largest when the number of layers N is equal to 3, and they decrease in proportion to $1/N$ as N increases. Thus, for symmetric angle-ply laminates with many layers, the values of A_{16}, A_{26}, D_{16}, and D_{26} can be quite small compared to other A_{ij} and D_{ij}.

A laminate composed of multiple anisotropic layers that are symmetrically disposed about the midplane of the laminate do not have any stiffness simplification other than $B_{ij} = 0$, which holds for *all* symmetric laminates. Stiffnesses A_{16}, A_{26}, D_{16}, and D_{26} are not zero, and they do not necessarily go to zero as the number of layers is increased.

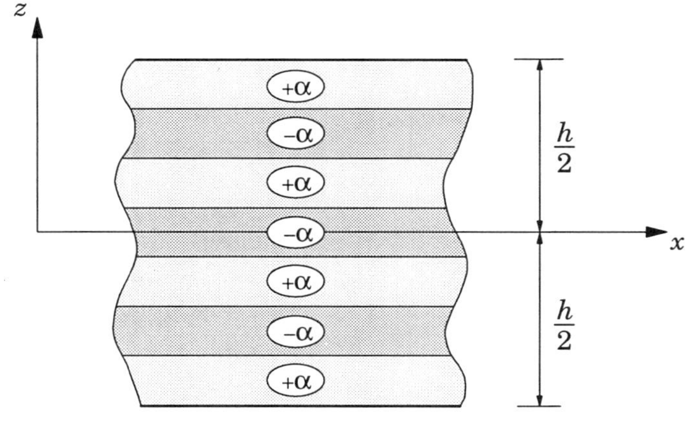

Figure 5.5–6: A symmetric angle-ply laminate.

In general, symmetric laminates are preferred wherever they meet the application requirements. Symmetric laminates are much easier to analyze than general or unsymmetric laminates. Further, symmetric laminates do not have a tendency to twist due to thermally induced contractions that occur during cooling following the curing process.

Example 5.5–2:

A general symmetric laminate $(30/0/90/-45)_s$ of total thickness 1 in. and made of boron–epoxy layers [see Eqs. (19) and (20) for material properties] has the following laminate stiffnesses:

$$[A] = \begin{bmatrix} 15.491 & 3.565 & 0.468 \\ 3.565 & 12.095 & -0.923 \\ 0.468 & -0.923 & 4.311 \end{bmatrix} \text{msi-in.}, \quad [D] = \begin{bmatrix} 1.683 & 0.303 & 0.409 \\ 0.303 & 0.604 & 0.141 \\ 0.409 & 0.141 & 0.366 \end{bmatrix} \text{msi-in.}^3$$

The transverse shear stiffnesses are

$$A_{44} = 0.9938 \text{ msi-in.}, \quad A_{45} = -0.0151 \text{ msi-in.}, \quad A_{55} = 1.1063 \text{ msi-in.}$$

The thermal stress resultants are $(T_0 \neq 0, T_1 = 0)$

$$\begin{Bmatrix} N_{xx}^T \\ N_{yy}^T \\ N_{xy}^T \end{Bmatrix} = \begin{Bmatrix} 57.241 \\ 50.307 \\ -0.929 \end{Bmatrix} T_0 \text{ msi/in.}, \quad \begin{Bmatrix} M_{xx}^T \\ M_{yy}^T \\ M_{xy}^T \end{Bmatrix} = \begin{Bmatrix} 0 \\ 0 \\ 0 \end{Bmatrix} \text{lb–in.}^2$$

A symmetric cross-ply laminate $(0/90/0/90)_s$ of boron–epoxy layers has the stiffnesses

$$[A] = \begin{bmatrix} 16.604 & 0.755 & 0 \\ 0.755 & 16.604 & 0 \\ 0 & 0 & 1.5 \end{bmatrix} \text{msi-in.}, \quad [D] = \begin{bmatrix} 1.808 & 0.063 & 0 \\ 0.063 & 0.959 & 0 \\ 0 & 0 & 0.125 \end{bmatrix} \text{msi-in.}^3$$

The transverse shear stiffnesses are

$$A_{44} = 1.05 \text{ msi-in.}, \quad A_{45} = 0.0 \text{ msi-in.}, \quad A_{55} = 1.05 \text{ msi-in.}$$

Note that the cross-ply laminate considered here is equivalent to (0/90/0/90/0/90/0) with all layers except the middle layer have a thickness of $h/8$ and the middle layer has a thickness of $h/4$, where h is the total thickness of the laminate.

A symmetric angle-ply laminate $(30/-30/45/-45)_s$ of boron–epoxy layers has the stiffnesses

$$[A] = \begin{bmatrix} 14.379 & 6.376 & 0 \\ 6.376 & 7.586 & 0 \\ 0 & 0 & 7.122 \end{bmatrix} \text{msi-in.}, \quad [D] = \begin{bmatrix} 1.461 & 0.481 & 0.256 \\ 0.481 & 0.470 & 0.126 \\ 0.256 & 0.126 & 0.543 \end{bmatrix} \text{msi-in.}^3$$

The transverse shear stiffnesses are

$$A_{44} = 0.9375 \text{ msi-in.}, \quad A_{45} = 0.0 \text{ msi-in.}, \quad A_{55} = 1.1625 \text{ msi-in.}$$

Example 5.5–3:

Consider a symmetric laminate $(0/90)_s$ made of boron–epoxy layers of thickness 0.005 in. Suppose that the laminate is subjected to loads such that it experiences only nonzero strain of $\varepsilon_{xx}^{0} = 10^3\mu$ in./in. We wish to determine the forces and moment resultants.

The only nonzero strain is $\varepsilon_{xx} = \varepsilon_{xx}^{(0)}$. Hence the force resultants in the laminate are given by

$$\begin{Bmatrix} N_{xx} \\ N_{yy} \\ N_{xy} \end{Bmatrix} = \begin{bmatrix} A_{11} & A_{12} & 0 \\ A_{12} & A_{22} & 0 \\ 0 & 0 & A_{66} \end{bmatrix} \begin{Bmatrix} \varepsilon_{xx}^{(0)} \\ \varepsilon_{yy}^{(0)} \\ \gamma_{xy}^{(0)} \end{Bmatrix}$$
$$= \begin{bmatrix} 0.3321 & 0.0151 & 0 \\ 0.0151 & 0.3321 & 0 \\ 0 & 0 & 0.03 \end{bmatrix} \begin{Bmatrix} 1{,}000 \\ 0 \\ 0 \end{Bmatrix} = \begin{Bmatrix} 332.1 \\ 15.1 \\ 0 \end{Bmatrix} \text{ lbs/in.}$$

All moments will be zero on account of the fact that there are no bending strains and the coupling stiffnesses B_{ij} are zero.

Now suppose that the laminate is subjected to loads such that it experiences only nonzero strain of $\varepsilon_{xx}^{(1)} = 0.1$. Hence, the only nonzero strain is $\varepsilon_{xx} = \varepsilon_{xx}^{(1)} z$. Then the force resultants are zero, and the moment resultants are given by

$$\begin{Bmatrix} M_{xx} \\ M_{yy} \\ M_{xy} \end{Bmatrix} = \begin{bmatrix} D_{11} & D_{12} & 0 \\ D_{12} & D_{22} & 0 \\ 0 & 0 & D_{66} \end{bmatrix} \begin{Bmatrix} \varepsilon_{xx}^{(1)} \\ \varepsilon_{yy}^{(1)} \\ \gamma_{xy}^{(1)} \end{Bmatrix}$$
$$= \begin{bmatrix} 17.862 & 0.503 & 0 \\ 0.503 & 4.277 & 0 \\ 0 & 0 & 1 \end{bmatrix} \begin{Bmatrix} 0.1 \\ 0 \\ 0 \end{Bmatrix} = \begin{Bmatrix} 1.7862 \\ 0.0503 \\ 0 \end{Bmatrix} \text{ lb-in.}$$

5.5–4 Antisymmetric Laminates

Although symmetric laminates are more desirable from an analysis standpoint, they may not meet the design requirements in some applications. For example, a heat shield receives heat from one side, and thus requires nonsymmetric laminates to effectively shield the heat. Another example that requires coupling is provided by turbine blades with pretwist. Moreover, the shear stiffness of laminates can be increased by orienting the layers at angle to the laminate coordinates.

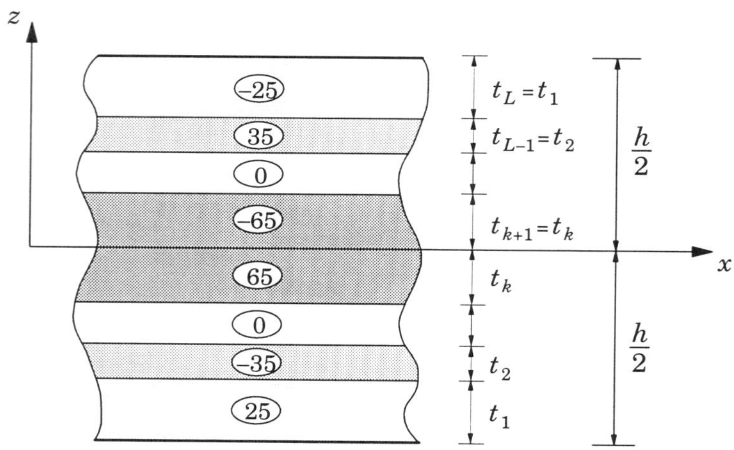

Figure 5.5–7: An antisymmetric laminate.

The general class of antisymmetric laminates must have an *even* number of orthotropic laminae if adjacent laminae have equal thicknesses and alternating orientations: $(\theta/-\theta)$, $0° \leq \theta \leq 90°$. Due to the antisymmetry of the lamination scheme (see Figure 5.5–7) but symmetry of the thicknesses of each pair of layers, this class of antisymmetric laminates has the feature that $A_{16} = A_{26} = D_{16} = D_{26} = 0$. The coupling stiffnesses B_{ij} are not all zero; they go to zero as the number of layers is increased. For a general antisymmetric laminate, the relations between the stress resultants and the strains are given by

$$\left\{\begin{array}{c} N_{xx} \\ N_{yy} \\ N_{xy} \end{array}\right\} = \begin{bmatrix} A_{11} & A_{12} & 0 \\ A_{12} & A_{22} & 0 \\ 0 & 0 & A_{66} \end{bmatrix} \left\{\begin{array}{c} \varepsilon_{xx}^{(0)} \\ \varepsilon_{yy}^{(0)} \\ \gamma_{xy}^{(0)} \end{array}\right\} + \begin{bmatrix} B_{11} & B_{12} & B_{16} \\ B_{12} & B_{22} & B_{26} \\ B_{16} & B_{26} & B_{66} \end{bmatrix} \left\{\begin{array}{c} \varepsilon_{xx}^{(1)} \\ \varepsilon_{yy}^{(1)} \\ \gamma_{xy}^{(1)} \end{array}\right\} \tag{5.5 – 26a}$$

$$\left\{\begin{array}{c} M_{xx} \\ M_{yy} \\ M_{xy} \end{array}\right\} = \begin{bmatrix} B_{11} & B_{12} & B_{16} \\ B_{12} & B_{22} & B_{26} \\ B_{16} & B_{26} & B_{66} \end{bmatrix} \left\{\begin{array}{c} \varepsilon_{xx}^{(0)} \\ \varepsilon_{yy}^{(0)} \\ \gamma_{xy}^{(0)} \end{array}\right\} + \begin{bmatrix} D_{11} & D_{12} & 0 \\ D_{12} & D_{22} & 0 \\ 0 & 0 & D_{66} \end{bmatrix} \left\{\begin{array}{c} \varepsilon_{xx}^{(1)} \\ \varepsilon_{yy}^{(1)} \\ \gamma_{xy}^{(1)} \end{array}\right\} \tag{5.5 – 26b}$$

$$\left\{\begin{array}{c} Q_y \\ Q_x \end{array}\right\} = K \begin{bmatrix} A_{44} & 0 \\ 0 & A_{55} \end{bmatrix} \left\{\begin{array}{c} \gamma_{yz}^{(0)} \\ \gamma_{xz}^{(0)} \end{array}\right\} \tag{5.5 – 26c}$$

The thermal force resultants are given by

$$\left\{\begin{array}{c} N_{xx}^T \\ N_{yy}^T \\ N_{xy}^T \end{array}\right\} = \sum_{k=1}^{L} \int_{z_k}^{z_{k+1}} \begin{bmatrix} \bar{Q}_{11} & \bar{Q}_{12} & \bar{Q}_{16} \\ \bar{Q}_{12} & \bar{Q}_{22} & \bar{Q}_{26} \\ \bar{Q}_{16} & \bar{Q}_{26} & \bar{Q}_{66} \end{bmatrix}^{(k)} \left\{\begin{array}{c} \alpha_{xx} \\ \alpha_{yy} \\ 2\alpha_{xy} \end{array}\right\}^{(k)} \Delta T dz \qquad (5.5-27)$$

Similar expression holds for $\{M^T\}$.

In the following pages, we discuss some special cases of the class of antisymmetric laminates described above (*i.e.*, laminates that have an even number of orthotropic laminae, each pair having equal thicknesses and alternating orientations).

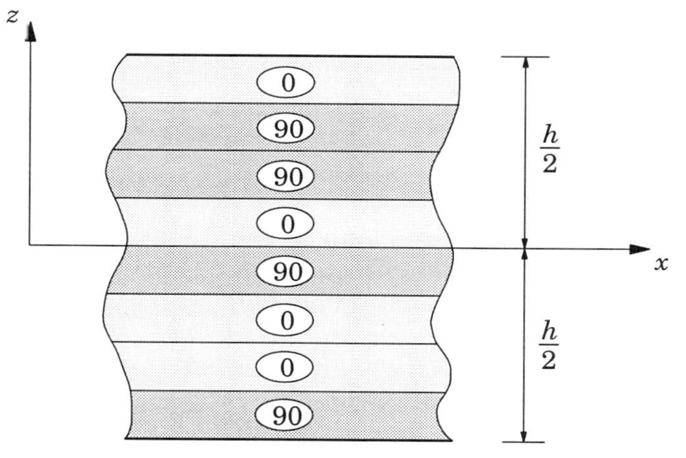

Figure 5.5–8: An antisymmetric cross-ply laminate.

Antisymmetric cross-ply Laminates

A special case of antisymmetric laminates are those which have an even number of orthotropic layers with principal material directions alternating at 0° to 90° to the laminate axes. Such laminates are called *antisymmetric cross-ply laminates.* Examples of antisymmetric cross-ply laminates are (0/90/0/90/ ···) with all layers of the same thickness, and (0/90/90/0/0/90) with layers of the thicknesses ($h_1/h_2/h_3/h_3/h_2/h_1$). Note that for every 0° layer of a given thickness and location, there is a 90° layer of the same thickness and location on the other side of the midplane (see Figure 5.5–8). For these laminates, the coupling stiffnesses B_{ij} have the properties

$$B_{22} = -B_{11}, \quad \text{and all other} \quad B_{ij} = 0 \qquad (5.5-28)$$

The relations between the stress resultants and the strains are

$$\begin{Bmatrix} N_{xx} \\ N_{yy} \\ N_{xy} \end{Bmatrix} = \begin{bmatrix} A_{11} & A_{12} & 0 \\ A_{12} & A_{22} & 0 \\ 0 & 0 & A_{66} \end{bmatrix} \begin{Bmatrix} \varepsilon_{xx}^{(0)} \\ \varepsilon_{yy}^{(0)} \\ \gamma_{xy}^{(0)} \end{Bmatrix} + \begin{bmatrix} B_{11} & 0 & 0 \\ 0 & -B_{11} & 0 \\ 0 & 0 & 0 \end{bmatrix} \begin{Bmatrix} \varepsilon_{xx}^{(1)} \\ \varepsilon_{yy}^{(1)} \\ \gamma_{xy}^{(1)} \end{Bmatrix} \tag{5.5-29a}$$

$$\begin{Bmatrix} M_{xx} \\ M_{yy} \\ M_{xy} \end{Bmatrix} = \begin{bmatrix} B_{11} & 0 & 0 \\ 0 & -B_{11} & 0 \\ 0 & 0 & 0 \end{bmatrix} \begin{Bmatrix} \varepsilon_{xx}^{(0)} \\ \varepsilon_{yy}^{(0)} \\ \gamma_{xy}^{(0)} \end{Bmatrix} + \begin{bmatrix} D_{11} & D_{12} & 0 \\ D_{12} & D_{22} & 0 \\ 0 & 0 & D_{66} \end{bmatrix} \begin{Bmatrix} \varepsilon_{xx}^{(1)} \\ \varepsilon_{yy}^{(1)} \\ \gamma_{xy}^{(1)} \end{Bmatrix} \tag{5.5-29b}$$

$$\begin{Bmatrix} Q_y \\ Q_x \end{Bmatrix} = K \begin{bmatrix} A_{44} & 0 \\ 0 & A_{55} \end{bmatrix} \begin{Bmatrix} \gamma_{yz}^{(0)} \\ \gamma_{xz}^{(0)} \end{Bmatrix} \tag{5.5-30}$$

A *regular antisymmetric cross-ply laminate* is one that has an even number of layers of equal thickness and the same material properties and which have alternating 0° and 90° orientations. For these laminates, the coupling coefficient B_{11} approaches zero as the number of layers is increased.

Antisymmetric Angle-ply Laminates

An antisymmetric angle-ply laminate has an even number of orthotropic layers with principal material directions alternating at θ degrees to the laminate axes on one side of the midplane and corresponding equal thickness laminae oriented at $-\theta$ degrees on the other side. When $\theta = 0$, $-\theta$ should be interpreted as 90 or vice versa. A *regular antisymmetric angle-ply laminate* is one that has an even number of layers of equal thickness and material properties. An example is given by (–45/40/–15/15/–40/45).

For antisymmetric angle-ply laminates without 90° layers, the stiffnesses can be simplified as

$$A_{16} = A_{26} = D_{16} = D_{26} = 0; \quad B_{11} = B_{22} = B_{12} = B_{66} = 0 \tag{5.5-31}$$

The relations between the stress resultants and the strains are

$$\begin{Bmatrix} N_{xx} \\ N_{yy} \\ N_{xy} \end{Bmatrix} = \begin{bmatrix} A_{11} & A_{12} & 0 \\ A_{12} & A_{22} & 0 \\ 0 & 0 & A_{66} \end{bmatrix} \begin{Bmatrix} \varepsilon_{xx}^{(0)} \\ \varepsilon_{yy}^{(0)} \\ \gamma_{xy}^{(0)} \end{Bmatrix} + \begin{bmatrix} 0 & 0 & B_{16} \\ 0 & 0 & B_{26} \\ B_{16} & B_{26} & 0 \end{bmatrix} \begin{Bmatrix} \varepsilon_{xx}^{(1)} \\ \varepsilon_{yy}^{(1)} \\ \gamma_{xy}^{(1)} \end{Bmatrix} \tag{5.5-32}$$

$$\begin{Bmatrix} M_{xx} \\ M_{yy} \\ M_{xy} \end{Bmatrix} = \begin{bmatrix} 0 & 0 & B_{16} \\ 0 & 0 & B_{26} \\ B_{16} & B_{26} & 0 \end{bmatrix} \begin{Bmatrix} \varepsilon_{xx}^{(0)} \\ \varepsilon_{yy}^{(0)} \\ \gamma_{xy}^{(0)} \end{Bmatrix} + \begin{bmatrix} D_{11} & D_{12} & 0 \\ D_{12} & D_{22} & 0 \\ 0 & 0 & D_{66} \end{bmatrix} \begin{Bmatrix} \varepsilon_{xx}^{(1)} \\ \varepsilon_{yy}^{(1)} \\ \gamma_{xy}^{(1)} \end{Bmatrix} \tag{5.5-33}$$

$$\begin{Bmatrix} Q_y \\ Q_x \end{Bmatrix} = K \begin{bmatrix} A_{44} & 0 \\ 0 & A_{55} \end{bmatrix} \begin{Bmatrix} \gamma_{yz}^{(0)} \\ \gamma_{xz}^{(0)} \end{Bmatrix} \tag{5.5-34}$$

For a fixed laminate thickness, the stiffnesses B_{16} and B_{26} go to zero as the number of layers in the laminate increases.

Example 5.5–4:

A regular antisymmetric cross-ply laminate (0/90/0/90/0/90/0/90) of boron–epoxy layers has the laminate stiffnesses

$$[A] = \begin{bmatrix} 16.604 & 0.755 & 0 \\ 0.755 & 16.604 & 0 \\ 0 & 0 & 1.5 \end{bmatrix} \text{ msi-in.}, \quad [B] = \begin{bmatrix} -0.849 & 0 & 0 \\ 0 & 0.849 & 0 \\ 0 & 0 & 0 \end{bmatrix} \text{ msi-in.}^2$$

$$[D] = \begin{bmatrix} 1.384 & 0.063 & 0 \\ 0.063 & 1.384 & 0 \\ 0 & 0 & 0.125 \end{bmatrix} \text{ msi-in.}^3, \quad \begin{Bmatrix} A_{44} \\ A_{45} \\ A_{55} \end{Bmatrix} = \begin{Bmatrix} 1.050 \\ 0 \\ 1.050 \end{Bmatrix} \text{ msi-in.}$$

Note that if the same 0° and 90° layers are positioned differently, say (0/90/90/0/90/0/0/90), then the coefficients B_{ij} would vanish (why?).

An antisymmetric angle-ply laminate (–45/45/30/0/0/–30/–45/45) of boron–epoxy layers has the laminate stiffnesses (units are obvious)

$$[A] = \begin{bmatrix} 17.281 & 5.172 & 0 \\ 5.172 & 7.093 & 0 \\ 0 & 0 & 5.917 \end{bmatrix}, \quad [B] = \begin{bmatrix} 0 & 0 & -0.194 \\ 0 & 0 & 0.067 \\ -0.194 & 0.067 & 0 \end{bmatrix}$$

$$[D] = \begin{bmatrix} 0.951 & 0.575 & 0 \\ 0.575 & 0.792 & 0 \\ 0 & 0 & 0.637 \end{bmatrix}, \quad \begin{Bmatrix} A_{44} \\ A_{45} \\ A_{55} \end{Bmatrix} = \begin{Bmatrix} 0.881 \\ 0 \\ 1.219 \end{Bmatrix}$$

A general antisymmetric laminate $(30/0/90/{-45})_{as} \equiv (30/0/90/{-45}/45/0/90/{-30})$ of total thickness 1 in. and composed of boron–epoxy layers has the following laminate stiffnesses and thermal resultants:

$$[A] = \begin{bmatrix} 15.491 & 3.565 & 0 \\ 3.565 & 12.095 & 0 \\ 0 & 0 & 4.311 \end{bmatrix}, \quad [B] = \begin{bmatrix} -0.425 & 0 & -0.842 \\ 0 & 0.425 & -0.233 \\ -0.842 & -0.233 & 4.311 \end{bmatrix}$$

$$[D] = \begin{bmatrix} 1.470 & 0.303 & 0 \\ 0.303 & 0.816 & 0 \\ 0 & 0 & 0.366 \end{bmatrix}, \quad \begin{Bmatrix} A_{44} \\ A_{45} \\ A_{55} \end{Bmatrix} = \begin{Bmatrix} 0.9938 \\ 0 \\ 1.1063 \end{Bmatrix}$$

$$\begin{Bmatrix} N_{xx}^T \\ N_{yy}^T \\ N_{xy}^T \end{Bmatrix} = \begin{Bmatrix} 5573.6 \\ 5958.5 \\ 0 \end{Bmatrix} \text{ msi–in.}, \quad \begin{Bmatrix} M_{xx}^T \\ M_{yy}^T \\ M_{xy}^T \end{Bmatrix} = \begin{Bmatrix} 48.113 \\ -48.113 \\ 121.78 \end{Bmatrix} \text{ msi–in.}^2$$

5.5–5 Balanced and Quasi-isotropic Laminates

A laminate is said to be *balanced* if for every layer in the laminate there exists, somewhere in the laminate, another layer with identical material and thickness but *opposite* fiber orientation. The two layers are not necessarily symmetrically located with respect to the midplane. Thus, the unsymmetric laminate $(\pm 35/0)_T$=(35/–35/0) as well as the symmetric laminate $(\pm 35/0)_s$ are balanced laminates. The characteristic feature of any balanced laminate is that the in-plane shear stiffnesses A_{16} and A_{26} are zero. The reason is that Q_{16} and Q_{26} from opposite orientations of the pair of layers are of opposite sign and therefore the net contribution from the pair to A_{16} and A_{26} is zero:

$$(Q_{16})_\theta = -(Q_{16})_{(-\theta)}, \quad (Q_{26})_\theta = -(Q_{26})_{(-\theta)}$$

For a general balanced laminate, the laminate constitutive relations are not that much simpler than for a general laminate. However, for a symmetric balanced laminate they are given by Eqs. (13)–(15) with $A_{16} = A_{26} = 0$.

Laminates consisting of three or more orthotropic laminae of identical material and thickness which are oriented at the same angle relative to adjacent laminae exhibit in-plane isotropy in the sense that $A_{11} = A_{22}$, $A_{66} = (A_{11} - A_{12})/2$, and $A_{16} = A_{26} = 0$. Such laminates are called *quasi–isotropic* laminates. Examples of quasi–isotropic laminates are provided by (90/45/0/–45) and (60/0/–60) (see Example 5.3–2). When the bending–stretching coupling coefficients are zero, the relations between force resultants and membrane strains are the same as those for isotropic plates. The stress resultants are given by

$$\begin{Bmatrix} N_{xx} \\ N_{yy} \\ N_{xy} \end{Bmatrix} = \begin{bmatrix} A_{11} & A_{12} & 0 \\ A_{12} & A_{11} & 0 \\ 0 & 0 & (A_{11} - A_{12})/2 \end{bmatrix} \begin{Bmatrix} \varepsilon_{xx}^{(0)} \\ \varepsilon_{yy}^{(0)} \\ \gamma_{xy}^{(0)} \end{Bmatrix}$$

Exercises

5.1 Suppose that the displacements (u, v, w) along the three coordinate axes (x, y, z) in a laminated beam can be expressed as

$$\begin{aligned} u(x,z) &= u_0(x) + z\left[c_0 \frac{dw_0}{dx} + c_1\phi(x)\right] \\ v(x,z) &= 0 \\ w(x,z) &= w_0(x) \end{aligned} \tag{1}$$

where (u_0, w_0) denote the displacements of a point $(x, y, 0)$ along the x and z directions, respectively, and ϕ denotes the rotation of a transverse normal about the $y-$axis. Show that the nonzero linear strains are given by

$$\varepsilon_{xx} = \varepsilon_{xx}^{(0)} + z\varepsilon_{xx}^{(1)}, \quad 2\varepsilon_{xz} = 2\varepsilon_{xz}^{(0)} \tag{2a}$$

where

$$\varepsilon_{xx}^{(0)} = \frac{du_0}{dx}\ ,\ \varepsilon_{x}^{(1)} = c_0\frac{d^2w_0}{dx^2} + c_1\frac{d\phi}{dx}\ ,\ 2\varepsilon_{xz}^{(0)} = (1+c_0)\frac{dw_0}{dx} + c_1\phi \tag{2b}$$

5.2 (*Continuation of Exercise 1*) Use the principle of virtual displacements to derive the equations of equilibrium and the natural and essential boundary conditions associated with the displacement field of Exercise 5.1, when the beam is subjected to axial distributed load $p(x)$ and transverse distributed load $q(x)$. In particular, show that

$$\begin{aligned} \delta u_0 :&\quad \frac{dN_{xx}}{dx} + p = 0 \\ \delta\phi :&\quad \frac{d}{dx}(c_1 M_{xx}) - c_1 Q_x = 0 \\ \delta w_0 :&\quad \frac{d^2}{dx^2}(c_0 M_{xx}) - (1+c_0)\frac{dQ_x}{dx} - q = 0 \end{aligned} \tag{3}$$

and the boundary conditions are of the form

$$\begin{aligned} N_{xx} &\quad \text{or} \quad u_0 \\ c_1 M_{xx} &\quad \text{or} \quad \phi \\ -\frac{d}{dx}(c_0 M_{xx}) + (1+c_0)Q_x &\quad \text{or} \quad w_0 \\ c_0 M_{xx} &\quad \text{or} \quad \frac{dw_0}{dx} \end{aligned} \tag{4}$$

where

$$N_{xx} = \int_A \sigma_{xx}\, dA,\quad M_{xx} = \int_A \sigma_{xx} z\, dA,\quad Q_x = \int_A \sigma_{xz}\, dA \tag{5}$$

Note that the displacement field (1), hence the equations of equilibrium (3), contain those of the classical (Euler–Bernoulli) beam theory ($c_0 = -1,\ c_1 = 0$) and the first-order (Timoshenko) beam theory ($c_0 = 0,\ c_1 = 1$).

5.3 (*Continuation of Exercise 1*) Assume linear elastic constitutive behavior and show that the laminated beam's constitutive equations are given by

$$\left\{\begin{matrix} N_{xx} \\ \\ M_{xx} \end{matrix}\right\} = \begin{bmatrix} A_{11} & B_{11} \\ B_{11} & D_{11} \end{bmatrix} \left\{\begin{matrix} \varepsilon_{xx}^{(0)} \\ \\ \varepsilon_{xx}^{(1)} \end{matrix}\right\},\quad Q_x = 2A_{55}\varepsilon_{xz}^{(0)} \tag{6a}$$

where

$$(A_{11}, B_{11}, D_{11}) = \int_A E_1(1, z, z^2)\, dA,\quad A_{55} = \int_A G_{13}\, dA \tag{6b}$$

5.4 The 3-D equilibrium equations of a kth layer, in the absence of body forces, can be expressed in index notation as

$$\frac{\partial \sigma_{\alpha\beta}^k}{\partial x_\beta} + \frac{\partial \sigma_{\alpha 3}^k}{\partial x_3} = 0 \tag{1}$$

$$\frac{\partial \sigma_{\alpha 3}^k}{\partial x_\alpha} + \frac{\partial \sigma_{33}^k}{\partial x_3} = 0 \tag{2}$$

where summation on repeated subscripts ($\alpha, \beta = 1, 2$) is implied. Integrate the equations over the thickness (z_k, z_{k+1}) with respect to $z = x_3$ to obtain:

$$\frac{\partial N_{\alpha\beta}^{(k)}}{\partial x_\beta} + \sigma_{\alpha 3}^{(k+1)} - \sigma_{\alpha 3}^{(k)} = 0 \tag{3}$$

$$\frac{\partial Q_\alpha^{(k)}}{\partial x_\alpha} + \sigma_{33}^{(k+1)} - \sigma_{33}^{(k)} = 0 \tag{4}$$

for $k = 1, 2, \cdots, N$ and $\alpha, \beta = 1, 2$ $(x_1 = x, x_2 = y, x_3 = z)$, where N is the total number of layers, and

$$\left(N_{\alpha\beta}^{(k)}, M_{\alpha\beta}^{(k)}\right) = \int_{z_k}^{z_{k+1}} (1, z)\sigma_{\alpha\beta}^{(k)}\, dz\ , \quad Q_\alpha^{(k)} = \int_{z_k}^{z_{k+1}} \sigma_{\alpha 3}^{(k)}\, dz \tag{5}$$

$$\sigma_{ij}^{(k)} = \sigma_{ij}(x_\beta, z_k) \tag{6}$$

5.5 (*Continuation of Exercise 5.4*) Multiply the equilibrium equations

$$\frac{\partial \sigma_{\alpha\beta}^k}{\partial x_\beta} + \frac{\partial \sigma_{\alpha 3}^k}{\partial x_3} = 0 \tag{1}$$

with z and integrate over the lamina thickness to obtain the third equation

$$\frac{\partial M_{\alpha\beta}^{(k)}}{\partial x_\beta} + \sigma_{\alpha 3}^{(k+1)} z_{k+1} - \sigma_{\alpha 3}^{(k)} z_k - Q_\alpha^k = 0 \tag{2}$$

5.6 Starting with a linear distribution of the displacements through the laminate thickness in terms of unknown functions $(u_0, v_0, w_0, F_1, F_2, F_3)$

$$\begin{aligned} u(x,y,z,t) &= u_0(x,y,t) + zF_1(x,y,t) \\ v(x,y,z,t) &= v_0(x,y,t) + zF_2(x,y,t) \\ w(x,y,z,t) &= w_0(x,y,t) + zF_3(x,y,t) \end{aligned}$$

determine the functions (F_1, F_2, F_3) such that the Kirchhoff hypothesis holds:

$$\frac{\partial w}{\partial z} = 0 \, , \quad \frac{\partial u}{\partial z} = -\frac{\partial w}{\partial x} \, , \quad \frac{\partial v}{\partial z} = -\frac{\partial w}{\partial y}$$

5.7 Consider a single, orthotropic layer plate ($Q_{45} = 0$), and assume that the material coordinates coincide with the plate coordinates. Compute the stresses $(\sigma_{xx}, \sigma_{yy}, \sigma_{xy})$ using the constitutive equations of the first-order plate theory, and then use the equilibrium equations of the three-dimensional elasticity theory to determine the transverse stresses $(\sigma_{xz}, \sigma_{yz}, \sigma_{zz})$ as a function of the thickness coordinate.

5.8 Consider a single, orthotropic layer plate ($Q_{45} = 0$), and assume that the material coordinates coincide with the plate coordinates. According to the first-order theory, the strain energy due to transverse shear stresses is given by

$$\begin{aligned} U_s &= \frac{1}{2}\int_{\Omega_0}\int_{-\frac{h}{2}}^{\frac{h}{2}} \left(\sigma_{xz}\gamma_{xz}^{(0)} + \sigma_{yz}\gamma_{yz}^{(0)}\right) dzdxdy \\ &= \frac{1}{2}\int_{\Omega_0}\int_{-\frac{h}{2}}^{\frac{h}{2}} \left[Q_{55}\left(\gamma_{xz}^{(0)}\right)^2 + Q_{44}\left(\gamma_{yz}^{(0)}\right)^2\right] dzdxdy \\ &= \frac{1}{2K}\int_{\Omega_0} \left(\frac{Q_x^2}{A_{55}} + \frac{Q_y^2}{A_{44}}\right) dxdy \end{aligned}$$

Compute $\bar{U}_s$ using the transverse shear stresses obtained in Exercise 5.7 from the three–dimensional elasticity, and equate it with U_s to determine the shear correction coefficient, K.

5.9 Consider the equations of motion of 3-D elasticity [see Eq. (2.6–2)] in the absence of body forces:

$$\frac{\partial \sigma_{xx}}{\partial x} + \frac{\partial \sigma_{xy}}{\partial y} + \frac{\partial \sigma_{xz}}{\partial z} = \rho_0 \frac{\partial^2 u}{\partial t^2}$$

$$\frac{\partial \sigma_{xy}}{\partial x} + \frac{\partial \sigma_{yy}}{\partial y} + \frac{\partial \sigma_{yz}}{\partial z} = \rho_0 \frac{\partial^2 v}{\partial t^2}$$

$$\frac{\partial \sigma_{xz}}{\partial x} + \frac{\partial \sigma_{yz}}{\partial y} + \frac{\partial \sigma_{zz}}{\partial z} = \rho_0 \frac{\partial^2 w}{\partial t^2}$$

Integrate the above equations with respect to z over the interval $(-h/2, h/2)$ and express the results in terms of the force resultants defined in Eq. (5.3–20a). Use the following boundary conditions:

$$\sigma_{xz}(x, y, -\frac{h}{2}) = 0, \ \sigma_{xz}(x, y, \frac{h}{2}) = 0, \ \sigma_{yz}(x, y, -\frac{h}{2}) = 0, \ \sigma_{yz}(x, y, \frac{h}{2}) = 0$$

$$\sigma_{zz}(x,y,-\frac{h}{2}) = q_b, \ \sigma_{zz}(x,y,\frac{h}{2}) = q_t$$

Next, multiply the equations of motion with z and integrate with respect to z over the interval $(-h/2, h/2)$ and express the results in terms of the moment resultants defined in Eq. (5.3–20a).

5.10 Show that the membrane strains $\{\varepsilon^0\}$ and the moment resultants $\{M\}$ in the classical or first-order laminated plate theory can be expressed in terms of force resultants $\{N\}$ and bending strains $\{\varepsilon^1\}$ as

$$\{\varepsilon^0\} = [A]^{-1}\left(\{N\} - [B]\{\varepsilon^1\}\right)$$
$$\{M\} = \left([B][A]^{-1}\right)\{N\} - \left([B][A]^{-1}[B] - [D]\right)\{\varepsilon^1\}$$

These equations bring out the bending-extensional coupling for laminates with nonzero $[B]$. For example, when the bending strains are zero, the applied in-plane forces induce bending moments for laminates with nonzero coupling coefficients $[B]$.

5.11 Show that if $B_{ij} = 0$ (*e.g.*, for symmetric laminates), the equation of motion governing the transverse deflection w_0 in the classical laminate theory is

$$D_{11}\frac{\partial^4 w_0}{\partial x^4} + 4D_{16}\frac{\partial^4 w_0}{\partial x^3 \partial y} + 2\left(D_{12} + 2D_{66}\right)\frac{\partial^4 w_0}{\partial x^2 \partial y^2} + 4D_{26}\frac{\partial^4 w_0}{\partial x \partial y^3} + D_{22}\frac{\partial^4 w_0}{\partial y^4}$$
$$-\mathcal{N}(w) = q + I_0\frac{\partial^2 w_0}{\partial t^2} - I_2\frac{\partial^2}{\partial t^2}\left(\frac{\partial^2 w_0}{\partial x^2} + \frac{\partial^2 w_0}{\partial y^2}\right) + I_1\frac{\partial^2}{\partial t^2}\left(\frac{\partial u_0}{\partial x} + \frac{\partial v_0}{\partial y}\right)$$

5.12 Show that for a general laminate composed of multiple isotropic layers, the laminate stiffness $A_{16}, A_{26}, B_{16}, B_{26}, D_{16}$, and D_{26} are zero, and that $A_{22} = A_{11}, B_{22} = B_{11}$, and $D_{22} = D_{11}$.

5.13 Show that for a general laminate composed of multiple specially orthotropic layers, the laminate stiffness $A_{16}, A_{26}, B_{16}, B_{26}, D_{16}$, and D_{26} are zero.

5.14 Show that for antisymmetric laminates the stiffnesses, A_{16}, A_{26}, D_{16}, and D_{26} are zero, and the coupling stiffnesses B_{ij} are not zero.

5.15 Show that for antisymmetric cross-ply laminates, the coupling stiffnesses B_{ij} have the properties: $B_{22} = -B_{11}$ and all other $B_{ij} = 0$.

5.16 Show that for antisymmetric angle-ply laminated plates, the following stiffnesses are zero: $A_{16}, A_{26}, D_{16}, D_{26}, B_{11}, B_{22}, B_{12}$, and B_{66}.

5.17 Show that for laminates $[\alpha/\beta/\beta/\alpha/\beta/\alpha/\alpha/\beta]$ where $-90° \leq \alpha \leq 90°$ and $-90° \leq \beta \leq 90°$, coefficients B_{ij} are zero.

5.18 The material properties of AS/3501 graphite–epoxy material layers are:

$$E_1 = 140 \times 10^3 \text{ MPa}, \ E_2 = 10 \times 10^3 \text{ MPa}, \ G_{12} = 7 \times 10^3 \text{ MPa}$$

$$G_{13} = 7 \times 10^3 \text{ MPa}, \ G_{23} = 7 \times 10^3 \text{ MPa}, \ \nu_{12} = 0.3$$

$$\alpha_1 = -0.3 \times 10^{-6} \text{ m/m/}^\circ \text{ K}, \ \alpha_2 = 28 \times 10^{-6} \text{ m/m/}^\circ \text{ K}$$

Determine the stiffnesses $[A], [B]$, and $[D]$ for the antisymmetric laminate $[0/90]$ composed of equal thickness (0.5 mm) layers.

5.19 Determine the stiffnesses $[A], [B]$, and $[D]$ for an antisymmetric laminate (–45/45) composed of equal thickness (0.5 mm) layers of AS/3501 graphite–epoxy layers (see Exercise 5.18 for the material properties).

5.20 If the laminate of Exercise 5.18 is heated from 20° to 90°, determine the thermal forces and moments generated in the laminate, if it were restrained from free expansion.

5.21 If the laminate in Exercise 19 is made of four layers (–45/45/–45/45) of thickness 0.25 mm each, show that the stiffnesses $[A]$ and $[D]$ remain unchanged. Compare the stiffnesses B_{ij} for the two laminates (do they increase or decrease in values?).

5.22 Suppose that a four-layer $(0/90)_s$ symmetric laminate is subjected to loads such that the only nonzero strain at a point (x, y) is $\varepsilon_{xx}^{(0)} = 10^3 \mu$ in./in. The material properties of a lamina are (typical of a graphite–epoxy material) $E_1 = 20$ msi, $E_2 = 1.30$ msi, $G_{12} = 1.03$ msi, $\nu_{12} = 0.3$. Assume that each layer is of thickness 0.005 in. Determine the state of stress $(\sigma_{xx}, \sigma_{yy}, \sigma_{xy})$ with respect to the laminate coordinates in each layer. Interpret the results you obtain in light of the assumed strains.

5.23 Compute the stains and stresses in the principal material coordinate system of each layer for the problem in Exercise 22.

5.24 Compute the stress resultants N's and M's for the problem in Exercise 22.

5.25 Repeat Exercise 22 for the case in which the laminate is subjected to loads such that the only nonzero strain at a point (x, y) is $\varepsilon_{xx}^{(1)} = (1/12)$ /in.

5.26 Compute the stains and stresses in the principal material coordinate system of each layer for the problem in Exercise 25.

5.27 Compute the stress resultants N's and M's for the problem in Exercise 25.

5.28 Determine the displacement associated with the assumed strain field in Exercise 25.

5.29 Suppose that a six-layer $(\pm 45/0)_s$ symmetric laminate is subjected to loads such that the only nonzero strain at a point (x, y) is $\varepsilon_{xx}^{(0)} = 10^3 \mu$ in./in. The thickness and material properties of a lamina are the same as those listed in Exercise 22. Determine the state of stress $(\sigma_{xx}, \sigma_{yy}, \sigma_{xy})$ and force resultants.

5.30 Repeat Exercise 29 for the case in which the laminate is subjected to loads such that the only nonzero strain at a point (x, y) is $\varepsilon_{xx}^{(1)} = (1/12)$ /in.

5.31 Suppose that a three-layer $(\pm 45/0)$ unsymmetric laminate is subjected to loads such that the only nonzero strain at a point (x, y) is $\varepsilon_{xx}^{(0)} = 10^{-3}$ in./in. The thickness and material properties of a lamina are the same as those listed in Exercise 22. Determine the state of stress $(\sigma_{xx}, \sigma_{yy}, \sigma_{xy})$ and stress resultants.

References for Additional Reading

1. Cauchy, A. L., "Sur l'equilibre et le mouvement d'une plaque solide," *Exercises de Mathematique,* **3**, 328–355 (1828).
2. Poisson, S. D., "Memoire sur l'equilibre et le mouvement des corps elastique," *Mem. Acad. Sci.,* **8**(2), 357–570 (1829).
3. Kirchhoff, G., "Uber das Gleichgwich und die Bewegung einer Elastischen Scheibe," *J. Angew. Math.,* **40**, 51–88 (1850).
4. Basset, A. B., "On the Extension and Flexure of Cylindrical and Spherical Thin Elastic Shells," *Phil. Trans. Royal Soc.,* (London) Ser. A, **181** (6), 433–480 (1890).
5. Goodier, J. N., "On the Problem of the Beam and the Plate in the Theory of Elasticity," *Transactions of the Royal Society of Canada,* **32,** 65–88 (1938).
6. Reissner, E., "On the Theory of Bending of Elastic Plates," *Journal of Mathematical Physics,* **23,** 184–191 (1944).
7. Reissner, E., "The Effect of Transverse Shear Deformation on the Bending of Elastic Plates," *Journal of Applied Mechanics,* **12,** 69–77 (1945).
8. Reissner, E., "Reflections on the Theory of Elastic Plates," *Applied Mechanics Reviews,* **38**(11), 1453–1464 (1985).
9. Bollé, E., "Contribution au Probleme Lineare de Flexion d'une Plaque Elastique," *Bull. Tech. Suisse. Romande.,* **73,** 281–285 and 293–298 (1947).
10. Hencky, H., "Uber die Berucksichtigung der Schubverzerrung in ebenen Platten," *Ing. Arch.,* **16**, 72–76 (1947).
11. Hildebrand, F. B., Reissner, E., and Thomas, G. B., "Notes on the Foundations of the Theory of Small Displacements of Orthotropic Shells," NACA TN-1833, Washington, D.C. (1949).
12. Mindlin, R. D., "Influence of Rotatory Inertia and Shear on Flexural Motions of Isotropic, Elastic Plates," *Journal of Applied Mechanics, Transactions of ASME*, **18**, 31–38 (1951).
13. Vlasov, B. F., "Ob uravneniyakh teovii isgiba plastinok (On the Equations of the Theory of Bending of Plates)," *Izv. Akd. Nauk SSR*, OTN, **4**, 102–109 (1958).
14. Panc, V., *Theories of Elastic Plates,* Noordhoff, Leyden, Netherlands, 1975.
15. Reissner, E. and Stavsky, Y., "Bending and Stretching of Certain Types of Aeolotropic Elastic Plates," *Journal of Applied Mechanics,* **28**, 402–408 (1961).
16. Stavsky, Y., "Bending and Stretching of Laminated Aeolotropic Plates," *Journal of Engineering Mechanics, ASCE*, **87** (EM6), 31–56 (1961).

17. Dong, S. B., Pister, K. S., and Taylor, R. L., "On the Theory of Laminated Anisotropic Shells and Plates," *Journal of Aeronautical Science,* **29**(8), 969–975 (1962).

18. Yang, P. C., Norris, C. H., and Stavsky, Y., "Elastic Wave Propagation in Heterogeneous Plates," *Int. J. Solids and Structures*, **2**, 665–684 (1966).

19. Ambartsumyan, S. A., *Theory of Anisotropic Plates,* translated from Russian by T. Cheron, Technomic, Stamford, CT, 1969.

20. Whitney, J. M. and Leissa, A. W., "Analysis of Heterogeneous Anisotropic Plates," *Journal of Applied Mechanics,* **36**(2), 261–266 (1969).

21. Whitney, J. M., "The Effect of Transverse Shear Deformation in the Bending of Laminated Plates," *Journal of Composite Materials*, **3**, 534–547 (1969).

22. Whitney, J. M. and Pagano, N. J., "Shear Deformation in Heterogeneous Anisotropic Plates," *Journal of Applied Mechanics*, **37** (4), 1031–1036 (1970).

23. Reissner, E., "A Consistent Treatment of Transverse Shear Deformations in Laminated Anisotropic Plates," *AIAA J.,* **10**(5), 716–718 (1972).

24. Librescu, L., *Elastostatics and Kinetics of Anisotropic and Heterogeneous Shell–type Structures,* Noordhoff, Leyden, Netherlands, 1975.

25. Reissner, E., "Note on the Effect of Transverse Shear Deformation in Laminated Anisotropic Plates," *Computer Methods in Applied Mechanics and Engineering*, **20**, 203–209 (1979).

26. Reddy, J. N., *Energy and Variational Methods in Applied Mechanics,* John Wiley and Sons, New York, 1984.

27. Librescu, L. and Reddy, J. N., "A Critical Review and Generalization of Transverse Shear Deformable Anisotropic Plate Theories," Euromech Colloquium 219, Kassel, Germany, Sept. 1986, *Refined Dynamical Theories of Beams, Plates and Shells and Their Applications,* I. Elishakoff and H. Irretier (eds.), Springer–Verlag, Berlin, 1987, pp. 32–43.

28. Whitney, J. M., "Shear Correction Factors for Orthotropic Laminates Under Static Load," *Journal of Applied Mechanics*, **40** (1), 302–304 (1973).

29. Bert, C. W., "Simplified Analysis of Static Shear Correction Factors for Beams of Non–Homogeneous Cross Section," *Journal of Composite Materials,* **7**, 525–529 (1973).

30. Chow, T. S., "On the Propagation of Flexural Waves in an Orthotropic Laminated Plate and Its Response to an Impulsive Load," *Journal of Composite Materials,* **5**, 306–319 (1971).

31. Srinivas, S. R., Joga Rao, C. V., and Rao, A. K., "An Exact Analysis for Vibration of Simply–Supported Homogeneous and Laminated Thick Rectangular Plates," *Journal of Sound and Vibration,* **12**, 187–199 (1970).

32. Wittrick, W. H., "Analytical Three-Dimensional Elasticity Solutions to Some Plate Problems and Some Observations on Mindlin's Plate Theory," *International Journal of Solids and Structures,* **23**, 441–464 (1987).

33. Whitney, J. M. and Sun, C. T., "A Higher Order Theory for Extensional Motion of Laminated Composites," *Journal of Sound and Vibration,* **30**, 85–97 (1973).
34. Sun, C. T. and Whitney, J. M., "Theories for the Dynamic Response of Laminated Plates," *AIAA Journal,* **11**(2), 178–183 (1973).
35. Lo, K. H., Christensen, R. M., and Wu, E. M., "A Higher Order Theory of Plate Deformation, Part 2; Laminated Plates," *Journal of Applied Mechanics,* **44**, 669–676 (1977).
36. Krishna Murty, A. V., "Higher Order Theory for Vibration of Thick Plates," *AIAA Journal,* **15**(12), 1823–1824 (1977).
37. Murthy, M. V. V., "An Improved Transverse Shear Deformation Theory for Laminated Anisotropic Plates," NASA Technical Paper 1903, 1–37 (1981).
38. Reddy, J. N., "A Simple Higher-Order Theory for Laminated Composite Plates," *Journal of Applied Mechanics,* **51**, 745–752 (1984).
39. Reddy, J. N., "A General Non-Linear Third-Order Theory of Plates with Moderate Thickness," *International Journal of Non-Linear Mechanics,* **25**(6), 677–686 (1990).
40. Jones, R. M., *Mechanics of Composite Materials*, Scripta Book Co., Washington, D.C., 1975.
41. Timoshenko, S. P. and Woinowsky–Krieger, S., *Theory of Plates and Shells,* McGraw–Hill, New York, 1959.
42. Lekhnitskii, S. G., *Anisotropic Plates,* Translated from Russian by S. W. Tsai and T. Cheron, Gordon and Breach, Newark, New Jersey, 1968.
43. Ashton, J. E. and Whitney, J. M., *Theory of Laminated Plates,* Technomic, Stamford, Connecticut, 1970.
44. Szilard, R., *Theory and Analysis of Plates, Classical and Numerical Methods,* Prentice–Hall, Englewood Cliffs, New Jersey, 1974.
45. Vinson, J. R. and Sierakowski, R. L., *The Behavior of Structures Composed of Composite Materials*, Kluwer, Netherlands, 1986.
46. Whitney, J. M., *Structural Analysis of Laminated Anisotropic Plates,* Technomic, Lancaster, Pennsylvania, 1987.
47. Vasiliev, V. V., *Mechanics of Composite Structures*, Translated from Russian by L. I. Man, Taylor & Francis, Washington, DC, 1988.
48. Ochoa, O. O. and Reddy, J. N., *Finite Element Analysis of Composite Laminates*, Kluwer, Netherlands, 1992.
49. Reddy, J. N. (ed.), *Mechanics of Composite Materials. Selected Works of Nicholas J. Pagano*, Kluwer, Netherlands, 1994.
50. Reddy, J. N. and Miravete, A., *Practical Analysis of Composite Laminates,* CRC Press, Boca Raton, Florida, 1995.

Chapter Six

One-Dimensional Analysis of Laminated Plates

6.1 Introduction

There are two cases of laminated plates that can be treated as one-dimensional problems: (1) laminated beams, and (2) cylindrical bending of laminated plate strips. When the width b (*i.e.*, length along the y−axis) of a laminated plate is very small compared to the length along the x−axis, it is treated as a laminated beam (see Figure 6.1–1). In cylindrical bending, the laminated plate is assumed to be a plate strip that is very long along the y−axis and has a finite dimension a along the x−axis (see Figure 6.1–2). The transverse load $q(x)$ is assumed to be uniform at any section parallel to the x−axis. In such a case, the deflection w_0 and displacements (u_0, v_0) of the plate are functions of only x, and all derivatives with respect to y are zero. The cylindrical bending problem is a *plane strain* problem, whereas the beam problem is a *plane stress* problem.

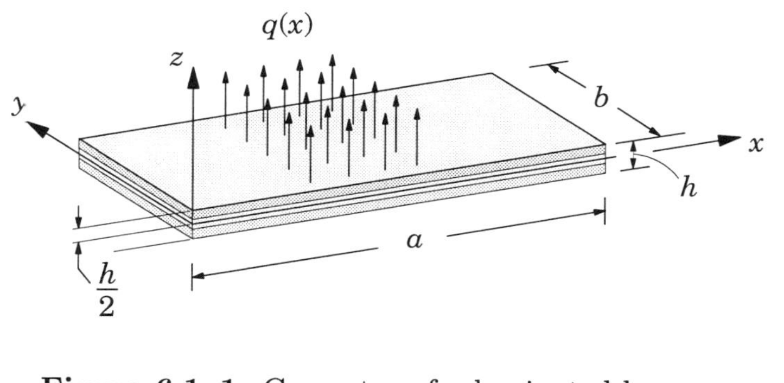

Figure 6.1–1: Geometry of a laminated beam.

In this chapter we develop exact analytical solutions for the two classes of problems. An *exact* solution of a problem is one which satisfies the governing equations at every point of the domain and the boundary and initial conditions of the problem. A numerical solution is one that is obtained by satisfying the governing equations and boundary conditions of the problem in an approximate sense. The solutions obtained with any of the variational methods (see Chapter 3) and numerical methods, such as the finite difference, finite element, and boundary element methods, are termed numerical solutions. An exact solution can be either closed-form or an infinite series. *Closed-form* solutions are those which can be expressed in terms of a finite number of terms. For example, $u(x) = 2 - x + 3x^2 + 4\sin n\pi x$ is a closed-form solution, whereas a solution in the form of a convergent series

$$u(x) = \sum_{n=1}^{\infty} a_n \sin n\pi x \tag{6.1-1}$$

where a_n are real numbers, is *not* a closed-form solution because the number of terms in the series is not finite. Since the series solution, in reality, is evaluated for a finite number of terms, it is, in a sense, approximate. The finite-sum series solution

$$u_N(x) = \sum_{n=1}^{N} a_n \sin n\pi x \tag{6.1-2}$$

will be termed analytical solution, although it is approximate because not all terms of the series (1) are included in (2). For all practical purposes, it is 'exact.'

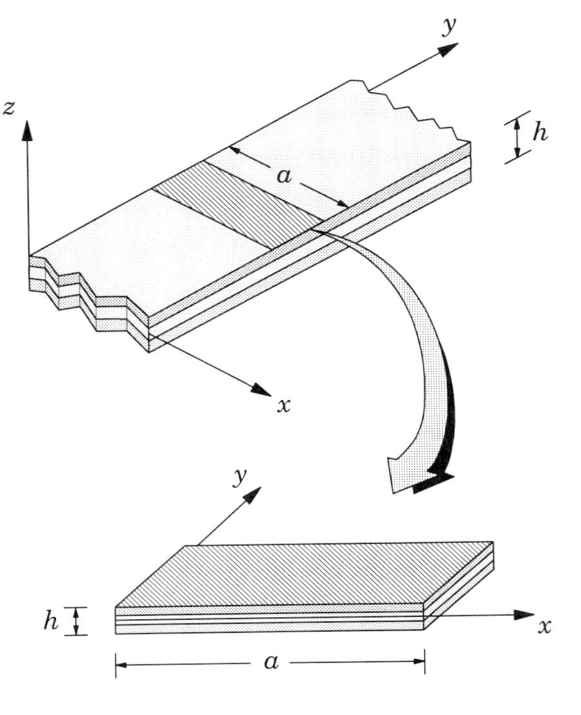

Figure 6.1–2: Geometry of a plate strip in cylindrical bending.

Due to their one-dimensional nature, analytical – exact as well as numerical – solutions can be developed for a number of laminated beams and plate strips. The analytical solutions presented here for simple problems serve as a basis of understanding the response. In addition, the results can serve as a reference for comparison for computational methods designed to analyze more complicated problems.

6.2 Analysis of Laminated Beams Using CLPT

6.2–1 Governing Equations

Here we consider the bending of symmetrically laminated beams according to CLPT. For symmetric laminates, the equations for bending deflection are uncoupled from those of the stretching displacements. If the in-plane forces are zero, the in-plane displacements (u_0, v_0) are zero, and the problem is reduced to one of solving for bending deflection and stresses.

In deriving the laminated beam theory we assume that

$$M_{yy} = M_{xy} = 0 \qquad (6.2-1)$$

everywhere in the beam. The classical laminated plate theory constitutive equations for symmetric laminates, in the absence of in-plane forces, are given by [see Eqs. (5.3–44)]

$$\begin{Bmatrix} M_{xx} \\ M_{yy} \\ M_{xy} \end{Bmatrix} = - \begin{bmatrix} D_{11} & D_{12} & D_{16} \\ D_{12} & D_{22} & D_{26} \\ D_{16} & D_{26} & D_{66} \end{bmatrix} \begin{Bmatrix} \frac{\partial^2 w_0}{\partial x^2} \\ \frac{\partial^2 w_0}{\partial y^2} \\ 2\frac{\partial^2 w_0}{\partial x \partial y} \end{Bmatrix} \qquad (6.2-2a)$$

or, in inverse form, we have

$$\begin{Bmatrix} \frac{\partial^2 w_0}{\partial x^2} \\ \frac{\partial^2 w_0}{\partial y^2} \\ 2\frac{\partial^2 w_0}{\partial x \partial y} \end{Bmatrix} = - \begin{bmatrix} D_{11}^* & D_{12}^* & D_{16}^* \\ D_{12}^* & D_{22}^* & D_{26}^* \\ D_{16}^* & D_{26}^* & D_{66}^* \end{bmatrix} \begin{Bmatrix} M_{xx} \\ M_{yy} \\ M_{xy} \end{Bmatrix} \qquad (6.2-2b)$$

where D_{ij}^* denote the elements of the inverse matrix of D_{ij}. In view of the assumption (1), we have

$$\frac{\partial^2 w_0}{\partial x^2} = -D_{11}^* M_{xx}, \quad \frac{\partial^2 w_0}{\partial y^2} = -D_{12}^* M_{xx}, \quad 2\frac{\partial^2 w_0}{\partial x \partial y} = -D_{16}^* M_{xx} \qquad (6.2-3a)$$

where

$$\begin{aligned} D_{11}^* &= (D_{22}D_{66} - D_{26}D_{26})/D^* \\ D_{12}^* &= (D_{16}D_{26} - D_{12}D_{66})/D^* \\ D_{16}^* &= (D_{12}D_{26} - D_{22}D_{16})/D^* \\ D^* &= D_{11}D_1 + D_{12}D_2 + D_{16}D_3 \ , \quad D_1 = D_{22}D_{66} - D_{26}D_{26} \\ D_2 &= D_{16}D_{26} - D_{12}D_{66} \ , \quad D_3 = D_{12}D_{26} - D_{22}D_{16} \end{aligned} \qquad (6.2-3b)$$

Equations (3) indicate that the transverse deflection w_0 cannot be independent of the coordinate y due to the Poisson effect (D_{12}^*) and anisotropic shear coupling (D_{16}^*). These effects can be neglected only for long beams (*i.e.*, when the length-to-width ratio is large). The length-to-width ratio for which the transverse deflection can be assumed to be independent of y is a function of the lamination scheme. For angle-ply laminates this ratio must be rather large to make the twisting curvature negligible.

In the following derivations we assume that the laminated beam under consideration is long enough to make the effects of the Poisson ratio and shear coupling on the deflection negligible. Then the transverse deflection can be treated only as a function of coordinate x (along the length of the beam) and time t:

$$w_0 = w_0(x,t) \tag{6.2-4}$$

Then we can write

$$\frac{\partial^2 w_0}{\partial x^2} = -D_{11}^* M_{xx} \tag{6.2-5}$$

In order to cast Eq. (5) in the familiar form used in the classical Euler–Bernoulli beam theory, we introduce the quantities

$$M = bM_{xx}, \quad Q = bQ_x, \quad E_{xx}^b = \frac{12}{h^3 D_{11}^*} = \frac{b}{I_{yy} D_{11}^*}\ , \quad I_{yy} = \frac{bh^3}{12} \tag{6.2-6}$$

and write Eq. (5) as

$$\frac{\partial^2 w_0}{\partial x^2} = -\frac{M}{E_{xx}^b I_{yy}} \quad \text{or} \quad M(x) = -E_{xx}^b I_{yy} \frac{\partial^2 w_0}{\partial x^2} \tag{6.2-7a}$$

and the shear force and bending moments are related by

$$Q_x = \frac{\partial M_{xx}}{\partial x} \quad \text{or} \quad Q = \frac{\partial M}{\partial x} \tag{6.2-7b}$$

where b is the width and h is the total thickness of the laminate.

The equation of motion of laminated beams can be obtained from Eq. (5.3–25c) by setting all terms involving differentiation with respect to y to zero:

$$\frac{\partial^2 M_{xx}}{\partial x^2} + \hat{N}_{xx} \frac{\partial^2 w_0}{\partial x^2} + q = I_0 \frac{\partial^2 w_0}{\partial t^2} - I_2 \frac{\partial^4 w_0}{\partial x^2 \partial t^2} \tag{6.2-8a}$$

or, for symmetrically laminated long beams, we have

$$-\frac{\partial^2}{\partial x^2}\left(E_{xx}^b I_{yy} \frac{\partial^2 w_0}{\partial x^2}\right) + b\hat{N}_{xx} \frac{\partial^2 w_0}{\partial x^2} + \hat{q} = \hat{I}_0 \frac{\partial^2 w_0}{\partial t^2} - \hat{I}_2 \frac{\partial^4 w_0}{\partial x^2 \partial t^2} \tag{6.2-8b}$$

where $\hat{N}_{xx}$ is the applied axial load, and

$$\hat{q} = bq, \quad \hat{I}_0 = bI_0, \quad \hat{I}_2 = bI_2, \quad I_i = b\int_{-\frac{h}{2}}^{\frac{h}{2}} \rho(z)^i \, dz \quad (i = 0, 1, 2) \qquad (6.2-8c)$$

The boundary conditions are of the form

$$\text{Geometric :} \qquad \text{specify} \quad w_0 \,, \quad \frac{\partial w_0}{\partial x} \qquad (6.2-9a)$$

$$\text{Force :} \qquad \text{specify} \quad Q \equiv \frac{\partial M}{\partial x} \,, \quad M \qquad (6.2-9b)$$

Equations (7)–(9) are identical, in form, to those of the Euler–Bernoulli beam theory of homogeneous, isotropic beams. Hence, the solutions available for deflections of isotropic beams under various boundary conditions can be readily used for laminated beams by replacing the modulus E with E^b_{xx} and multiplying loads and inertias with b. Note that the rotary (or rotatory) inertia I_2 is not neglected in Eq. (8).

6.2–2 Bending

For static bending without the axial force, $\hat{N}_{xx} = 0$, Eqs. (7) and (8b) take the form [*cf.* Eqs. (3.4–23) and (3.4–32)]

$$\frac{d^2 w_0}{dx^2} = -\frac{M}{E^b_{xx} I_{yy}} \,, \qquad E^b_{xx} I_{yy} \frac{d^4 w_0}{dx^4} = \hat{q} \qquad (6.2-10a,b)$$

where $\hat{q} = bq$. Equation (10a) is the most convenient when it is possible to express the bending moment M in terms of the applied loads. For indeterminate beams, use of Eq. (10b) is more convenient.

General Solutions

The general solutions of Eqs. (10a,b) are obtained by direct integration. We obtain from Eq. (10a)

$$E^b_{xx} I_{yy} w_0(x) = -\int^x \left[\int^\eta M(\xi) d\xi\right] d\eta + b_1 x + b_2 \qquad (6.2-11a)$$

and from Eq. (10b)

$$E^b_{xx} I_{yy} w_0(x) = \int^x \left\{\int^\xi \left[\int^\eta \left(\int^\zeta \hat{q}(\mu) d\mu\right) d\zeta\right] d\eta\right\} d\xi + c_1 \frac{x^3}{6} + c_2 \frac{x^2}{2} + c_3 x + c_4 \qquad (6.2-11b)$$

The constants of integration, b_1, b_2, and c_1 through c_4, can be determined using the boundary conditions of the problem. The boundary conditions for various types of supports are defined below:

$$\begin{array}{rl} \text{Free :} & Q \equiv \dfrac{dM}{dx} = 0\ , \quad M = 0 \\ \text{Simply Supported :} & w_0 = 0\ , \quad M = 0 \\ \text{Clamped :} & w_0 = 0\ , \quad \dfrac{dw_0}{dx} = 0 \end{array} \tag{6.2 – 11c}$$

Calculation of Stresses

The in-plane stresses in the kth layer can be computed from the equations [see Eqs. (5.3–12a) and (2b)]

$$\begin{aligned} \begin{Bmatrix} \sigma_{xx} \\ \sigma_{yy} \\ \sigma_{xy} \end{Bmatrix}^{(k)} &= z \begin{bmatrix} \bar{Q}_{11} & \bar{Q}_{12} & \bar{Q}_{16} \\ \bar{Q}_{12} & \bar{Q}_{22} & \bar{Q}_{26} \\ \bar{Q}_{16} & \bar{Q}_{26} & \bar{Q}_{66} \end{bmatrix}^{(k)} \begin{Bmatrix} -\frac{\partial^2 w_0}{\partial x^2} \\ -\frac{\partial^2 w_0}{\partial y^2} \\ -2\frac{\partial^2 w_0}{\partial x \partial y} \end{Bmatrix} \\ &= \frac{z}{b} \begin{bmatrix} \bar{Q}_{11} & \bar{Q}_{12} & \bar{Q}_{16} \\ \bar{Q}_{12} & \bar{Q}_{22} & \bar{Q}_{26} \\ \bar{Q}_{16} & \bar{Q}_{26} & \bar{Q}_{66} \end{bmatrix}^{(k)} \begin{bmatrix} D^*_{11} & D^*_{12} & D^*_{16} \\ D^*_{12} & D^*_{22} & D^*_{26} \\ D^*_{16} & D^*_{26} & D^*_{66} \end{bmatrix} \begin{Bmatrix} M \\ 0 \\ 0 \end{Bmatrix} \end{aligned} \tag{6.2 – 12a}$$

or

$$\begin{aligned} \sigma^{(k)}_{xx}(x,z) &= \frac{M(x)z}{b} \left(\bar{Q}^{(k)}_{11} D^*_{11} + \bar{Q}^{(k)}_{12} D^*_{12} + \bar{Q}^{(k)}_{16} D^*_{16} \right) \\ \sigma^{(k)}_{yy}(x,z) &= \frac{M(x)z}{b} \left(\bar{Q}^{(k)}_{12} D^*_{11} + \bar{Q}^{(k)}_{22} D^*_{12} + \bar{Q}^{(k)}_{26} D^*_{16} \right) \\ \sigma^{(k)}_{xy}(x,z) &= \frac{M(x)z}{b} \left(\bar{Q}^{(k)}_{16} D^*_{11} + \bar{Q}^{(k)}_{26} D^*_{12} + \bar{Q}^{(k)}_{66} D^*_{16} \right) \end{aligned} \tag{6.2 – 12b}$$

In general, the maximum stress does not occur at the top or bottom of a laminated beam. The maximum stress location through the beam thickness depends on the lamination scheme. As will be seen later in this section, the 0° layers take the most axial stress.

The stresses given by Eq. (12b) are approximate for the purpose of analyzing laminated beams. They are not valid especially in the free-edge zone, where the stress state is three dimensional. The width of the edge zone is about the order of the thickness of the beam.

In the classical beam theory, the interlaminar stresses $(\sigma_{xz}, \sigma_{zz})$ are identically zero when computed using the constitutive equations. However, these stresses do exist in reality and they can be responsible for failures in composite laminates because of the relatively low shear and transverse normal strengths of materials used. Interlaminar stresses may be computed using the equilibrium equations of 3-D elasticity [see Eq. (2.6–2)]:

$$\begin{aligned}
0 &= \frac{\partial \sigma_{xx}}{\partial x} + \frac{\partial \sigma_{xy}}{\partial y} + \frac{\partial \sigma_{xz}}{\partial z}\\
0 &= \frac{\partial \sigma_{xy}}{\partial x} + \frac{\partial \sigma_{yy}}{\partial y} + \frac{\partial \sigma_{yz}}{\partial z}\\
0 &= \frac{\partial \sigma_{xz}}{\partial x} + \frac{\partial \sigma_{yz}}{\partial y} + \frac{\partial \sigma_{zz}}{\partial z}
\end{aligned} \tag{6.2-13}$$

For each layer, these equations may be integrated with respect to z to obtain the interlaminar stresses within each layer $(z_k \leq z \leq z_{k+1})$:

$$\sigma_{xz}^{(k)} = -\int_{z_k}^{z} \left(\frac{\partial \sigma_{xx}^{(k)}}{\partial x} + \frac{\partial \sigma_{xy}^{(k)}}{\partial y} \right) dz + G^{(k)} \tag{6.2-14a}$$

$$\sigma_{yz}^{(k)} = -\int_{z_k}^{z} \left(\frac{\partial \sigma_{xy}^{(k)}}{\partial x} + \frac{\partial \sigma_{yy}^{(k)}}{\partial y} \right) dz + F^{(k)} \tag{6.2-14b}$$

$$\sigma_{zz}^{(k)} = -\int_{z_k}^{z} \left(\frac{\partial \sigma_{xz}^{(k)}}{\partial x} + \frac{\partial \sigma_{yz}^{(k)}}{\partial y} \right) dz + H^{(k)} \tag{6.2-14c}$$

where $(\sigma_{xx}^{(k)}, \sigma_{xy}^{(k)}, \sigma_{yy}^{(k)})$ are known from Eq. (12), and $G^{(k)}$, $F^{(k)}$, and $H^{(k)}$ are constants.

For beams, all variables are independent of y and $v = 0$. Hence, derivatives with respect to y are zero. For example, from Eqs. (14a,c) and (12b), we obtain

$$\boxed{\begin{aligned}
\sigma_{xz}^{(k)}(x,z) &= -Q_x(x)\left(\bar{Q}_{11}^{(k)} D_{11}^* + \bar{Q}_{12}^{(k)} D_{12}^* + \bar{Q}_{16}^{(k)} D_{16}^*\right)\left(\frac{z^2 - z_k^2}{2}\right) + G^{(k)}\\
\sigma_{zz}^{(k)}(x,z) &= -\frac{dQ_x}{dx}\left(\bar{Q}_{11}^{(k)} D_{11}^* + \bar{Q}_{12}^{(k)} D_{12}^* + \bar{Q}_{16}^{(k)} D_{16}^*\right)\left(\frac{z^3 - z_k^3}{6}\right) + H^{(k)}
\end{aligned}} \tag{6.2-15a,b}$$

where Eqs. (6) and (7b) are used to replace dM/dx with $Q = bQ_x$, and $G^{(k)}$ and $H^{(k)}$ are the integration constants, which are evaluated using the boundary and interface continuity conditions. For layer 1, the constants should be such that σ_{xz} and σ_{zz} equal the shear and normal stresses at the bottom face of the laminate. For example, if the laminate bottom is stress free, we have $G^{(1)} = 0$ and $H^{(1)} = 0$. The constants $G^{(k)}$ and $H^{(k)}$ for $k = 2, 3, \cdots$, are determined by requiring that $\sigma_{xz}^{(k)}$ and $\sigma_{zz}^{(k)}$ be continuous at the layer interfaces (see Figure 6.2–1):

$$\sigma_{xz}^{(k)}(x, z_{k+1}) = \sigma_{xz}^{(k+1)}(x, z_{k+1}), \quad \sigma_{zz}^{(k)}(x, z_{k+1}) = \sigma_{zz}^{(k+1)}(x, z_{k+1})$$

This gives, for $k = 1, 2, \cdots$, the result

$$\begin{aligned} G^{(k+1)} &= -Q_x(x)\left(\bar{Q}_{11}^{(k)} D_{11}^* + \bar{Q}_{12}^{(k)} D_{12}^* + \bar{Q}_{16}^{(k)} D_{16}^*\right)\left(\frac{z_{k+1}^2 - z_k^2}{2}\right) + G^{(k)} \\ &= \sigma_{xz}^{(k)}(x, z_{k+1}) \end{aligned} \quad (6.2-16a)$$

$$\begin{aligned} H^{(k+1)} &= -\frac{dQ_x}{dx}\left(\bar{Q}_{11}^{(k)} D_{11}^* + \bar{Q}_{12}^{(k)} D_{12}^* + \bar{Q}_{16}^{(k)} D_{16}^*\right)\left(\frac{z_{k+1}^3 - z_k^3}{6}\right) + H^{(k)} \\ &= \sigma_{zz}^{(k)}(x, z_{k+1}) \end{aligned} \quad (6.2-16b)$$

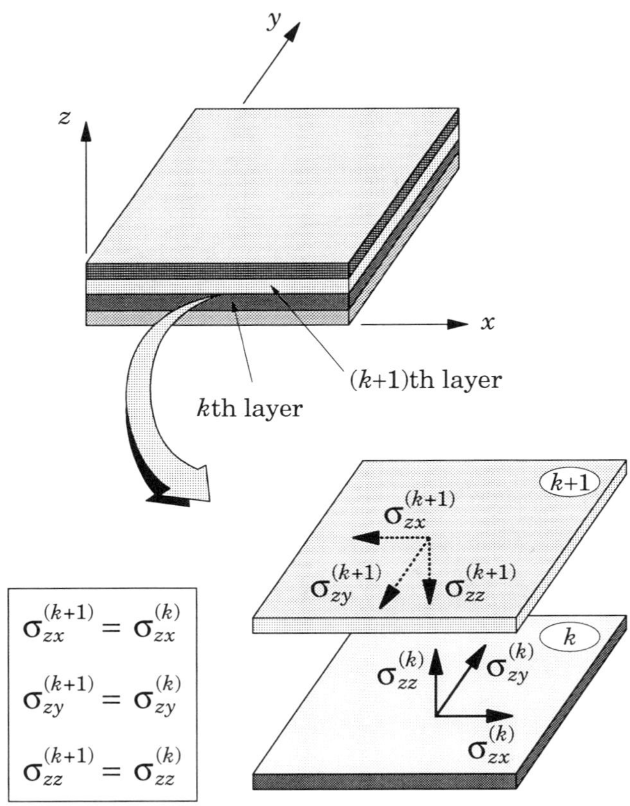

Figure 6.2–1: Equilibrium of interlaminar stresses in a laminated plate.

Note from Eqs. (15a,b) that the transverse shear stress σ_{xz} is quadratic and normal stress σ_{zz} is cubic through the thickness of each lamina. The distributions are described by different functions in different layers but they are continuous across layers (the so-called C^0–continuity).

Example 6.2–1 (*Simply supported beam*):

Consider a simply supported beam with a center point load (see Figure 6.2–2). This case is known as the *three-point bending*. The deflection is symmetric about the point $x = a/2$. The expression for the bending moment is

$$M(x) = -\frac{(F_0 b)x}{2}\,, \quad \text{for} \quad 0 \leq x \leq \frac{a}{2} \qquad (6.2-17)$$

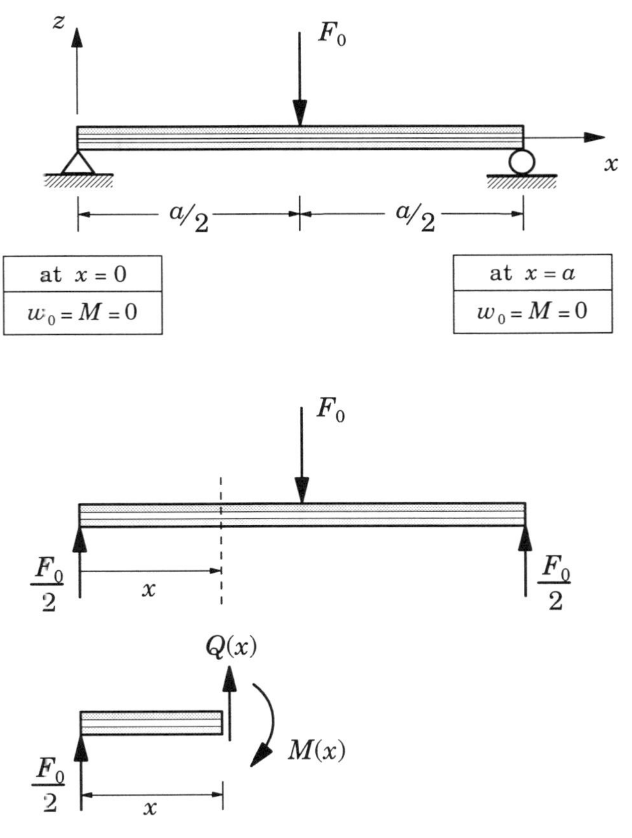

Figure 6.2–2: Three-point bending of a laminated beam.

Substituting this expression into Eq. (11a) and evaluating the integrals, we obtain

$$E^b_{xx} I_{yy} w_0(x) = \frac{F_0 b x^3}{12} + c_1 x + c_2$$

The constants c_1 and c_2 are evaluated using the boundary conditions of the problem

$$w_0(0) = 0, \quad \frac{dw_0}{dx}(a/2) = 0$$

We obtain $(c_1 = -F_0 b a^2/16, c_2 = 0)$

$$w_0(x) = -\frac{F_0 b a^3}{48 E^b_{xx} I_{yy}} \left[3\left(\frac{x}{a}\right) - 4\left(\frac{x}{a}\right)^3 \right] \tag{6.2 – 18}$$

The deflection is the maximum at $x = a/2$, which is given by

$$w_{max} = -\frac{F_0 b a^3}{48 E^b_{xx} I_{yy}} \equiv -w_c \tag{6.2 – 19}$$

This expression can be used to determine the modulus of the material in terms of the measured center deflection w_c, applied load F_0, and the geometric parameters of the laminated beam in a three-point bend test:

$$E^b_{xx} = \frac{F_0 b a^3}{4 b h^3 w_c} \tag{6.2 – 20}$$

The maximum in-plane stress σ_{xx} occurs at $x = a/2$ $(M(a/2) = -F_0 b a/4)$

$$\sigma^{(k)}_{xx}(a/2, z) = -\frac{F_0 a z}{4} \left(\bar{Q}^{(k)}_{11} D^*_{11} + \bar{Q}^{(k)}_{12} D^*_{12} + \bar{Q}^{(k)}_{16} D^*_{16} \right) \tag{6.2 – 21}$$

Example 6.2–2 (*Clamped beam*):

Consider a laminated beam, clamped at both ends, and subjected to uniformly distributed load acting downward, $q = -q_0$ (see Figure 6.2–3). The deflection is symmetric about the point $x = a/2$. We have from Eq. (11b) the result

$$E^b_{xx} I_{yy} w_0(x) = -\frac{q_0 b x^4}{24} + c_1 \frac{x^3}{6} + c_2 \frac{x^2}{2} + c_3 x + c_4$$

The constants c_1 through c_4 are evaluated using the boundary conditions of the half (because of the symmetry) or full beam. For the full beam case we have

$$w_0(0) = 0, \quad w_0(a) = 0, \quad \frac{dw_0}{dx}(0) = 0, \quad \frac{dw_0}{dx}(a) = 0$$

and for the half beam model we have

$$w_0(0) = 0, \quad \frac{dw_0}{dx}(0) = 0, \quad \frac{dw_0}{dx}(\frac{a}{2}) = 0, \quad Q(\frac{a}{2}) = \frac{dM}{dx} = -E_{xx}I_{yy}\frac{d^3w_0}{dx^3}(\frac{a}{2}) = 0$$

Either set of boundary conditions will yield the same solution. We obtain ($c_1 = q_0ba/2,\ c_2 = -q_0ba^2/12,\ c_3 = c_4 = 0$)

$$w_0(x) = -\frac{q_0ba^4}{24E^b_{xx}I_{yy}}\left[\left(\frac{x}{a}\right)^2 - \left(\frac{x}{a}\right)\right]^2 \qquad (6.2-22)$$

The deflection is the maximum at $x = a/2$, which is given by

$$w_{max} = -\frac{q_0ba^4}{384E^b_{xx}I_{yy}} \qquad (6.2-23)$$

The maximum bending moment, and hence the maximum in-plane stress σ_{xx}, occurs at $x = 0, a$:

$$M(x) = \frac{q_0ba^2}{12}\left[1 - 6\left(\frac{x}{a}\right) + 6\left(\frac{x}{a}\right)^2\right], \quad M_{max} = \frac{q_0ba^2}{12}$$

$$\sigma^{(k)}_{xx}(0, z) = \frac{q_0a^2z}{12}\left(Q^{(k)}_{11}D^*_{11} + Q^{(k)}_{12}D^*_{12} + Q^{(k)}_{16}D^*_{16}\right) \qquad (6.2-24)$$

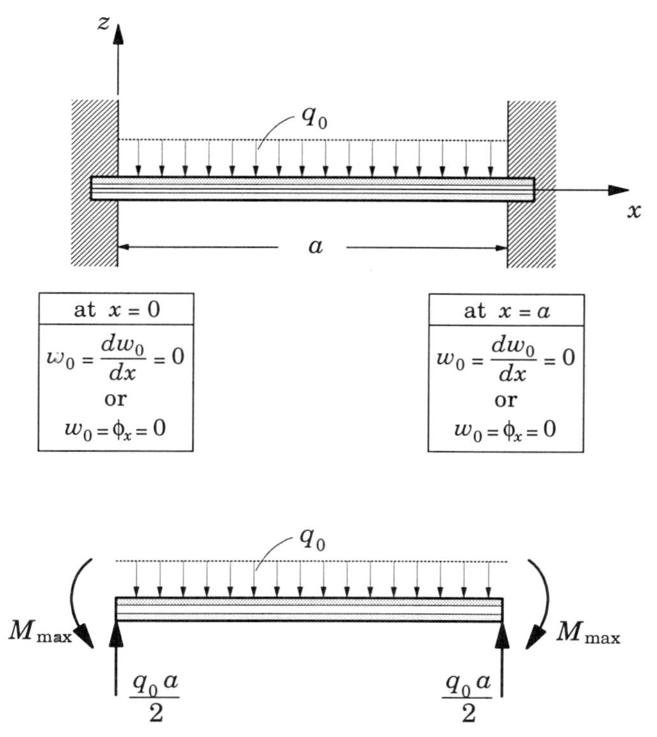

Figure 6.2–3: Clamped beam under uniformly distributed load.

Expressions for the transverse deflection of laminated beams with simple supports, clamped edges, and clamped-free (cantilever) supports and subjected to a transverse point load or uniformly distributed load are presented in Table 6.2–1. The maximum deflections and bending moments are also listed (note that the loads are assumed to be applied in the downward direction). Recall that $w_0(x)$ is taken positive upward and $M(x)$ is positive clockwise on the right end. When both point load and uniformly distributed load are applied simultaneously, the solution can be obtained by superposing (*i.e.*, adding) the expressions corresponding to each load. Expressions for other boundary conditions can be found in textbooks on a first course in deformable solids. The effects of material properties and stacking sequence are accounted for through the bending stiffness $E_{xx}^b I_{yy} = b/D_{11}^*$, as can be seen from Eqs. (6) and (3b).

Figures 6.2–4 and 6.2–5 show the maximum normal stress distribution, as predicted by Eq. (12b), through the thickness of $(0/45/{-45}/90)_s$ (0° corresponds to outer layers) and $(90/45/{-45}/0)_s$ (90° corresponds to outer layers) laminated beams, respectively, subjected to three-point bending ($F_0 = 1.0, b = 0.2, a = 1.0, h = 0.1$). The following layer material properties are used ($E_2 = 1$ msi):

$$\frac{E_1}{E_2} = 25, \quad G_{12} = G_{13} = 0.5E_2, \quad G_{23} = 0.2E_2, \quad \nu = 0.25 \qquad (6.2-25)$$

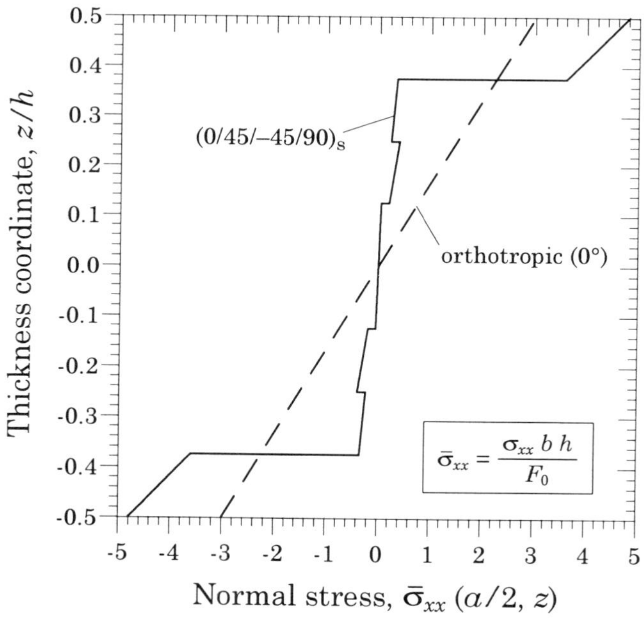

Figure 6.2–4: Maximum normal stress, $\sigma_{xx}(a/2, z)$, distribution through the thickness of a symmetrically laminated $(0/\pm 45/90)_s$ beam.

Table 6.2–1: Transverse deflections of laminated composite beams with various boundary conditions and subjected to point load or uniformly distributed load (acting downward) according to the classical beam theory.

Laminated Beam	Deflection, $w_0(x)$	w_{max} and M_{max}
• *Hinged–Hinged*		
Central point load	$\frac{c_1}{48}\left[3\left(\frac{x}{a}\right)-4\left(\frac{x}{a}\right)^3\right]$	$w^c_{max}=\frac{1}{48}c_1$ $M^c_{max}=-\frac{1}{4}c_3$
Uniform load	$\frac{c_2}{24}\left[\left(\frac{x}{a}\right)-2\left(\frac{x}{a}\right)^3+\left(\frac{x}{a}\right)^4\right]$	$w^c_{max}=\frac{5}{384}c_2$ $M^c_{max}=-\frac{1}{8}c_4$
• *Fixed–Fixed*		
Central point load	$\frac{c_1}{48}\left[3\left(\frac{x}{a}\right)^2-4\left(\frac{x}{a}\right)^3\right]$	$w^c_{max}=\frac{1}{192}c_1$ $M^0_{max}=\frac{1}{8}c_3$
Uniform load	$\frac{c_2}{24}\left[\left(\frac{x}{a}\right)^2-\left(\frac{x}{a}\right)\right]^2$	$w^c_{max}=\frac{1}{384}c_2$ $M^0_{max}=\frac{1}{12}c_4$
• *Fixed–Free*		
Point load at free end	$\frac{c_1}{6}\left[3\left(\frac{x}{a}\right)^2-\left(\frac{x}{a}\right)^3\right]$	$w^a_{max}=\frac{1}{3}c_1$ $M^0_{max}=c_3$
Uniform load	$\frac{c_2}{24}\left[6\left(\frac{x}{a}\right)^2-4\left(\frac{x}{a}\right)^3+\left(\frac{x}{a}\right)^4\right]$	$w^a_{max}=\frac{1}{8}c_2$ $M^0_{max}=\frac{1}{2}c_4$

Superscript 'c' refers to the center (at $x = a/2$), 'a' to the end $x = a$, and '0' refers to $x = 0$. The constants in the expressions for the deflection are defined as

$$c_1 = -\frac{F_0 ba^3}{E^b_{xx}I_{yy}}\,, \quad c_2 = -\frac{q_0 ba^4}{E^b_{xx}I_{yy}}\,, \quad c_3 = F_0 ba, \quad c_4 = F_0 ba^2$$

The maximum normal stress distribution in an orthotropic beam (with eight 0° layers) is shown in the figures by dashed lines. It is clear the 0° layer carries the most axial stress while the 90° layer carries the least axial stress, in proportion to their axial stiffness.

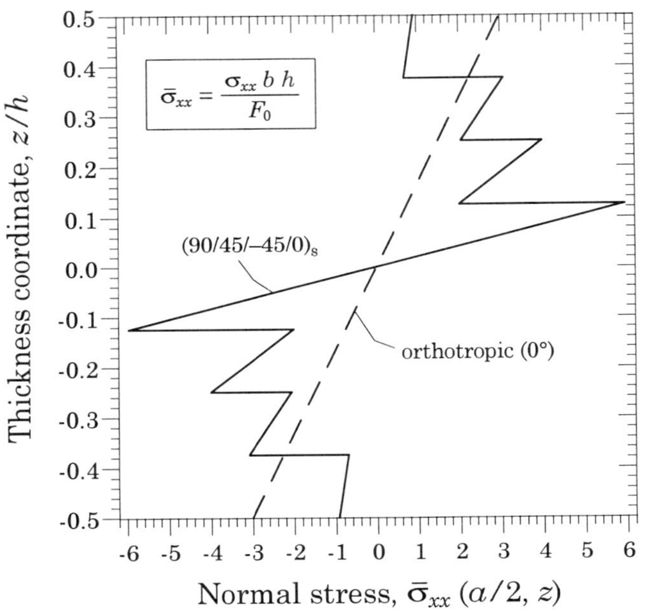

Figure 6.2–5: Maximum normal stress, $\sigma_{xx}(a/2, z)$, distribution through the thickness of a symmetrically laminated $(90/\pm 45/0)_s$ beam.

Figures 6.2–6 and 6.2–7 show the effect of stacking sequence on maximum transverse shear stress, as predicted by Eq. (15a), for laminates $(0/45/\text{–}45/90)_s$ and $(90/45/\text{–}45/0)_s$, respectively ($F_0 = 1.0, b = 0.2, a = 1.0, h = 0.1$). The parabolic distribution of transverse shear stress through an orthotropic beam is shown in dashed lines for comparison. The maximum stress value is dependent on the stacking sequence and considerably different from that in a homogeneous beam.

6.2–3 Buckling

A beam subjected to axial compressive load $\hat{N}_{xx} = -N_{xx}^0$ remains straight but shortens as the load increases from zero to a certain magnitude. If a small additional axial or lateral disturbance applied to the beam keeps it in equilibrium, then the beam is said to be *stable.* If the small additional disturbance results in a large response and the beam does not return to its original equilibrium configuration, the beam is said be *unstable.* The onset of instability is called *buckling* (see Figure 6.2–8). The magnitude of the compressive axial load at which the beam becomes unstable is termed the

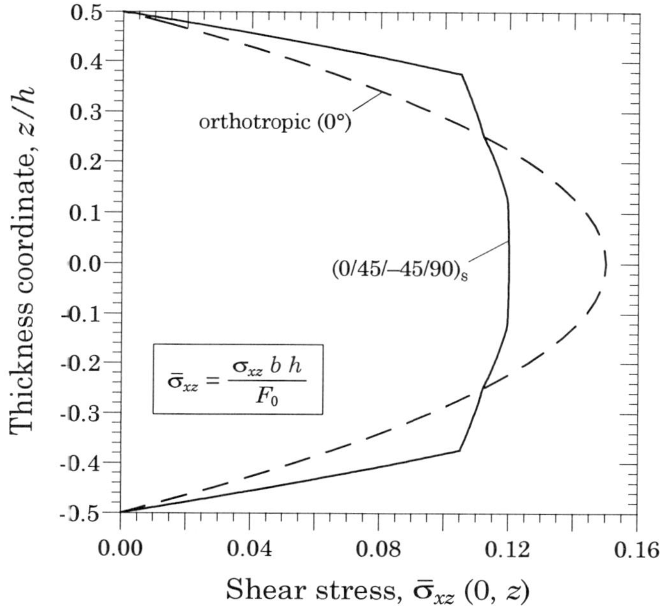

Figure 6.2–6: Variation of transverse shear stress (σ_{xz}) through the thickness of a symmetrically laminated $(0/\pm 45/90)_s$ beam subjected to three-point bending ($F_0 = 1.0, b = 0.2, a = 1.0, h = 0.1$).

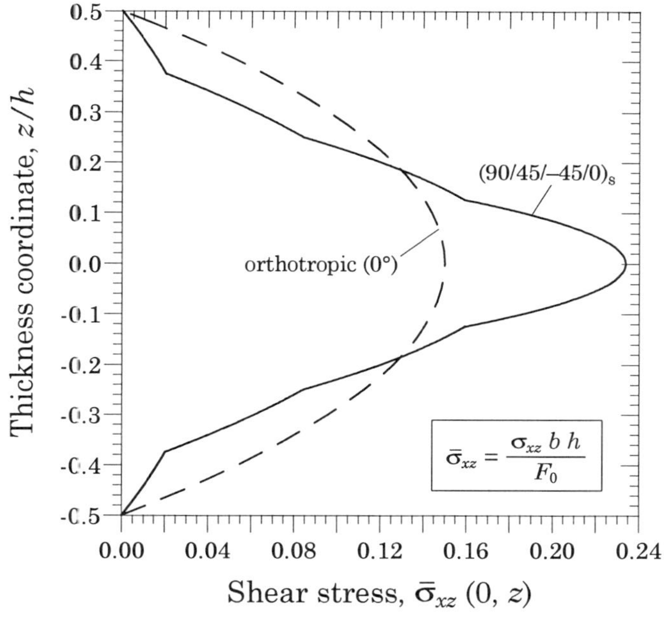

Figure 6.2–7: Variation of transverse shear stress (σ_{xz}) through the thickness of a symmetrically laminated $(90/\pm 45/0)_s$ beam subjected to three-point bending (see Figure 6.2–6 for data).

critical buckling load. If the load is increased beyond this critical buckling load, it results in a large deflection and the beam seeks another equilibrium configuration. Thus, the load at which a beam becomes unstable is of practical importance in the design of structural elements. Here we determine critical buckling loads for laminated straight beams. The equation governing buckling of laminated beams is also given by Eq. (8b), wherein the applied transverse load and inertia terms are set to zero, and axial force is assumed to be unknown. In addition, the deflection is measured from onset of buckling, and it is termed *buckling deflection.*

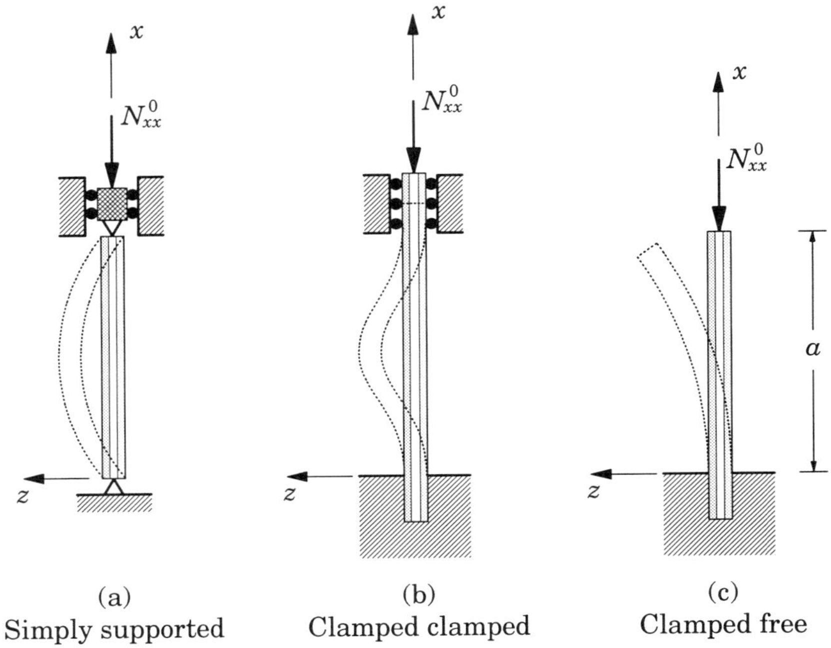

Figure 6.2–8: Buckling of laminated beams under various edge conditions.

Setting $\hat{N}_{xx} = -N^0_{xx}$, $\hat{q} = 0$, and all inertia terms to zero in Eqs. (8b), we obtain the equation

$$\boxed{\frac{d^4W}{dx^4} + \frac{bN^0_{xx}}{E^b_{xx}I_{yy}}\frac{d^2W}{dx^2} = 0} \tag{6.2-26}$$

where W denotes the buckling deflection. Equation (26) is obtained from the nonlinear equilibrium equation

$$E^b_{xx}I_{yy}\frac{d^4w_0}{dx^4} + bN_{xx}(w_0)\frac{d^2w_0}{dx^2} = 0$$

by substituting $w_0 = w_0^e + W$, where w_0^e is the original equilibrium (prebuckling) deflection and W is the buckling deflection. Note that w_0^e satisfies the equation

$$E_{xx}^b I_{yy} \frac{d^4 w_0^e}{dx^4} + bN_{xx}(w_0^e)\frac{d^2 w_0^e}{dx^2} = 0$$

[The reader is asked to verify the result in Eq. (26).]

Integrating Eq. (26) twice with respect to x, we obtain

$$\frac{d^2W}{dx^2} + \frac{bN_{xx}^0}{E_{xx}^b I_{yy}} W = K_1 x + K_2 \tag{6.2 – 27}$$

The general solution of Eq. (26) is

$$\boxed{W(x) = c_1 \sin \lambda_b x + c_2 \cos \lambda_b x + c_3 x + c_4} \tag{6.2 – 28}$$

where

$$\lambda_b^2 = \frac{bN_{xx}^0}{E_{xx}^b I_{yy}}\,, \quad c_3 = \frac{K_1}{\lambda_b^2}\,,\; c_4 = \frac{K_2}{\lambda_b^2} \tag{6.2 – 29}$$

and the constants c_1, c_2, c_3, and c_4 can be determined using the boundary conditions of the beam.

We are interested in determining the values of λ_b for which there exists a nonzero solution $W(x)$, *i.e.*, when beam experiences deflection. Once such a λ_b is known (often there will be many), the buckling load is determined from Eq. (29):

$$N_{xx}^0 = \left(\frac{E_{xx}^b I_{yy}}{b}\right) \lambda_b^2 \tag{6.2 – 30}$$

The smallest value of N_{xx}^0, which is given by the smallest value of λ_b, is the critical buckling load. The buckling shape (or mode) is given by $W(x)$. In the following, we consider beams with different boundary conditions to determine λ_b and then the critical buckling load for each beam.

Example 6.2–3 (*Simply supported beam*):

For a simply supported beam, the boundary conditions are

$$w_0(0) = 0, \quad w_0(a) = 0, \quad M_{xx}(0) = 0, \quad M_{xx}(a) = 0 \tag{6.2 – 31a}$$

These boundary conditions imply

$$W(0) = 0, \quad W(a) = 0, \quad \frac{d^2W}{dx^2}(0) = 0, \quad \frac{d^2W}{dx^2}(a) = 0 \tag{6.2 – 31b}$$

We have

$$
\begin{aligned}
W(0) = 0: &\quad c_2 + c_4 = 0 \\
W''(0) = 0: &\quad -c_2\lambda_b^2 = 0 \ \text{ which implies } \ c_2 = 0, \ c_4 = 0 \\
W(a) = 0: &\quad c_1 \sin\lambda_b a + c_3 a = 0 \\
W''(a) = 0: &\quad c_1 \sin\lambda_b a = 0 \ \text{ which implies } \ c_3 = 0
\end{aligned} \tag{6.2-32}
$$

For a nontrivial solution, the condition

$$
c_1 \sin\lambda_b a = 0 \ \text{ implies that } \ \lambda_b a = n\pi, \ n = 1, 2, \cdots \tag{6.2-33}
$$

and the buckling load is given by

$$
bN_{xx}^0 = E_{xx}^b I_{yy} \left(\frac{n\pi}{a}\right)^2 \tag{6.2-34a}
$$

The buckling mode is

$$
W(x) = c_1 \sin\frac{n\pi x}{a}, \quad c_1 \neq 0 \tag{6.2-34b}
$$

The critical buckling load becomes $(n = 1)$

$$
N_{cr} = \left(\frac{\pi}{a}\right)^2 \frac{E_{xx}^b I_{yy}}{b} = \left(\frac{\pi^2}{12}\right) \frac{E_{xx}^b h^3}{a^2} \tag{6.2-35}
$$

and the buckling mode (eigenfunction) associated with it is

$$
W(x) = c_1 \sin\frac{\pi x}{a}
$$

Example 6.2–4 (*Clamped beam*):

When the beam is fixed at both ends, the boundary conditions are

$$
w_0(0) = 0, \quad \frac{dw_0}{dx}(0) = 0, \quad w_0(a) = 0, \quad \frac{dw_0}{dx}(a) = 0 \tag{6.2-36a}
$$

which can be expressed as

$$
W(0) = 0, \quad \frac{dW}{dx}(0) = 0, \quad W(a) = 0, \quad \frac{dW}{dx}(a) = 0 \tag{6.2-36b}
$$

We have

$$
\begin{aligned}
W(0) = 0: &\quad c_2 + c_4 = 0 \\
W'(0) = 0: &\quad c_1\lambda_b + c_3 = 0 \\
W(a) = 0: &\quad c_1 \sin\lambda_b a + c_2 \cos\lambda_b a + c_3 a + c_4 = 0 \\
W'(a) = 0: &\quad c_1\lambda_b \cos\lambda_b a - c_2\lambda_b \sin\lambda_b a + c_3 = 0
\end{aligned} \tag{6.2-37}
$$

Expressing these equations in terms of constants c_1 and c_2, we obtain

$$\begin{aligned} c_1(\sin\lambda_b a - \lambda_b a) + c_2(\cos\lambda_b a - 1) &= 0 \\ c_1(\cos\lambda_b a - 1) - c_2\sin\lambda_b a &= 0 \end{aligned} \tag{6.2-38a}$$

For a nontrivial solution, the determinant of the coefficient matrix of the above two equations must be zero (eigenvalue problem):

$$\begin{aligned} 0 &= \begin{vmatrix} \sin\lambda_b a - \lambda_b a & \cos\lambda_b a - 1 \\ \cos\lambda_b a - 1 & -\sin\lambda_b a \end{vmatrix} \\ &= \lambda_b a\ \sin\lambda_b a + 2\cos\lambda_b a - 2 \end{aligned} \tag{6.2-38b}$$

The solution of the transcendental equation (38b), known as the *characteristic equation*, gives the eigenvalues $\lambda_b a$, and the buckling load is calculated from Eq. (30). A plot of the function $f(x) = x\ \sin x + 2\cos x - 2$ against x shows that $f(x)$ is zero at $x = 0, 6.2832(=2\pi), 8.9868, 12.5664(=4\pi), 15.4505, 6\pi, \cdots (\lambda_{2n-1}a = 2n\pi)$. Hence, the critical buckling load is

$$N_{cr} = \left(\frac{2\pi}{a}\right)^2 \frac{E^b_{xx} I_{yy}}{b} = \left(\frac{\pi^2}{3}\right)\frac{E^b_{xx}h^3}{a^2} \tag{6.2-39}$$

Table 6.2–2 contains governing equations for λ_b, some typical values of it, and values of the constants c_1, c_2, c_3, and c_4 for several combinations of simply supported (hinged), clamped (fixed), and free-edge conditions. For example, the critical buckling load of a cantilever (*i.e.*, fixed at one end and free at the other end) beam, the boundary conditions are

$$w_0(0) = 0, \quad \frac{dw_0}{dx}(0) = 0, \quad Q_x(a) = 0, \quad M_{xx}(a) = 0 \tag{6.2-40a}$$

which are equivalent to

$$\begin{aligned} &W = 0, \quad \frac{dW}{dx} = 0 \quad \text{at } x = 0 \\ &\left(\frac{d^3W}{dx^3} + \lambda_b^2\frac{dW}{dx}\right) = 0, \quad \frac{d^2W}{dx^2} = 0 \quad \text{at } x = a \end{aligned} \tag{6.2-40b}$$

The critical buckling load is given by

$$N_{cr} = \left(\frac{\pi}{2a}\right)^2 \frac{E^b_{xx} I_{yy}}{b} = \frac{\pi^2 h^3 E^b_{xx}}{48a^2} \tag{6.2-41}$$

Table 6.2–2: Values of the constants and eigenvalues for buckling of laminated composite beams with various boundary conditions ($\lambda^2 \equiv bN^0_{xx}/E^b_{xx}I_{yy} = (e_n/a)^2$). The classical laminate theory is used.

End conditions at $x=0$ and $x=a$	Constants†	Characteristic equation and values* of $e_n \equiv \lambda_n a$
• Hinged–Hinged	$c_1 \neq 0,\ c_2 = c_3 = c_4 = 0$	$\sin e_n = 0$ $e_n = n\pi$
• Fixed–Fixed	$c_1 = 1/(\sin e_n - e_n)$ $c_3 = -1/\lambda_n$ $c_2 = -c_4 = 1/(\cos e_n - 1)$	$e_n \sin e_n = 2(1-\cos e_n)$ $e_n = 2\pi, 8.987, 4\pi, \cdots$
• Fixed–Free	$c_1 = c_3 = 0$ $c_2 = -c_4 \neq 0$	$\cos e_n = 0$ $e_n = (2n-1)\pi/2$
• Free–Free	$c_1 = c_3 = 0$ $c_2 \neq 0,\ c_4 \neq 0$	$\sin e_n = 0$ $e_n = n\pi$
• Hinged–Fixed	$c_1 = 1/e_n \cos e_n,\ c_3 = -1$ $c_2 = c_4 = 0$	$\tan e_n = e_n$ $e_n = 4.493, 7.725, \cdots$

† See Eq. (6.2–28): $W(x) = c_1 \sin\lambda_b x + c_2\cos\lambda_b x + c_3 x + c_4$.

*For critical buckling load, only the first (minimum) value of $e = \lambda a$ is needed.

6.2–4 Vibration

For natural vibration, the solution is assumed to be periodic

$$w_0(x,t) = W(x)e^{i\omega t}, \quad i = \sqrt{-1} \tag{6.2-42}$$

In the absence of applied transverse load q, the governing equation (8b) reduces to

$$\boxed{E^b_{xx}I_{yy}\frac{d^4W}{dx^4} - b\hat{N}_{xx}\frac{d^2W}{dx^2} = \omega^2\hat{I}_0 W - \omega^2\hat{I}_2\frac{d^2W}{dx^2}} \tag{6.2-43}$$

Equation (43) has the general form

$$p\frac{d^4W}{dx^4} + q\frac{d^2W}{dx^2} - rW = 0 \tag{6.2-44}$$

where

$$p = E^b_{xx} I_{yy}, \quad q = \omega^2 \hat{I}_2 - b\hat{N}_{xx}, \quad r = \omega^2 \hat{I}_0 \tag{6.2 – 45}$$

The general solution of Eq. (44) is

$$W(x) = c_1 \sin \lambda x + c_2 \cos \lambda x + c_3 \sinh \mu x + c_4 \cosh \mu x$$

$$\lambda = \sqrt{\frac{1}{2p}\left(q + \sqrt{q^2 + 4pr}\right)}, \quad \mu = \sqrt{\frac{1}{2p}\left(-q + \sqrt{q^2 + 4pr}\right)} \tag{6.2 – 46a, b}$$

and c_1, c_2, c_3, and c_4 are constants, which are to be determined using the boundary conditions.

From Eq. (46b), we have

$$\left(2p\lambda^2 - q\right)^2 = q^2 + 4pr \quad \text{or} \quad p\lambda^4 - q\lambda^2 - r = 0 \tag{6.2 – 47a}$$

$$\left(2p\mu^2 + q\right)^2 = q^2 + 4pr \quad \text{or} \quad p\mu^4 + q\mu^2 - r = 0 \tag{6.2 – 47b}$$

Substituting for p, q, and r from Eq. (45) into Eq. (47a,b) and solving for ω^2, we obtain

$$\omega^2 = \lambda^4 \left(\frac{E^b_{xx} I_{yy}}{\hat{I}_0}\right)\left(\frac{1 + P_1}{1 + R_1}\right), \quad P_1 = \frac{b\hat{N}_{xx}}{E^b_{xx} I_{yy} \lambda^2}, \quad R_1 = \frac{\hat{I}_2}{\hat{I}_0}\lambda^2 \tag{6.2 – 48a}$$

$$\omega^2 = \mu^4 \left(\frac{E^b_{xx} I_{yy}}{\hat{I}_0}\right)\left(\frac{1 - P_2}{1 - R_2}\right), \quad P_2 = \frac{b\hat{N}_{xx}}{E^b_{xx} I_{yy} \mu^2}, \quad R_2 = \frac{\hat{I}_2}{\hat{I}_0}\mu^2 \tag{6.2 – 48b}$$

The two expressions for ω in Eqs. (48a,b) are the same and hence either one can be used to calculate the frequency once λ is known.

When the applied axial load is zero, the frequency of vibration can be calculated from

$$\omega^2 = \lambda^4 \frac{E^b_{xx} I_{yy}}{\hat{I}_0}\left(1 - \frac{\hat{I}_2 \lambda^2}{\hat{I}_0 + \hat{I}_2 \lambda^2}\right) = \mu^4 \frac{E^b_{xx} I_{yy}}{\hat{I}_0}\left(1 + \frac{\hat{I}_2 \mu^2}{\hat{I}_0 - \hat{I}_2 \mu^2}\right) \tag{6.2 – 49}$$

It is clear from the first expression that rotary inertia decreases the frequency of natural vibration. If the rotary inertia is neglected, we have $\lambda = \mu$ and

$$\omega = \lambda^2 a_0, \quad a_0 = \sqrt{\frac{E^b_{xx} I_{yy}}{\hat{I}_0}} \tag{6.2 – 50}$$

In the following discussion beams with both ends simply supported or clamped are considered to illustrate the procedure to evaluate the constants c_1 through c_4, and more importantly, to determine λ so that Eqs. (46)–(48) can be used to find ω. The smallest frequency is known as the *fundamental frequency*. For other boundary conditions, the reader is referred to Table 6.2–3.

Table 6.2–3: Values of the constants and eigenvalues for natural vibration of laminated composite beams with various boundary conditions ($\lambda_n^4 \equiv \omega_n^2 I_0 / E_{xx}^b I_{yy} = (e_n/a)^4$). The classical laminate theory *without* rotary inertia is used.

End conditions at $x = 0$ and $x = a$	Constants†	Characterstic equation and values of $e_n \equiv \lambda_n a$
• Hinged–Hinged	$c_1 \neq 0,\ c_2 = c_3 = c_4 = 0$	$\sin e_n = 0$ $e_n = n\pi$
• Fixed–Fixed	$c_1 = -c_3 = 1/(\sin e_n - \sinh e_n)$ $-c_2 = c_4 = 1/(\cos e_n - \cosh e_n)$	$\cos e_n \cosh e_n - 1 = 0$ $e_n = 4.730, 7.853, \cdots$
• Fixed–Free	$c_1 = -c_3 = 1/(\sin e_n + \sinh e_n)$ $-c_2 = c_4 = 1/(\cos e_n + \cosh e_n)$	$\cos e_n \cosh e_n + 1 = 0$ $e_n = 1.875, 4.694, \cdots$
• Free–Free	$c_1 = c_3 = 1/(\sin e_n - \sinh e_n)$ $c_2 = c_4 = -1/(\cos e_n - \cosh e_n)$	$\cos e_n \cosh e_n - 1 = 0$ $e_n = 4.730, 7.853, \cdots$
• Hinged–Fixed	$c_1 = 1/\sin e_n,\ c_3 = 1/\sinh e_n$ $c_2 = c_4 = 0$	$\tan e_n = \tanh e_n$ $e_n = 3.927, 7.069, \cdots$
• Hinged–Free	$c_1 = 1/\sin e_n,\ c_3 = -1/\sinh e_n$ $c_2 = c_4 = 0$	$\tan e_n = \tanh e_n$ $e_n = 3.927, 7.069, \cdots$

† See Eq. (6.2–46a): $W(x) = c_1 \sin \lambda x + c_2 \cos \lambda x + c_3 \sinh \mu x + c_4 \cosh \mu x$.

Example 6.2–5 (*Simply supported beam*):

For a simply supported beam, the boundary conditions in Eq. (31b) give

$$c_2 = c_3 = c_4 = 0 \tag{6.2-51}$$

$$c_1 \sin \lambda a = 0, \text{ which implies } \lambda = \frac{n\pi}{a} \tag{6.2-52}$$

Substituting for λ from Eqs. (45) and (46a) into Eq. (48a), we obtain

$$\omega_n = \left(\frac{n\pi}{a}\right)^2 a_0 \sqrt{1 + \frac{b\hat{N}_{xx}}{(\frac{n\pi}{a})^2 E_{xx}^b I_{yy}}} \sqrt{\frac{1}{1 + (\frac{n\pi}{a})^2 \frac{\hat{I}_2}{\hat{I}_0}}} \qquad (6.2-53a)$$

If the rotary inertia is neglected, we obtain

$$\omega_n = \left(\frac{n\pi}{a}\right)^2 a_0 \sqrt{1 + \frac{b\hat{N}_{xx}}{(\frac{n\pi}{a})^2 E_{xx}^b I_{yy}}} \qquad (6.2-53b)$$

Thus the effect of the axial tensile force $\hat{N}_{xx}$ is to increase the natural frequencies. If we have a very flexible beam, say a cable under large tension, the second term under the radical in Eq. (53b) becomes very large in comparison with unity; if n is not large, we have

$$\omega_n \approx \frac{n\pi}{a} \sqrt{\frac{\hat{N}_{xx}}{I_0}} \qquad (6.2-53c)$$

which are natural frequencies of a stretched laminated cable. We also note from Eq. (53b) that frequencies of natural vibration decrease when a compressive force instead of a tensile force is acting on the beam.

When $\hat{N}_{xx} = 0$, we obtain from Eq. (53a)

$$\begin{aligned} \omega_n &= \left(\frac{n\pi}{a}\right)^2 a_0 \sqrt{\frac{1}{1 + (\frac{n\pi}{a})^2 \frac{\hat{I}_2}{\hat{I}_0}}} \\ &= \left(\frac{n\pi}{a}\right)^2 \sqrt{\frac{E_{xx}^b I_{yy}}{\hat{I}_0}} \sqrt{1 - \frac{\hat{I}_2 (\frac{n\pi}{a})^2}{(\frac{n\pi}{a})^2 \hat{I}_2 + \hat{I}_0}} \end{aligned} \qquad (6.2-54)$$

Thus, rotatory inertia decreases frequencies of natural vibration. If the rotatory inertia is neglected, we obtain

$$\omega_n = \left(\frac{n\pi}{a}\right)^2 \sqrt{\frac{E_{xx}^b I_{yy}}{\hat{I}_0}} \qquad (6.2-55)$$

Example 6.2–6 (*Clamped beam*): ______________________________

For a beam clamped at both ends, the boundary conditions in Eq. (36) lead to

$$c_2 + c_4 = 0, \quad \lambda c_1 + \mu c_3 = 0 \qquad (6.2-56)$$

and the eigenvalue problem

$$\begin{bmatrix} \sin\lambda a - \left(\frac{\lambda}{\mu}\right)\sinh\mu a & \cos\lambda a - \cosh\mu a \\ \cos\lambda a - \cosh\mu a & -\sin\lambda a - \left(\frac{\mu}{\lambda}\right)\sinh\mu a \end{bmatrix} \begin{Bmatrix} c_1 \\ c_2 \end{Bmatrix} = \begin{Bmatrix} 0 \\ 0 \end{Bmatrix} \qquad (6.2-57)$$

where relations (56) are used to eliminate c_3 and c_4. For nonzero c_1 and c_2, we require the determinant of the coefficient matrix of the above equations to vanish, which yields the characteristic polynomial

$$-2 + 2\cos\lambda a \cosh\mu a + \left(\frac{\lambda}{\mu} - \frac{\mu}{\lambda}\right)\sin\lambda a \sinh\mu a = 0 \qquad (6.2-58)$$

The solution of this nonlinear equation gives λ and μ. Then the natural frequency of vibration can be calculated from Eq. (48a) or (48b); if the applied axial force is zero, Eq. (49) can be used to calculate the frequency of vibration.

For natural vibration without rotatory inertia and applied in-plane force (*i.e.*, $q = 0$ in Eq. (46a,b) and $\lambda = \mu$), Eq. (58) takes the simpler form

$$\cos\lambda a \cosh\lambda a - 1 = 0 \qquad (6.2-59)$$

Equation (59) is satisfied for the following values of λ:

$$\lambda_1 a = 4.730,\ \lambda_2 a = 7.853,\ \cdots,\ \lambda_n a = (n + \frac{1}{2})\pi \qquad (6.2-60)$$

Maximum transverse deflections, critical buckling loads, and fundamental natural frequencies of various laminated beams, according to the classical beam theory, are presented in Table 6.2–4 for simply supported (hinged–hinged), clamped (fixed–fixed), and cantilever (clamped–free) boundary conditions. In the case of bending, the point load is $F_0 b$, where F_0 is the line load across the width of the beam (force/unit length), and the distributed line load along the length is $q_0 b$, where q_0 is the intensity of the distributed load (force/unit square area). In Table 6.2–4, the first row corresponds to deflections due to point load F_0, and the second row corresponds to deflections due to uniformly distributed load q_0. Also, on the second and third rows, frequencies corresponding to $a/h = 100$ and $a/h = 10$ are listed when rotary inertia is included. All other frequencies were computed by neglecting the rotary inertia. The following nondimensionalizations are used:

$$\bar{w} = -w_{max} E_2 h^3 \times 10^2 / q_0 a^4 \quad (F_0 = q_0 a)$$

$$\bar{N} = N_{xx}^0 a^2 / E_2 h^3,\quad \bar{\omega} = \omega_1 a^2 \sqrt{I_0 / E_2 h^3} \qquad (6.2-61)$$

The stiffness in a laminate is largest in the fiber direction because $E_1 > E_2$. Also, the bending stiffness increases with (cube of) the distance of the 0° layers from the midplane. Thus, the 0°–laminated beam is stiffer in bending than the 90°–laminated beam, and therefore, 0° beam has smaller deflection and larger buckling load and natural frequencies when compared to the 90° beam. Since the 0° laminae are placed farther from the midplane in $(0/90)_s$ laminate, it has smaller deflection and larger buckling load and natural frequencies when compared to the $(90/0)_s$ beams. Similarly, due to the placement of the 0° layers, laminate A is stiffer than laminate B, and laminate B is stiffer than laminate C. Symmetric angle-ply laminated beams $(\theta/-\theta)_s$ have the same stiffness characteristics as $(-\theta/\theta)_s$, and they are less stiff compared to the symmetric cross-ply laminated beams.

Table 6.2–4: Maximum transverse deflections, critical buckling loads, and fundamental frequencies of laminated beams according to the classical beam theory ($h_n = h/n$, $E_1/E_2 = 25$, $G_{12} = G_{13} = 0.5E_2$, $G_{23} = 0.2E_2$, $\nu_{12} = 0.25$).

	Hinged–Hinged			Clamped–Clamped			Clamped–Free		
Laminate	$\bar{w}$	$\bar{N}$	$\bar{\omega}$	$\bar{w}$	$\bar{N}$	$\bar{\omega}$	$\bar{w}$	$\bar{N}$	$\bar{\omega}$
0	1.000	20.562	14.246	0.250	82.247	32.292	16.000	5.140	5.074
	0.625		14.245	0.125		32.291	6.000		5.074
			14.187			32.129			5.071
90	25.000	0.822	2.849	6.250	3.290	6.458	400.00	0.205	1.015
	15.625			3.125			150.00		
$(0/90)_s$	1.134	18.127	13.375	0.283	72.507	30.320	18.149	4.532	4.764
	0.709			0.142			6.806		
$(90/0)_s$	6.239	3.296	5.703	1.560	13.183	12.929	99.821	0.824	2.032
	3.899			0.780			37.433		
$(45/-45)_s$	14.308	1.437	3.766	3.577	5.748	8.537	228.93	0.359	1.341
	8.942			1.788			85.847		
Laminate A	1.607	12.790	11.236	0.402	51.162	25.469	25.721	3.197	4.002
	1.005			0.201			9.645		
Laminate B	2.801	7.341	8.512	0.700	29.366	19.296	44.813	1.835	3.032
	1.751			0.350			16.805		
Laminate C	7.945	2.588	5.054	1.986	10.351	11.456	127.13	0.647	1.800
	4.966			0.993			47.673		

Laminate A = $(0/\pm 45/90)_s$, Laminate B = $(45/0/-45/90)_s$,

Laminate C = $(90/\pm 45/0)_s$.

We note that for clamped–clamped and clamped–free beams, the calculation of natural frequencies require the solutions of transcendental equations (58) and (61) for λ. For the case where rotary inertia is negligible, the roots of these equations are given in Table 6.2–3. To see the effect of rotary inertia, Eqs. (58) and (61) were solved for λ and the frequencies were calculated. From the frequencies listed in rows 2 and 3 of Table 6.2–4, it is clear that the effect of rotary inertia on fundamental frequencies is negligible for small length-to-height ratios. Except for second and third rows, all other frequencies listed in the table were calculated by neglecting the rotary inertia, in which case the values of λ_1 given in Table 6.2–3 are applicable.

6.3 Analysis of Laminated Beams Using FSDT

6.3–1 Governing Equations

Here we consider bending of symmetrically laminated beams using the first-order shear deformation theory. When applied to beams, FSDT is known as the *Timoshenko beam theory.* The governing equations can be readily obtained from the results of Section 5.4.

The laminate constitutive equations for symmetric laminates, in the absence of in-plane forces, are given by [see Eqs. (5.4–21) and (5.4–22)]

$$\begin{Bmatrix} M_{xx} \\ M_{yy} \\ M_{xy} \end{Bmatrix} = \begin{bmatrix} D_{11} & D_{12} & D_{16} \\ D_{12} & D_{22} & D_{26} \\ D_{16} & D_{26} & D_{66} \end{bmatrix} \begin{Bmatrix} \frac{\partial \phi_x}{\partial x} \\ \frac{\partial \phi_y}{\partial y} \\ \frac{\partial \phi_x}{\partial y} + \frac{\partial \phi_y}{\partial x} \end{Bmatrix} \tag{6.3 - 1a}$$

$$\begin{Bmatrix} Q_y \\ Q_x \end{Bmatrix} = K \begin{bmatrix} A_{44} & A_{45} \\ A_{45} & A_{55} \end{bmatrix} \begin{Bmatrix} \frac{\partial w_0}{\partial y} + \phi_y \\ \frac{\partial w_0}{\partial x} + \phi_x \end{Bmatrix} \tag{6.3 - 1b}$$

or, in inverse form, we have

$$\begin{Bmatrix} \frac{\partial \phi_x}{\partial x} \\ \frac{\partial \phi_y}{\partial y} \\ \frac{\partial \phi_x}{\partial y} + \frac{\partial \phi_y}{\partial x} \end{Bmatrix} = \begin{bmatrix} D^*_{11} & D^*_{12} & D^*_{16} \\ D^*_{12} & D^*_{22} & D^*_{26} \\ D^*_{16} & D^*_{26} & D^*_{66} \end{bmatrix} \begin{Bmatrix} M_{xx} \\ M_{yy} \\ M_{xy} \end{Bmatrix} \tag{6.3 - 2a}$$

$$\begin{Bmatrix} \frac{\partial w_0}{\partial y} + \phi_y \\ \frac{\partial w_0}{\partial x} + \phi_x \end{Bmatrix} = \frac{1}{K} \begin{bmatrix} A^*_{44} & A^*_{45} \\ A^*_{45} & A^*_{55} \end{bmatrix} \begin{Bmatrix} Q_y \\ Q_x \end{Bmatrix} \tag{6.3 - 2b}$$

where K is the shear correction coefficient, D^*_{ij}, $(i, j = 1, 2, 6)$ denote the elements of the inverse of $[D]$, and A^*_{ij}, $(i, j = 4, 5)$ denote elements of the inverse of $[A]$:

$$A^*_{44} = \frac{A_{55}}{A}\,, \quad A^*_{55} = \frac{A_{44}}{A}\,, \quad A^*_{45} = -\frac{A_{45}}{A}\,, \quad A = A_{44}A_{55} - A_{45}A_{45} \tag{6.3 - 3}$$

As in Section 6.2, we assume that $M_{yy} = M_{xy} = Q_y = \phi_y = 0$ and both w_0 and ϕ_x are functions of only x and t:

$$w_0 = w_0(x,t), \quad \phi_x = \phi_x(x,t) \tag{6.3-4}$$

From Eq. (5.4–1) the displacement field takes the form (when the in-plane displacements u_0 and v_0 are zero)

$$u(x,z) = z\phi_x(x), \quad w(x,z) = w_0(x) \tag{6.3-5a}$$

and the linear strain–displacement relations give

$$\varepsilon_{xx} = z\frac{\partial \phi_x}{\partial x}\ , \ 2\varepsilon_{xz} = \frac{\partial w_0}{\partial x} + \phi_x \tag{6.3-5b}$$

From Eqs. (2a) and (2b) we have

$$\frac{\partial \phi_x}{\partial x} = D_{11}^* M_{xx}, \quad \frac{\partial w_0}{\partial x} + \phi_x = \frac{A_{55}^*}{K} Q_x \tag{6.3-6}$$

or

$$E_{xx}^b I_{yy} \frac{\partial \phi_x}{\partial x} = M(x), \quad M(x) = bM_{xx}, \quad E_{xx}^b = \frac{12}{D_{11}^* h^3} \tag{6.3-7a}$$

$$KG_{xz}^b bh\left(\frac{\partial w_0}{\partial x} + \phi_x\right) = Q(x), \quad Q(x) = bQ_x, \quad G_{xz}^b = \frac{1}{A_{55}^* h} \tag{6.3-7b}$$

where K denotes the the shear correction coefficient. The equations of motion from Eq. (5.4–13) are

$$\frac{\partial Q_x}{\partial x} + \hat{N}_{xx}\frac{\partial^2 w_0}{\partial x^2} + q = I_0 \frac{\partial^2 w_0}{\partial t^2} \tag{6.3-8a}$$

$$\frac{\partial M_{xx}}{\partial x} - Q_x = I_2 \frac{\partial^2 \phi_x}{\partial t^2} \tag{6.3-8b}$$

Using Eq. (7) in Eq. (8), the equations of motion can be recast in terms of the displacement functions:

$$\boxed{\begin{aligned} KG_{xz}^b bh\left(\frac{\partial^2 w_0}{\partial x^2} + \frac{\partial \phi_x}{\partial x}\right) + b\hat{N}_{xx}\frac{\partial^2 w_0}{\partial x^2} + \hat{q} &= \hat{I}_0 \frac{\partial^2 w_0}{\partial t^2} \\ E_{xx}^b I_{yy}\frac{\partial^2 \phi_x}{\partial x^2} - KG_{xz}^b bh\left(\frac{\partial w_0}{\partial x} + \phi_x\right) &= \hat{I}_2 \frac{\partial^2 \phi_x}{\partial t^2} \end{aligned}} \tag{6.3-9a,b}$$

where

$$\hat{q} = bq, \quad \hat{I}_0 = bI_0, \quad \hat{I}_2 = bI_2 \tag{6.3-9c}$$

6.3–2 Bending

Note that when the laminated beam problem is such that the bending moment $M(x)$ and $Q(x)$ can be written readily in terms of known applied loads (like in statically determinate beam problems), Eq. (7a) can be utilized to determine ϕ_x, and then w_0 can be determined using Eq. (7b). When $M(x)$ and $Q(x)$ cannot be expressed in terms of known loads, Eqs. (9a,b) are used to determine $w_0(x)$ and $\phi_x(x)$. In the latter case, the following relations prove to be useful.

For bending analysis, Eqs. (9a,b) reduce to

$$KG_{xz}^b bh\left(\frac{d^2w_0}{dx^2}+\frac{d\phi_x}{dx}\right)+\hat{q}=0$$
$$E_{xx}^b I_{yy}\frac{d^2\phi_x}{dx^2}-KG_{xz}^b bh\left(\frac{dw_0}{dx}+\phi_x\right)=0 \qquad (6.3-10a,b)$$

Integrating Eq. (10a) with respect to x, we obtain

$$KG_{xz}^b bh\left(\frac{dw_0}{dx}+\phi_x\right)=-\int^x \hat{q}(\xi)d\xi+c_1 \qquad (6.3-11)$$

Substituting the result into Eq. (10b) and integrating with respect to x, we obtain

$$E_{xx}^b I_{yy}\frac{d\phi_x}{dx}=-\int^x\int^\eta \hat{q}(\xi)d\xi d\eta+c_1x+c_2 \qquad (6.3-12a)$$

$$E_{xx}^b I_{yy}\phi_x(x)=-\int^x\int^\zeta\int^\eta \hat{q}(\xi)d\xi d\eta d\zeta+c_1\frac{x^2}{2}+c_2x+c_3 \qquad (6.3-12b)$$

Substituting for $\phi(x)$ from Eq. (12b) into Eq. (11), we arrive at

$$\frac{dw_0}{dx}=-\frac{1}{E_{xx}^b I_{yy}}\left[-\int^x\int^\zeta\int^\eta \hat{q}(\xi)d\xi d\eta d\zeta+c_1\frac{x^2}{2}+c_2x+c_3\right]$$
$$+\frac{1}{KG_{xz}^b bh}\left[-\int^x \hat{q}(\xi)d\xi+c_1\right] \qquad (6.3-13a)$$

$$
\boxed{\begin{aligned} w_0(x) = &- \frac{1}{E^b_{xx} I_{yy}} \left[-\int^x \int^\xi \int^\eta \int^\mu \hat{q}(\zeta) d\zeta d\mu d\eta d\xi + c_1 \frac{x^3}{6} + c_2 \frac{x^2}{2} + c_3 x + c_4 \right] \\ &+ \frac{1}{KG^b_{xz} bh} \left[-\int^x \int^\xi \hat{q}(\zeta) d\zeta d\xi + c_1 x \right] \end{aligned}}
\tag{6.3 – 13b}
$$

where the constants of integration c_1 through c_4 can be determined using the boundary conditions of the beam.

It is informative to note from Eq. (13) that the transverse deflection of the Timoshenko beam theory consists of two parts, one due to pure bending and other due to transverse shear:

$$
w_0(x) = w^b_0(x) + w^s_0(x) \tag{6.3 – 14a}
$$

where

$$
\begin{aligned} w^b_0(x) &= \frac{1}{E^b_{xx} I_{yy}} \left[\int^x \int^\xi \int^\eta \int^\mu \hat{q}(\zeta) d\zeta d\mu d\eta d\xi - c_1 \frac{x^3}{6} - c_2 \frac{x^2}{2} - c_3 x - c_4 \right] \\ w^s_0(x) &= \frac{1}{KG^o_{xz} bh} \left[-\int^x \int^\xi \hat{q}(\zeta) d\zeta d\xi + c_1 x \right] \end{aligned} \tag{6.3 – 14b}
$$

The pure bending deflection $w^b_0(x)$ is the same as that derived in the classical beam theory [*cf.* Eq. (6.2–11b)]. When the transverse shear stiffness is infinite, the shear deflection $w^s_0(x)$ goes to zero, and the Timoshenko beam theory solutions reduce to those of the classical beam theory. In fact, one can establish exact relationships between the solutions of the Euler–Bernoulli beam solutions and Timoshenko beam solutions (see Wang [27] and Reddy *et al.* [28]). These relationships enable one to obtain the Timoshenko beam solutions from known classical beam solutions for any set of boundary conditions (see Exercises 6.39 and 6.40).

The expressions for in-plane stresses of the Timoshenko beam theory remain the same as those in the classical beam theory [see Eq. (6.2–12b)]. The expressions given in Eqs. (6.2–15a,b) for transverse shear stresses derived from 3–D equilibrium are also valid for the present case.

The transverse shear stress can also be computed via constitutive equation in the Timoshenko beam theory. We have

$$
\sigma^{(k)}_{xz}(x,z) = \bar{Q}^{(k)}_{55} A^*_{55} \frac{Q(x)}{b} \tag{6.3 – 15}
$$

Example 6.3–1 (*Simply supported beam*):

Here we consider the three-point bending problem of Section 6.2 (see Figure 6.2–2). For this case, the bending moment [see Eq. (6.2–17)] and shear forces are

$$M(x) = -\frac{F_0 bx}{2}\ , \quad Q(x) = \frac{dM}{dx} = -\frac{F_0 b}{2}\ , \quad 0 \le x \le \frac{a}{2} \tag{6.3 – 16}$$

Using Eq. (16) for M in Eq. (7a) and integrating with respect to x, we obtain

$$\phi_x(x) = -\frac{F_0 b}{4E_{xx}^b I_{yy}} x^2 + c_1$$

By symmetry, $u_1 = u_0 + z\phi_x$ is zero at $x = a/2$. This implies that $\phi_x(a/2) = 0$. Hence

$$c_1 = \frac{F_0 ba^2}{16E_{xx}^b I_{yy}}$$

and the solution becomes

$$\phi_x(x) = \frac{F_0 ba^2}{16E_{xx}^b I_{yy}}\left[1 - 4\left(\frac{x}{a}\right)^2\right], \quad 0 \le x \le \frac{a}{2} \tag{6.3 – 17}$$

It is interesting to note from Eq. (17) that the rotation function $\phi_x(x)$ is the same as the slope $-dw_0/dx$ from the Euler–Bernoulli beam theory (*i.e.*, ϕ_x is independent of transverse shear stiffness). Consequently, the bending moment [see Eq. (7a)], and therefore the axial stress, is independent of shear deformation. In fact, ϕ_x is independent of shear deformation for all statically determinate beams and indeterminate beams with symmetric boundary conditions and loading (see Wang [27]). However, for general statically indeterminate beams, the rotation ϕ_x will depend on the shear stiffness $KG_{xz}^b bh$ (see Exercise 6.11).

Substituting for ϕ_x into Eq. (7b), we obtain

$$\frac{dw_0}{dx} = -\frac{F_0 ba^2}{16E_{xx}^b I_{yy}}\left[1 - 4\left(\frac{x}{a}\right)^2\right] - \frac{F_0 b}{2KG_{xz}^b bh} \tag{6.3 – 18a}$$

Let us denote the first expression in (18a) by

$$\frac{dw_0^b}{dx} = -\frac{F_0 ba^2}{16E_{xx}^b I_{yy}}\left[1 - 4\left(\frac{x}{a}\right)^2\right] = -\phi_x(x) \tag{6.3 – 18b}$$

In light of Eq. (14a), the first part of Eq. (18a) can be viewed as the slope (or rotation) due to bending and the second one due to transverse shear strain:

$$\frac{dw_0}{dx} \equiv \frac{dw_0^b}{dx} + \frac{dw_0^s}{dx} \tag{6.3 – 18c}$$

Indeed, dw_0^s/dx can be interpreted as the transverse shear strain [*cf.* Eq. (5b)]

$$\frac{dw_0^s}{dx} = \frac{dw_0}{dx} - \frac{dw_0^b}{dx} = \frac{dw_0}{dx} + \phi_x(x) \equiv \gamma_{xz} \tag{6.3-19}$$

Note from Eq. (18a) that, in contrast to the classical beam theory, the slope dw_0/dx at the center of the beam in the Timoshenko beam theory is nonzero. We have ($I_{yy} = bh^3/12$)

$$\frac{dw_0}{dx}(\frac{a}{2}) = -\frac{F_0 b}{2KG_{xz}^b bh} \tag{6.3-20}$$

However, $dw_0^b/dx = -\phi_x$ is zero at $x = a/2$. Integrating Eq. (18a) with respect to x, we arrive at the expression

$$w_0(x) = -\frac{F_0 ba^3}{48E_{xx}^b I_{yy}}\left[3\left(\frac{x}{a}\right) - 4\left(\frac{x}{a}\right)^3\right] - \frac{F_0 ba}{2KG_{xz}^b bh}\left(\frac{x}{a}\right) \tag{6.3-21}$$

where the constant of integration is found to be zero on account of the boundary condition $w_0(0) = 0$. Note that the first part (w_0^b) is the same as that obtained in the classical beam theory [*cf.* Eq. (6.2–18)].

The maximum deflection occurs at $x = a/2$ and it is given by

$$\begin{aligned} w_{max} &= -\frac{F_0 ba^3}{48E_{xx}^b I_{yy}} - \frac{F_0 ba}{4KG_{xz}^b bh} \\ &= -\frac{F_0 ba^3}{48E_{xx}^b I_{yy}}\left[1 + \left(\frac{E_{xx}^b}{KG_{xz}^b}\right)\left(\frac{h}{a}\right)^2\right] \end{aligned} \tag{6.3-22}$$

Equation (22) shows that the effect of shear deformation is to increase the deflection. The contribution due to shear deformation to the deflection depends on the modulus ratio E_{xx}^b/G_{xz}^b as well as the ratio of thickness to length h/a. The effect of shear deformation is negligible for thin and long beams.

Example 6.3–2 (*Clamped beam*):

Consider a laminated beam fixed at both ends and subjected to uniformly distributed transverse load $q_0 b$ as well as a point load $F_0 b$ at the center, both acting downward. For this case, the boundary conditions are

$$u(0,z) = 0, \quad u(\frac{a}{2},z) = 0, \quad w(0,z) = 0, \quad Q_x(\frac{a}{2},z) = -\frac{F_0 b}{2} \tag{6.3-23}$$

which in turn imply that

$$\phi_x(0) = 0, \quad \phi_x(\frac{a}{2}) = 0, \quad w_0(0) = 0, \quad KG_{xz}^b bh\left(\frac{dw_0}{dx} + \phi_x\right)(\frac{a}{2}) = -\frac{F_0 b}{2} \tag{6.3-24}$$

The the solution is

$$\phi_x(x) = -\frac{q_0 b a^3}{12E^b_{xx}I_{yy}}\left[-2\left(\frac{x}{a}\right)^3 + 3\left(\frac{x}{a}\right)^2 - \left(\frac{x}{a}\right)\right] - \frac{F_0 b a^2}{8E^b_{xx}I_{yy}}\left[-2\left(\frac{x}{a}\right)^2 + \left(\frac{x}{a}\right)\right] \qquad (6.3-25)$$

$$w_0(x) = -\frac{q_0 b a^4}{24E^b_{xx}I_{yy}}\left[\left(\frac{x}{a}\right)^2 - \left(\frac{x}{a}\right)\right]^2 - \frac{q_0 b a^2}{2KG^b_{xz}bh}\left[\left(\frac{x}{a}\right) - \left(\frac{x}{a}\right)^2\right] - \frac{F_0 b a^3}{48E^b_{xx}I_{yy}}\left[3\left(\frac{x}{a}\right)^2 - 4\left(\frac{x}{a}\right)^3\right] - \frac{F_0 b a}{2KG^b_{xz}bh}\left(\frac{x}{a}\right) \qquad (6.3-26)$$

The maximum deflection is at $x = a/2$ and is given by [*cf.* Eq. (6.2–23)]

$$w_{max} = -\frac{q_0 b a^4}{384E^b_{xx}I_{yy}} - \frac{q_0 a^2}{8KG^b_{xz}h} - \frac{F_0 b a^3}{192E^b_{xx}I_{yy}} - \frac{F_0 a}{4KG^b_{xz}h} = -\left[\frac{q_0 b a^4}{384E^b_{xx}I_{yy}} + \frac{F_0 b a^3}{192E^b_{xx}I_{yy}}\right](1+S) \qquad (6.3-27a)$$

where S is the positive parameter that characterizes the contribution due to the transverse shear strain to the displacement field

$$S = 4\left(\frac{E^b_{xx}}{KG^b_{xz}}\right)\left(\frac{h}{a}\right)^2 \qquad (6.3-27b)$$

Table 6.3–1 contains expressions for transverse deflections and maximum transverse deflections of laminated beams according to the first-order shear deformation theory. By comparison to the classical theory (see Table 6.2–1), it is clear that the shear deformation increases the deflection.

Table 6.3–2 contains maximum transverse deflections $\hat{w}$ of various laminated beams according to the Timoshenko shear deformation beam theory. The effect of length-to-height (or thickness) ratios of the beam on the deflections can be seen from the results. Thin or long beams do not experience transverse shear strains. Clamped beams show the most difference in deflections due to transverse shear deformation (*i.e.*, accounting for the transverse shear strain). The effect of shear deformation on maximum deflection can be seen from Figures 6.3–1 and 6.3–2, where the nondimensionalized maximum deflection, $\bar{w} = -w_{max}E_2h^3/q_0a^4$ $(F_0 = q_0 a)$, of a simply supported beam is plotted as a function of length-to-height ratio a/h for various laminated beams under a point load and uniformly distributed load, respectively. The material properties of a lamina are taken to be those in Eq. (6.2–25). The effect of shear deformation is more significant for beams with length-to-thickness ratios smaller than 10.

Table 6.3–1: Transverse deflections of laminated composite beams with various boundary conditions and subjected to point load or uniformly distributed load (acting downward) according to the shear deformation theory.

$$c_1 = -\frac{F_0 ba^3}{E^b_{xx} I_{yy}}, \quad c_2 = -\frac{q_0 ba^4}{E^b_{xx} I_{yy}}, \quad s_1 = -\frac{F_0 ba}{K G^b_{xz} bh}, \quad s_2 = -\frac{q_0 ba^2}{G^b_{xz} bh}$$

Laminated Beam	Deflection, $w_0(x)$	Max. Deflection
• *Hinged–Hinged*		
Central point load	$\frac{c_1}{48}\left[3\left(\frac{x}{a}\right) - 4\left(\frac{x}{a}\right)^3\right]$ $+\frac{s_1}{2}\left(\frac{x}{a}\right)$	$\frac{1}{48}c_1 + \frac{1}{4}s_1$ at $x = \frac{a}{2}$
Uniform load	$\frac{c_2}{24}\left[\left(\frac{x}{a}\right) - 2\left(\frac{x}{a}\right)^3 + \left(\frac{x}{a}\right)^4\right]$ $+\frac{s_2}{2}\left[\left(\frac{x}{a}\right) - \left(\frac{x}{a}\right)^2\right]$	$\frac{5}{384}c_2 + \frac{1}{8}s_2$ at $x = \frac{a}{2}$
• *Fixed–Fixed*		
Central point load	$\frac{c_1}{48}\left[3\left(\frac{x}{a}\right)^2 - 4\left(\frac{x}{a}\right)^3\right]$ $+\frac{s_1}{2}\left(\frac{x}{a}\right)$	$\frac{1}{192}c_1 + \frac{1}{4}s_1$ at $x = \frac{a}{2}$
Uniform load	$\frac{c_2}{24}\left[\left(\frac{x}{a}\right)^2 - \left(\frac{x}{a}\right)\right]^2$ $+\frac{s_2}{2}\left[\left(\frac{x}{a}\right) - \left(\frac{x}{a}\right)^2\right]$	$\frac{1}{384}c_2 + \frac{1}{8}s_2$ at $x = \frac{a}{2}$
• *Fixed–Free*		
Point load at free end	$\frac{c_1}{6}\left[3\left(\frac{x}{a}\right)^2 - \left(\frac{x}{a}\right)^3\right]$ $+s_1\left(\frac{x}{a}\right)$	$\frac{1}{3}c_1 + s_1$ at $x = a$
Uniform load	$\frac{c_2}{24}\left[6\left(\frac{x}{a}\right)^2 - 4\left(\frac{x}{a}\right)^3 + \left(\frac{x}{a}\right)^4\right]$ $+\frac{s_2}{2}\left[\left(2\frac{x}{a}\right) - \left(\frac{x}{a}\right)^2\right]$	$\frac{1}{8}c_2 + \frac{1}{2}s_2$ at $x = a$

Table 6.3–2: Maximum transverse deflections of laminated beams according to the Timoshenko beam theory† ($h_n = h/n$, $E_1/E_2 = 25$, $G_{12} = G_{13} = 0.5E_2$, $G_{23} = 0.2E_2$, $\nu_{12} = 0.25$).

Laminate	Hinged–Hinged			Clamped–Clamped			Clamped–Free		
$\frac{a}{h}$	100	20	10	100	20	10	100	20	10
0	1.001	1.150	1.600	0.256	0.400	0.850	16.024	16.600	18.400
	0.628	0.700	0.925	0.128	0.200	0.425	6.012	6.300	7.200
90	25.015	25.375	26.500	6.265	6.625	7.750	400.00	401.50	406.00
	15.633	15.813	16.375	3.132	3.312	3.875	150.00	150.75	153.00
$(90/0)_s$	1.143	1.348	1.991	0.292	0.498	1.141	18.184	19.006	21.578
	0.713	0.816	1.137	0.146	0.249	0.570	6.823	7.234	8.520
$(45/-45)_s$	14.316	14.522	15.165	3.585	3.791	4.434	228.96	229.78	232.35
	8.947	9.049	9.371	1.793	1.895	2.217	85.864	86.276	87.561

†The first row of each laminate refers to nondimensionalized maximum deflections under point load ($F_0 b$) and the second one refers to maximum deflections under uniformly distributed load ($q_0 b$). The deflection is nondimensionalized as $\hat{w} = -w_{max}(E_2 h^3/q_0 a^4) \times 10^2$ ($F_0 = q_0 a$).

6.3–3 Buckling

For buckling analysis, the inertia terms and the applied transverse load q in Eqs. (9a,b) are set to zero to obtain the governing equations of buckling under compressive edge load $\hat{N}_{xx} = -N^0_{xx}$:

$$KG^b_{xz}bh\left(\frac{d^2W}{dx^2} + \frac{d\mathcal{X}}{dx}\right) + b\hat{N}_{xx}\frac{d^2W}{dx^2} = 0 \qquad (6.3-28a)$$

$$E^b_{xx}I_{yy}\frac{d^2\mathcal{X}}{dx^2} - KG^b_{xz}bh\left(\frac{dW}{dx} + \mathcal{X}\right) = 0 \qquad (6.3-28b)$$

Solving Eq. (28a) for $d\mathcal{X}/dx$ one obtains

$$KG^b_{xz}bh\frac{d\mathcal{X}}{dx} = -\left(KG^b_{xz}bh - bN^0_{xx}\right)\frac{d^2W}{dx^2} \qquad (6.3-29)$$

Integration with respect to x yields

$$KG^b_{xz}bh\mathcal{X}(x) = -\left(KG^b_{xz}bh - bN^0_{xx}\right)\frac{dW}{dx} + K_1 \qquad (6.3-30)$$

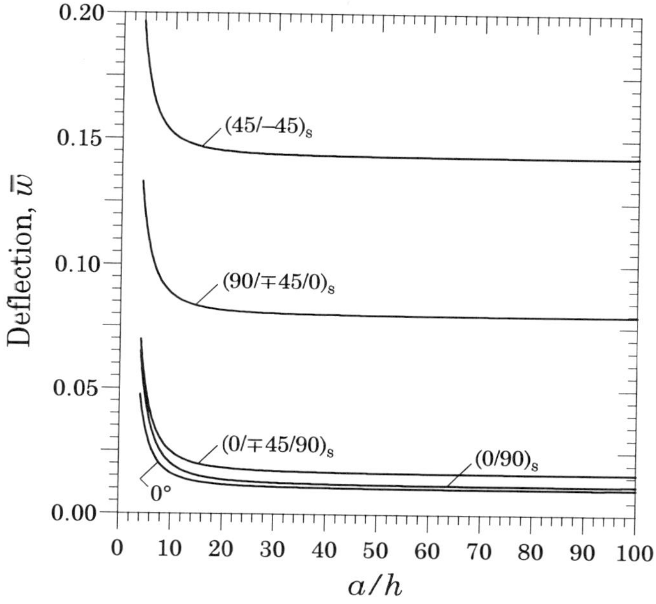

Figure 6.3–1: Transverse deflection ($\bar{w}$) versus length-to-thickness ratio (a/h) of simply supported beams under center point load.

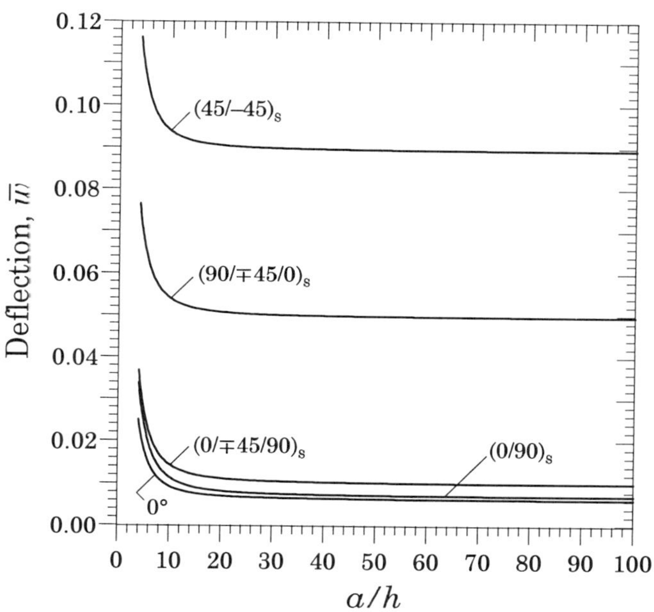

Figure 6.3–2: Transverse deflection ($\bar{w}$) versus length-to-thickness ratio (a/h) of simply supported beams under uniformly distributed load.

Next differentiate Eq. (28b) with respect to x

$$E_{xx}^b I_{yy}\frac{d^3\mathcal{X}}{dx^3} - KG_{xz}^b bh\left(\frac{d^2W}{dx^2} + \frac{d\mathcal{X}}{dx}\right) = 0$$

and substitute for $d\mathcal{X}/dx$ from Eq. (29) to obtain the result

$$-E_{xx}^b I_{yy}\frac{d^2}{dx^2}\left[\left(1 - \frac{bN_{xx}^0}{KG_{xz}^b bh}\right)\frac{d^2W}{dx^2}\right]$$

$$-KG_{xz}^b bh\left[\frac{d^2W}{dx^2} - \left(1 - \frac{bN_{xx}^0}{KG_{xz}^b bh}\right)\frac{d^2W}{dx^2}\right] = 0$$

or

$$E_{xx}^b I_{yy}\left(1 - \frac{bN_{xx}^0}{KG_{xz}^b bh}\right)\frac{d^4W}{dx^4} + bN_{xx}^0\frac{d^2W}{dx^2} = 0 \tag{6.3-31}$$

The general solution of Eq. (31) is

$$W(x) = c_1\sin\lambda x + c_2\cos\lambda x + c_3 x + c_4 \tag{6.3-32}$$

where

$$\lambda^2 = \frac{bN_{xx}^0}{\left(1 - \frac{bN_{xx}^0}{KG_{xz}^b bh}\right)E_{xx}^b I_{yy}} \quad \text{or} \quad bN_{xx}^0 = \frac{\lambda^2 E_{xx}^b I_{yy}}{\left(1 + \frac{\lambda^2 E_{xx}^b I_{yy}}{KG_{xz}^b bh}\right)} \tag{6.3-33}$$

and c_1 through c_4 are constants of integration, which must be evaluated using the boundary conditions.

Example 6.3–3 (*Simply supported beam*):

For a simply supported beam, the boundary conditions are [see Eq. (6.2–31a)]

$$W(0) = 0, \quad W(a) = 0, \quad \frac{d\mathcal{X}}{dx}(0) = 0, \quad \frac{d\mathcal{X}}{dx}(a) = 0 \tag{6.3-34a}$$

In view of Eq. (29), the above conditions are equivalent to

$$W(0) = 0, \quad W(a) = 0, \quad \frac{d^2W}{dx^2}(0) = 0, \quad \frac{d^2W}{dx^2}(a) = 0 \tag{6.3-34b}$$

The boundary conditions in Eq. (34b) lead to the result $c_2 = c_3 = c_4 = 0$, and for $c_1 \neq 0$ the requirement

$$\sin\lambda a = 0 \;\text{ implies }\; \lambda a = n\pi \tag{6.3-35}$$

Substituting for λ from Eq. (35) into Eq. (33) for N_{xx}^0, we obtain

$$bN_{xx}^0 = E_{xx}^b I_{yy}\left(\frac{n\pi}{a}\right)^2\left[\frac{KG_{xz}^b bh}{KG_{xz}^b bh + E_{xx}^b I_{yy}\left(\frac{n\pi}{a}\right)^2}\right]$$
$$= E_{xx}^b I_{yy}\left(\frac{n\pi}{a}\right)^2\left[1 - \frac{E_{xx}^b I_{yy}\left(\frac{n\pi}{a}\right)^2}{KG_{xz}^b bh + E_{xx}^b I_{yy}\left(\frac{n\pi}{a}\right)^2}\right] \qquad (6.3-36)$$

The critical buckling load is given by the minimum ($n = 1$)

$$bN_{cr} = E_{xx}^b I_{yy}\left(\frac{\pi}{a}\right)^2\left[1 - \frac{E_{xx}^b I_{yy}\left(\frac{\pi}{a}\right)^2}{KG_{xz}^b bh + E_{xx}^b I_{yy}\left(\frac{\pi}{a}\right)^2}\right] \qquad (6.3-37)$$

It is clear from the result in Eq. (37) that shear deformation has the effect of decreasing the buckling load [*cf.* Eq. (6.2–35)].

Example 6.3–4 (*Clamped beam*):

For a beam fixed at both ends, the boundary conditions are

$$W(0) = 0, \quad W(a) = 0, \quad \mathcal{X}(0) = 0, \quad \mathcal{X}(a) = 0 \qquad (6.3-38)$$

In order to impose the boundary conditions on $\mathcal{X}$, we use Eq. (30). The constant K_1 appearing in Eq. (30) can be shown (see Exercise 6.10) to be equal to $K_1 = -c_3(bN_{xx}^0)$. The boundary conditions yield

$$c_2 + c_4 = 0, \quad c_1 \sin\lambda a + c_2\cos\lambda a + c_3 a + c_4 = 0$$
$$-\left(1 - \frac{bN_{xx}^0}{KG_{xz}^b bh}\right)\lambda c_1 - c_3 = 0$$
$$-\left(1 - \frac{bN_{xx}^0}{KG_{xz}^b bh}\right)(\lambda c_1\cos\lambda a - \lambda c_2\sin\lambda a) - c_3 = 0$$

Expressing c_1 and c_2 in terms of c_3 and c_4, noting that

$$1 - \frac{bN_{xx}^0}{KG_{xz}^b bh} = \frac{1}{1 + \frac{\lambda^2 E_{xx}^b I_{yy}}{KG_{xz}^b bh}}$$

and then setting the determinant of the resulting algebraic equations among c_1 and c_2 to zero, we obtain

$$2(\cos\lambda a - 1)\left(1 + \frac{\lambda^2 E_{xx}^b I_{yy}}{KG_{xz}^b bh}\right) + \lambda a\ \sin\lambda a = 0 \qquad (6.3-39)$$

Once the value of λa is determined by solving the nonlinear equation (39), the buckling load can be readily determined from Eq. (33).

6.3–4 Vibration

For natural vibration, we assume that the applied axial force and transverse load are zero and that the motion is periodic. Equations (9a,b) take the form

$$KG_{xz}^b bh\left(\frac{d^2W}{dx^2}+\frac{d\mathcal{X}}{dx}\right)+\omega^2\hat{I}_0W=0 \qquad (6.3-40a)$$

$$E_{xx}^b I_{yy}\frac{d^2\mathcal{X}}{dx^2}-KG_{xz}^b bh\left(\frac{dW}{dx}+\mathcal{X}\right)+\omega^2\hat{I}_2\mathcal{X}=0 \qquad (6.3-40b)$$

We use the same procedure as before to eliminate $\mathcal{X}$ from Eqs. (40a,b). From Eq. (40a), we have

$$KG_{xz}^b bh\frac{d\mathcal{X}}{dx}=-\hat{I}_0\omega^2W-KG_{xz}^b bh\frac{d^2W}{dx^2} \qquad (6.3-41)$$

Substitute the above result into the derivative of Eq. (40b) for $d\mathcal{X}/dx$ and obtain the result

$$E_{xx}^b I_{yy}\frac{d^4W}{dx^4}+\left(\frac{E_{xx}^b I_{yy}\hat{I}_0}{KG_{xz}^b bh}+\hat{I}_2\right)\omega^2\frac{d^2W}{dx^2}-\left(1-\frac{\omega^2\hat{I}_2}{KG_{xz}^b bh}\right)\hat{I}_0\omega^2W=0 \quad (6.3-42a)$$

or

$$p\frac{d^4W}{dx^4}+q\frac{d^2W}{dx^2}-rW=0 \qquad (6.3-42b)$$

where

$$p=E_{xx}^b I_{yy},\quad q=\left(\frac{E_{xx}^b I_{yy}}{KG_{xz}^b bh}+\frac{\hat{I}_2}{\hat{I}_0}\right)\hat{I}_0\omega^2,\quad r=\left(1-\frac{\omega^2\hat{I}_2}{KG_{xz}^b bh}\right)\hat{I}_0\omega^2 \quad (6.3-42c)$$

The general solution of Eq. (42b) is

$$W(x)=c_1\sin\lambda x+c_2\cos\lambda x+c_3\sinh\mu x+c_4\cosh\mu x \qquad (6.3-43a)$$

where

$$\lambda=\sqrt{\frac{1}{2p}\left(q+\sqrt{q^2+4pr}\right)},\quad \mu=\sqrt{\frac{1}{2p}\left(-q+\sqrt{q^2+4pr}\right)} \qquad (6.3-43b)$$

and c_1, c_2, c_3, and c_4 are constants, which are to be determined using the boundary conditions. Note that we have

$$\left(2\lambda^2p-q\right)^2=q^2+4pr \quad \text{or} \quad p\lambda^4-q\lambda^2-r=0 \qquad (6.3-44)$$

Alternatively, Eq. (42a) can be written, with W given by Eq. (43), in terms of ω as

$$P\omega^4 - Q\omega^2 + R = 0 \tag{6.3-45a}$$

where

$$P = \frac{\hat{I}_2}{KG^b_{xz}bh}\,, \quad Q = \left[1 + \left(\frac{E^b_{xx}I_{yy}}{KG^b_{xz}bh} + \frac{\hat{I}_2}{\hat{I}_0}\right)\lambda^2\right], \quad R = \left(\frac{E^b_{xx}I_{yy}}{\hat{I}_0}\right)\lambda^4 \tag{6.3-45b}$$

Hence, there are two (sets of) roots of this equation (when $\hat{I}_2 \neq 0$)

$$(\omega^2)_1 = \frac{1}{2P}\left(Q - \sqrt{Q^2 - 4PR}\right), \quad (\omega^2)_2 = \frac{1}{2P}\left(Q + \sqrt{Q^2 - 4PR}\right) \tag{6.3-46}$$

It can be shown that $Q^2 - 4PR > 0$ (and $PQ > 0$), and therefore the frequency given by the first equation is the smaller of the two values. When the rotary inertia is neglected, we have $P = 0$ and the frequency is given by

$$\omega^2 = \frac{R}{\bar{Q}}\,, \quad \bar{Q} = \left[1 + \left(\frac{E^b_{xx}I_{yy}}{KG^b_{xz}bh}\right)\lambda^2\right], \quad R = \left(\frac{E^b_{xx}I_{yy}}{\hat{I}_0}\right)\lambda^4 \tag{6.3-47}$$

Example 6.3–5 (*Simply supported beam*):

For a simply supported beam, the boundary conditions in Eq. (34b) yield $c_2 = c_3 = c_4 = 0$ and

$$c_1 \sin \lambda a = 0, \text{ which implies } \lambda_n = \frac{n\pi}{a} \tag{6.3-48}$$

Substitution of λ from Eq. (48) into Eq. (47) and the result into Eq. (46a,b) gives two frequencies for each value of λ. The fundamental frequency will come from Eq. (46a).

When the rotatory inertia is neglected, we obtain from Eq. (47) the result

$$\omega_n = \left(\frac{n\pi}{a}\right)^2 \sqrt{\frac{E^b_{xx}I_{yy}}{\hat{I}_0}} \sqrt{1 - \frac{(\frac{n\pi}{a})^2 E^b_{xx}I_{yy}}{KG^b_{xz}bh + (\frac{n\pi}{a})^2 E^b_{xx}I_{yy}}} \tag{6.3-49}$$

Thus, shear deformation decreases frequencies of natural vibration [see Eq. (6.2–55)].

Example 6.3–6 (*Clamped beam*):

Using Eq. (40a) and expression (43a) for $W(x)$, $d\mathcal{X}/dx$ can be determined in terms of the constants c_1 through c_4, which then can be integrated with respect to x to obtain an expression for $\mathcal{X}$. Using the boundary conditions in Eq. (38), we obtain

$$c_2 + c_4 = 0, \quad c_1 \sin\lambda a + c_2 \cos\lambda a + c_3 \sinh\mu a + c_4 \cosh\mu a = 0$$

$$S_{11}c_1 - S_{22}c_3 = 0, \quad S_{11}c_1 - S_{11}c_2 - S_{22}c_3 - S_{22}c_4 = 0 \qquad (6.3-50a)$$

where

$$S_{11} = \mu\left(\hat{I}_0\omega^2 - \lambda^2 K G^b_{xz} bh\right), \quad S_{22} = \lambda\left(\hat{I}_0\omega^2 + \mu^2 K G^b_{xz} bh\right) \qquad (6.3-50b)$$

Eliminating c_2 and c_4 from the above equations, and setting the determinant of the resulting equations among c_1 and c_2 to zero (for a nontrivial solution), we obtain

$$-2 + 2\cos\lambda a \cosh\mu a + \sin\lambda a \sinh\mu a \left(\frac{S_{22}}{S_{11}} - \frac{S_{11}}{S_{22}}\right) = 0 \qquad (6.3-51)$$

Table 6.3–3 contains critical buckling loads and fundamental frequencies of various laminated beams according to the Timoshenko beam theory. The first row of each laminate refers to the nondimensionalized critical buckling load, the second row refers to nondimensionalized fundamental frequencies with rotary inertia, and the fourth row refers to fundamental frequencies without rotary inertia. The numbers in rows 3 and 5 refer to the fundamental frequencies calculated using the frequency equations of the classical laminate theory (for the simply supported boundary conditions, the frequency equations are the same in both theories). The following nondimensionalizations are used:

$$\bar{N} = N^0_{xx}(a^2/E_2h^3), \quad \bar{\omega} = \omega_1 a^2\sqrt{I_0/E_2h^3}$$

The frequency equations (51) of the Timoshenko theory depend, for clamped–clamped and clamped–free boundary conditions, on the lamination scheme and geometric parameters (through S_{ij}), whereas those of the classical laminate theory [see Eqs. (6.2–58) and (6.2–59)] are independent of the beam geometry or material properties. Thus, there are two different things that influence the frequencies in the Timoshenko theory: (i) the effect of transverse shear deformation [see Eqs. (47) and (49)], and (ii) the values of λ, which are governed by different equations than those of the classical theory (for clamped–clamped and clamped–free beams). The second effect is not significant, as can be seen from rows 3 and 5 of Table 6.3–3. Also, for clamped–clamped and clamped–free boundary conditions, the effect of rotary inertia on the frequencies is not as obvious as it was in the case of simply supported beams, where the rotary inertia would decrease the frequencies. From the results presented in Table 6.3–3, it appears that rotary inertia may actually increase the frequencies slightly.

Table 6.3–3: Critical buckling loads ($\bar{N}$) and fundamental frequencies ($\bar{\omega}$) of laminated beams according to the Timoshenko beam theory ($E_1/E_2 = 25$, $G_{12} = G_{13} = 0.5E_2$, $G_{23} = 0.2E_2$, $\nu_{12} = 0.25$).

Laminate		Hinged–Hinged			Clamped–Clamped			Clamped–Free		
$\frac{a}{h}$		100	20	10	100	20	10	100	20	10
0	$\bar{N}$	20.461	18.304	13.768	80.655	55.070	27.656	5.134	4.987	4.576
	$\bar{\omega}(\hat{I}_2 \neq 0)$	14.210	13.430	11.635	31.899	25.327	17.212	5.070	4.930	4.528
		14.210	13.430	11.635	32.110	28.506	22.140	5.010	4.965	4.675
	$\bar{\omega}(\hat{I}_2 = 0)$	14.211	13.441	11.657	31.824	24.636	16.680	5.063	4.813	4.229
		14.211	13.441	11.657	32.113	28.547	22.186	5.070	4.966	4.680
90		0.822	0.812	0.784	3.283	3.135	2.747	0.205	0.205	0.203
		2.848	2.829	2.771	6.450	6.260	5.761	1.015	1.012	1.004
		2.848	2.829	2.771	6.454	6.356	6.079	1.015	1.012	1.005
		2.848	2.832	2.781	6.449	6.232	5.681	1.015	1.009	0.993
		2.848	2.832	2.781	6.455	6.370	6.125	1.015	1.013	1.006
$(90/0)_s$		18.015	15.689	11.179	70.748	44.716	20.800	4.525	4.362	3.922
		13.334	12.434	10.488	29.857	22.672	14.837	4.758	4.594	4.132
		13.334	12.434	10.488	30.106	26.041	19.504	4.759	4.636	4.307
$(45/\text{–}45)_s$		1.436	1.419	1.369	5.737	5.478	4.802	0.359	0.358	0.355
		3.765	3.739	3.663	8.526	8.275	7.616	1.341	1.338	1.326
		3.765	3.739	3.663	8.531	8.402	8.036	1.341	1.338	1.328

The effect of length-to-height (or thickness) ratios of the beam on critical buckling loads $\bar{N}$ and fundamental frequencies $\bar{\omega}$ is shown in Figures 6.3–3 and 6.3–4, respectively, for various lamination schemes. The material properties used are those listed in Eq. (6.2–25). Transverse shear deformation has the effect of decreasing both buckling loads and natural frequencies. Thus, the classical laminate theory overpredicts buckling loads and natural frequencies. This is primarily due to the assumed infinite rigidity of the transverse normals in the classical laminate theory. Note that the assumption does not yield a conservative result; *i.e.*, if one designs a beam for buckling load based on the classical laminate theory and if no safety factor is used, it will fail for a working load smaller than the critical buckling load.

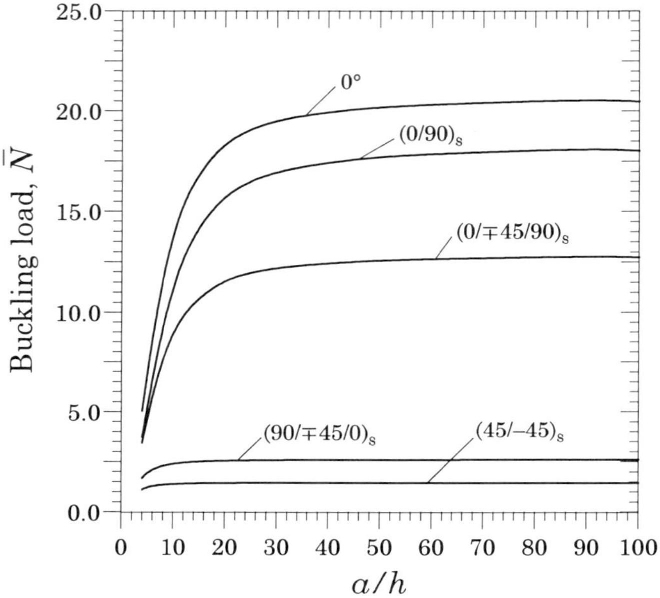

Figure 6.3–3: Nondimensionalized critical buckling load ($\bar{N}$) versus length-to-thickness ratio (a/h) of simply supported beams.

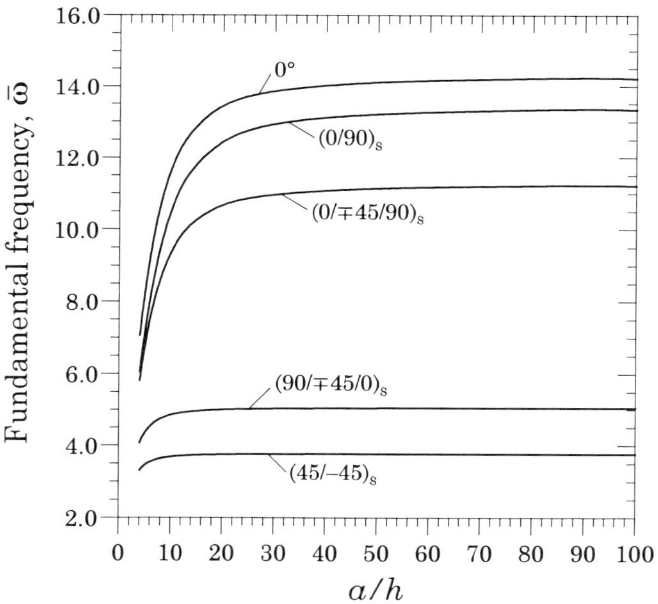

Figure 6.3–4: Nondimensionalized fundamental frequency ($\bar{\omega}$) versus length-to-thickness ratio (a/h) of simply supported beams.

6.4 Cylindrical Bending Using CLPT

6.4–1 Governing Equations

Consider a laminated rectangular plate strip, and let the x and y coordinates be parallel to the edges of the strip. Suppose that the plate is long in the y–direction and has a finite dimension along the x–direction, and subjected to a transverse load $q(x)$ that is uniform at any section parallel to the x–axis. In such a case, the deflection w_0 and displacements (u_0, v_0) of the plate are functions of only x. Therefore, all derivatives with respect to y are zero, and the plate bends into a cylindrical surface. For this *cylindrical bending* problem (see Figure 6.1–2), the governing equations of motion according to the linear classical laminate plate theory (CLPT) are given by [see Example 5.3–1; Eqs. (5.3–48)]

$$A_{11}\frac{\partial^2 u_0}{\partial x^2} + A_{16}\frac{\partial^2 v_0}{\partial x^2} - B_{11}\frac{\partial^3 w_0}{\partial x^3} - \frac{\partial N_{xx}^T}{\partial x} = I_0\frac{\partial^2 u_0}{\partial t^2} - I_1\frac{\partial^3 w_0}{\partial x \partial t^2} \qquad (6.4-1a)$$

$$A_{16}\frac{\partial^2 u_0}{\partial x^2} + A_{66}\frac{\partial^2 v_0}{\partial x^2} - B_{16}\frac{\partial^3 w_0}{\partial x^3} - \frac{\partial N_{xy}^T}{\partial x} = I_0\frac{\partial^2 v_0}{\partial t^2} \qquad (6.4-1b)$$

$$\begin{aligned} B_{11}\frac{\partial^3 u_0}{\partial x^3} + B_{16}\frac{\partial^3 v_0}{\partial x^3} - D_{11}\frac{\partial^4 w_0}{\partial x^4} + \frac{\partial}{\partial x}\left(\hat{N}_{xx}\frac{\partial w_0}{\partial x}\right) - \frac{\partial^2 M_{xx}^T}{\partial x^2} + q \\ = I_0\frac{\partial^2 w_0}{\partial t^2} - I_2\frac{\partial^4 w_0}{\partial x^2 \partial t^2} + I_1\frac{\partial^3 u_0}{\partial x \partial t^2} \end{aligned} \qquad (6.4-1c)$$

where $\hat{N}_{xx}$ is an applied axial load, and

$$(I_0, I_1, I_2) = \sum_{k=1}^{L}\int_{z_k}^{z_{k+1}} (1, z, z^2)\rho_0^{(k)}\, dz \qquad (6.4-1d)$$

For a general lamination scheme, the three equations are fully coupled. In the case of cross-ply laminates, the second equation becomes uncoupled from the rest. In the general case, Eqs. (1a–d) can be expressed in an alternative form by solving the first two equations for u'' and v'' and substituting the results into the third equation

$$\begin{aligned} A\frac{\partial^2 u_0}{\partial x^2} =& B\frac{\partial^3 w_0}{\partial x^3} + A_{66}\frac{\partial N_{xx}^T}{\partial x} - A_{16}\frac{\partial N_{xy}^T}{\partial x} + A_{66}I_0\frac{\partial^2 u_0}{\partial t^2} - A_{16}I_0\frac{\partial^2 v_0}{\partial t^2} \\ &- A_{66}I_1\frac{\partial^3 w_0}{\partial x \partial t^2} \end{aligned} \qquad (6.4-2a)$$

$$A\frac{\partial^2 v_0}{\partial x^2} = C\frac{\partial^3 w_0}{\partial x^3} + A_{11}\frac{\partial N_{xy}^T}{\partial x} - A_{16}\frac{\partial N_{xx}^T}{\partial x} + A_{11}I_0\frac{\partial^2 v_0}{\partial t^2} - A_{16}I_0\frac{\partial^2 u_0}{\partial t^2}$$

$$+ A_{16} I_1 \frac{\partial^3 w_0}{\partial x \partial t^2} \tag{6.4-2b}$$

$$D\frac{\partial^4 w_0}{\partial x^4} = \bar{B}\frac{\partial^2 N_{xx}^T}{\partial x^2} + \bar{C}\frac{\partial^2 N_{xy}^T}{\partial x^2} - \left(I_1 - \bar{B}I_0\right)\frac{\partial^3 u_0}{\partial x \partial t^2} + \bar{C}I_0\frac{\partial^3 v_0}{\partial x \partial t^2} - I_0\frac{\partial^2 w_0}{\partial t^2}$$
$$+ \left(I_2 - \bar{B}I_1\right)\frac{\partial^4 w_0}{\partial x^2 \partial t^2} - \frac{\partial^2 M_{xx}^T}{\partial x^2} + \frac{\partial}{\partial x}\left(\hat{N}_{xx}\frac{\partial w_0}{\partial x}\right) + q \tag{6.4-2c}$$

where

$$A = A_{11}A_{66} - A_{16}A_{16}, \quad B = B_{11}A_{66} - B_{16}A_{16}, \quad C = A_{11}B_{16} - A_{16}B_{11}$$

$$D = D_{11} - B_{11}\bar{B} - B_{16}\bar{C}, \quad \bar{B} = \frac{B}{A}, \quad \bar{C} = \frac{C}{A} \tag{6.4-2d}$$

Note that $C = 0$ for a cross-ply laminate ($A_{16} = B_{16} = D_{16} = 0$), and v is identically zero unless N_{xy}^T is at least a linear function of x.

If the in-plane inertias are neglected, Eq. (2c) for w_0 is uncoupled from those of u_0 and v_0. In the absence of thermal forces and axial loads, Eq. (2c) will have the same form as Eq. (6.2–8b). Therefore, the solutions developed in Sections 6.2–2 through 6.2–4 are also valid for cylindrical bending with appropriate change of the coefficients.

6.4–2 Bending

For static bending analysis, Eqs. $(2a, b, c)$ reduce to

$$A\frac{d^2 u_0}{dx^2} = B\frac{d^3 w_0}{dx^3} + A_{66}\frac{dN_{xx}^T}{dx} - A_{16}\frac{dN_{xy}^T}{dx} \tag{6.4-3a}$$
$$A\frac{d^2 v_0}{dx^2} = C\frac{d^3 w_0}{dx^3} + A_{11}\frac{dN_{xy}^T}{dx} - A_{16}\frac{dN_{xx}^T}{dx} \tag{6.4-3b}$$
$$D\frac{d^4 w_0}{dx^4} = \bar{B}\frac{d^2 N_{xx}^T}{dx^2} + \bar{C}\frac{d^2 N_{xy}^T}{dx^2} - \frac{d^2 M_{xx}^T}{dx^2} + q \tag{6.4-3c}$$

Equation (3c) governing w_0 is uncoupled from those governing (u_0, v_0). Equation (3c) closely resembles that for symmetrically laminated beams [see Eq. (6.2–10b)]. While Eq. (3c) is valid for more general laminates (symmetric as well as nonsymmetric), it differs from Eq. (6.2–10b) mainly in the bending stiffness term. Hence, much of the discussion presented in Section 6.2 on exact solutions applies to Eq. (3c). The limitation on the lamination scheme in cylindrical bending comes from the boundary conditions on all three displacements of the problem. When both edges are simply supported or clamped, exact solutions can be developed without any restrictions on the lamination scheme. For clamped–free laminated plate strips, satisfaction of the boundary conditions places a restriction on the lamination scheme, as will be seen shortly.

Since Eq. (3c) is uncoupled from Eqs. (3a,b), it can be integrated, for given thermal and mechanical loads, to obtain $w_0(x)$, and the result can be used in Eqs. (3a) and (3b) to determine $u_0(x)$ and $v_0(x)$:

$$D\frac{d^3w_0}{dx^3} = \bar{B}\frac{dN_{xx}^T}{dx} + \bar{C}\frac{dN_{xy}^T}{dx} - \frac{dM_{xx}^T}{dx} + \int^x q(\xi)\ d\xi + c_1 \qquad (6.4-4a)$$

$$A\frac{d^2u_0}{dx^2} = \hat{B}\int^x q(\xi)\ d\xi + G_1\frac{dN_{xx}^T}{dx} + F_1\frac{dN_{xy}^T}{dx} - \hat{B}\frac{dM_{xx}^T}{dx} + a_1 \qquad (6.4-4b)$$

$$A\frac{d^2v_0}{dx^2} = \hat{C}\int^x q(\xi)\ d\xi + G_2\frac{dN_{xx}^T}{dx} + F_2\frac{dN_{xy}^T}{dx} - \hat{C}\frac{dM_{xx}^T}{dx} + b_1 \qquad (6.4-4c)$$

where

$$G_1 = \frac{\bar{B}B}{D} + A_{66}, \quad F_1 = \frac{\bar{B}B}{D} - A_{16}, \quad \hat{B} = \frac{B}{D}$$

$$G_2 = \frac{\bar{B}C}{D} - A_{16}, \quad F_2 = \frac{\bar{B}C}{D} + A_{11}, \quad \hat{C} = \frac{C}{D}$$

$$a_1 = \hat{B}c_1, \quad b_1 = \hat{C}c_1 \qquad (6.4-5)$$

Further integrations lead to

$$A\frac{du_0}{dx} = \hat{B}\int^x\left(\int^\xi q(\eta)\ d\eta\right)d\xi + G_1N_{xx}^T + F_1N_{xy}^T - \hat{B}M_{xx}^T + a_1x + a_2 \qquad (6.4-6a)$$

$$A\frac{dv_0}{dx} = \hat{C}\int^x\left(\int^\xi q(\eta)\ d\eta\right)d\xi + G_2N_{xx}^T + F_2N_{xy}^T - \hat{C}M_{xx}^T + b_1x + b_2 \qquad (6.4-6b)$$

$$D\frac{d^2w_0}{dx^2} = \bar{B}N_{xx}^T + \bar{C}N_{xy}^T - M_{xx}^T + \int^x\left(\int^\xi q(\eta)\ d\eta\right)d\xi + c_1x + c_2 \qquad (6.4-6c)$$

and

$$\begin{aligned} Au_0(x) =&\hat{B}\int^x\left[\int^\xi\left(\int^\eta q(\zeta)\ d\zeta\right)\ d\eta\right]d\xi + G_1\int^x N_{xx}^T(\xi)d\xi + F_1\int^x N_{xy}^T(\xi)d\xi \\ &- \hat{B}\int^x M_{xx}^T(\xi)d\xi + a_1\frac{x^2}{2} + a_2x + a_3 \end{aligned} \qquad (6.4-7a)$$

$$\begin{aligned} Av_0(x) =&\hat{C}\int^x\left[\int^\xi\left(\int^\eta q(\zeta)\ d\zeta\right)\ d\eta\right]d\xi + G_2\int^x N_{xx}^T(\xi)d\xi + F_2\int^x N_{xy}^T(\xi)d\xi \\ &- \hat{C}\int^x M_{xx}^T(\xi)d\xi + b_1\frac{x^2}{2} + b_2x + b_3 \end{aligned} \qquad (6.4-7b)$$

$$D\frac{dw_0}{dx} = \int^x \left[\int^\xi \left(\int^\eta q(\zeta)\, d\zeta\right)\, d\eta\right] d\xi + \bar{B}\int^x N_{xx}^T(\xi) d\xi + \bar{C}\int^x N_{xy}^T(\xi) d\xi$$
$$- \int^x M_{xx}^T(\xi) d\xi + c_1\frac{x^2}{2} + c_2 x + c_3 \qquad (6.4-7c)$$

$$Dw_0(x) = \int^x \left\{\int^\xi \left[\int^\eta \left(\int^\zeta q(\mu) d\mu\right)\, d\zeta\right]\, d\eta\right\} d\xi + \bar{B}\int^x \left(\int^\xi N_{xx}^T(\eta) d\eta\right) d\xi$$
$$+ \bar{C}\int^x \left(\int^\xi N_{xy}^T(\eta) d\eta\right) d\xi - \int^x \left(\int^\xi M_{xx}^T(\eta) d\eta\right) d\xi$$
$$+ c_1\frac{x^3}{6} + c_2\frac{x^2}{2} + c_3 x + c_4 \qquad (6.4-7d)$$

If the temperature distribution in the laminate is of the form

$$\Delta T(x,z) = T_0 + zT_1 \qquad (6.4-8)$$

where T_0 and T_1 are constants, then we have

$$N_{xx}^T = \sum_{k=1}^{L} \int_{z_k}^{z_{k+1}} \left(\bar{Q}_{11}\alpha_{xx} + \bar{Q}_{12}\alpha_{yy} + 2\bar{Q}_{16}\alpha_{xy}\right)^{(k)} \Delta T\ dz$$
$$= A_1^T T_0 + B_1^T T_1 \qquad (6.4-9a)$$

$$N_{xy}^T = \sum_{k=1}^{L} \int_{z_k}^{z_{k+1}} \left(\bar{Q}_{16}\alpha_{xx} + \bar{Q}_{26}\alpha_{yy} + 2\bar{Q}_{66}\alpha_{xy}\right)^{(k)} \Delta T\ dz$$
$$= A_6^T T_0 + B_6^T T_1 \qquad (6.4-9b)$$

$$M_{xx}^T = \sum_{k=1}^{L} \int_{z_k}^{z_{k+1}} \left(\bar{Q}_{11}\alpha_{xx} + \bar{Q}_{12}\alpha_{yy} + 2\bar{Q}_{16}\alpha_{xy}\right)^{(k)} \Delta T z\ dz$$
$$= B_1^T T_0 + D_1^T T_1 \qquad (6.4-9c)$$

where

$$A_j^T = \sum_{k=1}^{L} \int_{z_k}^{z_{k+1}} \left(\bar{Q}_{j1}\alpha_{xx} + \bar{Q}_{j2}\alpha_{yy} + 2\bar{Q}_{j6}\alpha_{xy}\right)^{(k)}\ dz$$
$$B_j^T = \sum_{k=1}^{L} \int_{z_k}^{z_{k+1}} \left(\bar{Q}_{j1}\alpha_{xx} + \bar{Q}_{j2}\alpha_{yy} + 2\bar{Q}_{j6}\alpha_{xy}\right)^{(k)} z\ dz$$
$$D_j^T = \sum_{k=1}^{L} \int_{z_k}^{z_{k+1}} \left(\bar{Q}_{j1}\alpha_{xx} + \bar{Q}_{j2}\alpha_{yy} + 2\bar{Q}_{j6}\alpha_{xy}\right)^{(k)} z^2\ dz \qquad (6.4-10)$$

In addition, if $q = -q_0$ (uniformly distributed load applied downward), expressions in Eqs. (7) become

$$
\begin{aligned}
Au_0(x) =& -\hat{B}q_0\frac{x^3}{6} + G_1\left(A_1^T T_0 + B_1^T T_1\right)x + F_1\left(A_6^T T_0 + B_6^T T_1\right)x \\
& - \hat{B}\left(B_1^T T_0 + D_1^T T_1\right)x + a_1\frac{x^2}{2} + a_2 x + a_3 \\
\equiv& -\hat{B}q_0\frac{x^3}{6} + a_1\frac{x^2}{2} + \hat{a}_2 x + a_3 \qquad (6.4-11a)\\
Av_0(x) =& -\hat{C}q_0\frac{x^3}{6} + G_2\left(A_1^T T_0 + B_1^T T_1\right)x + F_2\left(A_6^T T_0 + B_6^T T_1\right)x \\
& - \hat{C}\left(B_1^T T_0 + D_1^T T_1\right)x + b_1\frac{x^2}{2} + b_2 x + b_3 \\
\equiv& -\hat{C}q_0\frac{x^3}{6} + b_1\frac{x^2}{2} + \hat{b}_2 x + b_3 \qquad (6.4-11b)\\
Dw_0(x) =& -q_0\frac{x^4}{24} + \bar{B}\left(A_1^T T_0 + B_1^T T_1\right)\frac{x^2}{2} + \bar{C}\left(A_6^T T_0 + B_6^T T_1\right)\frac{x^2}{2} \\
& - \left(B_1^T T_0 + D_1^T T_1\right)\frac{x^2}{2} + c_1\frac{x^3}{6} + c_2\frac{x^2}{2} + c_3 x + c_4 \\
\equiv& -q_0\frac{x^4}{24} + c_1\frac{x^3}{6} + \hat{c}_2\frac{x^2}{2} + c_3 x + c_4 \qquad (6.4-11c)
\end{aligned}
$$

The constants of integration a_i, b_i, and c_i can be determined using the boundary conditions.

The in-plane stresses in each layer can be computed using the constitutive equations, and the transverse stresses can be determined using equilibrium equations of 3-D elasticity [see Eqs. (6.2–13) and (6.2–14)]. For a cross-ply laminate the only nonzero strain is ε_{xx}.

Example 6.4–1 (*Simply supported beam*):

For a plate strip with simply supported edges at $x = 0$ and $x = a$, the boundary conditions are (see Table 6.4–1)

$$N_{xx} = 0,\ w_0 = 0,\ M_{xx} = 0 \qquad (6.4-12)$$

where

$$N_{xx} = A_{11}\frac{du_0}{dx} + A_{16}\frac{dv_0}{dx} - B_{11}\frac{d^2w_0}{dx^2} - N_{xx}^T \qquad (6.4-13a)$$

$$N_{yy} = A_{12}\frac{du_0}{dx} + A_{26}\frac{dv_0}{dx} - B_{12}\frac{d^2w_0}{dx^2} - N_{yy}^T \qquad (6.4-13b)$$

$$N_{xy} = A_{16}\frac{du_0}{dx} + A_{66}\frac{dv_0}{dx} - B_{16}\frac{d^2w_0}{dx^2} - N_{xy}^T \qquad (6.4-13c)$$

$$M_{xx} = B_{11}\frac{du_0}{dx} + B_{16}\frac{dv_0}{dx} - D_{11}\frac{d^2w_0}{dx^2} - M_{xx}^T \tag{6.4-14a}$$

$$M_{yy} = B_{12}\frac{du_0}{dx} + B_{26}\frac{dv_0}{dx} - D_{12}\frac{d^2w_0}{dx^2} - M_{yy}^T \tag{6.4-14b}$$

$$M_{xy} = B_{16}\frac{du_0}{dx} + B_{66}\frac{dv_0}{dx} - D_{16}\frac{d^2w_0}{dx^2} - M_{xy}^T \tag{6.4-14c}$$

From Eqs. (12), (13a), and (14a) it follows that, for an arbitrary lamination scheme and $dv_0/dx = 0$, we must have at $x = 0, a$

$$w_0 = 0, \quad \frac{du_0}{dx} = \hat{N}_{xx}^T, \quad \frac{dv_0}{dx} = 0, \quad \frac{d^2w_0}{dx^2} = \hat{M}_{xx}^T \tag{6.4-15a}$$

$$\hat{N}_{xx}^T = \frac{D_{11}N_{xx}^T - B_{11}M_{xx}^T}{A_{11}D_{11} - B_{11}B_{11}}, \quad \hat{M}_{xx}^T = \frac{B_{11}N_{xx}^T - A_{11}M_{xx}^T}{A_{11}D_{11} - B_{11}B_{11}} \tag{6.4-15b}$$

Since only the derivatives of u_0 and v_0 are specified at the boundary points, the solution for u_0 and v_0 can be determined only with an arbitrary constant (*i.e.*, rigid body motion is not eliminated).

Using boundary conditions (15) in Eq. (6.4–11), we obtain

$$u_0(x) = -\frac{B}{AD}\frac{q_0a^3}{12}\left[2\left(\frac{x}{a}\right)^3 - 3\left(\frac{x}{a}\right)^2\right] + \hat{N}_{xx}^T\, x + a_3 \tag{6.4-16a}$$

$$v_0(x) = -\frac{C}{AD}\frac{q_0a^3}{12}\left[2\left(\frac{x}{a}\right)^3 - 3\left(\frac{x}{a}\right)^2\right] + b_3 \tag{6.4-16b}$$

$$\begin{aligned} w_0(x) = &-\frac{q_0a^4}{24D}\left[\left(\frac{x}{a}\right)^4 - 2\left(\frac{x}{a}\right)^3 + \left(\frac{x}{a}\right)\right] \\ &+ \frac{\hat{M}_{xx}^T a^2}{2}\left[\left(\frac{x}{a}\right)^2 - \left(\frac{x}{a}\right)\right] \end{aligned} \tag{6.4-16c}$$

where the constants a_3 and b_3 can be interpreted as rigid body displacements. The constants can be determined by setting $u_0(0) = 0$ and $v_0(0) = 0$, which give $a_3 = b_3 = 0$.

The stress resultants for any x are then given by substituting Eqs. (16) into Eqs. (13) and (14):

$$N_{xx} = 0, \quad N_{xy} = A_{16}\hat{N}_{xx}^T - B_{16}\hat{M}_{xx}^T - N_{xy}^T \tag{6.4-17a}$$

$$\begin{aligned} N_{yy} = &-\frac{q_0a^2}{2AD}\left(A_{12}B + A_{26}C - B_{12}A\right)\left[\left(\frac{x}{a}\right)^2 - \left(\frac{x}{a}\right)\right] \\ &+ A_{16}\hat{N}_{xx}^T - B_{16}\hat{M}_{xx}^T \end{aligned} \tag{6.4-17b}$$

$$M_{xx} = \frac{q_0 a^2}{2}\left[\left(\frac{x}{a}\right)^2 - \left(\frac{x}{a}\right)\right] \tag{6.4 – 17c}$$

$$M_{xy} = -\frac{q_0 a^2}{2AD}\left(B_{16}B + B_{66}C - D_{16}A\right)\left[\left(\frac{x}{a}\right)^2 - \left(\frac{x}{a}\right)\right] + B_{16}\hat{N}^T_{xx} - D_{16}\hat{M}^T_{xx} - M^T_{xy} \tag{6.4 – 17d}$$

$$M_{yy} = -\frac{q_0 a^2}{2AD}\left(B_{12}B + B_{26}C - D_{12}A\right)\left[\left(\frac{x}{a}\right)^2 - \left(\frac{x}{a}\right)\right] + B_{12}\hat{N}^T_{xx} - D_{12}\hat{M}^T_{xx} - M^T_{yy} \tag{6.4 – 17e}$$

The maximum transverse deflection occurs at $x = a/2$, and it is given by

$$w_{max} = -\frac{5q_0 a^4}{384D} - \frac{\hat{M}^T_{xx} a^2}{8} \tag{6.4 – 18}$$

In order to see the effect of the bending–stretching coupling on the transverse deflection, the reciprocal of the bending stiffness D [see Eq. (2d)] is expressed as

$$\frac{1}{D} = \frac{1}{D_{11}}\left(\frac{D_{11}}{D}\right) = \frac{1}{D_{11}}\left(\frac{D + B_{11}\bar{B} + B_{16}\bar{C}}{D}\right)$$

Hence, the maximum deflection can be expressed in the form

$$w_{max} = -\frac{5q_0 a^4}{384D_{11}}\left(1 + \frac{B_{11}\bar{B} + B_{16}\bar{C}}{D}\right) - \frac{\hat{M}^T_{xx} a^2}{8} \tag{6.4 – 19}$$

For symmetric laminates the coupling terms are zero, and the maximum deflection is given by

$$w_{max} = -\frac{5q_0 a^4}{384D_{11}} - \frac{\hat{M}^T_{xx} a^2}{8} \tag{6.4 – 20}$$

It can be shown that the expression $B_{11}\bar{B} + B_{16}\bar{C}$ is always positive. Therefore, it follows that the effect of the coupling is to increase the maximum transverse deflection of the plate strip. For example, for antisymmetric cross-ply laminates, we have $A_{16} = A_{26} = B_{16} = B_{26} = D_{16} = D_{26} = 0$, $\bar{B} = B_{11}/A_{11}$, $\bar{C} = 0$, and $D = D_{11} - B_{11}^2/A_{11}$. Thus the maximum deflection becomes

$$w_{max} = -\frac{5q_0 a^4}{384D_{11}}\left(1 + \frac{B_{11}^2}{A_{11}D_{11} - B_{11}^2}\right) - \frac{\hat{M}^T_{xx} a^2}{8} \tag{6.4 – 21}$$

In the case of antisymmetric angle-ply laminates, we have $A_{16} = A_{26} = B_{11} = B_{22} = B_{12} = B_{66} = D_{16} = D_{26} = 0$, $\bar{B} = 0$, $\bar{C} = B_{16}/A_{66}$, and $D = D_{11} - B_{16}^2/A_{66}$. The maximum deflection becomes

$$w_{max} = -\frac{5q_0 a^4}{384D_{11}}\left(1 + \frac{B_{16}^2}{A_{66}D_{11} - B_{16}^2}\right) - \frac{\hat{M}^T_{xx} a^2}{8} \tag{6.4 – 22}$$

Note that when the bending–stretching coupling terms are zero (*e.g.*, for symmetric laminates), the cylindrical bending and laminated beam solutions have the same form. The difference is only in the bending stiffness term. The bending stiffness D_{11} used in cylindrical bending is given by

$$D_{11} = \frac{E_{xx}^b h^3}{12(1 - \nu_{xy}^b \nu_{yx}^b)} = \frac{E_{xx}^b h^3}{12[1 - (\nu_{xy}^b)^2 (E_{xx}^b / E_{yy}^b)]} \tag{6.4 – 23}$$

whereas the bending stiffness used in the beam theory is $E_{xx}^b I_{yy} = E_{xx}^b bh^3/12$. Thus, the difference is in the expression containing Poisson's ratios, which is due to the plane strain assumption used in cylindrical bending compared to the plane stress assumption used in the beam theory. The difference between the two solutions will be the most for laminates containing angle-ply layers, where ν_{xy}^b can be very large.

Table 6.4–1: Boundary conditions in the classical (CLPT) and first-order shear deformation (FSDT) theories of beams and plate strips. The boundary conditions on u_0 and v_0 are only for laminated strips in cylindrical bending.

Edge Condition	CLPT		FSDT	
free 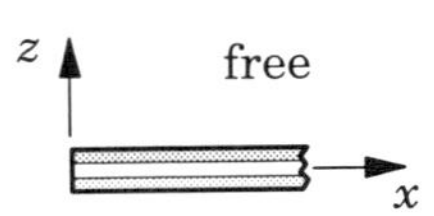	$N_{xx}=0$	$N_{xy}=0$	$N_{xx}=0$	$N_{xy}=0$
	$M_{xx}=0$	$\frac{dM_{xx}}{dx}=0$	$M_{xx}=0$	$Q_x=0$
roller 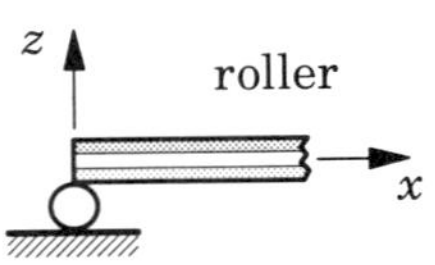	$w_0=0$	$\frac{dv_0}{dx}=0$	$w_0=0$	$\frac{dv_0}{dx}=0$
	$N_{xx}=0$	$M_{xx}=0$	$N_{xx}=0$	$M_{xx}=0$
simple support 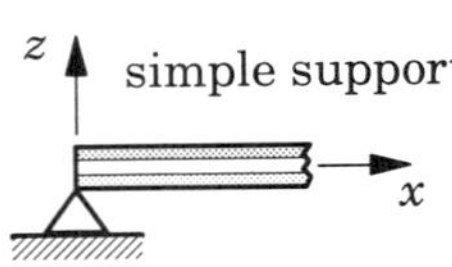	$u_0=0$	$w_0=0$	$u_0=0$	$w_0=0$
	$\frac{dv_0}{dx}=0$	$M_{xx}=0$	$\frac{dv_0}{dx}=0$	$M_{xx}=0$
clamped 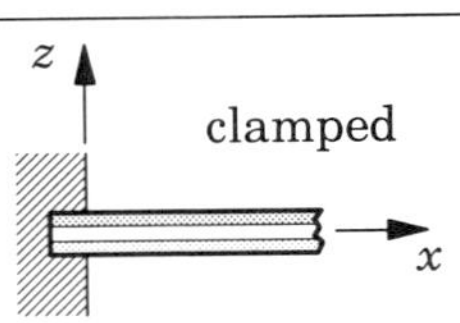	$u_0=0$	$v_0=0$	$u_0=0$	$v_0=0$
	$w_0=0$	$\frac{dw_0}{dx}=0$	$w_0=0$	$\phi_x=0$

6.4–3 Buckling

The equilibrium of the plate strip under the applied in-plane compressive load $\hat{N}_{xx} = -N_{xx}^0$ can be obtained from Eqs. $(2a, b, c)$ by omitting the inertia terms and thermal resultants

$$A\frac{d^2U}{dx^2} = B\frac{d^3W}{dx^3} \qquad (6.4-24)$$

$$A\frac{d^2V}{dx^2} = C\frac{d^3W}{dx^3} \qquad (6.4-25)$$

$$D\frac{d^4W}{dx^4} = -N_{xx}^0\frac{d^2W}{dx^2} \qquad (6.4-26)$$

where (U, V, W) denote the displacements measured from the prebuckling equilibrium state.

Equation (26), which is uncoupled from (24) and (25), can be integrated twice with respect to x to obtain

$$D\frac{d^2W}{dx^2} + N_{xx}^0 W = K_1 x + K_2 \qquad (6.4-27)$$

where K_1 and K_2 are constants. The general solution of Eq. (27) is

$$W(x) = c_1 \sin\lambda x + c_2 \cos\lambda x + c_3 x + c_4 \qquad (6.4-28)$$

where $c_3 = K_1/\lambda^2, c_4 = K_2/\lambda^2$, and

$$\lambda^2 = \frac{N_{xx}^0}{D} \quad \text{or} \quad N_{xx}^0 = D\lambda^2 \qquad (6.4-29)$$

The three of the four constants c_1, c_2, c_3, c_4, and λ are determined using (four) boundary conditions of the problem. Once λ is known, the buckling load can be determined using Eq. (29). The results of Section 6.2–3 are applicable here with $b = 1$ and $E_{xx}^b I_{yy} = D$. Here we consider only the case of simply supported boundary conditions for illustrative purposes.

Example 6.4–2:

When the plate strip is simply supported at $x = 0, a$, from Eq. (15a) we have

$$W = 0, \quad \frac{dU}{dx} = 0, \quad \frac{dV}{dx} = 0, \quad \frac{d^2W}{dx^2} = 0 \qquad (6.4-30)$$

Use of the boundary conditions on W gives $c_1 = c_2 = c_3 = 0$ and the result

$$\sin\lambda a \equiv \sin\left(\sqrt{\frac{N_{xx}^0}{D}}\right) = 0, \quad \text{or} \quad N_{xx}^0 = D\left(\frac{n\pi}{a}\right)^2 \qquad (6.4-31)$$

The critical buckling load N_{cr} is given by $(n = 1)$

$$N_{cr} = D_{11}\frac{\pi^2}{a^2}\left(1 - \frac{B_{11}\bar{B} + B_{16}\bar{C}}{D_{11}A}\right) \tag{6.4-32}$$

Thus the effect of the bending–extensional coupling is to decrease the critical buckling load.

Recall from Section 6.2–3 that when both edges are clamped, λ is determined by solving the equation

$$\lambda a \;\sin\lambda a + 2\cos\lambda a - 2 = 0 \tag{6.4-33}$$

The smallest root of this equation is $\lambda = 2\pi$, and the critical buckling load becomes

$$N_{cr} = D_{11}\frac{4\pi^2}{a^2}\left(1 - \frac{B_{11}\bar{B} + B_{16}\bar{C}}{D_{11}A}\right) \tag{6.4-34}$$

6.4–4 Vibration

For vibration in the absence of in-plane inertias, thermal forces, and transverse load, Eq. (2c) is reduced to

$$D\frac{\partial^4 w_0}{\partial x^4} = \bar{I}_2\frac{\partial^4 w_0}{\partial x^2 \partial t^2} - I_0\frac{\partial^2 w_0}{\partial t^2} + \hat{N}_{xx}\frac{\partial^2 w_0}{\partial x^2} \tag{6.4-35}$$

where $\bar{I}_2 = I_2 - \bar{B}I_1$. For a periodic motion, we assume

$$w_0(x,t) = W(x)e^{i\omega t}, \quad i = \sqrt{-1} \tag{6.4-36}$$

where ω is the natural frequency of vibration. Then Eq. (35) becomes

$$D\frac{d^4W}{dx^4} - \hat{N}_{xx}\frac{d^2W}{dx^2} = I_0\omega^2 W - \bar{I}_2\omega^2\frac{d^2W}{dx^2} \tag{6.4-37}$$

Equation (35) has the same form as Eq. (6.2–43). Hence, all of the results of Section 6.2–4 are applicable here with $b = 1$ ($\hat{I}_0 = I_0, \hat{I}_2 = \bar{I}_2$) and $E^b_{xx}I_{yy} = D$. We summarize the results here for completeness.

The general solution of Eq. (37) is

$$W(x) = c_1\sin\lambda x + c_2\cos\lambda x + c_3\sinh\mu x + c_4\cosh\mu x \tag{6.4-38}$$

where

$$\lambda = \sqrt{\frac{1}{2r}\left(q + \sqrt{q^2 + 4pr}\right)}, \; \mu = \sqrt{\frac{1}{2r}\left(-q + \sqrt{q^2 + 4pr}\right)} \tag{6.4-39}$$

$$p = D, \quad q = \bar{I}_2\omega^2 - \hat{N}_{xx}, \quad r = I_0\omega^2 \tag{6.4-40}$$

and c_1, c_2, c_3, and c_4 are integration constants, which are determined using the boundary conditions. For natural vibration without rotary inertia and applied axial load, the equation for $\lambda = \mu$ reduces to

$$\lambda^2 = \sqrt{r/p} \tag{6.4-41}$$

If the applied axial force is zero, the natural frequency of vibration, with rotary inertia included, is given by

$$\omega^2 = \lambda^4 \frac{D}{I_0}\left(1 - \frac{\bar{I}_2\lambda^2}{I_0 + \bar{I}_2\lambda^2}\right) \tag{6.4-42}$$

When rotary inertia is neglected, we have

$$\omega = \lambda^2\sqrt{\frac{D}{I_0}} \tag{6.4-43}$$

Example 6.4–3:

For a simply supported plate strip, λ_n is given by $\lambda_n = \frac{n\pi}{a}$ and from Eq. (42) it follows that

$$\omega_n = \left(\frac{n\pi}{a}\right)^2 \sqrt{\frac{D}{I_0}} \sqrt{\frac{1}{1 + (\frac{n\pi}{a})^2(\bar{I}_2/I_0)}} \tag{6.4-44}$$

Note that the rotatory inertia has the effect of decreasing the natural frequency. When the rotatory inertia is zero, we have

$$\omega_n = \left(\frac{n\pi}{a}\right)^2 \sqrt{\frac{D}{I_0}} \tag{6.4-45}$$

For a plate strip clamped at both ends, λ must be determined from [see Eqs. (6.2–56)–(6.2–60)]

$$-2 + 2\cos\lambda a \cosh\mu a + \left(\frac{\lambda}{\mu} - \frac{\mu}{\lambda}\right)\sin\lambda a \sinh\mu a = 0 \tag{6.4-46}$$

For natural vibration without rotatory inertia, Eq. (46) takes the simpler form

$$\cos\lambda a \cosh\lambda a - 1 = 0 \tag{6.4-47}$$

The roots of Eq. (47) are

$$\lambda_1 a = 4.730, \quad \lambda_2 a = 7.853, \quad \lambda_3 a = 10.996, \quad \cdots, \quad \lambda_n a \approx (n + \frac{1}{2})\pi \tag{6.4-48}$$

In general, the roots of Eq. (46) are not the same as those of Eq. (47). If one approximates Eq. (46) as (47) (*i.e.*, $\lambda \approx \mu$), the roots in Eq. (48) can be used to determine the natural frequencies of vibration *with rotary inertia* from Eq. (42). When rotary inertia is neglected, the frequencies are given by Eq. (43) with λ as given in Eq. (48). The frequencies obtained from Eq. (42) with the values of λ from Eq. (48) are only an approximation of the frequencies with rotary inertia.

Figure 6.4–1 contains a plot of the nondimensionalized fundamental frequency $\bar{\omega} = \omega a^2\sqrt{I_0/E_2h^3}$ of a simply supported plate strip with rotary inertia versus length-to-thickness ratio, a/h. For small values of a/h, rotary inertia is more significant in reducing the frequency than for thin and long plate strips.

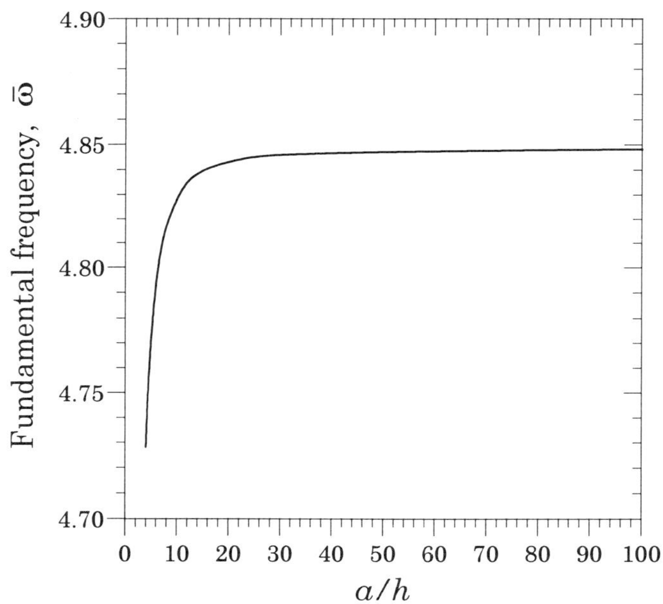

Figure 6.4–1: Effect of rotary inertia on nondimensionalized fundamental frequency of a simply supported (−45/45) laminated plate strip.

Table 6.4–2 contains nondimensionalized maximum deflections, critical buckling loads, and fundamental natural frequencies of simply supported and clamped (at both ends) laminated plate strips with various lamination schemes. Compared to laminated beams (see Table 6.2–4), laminated plates in cylindrical bending undergo smaller displacements and have larger buckling loads and frequencies. This is due to the Poisson effect discussed earlier. All of the frequencies listed in Table 6.4–2 are for the case where rotary inertia is included and a/h = 10. The (0/90/0) laminates have larger bending stiffness as well as axial stiffness compared to the (90/0/90) laminates. This is due to the fact there are two 0° layers and they are placed farther from the midplane in the first laminate than in the second laminate. Hence, (0/90/0) laminates undergo smaller deflections, and have larger buckling loads and natural frequencies. The $(0/90)_s$ laminates have larger bending stiffness than the $(90/0)_s$ laminates; both have the same axial stiffness. The antisymmetric laminates have some of the $B_{ij} \neq 0$ and thus are relatively flexible when compared to symmetric laminates.

Table 6.4–2: Maximum deflections ($\hat{w}$) under uniform load, critical buckling loads ($\bar{N}$), and fundamental frequencies ($\bar{\omega}$) of laminated plate strips according to the classical laminate theory ($E_1/E_2 = 25$, $G_{12} = G_{13} = 0.5E_2$, $G_{23} = 0.2E_2$, $\nu_{12} = 0.25$).

Laminate	Hinged–Hinged			Clamped–Clamped		
	$\hat{w}$	$\bar{N}$	$\bar{\omega}$	$\hat{w}$	$\bar{N}$	$\bar{\omega}$
0	0.623	20.613	14.205	0.125	82.453	32.169
90	15.586	0.824	2.841	3.117	3.298	6.434
(0/90/0)	0.646	19.880	13.950	0.129	79.521	31.592
(90/0/90)	8.251	1.557	3.905	1.650	6.230	8.842
(0/90)	3.321	3.869	6.154	0.664	15.476	13.937
$(0/90)_{as}$	1.427	9.006	9.389	0.285	36.026	21.264
$(0/90)_s$	0.708	18.140	13.326	0.142	72.558	30.177
$(90/0)_s$	3.896	3.298	5.682	0.779	13.192	12.868
(–45/45)	5.396	2.382	4.828	1.079	9.526	10.935
$(-45/45)_{as}$	2.570	5.000	6.996	0.514	20.003	15.845
$(45/-45)_s$	2.188	5.873	7.583	0.437	23.495	17.172
Laminate A	4.035	3.185	5.584	0.807	12.740	12.645
Laminate B	0.897	14.316	11.838	0.179	57.264	26.809

$(\cdot,\cdot)_s$=symmetric, $(\cdot,\cdot)_{as}$=antisymmetric (four layers).

Laminate A: $(90/\pm 45/0)_s$; Laminate B: $(0/\pm 45/90)_s$

$\hat{w} = -w_{max}(E_2h^3/q_0a^4) \times 10^2, \quad \bar{N} = N_{xx}^0(a^2/E_2h^3), \quad \bar{\omega} = \omega a^2\sqrt{I_0/E_2h^3}$.

Figures 6.4–2 and 6.4–3 show the effect of lamination angle on maximum deflections $\bar{w} = -w_{max}(E_2h^3/q_0a^4)$, critical buckling load $\bar{N}$, and fundamental frequency $\bar{\omega}$ of two-layer antisymmetric angle-ply $(-\theta/\theta)$ plates. It should be noted that antisymmetric angle-ply laminates with more than two plies are stiffer, *i.e.*, deflect less and carry more buckling load.

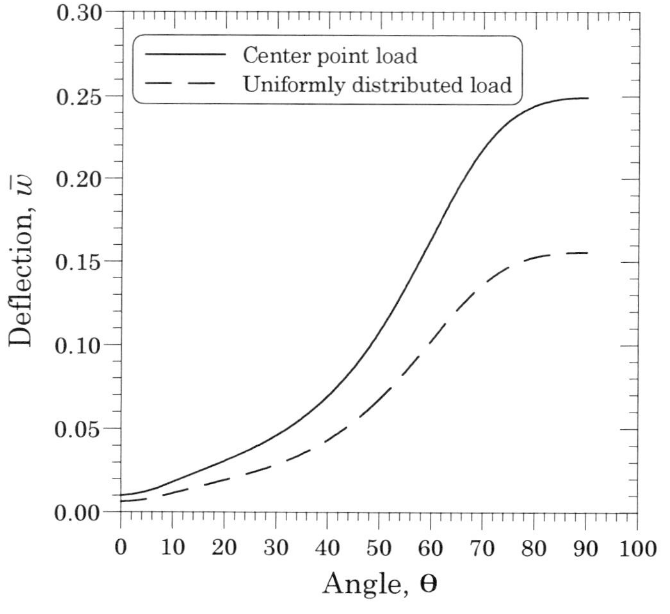

Figure 6.4–2: Nondimensionalized maximum transverse deflection ($\bar{w}$) versus lamination angle (θ) of a simply supported ($-\theta/\theta$) laminated plate strip in cylindrical bending (CLPT).

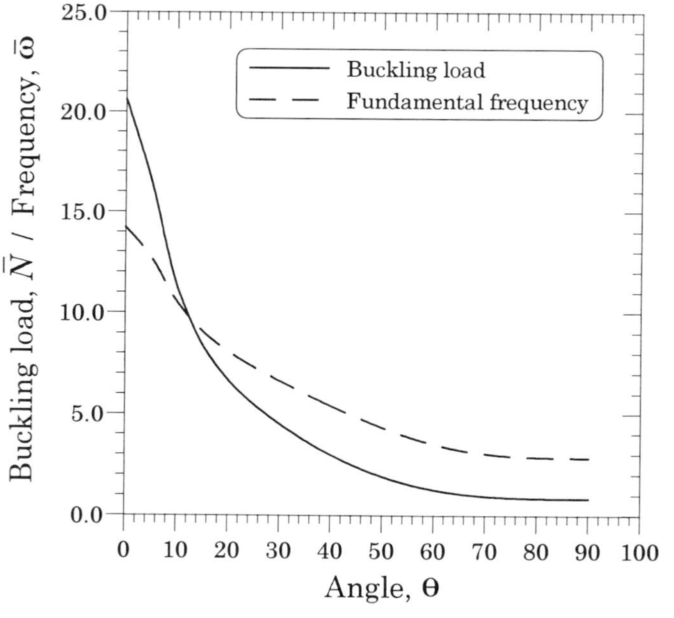

Figure 6.4–3: Nondimensionalized critical buckling load ($\bar{N}$) and fundamental frequency ($\bar{\omega}$) versus lamination angle (θ) of a simply supported ($-\theta/\theta$) laminated plate strip in cylindrical bending (CLPT).

6.5 Cylindrical Bending Using FSDT

6.5–1 Governing Equations

In order to see the effect of shear deformation on bending deflections and buckling loads, we consider the equations of motion for cylindrical bending according to the first-order shear deformation theory (FSDT) [see Eqs. (5.4–23)–(5.4–27)]:

$$A_{11}\frac{\partial^2 u_0}{\partial x^2}+A_{16}\frac{\partial^2 v_0}{\partial x^2}+B_{11}\frac{\partial^2 \phi_x}{\partial x^2}+B_{16}\frac{\partial^2 \phi_y}{\partial x^2}-\frac{\partial N_{xx}^T}{\partial x}=I_0\frac{\partial^2 u_0}{\partial t^2}+I_1\frac{\partial^2 \phi_x}{\partial t^2} \quad (6.5-1a)$$

$$A_{16}\frac{\partial^2 u_0}{\partial x^2}+A_{66}\frac{\partial^2 v_0}{\partial x^2}+B_{16}\frac{\partial^2 \phi_x}{\partial x^2}+B_{66}\frac{\partial^2 \phi_y}{\partial x^2}-\frac{\partial N_{xy}^T}{\partial x}=I_0\frac{\partial^2 v_0}{\partial t^2}+I_1\frac{\partial^2 \phi_y}{\partial t^2} \quad (6.5-1b)$$

$$\begin{aligned}B_{11}\frac{\partial^2 u_0}{\partial x^2}+B_{16}\frac{\partial^2 v_0}{\partial x^2}+D_{11}\frac{\partial^2 \phi_x}{\partial x^2}+D_{16}\frac{\partial^2 \phi_y}{\partial x^2}-KA_{55}\left(\frac{\partial w_0}{\partial x}+\phi_x\right)\\ -KA_{45}\phi_y-\frac{\partial M_{xx}^T}{\partial x}=I_2\frac{\partial^2 \phi_x}{\partial t^2}+I_1\frac{\partial^2 u_0}{\partial t^2}\end{aligned} \quad (6.5-1c)$$

$$\begin{aligned}B_{16}\frac{\partial^2 u_0}{\partial x^2}+B_{66}\frac{\partial^2 v_0}{\partial x^2}+D_{16}\frac{\partial^2 \phi_x}{\partial x^2}+D_{66}\frac{\partial^2 \phi_y}{\partial x^2}-KA_{44}\phi_y-KA_{45}\left(\frac{\partial w_0}{\partial x}+\phi_x\right)\\ -\frac{\partial M_{xy}^T}{\partial x}=I_2\frac{\partial^2 \phi_y}{\partial t^2}+I_1\frac{\partial^2 v_0}{\partial t^2}\end{aligned} \quad (6.5-1d)$$

$$KA_{55}\left(\frac{\partial^2 w_0}{\partial x^2}+\frac{\partial \phi_x}{\partial x}\right)+KA_{45}\frac{\partial \phi_y}{\partial x}+\frac{\partial}{\partial x}\left(N_{xx}\frac{\partial w_0}{\partial x}\right)+q=I_0\frac{\partial^2 w_0}{\partial t^2} \quad (6.5-1e)$$

For cylindrical bending we further assume that $\phi_y = 0$ everywhere, and omit Eq. (1d) from further consideration. For the purpose of developing analytical solutions, we neglect the in-plane inertia terms and assume that there are no thermal effects. Then Eqs. (1a–e) are simplified to

$$A_{11}\frac{\partial^2 u_0}{\partial x^2}+A_{16}\frac{\partial^2 v_0}{\partial x^2}+B_{11}\frac{\partial^2 \phi_x}{\partial x^2}=I_1\frac{\partial^2 \phi_x}{\partial t^2} \quad (6.5-2a)$$

$$A_{16}\frac{\partial^2 u_0}{\partial x^2}+A_{66}\frac{\partial^2 v_0}{\partial x^2}+B_{16}\frac{\partial^2 \phi_x}{\partial x^2}=0 \quad (6.5-2b)$$

$$B_{11}\frac{\partial^2 u_0}{\partial x^2}+B_{16}\frac{\partial^2 v_0}{\partial x^2}+D_{11}\frac{\partial^2 \phi_x}{\partial x^2}-KA_{55}\left(\frac{\partial w_0}{\partial x}+\phi_x\right)=I_2\frac{\partial^2 \phi_x}{\partial t^2} \quad (6.5-2c)$$

$$KA_{55}\left(\frac{\partial^2 w_0}{\partial x^2}+\frac{\partial \phi_x}{\partial x}\right)+\frac{\partial}{\partial x}\left(N_{xx}\frac{\partial w_0}{\partial x}\right)+q=I_0\frac{\partial^2 w_0}{\partial t^2} \quad (6.5-2d)$$

Next, we eliminate u_0 and v_0 from Eqs. (2a–c) by solving (2a) and (2b) for u_0 and v_0 in terms of ϕ_x and substituting the result into Eq. (2c):

$$KA_{55}\left(\frac{\partial^2 w_0}{\partial x^2}+\frac{\partial \phi_x}{\partial x}\right)+\hat{N}_{xx}\frac{\partial^2 w_0}{\partial x^2}+q=I_0\frac{\partial^2 w_0}{\partial t^2} \tag{6.5-3}$$

$$D\frac{\partial^2 \phi_x}{\partial x^2}-KA_{55}\left(\frac{\partial w_0}{\partial x}+\phi_x\right)=I_2\frac{\partial^2 \phi_x}{\partial t^2} \tag{6.5-4}$$

Equations (3) and (4) are similar to Eqs. (6.3–9a,b) for laminated beams, and therefore all developments of Section 6.3 would apply here.

6.5–2 Bending

For static analysis, Eqs. (3) and (4) reduce to

$$KA_{55}\left(\frac{d^2 w_0}{dx^2}+\frac{d\phi_x}{dx}\right)+q=0 \tag{6.5-5a}$$

$$D\frac{d^2 \phi_x}{dx^2}-KA_{55}\left(\frac{dw_0}{dx}+\phi_x\right)=0 \tag{6.5-5b}$$

Following the procedure of Section 6.3–2, we obtain [see Eqs. (6.3–12)–(6.3–14)] the general solution for the rotation

$$\boxed{\phi_x(x)=\frac{1}{D}\left[-\int^x\int^\xi\int^\eta q(\zeta)\ d\zeta d\eta d\xi+c_1\frac{x^2}{2}+c_2x+c_3\right]} \tag{6.5-6}$$

and transverse deflection

$$\boxed{\begin{aligned} w_0(x)=&-\frac{1}{D}\left[-\int^x\int^\xi\int^\eta\int^\mu q(\zeta)\ d\zeta d\mu d\eta d\xi+c_1\frac{x^3}{6}+c_2\frac{x^2}{2}+c_3x+c_4\right]\\ &+\frac{1}{KA_{55}}\left[-\int^x\int^\xi q(\zeta)d\zeta d\xi+c_1x\right]\end{aligned}} \tag{6.5-7}$$

where the constants of integration c_1 through c_4 can be determined using the boundary conditions. The solutions developed are general in the sense that they are applicable to any symmetrically laminated beams. Next we illustrate the procedure to determine the constants for beams with both edges simply supported or clamped.

Example 6.5–1 (*Simply supported beam*):

For a plate strip simply supported at both ends and subjected to uniformly distributed load $q = -q_0$ as well as a downward point load F_0 at the center, we obtain

$$\phi_x(x) = \frac{q_0 a^3}{24D}\left[4\left(\frac{x}{a}\right)^3 - 6\left(\frac{x}{a}\right)^2 + 1\right] - \frac{F_0 a^2}{16D}\left[1 - 4\left(\frac{x}{a}\right)^2\right] \tag{6.5-8}$$

$$\begin{aligned} w_0(x) = &- \frac{q_0 a^4}{24D}\left[\left(\frac{x}{a}\right)^4 - 2\left(\frac{x}{a}\right)^3 + \left(\frac{x}{a}\right)\right] - \frac{q_0 a^2}{2KA_{55}}\left[\left(\frac{x}{a}\right) - \left(\frac{x}{a}\right)^2\right] \\ &- \frac{F_0 a^3}{48D}\left[3\left(\frac{x}{a}\right) - 4\left(\frac{x}{a}\right)^3\right] - \frac{F_0 a}{2KA_{55}}\left(\frac{x}{a}\right) \end{aligned} \tag{6.5-9}$$

The maximum deflection occurs at $x = a/2$ and it is given by

$$w_{max} = -\left(\frac{5q_0 a^4}{384D} + \frac{q_0 a^2}{8KA_{55}} + \frac{F_0 a^3}{48D} + \frac{F_0 a}{4KA_{55}}\right) \tag{6.5-10}$$

Example 6.5–2 (*Clamped beam*):

Consider a laminated plate strip fixed at both ends and subjected to uniformly distributed transverse load q_0 and a point load F_0 at the center, both acting downward. For this case, the solution is given by

$$\begin{aligned} \phi_x(x) = &\frac{q_0 a^3}{12D}\left[2\left(\frac{x}{a}\right)^3 - 3\left(\frac{x}{a}\right)^2 + \left(\frac{x}{a}\right)\right] \\ &+ \frac{F_0 a^2}{8D}\left[2\left(\frac{x}{a}\right)^2 - \left(\frac{x}{a}\right)\right] \end{aligned} \tag{6.5-11}$$

$$\begin{aligned} w_0(x) = &- \frac{q_0 a^4}{24D}\left[\left(\frac{x}{a}\right)^2 - \left(\frac{x}{a}\right)\right]^2 - \frac{q_0 a^2}{2KA_{55}}\left[\left(\frac{x}{a}\right) - \left(\frac{x}{a}\right)^2\right] \\ &- \frac{F_0 a^3}{48D}\left[3\left(\frac{x}{a}\right)^2 - 4\left(\frac{x}{a}\right)^3\right] - \frac{F_0 a}{2KA_{55}}\left(\frac{x}{a}\right) \end{aligned} \tag{6.5-12}$$

The maximum deflection is given by

$$w_{max} = -\left(\frac{q_0 a^4}{384D} + \frac{q_0 a^2}{8KA_{55}} + \frac{F_0 a^3}{192D} + \frac{F_0 a}{4KA_{55}}\right) \tag{6.5-13}$$

The determination of the shear correction coefficient K for laminated structures is still an unresolved issue. Values of K for various special cases are available in the literature (see [4–8]). The most commonly used value of $K = 5/6$ is based on homogeneous, isotropic plates (see Section 5.4), although K depends, in general, on the lamination scheme, geometry, and material properties.

Figure 6.5–1 shows the effect of shear deformation, shear correction coefficient, and lamination scheme on nondimensionalized deflections $\bar{w} = -w_{max}(E_2 h^3/q_0 a^4)$ of simply supported, cross-ply (0/90) and angle-ply (45/–45) laminates under uniformly distributed load. The shear correction factor has little influence on the global response for the antisymmetric laminates analyzed. The effect of shear deformation is to increase the deflections, especially for $a/h \leq 10$. Antisymmetric angle-ply laminates are relatively more flexible than antisymmetric cross-ply laminates.

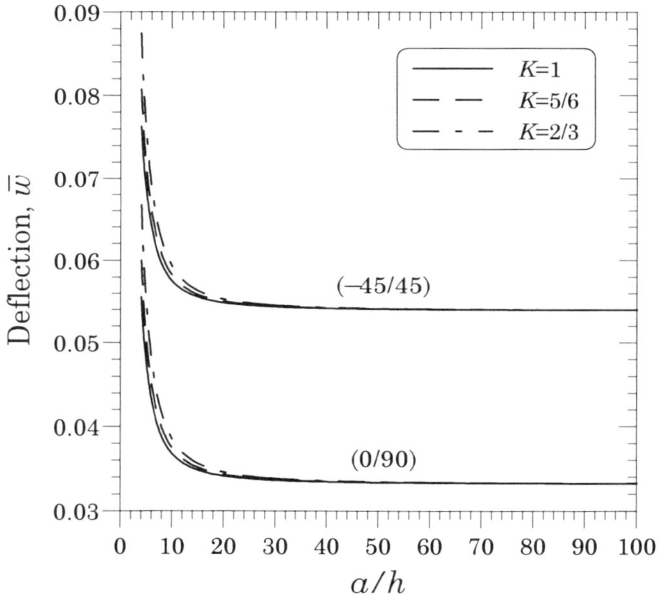

Figure 6.5–1: Transverse deflection ($\bar{w}$) versus length-to-thickness ratio (a/h) of a simply supported plate strips ($K = 1.0, 5/6, 2/3$).

Figure 6.5–2 contains plots of nondimensionalized maximum deflection versus length-to-height ratio for two-layer antisymmetric cross-ply (0/90) and angle-ply (45/–45) laminates ($K = 5/6$) under uniformly distributed load and with simply supported edges as well as for clamped edges. For clamped boundary conditions, shear deformation is relatively more significant for $a/h \leq 10$. The effect of orthotropy on deflections is shown in Figure 6.5–3 ($G_{12} = G_{13} = 0.5E_2,\ G_{23} = 0.2E_2,\ \nu_{12} = 0.25$, and $K = 5/6$).

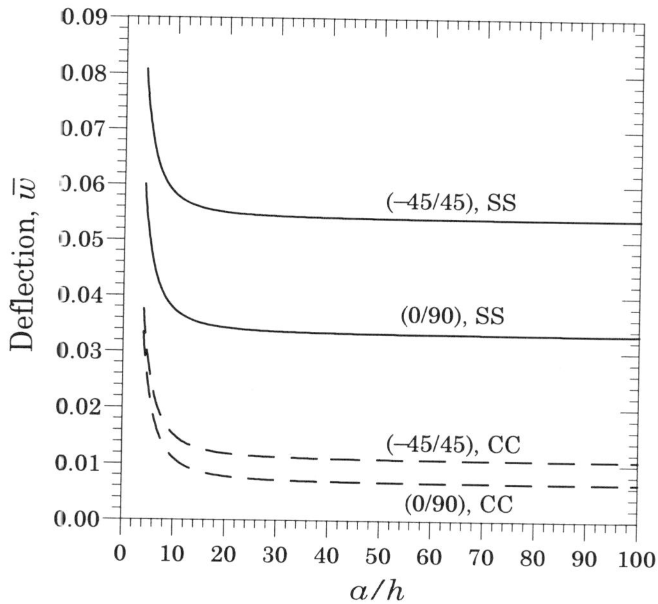

Figure 6.5–2: Transverse deflection ($\bar{w}$) versus length-to-thickness ratio (a/h) of simply supported (SS) and clamped (CC) plate strips.

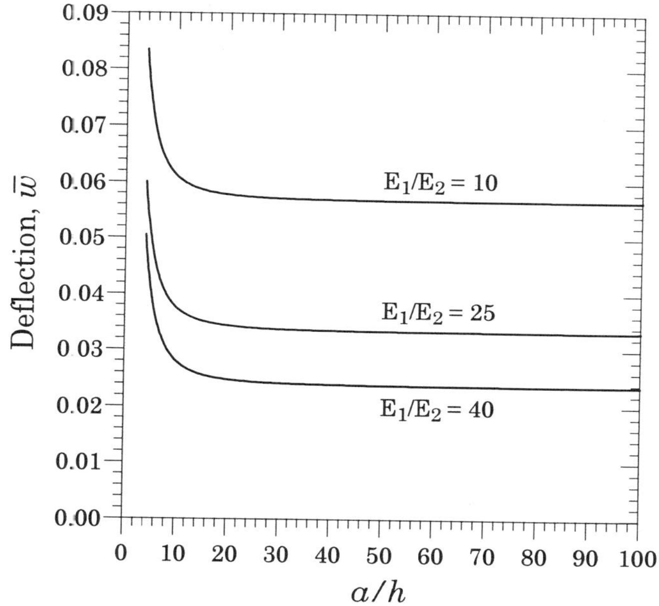

Figure 6.5–3: The effect of material orthotropy and shear deformation on transverse deflections $\bar{w}$ of simply supported cross-ply (0/90) laminated plate strips under uniformly distributed load.

6.5–3 Buckling

For stability analysis, we set $q = 0$, $\hat{N}_{xx} = -N_{xx}^0$, and $I_0 = I_2 = 0$ in Eqs. (3) and (4):

$$KA_{55}\left(\frac{d^2W}{dx^2} + \frac{d\mathcal{X}}{dx}\right) + \hat{N}_{xx}\frac{d^2W}{dx^2} = 0 \tag{6.5-14a}$$

$$D\frac{d^2\mathcal{X}}{dx^2} - KA_{55}\left(\frac{dW}{dx} + \mathcal{X}\right) = 0 \tag{6.5-14b}$$

Following the procedure of Section 6.3–3, we obtain

$$\frac{d\mathcal{X}}{dx} = -\left(1 - \frac{N_{xx}^0}{KA_{55}}\right)\frac{d^2W}{dx^2} \tag{6.5-15}$$

$$\mathcal{X}(x) = -\left(1 - \frac{N_{xx}^0}{KA_{55}}\right)\frac{dW}{dx} + K_1 \tag{6.5-16}$$

$$D\left(1 - \frac{N_{xx}^0}{KA_{55}}\right)\frac{d^4W}{dx^4} + N_{xx}^0\frac{d^2W}{dx^2} = 0 \tag{6.5-17}$$

The general solution of Eq. (17) is

$$W(x) = c_1 \sin\lambda x + c_2\cos\lambda x + c_3 x + c_4 \tag{6.5-18}$$

where

$$\lambda^2 = \frac{N_{xx}^0}{\left(1 - \frac{N_{xx}^0}{KA_{55}}\right)D} \quad \text{or} \quad N_{xx}^0 = \frac{\lambda^2 D}{\left(1 + \frac{\lambda^2 D}{KA_{55}}\right)} \tag{6.5-19}$$

and c_1 through c_4 are constants of integration, which are evaluated using the boundary conditions.

Example 6.5–3:

For a simply supported plate strip, the critical buckling load is given by

$$N_{cr} = \left(\frac{\pi}{a}\right)^2 D\left[1 - \frac{D\left(\frac{\pi}{a}\right)^2}{KA_{55} + D\left(\frac{\pi}{a}\right)^2}\right] \tag{6.5-20}$$

Thus, the effect of the transverse shear deformation is to decrease the buckling load. Omission of the transverse shear deformation in the classical theory amounts to assuming infinite rigidity in the transverse direction (*i.e.*, $A_{55} = G_{13} = \infty$); hence, classical laminate theory models the structure stiffer than it is.

For a plate strip fixed at both ends, λ is governed by the equation

$$2(\cos\lambda a - 1)\left(1 + \frac{\lambda^2 D}{KA_{55}}\right) + \lambda a\,\sin\lambda a = 0 \tag{6.5-21}$$

The roots of the equation are approximately the same as for the case in which shear deformation is neglected [see Eq. (6.2–38b)]. The first root of the equation is $\lambda_1 = 2\pi$. Hence, the critical buckling load is given by

$$N_{cr} = \left(\frac{2\pi}{a}\right)^2 D \left[1 - \frac{D\left(\frac{2\pi}{a}\right)^2}{KA_{55} + D\left(\frac{2\pi}{a}\right)^2}\right] \tag{6.5 – 22}$$

Figures 6.5–4 and 6.5–5 show the effect of shear deformation and modulus ratio on nondimensionalized critical buckling loads $\bar{N} = N_{xx}^0(a^2/E_2h^3)$ of two-layer antisymmetric angle-ply (–45/45) and cross-ply (0/90) plate strips ($E_1/E_2 = 25$, $G_{12} = G_{13} = 0.5E_2$, $G_{23} = 0.2E_2$, $\nu = 0.25$, $K = 5/6$). In Figure 6.5–4 results are presented for simply supported as well as clamped boundary conditions. The effect of shear deformation is significant for $a/h \leq 10$ in the case of simply supported boundary conditions, and $a/h \leq 20$ in the case of clamped boundary conditions. The effect of shear deformation is more for materials with larger modulus ratios (see Figure 6.5–5).

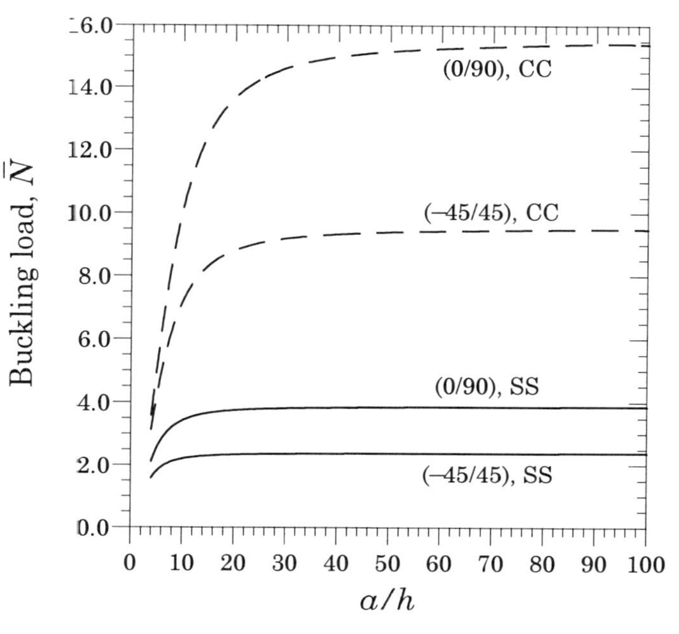

Figure 6.5–4: The effect of shear deformation on the critical buckling loads of simply supported (SS) and clamped (CC) cross-ply and angle-ply plate strips.

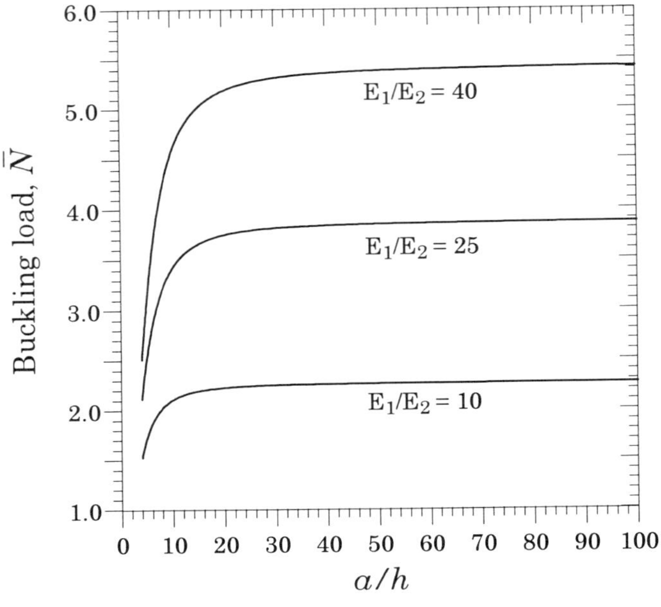

Figure 6.5–5: The effect of material orthotropy and shear deformation on critical buckling loads of simply supported cross-ply (0/90) laminated plate strips.

6.5–4 Vibration

For a periodic motion, we assume solution in the form

$$w_0(x,t) = W(x)e^{i\omega t}, \quad \phi_x(x,t) = \mathcal{X}(x)e^{i\omega t}, \quad i = \sqrt{-1}$$

where ω is the natural frequency of vibration, and Eqs. (3) and (4) become [*cf.* Eq. (6.3–40a,b)]

$$D\frac{d^2\mathcal{X}}{dx^2} - KA_{55}\left(\frac{dW}{dx} + \mathcal{X}\right) + I_2\omega^2\mathcal{X} = 0 \qquad (6.5-23a)$$

$$KA_{55}\left(\frac{d^2W}{dx^2} + \frac{d\mathcal{X}}{dx}\right) + I_0\omega^2 W = 0 \qquad (6.5-23b)$$

Following the results of Section 6.3–4, we obtain

$$p\frac{d^4W}{dx^4} + q\frac{d^2W}{dx^2} - rW = 0 \qquad (6.5-24a)$$

where

$$p = D, \quad q = \frac{I_0 D}{KA_{55}}\omega^2, \quad r = I_0\omega^2 \qquad (6.5-24b)$$

The general solution of Eq. (24a) is

$$W(x) = c_1 \sin \lambda x + c_2 \cos \lambda x + c_3 \sinh \mu x + c_4 \cosh \mu x \qquad (6.5-25a)$$

where

$$\lambda = \sqrt{\frac{1}{2p}\left(q + \sqrt{q^2 + 4pr}\right)}, \quad \mu = \sqrt{\frac{1}{2p}\left(-q + \sqrt{q^2 + 4pr}\right)} \qquad (6.5-25b)$$

and c_1, c_2, c_3, and c_4 are integration constants. Use of the boundary conditions lead to the determination of three of the four constants, the fourth one being arbitrary, and an equation governing λ and μ (see Section 6.3–4 for details). The frequencies ω can be determined from

$$(\omega^2)_1 = \frac{1}{2P}\left(Q - \sqrt{Q^2 - 4PR}\right) \qquad (6.5-26a)$$

$$(\omega^2)_2 = \frac{1}{2P}\left(Q + \sqrt{Q^2 - 4PR}\right) \qquad (6.5-26b)$$

where

$$P = \frac{I_2}{KA_{55}}, \quad Q = \left[1 + \left(\frac{D}{KA_{55}} + \frac{I_2}{I_0}\right)\lambda^2\right], \quad R = \left(\frac{D}{I_0}\right)\lambda^4 \qquad (6.5-26c)$$

When the rotary inertia is neglected, we have $P = 0$ and the frequency is given by

$$\omega^2 = \frac{\bar{Q}}{R}, \quad \bar{Q} = \left[1 + \left(\frac{D}{KA_{55}}\right)\lambda^2\right], \quad R = \left(\frac{D}{I_0}\right)\lambda^4 \qquad (6.5-27)$$

Example 6.5–4:

For a simply supported plate strip, the boundary conditions give $c_2 = c_3 = c_4 = 0$, and

$$\sin \lambda a = 0, \text{ or } \lambda_n = \frac{n\pi}{a} \qquad (6.5-28)$$

Substitution of λ from Eq. (28) into Eq. (26a,b) gives two frequencies for each value of λ. The fundamental frequency will come from Eq. (26a). When the rotatory inertia is neglected, we obtain from Eq. (27) the result

$$\omega_n = \left(\frac{n\pi}{a}\right)^2 \sqrt{\frac{D}{I_0}} \sqrt{\frac{KA_{55}}{KA_{55} + (\frac{n\pi}{a})^2 D}} \qquad (6.5-29)$$

By neglecting the shear deformation (*i.e.*, $A_{55} = G_{13} = \infty$) we obtain the result

$$\omega_n = \left(\frac{n\pi}{a}\right)^2 \sqrt{\frac{D}{I_0}} \qquad (6.5-30)$$

which is the same as in Eq. (6.4–45). Thus, the effect of shear deformation is to reduce the frequency of natural vibration.

For a laminated strip with clamped edges, the following equation governs λ:

$$-2 + 2\cos\lambda a \cosh\mu a + \sin\lambda a \sinh\mu a \left(\frac{S_{22}}{S_{11}} - \frac{S_{11}}{S_{22}}\right) = 0 \qquad (6.5-31a)$$

$$S_{11} = \mu\left(I_0\omega^2 - \lambda^2 K A_{55}\right), \quad S_{22} = \lambda\left(I_0\omega^2 + \mu^2 K A_{55}\right) \qquad (6.5-31b)$$

Once the value of λ is known, frequencies of vibration can be determined from Eqs. (26a,b).

Figures 6.5–6 and 6.5–7 show the effect of shear deformation and modulus ratio (E_1/E_2) on nondimensionalized fundamental frequencies $\bar{\omega} = \omega a^2\sqrt{I_0/E_2h^3}$ of two-layer antisymmetric angle-ply (–45/45) and cross-ply (0/90) plate strips ($K = 5/6$, $E_1/E_2 = 25$, $G_{12} = G_{13} = 0.5E_2$, $G_{23} = 0.2E_2$, $\nu_{12} = 0.25$). From Figure 6.5–6 it is clear that shear deformation effect in decreasing frequencies is felt for $a/h \leq 10$ for simply supported boundary conditions, whereas for clamped boundary conditions the effect is felt for $a/h \leq 15$. Also, the effect of shear deformation is more for materials with larger modulus ratio, as can be seen from the results of Figure 6.5–7.

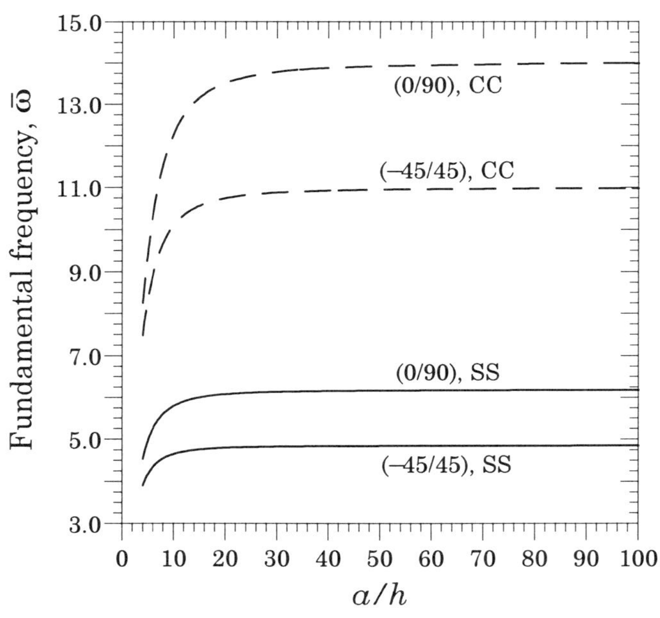

Figure 6.5–6: The effect of shear deformation on the fundamental frequencies of simply supported and clamped cross-ply and angle-ply plate strips.

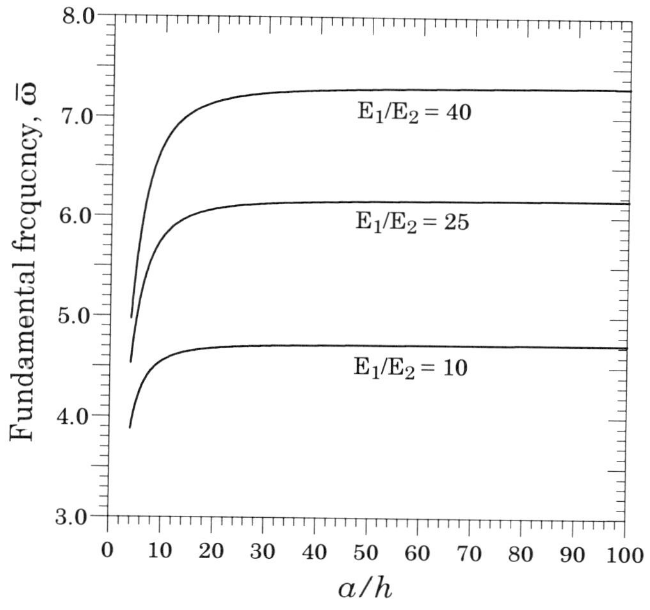

Figure 6.5–7: The effect of material orthotropy and shear deformation on fundamental frequencies of simply supported cross-ply (0/90) laminated plate strips.

6.6 Closing Remarks

In this chapter analytical solutions are developed for laminated beams and plate strips in cylindrical bending using the classical and first-order shear deformation theories. Analytical solutions are presented for static bending, natural vibration, and buckling problems under a number of boundary conditions.

When closed-form solutions can be derived, they are preferred over the series solutions. However, when exact closed-form solutions cannot be developed, the series solutions are the best alternative. For example, closed-form analytical solutions cannot be derived for laminated rectangular plates but the Navier solutions, which are infinite series except for a sinusoidally distributed load, can be derived, as discussed in the forthcoming chapters. When analytical solutions cannot be derived at all, numerical solutions based on the finite element method (see Chapter 10) can be used to determine the solutions.

Exercises

6.1 Consider a simply supported laminated beam under point loads F_0 at $x = a/4$ and $x = 3a/4$ (the so-called *four-point bending*). Use the symmetry about $x = a/2$ to determine the deflection $w_0(x)$ using the classical beam theory. (*Ans:* The maximum deflection is $w_{max} = -11F_0a^3/384E^b_{xx}I_{yy}$.)

6.2 Determine the static deflection of a clamped laminated beam under uniformly distributed load q_0 and a point load F_0 at the midspan using the classical beam theory.

6.3 Show that the critical buckling load of a clamped-free laminated beam using the classical beam theory is given by

$$N_{cr} = \left(\frac{E^b_{xx}I_{yy}}{b}\right)\left(\frac{\pi}{2a}\right)^2$$

6.4 Show that the characteristic equation governing buckling of a clamped-hinged laminated beam using the classical beam theory is given by

$$\sin \lambda a - \lambda a \cos \lambda a = 0$$

6.5 Show that the characteristic equation governing natural vibration of a clamped-free laminated beam using the classical beam theory is given by

$$\cos \lambda a \cosh \lambda a + 1 = 0$$

6.6 Show that the characteristic equation governing natural vibration of a clamped-hinged laminated beam using the classical beam theory, when rotary inertia is neglected, is given by

$$\sin \lambda a \cosh \lambda a - \cos \lambda a \sinh \lambda a = 0$$

6.7 Show that the characteristic equation governing natural vibration of a hinged-free laminated beam using the classical beam theory, when rotary inertia is neglected, is the same as that for a clamped-hinged beam.

6.8 Show that the characteristic equation governing natural vibration of a clamped-hinged laminated beam using the classical beam theory, when rotary inertia is *not* neglected, is given by Eq. (6.2–58).

6.9 Show that Eqs. (6.3-10a,b) can be reduced to the single equation

$$E^b_{xx}I_{yy}\frac{d^4w_0}{dx^4} = \hat{q}(x) - \left(\frac{E^b_{xx}I_{yy}}{KG^b_{xz}bh}\right)\frac{d^2\hat{q}}{dx^2}$$

This equation shows that the deflection of the Timoshenko beam theory can be obtained from that of the classical beam theory by replacing the load $\hat{q}$ [see Eq. (6.2–10b)] with a equivalent load given by the right-hand side of the above equation. Although the effect of shear deformation is zero when the load variation is linear or less, this effect will come through the boundary conditions.

6.10 Show that the equations governing the stability of a laminated beam according to the Timoshenko theory can be expressed as

$$KG_{xz}^b bh\mathcal{X} + \left(KG_{xz}^b bh - bN_{xx}^0\right)\frac{dW}{dx} = K_1 \tag{1}$$

$$EI\frac{d\mathcal{X}}{dx} - bN_{xx}^0 W = K_1 x + K_2 \tag{2}$$

Combine the above two equations to arrive at

$$\frac{d^2W}{dx^2} + \lambda^2 W + \frac{\lambda^2}{bN^0xx}(K_1 x + K_2) = 0 \tag{3}$$

Show that the general solution of Eq. (3) is

$$W(x) = c_1 \sin\lambda x + c_2\cos\lambda x + c_3 x + c_4 \tag{4a}$$

$$c_3 = -\frac{1}{bN_{xx}^0}K_1\ , \quad c_4 = -\frac{1}{bN_{xx}^0}K_2 \tag{4b}$$

6.11 Show that the solution to the equations governing the bending of a hinged-fixed beam according to the Timoshenko beam theory, under uniformly distributed transverse load, is given by

$$\begin{aligned} w_0(x) = &\frac{q_0a^4}{48E_{xx}^bI_{yy}}\left[2\left(\frac{x}{a}\right)^4 - \left(\frac{5+12\alpha}{1+3\alpha}\right)\left(\frac{x}{a}\right)^3 + 3\left(\frac{1}{1+3\alpha}\right)\left(\frac{x}{a}\right)^2\right]\\ &+\frac{q_0a^2}{8KG_{xz}^bbh}\left[\left(\frac{5+12\alpha}{1+3\alpha}\right)\left(\frac{x}{a}\right) - 4\left(\frac{x}{a}\right)^2\right] \end{aligned} \tag{a}$$

$$\phi_x(x) = -\frac{q_0a^3}{48E_{xx}^bI_{yy}}\left[8\left(\frac{x}{a}\right)^3 - 3\left(\frac{5+12\alpha}{1+3\alpha}\right)\left(\frac{x}{a}\right)^2 + 6\left(\frac{1}{1+3\alpha}\right)\left(\frac{x}{a}\right)\right] \tag{b}$$

where $\alpha = E_{xx}^bI_{yy}/KG_{xz}^bbha^2$.

6.12 Show that the characteristic equation governing the buckling load of a hinged-fixed beam according to the Timoshenko beam theory is given by

$$\lambda a\ \cos\lambda a - \sin\lambda a\left(1 + \frac{\lambda^2E_{xx}^bI_{yy}}{KG_{xz}^bbh}\right) = 0$$

Ans: The boundary conditions give

$$w_0(0) = 0 \text{ gives } c_2 + c_4 = 0 \tag{a}$$

$$\phi_x(0) = 0 \text{ gives } \left(1 - \frac{bN_{xx}^0}{KG_{xz}^b bh}\right)\lambda c_1 + c_3 = 0 \tag{b}$$

$$w_0(a) = 0 \text{ gives } c_1 \sin\lambda a + c_2 \cos\lambda a + c_3 a + c_4 = 0 \tag{c}$$

$$\frac{d\phi_x}{dx}(a) = 0 \text{ gives } \lambda^2\left(1 - \frac{bN_{xx}^0}{KG_{xz}^b bh}\right)(c_1 \sin\lambda a + c_2 \cos\lambda a) = 0 \tag{d}$$

In addition, note that

$$1 - \frac{bN_{xx}^0}{KG_{xz}^b bh} = \frac{1}{1 + \frac{\lambda^2 E_{xx}^b I_{yy}}{KG_{xz}^b bh}} \tag{e}$$

6.13 Determine the critical buckling load of a clamped-free laminated beam using the Timoshenko beam theory.

6.14 Show that the characteristic equation governing natural vibrations of a clamped-free beam according to the Timoshenko beam theory is given by

$$\left(\lambda^2 - \mu^2\frac{S_2}{S_1}\right) + \left(\mu^2 - \lambda^2\frac{S_2}{S_1}\right)\cos\lambda a \cosh\mu a + \mu\lambda\left(1 + \frac{S_2}{S_1}\right)\sin\lambda a \sinh\mu a = 0 \tag{1}$$

where

$$S_1 = \hat{I}_0\omega^2 - \lambda^2 KG_{xz}^b bh\ , \quad S_2 = \hat{I}_0\omega^2 + \mu^2 KG_{xz}^b bh \tag{2}$$

6.15 Show that the characteristic equation governing natural vibrations of a clamped-hinged beam according to the Timoshenko beam theory is given by

$$S_{11}\cos\lambda a \sinh\mu a + S_{22}\sin\lambda a \cosh\mu a = 0 \tag{1}$$

where

$$S_{11} = \mu\left(\hat{I}_0\omega^2 - \lambda^2 KG_{xz}^b bh\right)\ , \quad S_{22} = \lambda\left(\hat{I}_0\omega^2 + \mu^2 KG_{xz}^b bh\right) \tag{2}$$

6.16 Derive the equations of equilibrium for cylindrical bending using the principle of virtual displacements, $\delta W = 0$, where

$$\delta W = \int_0^a \left\{ N_{xx}\,\delta\left[\frac{du_0}{dx} + \frac{1}{2}\left(\frac{dw_0}{dx}\right)^2\right] + N_{xy}\,\delta\left(\frac{dv_0}{dx}\right) + M_{xx}\,\delta\left(-\frac{d^2w_0}{dx^2}\right) - q\delta w_0\right\} dx$$

Use the laminate constitutive equations (6.4–13a), (6.4–13c), and (6.4–14a) to express the resulting Euler–Lagrange equations in terms of the displacements and the thermal stress resultants. These equations are a static version of those in Eqs. (6.4–1).

6.17 Consider the equations of equilibrium of cross-ply laminates in cylindrical bending in the absence of thermal effects:

$$A_{11}\frac{d^2u_0}{dx^2} - B_{11}\frac{d^3w_0}{dx^3} = 0$$

$$-B_{11}\frac{d^3u_0}{dx^3} + D_{11}\frac{d^4w_0}{dx^4} = q$$

Show that the Navier solution of these equations is given by

$$u_0(x) = \frac{B_{11}}{D}\sum_{m=1}\frac{Q_m}{\alpha_m^3}\cos\alpha_m x\ , \quad w_0(x) = \frac{A_{11}}{D}\sum_{m=1}\frac{Q_m}{\alpha_m^4}\sin\alpha_m x$$

where $D = A_{11}D_{11} - B_{11}^2$ and $\alpha_m = \frac{m\pi}{a}$. The load $q(x)$ is also expanded in sine series with coefficient Q_m.

6.18 For the cylindrical bending problem of cross-ply plates (see Exercise 6.19), show that (a) the stresses in the kth layer are given by

$$\sigma_{xx}^k = \frac{(A_{11}z - B_{11})\,Q_{11}^k}{D}\sum_{m=1}\frac{Q_m}{\alpha_m^2}\sin\alpha_m x$$

$$\sigma_{yy}^k = \frac{(A_{11}z - B_{11})\,Q_{12}^k}{D}\sum_{m=1}\frac{Q_m}{\alpha_m^2}\sin\alpha_m x$$

and (b) the transverse stresses from the 3-D equations of equilibrium are given by

$$\sigma_{xz}^k = -\frac{Q_{11}^k}{D}\left(A_{11}\frac{z^2}{2} - B_{11}z + G_i^k\right)\sum_{m=1}\frac{Q_m}{\alpha_m}\cos\alpha_m x$$

$$\sigma_{zz}^k = -\frac{Q_{11}^k}{D}\left(A_{11}\frac{z^3}{6} - B_{11}\frac{z^2}{2} + G^k z + H^k\right)\sum_{m=1}\frac{Q_m}{\alpha_m}\sin\alpha_m x$$

where G^k and H^k are constants to be determined such that the stress boundary conditions on σ_{xz} and σ_{zz} at $z = \pm h/2$ and the stress continuity conditions at the interfaces are satisfied.

6.19 Use the total potential energy functional

$$\Pi(w_0) = \frac{1}{2}\int_0^a \left[E_{xx}^b I_{yy} \left(\frac{d^2 w_0}{dx^2}\right)^2 - bN_{xx}^0 \left(\frac{dw_0}{dx}\right)^2 - I_0 b\omega^2 w_0^2 \right] dx$$

to construct a one-parameter Rayleigh–Ritz solution to determine the natural frequency of vibration, ω, of a *simply supported* laminated beam with compressive load N_{xx}^0. Use algebraic polynomials for the approximate functions. (*Ans:* $\omega = (1/a)\sqrt{(10/I_0)[(12E_{xx}^b I_{yy}/a^2 b) - N_{xx}^0]}$.)

6.20 Repeat Exercise 6.19 for a laminated beam with clamped boundary condition at $x = 0$ and free at $x = a$ (*i.e., cantilever beam*). (*Ans:* $\omega = (1/a)\sqrt{(5/3I_0)[(12E_{xx}^b I_{yy}/a^2 b) - 4N_{xx}^0]}$.)

6.21 Use the total potential energy functional

$$\Pi(u_0, v_0, w_0) = \int_0^a \left[\frac{A_{11}}{2}\left(\frac{du_0}{dx}\right)^2 + A_{16}\frac{du_0}{dx}\frac{dv_0}{dx} + \frac{A_{66}}{2}\left(\frac{dv_0}{dx}\right)^2 \right.$$
$$\left. - \frac{d^2 w_0}{dx^2}\left(B_{11}\frac{du_0}{dx} + B_{16}\frac{dv_0}{dx}\right) + \frac{D_{11}}{2}\left(\frac{d^2 w_0}{dx^2}\right)^2 - q w_0 \right] dx$$

to construct a one-parameter (for each variable) Ritz solution of (u_0, v_0, w_0) for a simply supported plate strip. Use algebraic polynomials for the approximate functions. (*Ans:* $a_1 = -Bq_0 a^2/12AD,\ b_1 = -Cq_0 a^2/12AD,\ c_1 = q_0 a^2/24D$.)

6.22 Repeat Exercise 6.21 for a plate strip with clamped boundary conditions at $x = 0$ and free boundary conditions at $x = a$. (*Ans:* $a_1 = Bq_0 a^2/6AD,\ b_1 = Cq_0 a^2/6AD,\ c_1 = q_0 a^2/12D$.)

6.23 Use the total potential energy functional

$$\Pi(u_0, v_0, w_0) = \int_0^a \left[\frac{A_{11}}{2}\left(\frac{du_0}{dx}\right)^2 + A_{16}\frac{du_0}{dx}\frac{dv_0}{dx} + \frac{A_{66}}{2}\left(\frac{dv_0}{dx}\right)^2 - \frac{N_{xx}^0}{2}\left(\frac{dw_0}{dx}\right)^2 \right.$$
$$\left. - \frac{d^2 w_0}{dx^2}\left(B_{11}\frac{du_0}{dx} + B_{16}\frac{dv_0}{dx}\right) + \frac{D_{11}}{2}\left(\frac{d^2 w_0}{dx^2}\right)^2 \right] dx$$

to construct a one–parameter (for each variable) Ritz solution to determine the critical buckling load N_{cr} of a plate strip with clamped boundary conditions at $x = 0$ and free boundary conditions at $x = a$. Use algebraic polynomials for the approximate functions. (*Ans:* $P_{cr} = 3D/a^2$.)

6.24 Use the total potential energy functional

$$\Pi(u_0, v_0, w_0) = \int_0^a \left[\frac{A_{11}}{2}\left(\frac{du_0}{dx}\right)^2 + A_{16}\frac{du_0}{dx}\frac{dv_0}{dx} + \frac{A_{66}}{2}\left(\frac{dv_0}{dx}\right)^2 - \frac{N_{xx}^0}{2}\left(\frac{dw_0}{dx}\right)^2 \right.$$
$$\left. - \frac{d^2 w_0}{dx^2}\left(B_{11}\frac{du_0}{dx} + B_{16}\frac{dv_0}{dx}\right) + \frac{D_{11}}{2}\left(\frac{d^2 w_0}{dx^2}\right)^2 - \frac{I_0\omega^2}{2} w_0^2 \right] dx$$

to construct a one-parameter (for each variable) Ritz solution to determine the natural frequency of vibration, ω, of a simply supported plate strip with edge compressive load N_{xx}^0. Use algebraic polynomials for the approximate functions. (*Ans:* $\omega = (1/a)\sqrt{(10/I_0)[(12D/a^2) - N_{xx}^0]}$.)

6.25 Repeat Exercise 6.24 for a plate strip with clamped boundary condition at $x = 0$ and free at $x = a$. (*Ans:* $\omega = (1/a)\sqrt{(20/3I_0)[(3D/a^2) - N_{xx}^0]}$.)

6.26 Repeat Exercise 6.25 for cylindrical bending of a plate strip using the first-order shear deformation theory but neglecting rotary inertia.

6.27 Consider the buckling of a uniform beam according to the Timoshenko beam theory. The total potential energy functional for the problem can be written as

$$\Pi(w_0, \phi_x) = \frac{1}{2}\int_0^a \left[D\left(\frac{d\phi_x}{dx}\right)^2 + S\left(\frac{dw_0}{dx} - \phi_x\right)^2 - N_{xx}^0\left(\frac{dw_0}{dx}\right)^2 \right] dx$$

where $w_0(x)$ is the transverse deflection, ϕ_x is the rotation, D is the flexural stiffness, S is the shear stiffness, and N_{xx}^0 is the axial compressive load. Determine the critical buckling load of a beam clamped at one end and simply supported at the other end. Use one-parameter Rayleigh–Ritz approximation for each variable.

6.28 Consider a laminated beam of length L, flexural stiffness EI=constant, and subjected to uniformly distributed transverse load $q(x) = q_0$. Suppose that the beam is subjected to the following geometric boundary conditions

$$w_0(0) = u_1, \quad -\frac{dw_0}{dx}(0) = u_2, \quad w_0(L) = u_3, \quad -\frac{dw_0}{dx}(L) = u_4 \qquad (a)$$

and force boundary conditions

$$\frac{d}{dx}\left(EI\frac{d^2 w_0}{dx}\right)_{x=0} = Q_1, \quad \left(EI\frac{d^2 w_0}{dx}\right)_{x=0} = Q_2$$
$$-\frac{d}{dx}\left(EI\frac{d^2 w_0}{dx}\right)_{x=L} = Q_3, \quad -\left(EI\frac{d^2 w_0}{dx}\right)_{x=L} = Q_4 \qquad (b)$$

Here (u_1, u_2) and (u_3, u_4) denote the transverse deflections and rotations (clockwise) at the left and right ends, respectively, and (Q_1, Q_3) and (Q_2, Q_4) are the associated shear forces and bending moments at the same points. Note that u_i and Q_i are introduced into the formulation to have the convenience of specifying a geometric or force boundary condition.

Assume Rayleigh–Ritz approximation of the form (the exact solution of the homogeneous equation, $EId^4w_0/dx^4 = 0$ suggests this polynomial)

$$w_0(x) = c_1 + c_2 x + c_3 x^2 + c_4 x^3 \qquad (c)$$

and express the constants c_1, c_2, c_3, and c_4 in terms of u_1, u_2, u_3, and u_4 using the geometric boundary conditions (a) and rewrite (c) in the form

$$w_0(x) = u_1\varphi_1(x) + u_2\varphi_2(x) + u_3\varphi_3(x) + u_4\varphi_4(x) \qquad (d)$$

Define the functions $\varphi_i(x)$ $(i = 1, 2, 3, 4)$ that you derived. These functions can serve as the approximation functions for the Rayleigh–Ritz method (see the next exercise). (*Ans:* φ_i are the same as the Hermite cubic interpolation functions given in Section 10.2.)

6.29 (*Continuation of Exercise 6.28*) Substitute the approximation

$$w_0(x) = \sum_{j=1}^{4} u_j \varphi_j(x) \qquad (a)$$

into the total potential energy functional associated with the Euler–Bernoulli beam theory

$$\Pi(w_0) = \frac{1}{2}\int_0^L EI\left(\frac{d^2 w_0}{dx^2}\right)^2 dx - \int_0^L q(x) w_0(x)\, dx - \sum_{j=1}^{4} Q_j u_j \qquad (b)$$

and express it in the form

$$\Pi(u_1, u_2, u_3, u_4) = \sum_{i=1}^{4}\sum_{j=1}^{4} \frac{1}{2} K_{ij} u_i u_j - \sum_{j=1}^{4} (q_i u_i + Q_j u_j) \qquad (c)$$

(a) Define and evaluate the coefficients K_{ij} of the stiffness matrix and q_i of the force vector when EI = constant and $q(x) = q_0$, a constant, and (b) use the total potential energy principle to determine the four-parameter Ritz solution for the problem. In particular, show that

$$[K]\{u\} = \{q\} + \{Q\} \qquad (d)$$

(*Ans:* The stiffness matrix $[K]$ and force vector $\{q\}$ are the same as those given in Section 10.2 for the Euler–Bernoulli beam element.)

6.30 Since Eq. (d) of Exercise 6.29 is valid for any boundary conditions, it can be used to determine solutions (which turn out to be exact) even for indeterminate beams. In particular, determine the displacement in the spring that supports the right end of a beam when the left end is fixed and the beam is subjected to uniformly distributed transverse load q_0.

6.31 Equations (6.3–12b) and (6.3–13b) for ϕ_x and w_0 of the Timoshenko beam theory suggest that they can be approximated with quadratic and cubic polynomials

$$\phi_x(x) = a_1 + a_2 x + a_3 x^2, \quad w_0(x) = c_1 + c_2 x + c_3 x^2 + c_3 x^3$$

Rewrite the constants a_i in terms of the values of ϕ_x at $x = 0,\ x = 0.5L$, and $x = L$ and obtain

$$\phi_x(x) = \Phi_1\psi_1(x) + \Phi_2\psi_2(x) + \Phi_3\psi_3(x) \tag{a}$$

where $\Phi_1 = \phi_x(0)$ etc. Show that $\psi_i(x)$ $(i = 1,2,3)$ are the quadratic Lagrange interpolation functions derived in Section 10.3.

6.32 Use Eq. (a) of Exercise 6.31 and Eq. (a) of Exercise 6.29 to express the total potential energy functional in terms of u_j and Φ_j:

$$\begin{aligned}\Pi(w_0, \phi_x) =& \frac{1}{2}\int_0^a \left[E^b_{xx} I_{yy}\left(\frac{d\phi_x}{dx}\right)^2 + KG^b_{xz} bh\left(\frac{dw_0}{dx} - \phi_x\right)^2\right] dx \\ & - \int_0^a \hat{q} w_0 dx - \sum_{j=1}^{4} Q_j u_j - \sum_{j=1}^{3} P_j \Phi_j \end{aligned} \tag{a}$$

where P_j $(j = 1,2,3)$ are the moments corresponding to the rotations Φ_j. Then use the total potential energy principle to derive the Rayleigh–Ritz equations for the problem.

6.33 The deflection, bending moment, and shear force of the Timoshenko beam theory can be expressed in terms of the corresponding quantities of the Euler–Bernoulli beam theory (see [27,28]). In order to establish these relationships, we use the following equations of the two theories:

$$-D_{xx}\frac{d^4 w_0^E}{dx^4} = -q(x), \quad D_{xx}\frac{d^3\phi^T}{dx^3} = -q(x), \quad A_{xz}K_s\left(\frac{d\phi^T}{dx} + \frac{d^2 w_0^T}{dx^2}\right) = -q(x)$$

where K_s is the shear correction coefficient, and superscripts E and T on variables refer to the Euler–Bernoulli and Timoshenko beam theories. Show that

$$
\begin{aligned}
D_{xx}w_0^T(x) =& D_{xx}w_0^E(x) + \frac{D_{xx}}{A_{xz}K_s}M_{xx}^E(x) + C_1\left(\frac{D_{xx}}{A_{xz}K_s}x - \frac{x^3}{6}\right) \\
& - C_2\frac{x^2}{2} - C_3x - C_4 \qquad (1) \\
D_{xx}\phi^T(x) =& - D_{xx}\frac{dw_0^E}{dx} + C_1\frac{x^2}{2} + C_2x + C_3 \qquad (2) \\
M_{xx}^T(x) =& M_{xx}^E(x) + C_1x + C_2 \qquad (3) \\
Q_x^T(x) =& Q_x^E(x) + C_1 \qquad (4)
\end{aligned}
$$

where C_1, C_2, C_3, and C_4 are constants of integration, which are to be determined using the boundary conditions of the particular beam.

6.34 Show that for simply supported beams all C_i of Exercise 6.33 are zero.

6.35 Show that for cantilevered beams all C_i except $C_4 = M_{xx}^E(0)/(D_{xx}A_{xz}K_s)$ of Exercise 6.33 are zero.

6.36 Consider bending of a beam of length L, clamped (or fixed) at the left end and simply supported at the right, and subjected to a uniformly distributed transverse load q_0. The boundary conditions of the Euler–Bernoulli and Timoshenko beam theories for the problem are as follows:

$$
\begin{aligned}
\text{EBT}: &\quad w_0^E(0) = w_0^E(L) = \frac{dw_0^E}{dx}(0) = M_{xx}^E(L) = 0 \qquad (1) \\
\text{TBT}: &\quad w_0^T(0) = w_0^T(L) = \phi^T(0) = M_{xx}^T(L) = 0 \qquad (2)
\end{aligned}
$$

Show that the constants of integration in Exercise 6.33 are given by

$$
C_1 = \frac{3\Omega}{(1+3\Omega)L}M_{xx}^E(0), \quad C_2 = -C_1L, \quad C_3 = 0, \quad C_4 = \Omega M_{xx}^E(0)L^2 \qquad (3)
$$

where $\Omega = D_{xx}/(A_{xz}K_sL^2)$.

6.37 The governing equations of the Reddy third-order beam theory are given in Exercie 3.17. For static case, show that

$$
\frac{d^2M_{xx}^R}{dx^2} = -q(x), \quad M_{xx}^R(x) = M_{xx}^E(x) + C_1x + C_2
$$

where C_1 and C_2 are constants of integration and the superscripts E and R on variables refer to the Euler–Bernoulli and Reddy beam theories.

References for Additional Reading

1. Reddy, J. N., *Energy and Variational Methods in Applied Mechanics,* John Wiley, New York, 1984.
2. Reddy, J. N. (ed.), *Mechanics of Composite Materials. Selected Works of Nicholas J. Pagano*, Kluwer, Netherlands, 1994.
3. Whitney, J. M. , *Structural Analysis of Laminated Anisotropic Plates,* Technomic, Lancaster, PA, 1987.
4. Reissner, E., "The Effect of Transverse Shear Deformation on the Bending of Elastic Plates," *Journal of Applied Mechanics,* **12**, 69–77 (1945).
5. Mindlin, R. D., "Influence of Rotatory Inertia and Shear on Flexural Motions of Isotropic Elastic Plates," *Journal of Applied Mechanics, Transactions of ASME*, **18**, 31–38 (1951).
6. Uflyand, Ya. S., "The Propagation of Waves in the Transverse Vibrations of Bars and Plates," Akad. Nauk SSSR. *Prikl Mat. Mekh.,* **12**, 287–300 (1948) (in Russian).
7. Yang, P. C., Norris, C. H., and Stavsky, Y., "Elastic Wave Propagation in Heterogeneous Plates," *International Journal of Solids and Structures*, **2**, 665–684 (1966).
8. Whitney, J. M., "Shear Correction Factors for Orthotropic Laminates Under Static Load," *Journal of Applied Mechanics*, **40**(1), 302–304 (1973).
9. Brogan, W. L., *Modern Control Theory,* Prentice–Hall, Englewood Cliffs, New Jersey, 1985.
10. Franklin, J. N., *Matrix Theory,* Prentice–Hall, Englewood Cliffs, New Jersey, 1968.
11. Goldberg, J. L. and Schwartz, A. J., *Systems of Ordinary Differential Equations, An Introduction,* Harper and Row, New York, 1972.
12. Khdeir, A. A. and Reddy, J. N., "Free Vibration of Cross-Ply Laminated Beams with Arbitrary Boundary Conditions," *International Journal of Engineering Science,* **32** (12), 1971–1980 (1994).
13. Pagano, N. J., "Exact Solutions for Composite Laminates in Cylindrical Bending," *Journal of Composite Materials,* **3**, 398–411 (1969).
14. Pagano, N. J., "Influence of Shear Coupling in Cylindrical Bending of Anisotropic Laminates," *Journal of Composite Materials*, **4**, 330–343 (1970).
15. Pagano, N. J., and Wang, A. S. D., "Further Study of Composite Laminates Under Cylindrical Bending," *Journal of Composite Materials*, **5**, 521–528 (1971).
16. Weaver, W., Jr., Timoshenko, S. P., and Young, D. H., *Vibration Problems in Engineering*, Fifth Edition, John Wiley, New York, 1990.
17. Clough, R. W. and Penzien, J., *Dynamics of Structures*, McGraw–Hill, New York, 1975.

18. Pipes, L. A. and Harvill, L. R., *Applied Mathematics for Engineers and Physicists*, Third Edition, McGraw–Hill, New York, 1970.

19. Timoshenko, S. P. and Gere, J. P., *Theory of Elastic Stability,* Second Edition, McGraw–Hill, New York, 1959.

20. Timoshenko, S. P., "On the Correction for Shear of the Differential Equation for Transverse Vibrations of Prismatic Bars," *Philosophical Magazine,* **41**, 744–746 (1921).

21. Timoshenko, S. P., "On the Transverse Vibrations of Bars of Uniform Cross Section," *Philosophical Magazine,* **43**, 125–131 (1922).

22. Timoshenko, S. P. and Woinowsky–Krieger, S., *Theory of Plates and Shells,* McGraw–Hill, Singapore (1970).

23. Levinson, M., "A New Rectangular Beam Theory," *Journal of Sound and Vibration,* **74**, 81–87 (1981).

24. Bickford, W. B., "A Consistent Higher Order Beam Theory," *Developments in Theoretical and Applied Mechanics,* **11**, 137–150 (1982).

25. Reddy, J. N., "A Simple Higher-Order Theory for Laminated Composite Plates," *Journal of Applied Mechanics,* **51**, 745–752 (1984).

26. Heyliger, P. R. and Reddy, J. N., "A Higher-Order Beam Finite Element for Bending and Vibration Problems," *Journal of Sound and Vibration,* **126**(2), 309–326 (1988).

27. Wang, C. M., "Timoshenko Beam-Bending Solutions in Terms of Euler–Bernoulli Solutions, *Journal of Engineering Mechanics,* ASCE, **121**(6), 763–765 (1995).

28. Reddy, J. N., Wang, C. M., and Lee, K. H., "Relationships Between Bending Solutions of Classical and Shear Deformation Beam Theories," *International Journal of Solids and Structures,* in press.

Chapter Seven

Analysis of Specially Orthotropic Plates Using CLPT

7.1 Introduction

The governing equations of composite laminates according to various laminate theories were developed in Chapter 5. These equations can be solved either analytically or numerically for the generalized displacements and strains. Stresses can be determined using either the constitutive equations or the 3-D equilibrium equations expressed in terms of stresses. Analytical solutions can be developed for certain one-dimensional problems, as was done in Chapter 6 for beams and cylindrical bending problems, and for rectangular laminates with certain lamination schemes and boundary conditions.

In this chapter we develop analytical solutions of *specially orthotropic plates, i.e.*, plates for which the bending–stretching coupling coefficients B_{ij} and bending–twisting coefficients D_{16} and D_{26} are zero, using the classical laminate theory. The solution methods used here are the Navier method, the Lévy method with the state–space approach, and the Rayleigh–Ritz method. The Navier solutions can be developed for a rectangular laminate when all four edges of the laminate are simply supported. The Lévy solutions can be developed for plates with two opposite edges simply supported and the remaining two edges having any possible combination of boundary conditions: free, simple support, or fixed support. The Rayleigh–Ritz method can be used to determine approximate solutions for more general boundary conditions.

The analysis of specially orthotropic laminates is greatly simplified due to the uncoupling of bending deformation from extensional deformation and the fact that $D_{16} = D_{26} = 0$. This class of laminates will be used to gain a basic understanding of the response. Although most laminates of practical interest do not qualify as specially orthotropic plates because of the presence of bending–twisting coupling terms D_{16} and D_{26}, they may represent reasonable approximations to more complex laminates. In the subsequent chapters, the solutions obtained for more complicated laminates will be compared with those of the specially orthotropic plates to assess their behavior.

The equation of motion governing bending deflection w_0 of a specially orthotropic plate can be deduced from Eq. (5.3–47) by omitting the nonlinear terms, bending–stretching terms, and bending–twisting terms. We have

$$
\begin{aligned}
&-\left[D_{11}\frac{\partial^4 w_0}{\partial x^4}+2\left(D_{12}+2D_{66}\right)\frac{\partial^4 w_0}{\partial x^2\partial y^2}+D_{22}\frac{\partial^4 w_0}{\partial y^4}\right]+q\\
&-\left(\frac{\partial^2 M_{xx}^T}{\partial x^2}+2\frac{\partial^2 M_{xy}^T}{\partial x\partial y}+\frac{\partial^2 M_{yy}^T}{\partial y^2}\right)+\hat{N}_{xx}\frac{\partial^2 w_0}{\partial x^2}+2\hat{N}_{xy}\frac{\partial^2 w_0}{\partial x\partial y}+\hat{N}_{yy}\frac{\partial^2 w_0}{\partial y^2}\\
&=I_0\ddot{w}_0-I_2\left(\frac{\partial^2 \ddot{w}_0}{\partial x^2}+\frac{\partial^2 \ddot{w}_0}{\partial y^2}\right)
\end{aligned}
\qquad (7.1-1)
$$

Equation (1) must be solved, in conjunction with appropriate boundary conditions [see Eq. (5.3–34)] and initial conditions of the problem, for the desired response. The boundary conditions at any point on the boundary are of the form: specify

$$w_0 \quad \text{or} \quad Q_n \equiv Q_x n_x + Q_y n_y \qquad (7.1-2)$$

and

$$\frac{\partial w_0}{\partial n} \quad \text{or} \quad M_{nn} \qquad (7.1-3)$$

In this chapter, we wish to determine static deflections and stresses, frequencies of natural vibration, and buckling loads under in-plane compressive or shear loads of specially orthotropic plates. We seek exact solutions whenever possible, and approximate solutions using the Rayleigh–Ritz method when exact solutions cannot be developed.

7.2 Bending of Simply Supported Rectangular Plates

7.2–1 Governing Equations

Here we consider the static bending in the absence of thermal effects and in-plane forces. Equation (7.1–1) for this case reduces to

$$D_{11}\frac{\partial^4 w_0}{\partial x^4}+2\left(D_{12}+2D_{66}\right)\frac{\partial^4 w_0}{\partial x^2\partial y^2}+D_{22}\frac{\partial^4 w_0}{\partial y^4}=q \qquad (7.2-1)$$

The simply supported boundary conditions on all four edges of the rectangular plate (see Figure 7.2–1) can be expressed as

$$w_0(x,0)=0,\quad w_0(x,b)=0,\quad w_0(0,y)=0,\quad w_0(a,y)=0 \qquad (7.2-2a)$$

$$M_{xx}(0,y)=0,\quad M_{xx}(a,y)=0,\quad M_{yy}(x,0)=0,\quad M_{yy}(x,b)=0 \qquad (7.2-2b)$$

where the bending moments are related to the transverse deflection by the equations

$$
\begin{aligned}
M_{xx} &= -\left(D_{11}\frac{\partial^2 w_0}{\partial x^2} + D_{12}\frac{\partial^2 w_0}{\partial y^2}\right) \\
M_{yy} &= -\left(D_{12}\frac{\partial^2 w_0}{\partial x^2} + D_{22}\frac{\partial^2 w_0}{\partial y^2}\right) \\
M_{xy} &= -2D_{66}\frac{\partial^2 w_0}{\partial x \partial y}
\end{aligned} \tag{7.2 – 3}
$$

and a and b denote the in-plane dimensions along the $x-$ and $y-$coordinate directions of the rectangular laminate. The origin of the coordinate system is taken at the lower left corner of the midplane (see Figure 7.2–1).

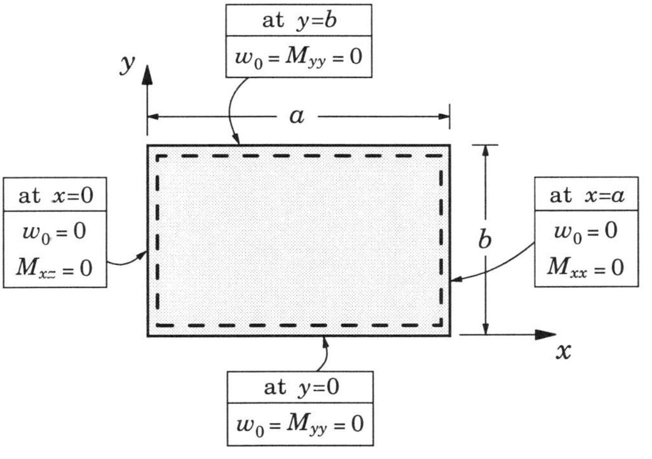

Figure 7.2–1: Geometry, coordinate system, and simply supported boundary conditions for a rectangular plate.

7.2–2 The Navier Solution

In the Navier method the displacement w_0 is expanded in a double trigonometric (Fourier) series in terms of unknown parameters. The choice of the trigonometric functions in the series is restricted to those which satisfy the boundary conditions of the problem. The load $q(x, y)$ is also expanded in double trigonometric series. Substitution of the displacement and load expansions into the governing equation should result in an invertible set of algebraic equations among the parameters of the displacement expansion. Otherwise, the Navier solution cannot be developed for the problem. The simply supported boundary conditions in Eq. (2) admit the Navier solution.

The boundary conditions in Eq. (2) are satisfied by the following form of the transverse deflection

$$w_0(x,y) = \sum_{n=1}^{\infty}\sum_{m=1}^{\infty} W_{mn} \; \sin\alpha x \; \sin\beta y \qquad (7.2-4)$$

where $\alpha = m\pi/a$ and $\beta = n\pi/b$, and W_{mn} are coefficients to be determined such that the governing equation (1) is satisfied everywhere in the domain of the plate. We assume that the load can also be expanded in the series form as

$$q(x,y) = \sum_{n=1}^{\infty}\sum_{m=1}^{\infty} Q_{mn} \sin\alpha x \; \sin\beta y \qquad (7.2-5a)$$

where

$$Q_{mn} = \frac{4}{ab}\int_0^b \int_0^a q(x,y)\; \sin\frac{m\pi x}{a} \; \sin\frac{n\pi y}{b}\; dxdy \qquad (7.2-5b)$$

Substitution of the expansions (4) and (5) into Eq. (1) yields

$$\sum_{n=1}^{\infty}\sum_{m=1}^{\infty}\left\{-W_{mn}\left[D_{11}\alpha^4 + 2(D_{12}+2D_{66})\alpha^2\beta^2 + D_{22}\beta^4\right] + Q_{mn}\right\}\sin\alpha x \; \sin\beta y = 0 \qquad (7.2-6)$$

Since the equation must hold for every point (x,y) of the domain $0 < x < a$ and $0 < y < b$, the expression inside the curl brackets (or braces) should be zero for every m and n. This yields

$$W_{mn} = \frac{Q_{mn}}{d_{mn}} \qquad (7.2-7a)$$

$$d_{mn} = \frac{\pi^4}{b^4}\left[D_{11}m^4s^4 + 2(D_{12}+2D_{66})m^2n^2s^2 + D_{22}n^4\right] \qquad (7.2-7b)$$

where s denotes the plate aspect ratio, $s = b/a$. Then the solution in Eq. (4) becomes

$$w_0(x,y) = \sum_{n=1}^{\infty}\sum_{m=1}^{\infty} \frac{Q_{mn}}{d_{mn}} \; \sin\alpha x \; \sin\beta y \qquad (7.2-8)$$

The effect of thermal moments can be easily incorporated into the calculation.

The load coefficients Q_{mn} for various types of loading [see Eq. (5b)] are listed in Table 7.2–1. The Navier solution for a sinusoidally distributed transverse load

$$q(x,y) = q_0 \sin\frac{\pi x}{a} \; \sin\frac{\pi y}{b} \qquad (7.2-9)$$

Table 7.2–1: Coefficients in the double trigonometric series expansion of loads in the Navier method.

Load $q(x,y)$	Coefficients Q_{mn}
Uniform load, $q = -q_0$ 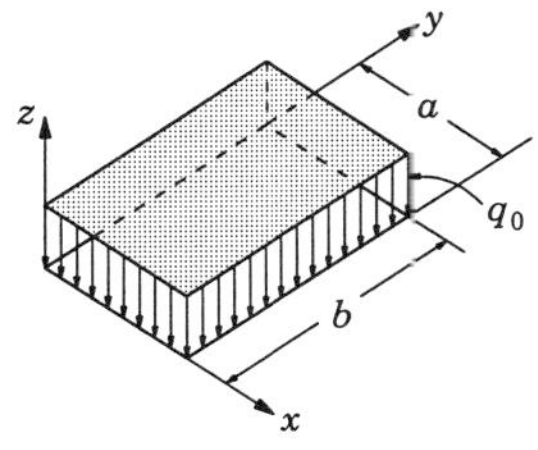	$Q_{mn} = -\frac{16q_0}{\pi^2 mn}$ $(m,n = 1,3,5,\cdots)$
Hydrostatic load, $q(x,y) = -q_0\frac{y}{b}$ 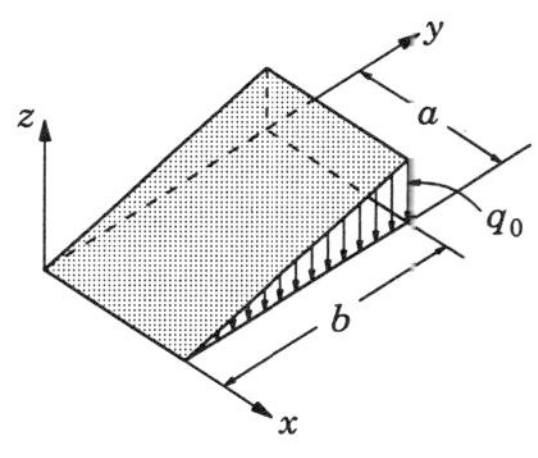	$Q_{mn} = -\frac{8q_0 \cos n\pi}{\pi^2 mn}$ $(m,n = 1,3,5,\cdots)$
Point load, $q(x,y) = -Q_0$ at (x_0, y_0) 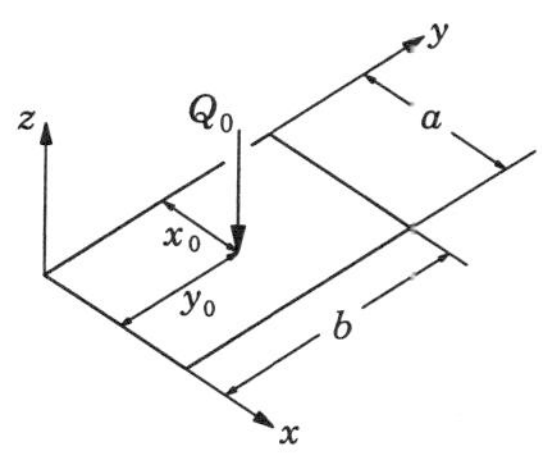	$Q_{mn} = -\frac{4Q_0}{ab} \sin \frac{m\pi x_0}{a} \sin \frac{n\pi y_0}{b}$ $(m,n = 1,2,3,\cdots)$
Line load, $q(x,y) = -q_0$ at $y = y_0$ 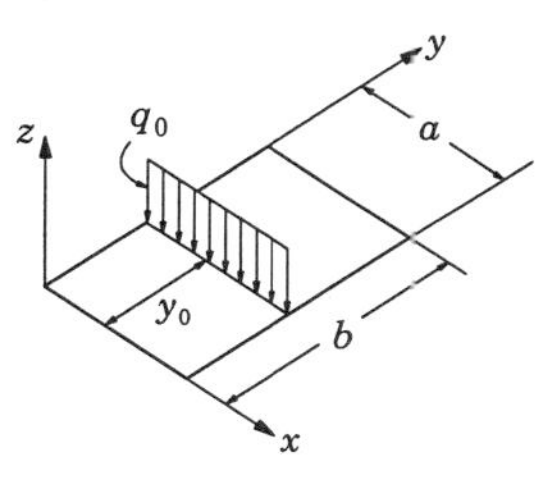	$Q_{mn} = -\frac{4q_0}{\pi bm} \sin \frac{n\pi y_0}{b}$ $(m,n = 1,2,3,\cdots)$

is a one-term solution ($Q_{mn} = q_0$ and $m = n = 1$), and therefore it is a *closed-form solution.* For other types of loads, the Navier solution is a series solution, which can be evaluated for sufficient number of terms in the series. In particular, for uniformly distributed load $q(x,y) = q_0$, a constant, we have

$$Q_{mn} = \frac{16q_0}{\pi^2 mn} \quad \text{for } m, n, \text{ odd} \tag{7.2 – 10}$$

For a point load Q_0 located at (x_0, y_0), the load coefficients are given by $[q(x,y) = Q_0\delta(x - x_0, y - y_0)]$

$$Q_{mn} = \frac{4Q_0}{ab}\sin\frac{m\pi x_0}{a}\,\sin\frac{n\pi y_0}{b} \tag{7.2 – 11}$$

The bending moments can be calculated from

$$M_{xx} = \sum_{n=1}^{\infty}\sum_{m=1}^{\infty}\left(D_{11}\alpha^2 + D_{12}\beta^2\right)W_{mn}\,\sin\frac{m\pi x}{a}\,\sin\frac{n\pi y}{b} \tag{7.2 – 12a}$$

$$M_{yy} = \sum_{n=1}^{\infty}\sum_{m=1}^{\infty}\left(D_{12}\alpha^2 + D_{22}\beta^2\right)W_{mn}\,\sin\frac{m\pi x}{a}\,\sin\frac{n\pi y}{b} \tag{7.2 – 12b}$$

$$M_{xy} = -2\sum_{n=1}^{\infty}\sum_{m=1}^{\infty}\alpha\beta D_{66}W_{mn}\,\cos\frac{m\pi x}{a}\,\cos\frac{n\pi y}{b} \tag{7.2 – 12c}$$

The in-plane stresses can be computed from Eqs. (5.3–12a)

$$\begin{Bmatrix}\sigma_{xx}\\ \sigma_{yy}\\ \sigma_{xy}\end{Bmatrix}^{(k)} = -z\begin{bmatrix}\bar{Q}_{11} & \bar{Q}_{12} & 0\\ \bar{Q}_{12} & \bar{Q}_{22} & 0\\ 0 & 0 & \bar{Q}_{66}\end{bmatrix}^{(k)}\begin{Bmatrix}\frac{\partial^2 w_0}{\partial x^2}\\ \frac{\partial^2 w_0}{\partial y^2}\\ 2\frac{\partial^2 w_0}{\partial x\partial y}\end{Bmatrix}$$

$$= z\sum_{n=1}^{\infty}\sum_{m=1}^{\infty}W_{mn}\begin{Bmatrix}\left(\bar{Q}_{11}^{(k)}\alpha^2 + \bar{Q}_{12}^{(k)}\beta^2\right)\sin\frac{m\pi x}{a}\,\sin\frac{n\pi y}{b}\\ \left(\bar{Q}_{12}^{(k)}\alpha^2 + \bar{Q}_{22}^{(k)}\beta^2\right)\sin\frac{m\pi x}{a}\,\sin\frac{n\pi y}{b}\\ -2\bar{Q}_{66}^{(k)}\alpha\beta\,\cos\frac{m\pi x}{a}\,\cos\frac{n\pi y}{b}\end{Bmatrix} \tag{7.2 – 13}$$

The maximum normal stresses occur at $(x, y, z) = (a/2, b/2, h/2)$, and the shear stress is maximum at $(x, y, z) = (a, b, -h/2)$.

The interlaminar stresses are identically zero when computed from the constitutive equations in the classical laminate theory. However, they can be computed using the 3-D stress equilibrium equations [see Eqs. (6.2–13)] for any $-h/2 \le z \le z_{k+1}$:

$$\sigma_{xz}^{(k)} = -\int_{z_k}^{z}\left(\frac{\partial\sigma_{xx}^{(k)}}{\partial x} + \frac{\partial\sigma_{xy}^{(k)}}{\partial y}\right)dz + C_1^{(k)}(x,y)$$

$$\sigma_{yz}^{(k)} = -\int_{z_k}^{z}\left(\frac{\partial\sigma_{xy}^{(k)}}{\partial x} + \frac{\partial\sigma_{yy}^{(k)}}{\partial y}\right)dz + C_2^{(k)}(x,y)$$

$$\sigma_{zz}^{(k)} = -\int_{z_k}^{z}\left(\frac{\partial\sigma_{xz}^{(k)}}{\partial x} + \frac{\partial\sigma_{yz}^{(k)}}{\partial y}\right)dz + C_3^{(k)}(x,y) \tag{7.2 – 14}$$

where the stresses $\sigma_{xx}^{(k)}, \sigma_{xy}^{(k)}$, and $\sigma_{yy}^{(k)}$ are known from Eq. (13), and $C_i^{(k)}$ are functions to be determined using the boundary conditions, $\sigma_{xz}(x, y, -h/2) = \sigma_{yz}(x, y, -h/2) = \sigma_{zz}(x, y, -h/2) = 0$ and continuity of stresses at layer interfaces. We obtain

$$
\begin{aligned}
\sigma_{xz}^{(k)} &= -\mathcal{X}(z) \sum_{n=1}^{\infty} \sum_{m=1}^{\infty} T_{12}^{(k)} W_{mn} \cos \frac{m\pi x}{a} \sin \frac{n\pi y}{b} \\
\sigma_{yz}^{(k)} &= -\mathcal{Y}(z) \sum_{n=1}^{\infty} \sum_{m=1}^{\infty} T_{13}^{(k)} W_{mn} \sin \frac{m\pi x}{a} \cos \frac{n\pi y}{b} \\
\sigma_{zz}^{(k)} &= -\mathcal{Z}(z) \sum_{n=1}^{\infty} \sum_{m=1}^{\infty} T_{33}^{(k)} W_{mn} \sin \frac{m\pi x}{a} \sin \frac{n\pi y}{b}
\end{aligned}
\qquad (7.2-15a)
$$

where

$$
\begin{aligned}
\mathcal{X}(z) = \mathcal{Y}(z) &= \left(\frac{z^2 - z_k^2}{2}\right) \\
\mathcal{Z}(z) &= \left[-\frac{h^3}{24} + \frac{z}{6}\left(z^2 - 3z_k^2\right)\right]
\end{aligned}
\qquad (7.2-15b)
$$

and $(T_{33}^{(k)} = \alpha T_{12} + \beta T_{13})$

$$
\begin{aligned}
T_{12}^{(k)} &= \alpha^3 \bar{Q}_{11}^{(k)} + \alpha\beta^2 \left(2\bar{Q}_{66}^{(k)} + \bar{Q}_{12}^{(k)}\right) \\
T_{13}^{(k)} &= \beta^3 \bar{Q}_{22}^{(k)} + \alpha^2\beta \left(2\bar{Q}_{66}^{(k)} + \bar{Q}_{12}^{(k)}\right) \\
T_{33}^{(k)} &= \alpha^4 \bar{Q}_{11}^{(k)} + 2\alpha^2\beta^2 \left(2\bar{Q}_{66}^{(k)} + \bar{Q}_{12}^{(k)}\right) + \beta^4 \bar{Q}_{22}^{(k)}
\end{aligned}
\qquad (7.2-15c)
$$

For single-layer plates, the expressions in Eq. (15a) can be simplified to

$$
\begin{aligned}
\sigma_{xz} &= \frac{h^2}{8}\left[1 - \left(\frac{2z}{h}\right)^2\right] \sum_{n=1}^{\infty} \sum_{m=1}^{\infty} T_{12}^{(1)} W_{mn} \cos \frac{m\pi x}{a} \sin \frac{n\pi y}{b} \\
\sigma_{yz} &= \frac{h^2}{8}\left[1 - \left(\frac{2z}{h}\right)^2\right] \sum_{n=1}^{\infty} \sum_{m=1}^{\infty} T_{13}^{(1)} W_{mn} \sin \frac{m\pi x}{a} \cos \frac{n\pi y}{b} \\
\sigma_{zz} &= -\frac{h^3}{48}\left\{\left[1 + \left(\frac{2z}{h}\right)^3\right] - 3\left[1 + \left(\frac{2z}{h}\right)\right]\right\} \times \\
&\quad \sum_{n=1}^{\infty} \sum_{m=1}^{\infty} T_{11}^{(1)} W_{mn} \sin \frac{m\pi x}{a} \sin \frac{n\pi y}{b}
\end{aligned}
\qquad (7.2-16)
$$

In integrating the stress-equilibrium equations it is assumed that the stresses $(\sigma_{xz}, \sigma_{yz}, \sigma_{zz})$ are zero at the bottom surface of the plate $(z = -h/2)$. Because of the assumptions of the laminate plate theory, they will also be zero at $z = h/2$. The transverse shear stress σ_{xz} is the maximum at $(x, y, z) = (0, b/2, 0)$, σ_{yz} is the maximum at $(x, y, z) = (a/2, 0, 0)$, and the transverse normal stress σ_{zz} is the maximum at $(x, y, z) = (a/2, b/2, h/2)$.

Table 7.2–2: Transverse deflections and stresses in specially orthotropic square laminates subjected to various types of mechanical and thermal loads ($h_n = h/n$, $E_1/E_2 = 25$, $G_{12} = G_{13} = 0.5E_2$, $G_{23} = 0.2E_2$, $\nu_{12} = 0.25$, $\alpha_1 = 3\alpha_2$, $T_0 = 0$).

	Mechanical				Thermal			
Laminate	$\hat{w}$	$\bar{\sigma}_{xx}$	$\bar{\sigma}_{yy}$	$\bar{\sigma}_{xy}$	$\hat{w}$	$\bar{\sigma}_{xx}$	$\bar{\sigma}_{yy}$	$\bar{\sigma}_{xy}$
				SSL				
0°	0.4312	0.5387	0.0267	0.0213	10.312	–0.003	–0.099	0.0509
(0°/90°/0°)	0.4312	0.5387	0.0267	0.0213	10.312	–0.003	–0.099	0.0509
				UDL				
0°	0.6497	0.7866	0.0244	0.0463	14.331	1.4343	0.0246	0.2563
(0°/90°/0°)	0.6660	0.8075	0.0306	0.0425	14.860	1.5040	0.0439	0.2281
				CPL				
0°	2.3170	5.4330	0.7520	0.0410	6.5680	0.5305	–0.065	4.0548
(0°/90°/0°)	2.1257	5.0052	0.5322	0.0559	7.4458	0.6557	–0.030	3.6735

SSL=Sinusoidal load; UDL=Uniformly distributed load; CPL=Central point load.

Table 7.2–2 contains the nondimensionalized maximum transverse deflections and stresses of square laminates under various types of loads. For the case of mechanical loading, the deflection and stresses are nondimensionalized as follows:

$$\hat{w} = w_0(0,0)\left(\frac{E_2 h^3}{a^4 q_0}\right) \times 10^2; \quad \bar{\sigma}_{xx} = \sigma_{xx}(a/2, b/2, h/2)\left(\frac{h^2}{a^2 q_0}\right)$$

$$\bar{\sigma}_{yy} = \sigma_{yy}(a/2, b/2, h/2)\left(\frac{h^2}{a^2 q_0}\right); \quad \bar{\sigma}_{xy} = \sigma_{xy}(a, b, -h/2)\left(\frac{h^2}{a^2 q_0}\right)$$

$$\bar{\sigma}_{xz} = \sigma_{yy}(0, b/2, 0)\left(\frac{h}{a q_0}\right); \quad \bar{\sigma}_{yz} = \sigma_{xy}(a/2, 0, 0)\left(\frac{h}{a q_0}\right) \qquad (7.2-17)$$

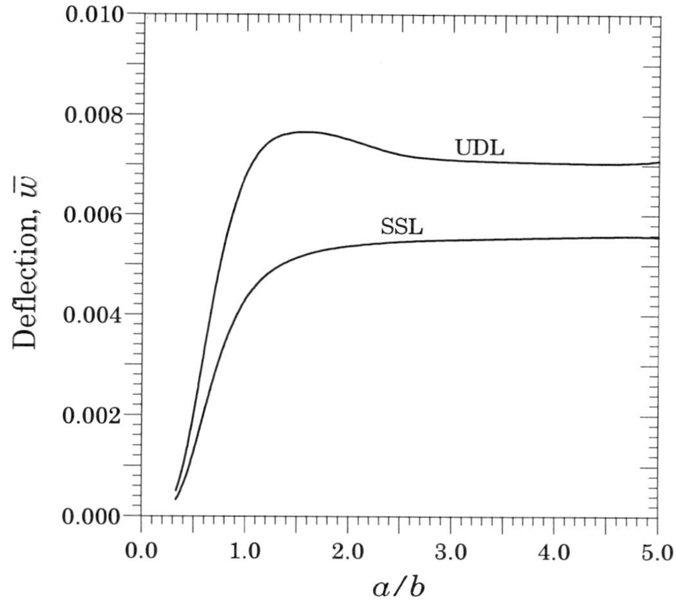

Figure 7.2–2: Nondimensionalized maximum transverse displacement $\bar{w} = w_0(E_2h^3/a^4q_0)$ versus plate aspect ratio (a/b) of symmetric cross-ply (0/90/90/0) laminates.

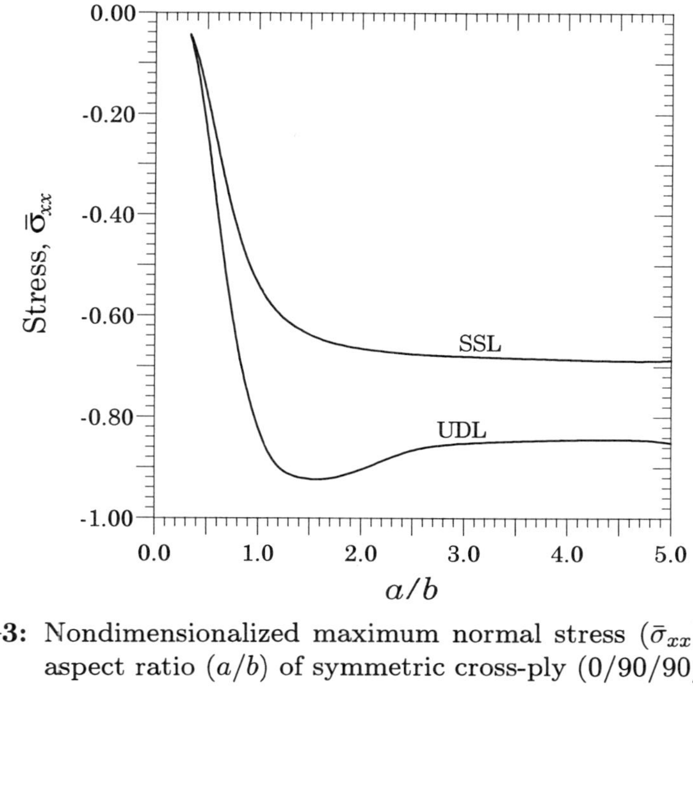

Figure 7.2–3: Nondimensionalized maximum normal stress $(\bar{\sigma}_{xx})$ versus plate aspect ratio (a/b) of symmetric cross-ply (0/90/90/0) laminates.

For the thermal load case, the nondimensionalized quantities are defined as

$$\hat{w} = w_0(0,0)\beta \times 10^2; \; \bar{\sigma}_{xx} = \sigma_{xx}(a/2, b/2, h/2)\left(\frac{\beta}{E_2}\right); \; \beta = \frac{1}{\alpha_1 T_1 a^2}$$

$$\bar{\sigma}_{yy} = \sigma_{yy}(a/2, b/2, h/2)\left(\frac{\beta}{E_2}\right); \; \bar{\sigma}_{xy} = \sigma_{xy}(a, b, -h/2)\left(\frac{\beta}{E_2}\right) \qquad (7.2-18)$$

The mechanical load consists of only the transverse load $q(x, y)$, and the thermal load consists of linear temperature distribution through the laminate thickness, $\Delta T = zT_1(x, y)$. Both q and T_1 are assumed to be sinusoidal, uniform, or point functions. In the case of uniform and point source distribution, the first ten terms of the double trigonometric series are evaluated.

Plots of nondimensionalized maximum transverse deflection $\bar{w}$ and normal stress $\bar{\sigma}_{xx}$ as a function of the plate aspect ratio a/b are shown in Figures 7.2–2 and 7.2–3, respectively, for symmetric cross-ply (0/90/90/0) laminates under uniformly distributed (UDL) and sinusoidally distributed (SSL) loads. The material properties of the lamina are taken to be: $E_1/E_2 = 25$, $G_{12} = G_{13} = 0.5E_2$, and $\nu_{12} = 0.25$. For uniformly distributed load, the maximum deflection and (negative) stress occur for an aspect ratio around 1.5, whereas for sinusoidally distributed load the maxima are reached around $a/b = 2.5$.

Figures 7.2–4 and 7.2–5 show the distributions of the maximum in-plane normal stresses $\bar{\sigma}_{xx}$ and $\bar{\sigma}_{yy}$, respectively, through the thickness for laminates (0/90/0) and (0/90/90/0) under sinusoidally distributed transverse load, and Figure 7.2–6 shows the distribution of the maximum transverse shear stresses through the thickness for the two laminates ($a/b = 1$, $E_1 = 25E_2$, $G_{12} = G_{13} = 0.5E_2$, $\nu_{12} = 0.25$).

7.3 Bending of Plates with Two Opposite Edges Simply Supported

7.3–1 The Lévy Solution Procedure

Consider a rectangular plate with simply supported edges along $y = 0, b$ and subjected to a transverse load q. The other two edges at $x = 0, a$, can each be free, simply supported, or clamped, independent of the other. For such problems, the Navier solution cannot be developed. However, the idea of the Navier method can be applied with respect to the simply supported boundary conditions at $y = 0, b$ to reduce the partial differential equation (7.2–1) to an ordinary differential equation with respect to the coordinate x, which may then be solved exactly or approximately. This procedure is known as the Lévy method.

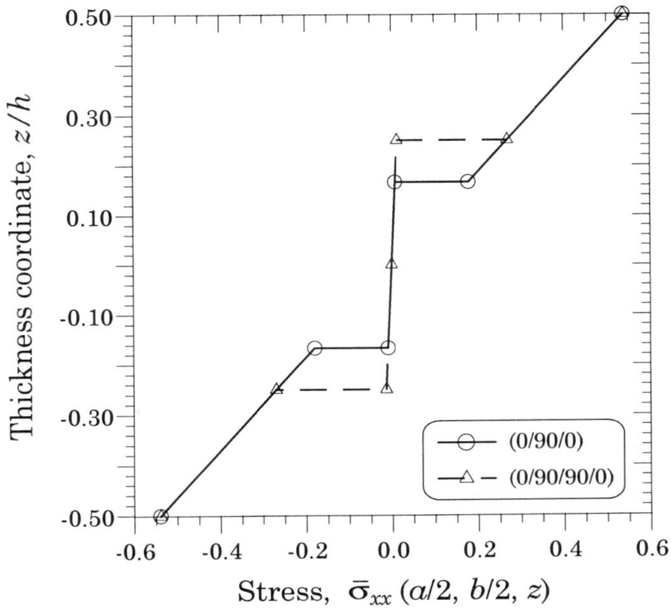

Figure 7.2–4: Variation of nondimensionalized maximum normal stress ($\bar{\sigma}_{xx}$) through the thickness (z/h) of square cross-ply laminates under sinusoidally distributed load.

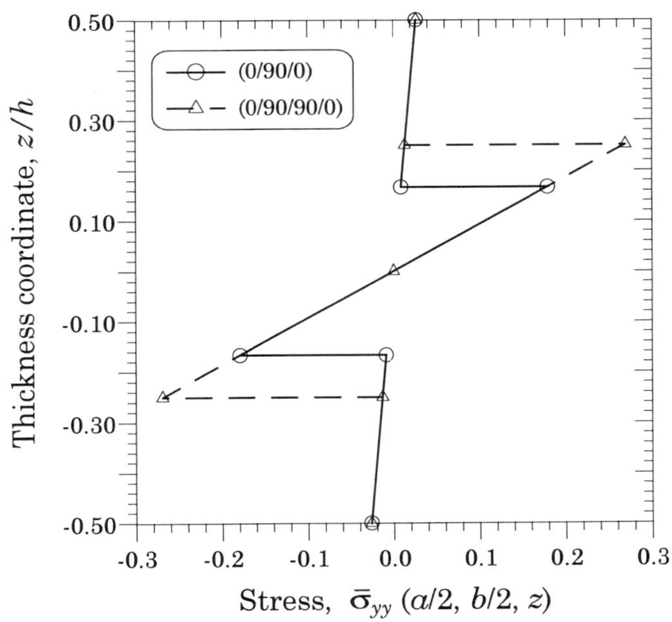

Figure 7.2–5: Variation of nondimensionalized maximum normal stress ($\bar{\sigma}_{yy}$) through the thickness (z/h) of square cross-ply laminates under sinusoidally distributed load.

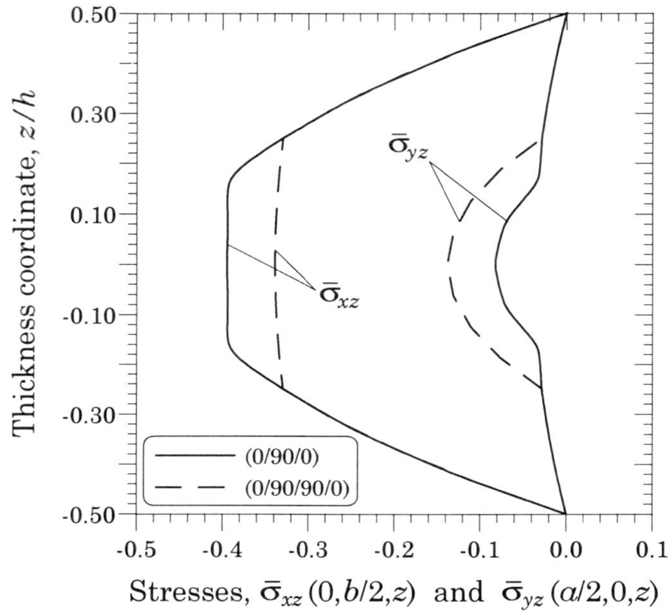

Figure 7.2–6: Variation of nondimensionalized maximum transverse shear stresses, $\bar{\sigma}_{yz}$ and $\bar{\sigma}_{xz}$, through the thickness (z/h) of square cross-ply laminates. The stresses are the same in both laminates.

The solution to the problem of a rectangular plate with two opposite edges simply supported and the other two edges having arbitrary boundary conditions can be represented in terms of single Fourier series as

$$w_0(x,y) = \sum_{n=1}^{\infty} W_n(x) \ \sin \beta y \tag{7.3 – 1}$$

Similarly, the load is represented as

$$q(x,y) = \sum_{n=1}^{\infty} Q_n(x) \ \sin \beta y \tag{7.3 – 2a}$$

where $Q_n(x)$ are given by (see Table 7.3–1)

$$Q_n(x) = \frac{2}{b}\int_0^b q(x,y) \ \sin \beta y \tag{7.3 – 2b}$$

The assumed solution in Eq. (1) satisfies the simply supported boundary conditions on edges $y = 0, a$. In the case of uniformly distributed load of intensity q_0, the coefficients Q_n are given by

$$Q_n(x) = \frac{4q_0}{n\pi}\ , \quad n = 1, 3, \cdots \tag{7.3 – 3}$$

Table 7.3–1: Coefficients in the single trigonometric series expansion of loads in the Lévy method.

Load $q(x)$	Coefficients Q_m
Uniform load, $q = -q_0$ 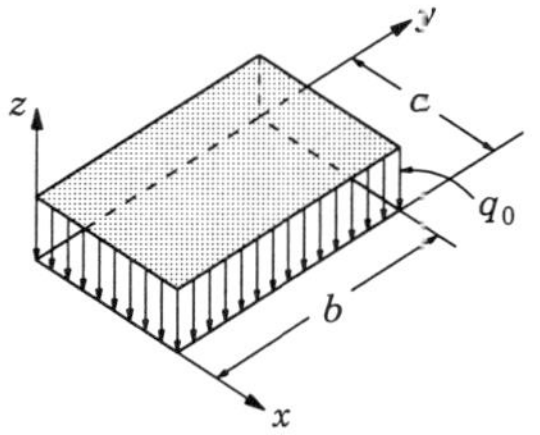 	$Q_m = -\frac{4q_0}{\pi m}$ $(m = 1, 3, 5, \cdots)$
Hydrostatic load, $q(x) = -q_0 \frac{y}{b}$ 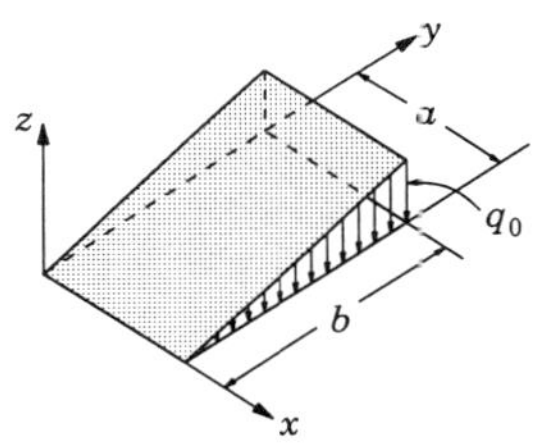 	$Q_m = -\frac{2q_0}{n\pi}$ $(n = 1, 2, 3 \cdots)$
Point load, $q(x) = -Q_0$ at (x_0, y_0) 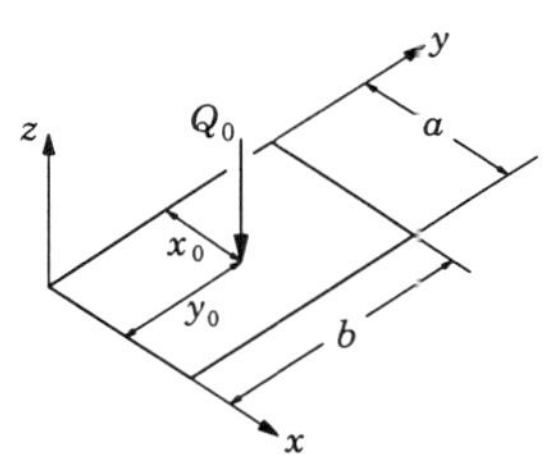 	$Q_m = -\frac{2Q_0}{b} \sin \frac{n\pi y_0}{b}$ $(n = 1, 2, 3, \cdots)$
Line load, $q(x) = -q_0$ at $y = y_0$ 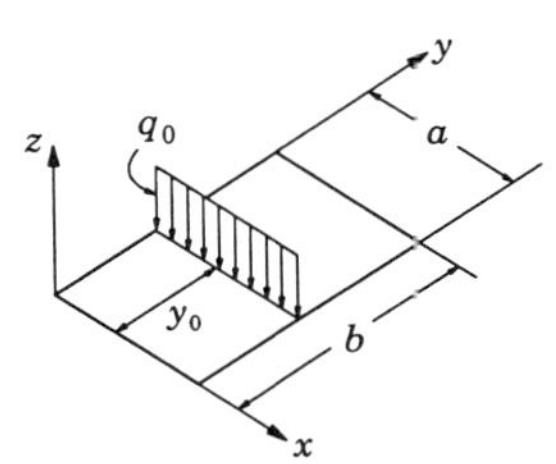 	$Q_m = -\frac{2q_0}{b} \sin \frac{n\pi y_0}{b}$ $(n = 1, 2, 3, \cdots)$

Substituting Eqs. (7.2–5) and (7.3–1) into Eq. (7.2–1), we obtain

$$\sum_{n=1}^{\infty}\left[D_{11}\frac{d^4W_n}{dx^4}-2\left(D_{12}+2D_{66}\right)\beta^2\frac{d^2W_n}{dx^2}+D_{22}\beta^4W_n-Q_n\right]\sin\beta y=0 \quad (7.3-4)$$

Since the result must hold for any y, it follows that the expression in the square brackets must be zero:

$$D_{11}\frac{d^4W_n}{dx^4}-2\left(D_{12}+2D_{66}\right)\beta^2\frac{d^2W_n}{dx^2}+D_{22}\beta^4W_n=Q_n \quad (7.3-5)$$

The ordinary fourth-order differential equation (5) can be solved either analytically or by an approximate method. Analytically, Eq. (5) can be solved directly or by the so-called state–space approach used in control theory (see [11,12]). As for approximate methods, the Rayleigh–Ritz, finite difference, and finite element methods are good candidates. Here we discuss direct analytical solution, analytical solution by the state–space approach, and approximate solution by the Rayleigh–Ritz method.

7.3–2 Analytical Solutions

The general form of the analytical (exact) solution to the fourth-order differential equation (5) consists of two parts: homogeneous and nonhomogeneous (or particular) solutions. The homogeneous solution is of the form

$$W_n^h(x)=C\exp\left(\lambda x\right) \quad (7.3-6)$$

where λ denotes a root of the algebraic equation

$$D_{11}\lambda^4-2\left(D_{12}+2D_{66}\right)\beta^2\lambda^2+D_{22}\beta^4=0 \quad (7.3-7)$$

Since there are four roots, the solution (5) can be written as a linear combination of functions of these four roots. The true form of the solution depends on the nature of the roots, *i.e.*, real or complex and equal or distinct. We consider three cases.

Case 1: Roots are real and distinct

When $(D_{12}+2D_{66})^2>D_{11}D_{22}$, the roots are real and unequal:

$$(\lambda_1)^2=(-\lambda_2)^2=\frac{\beta^2}{D_{11}}\left[D_{12}+2D_{66}-\sqrt{\left(D_{12}+2D_{66}\right)^2-D_{11}D_{22}}\right]$$

$$(\lambda_3)^2=(-\lambda_4)^2=\frac{\beta^2}{D_{11}}\left[D_{12}+2D_{66}+\sqrt{\left(D_{12}+2D_{66}\right)^2-D_{11}D_{22}}\right] \quad (7.3-8)$$

The homogeneous part of the solution is of the form

$$W_n^h(x) = A_n \cosh \lambda_1 x + B_n \sinh \lambda_1 x + C_n \cosh \lambda_3 x + D_n \sinh \lambda_3 x \qquad (7.3-9)$$

Case 2: Roots are real and equal

When $(D_{12} + 2D_{66})^2 = D_{11}D_{22}$, the roots are real but equal

$$\lambda_1 = \lambda_2 = -\lambda_3 = -\lambda_4 \equiv \lambda, \quad \lambda^2 = \frac{\beta^2}{D_{11}}(D_{12} + 2D_{66}) \qquad (7.3-10)$$

and the homogeneous part of the solution is of the form

$$W_n^h(x) = (A_n + B_n x) \cosh \lambda x + (C_n + D_n x) \sinh \lambda x \qquad (7.3-11)$$

Case 3: Roots are complex

When $(D_{12} + 2D_{66})^2 < D_{11}D_{22}$, the roots are complex and they appear in complex conjugate pairs $\lambda_1 \pm i\lambda_2$ and $-\lambda_1 \pm i\lambda_2$ $(i = \sqrt{-1}, \lambda_1 > 0, \lambda_2 > 0)$:

$$\begin{aligned}
(\lambda_1)^2 &= \frac{\beta^2}{2D_{11}}\left[\sqrt{D_{11}D_{22}} + (D_{12} + 2D_{66})\right] \\
(\lambda_2)^2 &= \frac{\beta^2}{2D_{11}}\left[\sqrt{D_{11}D_{22}} - (D_{12} + 2D_{66})\right]
\end{aligned} \qquad (7.3-12)$$

The homogeneous part of the solution is of the form

$$\begin{aligned}
W_n^h(x) = &(A_n \cos \lambda_2 x + B_n \sin \lambda_2 x) \cosh \lambda_1 x \\
&+ (C_n \cos \lambda_2 x + D_n \sin \lambda_2 x) \sinh \lambda_1 x
\end{aligned} \qquad (7.3-13)$$

The particular solution of the fourth-order differential equation (5) in the general case in which Q_n is a function of x can be determined using the method of undetermined coefficients (see Pipes and Harvill [10]). When Q_n is a constant, the particular solution is a constant and is determined by substituting it into Eq. (5). We obtain $kD_{22}\beta^4 = Q_n$. Hence, the particular solution becomes

$$W_n^p(x) = \frac{Q_n}{D_{22}\beta^4} \equiv \hat{Q}_n \qquad (7.3-14)$$

The four constants A_n, B_n, C_n, and D_n in Eqs. (9), (11), and (13) can be determined using the four boundary conditions associated with the edges $x = 0, a$ (in addition to the simply supported boundary conditions on the edges $y = 0, b$). Note that the particular case (*i.e.*, Case 1, Case 2, or Case 3) in a problem is dictated by the plate stiffnesses, D_{ij}. Here we illustrate the procedure for simply supported and clamped boundary conditions in the case of real and distinct roots. The solution in this case is given by

$$w_0(x,y) = \sum_{n=1}^{\infty} (A_n \cosh \lambda_1 x + B_n \sinh \lambda_1 x + C_n \cosh \lambda_3 x + D_n \sinh \lambda_3 x + W_n^p) \sin \beta_n y \tag{7.3 – 15}$$

In the following discussion we assume that the applied transverse load is uniformly distributed.

Simply supported plate

The simply supported boundary conditions on edges $x = 0, a$ (and $y = 0, b$) are

$$w_0 = 0, \quad M_{xx} \equiv -\left(D_{11}\frac{\partial^2 w_0}{\partial x^2} + D_{12}\frac{\partial^2 w_0}{\partial y^2}\right) = 0 \tag{7.3 – 16}$$

Using (15) in (16), we obtain

$$A_n + C_n + \hat{Q}_n = 0$$

$$A_n \cosh \lambda_1 a + B_n \sinh \lambda_1 a + C_n \cosh \lambda_3 a + D_n \sinh \lambda_3 a + \hat{Q}_n = 0$$

$$D_{11}\left(\lambda_1^2 A_n + \lambda_3^2 C_n\right) - D_{12}\left(A_n + C_n + \hat{Q}_n\right)\beta^2 = 0$$

$$D_{11}\left(A_n\lambda_1^2 \cosh \lambda_1 a + B_n\lambda_1^2 \sinh \lambda_1 a + C_n\lambda_3^2 \cosh \lambda_3 a + D_n\lambda_3^2 \sinh \lambda_3 a\right)$$
$$-D_{12}\left(A_n \cosh \lambda_1 a + B_n \sinh \lambda_1 a + C_n \cosh \lambda_3 a + D_n \sinh \lambda_3 a + \hat{Q}_n\right)\beta^2 = 0$$

where $\hat{Q}_n = Q_n/\beta^4 D_{22}$. By virtue of the first two equations, the coefficients of D_{12} in the last two equations are identically zero. The four equations can be expressed in matrix form as

$$\begin{bmatrix} 1 & 0 & 1 & 0 \\ \cosh \lambda_1 a & \sinh \lambda_1 a & \cosh \lambda_3 a & \sinh \lambda_3 a \\ \lambda_1^2 & 0 & \lambda_3^2 & 0 \\ \lambda_1^2 \cosh \lambda_1 a & \lambda_1^2 \sinh \lambda_1 a & \lambda_3^2 \cosh \lambda_3 a & \lambda_3^2 \sinh \lambda_3 a \end{bmatrix} \begin{Bmatrix} A_n \\ B_n \\ C_n \\ D_n \end{Bmatrix} = -\begin{Bmatrix} \hat{Q}_n \\ \hat{Q}_n \\ 0 \\ 0 \end{Bmatrix} \tag{7.3 – 17}$$

The determinant of the 4×4 coefficient matrix in Eq. (17) is $(\lambda_3^2 - \lambda_3^2)^2 \sinh \lambda_1 a \sinh \lambda_3 a$. The solution of the matrix equation yields

$$A_n = -\hat{Q}_n \frac{\lambda_3^2}{(\lambda_3^2 - \lambda_1^2)}\,, \quad B_n = -\hat{Q}_n \frac{\lambda_3^2\,(1 - \cosh \lambda_1)}{\sinh \lambda_1 a\,(\lambda_3^2 - \lambda_1^2)}$$
$$C_n = \hat{Q}_n \frac{\lambda_3^2}{(\lambda_3^2 - \lambda_1^2)}\,, \quad D_n = \hat{Q}_n \frac{\lambda_3^2\,(1 - \cosh \lambda_2)}{\sinh \lambda_2 a\,(\lambda_3^2 - \lambda_1^2)} \tag{7.3 – 18}$$

Simply supported at y=0,b and clamped at x=0,a

For clamped boundary conditions on edges $x = 0, a$, we require

$$w_0 = 0, \quad \frac{\partial w_0}{\partial x} = 0 \tag{7.3 – 19}$$

which yield

$$\begin{bmatrix} 1 & 0 & 1 & 0 \\ \cosh\lambda_1 a & \sinh\lambda_1 a & \cosh\lambda_3 a & \sinh\lambda_3 a \\ 0 & \lambda_1 & 0 & \lambda_3 \\ \lambda_1\sinh\lambda_1 a & \lambda_1\cosh\lambda_1 a & \lambda_3\sinh\lambda_3 a & \lambda_3\cosh\lambda_3 a \end{bmatrix} \begin{Bmatrix} A_n \\ B_n \\ C_n \\ D_n \end{Bmatrix} = - \begin{Bmatrix} \hat{Q}_n \\ \hat{Q}_n \\ 0 \\ 0 \end{Bmatrix} \tag{7.3 – 20}$$

The solution of the matrix equation (20) is

$$\begin{aligned} A_n =& \frac{\hat{Q}_n\lambda_3}{E_n}\left[(\lambda_1\sinh\lambda_3 a - \lambda_3\sinh\lambda_1 a)\sinh\lambda_3 a \right.\\ &\left. + \lambda_1(\cosh\lambda_1 a - \cosh\lambda_3 a)(\cosh\lambda_3 a - 1)\right] \\ B_n =& \frac{\hat{Q}_n\lambda_3}{E_n}\left[\lambda_3\sinh\lambda_3 a\,(\cosh\lambda_1 a - 1) + \lambda_1\sinh\lambda_1 a\,(1 - \cosh\lambda_3 a)\right] \\ C_n =& -(A_n + Q_n), \quad D_n = -\frac{\lambda_1}{\lambda_3}B_n \end{aligned} \tag{7.3 – 21}$$

where E_n is the determinant of the coefficient matrix

$$\begin{aligned} E_n =& -(\lambda_3\sinh\lambda_1 a - \lambda_1\sinh\lambda_3 a)(\lambda_1\sinh\lambda_1 a - \lambda_3\sinh\lambda_3 a) \\ &+\lambda_1\lambda_3(\cosh\lambda_3 a - \cosh\lambda_1 a)^2 \end{aligned} \tag{7.3 – 22}$$

An alternative method of solving Eq. (5) is provided by the state–space approach [12]. The approach involves writing a higher-order ordinary differential equation as a first-order matrix equation, and its solution is obtained using matrix methods in terms of the eigenvalues of the matrix operator. In the present case, the linear ordinary differential equation in (5) with constant coefficients can be expressed in the form of a single, first-order matrix differential equation

$$\{Z'\} = [T]\{Z\} + \{F\} \tag{7.3 – 23}$$

$$\{Z\} = \begin{Bmatrix} W_n \\ W_n' \\ W_n'' \\ W_n''' \end{Bmatrix}, \quad [T] = \begin{bmatrix} 0 & 1 & 0 & 0 \\ 0 & 0 & 1 & 0 \\ 0 & 0 & 0 & 1 \\ C_1 & 0 & C_2 & 0 \end{bmatrix}, \quad \{F\} = \begin{Bmatrix} 0 \\ 0 \\ 0 \\ \hat{Q}_n \end{Bmatrix} \tag{7.3 – 24}$$

$$C_1 = -\frac{D_{22}}{D_{11}}\beta^4, \quad C_2 = 2\frac{(D_{12} + 2D_{66})}{D_{11}}\beta^2, \quad \hat{Q}_n = \frac{Q_n}{D_{11}} \tag{7.3 – 25}$$

The general solution of Eq. (23) is given by

$$\begin{aligned}\mathbf{Z}(x) &= e^{\mathbf{T}x}\left(\mathbf{K} + \int^{x} e^{-\mathbf{T}\xi}\ \mathbf{F}(\xi)\ d\xi\right)\\ &\equiv \mathbf{G}(x)\mathbf{K} + \mathbf{H}(x)\end{aligned} \tag{7.3-26}$$

Here $e^{\mathbf{T}x}$ denotes the matrix product

$$e^{\mathbf{T}x} = [E]\begin{bmatrix} e^{\lambda_1 x} & & 0 \\ & \ddots & \\ 0 & & e^{\lambda_4 x}\end{bmatrix}[E]^{-1} \tag{7.3-27}$$

Here $[E]$ is the matrix of distinct eigenvectors of matrix $[T]$, $[E]^{-1}$ denotes its inverse, λ_j $(j = 1, 2, 3, 4)$ are the eigenvalues associated with matrix $[T]$, and $\{K\}$ is a vector of constants to be determined using the boundary conditions of the problem.

As an example, consider the case of simply supported boundary condition at $x = 0$ and clamped boundary condition at $x = a$. The simply supported boundary conditions (7.2–2a,b) at $x = 0$ imply [see Eq. (1)]

$$W_n(0) = 0, \quad D_{11}W_n''(0) - D_{12}\beta^2 W_n(0) = 0 \tag{7.3-28a}$$

The clamped boundary conditions (20) at $x = a$ imply

$$W_n(a) = 0, \quad W_n'(a) = 0 \tag{7.3-28b}$$

These four conditions in turn yield, in view of Eq. (26), the following four nonhomogeneous algebraic equations among K_i $(i = 1, 2, 3, 4)$:

$$\sum_{j=1}^{4} G_{1j}(0)K_j + H_1(0) = 0$$

$$\sum_{j=1}^{4}\left(D_{11}G_{3j}(0) - \beta^2 D_{12}G_{1j}(0)\right)K_j + D_{11}H_3(0) - D_{12}\beta^2 H_1(0) = 0$$

$$\sum_{j=1}^{4} G_{1j}(a)K_j + H_1(a) = 0, \quad \sum_{j=1}^{4} G_{2j}(a)K_j + H_2(a) = 0 \tag{7.3-29}$$

These equations can be solved for the four constants. In general, the procedure is algebraically complicated, and therefore all calculations, *i.e.*, matrix multiplication, determination of eigenvalues and constants K_i, and evaluation of the solution, are made using a computer.

Table 7.3–2 contains numerical results for three-layer, cross-ply (0°/90°/0°), square laminates under uniformly distributed transverse load.

The lamina material properties used are $E_1 = 19.2$ msi, $E_2 = 1.56$ msi, $G_{12} = 0.82$ msi, and $\nu_{12} = 0.24$. The transverse deflection and stresses are nondimensionalized as follows:

$$\hat{w} = w_0(0,0)\left(\frac{E_2 h^3}{a^4 q_0}\right) \times 10^2; \quad \bar{\sigma}_{xx} = \sigma_{xx}(a/2, b/2, h/2)\left(\frac{h^2}{a^2 q_0}\right) \times 10$$

$$\bar{\sigma}_{yy} = \sigma_{yy}(a/2, b/2, h/6)\left(\frac{h^2}{a^2 q_0}\right) \times 10 \qquad (7.3-30)$$

The notation SF, for example, is used to denote a plate with edge $x = 0$ is simply supported (S) and edge $x = a$ is free (F); of course, edges $y = 0, b$ are simply supported.

Table 7.3–2: Nondimensional center deflections ($\hat{w}$) and in-plane normal stresses ($\bar{\sigma}_{xx}$ and $\bar{\sigma}_{yy}$) of symmetric cross-ply (0°/90°/0°) square plates subjected to uniform distribution of transverse load and for various boundary conditions.

Variable	SS	SC	CC	FF	FS	FC
$\hat{w}$	1.206	0.544	0.280	10.920	5.992	2.376
$\bar{\sigma}_{xx}$	7.251	4.082	2.787	0.195	3.778	1.685
$\bar{\sigma}_{yy}$	1.938	0.651	0.185	21.597	11.621	4.313

7.3–3 Rayleigh–Ritz Solution

Equation (5) can also be solved using the Rayleigh–Ritz method. In the Rayleigh–Ritz method, we seek solution of (5) in the form

$$W_n(x) \approx \sum_{j=1}^{N} c_j \varphi_j(x) \qquad (7.3-31)$$

where $\varphi_j(x)$ are approximation functions that must meet the continuity and completeness conditions and satisfy the homogeneous form of the geometric boundary conditions [see Eq. (3.5–2)]. The parameters c_j are then determined by requiring that the weak form of Eq. (5) be satisfied:

$$0 = \int_0^a \left[D_{11}\frac{d^2 W_n}{dx^2}\frac{d^2 \delta W_n}{dx^2} + 2\left(D_{12} + 2D_{66}\right)\beta^2 \frac{dW_n}{dx}\frac{d\delta W_n}{dx} + D_{22}\beta^4 W_n \delta W_n - Q_n \delta W_n\right] dx \qquad (7.3-32)$$

where δW_n denotes the virtual variation in W_n

$$\delta W_n(x) \approx \sum_{i=1}^{N} \delta c_i \; \varphi_i(x) \tag{7.3-33}$$

Substituting (31) and (33) into (32), we obtain

$$0 = \sum_{i=1}^{N} \left\{ \sum_{j=1}^{N} c_j \int_0^a \left[D_{11} \frac{d^2\varphi_j}{dx^2}\frac{d^2\varphi_i}{dx^2} + 2\left(D_{12} + 2D_{66}\right)\beta^2 \frac{d\varphi_j}{dx}\frac{d\varphi_i}{dx} \right.\right.$$
$$\left.\left. + D_{22}\beta^4 \varphi_j \varphi_i \right] dx - \int_0^a Q_n \varphi_i dx \right\} \delta c_i \tag{7.3-34}$$

Since the above expression must hold for all arbitrary values of δc_i, it follows that the expression in the curly bracket must be zero. We have

$$0 = \sum_{j=1}^{N} A_{ij} c_j - F_i \quad \text{or} \quad [A]\{c\} = \{F\} \tag{7.3-35a}$$

where

$$A_{ij} = \int_0^a \left[D_{11} \frac{d^2\varphi_j}{dx^2}\frac{d^2\varphi_i}{dx^2} + 2\left(D_{12} + 2D_{66}\right)\beta^2 \frac{d\varphi_j}{dx}\frac{d\varphi_i}{dx} + D_{22}\beta^4 \varphi_j \varphi_i \right] dx$$
$$F_i = \int_0^a Q_n \varphi_i \; dx \tag{7.3-35b}$$

Equation (35a) represents a set of N algebraic equations among c_i.

As an example, we consider the case in which the edges $x = 0, a$ are clamped. The geometric boundary conditions are given by Eq. (20):

$$w_0 = 0, \quad \frac{\partial w_0}{\partial x} = 0 \tag{7.3-36}$$

Hence, the approximation functions φ_i must be selected such that $\varphi_i = 0$ and $(d\varphi_i/dx)$ are zero at$x = 0, a$.

If an algebraic polynomial is to be selected, one may begin with the five-term complete polynomial

$$\varphi_1(x) = K_0 + K_1 x + K_2 x^2 + K_3 x^3 + K_4 x^4$$

and determine four of the five constants K_i in terms of the remaining constant using the four boundary conditions. The constant is arbitrary and may be set to unity. We obtain

$$\varphi_1(x) = \left(\frac{x}{a}\right)^2 \left(1 - \frac{x}{a}\right)^2$$

The ith function can be written as

$$\varphi_i(x) = \left(\frac{x}{a}\right)^{i+1}\left(1-\frac{x}{a}\right)^2, \quad i = 1, 2, \cdots, n \tag{7.3 – 37}$$

For the choice of $\varphi_i(x)$ in (37), we have

$$\begin{aligned}
\int_0^a \varphi_i \, dx =& a\left(\frac{1}{i+2} - \frac{2}{i+3} + \frac{1}{i+4}\right)\\
\int_0^a \varphi_i\varphi_j \, dx =& a\left(\frac{1}{i+j+3} - \frac{4}{i+j+4} + \frac{6}{i+j+5} - \frac{4}{i+j+6} + \frac{1}{i+j+7}\right)\\
\int_0^a \frac{d\varphi_i}{dx}\frac{d\varphi_j}{dx} \, dx =& \frac{1}{a}\left[\frac{(i+1)(j+1)}{i+j+1} - 2\frac{(i+1)(j+2)+(j+1)(i+2)}{i+j+2}\right.\\
&+ \frac{(i+1)(j+3)+4(i+2)(j+2)+(i+3)(j+1)}{i+j+3}\\
&\left.-2\frac{(i+2)(j+3)+(i+3)(j+2)}{i+j+4} + \frac{(i+3)(j+3)}{i+j+5}\right]\\
\int_0^a \frac{d^2\varphi_i}{dx^2}\frac{d^2\varphi_j}{dx^2} \, dx =& \frac{1}{a^3}\left[\frac{ij(i+1)(j+1)}{i+j-1} - 2(i+1)(j+1)\frac{i(j+2)+(i+2)j}{i+j}\right.\\
&+ \frac{i(i+1)(j+2)(j+3)+4(i+1)(i+2)(j+1)(j+2)}{i+j+1}\\
&+ \frac{j(i+2)(i+3)(j+1)}{i+j+1} + \frac{(i+2)(i+3)(j+2)(j+3)}{i+j+3}\\
&\left.-2(i+2)(j+2)\frac{(i+1)(j+3)+(i+3)(j+1)}{i+j+2}\right]
\end{aligned} \tag{7.3 – 38}$$

For $N = 1$, Eq. (35a) gives

$$\left[\frac{4}{5a^3}D_{11} + \frac{4}{105a}\left(D_{12}+2D_{66}\right)\beta^2 + \frac{a}{630}D_{22}\beta^4\right]c_1 = \frac{a}{30}Q_n$$

and the solution (1) becomes

$$w_0(x,y) \approx \left[\left(\frac{x}{a}\right) - \left(\frac{x}{a}\right)^2\right]^2 \sum_{n=1}^{\infty} c_1(n)\sin\beta y \tag{7.3 – 39a}$$

with $\beta = n\pi/b$ and

$$c_1(n) = \frac{1}{30}Q_n a^4\left[\frac{4}{5}D_{11} + \frac{4}{105}\left(D_{12}+2D_{66}\right)\beta^2 a^2 + \frac{1}{630}D_{22}\beta^4 a^4\right]^{-1} \tag{7.3 – 39b}$$

The center deflection is given by

$$\begin{aligned} w_0(a/2,b/2) \approx & \frac{1}{16}\sum_{n=1}^{\infty} c_1(n)\sin\frac{n\pi}{2} \\ = & \frac{1}{16}\left[c_1(1)-c_1(3)+c_1(5)-\cdots\right] \end{aligned} \qquad (7.3-40)$$

For a symmetric cross-ply laminate (0/90/0) with ply properties $E_1 = 19.2$ msi, $E_2 = 1.56$ msi, $G_{12} = 0.82$ msi, and $\nu_{12} = 0.25$, the bending stiffnesses, for $h = 0.01$, are $D_{11} = 1.5528$, $D_{12} = 0.031347$, $D_{22} = 0.18531$, and $D_{66} = 0.068333$ lb-in. For uniformly distributed load q_0, we have ($s = a/b$)

$$c_1(n) = \frac{4q_0}{30n\pi\left(1.24224 + 0.06317n^2s^2 + 0.02865n^4s^4\right)}$$

for $n = 1, 3, 5, \cdots$. For a square plate, the maximum deflection becomes

$$w_0(a/2,b/2) \approx \frac{1}{16}\left(0.03181 - 0.003424 + 0.001509 - \cdots\right)$$

The series converges slowly unless we also increase the number of parameters in the x–coordinate [see Eq. (31)]. The "exact" solution for a square laminate under uniformly distributed load is

$$w_0(a/2,b/2) = 0.001795q_0a^4 \qquad (7.3-41)$$

whereas the one-term ($n = 1$ and $N = 1$) solution predicted by Eq. (40) is $0.001988q_0a^4$. The two-term solution ($n = 1,3$ and $N = 1$) is $0.001774q_0a^4$.

Other choices of $\varphi_i(x)$ are provided by the eigenfunctions $W(x)$ of beams developed in Chapter 6 [see Eq. (6.2–46a)]. For example, for clamped boundary conditions, we use the eigenfunctions of a beam with clamped ends. From Eq. (6.2–46a) and Table 6.2–3, we have

$$\varphi_i(x) = \sin\lambda_i x - \sinh\lambda_i x + \alpha_i\left(\cosh\lambda_i x - \cos\lambda_i x\right) \qquad (7.3-42a)$$

$$\frac{d\varphi_i}{dx} = \lambda_i\left[\cos\lambda_i x - \cosh\lambda_i x + \alpha_i\left(\sinh\lambda_i x + \sin\lambda_i x\right)\right] \qquad (7.3-42b)$$

where λ_i are the roots of the characteristic equation (6.2–59)

$$\cos\lambda_i a\cosh\lambda_i a - 1 = 0 \qquad (7.3-43)$$

and

$$\alpha_i = \frac{\sinh\lambda_i a - \sin\lambda_i a}{\cosh\lambda_i a - \cos\lambda_i a} = \frac{\cosh\lambda_i a - \cos\lambda_i a}{\sinh\lambda_i a + \sin\lambda_i a} \qquad (7.3-44)$$

Clearly, φ_i and $(d\varphi_i/dx)$ are zero at $x = 0, a$. Recall from Table 6.2–3 [also see Eq. (6.2–60)] that the roots λ_i of the characteristic equation (43) are given by

$$\lambda_1 a = 4.730,\ \lambda_2 a = 7.853,\ \cdots\ ,\ \lambda_i a \approx (2i+1)\frac{\pi}{2} \tag{7.3 – 45}$$

The corresponding values of α_i are

$$\alpha_1 = 1.0178,\ \alpha_2 = 0.99922,\quad \alpha_i = 1 \text{ for } i > 2 \tag{7.3 – 46}$$

Hence, the first two eigenfunctions are

$$\begin{aligned}\varphi_1(x) &= \sin\frac{4.73x}{a} - \sinh\frac{4.73x}{a} + 1.0178\left(\cosh\frac{4.73x}{a} - \cos\frac{4.73x}{a}\right)\\ \varphi_2(x) &= \sin\frac{7.853x}{a} - \sinh\frac{7.853x}{a} + 0.9992\left(\cosh\frac{7.853x}{a} - \cos\frac{7.853x}{a}\right)\end{aligned} \tag{7.3 – 47}$$

For $N = 1$, Eq. (35a) yields

$$\begin{aligned}c_1(n) =& 0.84555 Q_n a^4 \Big[518.53135 D_{11} + 12.7442\,(D_{12} + 2D_{66})\,(n\pi)^2 s^2\\ &+ 1.035965 D_{22}(n\pi)^4 s^4\Big]^{-1}\end{aligned} \tag{7.3 – 48}$$

For the symmetric cross-ply laminate considered above, the center deflection $(X_1(a/2) = 1.61637)$ predicted for $n = 1$ is $0.002009 q_0 a^4$ compared to the exact solution of $0.001795 q_0 a^4$.

7.4 Bending of Rectangular Plates with Various Boundary Conditions

7.4–1 Virtual Work Statements

The Navier and Lévy type solutions do not exist for rectangular plates with all four edges clamped or when two parallel edges are not simply supported. Therefore, an approximate method must be utilized to determine solutions of these plates. In this section, we discuss applications of the Rayleigh–Ritz method to determine the bending deflections of specially orthotropic rectangular plates with various boundary conditions.

The virtual work statement (or weak form) and the total potential energy expressions for a specially orthotropic rectangular plate are [see Eq. (5.3–19)]

$$\begin{aligned}0 &= \int_0^b \int_0^a \left(M_{xx}\delta\varepsilon_{xx}^{(1)} + M_{yy}\delta\varepsilon_{yy}^{(1)} + M_{xy}\delta\varepsilon_{xy}^{(1)} - q\delta w_0\right) dxdy\\ &= \int_0^b \int_0^a \Big[D_{11}\frac{\partial^2 w_0}{\partial x^2}\frac{\partial^2 \delta w_0}{\partial x^2} + D_{12}\left(\frac{\partial^2 w_0}{\partial y^2}\frac{\partial^2 \delta w_0}{\partial x^2} + \frac{\partial^2 w_0}{\partial x^2}\frac{\partial^2 \delta w_0}{\partial y^2}\right)\\ &\quad + 4D_{66}\frac{\partial^2 w_0}{\partial x\partial y}\frac{\partial^2 \delta w_0}{\partial x\partial y} + D_{22}\frac{\partial^2 w_0}{\partial y^2}\frac{\partial^2 \delta w_0}{\partial y^2} - q\delta w_0\Big] dxdy\end{aligned} \tag{7.4 – 1}$$

and

$$\Pi(w_0) = \frac{1}{2}\int_0^b \int_0^a \left[D_{11}\left(\frac{\partial^2 w_0}{\partial x^2}\right)^2 + 2D_{12}\frac{\partial^2 w_0}{\partial x^2}\frac{\partial^2 w_0}{\partial y^2} + 4D_{66}\left(\frac{\partial^2 w_0}{\partial x \partial y}\right)^2 + D_{22}\left(\frac{\partial^2 w_0}{\partial y^2}\right)^2 - 2qw_0 \right] dxdy \qquad (7.4-2)$$

The above expressions should be appended with appropriate terms due to any additional applied edge forces and moments.

7.4–2 Clamped Plates

Consider a rectangular plate with all edges clamped and subjected to distributed transverse load $q(x, y)$. The boundary conditions associated with the clamped plate are

$$w_0 = 0 \text{ and } \frac{\partial w_0}{\partial x} = 0 \quad \text{at } x = 0, a \qquad (7.4-3a)$$

$$w_0 = 0 \text{ and } \frac{\partial w_0}{\partial y} = 0 \quad \text{at } y = 0, b \qquad (7.4-3b)$$

We assume the Rayleigh–Ritz approximation in the form

$$w_0(x, y) \approx W_{mn}(x, y) = \sum_i^m \sum_j^n c_{ij}\ \varphi_{ij}(x, y) \qquad (7.4-4)$$

where the approximation functions φ_{ij} satisfy all the (homogeneous) geometric boundary conditions in Eqs. (3a,b). For this problem, therefore, both the Galerkin and Rayleigh–Ritz methods give the same solution for the same choice of approximation functions.

In view of the rectangular geometry and clamped boundary conditions, the approximation functions $\varphi_{ij}(x, y)$ can be expressed as a tensor product of the one-dimensional functions given in Eq. (7.3–37) or (7.3–42a):

$$\varphi_{ij}(x, y) = X_i(x)Y_j(y) \qquad (7.4-5)$$

where

$$X_i(x) = \left(\frac{x}{a}\right)^{i+1}\left(1 - \frac{x}{a}\right)^2$$
$$Y_j(y) = \left(\frac{y}{b}\right)^{j+1}\left(1 - \frac{y}{b}\right)^2 \qquad (7.4-6)$$

or

$$
\begin{aligned}
X_i(x) &= \sin\lambda_i x - \sinh\lambda_i x + \alpha_i\left(\cosh\lambda_i x - \cos\lambda_i x\right)\\
Y_j(y) &= \sin\lambda_j y - \sinh\lambda_j y + \alpha_j\left(\cosh\lambda_j y - \cos\lambda_j y\right)
\end{aligned} \tag{7.4 - 7}
$$

for $i = 1, 2, \cdots, m$; $j = 1, 2, \cdots, n$. The parameters λ_i and α_i are defined in Eq. (7.3–43) and (7.3–44), respectively.

Substituting Eq. (4), with φ_{ij} given by Eq. (5), and

$$
\delta w_0 = \sum_p^m \sum_q^n \delta c_{pq} \varphi_{pq}
$$

into Eq. (1), we obtain

$$
\begin{aligned}
0 = \sum_p^m \sum_q^n \Bigg\{ \sum_i^m \sum_j^n c_{ij} \int_0^b \int_0^a \Bigg[& D_{11} \frac{d^2X_i}{dx^2} Y_j \frac{d^2X_p}{dx^2} Y_q + 4D_{66} \frac{dX_i}{dx}\frac{dY_j}{dy}\frac{dX_p}{dx}\frac{dY_q}{dy} \\
& + D_{12}\left(X_i \frac{d^2Y_j}{dy^2}\frac{d^2X_p}{dx^2} Y_q + \frac{d^2X_i}{dx^2} Y_j X_p \frac{d^2Y_q}{dy^2} \right) \\
& + D_{22} X_i \frac{d^2Y_j}{dy^2} X_p \frac{d^2Y_q}{dy^2} \Bigg] dxdy \\
& - \int_0^b \int_0^a q X_p Y_q \, dxdy \Bigg\} \delta c_{pq}
\end{aligned} \tag{7.4 - 8a}
$$

Since the statement should hold for any arbitrary variations δc_{pq}, the expression inside the curly bracket should be zero for all $p, q = 1, 2, \cdots$:

$$
\begin{aligned}
0 = \sum_i^m \sum_j^n \Bigg\{ \int_0^b \int_0^a \Bigg[& D_{11} \frac{d^2X_i}{dx^2} Y_j \frac{d^2X_p}{dx^2} Y_q + 4D_{66} \frac{dX_i}{dx}\frac{dY_j}{dy}\frac{dX_p}{dx}\frac{dY_q}{dy} \\
& + D_{12}\left(X_i \frac{d^2Y_j}{dy^2}\frac{d^2X_p}{dx^2} Y_q + \frac{d^2X_i}{dx^2} Y_j X_p \frac{d^2Y_q}{dy^2} \right) \\
& + D_{22} X_i \frac{d^2Y_j}{dy^2} X_p \frac{d^2Y_q}{dy^2} \Bigg] dxdy \Bigg\} c_{ij} \\
& - \int_0^b \int_0^a q X_p Y_q \, dxdy
\end{aligned} \tag{7.4 - 8b}
$$

Equation (8b) represents $m \times n$ algebraic equations among the coefficients c_{ij}. Note that all integrals in (8b) are line integrals, and they involve evaluating five different integrals

$$
\int_0^a X_i \, dx, \quad \int_0^a X_i X_p \, dx, \quad \int_0^a \frac{dX_i}{dx}\frac{dX_p}{dx} \, dx
$$

$$\int_0^a X_i \frac{d^2 X_p}{dx^2}\, dx, \quad \int_0^a \frac{d^2 X_i}{dx^2}\frac{d^2 X_p}{dx^2}\, dx \tag{7.4-9a}$$

As an example we consider the algebraic functions in (6) with $m = n = 1$ and $q = q_0$ (uniformly distributed load). The integrals in Eq. (9a) for this case are given by

$$\int_0^a X_1\, dx = \frac{a}{30}\,, \quad \int_0^a X_1 X_1\, dx = \frac{a}{630}\,, \quad \int_0^a \frac{dX_1}{dx}\frac{dX_1}{dx}\, dx = \frac{2}{105a}$$

$$\int_0^a X_1 \frac{d^2 X_1}{dx^2}\, dx = -\frac{2}{105a}\,, \quad \int_0^a \frac{d^2 X_1}{dx^2}\frac{d^2 X_1}{dx^2}\, dx = \frac{4}{5a^3} \tag{7.4-9b}$$

Substituting the integral values into (8b), we obtain

$$\begin{aligned} 0 = &\left[\left(\frac{4}{5a^3}\right)\left(\frac{b}{630}\right) D_{11} + 4D_{66}\left(\frac{2}{105a}\right)\left(\frac{2}{105b}\right) + 2D_{12}\left(-\frac{2}{105a}\right)\left(-\frac{2}{105b}\right)\right. \\ &\left. + \left(\frac{a}{630}\right)\left(\frac{4}{5b^3}\right) D_{22}\right] c_{11} - \left(\frac{a}{30}\right)\left(\frac{b}{30}\right) q_0 \end{aligned}$$

or

$$\left[\frac{7}{a^4} D_{11} + \frac{4}{a^2 b^2}(D_{12} + 2D_{66}) + \frac{7}{b^4} D_{22}\right] c_{11} = \frac{49}{8} q_0 \tag{7.4-10}$$

and the one–parameter solution becomes

$$W_{11}(x,y) = \left(\frac{49}{8}\right) \frac{q_0 a^4 \left[\frac{x}{a} - \left(\frac{x}{a}\right)^2\right]^2 \left[\frac{y}{b} - \left(\frac{y}{b}\right)^2\right]^2}{7D_{11} + 4(D_{12} + 2D_{66}) s^2 + 7D_{22} s^4} \tag{7.4-11}$$

where $s = a/b$ denotes the plate aspect ratio. The maximum deflection occurs at $x = a/2$ and $y = b/2$:

$$W_{11}\left(\frac{a}{2}, \frac{b}{2}\right) = 0.00342 \frac{q_0 a^4}{D_{11} + 0.5714(D_{12} + 2D_{66}) s^2 + D_{22} s^4} \tag{7.4-12}$$

The algebra involved in evaluating the integrals in Eq. (9a) is quite tedious for the choice of approximation functions in (7). An algebraic manipulator (*e.g., Maple* or *Mathematica*) may be used to evaluate them. For $m = n = 1$, the functions in (7) are given by

$$\begin{aligned} X_1(x) &= \sin\frac{4.73x}{a} - \sinh\frac{4.73x}{a} + 1.0178\left(\cosh\frac{4.73x}{a} - \cos\frac{4.73x}{a}\right) \\ Y_1(y) &= \sin\frac{4.73y}{b} - \sinh\frac{4.73y}{b} + 1.0178\left(\cosh\frac{4.73y}{b} - \cos\frac{4.73y}{b}\right) \end{aligned} \tag{7.4-13a}$$

and substitution into (9b) gives

$$\int_0^a X_1\ dx = 0.84555a, \quad \int_0^a X_1 X_1\ dx = 1.035966a, \quad \int_0^a \frac{dX_1}{dx}\frac{dX_1}{dx}\ dx = \frac{12.7442}{a}$$

$$\int_0^a X_1 \frac{d^2 X_1}{dx^2}\ dx = -\frac{12.7442}{a}, \quad \int_0^a \frac{d^2 X_1}{dx^2}\frac{d^2 X_1}{dx^2}\ dx = \frac{518.531348}{a^3} \qquad (7.4-13b)$$

Then Eq. (8b) becomes

$$\left[\frac{537.181b}{a^3} D_{11} + \frac{324.829}{ab}(D_{12} + 2D_{66}) + \frac{537.181a}{b^3} D_{22}\right] c_{11} = 0.715 q_0 ab \qquad (7.4-14)$$

The maximum deflection is given by $(X_1(a/2) = Y_1(b/2) = 1.6164)$

$$W_{11}(\frac{a}{2}, \frac{b}{2}) = 0.00348 \frac{q_0 a^4}{D_{11} b^4 + 0.6047(D_{12} + 2D_{66})s^2 + D_{22} s^4} \qquad (7.4-15)$$

where $s = a/b$ denotes the plate aspect ratio. For an isotropic square plate $(a/b = 1, D_{11} = D_{22} = D_{12}+2D_{66} = D)$, the maximum deflection (15) becomes

$$W_{11}(\frac{a}{2}, \frac{b}{2}) = 0.00134 \frac{q_0 a^4}{D}$$

whereas Eq. (12) gives

$$W_{11}(\frac{a}{2}, \frac{b}{2}) = 0.00133 \frac{q_0 a^4}{D}$$

The "exact" solution (see Timoshenko and Woinowsky–Krieger [6]) is

$$W_{11}(\frac{a}{2}, \frac{b}{2}) = 0.00126 \frac{q_0 a^4}{D}$$

7.4–3 Approximation Functions for Other Boundary Conditions

Here we discuss the approximation functions $\varphi_{ij} = X_i(x)Y_j(y)$ required in the Rayleigh–Ritz approximation (4) of specially orthotropic rectangular plates with a variety of boundary conditions (see Hearman [8]). The choice is restricted to the tensor products of eigenfunctions (see Table 6.2–3) of beams with corresponding boundary conditions.

Clamped at $x = 0, a$ and simply supported at $y = 0, b$

$$X_i(x) = \sin \lambda_i x - \sinh \lambda_i x + \alpha_i (\cosh \lambda_i x - \cos \lambda_i x)$$
$$Y_j(y) = \sin \frac{n\pi y}{b} \qquad (7.4-16a)$$

where λ_i are the roots of the characteristic equation

$$\cos\lambda_i a \;\; \cosh\lambda_i a - 1 = 0 \qquad (7.4-16b)$$

and

$$\alpha_i = \frac{\sinh\lambda_i a - \sin\lambda_i a}{\cosh\lambda_i a - \cos\lambda_i a} \qquad (7.4-16c)$$

Clamped at $x = 0$, free at $x = a$, and simply supported at $y = 0, b$

$$\begin{aligned} X_i(x) &= \sin\lambda_i x - \sinh\lambda_i x + \alpha_i\left(\cosh\lambda_i x - \cos\lambda_i x\right) \\ Y_j(y) &= \sin\frac{n\pi y}{b} \end{aligned} \qquad (7.4-17a)$$

where λ_i are the roots of the characteristic equation

$$\cos\lambda_i a \;\; \cosh\lambda_i a + 1 = 0 \qquad (7.4-17b)$$

and

$$\alpha_i = \frac{\sinh\lambda_i a + \sin\lambda_i a}{\cosh\lambda_i a + \cos\lambda_i a} \qquad (7.4-17c)$$

Free at $x = 0, a$ and simply supported at $y = 0, b$

$$\begin{aligned} X_i(x) &= \sin\lambda_i x + \sinh\lambda_i x - \alpha_i\left(\cosh\lambda_i x + \cos\lambda_i x\right) \\ Y_j(y) &= \sin\frac{n\pi y}{b} \end{aligned} \qquad (7.4-18a)$$

where λ_i are the roots of the characteristic equation

$$\cos\lambda_i a \;\; \cosh\lambda_i a - 1 = 0 \qquad (7.4-18b)$$

and

$$\alpha_i = \frac{\sinh\lambda_i a - \sin\lambda_i a}{\cosh\lambda_i a - \cos\lambda_i a} \qquad (7.4-18c)$$

Simply supported at $x = 0$ and $y = 0, b$, and Clamped at $x = a$

$$\begin{aligned} X_i(x) &= \sinh\lambda_i a \sin\lambda_i x + \sin\lambda_i a \sinh\lambda_i x \\ Y_j(y) &= \sin\frac{n\pi y}{b} \end{aligned} \qquad (7.4-19a)$$

where λ_i are the roots of the characteristic equation

$$\tan\lambda_i a - \tanh\lambda_i a = 0 \qquad (7.4-19b)$$

Simply supported at $x = 0$ and $y = 0, b$, and free at $x = a$

$$
\begin{aligned}
X_i(x) &= \sinh \lambda_i a \;\; \sin \lambda_i x - \sin \lambda_i a \;\; \sinh \lambda_i x \\
Y_j(y) &= \sin \frac{n\pi y}{b}
\end{aligned}
\tag{7.4 – 20a}
$$

where λ_i are the roots of the characteristic equation

$$
\tan \lambda_i a - \tanh \lambda_i a = 0 \tag{7.4 – 20b}
$$

Clamped at $x = 0$, and free at $x = a$ and $y = 0, b$

$$
\begin{aligned}
X_i(x) &= \sin \lambda_i x - \sinh \lambda_i x + \alpha_i \left(\cosh \lambda_i x - \cos \lambda_i x\right) \\
Y_j(y) &= \sin \mu_j y + \sinh \mu_j y - \beta_i \left(\cosh \mu_j y + \cos \mu_j y\right)
\end{aligned}
\tag{7.4 – 21a}
$$

where λ_i and μ_j are the roots of the characteristic equations

$$
\cos \lambda_i a \;\; \cosh \lambda_i a + 1 = 0, \quad \cos \mu_j b \cosh \mu_j b - 1 = 0 \tag{7.4 – 21b}
$$

and

$$
\alpha_i = \frac{\sinh \lambda_i a + \sin \lambda_i a}{\cosh \lambda_i a + \cos \lambda_i a}, \quad \beta_j = \frac{\sinh \mu_j b - \sin \mu_i b}{\cosh \mu_j b - \cos \mu_j b} \tag{7.4 – 21c}
$$

Clamped at $x = 0$, simply supported at $y = 0$, and free at $x = a$ and $y = b$

$$
\begin{aligned}
X_i(x) &= \sin \lambda_i x - \sinh \lambda_i x + \alpha_i \left(\cosh \lambda_i x - \cos \lambda_i x\right) \\
Y_j(y) &= \sinh \mu_j b \;\; \sin \mu_j y - \sin \mu_j b \;\; \sinh \mu_j y
\end{aligned}
\tag{7.4 – 22a}
$$

where λ_i and μ_j are the roots of the characteristic equations

$$
\cos \lambda_i a \;\; \cosh \lambda_i a + 1 = 0, \quad \tan \mu_j b - \tanh \mu_j b = 0 \tag{7.4 – 22b}
$$

and

$$
\alpha_i = \frac{\sinh \lambda_i a + \sin \lambda_i a}{\cosh \lambda_i a + \cos \lambda_i a} \tag{7.4 – 22c}
$$

Similarly, one can construct the approximation functions for any combination of fixed, hinged, and free boundary conditions on the four edges of a rectangular plate. Of course, the most difficult part is to evaluate the integrals of these functions as required in Eq. (8b). One may use a symbolic manipulator, such as Mathematica or Maple, to evaluate the integrals. When general laminated plates are considered, tensor products of the beam eigenfunctions can still be used for the approximation of the transverse deflection with appropriate functions for the in-plane displacements. In general, the Rayleigh–Ritz method for general rectangular laminates with arbitrary boundary conditions is algebraically more complicated than a numerical method, such as the finite element method.

7.5 Buckling of Simply Supported Plates Under Compressive Loads

7.5–1 Governing Equations

When a plate is subjected to in-plane compressive forces, $\hat{N}_{xx} < 0$ and $\hat{N}_{yy} < 0$, $\hat{N}_{xy} = 0$, and if the forces are sufficiently small, the equilibrium of the plate is stable (see Figure 7.5–1). The plate remains flat until a certain load is reached. At that load, called the buckling load, the stable state of the plate is disturbed and the plate seeks an alternative equilibrium configuration accompanied by a change in the load–deflection behavior. The phenomenon of changing the equilibrium configuration at the same load and without drastic changes in deformation is termed *bifurcation*. The load–deflection curve for buckled plates is often bilinear. The magnitude of the buckling load depends, as will be shown shortly, on geometry, material properties, as well as on the buckling mode shape. Here we determine the critical buckling loads of simply supported specially orthotropic plates using the Navier method.

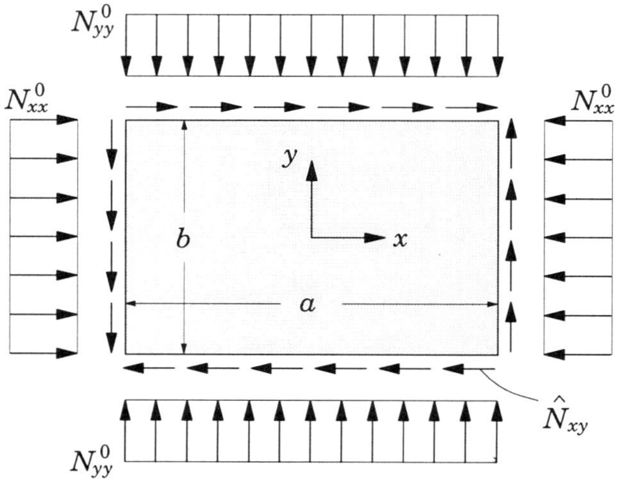

Figure 7.5–1: Buckling of a plate under in-plane compressive edge forces ($\hat{N}_{xx} = -N_{xx}^0$, $\hat{N}_{yy} = -N_{yy}^0$).

For the buckling analysis, we assume that the only applied loads are the in-plane forces and all other mechanical and thermal loads are zero. Since the prebuckling deformation w_0 is that of an equilibrium configuration, it satisfies the equilibrium equations, and the equation governing buckling deflection w_0^b is given by (see Section 6.2–3)

$$D_{11}\frac{\partial^4 w_0^b}{\partial x^4} + 2\left(D_{12} + 2D_{66}\right)\frac{\partial^4 w_0^b}{\partial x^2 \partial y^2} + D_{22}\frac{\partial^4 w_0^b}{\partial y^4} = \hat{N}_{xx}\frac{\partial^2 w_0^b}{\partial x^2} + \hat{N}_{yy}\frac{\partial^2 w_0^b}{\partial y^2} \quad (7.5-1)$$

For simplicity, we will omit the superscript 'b' on buckling deflection w_0^b. We wish to determine a nonzero deflection w_0 that satisfies Eq. (1) when the in-plane forces are

$$\hat{N}_{xx} = -N_0, \quad \hat{N}_{yy} = -kN_0, \quad k = \frac{\hat{N}_{yy}}{\hat{N}_{xx}} \tag{7.5 – 2}$$

and the edges are simply supported.

7.5–2 The Navier Solution

As in the case of bending, we select an expansion for w_0 that satisfies the boundary conditions in Eq. (7.2–2)

$$w_0(x,y) = W_{mn} \sin \alpha x \ \sin \beta y \tag{7.5 – 3}$$

Substituting Eq. (3) into Eq. (1), we obtain (for any m and n)

$$\begin{aligned} 0 =& \Big\{ - \left[D_{11}\alpha^4 + 2(D_{12} + 2D_{66})\alpha^2\beta^2 + D_{22}\beta^4 \right] + (\alpha^2 + k\beta^2)N_0 \Big\} \\ &\times W_{mn} \ \sin \alpha x \ \sin \beta y \end{aligned} \tag{7.5 – 4}$$

Since the equation must hold for every point (x, y) of the domain for nontrivial w_0 (*i.e.*, $W_{mn} \neq 0$), the expression inside the curly brackets should be zero for every m and n. This yields

$$N_0(m,n) = \frac{d_{mn}}{(\alpha^2 + k\beta^2)} \tag{7.5 – 5a}$$

where

$$d_{mn} = D_{11}\alpha^4 + 2(D_{12} + 2D_{66})\alpha^2\beta^2 + D_{22}\beta^4, \ \alpha = \frac{m\pi}{a}, \quad \beta = \frac{n\pi}{b} \tag{7.5 – 5b}$$

Thus, for each choice of m and n there corresponds a unique value of N_0. The *critical buckling load* is the smallest of $N_0(m, n)$. For a given laminate this value is dictated by a particular combination of the values of m and n. We investigate critical buckling loads of various laminates next.

7.5–3 Biaxial Compression of a Square Laminate ($k = 1$)

For a square laminate subjected to the same magnitude of compressive load on both edges (*i.e.*, biaxial compression with $k = 1$), Eq. (5a) yields

$$N_0(m,n) = \left(\frac{\pi^2}{a^2}\right) \frac{[D_{11}m^4 + 2(D_{12} + 2D_{66})m^2n^2 + D_{22}n^4]}{m^2 + n^2} \tag{7.5 – 6}$$

Now suppose that $D_{11} \geq D_{22}$. Then $D_{11}m^2$ increases more rapidly than the decrease in D_{22}/m^2 with an increase of m. Thus, the minimum of N_0 occurs when $m = 1$:

$$N_0(1,n) = \left(\frac{\pi^2}{a^2}\right) \frac{\left[D_{11} + 2(D_{12} + 2D_{66})n^2 + D_{22}n^4\right]}{1+n^2} \qquad (7.5-7)$$

The buckling load is a minimum when n is the nearest integer to the real number R

$$R^2 = -1 + \sqrt{1 + M_1 - 2M_2}\ , \quad M_1 = \frac{D_{11}}{D_{22}}\ , \quad M_2 = \frac{D_{12} + 2D_{22}}{D_{22}} \qquad (7.5-8)$$

For example, for modulus ratios of $M_1 = 10$ and $M_2 = 1$, we obtain $R = \sqrt{2}$ or $n = 1$. Hence, the critical buckling load becomes

$$N_{cr} \equiv N_0(1,1) = 6.5\left(\frac{\pi^2 D_{22}}{a^2}\right) \qquad (7.5-9)$$

For modulus ratios of $M_1 = 12$ and $M_2 = 1$, we obtain $R = 1.52$ or $n = 2$, and the critical buckling load becomes

$$N_{cr} = 7.2\left(\frac{\pi^2 D_{22}}{a^2}\right) \qquad (7.5-10)$$

For an isotropic ($D_{11} = D_{22} = D,\ D_{12} = \nu D$, and $2D_{66} = (1-\nu)D$) square plate under biaxial compression, the buckling can be calculated from Eq. (6):

$$N_0(m,n) = (m^2 + n^2)\left(\frac{\pi^2 D}{a^2}\right) \qquad (7.5-11)$$

and the critical buckling load occurs at $m = n = 1$, and it is equal to

$$N_{cr} \equiv N_0(1,1) = 2\left(\frac{\pi^2 D}{a^2}\right) \qquad (7.5-12)$$

7.5–4 Biaxial Loading of a Square Laminate

When the edges $x = 0, a$ are subjected to compressive load $\hat{N}_{xx} = -N_0$ and the edges $y = 0, b$ are subjected to tensile load $\hat{N}_{yy} = kN_0$, Eq. (5a) becomes

$$N_0(m,n) = \frac{\pi^2}{a^2} \frac{\left[D_{11}m^4 + 2(D_{12} + 2D_{66})m^2n^2 + D_{22}n^4\right]}{m^2 - kn^2} \qquad (7.5-13)$$

for $n^2 < m^2/k$. For example, when $k = 0.5$, the minimum buckling load occurs at $m = 1$ and $n = 1$. For the isotropic material properties used in Section 7.5–3, we have

$$N_{cr} = 26\left(\frac{\pi^2 D}{a^2}\right) \qquad (7.5-14)$$

7.5–5 Uniaxial Compression of a Rectangular Laminate ($k = 0$)

When $k = 0$ ($\hat{N}_{yy} = 0$), we have

$$N_0(m,n) = \frac{a^2}{m^2\pi^2}\left[D_{11}\left(\frac{m\pi}{a}\right)^4 + 2(D_{12}+2D_{66})\left(\frac{m\pi}{a}\right)^2\left(\frac{n\pi}{b}\right)^2 + D_{22}\left(\frac{n\pi}{b}\right)^4\right]$$
$$= \frac{\pi^2}{m^2b^2}\left[D_{11}m^4\left(\frac{b}{a}\right)^2 + 2(D_{12}+2D_{66})m^2n^2 + D_{22}n^4\left(\frac{a}{b}\right)^2\right] \qquad (7.5-15)$$

An examination of the expression in Eq. (15) shows that the smallest value of N_0, for any m, occurs for $n = 1$:

$$N_0(m,1) = \frac{\pi^2 D_{22}}{b^2}\left[m^2\frac{D_{11}}{D_{22}}\left(\frac{b}{a}\right)^2 + 2\frac{(D_{12}+2D_{66})}{D_{22}} + \frac{1}{m^2}\left(\frac{a}{b}\right)^2\right] \qquad (7.5-16)$$

The critical buckling load is then determined by finding the minimum of $N_0 = N_0(m)$ in Eq. (16) with respect to m. We have

$$\frac{dN_0}{dm} = 0 \quad \text{gives} \quad m^4 = \frac{D_{22}}{D_{11}}\left(\frac{a}{b}\right)^4 \qquad (7.5-17)$$

The second derivative of N_0 with respect to m can be shown to be positive. Since the value of m from Eq. (17) is not always an integer, the minimum buckling load cannot be predicted by substituting the value of m from Eq. (17) into Eq. (16). The minimum value of N_0 is given by Eq. (16) when m is the nearest integer value given by Eq. (17). Since the value of m depends on the ratio of the principal bending stiffnesses as well as plate aspect ratio, we must investigate the variation of N_0 with aspect ratio a/b for different values of m for a given laminate.

As an example, consider a laminate with $D_{11}/D_{22} = 10$ and $a/b = 1.778$. Then we have

$$N_0(m) = \frac{\pi^2 D_{22}}{b^2}\left[10m^2\left(\frac{b}{a}\right)^2 + 2 + \frac{1}{m^2}\left(\frac{a}{b}\right)^2\right] \qquad (7.5-18a)$$

with

$$m^4 = \frac{D_{22}}{D_{11}}\left(\frac{a}{b}\right)^4 = 0.1\times(1.778)^4 = 0.9994 \approx 1 \qquad (7.5-18b)$$

In fact, for aspect ratios (a/b) less than 2.66, we have

$$m^4 = \frac{D_{22}}{D_{11}}\left(\frac{a}{b}\right)^4 = 0.1 \times (2.66)^4 \quad \text{or} \quad m = 1.496 \tag{7.5 – 19}$$

Thus the closest integer is $m = 1$. The critical buckling load of a laminate with

$$\frac{D_{11}}{D_{22}} = 10, \quad (D_{12} + 2D_{66}) = D_{22}, \quad \frac{a}{b} < 2.66 \tag{7.5 – 20}$$

is given by

$$N_{cr} \equiv N_0(1,1) = \pi^2\left[10\left(\frac{b}{a}\right)^2 + 2 + \left(\frac{a}{b}\right)^2\right]\frac{D_{22}}{b^2} \tag{7.5 – 21}$$

For various aspect ratios, we have

$$\frac{a}{b} = 1: \quad N_{cr} = 13\frac{\pi^2 D_{22}}{b^2}\ ; \quad \frac{a}{b} = 1.5: \quad N_{cr} = 8.69\frac{\pi^2 D_{22}}{b^2}$$

$$\frac{a}{b} = 2: \quad N_{cr} = 8.5\frac{\pi^2 D_{22}}{b^2}\ ; \quad \frac{a}{b} = 2.5: \quad N_{cr} = 9.85\frac{\pi^2 D_{22}}{b^2}$$

It can be shown that if the laminate aspect ratio a/b is greater than 2.66 but less than 4.44, the buckling load is the minimum for $n = 1$ and $m = 2$ [using Eq. (17)]. For example, for $a/b = 3$, we have from Eq. (21)

$$N_0(1,1) = \frac{109}{9}\frac{\pi^2 D_{22}}{b^2} \approx 12.11\frac{\pi^2 D_{22}}{b^2}$$

$$N_0(1,2) = \frac{313}{36}\frac{\pi^2 D_{22}}{b^2} \approx 8.69\frac{\pi^2 D_{22}}{b^2} = N_{cr}$$

$$N_0(1,3) = 13\frac{\pi^2 D_{22}}{b^2}$$

Thus, for aspect ratios between 2.66 and 4.44, the plate buckles into two half–waves in the x–direction (and one half–wave in the y–direction). Thus larger aspect ratios lead to higher modes of buckling. Figure 7.5–2 contains a plot of the nondimensionalized buckling load $\bar{N}_0 = N_0 b^2/(\pi^2 D_{22})$ versus plate aspect ratio a/b for laminates whose material properties are $D_{11}/D_{22} = 10$, $(D_{12} + 2D_{66}) = D_{22}$. For aspect ratios less than 2.5, the plate buckles into a single half–wave in the x–direction. As the aspect ratio increases, the plate buckles into more and more half–waves in the x–direction. Note that intersections of two consecutive modes correspond to certain aspect ratios (see Figure 7.5-3). Thus, for each of these aspect ratios, there are two possible buckled mode shapes. The $\bar{N}_0$ versus a/b curve gets flatter with the increasing aspect ratio, and it approaches the value

$$N_{cr} = 2\left(\frac{\pi^2 D_{22}}{b^2}\right)\left[\sqrt{\frac{D_{11}}{D_{22}}} + (D_{12} + 2D_{66})\right] \tag{7.5 – 22}$$

which is obtained from Eq. (16) after substituting for m^2 from Eq. (17). For the data in Eq. (20), this limiting value of the critical buckling load is

$$N_{cr} = 8.32456\left(\frac{\pi^2 D_{22}}{b^2}\right) \tag{7.5 – 23}$$

For a square isotropic plate ($D_{11} = D_{22} = D,\ D_{12} = \nu D$, and $2D_{66} = (1-\nu)D$), we have $m = 1$ [from Eq. (17)], and the critical buckling load from Eq. (16) is

$$N_{cr} = 4\left(\frac{\pi^2 D}{a^2}\right) \tag{7.5 – 24}$$

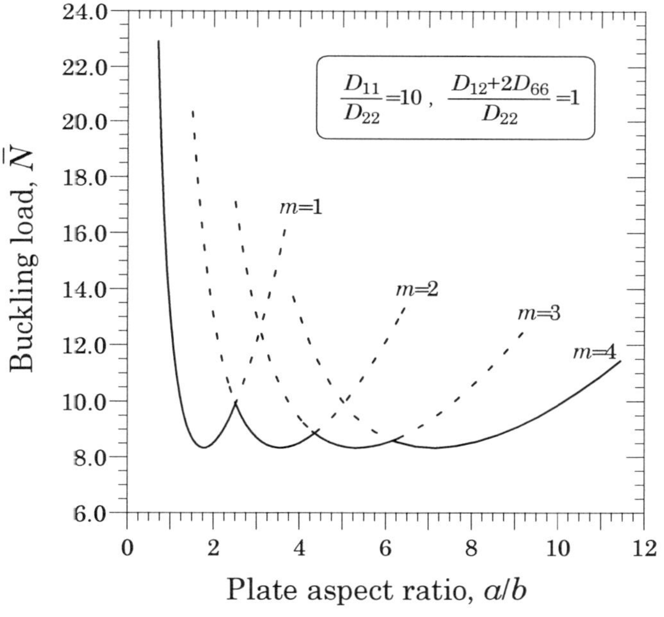

Figure 7.5–2: Nondimensionalized buckling load, $\bar{N} = N_0 b^2/(\pi^2 D_{22})$, versus plate aspect ratio a/b.

Table 7.5–1 shows the effect of plate aspect ratio and modulus ratio (anisotropy) on the critical buckling loads $\bar{N} = N_0 b^2/(\pi^2 D_{22})$ of rectangular laminates $(0/90)_s$ under uniform compression ($k = 0$) and biaxial compression ($k = 1$). In all cases the critical buckling mode is $(m, n) = (1, 1)$, except for $a/b = 0.5$ and $k = 1$, for which case the modes are (1,1), (1,2), (1,2), (1,2), and (1,3) for modulus ratios 5, 10, 20, 25, and 40, respectively. The nondimensionalized buckling load increases as the modulus ratio increases.

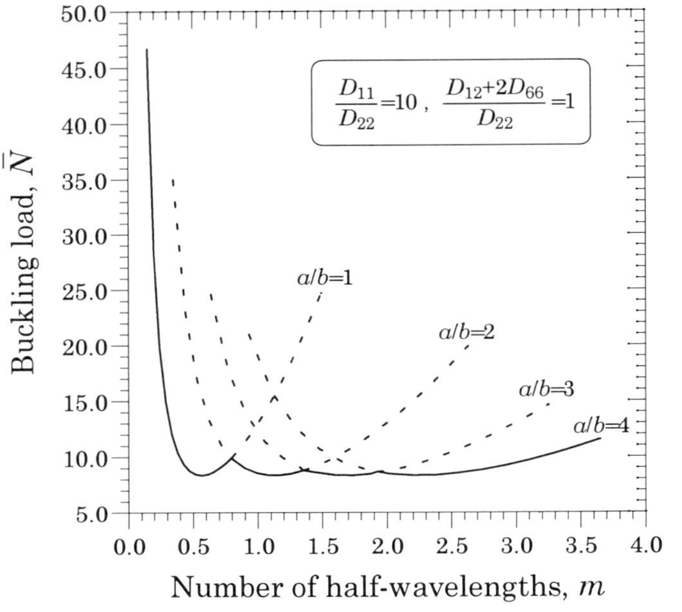

Figure 7.5–3: Nondimensionalized buckling load, $\bar{N}_0$, versus number of half-wavelengths m in the x−direction.

Table 7.5–1: Effect of plate aspect ratio and modulus ratio on the nondimensionalized buckling loads $\bar{N}$ of rectangular laminates $(0/90)_s$ under uniform compression ($k = 0$) and biaxial compression ($k = 1$) (E_1/E_2 varied, $G_{12} = G_{13} = 0.5E_2$, $G_{23} = 0.2E_2$, $\nu_{12} = 0.25$; all layers of equal thickness).

k	$\frac{a}{b}$	$\frac{E_1}{E_2} = 5$	10	20	25	40
	0.5	13.900	18.126	21.878	22.874	24.590
0	1.0	5.6500	6.3471	6.9611	7.1238	7.4037
	1.5	5.2333	5.2768	5.3099	5.3182	5.3322
	0.5	11.120	12.694	13.922	14.248	14.766
1	1.0	2.8250	3.1735	3.4806	3.5619	3.7019
	1.5	1.6103	1.6236	1.6338	1.6364	1.6407

Figure 7.5–4 shows plots of nondimensionalized critical buckling load $\bar{N} = N_0 b^2/(\pi^2 D_{22})$ as a function of the plate aspect ratio, a/b, for two different materials:

$$\text{Material 1:} \quad E_1 = 25E_2, \quad G_{12} = G_{13} = 0.5E_2, \quad \nu_{12} = 0.25$$

$$\text{Material 2:} \quad E_1 = 40E_2, \quad G_{12} = G_{13} = 0.6E_2, \quad \nu_{12} = 0.25$$

There is a mode change around $a/b > 2.2$ from (1,1) to (1,2).

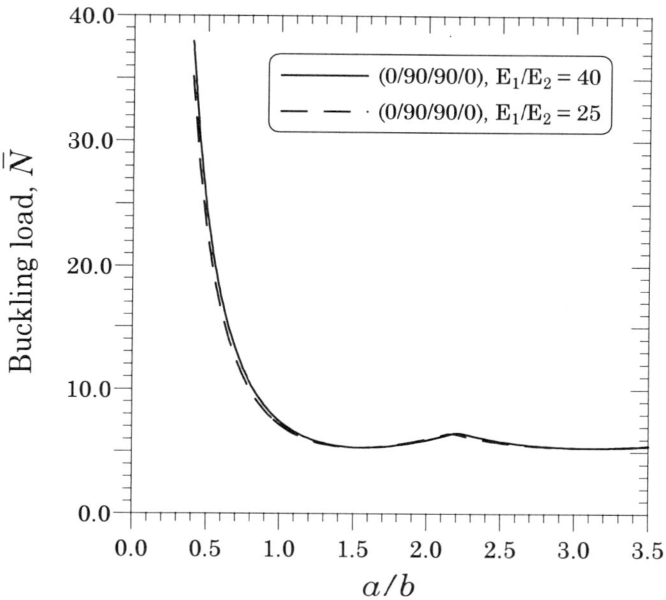

Figure 7.5–4: Nondimensionalized uniaxial critical buckling load ($\bar{N}$) versus plate aspect ratio (a/b) of symmetric cross-ply laminate $(0/90)_s$ for two different modular ratios.

7.6 Buckling of Rectangular Plates Under In-plane Shear Load

7.6–1 Governing Equation

In this section we consider buckling of specially orthotropic rectangular plates under in-plane shear load, N_{xy}^0. The problem does not permit the Navier solution; therefore, we use a variational method to solve the problem. When $\hat{N}_{xx} = \hat{N}_{yy} = 0$ and $\hat{N}_{xy} = N_{xy}^0$ (see Figure 7.5–1), the governing equation (7.5–1) takes the form

$$D_{11}\frac{\partial^4 w_0}{\partial x^4} + 2\left(D_{12} + 2D_{66}\right)\frac{\partial^4 w_0}{\partial x^2 \partial y^2} + D_{22}\frac{\partial^4 w_0}{\partial y^4} = 2N_{xy}^0\,\frac{\partial^2 w_0}{\partial x \partial y} \tag{7.6 – 1}$$

7.6–2 Simply Supported Plates

When the plate is simply supported on all its edges and subjected to in-plane shear, the Navier solution does not exist because the cross derivative term involving N_{xy}^0 will have a different coefficient $(\cos\alpha x \cos\beta y)$ than the rest of the expression in Eq. (1). Hence, we will seek the solution by a variational method.

Since the expression given in Eq. (7.5–3) for w_0 satisfies the geometric boundary conditions of the problem, the same functions are admissible in the Rayleigh–Ritz method:

$$w_0(x,y) \approx W_{MN} = \sum_{n=1}^{N}\sum_{m=1}^{M} c_{mn}\ \sin\alpha x\ \sin\beta y \qquad (7.6-2)$$

where $\alpha = m\pi/a$ and $\beta = n\pi/b$. Since the approximation functions

$$\varphi_{mn}(x,y) = \sin\frac{m\pi x}{a}\ \sin\frac{n\pi y}{b} \qquad (7.6-3)$$

also satisfy the natural boundary conditions of the problem, the Rayleigh–Ritz and Galerkin solutions are the same. Thus, substitution of Eq. (2) in the total potential energy functional for the Rayleigh–Ritz method

$$\Pi(w_0) = \frac{1}{2}\int_0^b\int_0^a\Bigg[D_{11}\left(\frac{\partial^2 w_0}{\partial x^2}\right)^2 + 2D_{12}\frac{\partial^2 w_0}{\partial x^2}\frac{\partial^2 w_0}{\partial y^2} + 4D_{66}\left(\frac{\partial^2 w_0}{\partial x\partial y}\right)^2 + D_{22}\left(\frac{\partial^2 w_0}{\partial y^2}\right)^2 - 2N_{xy}^0\left(\frac{\partial w_0}{\partial x}\frac{\partial w_0}{\partial y}\right)\Bigg]dxdy \qquad (7.6-4)$$

or the weighted–integral statement for the Galerkin method

$$0 = \int_0^b\int_0^a\left[D_{11}\frac{\partial^4 w_0}{\partial x^4} + 2\left(D_{12}+2D_{66}\right)\frac{\partial^4 w_0}{\partial x^2\partial y^2} + D_{22}\frac{\partial^4 w_0}{\partial y^4} - 2N_{xy}^0\frac{\partial^2 w_0}{\partial x\partial y}\right]\varphi_{pq}\,dxdy \qquad (7.6-5)$$

would lead to the same equations for the coefficients c_{mn}. Using the Galerkin method, we obtain

$$0 = \sum_{n=1}^{N}\sum_{m=1}^{M}\int_0^b\int_0^a\Big\{\left[D_{11}\alpha^4 + \left(D_{12}+2D_{66}\right)\alpha^2\beta^2 + D_{22}\beta^4\right]\varphi_{mn} - 2\alpha\beta N_{xy}^0\cos\alpha x\cos\beta y\Big\}c_{mn}\varphi_{pq}\ dxdy \qquad (7.6-6)$$

Using the identities

$$\int\sin\frac{m\pi x}{a}\ \sin\frac{n\pi x}{a}\ dx = \begin{cases} 0, & m\neq n \\ \frac{x}{2}, & m=n \end{cases}$$

$$\int\sin\lambda x\ \cos\mu x\ dx = -\left[\frac{\cos(\lambda-\mu)x}{(\lambda-\mu)} + \frac{\cos(\lambda+\mu)x}{(\lambda+\mu)}\right] \qquad (7.6-7)$$

we arrive at

$$0 = \frac{ab}{4}\left[D_{11}\left(\frac{p\pi}{a}\right)^4 + (D_{12}+2D_{66})\left(\frac{p\pi}{a}\right)^2\left(\frac{q\pi}{b}\right)^2 + D_{22}\left(\frac{q\pi}{b}\right)^4\right] c_{pq}$$

$$-2N_{xy}^0 \sum_{m=1}^{N}\sum_{n=1}^{M}\left(\frac{m\pi}{a}\right)\left(\frac{n\pi}{b}\right) S_{(mn)(pq)}\, c_{mn} \tag{7.6-8a}$$

where

$$S_{(mn)(pq)} = \int_0^b\int_0^a \cos\frac{m\pi x}{a}\cos\frac{n\pi y}{b}\sin\frac{p\pi x}{a}\sin\frac{q\pi y}{b}\,dxdy$$

$$= \left(\frac{4ab}{\pi^2}\right)\frac{pq}{(p^2-m^2)(q^2-n^2)} \quad \text{for } p^2 \neq m^2 \text{ and } q^2 \neq n^2 \tag{7.6-8b}$$

and the coefficients are zero when $p = m, p \pm m$ even, or when $q = n, q \pm n$ even.

The set of mn homogeneous equations (8a) define an eigenvalue problem

$$\left([A] - N_{xs}^0[S]\right)\{c\} = \{0\} \tag{7.6-9}$$

which has a nontrivial solution ($i.e., c_{mn} \neq 0$) when the determinant of the coefficient matrix is zero. Note that $[A]$ is a diagonal matrix while $[S]$ is a nonpositive-definite matrix; hence, the solution of (9) requires an eigenvalue routine that is suitable for nonpositive-definite matrices. It is found that the solution of (9) converges very slowly with increasing values of M and N (see [3,7]).

7.6–3 Clamped Plates

The total potential energy expression for the clamped rectangular plate under in-plane shear load N_{xy}^0 is

$$\Pi(w_0) = \frac{1}{2}\int_0^b\int_0^a \left[D_{11}\left(\frac{\partial^2 w_0}{\partial x^2}\right)^2 + 2D_{12}\frac{\partial^2 w_0}{\partial x^2}\frac{\partial^2 w_0}{\partial y^2} + 4D_{66}\left(\frac{\partial^2 w_0}{\partial x\partial y}\right)^2\right.$$

$$\left. + D_{22}\left(\frac{\partial^2 w_0}{\partial y^2}\right)^2 - 2N_{xy}^0\left(\frac{\partial w_0}{\partial x}\frac{\partial w_0}{\partial y}\right)\right]dxdy \tag{7.6-10a}$$

The minimum total potential energy principle requires that $\delta\Pi = 0$. We have

$$0 = \int_0^b\int_0^a \left(M_{xx}\delta\varepsilon_{xx}^{(1)} + M_{yy}\delta\varepsilon_{yy}^{(1)} + M_{xy}\delta\varepsilon_{xy}^{(1)} - q\delta w_0\right)dxdy$$

$$
\begin{aligned}
&= \int_0^b \int_0^a \bigg[D_{11} \frac{\partial^2 w_0}{\partial x^2} \frac{\partial^2 \delta w_0}{\partial x^2} + D_{12} \left(\frac{\partial^2 w_0}{\partial y^2} \frac{\partial^2 \delta w_0}{\partial x^2} + \frac{\partial^2 w_0}{\partial x^2} \frac{\partial^2 \delta w_0}{\partial y^2} \right) \\
&\quad + 4D_{66} \frac{\partial^2 w_0}{\partial x \partial y} \frac{\partial^2 \delta w_0}{\partial x \partial y} + D_{22} \frac{\partial^2 w_0}{\partial y^2} \frac{\partial^2 \delta w_0}{\partial y^2} \\
&\quad - N_{xy}^0 \left(\frac{\partial \delta w_0}{\partial x} \frac{\partial w_0}{\partial y} + \frac{\partial w_0}{\partial x} \frac{\partial \delta w_0}{\partial y} \right) \bigg] dx dy \qquad (7.6-10b)
\end{aligned}
$$

We assume Rayleigh–Ritz approximation of the form

$$
w_0(x,y) \approx W_{mn}(x,y) = \sum_{i=1}^{m} \sum_{j=1}^{n} c_{ij} \; \varphi_{ij}(x,y) \qquad (7.6-11a)
$$

where

$$
\varphi_{ij}(x,y) = X_i(x) Y_j(y) \qquad (7.6-11b)
$$

with

$$
X_i(x) = \left(\frac{x}{a} \right)^{i+1} \left(1 - \frac{x}{a} \right)^2, \quad Y_j(y) = \left(\frac{y}{b} \right)^{j+1} \left(1 - \frac{y}{b} \right)^2 \qquad (7.6-12)
$$

or

$$
\begin{aligned}
X_i(x) &= \sin \lambda_i x - \sinh \lambda_i x + \alpha_i \left(\cosh \lambda_i x - \cos \lambda_i x \right) \\
Y_j(y) &= \sin \lambda_j y - \sinh \lambda_j y + \alpha_j \left(\cosh \lambda_j y - \cos \lambda_j y \right) \qquad (7.6-13)
\end{aligned}
$$

for $i = 1, 2, \cdots, m$; $j = 1, 2, \cdots, n$. The parameters λ_i and α_i of Eq. (13) are defined in Eq. (7.3–45) and (7.3–46), respectively. Substituting Eq. (11) into Eq. (10b) we obtain

$$
\begin{aligned}
0 = \bigg\{ \int_0^b \int_0^a \bigg[& D_{11} \frac{d^2 X_i}{dx^2} Y_j \frac{d^2 X_p}{dx^2} Y_q + 2 \left(D_{12} + 2D_{66} \right) \frac{dX_i}{dx} \frac{dY_j}{dy} \frac{dX_p}{dx} \frac{dY_q}{dy} \\
& + D_{22} X_i \frac{d^2 Y_j}{dy^2} X_p \frac{d^2 Y_q}{dy^2} - N_{xy}^0 \left(\frac{dX_i}{dx} Y_j X_p \frac{dY_q}{dy} + X_i \frac{dY_j}{dy} \frac{dX_p}{dx} Y_q \right) \bigg] dx dy \bigg\} c_{pq} \\
& \qquad (7.6-14)
\end{aligned}
$$

When functions in Eq. (13) are used, at least two terms should be used because the coefficient of N_{xy}^0 is zero for $m = n = 1$; other coefficients are zero for $m = 1, n = 2$ and $m = 2, n = 1$. Using the approximation

$$
w_0(x,y) \approx c_{11} X_1(x) Y_1(y) + c_{22} X_2(x) Y_2(y) \qquad (7.6-15a)
$$

with [see Eq. (7.3–47)]

$$
\begin{aligned}
X_1(x) &= \sin\frac{4.73x}{a} - \sinh\frac{4.73x}{a} + 1.0178\left(\cosh\frac{4.73x}{a} - \cos\frac{4.73x}{a}\right)\\
X_2(x) &= \sin\frac{7.853x}{a} - \sinh\frac{7.853x}{a} + 0.9992\left(\cosh\frac{7.853x}{a} - \cos\frac{7.853x}{a}\right)\\
Y_1(y) &= \sin\frac{4.73y}{b} - \sinh\frac{4.73y}{b} + 1.0178\left(\cosh\frac{4.73y}{b} - \cos\frac{4.73y}{b}\right)\\
Y_2(y) &= \sin\frac{7.853y}{b} - \sinh\frac{7.853y}{b} + 0.9992\left(\cosh\frac{7.853y}{b} - \cos\frac{7.853y}{b}\right)
\end{aligned}
\tag{7.6 – 15b}
$$

we obtain

$$\left[\frac{537.181b}{a^3}D_{11} + \frac{324.829}{ab}(D_{12}+2D_{66}) + \frac{537.181a}{b^3}D_{22}\right]c_{11} - 23.107N_{xy}^0 c_{22} = 0$$

$$\left[\frac{3791.532b}{a^3}D_{11} + \frac{4227.255}{ab}(D_{12}+2D_{66}) + \frac{3791.532a}{b^3}D_{22}\right]c_{22} - 23.107N_{xy}^0 c_{11} = 0$$

or in matrix form

$$\begin{bmatrix} a_{11} & N_{xy}^0 a_{12} \\ N_{xy}^0 a_{12} & a_{22}\end{bmatrix}\begin{Bmatrix} c_{11} \\ c_{22}\end{Bmatrix} = -\begin{Bmatrix} 0 \\ 0\end{Bmatrix} \tag{7.6 – 16a}$$

where

$$a_{11} = \frac{537.181}{a^4}D_{11} + \frac{324.829}{a^2b^2}(D_{12}+2D_{66}) + \frac{537.181}{b^4}D_{22}, \quad a_{12} = \frac{23.107}{ab}$$

$$a_{22} = \frac{3791.532}{a^4}D_{11} + \frac{4227.255}{a^2b^2}(D_{12}+2D_{66}) + \frac{3791.532}{b^4}D_{22} \tag{7.6 – 16b}$$

For a nontrivial solution, the determinant of the coefficient matrix should be zero, $a_{11}a_{22} - a_{12}a_{12}(N_{xy}^0)^2 = 0$. Solving for the buckling load N_{xy}^0, we obtain

$$N_{xy}^0 = \pm\frac{1}{a_{12}}\sqrt{a_{11}a_{22}} \tag{7.6 – 17}$$

The $\pm$ sign indicates that the shear buckling load may be either positive or negative.

For an isotropic square plate, we have $a = b$ and $D_{11} = D_{22} = (D_{12} + 2D_{66}) = D$, and the shear buckling load predicted by Eq. (17) is

$$N_{xy}^0 = \pm 176\frac{D}{a^2} \tag{7.6 – 18}$$

whereas the "exact" critical buckling load is

$$N_{xy}^0 = \pm 145 \frac{D}{a^2} \tag{7.6 – 19}$$

The two-term Rayleigh–Ritz solution (18) is over 21% in error.

This concludes the discussion of shear buckling of rectangular plates. The variational solutions presented here for buckling under in-plane shear are only for illustrative purposes. More than two-term variational approximations are required to obtain accurate buckling loads. Once again, a symbolic manipulator proves to be effective in evaluating the integrals in the variational methods.

7.7 Vibration of Simply Supported Plates

7.7–1 Governing Equations

For natural vibration, all applied loads and the in-plane forces are set to zero in Eq. (7.1–1)

$$\begin{aligned} D_{11}\frac{\partial^4 w_0}{\partial x^4} + 2\left(D_{12} + 2D_{66}\right)\frac{\partial^4 w_0}{\partial x^2 \partial y^2} + D_{22}\frac{\partial^4 w_0}{\partial y^4} \\ + I_0 \ddot{w}_0 - I_2\left(\frac{\partial^2 \ddot{w}_0}{\partial x^2} + \frac{\partial^2 \ddot{w}_0}{\partial y^2}\right) = 0 \end{aligned} \tag{7.7 – 1a}$$

where

$$I_0 = \sum_{k=1}^{L} \rho_0^{(k)} \left(z_{k+1} - z_k\right), \quad I_2 = \frac{1}{3}\sum_{k=1}^{L} \rho_0^{(k)} \left(z_{k+1}^3 - z_k^3\right) \tag{7.7 – 1b}$$

7.7–2 Solution

We assume a periodic solution of the form

$$W_{mn}(t) = W_{mn}^0 e^{i\omega t} \tag{7.7 – 2}$$

where $i = \sqrt{-1}$ and ω is the frequency of natural vibration. Substituting (2) in (1), we obtain (for any m and n)

$$\begin{aligned} &\left\{D_{11}\alpha^4 + 2(D_{12} + 2D_{66})\alpha^2\beta^2 + D_{22}\beta^4 - \omega^2\left[I_0 + \left(\alpha^2 + \beta^2\right) I_2\right]\right\} \\ &\times W_{mn} \sin \alpha x \ \sin \beta y = 0 \end{aligned} \tag{7.7 – 3}$$

Since the equation must hold for every point (x, y) of the domain $0 < x < a$ and $0 < y < b$, the expression inside the braces should be zero for every m and n. This yields

$$\omega_{mn}^2 = \frac{\pi^4}{\tilde{I}_0 b^4}\left[D_{11}m^4\left(\frac{b}{a}\right)^4 + 2(D_{12}+2D_{66})m^2n^2\left(\frac{b}{a}\right)^2 + D_{22}n^4\right] \quad (7.7-4)$$

where

$$\tilde{I}_0 = I_0 + I_2\left[\left(\frac{m\pi}{a}\right)^2 + \left(\frac{n\pi}{b}\right)^2\right] \quad (7.7-5)$$

and L denotes the total number of layers in the laminate. For different values of m and n there corresponds a unique frequency ω_{mn} and a corresponding mode shape

$$w_0(x, y) = W_{mn}^0 \sin\frac{m\pi x}{a}\ \sin\frac{n\pi y}{b} \quad (7.7-6)$$

where W_{mn}^0 is the amplitude of the vibration mode (m, n). For square laminates, Eq. (4) reduces to

$$\omega_{mn}^2 = \left(\frac{\pi}{a}\right)^4 \frac{\left[D_{11}m^4 + 2(D_{12}+2D_{66})m^2n^2 + D_{22}n^4\right]}{\left[I_0 + I_2\left(\frac{\pi}{a}\right)^2(m^2+n^2)\right]} \quad (7.7-7)$$

When the rotatory inertia I_2 is not zero, it is not simple to find the lowest natural frequency (fundamental frequency). The rotatory (or rotary) inertia has the effect of reducing the frequency for any m and n.

When the rotatory inertia I_2 is neglected, the frequency of a rectangular specially orthotropic laminate reduces to

$$\omega_{mn}^2 = \frac{\pi^4}{I_0 b^4}\left[D_{11}m^4\left(\frac{b}{a}\right)^4 + 2(D_{12}+2D_{66})m^2n^2\left(\frac{b}{a}\right)^2 + D_{22}n^4\right] \quad (7.7-8)$$

and for a square plate we have

$$\omega_{mn}^2 = \frac{\pi^4}{I_0 b^4}\left[D_{11}m^4 + 2(D_{12}+2D_{66})m^2n^2 + D_{22}n^4\right] \quad (7.7-9)$$

The fundamental frequency occurs at $m = n = 1$:

$$\omega_{11}^2 = \frac{\pi^4}{I_0 b^4}\left[D_{11}\left(\frac{b}{a}\right)^4 + 2(D_{12}+2D_{66})\left(\frac{b}{a}\right)^2 + D_{22}\right] \quad (7.7-10)$$

For a rectangular isotropic plate, when the rotatory inertia is neglected, the frequency equation (8) becomes

$$\omega_{mn}^2 = \frac{D\pi^4}{I_0 b^4}\left[m^4\left(\frac{b}{a}\right)^4 + 2m^2n^2\left(\frac{b}{a}\right)^2 + n^4\right] = \frac{D\pi^4}{I_0 b^4}\left[m^2\left(\frac{b}{a}\right)^2 + n^2\right]^2 \quad (7.7-11)$$

and the fundamental frequency is given by

$$\omega_{11} = \frac{\pi^2}{b^2}\sqrt{\frac{D}{I_0}}\left[\left(\frac{b}{a}\right)^2 + 1\right] \tag{7.7-12}$$

Nondimensionalized frequencies, $\bar{\omega}_{mn} = \omega_{mn}(b^2/\pi^2)\sqrt{\rho h/D_{22}}$, of specially orthotropic square laminates are presented in Tables 7.7–1 and 7.7–2 for modulus ratios $E_1/E_2 = 10, 20$ ($G_{12} = G_{13} = 0.5E_2$, $G_{23} = 0.2E_2$, $\nu_{12} = 0.25$). The results presented in Table 7.7–1 are for $m, n = 1, 2, 3$, and for the case in which the rotary inertia is neglected. The fundamental frequency increases with modular ratio. As noted earlier, the effect of including rotary inertia is to decrease the frequency of vibration. The first four frequencies for an orthotropic (0°) plate correspond to the modes, (m, n)=(1,1), (1,2), (1,3), and (2,1), whereas for symmetric cross-ply plates the first four frequencies are provided by the modes: (m, n)=(1,1), (1,2), (2,1), and (1,3).

Table 7.7–1: Nondimensionalized fundamental frequencies of symmetric cross-ply laminates according to the classical plate theory ($\bar{\omega}_{mn} = \omega_{mn}(b^2/\pi^2)\sqrt{\rho h/D_{22}}$).

$\frac{E_1}{E_2}$	m	n	0°	$(0°/90°)_s$
	1	1	3.672	2.519
	1	2	5.996	4.986
	1	3	10.648	9.783
	2	1	13.075	8.515
10	2	2	14.690	10.077
	2	3	18.181	13.783
	3	1	28.868	18.704
	3	2	30.258	19.911
	3	3	33.053	22.674
	1	1	4.847	2.638
	1	2	6.781	4.917
	1	3	11.111	9.637
	2	1	18.193	9.354
20	2	2	19.388	10.554
	2	3	22.153	13.826
	3	1	40.539	20.752
	3	2	41.542	21.578
	3	3	43.623	13.826

Table 7.7–2: Nondimensionalized fundamental frequencies $\bar{\omega}$ of symmetric cross-ply laminates $(0/90)_s$ according to the classical plate theory.

	Without Rotary Inertia				With Rotary Inertia			
a/b	10*	20	30	40	10	20	30	40
0.5	27.486	36.234	43.237	49.252	26.941	35.516	42.381	48.277
1.0	10.068	12.931	15.264	17.285	9.986	12.826	15.141	17.145
1.5	7.562	9.832	11.666	13.248	7.518	9.774	11.598	13.170
2.0	6.887	9.082	10.839	12.348	6.852	9.036	10.784	12.348
2.5	6.627	8.819	10.563	12.056	6.595	8.777	10.513	11.999
3.0	6.502	8.701	10.443	11.933	6.472	8.661	10.396	11.879

*Modulus ratio E_1/E_2.

Figure 7.7–1 shows a plot of nondimensionalized fundamental frequency $\bar{\omega} = \omega(b^2/\pi^2)\sqrt{\rho h/D_{22}}$ as a function of plate aspect ratio for symmetric $(0/90)_s$ graphite–epoxy laminate ($E_1/E_2 = 40$, $G_{12} = G_{13} = 0.5E_2$, $\nu_{12} = 0.25$).

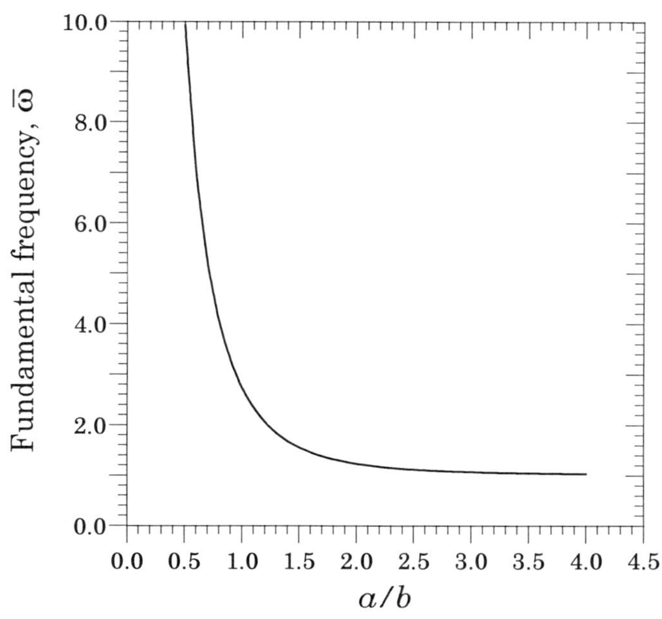

Figure 7.7–1: Nondimensionalized fundamental frequency $\bar{\omega}$ as a function of plate aspect ratio a/b for symmetric $(0/90)_s$ laminate.

7.8 Buckling and Vibration of Plates with Two Parallel Edges Simply Supported

7.8–1 Introduction

The Lévy method can be used to determine natural frequencies and critical buckling loads of rectangular laminates for which two (parallel) opposite edges are simply supported and the other two edges have any boundary conditions, as described in Section 7.3 for bending analysis. For other combinations of fixed, hinged, and free boundary conditions on the edges of rectangular plates, one may use the Rayleigh–Ritz method with the approximation functions suggested in Section 7.4–3.

Consider a rectangular laminate with in-plane dimensions a and b and total thickness h. The laminate coordinate system (x, y, z) is taken such that $-a/2 \leq x \leq a/2, 0 \leq y \leq b, -h/2 \leq z \leq h/2$ (see Figure 7.8–1). Here we assume that the edges $y = 0, b$ are simply supported, and the other two edges each have simply supported, clamped, or free boundary conditions. The equation governing buckling under in-plane normal forces and natural vibration of a specially orthotropic laminated plate is given by Eq. (7.1–1):

$$\begin{aligned} & D_{11}\frac{\partial^4 w_0}{\partial x^4} + 2\left(D_{12} + 2D_{66}\right)\frac{\partial^4 w_0}{\partial x^2 \partial y^2} + D_{22}\frac{\partial^4 w_0}{\partial y^4} \\ & - \hat{N}_{xx}\frac{\partial^2 w_0}{\partial x^2} - \hat{N}_{yy}\frac{\partial^2 w_0}{\partial y^2} + I_0\ddot{w}_0 - I_2\left(\frac{\partial^2 \ddot{w}_0}{\partial x^2} + \frac{\partial^2 \ddot{w}_0}{\partial y^2}\right) = 0 \qquad (7.8-1) \end{aligned}$$

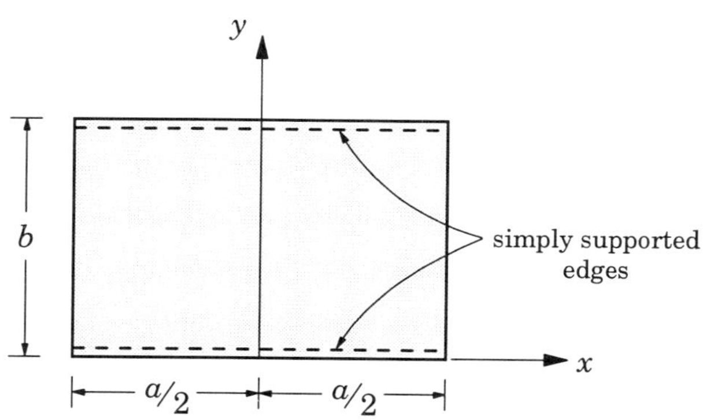

Figure 7.8–1: Geometry and coordinate system of a rectangular plate used in the Lévy method.

Recall that in the Lévy method the partial differential equation (1) is reduced to an ordinary differential equation in x by assuming solution in the form of a single Fourier series

$$w_0(x,y) = W_n(x)\ \sin\beta y, \quad \beta = \frac{n\pi}{b} \tag{7.8-2}$$

which satisfies the simply supported boundary conditions

$$w_0 = 0, \quad M_{xx} = -\left(D_{11}\frac{\partial^2 w_0}{\partial x^2} + D_{12}\frac{\partial^2 w_0}{\partial y^2}\right) = 0 \tag{7.8-3}$$

on edges $y = 0, b$. The ordinary differential equations obtained in the Lévy method can be solved either by direct integration or by means of the state–space approach. We discuss both procedures in the following sections.

7.8–2 Buckling by Direct Integration

Here we consider buckling under uniaxial compressive forces

$$\hat{N}_{xx} = 0, \quad \hat{N}_{yy} = -N_{yy}^0 \tag{7.8-4}$$

Substituting (2) and (4) into the governing equation (1) with the inertia terms zero, for any y, we obtain

$$D_{11}\frac{d^4W_n}{dx^4} - 2\left(D_{12} + 2D_{66}\right)\beta^2\frac{d^2W_n}{dx^2} + D_{22}\beta^4 W_n - N_{yy}^0\beta^2 W_n = 0 \tag{7.8-5}$$

We assume the general solution of Eq. (5) in the form

$$W_n(x) = A_n\cosh\lambda_1 x + B_n\sinh\lambda_1 x + C_n\cos\lambda_2 x + D_n\sin\lambda_2 x \tag{7.8-6}$$

where λ_i are the roots of the characteristic equation

$$D_{11}\lambda^4 - 2\left(D_{12} + 2D_{66}\right)\beta^2\lambda^2 + D_{22}\beta^4 - N_{yy}^0\beta^2\lambda = 0 \tag{7.8-7}$$

and they are given by

$$\begin{aligned}
(\lambda_1)^2 &= \frac{\beta^2}{D_{11}}\left[\sqrt{\left(D_{12} + 2D_{66}\right)^2 + D_{11}\left(\tilde{N}_{yy}^0 - D_{22}\right)} + \left(D_{12} + 2D_{66}\right)\right] \\
(\lambda_2)^2 &= \frac{\beta^2}{D_{11}}\left[\sqrt{\left(D_{12} + 2D_{66}\right)^2 + D_{11}\left(\tilde{N}_{yy}^0 - D_{22}\right)} - \left(D_{12} + 2D_{66}\right)\right]
\end{aligned} \tag{7.8-8}$$

where $\tilde{N}_{yy}^0 = N_{yy}^0/\beta^2$. The constants A_n, B_n, C_n, and D_n must be determined using the boundary conditions at $x = 0, a$.

For clamped boundary conditions on edges $x = 0, a$, for example, we require

$$W_n = 0, \quad \frac{dW_n}{dx} = 0 \tag{7.8-9}$$

which yield the eigenvalue problem

$$\begin{bmatrix} 1 & 0 & 1 & 0 \\ 0 & \lambda_1 & 0 & \lambda_2 \\ \cosh \lambda_1 a & \sinh \lambda_1 a & \cos \lambda_2 a & \sin \lambda_2 a \\ \lambda_1 \sinh \lambda_1 a & \lambda_1 \cosh \lambda_1 a & -\lambda_2 \sin \lambda_2 a & \lambda_2 \cos \lambda_2 a \end{bmatrix} \begin{Bmatrix} A_n \\ B_n \\ C_n \\ D_n \end{Bmatrix} = \begin{Bmatrix} 0 \\ 0 \\ 0 \\ 0 \end{Bmatrix} \tag{7.8-10}$$

For a nontrivial solution, $A_n \neq 0, B_n \neq 0, C_n \neq 0$, and $D_n \neq 0$, we set the determinant of the coefficient matrix in (10) to zero. We have [*cf.* Eq. 6.2–58)]

$$2\lambda_1\lambda_2 \left(1 - \cosh \lambda_1 a \ \cos \lambda_2 a\right) + (\lambda_1^2 - \lambda_2^2) \sin \lambda_1 a \ \sinh \lambda_2 a = 0 \tag{7.8-11}$$

Since λ_1 and λ_2 contain the buckling load N_{yy}^0, Eq. (11) can be used, in theory, to determine the critical buckling load of the plate. However, the complexity of (11) makes it less useful in readily computing the buckling loads of specially orthotropic plates. For specific plate stiffnesses, it is possible to numerically determine the critical buckling loads.

7.8–3 Vibration by Direct Integration

Here we consider natural vibration of a specially orthotropic plate. For periodic motion, we assume that

$$w_0(x, y, t) = \bar{w}_0(x, y) e^{i\omega t} \tag{7.8-12}$$

where $i = \sqrt{-1}$ and ω is the frequency of natural vibration. Then the amplitude of vibration $\bar{w}_0$ is approximated as in Eq. (2). Substituting (2) and (12) into the governing equation (1), with the in-plane forces zero, for any y and t we obtain

$$D_{11}\frac{d^4W_n}{dx^4} + \left[I_2\omega^2 - 2\left(D_{12} + 2D_{66}\right)\beta^2\right]\frac{d^2W_n}{dx^2} - \left[\omega^2\left(I_0 + I_2\beta^2\right) - D_{22}\beta^4\right]W_n = 0 \tag{7.8-13}$$

or

$$p\frac{d^4W_n}{dx^4} + q\frac{d^2W_n}{dx^2} - rW_n = 0 \tag{7.8-14a}$$

where

$$p = D_{11}, \quad q = I_2\omega^2 - 2\left(D_{12} + 2D_{66}\right)\beta^2, \quad r = \omega^2\left(I_0 + I_2\beta^2\right) - D_{22}\beta^4 \tag{7.8-14b}$$

Equation (14) is of the same form as Eq. (6.2–44), and the procedure described in Section 6.2–3 can be used to determine the natural frequencies for various boundary conditions on edges $x = 0, a$.

7.8–4 Buckling and Vibration by the State–Space Approach

As explained in Section 7.3, the governing differential equation in (1) can be reduced, with Eq. (2), to a system of a first-order matrix differential equation

$$\{Z'\} = [T]\{Z\} \tag{7.8 – 15}$$

where

$$\{Z\} = \begin{Bmatrix} W_m \\ W_m' \\ W_m'' \\ W_m''' \end{Bmatrix}, \quad [T] = \begin{bmatrix} 0 & 1 & 0 & 0 \\ 0 & 0 & 1 & 0 \\ 0 & 0 & 0 & 1 \\ C_1 & 0 & C_2 & 0 \end{bmatrix} \tag{7.8 – 16}$$

$\hat{N}_{xx} = -N_{xx}^0$, $\hat{N}_{yy} = -N_{yy}^0$, and

$$C_1 = -\frac{\left(\beta^4 D_{22} - \beta^2 \hat{N}_{yy}\right)}{D_{11}}, \quad C_2 = \frac{\left[2\beta^2(D_{12} + 2D_{66}) - \hat{N}_{xx}\right]}{D_{11}} \tag{7.8 – 17}$$

for buckling analysis and

$$C_1 = -\frac{\left(\beta^4 D_{22} - \hat{I}_0 \omega_m^2\right)}{D_{11}}, \quad C_2 = \frac{[2\beta^2(D_{12} + 2D_{66}) - I_2 \omega_m^2]}{D_{11}} \tag{7.8 – 18}$$

for free vibration analysis. Here ω_m denotes the frequency of vibration of the mth mode, $\hat{I}_0 = I_0 + \beta^2 I_2$, and $i = \sqrt{-1}$.

The solution of Eq. (15) is given by

$$\mathbf{Z}(x) = e^{\mathbf{A}x}\mathbf{K} \tag{7.8 – 19}$$

and the vector $\mathbf{K}$ of constants is to be determined from the boundary conditions. Substitution of Eq. (19) into the set of boundary conditions results in a homogeneous system of equations

$$[M]\{K\} = \{0\} \tag{7.8 – 20}$$

For a nontrivial solution, the determinant of the coefficient matrix in (20) should be zero:

$$|M_{ij}| = 0 \tag{7.8 – 21}$$

The roots of the above equation are the squares of the frequencies of natural vibration or, in the case of buckling, they denote the buckling loads.

The Lévy type solution procedure is used to evaluate the natural frequencies and critical buckling loads under uniaxial compression of specially orthotropic rectangular laminates. The lamina material properties used are $E_1/E_2 = 40$, $G_{12} = G_{13} = 0.6E_2$, and $\nu_{12} = 0.25$. Numerical results for the nondimensionalized fundamental frequencies ($\bar{\omega} = \omega(\frac{a^2}{h})\sqrt{\rho/E_2}$) and critical buckling loads under uniaxial compression ($\bar{N} = N_{xx}^0 b^2/E_2h^3$) of square, symmetric, cross-ply laminates are presented in Table 7.8–1 for various ratios of principal moduli of the material. Note that the nondimensionalized frequencies and buckling loads are the same for any odd number of layers $n = 3, 5, 7, \cdots$. Table 7.8–2 contains numerical results for various boundary conditions (see [16]). As before, the notation SF, for example, is used to indicate that edge $x = a/2$ is simply supported (S) and edge $x = -a/2$ is free (F).

Table 7.8–1: Nondimensionalized fundamental frequencies and critical buckling loads under uniaxial compression of simply supported symmetric cross-ply square plates as a function of the modulus ratio.

Laminate	$\frac{E_1}{E_2} = 3$	10	20	30	40
Fundamental Frequencies, $\bar{\omega} = \omega\left(\frac{a^2}{h}\right)\sqrt{\rho/E_2}$					
$(0/90/0)_n$	7.5357	10.650	13.948	16.605	18.891
(0/90/90/0)	7.5357	10.650	13.948	16.605	18.891
Uniaxial Critical Buckling Loads, $\bar{N} = N_{xx}^0 b^2/E_2h^3$					
$(0/90/0)_n$	5.754	11.492	19.712	27.936	36.160
(0/90/90/0)	5.754	11.492	19.712	27.936	36.160

7.9 Transient Analysis

7.9–1 Preliminary Comments

In this section we will develop transient solutions to specially orthotropic plates. Recall that in the static bending analysis of plates we developed the analytical solutions using the Navier method, the Lévy method, and the Rayleigh–Ritz method. The same methods can also be used to approximate the spatial variations of the transient solutions of plates. The resulting ordinary differential equations in time can be solved exactly when possible or numerically using a time–integration method. Here we consider simply supported plates to illustrate these ideas (see Reddy [21]).

Table 7.8–2: Nondimensionalized fundamental frequencies and critical buckling loads under uniaxial compression of symmetric cross-ply (0°/90°/0°) square plates for various boundary conditions and modulus ratios.

$\frac{E_1}{E_2}$	SS	SC	CC	FF	FS	FC
Fundamental Frequencies, $\bar{\omega} = \omega\left(\frac{a^2}{h}\right)\sqrt{\rho/E_2}$						
10	10.650	15.199	21.118	3.294	4.088	5.419
20	13.948	20.610	29.166	3.721	4.443	6.515
30	16.605	24.870	35.431	4.106	4.770	7.445
40(1)*	18.891	28.501	40.743	4.457	5.076	8.269
40(2)	26.938	34.533	45.233	17.827	18.473	19.789
40(3)	46.208	51.192	59.023	40.113	40.761	41.505
Biaxial Critical Buckling Loads, $\bar{N} = N_{xx}^0 b^2/E_2h^3$						
10	5.746	9.353(2)	13.468(2)	1.123	1.661	3.202
20	9.591(2)†	14.026(2)	21.709(3)	1.420	1.978	4.683
30	12.147(2)	18.703(2)	28.081(3)	1.722	2.288	6.142
40	14.704(2)	23.381(2)	34.454(3)	2.025	2.596	7.595

* Denotes the mode number m.

† Mode m in which the lowest buckling load occurs (otherwise, $m = 1$).

7.9–2 Spatial Variation of the Solution

The equation of motion governing bending deflection w_0 of a specially orthotropic plate, assuming no applied in-plane and thermal forces, is [see Eq. (7.1–1)]

$$-\left[D_{11}\frac{\partial^4 w_0}{\partial x^4} + 2\left(D_{12} + 2D_{66}\right)\frac{\partial^4 w_0}{\partial x^2 \partial y^2} + D_{22}\frac{\partial^4 w_0}{\partial y^4}\right] + q(x,y,t)$$
$$= I_0\ddot{w}_0 - I_2\left(\frac{\partial^2 \ddot{w}_0}{\partial x^2} + \frac{\partial^2 \ddot{w}_0}{\partial y^2}\right) \quad (7.9-1)$$

Suppose that the plate is simply supported with the boundary conditions

$$w_0(x,0,t) = 0, \quad w_0(x,b,t) = 0, \quad w_0(0,y,t) = 0, \quad w_0(a,y,t) = 0 \quad \text{for} \quad t \geq 0$$

$$M_{xx}(0,y,t) = 0, \quad M_{xx}(a,y,t) = 0, \quad M_{yy}(x,0,t) = 0, \quad M_{yy}(x,b,t) = 0 \quad \text{for} \quad t \geq 0 \quad (7.9-2)$$

and assume that the initial conditions are

$$w_0(x,y,0) = d_0(x,y), \quad \frac{\partial w_0}{\partial t}(x,y,0) = v_0(x,y) \quad \text{for all } x \text{ and } y \tag{7.9-3}$$

where d_0 and v_0 are the initial displacement and velocity, respectively.

We assume the following expansion of the transverse deflection to satisfy the boundary conditions (2) for any time $t \geq 0$

$$w_0(x,y,t) = \sum_{n=1}^{\infty}\sum_{m=1}^{\infty} W_{mn}(t) \, \sin \alpha x \, \sin \beta y \tag{7.9-4}$$

Similarly, we assume that the transverse load, initial displacement, and initial velocity can be expanded as

$$q(x,y,t) = \sum_{n=1}^{\infty}\sum_{m=1}^{\infty} Q_{mn}(t) \, \sin \alpha x \, \sin \beta y \tag{7.9-5}$$

$$d_0(x,y) = \sum_{n=1}^{\infty}\sum_{m=1}^{\infty} D_{mn} \, \sin \alpha x \, \sin \beta y \tag{7.9-6}$$

$$v_0(x,y) = \sum_{n=1}^{\infty}\sum_{m=1}^{\infty} V_{mn} \, \sin \alpha x \, \sin \beta y \tag{7.9-7}$$

where, for example, Q_{mn} are given by

$$Q_{mn}(t) = \frac{4}{ab}\int_0^b \int_0^a q(x,y,t) \, \sin \frac{m\pi x}{a} \, \sin \frac{n\pi y}{b} \, dxdy \tag{7.9-8}$$

Substituting the expansions (4) and (5) into Eq. (1), we obtain

$$\sum_{n=1}^{\infty}\sum_{m=1}^{\infty}\Big\{W_{mn}\left[D_{11}\alpha^4 + 2(D_{12}+2D_{66})\alpha^2\beta^2 + D_{22}\beta^4\right] + \left[I_0 + I_2\left(\alpha^2+\beta^2\right)\right]\ddot{W}_{mn} - Q_{mn}\Big\} \sin \alpha x \, \sin \beta y = 0 \tag{7.9-9}$$

Since the above expression must hold for all x and y, it follows that

$$W_{mn}\left[D_{11}\alpha^4 + 2(D_{12}+2D_{66})\alpha^2\beta^2 + D_{22}\beta^4\right] + \left[I_0 + I_2\left(\alpha^2+\beta^2\right)\right]\ddot{W}_{mn} - Q_{mn} = 0 \tag{7.9-10a}$$

or

$$K_{mn}W_{mn}(t) + M_{mn}\ddot{W}_{mn} = Q_{mn}(t) \tag{7.9-10b}$$

where

$$
\begin{aligned}
K_{mn} =& D_{11}\alpha^4 + 2(D_{12} + 2D_{66})\alpha^2\beta^2 + D_{22}\beta^4 \\
M_{mn} =& I_0 + I_2\left(\alpha^2 + \beta^2\right)
\end{aligned}
\qquad (7.9-10c)
$$

7.9–3 Time Integration

The ordinary differential equation (10a) can be solved either exactly or numerically. The numerical time integration methods will be discussed in the subsequent chapters. To solve it exactly, we first write Eq. (10a) in the form

$$
\frac{d^2W_{mn}}{dt^2} + \left(\frac{K_{mn}}{M_{mn}}\right)W_{mn} = \frac{1}{M_{mn}}Q_{mn}(t) \qquad (7.9-11)
$$

The solution of Eq. (11) is given by

$$
W_{mn}(t) = C_1e^{\lambda_1 t} + C_2e^{\lambda_2 t} + W^p_{mn}(t) \qquad (7.9-12)
$$

where C_1 and C_2 are constants to be determined using the initial conditions, $W^p_{mn}(t)$ is the particular solution

$$
W^p_{mn}(t) = \int^t \frac{r_1(\tau)r_2(t) - r_1(t)r_2(\tau)}{r_1(\tau)\dot{r}_2(\tau) - \dot{r}_1(\tau)r_2(\tau)}Q_{mn}(\tau)\, d\tau \qquad (7.9-13a)
$$

with $r_1(t) = e^{\lambda_1 t}$ and $r_2(t) = e^{\lambda_2 t}$, and λ_1 and λ_2 are the roots of the equation

$$
\lambda^2 + \frac{K_{mn}}{M_{mn}} = 0; \quad \lambda_1 = -i\mu,\ \lambda_2 = i\mu,\ i = \sqrt{-1},\ \mu = \sqrt{\frac{K_{mn}}{M_{mn}}} \qquad (7.9-13b)
$$

The solution becomes

$$
W_{mn}(t) = A\cos\mu t + B\sin\mu t + W^p_{mn}(t) \qquad (7.9-14a)
$$

$$
W^p_{mn}(t) = \frac{1}{2i\mu}\left(e^{i\mu t}\int^t e^{-i\mu\tau}Q_{mn}(\tau)d\tau - e^{-i\mu t}\int^t e^{i\mu\tau}Q_{mn}(\tau)d\tau\right) \qquad (7.9-14b)
$$

Once the load distribution, both spatially and with time, is known, the solution can be determined from Eq. (14a).

For a step loading, $Q_{mn}(t) = Q^0_{mn}H(t)$, where $H(t)$ denotes the Heavyside step function, Eq. (14a) takes the form

$$
W_{mn}(t) = A\cos\mu t + B\sin\mu t + \frac{1}{K_{mn}}Q^0_{mn} \qquad (7.9-15a)
$$

Using the initial conditions (3), we obtain

$$A = D_{mn} - \frac{1}{K_{mn}} Q^0_{mn}, \quad B = \frac{V_{mn}}{\mu} \tag{7.9 – 15b}$$

Thus the final solution (4) is given by

$$w_0(x,y,t) = \sum_{n=1}^{\infty}\sum_{m=1}^{\infty}\left[D_{mn}\cos\mu t + \frac{V_{mn}}{\mu}\sin\mu t + \frac{Q^0_{mn}}{K_{mn}}(1-\cos\mu t)\right]\sin\alpha x\ \sin\beta y \tag{7.9 – 16}$$

The coefficients Q^0_{mn} are given in Table 7.2–1 for various types of distributions. The same holds for D_{mn} and V_{mn}.

It should be noted that the procedure outlined above is valid irrespective of how one arrives at Eq. (10b); *e.g.*, Eq. (10b) could have been obtained using the Rayleigh–Ritz method or other methods. The exact solution of the differential equation (10b) can also be obtained using the Laplace transform method (see Table 6.6–1 for the Laplace transform pairs). Once the solution w_0 is known, stresses can be computed using Eqs. (7.2–13).

Figure 7.9–1 contains plots of the nondimensionalized center deflection $\bar{w} = w_0(E_2h^3/a^4q_0)$ as a function of time for a simply supported (SS–1) symmetric cross-ply (0/90/0) laminate ($h_1 = h_3 = h/4$, $h_2 = h/2$; $E_1/E_2 = 25$, $G_{12} = G_{13} = 0.5E_2$, $\nu_{12} = 0.25$; $a = b = 25$ cm, $h = 5$ cm) under a step loading that is sinusoidally distributed (SSL) or uniformly distributed (UDL) over the plate surface. It is assumed that the plate motion ensues from rest, *i.e.*, $d_0 = 0$ and $v_0 = 0$. The solution is plotted to show one complete wavelength. The dashed curve corresponds to the solution when the rotary inertia is neglected. The rotary inertia has the effect of increasing the wavelength slightly. Figure 7.9–2 contains plots of nondimensionalized center normal stress $\bar{\sigma}_{xx} = \sigma_{xx}(h^2/a^2q_0)$ as a function of time for the same laminates. Note that the stress variation for the uniformly distributed load case is not as smooth as for the sinusoidally distributed load case.

7.10 Closure

In this chapter analytical and Rayleigh–Ritz solutions for bending, buckling, natural vibration, and transient response of specially orthotropic plates are presented. In most cases, the numerical determination of actual solutions require evaluation of a series solution, solution of a transcendental equation, or determination of eigenvalues (in the state–space approach). Thus, even the "exact" solutions become approximate because of the truncation of an infinite series or round-off errors in the solution of nonlinear equations. The analytical solutions developed herein serve to help one understand, at least qualitatively, the behavior of laminated plates.

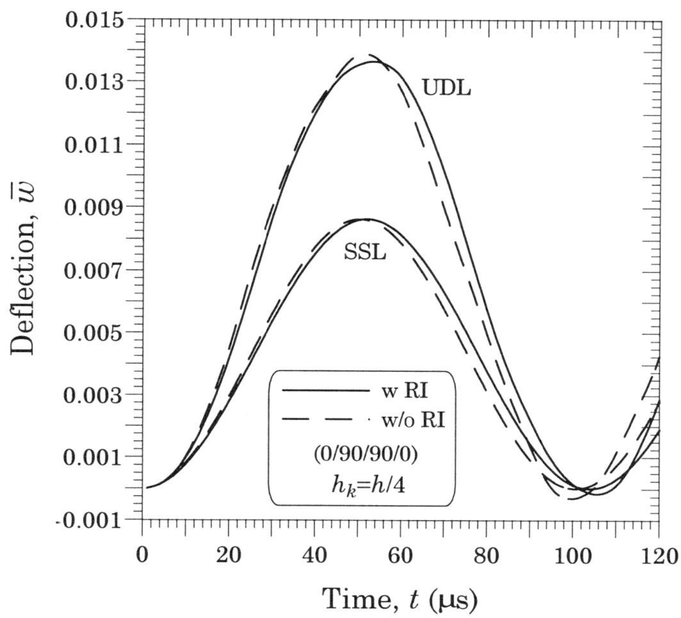

Figure 7.9–1: Nondimensionalized maximum transverse deflection ($\bar{w}$) versus time for a simply supported symmetric cross-ply $(0/90)_s$ laminate.

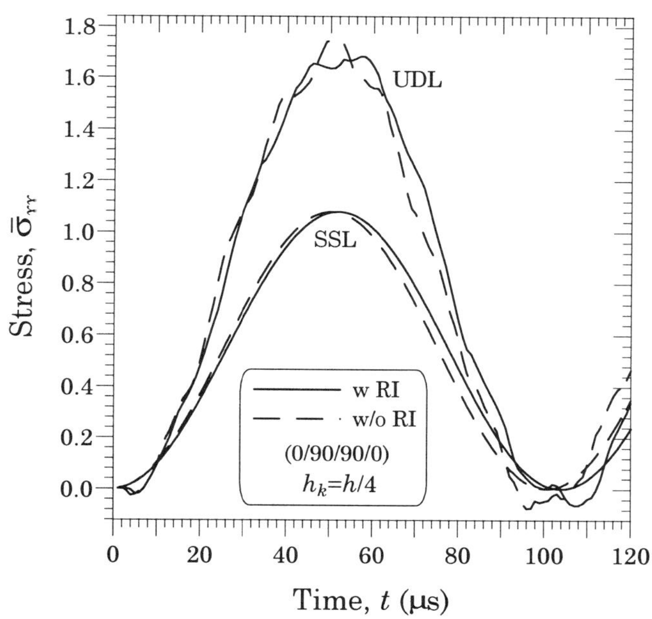

Figure 7.9–2: Nondimensionalized maximum normal stress ($\bar{\sigma}_{xx}$) versus time for a simply supported symmetric cross-ply $(0/90)_s$ laminate.

Exercises

7.1 Determine the displacement field of a simply supported plate strip under a concentrated (line) load F_0 at the center using the Navier solution method.

7.2 Derive the Navier solution of a simply supported rectangular plate under the following temperature distribution

$$T(x, y, z) = T_0(x, y) + zT_1(x, y)$$

where T_0 and T_1 are known functions of x and y only, which can be expanded in double Fourier series in the same way as the mechanical loading $q(x, y)$.

7.3 Derive the expressions for transverse shear stresses from 3-D equilibrium equations when the plate is subjected to the temperature distribution of the form given in Problem 7.2. Assume that T_0 and T_1 can be expanded in double sine series.

7.4 Determine the constants A_n, B_n, C_n, and D_n in the Lévy solution (7.3–15) of a specially orthotropic rectangular plate with simply supported edges at $y = 0, b$ and $x = a$, and clamped at $x = 0$. Assume uniformly distributed transverse load.

7.5 Determine the constants A_n, B_n, C_n, and D_n in the Lévy solution (7.3–15) of a specially orthotropic rectangular plate with simply supported edges at $y = 0, b$, clamped at $x = 0$, and free at $x = a$. Assume uniformly distributed transverse load. The boundary conditions for the free edge are [see Eq. (5.3–33b)]

$$M_{xx} = 0, \quad V_x \equiv Q_x + \frac{\partial M_{xy}}{\partial y} = 0$$

These boundary conditions can be expressed in terms of the transverse deflection as

$$D_{11}\frac{\partial^2 w_0}{\partial x^2} + D_{12}\frac{\partial^2 w_0}{\partial y^2} = 0, \quad D_{11}\frac{\partial^3 w_0}{\partial x^3} + D_{12}\frac{\partial^3 w_0}{\partial y^2 \partial x} + D_{66}\frac{\partial^3 w_0}{\partial y^2 \partial x} = 0$$

7.6 Determine the constants A_n, B_n, C_n, and D_n in the Lévy solution (7.3–15) of a specially orthotropic rectangular plate with simply supported edges at $y = 0, b$ and $x = a$, and free at $x = 0, a$. Assume uniformly distributed transverse load.

7.7 Use the following one–parameter Ritz approximation to determine the deflection of a simply supported rectangular plate:

$$w_0(x, y) \approx c_1 x(a - x)y(b - y)$$

Ans: The parameter c_1 is given by $c_1 = \frac{q_0 a^2 b^2}{36 R_{11}}$ with

$$R_{11} = 2\left[\frac{1}{15}D_{11}b^4 + \frac{1}{9}(D_{12} + 2D_{66})a^2b^2 + \frac{1}{15}D_{22}\right]$$

7.8 Show that the one–parameter Galerkin solution with the algebraic functions in Eq. (7.4–6) is also given by Eq. (7.4–11).

7.9 Use one–parameter Rayleigh–Ritz approximation of the form

$$w_0(x,y) \approx c_{11}\left(1-\cos\frac{2\pi x}{a}\right)\left(1-\cos\frac{2\pi y}{b}\right)$$

to determine the deflection of a rectangular plate with clamped edges and subjected to uniformly distributed transverse load.

7.10 Verify the result in Eq. (7.4–14).

7.11 Verify the result in Eq. (7.6–16).

7.12 Determine the critical buckling load of a rectangular orthotropic plate simply supported on edges $y = 0, b$ and clamped on edges $x = 0, a$ using the one–parameter Rayleigh–Ritz approximation of the form

$$w_0(x,y) = c_{11}\left(1-\cos\frac{2\pi x}{a}\right)\sin\frac{n\pi y}{b}$$

7.13 Determine the critical buckling load of a rectangular orthotropic plate simply supported on edges $y = 0, b$ and $x = 0$, and clamped on edge $x = a$ using the one–parameter Rayleigh–Ritz approximation of the form

$$w_0(x,y) = c_{11}\left(\frac{x}{a}\right)\sin\frac{n\pi y}{b}$$

7.14 Determine the transient response of simply supported specially orthotropic plate under transverse loading (a) $q(x,y,t) = q_0 H(t - t_0)$ and (b) $q(x,y,t) = q_0\delta(t - t_0)$, where $H(t)$ denotes the Heavyside step function and $\delta(t)$ is the Dirac delta function.

7.15 Solve Eq. (7.9–10) using the Laplace transform method.

References for Additional Reading

1. Reddy, J. N., *Energy and Variational Methods in Applied Mechanics,* John Wiley, New York, 1984.
2. Reddy, J. N. (ed.), *Mechanics of Composite Materials. Selected Works of Nicholas J. Pagano*, Kluwer, Netherlands, 1994.
3. Ashton, J. E. and Whitney, J. M., *Theory of Laminated Plates,* Technomic, Stamford, Connecticut, 1970.
4. Whitney, J. M., *Structural Analysis of Laminated Anisotropic Plates,* Technomic, Lancaster, Pennsylvania, 1987.
5. Szilard, R., *Theory and Analysis of Plates, Classical and Numerical Methods,* Prentice–Hall, Englewood Cliffs, New Jersey, 1974.

6. Timoshenko, S. P. and Woinowsky–Krieger, S., *Theory of Plates and Shells,* McGraw–Hill, New York, 1959.
7. Lekhnitskii, S. G., *Anisotropic Plates,* Translated from Russian by S. W. Tsai and T. Cheron, Gordon and Breach, Newark, New Jersey, 1968.
8. Hearman, R. F. S., "The Frequency of Flexural Vibration of Rectangular Orthotropic Plates with Clamped or Supported Edges," *Journal of Applied Mechanics,* **26**(4), 537–540 (1959).
9. Young, D., "Vibrations of Rectangular Plates by the Ritz Method," *Journal of Applied Mechanics,* **17**, 448–453 (1950).
10. Pipes, L. A. and Harvill, L. R., *Applied Mathematics for Engineers and Physicists,* Third Edition, McGraw–Hill, New York, 1970.
11. Franklin, J. N., *Matrix Theory,* Prentice–Hall, Englewood Cliffs, New Jersey, 1968.
12. Goldberg, J. L. and Schwartz, A. J., *Systems of Ordinary Differential Equations, An Introduction,* Harper and Row, New York, 1972.
13. Reddy, J. N., Khdeir, A. A., and Librescu, L., "Lévy Type Solutions for Symmetrically Laminated Rectangular Plates Using First-Order Shear Deformation Theory," *Journal of Applied Mechanics*, **54**, 740–742 (1987).
14. Khdeir, A. A., Reddy, J. N., and Librescu, L., "Analytical Solution of a Refined Shear Deformation Theory for Rectangular Composite Plates," *International Journal of Solids and Structures*, **23**, 1447–1463 (1987).
15. Khdeir, A. A. and Librescu, L., "Analysis of Symmetric Cross-Ply Laminated Elastic Plates Using a Higher-Order Theory: Part I: Stress and Displacement," *Composite Structures*, **9**, 189–213 (1988).
16. Khdeir, A. A. and Librescu, L., "Analysis of Symmetric Cross-Ply Laminated Elastic Plates Using a Higher-Order Theory: Part II: Buckling and Free Vibration," *Composite Structures*, **9**, 259–277 (1988).
17. Khdeir, A. A., "Free Vibration and Buckling of Symmetric Cross-Ply Laminated Plates by an Exact Method," *Journal of Sound and Vibration*, **126**(3), 447–461 (1988).
18. Wang, J. T.–S., "On the Solution of Plates of Composite Materials," *Journal of Composite Materials,* **3**, 590–592 (1969).
19. Ashton, J. E. and Waddoups, M. E., "Analysis of Anisotropic Plates," *Journal of Composite Materials,* **3**, 148–165 (1969).
20. Ashton, J. E., "Analysis of Anisotropic Plates II," *Journal of Composite Materials,* **3**, 470–479 (1969).
21. Reddy, J. N., "On the Solutions to Forced Motions of Rectangular Composite Plates," *Journal of Applied Mechanics,* **49**, 403–408 (1982).
22. Timoshenko, S. P. and Gere, J. P., *Theory of Elastic Stability,* Second Edition, McGraw–Hill, New York, 1959.

Chapter Eight

Analytical Solutions of Rectangular Laminates Using CLPT

8.1 Governing Equations in Terms of Displacements

In this chapter analytical solutions of antisymmetric cross-ply and angle-ply laminated plates based on the classical laminate theory are developed. The Navier method, the Lévy method with the state–space approach, and the Rayleigh–Ritz method are used, depending on the boundary conditions. In all cases considered in this chapter, the von Kármán nonlinear terms in the strain–displacement relations are omitted and the constitutive relations are assumed to be linear. Before we begin with the derivation of the exact solutions, it is useful to express the governing equations in terms of the generalized displacements of the theory.

The linear equations of motion of the classical laminated plate theory (CLPT) can be obtained from Eqs. (5.3–45)-(5.3–47) by setting the nonlinear terms to zero:

$$
\begin{aligned}
&A_{11}\frac{\partial^2 u_0}{\partial x^2} + 2A_{16}\frac{\partial^2 u_0}{\partial x\partial y} + A_{66}\frac{\partial^2 u_0}{\partial y^2} + A_{16}\frac{\partial^2 v_0}{\partial x^2} + (A_{12}+A_{66})\frac{\partial^2 v_0}{\partial x\partial y} + A_{26}\frac{\partial^2 v_0}{\partial y^2}\\
&-\left[B_{11}\frac{\partial^3 w_0}{\partial x^3} + 3B_{16}\frac{\partial^3 w_0}{\partial x^2\partial y} + (B_{12}+2B_{66})\frac{\partial^3 w_0}{\partial x\partial y^2} + B_{26}\frac{\partial^3 w_0}{\partial y^3}\right]\\
&-\left(\frac{\partial N_{xx}^T}{\partial x} + \frac{\partial N_{xy}^T}{\partial y}\right) = I_0\ddot{u}_0 - I_1\frac{\partial \ddot{w}_0}{\partial x}
\end{aligned}
\qquad (8.2-1)
$$

$$
\begin{aligned}
&A_{16}\frac{\partial^2 u_0}{\partial x^2} + (A_{12}+A_{66})\frac{\partial^2 u_0}{\partial x\partial y} + A_{26}\frac{\partial^2 u_0}{\partial y^2} + A_{66}\frac{\partial^2 v_0}{\partial x^2} + 2A_{26}\frac{\partial^2 v_0}{\partial x\partial y} + A_{22}\frac{\partial^2 v_0}{\partial y^2}\\
&-\left[B_{16}\frac{\partial^3 w_0}{\partial x^3} + (B_{12}+2B_{66})\frac{\partial^3 w_0}{\partial x^2\partial y} + 3B_{26}\frac{\partial^3 w_0}{\partial x\partial y^2} + B_{22}\frac{\partial^3 w_0}{\partial y^3}\right]\\
&-\left(\frac{\partial N_{xy}^T}{\partial x} + \frac{\partial N_{yy}^T}{\partial y}\right) = I_0\ddot{v}_0 - I_1\frac{\partial \ddot{w}_0}{\partial y}
\end{aligned}
\qquad (8.2-2)
$$

$$B_{11}\frac{\partial^3 u_0}{\partial x^3} + 3B_{16}\frac{\partial^3 u_0}{\partial x^2\partial y} + (B_{12}+2B_{66})\frac{\partial^3 u_0}{\partial x\partial y^2} + B_{26}\frac{\partial^3 u_0}{\partial y^3}$$
$$+ B_{16}\frac{\partial^3 v_0}{\partial x^3} + (B_{12}+2B_{66})\frac{\partial^3 v_0}{\partial x^2\partial y} + 3B_{26}\frac{\partial^3 v_0}{\partial x\partial y^2} + B_{22}\frac{\partial^3 v_0}{\partial y^3}$$
$$- \left[D_{11}\frac{\partial^4 w_0}{\partial x^4} + 4D_{16}\frac{\partial^4 w_0}{\partial x^3\partial y} + 2(D_{12}+2D_{66})\frac{\partial^4 w_0}{\partial x^2\partial y^2} + 4D_{26}\frac{\partial^4 w_0}{\partial x\partial y^3} + D_{22}\frac{\partial^4 w_0}{\partial y^4}\right]$$
$$- \left(\frac{\partial^2 M_{xx}^T}{\partial x^2} + 2\frac{\partial^2 M_{xy}^T}{\partial x\partial y} + \frac{\partial^2 M_{yy}^T}{\partial y^2}\right) + \hat{N}_{xx}\frac{\partial^2 w_0}{\partial x^2} + 2\hat{N}_{xy}\frac{\partial^2 w_0}{\partial x\partial y} + \hat{N}_{yy}\frac{\partial^2 w_0}{\partial y^2}$$
$$+ q = I_1\left(\frac{\partial \ddot{u}_0}{\partial x} + \frac{\partial \ddot{v}_0}{\partial y}\right) + I_0\ddot{w}_0 - I_2\left(\frac{\partial^2 \ddot{w}_0}{\partial x^2} + \frac{\partial^2 \ddot{w}_0}{\partial y^2}\right) \tag{8.2-3}$$

where N^T and M^T denote thermal resultants defined in Eq. (5.3–41), and $\hat{N}_{xx}$, $\hat{N}_{xy}$, and $\hat{N}_{yy}$ denote the applied edge forces (see Figure 8.2–1).

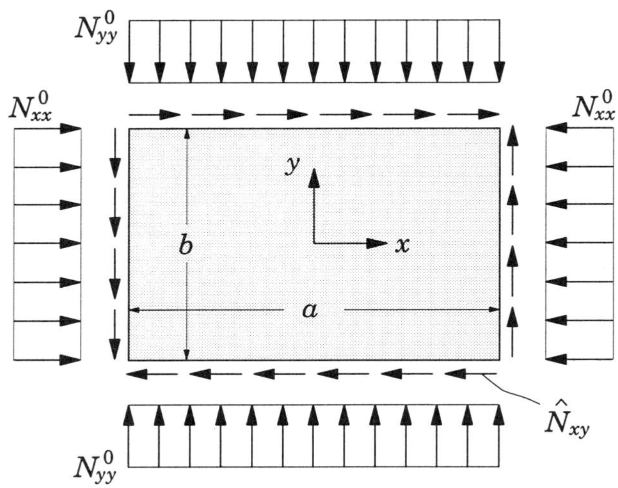

Figure 8.2–1: A plate with applied edge forces ($\hat{N}_{xx} = -N_{xx}^0$, $\hat{N}_{yy} = -N_{yy}^0$).

Equations (1)–(3) can be cast in differential operator form as[1]

$$\begin{bmatrix} c_{11} & c_{12} & c_{13} \\ c_{12} & c_{22} & c_{23} \\ c_{13} & c_{23} & c_{33} \end{bmatrix} \begin{Bmatrix} u_0 \\ v_0 \\ w_0 \end{Bmatrix} + \begin{bmatrix} m_{11} & 0 & m_{13} \\ 0 & m_{22} & m_{23} \\ m_{13} & m_{23} & m_{33} \end{bmatrix} \begin{Bmatrix} \ddot{u}_0 \\ \ddot{v}_0 \\ \ddot{w}_0 \end{Bmatrix} = \begin{Bmatrix} 0 \\ 0 \\ q \end{Bmatrix} + \begin{Bmatrix} f_1^T \\ f_2^T \\ f_3^T \end{Bmatrix} \tag{8.2-4}$$

[1] In order to make the coefficient matrices $[C]$ and $[M]$ symmetric, the third equation is multiplied with a negative sign.

where the coefficients c_{ij} and m_{ij} are

$$
\begin{aligned}
c_{11} &= A_{11}d_x^2 + 2A_{16}d_xd_y + A_{66}d_y^2\\
c_{12} &= A_{16}d_x^2 + (A_{12}+A_{66})d_xd_y + A_{26}d_y^2\\
c_{13} &= -\left[B_{11}d_x^3 + 3B_{16}d_x^2d_y + (B_{12}+2B_{66})d_xd_y^2 + B_{26}d_y^3\right]\\
c_{22} &= A_{66}d_x^2 + 2A_{26}d_xd_y + A_{22}d_y^2\\
c_{23} &= -\left[B_{16}d_x^3 + (B_{12}+2B_{66})d_x^2d_y + 3B_{26}d_xd_y^2 + B_{22}d_y^3\right]\\
c_{33} &= D_{11}d_x^4 + 4D_{16}d_x^3d_y + 2(D_{12}+2D_{66})d_x^2d_y^2 + 4D_{26}d_xd_y^3 + D_{22}d_y^4\\
&\quad - \left[\hat{N}_{xx}d_x^2 + 2\hat{N}_{xy}d_xd_y + \hat{N}_{yy}d_y^2\right]\\
m_{11} &= -I_0d_t^2\ ,\quad m_{13} = I_1d_xd_t^2\ ,\quad m_{22} = -I_0d_t^2\\
m_{23} &= I_1d_yd_t^2\ ,\quad m_{33} = I_0d_t^2 - I_2d_t^2(d_x^2+d_y^2)
\end{aligned}
\tag{8.2-5a}
$$

d_x^i, d_y^i, and d_t^i denote the differential operators

$$
d_x^i = \frac{\partial^i}{\partial x^i}\ ,\quad d_y^i = \frac{\partial^i}{\partial y^i}\ ,\quad d_t^i = \frac{\partial^i}{\partial t^i} \qquad (i=1,2,3,4) \tag{8.2-5b}
$$

and f_i^T are equivalent generalized thermal forces

$$
\begin{aligned}
f_1^T &= \frac{\partial N_{xx}^T}{\partial x} + \frac{\partial N_{xy}^T}{\partial y}\\
f_2^T &= \frac{\partial N_{xy}^T}{\partial x} + \frac{\partial N_{yy}^T}{\partial y}\\
f_3^T &= -\left(\frac{\partial^2 M_{xx}^T}{\partial x^2} + 2\frac{\partial^2 M_{xy}^T}{\partial y\partial x} + \frac{\partial^2 M_{yy}^T}{\partial y^2}\right)
\end{aligned}
\tag{8.2-5c}
$$

Note that the thermal forces and moments, $(N_{xx}^T, N_{yy}^T, N_{xy}^T)$ and $(M_{xx}^T, M_{yy}^T, M_{xy}^T)$, are known in terms of the temperature distribution and material coefficients as defined in Eqs. (5.3–41a,b).

8.2 Admissible Boundary Conditions for the Navier Solutions

In the Navier method the generalized displacements are expanded in a double trigonometric series in terms of unknown parameters (see Sections 6.6 and 7.2–2). The choice of the functions in the series is restricted to those which satisfy the boundary conditions of the problem. Substitution of the displacement expansions into the governing equations should result in a unique, invertible, set of algebraic equations among the parameters of the expansion. Otherwise, the Navier solution cannot be developed for the problem.

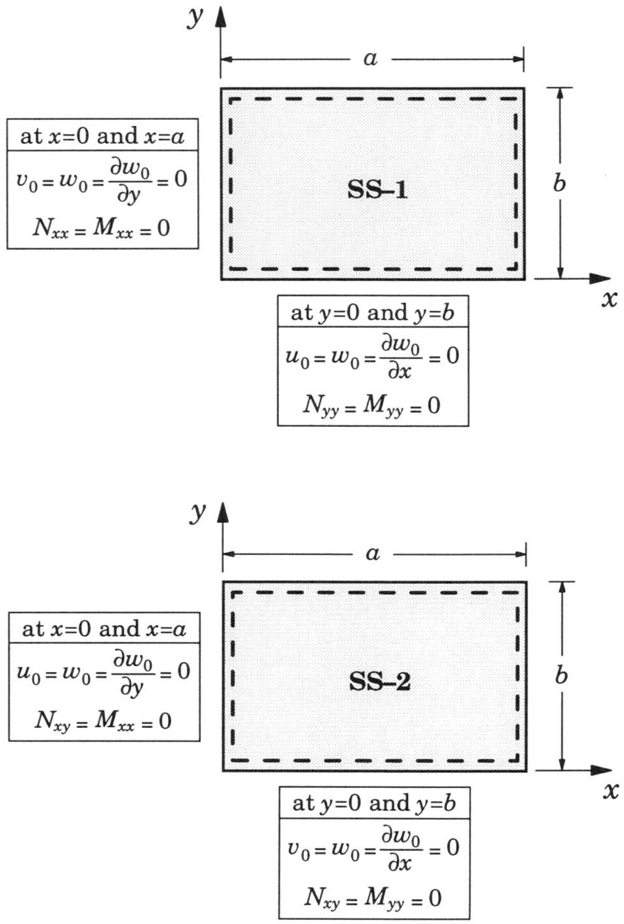

Figure 8.2–2: Two types of simply supported boundary conditions, SS–1 and SS–2, used in the analytical solutions of rectangular laminated plates in extension and bending.

The Navier solutions can be developed for rectangular laminates with two sets of simply supported boundary conditions. Even for these boundary conditions, not all laminates permit the Navier solution. We will determine which lamination schemes permit such solutions. The geometry, laminate coordinate system, and the two types of simply supported boundary conditions are shown in Figure 8.2–2. The two types of boundary conditions are given below.

Simply Supported–1 (SS–1):

$$u(x,0,0,t)=0,\quad u(x,b,0,t)=0,\quad v(0,y,0,t)=0,\quad v(a,y,0,t)=0$$

$$\frac{\partial u}{\partial z}(x,0,0,t)=0,\quad \frac{\partial u}{\partial z}(x,b,0,t)=0,\quad \frac{\partial v}{\partial z}(0,y,0,t)=0,\quad \frac{\partial v}{\partial z}(a,y,0,t)=0$$

$$w(x,0,0,t)=0,\quad w(x,b,0,t)=0,\quad w(0,y,0,t)=0,\quad w(a,y,0,t)=0 \quad (8.2-6a)$$

$$\int_{-\frac{h}{2}}^{\frac{h}{2}}\sigma_{xx}(0,y,z,t)\;dz=0,\quad \int_{-\frac{h}{2}}^{\frac{h}{2}}\sigma_{xx}(a,y,z,t)\;dz=0$$

$$\int_{-\frac{h}{2}}^{\frac{h}{2}}\sigma_{yy}(x,0,z,t)\;dz=0,\quad \int_{-\frac{h}{2}}^{\frac{h}{2}}\sigma_{yy}(x,b,z,t)\;dz=0$$

$$\int_{-\frac{h}{2}}^{\frac{h}{2}}\sigma_{xx}(0,y,z,t)\;z\;dz=0,\quad \int_{-\frac{h}{2}}^{\frac{h}{2}}\sigma_{xx}(a,y,z,t)\;z\;dz=0$$

$$\int_{-\frac{h}{2}}^{\frac{h}{2}}\sigma_{yy}(x,0,z,t)\;z\;dz=0,\quad \int_{-\frac{h}{2}}^{\frac{h}{2}}\sigma_{yy}(x,b,z,t)\;z\;dz=0 \quad (8.2-6b)$$

Simply Supported–2 (SS–2):

$$u(0,y,0,t)=0,\quad u(a,y,0,t)=0,\quad v(x,0,0,t)=0,\quad v(x,b,0,t)=0$$

$$\frac{\partial u}{\partial z}(x,0,0,t)=0,\quad \frac{\partial u}{\partial z}(x,b,0,t)=0,\quad \frac{\partial v}{\partial z}(0,y,0,t)=0,\quad \frac{\partial v}{\partial z}(a,y,0,t)=0$$

$$w(x,0,0,t)=0,\quad w(x,b,0,t)=0,\quad w(0,y,0,t)=0,\quad w(a,y,0,t)=0 \quad (8.2-7a)$$

$$\int_{-\frac{h}{2}}^{\frac{h}{2}}\sigma_{xy}(0,y,z,t)\;dz=0,\quad \int_{-\frac{h}{2}}^{\frac{h}{2}}\sigma_{xy}(a,y,z,t)\;dz=0$$

$$\int_{-\frac{h}{2}}^{\frac{h}{2}}\sigma_{xy}(x,0,z,t)\;dz=0,\quad \int_{-\frac{h}{2}}^{\frac{h}{2}}\sigma_{xy}(x,b,z,t)\;dz=0$$

$$\int_{-\frac{h}{2}}^{\frac{h}{2}}\sigma_{xx}(0,y,z,t)\;z\;dz=0,\quad \int_{-\frac{h}{2}}^{\frac{h}{2}}\sigma_{xx}(a,y,z,t)\;z\;dz=0$$

$$\int_{-\frac{h}{2}}^{\frac{h}{2}}\sigma_{yy}(x,0,z,t)\;z\;dz=0,\quad \int_{-\frac{h}{2}}^{\frac{h}{2}}\sigma_{yy}(x,b,z,t)\;z\;dz=0 \quad (8.2-7b)$$

Here a and b denote the in-plane dimensions along the x and y directions of a rectangular laminate. The origin of the coordinate system is taken at the lower left corner of the midplane (see Figure 8.2–2).

As will be shown in the following sections, the Navier solutions using SS–1 boundary conditions can be obtained only for laminates whose stiffnesses $A_{16}, A_{26}, B_{16}, B_{26}, D_{16}, D_{26}$, and A_{45} are zero. Thus, the Navier solutions for the SS–1 boundary conditions can be developed for laminates with a single generally orthotropic layer, symmetrically laminated plates with multiple specially orthotropic layers, and antisymmetric cross-ply laminated plates. Similarly, the Navier solutions using SS–2 boundary conditions can be obtained only for laminates whose stiffnesses $A_{16}, A_{26}, B_{11}, B_{12}, B_{22}, B_{66}, D_{16}, D_{26}$, and A_{45} are zero, *i.e.*, for laminates with a single generally orthotropic layer, symmetrically laminated plates with multiple specially orthotropic layers, and antisymmetric angle-ply laminated plates.

8.3 Navier Solutions of Antisymmetric Cross-Ply Laminates

8.3–1 Boundary Conditions

The conditions in Eqs. (8.2–6) imply, in view of Eq. (5.3–2), the following SS–1 boundary conditions on the generalized displacements and stress resultants of the classical laminate theory:

$$u_0(x,0,t) = 0, \quad \frac{\partial w_0}{\partial x}(x,0,t) = 0, \quad u_0(x,b,t) = 0, \quad \frac{\partial w_0}{\partial x}(x,b,t) = 0$$

$$v_0(0,y,t) = 0, \quad \frac{\partial w_0}{\partial y}(0,y,t) = 0, \quad v_0(a,y,t) = 0, \quad \frac{\partial w_0}{\partial y}(a,y,t) = 0$$

$$w_0(x,0,t) = 0, \quad w_0(x,b,t) = 0, \quad w_0(0,y,t) = 0, \quad w_0(a,y,t) = 0 \qquad (8.3-1)$$

$$N_{xx}(0,y,t) = 0, \quad N_{xx}(a,y,t) = 0, \quad N_{yy}(x,0,t) = 0, \quad N_{yy}(x,b,t) = 0$$

$$M_{xx}(0,y,t) = 0, \quad M_{xx}(a,y,t) = 0, \quad M_{yy}(x,0,t) = 0, \quad M_{yy}(x,b,t) = 0 \quad (8.3-2)$$

The displacement boundary conditions in (1) are satisfied by assuming the following form of the displacements

$$u_0(x,y,t) = \sum_{n=1}^{\infty}\sum_{m=1}^{\infty} U_{mn}(t)\cos\alpha x\ \sin\beta y \qquad (8.3-3a)$$

$$v_0(x,y,t) = \sum_{n=1}^{\infty}\sum_{m=1}^{\infty} V_{mn}(t)\sin\alpha x\ \cos\beta y \qquad (8.3-3b)$$

$$w_0(x,y,t) = \sum_{n=1}^{\infty}\sum_{m=1}^{\infty} W_{mn}(t)\sin\alpha x\ \sin\beta y \qquad (8.3-3c)$$

where $\alpha = m\pi/a$ and $\beta = n\pi/b$ and (U_{mn}, V_{mn}, W_{mn}) are coefficients to be determined. To see if the boundary conditions (2) on the stress resultants are also satisfied, we substitute expansions (3) into the expressions for N_{xx}, N_{yy} N_{xy}, M_{xx}, M_{yy}, and M_{xy} given in Eqs. (5.3–43) and (5.3–44):

$$\begin{aligned}
N_{xx} &= A_{11}\frac{\partial u_0}{\partial x} + A_{12}\frac{\partial v_0}{\partial y} + A_{16}\left(\frac{\partial u_0}{\partial y} + \frac{\partial v_0}{\partial x}\right) \\
&\quad - B_{11}\frac{\partial^2 w_0}{\partial x^2} - B_{12}\frac{\partial^2 w_0}{\partial y^2} - 2B_{16}\frac{\partial^2 w_0}{\partial x \partial y} - N_{xx}^T \\
&= \sum_{n=1}^{\infty}\sum_{m=1}^{\infty}\left[-\alpha A_{11}U_{mn} - \beta A_{12}V_{mn} + \left(B_{11}\alpha^2 + B_{12}\beta^2\right)W_{mn}\right] f(x,y) \\
&\quad + \sum_{n=1}^{\infty}\sum_{m=1}^{\infty}\left[A_{16}\left(\beta U_{mn} + \alpha V_{mn}\right) - 2\alpha\beta B_{16}W_{mn}\right] g(x,y) - N_{xx}^T \qquad (8.3-4a)
\end{aligned}$$

$$\begin{aligned}
N_{yy} &= A_{12}\frac{\partial u_0}{\partial x} + A_{22}\frac{\partial v_0}{\partial y} + A_{26}\left(\frac{\partial u_0}{\partial y} + \frac{\partial v_0}{\partial x}\right) \\
&\quad - B_{12}\frac{\partial^2 w_0}{\partial x^2} - B_{22}\frac{\partial^2 w_0}{\partial y^2} - 2B_{26}\frac{\partial^2 w_0}{\partial x \partial y} - N_{yy}^T \\
&= \sum_{n=1}^{\infty}\sum_{m=1}^{\infty}\left[-\alpha A_{12}U_{mn} - \beta A_{22}V_{mn} + \left(B_{12}\alpha^2 + B_{22}\beta^2\right)W_{mn}\right] f(x,y) \\
&\quad + \sum_{n=1}^{\infty}\sum_{m=1}^{\infty}\left[A_{26}\left(\beta U_{mn} + \alpha V_{mn}\right) - 2\alpha\beta B_{26}W_{mn}\right] g(x,y) - N_{yy}^T \qquad (8.3-4b)
\end{aligned}$$

$$\begin{aligned}
N_{xy} &= A_{16}\frac{\partial u_0}{\partial x} + A_{26}\frac{\partial v_0}{\partial y} + A_{66}\left(\frac{\partial u_0}{\partial y} + \frac{\partial v_0}{\partial x}\right) \\
&\quad - B_{16}\frac{\partial^2 w_0}{\partial x^2} - B_{26}\frac{\partial^2 w_0}{\partial y^2} - 2B_{66}\frac{\partial^2 w_0}{\partial x \partial y} - N_{xy}^T \\
&= \sum_{n=1}^{\infty}\sum_{m=1}^{\infty}\left[-\alpha A_{16}U_{mn} - \beta A_{26}V_{mn} + \left(B_{16}\alpha^2 + B_{26}\beta^2\right)W_{mn}\right] f(x,y) \\
&\quad + \sum_{n=1}^{\infty}\sum_{m=1}^{\infty}\left[A_{66}\left(\beta U_{mn} + \alpha V_{mn}\right) - 2\alpha\beta B_{66}W_{mn}\right] g(x,y) - N_{xy}^T \qquad (8.3-4c)
\end{aligned}$$

$$\begin{aligned}
M_{xx} &= B_{11}\frac{\partial u_0}{\partial x} + B_{12}\frac{\partial v_0}{\partial y} + B_{16}\left(\frac{\partial u_0}{\partial y} + \frac{\partial v_0}{\partial x}\right) \\
&\quad - D_{11}\frac{\partial^2 w_0}{\partial x^2} - D_{12}\frac{\partial^2 w_0}{\partial y^2} - 2D_{16}\frac{\partial^2 w_0}{\partial x \partial y} - M_{xx}^T \\
&= \sum_{n=1}^{\infty}\sum_{m=1}^{\infty}\left[-\alpha B_{11}U_{mn} - \beta B_{12}V_{mn} + \left(D_{11}\alpha^2 + D_{12}\beta^2\right)W_{mn}\right] f(x,y) \\
&\quad + \sum_{n=1}^{\infty}\sum_{m=1}^{\infty}\left[B_{16}\left(\beta U_{mn} + \alpha V_{mn}\right) - 2\alpha\beta D_{16}W_{mn}\right] g(x,y) - M_{xx}^T \qquad (8.3-5a)
\end{aligned}$$

$$
\begin{aligned}
M_{yy} &= B_{12}\frac{\partial u_0}{\partial x} + B_{22}\frac{\partial v_0}{\partial y} + B_{26}\left(\frac{\partial u_0}{\partial y} + \frac{\partial v_0}{\partial x}\right) \\
&- D_{12}\frac{\partial^2 w_0}{\partial x^2} - D_{22}\frac{\partial^2 w_0}{\partial y^2} - 2D_{26}\frac{\partial^2 w_0}{\partial x \partial y} - M_{yy}^T \\
&= \sum_{n=1}^{\infty}\sum_{m=1}^{\infty}\left[-\alpha B_{12}U_{mn} - \beta B_{22}V_{mn} + \left(D_{12}\alpha^2 + D_{22}\beta^2\right)W_{mn}\right] f(x,y) \\
&+ \sum_{n=1}^{\infty}\sum_{m=1}^{\infty}\left[B_{26}\left(\beta U_{mn} + \alpha V_{mn}\right) - 2\alpha\beta D_{26}W_{mn}\right] g(x,y) - M_{yy}^T \qquad (8.3-5b)
\end{aligned}
$$

$$
\begin{aligned}
M_{xy} &= B_{16}\frac{\partial u_0}{\partial x} + B_{26}\frac{\partial v_0}{\partial y} + B_{66}\left(\frac{\partial u_0}{\partial y} + \frac{\partial v_0}{\partial x}\right) \\
&- D_{16}\frac{\partial^2 w_0}{\partial x^2} - D_{26}\frac{\partial^2 w_0}{\partial y^2} - 2D_{66}\frac{\partial^2 w_0}{\partial x \partial y} - M_{xy}^T \\
&= \sum_{n=1}^{\infty}\sum_{m=1}^{\infty}\left[-\alpha B_{16}U_{mn} - \beta B_{26}V_{mn} + \left(D_{16}\alpha^2 + D_{26}\beta^2\right)W_{mn}\right] f(x,y) \\
&+ \sum_{n=1}^{\infty}\sum_{m=1}^{\infty}\left[B_{66}\left(\beta U_{mn} + \alpha V_{mn}\right) - 2\alpha\beta D_{66}W_{mn}\right] g(x,y) - M_{xy}^T \qquad (8.3-5c)
\end{aligned}
$$

where

$$
f(x,y) = \sin\alpha x \ \sin\beta y \ , \quad g(x,y) = \cos\alpha x \ \cos\beta y \qquad (8.3-6)
$$

Note that the boundary conditions in Eq. (2) on the stress and moment resultants N_{xx}, N_{yy}, M_{xx}, and M_{yy} can be satisfied only if the laminate stiffnesses $A_{16}, A_{26}, B_{16}, B_{26}, D_{16}, D_{26}$ are zero (because $g(x,y) \neq 0$ for $x = 0, a$ or $y = 0, b$); in addition, the thermal force and moment resultants must satisfy the boundary conditions in Eq. (2). Thus, the Navier solutions for rectangular laminated plates with SS–1 boundary conditions may exist only when the laminate stacking sequences are such that

$$
A_{16} = A_{26} = B_{16} = B_{26} = D_{16} = D_{26} = 0 \qquad (8.3-7a)
$$

$$
N_{xx}^T(0,y,t) = N_{xx}^T(a,y,t) = N_{yy}^T(x,0,t) = N_{yy}^T(x,b,t) = 0 \qquad (8.3-7b)
$$

$$
M_{xx}^T(0,y,t) = M_{xx}^T(a,y,t) = M_{yy}^T(x,0,t) = M_{yy}^T(x,b,t) = 0 \qquad (8.3-7c)
$$

From Section 5.5, it follows that plates with a single generally orthotropic layer, symmetrically laminated plates with multiple specially orthotropic layers, and antisymmetric cross-ply laminated plates, which include the former cases as special cases, admit the Navier solutions for the SS–1 boundary conditions. Although the Navier solutions cannot be developed for general laminates, *i.e.*, with no restrictions on laminate stiffnesses, approximate or numerical solutions may be constructed, as shown later in this chapter or in subsequent chapters.

8.3–2 Solution

Substitution of Eqs. (3) and (7) into Eqs. (8.2–1)–(8.2–3) yields

$$\sum_{n=1}^{\infty}\sum_{m=1}^{\infty}\Big[-\left(A_{11}\alpha^2+A_{66}\beta^2\right)U_{mn}-\left(A_{12}+A_{66}\right)\alpha\beta V_{mn}$$
$$+\left(B_{11}\alpha^3+\tilde{B}_{12}\alpha\beta^2\right)W_{mn}-I_0\ddot{U}_{mn}+I_1\alpha\ddot{W}_{mn}\Big]\cos\alpha x\sin\beta y$$
$$=\left(\frac{\partial N_{xx}^T}{\partial x}+\frac{\partial N_{xy}^T}{\partial y}\right) \qquad (8.3-8a)$$

$$\sum_{n=1}^{\infty}\sum_{m=1}^{\infty}\Big[-\left(A_{12}+A_{66}\right)\alpha\beta U_{mn}-\left(A_{66}\alpha^2+A_{22}\beta^2\right)V_{mn}$$
$$+\left(\tilde{B}_{12}\alpha^2\beta+B_{22}\beta^3\right)W_{mn}-I_0\ddot{V}_{mn}+I_1\beta\ddot{W}_{mn}\Big]\sin\alpha x\cos\beta y$$
$$=\left(\frac{\partial N_{xy}^T}{\partial x}+\frac{\partial N_{yy}^T}{\partial y}\right) \qquad (8.3-8b)$$

$$\sum_{n=1}^{\infty}\sum_{m=1}^{\infty}\Big[\left(B_{11}\alpha^3+\tilde{B}_{12}\alpha\beta^2\right)U_{mn}+\left(\tilde{B}_{12}\alpha^2\beta+B_{22}\beta^3\right)V_{mn}$$
$$-\left(D_{11}\alpha^4+2\tilde{D}_{12}\alpha^2\beta^2+D_{22}\beta^4\right)W_{mn}$$
$$-\left(\alpha^2\hat{N}_{xx}+\beta^2\hat{N}_{yy}\right)W_{mn}$$
$$+I_1\alpha\ddot{U}_{mn}+I_1\beta\ddot{V}_{mn}-\left(I_0+I_2(\alpha^2+\beta^2)\right)\ddot{W}_{mn}\Big]\sin\alpha x\ \sin\beta y$$
$$=\left(\frac{\partial^2 M_{xx}^T}{\partial x^2}+2\frac{\partial^2 M_{xy}^T}{\partial x\partial y}+\frac{\partial^2 M_{yy}^T}{\partial y^2}\right)-q(x,y) \qquad (8.3-8c)$$

where

$$\tilde{B}_{12}=B_{12}+2B_{66},\quad \tilde{D}_{12}=D_{12}+2D_{66} \qquad (8.3-9)$$

Note that the edge shear force $\hat{N}_{xy}$ is necessarily zero. In addition, for the class of laminates admissible here, inertia I_1 is zero.

An examination of Eqs. (8) shows that the mechanical force q and thermal forces and moments of Eqs. (8a–c) should also be expanded in the same form as their counterparts on the left side of the equality in Eqs. (8a,b,c). For example, the left side of Eq. (8c) has the form

$$\sum_{n=1}^{\infty}\sum_{m=1}^{\infty}C_{mn}^3\sin\alpha x\ \sin\beta y$$

where C^3_{mn} is the expression in the square bracket of Eq. (8c). Hence, the right side of Eq. (8c), which consists of the transverse load and thermal moments, should also be expanded in double sine series. Thus, $q(x, y, t)$ must be expanded as

$$q(x,y,t) = \sum_{n=1}^{\infty}\sum_{m=1}^{\infty} Q_{mn}(t) \sin \alpha x \ \sin \beta y \qquad (8.3-10a)$$

$$Q_{mn}(t) = \frac{4}{ab}\int_0^a\int_0^b q(x,y,t) \ \sin\frac{m\pi x}{a} \ \sin\frac{n\pi y}{b} \ dxdy \qquad (8.3-10b)$$

Since the thermal moments M^T_{xx}, M^T_{xy}, and M^T_{yy} are defined in terms of the same temperature increment $\Delta T(x, y, t)$ but they enter Eq. (8c) with different derivatives, it is expected that not all of them will contribute to the solution. Suppose that the temperature increment is expanded as

$$\Delta T(x,y,z,t) = \sum_{n=1}^{\infty}\sum_{m=1}^{\infty} T_{mn}(z,t) \sin \alpha x \ \sin \beta y \qquad (8.3-11a)$$

$$T_{mn}(z,t) = \frac{4}{ab}\int_0^a\int_0^b \Delta T(x,y,z,t) \ \sin\frac{m\pi x}{a} \ \sin\frac{n\pi y}{b} \ dxdy \qquad (8.3-11b)$$

Then we have from Eq. (5.3–41a,b)

$$\begin{Bmatrix} N^T_{xx} \\ N^T_{yy} \\ N^T_{xy} \end{Bmatrix} = \sum_{n=1}^{\infty}\sum_{m=1}^{\infty} \begin{Bmatrix} N^1_{mn}(t) \\ N^2_{mn}(t) \\ N^6_{mn}(t) \end{Bmatrix} \sin \alpha x \ \sin \beta y \qquad (8.3-12a)$$

$$\begin{Bmatrix} M^T_{xx} \\ M^T_{yy} \\ M^T_{xy} \end{Bmatrix} = \sum_{n=1}^{\infty}\sum_{m=1}^{\infty} \begin{Bmatrix} M^1_{mn}(t) \\ M^2_{mn}(t) \\ M^6_{mn}(t) \end{Bmatrix} \sin \alpha x \ \sin \beta y \qquad (8.3-12b)$$

$$\{N_{mn}(t)\} = \sum_{k=1}^{N}\int_{z_k}^{z_{k+1}} [\bar{Q}]^{(k)}\{\bar{\alpha}\}^{(k)} \ T_{mn}(z,t) \ dz \qquad (8.3-13a)$$

$$\{M_{mn}(t)\} = \sum_{k=1}^{N}\int_{z_k}^{z_{k+1}} [\bar{Q}]^{(k)}\{\bar{\alpha}\}^{(k)} \ T_{mn}(z,t) \ z \ dz \qquad (8.3-13b)$$

Thus we have

$$\begin{aligned} f^T_1 &= \frac{\partial N^T_{xx}}{\partial x} + \frac{\partial N^T_{xy}}{\partial y} \\ &= \sum_{n=1}^{\infty}\sum_{m=1}^{\infty}\left[\alpha N^1_{mn}(t)\cos\alpha x \ \sin\beta y + \beta N^6_{mn}(t)\sin\alpha x \ \cos\beta y\right] \end{aligned}$$

$$f^T_2 = \frac{\partial N^T_{xy}}{\partial x} + \frac{\partial N^T_{yy}}{\partial y}$$

$$= \sum_{n=1}^{\infty}\sum_{m=1}^{\infty}\left[\alpha N_{mn}^{6}(t)\cos\alpha x\ \sin\beta y + \beta N_{mn}^{2}(t)\sin\alpha x\ \cos\beta y\right]$$

$$f_3^T = \frac{\partial^2 M_{xx}^T}{\partial x^2} + 2\frac{\partial^2 M_{xy}^T}{\partial y \partial x} + \frac{\partial^2 M_{yy}^T}{\partial y^2}$$

$$= \sum_{n=1}^{\infty}\sum_{m=1}^{\infty}\left[-\left(\alpha^2 M_{mn}^{1}(t) + \beta^2 M_{mn}^{2}(t)\right)\sin\alpha x\ \sin\beta y \right.$$

$$\left. + 2\alpha\beta M_{mn}^{6}(t)\cos\alpha x\ \cos\beta y\right] \qquad (8.3-14)$$

This particular expansion of temperature distribution necessarily requires that N_{mn}^6 and M_{mn}^6 be zero because they must be of the form [see Eqs. (8a,b,c)]

$$f_1^T = \sum_{n=1}^{\infty}\sum_{m=1}^{\infty} f_{mn}^{1}(t)\cos\alpha x\ \sin\beta y$$

$$f_2^T = \sum_{n=1}^{\infty}\sum_{m=1}^{\infty} f_{mn}^{2}(t)\sin\alpha x\ \cos\beta y$$

$$f_3^T = \sum_{n=1}^{\infty}\sum_{m=1}^{\infty} f_{mn}^{3}(t)\sin\alpha x\ \sin\beta y \qquad (8.3-15)$$

This requirement places a restriction on the lamination scheme in order for the Navier solution to exist in the presence of temperature changes. The lamination scheme be such that

$$\sum_{k=1}^{N}\int_{z_k}^{z_{k+1}} [\bar{Q}]^{(k)}\{\bar{\alpha}\}^{(k)} T_{mn}(z,t)\ dz = \begin{Bmatrix} N_{mn}^1 \\ N_{mn}^2 \\ 0 \end{Bmatrix} \qquad (8.3-16a)$$

$$\sum_{k=1}^{N}\int_{z_k}^{z_{k+1}} [\bar{Q}]^{(k)}\{\bar{\alpha}\}^{(k)} T_{mn}(z,t)\ z\ dz = \begin{Bmatrix} M_{mn}^1 \\ M_{mn}^2 \\ 0 \end{Bmatrix} \qquad (8.3-16b)$$

For single-layer plates with a generally orthotropic layer, symmetrically laminated plates with multiple specially orthotropic layers, and antisymmetric cross-ply laminated plates, the conditions in Eq. (16a,b) are automatically satisfied. In order to include N_{mn}^6 and M_{mn}^6, the temperature distribution should be expanded in a double cosine series. Then N_{mn}^1, N_{mn}^2, M_{mn}^1, and M_{mn}^2 must be zero.

Substituting the expansions (10a) and (14) with $N_{mn}^6 = M_{mn}^6 = 0$ into Eq. (8), we obtain expressions of the form

$$\sum_{n=1}^{\infty}\sum_{m=1}^{\infty} a_{mn}(t)\cos\alpha x\ \sin\beta y = 0$$

$$\sum_{n=1}^{\infty}\sum_{m=1}^{\infty} b_{mn}(t)\sin\alpha x\ \cos\beta y = 0$$
$$\sum_{n=1}^{\infty}\sum_{m=1}^{\infty} c_{mn}(t)\sin\alpha x\ \sin\beta y = 0 \qquad (8.3-17)$$

where a_{mn}, b_{mn}, and c_{mn} are coefficients whose explicit form will be given shortly. Since Eqs. (17) must hold for any m, n, x, and y, it follows that $a_{mn} = 0$, $b_{mn} = 0$, and $c_{mn} = 0$ for every m and n. The explicit forms of the coefficients a_{mn}, b_{mn}, and c_{mn} are given by

$$\begin{aligned}
a_{mn} \equiv & -\left(A_{11}\alpha^2 + A_{66}\beta^2\right)U_{mn} - (A_{12}+A_{66})\,\alpha\beta V_{mn} \\
& + \left(B_{11}\alpha^3 + \tilde{B}_{12}\alpha\beta^2\right)W_{mn} - \alpha N^1_{mn} - I_0\ddot{U}_{mn} + I_1\alpha\ddot{W}_{mn} = 0 \quad (8.3-18a)\\
b_{mn} \equiv & -(A_{12}+A_{66})\,\alpha\beta U_{mn} - \left(A_{66}\alpha^2 + A_{22}\beta^2\right)V_{mn} \\
& + \left(\tilde{B}_{12}\alpha^2\beta + B_{22}\beta^3\right)W_{mn} - \beta N^2_{mn} - I_0\ddot{V}_{mn} + I_1\beta\ddot{W}_{mn} = 0 \quad (8.3-18b)\\
c_{mn} \equiv & \left(B_{11}\alpha^3 + \tilde{B}_{12}\alpha\beta^2\right)U_{mn} + \left(\tilde{B}_{12}\alpha^2\beta + B_{22}\beta^3\right)V_{mn} \\
& - \left(D_{11}\alpha^4 + 2\tilde{D}_{12}\alpha^2\beta^2 + D_{22}\beta^4\right)W_{mn} + Q_{mn} \\
& + \alpha^2 M^1_{mn} + \beta^2 M^2_{mn} - \left(\alpha^2\hat{N}_{xx} + \beta^2\hat{N}_{yy}\right)W_{mn} \\
& + I_1\alpha\ddot{U}_{mn} + I_1\beta\ddot{V}_{mn} - \left(I_0 + I_2(\alpha^2+\beta^2)\right)\ddot{W}_{mn} = 0 \quad (8.3-18c)
\end{aligned}$$

or in matrix form

$$\begin{bmatrix} \hat{c}_{11} & \hat{c}_{12} & \hat{c}_{13} \\ \hat{c}_{12} & \hat{c}_{22} & \hat{c}_{23} \\ \hat{c}_{13} & \hat{c}_{23} & \hat{c}_{33}+\tilde{s}_{33} \end{bmatrix} \begin{Bmatrix} U_{mn} \\ V_{mn} \\ W_{mn} \end{Bmatrix} + \begin{bmatrix} \hat{m}_{11} & 0 & -I_1\alpha \\ 0 & \hat{m}_{22} & -I_1\beta \\ -I_1\alpha & -I_1\beta & \hat{m}_{33} \end{bmatrix} \begin{Bmatrix} \ddot{U}_{mn} \\ \ddot{V}_{mn} \\ \ddot{W}_{mn} \end{Bmatrix}$$
$$= \begin{Bmatrix} 0 \\ 0 \\ Q_{mn} \end{Bmatrix} + \begin{Bmatrix} -\alpha N^1_{mn} \\ -\beta N^2_{mn} \\ \alpha^2 M^1_{mn} + \beta^2 M^2_{mn} \end{Bmatrix} \qquad (8.3-19)$$

where $\hat{c}_{ij}$ and $\hat{m}_{ij}$ are

$$\begin{aligned}
\hat{c}_{11} &= (A_{11}\alpha^2 + A_{66}\beta^2), \quad \hat{c}_{12} = (A_{12}+A_{66})\alpha\beta \\
\hat{c}_{13} &= -B_{11}\alpha^3 - (B_{12}+2B_{66})\alpha\beta^2, \quad \hat{c}_{22} = (A_{66}\alpha^2 + A_{22}\beta^2) \\
\hat{c}_{23} &= -(B_{12}+2B_{66})\alpha^2\beta - B_{22}\beta^3 \\
\hat{c}_{33} &= D_{11}\alpha^4 + 2(D_{12}+2D_{66})\alpha^2\beta^2 + D_{22}\beta^4 \\
\hat{m}_{11} &= I_0, \ \hat{m}_{22} = I_0, \ \hat{m}_{33} = \left(I_0 + I_2(\alpha^2+\beta^2)\right) \\
\tilde{s}_{33} &= \alpha^2\hat{N}_{xx} + \beta^2\hat{N}_{yy} \qquad (8.3-20)
\end{aligned}$$

and $\alpha = m\pi/a$ and $\beta = n\pi/b$.

Equations (19) provide three second-order differential equations in time among the three variables U_{mn}, V_{mn}, and W_{mn} for any fixed values of m and n. For transient (*i.e.*, time dependent) response, the differential equations in time can be solved either exactly or approximately.

8.3–3 Bending

The static solution can be obtained by solving the algebraic equations resulting from Eqs. (19) by setting the time derivative terms to zero:

$$\begin{bmatrix} \hat{c}_{11} & \hat{c}_{12} & \hat{c}_{13} \\ \hat{c}_{12} & \hat{c}_{22} & \hat{c}_{23} \\ \hat{c}_{13} & \hat{c}_{23} & \hat{c}_{33}+\tilde{s}_{33} \end{bmatrix} \begin{Bmatrix} U_{mn} \\ V_{mn} \\ W_{mn} \end{Bmatrix} = \begin{Bmatrix} 0 \\ 0 \\ Q_{mn} \end{Bmatrix} + \begin{Bmatrix} -\alpha N^1_{mn} \\ -\beta N^2_{mn} \\ \alpha^2 M^1_{mn} + \beta^2 M^2_{mn} \end{Bmatrix} \quad (8.3-21)$$

which can be solved for the coefficients U_{mn}, V_{mn}, and W_{mn} in terms of the coefficients Q_{mn}, N^1_{mn}, N^2_{mn}, M^1_{mn}, and M^2_{mn}. Then the final solution is given by Eqs. (3a,b,c).

Equations (21) can be solved using Cramer's rule or by the method of *static condensation*. The latter allows the elimination of a selected set of variables and retains a desired set of variables. The method is useful in later discussions of this book, and therefore it is described here. First, the column of unknowns is subdivided into two parts, $\{\Delta^1\}$ and $\{\Delta^2\}$, according to what is to be eliminated and what is to be retained. Suppose that we wish to eliminate the coefficients associated with the in-plane displacements and retain those associated with the transverse deflection. Then Eq. (21) can be written as

$$\begin{bmatrix} [K^{11}] & [K^{12}] \\ [K^{12}]^T & [K^{22}] \end{bmatrix} \begin{Bmatrix} \{\Delta^1\} \\ \{\Delta^2\} \end{Bmatrix} = \begin{Bmatrix} \{F^1\} \\ \{F^2\} \end{Bmatrix} \quad (8.3-22)$$

where

$$\{\Delta^1\} = \begin{Bmatrix} U_{mn} \\ V_{mn} \end{Bmatrix}, \ \{\Delta^2\} = W_{mn} \quad (8.3-23a)$$

$$[K^{11}] = \begin{bmatrix} \hat{c}_{11} & \hat{c}_{12} \\ \hat{c}_{12} & \hat{c}_{22} \end{bmatrix}, \ [K^{12}] = \begin{Bmatrix} \hat{c}_{13} \\ \hat{c}_{23} \end{Bmatrix}, \ [K^{22}] = \hat{c}_{33} + \tilde{s}_{33}$$

$$\{F^1\} = \begin{Bmatrix} -\alpha N^1_{mn} \\ -\beta N^2_{mn} \end{Bmatrix}, \ \{F^2\} = Q_{mn} + \alpha^2 M^1_{mn} + \beta^2 M^2_{mn} \quad (8.3-23b)$$

Equation (22) represents a pair of two matrix equations:

$$[K^{11}]\{\Delta^1\} + [K^{12}]\{\Delta^2\} = \{F^1\} \quad (8.3-24a)$$

$$[K^{12}]^T\{\Delta^1\} + [K^{22}]\{\Delta^2\} = \{F^2\} \quad (8.3-24b)$$

Solving Eq. (24a) for $\{\Delta^1\}$, which is to be eliminated, we obtain

$$\{\Delta^1\} = [K^{11}]^{-1}\left(\{F^1\} - [K^{12}]\{\Delta^2\}\right) \quad (8.3-25)$$

Then substituting the result into (24b), we obtain

$$\left([K^{22}] - [K^{12}]^T[K^{11}]^{-1}[K^{12}]\right)\{\Delta^2\} = \{F^2\} - [K^{12}]^T[K^{11}]^{-1}\{F^1\} \quad (8.3-26a)$$

or

$$[\hat{K}^{22}]\{\Delta^2\} = \{\hat{F}^2\} \quad (8.3-26b)$$

where

$$\begin{aligned}[\hat{K}^{22}] =& [K^{22}] - [K^{12}]^T[K^{11}]^{-1}[K^{12}] \\ \{\hat{F}^2\} =& \{F^2\} - [K^{12}]^T[K^{11}]^{-1}\{F^1\}\end{aligned} \quad (8.3-26c)$$

This procedure of eliminating (or condensing out) a subset of unknowns is known in structural mechanics as the method of *static condensation*. The calculation involves solving for $\{\Delta^2\}$ first, and then, if desired, solving for $\{\Delta^1\}$ next using Eq. (25).

Using the definitions (23) in (26) we obtain (when $\tilde{s}_{33} = 0$)

$$\begin{aligned}W_{mn} =& \frac{1}{a_{mn}}\left[Q_{mn} + \alpha^2 M^1_{mn} + \beta^2 M^2_{mn} - \frac{a_1}{a_0}\left(\alpha N^1_{mn}\right) - \frac{a_2}{a_0}\left(\beta N^2_{mn}\right)\right] \\ U_{mn} =& \frac{1}{a_0}\left[a_1 W_{mn} - \alpha N^1_{mn}\hat{c}_{22} + \beta N^2_{mn}\hat{c}_{12}\right] \\ V_{mn} =& \frac{1}{a_0}\left[a_2 W_{mn} + \alpha N^1_{mn}\hat{c}_{12} - \beta N^2_{mn}\hat{c}_{11}\right]\end{aligned} \quad (8.3-27a)$$

where

$$\begin{aligned}a_{mn} =& \hat{c}_{33} + \hat{c}_{13}\frac{a_1}{a_0} + \hat{c}_{23}\frac{a_2}{a_0} \\ a_0 =& \hat{c}_{11}\hat{c}_{22} - \hat{c}_{12}\hat{c}_{12} \\ a_1 =& \hat{c}_{12}\hat{c}_{23} - \hat{c}_{13}\hat{c}_{22} \\ a_2 =& \hat{c}_{13}\hat{c}_{12} - \hat{c}_{11}\hat{c}_{23}\end{aligned} \quad (8.3-27b)$$

Solution of Eq. (27a) for each $m, n = 1, 2, \ldots$ gives (U_{mn}, V_{mn}, W_{mn}), which can then be used to compute the solution (u_0, v_0, w_0) from Eq. (3). If there are no thermal loads, the solution becomes

$$W_{mn} = \frac{Q_{mn}}{a_{mn}}\ , \ U_{mn} = \frac{a_1 Q_{mn}}{a_0 a_{mn}}\ , \ V_{mn} = \frac{a_2 Q_{mn}}{a_0 a_{mn}} \quad (8.3-28)$$

Note that for antisymmetric cross-ply laminates, $B_{66} = 0$ and the coefficients in Eq. (27a) can be simplified.

8.3–4 Determination of Stresses

The in-plane stresses in each layer of a laminate are calculated from constitutive relations in Eq. (5.3–12a). Accounting for only mechanical and thermal effects, we obtain

$$\left\{\begin{array}{c}\sigma_{xx}\\ \sigma_{yy}\\ \sigma_{xy}\end{array}\right\}^{(k)} = \begin{bmatrix}\bar{Q}_{11} & \bar{Q}_{12} & \bar{Q}_{16}\\ \bar{Q}_{12} & \bar{Q}_{22} & \bar{Q}_{26}\\ \bar{Q}_{16} & \bar{Q}_{26} & \bar{Q}_{66}\end{bmatrix}^{(k)} \left(\left\{\begin{array}{c}\varepsilon_{xx}\\ \varepsilon_{yy}\\ 2\varepsilon_{xy}\end{array}\right\} - \left\{\begin{array}{c}\alpha_{xx}\\ \alpha_{yy}\\ 2\alpha_{xy}\end{array}\right\}\Delta T\right)$$

$$= \begin{bmatrix}\bar{Q}_{11} & \bar{Q}_{12} & \bar{Q}_{16}\\ \bar{Q}_{12} & \bar{Q}_{22} & \bar{Q}_{26}\\ \bar{Q}_{16} & \bar{Q}_{26} & \bar{Q}_{66}\end{bmatrix}^{k} \left\{\begin{array}{c}\varepsilon^0_{xx} - \alpha_{xx}\,T_0\\ \varepsilon^0_{yy} - \alpha_{yy}\,T_0\\ 2(\varepsilon^0_{xy} - \alpha_{xy}\,T_0)\end{array}\right\}$$

$$+\, z\begin{bmatrix}\bar{Q}_{11} & \bar{Q}_{12} & \bar{Q}_{16}\\ \bar{Q}_{12} & \bar{Q}_{22} & \bar{Q}_{26}\\ \bar{Q}_{16} & \bar{Q}_{26} & \bar{Q}_{66}\end{bmatrix}^{k} \left\{\begin{array}{c}\varepsilon^1_{xx} - \alpha_{xx}T_1\\ \varepsilon^1_{yy} - \alpha_{yy}T_1\\ 2(\varepsilon^1_{xy} - \alpha_{xy}T_1)\end{array}\right\} \qquad (8.3-29a)$$

where temperature increment ΔT is assumed to be of the form

$$\Delta T(x,y,z,t) = \sum_{m=1}^{\infty}\sum_{n=1}^{\infty}\left(T^0_{mn} + zT^1_{mn}\right)\sin\alpha x \sin\beta y \qquad (8.3-29b)$$

The in-plane stresses of a simply supported (SS–1) cross-ply laminate (*i.e.*, when $\bar{Q}_{16} = \bar{Q}_{26} = 0$ and $\alpha_{xy} = 0$) are then given by

$$\left\{\begin{array}{c}\sigma_{xx}\\ \sigma_{yy}\\ \sigma_{xy}\end{array}\right\}^{(k)} = \sum_{m=1}^{\infty}\sum_{n=1}^{\infty}\begin{bmatrix}\bar{Q}_{11} & \bar{Q}_{12} & 0\\ \bar{Q}_{12} & \bar{Q}_{22} & 0\\ 0 & 0 & \bar{Q}_{66}\end{bmatrix}^{(k)} \left\{\begin{array}{c}(R^{xx}_{mn} + zS^{xx}_{mn})\sin\alpha x\sin\beta y\\ (R^{yy}_{mn} + zS^{yy}_{mn})\sin\alpha x\sin\beta y\\ (R^{xy}_{mn} + zS^{xy}_{mn})\cos\alpha x\cos\beta y\end{array}\right\}$$

$$(8.3-30a)$$

$$\left\{\begin{array}{c}R^{xx}_{mn}\\ R^{yy}_{mn}\\ R^{xy}_{mn}\end{array}\right\} = \left\{\begin{array}{c}-\alpha U_{mn} - \alpha_{xx}T^0_{mn}\\ -\beta V_{mn} - \alpha_{yy}T^0_{mn}\\ \beta U_{mn} + \alpha V_{mn}\end{array}\right\},\quad \left\{\begin{array}{c}S^{xx}_{mn}\\ S^{yy}_{mn}\\ S^{xy}_{mn}\end{array}\right\} = \left\{\begin{array}{c}\alpha^2 W_{mn} - \alpha_{xx}T^1_{mn}\\ \beta^2 W_{mn} - \alpha_{yy}T^1_{mn}\\ -2\alpha\beta W_{mn}\end{array}\right\} \qquad (8.3-30b)$$

The maximum normal stresses occur at $(x,y,z) = (a/2, b/2, -h/2)$, and the shear stress is maximum at $(x,y,z) = (a,b,-h/2)$.

The transverse stresses in a laminate can be determined using the 3–D equilibrium equations [see Eqs. (7.2–14)] for any $-h/2 \le z \le z_{k+1}$

$$\sigma^{(k)}_{xz} = -\int_{z_k}^{z}\left(\frac{\partial\sigma^{(k)}_{xx}}{\partial x} + \frac{\partial\sigma^{(k)}_{xy}}{\partial y}\right)dz + C^{(k)}_1(x,y)$$

$$\sigma^{(k)}_{yz} = -\int_{z_k}^{z}\left(\frac{\partial\sigma^{(k)}_{xy}}{\partial x} + \frac{\partial\sigma^{(k)}_{yy}}{\partial y}\right)dz + C^{(k)}_2(x,y)$$

$$\sigma^{(k)}_{zz} = -\int_{z_k}^{z}\left(\frac{\partial\sigma^{(k)}_{xz}}{\partial x} + \frac{\partial\sigma^{(k)}_{yz}}{\partial y}\right)dz + C^{(k)}_3(x,y) \qquad (8.3-31)$$

where $\sigma_{xx}^{(k)}, \sigma_{xy}^{(k)}$, and $\sigma_{yy}^{(k)}$ are known from Eq. (29), and $C_i^{(k)}$ are functions to be determined using the boundary conditions

$$\sigma_{xz}(x,y,-h/2) = \sigma_{yz}(x,y,-h/2) = \sigma_{zz}(x,y,-h/2) = 0 \tag{8.3-32}$$

and continuity of stresses at layer interfaces:

$$\begin{aligned}
\sigma_{xz}^{(k)}(x,y,z_{k+1}) &= \sigma_{xz}^{(k+1)}(x,y,z_{k+1})\\
\sigma_{yz}^{(k)}(x,y,z_{k+1}) &= \sigma_{yz}^{(k+1)}(x,y,z_{k+1})\\
\sigma_{zz}^{(k)}(x,y,z_{k+1}) &= \sigma_{zz}^{(k+1)}(x,y,z_{k+1})
\end{aligned} \tag{8.3-33}$$

Substituting for the in-plane stresses from Eq. (30) into Eq. (31) and integrating with respect to z, we obtain

$$\left\{\begin{matrix}\sigma_{xz}\\ \sigma_{yz}\end{matrix}\right\}^{(k)} = (z-z_k)\left\{\begin{matrix}A_{xz}\\ A_{yz}\end{matrix}\right\}^{(k)} + \left(\frac{z^2-z_k^2}{2}\right)\left\{\begin{matrix}B_{xz}\\ B_{yz}\end{matrix}\right\}^{(k)} + \left\{\begin{matrix}C_1\\ C_2\end{matrix}\right\}^{(k)} \tag{8.3-34a}$$

where

$$\begin{aligned}
A_{xz}^{(k)} &= -\Big[Q_{11}^{(k)}(\varepsilon_{xx}^0-\alpha_{xx}T_0)_{,x} + Q_{12}^{(k)}(\varepsilon_{yy}^0-\alpha_{yy}T_0)_{,x} + 2Q_{16}^{(k)}(\varepsilon_{xy}^0-\alpha_{xy}T_0)_{,x}\\
&\quad + Q_{16}^{(k)}(\varepsilon_{xx}^0-\alpha_{xx}T_0)_{,y} + Q_{26}^{(k)}(\varepsilon_{yy}^0-\alpha_{yy}T_0)_{,y} + 2Q_{66}^{(k)}(\varepsilon_{xy}^0-\alpha_{xy}T_0)_{,y}\Big]\\
B_{xz}^{(k)} &= -\Big[Q_{11}^{(k)}(\varepsilon_{xx}^1-\alpha_{xx}T_1)_{,x} + Q_{12}^{(k)}(\varepsilon_{yy}^1-\alpha_{yy}T_1)_{,x} + 2Q_{16}^{(k)}(\varepsilon_{xy}^1-\alpha_{xy}T_1)_{,x}\\
&\quad + Q_{16}^{(k)}(\varepsilon_{xx}^1-\alpha_{xx}T_1)_{,y} + Q_{26}^{(k)}(\varepsilon_{yy}^1-\alpha_{yy}T_1)_{,y} + 2Q_{66}^{(k)}(\varepsilon_{xy}^1-\alpha_{xy}T_1)_{,y}\Big]\\
A_{yz}^{(k)} &= -\Big[Q_{16}^{(k)}(\varepsilon_{xx}^0-\alpha_{xx}T_0)_{,x} + Q_{26}^{(k)}(\varepsilon_{yy}^0-\alpha_{yy}T_0)_{,x} + 2Q_{66}^{(k)}(\varepsilon_{xy}^0-\alpha_{xy}T_0)_{,x}\\
&\quad + Q_{12}^{(k)}(\varepsilon_{xx}^0-\alpha_{xx}T_0)_{,y} + Q_{22}^{(k)}(\varepsilon_{yy}^0-\alpha_{yy}T_0)_{,y} + 2Q_{26}^{(k)}(\varepsilon_{xy}^0-\alpha_{xy}T_0)_{,y}\Big]\\
B_{yz}^{(k)} &= -\Big[Q_{16}^{(k)}(\varepsilon_{xx}^1-\alpha_{xx}T_1)_{,x} + Q_{26}^{(k)}(\varepsilon_{yy}^1-\alpha_{yy}T_1)_{,x} + 2Q_{66}^{(k)}(\varepsilon_{xy}^1-\alpha_{xy}T_1)_{,x}\\
&\quad + Q_{12}^{(k)}(\varepsilon_{xx}^1-\alpha_{xx}T_1)_{,y} + Q_{22}^{(k)}(\varepsilon_{yy}^1-\alpha_{yy}T_1)_{,y} + 2Q_{26}^{(k)}(\varepsilon_{xy}^1-\alpha_{xy}T_1)_{,y}\Big]
\end{aligned} \tag{8.3-34b}$$

and a comma followed by x or y denotes differentiation. The boundary conditions (32) yield $C_1^{(1)} = C_2^{(1)} = 0$. The interface continuity conditions (33) result in

$$C_1^{(k+1)} = \sigma_{xz}^{(k)}(x,y,z_{k+1}),\quad C_2^{(k+1)} = \sigma_{yz}^{(k)}(x,y,z_{k+1}) \tag{8.3-35}$$

for $k = 1, 2, \cdots, n$, where n denotes the number of layers.

Substituting the displacement and temperature expansions in Eqs. (3) and (30c), the following expressions for the interlaminar transverse shear stresses are obtained

$$
\begin{aligned}
\sigma_{xz}^{(k)}(x,y,z) &= \sum_{m=1}^{\infty}\sum_{n=1}^{\infty}\left[(z-z_k)\mathcal{A}_{mn}^{(k)} + \frac{1}{2}(z^2-z_k^2)\mathcal{B}_{mn}^{(k)}\right]\cos\alpha x\sin\beta y \\
&\quad + \sigma_{xz}^{(k-1)}(x,y,z_k) \\
\sigma_{yz}^{(k)}(x,y,z) &= \sum_{m=1}^{\infty}\sum_{n=1}^{\infty}\left[(z-z_k)\mathcal{C}_{mn}^{(k)} + \frac{1}{2}(z^2-z_k^2)\mathcal{D}_{mn}^{(k)}\right]\sin\alpha x\cos\beta y \\
&\quad + \sigma_{yz}^{(k-1)}(x,y,z_k) \\
\sigma_{xz}^{(0)}(x,y,z_1) &= 0, \quad \sigma_{yz}^{(0)}(x,y,z_1) = 0 \qquad (8.3-36a)
\end{aligned}
$$

where

$$
\begin{aligned}
\mathcal{A}_{mn}^{(k)} &= \left(\alpha^2\bar{Q}_{11}^{(k)} + \beta^2\bar{Q}_{66}^{(k)}\right)U_{mn} + \alpha\beta\left(\bar{Q}_{12}^{(k)} + \bar{Q}_{66}^{(k)}\right)V_{mn} \\
&\quad + \alpha\left(\bar{Q}_{11}\alpha_{xx} + \bar{Q}_{12}\alpha_{yy}\right)T_{mn}^{0} \\
\mathcal{B}_{mn}^{(k)} &= -\left[\alpha^3\bar{Q}_{11}^{(k)} + \alpha\beta^2\left(\bar{Q}_{12}^{(k)} + 2\bar{Q}_{66}^{(k)}\right)\right]W_{mn} \\
&\quad + \alpha\left(\bar{Q}_{11}\alpha_{xx} + \bar{Q}_{12}\alpha_{yy}\right)T_{mn}^{1} \\
\mathcal{C}_{mn}^{(k)} &= \alpha\beta\left(\bar{Q}_{12}^{(k)} + \bar{Q}_{66}^{(k)}\right)U_{mn} + \left(\alpha^2\bar{Q}_{66}^{(k)} + \beta^2\bar{Q}_{22}^{(k)}\right)V_{mn} \\
&\quad + \beta\left(\bar{Q}_{12}\alpha_{xx} + \bar{Q}_{22}\alpha_{yy}\right)T_{mn}^{0} \\
\mathcal{D}_{mn}^{(k)} &= -\left[\alpha^2\beta\left(\bar{Q}_{12}^{(k)} + 2\bar{Q}_{66}^{(k)}\right) + \beta^3\bar{Q}_{22}^{(k)}\right]W_{mn} \\
&\quad + \beta\left(\bar{Q}_{12}\alpha_{xx} + \bar{Q}_{22}\alpha_{yy}\right)T_{mn}^{1} \qquad (8.3-36b)
\end{aligned}
$$

The maximum of σ_{xz} occurs at $(x,y,z) = (0,b/2,0)$, and the maximum of σ_{yz} occurs at $(x,y,z) = (a/2,0,0)$.

The transverse normal stress is given by

$$
\begin{aligned}
\sigma_{zz}^{(k)}(x,y,z) = -\sum_{m=1}^{\infty}\sum_{n=1}^{\infty}\Bigg[&\left(\frac{z^2}{2} - zz_k\right)\left(\alpha\mathcal{A}_{mn}^{(k)} + \beta\mathcal{C}_{mn}^{(k)}\right) \\
&+ \left(\frac{z^3}{2} - \frac{zz_k^2}{2}\right)\left(\alpha\mathcal{B}_{mn}^{(k)} + \beta\mathcal{D}_{mn}^{(k)}\right) \\
&+ \mathcal{E}_{mn}^{(k)}\Bigg]\sin\alpha x\sin\beta y \qquad (8.3-37)
\end{aligned}
$$

where the functions $\mathcal{E}_{mn}^{(k)}$ are determined using the boundary and interface continuity conditions. We obtain

$$
\mathcal{E}_{mn}^{(1)} = -\frac{h^2}{4}\left(\alpha\mathcal{A}_{mn}^{(1)} + \beta\mathcal{C}_{mn}^{(1)}\right) + \frac{h^3}{24}\left(\alpha\mathcal{B}_{mn}^{(1)} + \beta\mathcal{D}_{mn}^{(1)}\right)
$$

$$\begin{aligned}\mathcal{E}_{mn}^{(k+1)} &= \mathcal{E}_{mn}^{(k)} - z_{k+1}\left(\frac{z_{k+1}}{2} - z_k\right)\left(\alpha\mathcal{A}_{mn}^{(k)} + \beta\mathcal{C}_{mn}^{(k)}\right)\\ &\quad - z_{k+1}\left(\frac{z_{k+1}^2}{6} - \frac{z_k^2}{2}\right)\left(\alpha\mathcal{B}_{mn}^{(k)} + \beta\mathcal{D}_{mn}^{(k)}\right)\\ &\quad - \frac{z_{k+1}^2}{2}\left(\alpha\mathcal{A}_{mn}^{(k+1)} + \beta\mathcal{C}_{mn}^{(k+1)}\right)\\ &\quad - \frac{z_{k+1}^3}{3}\left(\alpha\mathcal{B}_{mn}^{(k+1)} + \beta\mathcal{D}_{mn}^{(k+1)}\right)\end{aligned} \tag{8.3-38}$$

for $k = 1, 2, \cdots, n$.

The bending moments can be calculated from Eqs. (5a,b,c)

$$\begin{aligned}\begin{Bmatrix} M_{xx} \\ M_{yy} \\ M_{xy} \end{Bmatrix} &= \sum_{m=1}^{\infty}\sum_{n=1}^{\infty}\begin{bmatrix} B_{11} & B_{12} & 0 \\ B_{12} & B_{22} & 0 \\ 0 & 0 & B_{66} \end{bmatrix}\begin{Bmatrix} R_{mn}^{xx}\sin\alpha x\sin\beta y \\ R_{mn}^{yy}\sin\alpha x\sin\beta y \\ R_{mn}^{xy}\cos\alpha x\cos\beta y \end{Bmatrix}\\ &+ \sum_{m=1}^{\infty}\sum_{n=1}^{\infty}\begin{bmatrix} D_{11} & D_{12} & 0 \\ D_{12} & D_{22} & 0 \\ 0 & 0 & D_{66} \end{bmatrix}\begin{Bmatrix} S_{mn}^{xx}\sin\alpha x\sin\beta y \\ S_{mn}^{yy}\sin\alpha x\sin\beta y \\ S_{mn}^{xy}\cos\alpha x\cos\beta y \end{Bmatrix}\end{aligned} \tag{8.3-39}$$

For the definitions of R's and S's, see Eq. (30b). The maximum values of M_{xx} and M_{yy} occur at $(x, y) = (a/2, b/2)$, and the maximum of M_{xy} occurs at $(x, y) = (0, 0)$.

The following nondimensionalizations are used in presenting the numerical results:

$$\bar{w} = w_0(a/2, b/2)\left(\frac{E_2h^3}{a^4q_0}\right); \quad \bar{\sigma}_{xx} = \sigma_{xx}(a/2, b/2, z)\left(\frac{h^2}{a^2q_0}\right) \tag{8.3-40a}$$

$$\bar{\sigma}_{yy} = \sigma_{yy}(a/2, b/2, z)\left(\frac{h^2}{a^2q_0}\right); \quad \bar{\sigma}_{xy} = \sigma_{xy}(a, b, z)\left(\frac{h^2}{a^2q_0}\right) \tag{8.3-40b}$$

Tables 8.3–1 and 8.3–2 contain nondimensionalized deflections and stresses for antisymmetric cross-ply laminates (all B_{ij}, except for $B_{11} = -B_{22}$, are zero, and $A_{16} = A_{26} = D_{16} = D_{26} = 0$) under various types of mechanical loads. For comparison, results of symmetric laminates are also included. From these results it is clear that, for the same laminate thickness, antisymmetric cross-ply laminates with four or more layers are more desirable than two-layer laminates due to the reduction in deflections as well as stresses. The difference between two-layer and four- or eight-layer laminates is due to the bending–stretching coupling coefficients B_{ij}, which are dominant in the case of two layers. As the number of layers increase, the B_{ij} decrease and the laminate essentially behaves like a specially orthotropic plate.

Table 8.3–1: Transverse deflections and stresses in composite laminates subjected to sinusoidally distributed transverse load ($E_1/E_2 = 25,\ G_{12} = G_{13} = 0.5E_2,\ G_{23} = 0.2E_2,\ \nu_{12} = 0.25$).

	$(b/a = 1)^*$				$(b/a = 3)^\dagger$			
Laminate	$\bar{w} \times 10^2$	$\bar{\sigma}_{xx}$	$\bar{\sigma}_{yy}$	$\bar{\sigma}_{xy}$	$\bar{w} \times 10^2$	$\bar{\sigma}_{xx}$	$-\bar{\sigma}_{yy}$	$\bar{\sigma}_{xy}$
0	0.4312	0.5387	0.0267	0.0213	0.4859	0.6016	0.0087	0.0080
(0/90/0)	0.4312	0.5387	0.0267	0.0213	0.5034	0.6233	0.0090	0.0083
(0/90)	1.0636	0.0843	0.7157	0.0525	2.4628	1.6411	0.2065	0.0417
$(0/90)_2$	0.5065	0.0357	0.4868	0.0250	1.0850	1.0332	0.1267	0.0181
$(0/90)_4$	0.4479	0.0296	0.4950	0.0221	0.9519	1.0425	0.1269	0.0158

* $\bar{\sigma}_{xx}(a/2, b/2, h/2)$, $\bar{\sigma}_{yy}(a/2, b/2, h/2$, and $\bar{\sigma}_{xy}(a, b, -h/2) = -\bar{\sigma}_{xy}(a, b, h/2)$. For square antisymmetric cross-ply laminates, we have $\sigma_{xx}(a/2, b/2, \pm h/2) = \mp\sigma_{yy}(a/2, b/2, \mp h/2)$.

† $\bar{\sigma}_{xx}(a/2, b/2, -h/2)$, $\bar{\sigma}_{yy}(a/2, b/2, h/2)$, and $\bar{\sigma}_{xy}(a, b, -h/2)$.

Table 8.3–2: Transverse deflections and stresses in square laminates subjected to uniformly distributed transverse load (UDL) or central point load (CPL) ($E_1/E_2 = 25,\ G_{12} = G_{13} = 0.5E_2,\ G_{23} = 0.2E_2,\ \nu_{12} = 0.25;\ m, n = 1, 3, \cdots, 20$ are used to evaluate the series). See the foot note (*) of Table 8.3–1 for stress locations.

	UDL				CPL			
Laminate	$\bar{w} \times 10^2$	$\bar{\sigma}_{xx}$	$\bar{\sigma}_{yy}$	$\bar{\sigma}_{xy}$	$\bar{w} \times 10^2$	$\bar{\sigma}_{xx}$	$\bar{\sigma}_{yy}$	$\bar{\sigma}_{xy}$
(0/90)	1.6955	0.1268	1.0761	0.0933	4.6664	0.8019	6.8217	0.1932
(90/0)	1.6955	1.0761	0.1268	0.0933	4.6664	6.8217	0.8019	0.1932
$(0/90)_2$	0.8085	0.0541	0.7367	0.0442	2.2105	0.3298	4.4952	0.0932
$(0/90)_4$	0.7150	0.0449	0.7496	0.0391	1.9536	0.2728	4.5553	0.0825

The effect of stacking sequence on nondimensionalized maximum deflection $\bar{w} \times 10^2$ and in-plane normal stress $-\bar{\sigma}_{xx}(a/2, b/2, -h/2)$ of $(0/90)_k$ and $(0/90)_{ks}$ laminates under uniformly distributed transverse load can be seen from the results presented in Table 8.3–3. The following notation is used: $(0/90)_2 = (0/90/0/90)$ and $(0/90)_{2s} = (0/90/0/90)_s$. The material properties used are $G_{12} = G_{13} = 0.5E_2$, $G_{23} = 0.2E_2$, and $\nu_{12} = 0.25$, and the series is evaluated using $m, n = 1, 3, \cdots, 30$.

For the $(0/90)_k$ (antisymmetric cross-ply) laminates, both heterogeneity and anisotropy ratio influence the deflections, which decrease as the number of layers is increased. For $(0°/90°)_{ks}$ laminates, heterogeneity has little effect on deflections and stresses. The anisotropy ratio affects deflections and stresses; deflections decrease and stresses increase with increasing value of E_1/E_2. Also, it should be noted that deflections decrease and stresses increase with the number of layers for a fixed anisotropy ratio.

Table 8.3–3: Effect of lamination scheme on the deflections $\bar{w} \times 10^2$ (first row) and stresses $\bar{\sigma}_{xx}$ (second row) in square laminates subjected to uniformly distributed load (UDL).

	$[(0/90)_k]^*$					$[(0/90)_{ks}]^\dagger$				
k	$\frac{E_1}{E_2}=5$	10	15	20	25	$\frac{E_1}{E_2}=5$	10	15	20	25
1	3.071	2.529	2.169	1.902	1.695	2.273	1.435	1.047	0.824	0.680
	0.503	0.714	0.867	0.984	1.076	0.569	0.704	0.765	0.801	0.824
2	2.440	1.620	1.214	0.971	0.809	2.282	1.445	1.057	0.833	0.687
	0.500	0.619	0.678	0.713	0.737	0.568	0.704	0.767	0.803	0.827
4	2.321	1.487	1.094	0.865	0.715	2.283	1.447	1.058	0.834	0.688
	0.523	0.641	0.697	0.729	0.750	0.565	0.700	0.763	0.799	0.823

* $-\sigma_{xx}(z=-h/2)=\sigma_{yy}(z=h/2)$. † $-\sigma_{xx}(z=-h/2)=\sigma_{xx}(z=h/2)$.

Figures 8.3–1 through 8.3–4 show the effect of bending–stretching coupling and plate aspect ratio on the transverse deflection $\bar{w} = w_0E_2h^3/(q_0b^4)$ and normal stresses $\bar{\sigma} = [h^2/(q_0b^2)]\sigma(a/2, b/2, z_0)$ for a fixed $z = z_0$. The material properties used are $E_1/E_2 = 25$, $G_{12} = G_{13} = 0.5E_2$, and $\nu_{12} = 0.25$. The magnitude of deflections and stresses of symmetric laminates (0/90/90/0) are about two to three times that of antisymmetric (0/90/0/90) laminates for $a/b > 1$. For uniformly distributed load there corresponds an aspect ratio, around $a/b = 2.25$ for $(0/90)_2$ and $a/b = 3.5$ for $(0/90)_s$, for which the deflection is the maximum of all aspect ratios. The effect of coupling is to increase the deflections and stresses. The coupling coefficients B_{ij} decrease in magnitude (hence the effect of coupling decreases) with the increase in the number of layers (for the same total thickness of the plate) in antisymmetric cross-ply laminates. The nondimensionalized center deflection $\bar{w} = w_0E_2h^3/(q_0b^4)$ versus the aspect ratio a/b is shown in Figure 8.3–5 for $(0/90)_k$ $(k = 1, 2, 3, 4)$ laminates under sinusoidal transverse loading ($E_1 = 25E_2$, $G_{12} = G_{13} = 0.5E_2$, $G_{23} = 0.2E_2$, $\nu_{12} = 0.25$). The nondimensionalized deflections of the six-layer and eight-layer plates approach the limiting case of an orthotropic plate.

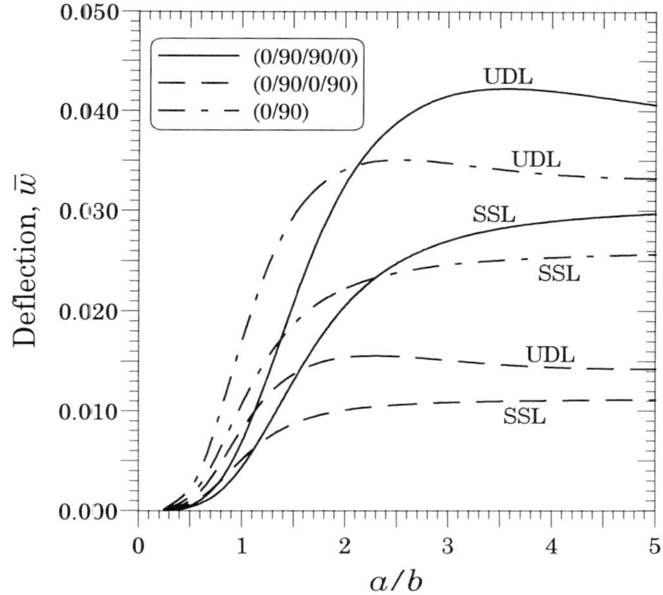

Figure 8.3–1: Nondimensionalized center transverse deflection ($\bar{w}$) versus plate aspect ratio (a/b) of simply supported (SS–1) laminates.

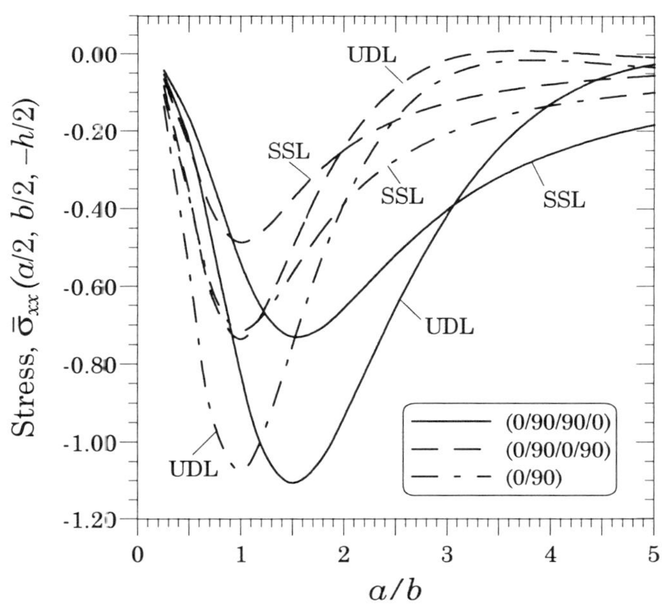

Figure 8.3–2: Nondimensionalized normal stress ($\bar{\sigma}_{xx}(a/2, b/2, -h/2)$) versus plate aspect ratio (a/b) for simply supported (SS–1) laminates.

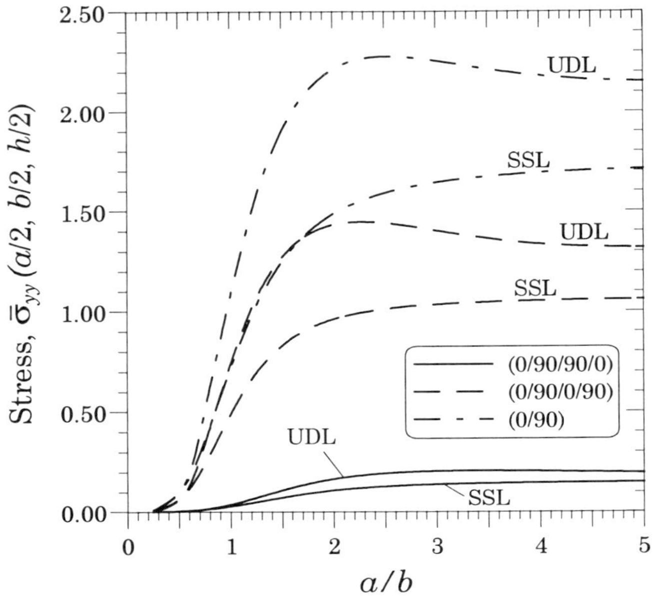

Figure 8.3–3: Nondimensionalized normal stress ($\bar{\sigma}_{yy}(a/2, b/2, h/2)$) versus plate aspect ratio (a/b) of simply supported (SS–1) laminates.

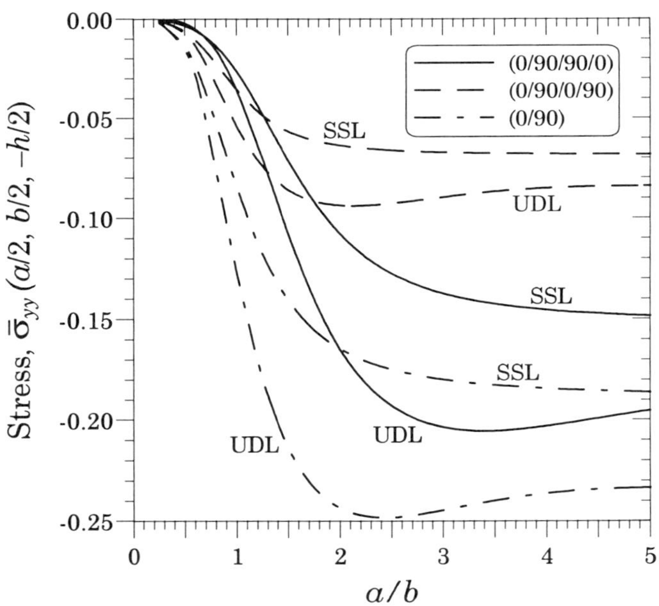

Figure 8.3–4: Nondimensionalized normal stress ($\bar{\sigma}_{yy}(a/2, b/2, -h/2)$) versus plate aspect ratio (a/b) for simply supported (SS–1) laminates.

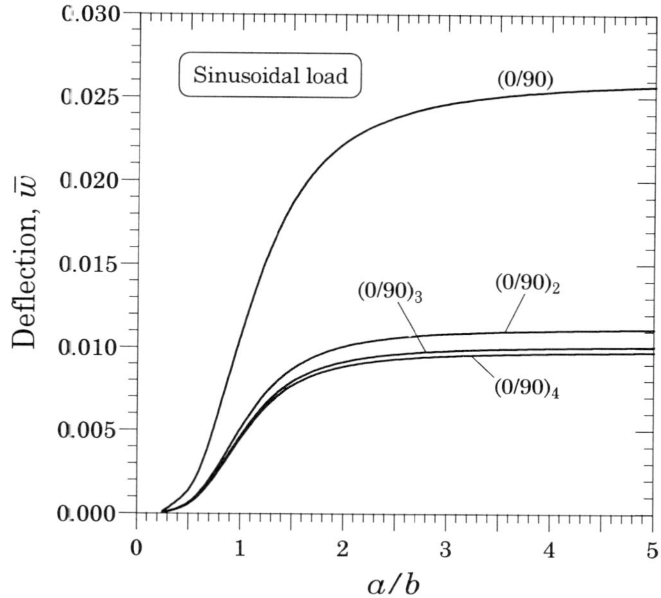

Figure 8.3–5: Nondimensionalized center transverse deflection versus aspect ratio for simply supported cross-ply laminates.

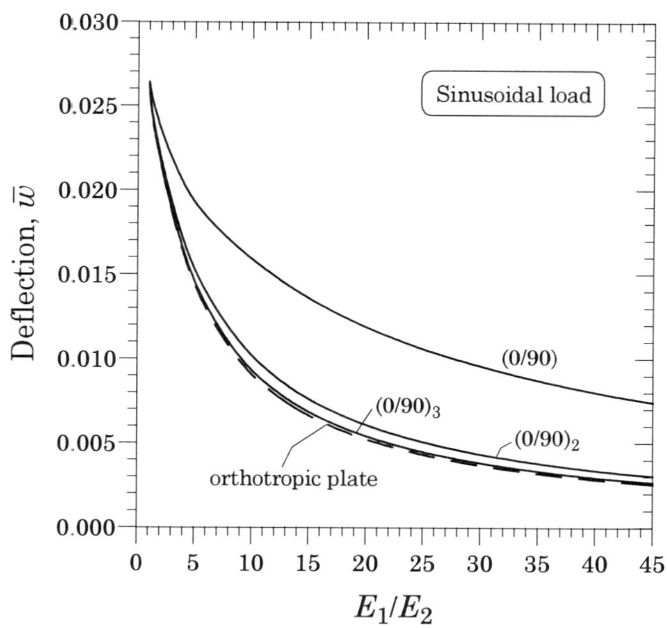

Figure 8.3–6: Nondimensionalized maximum transverse deflection versus modulus ratio for simply supported cross-ply laminates.

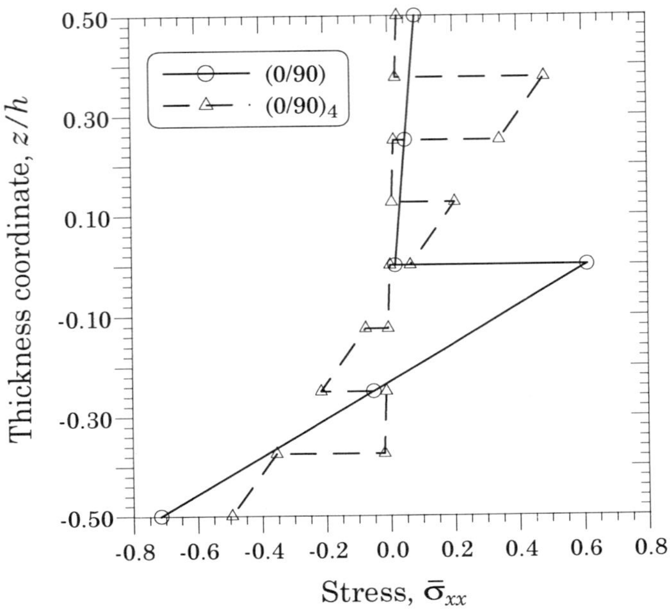

Figure 8.3–7: Nondimensionalized maximum normal stress ($\bar{\sigma}_{xx}$) versus plate thickness (z/h) for simply supported cross-ply laminates (SSL).

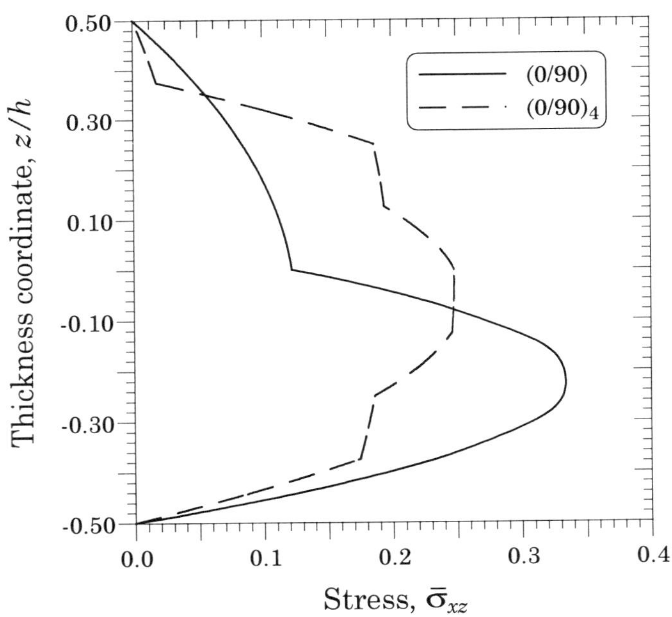

Figure 8.3–8: Nondimensionalized maximum shear stress ($\bar{\sigma}_{xz}$ obtained from equilibrium) versus plate thickness (z/h) for simply supported cross-ply laminates (SSL).

The dependence of the coupling effect on the modulus ratio is illustrated in Figure 8.3–5, where the maximum nondimensionalized deflection is plotted against the modulus ratio E_1/E_2 ($G_{12} = G_{13} = 0.5E_2$, and $\nu_{12} = 0.25$) for the sinusoidal load. The solution rapidly approaches that of an orthotropic plate for increasing number of layers.

Figures 8.3–7 and 8.3–8 show the distribution of the nondimensionalized maximum normal stress $\bar{\sigma}_{xx} = \sigma_{xx}(a/2, b/2, z)(h^2/q_0 b^2)$ and transverse shear stress $\bar{\sigma}_{xz} = \sigma_{xz}(a, b/2, z)(h/q_0 b)$, computed using the 3–D equilibrium equations, through the thickness of two-layer and eight-layer antisymmetric cross-ply laminates under sinusoidal loading ($a/b = 1$, $a/h = 100$, $E_1 = 25E_2$, $G_{12} = G_{13} = 0.5E_2$, $\nu_{12} = 0.25$). The two-layer plates experience larger stresses than eight-layer plates, and the stress concentration is reduced in the latter. Thus, the effect of the bending–stretching coupling present in two-layer plates on stresses is to increase the magnitude of stresses.

8.3–5 Buckling

For buckling analysis, we assume that the only applied loads are the in-plane forces

$$\hat{N}_{xx} = -N_0, \quad \hat{N}_{yy} = -kN_0, \quad k = \frac{\hat{N}_{yy}}{\hat{N}_{xx}} \tag{8.3-41}$$

and all other mechanical and thermal loads are zero. From Eq. (19) we have the eigenvalue problem

$$\begin{bmatrix} \hat{c}_{11} & \hat{c}_{12} & \hat{c}_{13} \\ \hat{c}_{12} & \hat{c}_{22} & \hat{c}_{23} \\ \hat{c}_{13} & \hat{c}_{23} & \hat{c}_{33} - N_0(\alpha^2 + k\beta^2) \end{bmatrix} \begin{Bmatrix} U_{mn} \\ V_{mn} \\ W_{mn} \end{Bmatrix} = \begin{Bmatrix} 0 \\ 0 \\ 0 \end{Bmatrix} \tag{8.3-42}$$

where $\hat{c}_{ij}$ are the coefficients defined in Eq. (20). For a nontrivial solution, $U_{mn} \neq 0$, $V_{mn} \neq 0$, and $W_{mn} \neq 0$, the determinant of the coefficient matrix in (42) should be zero:

$$\begin{vmatrix} \hat{c}_{11} & \hat{c}_{12} & \hat{c}_{13} \\ \hat{c}_{12} & \hat{c}_{22} & \hat{c}_{23} \\ \hat{c}_{13} & \hat{c}_{23} & \hat{c}_{33} - N_0(\alpha^2 + k\beta^2) \end{vmatrix} = 0 \tag{8.3-43}$$

or

$$N_0 = \frac{d_{mn}}{a_0(\alpha^2 + k\beta^2)}, \quad \alpha = \frac{m\pi}{a}, \quad \beta = \frac{n\pi}{b} \tag{8.3-44a}$$

where

$$d_{mn} = \begin{vmatrix} \hat{c}_{11} & \hat{c}_{12} & \hat{c}_{13} \\ \hat{c}_{12} & \hat{c}_{22} & \hat{c}_{23} \\ \hat{c}_{13} & \hat{c}_{23} & \hat{c}_{33} \end{vmatrix}, \quad a_0 = \hat{c}_{11}\hat{c}_{22} - \hat{c}_{12}\hat{c}_{12} \tag{8.3-44b}$$

Alternatively, using the static condensation procedure described in Eqs. (22)–(26), we obtain

$$\left[\hat{c}_{33} - N_0\left(\alpha^2 + k\beta^2\right) - \frac{\hat{c}_{13}\hat{c}_{22} - \hat{c}_{23}\hat{c}_{12}}{\hat{c}_{11}\hat{c}_{22} - \hat{c}_{12}\hat{c}_{12}}\hat{c}_{13} - \frac{\hat{c}_{11}\hat{c}_{23} - \hat{c}_{12}\hat{c}_{13}}{\hat{c}_{11}\hat{c}_{22} - \hat{c}_{12}\hat{c}_{12}}\hat{c}_{23}\right] W_{mn} = 0$$

Since $W_{mn} \neq 0$, we obtain

$$N_0 = \frac{1}{\alpha^2 + k\beta^2}\left(\hat{c}_{33} - \frac{\hat{c}_{13}\hat{c}_{22} - \hat{c}_{23}\hat{c}_{12}}{\hat{c}_{11}\hat{c}_{22} - \hat{c}_{12}\hat{c}_{12}}\hat{c}_{13} - \frac{\hat{c}_{11}\hat{c}_{23} - \hat{c}_{12}\hat{c}_{13}}{\hat{c}_{11}\hat{c}_{22} - \hat{c}_{12}\hat{c}_{12}}\hat{c}_{23}\right) \qquad (8.3-45)$$

Clearly, for each pair of m and n, there is a unique value of N_0. The critical buckling load is the smallest of all $N_0 = N_0(m,n)$:

$$N_{cr} = \min_{1 \leq m,n \leq \infty} \{N_0(m,n)\} \qquad (8.3-46)$$

Since $\hat{c}_{ij}$ depend on m and n, $N_0(m,n)$ is a complicated function of both m and n and no simple conclusions can be drawn about the mode (m,n) at buckling.

Antisymmetric cross-ply laminates have special stiffness characteristics given in Eq. (7a). Hence the buckling load for antisymmetric cross-ply laminates is given by Eq. (44a) or (45) with coefficients $\hat{c}_{ij}$ from Eq. (20).

For specially orthotropic plates, neither shear–twist coupling nor bending–extension coupling exists ($\hat{c}_{13} = \hat{c}_{23} = 0$), and therefore U_{mn} and V_{mn} are zero prior to onset of buckling. Therefore, we have (*cf.* Eq. (7.5–5a))

$$N_0 = \frac{\hat{c}_{33}}{(\alpha^2 + k\beta^2)} \qquad (8.3-47a)$$

$$\hat{c}_{33} = D_{11}\alpha^4 + 2(D_{12} + 2D_{66})\alpha^2\beta^2 + D_{22}\beta^4, \; \alpha = \frac{m\pi}{a}, \quad \beta = \frac{m\pi}{b} \qquad (8.3-47b)$$

Table 8.3–4 shows the effect of stacking sequence, plate aspect ratio, and modulus ratio on nondimensionalized critical buckling loads $\bar{N} = N_{cr}(b^2/\pi^2 D_{22})$ of rectangular laminates under uniform compression ($k = 0$) as well as biaxial compression ($k = 1$). The following material properties were used: *material 1:* $E_1/E_2 = 25$, $G_{12} = G_{13} = 0.5E_2$, $\nu_{12} = 0.25$ and *material 2:* $E_1/E_2 = 40$, $G_{12} = G_{13} = 0.6E_2$, $\nu_{12} = 0.25$. In all cases (also see Figures 8.3–9 through 8.3–11) the critical buckling mode is (m,n)=(1,1), except for the antisymmetric cross-ply laminate, with aspect ratio $a/b = 1.5$, in uniform compression. In the latter case, the mode is (2,1). The nondimensionalized buckling load increases for symmetric laminates whereas it decreases for antisymmetric laminates as the modulus ratio increases.

Table 8.3–4: Effect of lamination scheme, aspect ratio, and modulus ratio on the nondimensionalized buckling loads $\bar{N}$ of rectangular laminates under uniform compression and biaxial compression (material 1).

	$(0/90)_{as} \equiv (0°/90°/0°/90°)$					$(0/90)_s \equiv (0°/90°/90°/0°)$				
$\frac{a}{b}$	$\frac{E_1}{E_2} = 5$	10	20	25	40	$\frac{E_1}{E_2} = 5$	10	20	25	40
Uniaxial compression ($k = 0$)										
0.5	4.705	4.157	3.828	3.757	3.647	13.900	18.126	21.878	22.874	24.590
1.0	2.643	2.189	1.923	1.866	1.778	5.650	6.347	6.961	7.124	7.404
1.5	2.955	2.487	2.211	2.152	2.061	5.233	5.277	5.310	5.318	5.332
Biaxial compression ($k = 1$)										
0.5	3.764	3.325	3.062	3.005	2.917	11.120	12.694	13.922	14.248	14.766
1.0	1.322	1.095	0.962	0.933	0.889	2.825	3.174	3.481	3.562	3.702
1.5	1.009	0.860	0.773	0.754	0.725	1.610	1.624	1.634	1.636	1.641

The mode number is $(m, n) = (1, 1)$ for all cases, except for the following: (a) $(0°/90°)_2$, $a/b = 1.5$ and $k = 0$: mode is (2,1); (b) $(0°/90°)_s$, $a/b = 0.5$ and $k = 1$: modes are (1,1), (1,2), (1,2), (1,2), and (1,3) for modulus ratios 5, 10, 20, 25, and 40, respectively.

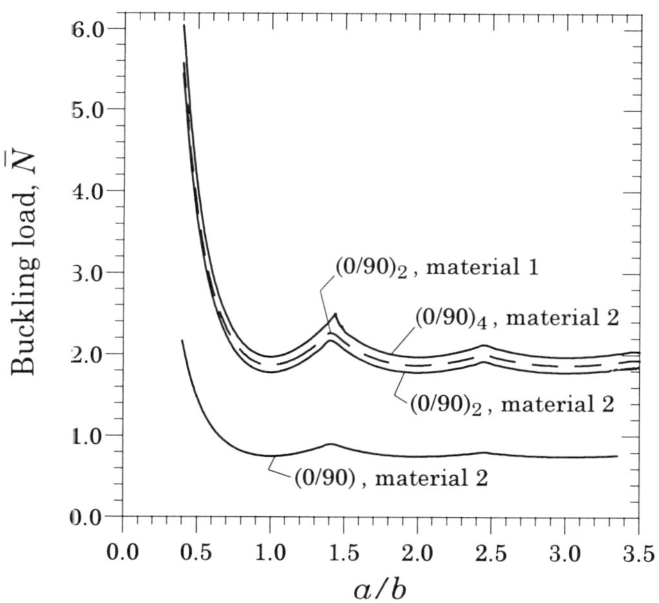

Figure 8.3–9: Nondimensionalized buckling ($\bar{N}$) load versus plate aspect ratio (a/b) for simply supported (SS–1) antisymmetric cross-ply laminates $(0/90)_n$ under uniaxial compression.

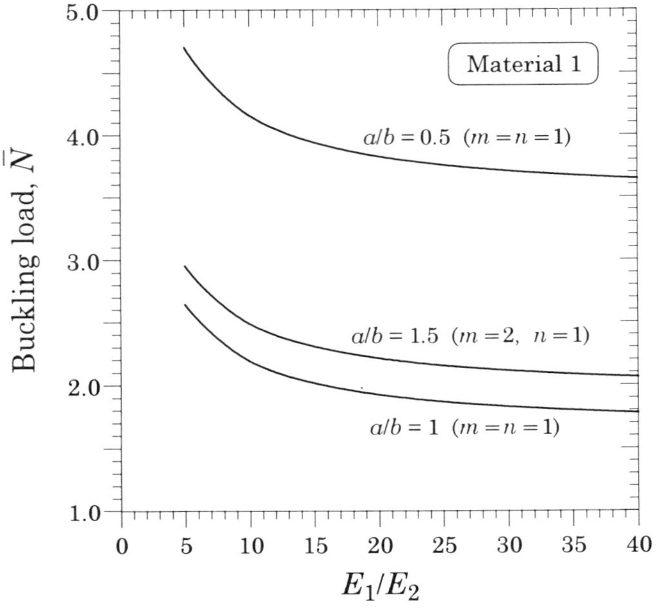

Figure 8.3–10: Nondimensionalized buckling load ($\bar{N}$) versus modulus ratio (E_1/E_2) for antisymmetric cross-ply laminates $(0/90)_2$ under uniaxial compression.

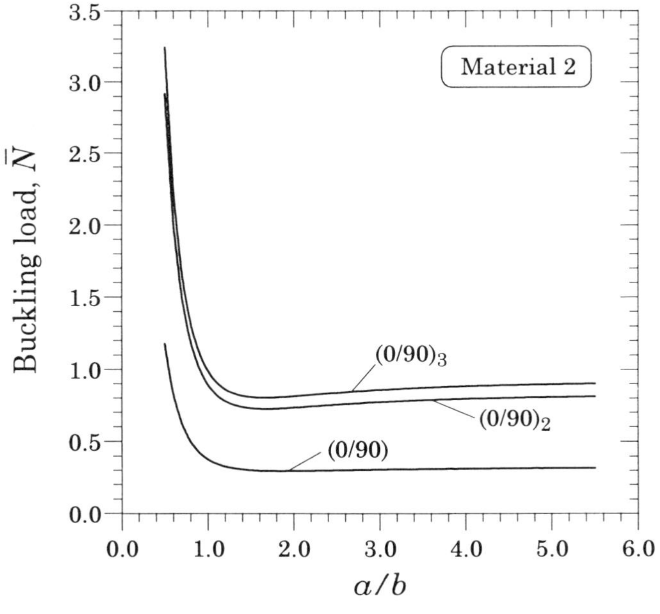

Figure 8.3–11: Nondimensionalized biaxial buckling load ($\bar{N}$) versus plate aspect ratio (a/b) for antisymmetric cross-ply laminates $(0/90)_n$ ($n = 1, 2, 3$) under biaxial compression ($k = 1$).

8.3–6 Vibration

For free vibration, all applied loads and the in-plane forces are set to zero, and we assume a periodic solution of the form

$$U_{mn}(t) = U_{mn}^0 e^{i\omega t}, \quad V_{mn}(t) = V_{mn}^0 e^{i\omega t}, \quad W_{mn}(t) = W_{mn}^0 e^{i\omega t} \tag{8.3-48}$$

where $i = \sqrt{-1}$ and ω is the frequency of natural vibration. Then Eq. (19) reduces to the eigenvalue problem

$$\left(\begin{bmatrix} \hat{c}_{11} & \hat{c}_{12} & \hat{c}_{13} \\ \hat{c}_{12} & \hat{c}_{22} & \hat{c}_{23} \\ \hat{c}_{13} & \hat{c}_{23} & \hat{c}_{33} \end{bmatrix} - \omega^2 \begin{bmatrix} \hat{m}_{11} & 0 & 0 \\ 0 & \hat{m}_{22} & 0 \\ 0 & 0 & \hat{m}_{33} \end{bmatrix} \right) \begin{Bmatrix} U_{mn}^0 \\ V_{mn}^0 \\ W_{mn}^0 \end{Bmatrix} = \begin{Bmatrix} 0 \\ 0 \\ 0 \end{Bmatrix} \tag{8.3-49}$$

For a nontrivial solution, $U_{mn}^0 \neq 0$, $V_{mn}^0 \neq 0$, $W_{mn}^0 \neq 0$, the determinant of the coefficient matrix in (49) should be zero, which yields the characteristic polynomial

$$-p\lambda^3 + q\lambda^2 - r\lambda + s = 0 \tag{8.3-50}$$

where $\lambda = \omega^2$ is the eigenvalue and

$$p = \begin{vmatrix} \hat{m}_{11} & 0 & 0 \\ 0 & \hat{m}_{22} & 0 \\ 0 & 0 & \hat{m}_{33} \end{vmatrix}, \quad s = \begin{vmatrix} \hat{c}_{11} & \hat{c}_{12} & \hat{c}_{13} \\ \hat{c}_{12} & \hat{c}_{22} & \hat{c}_{23} \\ \hat{c}_{13} & \hat{c}_{23} & \hat{c}_{33} \end{vmatrix}$$

$$q = \begin{vmatrix} \hat{c}_{11} & 0 & 0 \\ \hat{c}_{12} & \hat{m}_{22} & 0 \\ \hat{c}_{13} & 0 & \hat{m}_{33} \end{vmatrix} + \begin{vmatrix} \hat{m}_{11} & \hat{c}_{12} & 0 \\ \hat{0} & \hat{c}_{22} & 0 \\ 0 & \hat{c}_{23} & \hat{m}_{33} \end{vmatrix} + \begin{vmatrix} \hat{m}_{11} & 0 & \hat{c}_{13} \\ \hat{0} & \hat{m}_{22} & \hat{c}_{23} \\ 0 & 0 & \hat{c}_{33} \end{vmatrix}$$

$$r = \begin{vmatrix} \hat{c}_{11} & \hat{c}_{12} & 0 \\ \hat{c}_{12} & \hat{c}_{22} & 0 \\ \hat{c}_{13} & \hat{c}_{23} & \hat{m}_{33} \end{vmatrix} + \begin{vmatrix} \hat{c}_{11} & \hat{0} & \hat{c}_{13} \\ \hat{c}_{12} & \hat{m}_{22} & \hat{c}_{23} \\ \hat{c}_{13} & 0 & \hat{c}_{33} \end{vmatrix} + \begin{vmatrix} \hat{m}_{11} & \hat{c}_{12} & \hat{c}_{13} \\ \hat{0} & \hat{c}_{22} & \hat{c}_{23} \\ 0 & \hat{c}_{23} & \hat{c}_{33} \end{vmatrix} \tag{8.3-51}$$

The real positive roots of this cubic equation give the square of the natural frequency ω_{mn} associated with mode (m, n). The smallest of the frequencies is called the *fundamental frequency*. In general, ω_{11} is not the fundamental frequency; the smallest frequency might occur for values other than $m = n = 1$.

If the in-plane inertias are neglected (*i.e.*, $\hat{m}_{11} = \hat{m}_{22} = 0$), and irrespective of whether the rotary inertia is zero, Eq. (50) takes the same form as Eq. (43) with $N_0(\alpha^2 + k\beta^2)$ replaced by $\omega^2 \hat{m}_{33}$. Hence, from Eq. (43) we have

$$\omega^2 = \frac{1}{\hat{m}_{33}} \left(\hat{c}_{33} - \frac{\hat{c}_{13}\hat{c}_{22} - \hat{c}_{23}\hat{c}_{12}}{\hat{c}_{11}\hat{c}_{22} - \hat{c}_{12}\hat{c}_{12}} \hat{c}_{13} - \frac{\hat{c}_{11}\hat{c}_{23} - \hat{c}_{12}\hat{c}_{13}}{\hat{c}_{11}\hat{c}_{22} - \hat{c}_{12}\hat{c}_{12}} \hat{c}_{23} \right) \tag{8.3-52}$$

Note that if the in-plane inertias are not neglected, the eigenvalue problem cannot be simplified to a single equation even if the rotary inertia is zero.

Nondimensionalized frequencies, $\bar{\omega}_{mn} = \omega_{mn}(b^2/\pi^2)\sqrt{\rho h/D_{22}}$, of specially orthotropic and antisymmetric cross-ply square laminates are presented in Table 8.3–5 for modulus ratios E_1/E_2=10 and 20 ($G_{12} = G_{13} = 0.5E_2, G_{23} = 0.2E_2, \nu_{12} = 0.25$). All layers are of equal thickness. Results are presented for $m, n = 1, 2, 3$, and for the case in which the rotary inertia is neglected. The fundamental frequency increases with modular ratio. As noted earlier, the effect of including rotary inertia is to decrease the frequency of vibration. Note that the first four frequencies for an orthotropic (0°) plate correspond to the modes, (m, n)=(1,1), (1,2), (1,3), and (2,1), whereas for antisymmetric cross-ply plates they are (m, n)=(1,1), (1,2), (2,1), and (2,2). For symmetric cross-ply plates the first four frequencies are provided by the modes: (m, n)=(1,1), (1,2), (2,1), and (1,3). Also, we note that $\omega_{mn} = \omega_{nm}$ for antisymmetric laminates.

Table 8.3–5: Nondimensionalized fundamental frequencies $\bar{\omega}$ of cross-ply laminates according to the classical plate theory.

m	n	(0)	(0/90)	$(0/90)_2$	$(0/90)_4$	$(0/90)_s$
$E_1/E_2 = 10$						
1	1	3.672	1.183	1.479	1.545	2.519
1	2	5.996	3.174	4.077	4.274	4.986
1	3	10.648	6.666	8.698	9.136	9.783
2	1	13.075	3.174	4.077	4.274	8.515
2	2	14.690	4.733	5.918	6.179	10.077
2	3	18.181	7.927	10.034	10.494	13.783
3	1	28.868	6.666	8.698	9.136	18.704
3	2	30.258	7.927	10.034	10.494	19.911
3	3	33.053	10.650	13.317	13.904	22.674
$E_1/E_2 = 20$						
1	1	4.847	0.990	1.386	1.469	2.638
1	2	6.781	2.719	3.913	4.158	4.917
1	3	11.111	5.789	8.456	8.998	9.637
2	1	18.193	2.719	3.913	4.158	9.354
2	2	19.388	3.959	5.547	5.877	10.554
2	3	22.153	6.702	9.507	10.088	13.826
3	1	40.539	5.789	8.456	8.998	20.752
3	2	41.542	6.193	9.507	10.088	21.578
3	3	43.623	8.908	12.481	13.224	13.826

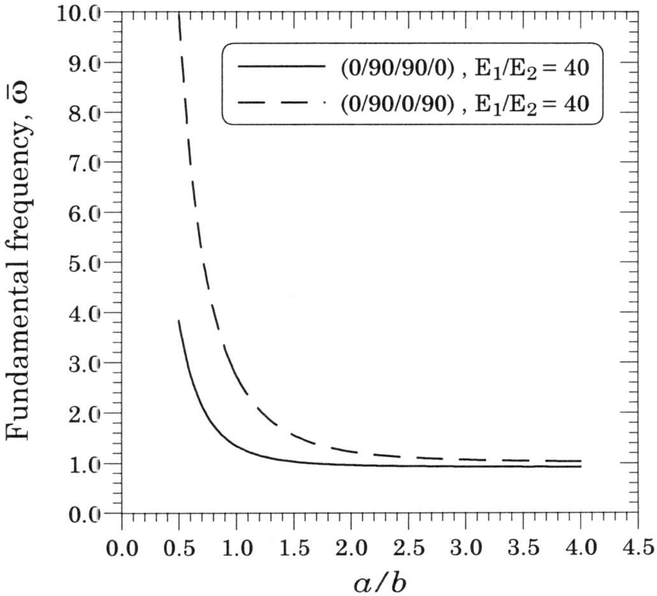

Figure 8.3–12: Nondimensionalized fundamental frequency ($\bar{\omega}$) versus plate aspect ratio (a/b) for cross-ply laminates.

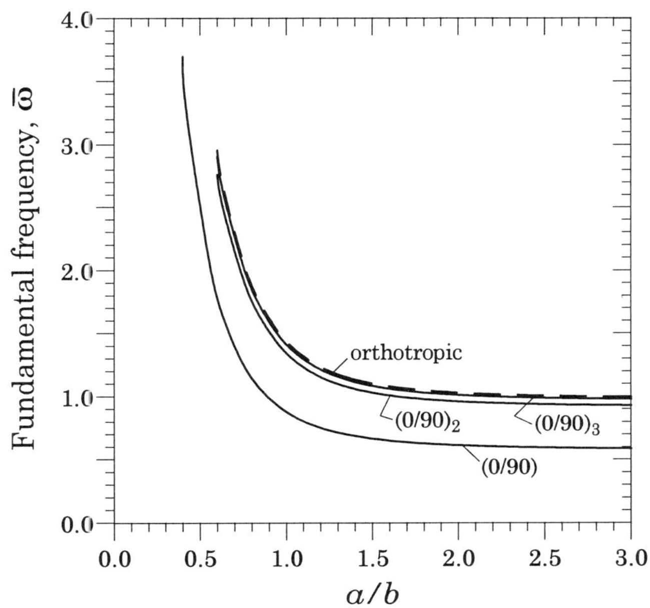

Figure 8.3–13: Nondimensionalized fundamental frequency ($\bar{\omega}$) versus plate aspect ratio (a/b) for antisymmetric cross-ply laminates.

Figure 8.3–12 shows a plot of fundamental frequency $\bar{\omega}$ versus aspect ratio a/b for symmetric $(0/90)_s$ cross-ply and antisymmetric $(0/90)_2$ cross-ply laminates. The material properties used are $E_1/E_2 = 40$, $G_{12} = G_{13} = 0.6E_2$, and $\nu_{12} = 0.25$. Figure 8.3–13 shows the effect of coupling between bending and extension on the fundamental frequencies of antisymmetric cross-ply laminates. The material properties used are $E_1/E_2 = 25$, $G_{12} = G_{13} = 0.5E_2$, and $\nu_{12} = 0.25$. With an increase in the number of layers, the frequencies approach those of the orthotropic plate. The bending–stretching coupling has the effect of lowering the vibration frequencies. For example, the two-layer plate has vibration frequencies about 40 percent lower than those of eight-layer antisymmetric laminate or orthotropic plate with the same total thickness.

8.4 Navier Solutions of Antisymmetric Angle-Ply Laminates

8.4–1 Boundary Conditions

The SS–2 boundary conditions in Eqs. (8.2–7) imply the conditions on the generalized displacements and resultants of the classical laminate theory:

$$u_0(0,y,t) = 0, \quad u_0(a,y,t) = 0, \quad v_0(x,0,t) = 0, \quad v_0(x,b,t) = 0$$

$$\frac{\partial w_0}{\partial x}(x,0,t) = 0, \quad \frac{\partial w_0}{\partial x}(x,b,t) = 0, \quad \frac{\partial w_0}{\partial y}(0,y,t) = 0, \quad \frac{\partial w_0}{\partial y}(a,y,t) = 0$$

$$w_0(x,0,t) = 0, \quad w_0(x,b,t) = 0, \quad w_0(0,y,t) = 0, \quad w_0(a,y,t) = 0 \qquad (8.4-1a)$$

$$N_{xy}(0,y,t) = 0, \quad N_{xy}(a,y,t) = 0, \quad N_{xy}(x,0,t) = 0, \quad N_{xy}(x,b,t) = 0$$

$$M_{xx}(0,y,t) = 0, \quad M_{xx}(a,y,t) = 0, \quad M_{yy}(x,0,t) = 0, \quad M_{yy}(x,b,t) = 0 \quad (8.4-1b)$$

The displacement boundary conditions in (1a) are satisfied by assuming the following form of the displacements

$$u_0(x,y,t) = \sum_{n=1}^{\infty}\sum_{m=1}^{\infty} U_{mn}(t)\ \sin\alpha x\ \cos\beta y \qquad (8.4-2a)$$

$$v_0(x,y,t) = \sum_{n=1}^{\infty}\sum_{m=1}^{\infty} V_{mn}(t)\ \cos\alpha x\ \sin\beta y \qquad (8.4-2b)$$

$$w_0(x,y,t) = \sum_{n=1}^{\infty}\sum_{m=1}^{\infty} W_{mn}(t)\ \sin\alpha x\ \sin\beta y \qquad (8.4-2c)$$

Substituting the expansions (2) into the expressions for N_{xx}, N_{yy}, N_{xy}, M_{xx}, M_{yy}, and M_{xy}, we obtain

$$\begin{aligned}
N_{xx} &= A_{11}\frac{\partial u_0}{\partial x} + A_{12}\frac{\partial v_0}{\partial y} + A_{16}\left(\frac{\partial u_0}{\partial y} + \frac{\partial v_0}{\partial x}\right) \\
&- B_{11}\frac{\partial^2 w_0}{\partial x^2} - B_{12}\frac{\partial^2 w_0}{\partial y^2} - 2B_{16}\frac{\partial^2 w_0}{\partial x \partial y} \\
&= \sum_{n=1}^{\infty}\sum_{m=1}^{\infty} \left(\alpha A_{11}U_{mn} + \beta A_{12}V_{mn} + 2B_{16}\alpha\beta W_{mn}\right)\cos\alpha x \cos\beta y \\
&+ \sum_{n=1}^{\infty}\sum_{m=1}^{\infty} \left[-A_{16}\left(\beta U_{mn} + \alpha V_{mn}\right) + \left(B_{11}\alpha^2 + B_{12}\beta^2\right)W_{mn}\right]\sin\alpha x \sin\beta y
\end{aligned}$$

$$\begin{aligned}
N_{yy} &= A_{12}\frac{\partial u_0}{\partial x} + A_{22}\frac{\partial v_0}{\partial y} + A_{26}\left(\frac{\partial u_0}{\partial y} + \frac{\partial v_0}{\partial x}\right) \\
&- B_{12}\frac{\partial^2 w_0}{\partial x^2} - B_{22}\frac{\partial^2 w_0}{\partial y^2} - 2B_{26}\frac{\partial^2 w_0}{\partial x \partial y} \\
&= \sum_{n=1}^{\infty}\sum_{m=1}^{\infty} \left(\alpha A_{12}U_{mn} + \beta A_{22}V_{mn} - 2B_{26}W_{mn}\right)\cos\alpha x \cos\beta y \\
&+ \sum_{n=1}^{\infty}\sum_{m=1}^{\infty} \left[-A_{26}\left(\beta U_{mn} + \alpha V_{mn}\right) + \left(B_{12}\alpha^2 + B_{22}\beta^2\right)W_{mn}\right]\sin\alpha x \sin\beta y
\end{aligned}$$

$$\begin{aligned}
N_{xy} &= A_{16}\frac{\partial u_0}{\partial x} + A_{26}\frac{\partial v_0}{\partial y} + A_{66}\left(\frac{\partial u_0}{\partial y} + \frac{\partial v_0}{\partial x}\right) \\
&- B_{16}\frac{\partial^2 w_0}{\partial x^2} - B_{26}\frac{\partial^2 w_0}{\partial y^2} - 2B_{66}\frac{\partial^2 w_0}{\partial x \partial y} \\
&= \sum_{n=1}^{\infty}\sum_{m=1}^{\infty} \left[\alpha A_{16}U_{mn} + \beta A_{26}V_{mn} - 2\alpha\beta B_{66}W_{mn}\right]\ \cos\alpha x\ \cos\beta y \\
&+ \sum_{n=1}^{\infty}\sum_{m=1}^{\infty} \left[-A_{66}\left(\beta U_{mn} + \alpha V_{mn}\right) + \left(B_{16}\alpha^2 + B_{26}\beta^2\right)W_{mn}\right]\ \sin\alpha x\ \sin\beta y
\end{aligned}$$

$$\begin{aligned}
M_{xx} &= B_{11}\frac{\partial u_0}{\partial x} + B_{12}\frac{\partial v_0}{\partial y} + B_{16}\left(\frac{\partial u_0}{\partial y} + \frac{\partial v_0}{\partial x}\right) \\
&- D_{11}\frac{\partial^2 w_0}{\partial x^2} - D_{12}\frac{\partial^2 w_0}{\partial y^2} - 2D_{16}\frac{\partial^2 w_0}{\partial x \partial y} \\
&= \sum_{n=1}^{\infty}\sum_{m=1}^{\infty} \left[\alpha B_{11}U_{mn} + \beta B_{12}V_{mn} - 2\alpha\beta D_{16}W_{mn}\right]\ \cos\alpha x\ \cos\beta y \\
&+ \sum_{n=1}^{\infty}\sum_{m=1}^{\infty} \left[-B_{16}\left(\beta U_{mn} + \alpha V_{mn}\right) + \left(D_{11}\alpha^2 + D_{12}\beta^2\right)W_{mn}\right]\ \sin\alpha x\ \sin\beta y
\end{aligned}$$

$$
\begin{aligned}
M_{yy} &= B_{12}\frac{\partial u_0}{\partial x} + B_{22}\frac{\partial v_0}{\partial y} + B_{26}\left(\frac{\partial u_0}{\partial y} + \frac{\partial v_0}{\partial x}\right) \\
&- D_{12}\frac{\partial^2 w_0}{\partial x^2} - D_{22}\frac{\partial^2 w_0}{\partial y^2} - 2D_{26}\frac{\partial^2 w_0}{\partial x \partial y} \\
&= \sum_{n=1}^{\infty}\sum_{m=1}^{\infty} \left[\alpha B_{12} U_{mn} + \beta B_{22} V_{mn} - 2\alpha\beta D_{26} W_{mn}\right] \cos\alpha x \ \cos\beta y \\
&+ \sum_{n=1}^{\infty}\sum_{m=1}^{\infty} \left[-B_{26}\left(\beta U_{mn} + \alpha V_{mn}\right) + \left(D_{12}\alpha^2 + D_{22}\beta^2\right) W_{mn}\right] \sin\alpha x \ \sin\beta y \\
M_{xy} &= B_{16}\frac{\partial u_0}{\partial x} + B_{26}\frac{\partial v_0}{\partial y} + B_{66}\left(\frac{\partial u_0}{\partial y} + \frac{\partial v_0}{\partial x}\right) \\
&- D_{16}\frac{\partial^2 w_0}{\partial x^2} - D_{26}\frac{\partial^2 w_0}{\partial y^2} - 2D_{66}\frac{\partial^2 w_0}{\partial x \partial y} \\
&= \sum_{n=1}^{\infty}\sum_{m=1}^{\infty} \left[\alpha B_{16} U_{mn} + \beta B_{26} V_{mn} - 2\alpha\beta D_{66} W_{mn}\right] \cos\alpha x \cos\beta y \\
&+ \sum_{n=1}^{\infty}\sum_{m=1}^{\infty} \left[-B_{66}\left(\beta U_{mn} + \alpha V_{mn}\right) + \left(D_{16}\alpha^2 + D_{26}\beta^2\right) W_{mn}\right] \sin\alpha x \sin\beta y
\end{aligned}
\tag{8.4-3}
$$

Note that the boundary conditions in (1b) on the stress and moment resultants N_{xy}, M_{xx}, and M_{yy} can be satisfied only if the laminate stiffnesses $A_{16}, A_{26}, B_{11}, B_{12}, B_{22}, B_{66}, D_{16}$, and D_{26} are zero. Thus, the Navier solutions for rectangular laminated plates with SS–2 boundary conditions exist only when the stacking sequences are such that

$$
A_{16} = A_{26} = B_{11} = B_{12} = B_{22} = B_{66} = D_{16} = D_{26} = 0 \tag{8.4-4}
$$

In addition, for dynamic problems, we must have $I_1 = 0$. From Section 5.5–4, it follows that plates with a single generally orthotropic layer, symmetrically laminated plates with multiple specially orthotropic layers, and antisymmetric angle-ply laminated plates admit the Navier solutions for the SS–2 boundary conditions.

8.4–2 Solution

Substitution of Eq. (2) into Eqs. (8.2–1)–(8.2–3) yields

$$
\begin{aligned}
\sum_{n=1}^{\infty}\sum_{m=1}^{\infty} \Big[&- \left(A_{11}\alpha^2 + A_{66}\beta^2\right) U_{mn} - \left(A_{12} + A_{66}\right)\alpha\beta V_{mn} + \left(3B_{16}\alpha^2\beta + B_{26}\beta^3\right) W_{mn} \\
&- I_0 \ddot{U}_{mn}\Big] \sin\alpha x \ \cos\beta y = \left(\frac{\partial N_{xx}^T}{\partial x} + \frac{\partial N_{xy}^T}{\partial y}\right)
\end{aligned}
\tag{8.4-5a}
$$

$$\sum_{n=1}^{\infty}\sum_{m=1}^{\infty}\Big[-(A_{12}+A_{66})\,\alpha\beta U_{mn}-\left(A_{66}\alpha^2+A_{22}\beta^2\right)V_{mn}+\left(B_{16}\alpha^3+3B_{26}\alpha\beta^2\right)W_{mn}$$

$$-I_0\ddot{V}_{mn}\Big]\cos\alpha x\ \sin\beta y=\left(\frac{\partial N_{xy}^T}{\partial x}+\frac{\partial N_{yy}^T}{\partial y}\right) \tag{8.4-5b}$$

$$\sum_{n=1}^{\infty}\sum_{m=1}^{\infty}\Big[\left(3B_{16}\alpha^2\beta+B_{26}\beta^3\right)U_{mn}+\left(B_{16}\alpha^3+3B_{26}\alpha\beta^2\right)V_{mn}$$

$$-\left(D_{11}\alpha^4+2(D_{12}+2D_{66})\alpha^2\beta^2+D_{22}\beta^4\right)W_{mn}$$

$$-\left(\alpha^2\hat{N}_{xx}+\beta^2\hat{N}_{yy}\right)W_{mn}-\left(I_0+I_2(\alpha^2+\beta^2)\right)\ddot{W}_{mn}\Big]\sin\alpha x\ \sin\beta y$$

$$=\left(\frac{\partial^2 M_{xx}^T}{\partial x^2}+2\frac{\partial^2 M_{xy}^T}{\partial x\partial y}+\frac{\partial^2 M_{yy}^T}{\partial y^2}\right)-q(x,y) \tag{8.4-5c}$$

Note that the edge shear force $\hat{N}_{xy}$ is necessarily zero. In addition, inertia I_1 is zero. If the transverse load and thermal resultants are expanded as before [see Eqs. (8.3–10)–(8.3–13)], then it follows from Eq. (8.3–14) that N_{mn}^1, N_{mn}^2, and M_{mn}^6 must be zero. If the temperature field is expanded in double cosine series, N_{mn}^6, M_{mn}^1, and M_{mn}^2 must be zero.

Equations (5) can be expressed in matrix form as

$$\begin{bmatrix}\hat{c}_{11} & \hat{c}_{12} & \hat{c}_{13}\\ \hat{c}_{12} & \hat{c}_{22} & \hat{c}_{23}\\ \hat{c}_{13} & \hat{c}_{23} & \hat{c}_{33}+\tilde{s}_{33}\end{bmatrix}\begin{Bmatrix}U_{mn}\\ V_{mn}\\ W_{mn}\end{Bmatrix}+\begin{bmatrix}\hat{m}_{11} & 0 & 0\\ 0 & \hat{m}_{22} & 0\\ 0 & 0 & \hat{m}_{33}\end{bmatrix}\begin{Bmatrix}\ddot{U}_{mn}\\ \ddot{V}_{mn}\\ \ddot{W}_{mn}\end{Bmatrix}$$

$$=\begin{Bmatrix}0\\ 0\\ Q_{mn}\end{Bmatrix}+\begin{Bmatrix}-\beta N_{mn}^6\\ -\alpha N_{mn}^6\\ \alpha^2 M_{mn}^1+\beta^2 M_{mn}^2\end{Bmatrix}\ \text{or}\ \begin{Bmatrix}\alpha N_{mn}^1\\ \beta N_{mn}^2\\ -2\alpha\beta M_{mn}^6\end{Bmatrix} \tag{8.4-6}$$

where

$$\hat{c}_{11}=A_{11}\alpha^2+A_{66}\beta^2,\quad \hat{c}_{12}=(A_{12}+A_{66})\alpha\beta$$
$$\hat{c}_{13}=-(3B_{16}\alpha^2+B_{26}\beta^2)\beta,\quad \hat{c}_{22}=A_{66}\alpha^2+A_{22}\beta^2$$
$$\hat{c}_{23}=-(B_{16}\alpha^2+3B_{26}\beta^2)\alpha$$
$$\hat{c}_{33}=D_{11}\alpha^4+2(D_{12}+2D_{66})\alpha^2\beta^2+D_{22}\beta^4 \tag{8.4-7a}$$
$$\tilde{s}_{33}=\alpha^2\hat{N}_{xx}+\beta^2\hat{N}_{yy} \tag{8.4-7b}$$
$$\hat{m}_{11}=I_0,\quad \hat{m}_{22}=I_0,\quad \hat{m}_{33}=I_0+I_2(\alpha^2+\beta^2) \tag{8.4-7c}$$

and $\alpha=m\pi/a$ and $\beta=n\pi/b$. The second column of thermal forces in Eq. (6) are valid for the case in which the temperature field is expanded in double cosine series.

8.4–3 Bending

The static solution can be obtained by setting the time derivative terms in Eq. (6) to zero:

$$\begin{bmatrix} \hat{c}_{11} & \hat{c}_{12} & \hat{c}_{13} \\ \hat{c}_{12} & \hat{c}_{22} & \hat{c}_{23} \\ \hat{c}_{13} & \hat{c}_{23} & \hat{c}_{33}+\tilde{s}_{33} \end{bmatrix} \begin{Bmatrix} U_{mn} \\ V_{mn} \\ W_{mn} \end{Bmatrix} = \begin{Bmatrix} 0 \\ 0 \\ Q_{mn} \end{Bmatrix} + \begin{Bmatrix} -\alpha N^6_{mn} \\ -\beta N^6_{mn} \\ \alpha^2 M^1_{mn} + \beta^2 M^2_{mn} \end{Bmatrix} \quad (8.4-8)$$

Using the static condensation procedure presented in Eqs. (8.3–22)–(8.3–26), we can determine the solution to Eq. (8) (when $\tilde{s}_{33} = 0$) as

$$\begin{aligned} W_{mn} =& \frac{1}{a_{mn}} \left[Q_{mn} + \alpha^2 M^1_{mn} + \beta^2 M^2_{mn} - \left(\frac{a_1}{a_0}\alpha + \frac{a_2}{a_0}\beta \right) N^6_{mn} \right] \\ U_{mn} =& \frac{1}{a_0} \left[a_1 W_{mn} - (\alpha \hat{c}_{22} - \beta \hat{c}_{12}) N^6_{mn} \right] \\ V_{mn} =& \frac{1}{a_0} \left[a_2 W_{mn} + (\alpha \hat{c}_{12} - \beta \hat{c}_{11}) N^6_{mn} \right] \end{aligned} \quad (8.4-9a)$$

where

$$\begin{aligned} a_{mn} =& \hat{c}_{33} + \hat{c}_{13}\frac{a_1}{a_0} + \hat{c}_{23}\frac{a_2}{a_0}, \; a_0 = \hat{c}_{11}\hat{c}_{22} - \hat{c}_{12}\hat{c}_{12} \\ a_1 =& \hat{c}_{12}\hat{c}_{23} - \hat{c}_{13}\hat{c}_{22}, \; a_2 = \hat{c}_{13}\hat{c}_{12} - \hat{c}_{11}\hat{c}_{23} \end{aligned} \quad (8.4-9b)$$

Solution of Eq. (9a) for each $m, n = 1, 2, \ldots$ gives (U_{mn}, V_{mn}, W_{mn}), which can then be used to compute the solution (u_0, v_0, w_0) from Eq. (2a–c). If there are no thermal loads, the solution becomes

$$W_{mn} = \frac{Q_{mn}}{a_{mn}}, \quad U_{mn} = \frac{a_1 Q_{mn}}{a_0 a_{mn}}, \quad V_{mn} = \frac{a_2 Q_{mn}}{a_0 a_{mn}} \quad (8.4-9c)$$

8.4–4 Determination of Stresses

The stresses in each layer of an antisymmetric angle-ply laminate can be calculated from [see Eq. (8.3–29)]

$$\begin{Bmatrix} \sigma_{xx} \\ \sigma_{yy} \\ \sigma_{xy} \end{Bmatrix}^{(k)} = \begin{bmatrix} \bar{Q}_{11} & \bar{Q}_{12} & \bar{Q}_{16} \\ \bar{Q}_{12} & \bar{Q}_{22} & \bar{Q}_{26} \\ \bar{Q}_{16} & \bar{Q}_{26} & \bar{Q}_{66} \end{bmatrix}^{(k)} \left(\begin{Bmatrix} \varepsilon_{xx} \\ \varepsilon_{yy} \\ 2\varepsilon_{xy} \end{Bmatrix} - \begin{Bmatrix} \alpha_{xx} \\ \alpha_{yy} \\ 2\alpha_{xy} \end{Bmatrix} \Delta T \right) \quad (8.4-10)$$

$$\begin{aligned} \left(\begin{Bmatrix} \varepsilon_{xx} \\ \varepsilon_{yy} \\ 2\varepsilon_{xy} \end{Bmatrix} - \begin{Bmatrix} \alpha_{xx} \\ \alpha_{yy} \\ 2\alpha_{xy} \end{Bmatrix} \Delta T \right) =& \sum_{m=1}^{\infty}\sum_{n=1}^{\infty} \begin{Bmatrix} \alpha U_{mn}\, g_{mn} - \alpha_{xx} T^0_{mn} f_{mn} \\ \beta V_{mn}\, g_{mn} - \alpha_{yy} T^0_{mn} f_{mn} \\ -\left(\beta U_{mn} + \alpha V_{mn} + 2\alpha_{xy} T^0_{mn}\right) f_{mn} \end{Bmatrix} \\ &+ \sum_{m=1}^{\infty}\sum_{n=1}^{\infty} z \begin{Bmatrix} \left(\alpha^2 W_{mn} - \alpha_{xx} T^1_{mn}\right) f_{mn} \\ \left(\beta^2 W_{mn} - \alpha_{yy} T^1_{mn}\right) f_{mn} \\ -2\alpha\beta W_{mn}\, g_{mn} - 2\alpha_{xy} T^1_{mn} f_{mn} \end{Bmatrix} \end{aligned} \quad (8.4-11a)$$

$$f_{mn} = \sin\frac{m\pi x}{a}\,\sin\frac{n\pi y}{b}\,,\quad g_{mn} = \cos\frac{m\pi x}{a}\,\cos\frac{n\pi y}{b} \tag{8.4-11b}$$

Note that the in-plane stresses in angle-ply laminates will have nonzero contributions from $\bar{Q}_{6i}$. Also, the maxima of $(\sigma_{xx}, \sigma_{yy})$ occur at $(a/2, b/2)$, and they have contributions from $\varepsilon^0_{xy}, \varepsilon^1_{xx}$, and ε^1_{yy}. The values of shear stress σ_{xy} at $(0,0)$ and $(a/2, b/2)$ may be comparable, and relative maximum depends on the specific laminate construction.

The transverse stresses are determined as described in Section 8.3–4. We obtain

$$\sigma^{(k)}_{xz}(x,y,z) = \sum_{m=1}^{\infty}\sum_{n=1}^{\infty}\left[(z-z_k)\bar{\mathcal{A}}^{(k)}_{mn} + \frac{1}{2}(z^2 - z_k^2)\bar{\mathcal{B}}^{(k)}_{mn}\right] + \sigma^{(k-1)}_{xz}(x,y,z_k)$$

$$\sigma^{(k)}_{yz}(x,y,z) = \sum_{m=1}^{\infty}\sum_{n=1}^{\infty}\left[(z-z_k)\bar{\mathcal{C}}^{(k)}_{mn} + \frac{1}{2}(z^2 - z_k^2)\bar{\mathcal{D}}^{(k)}_{mn}\right] + \sigma^{(k-1)}_{yz}(x,y,z_k) \tag{8.4-12a}$$

$$\sigma^{(0)}_{xz}(x,y,z_1) = 0,\quad \sigma^{(0)}_{yz}(x,y,z_1) = 0 \tag{8.4-12c}$$

where

$$\begin{aligned}
\bar{\mathcal{A}}^{(k)}_{mn} &= \left[\left(\alpha^2\bar{Q}^{(k)}_{11} + \beta^2\bar{Q}^{(k)}_{66}\right)U_{mn} + \alpha\beta\left(\bar{Q}^{(k)}_{12} + \bar{Q}^{(k)}_{66}\right)V_{mn}\right]\sin\alpha x\cos\beta y \\
&\quad + \left[2\alpha\beta\bar{Q}^{(k)}_{16}U_{mn} + \left(\alpha^2\bar{Q}^{(k)}_{16} + \beta^2\bar{Q}^{(k)}_{26}\right)V_{mn}\right]\cos\alpha x\sin\beta y \\
\bar{\mathcal{B}}^{(k)}_{mn} &= -\left[\alpha^3\bar{Q}^{(k)}_{11} + \alpha\beta^2\left(\bar{Q}^{(k)}_{12} + 2\bar{Q}^{(k)}_{66}\right)\right]W_{mn}\cos\alpha x\sin\beta y \\
&\quad - \left(3\bar{Q}^{(k)}_{16}\alpha^2\beta + \bar{Q}^{(k)}_{26}\beta^3\right)W_{mn}\sin\alpha x\cos\beta y \\
\bar{\mathcal{C}}^{(k)}_{mn} &= \left[\left(\alpha^2\bar{Q}^{(k)}_{16} + \beta^2\bar{Q}^{(k)}_{26}\right)U_{mn} + 2\alpha\beta\bar{Q}^{(k)}_{16}V_{mn}\right]\sin\alpha x\cos\beta y \\
&\quad + \left[\alpha\beta\left(\bar{Q}^{(k)}_{12} + \bar{Q}^{(k)}_{66}\right)U_{mn} + \left(\alpha^2\bar{Q}^{(k)}_{66} + \beta^2\bar{Q}^{(k)}_{22}\right)V_{mn}\right]\cos\alpha x\sin\beta y \\
\bar{\mathcal{D}}^{(k)}_{mn} &= -\left(\bar{Q}^{(k)}_{16}\alpha^3 + 3\bar{Q}^{(k)}_{26}\alpha\beta^2\right)W_{mn}\cos\alpha x\sin\beta y \\
&\quad - \left[\alpha^2\beta\left(\bar{Q}^{(k)}_{12} + 2\bar{Q}^{(k)}_{66}\right) + \beta^3\bar{Q}^{(k)}_{22}\right]W_{mn}\sin\alpha x\cos\beta y
\end{aligned} \tag{8.4-12b}$$

The maxima of σ_{xz} and σ_{yz} occur at $(x,y) = (0, b/2)$ and $(x,y) = (a/2, 0)$, respectively, although their values at $(x,y) = (a/2, 0)$ and $(x,y) = (0, b/2)$, respectively, are not zero. The location of the maximum through the thickness depends on the lamination scheme.

The bending moments can be calculated from Eq. (3)

$$\begin{aligned}
\begin{Bmatrix} M_{xx} \\ M_{yy} \\ M_{xy} \end{Bmatrix} &= \sum_{m=1}^{\infty}\sum_{n=1}^{\infty}\begin{bmatrix} 0 & 0 & B_{16} \\ 0 & 0 & B_{26} \\ B_{16} & B_{26} & 0 \end{bmatrix}\begin{Bmatrix} \alpha U_{mn}\cos\alpha x\cos\beta y \\ \beta V_{mn}\cos\alpha x\cos\beta y \\ -(\beta U_{mn} + \alpha V_{mn})\sin\alpha x\sin\beta y \end{Bmatrix} \\
&\quad + \sum_{m=1}^{\infty}\sum_{n=1}^{\infty}\begin{bmatrix} D_{11} & D_{12} & 0 \\ D_{12} & D_{22} & 0 \\ 0 & 0 & D_{66} \end{bmatrix}\begin{Bmatrix} \alpha^2 W_{mn}\sin\alpha x\sin\beta y \\ \beta^2 W_{mn}\sin\alpha x\sin\beta y \\ -2\alpha\beta W_{mn}\cos\alpha x\cos\beta y \end{Bmatrix}
\end{aligned} \tag{8.4-13}$$

Note that the locations of the maximum values of M_{xx}, M_{yy}, and M_{xy} cannot be determined in the general case. However, when the coupling coefficients are zero, maximum values of M_{xx} and M_{yy} occur at $(x,y) = (a/2, b/2)$, and the maximum of M_{xy} occurs at $(x,y) = (0,0)$.

The effect of bending–extension coupling and the dependence of the coupling on the modulus ratio can be seen from the deflections $\bar{w} = w_0(E_2h^3/q_0b^4)$ and stresses $\bar{\sigma}_{xx} = \sigma_{xx}(a/2, b/2, h/2)(h^2/q_0b^2)$ presented in Table 8.4–1 for antisymmetric angle-ply laminates $(-45/45)_k$ for $k = 2, 4$, and 8, and subjected to sinusoidal load (first line) and uniformly distributed load (second line). All layers are of equal thickness (E_1/E_2 varied, $G_{12} = G_{13} = 0.5E_2$, $G_{23} = 0.2E_2$, $\nu_{12} = 0.25$, $a/h = 10$). The series for uniform load is evaluated using $m, n = 1, 3, \cdots, 21$ terms. The effect of bending–stretching coupling is to increase the deflection and decrease the stress.

Table 8.4–1: Effect of lamination scheme on the transverse deflections $\bar{w}$ and stresses $\bar{\sigma}_{xx}$ in square angle-ply laminates $(-45/45)_k$.

Laminate	$\bar{w} \times 10^2$					$\bar{\sigma}_{xx}$				
	$\frac{E_1}{E_2} = 1$	10	20	30	40	$\frac{E_1}{E_2} = 1$	10	20	30	40
0°	2.639	0.908	0.523	0.367	0.283	0.174	0.462	0.524	0.549	0.563
	4.172	1.412	0.795	0.548	0.415	0.251	0.693	0.772	0.795	0.801
1†	2.887	1.117	0.757	0.577	0.467	0.190	0.217	0.242	0.256	0.265
	4.577	1.759	1.190	0.906	0.732	0.278	0.308	0.340	0.358	0.370
2	2.887	0.636	0.345	0.237	0.181	0.190	0.153	0.147	0.145	0.144
	4.577	0.999	0.542	0.372	0.283	0.278	0.214	0.205	0.201	0.199
4	2.887	0.574	0.304	0.207	0.157	0.190	0.151	0.146	0.144	0.143
	4.577	0.902	0.477	0.324	0.245	0.278	0.211	0.202	0.199	0.197

† Denotes k in the laminate $(-45/45)_k$.

Figure 8.4–1 contains a plot of the nondimensionalized deflection $\bar{w}$ versus plate aspect ratio for simply supported (SS–2) antisymmetric angle-ply laminates $(-45/45)_k$ under sinusoidal load. Figure 8.4–2 contains $\bar{w}$ as a function of the lamination angle θ for square laminates $(-\theta/\theta)_k$ under sinusoidal load. The material properties used are $E_1/E_2 = 25$, $G_{12} = G_{13} = 0.5E_2$, and $\nu_{12} = 0.25$. Clearly, the bending–extension coupling is quite significant for two-layered plates, but the coupling decreases very rapidly as the number of layers is increased. Lastly, nondimensionalized transverse deflections as a function of the modulus ratio for square laminates under sinusoidal transverse load are presented in Figure 8.4–3. The effect of coupling is significant for all modulus ratios except for those close to unity.

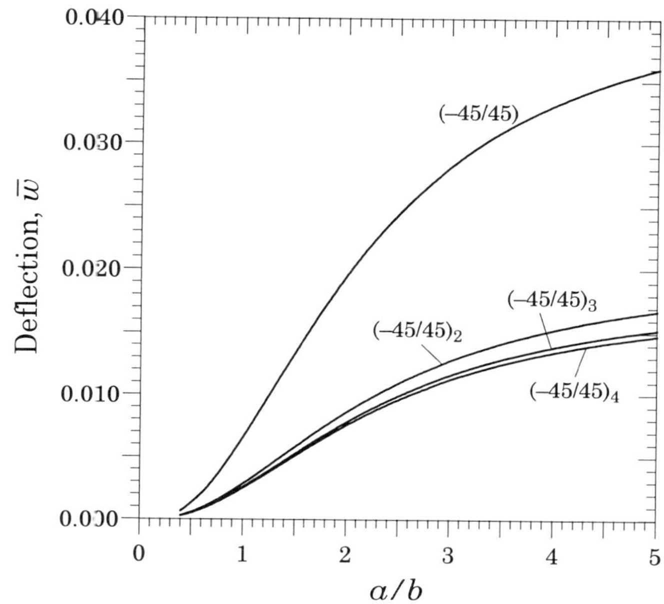

Figure 8.4–1: Nondimensionalized maximum transverse deflection ($\bar{w}$) versus plate aspect ratio (a/b) for antisymmetric angle-ply $(-45/45)_n$ ($n = 1, 2, 3, 4$) laminates under sinusoidal load.

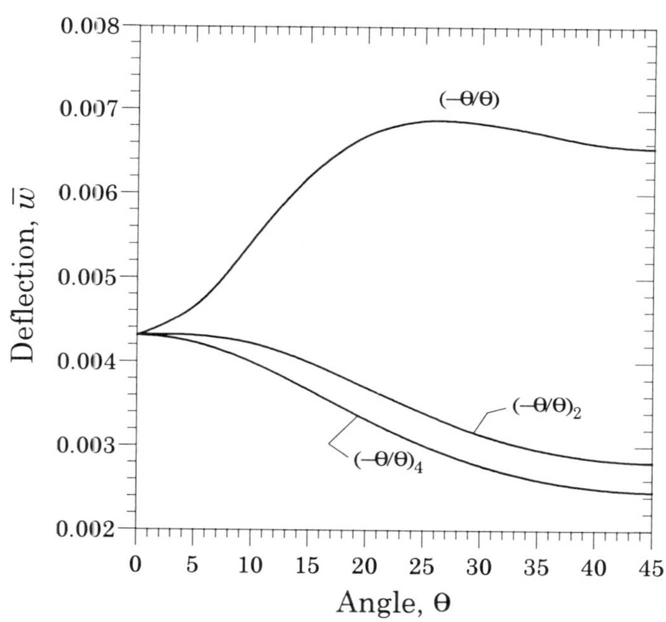

Figure 8.4–2: Nondimensionalized maximum transverse deflection ($\bar{w}$) versus lamination angle (θ) for antisymmetric angle-ply $(-\theta/\theta)_n$ ($n = 1, 2, 4$) laminates under sinusoidal load.

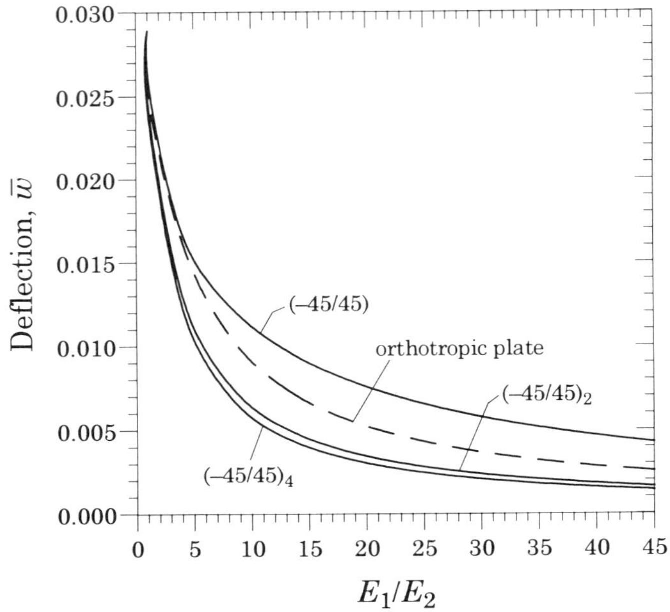

Figure 8.4–3: Nondimensionalized maximum transverse deflection ($\bar{w}$) versus modulus ratio (E_1/E_2) for antisymmetric angle-ply $(-45/45)_n$ $(n = 1, 2, 4)$ laminates under sinusoidal load.

Figures 8.4–4 and 8.4–5 show the plots of nondimensionalized transverse shear stresses $\bar{\sigma}_{xz}(0, b/2, z) = \bar{\sigma}_{yz}(a/2, 0, z)$ and $\bar{\sigma}_{xz}(a/2, 0, z) = \bar{\sigma}_{yz}(0, b/2, z)$, respectively, for two-layer and eight-layer antisymmetric angle-ply laminates $(-45/45/-45/\cdots)$ under sinusoidally distributed transverse load $(a/b = 1,\ E_1 = 25E_2,\ G_{12} = G_{13} = 0.5E_2,\ \nu_{12} = 0.25)$. Unlike in antisymmetric cross-ply laminates, the stress $\bar{\sigma}_{xz}$ is not zero at $(x, y) = (a/2, 0)$, although small in magnitude compared to that at $(x, y) = (0, b/2)$. Note that through-thickness variations are significantly altered when the number of layers are increased (for the same total laminate thickness). The parabolic type variation shown in Figure 8.4–4 is consistent with that of an orthotropic plate.

8.4–5 Buckling

For buckling analysis, we assume that the only applied loads are the in-plane forces and all other mechanical and thermal loads are zero:

$$\hat{N}_{xx} = -N_0\ , \quad \hat{N}_{yy} = -kN_0\ , \quad k = \frac{\hat{N}_{yy}}{\hat{N}_{xx}} \tag{8.4-14}$$

From Eq. (6) we have

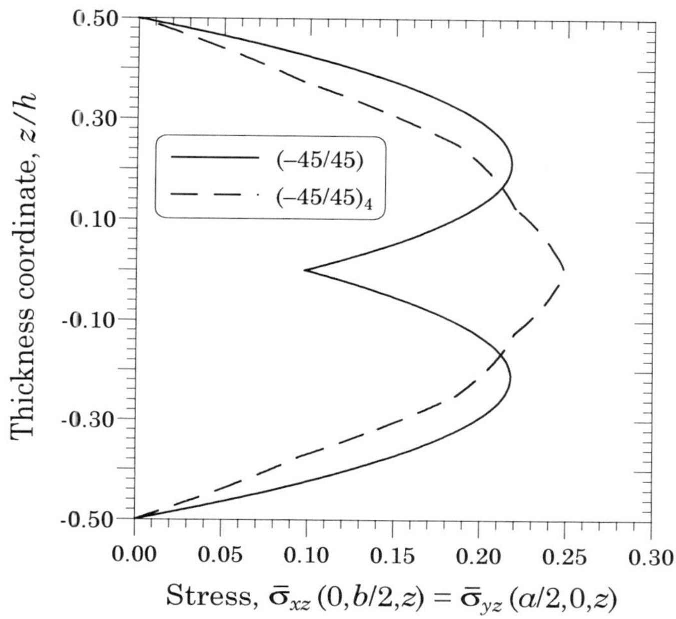

Figure 8.4–4: Variation of the nondimensionalized maximum transverse shear stresses through the thickness of antisymmetric angle-ply $(-45/45)_n$ laminates.

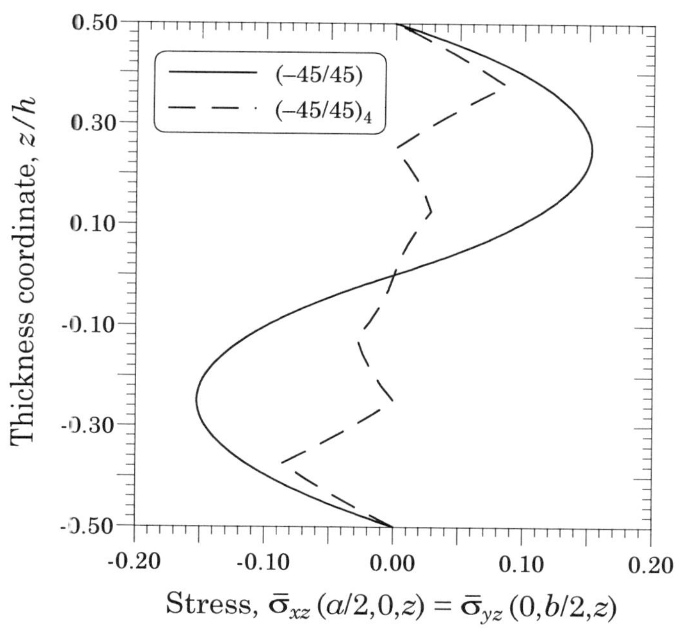

Figure 8.4–5: Variation of the nondimensionalized maximum transverse shear stresses through the thickness of antisymmetric angle-ply $(-45/45)_n$ laminates.

$$\begin{bmatrix} \hat{c}_{11} & \hat{c}_{12} & \hat{c}_{13} \\ \hat{c}_{12} & \hat{c}_{22} & \hat{c}_{23} \\ \hat{c}_{13} & \hat{c}_{23} & \hat{c}_{33} - N_0(\alpha^2 + k\beta^2) \end{bmatrix} \begin{Bmatrix} U_{mn} \\ V_{mn} \\ W_{mn} \end{Bmatrix} = \begin{Bmatrix} 0 \\ 0 \\ 0 \end{Bmatrix} \qquad (8.4-15)$$

where $\hat{c}_{ij}$ are the coefficients defined in Eq. (7a). Setting the determinant of the coefficient matrix in (15) to zero, we obtain

$$\begin{aligned} N_0 &= \frac{1}{\alpha^2 + k\beta^2}\left(\hat{c}_{33} - \frac{\hat{c}_{13}\hat{c}_{22} - \hat{c}_{23}\hat{c}_{12}}{\hat{c}_{11}\hat{c}_{22} - \hat{c}_{12}\hat{c}_{12}}\hat{c}_{13} - \frac{\hat{c}_{11}\hat{c}_{23} - \hat{c}_{12}\hat{c}_{13}}{\hat{c}_{11}\hat{c}_{22} - \hat{c}_{12}\hat{c}_{12}}\hat{c}_{23}\right) \\ &= \frac{1}{\alpha^2 + k\beta^2}\left(\hat{c}_{33} + \frac{2\hat{c}_{12}\hat{c}_{23}\hat{c}_{13} - \hat{c}_{22}\hat{c}_{13}^2 - \hat{c}_{11}\hat{c}_{23}^2}{\hat{c}_{11}\hat{c}_{22} - \hat{c}_{12}^2}\right) \end{aligned} \qquad (8.4-16)$$

Clearly, for each pair of m and n, there is a unique value of N_0. The critical buckling load is the smallest of all $N_0 = N_0(m,n)$. Since $\hat{c}_{ij}$ depend on m and n, $N_0(m,n)$ is a complicated function of both m and n and no conclusions can be drawn about the mode (m,n) at buckling.

For specially orthotropic laminates (*i.e.*, a plate with a single specially orthotropic layer or laminate consisting of specially orthotropic layers that are symmetrically arranged about the laminate middle surface), the only nonzero stiffnesses are $A_{11}, A_{12}, A_{22}, A_{66}, D_{11}, D_{12}, D_{22}$, and D_{66}. Thus, neither shear or twist coupling nor bending–extension coupling exists. For biaxial compressive in-plane loading, the buckling load is given by Eq. (8.3–45). The *specially orthotropic solution* for antisymmetric angle-ply laminates is the one that corresponds to the case in which $A_{16}, A_{26}, B_{16}, B_{26}, D_{16}$, and D_{26} are zero.

Table 8.4–2 contains nondimensionalized buckling loads ($\bar{N} = N_{cr}b^2/E_2h^3$) of antisymmetric angle-ply laminates under uniaxial and biaxial in-plane compressive loads. The material properties used for a typical lamina are: $G_{12} = 0.5E_2$, and $\nu_{12} = 0.25$. The buckling mode is (1,1), except for uniaxial compression with aspect ratio equal to 1.5.

Plots of nondimensionalized critical buckling loads versus plate aspect ratio (a/b) for simply supported (SS–2) angle-ply laminates $(45/-45)_k$ under uniaxial compressive in-plane loads are presented in Figure 8.4–6 for $E_1/E_2 = 40$, $G_{12} = G_{13} = 0.5E_2$, and $\nu_{12} = 0.25$. The buckling mode associated with the critical buckling load is $(m,n) = (1,1)$ for $a/b \leq 1.4$, $(m,n) = (2,1)$ for $1.5 \leq a/b \leq 2.4$, $(m,n) = (3,1)$ for $2.5 \leq a/b \leq 3.4$, $(m,n) = (4,1)$ for $3.5 \leq a/b \leq 4.4$, and $(m,n) = (5,1)$ for $4.5 \leq a/b \leq 5$. The effect of bending–stretching coupling is the most for two-layer laminates, and the orthotropic solution is rapidly approached as the number of plies is increased.

Table 8.4–2: Effect of coupling, plate aspect ratio, and modulus ratio on the nondimensionalized critical buckling load, $\bar{N} = N_{cr}\frac{b^2}{E_2h^3}$, of rectangular laminates under uniform compression and biaxial compression (E_1/E_2 varied, $G_{12} = G_{13} = 0.5E_2$, $\nu_{12} = 0.25$).

	(45/−45)				$(45/-45)_4$			
$\frac{a}{b}$	10†	20	25	40	10	20	25	40
Uniaxial compression ($k = 0$)								
0.5	1.280	1.838	2.110	2.919	2.406	4.442	5.460	8.513
1.0	0.918	1.355	1.568	2.200	1.787	3.376	4.171	6.554
1.5*	0.973	2.428	1.650	2.308	1.881	3.537	4.366	6.850
Biaxial compression ($k = 1$)								
0.5	1.025	1.471	1.688	2.335	1.925	3.554	4.368	6.811
1.0	0.459	0.678	0.784	1.100	0.893	1.688	2.085	3.277
1.5	0.317	0.914	0.534	0.745	0.608	1.140	1.406	2.203

† Modulus ratio.

* Mode is (2,1) for this row ($a/b = 1.5$); for all other cases, the mode is (1,1).

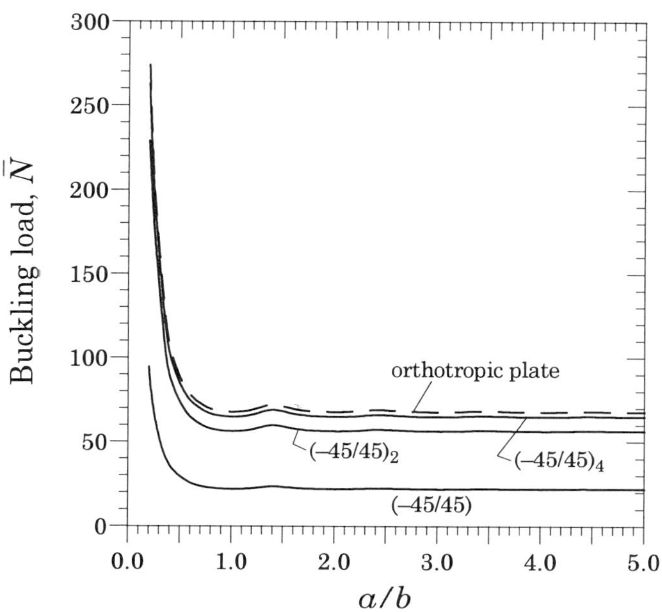

Figure 8.4–6: Nondimensionalized buckling load ($\bar{N}$) versus plate aspect ratio (a/b) of antisymmetric angle-ply square laminates under uniaxial compressive edge load.

Nondimensionalized critical buckling loads versus the lamination angle for uniaxial compression ($k = 0$) and biaxial compression ($k = 1$) antisymmetric angle-ply square laminates are shown in Figures 8.4–7 and 8.4–8, respectively, for $E_1 = 40E_2$, $G_{12} = 0.5E_2$, and $\nu_{12} = 0.25$. The plots shown in Figure 8.4–8 are symmetric about $\theta = 45°$. Note that, once again, the bending–stretching coupling severely reduces the buckling load for the two-layer plate. The effect is negligible for eight or more layers. The buckling load is the maximum for $\theta = 45°$.

8.4–6 Vibration

For free vibration Eq. (6) reduces to the eigenvalue problem

$$\left(\begin{bmatrix} \hat{c}_{11} & \hat{c}_{12} & \hat{c}_{13} \\ \hat{c}_{12} & \hat{c}_{22} & \hat{c}_{23} \\ \hat{c}_{13} & \hat{c}_{23} & \hat{c}_{33} \end{bmatrix} - \omega^2 \begin{bmatrix} \hat{m}_{11} & 0 & 0 \\ 0 & \hat{m}_{22} & 0 \\ 0 & 0 & \hat{m}_{33} \end{bmatrix}\right) \begin{Bmatrix} U_{mn}^0 \\ V_{mn}^0 \\ W_{mn}^0 \end{Bmatrix} = \begin{Bmatrix} 0 \\ 0 \\ 0 \end{Bmatrix} \tag{8.4 – 17}$$

where $\hat{c}_{ij}$ and $\hat{m}_{ij}$ are defined in Eq. (7). Setting the determinant of the coefficient matrix in (17) to zero, we obtain the cubic characteristic polynomial

$$-p\lambda^3 + q\lambda^2 - r\lambda + s = 0 \tag{8.4 – 18}$$

in the eigenvalue $\lambda = \omega^2$, where

$$p = \begin{vmatrix} \hat{m}_{11} & 0 & 0 \\ 0 & \hat{m}_{22} & 0 \\ 0 & 0 & \hat{m}_{33} \end{vmatrix}, \quad s = \begin{vmatrix} \hat{c}_{11} & \hat{c}_{12} & \hat{c}_{13} \\ \hat{c}_{12} & \hat{c}_{22} & \hat{c}_{23} \\ \hat{c}_{13} & \hat{c}_{23} & \hat{c}_{33} \end{vmatrix}$$

$$q = \begin{vmatrix} \hat{c}_{11} & 0 & 0 \\ \hat{c}_{12} & \hat{m}_{22} & 0 \\ \hat{c}_{13} & 0 & \hat{m}_{33} \end{vmatrix} + \begin{vmatrix} \hat{m}_{11} & \hat{c}_{12} & 0 \\ \hat{0} & \hat{c}_{22} & 0 \\ 0 & \hat{c}_{23} & \hat{m}_{33} \end{vmatrix} + \begin{vmatrix} \hat{m}_{11} & 0 & \hat{c}_{13} \\ \hat{0} & \hat{m}_{22} & \hat{c}_{23} \\ 0 & 0 & \hat{c}_{33} \end{vmatrix}$$

$$r = \begin{vmatrix} \hat{c}_{11} & \hat{c}_{12} & 0 \\ \hat{c}_{12} & \hat{c}_{22} & 0 \\ \hat{c}_{13} & \hat{c}_{23} & \hat{m}_{33} \end{vmatrix} + \begin{vmatrix} \hat{c}_{11} & \hat{0} & \hat{c}_{13} \\ \hat{c}_{12} & \hat{m}_{22} & \hat{c}_{23} \\ \hat{c}_{13} & 0 & \hat{c}_{33} \end{vmatrix} + \begin{vmatrix} \hat{m}_{11} & \hat{c}_{12} & \hat{c}_{13} \\ \hat{0} & \hat{c}_{22} & \hat{c}_{23} \\ 0 & \hat{c}_{23} & \hat{c}_{33} \end{vmatrix} \tag{8.4 – 19}$$

If the in-plane inertias are neglected (*i.e.*, $\hat{m}_{11} = \hat{m}_{22} = 0$), Eq. (17) yields

$$\omega^2 = \frac{1}{\hat{m}_{33}}\left(\hat{c}_{33} - \frac{\hat{c}_{13}\hat{c}_{22} - \hat{c}_{23}\hat{c}_{12}}{\hat{c}_{11}\hat{c}_{22} - \hat{c}_{12}\hat{c}_{12}}\hat{c}_{13} - \frac{\hat{c}_{11}\hat{c}_{23} - \hat{c}_{12}\hat{c}_{13}}{\hat{c}_{11}\hat{c}_{22} - \hat{c}_{12}\hat{c}_{12}}\hat{c}_{23}\right) \tag{8.4 – 20}$$

Note that ω is a function of the mode numbers (m, n) because the coefficients $\hat{c}_{ij}$ depend on m and n, as shown in Eq. (7a).

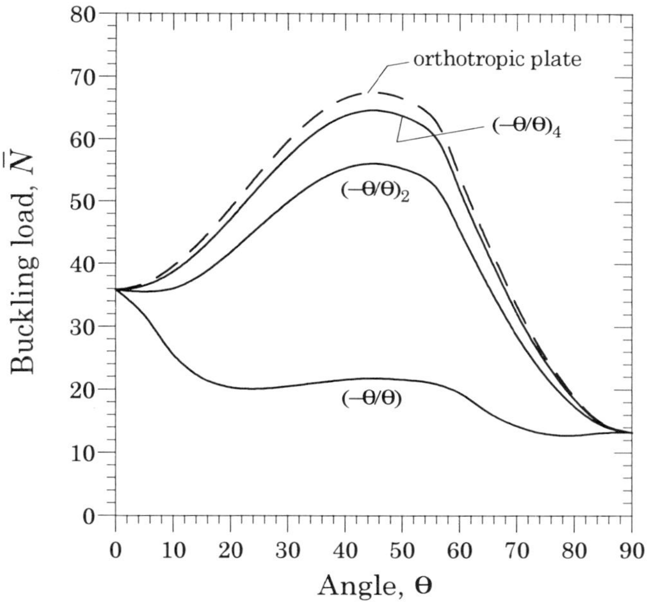

Figure 8.4–7: Nondimensionalized buckling load ($\bar{N}$) versus lamination angle (θ) of antisymmetric angle-ply square laminates under uniaxial compressive edge load.

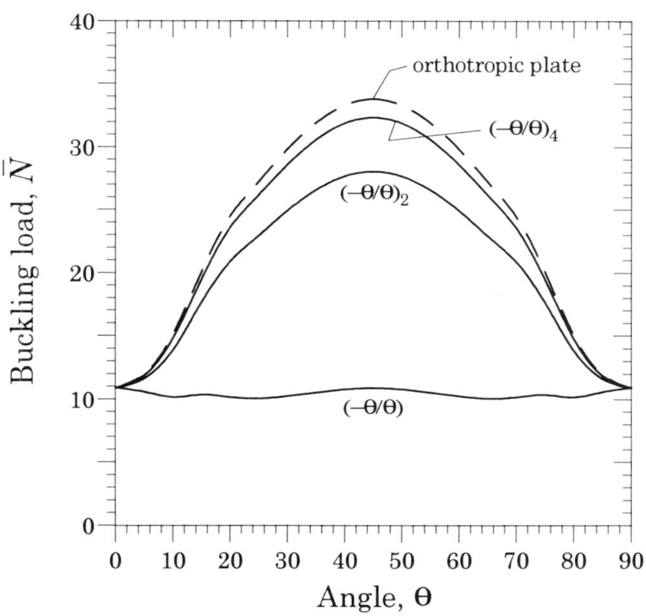

Figure 8.4–8: Nondimensionalized buckling load ($\bar{N}$) versus lamination angle (θ) of antisymmetric angle-ply square laminates under biaxial compressive edge loads.

Nondimensionalized fundamental frequencies $\bar{\omega} = \omega(b^2/\pi^2)\sqrt{\rho h/D_{22}}$ of graphite–epoxy composites with $E_1/E_2 = 40, G_{12}/E_2 = 0.5, \nu_{12} = 0.25$ and $a/b = 1$ are shown as a function of lamination angle in Figure 8.4–9. The bending–stretching coupling due to the presence of B_{16} and B_{26} has the effect of lowering the frequencies. The coupling is the maximum for two-layer plates, and it rapidly decreases with increasing number of layers. At $\theta = 45°$, the fundamental frequency of the two-layer plate is about 40 percent lower than that of the eight-layer laminate. The same conclusions hold for results presented in Figures 8.4–10 and 8.4–11. The effect of coupling is significant for all modulus ratios, and the difference between the two-layer solution and orthotropic solution increases with modulus ratio.

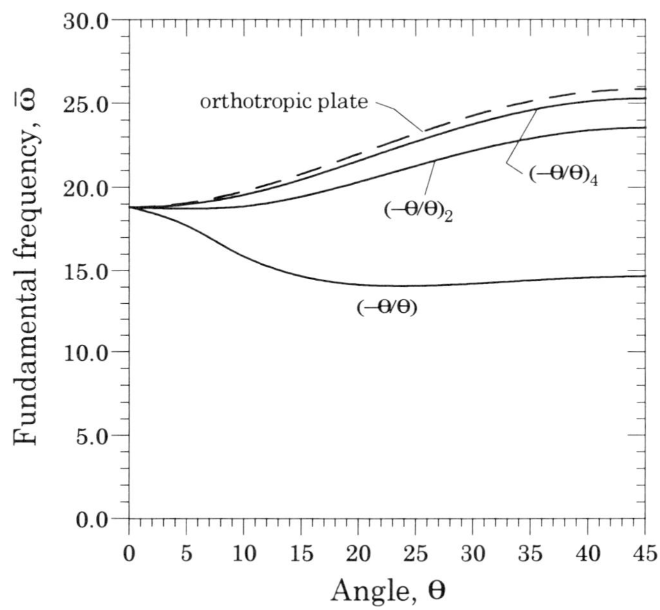

Figure 8.4–9: Nondimensionalized fundamental frequency versus lamination angle (θ) of antisymmetric angle-ply square laminates.

8.5 The Lévy Solutions

8.5–1 Introduction

The Lévy method can be used to solve the governing equations of various plate theories for rectangular laminates for which two (parallel) opposite edges are simply supported and the other two edges can have any boundary conditions. Here we describe the Lévy solution procedure for cross-ply and antisymmetric angle-ply laminates using the classical laminate plate theory (CLPT). However, details are presented for only cross-ply laminates.

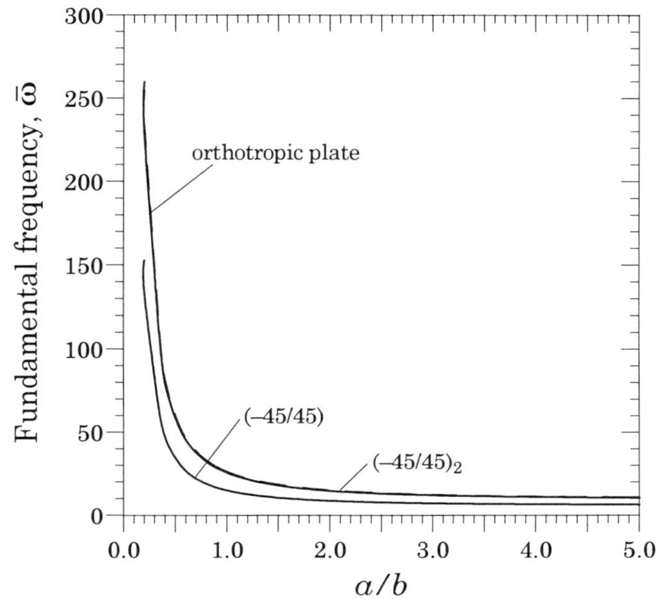

Figure 8.4–10: Nondimensionalized fundamental frequency ($\bar{\omega}$) versus plate aspect ratio (a/b) of antisymmetric angle-ply laminates.

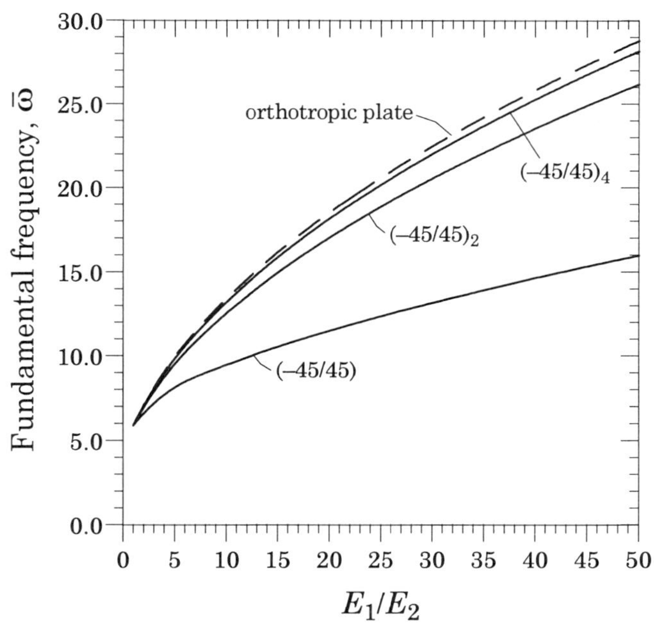

Figure 8.4–11: Nondimensionalized fundamental frequency ($\bar{\omega}$) versus modulus ratio E_1/E_2 of antisymmetric angle-ply square laminates.

Consider a rectangular laminate which has an even number of orthotropic layers with principal material directions alternating at 0° and 90° to the laminate axes (*i.e.*, antisymmetric cross-ply laminate). The planar dimensions are taken to be a and b, and the total thickness h. The laminate coordinate system (x, y, z) is taken such that $-a/2 \leq x \leq a/2, 0 \leq y \leq b, -h/2 \leq z \leq h/2$, as shown in Figure 8.5–1. Here we assume that the edges $y = 0, b$ are simply supported, and the other two edges can each have arbitrary boundary conditions (*e.g.*, simply supported, clamped, or free). The type of the boundary conditions for the classical laminate plate theory were derived in Section 5.3 [see Eq. (5.3–34)]. Note that at least one and only one quantity in each of the following pairs should be specified on the boundary:

$$(u_n, N_{nn}), \ (u_s, N_{ns}), \ (w_0, V_n) \ , \ (\frac{\partial w_0}{\partial n}, M_{nn}) \tag{8.5 - 1}$$

where n refers to the normal and s to the tangential directions at the boundary point.

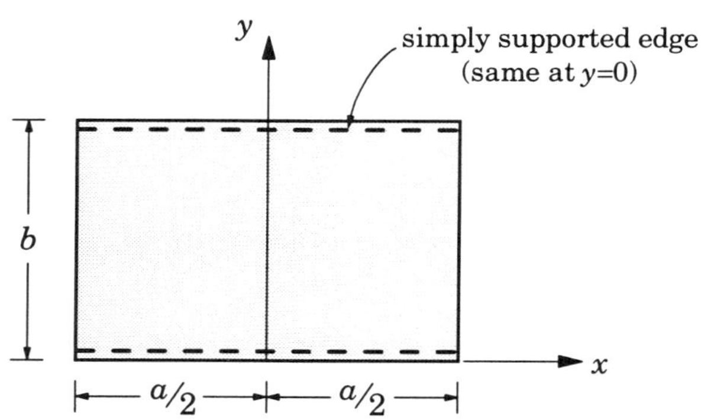

Figure 8.5–1: The coordinate system used for the Lévy solution of rectangular plates.

The simply supported boundary conditions on edges $y = 0, b$ $(n = \mp y, s = \mp x, u_n = v_0, u_s = u_0$, etc.) are expressed as follows:

$$u_0 = 0, \ N_{yy} = 0, \ w_0 = 0, \ M_{yy} = 0 \tag{8.5 - 2}$$

One of the following three types of boundary conditions may be used on the remaining two edges, $x = \mp\frac{a}{2}$ $(n = \mp x$ and $s = \mp y)$:

Simply supported (S): $N_{xx} = 0, \ v_0 = 0, \ w_0 = 0, \ M_{xx} = 0$ (8.5 – 3)

Clamped (C): $u_0 = 0, \ v_0 = 0, \ w_0 = 0, \ \frac{\partial w_0}{\partial x} = 0$ (8.5 – 4)

Free (F): $N_{xx} = 0, \ N_{xy} = 0, \ V_x \equiv \frac{\partial M_{xx}}{\partial x} + 2\frac{\partial M_{xy}}{\partial y} = 0, \ M_{xx} = 0$ (8.5 – 5)

The basic idea of the Lévy method is to seek a solution that satisfies the boundary conditions along the simply supported edges exactly, and thereby reduce the two-dimensional problem to a one-dimensional problem with respect to the coordinate x. This results in ordinary differential equations in x, which involve usually second- or higher-order derivatives of the unknown coefficients of the displacement expansion. These ordinary differential equations are then solved using the so-called *state–space approach* (see Brogan [2] or Franklin [3]).

For the case of antisymmetric cross-ply laminates, we have

$$B_{12} = B_{16} = B_{26} = B_{66} = A_{16} = A_{26} = D_{16} = D_{26} = 0 \qquad (8.5-6)$$

Consequently, from Eqs. (8.2–1)–(8.2–3), we have the following equations of motion of the classical laminate theory for the isothermal case:

$$\begin{aligned} &A_{11}\frac{\partial^2 u_0}{\partial x^2} + A_{66}\frac{\partial^2 u_0}{\partial y^2} + (A_{12}+A_{66})\frac{\partial^2 v_0}{\partial x \partial y} \\ &- \left[B_{11}\frac{\partial^3 w_0}{\partial x^3} + (B_{12}+2B_{66})\frac{\partial^3 w_0}{\partial x \partial y^2}\right] = I_0\ddot{u}_0 - I_1\frac{\partial \ddot{w}_0}{\partial x} \end{aligned} \qquad (8.5-7)$$

$$\begin{aligned} &(A_{12}+A_{66})\frac{\partial^2 u_0}{\partial x \partial y} + A_{66}\frac{\partial^2 v_0}{\partial x^2} + A_{22}\frac{\partial^2 v_0}{\partial y^2} \\ &- \left[(B_{12}+2B_{66})\frac{\partial^3 w_0}{\partial x^2 \partial y} + B_{22}\frac{\partial^3 w_0}{\partial y^3}\right] = I_0\ddot{v}_0 - I_1\frac{\partial \ddot{w}_0}{\partial y} \end{aligned} \qquad (8.5-8)$$

$$\begin{aligned} &B_{11}\frac{\partial^3 u_0}{\partial x^3} + (B_{12}+2B_{66})\left(\frac{\partial^3 u_0}{\partial x \partial y^2} + \frac{\partial^3 v_0}{\partial x^2 \partial y}\right) + B_{22}\frac{\partial^3 v_0}{\partial y^3} \\ &- \left[D_{11}\frac{\partial^4 w_0}{\partial x^4} + 2\left(D_{12}+2D_{66}\right)\frac{\partial^4 w_0}{\partial x^2 \partial y^2} + D_{22}\frac{\partial^4 w_0}{\partial y^4}\right] \\ &+ \hat{N}_{xx}\frac{\partial^2 w_0}{\partial x^2} + 2\hat{N}_{xy}\frac{\partial^2 w_0}{\partial x \partial y} + \hat{N}_{yy}\frac{\partial^2 w_0}{\partial y^2} + q \\ &= I_1\left(\frac{\partial \ddot{u}_0}{\partial x} + \frac{\partial \ddot{v}_0}{\partial y}\right) + I_0\ddot{w}_0 - I_2\left(\frac{\partial^2 \ddot{w}_0}{\partial x^2} + \frac{\partial^2 \ddot{w}_0}{\partial y^2}\right) \end{aligned} \qquad (8.5-9)$$

8.5–2 Solution Procedure

In the Lévy type procedure, we assume the following representation of the displacements:

$$u_0(x, y, t) = \sum_{m=1}^{\infty} U_m(x, t) \sin \beta y$$

$$v_0(x,y,t) = \sum_{m=1}^{\infty} V_m(x,t)\cos\beta y$$

$$w_0(x,y,t) = \sum_{m=1}^{\infty} W_m(x,t)\sin\beta y \qquad (8.5-10a)$$

The transverse load is expanded as

$$q(x,y,t) = \sum_{m=1}^{\infty} Q_m(x,t)\sin\beta y \qquad (8.5-10b)$$

where $\beta = \frac{m\pi}{b}$ and U_m, V_m, W_m, and Q_m denote amplitudes of u_0, v_0, w_0, and q, respectively. These expansions satisfy the simply supported (SS–1) boundary conditions (2) on edges $y = 0, b$.

The stress resultants derived from the displacement field (10a) are given by

$$N_{xx} = \sum_{m=1}^{\infty}\left(A_{11}U_m' - \beta A_{12}V_m + \beta^2 B_{12}W_m - B_{11}W_m''\right)\sin\beta y$$

$$N_{yy} = \sum_{m=1}^{\infty}\left(A_{12}U_m' - \beta A_{22}V_m + \beta^2 B_{22}W_m - B_{12}W_m''\right)\sin\beta y$$

$$N_{xy} = \sum_{m=1}^{\infty}\left(\beta A_{66}U_m + A_{66}V_m' - 2\beta B_{66}W_m'\right)\cos\beta y$$

$$M_{xx} = \sum_{m=1}^{\infty}\left(B_{11}U_m' - \beta B_{12}V_m + \beta^2 D_{12}W_m - D_{11}W_m''\right)\sin\beta y$$

$$M_{yy} = \sum_{m=1}^{\infty}\left(B_{12}U_m' - \beta B_{22}V_m + \beta^2 D_{22}W_m - D_{12}W_m''\right)\sin\beta y$$

$$M_{xy} = \sum_{m=1}^{\infty}\left(\beta B_{66}U_m + B_{66}V_m' - 2\beta D_{66}W_m'\right)\cos\beta y$$

$$\frac{\partial M_{xx}}{\partial x} = \sum_{m=1}^{\infty}\left(B_{11}U_m'' - \beta B_{12}V_m' + \beta^2 D_{12}W_m' - D_{11}W_m'''\right)\sin\beta y$$

$$\frac{\partial M_{xy}}{\partial y} = \sum_{m=1}^{\infty}\left(-\beta^2 B_{66}U_m - \beta B_{66}V_m' + 2\beta^2 D_{66}W_m'\right)\sin\beta y$$

$$\frac{\partial M_{yy}}{\partial y} = \sum_{m=1}^{\infty}\left(\beta B_{12}U_m' - \beta^2 B_{22}V_m - \beta D_{12}W_m'' + \beta^3 D_{22}W_m\right)\cos\beta y$$

$$\frac{\partial M_{xy}}{\partial x} = \sum_{m=1}^{\infty}\left(\beta B_{66}U_m' + B_{66}V_m'' - 2\beta D_{66}W_m''\right)\cos\beta y \qquad (8.5-11)$$

The boundary conditions in (3)–(5) on edges $x = \mp a/2$ require that U_m, V_m, W_m, and their derivatives with respect to x satisfy the following boundary conditions:

Simply supported $(A_{11}D_{11} - B_{11}^2 \neq 0)$:

$$U_m' = 0,\ V_m = 0,\ W_m = 0,\ W_m'' = 0 \qquad (8.5-12)$$

Clamped:

$$U_m = 0,\ V_m = 0,\ W_m = 0,\ W_m' = 0 \qquad (8.5-13)$$

Free:

$$A_{11}U_m' - \beta A_{12}V_m + \beta^2 B_{12}W_m - B_{11}W_m'' = 0 \qquad (8.5-14a)$$

$$\beta A_{66}U_m + A_{66}V_m' - 2\beta B_{66}W_m' = 0 \qquad (8.5-14b)$$

$$\begin{aligned}-2\beta^2 B_{66}U_m + B_{11}U_m'' - \beta\left(B_{12} + 2B_{66}\right)V_m'\\ +\beta^2\left(D_{12} + 4D_{66}\right)W_m' - D_{11}W_m''' &= 0\end{aligned} \qquad (8.5-14c)$$

$$B_{11}U_m' - \beta B_{12}V_m + \beta^2 D_{12}W_m - D_{11}W_m'' = 0 \qquad (8.5-14d)$$

Substituting Eq. (10) into Eqs. (7)–(9), expressing the results in terms of the highest derivatives U_m'', V_m'', and W_m'''', and substituting for U_m'' and V_m'' into the expression for W_m'''', we obtain (when $\hat{N}_{xy} = 0$)

$$U_m'' = C_1U_m + C_2V_m' - C_3W_m' + C_4W_m''' + D_1\ddot{U}_m - D_2\ddot{W}_m' \qquad (8.5-15a)$$

$$V_m'' = -C_5U_m' + C_6V_m - C_7W_m + C_8W_m'' + D_3\ddot{V}_m - D_4\ddot{W}_m \qquad (8.5-15b)$$

$$\begin{aligned}W_m'''' &= C_9U_m' + C_{10}V_m + C_{11}W_m + C_{12}W_m'' - C_0Q_m\\ &\quad + D_5\ddot{U}_m' + D_6\ddot{V}_m + D_7\ddot{W}_m + D_8\ddot{W}_m''\end{aligned} \qquad (8.5-15c)$$

The coefficients C_i appearing in Eqs. (15) are

$$\begin{aligned}
&C_1 = \frac{e_2}{e_1},\quad C_2 = \frac{e_3}{e_1},\quad C_3 = \frac{e_4}{e_1},\quad C_4 = \frac{e_5}{e_1}\\
&C_5 = \frac{e_3}{e_6},\quad C_6 = \frac{e_7}{e_6},\quad C_7 = \frac{e_8}{e_6},\quad C_8 = \frac{e_9}{e_6}\\
&C_9 = \left[-e_4 + e_5C_1 + (e_9 - e_5C_2)C_5\right]C_0\\
&C_{10} = \left[e_8 - (e_9 - e_5C_2)\right]C_0,\quad C_0 = (e_{13})^{-1}\\
&C_{11} = \left[-e_{11} - (e_9 - e_5C_2)C_7\right]C_0\\
&C_{12} = \left[-e_{12} - e_5C_3 - (e_9 - e_5C_2)C_8\right]C_0\\
&D_1 = \frac{d_1}{e_1},\quad D_2 = \frac{d_2}{e_1},\quad D_3 = \frac{d_1}{e_6},\quad D_4 = \frac{d_3}{e_6}\\
&D_5 = (e_5D_1 - d_2)C_0,\quad D_6 = \left[d_3 + (e_5C_2 - e_9)D_3\right]C_0\\
&D_7 = \left[-d_4(e_9 - e_5C_2)D_4\right]C_0,\quad D_8 = (d_5 - e_5D_2)C_0
\end{aligned} \qquad (8.5-16a)$$

where

$$\begin{aligned}
&e_1 = A_{11},\quad e_2 = \beta^2A_{66},\quad e_3 = \beta(A_{12} + A_{66}),\quad e_4 = \beta^2(B_{12} + 2B_{66}),\quad e_5 = B_{11}\\
&e_6 = A_{66},\quad e_7 = \beta^2A_{22},\quad e_8 = \beta^3B_{22},\quad e_9 = \beta(B_{12} + 2B_{66}),\quad e_{10} = D_{11}\\
&e_{11} = \left(\beta^4D_{22} + \beta^2\hat{N}_{yy}\right),\quad e_{12} = 2\beta^2(D_{12} + 2D_{66}) + \hat{N}_{xx},\quad e_{13} = e_{10} - e_5C_4\\
&d_1 = I_0,\quad d_2 = I_1,\quad d_3 = I_1\beta,\quad d_4 = I_0 + \beta^2I_2,\quad d_5 = I_2
\end{aligned} \qquad (8.5-16b)$$

The linear system of ordinary differential equations in (15) with constant coefficients can be expressed in the form of a single, first-order, matrix differential equation

$$\{Z'\} = [T]\{Z\} + \{F\} \tag{8.5 – 17}$$

where

$$\{Z\} = \begin{Bmatrix} U_m \\ U_m' \\ V_m \\ V_m' \\ W_m \\ W_m' \\ W_m'' \\ W_m''' \end{Bmatrix}, \quad \{b\} = \begin{Bmatrix} 0 \\ D_1\ddot{U}_m - D_2\ddot{W}_m' \\ 0 \\ D_3\ddot{V}_m - D_4\ddot{W}_m \\ 0 \\ 0 \\ 0 \\ \hat{Q}_m \end{Bmatrix} \tag{8.5 – 18a}$$

$$[T] = \begin{bmatrix} 0 & 1 & 0 & 0 & 0 & 0 & 0 & 0 \\ C_1 & 0 & 0 & C_2 & 0 & -C_3 & 0 & C_4 \\ 0 & 0 & 0 & 1 & 0 & 0 & 0 & 0 \\ 0 & -C_5 & C_6 & 0 & -C_7 & 0 & C_8 & 0 \\ 0 & 0 & 0 & 0 & 0 & 1 & 0 & 0 \\ 0 & 0 & 0 & 0 & 0 & 0 & 1 & 0 \\ 0 & 0 & 0 & 0 & 0 & 0 & 0 & 1 \\ 0 & C_9 & C_{10} & 0 & C_{11} & 0 & C_{12} & 0 \end{bmatrix} \tag{8.5 – 18b}$$

and

$$\hat{Q}_m = -C_0 Q_m + D_5\ddot{U}_m' + D_6\ddot{V}_m + D_7\ddot{W}_m + D_8\ddot{W}_m'' \tag{8.5 – 18c}$$

Next, we discuss the solution of Eq. (17) separately for bending, vibration, and buckling problems. In each case, we solve Eq. (17) or its special cases.

Bending

In the case of static bending, all variables are independent of time. Equations (17) and (18) hold with

$$\hat{Q}_m = C_0 Q_m \tag{8.5 – 19}$$

In addition, $\hat{N}_{xx}$ and $\hat{N}_{yy}$ appearing in the definition of the coefficients e_{12} and e_{11} are assumed to be zero.

The solution of Eq. (17), $\mathbf{Z}' = \mathbf{TZ} + \mathbf{F}$, when $\mathbf{T}$ is independent of x, is given by (see Franklin [3], Chapter 3)

$$\mathbf{Z}(x) = e^{\mathbf{T}x}\left(\mathbf{K} + \int^x e^{-\mathbf{T}\xi}\ \mathbf{F}(\xi)\ d\xi\right) \tag{8.5 – 20a}$$

where $e^{\mathbf{T}x}$ represents the matrix product

$$e^{\mathbf{T}x} = [E]\begin{bmatrix} e^{\lambda_1 x} & & & \mathbf{0} \\ & e^{\lambda_2 x} & & \\ & & \ddots & \\ \mathbf{0} & & & e^{\lambda_8 x} \end{bmatrix}[E]^{-1} \tag{8.5-20b}$$

$[E]$ denotes the matrix of distinct eigenvectors of the matrix $[T]$, $[E]^{-1}$ denotes its inverse, λ_i $(i = 1, 2, 3, \ldots 8)$ are the eigenvalues associated with matrix $[T]$, and $\mathbf{K}$ is a vector of constants to be determined from the boundary conditions (12)–(14).

Substitution of Eq. (20a) into any combination of boundary conditions (12)–(14) on edges $x = \mp a/2$ yields a nonhomogeneous system of equations

$$[M]\{K\} = \{R\} \tag{8.5-21}$$

which can be solved for the vector $\{K\}$. For example, consider the case in which the edge $x = -a/2$ is clamped and the edge $x = a/2$ is free. For uniformly distributed load (static bending case), the solution (20) can be written as

$$\begin{aligned} \{Z\} &= [E]\begin{bmatrix} e^{\lambda_1 x} & & & \mathbf{0} \\ & e^{\lambda_2 x} & & \\ & & \ddots & \\ \mathbf{0} & & & e^{\lambda_8 x} \end{bmatrix}[E]^{-1}\{K\} + [E]\begin{bmatrix} -\frac{1}{\lambda_1} & & & \mathbf{0} \\ & -\frac{1}{\lambda_2} & & \\ & & \ddots & \\ \mathbf{0} & & & -\frac{1}{\lambda_8} \end{bmatrix}[E]^{-1}\{F\} \\ &\equiv [G(x)]\{K\} + \{H(x)\} \end{aligned} \tag{8.5-22}$$

Now the components of the matrix $[M]$ and vector $\{R\}$ in Eq. (21) can be defined in terms of the coefficients G_{ij} and H_i, evaluated at $x = -a/2$ and $x = a/2$, as described below.

The *clamped* boundary condition at $x = -a/2$ requires [see Eq. (13)] that

$$M_{1j} = G_{1j}(-a/2),\ M_{2j} = G_{3j}(-a/2),\ M_{3j} = G_{5j}(-a/2),\ M_{4j} = G_{6j}(-a/2) \tag{8.5-23a}$$

The *free* boundary condition at $x = a/2$ requires [see Eqs. (14a,b,c)] that

$$\begin{aligned} M_{5j} &= A_{11}G_{2j}(a/2) - \beta A_{12}G_{3j}(a/2) + \beta^2 B_{12}G_{5j}(a/2) - B_{11}G_{7j}(a/2) \\ M_{6j} &= \beta A_{66}G_{1j}(a/2) + A_{66}G_{4j}(a/2) - 2\beta B_{66}G_{6j}(a/2) \\ M_{7j} &= \hat{C}_1 G_{1j}(a/2) + \hat{C}_2 G_{4j}(a/2) + \hat{C}_3 G_{6j}(a/2) + \hat{C}_4 G_{8j}(a/2) \\ M_{8j} &= B_{11}G_{2j}(a/2) - \beta B_{12}G_{3j}(a/2) + \beta^2 D_{12}G_{5j}(a/2) - D_{11}G_{7j}(a/2) \end{aligned} \tag{8.5-23b}$$

$$\begin{aligned} \hat{C}_1 &= C_1 B_{11} - \beta^2 B_{66},\ \hat{C}_2 = C_2 B_{11} - \beta(B_{12} + B_{66}) \\ \hat{C}_3 &= -C_3 B_{11} + \beta^2(D_{12} + 2D_{66}),\ \hat{C}_4 = C_4 B_{11} - D_{11} \end{aligned} \tag{8.5-23c}$$

Similarly, the coefficients R_i are defined by

$$\begin{aligned}
&R_1 = -H_1(-a/2),\ R_2 = -H_3(-a/2),\ R_3 = -H_5(-a/2),\ R_4 = -H_6(-a/2)\\
&R_5 = A_{11}H_2(a/2) - \beta A_{12}H_3(a/2) + \beta^2 B_{12}H_5(a/2) - B_{11}H_7(a/2)\\
&R_6 = \beta A_{66}H_1(a/2) + A_{66}H_4(a/2) - 2\beta B_{66}H_6(a/2)\\
&R_7 = \hat{C}_1 H_1(a/2) + \hat{C}_2 H_4(a/2) + \hat{C}_3 H_6(a/2) + \hat{C}_4 H_8(a/2)\\
&R_8 = B_{11}H_2(a/2) - \beta B_{12}H_3(a/2) + \beta^2 D_{12}H_5(a/2) - D_{11}H_7(a/2) \qquad (8.5-24)
\end{aligned}$$

Natural Vibration

In the case of natural vibration, the applied mechanical loads $(Q_m, \hat{N}_{xx}, \hat{N}_{yy})$ are assumed to be zero, and the solution is of the form

$$\begin{aligned}
u(x,y,t) &= U_m(x)\sin\beta y\ e^{i\omega_m t}\\
v(x,y,t) &= V_m(x)\cos\beta y\ e^{i\omega_m t}\\
w(x,y,t) &= W_m(x)\sin\beta y\ e^{i\omega_m t} \qquad (8.5-25)
\end{aligned}$$

where ω_m denotes the frequency of vibration of the mth mode, and $i = \sqrt{-1}$. Equation (15) becomes

$$\{Z'(x)\} = [A]\{Z(x)\} \qquad (8.5-26a)$$

with

$$[A] = \begin{bmatrix}
0 & 1 & 0 & 0 & 0 & 0 & 0 & 0\\
\hat{C}_1 & 0 & 0 & C_2 & 0 & -\hat{C}_3 & 0 & C_4\\
0 & 0 & 0 & 1 & 0 & 0 & 0 & 0\\
0 & -C_5 & \hat{C}_6 & 0 & -\hat{C}_7 & 0 & C_8 & 0\\
0 & 0 & 0 & 0 & 0 & 1 & 0 & 0\\
0 & 0 & 0 & 0 & 0 & 0 & 1 & 0\\
0 & 0 & 0 & 0 & 0 & 0 & 0 & 1\\
0 & \hat{C}_9 & \hat{C}_{10} & 0 & \hat{C}_{11} & 0 & \hat{C}_{12} & 0
\end{bmatrix} \qquad (8.5-26b)$$

where

$$\begin{aligned}
&\hat{C}_1 = C_1 - D_1\omega^2,\ \hat{C}_3 = C_3 - D_2\omega^2,\ \hat{C}_6 = C_6 - D_3\omega^2\\
&\hat{C}_7 = C_7 - D_4\omega^2,\ \hat{C}_9 = C_9 - D_5\omega^2,\ \hat{C}_{10} = C_{10} - D_6\omega^2\\
&\hat{C}_{11} = C_{11} - D_7\omega^2,\ \hat{C}_{12} = C_{12} - D_8\omega^2 \qquad (8.5-26c)
\end{aligned}$$

The solution of Eq. (26a) is given by

$$\mathbf{Z}(x) = e^{\mathbf{A}x}\mathbf{K} \qquad (8.5-27)$$

and the vector K of constants is determined from the boundary conditions. Substitution of Eq. (27) into the set of boundary conditions results in a homogeneous system of equations

$$[M]\{K\} = \{0\} \tag{8.5 – 28}$$

For a nontrivial solution, the determinant of the coefficient matrix in (28) should be zero:

$$|M_{ij}| = 0 \tag{8.5 – 29}$$

The roots of the above equation are (the squares of) the frequencies of natural vibration.

Buckling

In the case of buckling, the applied mechanical load Q_m is zero, and $\hat{N}_{xx}$ and $\hat{N}_{yy}$ are determined. The solution is assumed to be of the form

$$\begin{aligned} u(x, y) =& U_m(x) \sin \beta y \\ v(x, y) =& V_m(x) \cos \beta y \\ w(x, y) =& W_m(x) \sin \beta y \end{aligned} \tag{8.5 – 30}$$

The operator equation for this case is

$$\{Z^{'}(x)\} = [T]\{Z\} \tag{8.5 – 31}$$

where $[T]$ is the matrix defined in Eq. (18b). Note that the buckling loads enter the matrix through the coefficients C_{11} and C_{12}, which contain e_{11} and e_{12}, respectively [see Eqs. (16a) and (16b)]. The solution of Eq. (31) is given by

$$\mathbf{Z}(x) = e^{\mathbf{T}x}\mathbf{K} \tag{8.5 – 32}$$

and the vector **K** of constants is determined from the boundary conditions. Substitution of Eq. (32) into the set of boundary conditions results in a homogeneous system of equations

$$[M]\{K\} = \{0\} \tag{8.5 – 33}$$

For a nontrivial solution, the determinant of the coefficient matrix in (33) should be zero. The roots of this equation are the buckling loads, $\hat{N}_{xx}$ and $\hat{N}_{yy}$.

Computational Issues

Some comments are in order on the numerical solution of Eq. (22). Due to the sparse nature of matrix $[T]$ or $[A]$, the matrix $[M]$ appearing in Eqs. (21), (28), and (33) is often ill-conditioned and results in computer overflow or underflow. This can be overcome (see, for example, Nosier and Reddy [4]) by rewriting Eq. (22) as

$$\{Z(x)\} = [E]\begin{bmatrix} e^{\lambda_1 x} & & 0 \\ & e^{\lambda_2 x} & \\ & \ddots & \\ 0 & & e^{\lambda_8 x} \end{bmatrix}[E]^{-1}\{K\} + e^{\mathbf{T}x}\int^x e^{-\mathbf{T}\xi}\ \{F\}\ d\xi$$

$$= [E]\begin{bmatrix} e^{\lambda_1 x} & & 0 \\ & e^{\lambda_2 x} & \\ & \ddots & \\ 0 & & e^{\lambda_8 x} \end{bmatrix}\{\hat{K}\} + e^{\mathbf{T}x}\int^x e^{-\mathbf{T}\xi}\ \{F\}\ d\xi \quad (8.5-34a)$$

and

$$[\hat{M}]\{\hat{K}\} + \{F\} = \{0\}\ , \quad \{\hat{K}\} = [E]^{-1}\{K\} \qquad (8.5-34b)$$

The matrix $[\hat{M}]$ is not ill-conditioned and therefore can be easily inverted to solve for $\{\hat{K}\}$ and $\{K\} = [E]\{\hat{K}\}$. It should be noted that, while $\{K\}$ and $[M]$ are real-valued, $\{\hat{K}\}$ and $[\hat{M}]$ are complex-valued arrays.

Another source of difficulty in the numerical evaluation of the eigenvalues of the matrix $[T]$ or $[A]$ is due to the fact their diagonals have zero entries. This can be circumvented by adding a nonzero constant to all diagonal elements (*i.e.*, add $-c[I]$). The eigenvalues of the original matrix $[T]$ or $[A]$ are obtained from the eignevalues of the modified matrix by subtracting the same nonzero constant. The eigenvectors in both cases are the same.

8.5–3 Antisymmetric Cross-Ply Laminates

Here we present numerical results obtained with the Lévy method and the state–space solution approach. Khdeir and his colleagues developed solutions for static and dynamic (natural vibration as well as transient response) analyses and buckling of rectangular composite laminates with various lamination schemes and boundary conditions. The reader may consult the papers cited in the bibliography for detailed derivations and additional numerical results.

The notation used for rectangular laminates with different boundary conditions on edges $x = \pm a/2$ is as follows (see Figure 8.5–1): The notation SF, for example, is used to denote a plate for which edge $x = -a/2$ is simply supported (S) and edge $x = a/2$ is free (F). Since edges $y = 0, b$ are always simply supported, we also use the notation SSSF to denote SF. Thus SS is used in place of SSSS, SC in place of SSSC, CC in place of SSCC, and so on.

Bending

The following lamina properties, typical of graphite-epoxy material, are used in all numerical examples presented here:

$$E_1/E_2 = 25, \quad G_{12} = G_{13} = 0.5E_2, \quad G_{23} = 0.2E_2, \quad \nu_{12} = 0.25 \tag{8.5-35}$$

The loading, in all cases, is assumed to be sinusoidal

$$q(x,y) = q_0 \cos\frac{\pi x}{a} \sin\frac{\pi y}{b} \tag{8.5-36}$$

In the tables and figures, the results for deflections and stresses are presented in the following nondimensional form:

$$\bar{w} = w_0(0, b/2)\frac{E_2 h^3}{b^4 q_0} \times 10^2 \tag{8.5-37a}$$

$$\bar{\sigma}_{xx} = -\sigma_{xx}(0, \frac{b}{2}, -\frac{h}{2})\frac{h^2}{b^2 q_0} \times 10, \quad \bar{\sigma}_{yy} = \sigma_{yy}(0, \frac{b}{2}, \frac{h}{2})\frac{h^2}{b^2 q_0} \times 10 \tag{8.5-37b}$$

where h is the total thickness of the laminate and q_0 is the intensity of the distributed transverse load. For the coordinate system used in the nondimensionalization, one should refer to Figure 8.5–1.

Figures 8.5–2 and 8.5–3 contain plots of $\bar{w}$ versus E_1/E_2 for two-layer antisymmetric rectangular ($b/a = 2$) laminates ($G_{12} = G_{13} = 0.5E_2$, $\nu_{12} = 0.25$) under various boundary conditions on edges $x = \pm a/2$, showing the effect of material orthotropy on the deflections. The degree of orthotropy has less influence on the deflections for large ratios of E_1 to E_2. Table 8.5–1 contains numerical results of deflections and stresses for two- and ten-layer laminates.

Numerical results for deflections and stresses of cross-ply laminates subjected to sinusoidal distribution of temperature

$$T(x,y,z) = zT_1(x,y) = z\bar{T}_1 \cos\frac{\pi x}{a} \sin\frac{\pi y}{b} \tag{8.5-38}$$

are presented in Table 8.5–2. The material properties used are the same as those in Eq. (35), with $\alpha_2 = 3\alpha_1$. The following nondimensionalizations are used:

$$\bar{w} = w_0(0, b/2)\frac{1}{\alpha_1 \bar{T}_1 b^2} \times 10 \tag{8.5-39a}$$

$$\bar{\sigma}_{xx} = \sigma_{xx}(0, \frac{b}{2}, -\frac{h}{2})\frac{1}{\alpha_1 \bar{T}_1 b E_2} \times 10 \tag{8.5-39b}$$

$$\bar{\sigma}_{yy} = -\sigma_{yy}(0, \frac{b}{2}, \frac{h}{2})\frac{1}{\alpha_1 \bar{T}_1 b E_2} \times 10 \tag{8.5-39c}$$

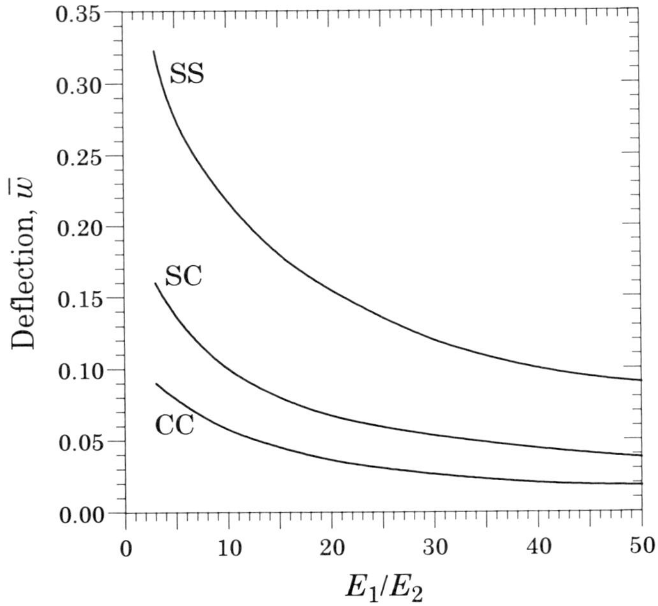

Figure 8.5–2: Nondimensionalized maximum transverse deflection ($\bar{w}$) versus modulus ratio (E_1/E_2) for antisymmetric cross-ply (0/90) laminates ($b/a = 2$) subjected to sinusoidal load.

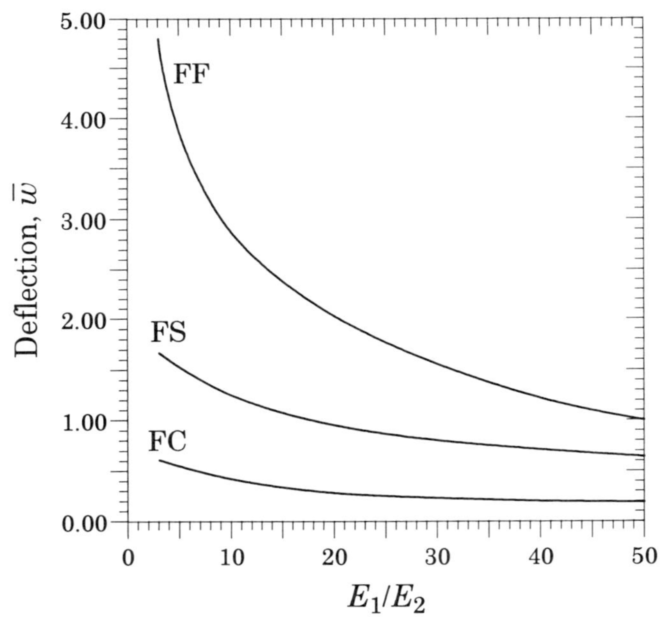

Figure 8.5–3: Nondimensionalized maximum transverse deflection ($\bar{w}$) versus modulus ratio (E_1/E_2) for antisymmetric cross-ply (0/90) laminates ($b/a = 2$) subjected to sinusoidal load.

Table 8.5–1: Nondimensionalized center deflections ($\bar{w}$) and in-plane normal stresses ($\bar{\sigma}_{xx}$ and $\bar{\sigma}_{yy}$) of antisymmetric cross-ply square plates subjected to sinusoidal distribution of transverse load and for various boundary conditions.

No. of Layers	Variable	SS	SC	CC	FF	FS	FC
2	$\bar{w}$	1.064	0.664	0.429	1.777	1.471	0.980
	$\bar{\sigma}_{xx}$	7.157	5.660	4.800	2.403	4.442	3.042
	$\bar{\sigma}_{yy}$	7.157	4.483	2.914	11.849	9.837	6.560
10	$\bar{w}$	0.442	0.266	0.167	0.665	0.579	0.380
	$\bar{\sigma}_{xx}$	5.009	3.829	3.167	1.725	2.986	1.865
	$\bar{\sigma}_{yy}$	5.009	3.025	1.911	7.480	6.531	4.284

Table 8.5–2: Nondimensionalized center deflections ($\bar{w}$) and in-plane normal stresses ($\bar{\sigma}_{xx}$ and $\bar{\sigma}_{yy}$) of cross-ply square plates subjected to sinusoidal distribution of temperature distribution and for various boundary conditions.

Laminate	Variable	SS	SC	CC	FF	FS
0	$\bar{w}$	1.0312	0.4543	0.2443	2.2935	1.6067
(0/90)	$\bar{w}$	1.1504	0.7183	0.4681	1.2639	1.2152
	$\bar{\sigma}_{yy}$	0.6148	5.1916	8.8393	2.1091	1.4684
$(0/90)_5$	$\bar{w}$	1.0331	0.6222	0.3914	1.0681	1.0546
(0/90/0)	$\bar{w}$	1.0312	0.4635	0.2512	1.6645	1.3800
	$\bar{\sigma}_{xx}$	0.0526	11.1264	15.2675	1.4489	0.8217

Vibration and Stability

The Lévy type solution procedure is used to evaluate the natural frequencies and critical buckling loads of antisymmetric cross-ply rectangular laminates. The following material properties are used in the analysis (*material 2*):

$$E_1/E_2 = 40, \quad G_{12} = G_{13} = 0.6E_2, \quad G_{23} = 0.5E_2, \quad \nu_{12} = 0.25 \qquad (8.5-40)$$

Table 8.5–3: Effect of degree of orthotropy of the individual layers on the dimensionless fundamental frequency, $\bar{\omega} = \omega(b^2/h)\sqrt{\rho/E_2}$, and critical buckling loads, $\bar{N} = N_{xx}(b^2/E_2h^3)$ $(k = 0)$, of simply supported antisymmetric square laminates (E_1/E_2 = varied, $G_{12} = G_{13} = 0.6E_2$, $G_{23} = 0.5E_2$, $\nu_{12} = 0.25$).

No. of Layers	E_1/E_2				
	3	10	20	30	40
Natural Frequencies $(\omega_{11})^{\dagger}$					
2	6.977	8.031	9.204	10.227	11.154
	7.034	8.097	9.278	10.310	11.244
4	7.353	9.987	12.826	15.141	17.145
	7.413	10.068	12.931	15.264	17.285
8	7.443	10.422	13.589	16.147	18.352
	7.505	10.507	13.701	16.279	18.502
10	7.455	10.473	13.678	16.264	18.492
	7.516	10.559	13.790	16.397	18.643
Critical Buckling Loads $(k = 0)$					
2	5.034	6.703	8.816	10.891	12.957
4	5.574	10.295	16.988	23.675	30.359
8	5.709	11.192	19.031	26.870	34.710
10	5.725	11.300	19.277	27.254	35.232

† Fundamental frequencies obtained with (first row) and without (second row) rotary inertia. When rotary inertia is included, the nondimensionalized frequencies depend on the ratio a/h; the frequencies are reported for $a/h = 10$.

Numerical results for the nondimensionalized fundamental frequencies of square, antisymmetric, cross-ply laminates (0/90/0/...) are presented in Table 8.5–3 for various boundary conditions, number of layers, and ratio of principal moduli of the material. The fundamental frequencies increase with increasing orthotropy E_1/E_2 as well as number of layers. Similar results for critical buckling loads are also presented in the same table. Results for fundamental frequencies and buckling loads are presented for various boundary conditions and aspect ratios in Table 8.5–4. The natural frequencies increase with an increase in the aspect ratio as well as the number of layers.

Table 8.5–4: Dimensionless fundamental frequencies, $\bar{\omega} = \omega(b^2/h)\sqrt{\rho/E_2}$, and uniaxial critical buckling loads, $\bar{N} = N_{xx}(b^2/E_2h^3)$, of antisymmetric cross-ply plates with various boundary conditions ($E_1 = 40E_2$, $G_{12} = G_{13} = 0.6E_2$, $G_{23} = 0.5E_2$, $\nu_{12} = 0.25$).

No. of Layers	b/a	FF	FS	FC	SS	SC	CC
Natural Frequencies $(\omega_{11})^\dagger$							
2	1	7.267	7.636	8.228	11.154	14.223	18.543
10	1	12.680	12.906	13.779	18.492	23.971	31.709
2	2	7.267	8.677	13.915	30.468	45.554	64.832
10	2	12.680	13.569	22.876	52.292	79.371	113.80
2	3	7.267	10.153	25.769	63.325	96.451	137.71
10	3	12.680	14.606	43.616	111.58	159.65	159.95
Critical Buckling Loads ($k = 0$)							
2	1	5.425	6.003	6.968	12.957	21.116	31.280
10	1	16.426	17.023	19.389	35.232	59.288	89.770

† Frequencies with rotary inertia included ($a/h = 10$).

8.5–4 Antisymmetric Angle-Ply Laminates

The Lévy solutions in conjunction with the state–space approach can also be obtained for antisymmetric angle-ply laminated plates. In this section numerical results of bending, free vibration, and in-plane compressive buckling of rectangular laminates are presented (see Khdeir [18]).

Bending

Nondimensionalized deflections, $\bar{w} = w_0(0, b/2)E_2h^3/q_0b^4 \times 10^2$, of square, antisymmetric angle-ply laminates (45/–45/45/–45) for various boundary conditions and uniformly distributed load of intensity q_0 are presented in Table 8.5–5. The material properties used are the same as those presented in Eq. (40). As one might expect, plates with a combination of free and simply supported boundary conditions deflect the most and those with simply supported and clamped boundary conditions deflect the least. Table 8.5–6 contains results for two- and ten-layer antisymmetric angle-ply laminates as a function of the lamination angle and for different boundary conditions. The material properties used in this case are

$$E_1 = 19.2 \times 10^6 \text{ psi (132.38 GPa)}, \quad E_2 = 1.56 \times 10^6 \text{ psi (10.76 GPa)}, \quad \nu_{12} = 0.24$$

$$G_{12} = G_{13} = 0.82 \times 10^6 \text{ psi } (5.65 \text{ GPa}), \quad G_{23} = 0.523 \times 10^6 \text{ psi } (3.61 \text{ GPa}) \qquad (8.5-41)$$

It is clear that the bending–stretching coupling is the most significant for two-layer laminates, and its effect is to make the laminate more flexible and hence deflects more than the ten-layer plates, for which the coupling is negligible.

Table 8.5–5: Effect of orthotropy on dimensionless deflections $\bar{w}$ of a (45/–45/45/–45) square laminated plate.

E_1/E_2	SS	SC	CC	FF	FS	FC
2	3.2142	2.2144	1.5308	10.470	6.2336	4.4460
10	1.0000	0.7467	0.5578	5.5710	2.3451	1.7473
20	0.5418	0.4120	0.3133	3.6574	1.3432	1.0104
30	0.3718	0.2847	0.2179	2.7376	0.9433	0.7121

Table 8.5–6: Effect of ply angle (θ) and number of layers (n) on dimensionless deflection $\bar{w}$ of a square plate [$(\theta/-\theta/\theta/\ldots/-\theta)$; material properties are as given in Eq. (41)].

θ	n	SS	SC	CC	FF	FS	FC
30	2	1.6185	1.2996	1.0352	4.3732	2.6983	2.2162
	10	0.8187	0.6898	0.5786	2.2621	1.3936	1.1913
45	2	1.5807	1.1675	0.8628	7.8765	3.5696	2.6432
	10	0.7391	0.5590	0.4229	4.7291	1.8069	1.3554
60	2	1.6184	1.0218	0.6657	12.3713	4.7491	2.8627
	10	0.8187	0.5188	0.3364	10.3035	2.7365	1.6391

Vibration and Buckling

Numerical results for nondimensionalized frequencies, $\bar{\omega} = \omega \frac{b^2}{h}\sqrt{\rho/E_2}$, and dimensionless uniaxial buckling loads, $\bar{N} = N_{xx}\frac{a^2}{E_2 h^3}$, are presented for various laminates in Tables 8.5–7 through 8.5–13. The material used in all these cases is assumed to be a high modulus graphite epoxy with the properties listed in Eq. (40). The fundamental frequencies presented are for the case in which rotary inertia is neglected. The parametric effects of the lamination angle, plate aspect ratio, and boundary conditions on frequencies and buckling loads can be seen from the results presented in these tables.

Table 8.5–7: Effect of in-plane orthotropy ratio on dimensioness fundamental frequency $\bar{\omega}$ of a (45/–45/45/–45) square laminated plate ($E_1/E_2 = 40,\ G_{12} = G_{13} = 0.6E_2,\ G_{23} = 0.5E_2,\ \nu_{12} = 0.25$).

E_1/E_2	SS	SC	CC	FF	FS	FC
2	7.02	8.39	10.24	3.44	4.24	4.62
10	12.54	14.43	16.90	4.65	6.78	7.38
20	17.02	19.43	22.53	5.72	8.92	9.70
30	20.53	23.37	27.00	6.60	10.62	11.55
40	23.53	26.73	30.83	7.37	12.08	13.14

Table 8.5–8: Effect of ply angle (θ) and number of layers (n) on dimensionless fundamental frequency $\bar{\omega}$ of a square laminate [$(\theta/{-\theta}/\theta/\ldots/{-\theta})$; $E_1/E_2 = 40,\ G_{12} = G_{13} = 0.6E_2,\ G_{23} = 0.5E_2,\ \nu_{12} = 0.25$].

θ	n	SS	SC	CC	FF	FS	FC
30	2	14.24	15.44	17.00	7.58	9.35	9.69
	10	23.95	25.59	27.58	12.37	15.38	15.84
45	2	14.64	16.75	19.48	5.12	7.79	8.48
	10	25.47	28.91	33.32	7.89	13.03	14.17
60	2	14.24	17.74	22.31	3.47	6.26	7.54
	10	23.95	29.86	37.62	4.32	9.92	11.96

Table 8.5–9: Effect of aspect ratio dimensionless fundamental frequency $\bar{\omega} = \omega\frac{a^2}{h}\sqrt{\rho/E_2}$ of a (45/–45/45/–45) square laminated plate ($E_1/E_2 = 40,\ G_{12} = G_{13} = 0.6E_2,\ G_{23} = 0.5E_2,\ \nu_{12} = 0.25$).

a/b	SS	SC	CC	FF	FS	FC
1	23.57	26.73	30.83	7.37	12.08	13.14
2	53.74	69.75	90.73	6.32	19.80	25.47
3	98.87	138.20	189.13	5.61	28.39	42.95

Table 8.5–10: Dimensionless frequency $\bar{\omega}_m$ for various mode numbers (m) of (45/–45/45/–45) square laminated plate ($E_1/E_2 = 40$, $G_{12} = G_{13} = 0.6E_2$, $G_{23} = 0.5E_2$, $\nu_{12} = 0.25$).

m	SS	SC	CC	FF	FS	FC
1	23.53	26.73	30.83	7.37	12.08	13.14
2	53.74	56.10	58.88	31.89	37.32	37.91
3	98.87	100.70	102.76	73.58	79.12	79.48
4	160.35	161.82	163.43	132.42	137.92	138.16
5	238.72	239.93	241.25	208.39	213.78	213.95

Table 8.5–11: Effect of plate aspect ratio (a/b) and number of layers (n) on uniaxial buckling load of simply supported angle-ply (45/–45/45/...) laminates ($E_1/E_2 = 40$, $G_{12} = G_{13} = 0.6E_2$, $G_{23} = 0.5E_2$, $\nu_{12} = 0.25$; $\bar{N} = N_{xx}(b^2/\pi^2 D_{22})$.

a/b	mode	$n = 2$	$n = 4$	$n = 6$	$n = 8$
0.5	(1,1)	3.2071	8.0934	8.9980	9.3150
1.0	(1,1)	2.4014	6.2045	6.9088	7.1552
1.5	(2,1)	2.5231	6.4901	7.2247	7.4819
2.0	(2,1)	2.4014	6.2045	6.9088	7.1552
2.5	(3,1)	2.4495	6.3173	7.0336	7.2843
3.0	(3,1)	2.4014	6.2045	6.9088	7.1552

Table 8.5–12: Effect of in-plane orthotropy ratio on dimensionless uniaxial buckling loads $\bar{N} = N_{cr}b^2/E_2h^3$ of a (45/–45/45/–45) square laminated plate ($E_1/E_2 = 40$, $G_{12} = G_{13} = 0.6E_2$, $G_{23} = 0.5E_2$, $\nu_{12} = 0.25$).

E_1/E_2	SS	SC	CC	FF	FS	FC
2	4.988	7.126	9.512	1.199	1.819	2.166
10	15.923	21.106	26.278	2.190	4.660	5.521
20	29.333	38.234	46.823	3.313	8.054	9.532
30	42.715	55.321	67.320	4.415	11.421	13.513
40	56.088	72.396	87.803	5.509	14.780	17.484

Table 8.5–13: Effect of ply angle (θ) and number of layers (n) on dimensionless uniaxial buckling loads $\bar{N} = N_{cr}b^2/E_2h^3$ of a square plate $[(\theta/-\theta/\theta/\ldots/-\theta);\ E_1/E_2 = 40,\ G_{12} = G_{13} = 0.6E_2,\ G_{23} = 0.5E_2,\ \nu_{12} = 0.25]$.

θ	n	SS	SC	CC	FF	FS	FC
30	2	20.543	24.158	29.269	5.822	8.857	9.520
	10	58.135	66.322	77.065	15.499	23.972	25.412
45	2	21.709	28.423	34.963	2.654	6.150	7.283
	10	65.714	84.707	102.596	6.300	17.189	20.332
60	2	19.564	23.834	29.547	1.221	3.975	5.756
	10	52.945	64.103	79.619	1.889	9.977	14.501

8.6 Analysis of Midplane Symmetric Laminates

8.6–1 Introduction

In the previous sections of this chapter we considered analytical solutions of bending, vibration, and buckling of antisymmetric cross-ply and angle-ply rectangular laminates. In these laminates, in general, the bending–stretching coupling stiffnesses B_{ij} were not zero but the bending–twisting coupling stiffnesses D_{16} and D_{26} were zero. In this section we consider laminates that are symmetric in *both* geometry and material properties about the middle plane. In such symmetric laminates, we have $B_{ij} = 0$ and D_{16} and D_{26} are not zero. The specially orthotropic plates considered in Chapter 7 are a special case of symmetric laminates. Laminates containing multiple generally orthotropic layers (*i.e.*, orthotropic layers whose principal material axes are not parallel to the plate axes) that are symmetrically placed about the midplane fall into the class of symmetric laminates.

An example of symmetric laminates is provided by the class of *regular symmetric angle-ply laminates,* $(\theta/-\theta/\theta)$, $0 \leq \theta \leq 90$ with equal thickness layers. The regular symmetric angle-ply laminates should contain an odd number of plies. A more general example of symmetric angle-ply laminate is provided by (30/–60/15/–60/30) with thicknesses $h_1 = h_5, h_2 = h_4$, and the midplane of the plate coincides with the midplane of the 15° ply. For symmetric angle-ply laminates the coupling terms A_{16}, A_{26}, D_{16}, and D_{26} are proportional to $1/N$, where N is the total number of layers in the laminate. Thus the coupling stiffnesses are the largest when $N = 3$ for symmetric angle-ply laminates, and they decrease with increasing N.

The symmetric angle-ply laminates, with $B_{ij} = 0$ and small A_{16}, A_{26}, D_{16}, and D_{26}, offer both analysis simplifications and practical advantages over more general laminates. For example, symmetric angle-ply laminates offer more shear stiffness than cross-ply laminates. Even when A_{16}, A_{26}, D_{16}, and D_{26} are small, they influence the laminate behavior significantly.

8.6–2 Governing Equations

The governing equations of motion of symmetric laminates according to the classical laminate theory can be obtained from (5.3–45)–(5.3–47) by setting $B_{ij} = 0$ and $I_1 = 0$. For linear analysis, we obtain

$$
\begin{aligned}
&A_{11}\frac{\partial^2 u_0}{\partial x^2} + A_{12}\frac{\partial^2 v_0}{\partial x \partial y} + A_{16}\left(\frac{\partial^2 u_0}{\partial x \partial y} + \frac{\partial^2 v_0}{\partial x^2}\right) \\
&+ A_{16}\frac{\partial^2 u_0}{\partial x \partial y} + A_{26}\frac{\partial^2 v_0}{\partial y^2} + A_{66}\left(\frac{\partial^2 u_0}{\partial y^2} + \frac{\partial^2 v_0}{\partial x \partial y}\right) \\
&- \left(\frac{\partial N_{xx}^T}{\partial x} + \frac{\partial N_{xy}^T}{\partial y}\right) = I_0\frac{\partial^2 u_0}{\partial t^2}
\end{aligned}
\tag{8.6 – 1}
$$

$$
\begin{aligned}
&A_{16}\frac{\partial^2 u_0}{\partial x^2} + A_{26}\frac{\partial^2 v_0}{\partial x \partial y} + A_{66}\left(\frac{\partial^2 u_0}{\partial x \partial y} + \frac{\partial^2 v_0}{\partial x^2}\right) \\
&+ A_{12}\frac{\partial^2 u_0}{\partial x \partial y} + A_{22}\frac{\partial^2 v_0}{\partial y^2} + A_{26}\left(\frac{\partial^2 u_0}{\partial y^2} + \frac{\partial^2 v_0}{\partial x \partial y}\right) \\
&- \left(\frac{\partial N_{xy}^T}{\partial x} + \frac{\partial N_{yy}^T}{\partial y}\right) = I_0\frac{\partial^2 v_0}{\partial t^2}
\end{aligned}
\tag{8.6 – 2}
$$

$$
\begin{aligned}
&- D_{11}\frac{\partial^4 w_0}{\partial x^4} - 2(D_{12} + 2D_{66})\frac{\partial^4 w_0}{\partial x^2 \partial y^2} - D_{22}\frac{\partial^4 w_0}{\partial y^4} \\
&- 4D_{16}\frac{\partial^4 w_0}{\partial x^3 \partial y} - 4D_{26}\frac{\partial^4 w_0}{\partial x \partial y^3} \\
&- \left(\frac{\partial^2 M_{xx}^T}{\partial x^2} + 2\frac{\partial^2 M_{xy}^T}{\partial y \partial x} + \frac{\partial^2 M_{yy}^T}{\partial y^2}\right) + q \\
&+ \frac{\partial}{\partial x}\left(\hat{N}_{xx}\frac{\partial w_0}{\partial x} + \hat{N}_{xy}\frac{\partial w_0}{\partial y}\right) + \frac{\partial}{\partial y}\left(\hat{N}_{xy}\frac{\partial w_0}{\partial x} + \hat{N}_{yy}\frac{\partial w_0}{\partial y}\right) \\
&= I_0\frac{\partial^2 w_0}{\partial t^2} - I_2\frac{\partial^2}{\partial t^2}\left(\frac{\partial^2 w_0}{\partial x^2} + \frac{\partial^2 w_0}{\partial y^2}\right)
\end{aligned}
\tag{8.6 – 3}
$$

where $\hat{N}_{xx}, \hat{N}_{yy}$, and $\hat{N}_{xy}$ are the applied edge forces.

Clearly the first two equations governing (u_0, v_0) are uncoupled from the third equation governing w_0. In the absence of any in-plane loads, the first two equations yield zero in-plane displacements everywhere. Because of the presence of the bending–twisting coupling stiffnesses, the Navier solutions of Eq. (3) cannot be developed, forcing us to use the Rayleigh–Ritz, Galerkin, or the finite element method. In the following sections we discuss the Rayleigh–Ritz solutions for symmetrically laminated plates.

8.6–3 Weak Forms

We can use the Rayleigh–Ritz method to determine an approximate solution to the bending, buckling, and natural vibrations of symmetric laminates. The weak form or the statement of the principle of minimum total potential energy for bending, buckling, and natural vibration problems is given below. For bending, the virtual work done to applied edge forces and moments should be added to the expression

$$
\begin{aligned}
0 = \int_0^b \int_0^a \Bigg\{ & D_{11} \frac{\partial^2 w_0}{\partial x^2} \frac{\partial^2 \delta w_0}{\partial x^2} + D_{12} \left(\frac{\partial^2 w_0}{\partial y^2} \frac{\partial^2 \delta w_0}{\partial x^2} + \frac{\partial^2 w_0}{\partial x^2} \frac{\partial^2 \delta w_0}{\partial y^2} \right) \\
& + D_{22} \frac{\partial^2 w_0}{\partial y^2} \frac{\partial^2 \delta w_0}{\partial y^2} + 4D_{66} \frac{\partial^2 w_0}{\partial x \partial y} \frac{\partial^2 \delta w_0}{\partial x \partial y} \\
& + 2D_{16} \left(\frac{\partial^2 w_0}{\partial x \partial y} \frac{\partial^2 \delta w_0}{\partial x^2} + \frac{\partial^2 w_0}{\partial x^2} \frac{\partial^2 \delta w_0}{\partial x \partial y} \right) \\
& + 2D_{26} \left(\frac{\partial^2 w_0}{\partial x \partial y} \frac{\partial^2 \delta w_0}{\partial y^2} + \frac{\partial^2 w_0}{\partial y^2} \frac{\partial^2 \delta w_0}{\partial x \partial y} \right) - q \delta w_0 \\
& + \hat{N}_{xx} \frac{\partial w_0}{\partial x} \frac{\partial \delta w_0}{\partial x} + \hat{N}_{xy} \left(\frac{\partial w_0}{\partial y} \frac{\partial \delta w_0}{\partial x} + \frac{\partial w_0}{\partial x} \frac{\partial \delta w_0}{\partial y} \right) + \hat{N}_{yy} \frac{\partial w_0}{\partial y} \frac{\partial \delta w_0}{\partial y} \\
& - \omega^2 \left[I_0 w_0 \delta w_0 + I_2 \left(\frac{\partial w_0}{\partial x} \frac{\partial \delta w_0}{\partial x} + \frac{\partial w_0}{\partial y} \frac{\partial \delta w_0}{\partial y} \right) \right] \Bigg\} dx dy \qquad (8.6-4)
\end{aligned}
$$

where ω denotes the frequency of natural vibration. For bending we set all terms involving the in-plane edge forces and frequency of vibration to zero. We set $q = 0$ and $\omega = 0$ for buckling analysis, and $q = 0$ and $\hat{N}_{xx} = \hat{N}_{yy} = \hat{N}_{xy} = 0$ for natural vibration.

8.6–4 The Rayleigh–Ritz Solution

We begin with the Rayleigh–Ritz approximation of the form

$$
w_0(x, y) \approx W_{MN}(x, y) = \sum_{i=1}^{M} \sum_{j=1}^{N} c_{ij} \ \varphi_{ij}(x, y) \qquad (8.6-5)
$$

where

$$
\varphi_{ij}(x, y) = X_i(x) Y_j(y) \qquad (8.6-6)
$$

and X_i and Y_j denote any admissible approximation functions for the problem. The choice is dictated by the essential (or geometric) boundary conditions of the problem. Substitution of (5) into Eq. (4) results in the following equations:

$$\begin{aligned}
0 = \sum_{i=1}^{M}\sum_{j=1}^{N}\Bigg\{\int_0^b\int_0^a\Bigg[&D_{11}\frac{d^2X_i}{dx^2}Y_j\frac{d^2X_p}{dx^2}Y_q + 4D_{66}\frac{dX_i}{dx}\frac{dY_j}{dy}\frac{dX_p}{dx}\frac{dY_q}{dy}\\
&+ D_{12}\left(X_i\frac{d^2Y_j}{dy^2}\frac{d^2X_p}{dx^2}Y_q + \frac{d^2X_i}{dx^2}Y_jX_p\frac{d^2Y_q}{dy^2}\right)\\
&+ D_{22}X_i\frac{d^2Y_j}{dy^2}X_p\frac{d^2Y_q}{dy^2}\\
&+ 2D_{16}\left(\frac{dX_i}{dx}\frac{dY_j}{dy}\frac{d^2X_p}{dx^2}Y_q + \frac{d^2X_i}{dx^2}Y_j\frac{dX_p}{dx}\frac{dY_q}{dy}\right)\\
&+ 2D_{26}\left(\frac{dX_i}{dx}\frac{dY_j}{dy}X_p\frac{d^2Y_q}{dy^2} + X_i\frac{d^2Y_j}{dy^2}\frac{dX_p}{dx}\frac{dY_q}{dy}\right)\Bigg]dxdy\Bigg\}c_{ij}\\
-\sum_{i=1}^{M}\sum_{j=1}^{N}\Bigg\{\int_0^b\int_0^a\Bigg[&\hat{N}_{xx}^0\frac{dX_i}{dx}Y_j\frac{dX_p}{dx}Y_q + \hat{N}_{yy}X_i\frac{dY_j}{dy}X_p\frac{dY_q}{dy}\\
&+ 2\hat{N}_{xy}^0\left(X_i\frac{dY_j}{dy}\frac{dX_p}{dx}Y_q + \frac{dX_i}{dx}Y_jX_p\frac{dY_q}{dy}\right)\Bigg]dxdy\Bigg\}c_{ij}\\
-\sum_{i=1}^{M}\sum_{j=1}^{N}\Bigg\{\int_0^b\int_0^a\omega^2\Bigg[&I_0X_iY_jX_pY_q + I_2\left(\frac{dX_i}{dx}Y_j\frac{dX_p}{dx}Y_q\right.\\
&\left.+X_i\frac{dY_j}{dy}X_p\frac{dY_q}{dy}\right)\Bigg]dxdy\Bigg\}c_{ij} - \int_0^b\int_0^a qX_pY_q\,dxdy \quad (8.6-7)
\end{aligned}$$

for $p = 1, 2, \cdots, M$ and $q = 1, 2, \cdots, N$.

8.6–5 Simply Supported Plates

Recall from Eq. (7.2–4) that the choice of the double sine series

$$\varphi_{ij}(x,y) = X_i(x)Y_j(y) \equiv \sin\frac{i\pi x}{a}\sin\frac{j\pi y}{b} \quad (8.6-8)$$

satisfies the simply supported (SS–1) boundary conditions. Substituting (8) into (7) we obtain

$$\begin{aligned}
0 = \sum_{i=1}^{M}\sum_{j=1}^{N}\Bigg[\int_0^b\int_0^a\Big(&A_{ijpq}\sin\alpha_ix\ \sin\beta_jy\ \sin\alpha_px\ \sin\beta_qy\\
&+ B_{ijpq}\cos\alpha_ix\ \cos\beta_jy\ \cos\alpha_px\ \cos\beta_qy
\end{aligned}$$

$$
\begin{aligned}
&- C_{ijpq} \cos\alpha_i x \ \cos\beta_j y \ \sin\alpha_p x \ \sin\beta_q y \\
&\left. - D_{ijpq} \sin\alpha_i x \ \sin\beta_j y \ \cos\alpha_p x \ \cos\beta_q y \Big) dxdy \right] c_{ij} \\
&- \int_0^b \int_0^a q(x,y) \ \sin\alpha_p x \ \sin\beta_q y \ dxdy \\
- \sum_{i=1}^{M} \sum_{j=1}^{N} \Bigg\{ \int_0^b \int_0^a \Big[&\hat{N}_{xx}^0 \cos\alpha_i x \ \sin\beta_j y \ \cos\alpha_p x \ \sin\beta_q y \\
&+ \hat{N}_{yy} \sin\alpha_i x \ \cos\beta_j y \ \sin\alpha_p x \ \cos\beta_q y \\
&+ 2\hat{N}_{xy}^0 (\sin\alpha_i x \ \cos\beta_j y \ \cos\alpha_p x \ \sin\beta_q y \\
&+ \cos\alpha_i x \ \sin\beta_j y \ \sin\alpha_p x \ \cos\beta_q y) \Big] dxdy \Bigg\} c_{ij} \\
- \sum_{i=1}^{M} \sum_{j=1}^{N} \Bigg\{ \int_0^b \int_0^a \omega^2 \big[&I_0 \sin\alpha_i x \ \sin\beta_j y \ \sin\alpha_p x \ \sin\beta_q y \\
&+ I_2 (\cos\alpha_i x \ \sin\beta_j y \ \cos\alpha_p x \ \sin\beta_q y \\
&+ \sin\alpha_i x \ \cos\beta_j y \ \sin\alpha_p x \ \cos\beta_q y) \big] \, dxdy \Bigg\} c_{ij} \qquad (8.6-9)
\end{aligned}
$$

where $\alpha_i = \frac{i\pi}{a}$, $\beta_j = \frac{j\pi}{b}$, and

$$
\begin{aligned}
A_{ijpq} &= D_{11}\alpha_i^2\alpha_p^2 + D_{12}\left(\beta_j^2\alpha_p^2 + \alpha_i^2\beta_q^2\right) + D_{22}\beta_j^2\beta_q^2 \\
B_{ijpq} &= 4D_{66}\alpha_i\beta_j\alpha_p\beta_q \\
C_{ijpq} &= 2\left(D_{16}\alpha_i\beta_j\alpha_p^2 + D_{26}\alpha_i\beta_j\beta_q^2\right) \\
D_{ijpq} &= 2\left(D_{16}\alpha_i^2\alpha_p\beta_q + D_{26}\beta_j^2\alpha_p\beta_q\right) \qquad (8.6-10)
\end{aligned}
$$

Suppose that the load $q(x,y)$ is also expanded in double sine series

$$
q(x,y) = \sum_{m}^{M} \sum_{n}^{N} Q_{mn} \sin\frac{m\pi x}{a} \sin\frac{n\pi y}{b} \qquad (8.6-11)
$$

In view of the integral identities

$$
\begin{aligned}
\int_0^a \sin\alpha_i x \ \sin\alpha_j x \ dx &= \begin{cases} 0, & i \neq j \\ \frac{a}{2}, & i = j \end{cases} \\
\int_0^a \cos\alpha_i x \ \cos\alpha_j x \ dx &= \begin{cases} 0, & i \neq j \\ \frac{a}{2}, & i = j \end{cases} \\
\int_0^a \cos\alpha_i x \ \sin\alpha_j x \ dx &= \begin{cases} 0, & i = j \ \text{ and } \ i+j \ \text{ even} \\ \frac{2j}{j^2-i^2}, & i \neq j \ \text{ and } \ i+j \ \text{ odd} \end{cases} \qquad (8.6-12)
\end{aligned}
$$

Eq. (9) can be simplified. In particular, when $D_{16} = D_{26} = 0$, Eq. (9) gives the Navier solutions presented in Chapter 7. When D_{16} and D_{26} are nonzero, the one-term Rayleigh–Ritz solution does not exist for a general symmetric laminate, because the solution does not contain the stiffness terms D_{16} and D_{26} due to the vanishing of the integrals. Thus the double sine series solution is incomplete, and it can only give an approximate solution to the symmetrically laminated plates when many terms in the series are used.

As reported by Ashton and Whitney [5], for a square plate with $D_{22} = 0.1D_{11}$, $D_{12} + 2D_{66} = 1.5D_{11}$, and $D_{16} = D_{26} = -0.5D_{11}$, the maximum deflection under uniformly distributed load, obtained with $M = N = 7$ in the series, is

$$w_0(a/2, a/2) = 0.00425\frac{q_0 a^4}{D_{11}} \tag{8.6 – 13}$$

For the same case, when D_{16} and D_{26} are neglected the maximum deflection is

$$w_0(a/2, a/2) = 0.00324\frac{q_0 a^4}{D_{11}} \tag{8.6 – 14}$$

Thus, the deflection is underpredicted by 23.76% when the bending–twisting coupling is neglected.

Similarly, it is found that the orthotropic plate solutions for buckling loads and natural frequencies of vibration are overpredicted in comparison to the solutions obtained with the bending–twisting coupling in place. In general, the task of computing the Rayleigh–Ritz solutions is algebraically complicated, and many terms have to be included to obtain accurate results.

8.6–6 Other Boundary Conditions

Equation (7) is also valid for other boundary conditions. Only the choice of the approximation functions X_i and Y_j is different for different boundary conditions. As discussed in Section 7.4–3, the eigenfunctions of the Euler–Bernoulli beams can be used for these functions (see Table 6.2–3; also see [22,23]). For example, for a symmetric laminate with all edges clamped, we can use the eigenfunctions of a beam with both ends clamped:

$$\begin{aligned} X_i(x) &= \sin\lambda_i x - \sinh\lambda_i x + \alpha_i\left(\cosh\lambda_i x - \cos\lambda_i x\right) \\ Y_j(y) &= \sin\lambda_j y - \sinh\lambda_j y + \alpha_j\left(\cosh\lambda_j y - \cos\lambda_j y\right) \end{aligned} \tag{8.6 – 15}$$

for $i = 1, 2, \cdots, M$; $j = 1, 2, \cdots, N$. The parameters λ_i are the roots of the characteristic equation

$$\cos\lambda_i a \cosh\lambda_i a - 1 = 0 \tag{8.6 – 16}$$

and

$$\alpha_i = \frac{\sinh \lambda_i a - \sin \lambda_i a}{\cosh \lambda_i a - \cos \lambda_i a} = \frac{\cosh \lambda_i a - \cos \lambda_i a}{\sinh \lambda_i a + \sin \lambda_i a} \tag{8.6 – 17}$$

We will not consider the topic of solving symmetrically laminated plates for bending deflections, buckling loads, and vibration frequencies by the Rayleigh–Ritz method further in this book. Interested readers may consult References 19–23.

8.7 Transient Analysis

8.7–1 Preliminary Comments

Here we discuss the procedures to determine the transient response of composite laminates. The equations of motion can be solved using analytical solution methods, such as the state–space approach (see Khdeir and Reddy [24–26]). Here we discuss a method which takes advantage of the static solution form for spatial variation and uses a numerical method to solve the resulting differential equations in time (see Reddy [27]).

As described in Section 7.9, there are two major steps in the solution process: (1) assume a spatial variation of the displacements and reduce the governing partial differential equations to a set of ordinary differential equations in time, and (2) solve the ordinary differential equations exactly if possible or numerically. The first step is amply illustrated in the preceding sections of this chapter. For example, the Navier solution method can be used to determine the spatial variation of the transient solution. The only difference is that the coefficients of the double Fourier series are assumed to be functions of time. Thus a typical dependent variable $\phi(x, y, t)$ is expanded as [see Eq. (8.3–3)]

$$\phi(x, y, t) = \sum_{m=1}^{\infty} \sum_{n=1}^{\infty} T_{mn}(t) F_{mn}(x, y)$$

where F_{mn} are suitable functions that satisfy the boundary conditions and T_{mn} are coefficients to be determined such that $\phi(x, y, t)$ satisfies its governing equation. The choice of a *separable* solution form as above implies that the general spatial variation is independent of time, and its amplitude may vary with time.

8.7–2 Equations of Motion

For simply supported cross-ply and antisymmetric angle-ply laminates, the Navier solution method can be used to reduce the governing equations of motion to differential equations in time. These are given by Eq. (8.3–19) for antisymmetric cross-ply laminates and by Eq. (8.4–6) for antisymmetric angle-ply laminates. In the absence of thermal effects and applied in-plane forces, these equations are of the form

$$\begin{bmatrix} \hat{c}_{11} & \hat{c}_{12} & \hat{c}_{13} \\ \hat{c}_{12} & \hat{c}_{22} & \hat{c}_{23} \\ \hat{c}_{13} & \hat{c}_{23} & \hat{c}_{33} \end{bmatrix} \begin{Bmatrix} U_{mn} \\ V_{mn} \\ W_{mn} \end{Bmatrix} + \begin{bmatrix} \hat{m}_{11} & 0 & 0 \\ 0 & \hat{m}_{22} & 0 \\ 0 & 0 & \hat{m}_{33} \end{bmatrix} \begin{Bmatrix} \ddot{U}_{mn} \\ \ddot{V}_{mn} \\ \ddot{W}_{mn} \end{Bmatrix} = \begin{Bmatrix} 0 \\ 0 \\ Q_{mn} \end{Bmatrix} \quad (8.7-1)$$

or

$$[M]\{\ddot{\Delta}\} + [K]\{\Delta\} = \{F\} \quad (8.7-2)$$

where the superposed dot denotes differentiation with respect to time, and

$$\{\Delta\} = \begin{Bmatrix} U_{mn} \\ V_{mn} \\ W_{mn} \end{Bmatrix}, \quad \{F\} = \begin{Bmatrix} 0 \\ 0 \\ Q_{mn} \end{Bmatrix} \quad (8.7-3)$$

The coefficients $\hat{c}_{ij}$ and $\hat{m}_{ij}$ of Eq. (1) are defined in Eqs. (8.3–20) and (8.4–7), respectively, for the two classes of laminates.

Equation (1) is subjected to the initial conditions

$$u_0(x,y,0) = d_0^1(x,y), \quad v_0(x,y,0) = d_0^2(x,y), \quad w_0(x,y,0) = d_0^3(x,y) \quad (8.7-4a)$$

$$\dot{u}_0(x,y,0) = v_0^1(x,y), \quad \dot{v}_0(x,y,0) = v_0^2(x,y), \quad \dot{w}_0(x,y,0) = v_0^3(x,y) \quad (8.7-4b)$$

We assume that the functions d^i and v^i $(i = 1,2,3)$ can also be expanded in the double Fourier series in the same way as the corresponding displacements. Then we have

$$\Delta_{mn}^1 = U_{mn}(0) = D_{mn}^1, \ \Delta_{mn}^2 = V_{mn}(0) = D_{mn}^2, \ \Delta_{mn}^3 = W_{mn}(0) = D_{mn}^3$$

$$\dot{\Delta}_{mn}^1 = \dot{U}_{mn}(0) = V_{mn}^1, \ \dot{\Delta}_{mn}^2 = \dot{V}_{mn}(0) = V_{mn}^2, \ \dot{\Delta}_{mn}^3 = \dot{W}_{mn}(0) = V_{mn}^3 \quad (8.7-5)$$

where D_{mn}^i and V_{mn}^i are the coefficients in the Fourier expansion of the ith initial displacement and velocity, respectively.

8.7–3 Numerical Time Integration

The set of three equations in (2), for any fixed m and n, can be solved exactly using either the Laplace transform method or the modal analysis methods. Both methods are algebraically complicated and require the determination of eigenvalues and eigenfunctions, as in the state–space method. Therefore we will not attempt them here. Alternatively, we seek numerical solutions to Eq. (2) using the well–known family of Newmark's integration schemes for second–order differential equations (see Reddy [27]). In this numerical integration method, the time derivatives are approximated using difference approximations (or truncated Taylor's series), and therefore solution is obtained only for discrete times and not as a continuous function of time.

In the Newmark method, the function (of time) and its first derivative are approximated using Taylor's series and only terms up to the second derivative are included:

$$\begin{aligned}
\{\Delta(t_{s+1})\} &= \{\Delta(t_s)\} + \delta t_s\{\dot{\Delta}(t_s)\} + \frac{1}{2}(\delta t_s)^2\{\ddot{\Delta}(t_{s+\gamma})\} \\
\{\dot{\Delta}(t_{s+1})\} &= \{\dot{\Delta}(t_s)\} + \delta t_s\{\ddot{\Delta}(t_{s+\alpha})\} \\
\{\ddot{\Delta}(t_{s+\alpha})\} &= (1-\alpha)\{\ddot{\Delta}(t_s)\} + \alpha\{\ddot{\Delta}(t_{s+1})\}, \quad 0 \le \alpha \le 1
\end{aligned} \tag{8.7-6}$$

where δt is the time increment, $\delta t_s = t_{s+1} - t_s$, and t_s is the current time and t_{s+1} is the next time at which we seek the solution. We assume that the solution at time t_s is known. Substituting the third equation into the first two in Eq. (6) and solving for $\{\ddot{\Delta}\}$, we obtain

$$\begin{aligned}
\{\dot{\Delta}\}_{s+1} &= \{\dot{\Delta}\}_s + a_1\{\ddot{\Delta}\}_s + a_2\{\ddot{\Delta}\}_{s+1} \\
\{\ddot{\Delta}\}_{s+1} &= a_3\left(\{\Delta\}_{s+1} - \{\Delta\}_s\right) - a_4\{\dot{\Delta}\}_s - a_5\{\ddot{\Delta}\}_s
\end{aligned} \tag{8.7-7}$$

where

$$a_1 = (1-\alpha)\delta t_s, \quad a_2 = \alpha\delta t_s, \quad a_3 = \frac{2}{\gamma(\delta t_s)^2}, \quad a_4 = a_3\delta t_s, \quad a_5 = \frac{(1-\gamma)}{\gamma} \tag{8.7-8}$$

and $\{\cdot\}_s$, for example, denotes the value of the enclosed vector at time t_s.

The parameters α and γ are selected such that the error introduced in the approximation (6) does not grow unboundedly as the scheme is applied at each time step to determine the solution at the next time. When the error introduced is bounded (hence the solution is bounded), such schemes are said to be numerically *stable* schemes. Sometimes, there is a restriction on the size of the time step that would make the error remain bounded. In such cases, the scheme is said to be *conditionally stable.* All schemes for which $\gamma \ge \alpha \ge 1/2$ are unconditionally stable. Schemes for which $\gamma < \alpha$ and $\alpha \ge 0.5$ are conditionally stable, and the stability condition is

$$\delta t \le \delta t_{cr} \equiv \frac{1}{\sqrt{2}\omega_{max}}(\alpha-\gamma)^{-\frac{1}{2}} \tag{8.7-9}$$

where ω_{max} denotes the maximum frequency of the discrete eigenvalue problem associate with Eq. (2):

$$\left([K] - \omega^2[M]\right)\{\Delta\} = \{0\} \tag{8.7-10}$$

The critical time step can also be expressed in terms of the *period of vibration,* $T = 2\pi/\omega$. It should be noted that the frequencies of vibration for different modes, axial, bending, torsional, and shear modes, are different. The critical time step for the element is the smallest of the critical time steps calculated using the maximum frequency of each mode of vibration.

The Newmark family contains several well–known schemes as special cases. The following choices of α and γ define some of the widely used schemes:

$$
\begin{aligned}
&\alpha = \frac{1}{2}\,, \quad \gamma = \frac{1}{2}\,, \quad \text{the constant average acceleration method (stable)}\\
&\alpha = \frac{1}{2}\,, \quad \gamma = \frac{1}{3}\,, \quad \text{the linear acceleration method (conditionally stable)}\\
&\alpha = \frac{1}{2}\,, \quad \gamma = \frac{1}{6}\,, \quad \text{the Fox–Goodwin scheme (conditionally stable)}\\
&\alpha = \frac{1}{2}\,, \quad \gamma = 0\,, \quad \text{the central difference method (conditionally stable)}\\
&\alpha = \frac{3}{2}\,, \quad \gamma = \frac{8}{5}\,, \quad \text{the Galerkin method (stable)}\\
&\alpha = \frac{3}{2}\,, \quad \gamma = 2\,, \quad \text{the backward difference method (stable)}
\end{aligned}
\tag{8.7 – 11}
$$

Premultiplying the second equation in (7) with $[M]_{s+1}$ and using Eq. (2) at $t = t_{s+1}$ to replace $[M]_{s+1}\{\ddot{\Delta}\}_{s+1}$, we obtain

$$
[\hat{K}]\{\Delta\}_{s+1} = \{\hat{F}\} \tag{8.7 – 12}
$$

$$
\begin{aligned}
[\hat{K}] &= [K]_{s+1} + a_3[M]_{s+1}\\
\{\hat{F}\} &= \{F\}_{s+1} + [M]_{s+1}\left(a_3\{\Delta\}_s + a_4\{\dot{\Delta}\}_s + a_5\{\ddot{\Delta}\}_s\right)
\end{aligned}
\tag{8.7 – 13}
$$

An alternative form of Eq. (12) is given by

$$
[\bar{K}]\{\ddot{\Delta}\}_{s+1} = \{\bar{F}\} \tag{8.7 – 14}
$$

$$
\begin{aligned}
[\bar{K}] &= [M]_{s+1} + \frac{1}{a_3}[K]_{s+1}\\
\{\bar{F}\} &= \{F\}_{s+1} - [K]_{s+1}\left(\{\Delta\}_s + \frac{a_4}{a_3}\{\dot{\Delta}\}_s + \frac{a_5}{a_3}\{\ddot{\Delta}\}_s\right)
\end{aligned}
\tag{8.7 – 15}
$$

Note that for the central difference scheme ($\gamma = 0$), it is necessary to use Eq. (14).

Equation (12) or (14) represents a system of algebraic equations among the (discrete) values of $\{\Delta(t)\}$ at time $t = t_{s+1}$ in terms of known values at time $t = t_s$. Thus the values $\Delta_1(t) = U_{mn}(t)$, $\Delta_2(t) = V_{mn}(t)$, and $\Delta_3(t) = W_{mn}(t)$ are determined at time $t = t_1, t_2, \cdots, t_s, \cdots$ by a repeated solution (or marching in time) of Eq. (12). At the first time step (*i.e.*, $s = 0$), the values $\{\Delta\}_0 = \{\Delta(0)\}$ and $\{\dot{\Delta}\}_0 = \{\dot{\Delta}(0)\}$ are known from the initial conditions (5) of the problem. However, $\{\ddot{\Delta}\}_0 = \{\ddot{\Delta}(0)\}$ is not known at time $t = 0$. Thus, the Newmark method is not a self-starting scheme. Although Eq. (2) is not valid for $t = 0$, it is used to determine $\{\ddot{\Delta}\}_0$ at $t = 0$:

$$\{\ddot{\Delta}\}_0 = [M]^{-1}\left\{\{F\} - [K]\{\Delta\}_0\right\}$$

The transient solution, for example, for the transverse deflection at time $t_s, s > 0$, is given by $(W_{mn}(t_s) = \Delta^3_{mn}(t_s))$

$$w_0(x, y, t_s) = \sum_{m=1}^{\infty}\sum_{n=1}^{\infty} W_{mn}(t_s)\ \sin\frac{m\pi x}{a}\sin\frac{m\pi y}{b} \qquad (8.7-16)$$

8.7–4 Numerical Results

Several examples of applications of the methodology described in this section are presented here. In all of the numerical examples, zero initial conditions were assumed. The following data (in dimensional form) were used in all of the computations:

$$\begin{aligned} a &= b = 25 \text{ cm},\ h = 1 \text{ cm}\ (a/b = 1,\ a/h = 25) \\ \rho &= 8 \times 10^{-6}\ \text{N-s}^2/\text{cm}^4,\ E_2 = 2.1 \times 10^6\ \text{N/cm}^2 \\ E_1 &= 25E_2,\ G_{12} = G_{13} = 0.5E_2,\ \nu_{12} = 0.25 \end{aligned} \qquad (8.7-17)$$

The values of α and γ in the Newmark integration scheme are taken to be 0.5, which correspond to constant-average acceleration method.

The effect of the time step on the accuracy of the solution was investigated using a simply supported antisymmetric cross-ply (0/90) laminate under uniformly distributed step loading. Table 8.7–1 shows the nondimensionalized center transverse deflection, $\bar{w} = w_0(E_2h^3/q_0a^4) \times 10^2$, at selective times for three different time steps: $\delta t = 5, 20$, and 50μs (μs$= 10^{-6}$ s). The effect of larger time step is to reduce the amplitude and increase the period. Plots of the nondimensionalized center deflection versus time for the same problem are shown in Figure 8.7–1. For all time steps below 10μs. the difference is not noticeable on the graphs. In all the following examples, $\delta t = 5\mu$s is used.

Table 8.7–1 Nondimensionalized center transverse deflections ($\bar{w}$) in simply supported (SS–1) cross-ply (0/90) laminates subjected to uniformly distributed transverse load (h = 1cm, $E_1/E_2 = 25$, $E_2 = 2.1 \times 10^6$ N/cm^2, $G_{12} = G_{13} = 0.5E_2$, $G_{23} = 0.2E_2$, $\nu_{12} =$ 0.25).

$\delta t^\dagger$	$t = 100^\dagger$	200	300	400	500	600	700	800	900
5	0.452	1.655	2.931	3.451	2.880	1.628	0.460	-0.003	0.511
20	0.372	1.568	2.862	3.404	2.943	1.796	0.530	-0.051	0.467
50	0.271	1.379	2.603	3.372	3.095	2.061	0.774	0.043	0.274

† Denotes time in microseconds (μs).

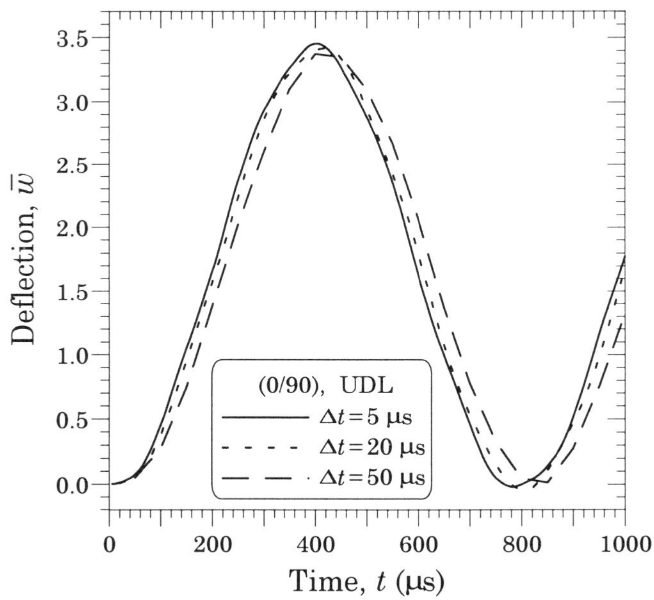

Figure 8.7–1: Nondimensionalized center transverse deflection ($\bar{w}$) versus time (t) for simply supported (SS–1) antisymmetric cross-ply (0/90) laminates subjected to uniformly distributed step loading; see Eq. (8.7–14) for the data.

Figures 8.7–2 through 8.7–5 contain nondimensionalized transverse deflections and normal and shear stresses in two-layer and eight-layer antisymmetric cross-ply (0/90/0/$\cdots$) square plates under suddenly applied transverse load. The nondimensionalizations used are the same as listed in Eq. (8.3–39), except that the nondimensionalized deflection plotted in the figures is $\bar{w} = w_0(E_2h^3/q_0a^4) \times 10^2$ (note the multiplicative factor). The normal stress $\bar{\sigma}_{xx} = \sigma_{xx}(h^2/q_0b^2)$ presented in Figure 8.7–4 is computed at the *bottom* (*i.e.*, at $z = -h/2$) of the laminate, which is larger than that at the *top* (*i.e.*, at $z = h/2$) of the laminate (see Figure 8.7–3). The effect of coupling on the transient response can be seen from the two-layer and eight-layer results. It has the effect of increasing the amplitude as well as the period. The maximum deflections and stresses for the static case are summarized below.

Laminate(0/90), SSL:

$$\bar{w} = 1.064, \quad \bar{\sigma}_{xx}(a/2, b/2, h/2) = 0.084$$

$$\bar{\sigma}_{xx}(a/2, b/2, -h/2) = -0.716, \quad \bar{\sigma}_{xy}(a, b, -h/2) = 0.053 \tag{8.7-18}$$

Laminate(0/90), UDL:

$$\bar{w} = 1.695, \quad \bar{\sigma}_{xx}(a/2, b/2, h/2) = 0.127$$

$$\bar{\sigma}_{xx}(a/2, b/2, -h/2) = -1.076, \quad \bar{\sigma}_{xy}(a, b, -h/2) = 0.093 \tag{8.7-19}$$

Laminate(0/90/0/$\cdots$), UDL:

$$\bar{w} = 0.715, \quad \bar{\sigma}_{xx}(a/2, b/2, h/2) = 0.045$$

$$\bar{\sigma}_{xx}(a/2, b/2, -h/2) = -0.749, \quad \bar{\sigma}_{xy}(a, b, -h/2) = 0.039 \tag{8.7-20}$$

Note that the maximum transient transverse deflection of (0/90) laminate under UDL, which occurs at $t = 400$ μs, is 2.035 times that of the static deflection. Similarly, the stresses are also about 2.035 times that of the static stresses.

Figures 8.7–6 through 8.7–8 contain nondimensionalized transverse deflections and shear and normal stresses in two-layer and eight-layer antisymmetric angle-ply (0/90/0/$\cdots$) square plates under suddenly applied transverse load. The same observations made for cross-ply laminates also apply for angle-ply plates. The angle-ply plates, for the same material and geometric dimensions, have smaller maximum deflections, stresses, and periods of oscillation. The maximum static deflections and stresses are given below.

Laminate(−45/45), UDL:

$$\bar{w} = 1.028, \quad \bar{\sigma}_{xx}(a/2, b/2, h/2) = 0.351, \quad \bar{\sigma}_{xy}(a, b, -h/2) = 0.442 \tag{8.7-21}$$

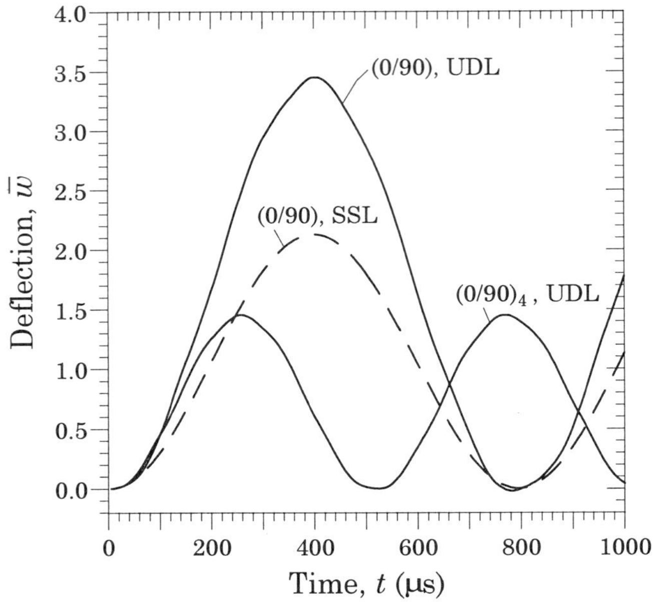

Figure 8.7–2: Nondimensionalized center transverse deflection ($\bar{w}$) versus time (t) for simply supported (SS–1) two-layer and eight-layer antisymmetric cross-ply laminates.

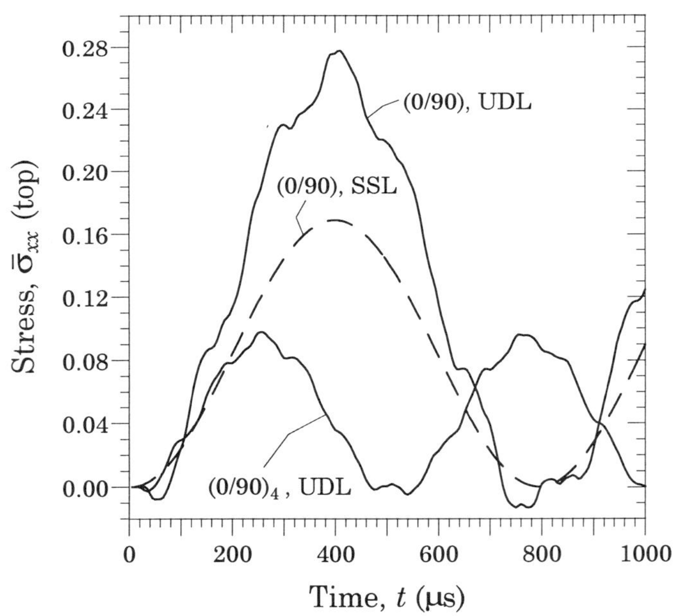

Figure 8.7–3: Nondimensionalized normal stress ($\bar{\sigma}_{xx}$) versus time (t) for simply supported (SS–1) two-layer and eight-layer antisymmetric cross-ply (0/90) laminates.

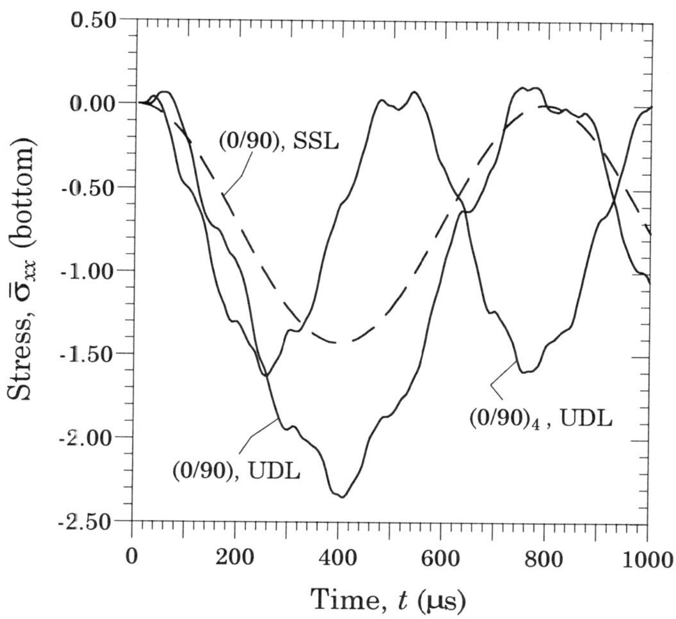

Figure 8.7–4: Nondimensionalized normal stress ($\bar{\sigma}_{xx}$ at the bottom of the laminate) versus time (t) for simply supported (SS–1) two-layer and eight-layer antisymmetric cross-ply (0/90) laminates.

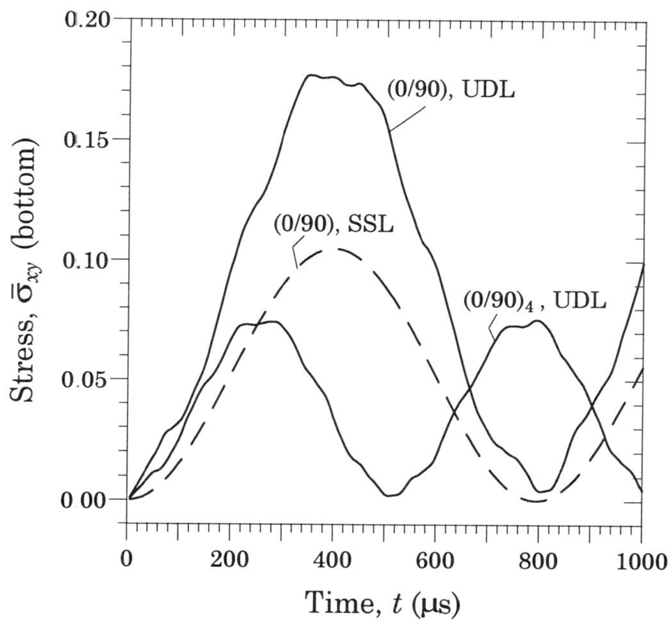

Figure 8.7–5: Nondimensionalized shear stress ($\bar{\sigma}_{xy}$) versus time (t) for simply supported (SS–1) two-layer and eight-layer antisymmetric cross-ply (0/90) laminates.

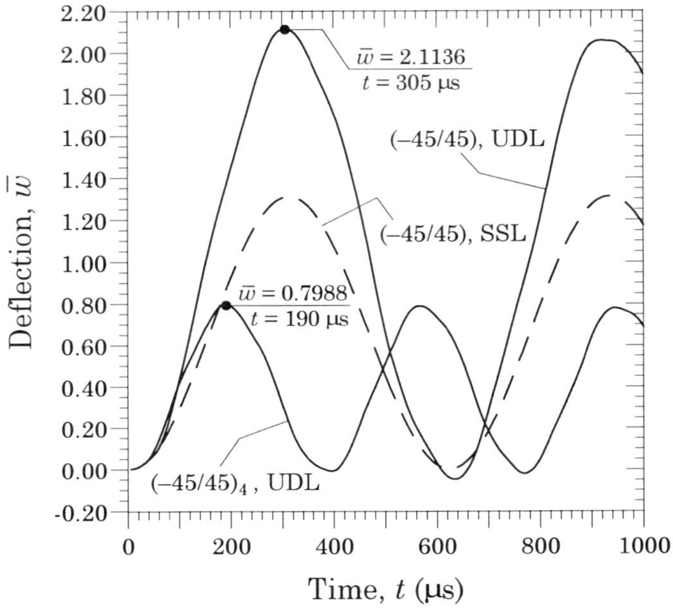

Figure 8.7–6: Nondimensionalized center transverse deflection ($\bar{w}$) versus time (t) for simply supported (SS–2) two-layer and eight-layer antisymmetric angle-ply $(-45/45)_n$ laminates.

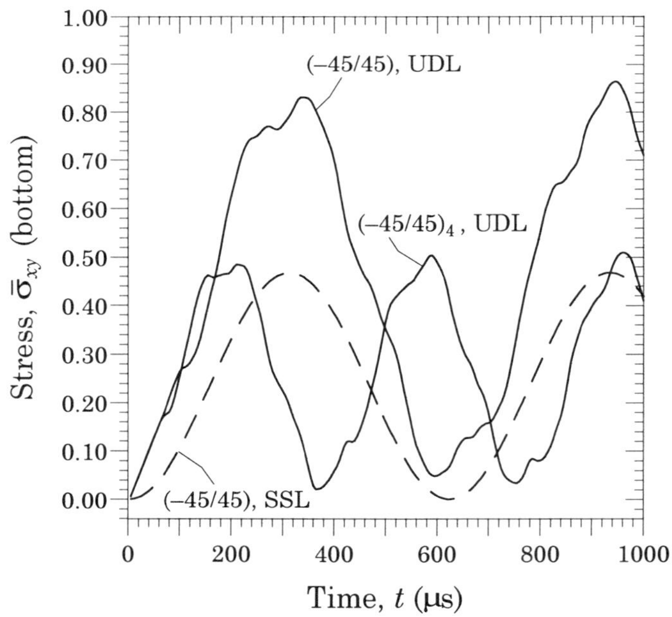

Figure 8.7–7: Nondimensionalized shear stress ($\bar{\sigma}_{xy}$) versus time (t) for simply supported (SS–2) two-layer and eight-layer antisymmetric angle-ply $(-45/45)_n$ laminates.

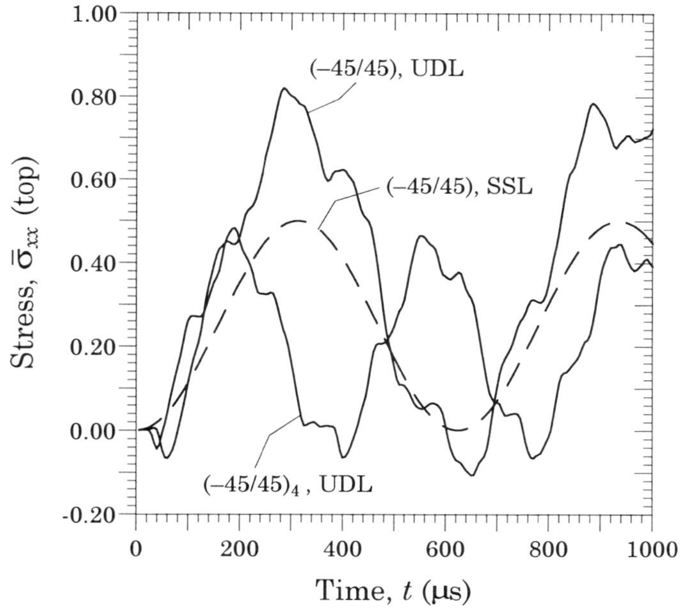

Figure 8.7–8: Nondimensionalized normal stress ($\bar{\sigma}_{xx}$) versus time (t) for simply supported (SS–2) two-layer and eight-layer antisymmetric angle-ply $(-45/45)_n$ laminates.

Laminate (–45/45/–45/$\cdots$), UDL:

$$\bar{w} = 0.386, \quad \bar{\sigma}_{xx}(a/2, b/2, h/2) = 0.201, \quad \bar{\sigma}_{xy}(a, b, -h/2) = 0.264 \qquad (8.7-22)$$

The maximum transient deflection for the two-layer plate is 2.114 and it occurs at $t = 305$ μs; it is about 2.056 times that of the static deflection. In the case of eight-layer laminate, the maximum transient deflection is 0.7988 and it occurs at $t = 190$ μs; it is 2.7 times that of the static deflection.

8.8 Summary

In this chapter analytical solutions for bending, buckling under in-plane compressive loads, and natural vibration of rectangular laminates with various boundary conditions are presented based on the classical laminate theory. The Navier solutions were developed for two classes of laminates: antisymmetric cross-ply laminates and antisymmetric angle-ply laminates, each for a specific type of simply supported boundary conditions, SS–1 and SS–2, respectively. The Lévy solutions with the state–space approach were developed for these classes of laminates when two opposite edges are simply supported with the other two edges having a variety of boundary conditions of choice. A discussion of symmetrically laminated plates, which are characterized by nonzero bending–twisting coupling terms, is also presented. For such laminates, one must use approximate methods, such as the Rayleigh–Ritz method or the finite element method because the Navier solutions do not exist for symmetric laminates. The Rayleigh–Ritz solutions for symmetric laminates are discussed in some detail. Lastly, a transient solution procedure for antisymmetric cross-ply and angle-ply laminates is presented. In this procedure, the solutions are assumed to be products of functions of spatial coordinates (x, y) only and functions of time t only (*i.e.*, separation of variables). The spatial functions are the same as those used in the static case, and the time variation is determined using the Newmark time integration scheme.

Numerical results were presented for static bending, buckling, natural vibration, and transient response of antisymmetric cross-ply and angle-ply laminates. The presence of bending–extensional coupling in a laminate generally reduces the effective stiffnesses and hence increases deflections and reduces buckling loads and natural frequencies. The coupling also increases the period of oscillation in the transient problems. The coupling is the most significant in two-layer laminates, and it decreases gradually as the number of layers is increased for fixed total thickness.

The presence of twist–curvature coupling in a laminate also has the effect of increasing deflections, decreasing buckling loads, and decreasing natural frequencies. The coupling dies out as the number of layers is increased for fixed total thickness.

The effects of bending–stretching coupling and twist–curvature coupling on deflections, buckling loads, and natural frequencies of general laminates, for example, unsymmetric laminates, can only be assessed by specific studies. Such laminates can be analyzed only with approximate methods of analysis.

In general, the bending–twisting coupling in symmetrically laminated plates has the effect of increasing deflections and decreasing buckling loads and natural frequencies of vibration. Analysis of such laminates by the Rayleigh–Ritz method is characterized by slow convergence.

Exercises

8.1 Verify Eq. (8.2–4) by casting Eqs. (8.2–1)–(8.2–3) in operator form.

8.2 Verify Eq. (8.3–19) by substituting expansions (8.3–3) into Eqs. (8.2–1)–(8.2–3) and assuming that conditions in Eqs. (8.3–7) hold.

8.3 Verify the solution in Eq. (8.3–27).

8.4 Derive the expressions for transverse shear stresses from 3–D equations of equilibrium for the isothermal case.

8.5 Derive the expressions for transverse shear stresses from 3–D equations of equilibrium for the case in which the temperature distribution in the plate is of the form

$$\Delta T(x,y,z) = T_0(x,y) + zT_1(x,y)$$

Assume that both T_0 and T_1 can be expanded in double sine series (similar to the mechanical load).

8.6 Verify Eq. (8.4–6) by substituting expansions (8.4–2) into Eqs. (8.2–1)–(8.2–3) and assuming that conditions in Eqs. (8.4–4) hold.

8.7 Verify the solution in Eq. (8.4–9).

8.8 Verify the expressions in Eq. (8.5–11) by substituting expansions (8.5–10) into the definitions of the resultants in Eqs. (5.3–43) and (5.3–44).

8.9 Verify Eqs. (8.5–15).

8.10 Consider antisymmetric angle-ply rectangular laminates with edges $x = 0$ and $x = a$ simply supported and the other two edges, $y = \pm b/2$, having arbitrary boundary conditions. Assume solution of the form

$$\begin{aligned} u_0(x,y,t) &= \sum_{m=1}^{\infty} U_m(y)\sin\alpha x \\ v_0(x,y,t) &= \sum_{m=1}^{\infty} V_m(y)\cos\alpha x \\ w_0(x,y,t) &= \sum_{m=1}^{\infty} W_m(y)\sin\alpha x \end{aligned} \qquad (a)$$

and load expansion in the form

$$q(x,y) = \sum_{m=1}^{\infty} Q_m(y)\sin\alpha x \qquad (b)$$

where $\alpha = m\pi/a$. Show that the equations of equilibrium of the classical laminated plate theory for such laminates (without any applied in-plane loading) can be reduced to the following ordinary differential equations

$$\begin{aligned} U''_m &= C_1U_m + C_2V'_m + C_3W'_m + C_4W'''_m \\ V''_m &= C_5U'_m + C_6V_m + C_7W_m + C_8W''_m \\ W''''_m &= C_9U'_m + C_{10}V_m + C_{11}W_m + C_{12}W''_m + C_{13}Q_m \end{aligned} \qquad (c)$$

where the primes indicate differentiation with respect to y, and the coefficients C_i are defined as

$$\begin{aligned} &C_1 = -e_1/e_2, \quad C_2 = -e_3/e_2, \quad C_3 = -e_4/e_2, \quad C_4 = -e_5/e_2 \\ &C_5 = -e_6/e_8, \quad C_6 = -e_7/e_8, \quad C_7 = -e_9/e_8, \quad C_8 = -e_{10}/e_8 \end{aligned}$$

$$C_9 = \left(e_{14} - \frac{e_1 e_{16}}{e_2} - \frac{e_6 e_{17}}{e_8} + \frac{e_3 e_6 e_{16}}{e_2 e_8}\right)\left(\frac{e_5 e_{16}}{e_2} - e_{13}\right)^{-1}$$
$$C_{10} = \left(e_{15} - \frac{e_7 e_{17}}{e_8} + \frac{e_3 e_7 e_{16}}{e_2 e_8}\right)\left(\frac{e_5 e_{16}}{e_2} - e_{13}\right)^{-1}$$
$$C_{11} = \left(e_{11} - \frac{e_9 e_{17}}{e_8} + \frac{e_3 e_9 e_{16}}{e_2 e_8}\right)\left(\frac{e_5 e_{16}}{e_2} - e_{13}\right)^{-1}$$
$$C_{12} = \left(e_{12} - \frac{e_4 e_{16}}{e_2} - \frac{e_{10} e_{17}}{e_8} + \frac{e_3 e_{10} e_{16}}{e_2 e_8}\right)\left(\frac{e_5 e_{16}}{e_2} - e_{13}\right)^{-1}$$
$$C_{13} = \left(e_{13} - \frac{e_5 e_{16}}{e_2}\right)^{-1} \qquad (d)$$

and the coefficients e_i are defined as

$$\begin{aligned} &c_1 = -\alpha^2 A_{11},\quad e_2 = A_{66},\quad e_3 = -\alpha(A_{12}+A_{66}),\quad e_4 = 3\alpha^2 B_{16}\\ &e_5 = -B_{26},\quad e_6 = -e_3,\quad e_7 = -\alpha^2 A_{66},\quad e_8 = A_{22}\\ &e_9 = \alpha^3 B_{16},\quad e_{10} = -3\alpha B_{26},\quad e_{11} = \alpha^4 D_{11},\quad e_{12} = -2\alpha^2(D_{12}+2D_{66})\\ &e_{13} = D_{22},\quad e_{14} = e_4,\quad e_{15} = -e_9,\quad e_{16} = e_5,\quad e_{17} = -e_{10} \end{aligned} \qquad (e)$$

8.11 Repeat Exercise 8.10 for the case of biaxial buckling. All definitions in Exercise 8.10 hold with exception of e_{11} and e_{12}, which are modified as

$$\begin{aligned} e_{11} &= \alpha^4 D_{11} - \alpha^2 N_{xx}^0\\ e_{12} &= -2\alpha^2(D_{12}+2D_{66}) + N_{yy}^0 \end{aligned}$$

where N_{xx}^0 and N_{yy}^0 are the in-plane compressive forces.

8.12 Repeat Exercise 8.10 for the case of free vibration. All definitions in Exercise 8.10 hold with exception of e_{11}, which is modified as

$$e_{11} = \alpha^4 D_{11} - I_0\omega_m^2,\quad I_0 = \int_{-h/2}^{h/2} \rho^{(m)} dz$$

where ω_m is the frequency of vibration associated with mode m.

8.13 Defining the state vector $\mathbf{Z}(\mathbf{y})$ as

$$Z_1 = U_m,\ Z_2 = U_m',\ Z_3 = V_m,\ Z_4 = V_m'$$

$$Z_5 = W_m,\ Z_6 = W_m',\ Z_7 = W_m'',\ Z_8 = W_m''' \qquad (a)$$

express Eqs. (c) of Exercise 8.10 as a first-order matrix equation of the form

$$\mathbf{Z}' = \mathbf{TZ} + \mathbf{F} \qquad (b)$$

where the matrix $\mathbf{T}$ and the column vector $\mathbf{F}$ are given by

$$\mathbf{T} = \begin{bmatrix} 0 & 1 & 0 & 0 & 0 & 0 & 0 & 0 \\ C_1 & 0 & 0 & C_2 & 0 & C_3 & 0 & C_4 \\ 0 & 0 & 0 & 1 & 0 & 0 & 0 & 0 \\ 0 & C_5 & C_6 & 0 & C_7 & 0 & C_8 & 0 \\ 0 & 0 & 0 & 0 & 0 & 1 & 0 & 0 \\ 0 & 0 & 0 & 0 & 0 & 0 & 1 & 0 \\ 0 & 0 & 0 & 0 & 0 & 0 & 0 & 1 \\ 0 & C_9 & C_{10} & 0 & C_{11} & 0 & C_{12} & 0 \end{bmatrix} \qquad (c)$$

$$\mathbf{F} = \{0,0,0,0,0,0,0,C_{13}\, Q_m\}^T \qquad (d)$$

8.14 Consider a symmetrically laminated rectangular plate under the transverse load $q(x,y)$. The governing equation for static bending analysis is given by

$$D_{11}\frac{\partial^4 w_0}{\partial x^4} + 4D_{16}\frac{\partial^4 w_0}{\partial x^3 \partial y} + 2\left(D_{12} + 2D_{66}\right)\frac{\partial^4 w_0}{\partial x^2 \partial y^2} + D_{22}\frac{\partial^4 w_0}{\partial y^4} + 4D_{26}\frac{\partial^4 w_0}{\partial x \partial y^3} = q$$

The weak form (or the virtual work statement) of the same equation is given by Eq. (8.6–6), without the in-plane force and inertial terms. Show that the Rayleigh–Ritz solution of the form

$$w_0(x,y) \approx \sum_{i=1}^{N} c_i \phi_i(x,y) \qquad (a)$$

requires the solution of the algebraic equation

$$[R]\{c\} = \{q\} \qquad (b)$$

where

$$\begin{aligned} R_{ij} = \int_0^b \int_0^a \Bigg[& D_{11}\frac{\partial^2 \phi_i}{\partial x^2}\frac{\partial^2 \phi_j}{\partial x^2} + D_{12}\left(\frac{\partial^2 \phi_i}{\partial y^2}\frac{\partial^2 \phi_j}{\partial x^2} + \frac{\partial^2 \phi_i}{\partial x^2}\frac{\partial^2 \phi_j}{\partial y^2}\right) \\ & + 4D_{66}\frac{\partial^2 \phi_i}{\partial x \partial y}\frac{\partial^2 \phi_j}{\partial x \partial y} + D_{22}\frac{\partial^2 \phi_i}{\partial y^2}\frac{\partial^2 \phi_j}{\partial y^2} \\ & + 2D_{16}\left(\frac{\partial^2 \phi_i}{\partial x^2}\frac{\partial^2 \phi_j}{\partial x \partial y} + \frac{\partial^2 \phi_i}{\partial x \partial y}\frac{\partial^2 \phi_j}{\partial x^2}\right) \\ & + 2D_{26}\left(\frac{\partial^2 \phi_i}{\partial y^2}\frac{\partial^2 \phi_j}{\partial x \partial y} + \frac{\partial^2 \phi_i}{\partial x \partial y}\frac{\partial^2 \phi_j}{\partial y^2}\right) \Bigg] dx dy \end{aligned}$$

$$q_i = \int_0^b \int_0^a q(x,y)\phi_i \, dx dy \qquad (c)$$

8.15 Consider a symmetrically laminated rectangular plate with simply supported edges. The boundary conditions are given by

$$w_0(x,0) = 0, \quad w_0(x,b) = 0, \quad w_0(0,y) = 0, \quad w_0(a,y) = 0$$

$$M_{xx}(0,y) = 0, \quad M_{xx}(a,y) = 0, \quad M_{yy}(x,0) = 0, \quad M_{yy}(x,b) = 0$$

where the bending moments are related to the transverse deflection by the equations

$$M_{xx} = -\left(D_{11}\frac{\partial^2 w_0}{\partial x^2} + 2D_{16}\frac{\partial^2 w_0}{\partial x \partial y} + D_{12}\frac{\partial^2 w_0}{\partial y^2}\right)$$
$$M_{yy} = -\left(D_{12}\frac{\partial^2 w_0}{\partial x^2} + 2D_{26}\frac{\partial^2 w_0}{\partial x \partial y} + D_{22}\frac{\partial^2 w_0}{\partial y^2}\right)$$

Find a two-parameter Rayleigh–Ritz approximation using algebraic polynomials. Note that the one-parameter approximation, $w_0(x,y) \approx c_1 xy(a-x)(b-y)$ does not give a solution for the case in which D_{16} and D_{26} are not zero.

Ans: For the approximation of the form

$$w_0(x,y) \approx c_1 xy(a-x)(b-y) + c_2 x^2 y^2 (a-x)(b-y)$$

the Ritz coefficients are given by

$$R_{11} = 2\left[D_{11}\frac{ab^5}{15} + 2(D_{12}+2D_{66})\frac{a^3b^3}{9} + D_{22}\frac{a^5 b}{15}\right]$$
$$R_{12} = \left[D_{11}\frac{a^2b^7}{60} + 2(D_{12}+2D_{66})\frac{a^4b^4}{36} + D_{22}\frac{a^7b^2}{60} - 2D_{16}\frac{a^3b^5}{60} - 2D_{26}\frac{a^5b^3}{60}\right]$$
$$R_{22} = 4\left[D_{11}\frac{a^3b^7}{105} + 2(D_{12}+2D_{66})\frac{a^5b^5}{225} + D_{22}\frac{a^7b^3}{105} + D_{16}\frac{a^4b^6}{10} + D_{26}\frac{a^6b^4}{10}\right]$$
$$q_1 = q_0\frac{a^3b^3}{36}, \quad q_2 = q_0\frac{a^7b^7}{(105)(105)}$$

References for Additional Reading

1. Reddy, J. N., *Energy and Variational Methods in Applied Mechanics,* John Wiley & Sons, New York, 1984.
2. Brogan, W. L., *Modern Control Theory,* Prentice–Hall, Englewood Cliffs, New Jersey, 1985.

3. Franklin, J. N., *Matrix Theory,* Prentice–Hall, Englewood Cliffs, New Jersey, 1968.
4. Nosier, A. and Reddy, J. N., "Vibration and Stability Analyses of Cross-Ply Laminated Circular Cylindrical Shells," *Journal of Sound and Vibration,* **157** (1), 139–159 (1992).
5. Ashton, J. E. and Whitney, J. M., *Theory of Laminated Plates,* Technomic, Stamford, CT, 1970.
6. Reddy, J. N. (ed.), *Mechanics of Composite Materials. Selected Works of Nicholas J. Pagano*, Kluwer, Netherlands, 1994.
7. Pagano, N. J., "Exact Solutions for Rectangular Bidirectional Composites and Sandwich Plates," *Journal of Composite Materials*, **4**(1), 20–34 (1970).
8. Pagano, N. J., and Hatfield, S. J., "Elastic Behavior of Multilayered Bidirectional Composites," *AIAA Journal*, **10**(7), 931–933 (1972).
9. Reddy, J. N. and Chao, W. C., "A Comparison of Closed Form and Finite Element Solutions of Thick Laminated Anisotropic Rectangular Plates," *Nuclear Engineering and Design*, **64**, 153–167 (1981).
10. Reddy, J. N., Khdeir, A. A., and Librescu, L., "Lévy Type Solutions for Symmetrically Laminated Rectangular Plates Using First–Order Shear Deformation Theory," *Journal of Applied Mechanics*, **54**, 740–742 (1987).
11. Khdeir, A. A., Reddy, J. N., and Librescu, L., "Analytical Solution of a Refined Shear Deformation Theory for Rectangular Composite Plates," *International Journal of Solids and Structures*, **23**, 1447–1463 (1987).
12. Khdeir, A. A. and Librescu, L., "Analysis of Symmetric Cross-Ply Laminated Elastic Plates Using a Higher-Order Theory: Part I: Stress and Displacement," *Composite Structures*, **9**, 189–213 (1988).
13. Khdeir, A. A. and Librescu, L., "Analysis of Symmetric Cross-Ply Laminated Elastic Plates Using a Higher–Order Theory: Part II: Buckling and Free Vibration," *Composite Structures*, **9**, 259–277 (1988).
14. Reddy, J. N. and Khdeir, A. A., "Buckling and Vibration of Laminated Composite Plates Using Various Plate Theories," *AIAA Journal*, **27**(12), 1808–1817 (1989).
15. Khdeir, A. A., "Free Vibration of Antisymmetric Angle-Ply Laminated Plates Including Various Boundary Conditions," *Journal of Sound and Vibration*, **122**(2), 377–388 (1988).
16. Khdeir, A. A., "Free Vibration and Buckling of Unsymmetric Cross-Ply Laminated Plates," *Journal of Sound and Vibration*, **128**, (3), 377–395 (1989).
17. Khdeir, A. A., "An Exact Approach to the Elastic State of Stress of Shear Deformable Antisymmetric Angle-Ply Laminated Plates," *Composite Structures*, **11**, 245–258 (1989).
18. Khdeir, A. A., "Comparison Between Shear Deformable and Kirchhoff Theories for Bending, Buckling, and Vibration of Antisymmetric Angle-Ply Laminated Plates," *Composite Structures*, **13,** 159–172 (1989)

19. Ashton, J. E. and Waddoups, M. E., "Analysis of Anisotropic Plates," *Journal of Composite Materials,* **3**, 148–165 (1969).
20. Ashton, J. E., "Analysis of Anisotropic Plates II," *Journal of Composite Materials,* **3**, 470–479 (1969).
21. Lekhnitskii, S. G., *Anisotropic Plates,* Translated from Russian by S. W. Tsai and T. Cheron, Gordon and Breach, Newark, New Jersey, 1968.
22. Hearman, R. F. S., "The Frequency of Flexural Vibration of Rectangular Orthotropic Plates with Clamped or Supported Edges," *Journal of Applied Mechanics,* **26**(4), 537–540 (1959).
23. Young, D. and Felgar, F. P., *Tables of Characteristic Functions Representing the Normal Modes of Vibration of a Beam,* University of Texas, Publication No. 4913, 1949.
24. Khdeir A. A., and Reddy, J. N. "Exact Solutions for the Transient Response of Symmetric Cross-Ply Laminates Using a Higher-Order Plate Theory," *Composites Science and Technology,* **34**, 205–224 (1989).
25. Khdeir A. A., and Reddy, J. N. "On the Forced Motions of Antisymmetric Cross-Ply Laminates," *International Journal of Mechanical Sciences,* **31**, 499–510 (1989).
26. Khdeir A. A., and Reddy, J. N. "Dynamic Response of Antisymmetric Angle-Ply Laminated Plates Subjected to Arbitrary Loading," *Journal of Sound and Vibration,* **126**, 437–445 (1988).
27. Reddy, J. N., "On the Solutions to Forced Motions of Rectangular Composite Plates," *Journal of Applied Mechanics,* **49**, 403–408 (1982).
28. Khdeir, A. A. and Reddy, J. N., "Analytical Solutions of Refined Plate Theories of Cross-Ply Composite Laminates," *Journal of Pressure Vessel Technology,* **113**(4), 570–578 (1991).

Chapter Nine

Analytical Solutions of Rectangular Laminates Using FSDT

9.1 Introduction

The classical laminate plate theory is based on the Kirchhoff assumptions, in which transverse normal and shear stresses are neglected. Although such stresses can be postcomputed through 3-D elasticity equilibrium equations, they are not always accurate. The equilibrium-derived transverse stress field is sufficiently accurate for homogeneous and thin plates; they are not accurate when plates are relatively thick (*i.e.*, $a/h < 20$). In the first-order shear deformation theory (FSDT), a constant state of transverse shear stresses is accounted for, and often the transverse normal stress is neglected. The FSDT allows the computation of interlaminar shear stresses through constitutive equations, which is quite simpler than deriving them through equilibrium equations. It should be noted that the interlaminar stresses derived from constitutive equations do not match, in general, those derived from equilibrium equations. In fact, the transverse shear stresses derived from the equilibrium equations are quadratic through lamina thickness, as was shown in Chapter 8 for CLPT, whereas those computed from constitutive equations are constant.

The more significant difference between the classical and first-order theories is the effect of including transverse shear deformation on the predicted deflections, frequencies, and buckling loads. As noted in Chapter 6, the classical laminate theory underpredicts deflections and overpredicts frequencies as well as buckling loads with plate side-to-thickness ratios of the order of 20 or less. For this reason alone it is necessary to use the first-order theory in the analysis of relatively thick laminated plates. In this chapter, we develop analytical solutions of rectangular laminates using the first-order shear deformation theory. The primary objective is to bring out the effect of shear deformation on deflections, stresses, frequencies, and buckling loads.

To discuss the Navier and other solutions, the equations of motion of the first-order plate theory, Eqs. (5.4–23) through (5.4–27), are expressed in terms of the generalized displacements (u_0, v_0, w_0, ϕ_x and ϕ_y) as

$$\begin{aligned}
&\frac{\partial}{\partial x}\left[A_{11}\frac{\partial u_0}{\partial x} + A_{12}\frac{\partial v_0}{\partial y} + A_{16}\left(\frac{\partial u_0}{\partial y} + \frac{\partial v_0}{\partial x}\right) + B_{11}\frac{\partial \phi_x}{\partial x} + B_{12}\frac{\partial \phi_y}{\partial y}\right.\\
&\quad\left. + B_{16}\left(\frac{\partial \phi_x}{\partial y} + \frac{\partial \phi_y}{\partial x}\right)\right] + \frac{\partial}{\partial y}\left[A_{16}\frac{\partial u_0}{\partial x} + A_{26}\frac{\partial v_0}{\partial y}\right.\\
&\quad\left. + A_{66}\left(\frac{\partial u_0}{\partial y} + \frac{\partial v_0}{\partial x}\right) + B_{16}\frac{\partial \phi_x}{\partial x} + B_{26}\frac{\partial \phi_y}{\partial y} + B_{66}\left(\frac{\partial \phi_x}{\partial y} + \frac{\partial \phi_y}{\partial x}\right)\right]\\
&\quad - \left(\frac{\partial N_{xx}^T}{\partial x} + \frac{\partial N_{xy}^T}{\partial y}\right) = I_0\frac{\partial^2 u_0}{\partial t^2} + I_1\frac{\partial^2 \phi_x}{\partial t^2}
\end{aligned} \tag{9.1-1}$$

$$\begin{aligned}
&\frac{\partial}{\partial x}\left[A_{16}\frac{\partial u_0}{\partial x} + A_{26}\frac{\partial v_0}{\partial y} + A_{66}\left(\frac{\partial u_0}{\partial y} + \frac{\partial v_0}{\partial x}\right) + B_{16}\frac{\partial \phi_x}{\partial x} + B_{26}\frac{\partial \phi_y}{\partial y}\right.\\
&\quad\left. + B_{66}\left(\frac{\partial \phi_x}{\partial y} + \frac{\partial \phi_y}{\partial x}\right)\right] + \frac{\partial}{\partial y}\left[A_{12}\frac{\partial u_0}{\partial x} + A_{22}\frac{\partial v_0}{\partial y}\right.\\
&\quad\left. + A_{26}\left(\frac{\partial u_0}{\partial y} + \frac{\partial v_0}{\partial x}\right) + B_{12}\frac{\partial \phi_x}{\partial x} + B_{22}\frac{\partial \phi_y}{\partial y} + B_{26}\left(\frac{\partial \phi_x}{\partial y} + \frac{\partial \phi_y}{\partial x}\right)\right]\\
&\quad - \left(\frac{\partial N_{xy}^T}{\partial x} + \frac{\partial N_{yy}^T}{\partial y}\right) = I_0\frac{\partial^2 v_0}{\partial t^2} + I_1\frac{\partial^2 \phi_y}{\partial t^2}
\end{aligned} \tag{9.1-2}$$

$$\begin{aligned}
&\frac{\partial}{\partial x}\left[KA_{45}\left(\frac{\partial w_0}{\partial y} + \phi_y\right) + KA_{55}\left(\frac{\partial w_0}{\partial x} + \phi_x\right)\right]\\
&\quad + \frac{\partial}{\partial y}\left[KA_{44}\left(\frac{\partial w_0}{\partial y} + \phi_y\right) + KA_{45}\left(\frac{\partial w_0}{\partial x} + \phi_x\right)\right]\\
&\quad + \hat{N}_{xx}\frac{\partial^2 w_0}{\partial x^2} + 2\hat{N}_{xy}\frac{\partial^2 w_0}{\partial x \partial y} + \hat{N}_{yy}\frac{\partial^2 w_0}{\partial y^2} + q = I_0\frac{\partial^2 w_0}{\partial t^2}
\end{aligned} \tag{9.1-3}$$

$$\begin{aligned}
&\frac{\partial}{\partial x}\left[B_{11}\frac{\partial u_0}{\partial x} + B_{12}\frac{\partial v_0}{\partial y} + B_{16}\left(\frac{\partial u_0}{\partial y} + \frac{\partial v_0}{\partial x}\right) + D_{11}\frac{\partial \phi_x}{\partial x} + D_{12}\frac{\partial \phi_y}{\partial y}\right.\\
&\quad\left. + D_{16}\left(\frac{\partial \phi_x}{\partial y} + \frac{\partial \phi_y}{\partial x}\right)\right] + \frac{\partial}{\partial y}\left[B_{16}\frac{\partial u_0}{\partial x} + B_{26}\frac{\partial v_0}{\partial y}\right.\\
&\quad\left. + B_{66}\left(\frac{\partial u_0}{\partial y} + \frac{\partial v_0}{\partial x}\right) + D_{16}\frac{\partial \phi_x}{\partial x} + D_{26}\frac{\partial \phi_y}{\partial y} + D_{66}\left(\frac{\partial \phi_x}{\partial y} + \frac{\partial \phi_y}{\partial x}\right)\right]\\
&\quad - \left[KA_{45}\left(\frac{\partial w_0}{\partial y} + \phi_y\right) + KA_{55}\left(\frac{\partial w_0}{\partial x} + \phi_x\right)\right]\\
&\quad - \left(\frac{\partial M_{xx}^T}{\partial x} + \frac{\partial M_{xy}^T}{\partial y}\right) = I_2\frac{\partial^2 \phi_x}{\partial t^2} + I_1\frac{\partial^2 u_0}{\partial t^2}
\end{aligned} \tag{9.1-4}$$

$$\frac{\partial}{\partial x}\left[B_{16}\frac{\partial u_0}{\partial x}+B_{26}\frac{\partial v_0}{\partial y}+B_{66}\left(\frac{\partial u_0}{\partial y}+\frac{\partial v_0}{\partial x}\right)+D_{16}\frac{\partial \phi_x}{\partial x}\right.$$
$$\left.+D_{26}\frac{\partial \phi_y}{\partial y}+D_{66}\left(\frac{\partial \phi_x}{\partial y}+\frac{\partial \phi_y}{\partial x}\right)\right]+\frac{\partial}{\partial y}\left[B_{12}\frac{\partial u_0}{\partial x}+B_{22}\frac{\partial v_0}{\partial y}\right.$$
$$\left.+B_{26}\left(\frac{\partial u_0}{\partial y}+\frac{\partial v_0}{\partial x}\right)+D_{12}\frac{\partial \phi_x}{\partial y}+D_{22}\frac{\partial \phi_y}{\partial y}+D_{26}\left(\frac{\partial \phi_x}{\partial y}+\frac{\partial \phi_y}{\partial x}\right)\right]$$
$$-\left[KA_{44}\left(\frac{\partial w_0}{\partial y}+\phi_y\right)+KA_{45}\left(\frac{\partial w_0}{\partial x}+\phi_x\right)\right]$$
$$-\left(\frac{\partial M_{xy}^T}{\partial x}+\frac{\partial M_{yy}^T}{\partial y}\right)=I_2\frac{\partial^2 \phi_y}{\partial t^2}+I_1\frac{\partial^2 v_0}{\partial t^2} \qquad (9.1-5)$$

where the thermal resultants, $(N_{xx}^T, N_{yy}^T, N_{xy}^T)$ and $(M_{xx}^T, M_{yy}^T, M_{xy}^T)$, are defined in Eqs. (5.3–41a,b).

9.2 Simply Supported Antisymmetric Cross-Ply Laminates

9.2–1 Solution for the General Case

The SS–1 boundary conditions for the first-order shear deformation plate theory (FSDT) are (Figure 9.2–1):

$$u_0(x,0,t)=0,\quad \phi_x(x,0,t)=0,\quad u_0(x,b,t)=0,\quad \phi_x(x,b,t)=0$$
$$v_0(0,y,t)=0,\quad \phi_y(0,y,t)=0,\quad v_0(a,y,t)=0,\quad \phi_y(a,y,t)=0$$
$$w_0(x,0,t)=0,\quad w_0(x,b,t)=0,\quad w_0(0,y,t)=0,\quad w_0(a,y,t)=0 \qquad (9.2-1a)$$
$$N_{xx}(0,y,t)=0,\quad N_{xx}(a,y,t)=0,\quad N_{yy}(x,0,t)=0,\quad N_{yy}(x,b,t)=0$$
$$M_{xx}(0,y,t)=0,\quad M_{xx}(a,y,t)=0,\quad M_{yy}(x,0,t)=0,\quad M_{yy}(x,b,t)=0 \qquad (9.2-1b)$$

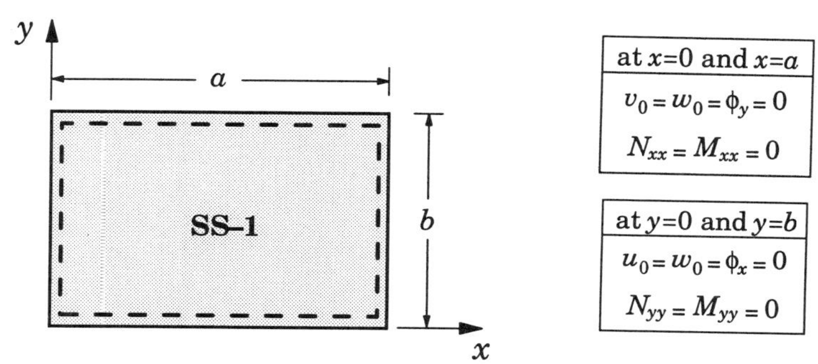

Figure 9.2–1: The simply supported boundary conditions for antisymmetric cross-ply laminates using the first-order shear deformation theory (SS–1).

The boundary conditions in (1b) are satisfied by the following expansions

$$u_0(x,y,t) = \sum_{n=1}^{\infty}\sum_{m=1}^{\infty} U_{mn}(t)\cos\alpha x \sin\beta y \tag{9.2 - 2a}$$

$$v_0(x,y,t) = \sum_{n=1}^{\infty}\sum_{m=1}^{\infty} V_{mn}(t)\sin\alpha x \cos\beta y \tag{9.2 - 2b}$$

$$w_0(x,y,t) = \sum_{n=1}^{\infty}\sum_{m=1}^{\infty} W_{mn}(t)\sin\alpha x \sin\beta y \tag{9.2 - 3}$$

$$\phi_x(x,y,t) = \sum_{n=1}^{\infty}\sum_{m=1}^{\infty} X_{mn}(t)\cos\alpha x \sin\beta y \tag{9.2 - 4a}$$

$$\phi_y(x,y,t) = \sum_{n=1}^{\infty}\sum_{m=1}^{\infty} Y_{mn}(t)\sin\alpha x \cos\beta y \tag{9.2 - 4b}$$

The mechanical and thermal loads are also expanded in double Fourier sine series

$$q(x,y,t) = \sum_{n=1}^{\infty}\sum_{m=1}^{\infty} Q_{mn}(t)\sin\alpha x \sin\beta y \tag{9.2 - 5a}$$

$$\Delta T(x,y,z,t) = \sum_{n=1}^{\infty}\sum_{m=1}^{\infty} T_{mn}(z,t)\sin\alpha x \;\sin\beta y \tag{9.2 - 5b}$$

where

$$Q_{mn}(z,t) = \frac{4}{ab}\int_0^a\int_0^b q(x,y,t)\ \sin\frac{m\pi x}{a}\ \sin\frac{n\pi y}{b}\ dxdy \tag{9.2 - 6a}$$

$$T_{mn}(z,t) = \frac{4}{ab}\int_0^a\int_0^b \Delta T(x,y,z,t)\ \sin\frac{m\pi x}{a}\ \sin\frac{n\pi y}{b}\ dxdy \tag{9.2 - 6b}$$

Substitution of Eqs. (2)–(5) into Eqs. (9.1–1)–(9.1–5) will show that the Navier solution exists only if

$$A_{16}=0,\ A_{26}=0,\ A_{45}=0,\ B_{16}=0,\ B_{26}=0,\ D_{16}=0,\ D_{26}=0,\ I_1=0$$

i.e., for the same laminates as those for the classical laminate theory. For such laminates the coefficients $(U_{mn}, V_{mn}, W_{mn}, X_{mn}, Y_{mn})$ of the Navier solution can be calculated from

$$\begin{bmatrix} \hat{s}_{11} & \hat{s}_{12} & 0 & \hat{s}_{14} & \hat{s}_{15} \\ \hat{s}_{12} & \hat{s}_{22} & 0 & \hat{s}_{24} & \hat{s}_{25} \\ 0 & 0 & \hat{s}_{33}+\tilde{s}_{33} & \hat{s}_{34} & \hat{s}_{35} \\ \hat{s}_{14} & \hat{s}_{24} & \hat{s}_{34} & \hat{s}_{44} & \hat{s}_{45} \\ \hat{s}_{15} & \hat{s}_{25} & \hat{s}_{35} & \hat{s}_{45} & \hat{s}_{55} \end{bmatrix} \begin{Bmatrix} U_{mn} \\ V_{mn} \\ W_{mn} \\ X_{mn} \\ Y_{mn} \end{Bmatrix} +$$

$$\begin{bmatrix} \hat{m}_{11} & 0 & 0 & 0 & 0 \\ 0 & \hat{m}_{22} & 0 & 0 & 0 \\ 0 & 0 & \hat{m}_{33} & 0 & 0 \\ 0 & 0 & 0 & \hat{m}_{44} & 0 \\ 0 & 0 & 0 & 0 & \hat{m}_{55} \end{bmatrix} \begin{Bmatrix} \ddot{U}_{mn} \\ \ddot{V}_{mn} \\ \ddot{W}_{mn} \\ \ddot{X}_{mn} \\ \ddot{Y}_{mn} \end{Bmatrix} = \begin{Bmatrix} 0 \\ 0 \\ Q_{mn} \\ 0 \\ 0 \end{Bmatrix} - \begin{Bmatrix} \alpha N^1_{mn} \\ \beta N^2_{mn} \\ 0 \\ \alpha M^1_{mn} \\ \beta M^2_{mn} \end{Bmatrix} \quad (9.2-7a)$$

where $\hat{s}_{ij}$ and $\hat{m}_{ij}$

$$\hat{s}_{11} = (A_{11}\alpha^2 + A_{66}\beta^2), \quad \hat{s}_{12} = (A_{12} + A_{66})\alpha\beta$$
$$\hat{s}_{14} = (B_{11}\alpha^2 + B_{66}\beta^2), \quad \hat{s}_{15} = (B_{12} + B_{66})\alpha\beta$$
$$\hat{s}_{22} = (A_{66}\alpha^2 + A_{22}\beta^2), \quad \hat{s}_{24} = \hat{s}_{15}$$
$$\hat{s}_{25} = (B_{66}\alpha^2 + B_{22}\beta^2), \quad \hat{s}_{33} = K(A_{55}\alpha^2 + A_{44}\beta^2)$$
$$\tilde{s}_{33} = \hat{N}_{xx}\alpha^2 + \hat{N}_{yy}\beta^2$$
$$\hat{s}_{34} = KA_{55}\alpha, \quad \hat{s}_{35} = KA_{44}\beta, \quad \hat{s}_{44} = (D_{11}\alpha^2 + D_{66}\beta^2 + KA_{55})$$
$$\hat{s}_{45} = (D_{12} + D_{66})\alpha\beta, \quad \hat{s}_{55} = (D_{66}\alpha^2 + D_{22}\beta^2 + KA_{44}) \quad (9.2-7b)$$

$$\hat{m}_{11} = I_0, \quad \hat{m}_{22} = I_0, \quad \hat{m}_{33} = I_0, \quad \hat{m}_{44} = I_2, \quad \hat{m}_{55} = I_2 \quad (9.2-7c)$$

where the thermal coefficients $N^1_{mn}, N^2_{mn}, M^1_{mn}$, and M^2_{mn} are defined in Eqs. (8.3–13a,b).

9.2–2 Bending

The static solution can be obtained from Eqs. (7) by setting the time derivative terms and edge forces to zero:

$$\begin{bmatrix} \hat{s}_{11} & \hat{s}_{12} & 0 & \hat{s}_{14} & \hat{s}_{15} \\ \hat{s}_{12} & \hat{s}_{22} & 0 & \hat{s}_{24} & \hat{s}_{25} \\ 0 & 0 & \hat{s}_{33} & \hat{s}_{34} & \hat{s}_{35} \\ \hat{s}_{14} & \hat{s}_{24} & \hat{s}_{34} & \hat{s}_{44} & \hat{s}_{45} \\ \hat{s}_{15} & \hat{s}_{25} & \hat{s}_{35} & \hat{s}_{45} & \hat{s}_{55} \end{bmatrix} \begin{Bmatrix} U_{mn} \\ V_{mn} \\ W_{mn} \\ X_{mn} \\ Y_{mn} \end{Bmatrix} = \begin{Bmatrix} 0 \\ 0 \\ Q_{mn} \\ 0 \\ 0 \end{Bmatrix} - \begin{Bmatrix} \alpha N^1_{mn} \\ \beta N^2_{mn} \\ 0 \\ \alpha M^1_{mn} \\ \beta M^2_{mn} \end{Bmatrix} \quad (9.2-8)$$

Solution of Eq. (8) for each $m, n = 1, 2, \ldots$ gives $(U_{mn}, V_{mn}, W_{mn}, X_{mn}, Y_{mn})$, which can then be used to compute the solution $(u_0, v_0, w_0, \phi_x, \phi_y)$ from Eqs. (2)–(4). Antisymmetric cross-ply laminates have the following additional stiffness characteristics [see Eqs. (5.5–29a,b)]:

$$B_{12} = 0, \quad B_{22} = -B_{11}, \quad B_{66} = 0 \quad (9.2-9)$$

Hence, the matrix coefficients in Eq. (7b) can be simplified.

The stresses in each layer can be computed using the constitutive equations (see Section 8.3–4). The in-plane stresses of a simply supported (SS–1) cross-ply laminated plate (*i.e.*, when $\bar{Q}_{16} = \bar{Q}_{26} = \bar{Q}_{45} = 0$ and $\alpha_{xy} = 0$) are then given by

$$\begin{Bmatrix} \sigma_{xx} \\ \sigma_{yy} \\ \sigma_{xy} \end{Bmatrix}^{(k)} = \sum_{m=1}^{\infty}\sum_{n=1}^{\infty} \begin{bmatrix} \bar{Q}_{11} & \bar{Q}_{12} & 0 \\ \bar{Q}_{12} & \bar{Q}_{22} & 0 \\ 0 & 0 & \bar{Q}_{66} \end{bmatrix}^{(k)} \begin{Bmatrix} (R^{xx}_{mn} + zS^{xx}_{mn}) \sin\alpha x \sin\beta y \\ (R^{yy}_{mn} + zS^{yy}_{mn}) \sin\alpha x \sin\beta y \\ (R^{xy}_{mn} + zS^{xy}_{mn}) \cos\alpha x \cos\beta y \end{Bmatrix} \quad (9.2-10a)$$

where

$$\left\{\begin{array}{c} R^{xx}_{mn} \\ R^{yy}_{mn} \\ R^{xy}_{mn} \end{array}\right\} = \left\{\begin{array}{c} -\alpha U_{mn} - \alpha_{xx}T^0_{mn} \\ -\beta V_{mn} - \alpha_{yy}T^0_{mn} \\ \beta U_{mn} + \alpha V_{mn} \end{array}\right\}, \quad \left\{\begin{array}{c} S^{xx}_{mn} \\ S^{yy}_{mn} \\ S^{xy}_{mn} \end{array}\right\} = \left\{\begin{array}{c} -\alpha X_{mn} - \alpha_{xx}T^1_{mn} \\ -\beta Y_{mn} - \alpha_{yy}T^1_{mn} \\ \beta X_{mn} + \alpha Y_{mn} \end{array}\right\} \tag{9.2 – 10b}$$

where temperature increment ΔT is assumed to be of the form

$$\Delta T(x, y, z, t) = \sum_{m=1}^{\infty}\sum_{n=1}^{\infty} \left(T^0_{mn} + zT^1_{mn}\right) \sin\alpha x \sin\beta y \tag{9.2 – 10c}$$

The transverse shear stresses from the constitutive equations are given by

$$\left\{\begin{array}{c} \sigma_{yz} \\ \sigma_{xz} \end{array}\right\}^{(k)} = \sum_{m=1}^{\infty}\sum_{n=1}^{\infty} \begin{bmatrix} \bar{Q}_{44} & 0 \\ 0 & \bar{Q}_{55} \end{bmatrix}^{(k)} \left\{\begin{array}{c} (Y_{mn} + \beta W_{mn}) \sin\alpha x \cos\beta y \\ (X_{mn} + \alpha W_{mn}) \cos\alpha x \sin\beta y \end{array}\right\} \tag{9.2 – 11}$$

Note that the stresses are layerwise constant through the thickness.

The bending moments are calculated from

$$\begin{aligned} \left\{\begin{array}{c} M_{xx} \\ M_{yy} \\ M_{xy} \end{array}\right\} &= \sum_{m=1}^{\infty}\sum_{n=1}^{\infty} \begin{bmatrix} B_{11} & B_{12} & 0 \\ B_{12} & B_{22} & 0 \\ 0 & 0 & B_{66} \end{bmatrix} \left\{\begin{array}{c} -\alpha U_{mn} \sin\alpha x \sin\beta y \\ -\beta V_{mn} \sin\alpha x \sin\beta y \\ (\beta U_{mn} + \alpha V_{mn}) \cos\alpha x \cos\beta y \end{array}\right\} \\ &+ \sum_{m=1}^{\infty}\sum_{n=1}^{\infty} \begin{bmatrix} D_{11} & D_{12} & 0 \\ D_{12} & D_{22} & 0 \\ 0 & 0 & D_{66} \end{bmatrix} \left\{\begin{array}{c} -\alpha X_{mn} \sin\alpha x \sin\beta y \\ -\beta Y_{mn} \sin\alpha x \sin\beta y \\ (\alpha Y_{mn} + \beta X_{mn}) \cos\alpha x \cos\beta y \end{array}\right\} \end{aligned} \tag{9.2 – 12}$$

As discussed in Chapter 8, the transverse stresses can also be determined using the equilibrium equations of 3-D elasticity. Following the procedure outlined in Eqs. (8.3–31)–(8.3–37), we obtain

$$\begin{aligned} \sigma^{(k)}_{xz}(x, y, z) &= \sum_{m=1}^{\infty}\sum_{n=1}^{\infty} \left[(z - z_k)\mathcal{A}^{(k)}_{mn} + \frac{1}{2}(z^2 - z_k^2)\mathcal{B}^{(k)}_{mn}\right] \cos\alpha x \sin\beta y \\ &+ \sigma^{(k-1)}_{xz}(x, y, z_k) \\ \sigma^{(k)}_{yz}(x, y, z) &= \sum_{m=1}^{\infty}\sum_{n=1}^{\infty} \left[(z - z_k)\mathcal{C}^{(k)}_{mn} + \frac{1}{2}(z^2 - z_k^2)\mathcal{D}^{(k)}_{mn}\right] \sin\alpha x \cos\beta y \\ &+ \sigma^{(k-1)}_{yz}(x, y, z_k) \end{aligned} \tag{9.2 – 13a}$$

where $\sigma^{(0)}_{xz}(x, y, z_1) = \sigma^{(0)}_{yz}(x, y, z_1) = 0$, and

$$\begin{aligned} \mathcal{A}^{(k)}_{mn} &= \left[\left(\alpha^2\bar{Q}^{(k)}_{11} + \beta^2\bar{Q}^{(k)}_{66}\right) U_{mn} + \alpha\beta\left(\bar{Q}^{(k)}_{12} + \bar{Q}^{(k)}_{66}\right) V_{mn}\right] \\ \mathcal{B}^{(k)}_{mn} &= \left[\left(\alpha^2\bar{Q}^{(k)}_{11} + \beta^2\bar{Q}^{(k)}_{66}\right) X_{mn} + \alpha\beta\left(\bar{Q}^{(k)}_{12} + \bar{Q}^{(k)}_{66}\right) Y_{mn}\right] \\ \mathcal{C}^{(k)}_{mn} &= \left[\alpha\beta\left(\bar{Q}^{(k)}_{12} + \bar{Q}^{(k)}_{66}\right) U_{mn} + \left(\alpha^2\bar{Q}^{(k)}_{66} + \beta^2\bar{Q}^{(k)}_{22}\right) V_{mn}\right] \\ \mathcal{D}^{(k)}_{mn} &= \left[\alpha\beta\left(\bar{Q}^{(k)}_{12} + \bar{Q}^{(k)}_{66}\right) X_{mn} + \left(\alpha^2\bar{Q}^{(k)}_{66} + \beta^2\bar{Q}^{(k)}_{22}\right) Y_{mn}\right] \end{aligned} \tag{9.2 – 13b}$$

The transverse normal stress can be computed using Eq. (8.3–37a,b) with the coefficients $\mathcal{A}_{mn}^{(k)}, \mathcal{B}_{mn}^{(k)}, \mathcal{C}_{mn}^{(k)}$, and $\mathcal{D}_{mn}^{(k)}$ defined in Eq. (9.2–13b).

Example 9.2–1:

Specially orthotropic plates differ from antisymmetric cross-ply laminates in that all B_{ij} are zero. Consequently, $\hat{s}_{14} = 0$, $\hat{s}_{15} = 0$, $\hat{s}_{24} = 0$, and $\hat{s}_{25} = 0$. It is clear from Eq. (8) that U_{mn} and V_{mn} are uncoupled from (W_{mn}, X_{mn}, Y_{mn}):

$$\begin{bmatrix} \hat{s}_{11} & \hat{s}_{12} & 0 & 0 & 0 \\ \hat{s}_{12} & \hat{s}_{22} & 0 & 0 & 0 \\ 0 & 0 & \hat{s}_{33} & \hat{s}_{34} & \hat{s}_{35} \\ 0 & 0 & \hat{s}_{34} & \hat{s}_{44} & \hat{s}_{45} \\ 0 & 0 & \hat{s}_{35} & \hat{s}_{45} & \hat{s}_{55} \end{bmatrix} \begin{Bmatrix} U_{mn} \\ V_{mn} \\ W_{mn} \\ X_{mn} \\ Y_{mn} \end{Bmatrix} = \begin{Bmatrix} 0 \\ 0 \\ Q_{mn} \\ 0 \\ 0 \end{Bmatrix} - \begin{Bmatrix} \alpha N_{mn}^1 \\ \beta N_{mn}^2 \\ 0 \\ \alpha M_{mn}^1 \\ \beta M_{mn}^2 \end{Bmatrix} \qquad (9.2-14)$$

or

$$\begin{bmatrix} \hat{s}_{11} & \hat{s}_{12} \\ \hat{s}_{12} & \hat{s}_{22} \end{bmatrix} \begin{Bmatrix} U_{mn} \\ V_{mn} \end{Bmatrix} = - \begin{Bmatrix} \alpha N_{mn}^1 \\ \beta N_{mn}^2 \end{Bmatrix} \qquad (9.2-15a)$$

and

$$\begin{bmatrix} \hat{s}_{33} & \hat{s}_{34} & \hat{s}_{35} \\ \hat{s}_{34} & \hat{s}_{44} & \hat{s}_{45} \\ \hat{s}_{35} & \hat{s}_{45} & \hat{s}_{55} \end{bmatrix} \begin{Bmatrix} W_{mn} \\ X_{mn} \\ Y_{mn} \end{Bmatrix} = \begin{Bmatrix} Q_{mn} \\ 0 \\ 0 \end{Bmatrix} - \begin{Bmatrix} 0 \\ \alpha M_{mn}^1 \\ \beta M_{mn}^2 \end{Bmatrix} \qquad (9.2-15b)$$

The solution of Eq. (15a) is given by

$$\begin{aligned} U_{mn} &= -\frac{1}{a_{mn}} \left(\hat{s}_{22} \alpha N_{mn}^1 - \hat{s}_{12} \beta N_{mn}^2 \right) \\ V_{mn} &= -\frac{1}{a_{mn}} \left(\hat{s}_{11} \beta N_{mn}^2 - \hat{s}_{12} \alpha N_{mn}^1 \right) \end{aligned} \qquad (9.2-16)$$

where $a_{mn} = \hat{s}_{11}\hat{s}_{22} - \hat{s}_{12}\hat{s}_{12}$. The in-plane deflections are identically zero when the thermal (and in-plane edge) forces are zero.

Equation (15b) can be solved either directly (by inverting the 3×3 coefficient matrix) or by using the static condensation procedure outlined in Chapter 8 [see Eqs. (8.3–22)–(8.3–26)]. Using the latter, we arrive at

$$\begin{aligned} W_{mn} &= \frac{1}{b_{mn}} \left[Q_{mn} + \frac{\hat{s}_{34}}{b_0} \left(\alpha M_{mn}^1 \hat{s}_{55} - \beta M_{mn}^2 \hat{s}_{45} \right) - \frac{\hat{s}_{35}}{b_0} \left(\alpha M_{mn}^1 \hat{s}_{45} - \beta M_{mn}^2 \hat{s}_{44} \right) \right] \\ X_{mn} &= \frac{1}{b_0} \left[b_1 W_{mn} - \left(\alpha M_{mn}^1 \hat{s}_{55} - \beta M_{mn}^2 \hat{s}_{45} \right) \right] \\ Y_{mn} &= \frac{1}{b_0} \left[b_2 W_{mn} + \left(\alpha M_{mn}^1 \hat{s}_{45} - \beta M_{mn}^2 \hat{s}_{44} \right) \right] \end{aligned} \qquad (9.2-17a)$$

where

$$\begin{aligned} b_{mn} &= \hat{s}_{33} + \hat{s}_{34} \frac{b_1}{b_0} + \hat{s}_{35} \frac{b_2}{b_0}, \quad b_0 = \hat{s}_{44}\hat{s}_{55} - \hat{s}_{45}\hat{s}_{45} \\ b_1 &= \hat{s}_{45}\hat{s}_{35} - \hat{s}_{34}\hat{s}_{55}, \quad b_2 = \hat{s}_{34}\hat{s}_{45} - \hat{s}_{44}\hat{s}_{35} \end{aligned} \qquad (9.2-17b)$$

When the thermal forces are zero, the bending deflections are given by

$$w_0(x,y) = \sum_{n=1}^{\infty}\sum_{m=1}^{\infty} W_{mn} \sin\frac{m\pi x}{a} \sin\frac{n\pi y}{b} \tag{9.2-18a}$$

$$\phi_x(x,y) = \sum_{n=1}^{\infty}\sum_{m=1}^{\infty} X_{mn} \cos\frac{m\pi x}{a} \sin\frac{n\pi y}{b} \tag{9.2-18b}$$

$$\phi_y(x,y) = \sum_{n=1}^{\infty}\sum_{m=1}^{\infty} Y_{mn} \sin\frac{m\pi x}{a} \cos\frac{n\pi y}{b} \tag{9.2-18c}$$

with

$$W_{mn} = \frac{1}{b_{mn}} Q_{mn},\ X_{mn} = \frac{b_1}{b_0 b_{mn}} Q_{mn},\ Y_{mn} = \frac{b_2}{b_0 b_{mn}} Q_{mn} \tag{9.2-19}$$

The bending moments are given by

$$M_{xx} = -\sum_{n=1}^{\infty}\sum_{m=1}^{\infty} \left(D_{11}\alpha X_{mn} + D_{12}\beta Y_{mn}\right) \sin\frac{m\pi x}{a} \sin\frac{n\pi y}{b} \tag{9.2-20a}$$

$$M_{yy} = -\sum_{n=1}^{\infty}\sum_{m=1}^{\infty} \left(D_{12}\alpha X_{mn} + D_{22}\beta Y_{mn}\right) \sin\frac{m\pi x}{a} \sin\frac{n\pi y}{b} \tag{9.2-20b}$$

$$M_{xy} = D_{66}\sum_{n=1}^{\infty}\sum_{m=1}^{\infty} \left(\beta X_{mn} + \alpha Y_{mn}\right) \cos\frac{m\pi x}{a} \cos\frac{n\pi y}{b} \tag{9.2-20c}$$

The in-plane stresses are given by

$$\left\{\begin{matrix}\sigma_{xx}\\ \sigma_{yy}\\ \sigma_{xy}\end{matrix}\right\}^{(k)} = -z\sum_{n=1}^{\infty}\sum_{m=1}^{\infty}\left\{\begin{matrix}\left(\bar{Q}_{11}^{(k)}\alpha X_{mn} + \bar{Q}_{12}^{(k)}\beta Y_{mn}\right)\sin\frac{m\pi x}{a}\ \sin\frac{n\pi y}{b}\\ \left(\bar{Q}_{12}^{(k)}\alpha X_{mn} + \bar{Q}_{22}^{(k)}\beta Y_{mn}\right)\sin\frac{m\pi x}{a}\ \sin\frac{n\pi y}{b}\\ -2\bar{Q}_{66}^{(k)}\left(\beta X_{mn} + \alpha Y_{mn}\right)\ \cos\frac{m\pi x}{a}\ \cos\frac{n\pi y}{b}\end{matrix}\right\} \tag{9.2-21}$$

and the transverse shear stresses are given by

$$\left\{\begin{matrix}\sigma_{yz}\\ \sigma_{xz}\end{matrix}\right\}^{(k)} = \sum_{n=1}^{\infty}\sum_{m=1}^{\infty}\left\{\begin{matrix}\bar{Q}_{44}^{(k)}\left(Y_{mn} + \beta W_{mn}\right)\sin\frac{m\pi x}{a}\ \cos\frac{n\pi y}{b}\\ \bar{Q}_{55}^{(k)}\left(X_{mn} + \alpha W_{mn}\right)\cos\frac{m\pi x}{a}\ \sin\frac{n\pi y}{b}\end{matrix}\right\} \tag{9.2-22}$$

The interlaminar stresses, computed using the 3-D stress equilibrium equations, are given by

$$\sigma_{xz}^{(k)} = \left(\frac{z^2 - z_k^2}{2}\right)\left(T_{11}^{(k)} X_{mn} + T_{12}^{(k)} Y_{mn}\right)\ \cos\frac{m\pi x}{a}\ \sin\frac{n\pi y}{b}$$

$$\sigma_{yz}^{(k)} = \left(\frac{z^2 - z_k^2}{2}\right)\left(T_{12}^{(k)} X_{mn} + T_{22}^{(k)} Y_{mn}\right)\ \sin\frac{m\pi x}{a}\ \cos\frac{n\pi y}{b}$$

$$\sigma_{zz}^{(k)} = \left[-\frac{h^3}{24} + \frac{z}{6}\left(z^2 - 3z^k\right)\right]\left(T_{31}^{(k)} X_{mn} + T_{32}^{(k)} Y_{mn}\right)\ \sin\frac{m\pi x}{a}\ \sin\frac{n\pi y}{b} \tag{9.2-23a}$$

where

$$T_{11}^{(k)} = \alpha^2 \bar{Q}_{11}^{(k)} + \beta^2 \bar{Q}_{66}^{(k)},\ T_{12}^{(k)} = \alpha\beta(\bar{Q}_{12}^{(k)} + \bar{Q}_{66}^{(k)}),\ T_{22}^{(k)} = \alpha^2 \bar{Q}_{66}^{(k)} + \beta^2 \bar{Q}_{22}^{(k)}$$

$$T_{31}^{(k)} = \alpha^3 \bar{Q}_{11}^{(k)} + \alpha\beta^2(2\bar{Q}_{66}^{(k)} + \bar{Q}_{12}^{(k)}),\ T_{32}^{(k)} = \alpha^2\beta(\bar{Q}_{12}^{(k)} + 2\bar{Q}_{66}^{(k)}) + \beta^3 \bar{Q}_{22}^{(k)} \quad (9.2-23b)$$

For single-layer plates, Eqs. (23a) reduce to

$$\begin{aligned}
\sigma_{xz} &= -\frac{h^2}{8}\left[1 - \left(\frac{2z}{h}\right)^2\right]\left(T_{11}^{(k)} X_{mn} + T_{12}^{(k)} Y_{mn}\right) \cos\alpha x\ \sin\beta y \\
\sigma_{yz} &= -\frac{h^2}{8}\left[1 - \left(\frac{2z}{h}\right)^2\right]\left(T_{12}^{(k)} X_{mn} + T_{22}^{(k)} Y_{mn}\right) \sin\alpha x\ \cos\beta y \\
\sigma_{zz} &= \frac{h^3}{48}\left\{\left[1 + \left(\frac{2z}{h}\right)^3\right] - 3\left[1 + \left(\frac{2z}{h}\right)\right]\right\}\left(T_{31}^{(k)} X_{mn} + T_{32}^{(k)} Y_{mn}\right) \sin\alpha x\ \sin\beta y
\end{aligned} \quad (9.2-24)$$

Numerical results for the maximum transverse deflection and stresses of symmetric laminates are discussed next. The following nondimensionalizations are used to present results in graphical and tabular forms:

$$\begin{aligned}
\bar{w} &= w_0\left(\frac{E_2 h^3}{b^4 q_0}\right), \quad \bar{\sigma}_{xx} = \sigma_{xx}\left(\frac{h^2}{b^2 q_0}\right), \quad \bar{\sigma}_{yy} = \sigma_{yy}\left(\frac{h^2}{b^2 q_0}\right) \\
\bar{\sigma}_{xy} &= \sigma_{xy}\left(\frac{h^2}{b^2 q_0}\right), \quad \bar{\sigma}_{xz} = \sigma_{xz}\left(\frac{h}{b q_0}\right), \quad \bar{\sigma}_{yz} = \sigma_{yz}\left(\frac{h}{b q_0}\right)
\end{aligned} \quad (9.2-25)$$

Table 9.2–1 contains the maximum nondimensionalized deflections and stresses of simply supported square symmetric laminates (0/90/90/0) and (0/90/0) under sinusoidally distributed load (SSL) as well as uniformly distributed load (UDL) and for different side-to-thickness ratios ($E_1 = 25E_2$, $G_{12} = G_{13} = 0.5E_2$, $G_{23} = 0.2E_2$, $\nu_{12} = 0.25$, $K = 5/6$). The membrane stresses were evaluated at the following locations: $\bar{\sigma}_{xx}(a/2, b/2, \frac{h}{2})$, $\bar{\sigma}_{yy}(a/2, b/2, \frac{h}{4})$, and $\bar{\sigma}_{xy}(a, b, -\frac{h}{2})$. The transverse shear stresses are calculated using the constitutive equations. For the (0/90/0) laminate, σ_{xz} is evaluated at $(x, y) = (0, b/2)$ in layers 1 and 3, and σ_{yz} is computed at $(x, y) = (a/2, 0)$ in layer 2.

The nondimensionalized quantities in the classical laminate theory are independent of the side-to-thickness ratio. The influence of transverse shear deformation is to increase the transverse deflection. The difference between the deflections predicted by the first-order shear deformation theory and classical plate theory increases with the ratio h/a. For example, for $a/h = 10$ and sinusoidal loading, the classical plate theory underpredicts deflections as much as about 35%, whereas it is only 12% for $a/h = 20$. Shear deformation has different effects on different stresses.

Table 9.2–1: Effect of transverse shear deformation on nondimensionalized maximum transverse deflections and stresses of simply supported (SS–1) symmetric cross-ply square plates.

a/h	Load	$\bar{w} \times 10^2$	$\bar{\sigma}_{xx}$	$\bar{\sigma}_{yy}$	$\bar{\sigma}_{xy}$	$\bar{\sigma}_{xz}$	$\bar{\sigma}_{yz}$
Orthotropic Plate [σ_{yy} is evaluated at $(x, y, z)=(a/2, b/2, h/2)$]							
10	SSL	0.6383	0.5248	0.0338	0.0246	0.3452	0.0367
						0.4315	0.0459†
	UDL	0.9519	0.7706	0.0352	0.0539	0.6147	0.1529
						0.7684	0.1911†
20	SSL	0.4836	0.5350	0.0286	0.0222	0.3501	0.0319
	UDL	0.7262	0.7828	0.0272	0.0487	0.6194	0.1466
100	SSL	0.4333	0.5385	0.0267	0.0213	0.3518	0.0302
	UDL	0.6528	0.7865	0.0245	0.0464	0.6206	0.1449
CLPT	SSL	0.4312	0.5387	0.0267	0.0213	0.4398	0.0377†
	UDL	0.6497	0.7866	0.0244	0.0463	0.7758	0.1811†
Symmetric Laminate, (0/90/90/0)							
10	SSL	0.6627	0.4989	0.3614	0.0241	0.4165	0.1292
						0.3181	0.1807†
	UDL	1.0250	0.7577	0.5006	0.0470	0.7986	0.3499
						0.6081	0.5091†
20	SSL	0.4912	0.5273	0.2956	0.0221	0.4370	0.1087
	UDL	0.7694	0.8045	0.3968	0.0420	0.8305	0.3228
100	SSL	0.4337	0.5382	0.2704	0.0213	0.4448	0.1008
	UDL	0.6833	0.8420	0.3558	0.0396	0.8420	0.3140
CLPT	SSL	0.4312	0.5387	0.2694	0.0213	0.3393	0.1382†
	UDL	0.6796	0.8236	0.3540	0.0395	0.6404	0.4548†
Symmetric Laminate, (0/90/0)							
10	SSL	0.6693	0.5134	0.2536	0.0252	0.4089	0.0914
						0.3806	0.1108†
	UDL	1.0219	0.7719	0.3072	0.0514	0.7548	0.3107
						0.7014	0.4156†
20	SSL	0.4921	0.5318	0.1997	0.0223	0.4205	0.0759
	UDL	0.7572	0.7983	0.2227	0.0453	0.7697	0.2902
100	SSL	0.4337	0.5384	0.1804	0.0213	0.4247	0.0703
	UDL	0.6697	0.8072	0.1925	0.0426	0.7744	0.2842
CLPT	SSL	0.4312	0.5387	0.1796	0.0213	0.3951	0.0823†
	UDL	0.6660	0.8075	0.1912	0.0425	0.7191	0.3791†

† σ_{xz} and σ_{yz} calculated from equilibrium equations (at $z = 0$).

Table 9.2–2: Effect of transverse shear deformation on nondimensionalized maximum transverse deflections and stresses of simply supported (SS–1) symmetric cross-ply square plates.

a/h	Load	$\bar{w}$	$\bar{\sigma}_{xx}$	$\bar{\sigma}_{yy}$	$\bar{\sigma}_{xy}$	$\bar{\sigma}_{xz}$	$\bar{\sigma}_{yz}$
Symmetric Laminate, (0/90/90/0/90/90/0)							
10	SSL	0.6213	0.5021	0.4107	0.0221	0.3459	0.1998
	UDL	0.9643	0.7605	0.6016	0.0422	0.6927	0.4630
20	SSL	0.4796	0.5276	0.3748	0.0215	0.3617	0.1840
	UDL	0.7575	0.8059	0.5475	0.0396	0.7212	0.4438
100	SSL	0.4332	0.5382	0.3598	0.0213	0.3683	0.1774
	UDL	0.6896	0.8260	0.5241	0.0381	0.7322	0.4365
CLPT	SSL	0.4312	0.5387	0.3591	0.0213	–	–
	UDL	0.6867	0.8270	0.5230	0.0380	–	–
Symmetric Laminate, (0/90/0/90/0)							
10	SSL	0.6277	0.5044	0.3852	0.0226	0.3535	0.1770
	UDL	0.9727	0.7649	0.5525	0.0436	0.6901	0.4410
20	SSL	0.4814	0.5285	0.3416	0.0217	0.3685	0.1591
	UDL	0.7581	0.8080	0.4844	0.0403	0.7166	0.4188
100	SSL	0.4333	0.5383	0.3240	0.0213	0.3746	0.1519
	UDL	0.6874	0.8264	0.4559	0.0386	0.7267	0.4108
CLPT	SSL	0.4312	0.5387	0.3232	0.0213	–	–
	UDL	0.6844	0.8272	0.4546	0.0385	–	–

Table 9.2–2 contains results for laminates (0/90/90/0/90/90/0) and (0/90/0/90/0). The material properties used are $E_1 = 25E_2$, $G_{12} = G_{13} = 0.5E_2$, $G_{23} = 0.2E_2$, $\nu_{12} = 0.25$, and $K = 5/6$. The same nondimensionalization as before is used except for the following quantities:

$$\bar{\sigma}_{yy} = \sigma_{yy}(a/2, b/2, \frac{h}{3})\frac{h^2}{b^2 q_0}$$

$$\bar{\sigma}_{xz} = \sigma_{xz}(0, b/2, k = 1,3,5)\frac{h}{bq_0}, \quad \bar{\sigma}_{yz} = \sigma_{yz}(a/2, 0, k = 2,4)\frac{h}{bq_0} \quad (9.2-26)$$

where k denotes the layer number. While the classical theory predictions are the same for (0/90/90/0) and (0/90/90/0/90/90/0), and for (0/90/0) and (0/90/0/90/0), the first-order theory results are slightly different. The influence of transverse shear deformation is less in the case of the laminates presented in Table 9.2–2. Thus, as the number of layers is increased, the effect of transverse shear strains on deflections and stresses decreases. Figure 9.2–2 clearly shows the diminishing effect of transverse shear deformation on deflections, the effect being negligible for side-to-thickness ratios larger than 20.

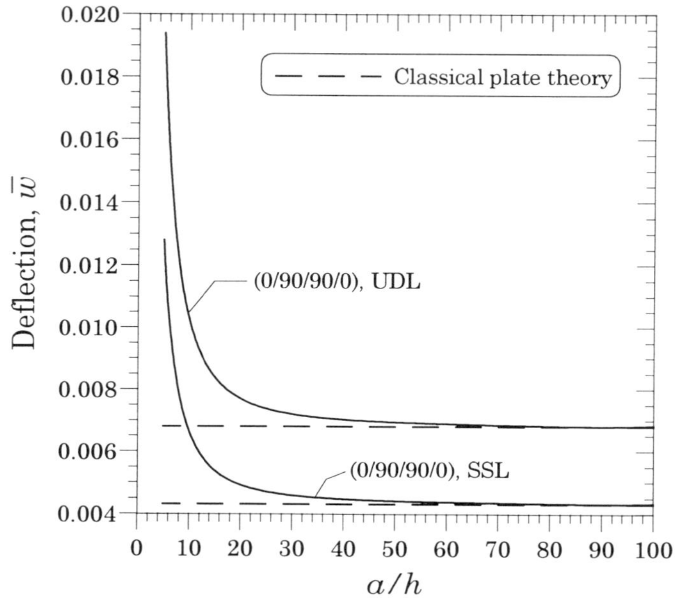

Figure 9.2–2: Nondimensionalized center transverse deflection ($\bar{w}$) versus side-to-thickness ratio (a/h) for simply supported (SS–1) symmetric cross-ply (0/90/90/0) square laminates subjected to uniformly or sinusoidally distributed transverse loading; dashed lines correspond to the classical plate theory (CLPT) solutions.

Table 9.2–3 contains numerical results of nondimensionalized transverse deflections and stresses $[\bar{\sigma}_{xx}(a/2, b/2, -h/2) = -\bar{\sigma}_{yy}(a/2, b/2, h/2)$ and $\bar{\sigma}_{xz} = \bar{\sigma}_{yz}]$ of antisymmetric cross-ply laminates subjected to sinusoidally and uniformly distributed transverse loads. The stresses are nondimensionalized as in Eq. (8.3–39). The locations of the maximum stresses, computed using the constitutive equations, are as follows:

$$\sigma_{xx}(a/2, b/2, -\frac{h}{2}),\ \sigma_{yy}(a/2, b/2, \frac{h}{2}),\ \sigma_{xy}(a, b, -\frac{h}{2})$$

$$\sigma_{xz}(0, b/2, k = 1),\ \sigma_{yz}(a/2, 0, k = n) \qquad (9.2-27)$$

We note that the two-layer laminate exhibits quite different behavior, due to bending–extensional coupling, from the eight-layer laminate, and the results for the eight-layer laminate are much the same as those of symmetric laminates in Tables 9.2–1 and 9.2–2. Figure 9.2–3 shows the effect of transverse shear deformation and bending–extensional coupling on deflections. The eight-layer antisymmetric cross-ply plate behaves much like an orthotropic plate (results are not shown in the figure).

Table 9.2–3: Effect of transverse shear deformation on nondimensionalized maximum transverse deflections and stresses of simply supported (SS–1) antisymmetric cross-ply square plates ($h_k = h/n, E_1 = 25E_2,\ G_{12} = G_{13} = 0.5E_2,\ G_{23} = 0.2E_2,\ \nu_{12} = 0.25,\ K = 5/6$).

a/h	Load	$\bar{w} \times 10^2$	$\bar{\sigma}_{yy}$	$\bar{\sigma}_{xy}$	$\bar{\sigma}_{xz}$	$\bar{\sigma}_{xz}^{\dagger}$
Antisymmetric Laminate, (0/90)						
10	SSL	1.2373	0.7157	0.0525	0.2728	0.3322
	UDL	1.9468	1.0715	0.0960	0.5772	0.7250
20	SSL	1.1070	0.7157	0.0525	0.2728	0.3322
	UDL	1.7582	1.0747	0.0943	0.5802	0.7285
100	SSL	1.0653	0.7157	0.0525	0.2728	0.3322
	UDL	1.6980	1.0761	0.0933	0.5813	0.7297
CLPT	SSL	1.0636	0.7157	0.0525	–	0.3322
	UDL	1.6955	1.0761	0.0933	–	0.7297
Antisymmetric Laminate, $(0/90)_4$						
10	SSL	0.6216	0.4950	0.0221	0.2728	0.2480
	UDL	0.9660	0.7415	0.0420	0.5787	0.5264
20	SSL	0.4913	0.4950	0.0221	0.2728	0.2480
	UDL	0.7776	0.7468	0.0402	0.5839	0.5311
100	SSL	0.4496	0.4950	0.0221	0.2728	0.2480
	UDL	0.7175	0.7494	0.0391	0.5857	0.5328
CLPT	SSL	0.4479	0.4950	0.0221	–	0.2479
	UDL	0.7150	0.7496	0.0391	–	0.5330

† Maximum stresses derived from equilibrium. The reported values are at $z = \pm h/4$ for (0/90) laminate, and at $z = 0$ for $(0/90)_4$ laminate.

Figures 9.2–4 through 9.2–7 show plots of maximum normal stresses, $\bar{\sigma}_{xx}(a/2, b/2, z)$ and $\bar{\sigma}_{yy}(a/2, b/2, z)$, and maximum transverse shear stresses, $\bar{\sigma}_{xz}(0, b/2, z)$ and $\bar{\sigma}_{yz}(a/2, 0, z)$, through the thickness of simply supported square laminates (0/90/90/0) under sinusoidally distributed transverse load. The material properties used are $E_1 = 25E_2$, $G_{12} = G_{13} = 0.5E_2$, $G_{23} = 0.2E_2$, $\nu_{12} = 0.25$, and $K = 5/6$. The dashed lines correspond to classical plate theory solutions. In Figures 9.2–6 and 9.2–7, stresses computed using the constitutive relations are also included. In the case of $\bar{\sigma}_{xz}$, the equilibrium equations predict a stress variation that is inconsistent with that predicted by constitutive relations; equilibrium equations predict the maximum stress to be at the midplane of the plate, while the constitutive equations predict maximum stress in the outer layers. It turns out that (see Pagano [6]) the constitutive equations yield, qualitatively, the correct stress variation.

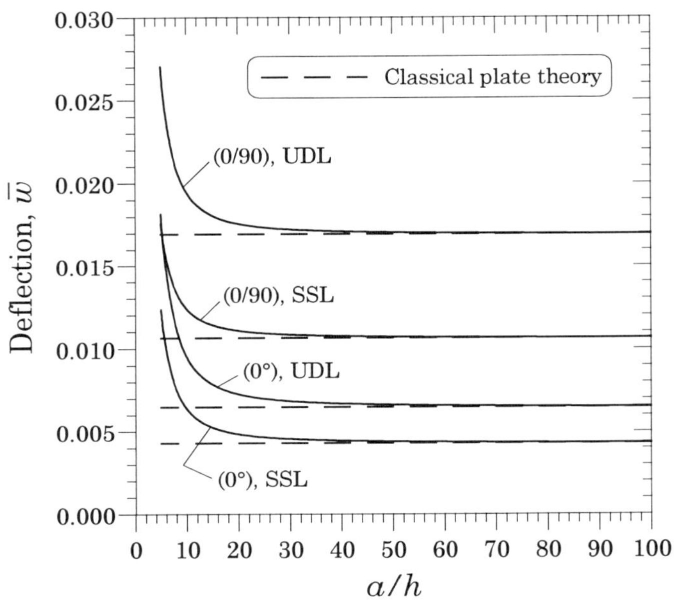

Figure 9.2–3: Nondimensionalized center transverse deflection ($\bar{w}$) versus side-to-thickness ratio (a/h) for simply supported (SS–1) orthotropic and antisymmetric cross-ply (0/90) laminates under sinusoidally distributed transverse loading.

Table 9.2–4 contains nondimensionalized deflections, $\bar{w} = w_0/(\alpha_1 T_1 b^2)$, of simply supported plates subjected to the temperature field of the form given in Eq. (10c). The material properties of orthotropic layers are assumed to be $E_1 = 25E_2$, $G_{12} = G_{13} = 0.5E_2$, $G_{23} = 0.2E_2$, $\nu_{12} = 0.25$, $K = 5/6$, and $\alpha_2 = 3\alpha_3$. The results in the table correspond to $T_0 = 0$ and $T_1 \neq 0$. We note that the effect of shear deformation on thermal deflections is negligible.

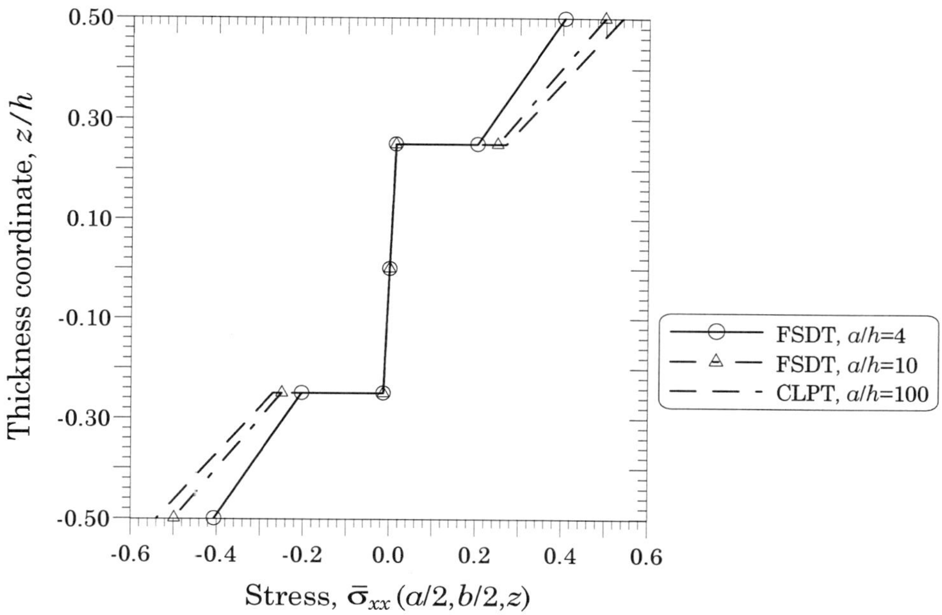

Figure 9.2–4: Nondimensionalized normal stress ($\bar{\sigma}_{xx}$) versus thickness (z/h) for simply supported (SS–1) symmetric cross-ply (0/90/90/0) laminates.

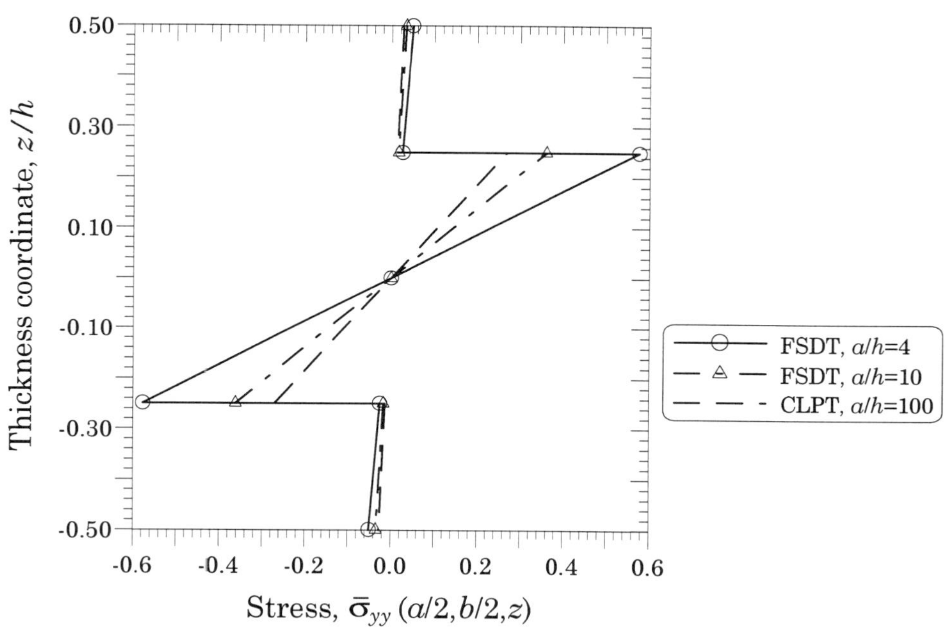

Figure 9.2–5: Nondimensionalized normal stress ($\bar{\sigma}_{yy}$) versus thickness (z/h) for simply supported (SS–1) symmetric cross-ply (0/90/90/0) laminates.

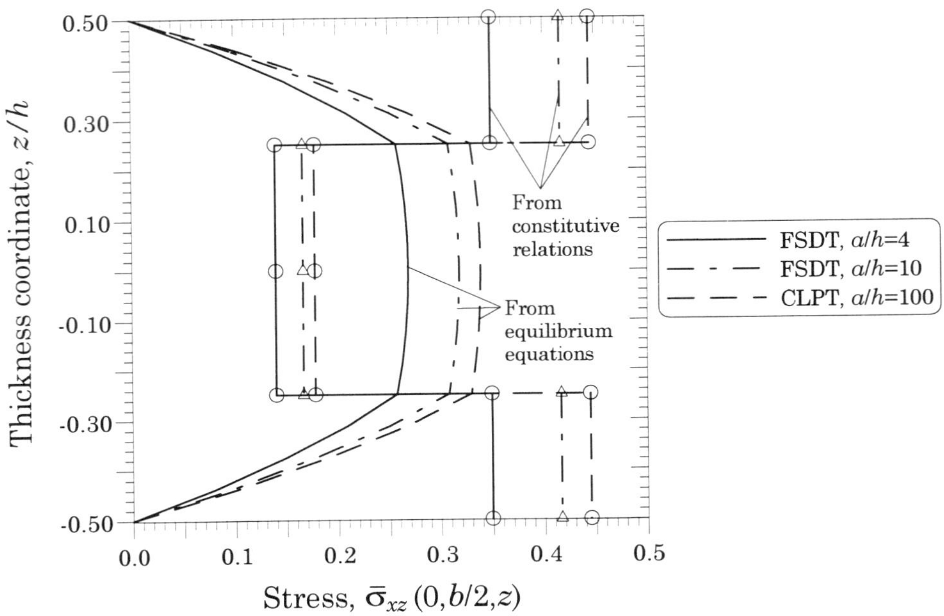

Figure 9.2–6: Nondimensionalized shear stress ($\bar{\sigma}_{xz}$) versus thickness (z/h) for simply supported (SS–1) symmetric cross-ply (0/90/90/0) laminates.

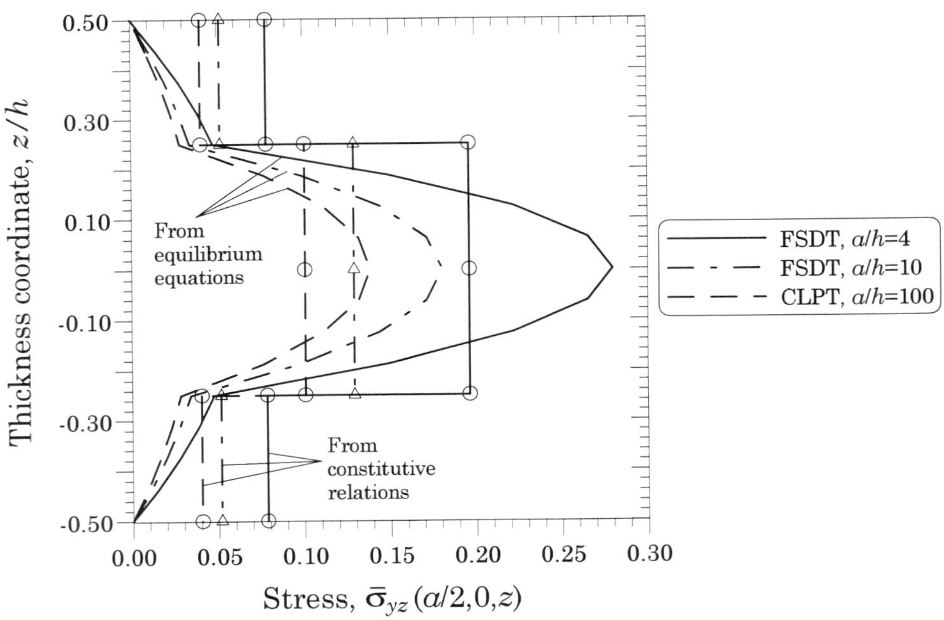

Figure 9.2–7: Nondimensionalized shear stress ($\bar{\sigma}_{yz}$) versus thickness (z/h) for simply supported (SS–1) symmetric cross-ply (0/90/90/0) laminates.

Table 9.2–4: Effect of the aspect ratio and side-to-thickness ratio on the deflection of simply supported (SS–1) plates subjected to temperature field that is uniform in the xy–plane and linearly varying through the thickness ($q_0 = 0, T_0 = 0, T_1 =$ constant).

Load	a/h	$a/b = 1$	$a/b = 1.5$	$a/b = 2$	$a/b = 2.5$	$a/b = 3$
Isotropic†						
SSL	10	0.6586	0.9119	1.0537	1.1355	1.1855
UDL	10	0.9575	1.3097	1.4798	1.5582	1.5938
Orthotropic						
SSL	10	1.0440	2.1129	3.0623	3.6394	3.8883
	20	1.0346	2.1128	3.0758	3.6560	3.9002
	100	1.0313	2.1127	3.0804	3.6617	3.9042
	CLPT	1.0312	2.1127	3.0806	3.6619	3.9044
UDL	10	1.4603	3.1321	4.5966	5.4269	5.6987
	20	1.4409	3.1339	4.6243	5.4609	5.7239
	100	1.4334	3.1343	4.6342	5.4729	5.7327
	CLPT	1.4331	3.1344	4.6346	5.4734	5.7330
Laminate,	(0/90)					
SSL	10	1.1504	1.4673	1.5186	1.5122	1.4984
	20	1.1504	1.4613	1.5091	1.5026	1.4898
	100	1.1504	1.4592	1.5058	1.4994	1.4869
	CLPT	1.1504	1.4591	1.5057	1.4993	1.4868
UDL	10	1.7213	2.1446	2.1100	1.9862	1.8796
	20	1.7269	2.1394	2.0965	1.9703	1.8649
	100	1.7293	2.1377	2.0918	1.9649	1.8600
	CLPT	1.7294	2.1376	2.0916	1.9647	1.8598
Laminate,	$(0/90)_4$					
SSL	10	1.0343	1.3000	1.3113	1.2781	1.2469
	20	1.0343	1.2837	1.2870	1.2555	1.2280
	100	1.0343	1.2776	1.2783	1.2477	1.2216
	CLPT	1.0343	1.2773	1.2779	1.2474	1.2214
UDL	10	1.5498	1.9026	1.8166	1.6603	1.5334
	20	1.5607	1.8862	1,7816	1.6224	1.5002
	100	1.5661	1.8801	1.7690	1.6091	1.4886
	CLPT	1.5664	1.8799	1.7685	1.6085	1.4881

Laminate, (0/90/90/0)

SSL	10	1.0421	1.7130	1.9680	1.9807	1.9227
	20	1.0343	1.7339	1.9858	1.9854	1.9193
	100	1.0313	1.7419	1.9923	1.9871	1.9181
	CLPT	1.0312	1.7422	1.9925	1.9872	1.9181
UDL	10	1.5452	2.5733	2.8961	2.8045	2.5921
	20	1.5357	2.6169	2.9352	2.8191	2.5877
	100	1.5318	2.6343	2.9500	2.8243	2.5859
	CLPT	1.5316	2.6350	2.9507	2.8245	2.5859

† $\nu = 0.3$; both CLPT and FSDT solutions are the same and independent of a/h.

9.2–3 Buckling

For buckling analysis, we assume that the only applied loads are the in-plane forces

$$\hat{N}_{xx} = -N_0, \quad \hat{N}_{yy} = -kN_0, \quad k = \frac{\hat{N}_{yy}}{\hat{N}_{xx}} \tag{9.2-28}$$

and all other mechanical and thermal loads are zero. From Eq. (7) we have

$$\begin{bmatrix} \hat{s}_{11} & \hat{s}_{12} & 0 & \hat{s}_{14} & \hat{s}_{15} \\ \hat{s}_{12} & \hat{s}_{22} & 0 & \hat{s}_{24} & \hat{s}_{25} \\ 0 & 0 & \hat{s}_{33} - N_0\left(\alpha^2 + k\beta^2\right) & \hat{s}_{34} & \hat{s}_{35} \\ \hat{s}_{14} & \hat{s}_{24} & \hat{s}_{34} & \hat{s}_{44} & \hat{s}_{45} \\ \hat{s}_{15} & \hat{s}_{25} & \hat{s}_{35} & \hat{s}_{45} & \hat{s}_{55} \end{bmatrix} \begin{Bmatrix} U_{mn} \\ V_{mn} \\ W_{mn} \\ X_{mn} \\ Y_{mn} \end{Bmatrix} = \begin{Bmatrix} 0 \\ 0 \\ 0 \\ 0 \\ 0 \end{Bmatrix} \tag{9.2-29}$$

Following the condensation of variables procedure to eliminate the in-plane displacements U_{mn} and V_{mn}, we obtain

$$\begin{bmatrix} \hat{s}_{33} - N_0\left(\alpha^2 + k\beta^2\right) & \hat{s}_{34} & \hat{s}_{35} \\ \hat{s}_{34} & \bar{s}_{44} & \bar{s}_{45} \\ \hat{s}_{35} & \bar{s}_{45} & \bar{s}_{55} \end{bmatrix} \begin{Bmatrix} W_{mn} \\ X_{mn} \\ Y_{mn} \end{Bmatrix} = \begin{Bmatrix} 0 \\ 0 \\ 0 \end{Bmatrix} \tag{9.2-30a}$$

$$\begin{aligned} \bar{s}_{44} &= \hat{s}_{44} - \hat{s}_{14}\frac{b_1}{b_0} - \hat{s}_{24}\frac{b_2}{b_0}, \quad \bar{s}_{45} = \hat{s}_{45} - \hat{s}_{15}\frac{b_1}{b_0} - \hat{s}_{25}\frac{b_2}{b_0} \\ \bar{s}_{55} &= \hat{s}_{55} - \hat{s}_{15}\frac{b_3}{b_0} - \hat{s}_{25}\frac{b_4}{b_0}, \quad b_0 = \hat{s}_{11}\hat{s}_{22} - \hat{s}_{12}\hat{s}_{12} \\ b_1 &= \hat{s}_{14}\hat{s}_{22} - \hat{s}_{12}\hat{s}_{24}, \quad b_2 = \hat{s}_{11}\hat{s}_{24} - \hat{s}_{12}\hat{s}_{14} \\ b_3 &= \hat{s}_{15}\hat{s}_{22} - \hat{s}_{12}\hat{s}_{25}, \quad b_4 = \hat{s}_{11}\hat{s}_{25} - \hat{s}_{12}\hat{s}_{15} \end{aligned} \tag{9.2-30b}$$

Repeating the procedure to eliminate X_{mn} and Y_{mn}, we obtain

$$N_0 = \frac{1}{\alpha^2 + \beta^2}\left(\hat{s}_{33} - \frac{\hat{s}_{34}\bar{s}_{55} - \hat{s}_{35}\bar{s}_{45}}{\bar{s}_{44}\bar{s}_{55} - \bar{s}_{45}\bar{s}_{45}}\hat{s}_{34} - \frac{\bar{s}_{44}\hat{s}_{35} - \bar{s}_{45}\hat{s}_{34}}{\bar{s}_{44}\bar{s}_{55} - \bar{s}_{45}\bar{s}_{45}}\hat{s}_{35}\right) \qquad (9.2-31)$$

Alternatively, we can eliminate X_{mn} and Y_{mn} first and then eliminate U_{mn} and V_{mn} to obtain an expression equivalent to the one given in Eq. (31); see next section for details.

Example 9.2–2 (*Specially orthotropic plates*):

For specially orthotropic plates, we have from Eq. (30b) $\hat{s}_{14} = \hat{s}_{15} = 0$ and $\hat{s}_{24} = \hat{s}_{25} = 0$; consequently, $b_1 = b_2 = b_3 = b_4 = 0$ and $\bar{s}_{44} = \hat{s}_{44}$, $\bar{s}_{45} = \hat{s}_{45}$, and $\bar{s}_{55} = \hat{s}_{55}$. Equation (31) takes the form

$$N_0 = \frac{1}{\alpha^2 + k\beta^2}\left(\hat{s}_{33} - \frac{\hat{s}_{34}\hat{s}_{55} - \hat{s}_{35}\hat{s}_{45}}{\hat{s}_{44}\hat{s}_{55} - \hat{s}_{45}\hat{s}_{45}}\hat{s}_{34} - \frac{\hat{s}_{44}\hat{s}_{35} - \hat{s}_{45}\hat{s}_{34}}{\hat{s}_{44}\hat{s}_{55} - \hat{s}_{45}\hat{s}_{45}}\hat{s}_{35}\right) \qquad (9.2-32)$$

Using the definitions of $\hat{s}_{ij}$ from Eq. (7), we can write

$$N_0 = \left(\frac{1}{\alpha^2 + k\beta^2}\right)\left(\frac{K^2 A_{44} A_{55}\hat{c}_{33} + (KA_{55}\alpha^2 + KA_{44}\beta^2)\,c_1}{c_1 + KA_{44}c_2 + KA_{55}c_3 + K^2 A_{44}A_{55}}\right) \qquad (9.2-33a)$$

$$= \left(\frac{1}{\alpha^2 + k\beta^2}\right)\left(\frac{\hat{c}_{33} + \left(\frac{\alpha^2}{KA_{44}} + \frac{\beta^2}{KA_{55}}\right)c_1}{1 + \frac{c_1}{K^2A_{44}A_{55}} + \frac{c_2}{KA_{55}} + \frac{c_3}{KA_{44}}}\right) \qquad (9.2-33b)$$

$$\hat{c}_{33} = D_{11}\alpha^4 + 2\,(D_{12} + 2D_{66})\,\alpha^2\beta^2 + D_{22}\beta^4, \quad c_1 = c_2c_3 - (c_4)^2 > 0$$
$$c_2 = D_{11}\alpha^2 + D_{66}\beta^2, \; c_3 = D_{66}\alpha^2 + D_{22}\beta^2, \; c_4 = (D_{12} + D_{66})\,\alpha\beta \quad (9.2-33c)$$

Clearly, when the effect of transverse shear deformation is neglected, Eq. (33b) yields the result (8.3–47a) obtained using the classical plate theory. The expression in (33b) is of the form

$$\frac{\hat{c}_{33} + k_1}{1 + k_2} \quad \text{where} \quad k_1 < k_2 \qquad (9.2-34a)$$

from which it follows that

$$\hat{c}_{33} \geq \frac{\hat{c}_{33} + k_1}{1 + k_2} \qquad (9.2-34b)$$

indicating that transverse shear deformation has the effect of reducing the buckling load (as long as $\hat{c}_{33} > 1$).

No conclusions can be drawn from the complicated expression of the buckling load concerning its minimum. Hence, a parametric study is carried out to determine the minimum buckling load, which occurs at $m = n = 1$. For an isotropic plate, the critical buckling load becomes

$$N_{cr} = 4D\left(\frac{\pi}{a}\right)^2 \frac{\left[1 + \frac{3(1-\nu^2)\pi^2(h/a)^2}{K}\right]}{\left[1 + \frac{72(1+\nu)(1-\nu^2)\pi^4(h/a)^4}{K^2} + \frac{6(1+\nu)(3-\nu)\pi^2(h/a)^2}{K}\right]} \qquad (9.2-35)$$

Table 9.2–5 contains nondimensionalized critical buckling loads of a square orthotropic plate and symmetric square laminates (0/90/0), (0/90/0/90/0), (0/90/0/90/0/90/0), and (0/90/0/90/0/90/0/90/0) under uniaxial and biaxial loadings. In these laminates the 0° layers and 90° layers have the same total thickness. For example, in the case of the nine-layer laminate the individual layer thicknesses are 0.1, 0.125, 0.1, 0.125, 0.1, 0.125, 0.1, 0.125, and 0.1, respectively. The critical buckling loads in all cases occurred in mode (1,1), except for orthotropic plates in biaxial compression, for which the mode is (2,1). For the side-to-thickness ratio of 10, for example, the classical laminate theory overpredicts the critical buckling loads by as much as 48% for orthotropic plates, and the error is less for thin plates. Figure 9.2–8 shows the effect of transverse shear deformation on critical buckling loads of symmetric (0/90/90/0) laminates under uniaxial and biaxial compression ($a/b = 1$; $E_1/E_2 = 25$, $G_{12} = G_{13} = 0.5E_2$, $G_{23} = 0.2E_2$, $\nu_{12} = 0.25$).

Table 9.2–5: Effect of shear deformation on nondimensionalized critical buckling loads, $\bar{N} = N_{cr}(a^2/E_2h^3)$, of simply supported (SS–1) symmetric cross-ply square plates ($E_1 = 25E_2$, $G_{12} = G_{13} = 0.5E_2$, $G_{23} = 0.2E_2$, $\nu_{12} = 0.25$, $K = 5/6$).

a/h	0°	Three-ply	Five-ply	Seven-ply	Nine-ply
Uniaxial Compression ($k = 0$)					
10	15.874	15.289	16.309	16.535	16.622
20	20.953	20.628	21.125	21.237	21.281
25	21.800	21.568	21.917	21.996	22.027
50	23.046	22.978	23.078	23.101	23.109
100	23.381	23.363	23.389	23.395	23.397
CLPT	23.495	23.495	23.495	23.495	23.495
Biaxial Compression ($k = 1$)					
10	5.837†	7.644	8.154	8.267	8.311
20	7.555	10.314	10.562	10.619	10.641
25	7.839	10.784	10.958	10.998	11.014
50	8.257	11.489	11.539	11.550	11.555
100	8.369	11.682	11.695	11.698	11.699
CLPT	8.407	11.747	11.747	11.747	11.747

† Mode for orthotropic plates in biaxial compression is $(m, n) = (2, 1)$.

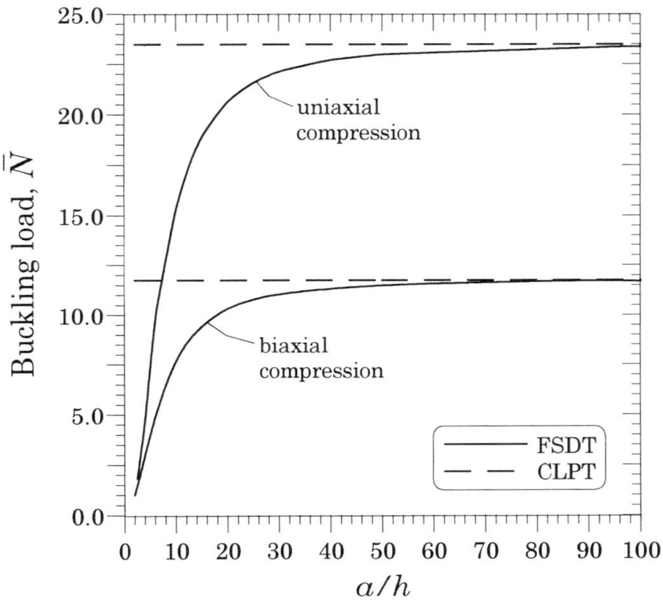

Figure 9.2–8: Nondimensionalized critical buckling load ($\bar{N}$) versus side-to-thickness ratio (a/h) for simply supported (SS–1) symmetric cross-ply (0/90/90/0) square laminates.

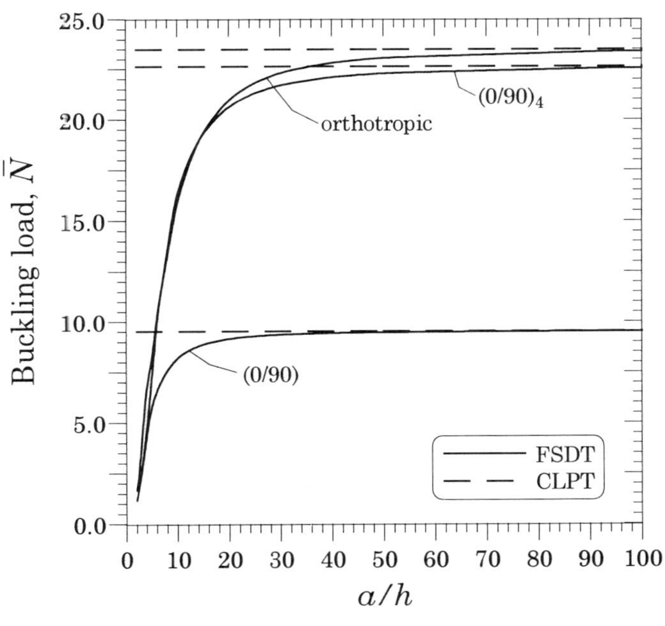

Figure 9.2–9: Nondimensionalized critical buckling load ($\bar{N}$) versus side-to-thickness ratio (a/h) for simply supported (SS–1) antisymmetric cross-ply $(0/90)_n$ square laminates.

Figure 9.2–9 shows the effect of transverse shear deformation and bending–extensional coupling on critical buckling loads ($a/b = 1$; $E_1/E_2 = 25$, $G_{12} = G_{13} = 0.5E_2$, $G_{23} = 0.2E_2$, $\nu_{12} = 0.25$). The eight-layer antisymmetric cross-ply plate behaves much like an orthotropic plate. Critical buckling loads of two-layer and eight-layer antisymmetric cross-ply laminated plates under uniaxial and biaxial loading are presented in Table 9.2–6 for modulus ratios E_1/E_2=10, 25, and 40. The effect of shear deformation on buckling loads is not as significant as for deflections. Note that the same critical buckling loads are valid for a rectangular laminate with aspect ratio $a/b = 3$, except that the mode at critical buckling is $(m, n) = (3, 1)$.

Table 9.2–6: Effect of shear deformation on nondimensionalized critical buckling loads, $\bar{N} = N_{cr}(a^2/E_2h^3)$, of simply supported (SS–1) antisymmetric cross-ply square plates ($G_{12} = G_{13} = 0.5E_2$, $G_{23} = 0.2E_2$, $\nu_{12} = 0.25$, $K = 5/6$).

	$E_1/E_2 = 10$		$E_1/E_2 = 25$		$E_1/E_2 = 40$	
b/h	(0/90)	$(0/90)_4$	(0/90)	$(0/90)_4$	(0/90)	$(0/90)_4$
Uniaxial Compression ($k = 0$); mode: (1,1)						
10	5.746	9.158	8.189	16.301	10.381	21.631
20	6.205	10.380	9.153	20.623	11.980	29.965
100	6.367	10.843	9.511	22.535	12.601	34.179
CLPT	6.374	10.864	9.526	22.622	12.628	34.381
Biaxial Compression ($k = 1$); mode: (1,1)						
10	2.873	4.579	4.094	8.150	5.190	10.816
20	3.102	5.190	4.576	10.311	5.990	14.983
100	3.184	5.422	4.755	11.267	6.300	17.090
CLPT	3.187	5.432	4.763	11.311	6.314	17.190

9.2–4 Vibration

For free vibration, we set the thermal and mechanical loads to zero, and substitute

$$U_{mn}(t) = U_{mn}^0 e^{i\omega t}, \quad V_{mn}(t) = V_{mn}^0 e^{i\omega t}$$

and so on in Eq. (7) and obtain

$$\left([\hat{S}] - \omega^2[\hat{M}]\right)\{\Delta\} = \{0\} \tag{9.2 - 36a}$$

where $\{\Delta\}^{\rm T} = \{U^0_{mn}\ V^0_{mn}\ W^0_{mn}\ X^0_{mn}\ Y^0_{mn}\}$

$$[\hat{S}] = \begin{bmatrix} \hat{s}_{11} & \hat{s}_{12} & 0 & \hat{s}_{14} & \hat{s}_{15} \\ \hat{s}_{12} & \hat{s}_{22} & 0 & \hat{s}_{24} & \hat{s}_{25} \\ 0 & 0 & \hat{s}_{33} & \hat{s}_{34} & \hat{s}_{35} \\ \hat{s}_{14} & \hat{s}_{24} & \hat{s}_{34} & \hat{s}_{44} & \hat{s}_{45} \\ \hat{s}_{15} & \hat{s}_{25} & \hat{s}_{35} & \hat{s}_{45} & \hat{s}_{55} \end{bmatrix}, \quad [\hat{M}] = \begin{bmatrix} \hat{m}_{11} & 0 & 0 & 0 & 0 \\ 0 & \hat{m}_{22} & 0 & 0 & 0 \\ 0 & 0 & \hat{m}_{33} & 0 & 0 \\ 0 & 0 & 0 & \hat{m}_{44} & 0 \\ 0 & 0 & 0 & 0 & \hat{m}_{55} \end{bmatrix} \tag{9.2 – 36b}$$

and the coefficients $\hat{s}_{ij}$ and $\hat{m}_{ij}$ are defined in Eqs. (7b,c).

When rotary inertia is omitted, Eq. (36) can be simplified by eliminating X_{mn} and Y_{mn} (say, using the static condensation method). We obtain the following 3×3 system of eigenvalue problem [*cf.* Eq. (8.3–49)]:

$$\left(\begin{bmatrix} \bar{s}_{11} & \bar{s}_{12} & \bar{s}_{13} \\ \bar{s}_{12} & \bar{s}_{22} & \bar{s}_{23} \\ \bar{s}_{13} & \bar{s}_{23} & \bar{s}_{33} \end{bmatrix} - \omega^2 \begin{bmatrix} \hat{m}_{11} & 0 & 0 \\ 0 & \hat{m}_{22} & 0 \\ 0 & 0 & \hat{m}_{33} \end{bmatrix} \right) \begin{Bmatrix} U^0_{mn} \\ V^0_{mn} \\ W^0_{mn} \end{Bmatrix} = \begin{Bmatrix} 0 \\ 0 \\ 0 \end{Bmatrix} \tag{9.2 – 37a}$$

where $(\bar{s}_{ij} = \bar{s}_{ji})$

$$\begin{aligned}
\bar{s}_{11} &= \hat{s}_{11} - (\hat{s}_{14}\hat{s}_{55} - \hat{s}_{15}\hat{s}_{45})\,\hat{s}_{14}/\hat{s}_{00} - (\hat{s}_{15}\hat{s}_{44} - \hat{s}_{14}\hat{s}_{45})\,\hat{s}_{15}/\hat{s}_{00} \\
\bar{s}_{12} &= \hat{s}_{12} - (\hat{s}_{24}\hat{s}_{55} - \hat{s}_{25}\hat{s}_{45})\,\hat{s}_{14}/\hat{s}_{00} - (\hat{s}_{25}\hat{s}_{44} - \hat{s}_{24}\hat{s}_{45})\,\hat{s}_{15}/\hat{s}_{00} \\
\bar{s}_{13} &= -(\hat{s}_{34}\hat{s}_{55} - \hat{s}_{35}\hat{s}_{45})\,\hat{s}_{14}/\hat{s}_{00} - (\hat{s}_{35}\hat{s}_{44} - \hat{s}_{34}\hat{s}_{45})\,\hat{s}_{15}/\hat{s}_{00} \\
\bar{s}_{22} &= \hat{s}_{22} - (\hat{s}_{24}\hat{s}_{55} - \hat{s}_{25}\hat{s}_{45})\,\hat{s}_{24}/\hat{s}_{00} - (\hat{s}_{25}\hat{s}_{44} - \hat{s}_{24}\hat{s}_{45})\,\hat{s}_{25}/\hat{s}_{00} \\
\bar{s}_{23} &= \hat{s}_{23} - (\hat{s}_{34}\hat{s}_{55} - \hat{s}_{35}\hat{s}_{45})\,\hat{s}_{24}/\hat{s}_{00} - (\hat{s}_{35}\hat{s}_{44} - \hat{s}_{34}\hat{s}_{45})\,\hat{s}_{25}/\hat{s}_{00} \\
\bar{s}_{33} &= \hat{s}_{33} - (\hat{s}_{34}\hat{s}_{55} - \hat{s}_{35}\hat{s}_{45})\,\hat{s}_{34}/\hat{s}_{00} - (\hat{s}_{35}\hat{s}_{44} - \hat{s}_{34}\hat{s}_{45})\,\hat{s}_{35}/\hat{s}_{00} \\
\bar{s}_{00} &= \hat{s}_{44}\hat{s}_{55} - \hat{s}_{45}\hat{s}_{45}
\end{aligned} \tag{9.2 – 37b}$$

If the in-plane and rotatory inertias are omitted (*i.e.*, $\hat{m}_{11} = \hat{m}_{22} = \hat{m}_{44} = \hat{m}_{55} = 0$), we have [*cf.* Eq. (8.3–52)]

$$\omega^2 = \frac{1}{\hat{m}_{33}} \left(\bar{s}_{33} - \frac{\bar{s}_{13}\bar{s}_{22} - \bar{s}_{23}\bar{s}_{12}}{\bar{s}_{11}\bar{s}_{22} - \bar{s}_{12}\bar{s}_{12}}\bar{s}_{13} - \frac{\bar{s}_{11}\bar{s}_{23} - \bar{s}_{12}\bar{s}_{13}}{\bar{s}_{11}\bar{s}_{22} - \bar{s}_{12}\bar{s}_{12}}\bar{s}_{23} \right) \tag{9.2 – 38}$$

If frequencies of in-plane vibration of specially orthotropic laminates or natural frequencies of flexural or in-plane vibration of antisymmetric laminates are required, one must use Eq. (37a).

Example 9.2–3 (*Specially orthotropic plates*):

For specially orthotropic plates, the in-plane displacements are uncoupled from the transverse deflection, and therefore the natural frequencies of vibration are given by Eq. (37); Eq. (38) gives the same frequencies of flexural vibration as Eq. (37a) for this case.

Tables 9.2–7 contains fundamental frequencies of isotropic plates. Similar results are presented in Tables 9.2–8 and 9.2–9 for symmetric cross-ply laminates. The effect of the shear correction factor is to decrease the frequencies, *i.e.*, the smaller the K, the smaller are the frequencies. The rotary inertia (RI) also has the effect of decreasing frequencies. Figure 9.2–10 shows the effect of transverse shear deformation and rotary inertia on fundamental natural frequencies of orthotropic and symmetric cross-ply (0/90/90/0) square plates ($E_1/E_2 = 25$, $G_{12} = G_{13} = 0.5E_2$, $G_{23} = 0.2E_2$, $\nu_{12} = 0.25$). The symmetric cross-ply plate behaves much like an orthotropic plate. The effect of rotary inertia is negligible in FSDT and therefore not shown in the figure.

Table 9.2–7: Effect of shear deformation, rotatory inertia, and shear correction coefficient on nondimensionalized natural frequencies of simply supported (SS–1) isotropic square plates ($\bar{\omega} = \omega(a^2/h)\sqrt{\rho/E}$; $\nu = 0.3, a/h = 10$).

m	n	CLPT† w/o RI	CLPT with RI	K	FSDT w/o RI	FSDT with RI
1	1	5.973	5.925	5/6	5.812	5.769
				2/3	5.773	5.732
				1.0	5.838	5.794
2	1	14.933	14.635	5/6	13.980	13.764
				2/3	13.769	13.568
				1.0	14.127	13.899
2	2	23.893	23.144	5/6	21.583	21.121
				2/3	21.103	20.688
				1.0	21.922	21.424
3	1	29.867	28.709	5/6	26.378	25.734
				2/3	25.682	25.115
				1.0	26.875	26.171
3	2	38.827	36.904	5/6	33.213	32.284
				2/3	32.153	31.357
				1.0	33.982	32.946
4	1	50.744	47.558	5/6	41.744	40.436
				2/3	40.150	39.063
				1.0	42.919	41.427
3	3	53.760	50.174	5/6	43.785	42.383
				2/3	42.051	40.895
				1.0	45.067	43.461

† w/o RI means without rotary inertia.

Table 9.2–8: Effect of shear deformation, rotatory inertia, and shear correction coefficient on nondimensionalized natural frequencies ($\bar{\omega} = \omega(a^2/h)\sqrt{\rho/E}$) of simply supported (SS–1) symmetric cross-ply (0/90/0) square plates ($h_k = h/3$; $E_1 = 25E_2$, $G_{12} = G_{13} = 0.5E_2$, $G_{23} = 0.2E_2$, $\nu_{12} = 0.25$).

a/h	m	n	CLPT w/o RI	CLPT with RI	FSDT w/o RI	FSDT with RI
10	1	1	15.228	15.104	12.593	12.527†
					12.223	12.163
	1	2	22.877	22.421	19.440	19.203
					18.942	18.729
	1	3	40.299	38.738	32.496	31.921
					31.421	30.932
	2	1	56.885	55.751	33.097	32.931
					31.131	30.991
	2	2	60.911	59.001	36.786	36.362
					34.794	34.434
	1	4	66.754	62.526	48.837	47.854
					46.714	45.923
	2	3	71.522	67.980	45.484	44.720
					43.212	42.585
100	1	1	15.228	15.227	15.192	15.191
					15.185	15.183
	1	2	22.877	22.873	22.831	22.827
					22.822	22.817
	1	3	40.299	40.283	40.190	40.174
					40.169	40.153
	2	1	56.885	56.874	56.330	56.319
					56.221	56.210
	2	2	60.911	60.891	60.342	60.322
					60.230	60.211
	1	4	66.754	66.708	66.466	66.421
					66.409	66.364
	2	3	71.522	71.484	70.919	70.882
					70.801	70.764

† The first line corresponds to shear correction coefficient of $K = 1.0$ and the second line corresponds to shear correction coefficient of $K = 5/6$.

Table 9.2–9: Effect of shear deformation on dimensionless natural frequencies of simply supported (SS–1) symmetric cross-ply plates ($\bar{\omega} = \omega(a^2/h)\sqrt{\rho/E}$; $E_1 = 25E_2$, $G_{12} = G_{13} = 0.5E_2$, $G_{23} = 0.2E_2$, $\nu_{12} = 0.25$, $K = 5/6$; rotary inertia is included; the total thickness of all 0° layers and all 90° layers is the same, $h/2$).

a/h	Theory	0°	Three-ply	Five-ply	Seven-ply	Nine-ply
5	FSDT	8.909	8.766	9.215	9.301	9.333
	CLPT	14.750	14.750	14.750	14.750	14.750
10	FSDT	12.452	12.227	12.633	12.720	12.754
	CLPT	15.104	15.104	15.104	15.104	15.104
20	FSDT	14.355	14.244	14.415	14.453	14.468
	CLPT	15.197	15.197	15.197	15.197	15.197
25	FSDT	14.651	14.573	14.690	14.717	14.727
	CLPT	15.208	15.208	15.208	15.208	15.208
50	FSDT	15.077	15.055	15.087	15.095	15.098
	CLPT	15.223	15.223	15.223	15.223	15.223
100	FSDT	15.190	15.184	15.192	15.194	15.195
	CLPT	15.227	15.227	15.227	15.227	15.227

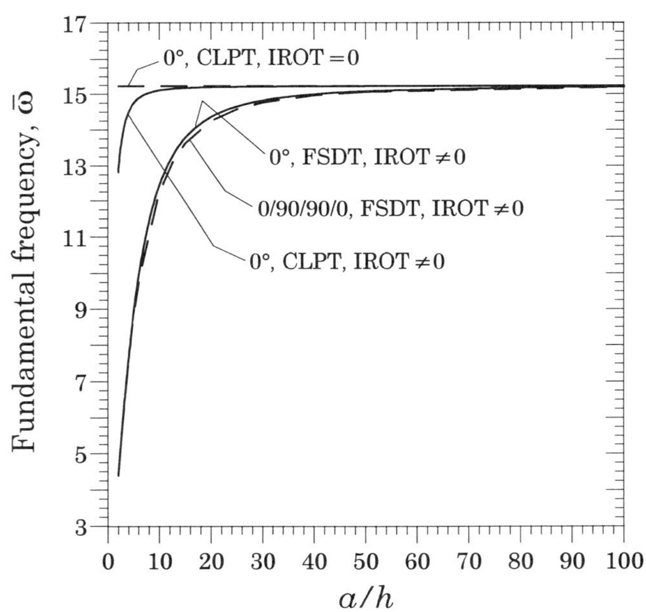

Figure 9.2–10: Nondimensionalized fundamental frequency ($\bar{\omega}$) versus side-to-thickness ratio (a/b) for simply supported (SS–1), orthotropic and symmetric cross-ply (0/90/90/0) laminates.

Table 9.2–10 contains numerical values of fundamental frequencies of antisymmetric cross-ply laminated plates. Results for both two-layer and eight-layer plates for various modulus ratios and two aspect ratios are presented. Figure 9.2–11 shows the effect of transverse shear deformation, bending–extensional coupling, and rotary inertia on fundamental natural frequencies ($E_1/E_2 = 25$, $G_{12} = G_{13} = 0.5E_2$, $G_{23} = 0.2E_2$, $\nu_{12} = 0.25$). The eight-layer antisymmetric cross-ply plate behaves much like an orthotropic plate. The effect of rotary inertia is negligible in FSDT and therefore not shown in the figure.

Table 9.2–10: Effect of shear deformation on nondimensionalized fundamental frequencies of simply supported (SS–1) antisymmetric cross-ply square plates ($G_{12} = G_{13} = 0.5E_2$, $G_{23} = 0.2E_2$, $\nu_{12} = 0.25$, $K = 5/6$).

		$E_1/E_2 = 10$		$E_1/E_2 = 25$		$E_1/E_2 = 40$	
b/h	Theory	(0/90)	$(0/90)_4$	(0/90)	$(0/90)_4$	(0/90)	$(0/90)_4$
Square Plate ($a/b = 1$)							
10	FSDT	7.454	9.450	8.900	12.628	10.027	14.562
	CLPT	7.832	10.268	9.566	14.816	11.011	18.265
20	FSDT	7.802	10.102	9.474	14.241	10.840	17.169
	CLPT	7.906	10.333	9.663	14.910	11.125	18.381
100	FSDT	7.926	10.344	9.687	14.912	11.150	18.365
	CLPT	7.931	10.354	9.695	14.941	11.163	18.419
Rectangular Plate ($a/b = 3$)							
10	FSDT	4.751	6.319	5.952	8.800	6.846	10.269
	CLPT	4.930	6.772	6.324	10.201	7.437	12.738
20	FSDT	4.908	6.592	6.258	9.819	7.313	11.990
	CLPT	4.956	6.796	6.360	10.238	7.480	12.783
100	FSDT	4.962	6.799	6.367	10.231	7.486	12.763
	CLPT	4.964	6.804	6.372	10.249	7.493	12.798

9.3 Simply Supported Antisymmetric Angle-Ply Laminates

9.3–1 Boundary Conditions

The boundary conditions in (8.2–7) imply the following SS–2 boundary conditions on the generalized displacements and resultants of the first-order laminate theory (see Figure 9.3–1):

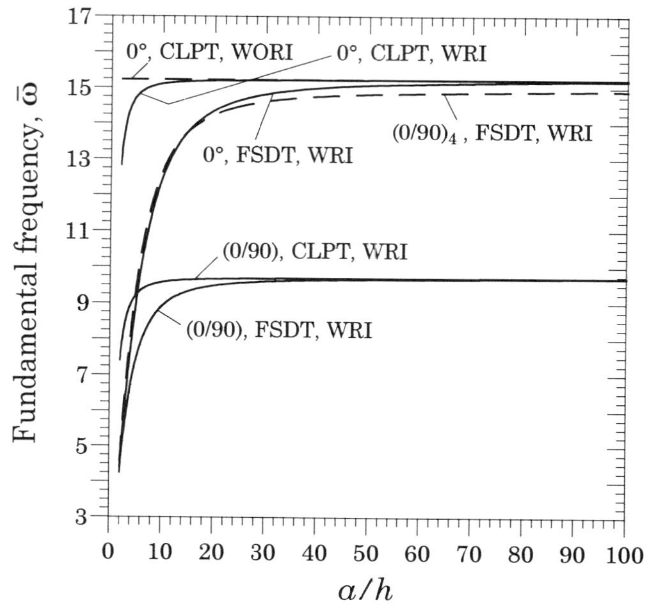

Figure 9.2–11: Nondimensionalized fundamental frequency ($\bar{\omega}$) versus side-to-thickness ratio (a/h) for simply supported (SS–1), antisymmetric cross-ply (0/90) laminates.

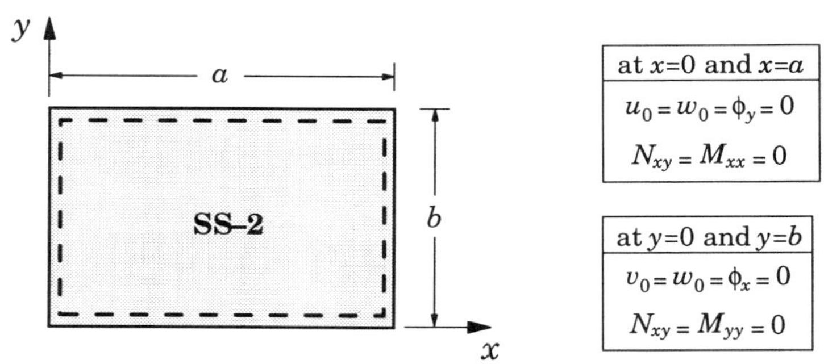

Figure 9.3–1: The simply supported boundary conditions for antisymmetric angle-ply laminates (SS–2).

$$u_0(0, y, t) = 0, \quad u_0(a, y, t) = 0, \quad v_0(x, 0, t) = 0, \quad v_0(x, b, t) = 0$$

$$\phi_x(x, 0, t) = 0, \quad \phi_x(x, b, t) = 0, \quad \phi_y(0, y, t) = 0, \quad \phi_y(a, y, t) = 0$$

$$w_0(x, 0, t) = 0, \quad w_0(x, b, t) = 0, \quad w_0(0, y, t) = 0, \quad w_0(a, y, t) = 0 \qquad (9.3-1a)$$

$$N_{xy}(0,y,t)=0,\quad N_{xy}(a,y,t)=0,\quad N_{xy}(x,0,t)=0,\quad N_{xy}(x,b,t)=0$$
$$M_{xx}(0,y,t)=0,\quad M_{xx}(a,y,t)=0,\quad M_{yy}(x,0,t)=0,\quad M_{yy}(x,b,t)=0 \quad (9.3-1b)$$

9.3–2 The Navier Solution

The boundary conditions in (1) are satisfied by the expansions

$$u_0(x,y,t)=\sum_{n=1}^{\infty}\sum_{m=1}^{\infty}U_{mn}(t)\sin\alpha x\cos\beta y \qquad (9.3-2a)$$

$$v_0(x,y,t)=\sum_{n=1}^{\infty}\sum_{m=1}^{\infty}V_{mn}(t)\cos\alpha x\sin\beta y \qquad (9.3-2b)$$

$$w_0(x,y,t)=\sum_{n=1}^{\infty}\sum_{m=1}^{\infty}W_{mn}(t)\sin\alpha x\sin\beta y \qquad (9.3-3)$$

$$\phi_x(x,y,t)=\sum_{n=1}^{\infty}\sum_{m=1}^{\infty}X_{mn}(t)\cos\alpha x\sin\beta y \qquad (9.3-4a)$$

$$\phi_y(x,y,t)=\sum_{n=1}^{\infty}\sum_{m=1}^{\infty}Y_{mn}(t)\sin\alpha x\cos\beta y \qquad (9.3-4b)$$

Substitution of Eqs. (2a,b), (3), and (4a,b) into Eqs. (9.1–1)–(9.1–5) shows that the Navier solution exists only if

$$A_{16}=0,\ A_{26}=0,\ A_{45}=0,\ B_{11}=0,\ B_{12}=0,\ B_{22}=0,\ B_{66}=0$$
$$D_{16}=0,\ D_{26}=0,\ I_1=0 \qquad (9.3-5)$$

i.e., for antisymmetric angle-ply laminates [see Eq. (5.5–31)].

The coefficients $U_{mn}, V_{mn}, W_{mn}, X_{mn}, Y_{mn}$ can be determined from the equations

$$\begin{bmatrix}\hat{s}_{11} & \hat{s}_{12} & 0 & \hat{s}_{14} & \hat{s}_{15}\\ \hat{s}_{12} & \hat{s}_{22} & 0 & \hat{s}_{24} & \hat{s}_{25}\\ 0 & 0 & \hat{s}_{33}+\tilde{s}_{33} & \hat{s}_{34} & \hat{s}_{35}\\ \hat{s}_{14} & \hat{s}_{24} & \hat{s}_{34} & \hat{s}_{44} & \hat{s}_{45}\\ \hat{s}_{15} & \hat{s}_{25} & \hat{s}_{35} & \hat{s}_{54} & \hat{s}_{55}\end{bmatrix}\begin{Bmatrix}U_{mn}\\ V_{mn}\\ W_{mn}\\ X_{mn}\\ Y_{mn}\end{Bmatrix}+$$
$$\begin{bmatrix}\hat{m}_{11} & 0 & 0 & 0 & 0\\ 0 & \hat{m}_{22} & 0 & 0 & 0\\ 0 & 0 & \hat{m}_{33} & 0 & 0\\ 0 & 0 & 0 & \hat{m}_{44} & 0\\ 0 & 0 & 0 & 0 & \hat{m}_{55}\end{bmatrix}\begin{Bmatrix}\ddot{U}_{mn}\\ \ddot{V}_{mn}\\ \ddot{W}_{mn}\\ \ddot{X}_{mn}\\ \ddot{Y}_{mn}\end{Bmatrix}=\begin{Bmatrix}0\\ 0\\ Q_{mn}\\ 0\\ 0\end{Bmatrix}+\begin{Bmatrix}-\beta N^6_{mn}\\ -\alpha N^6_{mn}\\ 0\\ -\alpha M^1_{mn}\\ -\beta M^2_{mn}\end{Bmatrix} \qquad (9.3-6)$$

where

$$\hat{s}_{11} = (A_{11}\alpha^2 + A_{66}\beta^2), \quad \hat{s}_{12} = (A_{12} + A_{66})\alpha\beta$$
$$\hat{s}_{13} = 0, \quad \hat{s}_{14} = 2B_{16}\alpha\beta, \quad \hat{s}_{15} = (B_{16}\alpha^2 + B_{26}\beta^2)$$
$$\hat{s}_{22} = (A_{66}\alpha^2 + A_{22}\beta^2), \quad \hat{s}_{23} = 0, \quad \hat{s}_{24} = \hat{s}_{15}$$
$$\hat{s}_{25} = 2B_{26}\alpha\beta, \quad \hat{s}_{33} = K(A_{55}\alpha^2 + A_{44}\beta^2)$$
$$\hat{s}_{34} = KA_{55}\alpha, \quad \hat{s}_{35} = KA_{44}\beta, \quad \hat{s}_{44} = (D_{11}\alpha^2 + D_{66}\beta^2 + KA_{55})$$
$$\hat{s}_{45} = (D_{12} + D_{66})\alpha\beta, \quad \hat{s}_{55} = (D_{66}\alpha^2 + D_{22}\beta^2 + KA_{44}) \qquad (9.3-7)$$

and the thermal coefficients are defined in Eqs. (8.3–11)–(8.3–16), and $\tilde{S}_{ij}$ and $\hat{m}_{ij}$ are defined in Eqs (9.2–7a,b). Equation (5) can be specialized for static analysis, buckling under in-plane compressive loads, and natural vibration, as was discussed for antisymmetric cross-ply laminates.

The in-plane stresses in each layer can be computed from the equations

$$\begin{Bmatrix} \sigma_{xx} \\ \sigma_{yy} \\ \sigma_{xy} \end{Bmatrix}^{(k)} = \begin{bmatrix} \bar{Q}_{11} & \bar{Q}_{12} & \bar{Q}_{16} \\ \bar{Q}_{12} & \bar{Q}_{22} & \bar{Q}_{26} \\ \bar{Q}_{16} & \bar{Q}_{26} & \bar{Q}_{66} \end{bmatrix}^{(k)} \left(\begin{Bmatrix} \varepsilon_{xx} \\ \varepsilon_{yy} \\ 2\varepsilon_{xy} \end{Bmatrix} - \begin{Bmatrix} \alpha_{xx} \\ \alpha_{yy} \\ 2\alpha_{xy} \end{Bmatrix} \Delta T \right) \qquad (9.3-8)$$

$$\left(\begin{Bmatrix} \varepsilon_{xx} \\ \varepsilon_{yy} \\ 2\varepsilon_{xy} \end{Bmatrix} - \begin{Bmatrix} \alpha_{xx} \\ \alpha_{yy} \\ 2\alpha_{xy} \end{Bmatrix} \Delta T \right) = \sum_{m=1}^{\infty}\sum_{n=1}^{\infty} \begin{Bmatrix} \alpha U_{mn}\, g_{mn} - \alpha_{xx} T^0_{mn} f_{mn} \\ \beta V_{mn}\, g_{mn} - \alpha_{yy} T^0_{mn} f_{mn} \\ -(\beta U_{mn} + \alpha V_{mn} - 2\alpha_{xy} T^0_{mn})\, f_{mn} \end{Bmatrix}$$
$$- \sum_{m=1}^{\infty}\sum_{n=1}^{\infty} z \begin{Bmatrix} (\alpha X_{mn} + \alpha_{xx} T^1_{mn})\, f_{mn} \\ (\beta Y_{mn} + \alpha_{yy} T^1_{mn})\, f_{mn} \\ -(\beta X_{mn} + \alpha Y_{mn})\, g_{mn} + 2\alpha_{xy} T^1_{mn} f_{mn} \end{Bmatrix} \qquad (9.3-9a)$$

$$f_{mn} = \sin\frac{m\pi x}{a}\sin\frac{n\pi y}{b}, \quad g_{mn} = \cos\frac{m\pi x}{a}\cos\frac{n\pi y}{b} \qquad (9.3-9b)$$

The transverse shear stresses from the constitutive equations are given by

$$\begin{Bmatrix} \sigma_{yz} \\ \sigma_{xz} \end{Bmatrix}^{(k)} = \sum_{m=1}^{\infty}\sum_{n=1}^{\infty} \begin{bmatrix} \bar{Q}_{44} & \bar{Q}_{45} \\ \bar{Q}_{45} & \bar{Q}_{55} \end{bmatrix}^{(k)} \begin{Bmatrix} (Y_{mn} + \beta W_{mn})\sin\alpha x\cos\beta y \\ (X_{mn} + \alpha W_{mn})\cos\alpha x\sin\beta y \end{Bmatrix} \qquad (9.3-10)$$

Note that the stresses are layerwise constant through the thickness.

The transverse stresses determined from the equilibrium equations of 3-D elasticity are

$$\sigma^{(k)}_{xz}(x,y,z) = \sum_{m=1}^{\infty}\sum_{n=1}^{\infty}\left[(z - z_k)\bar{\mathcal{A}}^{(k)}_{mn} + \frac{1}{2}(z^2 - z_k^2)\bar{\mathcal{B}}^{(k)}_{mn}\right] + \sigma^{(k-1)}_{xz}(x,y,z_k)$$
$$\sigma^{(k)}_{yz}(x,y,z) = \sum_{m=1}^{\infty}\sum_{n=1}^{\infty}\left[(z - z_k)\bar{\mathcal{C}}^{(k)}_{mn} + \frac{1}{2}(z^2 - z_k^2)\bar{\mathcal{D}}^{(k)}_{mn}\right] + \sigma^{(k-1)}_{yz}(x,y,z_k)$$
$$(9.3-11a)$$

where

$$
\begin{aligned}
\bar{\mathcal{A}}_{mn}^{(k)} &= \left[\left(\alpha^2 \bar{Q}_{11}^{(k)} + \beta^2 \bar{Q}_{66}^{(k)}\right) U_{mn} + \alpha\beta \left(\bar{Q}_{12}^{(k)} + \bar{Q}_{66}^{(k)}\right) V_{mn}\right] \sin \alpha x \cos \beta y \\
&\quad + \left[2\alpha\beta \bar{Q}_{16}^{(k)} U_{mn} + \left(\alpha^2 \bar{Q}_{16}^{(k)} + \beta^2 \bar{Q}_{26}^{(k)}\right) V_{mn}\right] \cos \alpha x \sin \beta y \\
\bar{\mathcal{B}}_{mn}^{(k)} &= \left[\left(\alpha^2 \bar{Q}_{11}^{(k)} + \beta^2 \bar{Q}_{66}^{(k)}\right) X_{mn} + \alpha\beta \left(\bar{Q}_{12}^{(k)} + \bar{Q}_{66}^{(k)}\right) Y_{mn}\right] \cos \alpha x \sin \beta y \\
&\quad + \left[2\alpha\beta \bar{Q}_{16}^{(k)} X_{mn} + \left(\alpha^2 \bar{Q}_{16}^{(k)} + \beta^2 \bar{Q}_{26}^{(k)}\right) Y_{mn}\right] \sin \alpha x \cos \beta y \\
\bar{\mathcal{C}}_{mn}^{(k)} &= \left[\left(\alpha^2 \bar{Q}_{16}^{(k)} + \beta^2 \bar{Q}_{26}^{(k)}\right) U_{mn} + 2\alpha\beta \bar{Q}_{16}^{(k)} V_{mn}\right] \sin \alpha x \cos \beta y \\
&\quad + \left[\alpha\beta \left(\bar{Q}_{12}^{(k)} + \bar{Q}_{66}^{(k)}\right) U_{mn} + \left(\alpha^2 \bar{Q}_{66}^{(k)} + \beta^2 \bar{Q}_{22}^{(k)}\right) V_{mn}\right] \cos \alpha x \sin \beta y \\
\bar{\mathcal{D}}_{mn}^{(k)} &= \left[\left(\alpha^2 \bar{Q}_{16}^{(k)} + \beta^2 \bar{Q}_{66}^{(k)}\right) X_{mn} + 2\alpha\beta \bar{Q}_{26}^{(k)} Y_{mn}\right] \cos \alpha x \sin \beta y \\
&\quad + \left[\alpha\beta \left(\bar{Q}_{12}^{(k)} + \bar{Q}_{66}^{(k)}\right) X_{mn} + \left(\alpha^2 \bar{Q}_{66}^{(k)} + \beta^2 \bar{Q}_{22}^{(k)}\right) Y_{mn}\right] \sin \alpha x \cos \beta y
\end{aligned}
\tag{9.3-11b}
$$

9.3–3 Bending

Table 9.3–1 contains numerical results of nondimensionalized maximum deflections and stresses of simply supported (SS–2), two-layer and eight-layer antisymmetric angle-ply square laminates, (–45/45/–45/⋯), subjected to sinusoidally and uniformly distributed loads. The nondimensionalizations and locations of maximum quantities are as follows:

$$
\begin{aligned}
\bar{w} &= w_0(a/2, a/2)\frac{E_2 h^3}{b^4 q_0}, \quad \bar{\sigma}_{xx} = \sigma_{xx}(a/2, b/2, \frac{h}{2})\frac{h^2}{b^2 q_0} \\
\bar{\sigma}_{yy} &= \sigma_{yy}(a/2, b/2, \frac{h}{2})\frac{h^2}{b^2 q_0}, \quad \bar{\sigma}_{xy} = \sigma_{xy}(0, 0, -\frac{h}{2})\frac{h^2}{b^2 q_0} \\
\bar{\sigma}_{xz} &= \sigma_{xz}(0, b/2, k)\frac{h}{b q_0}, \quad \bar{\sigma}_{yz} = \sigma_{yz}(a/2, 0, k)\frac{h}{b q_0}
\end{aligned}
\tag{9.3-12}
$$

where $k = 1, 2, \cdots, n$ denotes the ply number. Both constitutive and equilibrium based transverse shear stresses are included in the table. While the deflections are sensitive to the transverse shear deformation, stresses are not. Table 9.3–2 contains the maximum transverse deflection and in-plane normal stress as a function of the modulus ratio of simply supported (SS–2) square, antisymmetric, two-layer (–45/45) and eight-layer (–45/45/–45/⋯) angle-ply laminates ($a/h = 10$) subjected to uniformly distributed load.

Figure 9.3–2 contains plots of the nondimensionalized transverse deflection versus side-to-thickness ratio (a/h) of various angle-ply laminates subjected to uniformly or sinusoidally distributed transverse loading ($a/b = 1$; $E_1/E_2 = 25$, $G_{12} = G_{13} = 0.5E_2$, $\nu_{12} = 0.25$, $K = 5/6$). The effect of transverse shear deformation is negligible for all values of a/h greater than 10. For values of a/h less than 10, the effect is quite significant. Figure 9.3–3 contains plots of the nondimensionalized deflection as a function of the lamination angle for two- and eight-layer antisymmetric angle-ply laminates ($a/h = 10$), $(-\theta/\theta/-\theta/\cdots)$, subjected to sinusoidally distributed transverse load. The effect of bending–stretching is significant in two-layer laminates.

Table 9.3–1: Effect of transverse shear deformation on nondimensionalized maximum transverse deflections and stresses of simply supported (SS–2) antisymmetric angle-ply square plates ($h_i = h/n, E_1 = 25E_2,\ G_{12} = G_{13} = 0.5E_2,\ G_{23} = 0.2E_2,\ \nu_{12} = 0.25,\ K = 5/6$).

a/h	Load	$\bar{w} \times 10^2$	$\bar{\sigma}_{xx}$	$\bar{\sigma}_{xy}$	$\bar{\sigma}_{xz}$	$\bar{\sigma}_{xz}^{\dagger}$
Antisymmetric Laminate, $(-45/45)$						
10	SSL	0.8284	0.2498	0.2336	0.2728	0.2143
	UDL	1.2792	0.3476	0.4274	0.5072	0.4238
20	SSL	0.6981	0.2498	0.2336	0.2728	0.2143
	UDL	1.0907	0.3496	0.4357	0.5065	0.4205
100	SSL	0.6564	0.2498	0.2336	0.2728	0.2143
	UDL	1.0305	0.3504	0.4417	0.5068	0.4189
CLPT	SSL	0.6547	0.2498	0.2336	–	0.2143
	UDL	1.0280	0.3504	0.4421	–	0.4188
Antisymmetric Laminate, $(-45/45)_4$						
10	SSL	0.4198	0.1445	0.1384	0.2728	0.2487
	UDL	0.6366	0.1957	0.2463	0.5070	0.4960
20	SSL	0.2896	0.1445	0.1384	0.2728	0.2487
	UDL	0.4483	0.1988	0.2550	0.5050	0.4884
100	SSL	0.2479	0.1445	0.1384	0.2728	0.2487
	UDL	0.3883	0.2005	0.2630	0.5054	0.4841
CLPT	SSL	0.2462	0.1445	0.1384	–	0.2487
	UDL	0.3858	0.2006	0.2637	–	0.4838

† Maximum stress derived from equilibrium. The values reported are at $z = \pm h/4$ for $(-45/45)$ laminate, and at $z = 0$ for the $(-45/45)_4$ laminate; the shear stress derived from constitutive relations will have two values at each interface, and the larger of the two is reported.

Table 9.3–2 Effect of lamination scheme and shear deformation on the transverse deflections and stresses in square antisymmetric angle-ply laminates subjected to uniformly distributed transverse load ($h_n = h/n$, E_1/E_2 varied, $G_{12} = G_{13} = 0.5E_2$, $G_{23} = 0.2E_2$, $\nu_{12} = 0.25$, $a/h = 10$; $m, n = 1, 3, \cdots, 21$ in the series are used to calculate the solutions for uniform load).

Theory	$\hat{w}^*$					$\bar{\sigma}_{xx}^{\dagger}$				
	$\frac{E_1}{E_2} = 1$	10	20	30	40	$\frac{E_1}{E_2} = 1$	10	20	30	40
Orthotropic Plate										
FSDT	4.480	1.678	1.089	0.856	0.731	0.255	0.681	0.755	0.780	0.791
CLPT	4.172	1.412	0.795	0.548	0.415	0.251	0.693	0.772	0.795	0.801
Laminate, (–45/45)										
FSDT	4.829	2.010	1.441	1.157	0.983	0.278	0.307	0.338	0.355	0.365
CLPT	4.577	1.759	1.190	0.906	0.732	0.278	0.308	0.340	0.358	0.370
Laminate, $(–45/45)_2$										
FSDT	4.829	1.251	0.792	0.622	0.534	0.278	0.213	0.201	0.196	0.193
CLPT	4.577	0.999	0.542	0.372	0.283	0.278	0.214	0.205	0.201	0.199
Laminate, $(–45/45)_4$										
FSDT	4.829	1.153	0.727	0.575	0.496	0.278	0.210	0.198	0.194	0.191
CLPT	4.577	0.902	0.477	0.324	0.245	0.278	0.211	0.203	0.199	0.198

* $\hat{w} = w_0 \frac{E_2 h^3}{q_0 b^4} \times 10^2$. $^{\dagger}\bar{\sigma}_{xx} = \sigma_{xx} \frac{1}{q_0}$.

Figures 9.3–4 and 9.3–5 contain nondimensionalized maximum transverse shear stress distributions through laminate thickness for an eight-layer antisymmetric angle-ply square laminate (–45/30/–45/0/0/45/–30/45) subjected to uniformly or sinusoidally distributed transverse loading. The side-to-thickness ratio is taken to be $a/h = 10$. The material properties used are: $E_1 = 25E_2, G_{12} = G_{13} = 0.5E_2$, $G_{23} = 0.2E_2$, $\nu_{12} = 0.25$, $K = 5/6$. The effect of transverse shear deformation is negligible on the stresses.

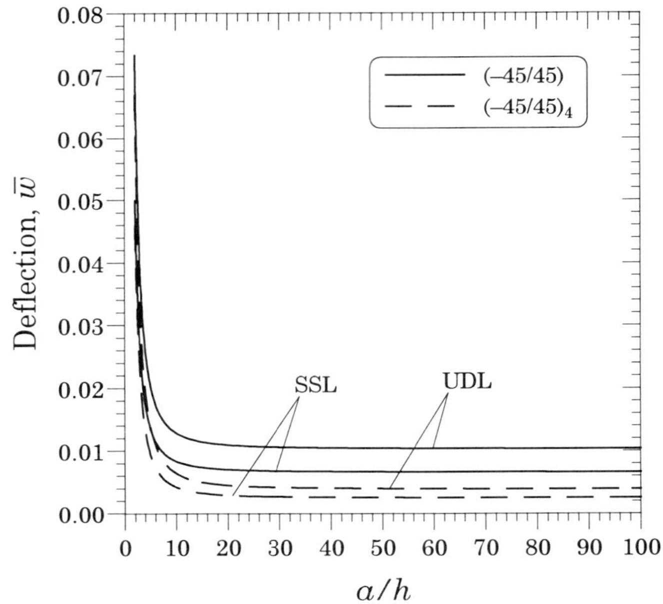

Figure 9.3–2: Nondimensionalized center transverse deflection $\bar{w}$ versus side-to-thickness ratio a/h for simply supported (SS–2) antisymmetric angle-ply $(-45/45)_n$ $(n = 1, 4)$ square laminates.

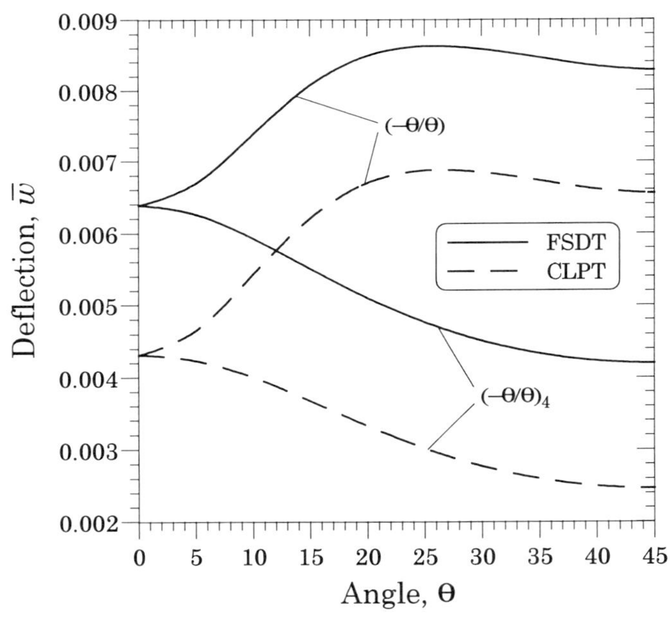

Figure 9.3–3: Nondimensionalized center transverse deflection $\bar{w}$ versus lamination angle θ for simply supported (SS–2) antisymmetric angle-ply $(-\theta/\theta)_n$ $(n = 1, 4)$ square laminates $(a/h = 10)$.

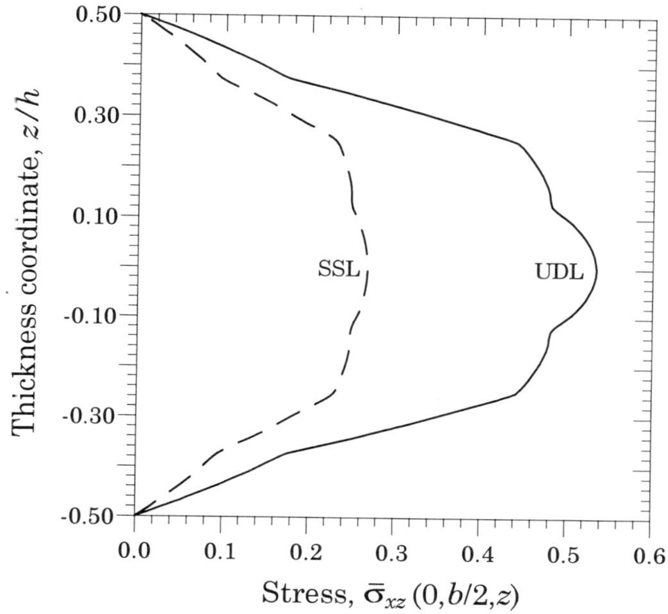

Figure 9.3–4: Distribution of transverse shear stress $\bar{\sigma}_{xz}(0, b/2, z)$ through thickness z/h for simply supported (SS–2) antisymmetric angle-ply (–45/30/–45/0/0/45/–30/45) square laminates.

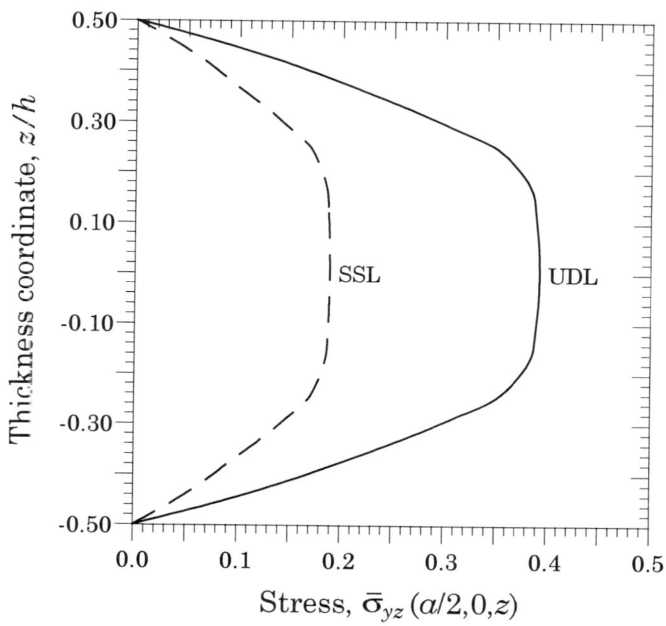

Figure 9.3–5: Distribution of transverse shear stress $\bar{\sigma}_{yz}(a/2, 0, z)$ through thickness z/h for simply supported (SS–2) antisymmetric angle-ply (–45/30/–45/0/0/45/–30/45) square laminates ($a/h = 10$).

9.3–4 Buckling

Table 9.3–3 contains critical buckling loads of uniaxially and biaxially compressed simply supported (SS–1) square, antisymmetric angle-ply laminates for various modulus ratios and two lamination schemes (–45/45) and (–45/45/–45/⋯). Note that for certain modulus ratios, side-to-thickness ratios, and lamination schemes, the shear deformation theory predicts buckling mode different from the classical laminate theory. Figures 9.3–6 and 9.3–7 show the effect of shear deformation, number of layers (bending–extensional coupling), and the lamination angle on critical buckling loads of antisymmetric angle-ply square laminates under uniaxial compressive loads ($a/b = 1$, $a/h = 10$, $E_1/E_2 = 25$, $G_{12} = G_{13} = 0.5E_2$, $\nu_{12} = 0.25$). The side-to-thickness ratio for the laminates in Figure 9.3–7 is taken to be $a/h = 10$.

Table 9.3–3: Effect of shear deformation on nondimensionalized critical buckling loads, $\bar{N} = N_{cr}(a^2/E_2h^3)$, of simply supported (SS–2) antisymmetric angle-ply, $(-45/45)_n$, square plates ($G_{12} = G_{13} = 0.5E_2$, $G_{23} = 0.2E_2$, $\nu_{12} = 0.25$, $K = 5/6$).

	$E_1/E_2 = 10$		$E_1/E_2 = 25$		$E_1/E_2 = 40$	
a/h	$n = 2$	$n = 8$	$n = 2$	$n = 8$	$n = 2$	$n = 8$
Uniaxial Compression ($k = 0$); mode: (1,1)						
10	7.847	13.542	12.231	21.082†	15.774†	24.514†
20	8.727	16.397	14.513	34.990	19.861	50.644
100	9.052	17.584	15.435	40.875	21.628	63.974
CLPT	9.066	17.637	15.476	41.163	21.709	64.683
Biaxial Compression ($k = 1$); mode: (1,1)						
10	3.923	6.771	6.115	12.067	7.910	15.336
20	4.364	8.199	7.257	17.495	9.930	25.322
100	4.526	8.792	7.717	20.437	10.810	31.987
CLPT	4.533	8.818	7.738	20.581	10.854	32.341

† Mode is (2,1).

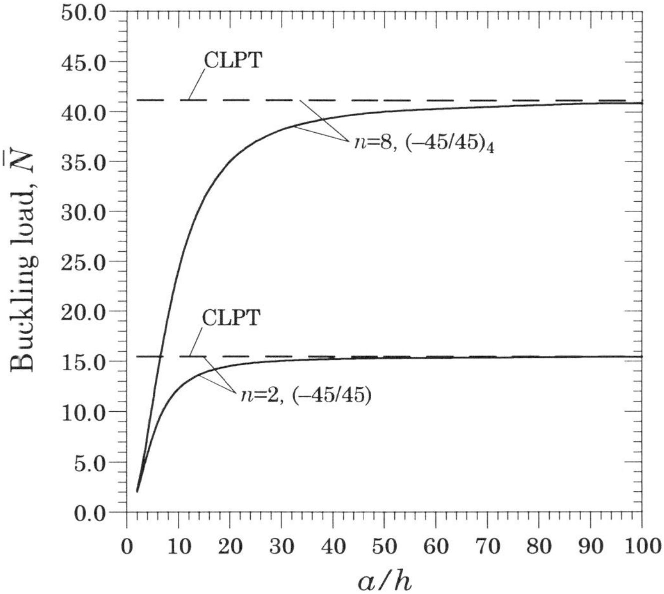

Figure 9.3–6: Critical buckling load ($\bar{N}$) versus side-to-thickness ratio (a/h) for simply supported (SS–2) antisymmetric angle-ply $(-45/45)_n$ square laminates.

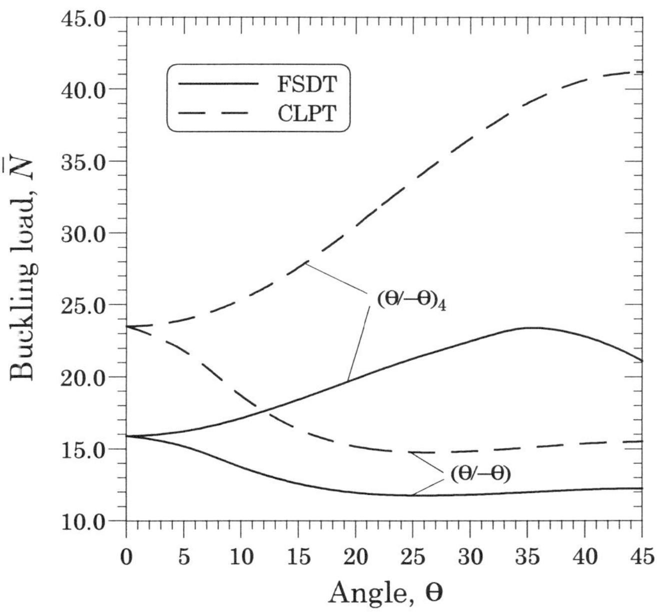

Figure 9.3–7: Critical buckling load ($\bar{N}$) versus lamination angle (θ) for simply supported (SS–2) antisymmetric angle-ply $(-\theta/\theta)_n$ laminates.

9.3–5 Vibration

Numerical results of nondimensionalized fundamental frequencies of antisymmetric angle-ply laminates (–45/45/–45/···) are presented in Table 9.3–4 for two different materials. Numerical results for two-layer (−45/45) and eight-layer $(-45/45)_4$ plates with $E_1/E_2 = 25$, $G_{12} = G_{13} = 0.5E_2$, $G_{23} = 0.2E_2$, $\nu_{12} = 0.25$, and $K = 5/6$ are given as a function of side-to-thickness ratio in Figure 9.3–8 and as a function of lamination angle in Figure 9.3–9. The effect of bending–stretching coupling ($i.e., B_{16}$ and B_{26}), transverse shear deformation ($i.e., \varepsilon_{xz} \neq 0$ and $\varepsilon_{yz} \neq 0$), and rotary inertia is to lower the fundamental frequencies. As the number of layers increases, the coupling decreases. The effect of shear deformation decreases with increasing values of a/h. This decrease is slower for eight-layer plates than two-layer plates. The effect of rotary inertia is negligible in FSDT, whereas it is significant in CLPT only for very thick plates.

Table 9.3–4: Effect of shear deformation on nondimensionalized natural frequencies of simply supported (SS–2) symmetric angle-ply (–45/45/–45/···) square plates [$\bar{\omega} = \omega(a^2/h)\sqrt{\rho/E_2}$, $K = 5/6$; rotary inertia is included; mode: (1,1)].

		Material 1		Material 2	
a/h	Theory	$n = 2$	$n = 8$	$n = 2$	$n = 8$
5	FSDT	8.498	10.285	10.799	12.893
	CLPT	11.737	15.708	14.508	17.207
10	FSDT	10.895	15.388	13.629	19.289
	CLPT	12.195	19.986	15.087	25.052
20	FSDT	11.933	18.555	14.815	23.259
	CLPT	12.317	20.113	15.242	25.212
100	FSDT	12.341	20.084	15.274	25.176
	CLPT	12.357	20.154	15.292	26.398

Material 1: $E_1 = 25E_2$, $G_{12} = G_{13} = 0.5E_2$, $G_{23} = 0.2E_2$, $\nu_{12} = 0.25$.

Material 2: $E_1 = 40E_2$, $G_{12} = G_{13} = 0.6E_2$, $G_{23} = 0.5E_2$, $\nu_{12} = 0.25$.

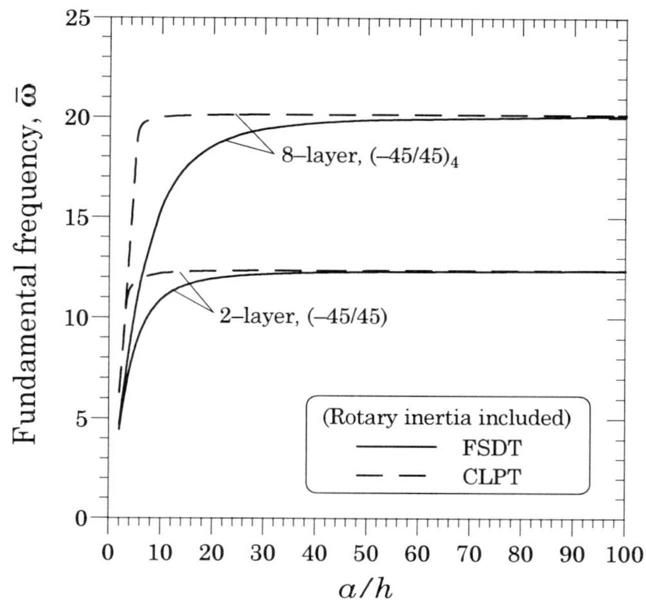

Figure 9.3–8: Nondimensionalized fundamental frequency $\bar{\omega}$ versus side-to-thickness ratio (a/h) for simply supported (SS–2) antisymmetric angle-ply $(-45/45)_n$ square laminates.

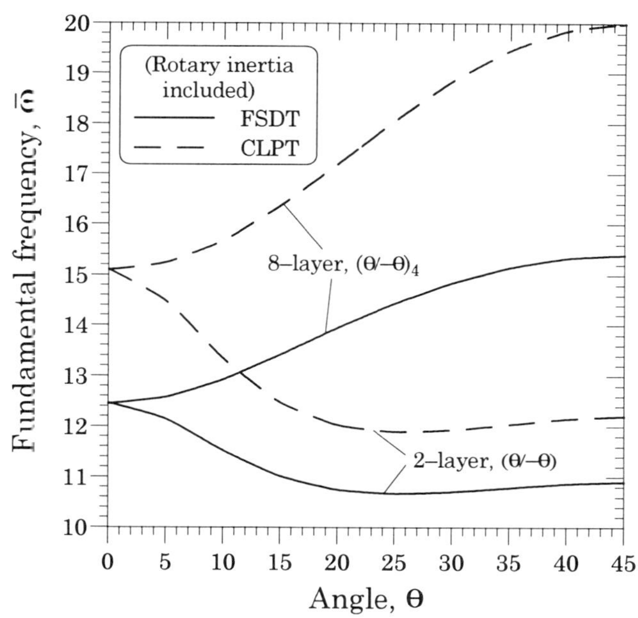

Figure 9.3–9: Nondimensionalized fundamental frequency $(\bar{\omega})$ versus lamination angle (θ) for simply supported (SS–2) antisymmetric angle-ply $(-\theta/\theta)_n$ square laminates.

9.4 Antisymmetric Cross-Ply Laminates with Two Opposite Edges Simply Supported

9.4–1 Introduction

In this section we present the Lévy type solutions for bending, natural vibration, and buckling of antisymmetric cross-ply laminates. In the interest of brevity, the discussion is limited to the bending case. For additional details and for a discussion of the free vibration and buckling analyses, the reader may consult References 8 and 23.

As described earlier, the Lévy solution technique involves choosing a solution form that satisfies the simply supported (SS–1) boundary conditions on two parallel edges of a rectangular laminate, and then the partial differential equations of equilibrium are reduced to ordinary differential equations in the coordinate parallel to the simply supported edges. The ordinary differential equations are then solved using the state–space approach.

Suppose that the edges $y = 0$ and $y = b$ are simply supported (SS–1), while the remaining edges $x = a/2$ and $x = -a/2$ have any combination of free, clamped, and simply supported boundary conditions (see Figure 9.4–1). We now proceed to describe the procedure for bending of cross-ply laminates. For additional details, the reader may consult the references at the end of the chapter.

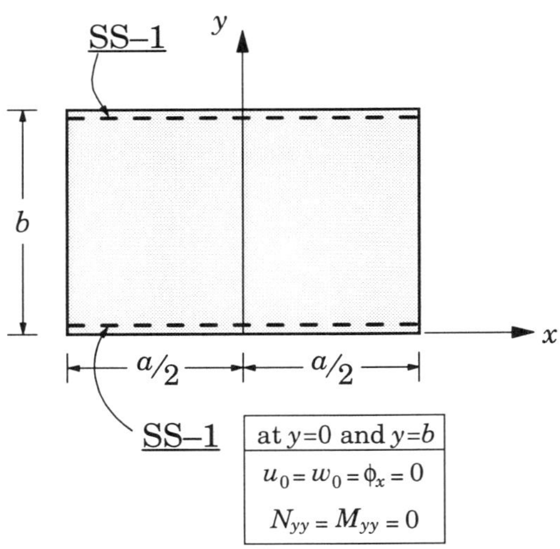

Figure 9.4–1: The coordinate system and boundary conditions used on the simply supported (SS–1) edges for the Lévy solutions of rectangular cross-ply laminates using the first-order shear deformation theory.

The equations of equilibrium appropriate for the antisymmetric cross-ply laminated plates, according to the first-order shear deformation plate theory, can be expressed in matrix form as

$$[L]\{\Delta\} = \{F\} \tag{9.4-1}$$

where $\{\Delta\}^{\mathrm{T}} = \{u_0, v_0, w_0, \phi_x, \phi_y\}$, $\{F\}^{\mathrm{T}} = \{0, 0, q, 0, 0\}$, and the coefficients $L_{ij} = L_{ji}$ are defined as

$$\begin{aligned}
&L_{11} = A_{11}d_x^2 + A_{66}d_y^2,\ L_{12} = (A_{12} + A_{66})d_xd_y,\ L_{13} = 0\\
&L_{14} = B_{11}d_x^2 + B_{66}d_y^2,\ L_{15} = (B_{12} + B_{66})d_xd_y,\ L_{22} = A_{66}d_x^2 + A_{22}d_y^2\\
&L_{23} = 0,\ L_{24} = L_{15},\ L_{25} = B_{22}d_y^2 + B_{66}d_x^2,\ L_{33} = -KA_{55}d_x^2 - KA_{44}d_y^2\\
&L_{34} = -KA_{55}d_x,\ L_{35} = -KA_{44}d_y,\ L_{44} = -KA_{55} + D_{11}d_x^2 + D_{66}d_y^2\\
&L_{45} = (D_{12} + D_{66})d_xd_y,\ L_{55} = -KA_{44} + D_{22}d_y^2 + D_{66}d_x^2
\end{aligned} \tag{9.4-2}$$

and

$$d_x^i = \frac{\partial^i}{\partial x^i},\quad d_y^i = \frac{\partial^i}{\partial y^i} \tag{9.4-3}$$

Note that the classical plate theory can be obtained as a special case of the first-order shear deformation theory by setting

$$\phi_x = -\frac{\partial w_0}{\partial x} \quad \text{and} \quad \phi_y = -\frac{\partial w_0}{\partial y} \tag{9.4-4}$$

9.4–2 The Lévy Type Solution

The Lévy method, in conjunction with the state–space concept, can be used to develop analytical solutions of Eq. (1) when the plate is simply supported on the edges $y = 0, b$ and the remaining edges $x = \pm a/2$ have any boundary conditions. The generalized displacements are expressed as products of undetermined functions and known trigonometric functions so as to satisfy the simply supported boundary conditions at $y = 0, b$ (see Figure 9.4–1):

$$u_0 = w_0 = \phi_x = N_{yy} = M_{yy} = 0 \tag{9.4-5}$$

The displacement field is represented as

$$\begin{Bmatrix} u_0(x,y)\\ v_0(x,y)\\ w_0(x,y)\\ \phi_x(x,y)\\ \phi_y(x,y) \end{Bmatrix} = \sum_{m=1}^{\infty} \begin{Bmatrix} U_m(x)\sin\beta y\\ V_m(x)\cos\beta y\\ W_m(x)\sin\beta y\\ X_m(x)\sin\beta y\\ Y_m(x)\cos\beta y \end{Bmatrix} \tag{9.4-6}$$

where $\beta = m\pi/b$. It can be easily verified that the boundary conditions (5) are satisfied by the displacement field in (6). The transverse load is also expanded as

$$q(x,y) = \sum_{m=1}^{\infty} Q_m \sin\frac{m\pi y}{b} \tag{9.4-7}$$

Substitution of the displacement field (6) into governing equations (1) results in five ordinary differential equations

$$\begin{aligned}
U''_m &= C_1 U_m + C_2 V'_m + C_3 W'_m + C_4 X_m + C_5 Y'_m \\
V''_m &= C_6 U'_m + C_7 V_m + C_8 W_m + C_9 X'_m + C_{10} Y_m \\
W''_m &= C_{11} W_m + C_{12} X'_m + C_{13} Y_m - Q_m/e_{13} \\
X''_m &= C_{14} U_m + C_{15} V'_m + C_{16} W'_m + C_{17} X_m + C_{18} Y'_m \\
Y''_m &= C_{19} U'_m + C_{20} V_m + C_{21} W_m + C_{22} X'_m + C_{23} Y_m
\end{aligned} \tag{9.4-8}$$

where the primes denote the derivative with respect to x. The coefficients in Eq. (8) are given by

$$\begin{aligned}
C_1 &= (e_3 e_{21} - e_5 e_{19})/e_0, \quad C_2 = (e_3 e_{18} - e_2 e_{19})/e_0 \\
C_3 &= e_3 e_{23}/e_0, \quad C_4 = (e_3 e_{22} - e_6 e_{19})/e_0 \\
C_5 &= (e_3 e_{20} - e_4 e_{19})/e_0, \quad C_6 = (e_8 e_{27} - e_{10} e_{25})/C_0 \\
C_7 &= (e_{11} e_{27} - e_{10} e_{28})/C_0, \quad C_8 = -e_{10} e_{30}/C_0 \\
C_9 &= (e_9 e_{27} - e_{10} e_{26})/C_0, \quad C_{10} = (e_{12} e_{27} - e_{10} e_{29})/C_0 \\
C_{11} &= -e_{15}/e_{13}, \quad C_{12} = -e_{14}/e_{13}, \quad C_{13} = -e_{16}/e_{13} \\
C_{14} &= (e_5 e_{17} - e_1 e_{21})/e_0, \quad C_{15} = (e_2 e_{17} - e_1 e_{18})/e_0, \quad C_{16} = -e_1 e_{23}/e_0 \\
C_{17} &= (e_6 e_{17} - e_1 e_{22})/e_0, \quad C_{18} = (e_4 e_{17} - e_1 e_{20})/e_0 \\
C_{19} &= (e_7 e_{25} - e_8 e_{24})/C_0, \quad C_{20} = (e_7 e_{28} - e_{11} e_{24})/C_0 \\
C_{21} &= e_7 e_{30}/C_0, \quad C_{22} = (e_7 e_{26} - e_9 e_{24})/C_0, \quad C_{23} = (e_7 e_{29} - e_{12} e_{24})/C_0 \\
e_0 &= e_1 e_{19} - e_3 e_{17}, \quad C_0 = e_{10} e_{24} - e_7 e_{27}
\end{aligned} \tag{9.4-9}$$

$$\begin{aligned}
e_1 &= A_{11}, \quad e_2 = -\beta(A_{12} + A_{66}), \quad e_3 = B_{11}, \quad e_4 = -\beta(B_{12} + B_{66}) \\
e_5 &= -\beta^2 A_{66}, \quad e_6 = -\beta^2 B_{66}, \quad e_7 = A_{66}, \quad e_8 = -e_2 \\
e_9 &= -e_4, \quad e_{10} = B_{66}, \quad e_{11} = -\beta^2 A_{22}, \quad e_{12} = -\beta^2 B_{22} \\
e_{13} &= KA_{55}, \quad e_{14} = KA_{55}, \quad e_{15} = -\beta^2 KA_{44} \\
e_{16} &= -\beta KA_{44}, \quad e_{17} = e_3, \quad e_{18} = e_4, \quad e_{19} = D_{11}, \quad e_{20} = -\beta(D_{11} + D_{66}) \\
e_{21} &= -\beta^2 B_{66}, \quad e_{22} = -\beta^2 D_{66} - KA_{55}, \quad e_{23} = -e_{14} \\
e_{24} &= e_{10}, \quad e_{25} = -e_4, \quad e_{26} = -e_{20}, \quad e_{27} = D_{66}, \quad e_{28} = -\beta^2 B_{22} \\
e_{29} &= -\beta^2 D_{22} - KA_{44}, \quad e_{30} = e_{16}
\end{aligned} \tag{9.4-10}$$

In order to reduce the system of equations (8) to a (matrix) system of first-order equations (*i.e.*, use the state–space approach), the components of the state vector $\mathbf{Z}(x)$ are defined as

$$Z_1 = U_m, \quad Z_2 = U'_m, \quad Z_3 = V_m, \quad Z_4 = V'_m, \quad Z_5 = W_m, \quad Z_6 = W'_m$$
$$Z_7 = X_m, \quad Z_8 = X'_m, \quad Z_9 = Y_m, \quad Z_{10} = Y'_m \tag{9.4-11}$$

Using the definitions (11), the systems of equations (8) may be converted to the form

$$\mathbf{Z}' = \mathbf{TZ} + \mathbf{r} \tag{9.4-12}$$

where the matrix $\mathbf{T}$ is the 10×10 matrix

$$[T] = \begin{bmatrix} 0 & 1 & 0 & 0 & 0 & 0 & 0 & 0 & 0 & 0 \\ C_1 & 0 & 0 & C_2 & 0 & C_3 & C_4 & 0 & 0 & C_5 \\ 0 & 0 & 0 & 1 & 0 & 0 & 0 & 0 & 0 & 0 \\ 0 & C_6 & C_7 & 0 & C_8 & 0 & 0 & C_9 & C_{10} & 0 \\ 0 & 0 & 0 & 0 & 0 & 1 & 0 & 0 & 0 & 0 \\ 0 & 0 & 0 & 0 & C_{11} & 0 & 0 & C_{12} & C_{13} & 0 \\ 0 & 0 & 0 & 0 & 0 & 0 & 0 & 1 & 0 & 0 \\ C_{14} & 0 & 0 & C_{15} & 0 & C_{16} & C_{17} & 0 & 0 & C_{18} \\ 0 & 0 & 0 & 0 & 0 & 0 & 0 & 0 & 0 & 1 \\ 0 & C_{19} & C_{20} & 0 & C_{21} & 0 & 0 & C_{22} & C_{23} & 0 \end{bmatrix} \tag{9.4-13a}$$

and the load vector $\mathbf{r}$ is defined as

$$\{r\} = \{0, 0, 0, 0, 0, -Q_m/e_{13}, 0, 0, 0, 0\}^{\mathrm{T}} \tag{9.4-13b}$$

$$Q_m(x) = \frac{2}{b}\int^b q(x,y)\sin\frac{m\pi y}{b}\,dy \tag{9.4-13c}$$

The solution to Eq. (12) is

$$\mathbf{Z} = e^{Tx}\{\mathbf{K} + \int^x e^{-T\zeta}\,\mathbf{r}\,d\zeta\} \tag{9.4-14}$$

Here $\mathbf{K}$ denotes constant column vector, which is to be determined from the boundary conditions on edges $x = \pm a/2$. The simply supported (S), clamped (C), and free (F) boundary conditions at the edges $x = \pm a/2$ can be expressed as

$$S: \; v_0 = w_0 = \phi_y = N_{xx} = M_{xx} = 0$$
$$C: \; u_0 = v_0 = w_0 = \phi_x = \phi_y = 0$$
$$F: \; N_{xx} = M_{xx} = Q_x = N_{xy} = M_{xy} = 0 \tag{9.4-15}$$

As described in Section 8.5, Eq. (12) can be used to obtain ten equations for the ten constants K_i by these boundary conditions.

The same procedure can be used to study natural vibration and buckling under in-plane compressive forces. The procedure was discussed earlier for the eigenvalue problems in Chapter 8 (see Section 8.5). Also, see References 8,9, and 15–27 at the end of the chapter.

9.4–3 Numerical Examples

Here we present numerical results for a number of example problems of bending, vibration, and buckling of rectangular, cross-ply laminated plates. For the purpose of comparison, the following two sets of lamina properties, typical of graphite–epoxy material, are used:

$$\textit{Material 1}: \quad E_1 = 25E_2, \quad G_{12} = G_{13} = 0.5E_2, \quad G_{23} = 0.2E_2, \quad \nu_{12} = 0.25 \tag{9.4-16}$$

$$\textit{Material 2}: \quad E_1 = 40E_2, \quad G_{12} = G_{13} = 0.6E_2, \quad G_{23} = 0.5E_2, \quad \nu_{12} = 0.25 \tag{9.4-17}$$

The shear correction coefficient for the first-order theory is taken to be $K = 5/6$. The notation SC, for example, refers to the boundary conditions used on the edges $x = \pm a/2$, while the other two edges (*i.e.*, $y = 0, b$), in all cases, are simply supported.

Bending

The loading in all cases considered here is assumed to be sinusoidal

$$q(x, y) = q_0 \cos \frac{\pi x}{a} \sin \frac{\pi y}{b} \tag{9.4-18}$$

In the tables and figures, the results for deflections and stresses are presented using the following nondimensional form (see Khdeir and Reddy [23]):

$$\bar{w} = w_0(0, b/2) \left(\frac{E_2 h^3}{b^4 q_0}\right) \times 10^2, \quad \bar{\sigma}_{xx} = -\sigma_{xx}(0, b/2, -h/2) \left(\frac{h^2}{b^2 q_0}\right) 10$$

$$\bar{\sigma}_{yy} = \sigma_{yy}(0, b/2, h/2) \left(\frac{h^2}{b^2 q_0}\right) 10, \quad \bar{\sigma}_{yz} = \sigma_{yz}(0, 0, 0) \left(\frac{h}{b q_0}\right) 10 \tag{9.4-19}$$

where h is the total thickness of the laminate.

Figures 9.4–2 and 9.4–3 contain plots of deflections versus side-to-thickness ratio b/h of two-layer and ten-layer antisymmetric $(0/90/\ldots)_{even}$ cross-ply laminates ($a = b/2$) with various boundary conditions (see Khdeir and Reddy [23]). The material properties used are those listed in Eq. (16) (*i.e.*, material 1). The classical laminate theory always underpredicts deflections because the plate is modeled as infinitely stiff through the thickness. Figures 9.4–4 and 9.4–5 contain plots of $\bar{w}$ vs. E_1/E_2 for the same load, $b/h = 10$, and $a = b/2$. As the degree of orthotropy increases, the difference between the deflections predicted by the classical and the first-order shear deformation theories increases, indicating that the shear deformation effect is more significant in anisotropic plates. Tables 9.4–1 through 9.4–4 contain numerical values of deflections and stresses in square plates obtained using the Lévy method.

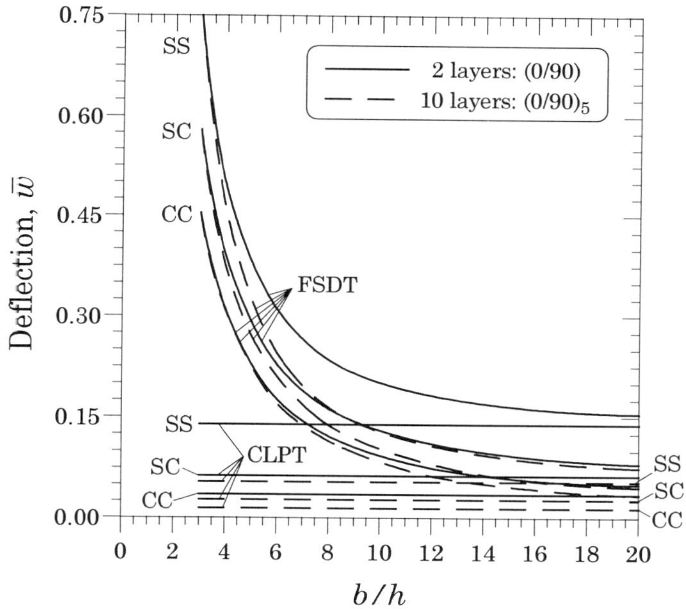

Figure 9.4–2: Nondimensionalized center transverse deflection ($\bar{w}$) versus side-to-thickness ratio (b/h) for antisymmetric cross-ply $(0/90)_n$ ($n = 1, 5$) laminates (Material 1, $b/a = 2$).

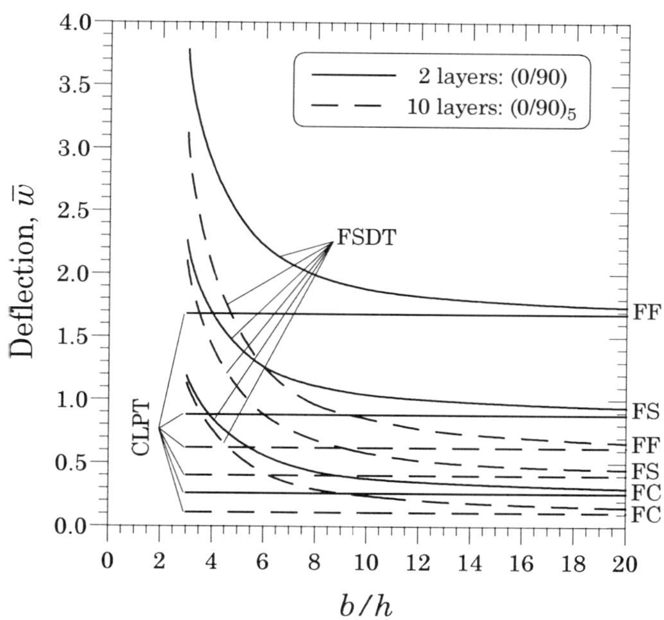

Figure 9.4–3: Nondimensionalized center transverse deflection ($\bar{w}$) versus side-to-thickness ratio (b/h) for antisymmetric cross-ply $(0/90)_n$ laminates (Material 1, $b/a = 2$).

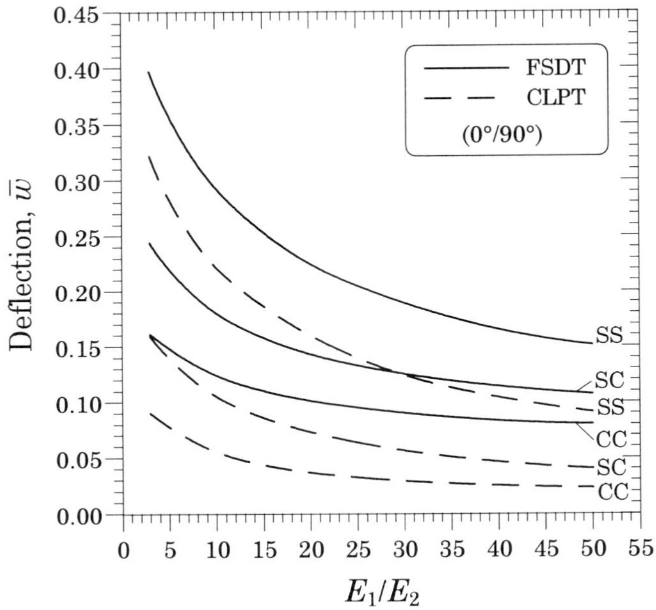

Figure 9.4–4: Nondimensionalized center transverse deflection ($\bar{w}$) versus modulus ratio (E_1/E_2) for antisymmetric cross-ply (0/90) laminates (Material 1, $b/h = 10$, $b/a = 2$).

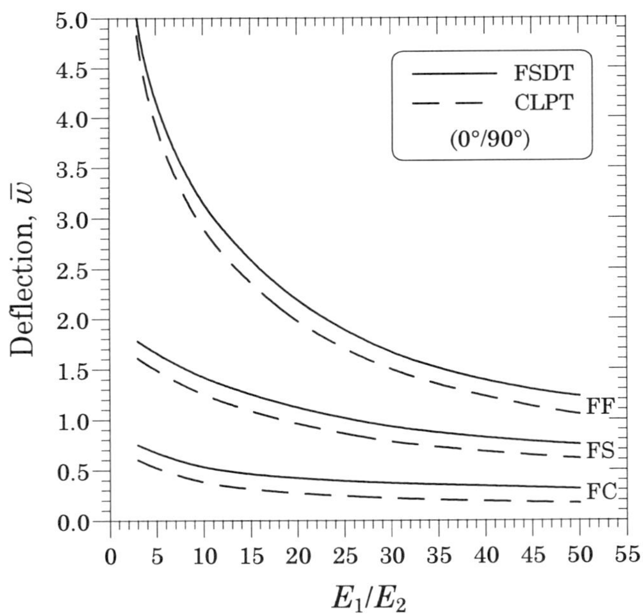

Figure 9.4–5: Nondimensionalized center transverse deflection ($\bar{w}$) versus modulus ratio (E_1/E_2) for antisymmetric cross-ply (0/90) laminates (Material 1, $b/h = 10$, $b/a = 2$).

Table 9.4–1: Nondimensionalized center deflection ($\bar{w}$) of antisymmetric cross-ply square plates with various boundary conditions (Material 1, $b/a = 1$).

Layers	$\frac{b}{h}$	Theory	SS	SC	CC	FF	FS	FC
2	5	FSDT	1.758	1.477	1.257	2.777	2.335	1.897
	10	FSDT	1.237	0.883	0.656	2.028	1.687	1.223
		CLPT†	1.064	0.664	0.429	1.777	1.471	0.980
10	5	FSDT	1.137	1.045	0.945	1.663	1.460	1.258
	10	FSDT	0.615	0.480	0.385	0.915	0.800	0.612
		CLPT†	0.442	0.266	0.167	0.665	0.579	0.380

† Results are independent of b/h.

Table 9.4–2: Nondimensionalized axial stress ($\bar{\sigma}_{xx}$) of antisymmetric cross-ply square plates with various boundary conditions (Material 1, $b/a = 1$).

Layers	$\frac{b}{h}$	Theory	SS	SC	CC	FF	FS	FC
2	5	FSDT	7.157	5.338	3.911	2.469	4.430	2.434
	10	FSDT	7.157	5.494	4.450	2.442	4.435	2.790
		CLPT†	7.157	5.660	4.800	2.403	4.442	3.042
10	5	FSDT	5.009	3.707	2.275	1.712	2.957	1.343
	10	FSDT	5.009	3.642	2.692	1.723	2.968	1.594
		CLPT†	5.009	3.829	3.167	1.725	2.986	1.865

† Results are independent of b/h.

Table 9.4–3: Nondimensionalized transverse shear stress ($\bar{\sigma}_{yz}$) of antisymmetric cross-ply square plates with various boundary conditions (Material 1, $b/a = 1$).

Layers	b/h	SS	SC	CC	FF	FS	FC
2	5	2.729	2.297	1.958	3.901	3.390	2.748
	10	2.729	1.993	1.523	3.882	3.383	2.449
10	5	2.729	2.498	2.248	3.883	3.437	2.951
	10	2.729	2.126	1.708	3.853	3.421	2.605

Table 9.4–4: Nondimensionalized axial stress ($\bar{\sigma}_{yy}$) of antisymmetric cross-ply square plates with various boundary conditions (Material 1, $b/a = 1$).

Layers	$\frac{b}{h}$	Theory	SS	SC	CC	FF	FS	FC
2	5	FSDT	7.157	6.034	5.153	11.907	9.848	8.047
	10	FSDT	7.157	5.109	3.799	11.884	9.847	7.150
		CLPT†	7.157	4.483	2.914	11.849	9.837	6.560
10	5	FSDT	5.009	4.628	4.212	7.583	6.590	5.706
	10	FSDT	5.009	3.904	3.135	7.533	6.566	5.029
		CLPT†	5.009	3.025	1.911	7.480	6.531	4.284

† Results are independent of b/h.

Vibration and Buckling

The effect of orthotropy and number of layers (*i.e.*, bending–stretching coupling) on the fundamental frequencies of simply supported, cross-ply, square laminates can be seen from the results presented in Table 9.4–5 (see Reddy and Khdeir [8]). The fundamental frequencies increase with an increase in degree of orthotropy and number of layers (or decrease of coupling). Similar results for critical buckling loads under uniaxial compression are included in Table 9.4–6.

Table 9.4–5: Effect of degree of orthotropy of the individual layers on the dimensionless fundamental frequency of simply supported antisymmetric cross-ply square laminates: $a/h = 5, \bar{\omega} = \omega\sqrt{\rho h^2/E_2}$ (Material 2).

Theory	Layers	$\frac{E_1}{E_2} = 3$	10	20	30	40
FSDT	2	0.24834	0.27757	0.30824	0.33284	0.35333
CLPT		0.27082	0.30968	0.35422	0.39335	0.42884
FSDT	4	0.26017	0.32898	0.38754	0.42479	0.45083
CLPT		0.28676	0.38877	0.49907	0.58900	0.66690
FSDT	6	0.26228	0.33673	0.39771	0.43531	0.46105
CLPT		0.28966	0.40215	0.52234	0.61963	0.70359
FSDT	10	0.26335	0.34053	0.40255	0.44023	0.46577
CLPT		0.29115	0.40888	0.53397	0.63489	0.72184

Table 9.4–6: Effect of degree of orthotropy of the individual layers on the dimensionless critical buckling loads, $\bar{N} = N_{cr}b^2/(E_2h^3)$, of simply supported (SS–1) antisymmetric cross-ply square laminates ($a/h = 10$) under uniaxial compression (Material 2).

Theory	Layers	$\frac{E_1}{E_2} = 3$	10	20	30	40
FSDT	2	4.7718	6.2465	8.0423	9.7347	11.353
CLPT		5.0338	6.7033	8.8158	10.891	12.957
FSDT	4	5.2543	9.2552	14.332	18.815	22.806
CLPT		5.5738	10.295	16.988	23.675	30.359
FSDT	6	5.343	9.7893	15.394	20.280	24.577
CLPT		5.674	10.960	18.502	26.042	33.582
FSDT	10	5.3884	10.060	15.927	21.008	25.450
CLPT		5.725	11.300	19.277	27.254	35.232

The effect of transverse shear deformation and boundary conditions on the fundamental frequencies of two-layer and ten-layer antisymmetric cross-ply laminates ($a/h = 10$) are examined in Table 9.4–7. The critical buckling loads for the same laminates under uniaxial compression are presented in Table 9.4–8. In all cases, the classical plate theory overpredicts frequencies and buckling loads.

Table 9.4–7: Effect of number of layers and transverse shear deformation on the dimensionless frequencies $\bar{\omega} = (\omega b^2/h)(\rho/E_2)^{1/2}$ of antisymmetric cross-ply square plates ($a/h = 10$) with various boundary conditions (Material 2).

Layers	Theory	FF	FS	FC	SS	SC	CC
2	FSDT	6.881	7.215	7.741	10.473	12.610	15.152
	CLPT	7.267	7.636	8.228	11.154	14.223	18.543
10	FSDT	10.900	11.079	11.862	15.779	18.044	20.471
	CLPT	12.680	12.906	13.779	18.492	23.971	31.709

Table 9.4–8: Effect of number of layers and transverse shear deformation on dimensionless critical buckling loads, $\bar{N} = N_{cr}b^2/E_2h^3$, of antisymmetric cross-ply square plates ($a/h = 10$) with various boundary conditions (Material 2).

Layers	Theory	FF	FS	FC	SS	SC	CC
2	FSDT	4.851	5.351	6.166	11.353	16.437	20.067
	CLPT	5.425	6.003	6.968	12.957	21.116	31.280
10	FSDT	12.092	12.524	14.358	25.450	32.614	34.837
	CLPT	16.426	17.023	19.389	35.232	59.288	89.770

9.5 Antisymmetric Angle-Ply Laminates with Two Opposite Edges Simply Supported

9.5–1 Introduction

As in the case of classical laminate theory, the Lévy–type solutions of the first-order theory can be developed for bending, buckling, and natural vibrations of antisymmetric angle-ply laminated rectangular plates with two opposite edges simply supported and the remaining ones subjected to a combination of clamped, simply supported, and free boundary conditions. In this section, we present the Lévy solution procedure for natural vibration and buckling analyses.

9.5–2 Governing Equations

Consider a rectangular laminated plate composed of an even number of identical layers having the principal material directions of orthotropy oriented at angles $+\theta$ and $-\theta$ with respect to the x–axis of the laminate (*i.e.*, antisymmetric angle-ply laminates). The laminates exhibit twisting–extensional coupling, and the differential equations (9.1–1)–(9.1–5) associated with the first-order theory take the form

$$\begin{aligned}
&\frac{\partial}{\partial x}\left[A_{11}\frac{\partial u_0}{\partial x} + A_{12}\frac{\partial v_0}{\partial y} + B_{16}\left(\frac{\partial \phi_x}{\partial y} + \frac{\partial \phi_y}{\partial x}\right)\right] + \\
&\frac{\partial}{\partial y}\left[A_{66}\left(\frac{\partial u_0}{\partial y} + \frac{\partial v_0}{\partial x}\right) + B_{16}\frac{\partial \phi_x}{\partial x} + B_{26}\frac{\partial \phi_y}{\partial y}\right] \\
&\quad - \left(\frac{\partial N_{xx}^T}{\partial x} + \frac{\partial N_{xy}^T}{\partial y}\right) = I_0\frac{\partial^2 u_0}{\partial t^2}
\end{aligned} \tag{9.5 – 1}$$

$$\frac{\partial}{\partial x}\left[A_{66}\left(\frac{\partial u_0}{\partial y}+\frac{\partial v_0}{\partial x}\right)+B_{16}\frac{\partial \phi_x}{\partial x}+B_{26}\frac{\partial \phi_y}{\partial y}\right]$$
$$+\frac{\partial}{\partial y}\left[A_{12}\frac{\partial u_0}{\partial x}+A_{22}\frac{\partial v_0}{\partial y}+B_{26}\left(\frac{\partial \phi_x}{\partial y}+\frac{\partial \phi_y}{\partial x}\right)\right]$$
$$-\left(\frac{\partial N_{xy}^T}{\partial x}+\frac{\partial N_{yy}^T}{\partial y}\right)=I_0\frac{\partial^2 v_0}{\partial t^2} \qquad (9.5-2)$$

$$\frac{\partial}{\partial x}\left[KA_{55}\left(\frac{\partial w_0}{\partial x}+\phi_x\right)\right]+\frac{\partial}{\partial y}\left[KA_{44}\left(\frac{\partial w_0}{\partial y}+\phi_y\right)\right]$$
$$+\hat{N}_{xx}\frac{\partial^2 w_0}{\partial x^2}+2\hat{N}_{xy}\frac{\partial^2 w_0}{\partial x\partial y}+\hat{N}_{yy}\frac{\partial^2 w_0}{\partial y^2}+q=I_0\frac{\partial^2 w_0}{\partial t^2} \qquad (9.5-3)$$

$$\frac{\partial}{\partial x}\left[B_{16}\left(\frac{\partial u_0}{\partial y}+\frac{\partial v_0}{\partial x}\right)+D_{11}\frac{\partial \phi_x}{\partial x}+D_{12}\frac{\partial \phi_y}{\partial y}\right]+$$
$$\frac{\partial}{\partial y}\left[B_{16}\frac{\partial u_0}{\partial x}+B_{26}\frac{\partial v_0}{\partial y}+D_{66}\left(\frac{\partial \phi_x}{\partial y}+\frac{\partial \phi_y}{\partial x}\right)\right]-$$
$$KA_{55}\left(\frac{\partial w_0}{\partial x}+\phi_x\right)-\left(\frac{\partial M_{xx}^T}{\partial x}+\frac{\partial M_{xy}^T}{\partial y}\right)=I_2\frac{\partial^2 \phi_x}{\partial t^2} \qquad (9.5-4)$$

$$\frac{\partial}{\partial x}\left[B_{16}\frac{\partial u_0}{\partial x}+B_{26}\frac{\partial v_0}{\partial y}+D_{66}\left(\frac{\partial \phi_x}{\partial y}+\frac{\partial \phi_y}{\partial x}\right)\right]+$$
$$\frac{\partial}{\partial y}\left[B_{26}\left(\frac{\partial u_0}{\partial y}+\frac{\partial v_0}{\partial x}\right)+D_{12}\frac{\partial \phi_x}{\partial y}+D_{22}\frac{\partial \phi_y}{\partial y}\right]-$$
$$KA_{44}\left(\frac{\partial w_0}{\partial y}+\phi_y\right)-\left(\frac{\partial M_{xy}^T}{\partial x}+\frac{\partial M_{yy}^T}{\partial y}\right)=I_2\frac{\partial^2 \phi_y}{\partial t^2} \qquad (9.5-5)$$

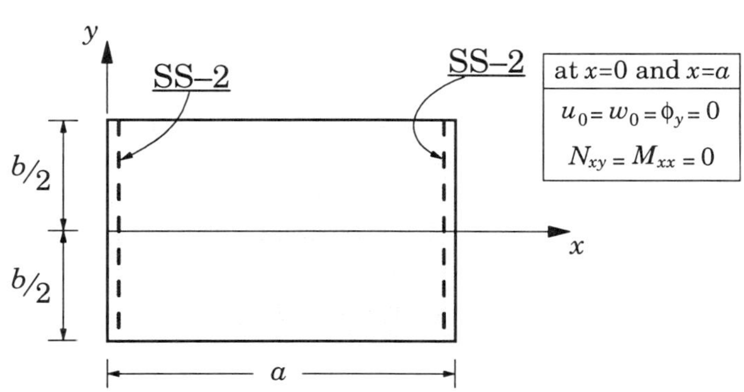

Figure 9.5–1: Boundary conditions used on simply supported edges for the Lévy solutions of rectangular angle-ply laminates (FSDT).

The following boundary conditions are considered (see Figure 9.5–1):

Simply supported (SS–1) at edges $x = 0, a$:

$$u_0 = w_0 = \phi_y = M_{xx} = N_{xy} = 0 \qquad (9.5-6a)$$

Simply supported (SS–1) at edges $y = \pm b/2$:

$$v_0 = w_0 = \phi_x = M_{yy} = N_{xy} = 0 \qquad (9.5-6b)$$

Clamped (C) at edges $y = \pm b/2$:

$$u_0 = v_0 = w_0 = \phi_x = \phi_y = 0 \qquad (9.5-7a)$$

Free (F) at edges $y = \pm b/2$:

$$M_{yy} = M_{xy} = N_{yy} = N_{xy} = Q_y = 0 \qquad (9.5-7b)$$

where the stress resultants can be expressed in terms of the generalized displacements as

$$N_{xx} = A_{11}\frac{\partial u_0}{\partial x} + A_{12}\frac{\partial v_0}{\partial y} + B_{16}(\frac{\partial \phi_x}{\partial y} + \frac{\partial \phi_y}{\partial x})$$

$$N_{yy} = A_{12}\frac{\partial u_0}{\partial x} + A_{22}\frac{\partial v_0}{\partial y} + B_{26}(\frac{\partial \phi_x}{\partial y} + \frac{\partial \phi_y}{\partial x})$$

$$N_{xy} = A_{66}(\frac{\partial u_0}{\partial y} + \frac{\partial v_0}{\partial x}) + B_{16}\frac{\partial \phi_x}{\partial x} + B_{26}\frac{\partial \phi_y}{\partial y}$$

$$M_{xx} = B_{16}(\frac{\partial u_0}{\partial y} + \frac{\partial v_0}{\partial x}) + D_{11}\frac{\partial \phi_x}{\partial x} + D_{12}\frac{\partial \phi_y}{\partial y}$$

$$M_{yy} = B_{26}(\frac{\partial u_0}{\partial y} + \frac{\partial v_0}{\partial x}) + D_{12}\frac{\partial \phi_x}{\partial x} + D_{22}\frac{\partial \phi_y}{\partial y}$$

$$M_{xy} = B_{16}\frac{\partial u_0}{\partial x} + B_{26}\frac{\partial v_0}{\partial y} + D_{66}(\frac{\partial \phi_x}{\partial y} + \frac{\partial \phi_y}{\partial x})$$

$$Q_x = KA_{55}(\frac{\partial w_0}{\partial x} + \phi_x)$$

$$Q_y = KA_{44}(\frac{\partial w_0}{\partial y} + \phi_y) \qquad (9.5-8)$$

9.5-3 The Lévy Solution

Here we present the Lévy type solution procedure in conjunction with the state–space concept to determine the compressive buckling loads of rectangular plates ($a \times b$). The edges $x = 0, a$ are assumed to be simply supported while the remaining edges, $y = \pm b/2$, having an arbitrary set of boundary conditions. The following representation of the displacement field is used:

$$\begin{aligned}
u_0(x,y) &= \sum_{n=1}^{\infty} U_m(y) \sin \alpha x \\
v_0(x,y) &= \sum_{n=1}^{\infty} V_m(y) \cos \alpha x \\
w_0(x,y) &= \sum_{n=1}^{\infty} W_m(y) \sin \alpha x \\
\phi_x(x,y) &= \sum_{n=1}^{\infty} X_m(y) \cos \alpha x \\
\phi_y(x,y) &= \sum_{n=1}^{\infty} Y_m(y) \sin \alpha x
\end{aligned} \tag{9.5 - 9}$$

where $\alpha = m\pi/a$. Substitution of Eq. (9) into Eqs. (1)–(5), after setting the inertia terms, q, and $\hat{N}_{xy}$ to zero, yields a system of ordinary differential equations in the $y-$coordinate. These equations, after some elementary algebraic manipulations, can be expressed as

$$\begin{aligned}
U''_m &= C_1 U_m + C_2 V'_m + C_3 W'_m + C_4 X'_m + C_5 Y_m \\
V''_m &= C_6 U'_m + C_7 V_m + C_8 W_m + C_9 X_m + C_{10} Y'_m \\
W''_m &= C_{11} W_m + C_{12} X_m + C_{13} Y'_m \\
X''_m &= C_{14} U'_m + C_{15} V_m + C_{16} W_m + C_{17} X_m + C_{18} Y'_m \\
Y''_m &= C_{19} U_m + C_{20} V'_m + C_{21} W'_m + C_{22} X'_m + C_{23} Y_m
\end{aligned} \tag{9.5 - 10a}$$

where the primes indicate differentiation with respect to y, and the constants C_i are defined as

$$\begin{aligned}
C_1 &= \alpha^2 \frac{A_{11} D_{22} - B_{16} B_{26}}{A_{66} D_{22} - B_{26}^2} \\
C_2 &= \alpha \frac{A_{12} D_{22} + A_{66} D_{22} - 2B_{26}^2}{A_{66} D_{22} - B_{26}^2} \\
C_3 &= -\frac{K A_{44} B_{26}}{A_{66} D_{22} - B_{26}^2} \\
C_4 &= \alpha \frac{2B_{16} D_{22} - B_{26} D_{12} - B_{26} D_{66}}{A_{66} D_{22} - B_{26}^2} \\
C_5 &= \frac{\alpha^2 B_{16} D_{22} - \alpha^2 B_{26} D_{66} - B_{26} K A_{44}}{A_{66} D_{22} - B_{26}^2} \\
C_6 &= \alpha \frac{2B_{16} B_{26} - A_{12} D_{66} - A_{66} D_{66}}{A_{22} D_{66} - B_{26}^2}
\end{aligned}$$

$$
\begin{aligned}
C_7 &= \alpha^2 \frac{A_{66}D_{66} - B_{16}B_{26}}{A_{22}D_{66} - B_{26}^2} \\
C_8 &= -\frac{\alpha K A_{55} B_{26}}{A_{22}D_{66} - B_{26}^2} \\
C_9 &= \frac{\alpha^2 B_{16}D_{66} - \alpha^2 B_{26}D_{11} - B_{26}KA_{55}}{A_{22}D_{66} - B_{26}^2} \\
C_{10} &= \alpha B_{26} \frac{D_{12} - D_{66}}{A_{22}D_{66} - B_{26}^2} \\
C_{11} &= \alpha^2 \frac{KA_{55} - \hat{N}_{xx}}{KA_{44} - \hat{N}_{yy}} \\
C_{12} &= \frac{\alpha K A_{55}}{KA_{44} - \hat{N}_{yy}} \\
C_{13} &= -\frac{\alpha K A_{44}}{KA_{44} - \hat{N}_{yy}} \\
C_{14} &= \alpha \frac{A_{12}B_{26} + A_{66}B_{26} - 2A_{22}B_{16}}{A_{22}D_{66} - B_{26}^2} \\
C_{15} &= \alpha^2 \frac{A_{22}B_{16} - A_{66}B_{26}}{A_{22}D_{66} - B_{26}^2} \\
C_{16} &= \frac{\alpha A_{22} K A_{55}}{A_{22}D_{66} - B_{26}^2} \\
C_{17} &= \frac{A_{22}KA_{55} + \alpha^2 A_{22}D_{11} - \alpha^2 B_{16}B_{26}}{A_{22}D_{66} - B_{26}^2} \\
C_{18} &= \alpha \frac{2B_{26}^2 - A_{22}D_{12} - A_{22}D_{66}}{A_{22}D_{66} - B_{26}^2} \\
C_{19} &= \alpha^2 \frac{A_{66}B_{16} - A_{11}B_{26}}{A_{66}D_{22} - B_{26}^2} \\
C_{20} &= \alpha B_{26} \frac{A_{66} - A_{12}}{A_{66}D_{22} - B_{26}^2} \\
C_{21} &= \frac{KA_{44}A_{66}}{A_{66}D_{22} - B_{26}^2} \\
C_{22} &= \alpha \frac{A_{66}D_{12} + A_{66}D_{66} - 2B_{16}B_{26}}{A_{66}D_{22} - B_{26}^2} \\
C_{23} &= \frac{KA_{44}A_{66} + \alpha^2 A_{66}D_{66} - \alpha^2 B_{16}B_{26}}{A_{66}D_{22} - B_{26}^2} \qquad (9.5-10b)
\end{aligned}
$$

Introducing the components of the state vector $\mathbf{Z} = \mathbf{Z}(y)$ as

$$
\begin{aligned}
&Z_1 = U_m, \quad Z_2 = U_m', \quad Z_3 = V_m, \quad Z_4 = V_m', \quad Z_5 = W_m \\
&Z_6 = W_m', \quad Z_7 = X_m, \quad Z_8 = X_m', \quad Z_9 = Y_m, \quad Z_{10} = Y_m' \qquad (9.5-11)
\end{aligned}
$$

Eq. (10a) may be reduced to the matrix form ($\mathbf{Z}' = \mathbf{TZ}$)

$$\{\frac{dZ}{dy}\} = [T]\{Z\} \tag{9.5-12a}$$

where the matrix $[T]$ is defined as

$$[T] = \begin{bmatrix} 0 & 1 & 0 & 0 & 0 & 0 & 0 & 0 & 0 & 0 \\ C_1 & 0 & 0 & C_2 & 0 & C_3 & 0 & C_4 & C_5 & 0 \\ 0 & 0 & 0 & 1 & 0 & 0 & 0 & 0 & 0 & 0 \\ 0 & C_6 & C_7 & 0 & C_8 & 0 & C_9 & 0 & 0 & C_{10} \\ 0 & 0 & 0 & 0 & 0 & 1 & 0 & 0 & 0 & 0 \\ 0 & 0 & 0 & 0 & C_{11} & 0 & C_{12} & 0 & 0 & C_{13} \\ 0 & 0 & 0 & 0 & 0 & 0 & 0 & 1 & 0 & 0 \\ 0 & C_{14} & C_{15} & 0 & C_{16} & 0 & C_{17} & 0 & 0 & C_{18} \\ 0 & 0 & 0 & 0 & 0 & 0 & 0 & 0 & 0 & 1 \\ C_{19} & 0 & 0 & C_{20} & 0 & C_{21} & 0 & C_{22} & C_{23} & 0 \end{bmatrix} \tag{9.5-12b}$$

A formal solution to Eq. (12a) is given by

$$\mathbf{Z}(y) = e^{\mathbf{T}y}\mathbf{K} \tag{9.5-13}$$

where $\mathbf{K}$ is a constant column vector to be determined using the boundary conditions. In the present case all eigenvalues of matrix $[T]$ are distinct. In the case of repeated eigenvalues, the Jordan canonical form must be used. Equation (13) in conjunction with the boundary conditions yields a homogeneous system of equations for the buckling problem

$$M_{ij}K_j = 0\ , \quad (i,j = 1,2,\cdots,10) \tag{9.5-14}$$

where the determinate equation

$$|M_{ij}| = 0 \tag{9.5-15}$$

allows determination of the buckling loads associated with the mth mode and with the considered edge conditions at $y = \pm b/2$.

The above solution procedure is also valid for the free vibration case, except that the elements of the operator $[T]$ should be modified to account for the inertia terms. In the case of static analysis, the elements of $[T]$ are modified by setting the in-plane force terms to zero, and the solution (12a) and equations for the determination of the constant vector $\mathbf{K}$ will be modified accordingly, as discussed in Chapter 8.

9.5–4 Numerical Examples

In the examples presented here the two sets of lamina properties given in Eqs. (9.4–16) and (9.4–17) are used. The shear correction coefficient for the first-order theory is taken to be $K = 5/6$. The loading in the case of bending is assumed to be sinusoidal. The deflections and stresses are nondimensionalized as given in Eqs. (9.4–19).

Bending

Table 9.5–1 contains maximum nondimensionalized deflections $\bar{w} \equiv w_0(a/2,0)E_2h^3 \times 10^2/(a^4q_0)$ of simply supported, two- and sixteen-layer angle-ply $(\theta/-\theta/\theta/-\theta,\ldots)$ square plates under sinusoidal loading (see Reddy and Chao [10]). The material properties used are those listed in Eq. (17) (*i.e.*, Material 2).

Table 9.5–1: Nondimensionalized deflection $\bar{w}$ as a function of number of layers, angle, and side-to-thickness ratio for simply supported (SS–2) angle-ply square plates under sinusoidally distributed transverse load (Material 2).

a/h	$\theta = 5°$		$\theta = 30°$		$\theta = 45°$	
	$n = 2$	$n = 16$	$n = 2$	$n = 16$	$n = 2$	$n = 16$
5	9.760	9.256	9.568	6.316	9.088	5.938
10	4.883	4.454	6.099	2.872	5.773	2.621
20	3.585	3.172	5.224	2.005	4.944	1.793
25	3.427	3.105	5.119	1.900	4.844	1.693
50	3.215	2.806	4.979	1.761	4.711	1.560
100	3.162	2.754	4.944	1.726	4.678	1.527

Table 9.5–2 contains nondimensionalized deflections $\bar{w}$ of angle-ply laminates subjected to uniformly distributed transverse load, under various boundary conditions (see Khdeir [21]), and with different values of E_1/E_2 (Material 2). Figure 9.5–2 shows the effect of side-to-thickness ratio (a/h) on the nondimensionalized center deflection of a square antisymmetric angle-ply laminate (45/–45/45/–45) under uniformly distributed transverse loading and for various boundary conditions. The material properties used are $E_1 = 19.2 \times 10^6$ psi (132.38 GPa), $E_2 = 1.56 \times 10^6$ psi (10.76 GPa), $G_{12} = G_{13} = 0.82 \times 10^6$ psi (5.65 GPa), $G_{23} = 0.523 \times 10^6$ psi (3.61 GPa), and $\nu_{12} = 0.25$. The effect of the ratio of principal moduli (Material 2) on the nondimensionalized center deflection is shown in Figure 9.5–3 for the same laminate.

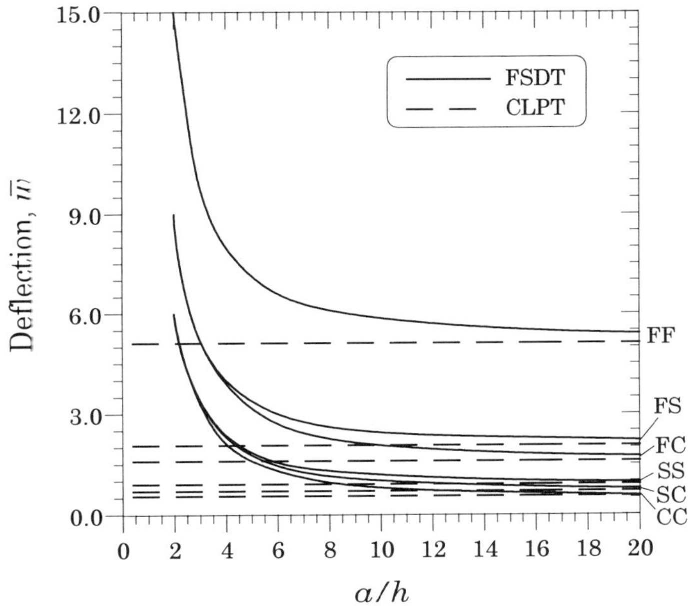

Figure 9.5–2: Nondimensionalized center transverse deflection ($\bar{w}$) versus side-to-thickness ratio (a/h) for square, antisymmetric angle-ply (45/–45/45/–45) laminates.

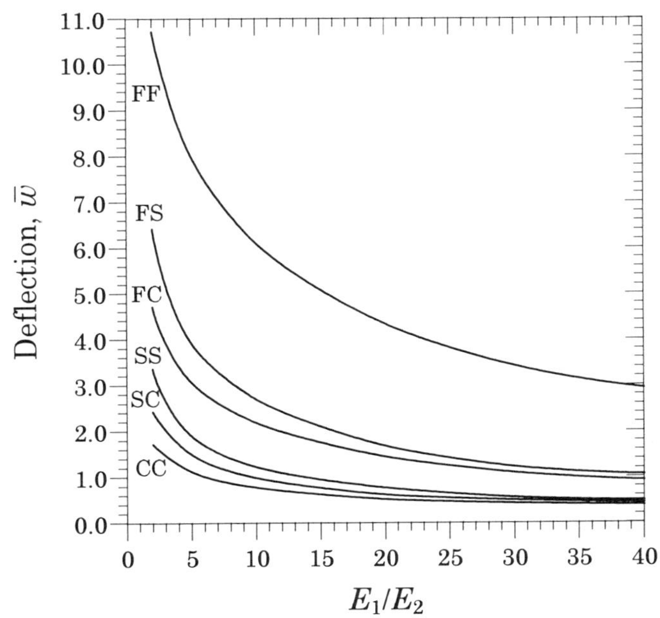

Figure 9.5–3: Nondimensionalized center transverse deflection ($\bar{w}$) versus modulus ratio (E_1/E_2) for square, antisymmetric angle-ply (45/–45/45/–45) laminates (Material 2, $a/h = 10$).

Table 9.5–2: Nondimensionalized deflections of simply supported (SS–2), four-layer antisymmetric angle-ply square plates (45/–45/45/–45) under uniformly distributed transverse load (Material 2, $a/h = 10$).

Theory	$\frac{E_1}{E_2}$	SS	SC	CC	FF	FS	FC
FSDT	2	3.375	2.423	1.753	10.735	6.447	4.743
CLPT		3.214	2.214	1.531	10.470	6.234	4.446
FSDT	10	1.160	0.944	0.771	6.049	2.611	2.109
CLPT		1.000	0.747	0.558	5.571	2.345	1.747
FSDT	20	0.701	0.602	0.518	4.284	1.623	1.379
CLPT		0.542	0.412	0.313	3.657	1.343	1.010
FSDT	30	0.531	0.471	0.417	3.422	1.225	1.075
CLPT		0.372	0.285	0.218	2.737	0.943	0.712

Table 9.5–3: Effect of ply angle (θ) and number of layers (n) on dimensionless fundamental frequency, $\bar{\omega}$, of antisymmetric angle-ply ($\theta°/-\theta°/\theta°/\ldots/-\theta°$) square plates (Material 2, $a/h = 10$).

$\theta°$	n	Theory	SS	SC	CC	FF	FS	FC
	2	FSDT	12.68	13.46	14.41	6.95	8.45	8.65
30		CLPT	14.24	15.44	17.00	7.58	9.35	9.69
	10	FSDT	18.51	19.11	19.81	10.11	12.33	12.48
		CLPT	23.95	25.59	27.58	12.37	15.38	15.84
	2	FSDT	13.04	14.23	15.63	4.76	7.13	7.52
45		CLPT	14.64	16.75	19.48	5.12	7.79	8.48
	10	FSDT	19.38	20.27	21.25	6.57	10.60	10.88
		CLPT	25.47	28.91	33.32	7.89	13.03	14.17
	2	FSDT	12.68	14.52	16.57	3.33	5.87	6.70
60		CLPT	14.24	17.74	22.31	3.47	6.26	7.54
	10	FSDT	18.51	19.82	21.21	3.82	8.53	9.22
		CLPT	23.95	29.86	37.62	4.32	9.92	11.96

Vibration and Buckling

The dimensionless frequencies, $\bar{\omega} \equiv \omega(a^2/h)\sqrt{(\rho/E_2)}$, of antisymmetric angle-ply laminates $(\theta°/-\theta°/\theta°/\cdots)$, under various boundary conditions and with different values of θ are presented in Table 9.5–3. Figure 9.5–4 shows the effect of side-to-thickness ratio (a/h) and Figure 9.5–5 shows the effect of the ratio of principal moduli on the nondimensionalized fundamental frequencies of a square antisymmetric angle-ply laminate (45/–45/45/–45) for various boundary conditions (see Khdeir [17, 21]). The material properties listed in Eq. (17) were used. The critical buckling loads, $\bar{N} = N_{cr}a^2/E_2h^3$, for the same laminates under uniaxial compressive load are included in Table 9.5–4.

Table 9.5–4: Effect of ply angle (θ) and number of layers (n) on dimensionless critical buckling load, $\bar{N}$, of antisymmetric angle-ply $(\theta°/-\theta°/\theta°/\ldots/-\theta°)$ square plates (Material 2, $a/h = 10$).

$\theta°$	n	Theory	SS	SC	CC	FF	FS	FC
	2	FSDT	16.613	18.718	21.447	4.991	7.372	7.739
30		CLPT	20.543	24.158	29.269	5.822	8.857	9.520
	10	FSDT	34.931	37.242	38.920	10.423	15.520	15.899
		CLPT	58.135	66.322	77.065	15.499	23.972	25.412
	2	FSDT	17.552	19.443	19.957	2.327	5.220	5.824
45		CLPT	21.709	28.423	34.963	2.654	6.150	7.283
	10	FSDT	33.173	33.263	33.356	4.401	11.484	12.092
		CLPT	65.714	84.707	102.59	6.300	17.189	20.332
	2	FSDT	13.863	14.722	15.598	1.136	3.543	4.605
60		CLPT	19.564	23.834	29.547	1.221	3.975	5.756
	10	FSDT	23.710	23.759	23.808	1.492	7.432	8.682
		CLPT	52.945	64.103	79.619	1.889	9.977	14.501

This completes the discussion of Lévy solutions of antisymmetric angle-ply rectangular plates with two opposite edges simply supported and the other two edges having a combination of boundary conditions. For additional theoretical as well as numerical results, the reader may consult the references at the end of the chapter.

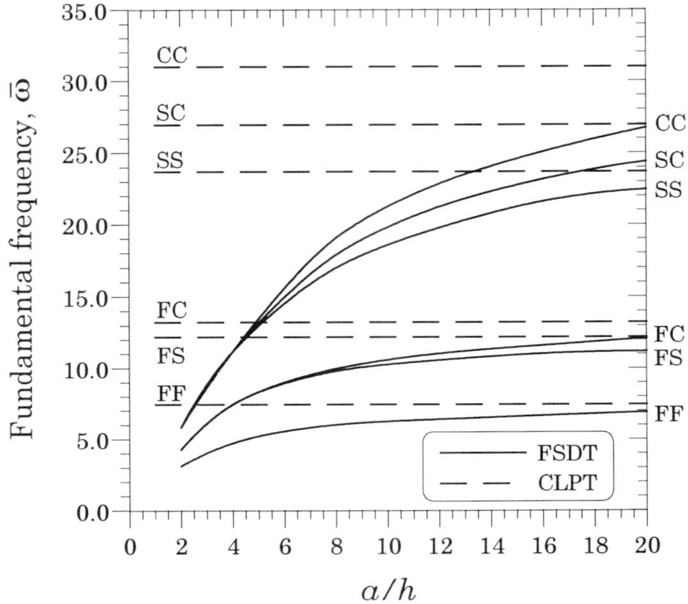

Figure 9.5–4: Nondimensionalized fundamental frequency ($\bar{\omega}$) versus side-to-thickness ratio (a/h) for square, antisymmetric angle-ply $(45/-45/45/-45)$ laminates (Material 2).

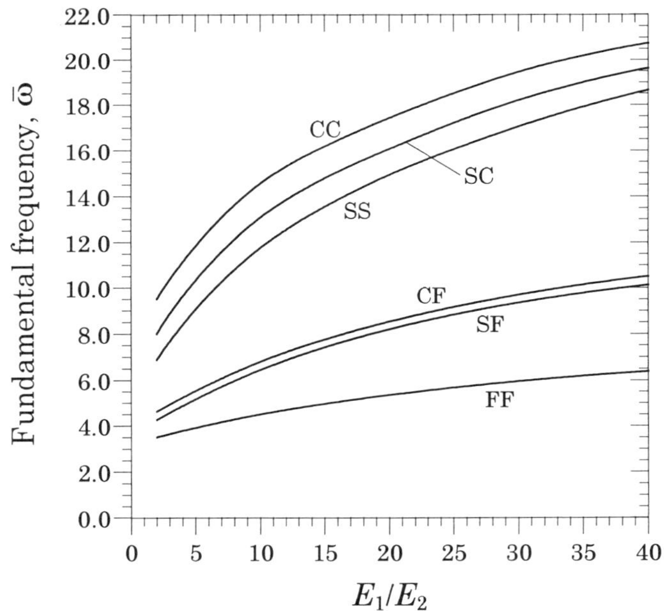

Figure 9.5–5: Nondimensionalized fundamental frequency ($\bar{\omega}$) versus modulus ratio (E_1/E_2) for square, antisymmetric angle-ply $(45/-45/45/-45)$ laminates ($a/h = 10$).

9.6 Transient Solutions

Transient solutions of antisymmetric cross-ply and angle-ply laminates using the first order theory can be developed either by the state–space approach or the combination of the Navier solution procedure and the Newmark time integration scheme as discussed in Section 8.7. The application of the state–space approach to the transient analysis of shear deformable theories can be found in the papers of Khdeir and Reddy [25–27].

The procedure of Section 8.7 is valid here when the coefficient matrices $[M]$ and $[K]$ in Eq. (8.7–2) are replaced with those in Eq. (9.2–7a) for cross-ply laminates and in Eq. (9.3–5) for angle-ply laminates. The solution vector $\{\Delta\}$ consists of the amplitudes of the five generalized displacements, $(u_0, v_0, w_0, \phi_x, \phi_y)$. Here we present numerical results based on this procedure.

Figure 9.6–1 shows plots of nondimensionalized center deflection, $\bar{w} = w_0[E_2h^3/(q_0a^4)]10^2$, versus time for antisymmetric cross-ply laminates (0/90) under sinusoidally distributed step loading (see Figure 9.3–1 for the coordinate system used here). The material properties used are

$$E_1 = 25E_2,\ E_2 = 2.1 \times 10^6 \text{ N/cm}^2,\ G_{12} = G_{13} = 0.5E_2,\ G_{23} = 0.2E_2,\ \nu_{12} = 0.25$$

$$\rho = 8 \times 10^{-6} \text{ N-s}^2/\text{cm}^4,\ a/b = 1, b = 25 \text{ cm} \qquad (9.6-1)$$

Results obtained with CLPT and FSDT are presented for two values of side-to-thickness ratios, $a/h = 10$ and 25. The plots in dashed lines correspond to CLPT. The effect of shear deformation is to increase the amplitude and period of the waves. Similar results are presented in Figure 9.6–2 for uniformly distributed step loading. A plot of the deflection under sinusoidal load (obtained with FSDT) is also included for comparison. Figures 9.6–3 and 9.6–4 contain the nondimensionalized center normal stress $\bar{\sigma}_{xx}(a/2, b/2, h/2)$ and shear stress $\bar{\sigma}_{xy}(a, b, -h/2)$ for the uniformly distributed load case ($a/h = 10$). Note that the effect of shear deformation on the amplitude of stresses is negligible; however, it still increases the period. Figures 9.6–5 and 9.6–6 contain plots of nondimensionalized center deflection for two-layer angle-ply (−45/45) laminates under sinusoidally and uniformly distributed step loadings, respectively, for $a/h = 10$. The same material properties as in Eq. (1) are used. Angle-ply laminates show larger increase in the period due to shear deformation. Figure 9.6–7 show a plot of the maximum in-plane displacement (u_0 or v_0) versus time, while Figures 9.6–8 through 9.6–10 show plots of normal stress at top and bottom of the laminate and shear stress at the bottom of the laminate for the uniform load case. Lastly, Figures 9.6–11 and 9.6–12 show plots of nondimensionalized transverse shear stress $\bar{\sigma}_{xz}$ in two-layer plates (0/90) and (−45/45) under uniformly and sinusoidally distributed transverse loads ($a/h = 10$).

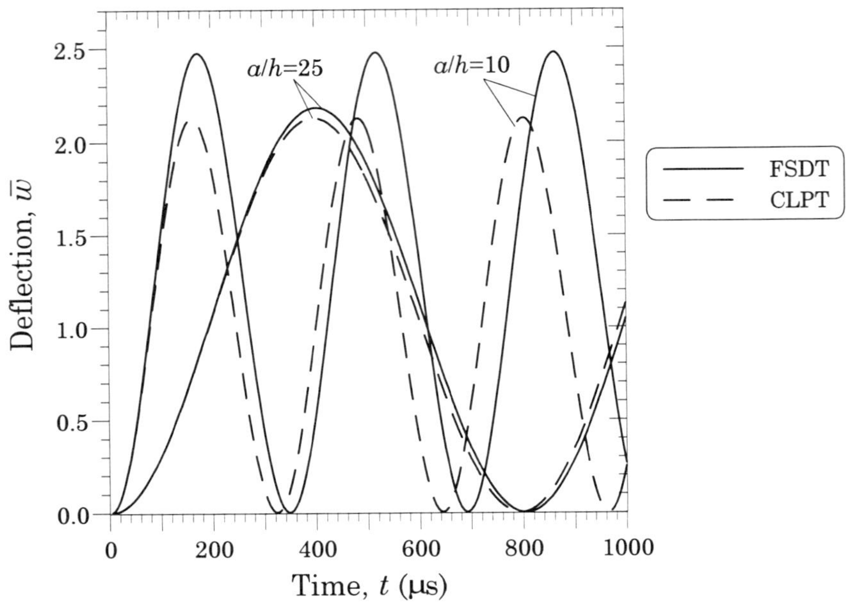

Figure 9.6–1: Nondimensionalized center transverse deflection ($\bar{w}$) versus time (t) for simply supported (SS–1) antisymmetric cross-ply (0/90) laminates.

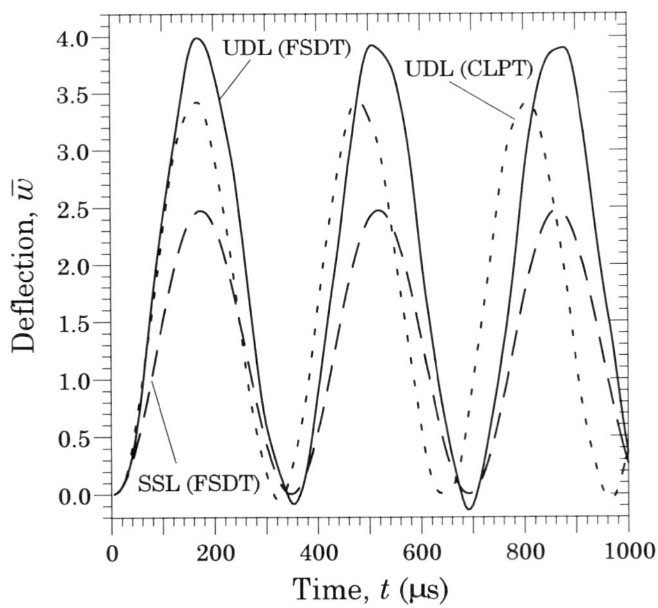

Figure 9.6–2: Nondimensionalized center transverse deflection ($\bar{w}$) versus time (t) for simply supported (SS–1) antisymmetric cross-ply (0/90) laminates.

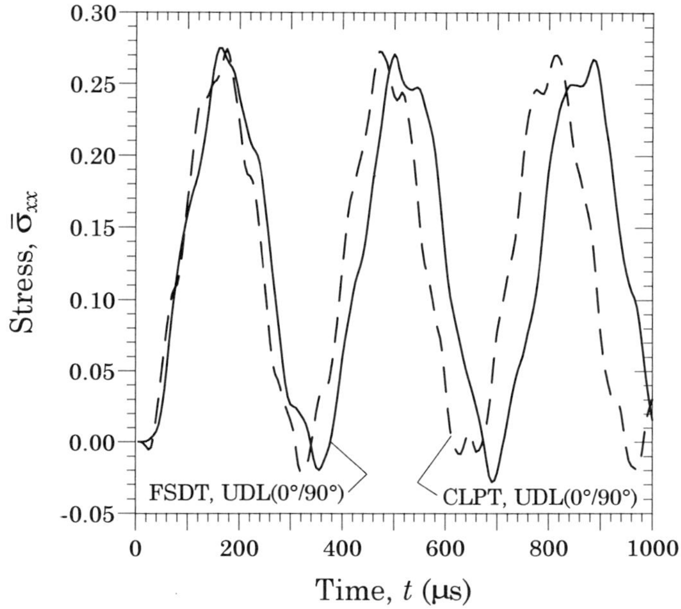

Figure 9.6–3: Nondimensionalized normal stress ($\bar{\sigma}_{xx}$) versus time (t) for simply supported (SS–1) antisymmetric cross-ply (0/90) laminates.

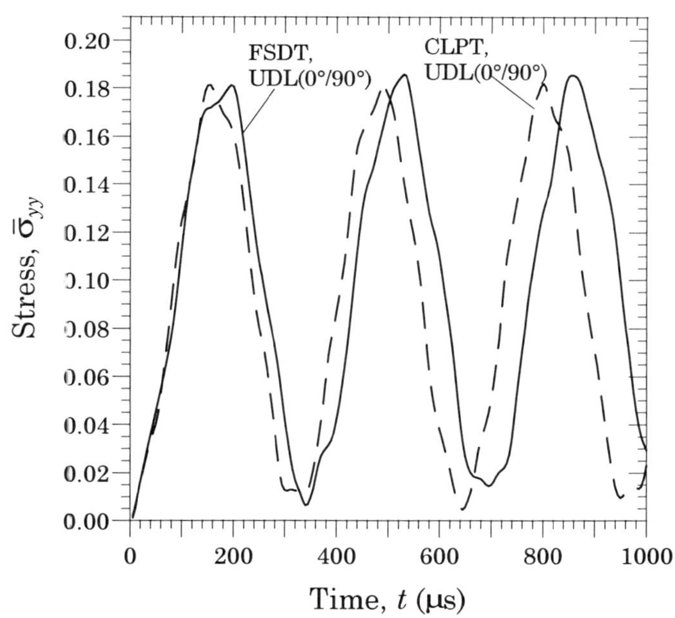

Figure 9.6–4: Nondimensionalized normal stress ($\bar{\sigma}_{yy}$) versus time (t) for simply supported (SS–1) antisymmetric cross-ply (0/90) laminates.

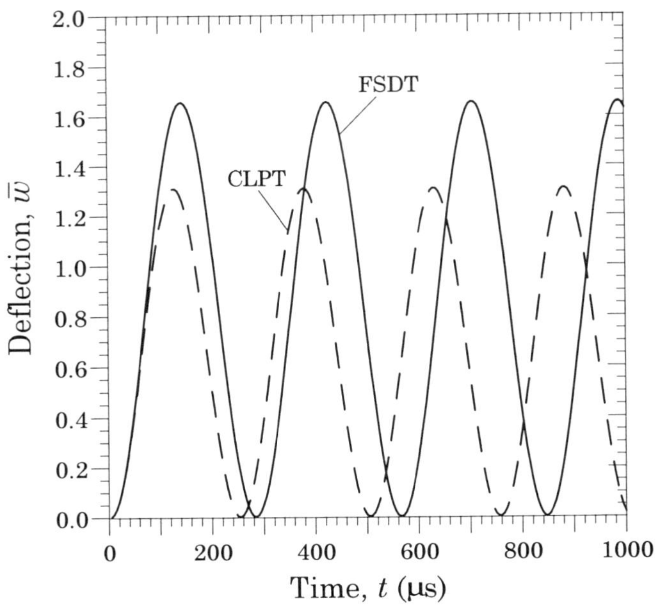

Figure 9.6–5: Nondimensionalized center transverse deflection ($\bar{w}$) versus time (t) for simply supported (SS–2) antisymmetric angle-ply (−45/45) laminates.

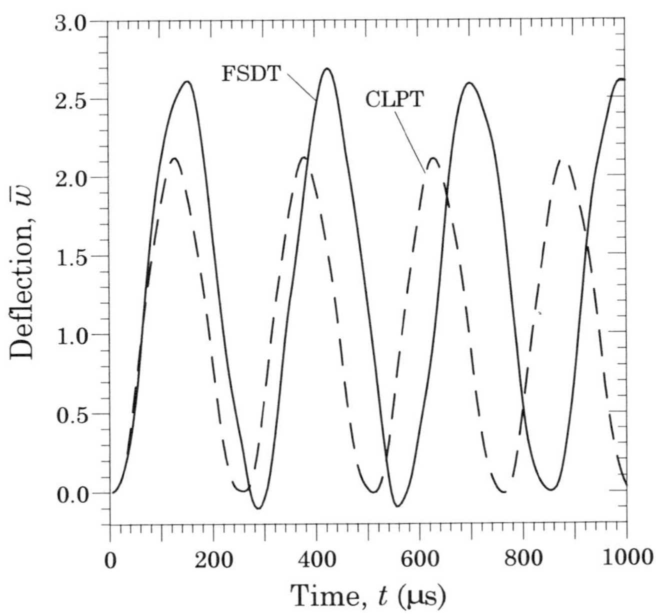

Figure 9.6–6: Nondimensionalized center transverse deflection ($\bar{w}$) versus time (t) for simply supported (SS–2) antisymmetric angle-ply (−45/45) laminates.

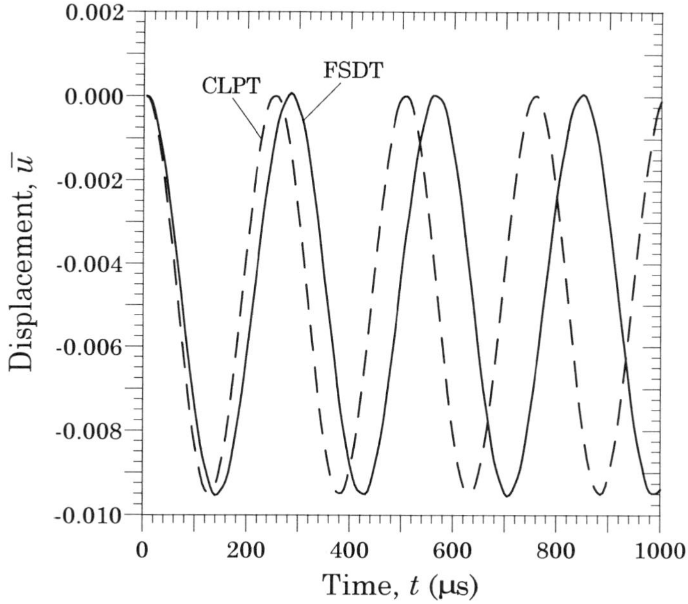

Figure 9.6–7: Nondimensionalized in-plane displacement ($\bar{u}$) versus time (t) for simply supported (SS–2) antisymmetric angle-ply ($-45/45$) laminates.

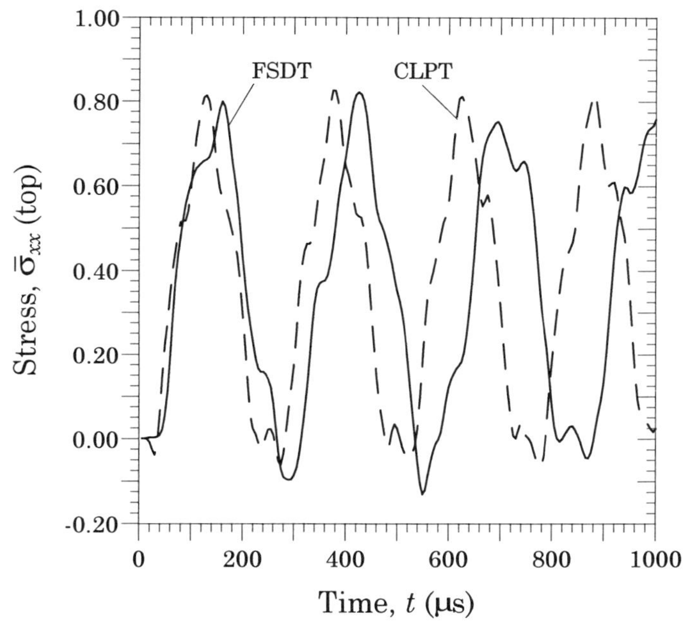

Figure 9.6–8: Nondimensionalized normal stress ($\bar{\sigma}_{xx}$ at the top of the laminate) versus time (t) for simply supported (SS–2) antisymmetric angle-ply ($-45/45$) laminates.

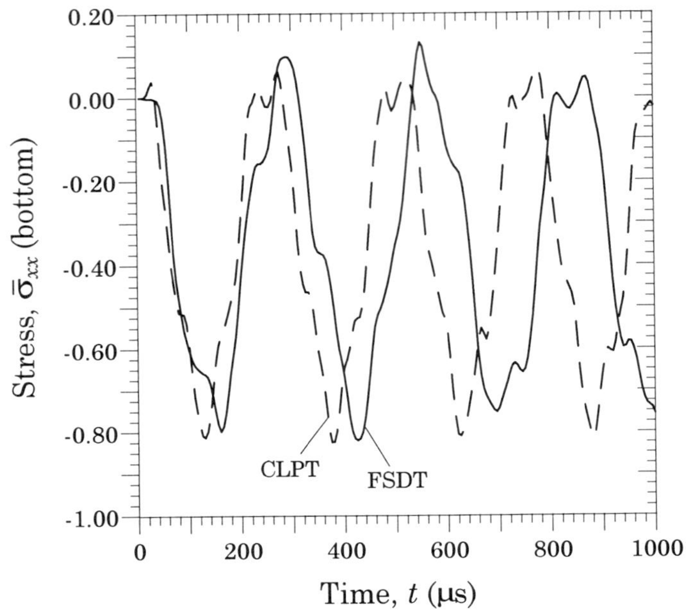

Figure 9.6–9: Nondimensionalized normal stress ($\bar{\sigma}_{xx}$ at the bottom of the laminate) versus time (t) for simply supported (SS–2) antisymmetric angle-ply (−45/45) laminates.

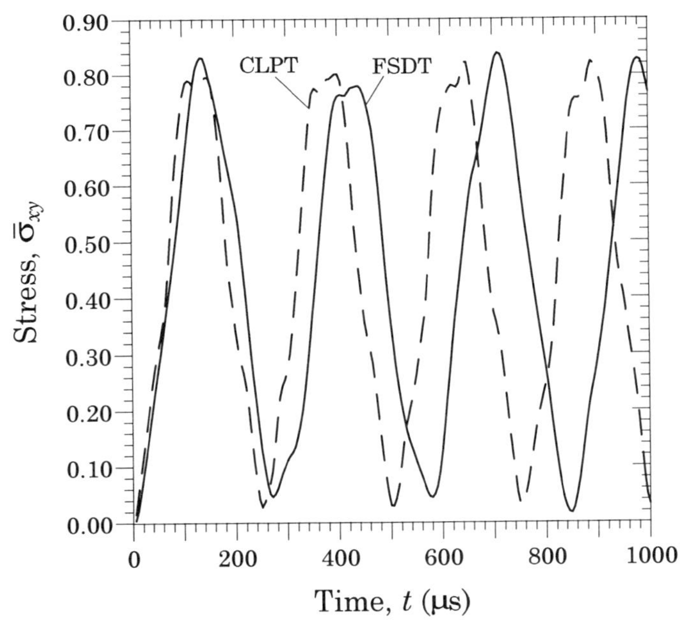

Figure 9.6–10: Nondimensionalized shear stress ($\bar{\sigma}_{xy}$) versus time (t) for simply supported (SS–2) antisymmetric angle-ply (−45/45) laminates.

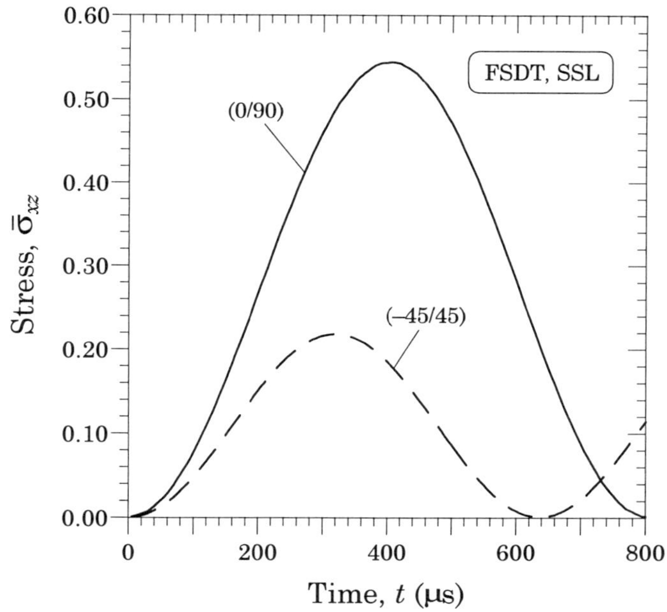

Figure 9.6–11: Nondimensionalized transverse shear stress ($\bar{\sigma}_{xz}$) versus time (t) for simply supported antisymmetric cross-ply (0/90) and angle-ply (−45/45) laminates.

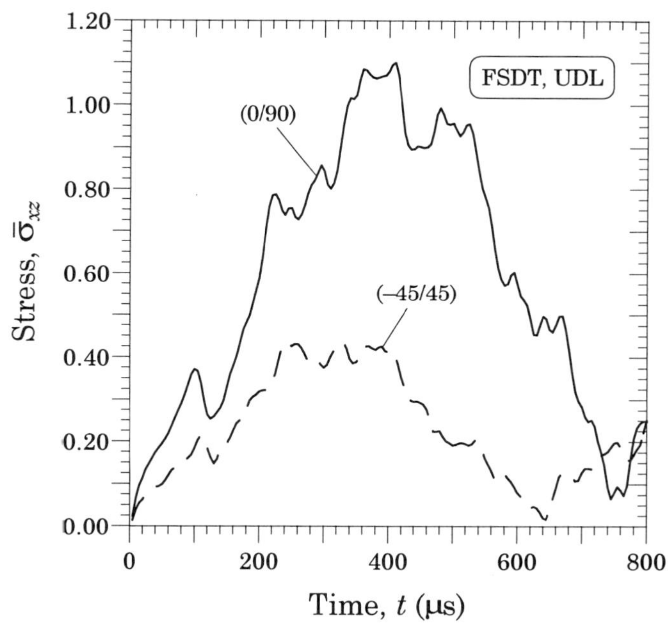

Figure 9.6–12: Nondimensionalized transverse shear stress ($\bar{\sigma}_{xz}$) versus time (t) for simply supported antisymmetric cross-ply (0/90) and angle-ply (−45/45) laminates.

For cross-ply laminates under uniformly distributed load, the ratio of maximum transient deflection to static deflection is found to be $\bar{w}_d/\bar{w}_s = 2.049$ for FSDT, whereas it is 2.017 for CLPT. From this, one may conclude that the effect of transverse shear is greater on dynamic response than static response in the two-layer cross-ply plates. For angle-ply laminates under uniformly distributed load, the ratio of maximum transient deflection to static deflection is found to be 2.041 for FSDT, whereas it is 2.062 for CLPT. This indicates that the effect of transverse shear is less on dynamic response than static response in the two-layer angle-ply plates. Table 9.6–1 contains the results of transient and static analysis.

Table 9.6–1: Nondimensionalized deflections and stresses of simply supported cross-ply (0/90) and angle-ply (–45/45) square plates under uniformly distributed transverse load ($E_1/E_2 = 25$, $E_2 = 2.1\times10^6$ psi, $G_{12} = G_{13} = 0.5E_2$, $G_{23} = 0.2E_2$, $\nu_{12} = 0.25$, $a/b = 1$, $a = 25$ cm, $a/h = 10$).

Theory	(0/90)			(−45/45)		
	$\bar{w}$	$\bar{\sigma}_{xx}$	$\bar{\sigma}_{xy}$	$\bar{w}$	$\bar{\sigma}_{xx}$	$\bar{\sigma}_{xy}$
FSDT†	3.990	0.275	0.181	2.611	0.799	0.832
	1.947	0.126	0.096	1.279	0.348	0.427
CLPT	3.421	0.274	0.181	2.120	0.813	0.795
	1.695	0.127	0.093	1.028	0.351	0.442

† The first line is the transient solution and the second line is the static solution.

9.7 Summary

Analytical solutions for bending, buckling under in-plane compressive loads, natural vibration, and transient response of rectangular laminates with various boundary conditions are presented based on the first-order shear deformation laminate theory. The Navier solutions were developed for two classes of laminates: antisymmetric cross-ply laminates and antisymmetric angle-ply laminates, each for a specific type of simply supported boundary conditions, SS–1 and SS–2, respectively. The Lévy solutions with the state–space approach were developed for these classes of laminates when two opposite edges are simply supported and other two edges having a variety of boundary conditions of choice.

Numerical results were presented for static bending, buckling, natural vibration, and transient response of antisymmetric cross-ply and angle-ply laminates. The bending–extensional coupling and transverse shear deformation in a laminate generally reduce the effective stiffnesses and hence increase deflections and reduce buckling loads and reduce natural frequencies. The effect of transverse shear deformation on transient response is to increase both amplitude and period of oscillation. The coupling is the most significant in two-layer laminates, and it decreases gradually as the number of layers is increased for fixed total thickness.

Exercises

9.1 Show that the solution of Eq. (9.2–15b) is given by

$$\begin{aligned} W_{mn} &= \frac{1}{b_{mn}}\left(a_1 Q_{mn} - a_2 \alpha M^1_{mn} - a_3 \beta M^2_{mn}\right)\\ X_{mn} &= \frac{1}{b_{mn}}\left(a_2 Q_{mn} - a_4 \alpha M^1_{mn} - a_5 \beta M^2_{mn}\right)\\ Y_{mn} &= \frac{1}{b_{mn}}\left(a_3 Q_{mn} - a_5 \alpha M^1_{mn} - a_6 \beta M^2_{mn}\right) \end{aligned} \tag{a}$$

where

$$b_{mn} = \begin{vmatrix} \hat{s}_{33} & \hat{s}_{34} & \hat{s}_{35} \\ \hat{s}_{34} & \hat{s}_{44} & \hat{s}_{45} \\ \hat{s}_{35} & \hat{s}_{45} & \hat{s}_{55} \end{vmatrix} = \hat{s}_{33}a_1 + \hat{s}_{34}a_2 + \hat{s}_{35}a_3$$

$$\begin{aligned} a_1 &= \hat{s}_{44}\hat{s}_{55} - \hat{s}_{45}\hat{s}_{45},\ a_2 = \hat{s}_{45}\hat{s}_{35} - \hat{s}_{34}\hat{s}_{55},\ a_3 = \hat{s}_{34}\hat{s}_{45} - \hat{s}_{44}\hat{s}_{35}\\ a_4 &= \hat{s}_{33}\hat{s}_{55} - \hat{s}_{35}\hat{s}_{35},\ a_5 = \hat{s}_{34}\hat{s}_{35} - \hat{s}_{33}\hat{s}_{45},\ a_6 = \hat{s}_{33}\hat{s}_{44} - \hat{s}_{34}\hat{s}_{34} \end{aligned} \tag{b}$$

9.2 Verify the expressions for transverse stresses presented in Eq. (9.2–23).

9.3 Formulate the Lévy type solution procedure for the natural vibration of antisymmetric cross-ply laminates. In particular, show that the operator $[T]$ in Eq. (9.4–13a) holds with

$$\begin{aligned} e_5 &= -\beta^2 A_{66} + I_0\omega_m^2,\quad e_6 = -\beta^2 B_{66} + I_1\omega_m^2\\ e_{11} &= -\beta^2 A_{22} + I_0\omega_m^2,\quad e_{12} = -\beta^2 B_{22} + I_1\omega_m^2\\ e_{15} &= -\beta^2 K A_{44} + I_0\omega_m^2\\ e_{21} &= -\beta^2 B_{66} + I_1\omega_m^2,\quad e_{22} = -\beta^2 D_{66} + I_2\omega_m^2\\ e_{28} &= -\beta^2 B_{22} + I_1\omega_m^2,\quad e_{29} = -\beta^2 D_{22} - K A_{44} + I_2\omega_m^2 \end{aligned}$$

9.4 Formulate the Lévy type solution procedure for the buckling of antisymmetric cross-ply laminates under in-plane compressive loads. In particular, show that the operator $[T]$ in Eq. (9.4–13a) holds with

$$e_{13} = K A_{55} + \hat{N}_{xx},\quad e_{15} = -\beta^2 K A_{44} - \beta^2 \hat{N}_{yy}$$

9.5 Verify the expressions for transverse stresses presented in Eq. (9.3–11).

9.6 Formulate the Lévy type solution procedure for the natural vibration of antisymmetric angle-ply laminates. In particular, show that the operator $[T]$ in Eq. (9.5-12b) holds with

$$C_1 = \frac{\alpha^2 (A_{11}D_{22} - B_{16}B_{26}) - I_0 D_{22}\omega_m^2}{A_{66}D_{22} - B_{26}^2}$$

$$C_5 = \frac{\alpha^2 (B_{16}D_{22} - B_{26}D_{66}) - B_{26}KA_{44} + I_2 B_{26}\omega_m^2}{A_{66}D_{22} - B_{26}^2}$$

$$C_7 = \frac{\alpha^2 (A_{66}D_{66} - B_{16}B_{26}) - I_0 D_{66}\omega_m^2}{A_{22}D_{66} - B_{26}^2}$$

$$C_9 = \frac{\alpha^2 (B_{16}D_{66} - D_{11}B_{26}) - KA_{55}B_{26} + I_2 B_{26}\omega_m^2}{A_{22}D_{66} - B_{26}^2}$$

$$C_{11} = \frac{\alpha^2 KA_{55} - I_0\omega_m^2}{KA_{44}}$$

$$C_{15} = \frac{\alpha^2 (A_{22}B_{16} - B_{26}A_{66}) + I_0 B_{26}\omega_m^2}{A_{22}D_{66} - B_{26}^2}$$

$$C_{17} = \frac{\alpha^2 (A_{22}D_{11} - B_{16}B_{26}) + A_{22}KA_{55} - I_2 A_{22}\omega_m^2}{A_{22}D_{66} - B_{26}^2}$$

$$C_{19} = \frac{\alpha^2 (A_{66}B_{16} - A_{11}B_{26}) + I_0 B_{26}\omega_m^2}{A_{22}D_{66} - B_{26}^2}$$

$$C_{23} = \frac{\alpha^2 (A_{66}D_{66} - B_{16}B_{26}) + A_{66}KA_{44} - I_2 A_{66}\omega_m^2}{A_{22}D_{66} - B_{26}^2}$$

and all other C_i are as defined in Eq. (9.5–10b) with $\hat{N}_{xx} = \hat{N}_{yy} = 0$.

References for Additional Reading

1. Khdeir, A. A., Reddy, J. N., and Librescu, L., "Lévy Type Solutions for Symmetrically Laminated Rectangular Plates Using First-Order Shear Deformation Theory," *Journal of Applied Mechanics*, **54**, 640–642 (1987).
2. Franklin, J. N., *Matrix Theory*, Prentice–Hall, Englewood Cliffs, New Jersey, 1968.
3. Brogan, W. L., *Modern Control Theory*, Prentice–Hall, Englewood Cliffs, New Jersey, 1985.
4. Reddy, J. N., *Energy and Variational Methods in Applied Mechanics*, John Wiley, New York, 1984.
5. Reddy, J. N. (ed.), *Mechanics of Composite Materials. Selected Works of Nicholas J. Pagano*, Kluwer, Netherlands, 1994.
6. Pagano, N. J., "Exact Solutions for Rectangular Bidirectional Composites and Sandwich Plates," *Journal of Composite Materials*, **4**(1), 20–34 (1970).

7. Pagano, N. J., and Hatfield, S. J., "Elastic Behavior of Multilayered Bidirectional Composites," *AIAA Journal*, **10**(7), 931–933 (1972).
8. Reddy, J. N. and Khdeir, A. A., "Buckling and Vibration of Laminated Composite Plates Using Various Plate Theories," *AIAA Journal*, **27** (12), 1808–1817 (1989).
9. Nosier, A. and Reddy, J. N., "On Vibration and Buckling of Symmetric Laminated Plates According to Shear Deformation Theories," *Acta Mechanica*, **94** (3–4), 123–170 (1992).
10. Reddy, J. N. and Chao, W. C., "A Comparison of Closed-Form and Finite Element Solutions of Thick Laminated Anisotropic Rectangular Plates," *Nuclear Engineering and Design*, **64**, 153–167 (1981).
11. Reddy, J. N. and Hsu, Y. S., "Effects of Shear Deformation and Anisotropy on the Thermal Bending of Layered Composite Plates," *Journal of Thermal Stresses*, **3**, 475–493 (1980).
12. Khdeir, A. A., Librescu, L., and Reddy, J. N., "Analytical Solution of a Refined Shear Deformation Theory for Rectangular Composite Plates," *International Journal of Solids and Structures*, **23**(10), 1447–1463 (1987).
13. Khdeir, A. A. and Reddy, J. N., "Dynamic Response of Antisymmetric Angle-Ply Laminated Plates Subjected to Arbitrary Loading," *Journal of Sound and Vibration*, **126**(3), 437–445 (1988).
14. Khdeir, A. A. and Librescu, L., "Analysis of Symmetric Cross-Ply Laminated Elastic Plates Using a Higher-Order Theory: Part I–Stress and Displacement," *Composite Structures*, **9**, 189–213 (1988).
15. Khdeir, A. A. and Librescu, L., "Analysis of Symmetric Cross-Ply Laminated Elastic Plates Using a Higher-Order Theory: Part II–Buckling and Free Vibration," *Composite Structures*, **9**, 259–277 (1988).
16. Khdeir, A. A., "Free Vibration and Buckling of Symmetric Cross-Ply Laminated Plates by an Exact Method," *Journal of Sound and Vibration*, **126**(3), 447–461 (1988).
17. Khdeir, A. A., "Free Vibration of Antisymmetric Angle-Ply Laminated Plates Including Various Boundary Conditions," *Journal of Sound and Vibration*, **122**(2), 377–388 (1988).
18. Sun, C. T. and Whitney, J. M., "On Theories for the Dynamic Response of Laminated Plates," *AIAA Journal*, **11**(2), 178–183 (1973).
19. Khdeir, A. A., "Free Vibration and Buckling of Unsymmetric Cross-Ply Laminated Plates Using a Refined Theory," *Journal of Sound and Vibration*, **128**(3), 377-395 (1989).
20. Khdeir, A. A., "An Exact Approach to the Elastic State of Stress of Shear Deformable Antisymmetric Angle-Ply Laminated Plates," *Composite Structures*, **11**, 245–258 (1989).
21. Khdeir, A. A., "Comparison Between Shear Deformable and Kirchhoff Theories for Bending, Buckling and Vibration of Antisymmetric Angle-Ply Laminated Plates," *Composite Structures*, **13**, 159–172 (1989).

22. Khdeir, A. A., "Stability of Antisymmetric Angle-Ply Laminated Plates," *ASCE Journal of Engineering Mechanics,* **115**, 952–962 (1989).
23. Khdeir, A. A. and Reddy, J. N., "Analytical Solutions of Refined Plate Theories of Cross-Ply Composite Laminates," *Journal of Pressure Vessel Technology,* **113**(4), 570–578 (1991).
24. Khdeir, A. A. and Reddy, J. N., "Thermal Stresses and Deflections of Cross-Ply Laminated Plates Using Refined Plate Theories," *Journal of Thermal Stresses,* **14**(4), 419–438 (1991).
25. Khdeir A. A. and Reddy, J. N. "Exact Solutions for the Transient Response of Symmetric Cross-Ply Laminates Using a Higher-Order Plate Theory," *Composites Science and Technology,* **34**, 205–224 (1989).
26. Khdeir A. A. and Reddy, J. N. "On the Forced Motions of Antisymmetric Cross-Ply Laminates," *International Journal of Mechanical Sciences,* **31**, 499–510 (1989).
27. Khdeir A. A. and Reddy, J. N. "Dynamic Response of Antisymmetric Angle-Ply Laminated Plates Subjected to Arbitrary Loading," *Journal of Sound and Vibration,* **126**, 437–445 (1988).
28. Reddy, J. N., "On the Solutions to Forced Motions of Rectangular Composite Plates," *Journal of Applied Mechanics,* **49**, 403–408 (1982).
29. Srinivas, S. and Rao, A. K., "Buckling of Thick Rectangular Plates," *AIAA Journal,* **7**, 1645 (1969).
30. Srinivas, S., Joga Rao, C. V., and Rao, A. K., "An Exact Analysis for Vibration of Simply Supported Homogeneous and Laminated Thick Rectangular Plates," *Journal of Sound and Vibration,* **12**, 187–199 (1970).
31. Srinivas, S., Joga Rao, C. V., and Rao, A. K., "Some Results from an Exact Analysis of Thick Laminates in Vibration and Buckling," *Journal of Applied Mechanics,* **37**, 868–870 (1970).
32. Srinivas, S. and Rao, A. K., "Bending, Vibration, and Buckling of Simply Supported Thick Orthotropic Rectangular Plates and Laminates," *International Journal of Solids and Structures,* **6**, 1463–1481 (1970).
33. Srinivas, S. and Rao, A. K., "A Three–Dimensional Solution for Plates and Laminates," *Journal of Franklin Institute,* **291,** 469–481 (1971).
34. Srinivas, S. and Rao, A. K., "Flexure of Thick Rectangular Plates," *Journal of Applied Mechanics,* **40,** 298–299 (1973).
35. Srinivas, S., "A Refined Analysis of Composite Laminates," *Journal of Sound and Vibration,* **30,** 495–507 (1973).

Chapter Ten

Finite Element Analysis of Composite Laminates

10.1 Introduction

In Chapters 6 through 9, the Navier, Lévy, and Rayleigh–Ritz solutions to the equations of composite beams and plates were presented for rectangular geometries. However, exact analytical or variational solutions to these problems cannot be developed when complex geometries, arbitrary boundary conditions, or nonlinearities are involved. Therefore, one must resort to approximate methods of analysis that are capable of solving such problems.

The *finite element method* is a powerful computational technique for the solution of differential and integral equations that arise in various fields of engineering and applied science. The method is a generalization of the classical variational (*i.e.*, the Rayleigh–Ritz) and weighted–residual (*e.g.*, Galerkin, least–squares, collocation, etc.) methods [1–5]. Since most real–world problems are defined on domains that are geometrically complex and may have different types of boundary conditions on different portions of the boundary of the domain, it is difficult to generate approximation functions required in the traditional variational methods. The basic idea of the finite element method is to view a given domain as an assemblage of simple geometric shapes, called *finite elements*, for which it is possible to systematically generate the approximation functions needed in the solution of differential equations by any of the variational and weighted–residual methods. The ability to represent domains with irregular geometries by a collection of finite elements makes the method a valuable practical tool for the solution of boundary, initial, and eigenvalue problems arising in various fields of engineering. The approximation functions are often constructed using ideas from interpolation theory, and hence they are also called *interpolation functions*. Thus the finite element method is a piecewise (or element wise) application of the variational and weighted–residual methods. For a given differential equation, it is possible to develop different finite element approximations (or *finite element models*), depending on the choice of a particular variational or weighted–residual method. For a detailed introduction to the finite element method, the reader is advised to consult References 1–5.

The major steps in the finite element analysis of a typical problem are (see Reddy [1,2])

1. Discretization of the domain into a set of finite elements (*mesh generation*).
2. Weighted–integral or weak formulation of the differential equation over a typical finite element (subdomain).
3. Development of the finite element model of the problem using its weighted–integral or weak form. The finite element model consists of a set of algebraic equations among the unknown parameters of the element.
4. Assembly of finite elements to obtain the global system (*i.e.*, for the total problem) of algebraic equations.
5. Imposition of boundary conditions.
6. Solution of equations.
7. Postcomputation of solution and quantities of interest.

The above steps of the finite element method make it a modular technique that can be implemented on a computer, independent of the shape of the domain and boundary conditions. In addition, the method allows coupling of various physical problems because finite elements based on different physical problems can be easily generated in the same computer program.

In this chapter, we develop finite element models of laminated composite beams and plates. The objective is to introduce the reader to the finite element formulations of laminated composite structures. While the coverage is not exhaustive in terms of solving complicated problems, for this is primarily a textbook, it helps the reader in gaining an understanding of the plate finite elements used in the analysis of practical problems.

It is important to note that any numerical or computational method is a means to analyze a practical engineering problem and that analysis is not an end in itself but rather an aid to design. *The value of the theory and analytical solutions presented in the preceding chapters to gain insight into the behavior of simple laminated beam and plate structures is immense in the numerical modeling of complicated problems by the finite element method or any numerical method.* Those who are quick to use a computer rather than *think* about the problem to be analyzed may find it difficult to interpret or explain the computer-generated results. Even to develop proper input data to a computer program requires a good understanding of the theory of the problem as well as the method on which the program is based.

10.2 Laminated Beams and Plate Strips by CLPT

10.2–1 Introduction

In this section we present the finite element models of laminated beams and plate strips (*i.e.*, cylindrical bending). Since the governing equations of symmetrically laminated beams and those of cylindrical bending are the same in form, we give a detailed development of the finite element model of laminated beams only. Those readers who have a good background in the finite element method may skip this section or browse through it. The development presented herein is fairly self-contained, and one need not consult other finite element books.

10.2–2 Governing Equation

The equation of motion governing the bending of *symmetrically laminated beams* according to the classical laminate theory is given by [see Eq. (6.2–8b)]

$$\frac{\partial^2}{\partial x^2}\left(E^b_{xx} I_{yy} \frac{\partial^2 w_0}{\partial x^2}\right) - b\hat{N}_{xx}\frac{\partial^2 w_0}{\partial x^2} - \hat{q} + \hat{I}_0 \frac{\partial^2 w_0}{\partial t^2} - \hat{I}_2 \frac{\partial^4 w_0}{\partial x^2 \partial t^2} = 0 \qquad (10.2-1a)$$

where $\hat{N}_{xx}$ is the axial load and

$$\hat{q} = bq, \quad \hat{I}_0 = bI_0, \quad \hat{I}_2 = bI_2 \qquad (10.2-1b)$$

and b is the width of the beam, $q(x,t)$ is the distributed transverse load, and I_0 and I_2 are mass inertias

$$I_0 = \int_{-\frac{h}{2}}^{\frac{h}{2}} \rho \; dz, \quad I_2 = \int_{-\frac{h}{2}}^{\frac{h}{2}} \rho z^2 \; dz \qquad (10.2-2)$$

We will follow the steps described in the previous section to develop the finite element model of Eq. (1a).

First, the domain of the beam, $0 < x < L$ (see Figure 10.2–1a), is divided into a set of subdomains. The division of a domain into a collection of subdomains facilitates accurate representation of the geometry, material properties, and the solution itself. Each subdomain is called a *beam finite element*, and the set of elements representing the total domain is called the *finite element mesh.* The interfaces of the elements and the ends $x = 0$ and $x = L$ are called *global nodes* of the mesh. A typical beam element occupies the domain $x_A < x < x_B$ (see Figure 10.2–1a), where x_A and $x = x_B$ denote the global coordinates of the ends of the typical element. Each element can have its own geometric properties: length, $h_e \equiv x_B - x_A$, cross-sectional area, A_e, and moment of inertia, I_e; material properties: E_e, $\hat{I}^e_0$, and $\hat{I}^e_2$; and applied load: $\hat{q}_e$.

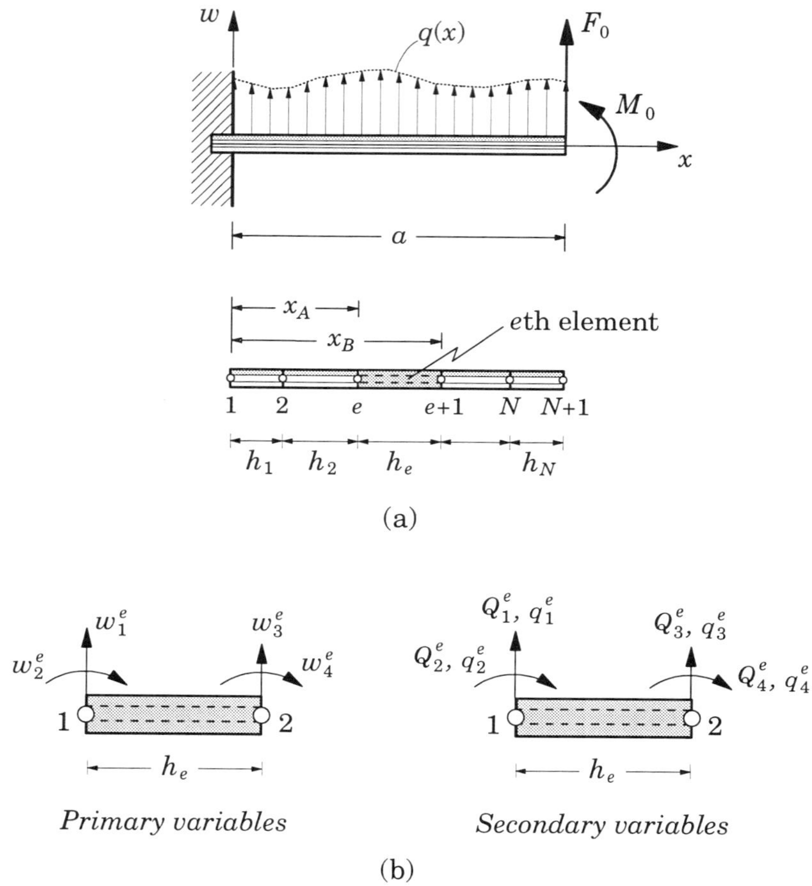

Figure 10.2–1: Domain, finite element subdivision, and typical finite element. (a) Laminated beam/plate strip. (b) A typical finite element with displacement and force degrees of freedom.

Next, the typical element is isolated from the mesh and its structural behavior is modeled using a variational method. In the present study, we use the Rayleigh–Ritz method (see Section 3.5–2) with algebraic polynomials for the approximation functions. In fact, the algebraic approximation functions are constructed systematically using the properties presented in Eq. (3.5–2). Recall that the Rayleigh–Ritz method uses either the total potential energy expression or the *weak form* (see Section 6.8–2) of the governing equation of the beam element. We use the weak form of the governing equation to develop the *finite element model*, which is a set of relations among the nodal displacements and forces (*i.e.*, primary and secondary variables).

When a portion of the body is isolated, we must show its free–body diagram (*i.e.*, show all its applied and reactive forces). For the beam element, the free–body diagram is shown in Figure 10.2–1b. The ends $x = x_A$ and $x = x_B$ are denoted as *element nodes* 1 and 2, respectively. The generalized forces Q_1^e and Q_3^e denote the transverse shear forces, whereas Q_2^e and Q_4^e denote the bending moments at nodes 1 and 2. Figure 10.2–1b defines the notation used for the generalized forces. Since the element is in equilibrium at every instant of time, it is expected that

$$Q_1^e + Q_3^e + \int_{x_A}^{x_B} \hat{q}dx = 0, \quad Q_2^e + Q_4^e - Q_3^e h_e - c\int_{x_A}^{x_B} \hat{q}x dx = 0$$

where c is a fraction of h_e and its value depends on the load distribution. These relations will be satisfied automatically if Q_i^e are determined from Eq. (1a) or its approximation (why?).

10.2–3 Weak Form

The weak form of Eq. (1a) over the typical element is obtained by the procedure described in Section 6.8–2 (also see Reddy [1], p. 145):

$$\begin{aligned}0 = &\int_{x_A}^{x_B} \left[E_{xx}^b I_{yy} \frac{d^2 v}{dx^2}\frac{\partial^2 w_0}{\partial x^2} + b\hat{N}_{xx}\frac{dv}{dx}\frac{\partial w_0}{\partial x} + \hat{I}_0 v\frac{\partial^2 w_0}{\partial t^2} + \hat{I}_2\frac{dv}{dx}\frac{\partial^3 w_0}{\partial x \partial t^2} - v\hat{q}\right] dx\\ &-Q_1^e v(x_A) - Q_3^e v(x_B) - Q_2^e\left(-\frac{dv}{dx}\right)_{x_A} - Q_4^e\left(-\frac{dv}{dx}\right)_{x_B}\end{aligned} \quad (10.2-3a)$$

where $v(x)$ denotes the weight function, and Q_i^e are related to the transverse deflection $w_0(x,t)$ by the relations

$$\begin{aligned}Q_1^e(t) &= \left[\frac{\partial}{\partial x}\left(E_{xx}^b I_{yy}\frac{\partial^2 w_0}{\partial x^2} - \hat{I}_2\frac{\partial^2 w_0}{\partial t^2}\right) - b\hat{N}_{xx}\frac{\partial w_0}{\partial x}\right]_{x=x_A}\\ Q_3^e(t) &= -\left[\frac{\partial}{\partial x}\left(E_{xx}^b I_{yy}\frac{\partial^2 w_0}{\partial x^2} - \hat{I}_2\frac{\partial^2 w_0}{\partial t^2}\right) - b\hat{N}_{xx}\frac{\partial w_0}{\partial x}\right]_{x=x_B}\\ Q_2^e(t) &= \left[E_{xx}^b I_{yy}\frac{\partial^2 w_0}{\partial x^2}\right]_{x=x_A}\\ Q_4^e(t) &= -\left[E_{xx}^b I_{yy}\frac{\partial^2 w_0}{\partial x^2}\right]_{x=x_B}\end{aligned} \quad (10.2-3b)$$

We note from the boundary terms that deflection and rotation, w_0 and $\theta \equiv -\partial w_0/\partial x$, are the *primary variables* of the problem (see Section 3.3), and they must be continuous at every point of the domain $0 < x < L$. The primary variables are the generalized displacements. The secondary variables are the generalized forces defined in Eq. (3b).

10.2–4 Semidiscrete Finite Element Model

The third step is to develop the finite element model based on the weak form (3a). The step amounts to using the Rayleigh–Ritz method to determine $w_0(x,t)$. To this end, we seek an approximation of w_0 over the element. Since both w_0 and θ must be continuous everywhere in the domain, they should be continuous at the interface of any two elements. We denote the generalized displacements at nodes 1 and 2 of the element as

$$w_0(x_A,t) = w_1^e(t), \quad \theta(x_A) \equiv -\frac{\partial w_0}{\partial x}|_{x=x_A} = w_2^e(t)$$
$$w_0(x_B,t) = w_3^e(t), \quad \theta(x_B) \equiv -\frac{\partial w_0}{\partial x}|_{x=x_B} = w_4^e(t) \qquad (10.2-4)$$

where w_1^e and w_3^e denote transverse deflections and w_2^e and w_4^e denote the slopes at nodes 1 and 2 (see Figure 10.2–1b). Then, continuity of the generalized variables at element interfaces can be enforced by simply equating the corresponding nodal values of elements connected at a node. We will return to this point when we reconnect (or assemble) the elements.

Suppose that we seek approximation of $w_0(x,t)$ in the form

$$w_0(x,t) \approx W^e(x,t) \equiv \sum_{j=1}^{n} w_j^e(t)\varphi_j^e(x) \qquad (10.2-5)$$

where the approximation functions $\varphi_j(x)$ must be such that the following requirements are met [see Eq. (3.5–2)]:

(1) $W^e(x,t)$ satisfies the end conditions in Eq. (4), and

(2) $W^e(x,t)$ has nonzero first and second derivatives with respect to x, as dictated by the weak form (3a).

(3) The approximation functions must be *complete*. The first condition is equivalent to satisfying the essential (or geometric) boundary conditions in the Rayleigh–Ritz method. Satisfaction of the first requirement often leads to the satisfaction of the second (*i.e.*, continuity requirement). The completeness property amounts to including all lower-order terms in the polynomial leading to the highest term included, and the property ensures convergence of the finite element solution to the true solution.

In order to satisfy the four end conditions in Eq. (4), we must select a four-parameter, *i.e.*, cubic, polynomial

$$W^e(x,t) = c_1^e(t) + c_2^e(t)x + c_3^e(t)x^2 + c_4^e(t)x^3 \qquad (10.2-6)$$

which clearly satisfies the continuity and completeness requirements. The parameters c_i^e are determined such that Eq. (6) satisfies the end conditions

(4). We have

$$
\begin{aligned}
w_1^e &= W^e(x_A,t) = c_1^e + c_2^e x_A + c_3^e x_A^2 + c_4^e x_A^3\\
w_2^e &= -\frac{\partial W^e}{\partial x}(x_A,t) = -c_2^e - 2c_3^e x_A - 3c_4^e x_A^2\\
w_3^e &= W^e(x_B,t) = c_1^e + c_2^e x_B + c_3^e x_B^2 + c_4^e x_B^3\\
w_4^e &= -\frac{\partial W^e}{\partial x}(x_B,t) = -c_2^e - 2c_3^e x_B - 3c_4^e x_B^2
\end{aligned}
\tag{10.2-7}
$$

Inverting the above equations and substituting the result into Eq. (6), we obtain

$$
W^e(x,t) = w_1^e(t)\varphi_1^e(x) + w_2^e(t)\varphi_2^e(x) + w_3^e(t)\varphi_3^e(x) + w_4^e(t)\varphi_4^e(x) \tag{10.2-8a}
$$

$$
\begin{aligned}
\varphi_1^e(x) &= 1 - 3\left(\frac{x-x_A}{h_e}\right)^2 + 2\left(\frac{x-x_A}{h_e}\right)^3\\
\varphi_2^e(x) &= -(x-x_A)\left(1 - \frac{x-x_A}{h_e}\right)^2\\
\varphi_3^e(x) &= 3\left(\frac{x-x_A}{h_e}\right)^2 - 2\left(\frac{x-x_A}{h_e}\right)^3\\
\varphi_4^e(x) &= -(x-x_A)\left[\left(\frac{x-x_A}{h_e}\right)^2 - \frac{x-x_A}{h_e}\right]
\end{aligned}
\tag{10.2-8b}
$$

The approximation functions φ_i $(i = 1,2,3,4)$ are called *Hermite cubic interpolation functions*, which are derived by interpolating the function and its derivatives at the nodes. In the interest of brevity, the element label 'e' will be omitted in the following discussion.

In the Rayleigh–Ritz method, w_0 is approximated as in Eq. (5), and the weak form (3a), which is equivalent to the governing equations (1a) and the end conditions (3b), is required to be satisfied for four different choices (because there are four undetermined parameters, w_i^e) of the weight function: $v(x) = \varphi_1(x), \varphi_2(x), \varphi_3(x)$, and $\varphi_4(x)$. Substituting for w_0 from Eq. (5) and $v = \varphi_i(x)$ into Eq. (3a), we obtain the ith equation of the finite element model:

$$
\begin{aligned}
0 = \int_{x_A}^{x_B} \Bigg[& E_{xx}^t I_{yy} \frac{d^2\varphi_i}{dx^2}\left(\sum_{j=1}^4 w_j^e(t)\frac{d^2\varphi_j}{dx^2}\right) + b\hat{N}_{xx}\frac{d\varphi_i}{dx}\left(\sum_{j=1}^4 w_j^e(t)\frac{d\varphi_j}{dx}\right)\\
& + \hat{I}_0\varphi_i\left(\sum_{j=1}^4 \varphi_j \frac{d^2 w_j^e}{dt^2}\right) + \hat{I}_2\frac{d\varphi_i}{dx}\left(\sum_{j=1}^4 \frac{d\varphi_j}{dx}\frac{d^2 w_j^e}{dt^2}\right) - \varphi_i\hat{q}\Bigg] dx\\
& - Q_1^e\varphi_i(x_A) - Q_3^e\varphi_i(x_B) - Q_2^e\left(-\frac{d\varphi_i}{dx}\right)_{x_A} - Q_4^e\left(-\frac{d\varphi_i}{dx}\right)_{x_B}
\end{aligned}
$$

$$
\begin{aligned}
&= \sum_{j=1}^{4} \int_{x_A}^{x_B} \left[\left(E_{xx}^b I_{yy} \frac{d^2\varphi_i}{dx^2}\frac{d^2\varphi_j}{dx^2} + b\hat{N}_{xx}\frac{d\varphi_i}{dx}\frac{d\varphi_j}{dx} \right) w_j^e \right. \\
&\left. + \left(\hat{I}_0 \varphi_i \varphi_j + \hat{I}_2 \frac{d\varphi_i}{dx}\frac{d\varphi_j}{dx} \right) \frac{d^2 w_j^e}{dt^2} - \varphi_i \hat{q} \right] dx - Q_1^e \varphi_i(x_A) - Q_3^e \varphi_i(x_B) \\
&- Q_2^e \left(-\frac{d\varphi_i}{dx} \right)_{x_A} - Q_4^e \left(-\frac{d\varphi_i}{dx} \right)_{x_B} \\
&\equiv \sum_{j=1}^{4} \left[\left(K_{ij}^e + G_{ij}^e \right) w_j^e + M_{ij}^e \frac{d^2 w_j^e}{dt^2} \right] - F_i^e
\end{aligned}
\tag{10.2-9a}
$$

where

$$
\begin{aligned}
K_{ij}^e &= \int_{x_A}^{x_B} E_{xx}^b I_{yy} \frac{d^2\varphi_i}{dx^2}\frac{d^2\varphi_j}{dx^2}\, dx \\
G_{ij}^e &= \int_{x_A}^{x_B} b\hat{N}_{xx} \frac{d\varphi_i}{dx}\frac{d\varphi_j}{dx}\, dx \\
M_{ij}^e &= \int_{x_A}^{x_B} \left(\hat{I}_0 \varphi_i \varphi_j + \hat{I}_2 \frac{d\varphi_i}{dx}\frac{d\varphi_j}{dx} \right) dx \\
F_i^e &= q_i^e + Q_1^e \varphi_i(x_A) + Q_3^e \varphi_i(x_B) \\
&\quad + Q_2^e \left(-\frac{d\varphi_i}{dx} \right)_{x_A} + Q_4^e \left(-\frac{d\varphi_i}{dx} \right)_{x_A} \\
q_i^e &= \int_{x_A}^{x_B} \varphi_i \hat{q}\, dx
\end{aligned}
\tag{10.2-9b}
$$

Equation (9a) can be written in matrix form as

$$
\left([K^e] + [G^e] \right) \{w^e\} + [M^e]\{\ddot{w}^e\} = \{F^e\}, \quad t > 0 \tag{10.2-10}
$$

where $[K^e]$ is known as the stiffness matrix, $[G^e]$ is the geometric stiffness matrix, $[M^e]$ is the mass matrix, and $\{F^e\}$ is the force vector. All of the matrices are symmetric: $[K^e] = [K^e]^T$ or $K_{ij}^e = K_{ji}^e$, etc. In view of the following properties of the approximation functions $\varphi_j^e(x)$, the expression for F_i^e can be simplified:

$$
\varphi_{2i-1}^e(\bar{x}_j) = \delta_{ij},\ \varphi_{2i}^e(\bar{x}_j) = 0,\ \frac{d\varphi_{2i-1}^e}{dx}|_{\bar{x}_j} = 0,\ \frac{d\varphi_{2i}^e}{dx}|_{\bar{x}_j} = \delta_{ij} \tag{10.2-11}
$$

where δ_{ij} denotes the Kronecker delta

$$
\delta_{ij} = \begin{cases} 1, & \text{when } i = j \\ 0, & \text{when } i \neq j \end{cases} \tag{10.2-12}
$$

The generalized force components become

$$
F_i^e = q_i^e + Q_i^e \tag{10.2-13}
$$

which are composed of nodal forces q_i^e due to the distributed load $\hat{q}$ and the nodal reactions Q_i^e. The latter are not known *a priori.*

Equation (10) is called the *semidiscrete finite element model* because it represents only the spatial approximation of the solution. The set of ordinary differential equations (10) in time must be further reduced to algebraic equations by means of a time-approximation scheme to complete the discretization.

The coefficient matrices $[K^e], [G^e]$, and $[M^e]$ and force components q_i^e can be computed by evaluating the integrals in Eq. (9b). For example, when the geometry and material properties are elementwise constant, we have

$$[K^e] = \frac{2E_{xx}^e I_{yy}^e}{h_e^3}\begin{bmatrix} 6 & -3h_e & -6 & -3h_e \\ -3h_e & 2h_e^2 & 3h_e & h_e^2 \\ -6 & 3h_e & 6 & 3h_e \\ -3h_e & h_e^2 & 3h_e & 2h_e^2 \end{bmatrix}, \quad \{q^e\} = \frac{q_0^e h_e}{12}\begin{Bmatrix} 6 \\ -h_e \\ 6 \\ h_e \end{Bmatrix}$$

$$[M^e] = \frac{\hat{I}_0^e}{420}\begin{bmatrix} 156 & -22h_e & 54 & 13h_e \\ -22h_e & 4h_e^2 & -13h_e & -3h_e^2 \\ 54 & -13h_e & 156 & 22h_e \\ -13h_e & -3h_e^2 & 22h_e & 4h_e^2 \end{bmatrix} + \frac{\hat{I}_2^e}{30h_e}[H^e]$$

$$[H^e] = \begin{bmatrix} 36 & -3h_e & -36 & -3h_e \\ -3h_e & 4h_e^2 & 3h_e & -h_e^2 \\ -36 & 3h_e & 36 & 3h_e \\ -3h_e & -h_e^2 & 3h_e & 4h_e^2 \end{bmatrix}, \quad [G^e] = \frac{b_e \hat{N}_{xx}^e}{30h_e}[H^e] \qquad (10.2-14)$$

If the problem is one of static bending under applied transverse load $\hat{q}$ and axial load $\hat{N}_{xx}$, then all variables are only functions of position and the time–derivative term is omitted in (10). The element equation (10) takes the form

$$([K^e] + [G^e])\{w^e\} = \{F^e\} \qquad (10.2-15)$$

If the problem is one of studying buckling under the axial compressive load, $\hat{N}_{xx} = -N_{xx}^0$, Eq. (13) reduces to

$$\left([K^e] - N_{xx}^0[\bar{G}^e]\right)\{w^e\} = \{Q^e\} \qquad (10.2-16)$$

where $[\bar{G}^e]$ is the same as $[G^e]$ but without $\hat{N}_{xx}$ [see Eq. (9b)].

In the case of natural vibrations of a beam, we assume $\hat{N}_{xx} = 0$ and that the nodal values $w_i^e(t)$ and reaction forces are periodic in time

$$w_j^e(t) = W_j^e e^{-i\omega t}, \; Q_j^e(t) = \bar{Q}_j^e e^{-i\omega t}, \quad i = \sqrt{-1} \qquad (10.2-17)$$

Then Eq. (10) reduces to

$$\left([K^e] - \omega^2[M^e]\right)\{W^e\} = \{\bar{Q}^e\} \qquad (10.2-18)$$

We note that both the buckling and natural vibration problems are eigenvalue problems in which we must determine the eigenvalues λ such that

$$([A] - \lambda[B])\{W\} = \{0\}$$

for any nontrivial (*i.e.*, nonzero) $\{W\}$. We will return to these topics later in this chapter.

10.2–5 Time Discretization

In the case of time dependent response of beams, Eq. (10) governs the motion, and the ordinary differential equations in time must be integrated to determine the nodal values $w_j^e(t)$ as functions of time. Although the matrix equation (10) can be solved, in principle, using the Laplace transforms for certain load cases, it is a tedious exercise and involves more computational effort than using a numerical time–integration scheme. Here we consider the Newmark family of time–integration schemes of Section 8.7–3 that are used widely in structural dynamics (see [58–65]). We repeat the essential elements of the method here for ready reference.

In the Newmark method, the first and second time derivatives are approximated as [see Eqs. (8.7–6) and (8.7–7)]

$$\begin{aligned}\{\dot{w}^e\}_{s+1} &= \{\dot{w}^e\}_s + a_1\{\ddot{w}^e\}_s + a_2\{\ddot{w}^e\}_{s+1}\\ \{\ddot{w}^e\}_{s+1} &= a_3\left(\{w^e\}_{s+1} - \{w^e\}_s\right) - a_4\{\dot{w}^e\}_s - a_5\{\ddot{w}^e\}_s \end{aligned} \qquad (10.2-19)$$

where the superposed dot denotes differentiation with respect to time, Δt is the time increment, $\Delta t_s = t_{s+1} - t_s$, and

$$a_1 = (1-\alpha)\Delta t_s, \quad a_2 = \alpha\Delta t_s, \quad a_3 = \frac{2}{\gamma(\Delta t_s)^2}$$

$$a_4 = \Delta t_s a_3, \quad a_5 = \frac{(1-\gamma)}{\gamma} \qquad (10.2-20)$$

and $\{\cdot\}_s$, for example, denotes the value of the enclosed vector at time t_s.

The parameters α and γ in the Newmark scheme are selected such that the scheme is either stable or conditionally stable; *i.e.*, the error introduced through the time approximation (19) does not grow unboundedly as we march in time. All schemes for which $\gamma \geq \alpha \geq 1/2$ are unconditionally stable. Schemes for which $\gamma < \alpha$ and $\alpha \geq 0.5$ are conditionally stable, and the stability condition is

$$\Delta t \leq \Delta t_{cr} \equiv \left[\frac{1}{2}\omega_{max}\left(\alpha - \gamma\right)\right]^{-\frac{1}{2}} \qquad (10.2-21)$$

where ω_{max} denotes the maximum frequency associated with the discrete system (18).

As listed in Eq. (8.7–11), the Newmark family contains the following widely used schemes:

$$
\begin{aligned}
&\alpha = \frac{1}{2}\,, \quad \gamma = \frac{1}{2}\,, \quad \text{the constant average acceleration method (stable)}\\
&\alpha = \frac{1}{2}\,, \quad \gamma = \frac{1}{3}\,, \quad \text{the linear acceleration method (conditionally stable)}\\
&\alpha = \frac{1}{2}\,, \quad \gamma = \frac{1}{6}\,, \quad \text{the Fox–Goodwin scheme (conditionally stable)}\\
&\alpha = \frac{1}{2}\,, \quad \gamma = 0\,, \quad \text{the central difference method (conditionally stable)}\\
&\alpha = \frac{3}{2}\,, \quad \gamma = \frac{8}{5}\,, \quad \text{the Galerkin method (stable)}\\
&\alpha = \frac{3}{2}\,, \quad \gamma = 2\,, \quad \text{the backward difference method (stable)}
\end{aligned}
\tag{10.2 – 22}
$$

Premultiplying the second equation in (19) with $[M^e]_{s+1}$ and using Eq. (10) at $t = t_{s+1}$ to replace $[M^e]_{s+1}\{w^e\}_{s+1}$, we obtain

$$
[\hat{K}^e]\{w^e\}_{s+1} = \{\hat{F}^e\} \tag{10.2 – 23}
$$

$$
\begin{aligned}
[\hat{K}^e] &= ([K^e]_{s+1} + [G^e]_{s+1}) + a_3[M^e]_{s+1}\\
\{\hat{F}^e\} &= \{F^e\}_{s+1} + [M^e]_{s+1}\left(a_3\{w^e\}_s + a_4\{\dot{w}^e\}_s + a_5\{\ddot{w}^e\}_s\right)
\end{aligned}
\tag{10.2 – 24}
$$

Note that for the central difference scheme ($\gamma = 0$), it is necessary to use an alternative form of Eq. (23) [see Eq. (8.7–14)].

Equation (23) represents a system of algebraic equations among the (discrete) values of $\{w^e(t)\}$ at time $t = t_{s+1}$ in terms of known values at time $t = t_s$. At the first time step (*i.e.*, $s = 0$), the values $\{w^e\}_0 = \{w^e(0)\}$ and $\{\dot{w}^e\}_0 = \{\dot{w}^e(0)\}$ are known from the initial conditions of the problem, and Eq. (10) is used to determine $\{\ddot{w}^e\}_0$ at $t = 0$:

$$
\{\ddot{w}^e\}_0 = [M^e]^{-1}\left\{\{F^e\} - ([K^e] + [G^e])\{w^e\}_0\right\} \tag{10.2 – 25}
$$

It is appropriate here to introduce the terminology of *explicit* and *implicit* numerical schemes. An explicit time integration scheme is one which yields a system of algebraic equations of the form

$$
\begin{bmatrix}
a_{11} & 0 & \cdots & 0\\
a_{21} & a_{22} & \cdots & 0\\
\cdots & \cdots & \cdots & \cdots\\
a_{n1} & \cdots & a_{n(n-1)} & a_{nn}
\end{bmatrix}
\begin{Bmatrix} x_1\\ x_2\\ \vdots\\ x_n \end{Bmatrix}
=
\begin{Bmatrix} b_1\\ b_2\\ \vdots\\ b_n \end{Bmatrix}
$$

which can be solved without inverting the matrix $[A]$. Clearly, a diagonal coefficient matrix would also meet the requirement. A scheme that leads to a system of equations of the form $[A]\{x\} = \{b\}$ whose solution can be determined only by inverting the matrix $[A]$ is termed an implicit scheme. A close examination of Eq. (23) or (24) shows that the coefficient matrix $[\hat{K}]$ or $[\bar{K}]$ is symmetric and banded but not diagonal, in general, for any of the schemes listed in Eq. (22). Therefore, by definition they are all implicit schemes. However, for the central difference scheme ($\gamma = 0$) we have $[\bar{K}] = [M]$, and if the mass matrix $[M]$ is diagonal, then it qualifies as an explicit scheme. The finite difference approximation of spatial variations of a problem often leads to a diagonal mass matrix, and therefore the central difference scheme is referred to as an implicit scheme. However, this is not the case with the finite element method. In general, the mass matrix derived [see Eq. (9b)] in the finite element method is not diagonal, and it is called a *consistent mass matrix* because its definition is consistent with the formulation of the problem and the spatial approximation used. In the interest of reducing the computational time of solving the assembled equations associated with (23) when the central difference method is used, one may diagonalize the mass matrix by some rational means, *e.g.*, dividing the total mass of the element equally among the element nodes. A diagonalized mass matrix of a finite element model is called the *lumped mass matrix* (see [1–5]). In summary, a time approximation scheme should be called explicit or implicit only in association with the model where it is used; the model dictates the implicit or explicit nature of the resulting system of equations.

10.2–6 Assembly of Element Equations

All of the development so far has been for a typical element. The element equations (15), (16), (18), or (23) cannot be solved at the element level because we have eight unknowns (w_j^e and Q_j^e) but only four equations. This is consistent with the fact one cannot solve the element equations independent of the total problem. To obtain exactly the same number of equations as there are unknowns in a problem, it is necessary to combine the equations of various elements in a meaningful way. First, we note that the primary variables, namely, the deflection and rotation, must be continuous everywhere, including at the element interfaces. For example, if the node 2 of element e is connected to node 1 of element f, then the continuity requirement can be expressed as (see Figure 10.2–2a)

$$w_3^e = w_1^f, \quad w_4^e = w_2^f \tag{10.2 - 26a}$$

Since elements e and f are connected at a node, the reaction forces from the two elements at the common node should be in equilibrium (see Figure 10.2–2b)

$$Q_3^e + Q_1^f = F_0, \quad Q_4^e + Q_2^f = M_0 \tag{10.2 - 26b}$$

where F_0 and M_0 are the applied transverse concentrated load (positive up) and bending moment (clockwise positive), respectively, at the node where the two elements are connected. If there is no applied load F_0 or bending moment M_0 at the node, their values in Eq. (26b) should be set to zero.

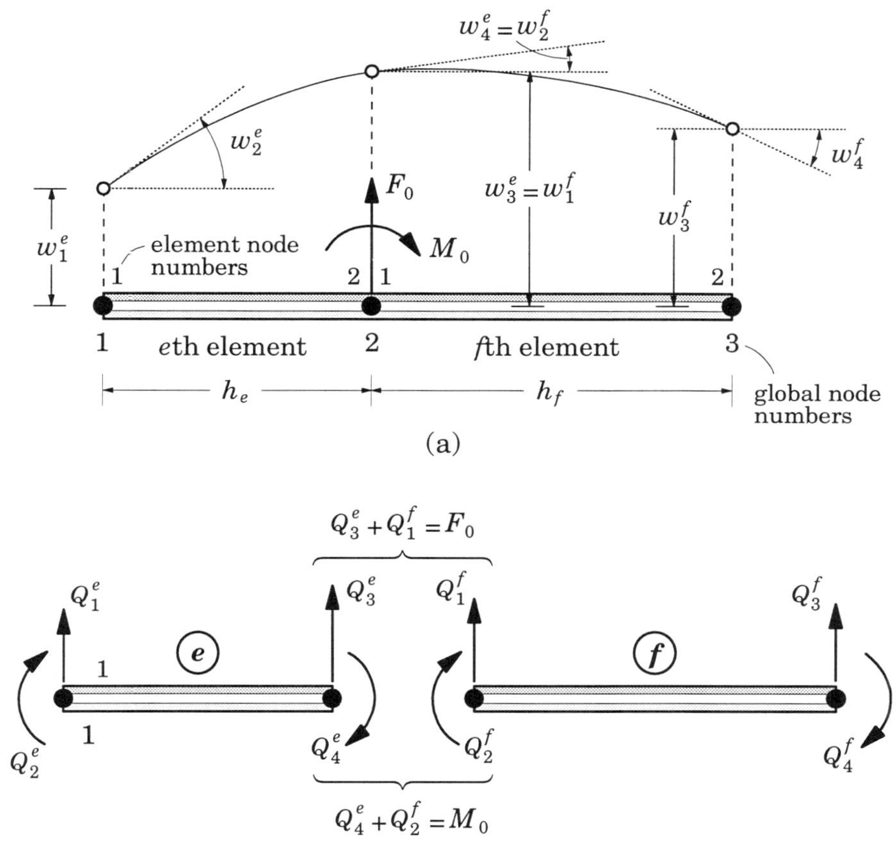

Figure 10.2–2: Assembly of two beam finite elements. (a) Continuity of generalized displacements. (b) Balance of generalized internal forces.

Thus, the assembly of all elements in a finite element mesh is based on the interelement continuity of primary variables and balance of secondary variables. Note that the external nodal forces q_i^e due to the distributed load do not enter Eq. (26b). The balance equations (26b) require that we must add the third equation of element e with the first equation of element f, and add the fourth equation of element e with the second equation of element f. Therefore, there will be only two equations for the node common to the two elements. If a beam is divided into N elements (see Figure 10.2–1a), there will be a total $N+1$ nodes, including $N-1$ internal nodes. Consequently, there are $2(N-1)$ equations from the internal nodes and two equations from each of the boundary nodes (node 1 and $N+1$) – a total of $2(N+1)$ equations. In view of the balance conditions (26b), generalized forces at the internal nodes are replaced by known applied forces. The generalized forces at node 1 and node $N+1$ will remain as a part of the $2(N+1)$ equations. In addition, there are $2(N+1)$ generalized displacements. Thus the total number of unknowns is equal to $2(N+1)+4$, whereas the number of equations is $2(N+1)$. The additional four equations are provided by the four boundary conditions of the problem. The boundary conditions involve specifying either a generalized displacement or the corresponding generalized force, but never both from the same pair.

As an example of the assembled equations, we consider a three-element mesh for static bending analysis. The same ideas apply for buckling, free vibration, and transient analyses. The assembled equations are

$$[K]\{U\} = \{F\} \tag{10.2-27a}$$

$$[K] = \begin{bmatrix} K_{11}^{(1)} & K_{12}^{(1)} & K_{13}^{(1)} & K_{14}^{(1)} & 0 & 0 & 0 & 0 \\ K_{21}^{(1)} & K_{22}^{(1)} & K_{23}^{(1)} & K_{24}^{(1)} & 0 & 0 & 0 & 0 \\ K_{31}^{(1)} & K_{32}^{(1)} & K_{33}^{(1)}+K_{11}^{(2)} & K_{34}^{(1)}+K_{12}^{(2)} & K_{13}^{(2)} & K_{14}^{(2)} & 0 & 0 \\ K_{41}^{(1)} & K_{42}^{(1)} & K_{43}^{(1)}+K_{21}^{(2)} & K_{44}^{(1)}+K_{22}^{(2)} & K_{23}^{(2)} & K_{24}^{(2)} & 0 & 0 \\ 0 & 0 & K_{31}^{(2)} & K_{32}^{(2)} & K_{33}^{(2)}+K_{11}^{(3)} & K_{34}^{(2)}+K_{12}^{(3)} & K_{13}^{(3)} & K_{14}^{(3)} \\ 0 & 0 & K_{41}^{(2)} & K_{42}^{(2)} & K_{43}^{(2)}+K_{21}^{(3)} & K_{44}^{(2)}+K_{22}^{(3)} & K_{23}^{(3)} & K_{24}^{(3)} \\ 0 & 0 & 0 & 0 & K_{31}^{(3)} & K_{32}^{(3)} & K_{33}^{(3)} & K_{34}^{(3)} \\ 0 & 0 & 0 & 0 & K_{41}^{(3)} & K_{42}^{(3)} & K_{43}^{(3)} & K_{44}^{(3)} \end{bmatrix}$$

$$\{U\} = \begin{Bmatrix} U_1 \\ U_2 \\ U_3 \\ U_4 \\ U_5 \\ U_6 \\ U_7 \\ U_8 \end{Bmatrix}, \quad \{F\} = \begin{Bmatrix} q_1^{(1)} \\ q_2^{(1)} \\ q_3^{(1)}+q_1^{(2)} \\ q_4^{(1)}+q_2^{(2)} \\ q_3^{(2)}+q_1^{(3)} \\ q_4^{(2)}+q_2^{(3)} \\ q_3^{(3)} \\ q_4^{(3)} \end{Bmatrix} + \begin{Bmatrix} Q_1^{(1)} \\ Q_2^{(1)} \\ Q_3^{(1)}+Q_1^{(2)} \\ Q_4^{(1)}+Q_2^{(2)} \\ Q_3^{(2)}+Q_1^{(3)} \\ Q_4^{(2)}+Q_2^{(3)} \\ Q_3^{(3)} \\ Q_4^{(3)} \end{Bmatrix} \tag{10.2-27b}$$

where U_I $(I = 1, 2, \cdots, 8)$ denote the generalized displacements associated with the global nodes. The Jth global node will have U_{2J-1} as the transverse deflection and U_{2J} as the rotation.

10.2–7 Imposition of Boundary Conditions

To illustrate the imposition of boundary conditions on the assembled equations, we consider static bending of a beam of length $L = 3h$, clamped at $x = 0$, and on a roller support at $x = 2h$. Suppose that a uniformly distributed load of intensity q_0 (per unit area) is distributed over the span $h \leq x \leq 2h$, and the free end $x = 3h$ is subjected to a point load F_0 and bending moment M_0 (see Figure 10.2–3). The beam must be modeled, because of the loading and boundary conditions, using three elements.

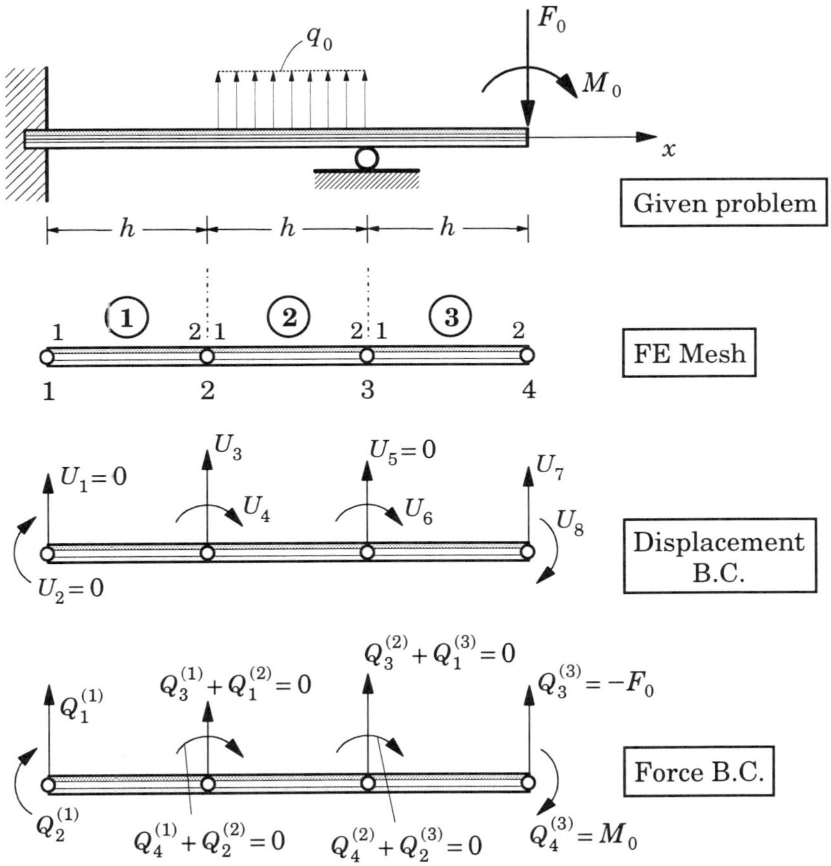

Figure 10.2–3: Bending of an indeterminate beam. Illustration of continuity of displacements, equilibrium of internal forces, and boundary conditions.

The geometric boundary conditions of the problem are

$$w_0(0) = 0, \quad \frac{dw_0}{dx}(0) = 0, \quad w_0(2h) = 0 \tag{10.2 - 28a}$$

Expressing the geometric conditions in terms of the global nodal variables, we have

$$U_1 = 0, \quad U_2 = 0, \quad U_5 = 0 \qquad (10.2-28b)$$

The force boundary conditions are

$$-\left[\frac{\partial}{\partial x}\left(E^b_{xx} I_{yy} \frac{\partial^2 w_0}{\partial x^2}\right)\right]_{x=3h} = -F_0, \quad \left[E^b_{xx} I_{yy} \frac{\partial^2 w_0}{\partial x^2}\right]_{x=3h} = M_0 \qquad (10.2-29a)$$

which can be expressed in terms of the nodal forces [see Eq. (3b)]

$$Q_3^{(3)} = -F_0, \quad Q_4^{(3)} = M_0 \qquad (10.2-29b)$$

The equilibrium of internal forces requires

$$Q_3^{(1)} + Q_1^{(2)} = 0, \quad Q_4^{(1)} + Q_2^{(2)} = 0, \quad Q_4^{(2)} + Q_2^{(3)} = 0 \qquad (10.2-29c)$$

The reaction force $Q_1^{(1)}$ and bending moment $Q_2^{(1)}$ at $x = 0$ and reaction force $Q_3^{(2)}$ at $x = 2h$ are unknown, and can be determined in the postcomputation.

Using the boundary conditions (28b) and (29b) and equilibrium conditions (29c) in the assembled equations, we obtain

$$\{U\} = \begin{Bmatrix} 0 \\ 0 \\ U_3 \\ U_4 \\ 0 \\ U_6 \\ U_7 \\ U_8 \end{Bmatrix}, \quad \{F\} = \frac{q_0 bh}{12}\begin{Bmatrix} 0 \\ 0 \\ 6 \\ -h \\ 6 \\ h \\ 0 \\ 0 \end{Bmatrix} + \begin{Bmatrix} Q_1^{(1)} \\ Q_2^{(1)} \\ 0 \\ 0 \\ Q_3^{(2)} + Q_1^{(3)} \\ 0 \\ -F_0 \\ M_0 \end{Bmatrix} \qquad (10.2-30)$$

There are five unknown generalized displacements and three unknown generalized forces to be determined from the eight equations. Often, the unknown generalized displacements are determined first, and the generalized forces are determined subsequently. To determine the unknown generalized displacements U_3, U_4, U_6, U_7, and U_8, we use the corresponding equations, equations 3, 4, 6, 7, and 8, which do not contain the unknown generalized forces. Since the specified displacements are zero, one can omit the rows and columns (1, 2, and 5) corresponding to the specified displacements. When the specified values are nonzero, the products of stiffnesses with the known displacements are moved to the right side of the equation. We have

$$\begin{bmatrix} K_{33}^{(1)} + K_{11}^{(2)} & K_{34}^{(1)} + K_{12}^{(2)} & K_{14}^{(2)} & 0 & 0 \\ K_{43}^{(1)} + K_{21}^{(2)} & K_{44}^{(1)} + K_{22}^{(2)} & K_{24}^{(2)} & 0 & 0 \\ K_{41}^{(2)} & K_{42}^{(2)} & K_{44}^{(2)} + K_{22}^{(3)} & K_{23}^{(3)} & K_{24}^{(3)} \\ 0 & 0 & K_{32}^{(3)} & K_{33}^{(3)} & K_{34}^{(3)} \\ 0 & 0 & K_{42}^{(3)} & K_{43}^{(3)} & K_{44}^{(3)} \end{bmatrix} \begin{Bmatrix} U_3 \\ U_4 \\ U_6 \\ U_7 \\ U_8 \end{Bmatrix}$$

$$
= \frac{q_0 bh}{12} \begin{Bmatrix} 6 \\ -h \\ h \\ 0 \\ 0 \end{Bmatrix} + \begin{Bmatrix} 0 \\ 0 \\ 0 \\ -F_0 \\ M_0 \end{Bmatrix} \tag{10.2-31a}
$$

The unknown generalized forces can be computed from equations 1, 2, and 5 of Eq. (27a):

$$
\begin{Bmatrix} Q_1^{(1)} \\ Q_2^{(1)} \\ Q_3^{(2)} + Q_1^{(3)} \end{Bmatrix} = \begin{bmatrix} K_{13}^{(1)} & K_{14}^{(1)} & 0 & 0 & 0 \\ K_{23}^{(1)} & K_{24}^{(1)} & 0 & 0 & 0 \\ K_{31}^{(2)} & K_{32}^{(2)} & K_{34}^{(2)} + K_{12}^{(3)} & K_{13}^{(3)} & K_{14}^{(3)} \end{bmatrix} \begin{Bmatrix} U_3 \\ U_4 \\ U_6 \\ U_7 \\ U_8 \end{Bmatrix} - \begin{Bmatrix} 0 \\ 0 \\ \frac{q_0 bh}{2} \end{Bmatrix} \tag{10.2-31b}
$$

10.2–8 Postprocessing

Once the nodal values of the generalized displacements are known, the transverse deflection and its derivatives up to and including order three can be determined from Eq. (8a) at any point within the element. Of course, only the deflection and its first derivatives will be continuous at the nodes, and second and third derivatives of w_0 will be discontinuous. If shear force and bending moments at any interior points are required, one may use Eqs. (3b). If these quantities are computed from Eq. (3b), they will be discontinuous, *i.e.*, the shear force and bending moments computed from two adjacent elements at the common node will be different. Equation (31b) gives unique and most accurate, if not exact, values of the shear force and bending moment at the nodes. The Hermite cubic beam finite element gives exact values of the deflection and slope at the nodes whenever the flexural stiffness is elementwise constant (see Reddy [1]).

This completes the finite element model development of laminated beams. Since the equation of motion governing the *cylindrical bending* of plate strips according to the classical laminate theory [see Eq. (6.4–2c)] is, in the absence of in-plane displacements, identical to that of symmetrically laminated beams, all of the discussion presented on finite element model development applies to cylindrical bending with the replacement of $E_{xx}^b I_{yy}$ with D [see Eq. (6.4–3c)].

10.3 Timoshenko Beam/Plate Theory

10.3–1 Governing Equations

The equations of motion governing symmetrically laminated beams according to the first–order shear deformation (*i.e.*, the Timoshenko) beam theory are [see Eq. (6.3–9)]

$$KG^b_{xz}bh\left(\frac{\partial^2 w_0}{\partial x^2}+\frac{\partial \phi_x}{\partial x}\right)+b\hat{N}_{xx}\frac{\partial^2 w_0}{\partial x^2}+\hat{q}=\hat{I}_0\frac{\partial^2 w_0}{\partial t^2} \qquad (10.3-1a)$$

$$E^b_{xx}I_{yy}\frac{\partial^2 \phi_x}{\partial x^2}-KG^b_{xz}bh\left(\frac{\partial w_0}{\partial x}+\phi_x\right)=\hat{I}_2\frac{\partial^2 \phi_x}{\partial t^2} \qquad (10.3-1b)$$

where

$$\hat{q}=bq, \quad \hat{I}_0=bI_0, \quad \hat{I}_2=bI_2 \qquad (10.3-1c)$$

We divide the total domain of the beam into a set of finite elements, isolate a typical finite element, construct weak forms of Eqs. (1a,b) over the typical element, assume interpolation of the primary variables, and develop the finite element model.

10.3–2 Weak Forms

The weak forms of Eqs. (1a,b) over the total domain $0 < x < L$ were constructed in Section 6.8 in connection with the Rayleigh–Ritz method. The same procedure can be used over the domain $x_A < x < x_B$ of a typical element to obtain the following weak forms:

$$0=\int_{x_A}^{x_B}\left[KG^b_{xz}bh\frac{dv_1}{dx}\left(\frac{\partial w_0}{\partial x}+\phi_x\right)+b\hat{N}_{xx}\frac{dv_1}{dx}\frac{\partial w_0}{\partial x}+v_1\hat{I}_0\frac{\partial^2 w_0}{\partial t^2}-v_1\hat{q}\right]dx$$
$$-Q^e_1 v_1(x_A)-Q^e_3 v_1(x_B) \qquad (10.3-2a)$$

$$0=\int_{x_A}^{x_B}\left[E^b_{xx}I_{yy}\frac{dv_2}{dx}\frac{\partial \phi_x}{\partial x}+KG^b_{xz}bh\ v_2\left(\frac{\partial w_0}{\partial x}+\phi_x\right)+\hat{I}_2 v_2\frac{\partial^2 \phi_x}{\partial t^2}\right]dx$$
$$-Q^e_2 v_2(x_A)-Q^e_4 v_2(x_B) \qquad (10.3-2b)$$

where v_1 and v_2 are weight functions, which have the same physical meaning (from energy considerations) as the spatial variations of w_0 and ϕ_x, respectively (see Reddy [1], p. 178). The Q^e_i have the same physical meaning as in the classical beam element [see Eq. (10.2–3)] but their relationship to the transverse deflection and rotation, w_0 and ϕ_x, is different:

$$Q^e_1(t)=-\left[KG^b_{xz}bh\left(\frac{\partial w_0}{\partial x}+\phi_x\right)+b\hat{N}_{xx}\frac{\partial w_0}{\partial x}\right]_{x=x_A}$$
$$Q^e_3(t)=\left[KG^b_{xz}bh\left(\frac{\partial w_0}{\partial x}+\phi_x\right)+b\hat{N}_{xx}\frac{\partial w_0}{\partial x}\right]_{x=x_B}$$
$$Q^e_2(t)=-\left[E^b_{xx}I_{yy}\frac{\partial \phi_x}{\partial x}\right]_{x=x_A}$$
$$Q^e_4(t)=\left[E^b_{xx}I_{yy}\frac{\partial \phi_x}{\partial x}\right]_{x=x_B} \qquad (10.3-3)$$

10.3–3 Semidiscrete Finite Element Model

is linear An examination of the weak form suggests that $w_0(x,t)$ and $\phi_x(x,t)$ are the primary variables and therefore must be carried as nodal variables. Suppose that

$$\begin{aligned} w_0(x_A,t) &= w_1^e(t), \quad w_0(x_B,t) = w_2^e(t) \\ \phi_x(x_A,t) &= \Phi_1^e(t), \quad \phi_x(x_B,t) = \Phi_2^e(t) \end{aligned} \tag{10.3-4}$$

where w_1^e and w_2^e denote transverse deflections and Φ_1^e and Φ_2^e denote the rotations at nodes 1 and 2, respectively. Thus, w_0 and ϕ_x each should satisfy the end conditions in (4) and be continuous with nonzero first derivatives. Purely from mathematical considerations, the lowest polynomial that satisfies these requirements for each of the variables:

$$w_0(x,t) \approx a_1 + a_2 x, \quad \phi_x(x,t) \approx b_1 + b_2 x \tag{10.3-5}$$

From physical considerations, this choice may not be consistent because deflection is a higher-order function than rotation. More discussion on this point is given later in this section.

Using Eqs. (4) and (5), the parameters a_1, a_2, b_1, and b_2 are expressed in terms of the nodal values of w_0 and ϕ_x:

$$\begin{aligned} w_1^e(t) &= a_1 + a_2 x_A, \quad w_2^e(t) = a_1 + a_2 x_B \\ \Phi_1^e(t) &= b_1 + b_2 x_A, \quad \Phi_2^e(t) = b_1 + b_2 x_B \end{aligned} \tag{10.3-6a}$$

or

$$\begin{aligned} a_1 &= \frac{w_1^e x_B - w_2^e x_A}{x_B - x_A}, \quad a_2 = \frac{w_2^e - w_1^e}{x_B - x_A} \\ b_1 &= \frac{\Phi_1^e x_B - \Phi_2^e x_A}{x_B - x_A}, \quad b_2 = \frac{\Phi_2^e - \Phi_1^e}{x_B - x_A} \end{aligned} \tag{10.3-6b}$$

Substituting (6b) into (6a) yields the finite element approximations

$$\begin{aligned} w^e(x,t) &\approx w_1^e(t)\psi_1^e(x) + w_2^e(t)\psi_2^e(x) \\ \phi^e(x,t) &\approx \Phi_1^e(t)\psi_1^e(x) + \Phi_2^e(t)\psi_2^e(x) \end{aligned} \tag{10.3-7a}$$

where

$$\psi_1^e(x) = \frac{x_B - x}{x_B - x_A}, \quad \psi_2^e(x) = \frac{x - x_A}{x_B - x_A} \tag{10.3-7b}$$

The $\psi_i^e(x)$ are called the *Lagrange interpolation functions,* which are derived by interpolating only the function and not its derivatives at the nodes. The Lagrange interpolation functions (of any degree) have the following (interpolation) properties:

$$\psi_j^e(x_i) = \delta_{ij}, \quad \sum_{j=1}^{n} \psi_j^e(x) = 1 \tag{10.3-8}$$

where x_i denotes the coordinate of the ith node.

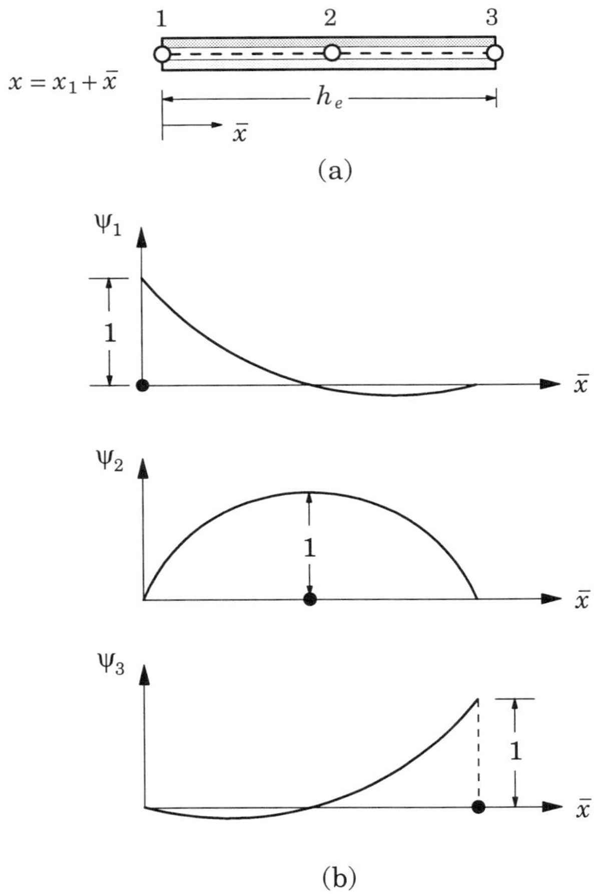

Figure 10.3–1: Quadratic Lagrange line element and interpolation functions.

Higher-order approximations of w_0 and ϕ_x can be developed using the same ideas. For example, quadratic approximation of $w_0(x,t)$ is given by

$$w_0(x,t) \approx a_1 + a_2 x + a_3 x^2 = \sum_{j=1}^{3} w_j^e(t)\psi_j^e(x) \tag{10.3 – 9}$$

where $w_j^e(t)$ denotes the value of $w(x,t)$ at the jth node of the element and ψ_j^e are the quadratic interpolation functions. These functions can be derived either using the procedure described for the linear interpolation functions or using the first property listed in Eq. (8). For example, $\psi_1^e(x)$ must be such that it is zero at nodes 2 and 3 (see Figure 10.3–1), and unity at its node, node 1. Thus, $\psi_j^e(x)$ is of the form

$$\psi_1^e(x) = c_1(x - x_2)(x - x_3)$$

where x_i is the coordinate of the ith node, and c_1 is a constant to be determined such that ψ_1^e is unity at $x = x_1$. We have

$$c_1 = \frac{1}{(x_1 - x_2)(x_1 - x_3)}$$

Hence

$$\psi_1^e(x) = \frac{(x - x_2)(x - x_3)}{(x_1 - x_2)(x_1 - x_3)} \tag{10.3 – 10a}$$

Similarly, we obtain

$$\psi_2^e(x) = \frac{(x - x_1)(x - x_3)}{(x_2 - x_1)(x_2 - x_3)}, \quad \psi_3^e(x) = \frac{(x - x_1)(x - x_2)}{(x_3 - x_1)(x_3 - x_2)} \tag{10.3 – 10b}$$

Note that except for the two end nodes, all other nodes are interior to the element. For a quadratic element, the midpoint is often chosen as the third node.

In general w_0 and ϕ_x need not be approximated by the same degree polynomials, although w_0 should be approximated using the same or greater degree polynomials than those used for ϕ_x. Suppose that

$$w_0(x,t) = \sum_{j=1}^{m} w_j^e(t)\psi_j^e(x), \quad \phi_x(x,t) = \sum_{j=1}^{n} \Phi_j^e(t)\Theta_j^e(x), \quad m \geq n \tag{10.3 – 11}$$

where $\psi_j^e(x)$ and $\Theta_j^e(x)$ are the Lagrange interpolation functions of degree $(m-1)$ and $(n-1)$, respectively. Substitution of (11) for w_0 and ϕ_x, and $v_1 = \psi_i^e$ and $v_2 = \Theta_j^e$, into the weak forms (2a,b) yields the semidiscrete finite element model

$$\begin{aligned} 0 &= \sum_{j=1}^{m} \left(K_{ij}^{11} + G_{ij}^e\right) w_j^e + \sum_{k=1}^{n} K_{ik}^{12}\Phi_k^e + \sum_{j=1}^{m} M_{ij}^{11}\frac{d^2 w_j^e}{dt^2} - F_i^1 \\ 0 &= \sum_{j=1}^{m} K_{jk}^{12} w_j^e + \sum_{j=1}^{n} K_{kj}^{22}\Phi_j^e + \sum_{j=1}^{n} M_{ij}^{22}\frac{d^2\Phi_j^e}{dt^2} - F_i^2 \end{aligned} \tag{10.3 – 12a}$$

where

$$\begin{aligned} K_{ij}^{11} &= \int_{x_A}^{x_B} KG_{xz}^b bh\frac{d\psi_i}{dx}\frac{d\psi_j}{dx}\, dx, \quad G_{ij}^e = \int_{x_A}^{x_B} b\hat{N}_{xx}\frac{d\psi_i}{dx}\frac{d\psi_j}{dx}\, dx \\ K_{ij}^{12} &= \int_{x_A}^{x_B} KG_{xz}^b bh\frac{d\psi_i}{dx}\Theta_j\, dx \end{aligned}$$

$$
\begin{aligned}
K_{ij}^{22} &= \int_{x_A}^{x_B} \left(E_{xx}^{b} I_{yy} \frac{d\Theta_i}{dx}\frac{d\Theta_j}{dx} + KG_{xz}^{b} bh \Theta_i \Theta_j \right) dx \\
M_{ij}^{11} &= \int_{x_A}^{x_B} \hat{I}_0 \psi_i \psi_j \, dx, \ M_{ij}^{22} = \int_{x_A}^{x_B} \hat{I}_2 \Theta_i \Theta_j \, dx \\
F_i^1 &= \int_{x_A}^{x_B} \psi_i \hat{q} \, dx + Q_1^e \psi_i(x_A) + Q_3^e \psi_i(x_B) \\
F_i^2 &= Q_2^e \Theta_i(x_A) + Q_4^e \Theta_i(x_B)
\end{aligned} \qquad (10.3-12b)
$$

The element equations (12a) can be expressed in matrix form as

$$
\begin{bmatrix} [K^{11}] & [K^{12}] \\ [K^{12}]^{\mathrm{T}} & [K^{22}] \end{bmatrix} \begin{Bmatrix} \{w\} \\ \{\Phi\} \end{Bmatrix} + \begin{bmatrix} [G] & [0] \\ [0] & [0] \end{bmatrix} \begin{Bmatrix} \{w\} \\ \{\Phi\} \end{Bmatrix} + \begin{bmatrix} [M^{11}] & [0] \\ [0] & [M^{22}] \end{bmatrix} \begin{Bmatrix} \{\ddot{w}\} \\ \{\ddot{\Phi}\} \end{Bmatrix} = \begin{Bmatrix} \{F^1\} \\ \{F^2\} \end{Bmatrix} \qquad (10.3-13)
$$

Shear Locking and Its Alleviation

When both w_0 and ϕ_x are approximated using linear interpolation functions, say $w_0 = a + bx$ and $\phi_x = c + dx$, the element becomes very stiff in the thin beam limit. Such behavior is known as *shear locking.* Note that when the beam is thin the transverse shear strain $(dw_0/dx) + \phi_x = a + c + dx$ should vanish for all x, requiring $a + c = 0$ and $d = 0$. On the other hand, the condition $d = 0$ (or $\phi_x = c$, a constant) makes the bending energy zero. Consequently, the element predicts zero solution in an attempt to satisfy the constraint, $(dw_0/dx) + \phi_x = 0$. To overcome the locking, one may use equal interpolation for both w_0 and ϕ_x but use a lower-order polynomial for the shear strain, $\gamma_{xz} = (dw_0/dx) + \phi_x$. This is often realized by using reduced integration to evaluate the stiffness coefficients associated with the transverse shear strain in the element stiffness matrix. All other coefficient matrices are evaluated using full integration. For this reduced integration element (RIE), the element matrices are given by

$$
[K^e] = \frac{KG_{xz}^b bh}{4L_e} \begin{bmatrix} 4 & -2L_e & -4 & -2L_e \\ -2L_e & L_e^2 + \mathcal{E} & 2L_e & L_e^2 + \mathcal{E} \\ -4 & 2L_e & 4 & 2L_e \\ -2L_e & L_e^2 - \mathcal{E} & 2L_e & L_e^2 + \mathcal{E} \end{bmatrix}, \quad \{q^e\} = \frac{q_0^e h_e}{2} \begin{Bmatrix} 1 \\ 0 \\ 1 \\ 0 \end{Bmatrix}
$$

$$
[M^e] = \frac{L_e}{6} \begin{bmatrix} 2\hat{I}_0 & 0 & \hat{I}_0 & 0 \\ 0 & 2\hat{I}_2 & 0 & \hat{I}_2 \\ \hat{I}_0 & 0 & 2\hat{I}_0 & 0 \\ 0 & \hat{I}_2 & 0 & 2\hat{I}_2 \end{bmatrix}, \quad [G^e] = \frac{b_e \hat{N}_{xx}^e}{L_e} \begin{bmatrix} 1 & 0 & -1 & 0 \\ 0 & 0 & 0 & 0 \\ -1 & 0 & 1 & 0 \\ 0 & 0 & 0 & 0 \end{bmatrix} \qquad (10.3-14)
$$

where L_e denotes the element length, $\mathcal{E} = 4E^b_{xx}I_{yy}/KG^b_{xz}bh$, and the elements of the nodal unknowns are rearranged in the order $(w^e_1, \Phi^e_1, w^e_2, \Phi^e_2)$.

Alternatively, one can use quadratic interpolation of w_0 and linear interpolation of ϕ_x, and use full integration to evaluate all stiffness coefficients, to overcome the locking. For such a consistent interpolation element (CIE) the stiffness matrix is of the order 5×5. If the deflection degree of freedom associated with the internal node is eliminated by static condensation, the resulting 4×4 stiffness matrix $[K^e]$ is identical to that for the reduced integration element (RIE) discussed above (see Reddy [1]). However, the force vector, for a uniformly distributed transverse load, is the same as that of the Hermite cubic element (see Eq. (10.2–14)). If (Hermite) cubic interpolation of w_0 and quadratic interpolation of ϕ_x is used, one may obtain a consistent interpolation Timoshenko finite element that predicts exact deflections and rotations at the nodes for all problems in which the flexural stiffness EI is constant (see Reddy *et al.* [90]).

10.3–4 Fully Discretized Finite Element Models

Bending

For static analysis, Eq. (13) takes the form

$$\begin{bmatrix} [K^{11}] & [K^{12}] \\ [K^{12}]^{\mathrm{T}} & [K^{22}] \end{bmatrix} \begin{Bmatrix} \{w\} \\ \{\Phi\} \end{Bmatrix} + \begin{bmatrix} [G] & [0] \\ [0] & [0] \end{bmatrix} \begin{Bmatrix} \{w\} \\ \{\Phi\} \end{Bmatrix} = \begin{Bmatrix} \{F^1\} \\ \{F^2\} \end{Bmatrix} \qquad (10.3-15)$$

which can be assembled and solved after imposing boundary conditions of the problem. In Eq. (15) it is assumed that $\hat{N}_{xx}$ appearing in the definition of G^e_{ij} is specified.

Buckling

For a study of buckling under compressive loads, $\hat{N}_{xx} = -N^0_{xx}$, Eq. (15) becomes

$$\begin{bmatrix} [K^{11}] & [K^{12}] \\ [K^{12}]^{\mathrm{T}} & [K^{22}] \end{bmatrix} \begin{Bmatrix} \{w\} \\ \{\Phi\} \end{Bmatrix} - bN^0_{xx} \begin{bmatrix} [\bar{G}] & [0] \\ [0] & [0] \end{bmatrix} \begin{Bmatrix} \{w\} \\ \{\Phi\} \end{Bmatrix} = \begin{Bmatrix} \{\bar{F}^1\} \\ \{F^2\} \end{Bmatrix} \qquad (10.3-16a)$$

where

$$\bar{G}^e_{ij} = \int_{x_A}^{x_B} \frac{d\psi_i}{dx}\frac{d\psi_j}{dx}\,dx, \quad \{\bar{F}^1\} = \{Q^e_1 \;\; Q^e_3\}^T \qquad (10.3-16b)$$

Since the coefficient matrix $[B]$ in the eigenvalue problem $[A]\{X\} - \lambda[B]\{X\} = \{0\}$ is singular, it is desirable to eliminate the nodal variables associated with the rotation by using the static condensation procedure discussed in Section 8.3–3. Solve the second matrix equation in (16a) for $\{\Phi\}$ in terms of $\{w\}$

$$\{\Phi\} = [K^{22}]^{-1}\left(\{F^2\} - [K^{21}]^{\mathrm{T}}\{w\}\right) \qquad (10.3-17)$$

and substitute into the first matrix equation of (16a) to obtain

$$[\hat{K}]\{w\} - bN_{xx}^0[\bar{G}]\{w\} = \{\hat{F}\} \qquad (10.3-18a)$$

where

$$\begin{aligned}[\hat{K}] =&[K^{11}] - [K^{12}][K^{22}]^{-1}[K^{12}]^{\mathrm{T}} \\ \{\hat{F}\} =&\{F^1\} - [K^{12}][K^{22}]^{-1}\{F^2\}\end{aligned} \qquad (10.3-18b)$$

Equation (18a) can be assembled and solved for the buckling loads after imposition of boundary conditions on the displacements only. To solve problems in which the rotations are also specified, the condensation procedure may be applied after assembly and imposition of boundary conditions.

Natural Vibration

For natural vibration, Eq. (13) reduces to

$$\left(\begin{bmatrix}[K^{11}]+[G] & [K^{12}]\\ [K^{12}]^{\mathrm{T}} & [K^{22}]\end{bmatrix} - \omega^2\begin{bmatrix}[M^{11}] & [0]\\ [0] & [M^{22}]\end{bmatrix}\right)\begin{Bmatrix}\{w\}\\ \{\Phi\}\end{Bmatrix} = \begin{Bmatrix}\{\bar{F}^1\}\\ \{F^2\}\end{Bmatrix} \quad (10.3-19)$$

where ω is the frequency of natural vibration. When the rotary inertia is assumed to be zero, the $[M^{22}] = [0]$, and once again we face the same problem as in the case of buckling. Thus, the neglect of rotary inertia does not simplify the problem, but complicates the analysis. Therefore, it is advised that rotary inertia be included in the finite element analysis.

Transient Analysis

In the case of transient analysis, Eq. (13) is reduced to a set of algebraic equations by using a time approximation scheme. Since the general form of Eq. (13) is the same as Eq. (10.2–10), all of the discussion in Section 10.2–5 applies. The fully discretized model for the transient analysis is given by

$$[\hat{K}^e]\{\Delta^e\}_{s+1} = \{\hat{F}^e\} \qquad (10.3-20a)$$

where

$$\begin{aligned}[\hat{K}^e] &= ([K^e]_{s+1} + [G^e]_{s+1}) + a_3[M^e]_{s+1} \\ \{\hat{F}^e\} &= \{F^e\}_{s+1} + [M^e]_{s+1}\left(a_3\{\Delta^e\}_s + a_4\{\dot{\Delta}^e\}_s + a_5\{\ddot{\Delta}^e\}_s\right)\end{aligned} \quad (10.3-20b)$$

and

$$[K^e] = \begin{bmatrix}[K^{11}] & [K^{12}]\\ [K^{12}]^{\mathrm{T}} & [K^{22}]\end{bmatrix}, \quad \{\Delta^e\} = \begin{Bmatrix}\{w^e\}\\ \{\Phi^e\}\end{Bmatrix}, \quad \{F^e\} = \begin{Bmatrix}\{F^1\}\\ \{F^2\}\end{Bmatrix} \quad (10.3-20c)$$

This completes the finite element model development of laminated beams according to the Timoshenko beam theory. The finite element models of cylindrical bending of laminated plate strips according to the first–order shear deformation plate theory can be deduced directly from those of laminated beams by the following change of the coefficients:

$$G_{xz}^b bh \longrightarrow A_{55}, \quad E_{xx}^b I_{yy} \longrightarrow D \tag{10.3-21}$$

In the next section numerical results obtained with the Euler–Bernoulli and Timoshenko beam elements are discussed.

The finite element models based on the classical theory of laminated beams and plate strips in cylindrical bending yield exact solutions for the displacements at the nodes irrespective of the number of elements. However, the linear and quadratic Timoshenko beam elements do not have this property. It is possible to develop a consistent interpolation Timoshenko beam finite element with cubic interpolation of w_0 and quadratic interpolation of ϕ_x, which is expected to yield exact nodal solutions. This should be clear from Eqs. (6.3–13b) and (6.3–12b). These equations also suggest an interdependent interpolation of w_0 and ϕ_x. For example, if $w_0(x)$ is interpolated by cubic polynomial

$$w_0(x) = c_1 \frac{x^3}{6} + c_2 \frac{x^2}{2} + c_3 x + c_4 \tag{10.3-22a}$$

then $\phi_x(x)$ can be interpolated as

$$\phi_x(x) = c_1 \frac{x^2}{2} + c_2 x + c_3 \tag{10.3-22b}$$

Of course, the coefficients $c_1 - c_4$ can be expressed in terms of the nodal values of w_0 and ϕ_x.

10.4 Numerical Results for Beams and Plate Strips

Although the Timoshenko beam element (RIE) with equal interpolation of w_0 and ϕ can be used to obtain results of both the classical beam theory and the Timoshenko beam theory, the classical beam theory is more accurate and reliable than the linear Timoshenko beam element in predicting the classical beam theory solutions for very thin beams ($a/h >> 100$). The quadratic Timoshenko beam element is accurate and reliable for both thick ($a/h \leq 100$) and thin beams ($a/h > 100$). The results presented in the following tables were obtained using the Timoshenko element with equal interpolation but with reduced integration of the transverse shear stiffnesses (see Reddy [1,2]).

Table 10.4–1 shows the nondimensionalized maximum deflections ($\bar{w} = -w_{max}E_2h^3 \times 10^2/q_0a^4$) under uniformly distributed load, critical buckling loads ($\bar{N} = N_{xx}^0 a^2/E_2h^3$), and fundamental natural frequencies ($\bar{\omega} = \omega_1 a^2\sqrt{I_0/E_2h^3}$) of simply supported beams. Table 10.4–2 contains similar results for plate strips in cylindrical bending. The finite element solution converges rapidly to the analytical solution as the mesh is refined.

Table 10.4–1: Maximum transverse deflections, critical buckling loads, and fundamental frequencies of simply supported laminated beams according to the classical and Timoshenko beam theories ($E_1/E_2 = 25$, $G_{12} = G_{13} = 0.5E_2$, $G_{23} = 0.2E_2$, $\nu_{12} = 0.25$, $a/h = 10$).

		CLPT			FSDT		
Laminate	$n^\dagger$	$\bar{w}$	$\bar{N}$	$\bar{\omega}$	$n^\ddagger$	$\bar{w}$	$\bar{\omega}$
0	1	0.6250	20.716	14.243	4L	0.9094	11.812
	2	0.6250	20.572	14.191	2Q	0.9250	11.642
	4	0.6250	20.562	14.188	4Q	0.9250	11.636
	Exact	0.6250	20.562	14.187	Exact	0.9250	11.635
$(0/90)_s$	1	0.7089	18.263	13.373	4L	1.1198	10.639
	2	0.7089	18.136	13.324	2Q	1.1375	10.494
	4	0.7089	18.127	13.321	4Q	1.1375	10.488
	Exact	0.7089	18.127	13.321	Exact	1.1375	10.488
$(45/-45)_s$	1	8.9424	1.448	3.765	4L	9.1474	3.732
	2	8.9424	1.438	3.752	2Q	9.3710	3.666
	4	8.9424	1.437	3.751	4Q	9.3710	3.663
	Exact	8.9424	1.437	3.751	Exact	9.3710	3.663

† Number of Hermite cubic elements; ‡ number of linear (L) or quadratic (Q) Lagrange elements.

There are two types of mesh refinements: (1) increasing the number of elements while keeping the type of element (*i.e.*, linear or quadratic) the same; and (2) increasing the order of the element (*i.e.*, degree of the polynomial) while keeping the number of elements the same. As a rule, the previous mesh should be a subset of the refined mesh. The first type of mesh refinement is termed an *h–refinement* and the second kind is called a *p–refinement*. Both types of mesh refinements were used to obtain the results in Tables 10.4–1 and 10.4–2.

Table 10.4–2: Maximum transverse deflections, critical buckling loads, and fundamental frequencies of simply supported laminated plate strips according to the classical and Timoshenko beam theories ($h_n = h/n$, $E_1/E_2 = 25$, $G_{12} = G_{13} = 0.5E_2$, $G_{23} = 0.2E_2$, $\nu_{12} = 0.25$, $a/h = 10$; n = number of layers).

		CLPT			FSDT		
Laminate	$N^\dagger$	$\bar{w}$	$\bar{N}$	$\bar{\omega}$	Mesh‡	$\bar{w}$	$\bar{\omega}$
0	1	0.6234	20.768	14.261	4L	0.9078	11.822
	2	0.6234	20.624	14.209	2Q	0.9234	11.652
	4	0.6234	20.614	14.205	4Q	0.9234	11.646
	Exact	0.6234	20.613	14.205	Exact	0.9234	11.645
(0/90)	1	3.3216	3.898	6.178	4L	3.6671	5.892
	2	3.3216	3.871	6.156	2Q	3.7502	5.792
	4	3.3216	3.869	6.154	4Q	3.7502	5.788
	Exact	3.3216	3.869	6.154	Exact	3.7502	5.788
(45/–45)	1	5.3959	2.399	4.847	4L	5.6896	4.731
	2	5.3959	2.383	4.830	2Q	5.8245	4.649
	4	5.3959	2.382	4.828	4Q	5.8245	4.645
	Exact	5.3959	2.382	4.828	Exact	5.8245	4.645
$(0/90)_s$	1	0.7084	18.276	13.378	4L	1.1193	10.642
	2	0.7084	18.149	13.329	2Q	1.1370	10.496
	4	0.7084	18.140	13.326	4Q	1.1370	10.490
	Exact	0.7084	18.140	13.326	Exact	1.1370	10.490
$(45/-45)_s$	1	2.1879	5.918	7.613	4L	2.5618	7.046
	2	2.1879	5.877	7.585	2Q	2.6165	6.931
	4	2.1879	5.874	7.583	4Q	2.6165	6.927
	Exact	2.1879	5.873	7.583	Exact	2.6165	6.927

† Number of Hermite cubic elements.

‡ number of linear (L) or quadratic (Q) Lagrange elements.

This completes the development of finite element models of symmetrically laminated beams and their applications to bending, buckling, and natural vibration. In the next section we discuss plate bending finite elements and present numerical results.

10.5 Finite Element Models of Laminated Plates (CLPT)

10.5–1 Weak Forms

In this section, finite element models of Eqs. (8.2–1)–(8.2–3) governing the motion of laminated plates according to the classical laminate theory are developed. For the sake of brevity, Eqs. (5.3–25a,b,c), which are expressed in terms of the stress resultants but equivalent to Eqs. (8.2–1)–(8.2–3), are used to develop the weak forms.

Multiplying Eqs. (5.3–25a,b,c) with δu_0, δv_0, and δw_0, respectively, and integrating over the element domain, we obtain

$$0=\int_{\Omega^e}\delta u_0\left[-\frac{\partial N_{xx}}{\partial x}-\frac{\partial N_{xy}}{\partial y}+I_0\frac{\partial^2 u_0}{\partial t^2}-I_1\frac{\partial^2}{\partial t^2}\left(\frac{\partial w_0}{\partial x}\right)\right]dxdy \qquad (10.5-1a)$$

$$0=\int_{\Omega^e}\delta v_0\left[-\frac{\partial N_{xy}}{\partial x}-\frac{\partial N_{yy}}{\partial y}+I_0\frac{\partial^2 v_0}{\partial t^2}-I_1\frac{\partial^2}{\partial t^2}\left(\frac{\partial w_0}{\partial y}\right)\right]dxdy \qquad (10.5-1b)$$

$$\begin{aligned}0=\int_{\Omega^e}\delta w_0\Bigg[&-\frac{\partial^2 M_{xx}}{\partial x^2}-2\frac{\partial^2 M_{xy}}{\partial y\partial x}-\frac{\partial^2 M_{yy}}{\partial y^2}-q\\&-\frac{\partial}{\partial x}\left(\hat{N}_{xx}\frac{\partial w_0}{\partial x}+\hat{N}_{xy}\frac{\partial w_0}{\partial y}\right)-\frac{\partial}{\partial y}\left(\hat{N}_{xy}\frac{\partial w_0}{\partial x}+\hat{N}_{yy}\frac{\partial w_0}{\partial y}\right)\\&+I_0\frac{\partial^2 w_0}{\partial t^2}-I_2\frac{\partial^2}{\partial t^2}\left(\frac{\partial^2 w_0}{\partial x^2}+\frac{\partial^2 w_0}{\partial y^2}\right)+I_1\frac{\partial^2}{\partial t^2}\left(\frac{\partial u_0}{\partial x}+\frac{\partial v_0}{\partial y}\right)\Bigg]dxdy\end{aligned} \qquad (10.5-1c)$$

where $\hat{N}_{xx}$, $\hat{N}_{xy}$, and $\hat{N}_{yy}$ are in-plane edge forces. The stress and moment resultants N_{xx}, M_{xx}, etc. are known in terms of the displacements (u_0, v_0, w_0) through Eq. (5.3–40). Note that the virtual displacements $(\delta u_0, \delta v_0, \delta w_0)$ take the role of weight functions in the development of weak forms. Integration by parts to weaken the differentiability of u_0, v_0, and w_0 results in the expressions

$$\begin{aligned}0=&\int_{\Omega^e}\left[\frac{\partial\delta u_0}{\partial x}N_{xx}+\frac{\partial\delta u_0}{\partial y}N_{xy}+I_0\delta u_0\frac{\partial^2 u_0}{\partial t^2}-I_1\delta u_0\frac{\partial^2}{\partial t^2}\left(\frac{\partial w_0}{\partial x}\right)\right]dxdy\\&-\oint_{\Gamma^e}\left(N_{xx}n_x+N_{xy}n_y\right)\delta u_0\,ds\end{aligned} \qquad (10.5-2a)$$

$$\begin{aligned}0=&\int_{\Omega^e}\left[\frac{\partial\delta v_0}{\partial x}N_{xy}+\frac{\partial\delta v_0}{\partial y}N_{yy}+I_0\delta v_0\frac{\partial^2 v_0}{\partial t^2}-I_1\delta v_0\frac{\partial^2}{\partial t^2}\left(\frac{\partial w_0}{\partial y}\right)\right]dxdy\\&-\oint_{\Gamma^e}\left(N_{xy}n_x+N_{yy}n_y\right)\delta v_0\,ds\end{aligned} \qquad (10.5-2b)$$

$$0=\int_{\Omega^e}\Bigg[-\frac{\partial^2\delta w_0}{\partial x^2}M_{xx}-2\frac{\partial^2\delta w_0}{\partial x\partial y}M_{xy}-\frac{\partial^2\delta w_0}{\partial y^2}M_{yy}-\delta w_0 q$$

$$
\begin{aligned}
&+ \frac{\partial \delta w_0}{\partial x}\left(\hat{N}_{xx}\frac{\partial w_0}{\partial x} + \hat{N}_{xy}\frac{\partial w_0}{\partial y}\right) + \frac{\partial \delta w_0}{\partial y}\left(\hat{N}_{xy}\frac{\partial w_0}{\partial x} + \hat{N}_{yy}\frac{\partial w_0}{\partial y}\right)\\
&+ I_0 \delta w_0 \frac{\partial^2 w_0}{\partial t^2} + I_2\left(\frac{\partial \delta w_0}{\partial x}\frac{\partial^3 w_0}{\partial x \partial t^2} + \frac{\partial \delta w_0}{\partial y}\frac{\partial^3 w_0}{\partial y \partial t^2}\right)\\
&- I_1\left(\frac{\partial \delta w_0}{\partial x}\frac{\partial^2 u_0}{\partial t^2} + \frac{\partial \delta w_0}{\partial y}\frac{\partial^2 v_0}{\partial t^2}\right)\Bigg] dxdy\\
&- \oint_{\Gamma^e}\left[\left(\frac{\partial M_{xx}}{\partial x} + \frac{\partial M_{xy}}{\partial y} + \hat{N}_{xx}\frac{\partial w_0}{\partial x} + \hat{N}_{xy}\frac{\partial w_0}{\partial y}\right) n_x\right.\\
&+ \left(\frac{\partial M_{xy}}{\partial x} + \frac{\partial M_{yy}}{\partial y} + \hat{N}_{xy}\frac{\partial w_0}{\partial x} + \hat{N}_{yy}\frac{\partial w_0}{\partial y}\right) n_y\\
&+ \left.\left(I_1\frac{\partial^2 u_0}{\partial t^2} - I_2\frac{\partial^3 w_0}{\partial x \partial t^2}\right) n_x + \left(I_1\frac{\partial^2 v_0}{\partial t^2} - I_2\frac{\partial^3 w_0}{\partial y \partial t^2}\right) n_y\right]\delta w_0 \, ds\\
&+ \oint_{\Gamma^e}\left[\frac{\partial \delta w_0}{\partial x}\left(M_{xx}n_x + M_{xy}n_y\right) + \frac{\partial \delta w_0}{\partial y}\left(M_{xy}n_x + M_{yy}n_y\right)\right] ds
\end{aligned}
\tag{10.5-2c}
$$

where (n_x, n_y) denote the direction cosines of the unit normal on the element boundary Γ^e. Integration by parts of the inertia terms in the last equation is necessitated by the symmetry considerations of the resulting weak form, which leads to symmetric mass matrix in the finite element model.

We note from the boundary terms in Eq. (2) that u_0, v_0, w_0, $\partial w_0/\partial x$, and $\partial w_0/\partial y$ are the primary variables (or generalized displacements), and

$$
\begin{aligned}
p_x &\equiv N_{xx}n_x + N_{xy}n_y, \quad p_y \equiv N_{xy}n_x + N_{yy}n_y\\
T_x &\equiv M_{xx}n_x + M_{xy}n_y, \quad T_y \equiv M_{xy}n_x + M_{yy}n_y\\
Q_n &\equiv \left(\frac{\partial M_{xx}}{\partial x} + \frac{\partial M_{xy}}{\partial y} + \hat{N}_{xx}\frac{\partial w_0}{\partial x} + \hat{N}_{xy}\frac{\partial w_0}{\partial y} + I_1\frac{\partial^2 u_0}{\partial t^2} - I_2\frac{\partial^3 w_0}{\partial x \partial t^2}\right) n_x\\
&+ \left(\frac{\partial M_{xy}}{\partial x} + \frac{\partial M_{yy}}{\partial y} + \hat{N}_{xy}\frac{\partial w_0}{\partial x} + \hat{N}_{yy}\frac{\partial w_0}{\partial y} + I_1\frac{\partial^2 v_0}{\partial t^2} - I_2\frac{\partial^3 w_0}{\partial y \partial t^2}\right) n_y
\end{aligned}
\tag{10.5-3}
$$

are the secondary degrees of freedom (or generalized forces). Thus, finite elements based on the classical plate theory require continuity of the transverse deflection and its normal derivative across element boundaries. Also, to satisfy the constant displacement (rigid body mode) and constant strain requirements, the polynomial expansion for w_0 should be a complete quadratic.

10.5–2 Spatial Approximations

First, we note that the stress and moment resultants contain first-order derivatives of (u_0, v_0) and second-order derivatives of w_0 with respect to the coordinates x and y. Second, the primary variables u_0, v_0, w_0, $\partial w_0/\partial x$, and $\partial w_0/\partial y$ must be carried as the nodal variables in order to enforce their interelement continuity. Thus, the displacements (u_0, v_0) must be approximated using the Lagrange interpolation functions, whereas w_0 should be approximated using Hermite interpolation functions over an element Ω^e. Let

$$
\begin{aligned}
u_0(x,y,t) &\approx \sum_{j=1}^{m} u_j^e(t)\psi_j^e(x,y)\\
v_0(x,y,t) &\approx \sum_{j=1}^{m} v_j^e(t)\psi_j^e(x,y)\\
w_0(x,y,t) &\approx \sum_{k=1}^{n} \Delta_k^e(t)\varphi_k^e(x,y) \qquad (10.5-4)
\end{aligned}
$$

where (u_j^e, v_j^e) denote the values of (u_0, v_0) at the jth node of the Lagrange elements, Δ_k^e denote the values of w_0 and its derivatives with respect to x and y at the k−th node, and (ψ_j^e, φ_k^e) are the Lagrange and Hermite interpolation functions, respectively.

Lagrange Interpolation Functions

The Lagrange interpolation functions $\psi_i^e(x,y)$ used for the in-plane displacements (u_0, v_0) can be derived as described for the one-dimensional functions (see Reddy [1], Chapter 9). The simplest Lagrange element in two dimensions is the triangular element with nodes at its vertices (see Figure 10.5–1), and its interpolation functions have the form

$$\psi_i^e(x,y) = a_i + b_i x + c_i y \qquad (10.5-5a)$$

The functions are linear in x and y, complete, and have nonzero first derivatives with respect to x and y. The linear triangular element (*i.e.*, element with linear variation of the dependent variables) can only represent a constant state of strains:

$$\varepsilon_{xx}^0 = \sum_{i=1}^{3} u_i^e \frac{\partial \psi_i^e}{\partial x} = \sum_{i=1}^{3} u_i^e b_i, \quad \varepsilon_{yy}^0 = \sum_{i=1}^{3} v_i^e \frac{\partial \psi_i^e}{\partial y} = \sum_{j=1}^{3} v_i^e c_i$$

$$2\varepsilon_{xy}^0 = \sum_{i=1}^{3}\left(u_i^e \frac{\partial \psi_i^e}{\partial y} + v_i^e \frac{\partial \psi_i^e}{\partial x}\right) = \sum_{i=1}^{3}(u_i^e c_i + v_i^e b_i) \qquad (10.5-5b)$$

For this reason the linear triangular element is known as the *constant strain triangle* (CST). A triangular element with quadratic variation of the dependent variables requires six nodes, because a complete quadratic polynomial in two dimensions has six coefficients:

$$\psi_i^e(x,y) = a_i + b_i x + c_i y + d_i xy + e_i x^2 + f_i y^2 \qquad (10.5-6a)$$

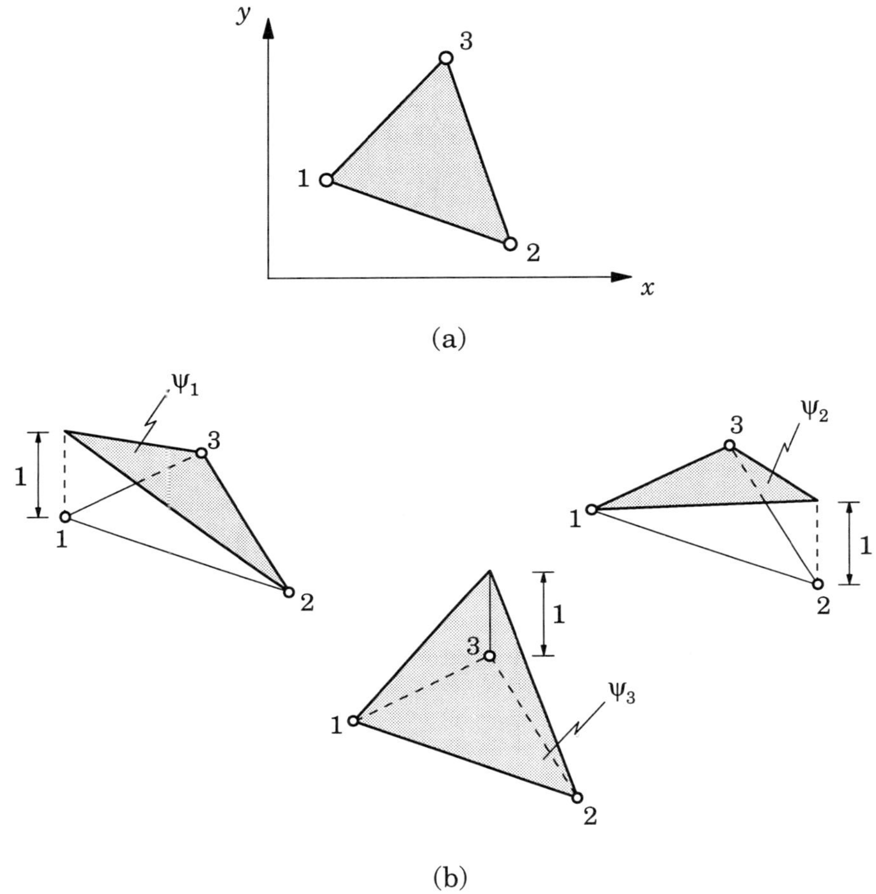

Figure 10.5–1: Linear Lagrange triangular element and its interpolation functions.

The three vertex nodes uniquely describe the geometry of the element (as in the linear element), and the other three nodes are placed at the midpoints of the sides (see Figure 10.5–2). The quadratic triangular element represents a state of linear strains:

$$\varepsilon_{xx}^0 = \sum_{i=1}^{6} u_i^e \left(b_i + 2e_i x + d_i y\right), \quad \varepsilon_{yy}^0 = \sum_{i=1}^{6} v_i^e \left(c_i + d_i x + 2f_i y\right)$$

$$2\varepsilon_{xy}^0 = \sum_{i=1}^{6} \left[u_i^e \left(c_i + d_i x + 2f_i y\right) + v_j^e \left(b_i + 2e_i x + d_i y\right)\right] \qquad (10.5-6b)$$

The interpolation functions for linear and quadratic triangular elements are presented below in terms of the area coordinates, L_i (see Figures 10.5–1 and

10.5–2):

$$\begin{Bmatrix} \psi_1^e \\ \psi_2^e \\ \psi_3^e \end{Bmatrix} = \begin{Bmatrix} L_1 \\ L_2 \\ L_3 \end{Bmatrix}, \quad \begin{Bmatrix} \psi_1^e \\ \psi_2^e \\ \psi_3^e \\ \psi_4^e \\ \psi_5^e \\ \psi_6^e \end{Bmatrix} = \begin{Bmatrix} L_1(2L_1-1) \\ L_2(2L_2-1) \\ L_3(2L_3-1) \\ 4L_1L_2 \\ 4L_2L_3 \\ 4L_3L_1 \end{Bmatrix} \qquad (10.5-7a)$$

where L_i are the area coordinates defined within an element

$$L_i = \frac{A_i}{A}, \quad A = \sum_{i=1}^{3} A_i \qquad (10.5-7b)$$

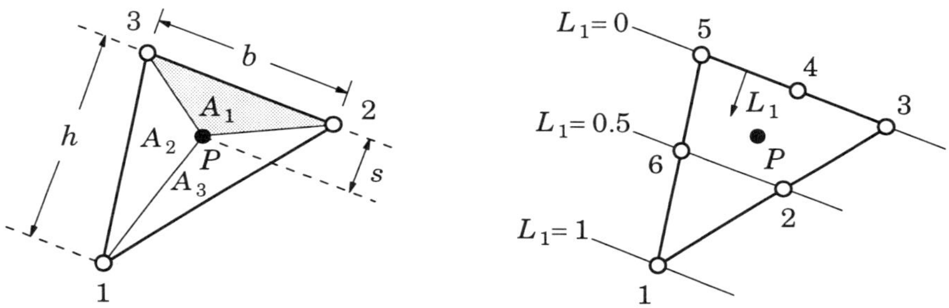

Figure 10.5–2: Quadratic Lagrange triangular element.

The simplest rectangular element has four nodes at vertices (see Figure 10.5–3), which define the geometry. The interpolation functions for this element have the form

$$\psi_i^e(x,y) = a_i + b_i x + c_i y + d_i xy \qquad (10.5-8a)$$

The strains in the linear rectangular element are partially linear (*i.e.*, at least linear in one coordinate)

$$\varepsilon_{xx}^0 = \sum_{i=1}^{4} u_i^e\,(b_i + d_i y), \quad \varepsilon_{yy}^0 = \sum_{j=1}^{4} v_i^e\,(c_i + d_i x)$$

$$2\varepsilon_{xy}^0 = \sum_{i=1}^{4} \left[u_i^e\,(c_i + d_i x) + v_i^e\,(b_i + d_i y)\right] \qquad (10.5-8b)$$

Note that the shear strain is represented as a bilinear function of the coordinates.

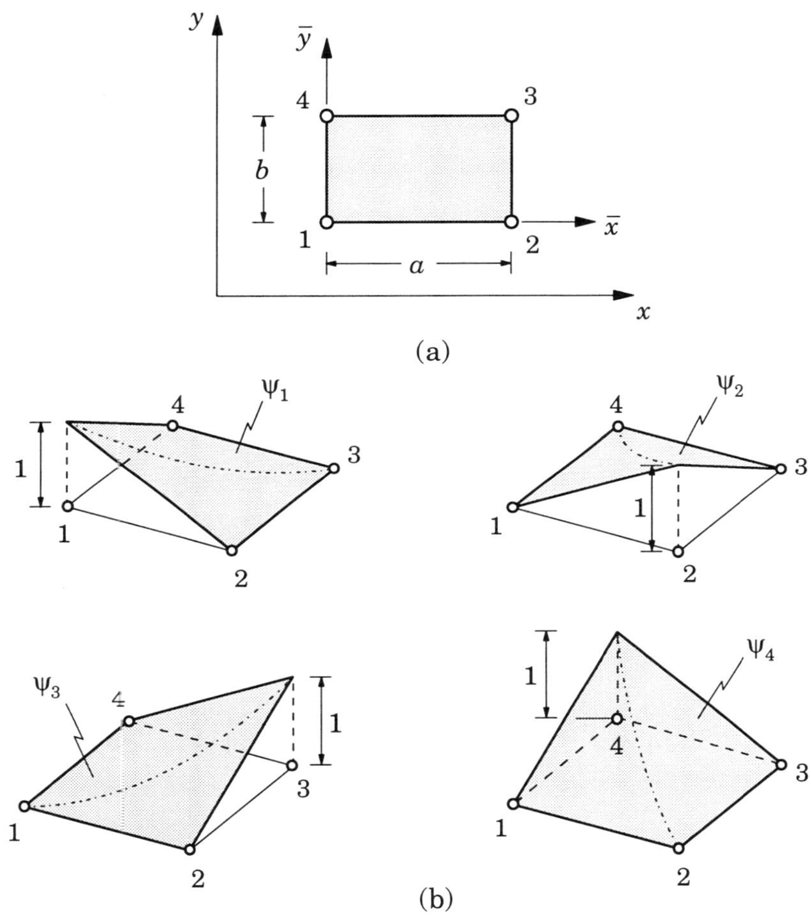

Figure 10.5–3: Linear Lagrange rectangular element and its interpolation functions.

A rectangular element with a complete quadratic polynomial representation

$$\psi_i^e(x,y) = a_i + b_i x + c_i y + d_i xy + e_i x^2 + f_i y^2 + g_i x^2 y + h_i xy^2 + k_i x^2 y^2 \tag{10.5 – 9a}$$

contains nine parameters and hence nine nodes (see Figure 10.5–4). In these elements the strains are represented at least as bilinear:

$$\varepsilon_{xx}^0 = \sum_{i=1}^{9} u_i^e \left(b_i + 2e_i x + d_i y + 2g_i xy + h_i y^2 + 2k_i xy^2\right)$$
$$\varepsilon_{yy}^0 = \sum_{j=1}^{9} v_i^e \left(c_i + d_i x + 2f_i y + 2h_i xy + g_i x^2 + 2k_i x^2 y\right) \qquad (10.5-9b)$$

and the shear strain is represented as a biquadratic function of the coordinates.

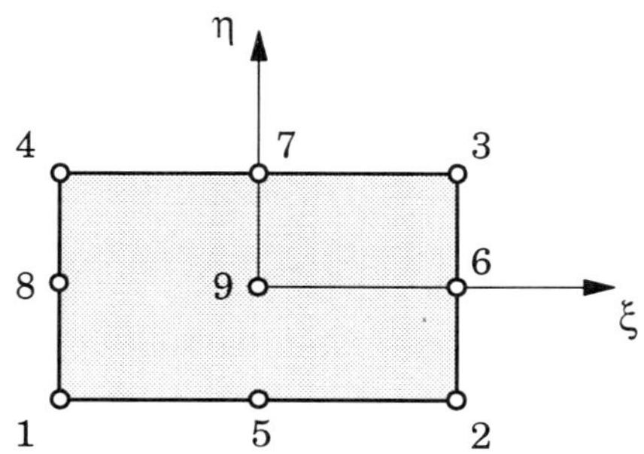

Figure 10.5–4: Nine-node quadratic Lagrange rectangular element.

The linear and quadratic Lagrange interpolation functions of rectangular elements are given below in terms of the element coordinates (ξ, η), called the *natural coordinates.*

$$\begin{Bmatrix} \psi_1^e \\ \psi_2^e \\ \psi_3^e \\ \psi_4^e \end{Bmatrix} = \frac{1}{4} \begin{Bmatrix} (1-\xi)(1-\eta) \\ (1+\xi)(1-\eta) \\ (1+\xi)(1+\eta) \\ (1-\xi)(1+\eta) \end{Bmatrix} \qquad (10.5-10)$$

$$\begin{Bmatrix} \psi_1^e \\ \psi_2^e \\ \psi_3^e \\ \psi_4^e \\ \psi_5^e \\ \psi_6^e \\ \psi_7^e \\ \psi_8^e \\ \psi_9^e \end{Bmatrix} = \frac{1}{4} \begin{Bmatrix} (1-\xi)(1-\eta)(-\xi-\eta-1) + (1-\xi^2)(1-\eta^2) \\ (1+\xi)(1-\eta)(\xi-\eta-1) + (1-\xi^2)(1-\eta^2) \\ (1+\xi)(1+\eta)(\xi+\eta-1) + (1-\xi^2)(1-\eta^2) \\ (1-\xi)(1+\eta)(-\xi+\eta-1) + (1-\xi^2)(1-\eta^2) \\ 2(1-\xi^2)(1-\eta) - (1-\xi^2)(1-\eta^2) \\ 2(1+\xi)(1-\eta^2) - (1-\xi^2)(1-\eta^2) \\ 2(1-\xi^2)(1+\eta) - (1-\xi^2)(1-\eta^2) \\ 2(1-\xi)(1-\eta^2) - (1-\xi^2)(1-\eta^2) \\ 4(1-\xi^2)(1-\eta^2) \end{Bmatrix} \qquad (10.5-11)$$

The *serendipity* family of Lagrange elements are those elements which have no interior nodes. Serendipity elements have fewer nodes compared to the higher-order Lagrange elements. The interpolation functions of the serendipity elements are not complete, and they cannot be obtained using tensor products of one-dimensional Lagrange interpolation functions. Instead, an alternative procedure must be employed, as discussed in Reference 1. The interpolation functions for the quadratic serendipity element are given in Eq. (12) below (also see Figure 10.5–5). Although the interpolation functions are not complete because the last term in Eq. (9a) is omitted, the serendipity elements have proven to be very effective in most practical applications (*serendipity* !).

$$\begin{Bmatrix} \psi_1^e \\ \psi_2^e \\ \psi_3^e \\ \psi_4^e \\ \psi_5^e \\ \psi_6^e \\ \psi_7^e \\ \psi_8^e \end{Bmatrix} = \frac{1}{4} \begin{Bmatrix} (1-\xi)(1-\eta)(-\xi-\eta-1) \\ (1+\xi)(1-\eta)(\xi-\eta-1) \\ (1+\xi)(1+\eta)(\xi+\eta-1) \\ (1-\xi)(1+\eta)(-\xi+\eta-1) \\ 2(1-\xi^2)(1-\eta) \\ 2(1+\xi)(1-\eta^2) \\ 2(1-\xi^2)(1+\eta) \\ 2(1-\xi)(1-\eta^2) \end{Bmatrix} \qquad (10.5-12)$$

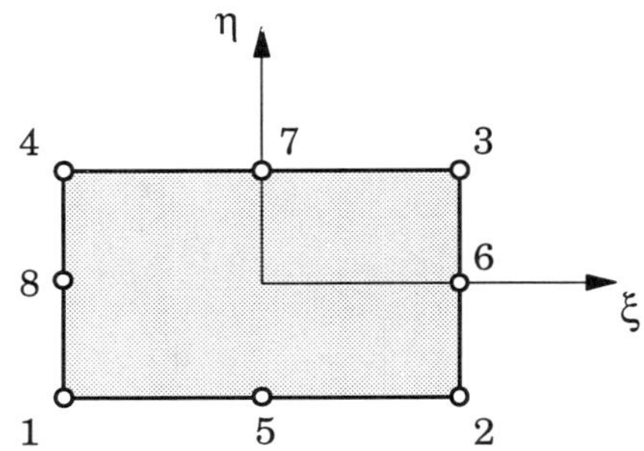

Figure 10.5–5: Eight-node quadratic serendipity rectangular element.

Hermite Interpolation Functions

There exists a vast literature on triangular and rectangular plate bending finite elements of isotropic or orthotropic plates based on the classical plate theory (*e.g.*, see References 6–27). Here we discuss triangular and rectangular C^1 plate bending elements.

There are two kinds of C^1 plate bending elements. A *conforming element* is one in which the interelement continuity of w_0, $\theta_x \equiv \partial w_0/\partial x$, and $\theta_y \equiv \partial w_0/\partial y$ (or $\partial w_0/\partial n$) are satisfied, and a *nonconforming element* is one in which the continuity of the normal slope, $\partial w_0/\partial n$, is not satisfied.

An effective nonconforming triangular element (the BCIZ triangle) was developed by Bazeley, Cheung, Irons, and Zienkiewicz [7], and it consists of three degrees of freedom (w_0, θ_x, θ_y) at the three vertex nodes (see Figure 10.5–6). The interpolation functions for this element can be expressed in terms of the area coordinates as

$$\begin{Bmatrix} \varphi_1^e \\ \varphi_2^e \\ \varphi_3^e \\ \varphi_4^e \\ \varphi_5^e \\ \varphi_6^e \\ \varphi_7^e \\ \varphi_8^e \\ \varphi_9^e \end{Bmatrix} = \begin{Bmatrix} L_1 + L_1^2 L_2 + L_1^2 L_3 - L_1 L_2^2 - L_1 L_3^2 \\ x_{31}(L_3 L_1^2 - f) - x_{12}(L_1^2 L_2 + f) \\ y_{31}(L_3 L_1^2 + f) - y_{12}(L_1^2 L_2 + f) \\ L_2 + L_2^2 L_3 + L_2^2 L_1 - L_2 L_3^2 - L_2 L_1^2 \\ x_{12}(L_1 L_2^2 - f) - x_{23}(L_2^2 L_3 + f) \\ y_{12}(L_1 L_2^2 + f) - y_{23}(L_2^2 L_3 + f) \\ L_3 + L_3^2 L_1 + L_3^2 L_2 - L_3 L_1^2 - L_3 L_2^2 \\ x_{23}(L_2 L_3^2 - f) - x_{31}(L_3^2 L_1 + f) \\ y_{23}(L_2 L_3^2 + f) - y_{31}(L_3^2 L_1 + f) \end{Bmatrix} \quad (10.5-13)$$

where $f = 0.5L_1L_2L_3$, $x_{ij} = x_i - x_j$, and $y_{ij} = y_i - y_j$, (x_i, y_i) being the global coordinates of the ith node.

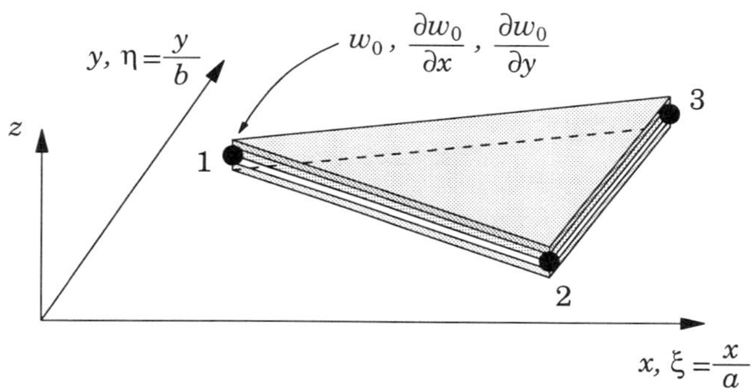

Figure 10.5–6: A nonconforming triangular element with three degrees of freedom ($w_0, \partial w_0/\partial x, \partial w_0/\partial y$) per node.

A conforming triangular element due to Clough and Tocher [22] is an assemblage of three triangles as shown in Figure 10.5–7. The normal slope continuity is enforced at the midside nodes between the subtriangles. In each subtriangle, the transverse deflection is represented by the polynomial ($i = 1, 2, 3$)

$$w_0^i(x,y) = a_i + b_i\xi + c_i\eta + d_i\xi\eta + e_i\xi^2 + f_i\eta^2 + g_i\xi^3 + h_i\xi^2\eta + k_i\xi\eta^2 + \ell_i\eta^3 \quad (10.5-14)$$

where (ξ, η) are the local coordinates, as shown in the Figure 10.5–7. The thirty coefficients are reduced to nine, three (w_0, $\partial w_0/\partial x$,$\partial w_0/\partial y$) at each vertex of the triangle, by equating the variables from the vertices of each subtriangle at the common points and normal slope between the midside points of subtriangles.

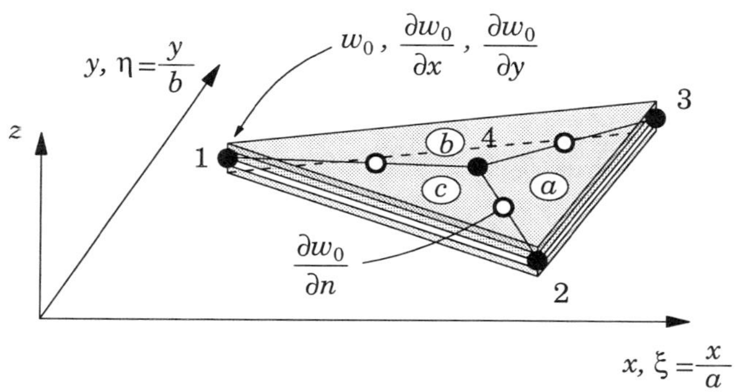

Figure 10.5–7: A conforming triangular element with three degrees of freedom.

A nonconforming rectangular element has w_0, θ_x, and θ_y as the nodal variables (see Figure 10.5–8). The element was developed by Melosh [18] and Zienkiewicz and Cheung [19]. The normal slope varies cubically along an edge whereas there are only two values of $\partial w_0/\partial n$ available on the edge. Therefore, the cubic polynomial for the normal derivative of w_0 is not the same on the edge common to two elements. The interpolation functions for this element can be expressed compactly as

$$\varphi_i^e = g_{i1} \ (i = 1, 4, 7, 10); \quad \varphi_i^e = g_{i2} \ (i = 2, 5, 8, 11)$$

$$\varphi_i^e = g_{i3} \ (i = 3, 6, 9, 12) \qquad (10.5 - 15a)$$

where

$$\begin{aligned}
g_{i1} =& \frac{1}{8}(1 + \xi_0)(1 + \eta_0)(2 + \xi_0 + \eta_0 - \xi^2 - \eta^2) \\
g_{i2} =& \frac{1}{8}\xi_i(\xi_0 - 1)(1 + \eta_0)(1 + \xi_0)^2 \\
g_{i3} =& \frac{1}{8}\eta_i(\eta_0 - 1)(1 + \xi_0)(1 + \eta_0)^2 \\
\xi =& (x - x_c)/2, \ \eta = (y - y_c)/b, \ \ \xi_0 = \xi\xi_i, \ \eta_0 = \eta\eta_i \qquad (10.5 - 15b)
\end{aligned}$$

where $2a$ and $2b$ are the sides of the rectangle, and (x_c, y_c) are the global coordinates of the center of the rectangle.

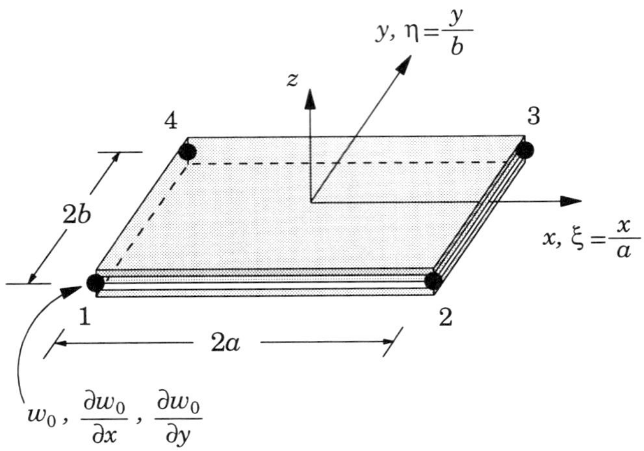

Figure 10.5–8: A nonconforming rectangular element with three degrees of freedom $(w_0, \partial w_0/\partial x, \partial w_0/\partial y)$ per node.

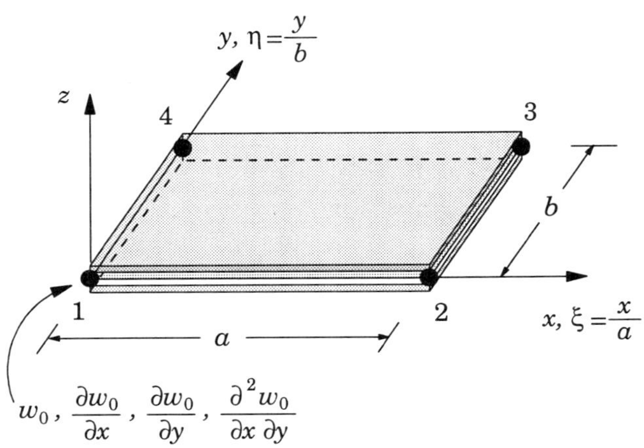

Figure 10.5–9: A conforming rectangular element with four degrees of freedom $(w_0, \partial w_0/\partial x, \partial w_0/\partial y, \partial^2 w_0/\partial x\partial y)$ per node.

A conforming rectangular element with w_0, $\partial w_0/\partial x$, $\partial w_0/\partial y$, and $\partial^2 w_0/\partial x\partial y$ as the nodal variables was developed by Bogner, Fox, and Schmidt [20]. The interpolation functions for this element (see Figure 10.5–9) are

$$\varphi_i^e = g_{i1}\ (i = 1, 5, 9, 13);\quad \varphi_i^e = g_{i2}\ (i = 2, 6, 10, 14)$$
$$\varphi_i^e = g_{i3}\ (i = 3, 7, 11, 15);\quad \varphi_i^e = g_{i4}\ (i = 4, 8, 12, 16) \qquad (10.5-16a)$$

$$
\begin{aligned}
g_{i1} =& \frac{1}{16}(\xi+\xi_i)^2(\xi_0-2)(\eta+\eta_i)^2(\eta_0-2)\\
g_{i2} =& \frac{1}{16}\xi_i(\xi+\xi_i)^2(1-\xi_0)(\eta+\eta_i)^2(\eta_0-2)\\
g_{i3} =& \frac{1}{16}\eta_i(\xi+\xi_i)^2(\xi_0-2)(\eta+\eta_i)^2(1-\eta_0)\\
g_{i4} =& \frac{1}{16}\xi_i\eta_i(\xi+\xi_i)^2(1-\xi_0)(\eta+\eta_i)^2(1-\eta_0) \qquad (10.5-16b)
\end{aligned}
$$

In this book we will use the Lagrange linear rectangular element for in-plane displacements and the conforming and nonconforming rectangular elements for bending deflections to present numerical results. The combined conforming element has a total of six degrees of freedom per node, whereas the nonconforming element has a total of five degrees of freedom per node. For the conforming rectangular element ($m = 4$ and $n = 12$) the total number of nodal degrees of freedom per element is 24, and the nonconforming element the total number of degrees of freedom per element is 12.

10.5–3 Semidiscrete Finite Element Model

Substituting approximations (4) for the displacements and the ith interpolation function for the virtual displacement ($\delta u_0 \sim \psi_i, \delta v_0 \sim \psi_i, \delta w_0 \sim \varphi_i$) into the weak forms, we obtain the ith equation associated with each weak form

$$
\begin{aligned}
0 &= \sum_{j=1}^{m}\left(K_{ij}^{11}u_j^e + K_{ij}^{12}v_j^e + M_{ij}^{11}\ddot{u}_j^e\right) + \sum_{k=1}^{n}\left(K_{ik}^{13}\Delta_k^e + M_{ik}^{13}\ddot{\Delta}_k^e\right) - F_i^1 - F_i^{T1}\\
0 &= \sum_{j=1}^{m}\left(K_{ij}^{21}u_j^e + K_{ij}^{22}v_j^e + M_{ij}^{22}\ddot{v}_j^e\right) + \sum_{k=1}^{n}\left(K_{ik}^{23}\Delta_k^e + M_{ik}^{23}\ddot{\Delta}_k^e\right) - F_i^2 - F_i^{T2}\\
0 &= \sum_{j=1}^{m}\left(K_{kj}^{31}u_j^e + K_{kj}^{32}v_j^e + M_{kj}^{31}\ddot{u}_j^e + M_{kj}^{32}\ddot{v}_j^e\right)\\
&\quad + \sum_{\ell=1}^{n}\left[\left(K_{k\ell}^{33} + G_{k\ell}^e\right)\Delta_\ell^e + M_{k\ell}^{33}\ddot{\Delta}_\ell^e\right] - F_k^3 - F_k^{T3} \qquad (10.5-17)
\end{aligned}
$$

where $i = 1, 2, \cdots, m; k = 1, 2, \cdots, n$. The coefficients of the stiffness matrix $K_{ij}^{\alpha\beta} = K_{ji}^{\beta\alpha}$, mass matrix $M_{ij}^{\alpha\beta} = M_{ji}^{\beta\alpha}$ (symmetric), and force vectors F_i^α and $F_i^{T\alpha}$ are defined as follows:

$$
\begin{aligned}
K_{ij}^{11} &= A_{11}S_{ij}^{xx} + A_{16}\left(S_{ij}^{xy} + S_{ij}^{yx}\right) + A_{66}S_{ij}^{yy}\\
K_{ij}^{12} &= A_{12}S_{ij}^{xy} + A_{16}S_{ij}^{xx} + A_{26}S_{ij}^{yy} + A_{66}S_{ij}^{yx}\\
K_{ij}^{22} &= A_{66}S_{ij}^{xx} + A_{26}\left(S_{ij}^{xy} + S_{ij}^{yx}\right) + A_{22}S_{ij}^{yy}
\end{aligned}
$$

$$
\begin{aligned}
K_{ik}^{13} &= -B_{11}R_{ik}^{xxx} - B_{12}R_{ik}^{xyy} - 2B_{26}R_{ik}^{xxy} - B_{16}R_{ik}^{yxx} - B_{26}R_{ik}^{yyy} - 2B_{66}R_{ik}^{yxy}\\
K_{ik}^{23} &= -B_{16}R_{ik}^{xxx} - B_{26}R_{ik}^{xyy} - 2B_{66}R_{ik}^{xxy} - B_{12}R_{ik}^{yxx} - B_{22}R_{ik}^{yyy} - 2B_{26}R_{ik}^{yxy}\\
K_{k\ell}^{33} &= D_{11}T_{k\ell}^{xxxx} + D_{12}\left(T_{k\ell}^{xxyy} + T_{k\ell}^{yyxx}\right) + 2D_{16}\left(T_{k\ell}^{xxxy} + T_{k\ell}^{xyxx}\right)\\
&\quad + 2D_{26}\left(T_{k\ell}^{xyyy} + T_{k\ell}^{yyxy}\right) + 4D_{66}T_{k\ell}^{xyxy} + D_{22}T_{k\ell}^{yyyy}\\
G_{ij}^{e} &= \int_{\Omega^e}\left[\hat{N}_{xx}S_{ij}^{xx} + \hat{N}_{xy}\left(S_{ij}^{xy} + S_{ij}^{yx}\right) + \hat{N}_{yy}S_{ij}^{yy}\right] dxdy\\
M_{ij}^{11} &= \int_{\Omega^e} I_0\psi_i^e\psi_j^e\, dxdy, \quad M_{ik}^{13} = -\int_{\Omega^e} I_1\psi_i^e\frac{\partial\varphi_k^e}{\partial x}\, dxdy\\
M_{ij}^{22} &= \int_{\Omega^e} I_0\psi_i^e\psi_j^e\, dxdy, \quad M_{ik}^{23} = -\int_{\Omega^e} I_1\psi_i^e\frac{\partial\varphi_k^e}{\partial y}\, dxdy\\
M_{k\ell}^{33} &= \int_{\Omega^e}\left[I_0\varphi_k^e\varphi_\ell^e + I_2\left(\frac{\partial\varphi_k^e}{\partial x}\frac{\partial\varphi_\ell^e}{\partial x} + \frac{\partial\varphi_k^e}{\partial y}\frac{\partial\varphi_\ell^e}{\partial y}\right)\right] dxdy\\
F_i^1 &= \oint_{\Gamma^e} p_x\psi_i^e\, ds, \quad F_i^2 = \int_{\Gamma_e} p_y\psi_i^e\, ds\\
F_k^3 &= \int_{\Omega^e} q\varphi_k^e\, dxdy + \oint_{\Gamma^e}\left(Q_n\varphi_k^e + T_x\frac{\partial\varphi_k^e}{\partial x} + T_y\frac{\partial\varphi_k^e}{\partial y}\right) ds\\
F_i^{T1} &= \int_{\Omega^e}\left(\frac{\partial\psi_i^e}{\partial x}N_{xx}^T + \frac{\partial\psi_i^e}{\partial y}N_{xy}^T\right) dxdy\\
F_i^{T2} &= \int_{\Omega^e}\left(\frac{\partial\psi_i^e}{\partial x}N_{xy}^T + \frac{\partial\psi_i^e}{\partial y}N_{yy}^T\right) dxdy\\
F_k^{T3} &= \int_{\Omega^e}\left(\frac{\partial^2\varphi_k^e}{\partial x^2}M_{xx}^T + 2\frac{\partial^2\varphi_k^e}{\partial x\partial y}M_{xy}^T + \frac{\partial^2\varphi_k^e}{\partial y^2}M_{yy}^T\right) dxdy \qquad (10.5-18a)
\end{aligned}
$$

where N_{xx}^T, M_{xx}^T, etc. are the thermal force and moment resultants [see Eq. (5.3–41)], and

$$
\begin{aligned}
S_{ij}^{\xi\eta} &= \int_{\Omega^e}\frac{\partial\psi_i^e}{\partial\xi}\frac{\partial\psi_j^e}{\partial\eta}\, dxdy\\
R_{ik}^{\xi\eta\zeta} &= \int_{\Omega^e}\frac{\partial\psi_i^e}{\partial\xi}\frac{\partial^2\varphi_k^e}{\partial\eta\partial\zeta}\, dxdy\\
T_{k\ell}^{\xi\eta\zeta\mu} &= \int_{\Omega^e}\frac{\partial^2\varphi_k^e}{\partial\xi\partial\eta}\frac{\partial^2\varphi_\ell^e}{\partial\zeta\partial\mu}\, dxdy \qquad (10.5-18b)
\end{aligned}
$$

and ξ, η, ζ, and μ can be equal to x or y. In matrix notation, Eq. (17) can be expressed as

$$
\begin{bmatrix} [K^{11}] & [K^{12}] & [K^{13}]\\ [K^{12}]^{\mathrm{T}} & [K^{22}] & [K^{23}]\\ [K^{13}]^{\mathrm{T}} & [K^{23}]^{\mathrm{T}} & [K^{33}] \end{bmatrix}
\begin{Bmatrix} \{u^e\}\\ \{v^e\}\\ \{\Delta^e\} \end{Bmatrix}
+ \begin{bmatrix} [0] & [0] & [0]\\ [0] & [0] & [0]\\ [0] & [0] & [G] \end{bmatrix}
\begin{Bmatrix} \{u^e\}\\ \{v^e\}\\ \{\Delta^e\} \end{Bmatrix}
$$
$$
+ \begin{bmatrix} [M^{11}] & [0] & [M^{13}]\\ [0] & [M^{22}] & [M^{23}]\\ [M^{13}]^{\mathrm{T}} & [M^{23}]^{\mathrm{T}} & [M^{33}] \end{bmatrix}
\begin{Bmatrix} \{\ddot{u}^e\}\\ \{\ddot{v}^e\}\\ \{\ddot{\Delta}^e\} \end{Bmatrix}
= \begin{Bmatrix} \{F^1\} + \{F^{T1}\}\\ \{F^2\} + \{F^{T2}\}\\ \{F^3\} + \{F^{T3}\} \end{Bmatrix} \qquad (10.5-19)
$$

This completes the finite element model development of the classical laminate theory. The finite element model in Eq. (17) or (19) is called a *displacement finite element model* because it is based on equations of motion expressed in terms of the displacements, and the generalized displacements are the primary nodal degrees of freedom.

It should be noted that the contributions of the internal forces defined in Eq. (3) to the force vector will cancel when element equations are assembled. They will remain in the force vector only when the element boundary coincides with the boundary of the domain being modeled (see Reddy [1], pp. 313–318). Of course, the contributions of the applied loads (*i.e.*, $q(x,y)$ and $\Delta T(x,y)$) to a node will add up from elements connected at the node and remain as a part of the force vector.

10.5–4 Fully Discretized Finite Element Models

Static Bending

In the case of static bending under applied mechanical and thermal loads, Eq. (19) reduces to

$$\begin{bmatrix} [K^{11}] & [K^{12}] & [K^{13}] \\ [K^{12}]^{\mathrm{T}} & [K^{22}] & [K^{23}] \\ [K^{13}]^{\mathrm{T}} & [K^{23}]^{\mathrm{T}} & [K^{33}]+[G] \end{bmatrix} \begin{Bmatrix} \{u^e\} \\ \{v^e\} \\ \{\Delta^e\} \end{Bmatrix} = \begin{Bmatrix} \{F^1\}+\{F^{T1}\} \\ \{F^2\}+\{F^{T2}\} \\ \{F^3\}+\{F^{T3}\} \end{Bmatrix} \tag{10.5 – 20}$$

where it is understood that all time–derivative terms are zero in the definitions (3).

Buckling

In the case of buckling under applied in-plane compressive and shear edge loads, Eq. (19) reduces to

$$\left(\begin{bmatrix} [K^{11}] & [K^{12}] & [K^{13}] \\ [K^{12}]^{\mathrm{T}} & [K^{22}] & [K^{23}] \\ [K^{13}]^{\mathrm{T}} & [K^{23}]^{\mathrm{T}} & [K^{33}] \end{bmatrix} - \lambda \begin{bmatrix} [0] & [0] & [0] \\ [0] & [0] & [0] \\ [0] & [0] & [\bar{G}] \end{bmatrix} \right) \begin{Bmatrix} \{u^e\} \\ \{v^e\} \\ \{\Delta^e\} \end{Bmatrix} = \begin{Bmatrix} \{F^1\} \\ \{F^2\} \\ \{\bar{F}^3\} \end{Bmatrix} \tag{10.5 – 21}$$

$$\begin{aligned} \lambda &= -\hat{N}_{xx}/\hat{N}_{xx}^0 = -\hat{N}_{yy}/\hat{N}_{yy}^0 = -\hat{N}_{xy}/\hat{N}_{xy}^0 \\ \bar{G}_{ij}^e &= \int_{\Omega^e} \left[\hat{N}_{xx}^0 S_{ij}^{xx} + \hat{N}_{xy}^0 \left(S_{ij}^{xy} + S_{ij}^{yx} \right) + \hat{N}_{yy}^0 S_{ij}^{yy} \right] dxdy \\ \bar{F}_k^3 &= \oint_{\Gamma^e} \left(Q_n \varphi_k^e + T_x \frac{\partial \varphi_k^e}{\partial x} + T_y \frac{\partial \varphi_k^e}{\partial y} \right) ds \end{aligned} \tag{10.5 – 22}$$

and all time-derivative terms in Eq. (3) are zero.

Natural Vibration

In the case of natural vibration, the response of the plate is assumed to be periodic. Equation (19) becomes

$$\left(\begin{bmatrix}[K^{11}] & [K^{12}] & [K^{13}]\\ [K^{12}]^{\mathrm{T}} & [K^{22}] & [K^{23}]\\ [K^{13}]^{\mathrm{T}} & [K^{23}]^{\mathrm{T}} & [K^{33}]\end{bmatrix} - \omega^2\begin{bmatrix}[M^{11}] & [0] & [M^{13}]\\ [0] & [M^{22}] & [M^{23}]\\ [M^{13}]^{\mathrm{T}} & [M^{23}]^{\mathrm{T}} & [M^{33}]\end{bmatrix}\right)\begin{Bmatrix}\{u^e\}\\ \{v^e\}\\ \{\Delta^e\}\end{Bmatrix}$$

$$= \begin{Bmatrix}\{F^1\}\\ \{F^2\}\\ \{\bar{F}^3\}\end{Bmatrix} \qquad (10.5-23)$$

where ω denotes the frequency of natural vibration.

Transient Analysis

For transient analysis, Eq. (19) can be written symbolically as

$$[K^e]\{\bar{\Delta}^e\} + [M^e]\{\ddot{\bar{\Delta}}^e\} = \{F^e\} \qquad (10.5-24)$$

where $[K^e]$ (which may contain $[G^e]$) and $[M^e]$ are the stiffness and mass matrices appearing in Eq. (19), and

$$\{\bar{\Delta}^e\} = \begin{Bmatrix}\{u^e\}\\ \{v^e\}\\ \{\Delta^e\}\end{Bmatrix} \qquad (10.5-25)$$

Since Eq. (24) is identical in form to Eq. (10.2–10), the time approximations discussed for Eq. (10.2–10) are also valid for Eq. (24). We have

$$[\hat{K}^e]_{s+1}\{\bar{\Delta}^e\}_{s+1} = \{\hat{F}^e\} \qquad (10.5-26a)$$

$$[\hat{K}^e]_{s+1} = ([K^e]_{s+1} + [G^e]_{s+1}) + a_3[M^e]_{s+1}$$

$$\{\hat{F}^e\} = \{F^e\}_{s+1} + [M^e]_{s+1}\{\ddot{\bar{\Delta}}^e\} \qquad (10.5-26b)$$

$$\{\ddot{\bar{\Delta}}^e\} = \left(a_3\{\bar{\Delta}^e\}_s + a_4\{\dot{\bar{\Delta}}^e\}_s + a_5\{\ddot{\bar{\Delta}}^e\}_s\right) \qquad (10.5-26c)$$

$$a_1 = (1-\alpha)\delta t_s, \quad a_2 = \alpha\delta t_s, \quad a_3 = \frac{2}{\gamma(\delta t_s)^2}$$

$$a_4 = \delta t_s a_3, \quad a_5 = \frac{(1-\gamma)}{\gamma} \qquad (10.5-26d)$$

and $\delta t_s = t_{s+1} - t_s$ is the time step; the notation δt in place of Δt for time step should be obvious in the context.

10.5–5 Quadrilateral Elements and Numerical Integration

Introduction

An accurate representation of irregular domains (*i.e.*, domains with curved boundaries) can be accomplished by the use of refined meshes and/or irregularly shaped elements. For example, a nonrectangular region cannot be represented using all rectangular elements; however, it can be represented by triangular and quadrilateral elements. However, it is easy to derive the interpolation functions for a rectangular element, and it is easier to evaluate integrals over rectangular geometries than over irregular geometries. Therefore, it is practical to use quadrilateral elements with straight or curved sides but have a means to generate interpolation functions and evaluate their integrals over the quadrilateral elements. A coordinate transformation between the coordinates (x, y) used in the formulation of the problem, called *global coordinates*, and the element coordinates $(\bar{x}, \bar{y})$ used to derive the interpolation functions of rectangular elements is introduced for this purpose. The transformation of the geometry and the variable coefficients of the differential equation from the problem coordinates (x,y) to the local coordinates $(\bar{x}, \bar{y})$ results in algebraically complex expressions, and they preclude analytical (*i.e.*, exact) evaluation of the integrals. Therefore, numerical integration is used to evaluate such complicated expressions.

While the element coordinate system, also called a *local coordinate system*, can be any convenient system that permits easy construction of the interpolation functions, it is useful to select one that is also convenient in the numerical evaluation of the integrals. Numerical integration schemes, such as the Gauss–Legendre numerical integration scheme, require the integral to be evaluated on a specific domain or with respect to a specific coordinate system. Gauss quadrature, for example, requires the integral to be expressed over a square region $\hat{\Omega}$ of dimension 2×2 and the coordinate system (ξ, η) be such that $-1 \leq (\xi, \eta) \leq 1$. The coordinates (ξ, η) are called *normalized* or *natural coordinates*. Thus, the transformation between (x, y) and (ξ, η) of a given integral expression defined over a quadrilateral element Ω^e to one on the domain $\hat{\Omega}$ facilitates the use of Gauss–Legendre quadrature to evaluate integrals. The element $\hat{\Omega}$ is called a *master element* (see Reddy [1], Chapter 9).

Coordinate Transformations

The transformation between Ω^e and $\hat{\Omega}$ is accomplished by a coordinate transformation of the form

$$x = \sum_{j=1}^{n} x_j^e \hat{\psi}_j^e(\xi, \eta) \; , \qquad y = \sum_{j=1}^{m} y_j^e \hat{\psi}_j^e(\xi, \eta) \qquad (10.5-27)$$

while a typical dependent variable $u(x, y)$ is approximated by

$$u(x,y) = \sum_{j=1}^{n} u_j^e \psi_j^e(x,y) = \sum_{j=1}^{n} u_j^e \psi_j^e(x(\xi,\eta), y(\xi,\eta)) \qquad (10.5-28)$$

where $\hat{\psi}_j^e$ denote the interpolation functions of the master element $\hat{\Omega}$ and ψ_j^e are interpolation functions of a typical element Ω^e over which u is approximated. Although the Lagrange interpolation of the geometry is implied by Eqs. (27) and (28), one can also use Hermite interpolation of the geometry and/or the solution as required. The transformation (27) maps a point (x, y) in a typical element Ω^e of the mesh to a point (ξ, η) in the master element $\hat{\Omega}$, and vice versa if the Jacobian of the transformation is positive-definite. The positive-definite requirement of the Jacobian dictates admissible geometries of elements in a mesh (see Reddy [1], pp. 421–448).

The interpolation functions ψ_j^e used for the approximation of the dependent variable are, in general, different from $\hat{\psi}_j^e$ used in the approximation of the geometry. Depending on the relative degree of approximations used for the geometry and the dependent variable(s), the finite element formulations are classified into three categories.

1. *Superparametric* ($m > n$): The polynomial degree of approximation used for the geometry is of higher order than that used for the dependent variable.
2. *Isoparametric* ($m = n$): Equal degree of approximation is used for both geometry and dependent variables.
3. *Subparametric* ($m < n$): Higher-order approximation of the dependent variable is used.

For example, in the finite element analysis of the Euler–Bernoulli beams, we may use linear Lagrange interpolation of the geometry

$$x = \sum_{j=1}^{2} x_j \hat{\psi}_j(\xi) \qquad (10.5-29)$$

whereas the Hermite cubic interpolation is used to approximate the transverse deflection

$$w_0(x) = \sum_{j=1}^{4} \Delta_j^e \varphi_j(x(\xi)) \qquad (10.5-30)$$

Then we say that subparametric formulation is used for the transverse deflection w_0. In the Timoshenko beam element we can use the same degree of interpolation for both geometry and dependent variables. Then we say that isoparametric formulation is used for the transverse deflection w_0 and rotation ϕ_x.

An example of the coordinate transformation in one dimension is provided by the linear transformation, which maps straight lines into straight lines

$$x = \sum_{i=1}^{2} x_i^e \hat{\psi}_i(\xi), \quad \hat{\psi}_1(\xi) = \frac{1}{2}(1-\xi), \quad \hat{\psi}_2(\xi) = \frac{1}{2}(1+\xi) \tag{10.5-31a}$$

where $x_1^e = x_e$ and $x_2^e = x_{e+1}$, x_i^e being the global coordinate of the ith node of the eth element, and x_I denotes the global coordinate of the Ith global node of the mesh. The transformation (31a) can be expressed directly in terms of x and ξ:

$$x = x_e + \frac{h_e}{2}(1+\xi) \tag{10.5-31b}$$

where $h_e = x_{e+1} - x_e$ is the element length. Note that the Lagrange and Hermite interpolation functions defined in Eqs. (10.3–7b) and (10.2–8b), respectively, can be written in terms of the natural coordinate ξ with the help of the linear coordinate transformation (31):

$$\psi_1 = \frac{1}{2}(1-\xi), \quad \psi_2 = \frac{1}{2}(1+\xi) \tag{10.5-32}$$

$$\begin{aligned} \varphi_1 &= \frac{1}{4}(2 - 3\xi^2 + \xi^3), \quad \varphi_2 = -\frac{h_e}{8}(1-\xi)(1-\xi^2) \\ \varphi_3 &= \frac{1}{4}(2 + 3\xi^2 - \xi^3), \quad \varphi_4 = \frac{h_e}{8}(1+\xi)(1-\xi^2) \end{aligned} \tag{10.5-33}$$

It should be noted that, once the approximations of geometry and solution are selected, the coordinate transformations have the sole purpose of numerically evaluating the integrals inside the computer program. *No transformation of the physical domain or the solution is involved in the finite element analysis.* The resulting algebraic equations of the finite element formulation are always among the nodal values of the physical domain and the nodal values are referred to the global coordinate system. Different elements of the finite element mesh can be generated from the same master element by assigning the global coordinates of the elements. Master elements of different order interpolation define different transformations and hence different collections of finite element meshes. Thus, with the help of an appropriate master element, any given element of a mesh can be generated. However, the transformations of a master element should be such that there exist no spurious gaps between elements and no element overlaps occur.

Numerical Integration: the Gauss Quadrature

Recall that a finite element model is a system of algebraic equations among the nodal values of the primary variables (generalized displacements) and secondary variables (generalized forces). The coefficients of these algebraic equations contain integrals of the physical parameters (*e.g.*, material properties) and functions used for the approximation of the primary variables. The integral expressions are, in general, complicated algebraically due to the spatial variation of the parameters or coordinate transformations. Therefore, numerical integration methods, known as numerical quadratures, are used to evaluate them. Here we discuss the Gauss quadrature, which is the most widely used method for master elements of rectangular or prismatic geometries.

We illustrate the essential elements of the Gauss quadrature by considering the following representative integral expression

$$K_{ij}^e = \int_{\Omega^e} \Big[a(x,y) \frac{\partial \psi_i^e}{\partial x}\frac{\partial \psi_j^e}{\partial x} + b(x,y)\left(\frac{\partial \psi_i^e}{\partial x}\frac{\partial \psi_j^e}{\partial y} + \frac{\partial \psi_i^e}{\partial x}\frac{\partial \psi_j^e}{\partial y}\right) + c(x,y)\frac{\partial \psi_i^e}{\partial y}\frac{\partial \psi_j^e}{\partial y} + d(x,y)\psi_i^e\psi_j^e\Big] dxdy \qquad (10.5-34)$$

We wish to transform the integral from Ω^e to the master element $\hat{\Omega} = \{(\xi,\eta) : -1 \leq \xi \leq 1,\ -1 \leq \eta \leq 1\}$ so that the Gauss quadrature can be used. Note that the integrand contains not only $\psi_i^e(x,y)$, but also their derivatives with respect to the global coordinates (x,y). The functions $\psi_i^e(x,y)$ can be easily expressed in terms of the local coordinates ξ and η by means of the transformation in Eq. (27), as was shown for one-dimensional Lagrange and Hermite functions in Eqs. (32) and (33). We must first develop relations $\partial\psi_i^e/\partial x$ and $\partial\psi_i^e/\partial y$ to $\partial\psi_i^e/\partial\xi$ and $\partial\psi_i^e/\partial\eta$ using the transformation (27).

By the chain rule of partial differentiation, we have

$$\begin{Bmatrix} \frac{\partial \psi_i^e}{\partial \xi} \\ \frac{\partial \psi_i^e}{\partial \eta} \end{Bmatrix} = \begin{bmatrix} \frac{\partial x}{\partial \xi} & \frac{\partial y}{\partial \xi} \\ \frac{\partial x}{\partial \eta} & \frac{\partial y}{\partial \eta} \end{bmatrix}^e \begin{Bmatrix} \frac{\partial \psi_i^e}{\partial x} \\ \frac{\partial \psi_i^e}{\partial y} \end{Bmatrix} \qquad (10.5-35)$$

which gives the relation between the derivatives of ψ_i^e with respect to the global and local coordinates. The coefficient matrix in Eq. (35) is called the *Jacobian matrix* of the transformation (27)

$$[J]^e = \begin{bmatrix} \frac{\partial x}{\partial \xi} & \frac{\partial y}{\partial \xi} \\ \frac{\partial x}{\partial \eta} & \frac{\partial y}{\partial \eta} \end{bmatrix}^e \qquad (10.5-36)$$

and its determinant $\mathcal{J}$ is called the *Jacobian*, which must be greater than zero in order to invert Eq. (35). Negative nonzero values of $\mathcal{J}$ imply that a right-hand coordinate system is transformed to a left-hand coordinate system, which should be avoided. Inverting Eq. (35), we obtain

$$\left\{\begin{array}{c}\frac{\partial\psi_i^e}{\partial x}\\ \frac{\partial\psi_i^e}{\partial y}\end{array}\right\} = [J]^{-1}\left\{\begin{array}{c}\frac{\partial\psi_i^e}{\partial \xi}\\ \frac{\partial\psi_i^e}{\partial \eta}\end{array}\right\} \tag{10.5-37}$$

This requires the Jacobian matrix $[\mathcal{J}]$ be nonsingular. The Jacobian can be determined using the transformation (27) in Eq. (36). We have

$$[J] = \begin{bmatrix}\frac{\partial x}{\partial\xi} & \frac{\partial y}{\partial\xi}\\ \frac{\partial x}{\partial\eta} & \frac{\partial y}{\partial\eta}\end{bmatrix} = \begin{bmatrix}\sum_{i=1}^m x_i\frac{\partial\hat\psi_i}{\partial\xi} & \sum_{i=1}^m y_i\frac{\partial\hat\psi_i}{\partial\xi}\\ \sum_{i=1}^m x_i\frac{\partial\hat\psi_i}{\partial\eta} & \sum_{i=1}^m y_i\frac{\partial\hat\psi_i}{\partial\eta}\end{bmatrix}$$

$$= \begin{bmatrix}\frac{\partial\hat\psi_1}{\partial\xi} & \frac{\partial\hat\psi_2}{\partial\xi} & \cdots & \frac{\partial\hat\psi_m}{\partial\xi}\\ \frac{\partial\hat\psi_1}{\partial\eta} & \frac{\partial\hat\psi_2}{\partial\eta} & \cdots & \frac{\partial\hat\psi_M}{\partial\eta}\end{bmatrix}\begin{bmatrix}x_1 & y_1\\ x_2 & y_2\\ \vdots & \vdots\\ x_m & y_m\end{bmatrix} \tag{10.5-38}$$

Thus, given the global coordinates (x_j, y_j) of element nodes and the interpolation functions $\hat\psi_j^e$ used for geometry, the Jacobian matrix can be evaluated using Eq. (38). Note that $\hat\psi_j^e$ are different, in general, from ψ_j^e used in the approximation of the dependent variables. The Jacobian is given by

$$\mathcal{J} = J_{11}J_{22} - J_{12}J_{21} \tag{10.5-39}$$

We have from Eq. (37)

$$\left\{\begin{array}{c}\frac{\partial\psi_i^e}{\partial x}\\ \frac{\partial\psi_i^e}{\partial y}\end{array}\right\} = [J]^{-1}\left\{\begin{array}{c}\frac{\partial\psi_i^e}{\partial \xi}\\ \frac{\partial\psi_i^e}{\partial \eta}\end{array}\right\} \equiv [J^*]\left\{\begin{array}{c}\frac{\partial\psi_i^e}{\partial \xi}\\ \frac{\partial\psi_i^e}{\partial \eta}\end{array}\right\} \tag{10.5-40a}$$

where

$$J_{11}^* = \frac{J_{22}}{\mathcal{J}}\,,\quad J_{12}^* = -\frac{J_{12}}{\mathcal{J}}\,,\quad J_{22}^* = \frac{J_{11}}{\mathcal{J}}\,,\quad J_{21}^* = -\frac{J_{21}}{\mathcal{J}} \tag{10.5-40b}$$

Returning the integral in Eq. (34), we can write it now in terms of the natural coordinates as

$$\begin{aligned}K_{ij}^e = \int_{\hat\Omega}\Bigg\{ & a(\xi,\eta)\left(J_{11}^*\frac{\partial\psi_i^e}{\partial\xi} + J_{12}^*\frac{\partial\psi_i^e}{\partial\eta}\right)\left(J_{11}^*\frac{\partial\psi_j^e}{\partial\xi} + J_{12}^*\frac{\partial\psi_j^e}{\partial\eta}\right)\\ & + b(\xi,\eta)\Bigg[\left(J_{11}^*\frac{\partial\psi_i^e}{\partial\xi} + J_{12}^*\frac{\partial\psi_i^e}{\partial\eta}\right)\left(J_{21}^*\frac{\partial\psi_j}{\partial\xi} + J_{22}^*\frac{\partial\psi_j}{\partial\eta}\right)\\ & + \left(J_{11}^*\frac{\partial\psi_i^e}{\partial\xi} + J_{12}^*\frac{\partial\psi_i^e}{\partial\eta}\right)\left(J_{21}^*\frac{\partial\psi_j}{\partial\xi} + J_{22}^*\frac{\partial\psi_j}{\partial\eta}\right)\Bigg]\\ & + c(\xi,\eta)\left(J_{21}^*\frac{\partial\psi_j}{\partial\xi} + J_{22}^*\frac{\partial\psi_j}{\partial\eta}\right)\left(J_{21}^*\frac{\partial\psi_j}{\partial\xi} + J_{22}^*\frac{\partial\psi_j}{\partial\eta}\right)\\ & + d(\xi,\eta)\psi_i^e\psi_j^e\Bigg\}\mathcal{J}\,d\xi d\eta \equiv \int_{\hat\Omega}F_{ij}(\xi,\eta)\,d\xi d\eta\end{aligned} \tag{10.5-41}$$

where the element area $dA = dxdy$ in element Ω^e is transformed to $dA \equiv dxdy = \mathcal{J} d\xi d\eta$ in the master element $\hat{\Omega}$.

Using the Gauss quadrature formulas for integrals defined over a rectangular master element $\hat{\Omega}$, which are the same as those for the one-dimensional quadrature, we obtain

$$\begin{aligned}\int_{\hat{\Omega}} F_{ij}(\xi,\eta)d\xi d\eta &= \int_{-1}^{1}\left[\int_{-1}^{1} F_{ij}(\xi,\eta)d\eta\right]d\xi \approx \int_{-1}^{1}\left[\sum_{J=1}^{N} F_{ij}(\xi,\eta_J)W_J\right]d\xi \\ &\approx \sum_{I=1}^{M}\sum_{J=1}^{N} F_{ij}(\xi_I,\eta_J)W_I W_J \end{aligned} \qquad (10.5-42)$$

where M and N denote the number of Gauss quadrature points in the ξ and η directions, (ξ_I, η_J) denote the Gauss points, and W_I and W_J denote the corresponding Gauss weights (see Table 10.5–1).

Table 10.5–1: Weights and points for the Gauss–Legendre quadrature in one coordinate direction.

N	Points ξ_I	Weights W_I
1	0.0000000000	2.0000000000
2	±0.5773502692	1.0000000000
3	0.0000000000	0.8888888889
	±0.7745966692	0.5555555555
4	±0.3399810435	0.6521451548
	±0.8611363116	0.3478548451
5	0.0000000000	0.5688888889
	±0.5384693101	0.4786286705
	±0.9061798459	0.2369268850
6	±0.2386191861	0.4679139346
	±0.6612093865	0.3607615730
	±0.9324695142	0.1713244924

The selection of the number of Gauss points required to evaluate the integrals accurately is based on the following rule: a polynomial of degree p is integrated exactly employing $N = \text{int}[\frac{1}{2}(p+1)]$; that is, the smallest integer greater than $\frac{1}{2}(p+1)$. In most cases, the interpolation functions are of the same degree in both ξ and η, and therefore one has $M = N$. When the integrand is of different degree in ξ and η, the number of Gauss points is selected on the basis of the largest-degree polynomial in one of the coordinates. The minimum allowable quadrature rule is one that computes the mass of the element exactly when the density is constant.

Table 10.5–2: Selection of the integration order and location of the Gauss points for linear, quadratic, and cubic quadrilateral elements (nodes not shown).

Element Type	Maximum Polynomial Degree	Order of Integration $(r \times r)$	Order of the Residual	Location of Integration Points† in Master Element
Linear $(r = 2)$	2	(2×2)	$O(h^4)$	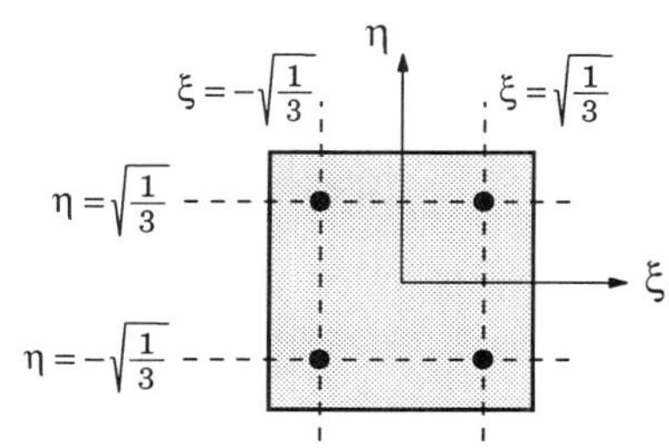
Quadratic $(r = 3)$	4	(3×3)	$O(h^6)$	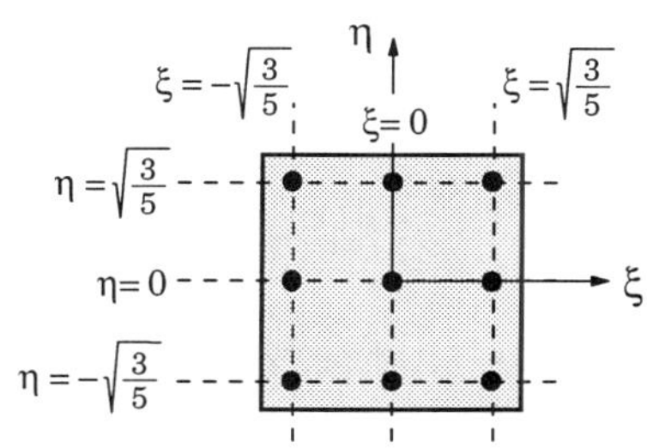
Cubic $(r = 4)$	6	(4×4)	$O(h^8)$	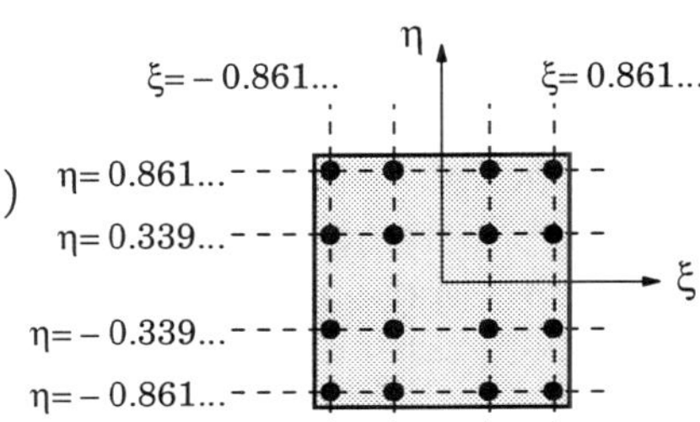

†See Table 10.5–1 for the integration points and weights for each coordinate direction.

Table 10.5–2 contains information on the selection of the integration order and the location of the Gauss points for linear, quadratic, and cubic elements. The maximum degree of the polynomial refers to the degree of the highest polynomial in ξ or η that is present in the integrand of the element matrices of the type in Eq. (32). Note that the polynomial degree of coefficients as well as J_{ij}^* and $\mathcal{J}$ should be accounted for in determining the total polynomial degree of the integrand. Of course, the coefficients a, b, and c and $\mathcal{J}_{ij}^*$ in general may not be polynomials. In those cases, their functional variations must be approximated by a suitable polynomial in order to determine the polynomial degree of the integrand. The $N \times N$ Gauss point locations are given by the *tensor product* of one-dimensional Gauss points ξ_I.

10.5–6 Postcomputation of Stresses

Once the generalized displacements at the nodes are determined, Eq.(4) can be used to determine the strains using the strain–displacement relations (5.3–10). For the case of small strains, displacements, and rotations, the strains at any point (x, y, z) in a typical element Ω^e can be computed from the equations

$$\left\{\begin{matrix} \varepsilon_{xx} \\ \varepsilon_{yy} \\ 2\varepsilon_{xy} \end{matrix}\right\}^e = \left\{\begin{matrix} \frac{\partial u_0}{\partial x} \\ \frac{\partial v_0}{\partial y} \\ \frac{\partial u_0}{\partial y} + \frac{\partial v_0}{\partial x} \end{matrix}\right\}^e - z \left\{\begin{matrix} \frac{\partial^2 w_0}{\partial x^2} \\ \frac{\partial^2 w_0}{\partial y^2} \\ 2\frac{\partial^2 w_0}{\partial x \partial y} \end{matrix}\right\}^e$$

$$= \sum_{j=1}^{m} \left\{\begin{matrix} u_j \frac{\partial \psi_j}{\partial x} \\ v_j \frac{\partial \psi_j}{\partial y} \\ u_j \frac{\partial \psi_j}{\partial y} + v_j \frac{\partial \psi_j}{\partial x} \end{matrix}\right\}^e - z \sum_{k=1}^{n} \Delta_k^e \left\{\begin{matrix} \frac{\partial^2 \varphi_k}{\partial x^2} \\ \frac{\partial^2 \varphi_k}{\partial y^2} \\ 2\frac{\partial^2 \varphi_k}{\partial x \partial y} \end{matrix}\right\}^e \quad (10.5-43)$$

Recall that only (u_0, v_0) and $(w_0, \partial w_0/\partial x, \partial w_0/\partial y)$ are continuous across element interfaces; the first derivatives of the in-plane displacements and the second derivatives of the transverse deflection are, in general, not continuous across element interfaces. In particular, the values of any strain component computed from different elements connected at a node are different.

The stresses at any point in the plate can be computed from the constitutive equations of a lamina, as given in Eq. (8.3–29). Since the strains are discontinuous, the stresses are also discontinuous across element interfaces, including nodes. It was shown by Barlow [28,29] that stresses computed at the Gauss points of an element are the most accurate. For the nonconforming and conforming rectangular plate elements used here, the strains and stresses are computed using the one-point Gauss rule, *i.e.*, at the center of the element.

10.5–7 Numerical Results

Here we use the conforming (C) and nonconforming (NC) rectangular finite elements to analyze laminated plates for bending and natural vibration. Additional numerical results will be presented in Section 10.6. Note should be made of the fact that the finite element model developed herein is not restricted to any particular lamination scheme, geometry, boundary conditions, or loading. Additional results are presented in Section 10.6–5.

The notation $m \times n$ mesh denotes m subdivisions along the $x-$axis and n subdivisions along the $y-$axis with the same type of elements. Solution symmetries available in a problem should be taken advantage of to identify the computational domain because they reduce computational effort. For example, a 2×2 mesh in a quadrant of the plate is the same as 4×4 mesh in the total plate, and the results obtained with the two meshes would be identical, within the round-off errors of the computation, if the solution exhibits biaxial symmetry. A solution is symmetric about a line only if (a) the geometry, including boundary conditions, (b) the material properties, and (c) the loading are symmetric about the line. The boundary conditions along a line of symmetry should be correctly identified and imposed in the finite element model. When one is not sure of the solution symmetry, it is advised that the whole plate be modeled.

Bending Analyses

For antisymmetric cross-ply and angle-ply rectangular laminates with their respective simply supported boundary conditions, a quadrant of the plate may be used as the computational domain of the finite element analysis. The boundary conditions along the symmetry lines for the cross-ply and angle-ply laminates are different, as shown in Figure 10.5–10. The symmetry boundary conditions can be identified from the Navier solutions for each case. When one is doubtful of the boundary conditions along the lines of symmetry, it is safe to use the full plate model. In the case of conforming element, it is necessary that the cross-derivative $\partial w_0/\partial x\partial y$ be also set to zero at the center of the plate when a quarter-plate model is used. Otherwise, the results will be less accurate.

Table 10.5–3 shows a comparison of finite element solutions with the analytical solutions of simply supported orthotropic and two-layer cross-ply and angle-ply (−45/45) square laminates under a uniformly distributed transverse load. In all cases, a quadrant of the plate was used in the finite element analysis. The stresses in the finite element analysis were computed at the Gauss points nearest to the locations at which the stresses were evaluated analytically [see Eq. (9.2–25) and (9.3–12)]. Stresses σ_{xx} and σ_{yy}

are computed at $(5a/8, 5b/8)$, $(9a/16, 9b/16)$, and $(17a/32, 17b/32)$ for uniform meshes 2×2, 4×4, and 8×8, respectively; the origin of the coordinate system is taken at the center of the laminate (see Figure 10.5–10); σ_{xy} is computed for the same meshes at $(3a/8, 3b/8)$, $(7a/16, 7b/16)$, and $(15a/32, 15b/32)$.

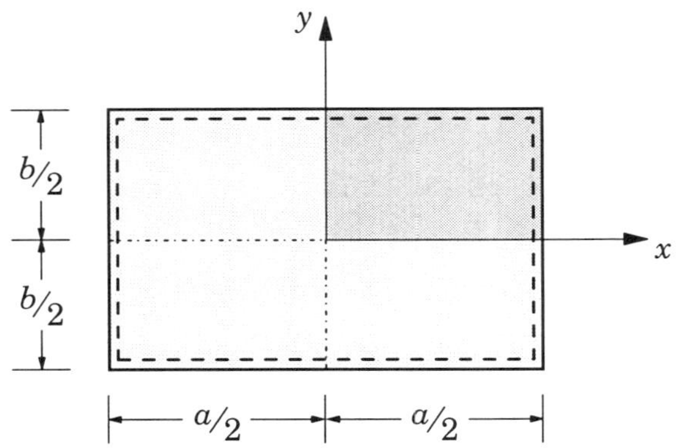

Theory	B.C.	$x = 0$		$y = 0$	
FSDT	SS–1	$u_0=0$	$\phi_x=0$	$v_0=0$	$\phi_y=0$
	SS–2	$v_0=0$	$\phi_x=0$	$u_0=0$	$\phi_y=0$
CLPT	SS–1	$u_0=0$	$\frac{\partial w_0}{\partial x}=0$	$v_0=0$	$\frac{\partial w_0}{\partial y}=0$
	SS–2	$v_0=0$	$\frac{\partial w_0}{\partial x}=0$	$u_0=0$	$\frac{\partial w_0}{\partial y}=0$

Figure 10.5–10: Symmetry boundary conditions for antisymmetric cross-ply and angle-ply laminates.

The conforming element ($\partial^2 w_0/\partial x\partial y = 0$ at the center of the plate) yields slightly better solutions than the nonconforming element, and both elements show good convergence. However, convergence of the displacements is always faster than stresses for the displacement-based finite elements, and the rate of convergence of stresses is two orders less than that of displacements for the CLPT-based element. Since the stresses in the finite element analysis are computed at locations different from the analytical solutions, they are expected to be different. Mesh refinement not only improves the accuracy of the solution, but the Gauss point locations also get closer to the node point locations (but never become the nodal locations), resulting in better agreement with the true solution.

Table 10.5–3: A comparison of the maximum transverse deflections and stresses† of simply supported square plates under uniformly distributed transverse load ($h_n = h/n$, $E_1/E_2 = 25$, $G_{12} = G_{13} = 0.5E_2$, $G_{23} = 0.2E_2$, $\nu_{12} = 0.25$; CLPT solutions).

Variable	Nonconforming			Conforming			Analytical
	2×2	4×4	8×8	2×2	4×4	8×8	Solution
Orthotropic Plate (SS–1)							
$\bar{w}$	0.7082	0.6635	0.6531	0.7532	0.6651	0.6517	0.6497
$\bar{\sigma}_{xx}$	0.7148	0.7709	0.7828	0.5772	0.7388	0.7759	0.7866
$\bar{\sigma}_{yy}$	0.0295	0.0253	0.0246	0.0283	0.0249	0.0246	0.0244
$\bar{\sigma}_{xy}$	0.0337	0.0421	0.0444	0.0369	0.0416	0.0448	0.0463
Cross-Ply (0/90) Plate (SS–1)							
$\bar{w}$	1.7937	1.7203	1.7017	1.7004	1.6909	1.6935	1.6955
$\bar{\sigma}_{xx}$	0.1109	0.1230	0.1259	0.1056	0.1215	0.1255	0.1268
$\bar{\sigma}_{yy}$	0.9436	1.0440	1.0683	0.8819	1.0263	1.0637	1.0761
$\bar{\sigma}_{xy}$	0.0751	0.0872	0.0914	0.0704	0.0853	0.0908	0.0933
Angle-Ply (−45/45) Plate (SS–2)							
$\bar{w}$	1.0524	1.0341	1.0296	1.0111	1.0222	1.0264	1.0280
$\bar{\sigma}_{xx} = \bar{\sigma}_{yy}$	0.2600	0.3279	0.3449	0.2639	0.3303	0.3445	0.3504
$\bar{\sigma}_{xy}$	0.3935	0.4264	0.4376	0.3852	0.4206	0.4356	0.4421
Angle-Ply $(-45/45)_4$ Plate (SS–2)							
$\bar{w}$	0.4045	0.3905	0.3870	0.3948	0.3870	0.3859	0.3858
$\bar{\sigma}_{xx} = \bar{\sigma}_{yy}$	0.1500	0.1880	0.1975	0.1406	0.1867	0.1973	0.2006
$\bar{\sigma}_{xy}$	0.2501	0.2612	0.2636	0.2495	0.2590	0.2627	0.2637

† The stresses are computed at the center of each finite element.

It is clear from the results presented in Table 10.5–3 that the convergence of the finite element results to the analytical solutions is very good. The slower convergence of stresses in two-layer angle-ply plates compared to the eight-layer laminate is due to the presence of bending stretching–coupling. Recall that in angle-ply laminates the analytical solution for stresses [compare Eq. (8.3–30a,b) with Eq. (8.4–10)] is the sum of two parts: one is a double sine series and the other is double cosine series. They are mutually exclusive at points $(a/2, b/2)$ and $(0, 0)$; However, the parts add up at the Gauss points (0.46875,0.46875) and (0.03125,0.03125). For example, the analytical results of stresses for the (−45/45) laminate are

$\bar{\sigma}_{xx} = 0.3486$ and $\bar{\sigma}_{xy} = 0.4312$ at the Gauss points (0.46875,0.46875) and (0.03125,0.03125), respectively, which shows better agreement between the analytical and finite element stress values.

Table 10.5–4: Maximum transverse deflections and stresses† of square laminates under uniformly distributed transverse load and for different boundary conditions ($h_n = h/n$, $E_1/E_2 = 25$, $G_{12} = G_{13} = 0.5E_2$, $G_{23} = 0.2E_2$, $\nu_{12} = 0.25$).

Variable	(0/90)			(−45/45)		
	SS–1	SS–2	CC	SS–1	SS–2	CC
$\bar{w}$	1.6839	0.8957	0.3814	0.6773	1.0208	0.3891
$\bar{\sigma}_{yy}$	1.0469	0.8939	0.5699	0.3392	0.3659	0.2202
$\bar{\sigma}_{xy}$	0.0851	0.0361	0.0181	0.3341	0.4201	0.1772

† The stresses are computed at the center of each finite element; 8×8 uniform mesh of conforming elements is used in the full plate.

The effect of the simply supported SS–1 ($u_s = w_0 = \theta_s = 0$) and SS–2 ($u_n = w_0 = \theta_s = 0$) and clamped ($u_n = u_s = w_0 = \theta_n = \theta_s = 0$) boundary conditions on two-layer cross-ply and angle-ply laminates is investigated using full plate models and 8×8 uniform mesh of conforming elements (no boundary condition on the cross–derivative was imposed) and the results are presented in Table 10.5–4. Recall that cross-ply laminates admit the Navier solutions for the SS–1 boundary conditions whereas antisymmetric angle-ply laminates admit for the SS–2 boundary conditions. Analytical solutions (*i.e.*, Navier or Lévy type solutions) are not available for cross-ply laminates with SS–2, antisymmetric angle-ply laminates with SS–1, and any laminate for clamped (CC) boundary conditions. The locations of maximum stresses are indicated below for various cases.

Cross-Ply SS–1 and SS–2 $[\sigma_{xx}(x_0, y_0, -h/2) = -\sigma_{yy}(x_0, y_0, h/2)]$:

$$\sigma_{yy}(0.5625a, 0.5625a, h/2) \ ; \quad \sigma_{xy}(0.9375a, 0.9375a, -h/2)$$

Cross-Ply Clamped $[\sigma_{xx}(x_0, y_0, -h/2) = -\sigma_{yy}(y_0, x_0, h/2)]$:

$$-\sigma_{yy}(0.4375a, 0.0625a, h/2) \ ; \quad \sigma_{xy}(0.8125a, 0.8125a, -h/2)$$

Angle-Ply $[\sigma_{xx}(x_0, y_0, h/2) = \sigma_{yy}(x_0, y_0, h/2)]$:

$$\text{SS–1:} \quad \sigma_{yy}(0.6875a, 0.3125a, h/2) \ ; \quad \sigma_{xy}(0.8125a, 0.1875a, -h/2)$$

SS–2: $\sigma_{yy}(0.8125a, 0.1875a, h/2)$; $-\sigma_{xy}(0.9375a, 0.9375a, h/2)$

Clamped: $-\sigma_{xx}(0.0625a, 0.4375a, h/2)$; $-\sigma_{xy}(0.0625a, 0.6875a, h/2)$

Table 10.5–5: A comparison of finite element (second row) and analytical (first row) solutions for nondimensional center deflections ($\bar{w}$) and in-plane normal stresses ($\bar{\sigma}_{xx}$ and $\bar{\sigma}_{yy}$) of antisymmetric cross-ply and angle-ply square plates subjected to sinusoidal distribution of transverse load and for various boundary conditions ($h_n = h/n$, $E_1/E_2 = 25$, $G_{12} = G_{13} = 0.5E_2$, $G_{23} = 0.2E_2$, $\nu_{12} = 0.25$).

No. of Layers	Variable	SS	SC	CC	FF	FS	FC
Cross-Ply Laminates (0/90/0 ···)							
2	$\bar{w}$	1.064	0.664	0.429	1.777	1.471	0.980
		1.055	0.652	0.415	1.761	1.445	0.966
	$\bar{\sigma}_{xx}$	7.157	5.660	4.800	2.403	4.442	3.042
		6.888^a	5.859^a	4.457^a	2.262^a	4.562^a	3.150^a
	$\bar{\sigma}_{yy}$	7.157	4.483	2.914	11.849	9.837	6.560
		6.888^a	4.496^a	2.722^a	11.516^a	10.029^b	7.142^a
10	$\bar{w}$	0.442	0.266	0.167	0.665	0.579	0.380
		0.442	0.266	0.168	0.665	0.579	0.381
	$\bar{\sigma}_{xx}$	5.009	3.829	3.167	1.725	2.986	1.865
		4.791^a	3.953^a	2.904^a	1.615^a	2.999^a	1.925^a
	$\bar{\sigma}_{yy}$	5.009	3.025	1.911	7.480	6.531	4.284
		4.791^a	3.046^a	1.811^a	7.258^a	6.711^b	4.694^a
Angle-Ply Laminates (–45/45/–45···) (FEM)							
2	$\bar{w}$	0.649	0.497	0.384	3.026	1.319	0.981
	$\bar{\sigma}_{xx}$	0.652^c	0.554^c	0.375^d	0.698^f	1.043^c	0.733^g
	$\bar{\sigma}_{yy}$	0.652^c	0.554^c	0.374^d	0.830^f	1.043^c	0.745^g
10	$\bar{w}$	0.243	0.191	0.429	1.426	0.524	0.397
	$\bar{\sigma}_{xx}$	0.188^c	0.163^c	0.116^e	0.357^f	0.318^c	0.249^g
	$\bar{\sigma}_{yy}$	0.188^c	0.163^c	0.115^e	0.427^f	0.317^c	0.254^g

a : (0.4375, 0.4375); b : (0.5625, 0.5625); c : (0.0625, 0.0625); d : (0.9375, 0.8125); e : (0.3125, 0.3125); f : (0.1875, 0.9375); g : (0.8125, 0.6875).

From the results (see Table 10.5–4) it is clear that SS–2 boundary conditions make the cross-ply laminate stiffer because they restrain the bidirectional composite from having normal in-plane displacements ($u_n = 0$). Similarly, SS–1 boundary conditions make the antisymmetric angle-ply stiffer by restraining the in-plane tangential displacements ($u_s = 0$). The clamped boundary conditions make both plates quite stiff compared to the simply supported boundary conditions.

A comparison of finite element solutions with the analytical solutions of antisymmetric cross-ply and angle-ply laminates under sinusoidal loading is presented in Table 10.5–5. The finite element solutions are obtained using 8×8 uniform mesh of conforming elements in the full plate. The stresses are computed at the Gauss points and the locations of the stresses are indicated in the footnote of the table.

Natural Vibration

The finite element solutions are compared with the analytical solutions of antisymmetric cross-ply and angle-ply laminates in Table 10.5–6. In all cases, a quarter-plate model with appropriate symmetry boundary conditions was used in the finite element analysis (see Reddy [71]). The finite element solutions show convergence to the analytical solutions with mesh refinements.

Table 10.5–6: A comparison of the natural frequencies,† $\bar{\omega} = \omega(b^2/h)\sqrt{\rho/E_2}$ of simply supported square plates ($h_n = h/n$, $E_1/E_2 = 40$, $G_{12} = G_{13} = 0.6E_2$, $G_{23} = 0.5E_2$, $\nu_{12} = 0.25$, $a/h = 10$).

$n^‡$	Nonconforming		Conforming		Analytical
	2×2	4×4	2×2	4×4	Solution
Cross-Ply $(0/90/0/\cdots)$ Plates (SS–1)					
2	10.882	11.080	11.215	11.186	11.154
4	16.461	16.960	16.745	17.101	17.145
8	17.559	18.140	17.795	18.279	18.352
Angle-Ply $(-45/45/-45/\cdots)$ Plates (SS–2)					
2	14.360	14.413	14.659	14.504	14.439
4	22.821	23.168	23.168	23.294	23.304
8	23.699	24.888	24.883	25.024	25.052

† The rotary inertia is included. ‡n = Number of layers in the laminate.

10.6 Finite Element Models of Laminated Plates (FSDT)

10.6–1 Weak Forms

Following the procedure described in Section 10.5–1, we can develop the weak forms of the equations governing the first-order shear deformation plate theory. We consider the linear equations of motion of FSDT from Eqs. (5.4–13), which are in terms of the stress resultants but equivalent to Eqs. (9.1–1) through (9.1–5). The generalized displacements of FSDT are $(u_0, v_0, w_0, \phi_x, \phi_y)$. The weak forms of the five equations in (5.4–13) are obtained by multiplying them with $\delta u_0, \delta v_0, \delta w_0, \delta\phi_x$, and $\delta\phi_y$, respectively, and integrating over the element domain. We obtain

$$0 = \int_{\Omega^e} \left(\frac{\partial \delta u_0}{\partial x} N_{xx} + \frac{\partial \delta u_0}{\partial y} N_{xy} + I_0 \delta u_0 \frac{\partial^2 u_0}{\partial t^2} + I_1 \delta u_0 \frac{\partial^2 \phi_x}{\partial t^2} \right) dxdy - \oint_{\Gamma^e} P_x \delta u_0 \, ds \tag{10.6 - 1a}$$

$$0 = \int_{\Omega^e} \left(\frac{\partial \delta v_0}{\partial x} N_{xy} + \frac{\partial \delta v_0}{\partial y} N_{yy} + I_0 \delta v_0 \frac{\partial^2 v_0}{\partial t^2} + I_1 \delta v_0 \frac{\partial^2 \phi_y}{\partial t^2} \right) dxdy - \oint_{\Gamma^e} P_y \delta v_0 \, ds \tag{10.6 - 1b}$$

$$\begin{aligned} 0 = \int_{\Omega^e} \Bigg[& \frac{\partial \delta w_0}{\partial x} Q_x + \frac{\partial \delta w_0}{\partial y} Q_y - \delta w_0 q + I_0 \delta w_0 \frac{\partial^2 w_0}{\partial t^2} \\ & + \frac{\partial \delta w_0}{\partial x} \left(\hat{N}_{xx} \frac{\partial w_0}{\partial x} + \hat{N}_{xy} \frac{\partial w_0}{\partial y} \right) + \frac{\partial \delta w_0}{\partial y} \left(\hat{N}_{xy} \frac{\partial w_0}{\partial x} + \hat{N}_{yy} \frac{\partial w_0}{\partial y} \right) \Bigg] dxdy \\ & - \oint_{\Gamma^e} \left[\left(Q_x + \hat{N}_{xx} \frac{\partial w_0}{\partial x} + \hat{N}_{xy} \frac{\partial w_0}{\partial y} \right) n_x + \left(Q_y + \hat{N}_{xy} \frac{\partial w_0}{\partial x} + \hat{N}_{yy} \frac{\partial w_0}{\partial y} \right) n_y \right] \delta w_0 \, ds \end{aligned} \tag{10.6 - 1c}$$

$$0 = \int_{\Omega^e} \left(\frac{\partial \delta \phi_x}{\partial x} M_{xx} + \frac{\partial \delta \phi_x}{\partial y} M_{xy} + \delta\phi_x Q_x + I_2 \delta\phi_x \frac{\partial^2 \phi_x}{\partial t^2} + I_1 \delta\phi_x \frac{\partial^2 u_0}{\partial t^2} \right) dxdy - \oint_{\Gamma^e} T_x \delta\phi_x \, ds \tag{10.6 - 1d}$$

$$0 = \int_{\Omega^e} \left(\frac{\partial \delta \phi_y}{\partial x} M_{xy} + \frac{\partial \delta \phi_y}{\partial y} M_{yy} + \delta\phi_y Q_y + I_2 \delta\phi_y \frac{\partial^2 \phi_y}{\partial t^2} + I_1 \delta\phi_y \frac{\partial^2 v_0}{\partial t^2} \right) dxdy - \oint_{\Gamma^e} T_y \delta\phi_y \, ds \tag{10.6 - 1e}$$

We note from the boundary terms in Eq. (1) that $(u_0, v_0, w_0, \phi_x, \phi_y)$ are the primary variables (or generalized displacements). Unlike in the classical plate theory, the rotations (ϕ_x, ϕ_y) are independent of w_0. Note also that no derivatives of w_0 are in the list of the primary variables. The secondary variables are

$$\begin{aligned}
P_x \equiv& N_{xx}n_x + N_{xy}n_y, \quad P_y \equiv N_{xy}n_x + N_{yy}n_y \\
T_x \equiv& M_{xx}n_x + M_{xy}n_y, \quad T_y \equiv M_{xy}n_x + M_{yy}n_y \\
Q_n \equiv& \left(Q_x + \hat{N}_{xx}\frac{\partial w_0}{\partial x} + \hat{N}_{xy}\frac{\partial w_0}{\partial y}\right) n_x \\
&+ \left(Q_y + \hat{N}_{xy}\frac{\partial w_0}{\partial x} + \hat{N}_{yy}\frac{\partial w_0}{\partial y}\right) n_y
\end{aligned} \tag{10.6 – 2}$$

10.6–2 Finite Element Model

The weak forms of the first-order theory contain, at the most, only the first derivatives of the dependent variables $(u_0, v_0, w_0, \phi_x, \phi_y)$. Therefore, they can all be approximated using the Lagrange interpolation functions. In principle, the sets (u_0, v_0), w_0, and (ϕ_x, ϕ_y) can be approximated with differing degrees of functions. For simplicity, we use the same interpolation for all variables. Let

$$u_0(x,y,t) = \sum_{j=1}^{m} u_j(t)\psi_j^e(x,y) \tag{10.6 – 3a}$$

$$v_0(x,y,t) = \sum_{j=1}^{m} v_j(t)\psi_j^e(x,y) \tag{10.6 – 3b}$$

$$w_0(x,y,t) = \sum_{j=1}^{n} w_j(t)\psi_j^e(x,y) \tag{10.6 – 4}$$

$$\phi_x(x,y,t) = \sum_{j=1}^{p} S_j^1(t)\psi_j^e(x,y) \tag{10.6 – 5a}$$

$$\phi_y(x,y,t) = \sum_{j=1}^{p} S_j^2(t)\psi_j^e(x,y) \tag{10.6 – 5b}$$

where ψ_j^e are Lagrange interpolation functions. In general, (u_0, v_0), w_0, and (ϕ_x, ϕ_y) may be interpolated with different degree of interpolation. One can use linear, quadratic, or higher-order interpolations of these sets.

Substituting Eqs. (3)–(5) for $(u_0, v_0, w_0, \phi_x, \phi_y)$ into the weak forms in Eq. (1), we obtain the semidiscrete finite element model of the first-order theory:

$$\left(\begin{bmatrix} [K^{11}] & [K^{12}] & [K^{13}] & [K^{14}] & [K^{15}] \\ [K^{12}]^T & [K^{22}] & [K^{23}] & [K^{24}] & [K^{25}] \\ [K^{13}]^T & [K^{23}]^T & [K^{33}] & [K^{34}] & [K^{35}] \\ [K^{14}]^T & [K^{24}]^T & [K^{34}]^T & [K^{44}] & [K^{45}] \\ [K^{15}]^T & [K^{25}]^T & [K^{35}]^T & [K^{45}]^T & [K^{55}] \end{bmatrix} + \begin{bmatrix} [0] & [0] & [0] & [0] & [0] \\ [0] & [0] & [0] & [0] & [0] \\ [0] & [0] & [G] & [0] & [0] \\ [0] & [0] & [0] & [0] & [0] \\ [0] & [0] & [0] & [0] & [0] \end{bmatrix}\right) \begin{Bmatrix} \{u^e\} \\ \{v^e\} \\ \{w^e\} \\ \{S^1\} \\ \{S^2\} \end{Bmatrix}$$

$$
+\begin{bmatrix} I_0[M] & [0] & [0] & I_1[M] & [0] \\ [0] & I_0[M] & [0] & [0] & I_1[M] \\ [0] & [0] & I_0[M] & [0] & [0] \\ I_1[M] & [0] & [0] & I_2[M] & [0] \\ [0] & I_1[M] & [0] & [0] & I_2[M] \end{bmatrix} \begin{Bmatrix} \{\ddot{u}^e\} \\ \{\ddot{v}^e\} \\ \{\ddot{w}^e\} \\ \{\ddot{S}^1\} \\ \{\ddot{S}^2\} \end{Bmatrix} = \begin{Bmatrix} \{F^1\} - \{F^{T1}\} \\ \{F^2\} - \{F^{T2}\} \\ \{F^3\} \\ \{F^4\} - \{F^{T4}\} \\ \{F^5\} - \{F^{T5}\} \end{Bmatrix} \tag{10.6-6a}
$$

or

$$
[K^e]\{\Delta^e\} + [M^e]\{\ddot{\Delta}^e\} = \{F^e\} \tag{10.6-6b}
$$

where the coefficients of the submatrices $[K^{\alpha\beta}]$ and $[M^{\alpha\beta}]$ and vectors $\{F^\alpha\}$ are defined for $(\alpha, \beta = 1, 2, ..., 5)$ by the expressions

$$
\begin{aligned}
K_{ij}^{1\alpha} &= \int_{\Omega^e} \left(\frac{\partial \psi_i^e}{\partial x} N_{1j}^\alpha + \frac{\partial \psi_i^e}{\partial y} N_{6j}^\alpha \right) dxdy \\
K_{ij}^{2\alpha} &= \int_{\Omega^e} \left(\frac{\partial \psi_i^e}{\partial x} N_{6j}^\alpha + \frac{\partial \psi_i^e}{\partial y} N_{2j}^\alpha \right) dxdy \\
K_{ij}^{3\alpha} &= \int_{\Omega^e} \left(\frac{\partial \psi_i^e}{\partial x} Q_{1j}^\alpha + \frac{\partial \psi_i^e}{\partial y} Q_{2j}^\alpha \right) dxdy \\
K_{ij}^{4\alpha} &= \int_{\Omega^e} \left(\frac{\partial \psi_i^e}{\partial x} M_{1j}^\alpha + \frac{\partial \psi_i^e}{\partial y} M_{6j}^\alpha + \psi_i^e Q_{1j}^\alpha \right) dxdy \\
K_{ij}^{5\alpha} &= \int_{\Omega^e} \left(\frac{\partial \psi_i^e}{\partial x} M_{6j}^\alpha + \frac{\partial \psi_i^e}{\partial y} M_{2j}^\alpha + \psi_i^e Q_{2j}^\alpha \right) dxdy \\
G_{ij} &= \int_{\Omega^e} \left[\hat{N}_{xx} \frac{\partial \psi_i^e}{\partial x} \frac{\partial \psi_j^e}{\partial y} + \hat{N}_{xy} \left(\frac{\partial \psi_i^e}{\partial x} \frac{\partial \psi_j^e}{\partial y} + \frac{\partial \psi_i^e}{\partial y} \frac{\partial \psi_j^e}{\partial x} \right) + \right. \\
&\qquad \left. \hat{N}_{yy} \frac{\partial \psi_i^e}{\partial y} \frac{\partial \psi_j^e}{\partial y} \right] dxdy \\
M_{ij} &= \int_{\Omega^e} \psi_i^e \psi_j^e \, dxdy
\end{aligned} \tag{10.6-7}
$$

The coefficients N_{Ij}^α, M_{Ij}^α, and Q_{Ij}^α for $\alpha = 1, 2, ..., 5$ and $I = 1, 2, 6$ are given by

$$
N_{1j}^1 = A_{11} \frac{\partial \psi_j^e}{\partial x} + A_{16} \frac{\partial \psi_j^e}{\partial y}, \quad N_{1j}^2 = A_{12} \frac{\partial \psi_j^e}{\partial y} + A_{16} \frac{\partial \psi_j^e}{\partial x}
$$

$$
N_{1j}^4 = B_{11} \frac{\partial \psi_j^e}{\partial x} + B_{16} \frac{\partial \psi_j^e}{\partial y}, \quad N_{1j}^5 = B_{12} \frac{\partial \psi_j^e}{\partial y} + B_{16} \frac{\partial \psi_j^e}{\partial x}
$$

$$
N_{2j}^1 = A_{12} \frac{\partial \psi_j^e}{\partial x} + A_{26} \frac{\partial \psi_j^e}{\partial y}, \quad N_{2j}^2 = A_{22} \frac{\partial \psi_j^e}{\partial y} + A_{26} \frac{\partial \psi_j^e}{\partial x}
$$

$$
N_{2j}^4 = B_{12} \frac{\partial \psi_j^e}{\partial x} + B_{26} \frac{\partial \psi_j^e}{\partial y}, \quad N_{2j}^5 = B_{22} \frac{\partial \psi_j^e}{\partial y} + B_{26} \frac{\partial \psi_j^e}{\partial x}
$$

$$N^1_{6j} = A_{16}\frac{\partial\psi^e_j}{\partial x} + A_{66}\frac{\partial\psi^e_j}{\partial y}\;,\quad N^2_{6j} = A_{26}\frac{\partial\psi^e_j}{\partial y} + A_{66}\frac{\partial\psi^e_j}{\partial x}$$

$$N^4_{6j} = B_{16}\frac{\partial\psi^e_j}{\partial x} + B_{66}\frac{\partial\psi^e_j}{\partial y}\;,\quad N^5_{6j} = B_{26}\frac{\partial\psi^e_j}{\partial y} + B_{66}\frac{\partial\psi^e_j}{\partial x}$$

$$M^1_{1j} = B_{11}\frac{\partial\psi^e_j}{\partial x} + B_{16}\frac{\partial\psi^e_j}{\partial y}\;,\quad M^2_{1j} = B_{12}\frac{\partial\psi^e_j}{\partial y} + B_{16}\frac{\partial\psi^e_j}{\partial x}$$

$$M^4_{1j} = D_{11}\frac{\partial\psi^e_j}{\partial x} + D_{16}\frac{\partial\psi^e_j}{\partial y}\;,\quad M^5_{1j} = D_{12}\frac{\partial\psi^e_j}{\partial y} + D_{16}\frac{\partial\psi^e_j}{\partial x}$$

$$M^1_{2j} = B_{12}\frac{\partial\psi^e_j}{\partial x} + B_{26}\frac{\partial\psi^e_j}{\partial y}\;,\quad M^2_{2j} = B_{22}\frac{\partial\psi^e_j}{\partial y} + B_{26}\frac{\partial\psi^e_j}{\partial x}$$

$$M^4_{2j} = D_{12}\frac{\partial\psi^e_j}{\partial x} + D_{26}\frac{\partial\psi^e_j}{\partial y}\;,\quad M^5_{2j} = D_{22}\frac{\partial\psi^e_j}{\partial y} + D_{26}\frac{\partial\psi^e_j}{\partial x}$$

$$M^1_{6j} = B_{16}\frac{\partial\psi^e_j}{\partial x} + B_{66}\frac{\partial\psi^e_j}{\partial y}\;,\quad M^2_{6j} = B_{26}\frac{\partial\psi^e_j}{\partial y} + B_{66}\frac{\partial\psi^e_j}{\partial x}$$

$$M^4_{6j} = D_{16}\frac{\partial\psi^e_j}{\partial x} + D_{66}\frac{\partial\psi^e_j}{\partial y}\;,\quad M^5_{6j} = D_{26}\frac{\partial\psi^e_j}{\partial y} + D_{66}\frac{\partial\psi^e_j}{\partial x}$$

$$N^3_{1j} = 0,\quad N^3_{2j} = 0,\quad N^3_{6j} = 0$$

$$M^3_{1j} = 0,\quad M^3_{2j} = 0,\quad M^3_{6j} = 0$$

$$Q^3_{1j} = 0,\quad Q^3_{2j} = 0 \qquad (10.6-8a)$$

$$F^1_i = \oint_{\Gamma^e} P_x\psi^e_i\; dxdy,\quad F^2_i = \int_{\Gamma_e} P_y\psi^e_i\; dxdy$$

$$F^3_i = \int_{\Omega^e} q\psi^e_i\; dxdy + \oint_{\Gamma^e} Q_n\psi^e_i\; ds$$

$$F^4_i = \oint_{\Gamma^e} T_x\psi^e_i\; dxdy,\quad F^5_i = \int_{\Gamma_e} T_y\psi^e_i\; dxdy$$

$$F^{T1}_i = \oint_{\Gamma^e}\left(\frac{\partial\psi^e_i}{\partial x}N^T_{xx} + \frac{\partial\psi^e_i}{\partial y}N^T_{xy}\right)\; dxdy$$

$$F^{T2}_i = \int_{\Gamma_e}\left(\frac{\partial\psi^e_i}{\partial x}N^T_{xy} + \frac{\partial\psi^e_i}{\partial y}N^T_{yy}\right)\; dxdy$$

$$F^{T4}_i = \oint_{\Gamma^e}\left(\frac{\partial\psi^e_i}{\partial x}M^T_{xx} + \frac{\partial\psi^e_i}{\partial y}M^T_{xy}\right)\; dxdy$$

$$F^{T5}_i = \int_{\Gamma_e}\left(\frac{\partial\psi^e_i}{\partial x}M^T_{xy} + \frac{\partial\psi^e_i}{\partial y}M^T_{yy}\right)\; dxdy \qquad (10.6-8b)$$

where N^T_{xx}, M^T_{xx}, etc. are the thermal force and moment resultants.

The displacement-based C^0 plate bending element of Eq. (6) is often referred to in the finite element literature as the *Mindlin plate element* due to the fact that it is based on the so-called Mindlin plate theory, which is labeled in this book as the first-order shear deformation plate theory. Any of the Lagrange interpolation functions presented in Eqs. (10.5–10), (10.5–11), and (10.5–12) may be used to approximate the displacement field $(u_0, v_0, w_0, \phi_x, \phi_y)$. When the bilinear rectangular element is used for all generalized displacements, the element stiffness matrices are of the order 20×20; and for the nine-node quadratic element they are 45×45 (see Figure 10.6–1).

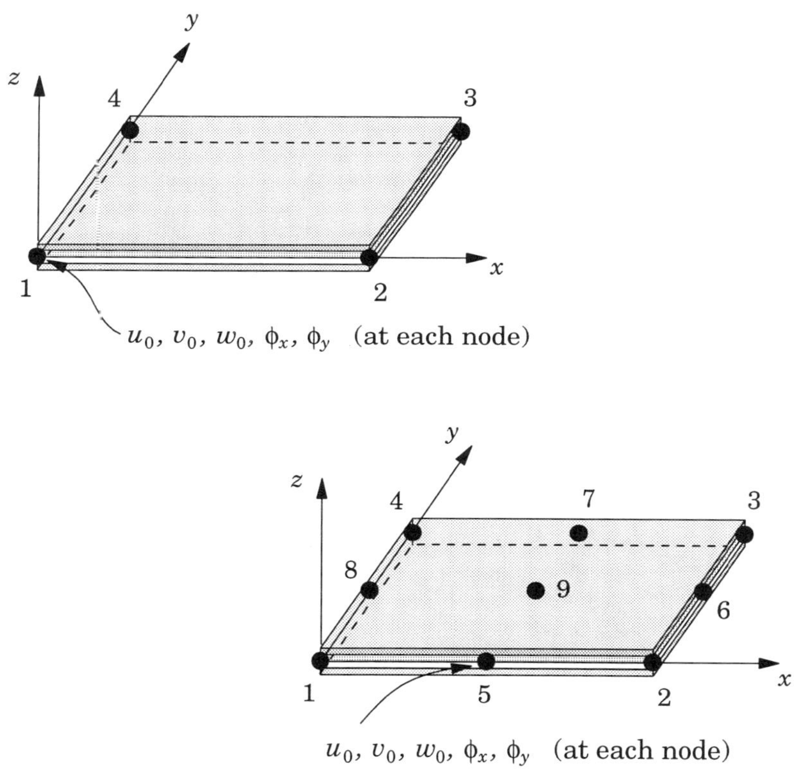

Figure 10.6–1: Linear and quadratic Lagrange rectangular elements for the first-order shear deformation theory.

Equation (6) can be simplified for static bending, buckling, natural vibration, and transient analyses, as described in Eqs. (10.5–20)–(10.5–24). The simplifications are obvious and therefore are not repeated for the FSDT element. However, numerical results for bending, buckling, natural vibration, and transient response will be discussed.

10.6–3 Penalty Function Formulation and Shear Locking

A finite element model equivalent to that in Eq. (6) can also be derived using the penalty function approach (see Reddy [30,31]) applied to the classical plate theory. For the sake of simplicity in discussion, consider the weak form of the classical plate theory without membrane strains, nonlinearity, and inertia terms [see Eq. (5.3–19)]:

$$0 = \int_{\Omega_0} \left(M_{xx}\delta\varepsilon_{xx}^{(1)} + M_{yy}\delta\varepsilon_{yy}^{(1)} + M_{xy}\delta\gamma_{xy}^{(1)} - q\delta w_0 \right) dxdy \qquad (10.6-9)$$

Next assume, again for the sake of simplicity, that the plate under consideration is orthotropic. Using the plate constitutive equations (5.3–44), we rewrite the weak form (9) in terms of the generalized displacements

$$\begin{aligned}
0 &= \int_{\Omega_0} \left[\left(D_{11}\frac{\partial^2 w_0}{\partial x^2} + D_{12}\frac{\partial^2 w_0}{\partial y^2} \right) \frac{\partial^2 \delta w_0}{\partial x^2} + \left(D_{12}\frac{\partial^2 w_0}{\partial x^2} + D_{22}\frac{\partial^2 w_0}{\partial y^2} \right) \frac{\partial^2 \delta w_0}{\partial y^2} \right. \\
&\quad \left. + 4D_{66}\frac{\partial^2 w_0}{\partial x \partial y}\frac{\partial^2 \delta w_0}{\partial x \partial y} - q\delta w_0 \right] dxdy \\
&= \delta \int_{\Omega_0} \frac{1}{2} \left[D_{11}\left(\frac{\partial^2 w_0}{\partial x^2}\right)^2 + 2D_{12}\frac{\partial^2 w_0}{\partial x^2}\frac{\partial^2 w_0}{\partial y^2} + D_{22}\left(\frac{\partial^2 w_0}{\partial y^2}\right)^2 \right. \\
&\quad \left. + 4D_{66}\left(\frac{\partial^2 w_0}{\partial x \partial y}\right)^2 - qw_0 \right] dxdy \equiv \delta\Pi_0(w_0) \qquad (10.6-10)
\end{aligned}$$

where Π_0 denotes the total potential energy functional

$$\begin{aligned}
\Pi_0(w_0) = \int_{\Omega_0} \frac{1}{2} &\left[D_{11}\left(\frac{\partial^2 w_0}{\partial x^2}\right)^2 + 2D_{12}\frac{\partial^2 w_0}{\partial x^2}\frac{\partial^2 w_0}{\partial y^2} + D_{22}\left(\frac{\partial^2 w_0}{\partial y^2}\right)^2 \right. \\
&\left. + 4D_{66}\left(\frac{\partial^2 w_0}{\partial x \partial y}\right)^2 - qw_0 \right] dxdy \qquad (10.6-11)
\end{aligned}$$

Equation (10) is a statement of the principle of the minimum potential energy, which is a special case of the principle of virtual displacements when the material of plate is assumed to obey Hooke's law. Introduce the variables θ_x and θ_y such that

$$\frac{\partial w_0}{\partial x} - \theta_x = 0, \quad \frac{\partial w_0}{\partial y} - \theta_y = 0 \qquad (10.6-12)$$

Then the potential energy functional takes the form

$$\begin{aligned}
\Pi(w_0, \theta_x, \theta_y) = \int_{\Omega_0} \frac{1}{2} &\left[D_{11}\left(\frac{\partial \theta_x}{\partial x}\right)^2 + 2D_{12}\frac{\partial \theta_x}{\partial x}\frac{\partial \theta_y}{\partial y} + D_{22}\left(\frac{\partial \theta_y}{\partial y}\right)^2 \right. \\
&\left. + D_{66}\left(\frac{\partial \theta_x}{\partial y} + \frac{\partial \theta_y}{\partial x}\right)^2 - qw_0 \right] dxdy \qquad (10.6-13)
\end{aligned}$$

The difference between $\Pi_0(w_0)$ and $\Pi(w_0, \theta_x, \theta_y)$ is that the latter contains at the most only the first derivatives of the dependent variables, and therefore will require C^0–interpolation in the finite element model. However, the potential energy functional $\Pi(w_0, \theta_x, \theta_y)$ in Eq. (13) does not include the fact that the dependent variables θ_x and θ_y are related to w_0 by Eqs. (12). Therefore, the principle of the minimum potential energy must be stated as one of minimizing the functional $\Pi(w_0, \theta_x, \theta_y)$ subjected to the (constraint) conditions in Eq. (12): minimize

$$\Pi(w_0, \theta_x, \theta_y) \tag{10.6 – 14a}$$

subjected to the constraints

$$\frac{\partial w_0}{\partial x} - \theta_x = 0, \quad \frac{\partial w_0}{\partial y} - \theta_y = 0 \tag{10.6 – 14b}$$

The constrained minimization problem (14) can be solved either using the Lagrange multiplier method or the penalty function method. In the Lagrange multiplier method we assume that there exist *Lagrange multipliers* λ_1 and λ_2 such that the constrained minimization problem is equivalent to $\delta\Pi_L(w_0, \theta_x, \theta_y, \lambda_1, \lambda_2) = 0$, where

$$\Pi_L \equiv \Pi(w_0, \theta_x, \theta_y) + \int_{\Omega_0} \left[\lambda_1 \left(\frac{\partial w_0}{\partial x} - \theta_x\right) + \lambda_2 \left(\frac{\partial w_0}{\partial y} - \theta_y\right)\right] dxdxy \tag{10.6 – 15}$$

The weak form, $\delta\Pi_L = 0$, can be used to construct a finite element model with C^0–interpolation of all dependent unknowns: $w_0, \theta_x, \theta_y, \lambda_1$, and λ_2. The Lagrange multipliers can be shown to have the meaning of shear forces $\lambda_1 \sim Q_x$ and $\lambda_2 \sim Q_y$. The finite element model based on the functional Π_L is called a *mixed finite element model*, because displacements are mixed with forces as element degrees of freedom.

In the penalty function method the constrained problem is posed as one of minimizing the functional $\Pi_p(w_0, \theta_x, \theta_y)$

$$\Pi_p \equiv \Pi(w_0, \theta_x, \theta_y) + \frac{1}{2}\int_{\Omega_0} \left[\gamma_1 \left(\frac{\partial w_0}{\partial x} - \theta_x\right)^2 + \gamma_2 \left(\frac{\partial w_0}{\partial y} - \theta_y\right)^2\right] dxdy \tag{10.6 – 16}$$

where γ_1 and γ_2 are called the *penalty parameters*, which are preselected positive functions of (x, y). In the penalty function method, the square of the error in each constraint is minimized along with the original functional, and the penalty parameters represent weights with which the constraints are minimized relative to the original functional. Thus, in the penalty function method the constraints are satisfied only approximately and the minimum character of the problem is penalized. The larger the values of the penalty parameters, the lesser the error in satisfying the constraint conditions. A desirable aspect of the penalty function method is that no new variables are introduced in addition to those in the original functional and the constraint equations.

At this juncture we should remind ourselves that the problem we are trying to formulate is the classical plate bending. A finite element model based on the penalty functional (16) is expected to give an approximate solution to the classical plate theory for sufficiently large values of the penalty parameters. However, there are computational problems arising from the finite element implementation of the model based on (16). Before we embark on the discussion of shear locking, it is useful to note the similarity between the functional Π_p and that of the first-order shear deformation plate theory.

The total potential energy functional for the first-order theory, omitting membrane effects and the von Kárman nonlinearity, and assuming orthotropic material behavior, can be derived from Eq. (5.4–9):

$$0 = \int_{\Omega_0} \left(M_{xx}\delta\varepsilon_{xx}^{(1)} + M_{yy}\delta\varepsilon_{yy}^{(1)} + M_{xy}\delta\gamma_{xy}^{(1)} + Q_x\delta\gamma_{xz}^{(0)} + Q_y\delta\gamma_{yz}^{(0)} \right.$$
$$\left. -q\delta w_0 \right) \, dxdy \tag{10.6 – 17}$$

Using the plate constitutive equations (5.4–21) and (5.4–22), we rewrite the weak form (17) in terms of the generalized displacements

$$0 = \int_{\Omega_0} \left[\left(D_{11}\frac{\partial \phi_x}{\partial x} + D_{12}\frac{\partial \phi_y}{\partial y} \right) \frac{\partial \delta\phi_x}{\partial x} + \left(D_{12}\frac{\partial \phi_x}{\partial x} + D_{22}\frac{\partial \phi_y}{\partial y} \right) \frac{\partial \delta\phi_y}{\partial y} \right.$$
$$+ 2D_{66}\left(\frac{\partial \phi_x}{\partial y} + \frac{\partial \phi_y}{\partial x} \right) \left(\frac{\partial \delta\phi_x}{\partial y} + \frac{\partial \delta\phi_y}{\partial x} \right) - q\delta w_0$$
$$\left. +A_{55}\left(\frac{\partial w_0}{\partial x} + \phi_x \right) \left(\frac{\partial \delta w_0}{\partial x} + \delta\phi_x \right) + A_{44}\left(\frac{\partial w_0}{\partial y} + \phi_y \right) \left(\frac{\partial \delta w_0}{\partial y} + \delta\phi_y \right) \right] \, dxdy$$
$$\tag{10.6 – 18}$$

which is the first variation of the functional

$$\bar{\Pi}(w_0, \phi_x, \phi_y) = \int_{\Omega_0} \frac{1}{2} \left[D_{11}\left(\frac{\partial \phi_x}{\partial x} \right)^2 + 2D_{12}\frac{\partial \phi_x}{\partial x}\frac{\partial \phi_y}{\partial y} + D_{22}\left(\frac{\partial \phi_y}{\partial y} \right)^2 \right.$$
$$+ D_{66}\left(\frac{\partial \phi_x}{\partial y} + \frac{\partial \phi_y}{\partial x} \right)^2 + A_{55}\left(\frac{\partial w_0}{\partial x} + \phi_x \right)^2$$
$$\left. +A_{44}\left(\frac{\partial w_0}{\partial y} + \phi_y \right)^2 - qw_0 \right] \, dxdy \tag{10.6 – 19}$$

We note the similarity between the functionals Π_p of (16) and $\bar{\Pi}$ in (19). They are the same with the following correspondence

$$\theta_x \sim -\phi_x, \quad \theta_y \sim -\phi_y, \quad \gamma_1 \sim A_{55}, \quad \gamma_2 \sim A_{44} \tag{10.6 – 20}$$

Thus, for a particular choice of the penalty parameters, we recover the first-order shear deformation theory from the penalty formulation of the classical plate theory; for large values of the penalty parameters, the classical plate theory is recovered. Indeed, use of the functional in (19) is more appropriate because it naturally gives rise to the classical plate theory as the plate thickness is reduced in relation to the plate in-plane dimensions. This is due to the fact that D_{ij} are proportional to h^3 whereas A_{ij} are proportional to h. Thus the penalty parameters are of the order h^{-2}. When $a/h = 100$, the penalty parameters A_{ij} are 10^4 times larger than D_{ij}, and hence the constraints (12) are satisfied accurately, *i.e.*, the classical plate theory is realized.

The C^0–plate bending elements based on the first-order shear deformation plate theory are among the simplest available in the literature. They are expected, in theory, to give the thin plate theory solution when the side-to-thickness ratio a/h is very large ($a/h \geq 100$). Unfortunately, when lower-order (quadratic or less) equal interpolation of the transverse deflection and rotations is used, the elements do not accurately represent the bending behavior as the side-to-thickness ratio of the element becomes large (*i.e.*, thin plate limit). For thin plates, the shearing strains ε_{xz} and ε_{yz} are required to vanish, and the plate elements based on the first-order theory become excessively stiff, yielding displacements that are too small compared to the true solution. This type of behavior is known as *shear locking*. There are a number of papers on the subject of shear locking and elements developed to alleviate the problem (see [30–57]).

Shear locking is due to the inability of shear deformable elements to accurately model the bending within an element under a state of zero transverse shearing strain. When thin plates are analyzed by the shear deformable elements, the energy due to transverse shear strains must vanish. Numerically this is equivalent to requiring the product of the shear stiffness matrix and the displacement vector be zero. Therefore, in order to obtain a nontrivial solution, the shear stiffness matrix must be singular. One way to achieve the singularity of the transverse shear stiffness matrix is to use an order of numerical integration lower than is necessary to evaluate the integrals exactly. Thus, *reduced integration* of transverse shear stiffnesses (*i.e.*, all coefficients in $K_{ij}^{\alpha\beta}$ that contain A_{44}, A_{45}, and A_{55}) is necessary. Higher-order elements or refined meshes of lower-order elements experience relatively less locking, but sometimes at the expense of rate of convergence.

In this chapter only rectangular or quadrilateral elements based on the first-order shear deformation theory are used. Equal interpolation of all generalized displacements is employed. Stiffness coefficients associated with the transverse shear deformation (*i.e.*, terms containing A_{44}, A_{45}, and A_{55}) are evaluated using reduced integration, and full integration is used for all other stiffness coefficients, mass coefficients, and force components.

With the suggested Gauss rule, highly distorted elements tend to have slower rates of convergence but they give sufficiently accurate results. Of course, one should avoid using highly distorted elements; most commercial codes issue warning messages when the element is highly distorted (*e.g.*, see Chapter 9, pp. 439–448 of the textbook by Reddy [1] for a discussion of modeling considerations).

10.6–4 Postcomputation of Stresses

Once the nodal values of generalized displacements $(u_0, v_0, w_0, \phi_x, \phi_y)$ have been obtained by solving the assembled equations of a problem, the strains are evaluated in each element by differentiating the displacements.

Since the displacements are only continuous across the element boundaries, strain continuity across the boundaries is not ensured. That is, along a boundary common to two elements, the strains and hence stresses take different values on the two sides of the interface. However, strains and hence stresses are continuous within an element.

As noted earlier, the strains and stresses are the most accurate if they are computed at the $(N-1)\times(N-1)$ Gauss points, where $N \times N$ is the Gauss quadrature rule used to evaluate the bending stiffness coefficients. For example, the linear rectangular plate bending element of the first-order theory requires 2×2 integration to evaluate the bending stiffnesses exactly. Then the one-point integration should be used to evaluate the transverse shear stiffness coefficients, strains, and stresses. Similarly, for a quadratic rectangular element the reduced integration rule is the 2×2 Gauss rule.

Since the displacements in the finite element models are referred to the global coordinates (x, y, z), the stresses are computed in the global coordinates using the relations (when thermal effects are not included) at the Barlow points (*i.e.*, reduced integration points)

$$\begin{Bmatrix} \sigma_{xx} \\ \sigma_{yy} \\ \sigma_{xy} \\ \sigma_{xz} \\ \sigma_{yz} \end{Bmatrix}^{(k)} = \begin{bmatrix} \bar{Q}_{11} & \bar{Q}_{12} & \bar{Q}_{16} & 0 & 0 \\ \bar{Q}_{12} & \bar{Q}_{22} & \bar{Q}_{26} & 0 & 0 \\ \bar{Q}_{16} & \bar{Q}_{26} & \bar{Q}_{66} & 0 & 0 \\ 0 & 0 & 0 & \bar{Q}_{55} & \bar{Q}_{45} \\ 0 & 0 & 0 & \bar{Q}_{45} & \bar{Q}_{44} \end{bmatrix}^{(k)} \begin{Bmatrix} \varepsilon_{xx} \\ \varepsilon_{yy} \\ \gamma_{xy} \\ \gamma_{xz} \\ \gamma_{yz} \end{Bmatrix} \tag{10.6 - 21}$$

where

$$\begin{Bmatrix} \varepsilon_{xx} \\ \varepsilon_{yy} \\ \gamma_{yz} \\ \gamma_{xz} \\ \gamma_{xy} \end{Bmatrix} = \begin{Bmatrix} \frac{\partial u_0}{\partial x} \\ \frac{\partial v_0}{\partial y} \\ \frac{\partial w_0}{\partial y} + \phi_y \\ \frac{\partial w_0}{\partial x} + \phi_x \\ \frac{\partial u_0}{\partial y} + \frac{\partial v_0}{\partial x} \end{Bmatrix} + z \begin{Bmatrix} \frac{\partial \phi_x}{\partial x} \\ \frac{\partial \phi_y}{\partial y} \\ 0 \\ 0 \\ \frac{\partial \phi_x}{\partial y} + \frac{\partial \phi_y}{\partial x} \end{Bmatrix} \tag{10.6 - 22}$$

If stresses and strains are required in the lamina principal material coordinates, for example, to check for failures, the strains and stresses of Eqs. (21) and (22) should be transformed to material coordinates associated with each layer using the transformation relations (4.3–14) and (4.3–10). Alternatively, the strains can be transformed using Eq. (4.3-14)

$$\begin{Bmatrix} \varepsilon_1 \\ \varepsilon_2 \\ \varepsilon_4 \\ \varepsilon_5 \\ \varepsilon_6 \end{Bmatrix}^{(k)} = [R]^{(k)} \begin{Bmatrix} \varepsilon_{xx} \\ \varepsilon_{yy} \\ \gamma_{yz} \\ \gamma_{xz} \\ \gamma_{xy} \end{Bmatrix} \tag{10.6-23a}$$

$$[R]^{(k)} = \begin{bmatrix} \cos^2\theta_k & \sin^2\theta_k & 0 & 0 & \sin\theta_k\cos\theta_k \\ \sin^2\theta_k & \cos^2\theta_k & 0 & 0 & -\sin\theta_k\cos\theta_k \\ 0 & 0 & \cos\theta_k & -\sin\theta_k & 0 \\ 0 & 0 & \sin\theta_k & \cos\theta_k & 0 \\ -2\sin\theta_k\cos\theta_k & 2\sin\theta_k\cos\theta_k & 0 & 0 & \cos^2\theta_k - \sin^2\theta_k \end{bmatrix} \tag{10.6-23b}$$

and then the lamina constitutive equations are used to compute the stresses:

$$\begin{Bmatrix} \sigma_1 \\ \sigma_2 \\ \sigma_4 \\ \sigma_5 \\ \sigma_6 \end{Bmatrix}^{(k)} = \begin{bmatrix} Q_{11} & Q_{12} & 0 & 0 & 0 \\ Q_{12} & Q_{22} & 0 & 0 & 0 \\ 0 & 0 & Q_{44} & 0 & 0 \\ 0 & 0 & 0 & Q_{55} & 0 \\ 0 & 0 & 0 & 0 & Q_{66} \end{bmatrix}^{(k)} \begin{Bmatrix} \varepsilon_1 \\ \varepsilon_2 \\ \varepsilon_4 \\ \varepsilon_5 \\ \varepsilon_6 \end{Bmatrix}^{(k)} \tag{10.6-24}$$

10.6–5 Numerical Results of Bending

First the effect of integration rule and the convergence characteristics of the C^0 finite element model based on equal interpolation is investigated using a simply supported (SS–1) cross-ply square laminate under sinusoidally distributed transverse load. The laminate consists of three plies (0/90/0) of thicknesses $h/4, h/2$, and $h/4$, where h denotes the total laminate thickness; it is equivalent to (0/90/90/0) laminate with equal thickness plies. For this problem, we have developed closed-form solutions in Chapters 7 and 9 (also see [66–85] for analytical solutions for bending, vibration, and stability of shear deformation plate theories), and Pagano [69,70] developed the 3–D elasticity solution for the problem. The material properties used are those typical of graphite–epoxy material (*Material 1*)

$$E_1 = 25E_2,\ G_{12} = G_{13} = 0.5E_2,\ G_{23} = 0.2E_2,\ \nu_{12} = 0.25 \tag{10.6-25}$$

and the shear correction factor is taken to be $K = 5/6$. The loading in all cases is assumed to be sinusoidal

$$q(x,y) = q_0 \cos\frac{\pi x}{a}\sin\frac{\pi y}{b} \tag{10.6-26}$$

The following nondimensionalizations of the quantities are used:

$$\bar{w} = w_0(0,0)\frac{E_2 h^3}{b^4 q_0}$$

$$\bar{\sigma}_{xx} = \sigma_{xx}(0,0,\frac{h}{2})\frac{h^2}{b^2 q_0} = -\sigma_{xx}(0,0,-\frac{h}{2})\frac{h^2}{b^2 q_0}$$

$$\bar{\sigma}_{yy} = \sigma_{yy}(0,0,\frac{h}{4})\frac{h^2}{b^2 q_0} = -\sigma_{yy}(0,0,-\frac{h}{4})\frac{h^2}{b^2 q_0}$$

$$\bar{\sigma}_{xy} = \sigma_{xy}(a/2,b/2,-\frac{h}{2})\frac{h^2}{b^2 q_0} = -\sigma_{xy}(a/2,b/2,\frac{h}{2})\frac{h^2}{b^2 q_0}$$

$$\bar{\sigma}_{xz} = \sigma_{xz}(a/2,0,k=1,3)\frac{h}{bq_0}\ ,\ \bar{\sigma}_{yz} = \sigma_{yz}(0,b/2,k=2)\frac{h}{bq_0} \qquad (10.6-27)$$

where the origin of the coordinate system is taken at the center of the plate, $-a/2 \leq x \leq a/2, -a/2 \leq y \leq a/2$, and $-h/2 \leq z \leq h/2$. As noted earlier, the stresses in the finite element analysis are computed at the reduced Gauss points, irrespective of the Gauss rule used for the evaluation of the element stiffness coefficients. The Gauss point locations differ for each mesh used. The Gauss point coordinates A and B are shown in Table 10.6–1. The notation nL stands for $n \times n$ uniform mesh of linear rectangular elements, nQ8 for $n \times n$ uniform mesh of eight-node quadratic elements, and nQ9 for $n \times n$ uniform mesh of nine-node quadratic elements in a quarter plate. The finite element solutions (FES) are compared with the 3–D elasticity solution (ELS) and the closed-form solutions (CFS) in Table 10.6–2 for three side-to-thickness ratios $a/h = 10, 20$, and 100. Various stresses in the finite element analysis are evaluated at the Gauss points as indicated below:

$$\sigma_{xx}(A,A,\frac{h}{2}),\ \sigma_{yy}(A,A,\frac{h}{4}),\ \sigma_{xy}(B,B,-\frac{h}{2})$$

$$\sigma_{xz}(B,A) \text{ in layers 1 and 3},\ \ \sigma_{yz}(A,B) \text{ in layer 2} \qquad (10.6-28)$$

Table 10.6–1: The Gauss point locations at which the stresses are computed in the finite element analysis.

Coordinate	2L	4L	8L	2Q8/2Q9	4Q8/4Q9
A	$0.125a$	$0.0625a$	$0.03125a$	$0.05283a$	$0.02642a$
B	$0.375a$	$0.4375a$	$0.46875a$	$0.44717a$	$0.47358a$

An examination of the numerical results presented in Table 10.6–2 shows that the FSDT finite element with equal interpolation of all generalized displacements does not experience shear locking for thick plates even when full integration rule is used. Shear locking is evident when the element is used to model thin plates ($a/h \geq 100$) with full integration rule (F). Also, higher-order elements show less locking but with slower convergence. The element behaves uniformly well for thin and thick plates when the reduced (R) or selectively reduced integration (S) rule is used.

Table 10.6–2: Effect of reduced integration on the nondimensionalized maximum deflections $\bar{w}$ and stresses $\bar{\sigma}$ of simply supported (SS–1) cross-ply (0/90/90/0) square plates subjected to sinusoidal loading (see Reddy and Chao [68]).

a/h	Source	$\bar{w} \times 10^2$	$\bar{\sigma}_{xx}$	$\bar{\sigma}_{yy}$	$\bar{\sigma}_{xy}$	$\bar{\sigma}_{xz}$	$\bar{\sigma}_{yz}$
Finite Element Solutions†							
10	2L–F	0.5901	0.3339	0.2454	0.0163	0.316	0.125
	2L–R	0.6508	0.3799	0.2838	0.0187	0.335	0.107
	2L–S	0.6655	0.3796	0.2882	0.0189	0.353	0.114
	4L–F	0.6427	0.4512	0.3280	0.0219	0.389	0.129
	4L–R	0.6599	0.4668	0.3406	0.0227	0.395	0.123
	4L–S	0.6632	0.4667	0.3419	0.0227	0.400	0.125
	2Q8-F	0.6605	0.4831	0.3492	0.0234	0.404	0.126
	2Q8-R	0.6615	0.4842	0.3509	0.0234	0.404	0.126
	2Q8-S	0.6613	0.4844	0.3509	0.0233	0.405	0.126
	2Q9-F	0.6551	0.4790	0.3400	0.0231	0.399	0.126
	2Q9-R	0.6633	0.4841	0.3508	0.0234	0.404	0.125
	2Q9-S	0.6631	0.4844	0.3509	0.0233	0.404	0.126
	8L–S	0.6628	0.4907	0.3565	0.0238	0.412	0.128
	4Q8-S	0.6626	0.4954	0.3589	0.0240	0.414	0.128
	4Q8-S	0.6627	0.4954	0.3589	0.0240	0.414	0.128
Analytical Solutions							
10	CFS	0.6627	0.4989	0.3614	0.0241	0.416 (0.318)	0.129 (0.181)‡
	ELS	0.7370	0.5590	0.4010	0.0276	0.301	0.196
Finite Element Solutions							
20	2L–F	0.3236	0.2645	0.1491	0.0111	0.303	0.139
	2L–R	0.4712	0.4036	0.2289	0.0170	0.353	0.089
	2L–S	0.4760	0.4043	0.2308	0.0171	0.373	0.094
	4L–F	0.4346	0.4365	0.2451	0.0183	0.395	0.123
	4L–R	0.4863	0.4940	0.2777	0.0207	0.415	0.103
	4L–S	0.4874	0.4942	0.2782	0.0207	0.420	0.105
	2Q8-F	0.4876	0.5082	0.2828	0.0214	0.424	0.106
	2Q8-R	0.4901	0.5117	0.2870	0.0214	0.424	0.106

	2Q8-S	0.4901	0.5120	0.2870	0.0214	0.424	0.106
	2Q9-F	0.4891	0.5083	0.2829	0.0214	0.424	0.106
	2Q9-R	0.4915	0.5118	0.2870	0.0213	0.424	0.106
	2Q9-S	0.4915	0.5120	0.2870	0.0213	0.424	0.105
	8L–S	0.4902	0.5189	0.2912	0.0218	0.433	0.108
	4Q8-S	0.4911	0.5236	0.2936	0.0219	0.434	0.108
	4Q9-S	0.4912	0.5236	0.2936	0.0219	0.434	0.108
Analytical Solutions							
20	CFS	0.4912	0.5273	0.2956	0.0221	0.437 (0.333)	0.109 (0.150)
	ELS	0.5128	0.5430	0.3080	0.0230	0.328	0.156
Finite Element Solutions							
100	2L–F	0.0315	0.0299	0.0151	0.0012	0.230	0.211
	2L–R	0.4107	0.4129	0.2076	0.0164	0.360	0.082
	2L–S	0.4120	0.4140	0.2082	0.0164	0.381	0.086
	4L–F	0.1034	0.1203	0.0604	0.0048	0.298	0.221
	4L–R	0.4281	0.5045	0.2535	0.0200	0.422	0.098
	4L–S	0.4284	0.5048	0.2537	0.0207	0.428	0.097
	2Q8-F	0.4143	0.4900	0.2435	0.0199	0.430	0.100
	2Q8-R	0.4319	0.5214	0.2621	0.0206	0.435	0.102
	2Q8-S	0.4319	0.5214	0.2620	0.0206	0.433	0.102
	2Q9-F	0.4193	0.4946	0.2420	0.0201	0.431	0.098
	2Q9-R	0.4339	0.5224	0.2625	0.0207	0.432	0.098
	2Q9-S	0.4339	0.5224	0.2550	0.0206	0.430	0.097
	8L–S	0.4324	0.5297	0.2662	0.0210	0.441	0.100
	4Q8-S	0.4336	0.5344	0.2685	0.0212	0.441	0.100
	4Q9-S	0.4337	0.5344	0.2685	0.0212	0.441	0.100
Analytical Solutions							
100	CFS	0.4337	0.5382	0.2704	0.0213	0.445 (0.339)	0.101 (0.139)
	ELS	0.4347	0.5390	0.2710	0.0214	0.339	0.139
	CLPT*	0.4313	0.5387	0.2667	0.0213	(0.339)	(0.138)

† F = full integration of all coefficients; R = reduced integration of all coefficients; S = selective integration: full integration of all except the transverse shear coefficients, which are evaluated using reduced integration rule.

‡ The values of transverse shear stresses in parentheses are obtained using the 3-D equilibrium equations.

* The CLPT solution is independent of side-to-thickness ratio, a/h.

Table 10.6–3: Comparison of nondimensionalized maximum deflections and stresses of simply supported (SS–1) five-layer (0/90/0/90/0) square plates under sinusoidal loading ($h_1 = h_3 = h_5 = h/6, h_2 = h_4 = h/4$, $E_1 = 25E_2$, $G_{12} = G_{13} = 0.5E_2$, $G_{23} = 0.2E_2$, $\nu_{12} = 0.25$, $K = 5/6$).

a/h	Source	$\bar{w} \times 10^2$	$\bar{\sigma}_{xx}$	$\bar{\sigma}_{yy}$	$\bar{\sigma}_{xy}$	$\bar{\sigma}_{xz}$	$\bar{\sigma}_{yz}$
2	ELS	5.2949	1.332	1.001	0.0634	0.227	0.186
	CFS	4.8346	0.4021	0.5517	0.0243	0.2838	0.2619
						(0.2112)	(0.3068)†
	FEM	4.8338	0.3993	0.5479	0.0241	0.2818	0.2601
4	ELS	1.8505	0.685	0.633	0.0384	0.238	0.229
	CFS	1.5623	0.4369	0.5026	0.0235	0.3054	0.2403
						(0.2267)	(0.2809)
	FEM	1.5620	0.4339	0.4991	0.0233	0.3033	0.2281
10	ELS	0.6771	0.545	0.430	0.0247	0.258	0.223
	CFS	0.6213	0.5021	0.4107	0.0221	0.3459	0.1998
						(0.2559)	(0.2324)
	FEM	0.6212	0.4986	0.4078	0.0219	0.3435	0.1984
20	ELS	0.4938	0.539	0.380	0.0222	0.268	0.212
	CFS	0.4796	0.5276	0.3748	0.0215	0.3617	0.1840
						(0.2673)	(0.2135)
	FEM	0.4796	0.5239	0.3722	0.0214	0.3592	0.1827
50	ELS	0.4412	0.539	0.363	0.0214	0.271	0.206
	CFS	0.4390	0.5368	0.3618	0.0213	0.3674	0.1783
						(0.2714)	(0.2066)
	FEM	0.4390	0.5331	0.3593	0.0212	0.3648	0.1770
100	ELS	0.4338	0.539	0.360	0.0213	0.272	0.205
	CFS	0.4332	0.5382	0.3598	0.0213	0.3683	0.1774
						(0.2720)	(0.2056)
	FEM	0.4331	0.5345	0.3573	0.0211	0.3655	0.1761
	CLPT	0.4313	0.5387	0.3591	0.0213	(0.2722)	(0.2052)†

† Stresses computed from 3-D equilibrium equations.

The finite element results are in excellent agreement with the closed-form solutions of the first-order shear deformation theory (FSDT). The displacements converge faster than stresses. This is expected with all displacement finite element models because the rate of convergence of gradients of the solution is one order less than the rate of convergence of the solution. Of course, the finite element solutions based on FSDT should not be expected to agree well with the 3-D elasticity solution; the finite element solutions should only converge to the closed-form solutions of FSDT.

Nondimensionalized maximum deflections and stresses in five-layer ($h_1 = h_3 = h_5 = h/6$, $h_2 = h_4 = h/4$) cross-ply (0/90/0/90/0) square laminates under sinusoidally distributed transverse load are compared in Table 10.6–3. The finite element results were obtained with 4×4 mesh of eight-node quadratic elements in a quarter plate are in excellent agreement with the closed-form solutions. Although the first-order shear deformation theory underpredicts deflections for small values of a/h, the stresses are in good agreement with those predicted by the 3-D elasticity theory; the error is relatively more for the five-layer case compared to the three-layer case shown in Table 10.6–2.

Table 10.6–4: Comparison of nondimensionalized maximum deflections and stresses of simply supported (SS–1) three-ply (0/90/0) square plates subjected to sinusoidal loading ($h_i = h/3$, $E_1 = 25E_2$, $G_{12} = G_{13} = 0.5E_2$, $G_{23} = 0.2E_2$, $\nu_{12} = 0.25$, $K = 5/6$).

a/h	Source		$\bar{w} \times 10^2$	$\bar{\sigma}_{xx}$	$\bar{\sigma}_{yy}$	$\bar{\sigma}_{xy}$	$\bar{\sigma}_{xz}$	$\bar{\sigma}_{yz}$
10	ELS		–	0.590	0.288	0.029	0.357	0.123
	CFS		0.6693	0.5134	0.2536	0.0252	0.4089	0.0915
							(0.3806)	(0.1108)†
	FEM	4Q8-S	0.6692	0.5098	0.2518	0.0250	0.4060	0.0908
20	ELS		–	0.552	0.210	0.0234	0.385	0.094
	CFS		0.4921	0.5318	0.1997	0.0223	0.4205	0.0759
							(0.3912)	(0.0901)
	FEM	4Q8-S	0.4921	0.5281	0.1983	0.0222	0.4176	0.0754
100	ELS		–	0.539	0.181	0.0213	0.395	0.083
	CFS		0.4337	0.5384	0.1804	0.0213	0.4247	0.0703
							(0.3950)	(0.0827)
	FEM	4Q8-S	0.4336	0.5346	0.1791	0.0212	0.4215	0.0699
	CLPT		0.4313	0.5387	0.1796	0.0213	(0.3951)	(0.0823)

† Values computed from equilibrium equations.

Table 10.6–4 shows a comparison of the elasticity solution of Pagano [70], with the closed-form and finite element solutions of a three-layer cross-ply (0/90/0) square plate under sinusoidally distributed transverse load. The layers are of equal thickness, with the material properties listed in Eq. (25). The same locations and nondimensionalizations as given in Eqs. (26) and (27) are used. The finite element results obtained with 4×4 mesh of eight-node quadratic elements in a quarter plate are in excellent agreement with the closed-form solutions.

While the classical laminate plate theory underpredicts deflections for small values of a/h, the stresses predicted are in general agreement with the first-order shear deformation theory and elasticity theory. Also, the transverse shear stresses predicted through equilibrium equations, for the laminates studied so far, are very close to those predicted by the elasticity theory.

Next we consider a sandwich plate subjected to sinusoidally distributed transverse loading. The face sheets (*i.e.*, layers 1 and 3) are assumed to be orthotropic with the following material properties

$$E_1 = 25E_2,\; E_2 = 10^6 \text{ psi},\; G_{12} = G_{13} = 0.5E_2,\; G_{23} = 0.2E_2,\; \nu_{12} = 0.25 \tag{10.6-29}$$

and the core material is transversely isotropic and is characterized by the following material properties:

$$E_1 = E_2 = 10^6 \text{ psi},\; G_{13} = G_{23} = 0.06 \times 10^6 \text{ psi},\; \nu_{12} = 0.25$$

$$G_{12} = \frac{E_1}{2(1+\nu_{12})} = 0.016 \times 10^6 \text{ psi} \tag{10.6-30}$$

Each face sheet is assumed to be one-tenth of the total thickness of the sandwich plate ($a = b$). The finite element results obtained with 4×4 mesh of eight-node quadratic elements with reduced integration (4Q8–R) are compared with the closed form solution and elasticity solution of Pagano [70] in Table 10.6–5. The stresses are nondimensionalized as before, and their locations with respect to a coordinate system whose origin is at the center of the plate are as follows:

$$\sigma_{xx}(0,0,\frac{h}{2}),\;\; \sigma_{yy}(0,0,\frac{h}{2}),\;\; \sigma_{xy}(a/2,b/2,-\frac{h}{2})$$

$$\sigma_{xz}(0,b/2,0),\;\; \sigma_{yz}(a/2,0,0) \tag{10.6-31}$$

The results indicate that the effect of shear deformation on deflections is significant in sandwich plates even at large values of a/h. The equilibrium-derived transverse shear stresses are surprisingly close to those predicted by the elasticity theory for $a/h \geq 10$, while those computed from constitutive equations are considerably underestimated for small side-to-thickness ratios. The transverse shear stress component σ_{yz} is significantly overestimated by CLPT. Figures 10.6–2 and 10.6–3 show the variation of the transverse shear stresses through the thickness of the sandwich plates for side-to-thickness ratios $a/h = 2, 10$, and 100.

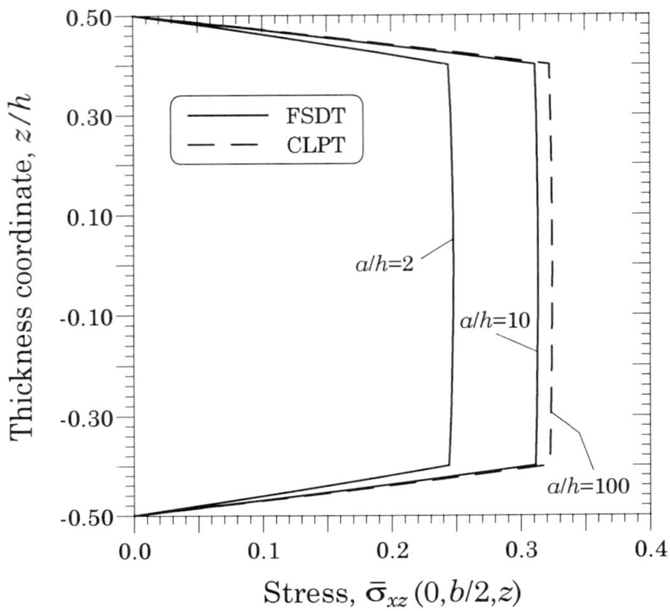

Figure 10.6–2: Distribution of transverse shear stress σ_{xz} through the thickness of a simply supported (SS–1) sandwich plate under sinusoidally distributed transverse load.

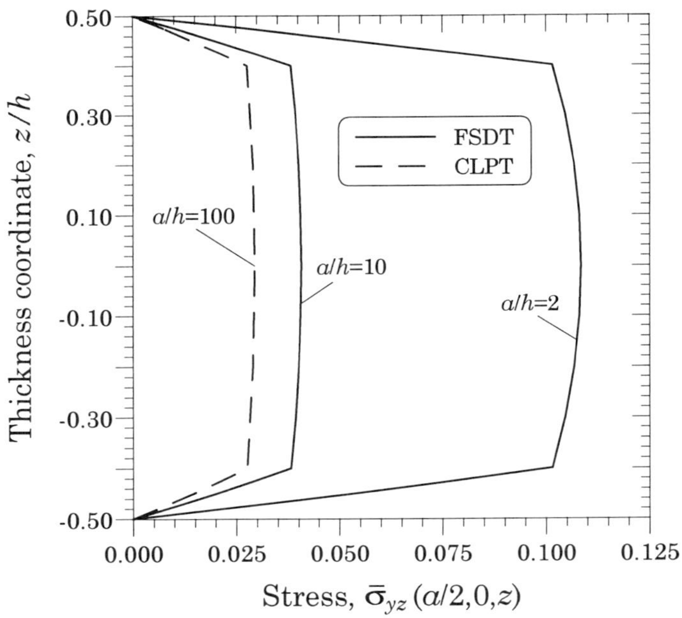

Figure 10.6–3: Distribution of transverse shear stress σ_{yz} through the thickness of a simply supported (SS–1) sandwich plate under sinusoidally distributed transverse load.

Table 10.6–5: Comparison of nondimensionalized maximum deflections and stresses in a simply supported (SS–1) sandwich plate subjected to sinusoidally varying transverse load ($h_1 = h_3 = 0.1h$, $h_2 = 0.8h$, $K = 5/6$).

a/h	Source	$\bar{w} \times 10^2$	$\bar{\sigma}_{xx}$	$\bar{\sigma}_{yy}$	$\bar{\sigma}_{xy}$	$\bar{\sigma}_{xz}$	$\bar{\sigma}_{yz}$
2	ELS	–	3.278	0.4517	0.2338	0.185	0.1399
	CFS	14.928	0.7378	0.2328	0.1263	0.1084	0.0781
						(0.2481)	(1085)†
	FEM	14.927	0.7326	0.2311	0.1254	0.1077	0.0776
4	ELS	–	1.556	0.2595	0.1481	0.239	0.1072
	CFS	4.7666	0.8918	0.1562	0.0907	0.1229	0.0537
						(0.2808)	(0.0746)
	FEM	4.7663	0.8856	0.1551	0.0901	0.1221	0.0534
10	ELS	–	1.153	0.1104	0.0717	0.300	0.0527
	CFS	1.5604	1.0457	0.0798	0.0552	0.1374	0.0293
						(0.3134)	(0.0408)
	FEM	1.5603	1.0384	0.0792	0.0548	0.1365	0.0278
20	ELS	–	1.110	0.0700	0.0511	0.317	0.0361
	CFS	1.0524	1.0831	0.0612	0.0466	0.1409	0.0234
						(0.3213)	(0.0325)
	FEM	1.0523	1.0755	0.0608	0.0462	0.1399	0.0233
50	ELS	–	1.099	0.0569	0.0446	0.323	0.0306
	CFS	0.9063	1.0947	0.0554	0.0439	0.1420	0.0216
						(0.3238)	(0.0300)
	FEM	0.9061	1.087	0.0551	0.0436	0.1410	0.0214
100	ELS	–	1.098	0.0550	0.0437	0.324	0.0297
	CFS	0.8852	1.0964	0.0546	0.0435	0.1422	0.0213
						(0.3242)	(0.0296)
	FEM	0.8851	1.0887	0.0542	0.0432	0.1412	0.0161
	CPT	0.8782	1.0970	0.0543	0.0433	(0.3243)	(0.0295)

† Values computed from equilibrium equations.

The same sandwich plate as discussed above is analyzed for simply supported and clamped boundary conditions when uniformly distributed load is used. Once again a quarter plate model is used with 4×4 mesh of quadratic FSDT elements and 8×8 mesh of CLPT conforming cubic elements. The results are presented in Table 10.6–6. The effect of shear deformation on the deflections is even more significant in clamped plates than in simply supported plates.

Table 10.6–6: Nondimensionalized maximum deflections and stresses in a square sandwich plate with simply supported (SS–1) and clamped boundary conditions ($h_1 = h_3 = 0.1h$, $h_2 = 0.8h$, $K = 5/6$).

a/h	Source	$\bar{w} \times 10^2$	$\bar{\sigma}_{xx}$	$\bar{\sigma}_{yy}$	$\bar{\sigma}_{xy}$	$\bar{\sigma}_{xz}$	$\bar{\sigma}_{yz}$
Simply supported plate under uniformly distributed load							
10	4Q8-R	2.3370	1.5430	0.0883	0.1136	0.2396	0.0991
50	4Q8-R	1.3671	1.5964	0.0526	0.0916	0.2433	0.0881
100	4Q8-R	1.3359	1.5978	0.0514	0.0906	0.2394	0.0880
CLPT	8CC-F‡	1.3296	1.5830	0.0509	0.0906	–	–
Clamped plate under uniformly distributed load							
10	4Q9-R†	1.2654	0.5018	0.0550	0.0120	0.2318	0.1445
50	4Q9-R	0.3111	0.5356	0.0108	0.0039	0.2406	0.1160
100	4Q9-R	0.2785	0.5347	0.0094	0.0030	0.2400	0.1148
CLPT	8CC-F‡	0.2951	0.5401	0.0145	0.0605	–	–

† The 4Q9-S element gives the same results as 4Q9-R.

‡ 8 × 8 mesh of conforming cubic elements with full integration for stiffness coefficient evaluation and one–point Gauss rule for stresses.

Next we turn our attention to angle-ply laminates. Table 10.6–7 shows maximum nondimensionalized deflections for two-layer and sixteen-layer angle-ply ($\theta/-\theta/\theta/-\theta,\ldots$) square plates under sinusoidal loading for $\theta = 5°, 30°$, and $45°$. The material properties of an individual layer are assumed to be (*Material 2*)

$$E_1 = 40E_2,\ G_{12} = G_{13} = 0.6E_2,\ G_{23} = 0.5E_2,\ \nu_{12} = 0.25 \qquad (10.6-32)$$

The finite element solutions, obtained with 4×4 mesh of nine-node elements in a quadrant and with reduced integration, are identical to the closed-form solutions (CFS) for all angles and side-to-thickness ratios, when the correct symmetry boundary conditions (SS–1) are used. Figure 10.6–4 shows the effect of side-to-thickness ratio (a/h), number of layers, and the lamination angle on the nondimensionalized maximum deflection of simply supported

(SS–1) antisymmetric angle-ply plates. The figure also shows the effect of using incorrect boundary conditions along the lines of symmetry; the symmetry conditions of SS–1 (see Figure 10.5–10) were used to obtain solutions of the two- and sixteen-layer laminates. It is clear from the results that when the lamination angle is small or the number of layers (n) is large, the error due to the incorrect symmetry boundary conditions is small. This is expected because for very small lamination angle, the laminate is close to being a cross-ply laminate; also, when the number of layers is large ($n \geq 8$), the laminate behaves like an orthotropic plate and hence the SS–1 boundary conditions are valid.

Table 10.6–7: Nondimensionalized deflections, $\bar{w} \times 10^2$, as a function of number of layers, angle, and side-to-thickness ratio for simply supported (SS–2) angle-ply square plates under sinusoidally distributed transverse load ($h_n = h/n$, $E_1 = 40E_2$, $G_{12} = G_{13} = 0.6E_2$, $G_{23} = 0.5E_2$, $\nu_{12} = 0.25$, $K = 5/6$, $n =$ Number of layers in the laminate).

a/h	Source	$\theta = 5°$		$\theta = 30°$		$\theta = 45°$	
		$n = 2$	$n = 10$	$n = 2$	$n = 10$	$n = 2$	$n = 10$
2	CFS	3.7849	3.6899	3.3330	2.9934	3.2300	2.9175
	FEM	3.7853	3.6903	3.3334	2.9937	3.2303	2.9178
4	CFS	1.3165	1.2611	1.2155	0.8904	1.1576	0.8450
	FEM	1.3166	1.2612	1.2156	0.8905	1.1576	0.8451
10	CFS	0.4883	0.4463	0.6099	0.2901	0.5773	0.2647
	FEM	0.4883	0.4464	0.6099	0.2901	0.5773	0.2647
20	CFS	0.3586	0.3181	0.5224	0.2033	0.4944	0.1818
	FEM	0.3586	0.3181	0.5224	0.2033	0.4944	0.1818
50	CFS	0.3216	0.2815	0.4979	0.1789	0.4712	0.1586
	FEM	0.3216	0.2815	0.4979	0.1789	0.4712	0.1586
100	CFS	0.3162	0.2763	0.4944	0.1755	0.4678	0.1553
	FEM	0.3162	0.2763	0.4944	0.1755	0.4678	0.1553
	CLPT	0.3145	0.2745	0.4932	0.1745	0.4667	0.1542

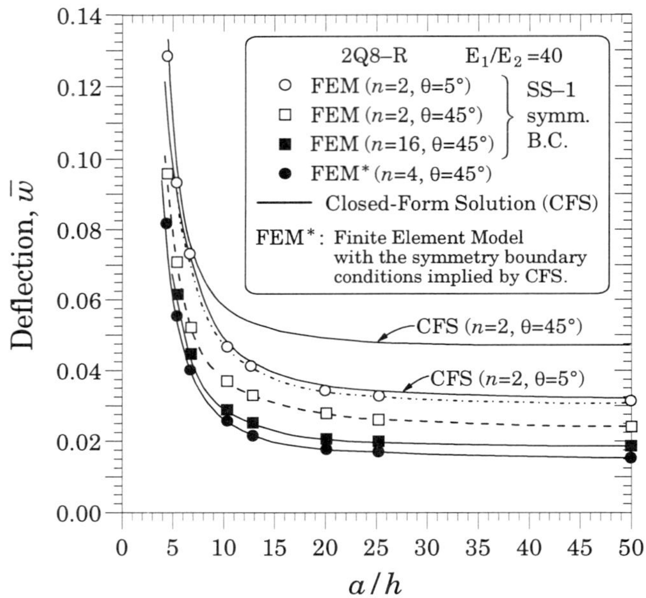

Figure 10.6–4: Effect of transverse shear deformation, lamination angle, number of layers, and symmetry boundary conditions on the nondimensionalized deflections of simply supported (SS–1) antisymmetric angle-ply laminates under sinusoidal load.

The stresses obtained with the same mesh as above for ten-layer antisymmetric angle-ply laminates are presented in Table 10.6–8. The stresses in the finite element analysis as well as in the closed–form solution are evaluated at the Gauss points:

$$(\bar{\sigma}_{xx}, \bar{\sigma}_{yy}) = (\sigma_{xx}, \sigma_{yy})(x_c, y_c, \frac{h}{2})\frac{h^2}{b^2 q_0}\ , \quad \bar{\sigma}_{xy} = \sigma_{xy}(x_s, y_s, -\frac{h}{2})\frac{h^2}{b^2 q_0}$$

$$x_c = y_c = 0.47358a, \quad x_s = y_s = 0.02642a \qquad (10.6-33)$$

The finite element stresses (second row) are in excellent agreement with the closed–form solution (first row). For $\theta = 45°$ it is found that the stresses are independent of side-to-thickness ratio.

Next, results for thermal bending are presented (see References 72 and 85). Figure 10.6–5 shows the effect of side-to-thickness ratio a/h on the nondimensionalized deflections and stresses of cross-ply and angle-ply plates subjected to temperature distribution that is linear through the thickness and varies sinusoidally in the plane of the plate ($q = 0, T_0 = 0, T_1 \neq 0$). The deflection and stresses are amplified to show the effect of thickness–shear strain. Clearly, the effect of shear deformation on thermal deflections and stresses is not as significant as in mechanically loaded plates.

Table 10.6–8: Nondimensionalized stresses for simply supported (SS–2) angle-ply ($-\theta/\theta/-\theta/\cdots$) square plates under sinusoidal transverse load ($h_n = h/n$, $n = 10$ $E_1 = 40E_2$, $G_{12} = G_{13} = 0.6E_2$, $G_{23} = 0.5E_2$, $\nu_{12} = 0.25$, $K = 5/6$, $4Q9 - R$, $n =$ number of layers).

a/h	$\theta = 5°$			$\theta = 30°$			$\theta^{\dagger} = 45°$	
	σ_{xx}	$\bar{\sigma}_{yy}$	$\bar{\sigma}_{xy}$	$\bar{\sigma}_{xx}$	$\bar{\sigma}_{yy}$	$\bar{\sigma}_{xy}$	$\bar{\sigma}_{xx}$	$\bar{\sigma}_{xy}$
2	0.3324	0.0964	0.0850	0.1530	0.0786	0.1688	0.1430	0.1392
	0.3280	0.0957	0.0846	0.1500	0.0774	0.1689	0.1421	0.1401
4	0.4423	0.0551	0.0511	0.2011	0.0890	0.1425	0.1430	0.1392
	0.4379	0.0547	0.0510	0.1982	0.0822	0.1423	0.1421	0.1401
10	0.5146	0.0279	0.0289	0.2328	0.0865	0.1252	0.1430	0.1392
	0.5103	0.0277	0.0290	0.2298	0.0854	0.1252	0.1421	0.1401
20	0.5286	0.0227	0.0246	0.2389	0.0871	0.1218	0.1430	0.1392
	0.5243	0.0225	0.0247	0.2359	0.0860	0.1219	0.1421	0.1401
50	0.5327	0.0211	0.0233	0.2407	0.0873	0.1208	0.1430	0.1392
	0.5284	0.0210	0.0234	0.2377	0.0862	0.1209	0.1421	0.1401
100	0.5333	0.0209	0.0231	0.2410	0.0873	0.1207	0.1430	0.1392
	0.5290	0.0207	0.0233	0.2380	0.0862	0.1208	0.1421	0.1401
CLPT	0.5290	0.0207	0.0233	0.2411	0.0873	0.1206	0.1440	0.1402

† The stresses are independent of side-to-thickness ratio.

Lastly, numerical results for transverse deflections and stresses of rectangular plates for a variety of boundary conditions are presented. The analytical results reported were obtained using the Lévy method with state-space approach (see Sections 9.4 and 9.5). For the finite element analysis, a 4×4 mesh is used for the classical plate theory, while 2×2 mesh of nine-node quadratic elements is used for the first-order theory when a quadrant of a plate is analyzed (for SS, CC, and FF boundary conditions), and equivalent 8×4 and 4×2 meshes were used for the half-plate models (in the case of SC, FS, and FC boundary conditions). The notation SC, for example, refers to simply supported boundary condition on the edge $x = -a/2$ and clamped boundary condition on the edge $x = a/2$, while the remaining two edges, $y = 0, b$, are simply supported. In all cases a sinusoidally varying transverse load is used, and the material properties of each lamina are assumed to be those of graphite–epoxy material with the properties listed in Eq. (25).

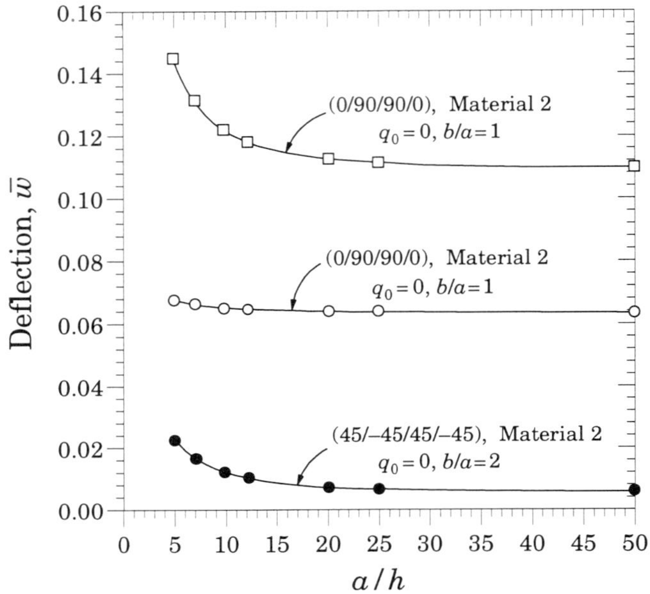

(a) Deflections

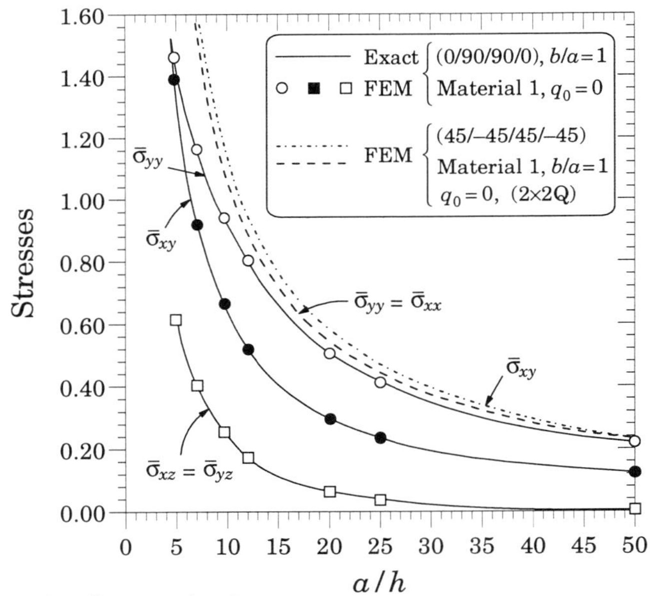

(b) Stresses (at Gauss points)

Figure 10.6–5: Effect of shear deformation on the nondimensionalized (a) deflections and (b) stresses of simply supported antisymmetric cross-ply and angle-ply laminates subjected to thermal loading.

The deflections and stresses are nondimensionalized as follows:

$$\bar{w} = w_0(0, b/2) \times \frac{E_2 h^3}{b^4 q_0}\ , \quad \bar{\sigma}_{xx} = \sigma_{xx}(0, \frac{b}{2}, -\frac{h}{2})\frac{h^2}{b^2 q_0} \times 10$$

$$\bar{\sigma}_{yy} = \sigma_{yy}(0, \frac{b}{2}, \frac{h}{2})\frac{h^2}{b^2 q_0} \times 10, \quad \bar{\sigma}_{yz} = \sigma_{yz}(0,0,0)\frac{h}{b q_0} \times 10 \qquad (10.6-34)$$

Tables 10.6–9 through 10.6–11 contain numerical values of deflections and stresses of antisymmetric cross-ply square plates $(0/90/0/\cdots)$ with various boundary conditions and sinusoidally distributed transverse load. The analytical (first row) and finite element solutions (second row) are presented. In all cases the finite element results are in good agreement with the analytical solutions.

Table 10.6–12 contains nondimensionalized deflections of angle-ply laminates subjected to uniformly distributed transverse load under various boundary conditions and with different values of E_1/E_2 ($G_{12} = G_{13} = 0.6E_2$, $G_{23} = 0.5E_2$, $\nu_{12} = 0.25$). The results were obtained using the Lévy method with state–space approach discussed in Section 9.5.

Table 10.6–9: Nondimensionalized center deflections $\bar{w} \times 10^2$ of antisymmetric, two-layer and ten-layer cross-ply square plates $(0/90)_n$.

n	$\frac{a}{h}$	Theory	SS	SC	CC	FF	FS	FC
	5	FSDT	1.758	1.477	1.257	2.777	2.335	1.897
			1.759	1.478	1.257	2.776	2.334	1.897†
		CLPT	1.064	0.664	0.429	1.777	1.471	0.980
1			1.043	0.648	0.417	1.786	1.465	0.977
	10	FSDT	1.237	0.883	0.656	2.028	1.687	1.223
			1.238	0.883	0.657	2.027	1.687	1.223
		CLPT	1.064	0.664	0.429	1.777	1.471	0.980
			1.043	0.648	0.417	1.786	1.465	0.977
	5	FSDT	1.137	1.045	0.945	1.663	1.460	1.258
			1.137	1.045	0.945	1.662	1.460	1.258
		CLPT	0.442	0.266	0.167	0.665	0.579	0.380
5			0.444	0.266	0.169	0.686	0.593	0.391
		FSDT	0.615	0.480	0.385	0.915	0.800	0.612
	5		0.616	0.480	0.386	0.914	0.800	0.612
		CLPT	0.442	0.266	0.167	0.665	0.579	0.380
			0.444	0.266	0.169	0.686	0.593	0.391

† The second row corresponds to finite element results.

Table 10.6–10: Nondimensionalized stress ($\bar{\sigma}_{xx}$) of antisymmetric cross-ply square plates (0/90/0/ $\cdots$); n denotes the total number of layers.

n	$\frac{a}{h}$	Theory	SS	SC	CC	FF	FS	FC
	5	FSDT	7.157	5.338	3.911	2.469	4.430	2.434
			6.948	5.465	3.707	2.359	4.479	2.542
		CLPT	7.157	5.660	4.800	2.403	4.442	3.042
2			6.659	5.782	4.348	2.034	4.288	2.991
	10	FSDT	7.157	5.494	4.450	2.442	4.435	2.790
			6.948	5.668	4.222	2.331	4.491	2.895
		CLPT	7.157	5.660	4.800	2.403	4.442	3.042
			6.659	5.782	4.348	2.034	4.288	2.991
	5	FSDT	5.009	3.707	2.275	1.712	2.957	1.343
			4.864	3.755	2.154	1.639	2.993	1.427
		CLPT	5.009	3.829	3.167	1.725	2.986	1.865
10			4.611	3.911	2.798	1.324	2.806	1.700
	10	FSDT	5.009	3.642	2.692	1.723	2.968	1.594
			4.863	3.754	2.550	1.648	3.010	1.674
		CLPT	5.009	3.829	3.167	1.725	2.986	1.865
			4.611	3.911	2.798	1.324	2.806	1.700

Table 10.6–11: Nondimensionalized stress ($\bar{\sigma}_{yy}$) of antisymmetric cross-ply (0/90/0/ $\cdots$) square plates.

n	$\frac{a}{h}$	Theory	SS	SC	CC	FF	FS	FC
	5	FSDT	7.157	6.034	5.153	11.907	9.848	8.047
			6.948	5.914	4.990	11.675	9.140	8.367
		CLPT	7.157	4.483	2.914	11.849	9.837	6.560
2			6.659	4.393	2.615	11.614	8.878	7.181
	10	FSDT	7.157	5.109	3.799	11.884	9.847	7.150
			6.948	5.082	3.661	11.654	9.120	7.610
		CLPT	7.157	4.483	2.914	11.849	9.837	6.560
			6.659	4.393	2.615	11.614	8.878	7.181
	5	FSDT	5.009	4.628	4.212	7.583	6.590	5.706
			4.864	4.511	4.086	7.429	6.141	5.844
		CLPT	5.009	3.025	1.911	7.480	6.531	4.284
10			4.611	2.942	1.694	7.395	5.935	4.782
	10	FSDT	5.009	3.904	3.135	7.533	6.566	5.029
			4.863	3.852	3.031	7.384	6.097	5.279
		CLPT	5.009	3.025	1.911	7.480	6.531	4.284
			4.611	2.942	1.694	7.395	5.935	4.782

Table 10.6–12: Nondimensionalized deflections $\bar{w} \equiv w_0(0, b/2)E_2h^3 \times 10^2/(a^4q_0)$ of simply supported (SS–2), four-layer antisymmetric angle-ply square plates [(45/–45/45/–45), $a/h = 10$].

Theory	$\frac{E_1}{E_2}$	SS	SC	CC	FF	FS	FC
FSDT	2	3.3749	2.4227	1.7530	10.7347	6.4469	4.7427
CLPT		3.2142	2.2144	1.6308	10.4702	6.2336	4.4460
FSDT	10	1.1598	0.9443	0.7708	6.0487	2.6109	2.1091
CLPT		1.000	0.7457	0.5578	5.5710	2.3451	1.7473
FSDT	20	0.7013	0.6024	0.5180	4.2843	1.6234	1.3787
CLPT		0.5418	0.4120	0.3133	3.6574	1.3432	1.0104
FSDT	30	0.5312	0.4706	0.4170	3.2417	1.2252	1.0749
CLPT		0.3718	0.2847	0.2179	2.7376	0.9433	0.7121

As noted in Section 8.6, the midplane symmetric plates are characterized by nonzero bending–twisting coupling coefficients D_{16} and D_{26}. Figure 10.6–6 contains plots of the nondimensionalized center deflection versus side-to-thickness ratio for three-layer (–45/45/–45) simply supported (SS–1) and clamped, square, symmetric laminates under uniformly distributed transverse load ($E_1 = 25E_2$, $G_{12} = G_{13} = 0.5E_2$, $G_{23} = 0.2E_2$, $\nu_{12} = 0.25$, $K = 5/6$, $a/h = 10$). The finite element results are obtained using 4×4 mesh of nine-node elements (*i.e.*, 4Q9–R). The solution obtained with omitting D_{16} and D_{26} is also shown in the figure. The bending–twisting coupling has the effect of increasing deflections. Also, the effect of shear deformation is more in the clamped plate than in the simply supported plate. Figure 10.6–7 shows the nondimensionalized center deflections obtained by the CLPT and FSDT ($a/h = 10$) as functions of the lamination angle for simply supported (SS–2) symmetric three-layer ($-\theta/\theta/-\theta$) plates under uniformly distributed load. The plots are symmetric about 45°.

Table 10.6–13: Effects of side-to-thickness ratio, integration, and type of element on the nondimensionalized fundamental frequency $\bar{\omega} = \omega(a^2/h)\sqrt{\rho/E_2}$ of simply supported (SS–1) cross-ply (0/90/90/0)square plates.

a/h	Serendipity Element			Lagrange Element			CFS
	F	R	S	F	R	S	
2	5.502	5.503	5.501	5.502	5.503	5.501	5.500
10	15.174	15.179	15.159	15.182	15.193	15.172	15.143
100	19.171	18.841	18.808	19.225	18.883	18.933	18.836

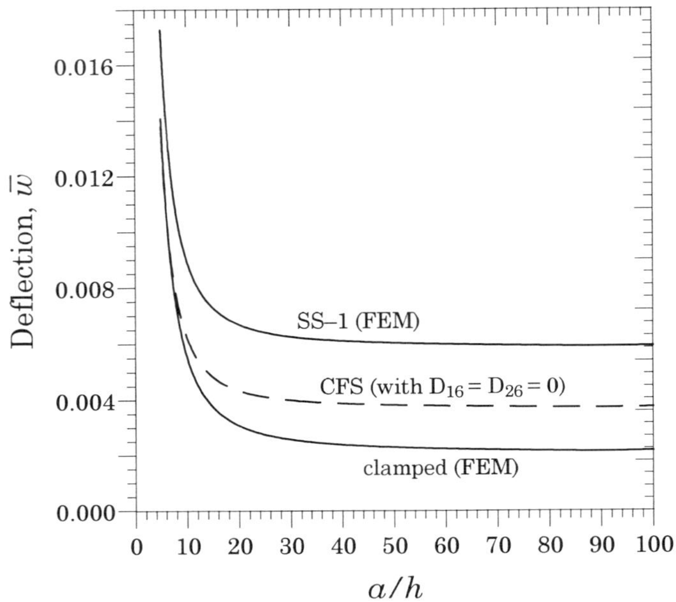

Figure 10.6–6: Nondimensionalized deflection versus side-to-thickness ratio for three-layer $(-45/45/-45)$, simply supported (SS–1) and clamped symmetric plates under uniform loading.

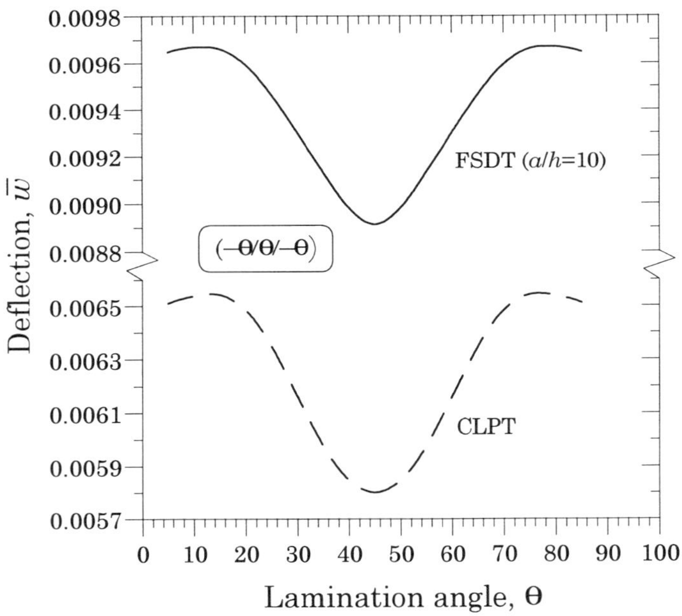

Figure 10.6–7: Nondimensionalized deflection versus lamination angle for simply supported (SS–1) symmetric three-layer $(-\theta/\theta/-\theta)$ plates under uniformly distributed load.

10.6–6 Numerical Results of Natural Vibration

The effect of reduced integration and the use of eight-node and nine-node elements on the accuracy of the natural frequencies are studied using three-layer cross-ply (0/90/0) laminate of Table 10.6–2. A 2×2 mesh in a quarter plate is used to obtain the results. Effects of side-to-thickness ratio, integration, and type of element on the nondimensionalized fundamental frequency $\bar{\omega}$ of simply supported (SS–1) cross-ply (0/90/90/0) square plates (Material 2; rotary inertia included) are presented in Table 10.6–13. From the results obtained, it is clear that both full (F) and selective (S) integrations give good results for thick plates ($a/h \leq 10$), whereas reduced integration (R) gives the best results for thin plates ($a/h \geq 100$). However, the reduced integration and selective integration rules both give good results for a wide range of side-to-thickness ratios.

Table 10.6–14 contains the lowest six natural frequencies of two- and four-layer cross-ply and angle-ply laminates with clamped edges. While the first two fundamental frequencies are very close, the higher frequencies are quite different for these laminates.

Table 10.6–14: The lowest six nondimensionalized frequencies $\bar{\omega} = \omega(a^2/h)\sqrt{\rho/E_2}$ of cross-ply and angle-ply square plates with clamped boundary conditions ($h_i = h/n$, $a/h = 10$, Material 2, 2Q8–R in full plate).

Laminate	a/h	ω_1	ω_2	ω_3	ω_4	ω_5	ω_6
(0/90)	2	5.900	10.137†	14.750†	18.123	18.245	20.759
	10	19.567	43.702†	128.43†	98.578	98.658	141.47
	100	25.677	244.77†	967.21†	1454.2†	1625.5	1687.1†
$(0/90)_2$	2	6.001	10.374†	23.477†	27.019	27.181	34.240†
	10	24.368	47.200†	101.35	101.69	141.23†	157.31
	100	40.506	259.23†	968.04	968.06	1455.6†	1627.2
(−45/45)	2	5.883	10.106†	14.75†	19.101	20.129	20.826
	10	18.979	41.482†	96.584	98.336	128.43†	160.64
	100	24.420	241.00†	966.94	967.16	1454.5†	1943.9
$(-45/45)_2$	2	5.992	10.359†	20.830	21.081	23.479†	31.771
	10	23.834	45.720†	99.174	101.37	141.26†	185.49
	100	38.201	249.30†	967.32	967.92	1455.8†	1947.0

† Occur in pairs.

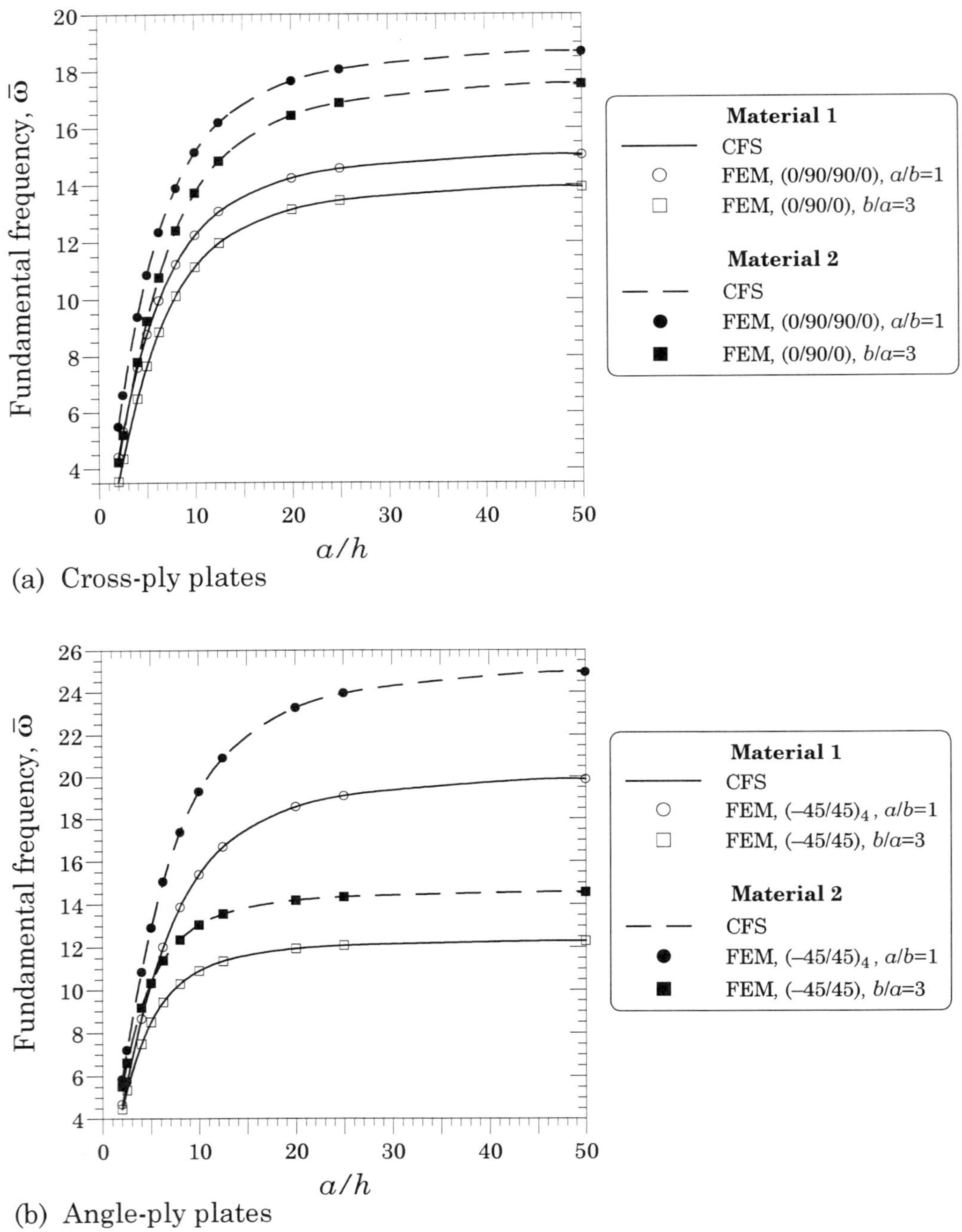

Figure 10.6–8: Nondimensionalized fundamental frequency versus side-to-thickness ratio of simply supported antisymmetric (a) cross-ply (SS–1) and (b) angle-ply (SS–2) plates.

Figure 10.6–8a shows the effect of side-to-thickness ratio on the nondimensionalized fundamental frequencies, $\bar{\omega} = \omega(a^2/h)\sqrt{\rho h/E_2}$, of cross-ply plates ($a/h = 10$, $E_1/E_2 = 25$, $G_{12} = G_{13} = 0.5E_2$, $G_{23} = 0.2E_2$, $\nu_{12} = 0.25$, $K = 5/6$). Similar results are presented for angle-ply plates in Figure 10.6–8b, which also contains a plot of the fundamental natural frequency versus the lamination angle for four-layer antisymmetric angle-ply laminates. The dashed line in Figure 10.6–8b corresponds to the case in which the SS–1 (incorrect) symmetry boundary conditions were used.

10.6–7 Transient Analysis Results

Here we present few results of transient analysis obtained using the shear deformable finite element. For additional results, the reader may consult References 86–88. The nine-node quadratic element with selective integration rule is used in the examples discussed here. In all cases initial conditions were assumed to be zero. The constant-average-acceleration scheme ($\alpha = \gamma = 0.5$) of Newmark is used for time integration.

Simply supported (SS–1) antisymmetric cross-ply square plates $(0/90)_n$ under suddenly applied sinusoidally (SSL) or uniformly (UDL) distributed transverse step load are analyzed. The boundary conditions for a quadrant are (see Figure 10.5–10):

$$\begin{aligned} x = 0: \quad u_0 = \phi_x = 0; \quad x = a/2: \quad v_0 = w_0 = \phi_y = 0 \\ y = 0: \quad v_0 = \phi_y = 0; \quad y = a/2: \quad u_0 = w_0 = \phi_x = 0 \end{aligned} \qquad (10.6-35)$$

The finite element results were obtained using a 2×2 mesh of nine-node quadratic FSDT finite elements in a quadrant. The following geometric and material properties were used:

$$a = b = 25 \text{ cm}, \quad h = 5 \text{ cm}, \quad q_0 = 10 \text{ N/cm}^2, \quad \Delta t = 5 \ \mu\text{s} \qquad (10.6-36)$$

$$\begin{aligned} \rho = 8 \times 10^{-6}\text{N-s}^2/\text{cm}^4, \quad E_1 = 25E_2 \text{ N/cm}^2, \quad E_2 = 2.1 \times 10^6 \text{ N/cm}^2 \\ G_{12} = G_{13} = 0.5E_2 \text{ N/cm}^2, \quad G_{23} = 0.2E_2 \text{ N/cm}^2 \end{aligned} \qquad (10.6-37)$$

The following nondimensionalizations are used:

$$\begin{aligned} \bar{w} = & w_0(a/2, a/2)\frac{E_2 h^3}{b^4 q_0} \times 10^2, \quad \bar{\sigma}_{xx} = \sigma_{xx}(a/2, b/2, \frac{h}{2})\frac{h^2}{b^2 q_0} \\ \bar{\sigma}_{yy} = & \sigma_{yy}(a/2, b/2, \frac{h}{2})\frac{h^2}{b^2 q_0}, \quad \bar{\sigma}_{xy} = \sigma_{xy}(0, 0, -\frac{h}{2})\frac{h^2}{b^2 q_0} \end{aligned} \qquad (10.6-38)$$

We note that

$$\sigma_{xx}(a/2,b/2,\frac{h}{2}) = -\sigma_{yy}(a/2,b/2,-\frac{h}{2})$$

$$\sigma_{yy}(a/2,b/2,\frac{h}{2}) = -\sigma_{xx}(a/2,b/2,-\frac{h}{2})$$

$$\sigma_{xy}(a/2,b/2,-\frac{h}{2}) = -\sigma_{xy}(a/2,b/2,\frac{h}{2}) \qquad (10.6-39)$$

Table 10.6–15 contains nondimensionalized center deflection and stresses for (0/90) laminate under sinusoidal load. The finite element results are compared with the the closed–form solutions (CFS) developed in Section 8.7 (also see Reddy [86]). The results are in good agreement with the analytical solutions.

Table 10.6–15: Comparison of transverse deflection and stresses obtained by the finite element method with closed–form solution of two-layer cross-ply square plate under suddenly applied sinusoidal load.

Time	$\bar{w}$		$\bar{\sigma}_{xx}$		$\bar{\sigma}_{xy}$	
(μs)	CFS†	FES	CFS	FES	CFS	FES
10	0.0514	0.0515	0.0080	0.0077	0.0058	0.0056
20	0.2486	0.2489	0.0777	0.0754	0.0525	0.0507
40	1.0152	1.0165	0.4962	0.4819	0.3073	0.2978
60	2.0523	2.0544	0.9679	0.9400	0.6039	0.5849
80	2.9782	2.9807	1.4649	1.4224	0.9075	0.8790
100	3.4784	3.4802	1.6724	1.6233	1.0404	1.0073
120	3.3620	3.3621	1.6544	1.6052	1.0251	0.9922
140	2.6836	2.6818	1.2807	1.2416	0.7979	0.7715
160	1.6705	1.6674	0.8195	0.7936	0.5080	0.4907
180	0.6952	0.6921	0.2997	0.2892	0.1902	0.1829
200	0.0902	0.0889	0.0394	0.0375	0.0248	0.0236

† Closed-form solution with Newmark's scheme for time integration.

Figures 10.6–9 through 10.6–12 show plots of center deflection $\bar{w}$ and maximum stresses $\bar{\sigma}_{xx}$, $\bar{\sigma}_{yy}$, and $\bar{\sigma}_{xy}$ of two-layer and eight-layer antisymmetric cross-ply square plates under suddenly applied sinusoidally or uniformly distributed transverse step load. The geometry and material properties are the same as listed in Eqs. (36) and (37). The finite element results are in excellent agreement with the analytical solutions of the first-order shear deformation plate theory.

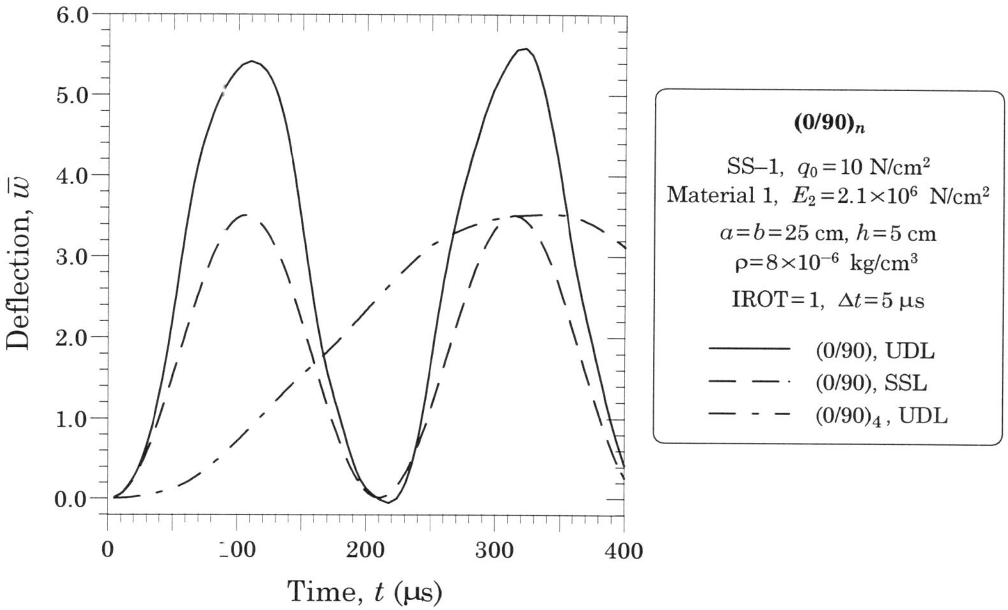

Figure 10.6–9: Plots of center deflection $\bar{w}$ versus time for a simply supported (SS–1) two-layer and eight-layer cross-ply square plates under sinusoidal or uniform step loading.

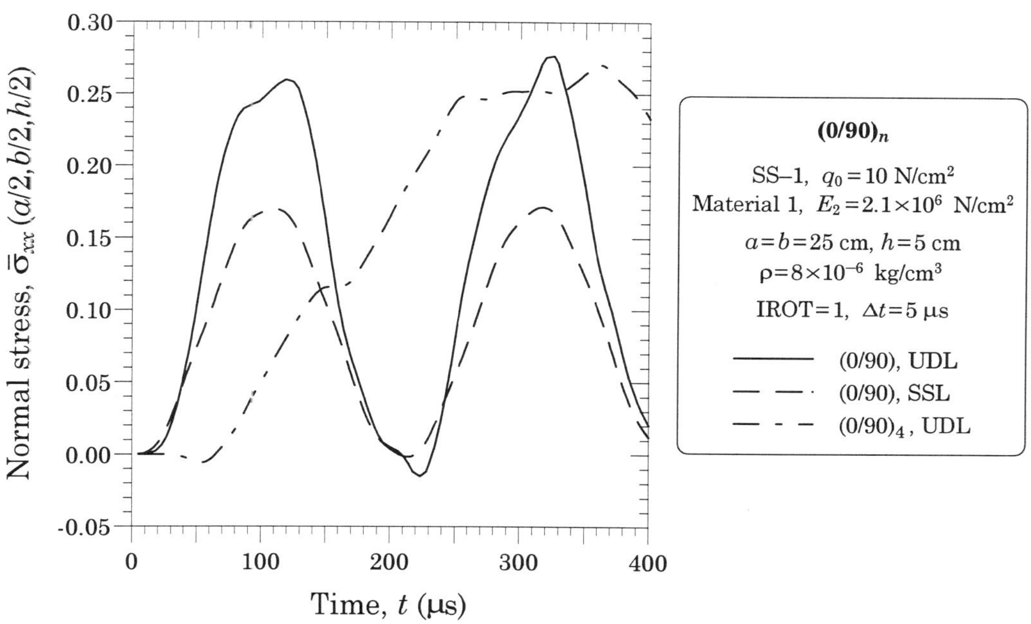

Figure 10.6–10: Plots of center normal stress $\bar{\sigma}_{xx}$ versus time for a simply supported (SS–1) two-layer and eight-layer cross-ply square plates under sinusoidal or uniform step loading.

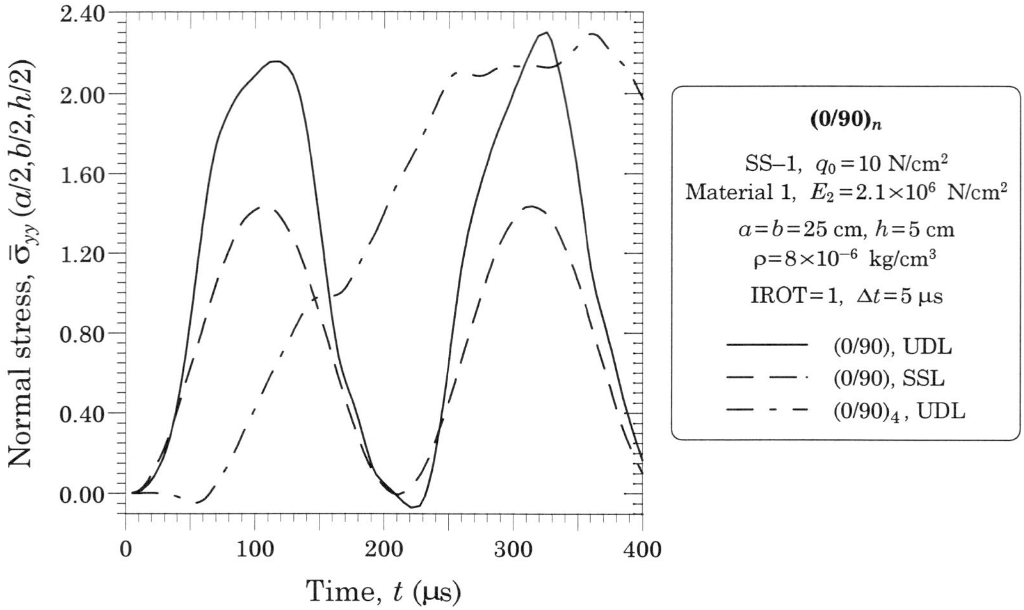

Figure 10.6–11: Plots of center normal stress $\bar{\sigma}_{yy}$ versus time for a simply supported (SS–1) two-layer and eight-layer cross-ply square plates under sinusoidal or uniform step loading.

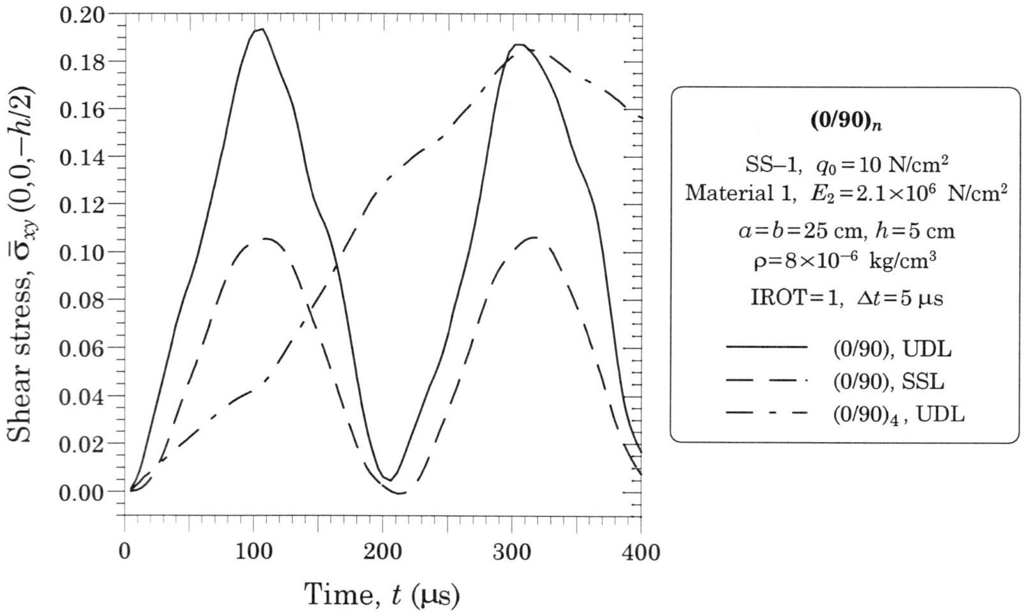

Figure 10.6–12: Plots of center normal stress $\bar{\sigma}_{xy}$ versus time for a simply supported (SS–1) two-layer and eight-layer cross-ply square plates under sinusoidal or uniform step loading.

10.7 Summary

In this chapter finite element models of laminated beams, plate strips and finite plates using the classical and first-order shear deformation plate theories are developed for static bending, buckling under in-plane compressive loads, natural vibration, and transient response. The finite element models are general in that they can be used for any lamination scheme, geometry, and boundary conditions. Numerical examples of bending, buckling, natural vibration, and transient response are presented in tabular and graphical form.

Exercises

10.1 (a) Derive finite element interpolation functions using w_0, $\theta = -dw_0/dx$, and d^2w_0/dx^2 as the nodal variables of an element with two (end) nodes, with a total of six degrees of freedom per element. Note that you must select a complete polynomial containing six parameters

$$w_0(x) = c_1 + c_2 x + c_3 x^2 + c_4 x^3 + c_5 x^4 + c_6 x^5$$

and derive the Hermite interpolation functions.

(b) Use the finite element approximation to compute the stiffness matrix $[K^e]$ of Eq. (10.2–9b).

Answer for Part (a):

$$\begin{aligned}
\varphi_1(x) =& 1 - 10\left(\frac{x}{h}\right)^3 + 15\left(\frac{x}{h}\right)^4 - 6\left(\frac{x}{h}\right)^5 \\
\varphi_2(x) =& h\left[-\left(\frac{x}{h}\right) + 6\left(\frac{x}{h}\right)^3 - 8\left(\frac{x}{h}\right)^4 + 3\left(\frac{x}{h}\right)^5\right] \\
\varphi_3(x) =& \frac{h^2}{2}\left[\left(\frac{x}{h}\right)^2 - 3\left(\frac{x}{h}\right)^3 + 3\left(\frac{x}{h}\right)^4 - \left(\frac{x}{h}\right)^5\right] \\
\varphi_4(x) =& -10\left(\frac{x}{h}\right)^3 + 15\left(\frac{x}{h}\right)^4 - 6\left(\frac{x}{h}\right)^5 \\
\varphi_5(x) =& h\left[-4\left(\frac{x}{h}\right)^3 + 7\left(\frac{x}{h}\right)^4 + 3\left(\frac{x}{h}\right)^5\right] \\
\varphi_6(x) =& \frac{h^2}{2}\left[\left(\frac{x}{h}\right)^3 - 2\left(\frac{x}{h}\right)^4 + \left(\frac{x}{h}\right)^5\right]
\end{aligned}$$

10.2 Derive finite element Hermite interpolation functions using w_0 and $\theta = -dw_0/dx$ as the nodal variables of an element with three nodes (two end nodes and the middle node), with a total of six degrees of freedom per element. As in Exercise 1, you must select a complete polynomial containing six parameters and derive the Hermite interpolation functions. *Ans:*

$$\varphi_1(x) =1 - 23\left(\frac{x}{h}\right)^2 + 66\left(\frac{x}{h}\right)^3 - 68\left(\frac{x}{h}\right)^4 + 24\left(\frac{x}{h}\right)^5$$
$$\varphi_2(x) =h\left[-\left(\frac{x}{h}\right) + 6\left(\frac{x}{h}\right)^2 - 13\left(\frac{x}{h}\right)^3 + 12\left(\frac{x}{h}\right)^4 - 4\left(\frac{x}{h}\right)^5\right]$$
$$\varphi_3(x) =16\left(\frac{x}{h}\right)^2 - 32\left(\frac{x}{h}\right)^3 + 16\left(\frac{x}{h}\right)^4$$
$$\varphi_4(x) =h\left[8\left(\frac{x}{h}\right)^2 - 32\left(\frac{x}{h}\right)^3 + 40\left(\frac{x}{h}\right)^4 - 16\left(\frac{x}{h}\right)^5\right]$$
$$\varphi_5(x) =7\left(\frac{x}{h}\right)^2 - 34\left(\frac{x}{h}\right)^3 + 52\left(\frac{x}{h}\right)^4 - 24\left(\frac{x}{h}\right)^5$$
$$\varphi_6(x) =h\left[\left(\frac{x}{h}\right)^2 - 5\left(\frac{x}{h}\right)^3 + 8\left(\frac{x}{h}\right)^4 - 4\left(\frac{x}{h}\right)^5\right]$$

10.3 Consider the spring–supported cantilever beam shown in Figure P10.3. Determine the deflection w_0 in the spring under uniformly distributed load of intensity q_0. Use the Hermite cubic beam finite element to solve the problem.

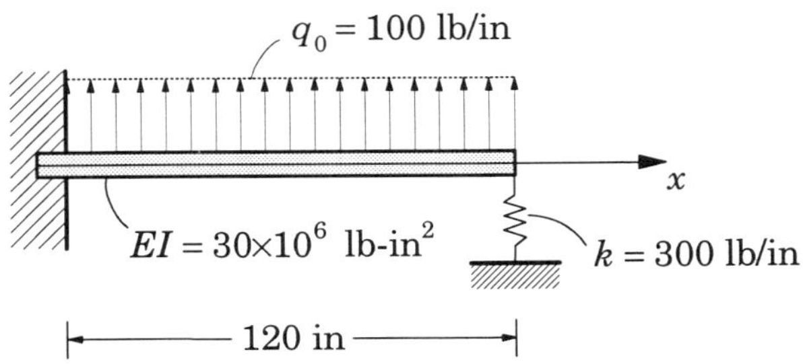

Figure P10.3

10.4 Determine the natural frequencies of vibration of a cantilever beam of length L using one Euler–Bernoulli beam element. Assume that the flexural stiffness EI and mass per unit length m are constant. The first two exact natural frequencies are $\omega_1 = 3.516\sqrt{EI/mL^2}$ and $\omega_2 = 22.03\sqrt{EI/mL^2}$. (*Ans:* $\omega_1 = 3.533\sqrt{EI/mL^2}$ and $\omega_2 = 34.81\sqrt{EI/mL^2}$.)

10.5 Determine the natural frequencies of vibration of a clamped–clamped beam of length L using one Euler–Bernoulli beam element. Assume that the flexural stiffness EI and mass per unit length m are constant.

10.6 Determine the natural frequencies of vibration of a cantilever beam of length L using one linear Timoshenko beam element. Assume that the flexural stiffness EI, shear stiffness KGA, and mass per unit length m are constant. Compare your answer with exact frequencies.

10.7 Develop a Timoshenko beam finite element based on the Hermite cubic interpolation of w_0 and Lagrange quadratic interpolation of ϕ_x. Eliminate (condense out) the midnode degree of freedom associated with ϕ_x and obtain 6×6 stiffness matrix and 6×1 force vector for the element.

10.8 (*Continuation of Exercise 10.7*) Condense out the slope $\theta \equiv -dw_0/dx$ degrees of freedom to reduce the element stiffness matrix from 6×6 to 4×4.

10.9 Develop a Timoshenko beam finite element based on the cubic interpolation of w_0 and interdependent quadratic interpolation of ϕ_x [see Eqs. (10.3–22a,b)]. *Ans:* The 4×4 stiffness matrix of the element is given by (see Reference 90 for details)

$$[K^e] = \mu^e \begin{bmatrix} \frac{12D_{xx}^e}{h_e^3} & -\frac{6D_{xx}^e}{h_e^2} & -\frac{12D_{xx}^e}{h_e^3} & -\frac{6D_{xx}^e}{h_e^2} \\ -\frac{6D_{xx}^e}{h_e^2} & \frac{4D_{xx}^e}{h_e} & \frac{6D_{xx}^e}{h_e^2} & \frac{2D_{xx}^e}{h_e} \\ -\frac{12D_{xx}^e}{h_e^3} & \frac{6D_{xx}^e}{h_e^2} & \frac{12D_{xx}^e}{h_e^3} & \frac{6D_{xx}^e}{h_e^2} \\ -\frac{6D_{xx}^e}{h_e^2} & \frac{2D_{xx}^e}{h_e} & \frac{6D_{xx}^e}{h_e^2} & \frac{4D_{xx}^e}{h_e} \end{bmatrix}$$

where

$$\mu^e = \frac{1}{1+12\Omega^e}\,, \qquad \Omega^e = \frac{D_{xx}^e}{A_{xz}^e K_s h_e^2}$$

10.10 Determine the transient solution $w_0(L,t)$ of a cantilevered beam subjected to a uniformly distributed step load. Assume that the initial conditions are zero.

10.11 Consider a rectangular classical plate theory element of dimension $2a \times 2b$. The displacement w_0 and slopes $\theta_x \equiv \partial w_0/\partial x$ and $\theta_y \equiv \partial w_0/\partial y$ are taken as the nodal degrees of freedom. Suppose that $w_0(x,y)$ is represented by the polynomial

$$\begin{aligned} w_0(x,y) =& c_1 + c_2 x + c_3 y + c_4 x^2 + c_5 xy + c_6 y^2 + c_7 x^3 + c_8 x^2 y \\ & c_9 xy^2 + c_{10} y^3 + c_{11} x^3 y + c_{12} xy^3 \end{aligned}$$

Show that the interelement continuity of the displacement is guaranteed. Check if the continuity of the normal derivative (for example, θ_x on side $x = 0$) is guaranteed.

10.12 For the plate element in Exercise 10.11, the deflection can be rewritten as

$$w_0(x,y) = \varphi_1(x,y)w_1 + \varphi_2(x,y)\theta_{x1} + \varphi_3(x,y)\theta_{y1} + \cdots + \varphi_{12}(x,y)\theta_{x4}$$

Take the coordinate system at the center of the element such that $-a < x < a$ and $-b < y < b$, and derive the first three interpolation functions using the following interpolation properties:

(i) $\varphi_1 = 1$ at node 1 and zero at all other nodes; $\partial\varphi_1/\partial x = 0$ and $\partial\varphi_1/\partial y = 0$ at all nodes.

(ii) $\varphi_2 = 0$ and $\partial\varphi_2/\partial x = 0$ at all nodes; and $\partial\varphi_2/\partial y = 1$ at node 1, and zero at all other nodes.

(iii) $\varphi_3 = 0$ and $\partial\varphi_3/\partial y = 0$ at all nodes; and $\partial\varphi_2/\partial x = -1$ at node 1, and zero at all other nodes. *Ans:*

$$\begin{aligned}
\varphi_1(x,y) =& \frac{1}{8}\left(1-\frac{x}{a}\right)\left(1-\frac{y}{b}\right)\left[2-\frac{x}{a}-\frac{y}{b}-\left(\frac{x}{a}\right)^2-\left(\frac{y}{b}\right)^2\right] \\
\varphi_2(x,y) =& \frac{b}{8}\left(1-\frac{x}{a}\right)\left(1-\frac{y}{b}\right)^2\left(1+\frac{y}{b}\right) \\
\varphi_3(x,y) =& \frac{a}{8}\left(1-\frac{x}{a}\right)^2\left(1-\frac{y}{b}\right)\left(1+\frac{x}{a}\right)
\end{aligned}$$

10.13 For a rectangular plate element, $-a < x < a$ and $-b < y < b$, derive the coefficients of the element stiffness matrix when w_0, ϕ_x, and ϕ_y are interpolated using linear Lagrange interpolation functions. Use exact integration to evaluate the integrals. (*Ans:* various submatrices involved in the element stiffness matrix can be found in Reference 1.)

10.14 Repeat Exercise 10.13 using reduced, one–point, integration to evaluate the shear stiffness coefficients.

10.15 Use the element stiffness matrix derived in Exercise 10.13 to study the bending of a cantilevered plate (say, $x = 0$ edge is clamped).

10.16 Use the element stiffness matrix derived in Exercise 10.14 to study the bending of a cantilevered plate (say, $x = 0$ edge is clamped).

References for Additional Reading

1. Reddy, J. N., *An Introduction to the Finite Element Method*, Second Edition, McGraw–Hill, New York, 1993.
2. Reddy, J. N., *Energy and Variational Methods in Applied Mechanics,* John Wiley, New York, 1984.
3. Burnett, D. S., *Finite Element Analysis*, Addison–Wesley, Reading, Massachusetts, 1987.
4. Zienkiewicz, O. C. and Taylor, R. L., *The Finite Element Method, Vol. 1: Linear Problems*, McGraw–Hill, New York, 1989.
5. Hughes, T. J. R., *The Finite Element Method, Linear Static and Dynamic Finite Element Analysis*, Prentice–Hall, Englewood Cliffs, New Jersey, 1987.
6. Hrabok, M. M. and Hrudey, T. M., "A Review and Catalog of Plate Bending Finite Elements," *Computers and Structures* **19**(3), 479–495 (1984).

7. Bazeley, G. P., Cheung, Y. K., Irons, B. M., and Zienkiewicz, O. C., "Triangular Elements in Bending – Conforming and Non-Conforming Solutions," *Proceedings of the Conference on Matrix Methods in Structural Mechanics,* Air Force Institute of Technology, Wright–Patterson Air Force Base, Ohio, 547–576 (1965).

8. Fraeijis de Veubeke, B., "A Conforming Finite Element for Plate Bending," *International Journal of Solids and Structures*, **4**(1), 95–108 (1968).

9. Argyris, J. H., Fried, I., and Scharpf, D. W., "The TUBA Family of Plate Bending Elements for the Matrix Displacement Method," *Journal of the Royal Aeronautical Society,* **72**, 701–709 (1969).

10. Bell, K., "A Refined Triangular Plate Bending Finite Element," *International Journal for Numerical Methods in Engineering*, **1**, 101–122 (1969).

11. Irons, B. M., "A Conforming Quartic Triangular Element for Plate Bending," *International Journal for Numerical Methods in Engineering*, **1**, 29–45 (1969).

12. Stricklin, J. A., Haisler, W., Tisdale, P., and Gunderson, R., "A Rapidly Converging Triangular Plate Element," *AIAA Journal*, **7**(1), 180–181 (1969).

13. Batoz, J. L., Bathe, K. J., and Ho, L. W., "A Study of Three-Node Triangular Plate Bending Elements," *International Journal for Numerical Methods in Engineering*, **15**(12), 1771–1812 (1980).

14. Batoz, J. L., "An Explicit Formulation for an Efficient Triangular Plate Bending Element," *International Journal for Numerical Methods in Engineering*, **18** (7), 1077–1089 (1985).

15. Batoz, J. L. and Tahar, M. B., "Evaluation of a New Quadrilateral Thin Plate Bending Element," *International Journal for Numerical Methods in Engineering*, **18**, 1655–1677 (1982).

16. Jeyachandrabose, C. and Kirkhope, J., "An Alternative Formulation for DKT Plate Bending Element," *International Journal for Numerical Methods in Engineering*, **21**(7), 1289–1293 (1985).

17. Hellan, K., "Analysis of Plates in Flexure by a Simplified Finite Element Method," *Acta Polytechnica Scandinavia*, Civil Engineering Series 46, Trondheim (1967).

18. Melosh, R. J., "Basis of Derivation of Matrices for the Direct Stiffness Method," *AIAA Journal,* **1**, 1631–1637 (1963).

19. Zienkiewicz, O. C. and Cheung, Y. K., "The Finite Element Method for Analysis of Elastic Isotropic and Orthotropic Slabs," *Proceeding of the Institute of Civil Engineers*, London, **28**, 471–488 (1964).

20. Bogner, F. K., Fox, R. L., and Schmidt, L. A., Jr., "The Generation of Interelement-Compatible Stiffness and Mass Matrices by the Use of Interpolation Formulas," *Proceedings of the Conference on Matrix Methods*

in Structural Mechanics, Air Force Institute of Technology, Wright–Patterson Air Force Base, Ohio, 397–443 (1965).

21. Anderson, R. G., Irons, B. M., and Zienkiewicz, O. C., "Vibration and Stability of Plates Using Finite Elements," *International Journal of Solids and Structures,* **4**(10), 1031–1055 (1968).
22. R. W. Clough and J. L. Tocher, "Finite Element Stiffness Matrices for Analysis of Plates in Bending," *Proceedings of the Conference on Matrix Methods in Structural Mechanics,* Air Force Institute of Technology, Wright–Patterson Air Force Base, Ohio, 515–545 (1965).
23. Clough, R. W. and Felippa, C. A., "A Refined Quadrilateral Element for Analysis of Plate Bending," *Proceedings of the Second Conference on Matrix Methods in Structural Mechanics,* Air Force Institute of Technology, Wright–Patterson Air Force Base, Ohio, 399–440 (1968).
24. Ahmad, S., Irons, B. M., and Zienkiewicz, O. C., "Analysis of Thick and Thin Shell Structures by Curved Finite Elements," *International Journal for Numerical Methods in Engineering,* **13**(4), 575–586 (1971).
25. Morley, L. S. D., "A Triangular Equilibrium Element with Linearly Varying Bending Moments for Plate Bending Problems," Technical Note, *Journal of the Aeronautical Society,* **71** (1967).
26. Morley, L. S. D., "The Triangular Equilibrium Element in the Solution of Plate Bending Problems," *Journal of the Aeronautical Society,* **72** (1968).
27. Morley, L. S. D., "The Constant-Moment Plate Bending Element," *Journal of Strain Analysis,* **6**(1), 20–24 (1971).
28. Barlow, J., "Optimal Stress Location in Finite Element Models," *International Journal for Numerical Methods in Engineering,* **10**, 243–251 (1976).
29. Barlow, J., "More on Optimal Stress Points — Reduced Integration Element Distortions and Error Estimation," *International Journal for Numerical Methods in Engineering,* **28**, 1486–1504 (1989).
30. Reddy, J. N., "Simple Finite Elements with Relaxed Continuity for Non–Linear Analysis of Plates," *Proceedings of the Third International Conference in Australia on Finite Element Methods,* University of New South Wales, Sydney, July 2–6, 1979.
31. Reddy, J. N., "A Penalty Plate-Bending Element for the Analysis of Laminated Anisotropic Composite Plates," *International Journal for Numerical Methods in Engineering,* **15**(8), 1187–1206 (1980).
32. Averill, R. C. and Reddy, J. N., "Behavior of Plate Elements Based on the First-Order Shear Deformation Theory," *Engineering Computations,* **7**, 57–74 (1990).
33. Reddy, J. N. and Averill, R. C., "Advances in the Modeling of Laminated Plates," *Computing Systems in Engineering,* **2**(5/6), 541–555 (1991).
34. Belytschko, T. and Tsay, C. S., "A Stabilization Procedure for the Quadrilateral Plate Bending Element with One-point Quadrature,"

International Journal for Numerical Methods in Engineering, **19** 405–420 (1983).

35. Park, K. C. and Flaggs, D. L., "A Symbolic Fourier Synthesis of a One-point Integrated Quadrilateral Plate Element," *Computer Methods in Applied Mechanics and Engineering*, **48**(2), 203–236 (1985).
36. Belytschko, T., Ong, J. S. -J., and Liu, W. K., "A Consistent Control of Spurious Singular Modes in the 9-node Lagrange Element for the Laplace and Mindlin Plate Equation," *Computer Methods in Applied Mechanics and Engineering*, **44**, 269–295 (1984).
37. Crisfield, M. A., "A Quadratic Mindlin Element Using Shear Constraints," *Computers and Structures*, **18**, 833–852 (1984).
38. Huang, H. C. and Hinton, E., "A Nine-Node Lagrangian Plate Element with Enhanced Shear Interpolation," *Engineering Computations*, **1**, 369–379 (1984).
39. Tessler, A. and Dong, S. B., "On a Hierarchy of Conforming Timoshenko Beam Elements," *Computers and Structures* **14**, 335–344 (1981).
40. Prathap, G. and Babu, C. R., "An Isoparametric Quadratic Thick Curved Beam Element," *International Journal for Numerical Methods in Engineering*, **23**(9), 1583–1600 (1986).
41. Stolarski, H. and Belytschko, T., "Membrane Locking and Reduced Integration for Curved Element," *Journal of Applied Mechanics*, **49** (1), 172–176 (1982).
42. Zienkiewicz, O. C., Too, J. J. M., and Taylor, R. L., "Reduced Integration Technique in General Analysis of Plates and Shells," *International Journal for Numerical Methods in Engineering*, **3**, 275–290 (1971).
43. Hughes, T. J. R. and Cohen, M., "The 'Heterosis' Finite Element for Plate Bending," *Journal of Computers and Structures,* **9**(5), 445–450 (1978).
44. Hinton, E. and Huang, H. C., "A Family of Quadrilateral Mindlin Plate Elements with Substitute Shear Strain Fields," *Computers and Structures* **23**(3), 409–431 (1986).
45. Hughes, T. J. R., Taylor, R. L., and Kanoknukulchai, W., "A Simple and Efficient Element for Plate Bending," *International Journal for Numerical Methods in Engineering*, **11**(10), 1529–1543 (1977).
46. Pugh, E. D. L., Hinton, E., and Zienkiewicz, O. C., "A Study of Quadrilateral Plate Bending Elements with 'Reduced' Integration," *International Journal for Numerical Methods in Engineering*, **12**(7), 1059–1079 (1978).
47. Malkus, D. S. and Hughes, T. J. R., "Mixed Finite Methods–Reduced and Selective Integration Techniques: A Unification of Concepts," *Computer Methods in Applied Mechanics and Engineering*, **15**(1), 63–81 (1978).
48. Hughes, T. J. R., Cohen, M., and Haroun, M., "Reduced and Selective Integration Techniques in the Finite Element Analysis of Plates," *Nuclear Engineering and Design*, **46**, 203–222 (1981).

49. Belytschko, T., Tsay, C. S., and Liu, W. K., "Stabilization Matrix for the Bilinear Mindlin Plate Element," *Computer Methods in Applied Mechanics and Engineering*, **29**, 313–327 (1981).
50. Hughes, T. J. R. and Tezduyar, T. E., "Finite Elements Based Upon Mindlin Plate Theory with Particular Reference to the Four-Node Bilinear Isoparametric Element," *Journal of Applied Mechanics*, **48**(3), 587–596 (1981).
51. Spilker, R. L. and Munir, N. I., "The Hybrid-Stress Model for Thin Plates," *International Journal for Numerical Methods in Engineering*, **15**(8), 1239–1260 (1980).
52. Spilker, R. L. and Munir, N. I., "A Serendipity Cubic-Displacement Hybrid-Stress Element for Thin and Moderately Thick Plates," *International Journal for Numerical Methods in Engineering*, **15**(8), 1261–1278 (1980).
53. Spilker, R. L. and Munir, N. I., "A Hybrid-Stress Quadratic Serendipity Displacement Mindlin Plate Bending Element," *Computers and Structures*, **12**, 11–21 (1980).
54. Crisfield, M. A., "A Four-Noded Plate Bending Element Using Shear Constraints; A Modified Version of Lyons' Element," *Computer Methods in Applied Mechanics and Engineering*, **38**, 93–120 (1983).
55. Tessler, A. and Hughes, T. J. R., "An Improved Treatment of Transverse Shear in the Mindlin-Type Four-Node Quadrilateral Element," *Computer Methods in Applied Mechanics and Engineering*, **39**, 311–335 (1983).
56. Tessler, A. and Hughes, T. J. R., "Three-Node Mindlin Plate Element with Improved Transverse Shear," *Computer Methods in Applied Mechanics and Engineering*, **50**, 71–101 (1985).
57. Bathe, K. J. and Dvorkin, E. N., "A Four-Node Plate Bending Element Based on Mindlin/Reissner Plate Theory and Mixed Interpolation," *International Journal for Numerical Methods in Engineering*, **21**, 367–383 (1985).
58. Newmark, N. M., "A Method of Computation for Structural Dynamics," *Journal of the Engineering Mechanics Division, ASCE*, 67–94 (1959).
59. Goudreau, G. L. and Taylor, R. L., "Evaluation of Numerical Integration Methods in Elastodynamics," *Computer Methods in Applied Mechanics and Engineering*, **2**, 69–97 (1972).
60. Hilber, H. M., "Analysis and Design of Numerical Integration Methods in Structural Dynamics," EERC Report No. 76-29, Earthquake Engineering Research Center, University of California, Berkeley, California, November 1976.
61. Hughes, T. J. R., "Analysis of Transient Algorithms with Particular Reference to Stability Behavior," *Computational Methods for Transient Analysis*, **1**, 67–155 (1983).

62. T. Belytschko, "An Overview of Semidiscretization and Time Integration Procedures," *Computational Methods for Transient Analysis*, **1**, 1–65 (1983).

63. Houbolt, J. C., "A Recurrence Matrix Solution for the Dynamic Response of Elastic Aircraft," *Journal of the Aeronautical Sciences*, **17**, 540–550 (1950).

64. Goudreau, G. L. and Taylor, R. L., "Evaluation of Numerical Methods in Elastodynamics," *Journal of Computer Methods in Applied Mechanics and Engineering*, **2**, 69–97 (1973).

65. Belytschko, T. and Hughes, T. J. R., (Eds.), *Computational Methods for Transient Analysis*, **1**, North–Holland, Amsterdam, Netherlands, 1983.

66. Reddy, J. N. and Khdeir, A. A., "Buckling and Vibration of Laminated Composite Plates Using Various Plate Theories," *AIAA Journal,* **27** (12), 1808–1817 (1989).

67. Nosier, A. and Reddy, J. N., "On Vibration and Buckling of Symmetric Laminated Plates According to Shear Deformation Theories," *Acta Mechanica*, **94** (3–4), 123–170 (1992).

68. Reddy, J. N. and Chao, W. C., "A Comparison of Closed-Form and Finite Element Solutions of Thick Laminated Anisotropic Rectangular Plates," *Nuclear Engineering and Design*, **64**, 153–167 (1981).

69. Pagano, N. J., "Exact Solutions for Composite Laminates in Cylindrical Bending," *Journal of Composite Materials*, **3**, 398–411 (1967).

70. Pagano, N. J., "Exact Solutions for Rectangular Bidirectional Composites and Sandwich Plates," *Journal of Composite Materials*, **4**, 20–34 (1970).

71. Reddy, J. N., "Free Vibration of Antisymmetric, Angle-Ply Laminated Plates Including Transverse Shear Deformation by the Finite Element Method," *Journal of Sound and Vibration,* **66**(4), 565–576 (1979).

72. Reddy, J. N. and Hsu, Y. S., "Effects of Shear Deformation and Anisotropy on the Thermal Bending of Layered Composite Plates," *Journal of Thermal Stresses*, **3,** 475–493 (1980).

73. Khdeir, A. A., Librescu, L., and Reddy, J. N., "Analytical Solution of a Refined Shear Deformation Theory for Rectangular Composite Plates," *International Journal of Solids and Structures*, **23**(10), 1447–1463 (1987).

74. Khdeir, A. A. and Reddy, J. N., "Dynamic Response of Antisymmetric Angle-Ply Laminated Plates Subjected to Arbitrary Loading," *Journal of Sound and Vibration*, **126**(3), 437–445 (1988).

75. Khdeir, A. A. and Librescu, L., "Analysis of Symmetric Cross-Ply Laminated Elastic Plates Using a Higher-Order Theory: Part I–Stress and Displacement," *Composite Structures,* **9**, 189–213 (1988).

76. Khdeir, A. A. and Librescu, L., "Analysis of Symmetric Cross-Ply Laminated Elastic Plates Using a Higher-Order Theory: Part II–Buckling and Free Vibration," *Composite Structures,* **9**, 259–277 (1988).

77. Khdeir, A. A., "Free Vibration and Buckling of Symmetric Cross-Ply Laminated Plates by an Exact Method," *Journal of Sound and Vibration,* **126**(3), 447–461 (1988).
78. Khdeir, A. A., "Free Vibration of Antisymmetric Angle-Ply Laminated Plates Including Various Boundary Conditions," *Journal of Sound and Vibration,* **122**(2), 377–388 (1988).
79. Khdeir, A. A., "Free Vibration and Buckling of Unsymmetric Cross-Ply Laminated Plates Using a Refined Theory," *Journal of Sound and Vibration,* **128**(3), 377–395 (1989).
80. Khdeir, A. A. and Reddy, J. N., "Exact Solutions for the Transient Response of Symmetric Cross-Ply Laminates Using a Higher-Order Plate Theory," *Composite Science and Technology,* **34**, 205–224 (1989).
81. Khdeir, A. A., "An Exact Approach to the Elastic State of Stress of Shear Deformable Antisymmetric Angle-Ply Laminated Plates," *Composite Structures,* **11**, 245–258 (1989).
82. Khdeir, A. A., "Comparison Between Shear Deformable and Kirchhoff Theories for Bending, Buckling and Vibration of Antisymmetric Angle-Ply Laminated Plates," *Composite Structures,* **13**, 159–172 (1989).
83. Khdeir, A. A., "Stability of Antisymmetric Angle-Ply Laminated Plates," *ASCE Journal of Engineering Mechanics,* **115**, 952–962 (1989).
84. Khdeir, A. A. and Reddy, J. N., "Analytical Solutions of Refined Plate Theories of Cross-Ply Composite Laminates," *Journal of Pressure Vessel Technology,* **113**(4), 570–578 (1991).
85. Khdeir, A. A. and Reddy, J. N., "Thermal Stresses and Deflections of Cross-Ply Laminated Plates Using Refined Plate Theories," *Journal of Thermal Stresses*, **14**(4), 419–438 (1991).
86. Reddy, J. N., "On the Solutions to Forced Motions of Rectangular Composite Plates," *Journal of Applied Mechanics,* **49**, 403–408 (1982).
87. Reddy, J. N., "Dynamic (Transient) Analysis of Layered Anisotropic Composite-Material Plates," *International Journal for Numerical Methods in Engineering,* **19**, 237–255 (1983).
88. Reddy, J. N., "Geometrically Nonlinear Transient Analysis of Laminated Composite Plates," *AIAA Journal,* **21**(4), 621–629 (1983).
89. Heyliger P. R. and Reddy, J. N., "A Higher-Order Beam Finite Element for Bending and Vibration Problems," *Journal of Sound and Vibration,* **126**(2), 309–326 (1988).
90. Reddy, J. N., Wang, C. M., and Lam, K. Y., "A Unified Beam Finite Element for the Euler–Bernoulli and Shear Deformation Beam Theories," *Communications in Numerical Methods in Engineering,* to appear.

Chapter Eleven

Refined Theories of Laminated Composite Plates

11.1 Introduction

The classical laminate plate theory and the first-order shear deformation theory are the simplest equivalent single-layer theories, and they adequately describe the kinematic behavior of most laminates. Higher-order theories can represent the kinematics better, may not require shear correction factors, and can yield more accurate interlaminar stress distributions. However, they involve considerably more computational effort. Therefore, such theories should be used only when necessary.

In principle, it is possible to expand the displacement field in terms of the thickness coordinate up to any desired degree. However, due to the algebraic complexity and computational effort involved with higher-order theories in return for marginal gain in accuracy, theories higher than third order have not been attempted. The reason for expanding the displacements up to the cubic term in the thickness coordinate is to have quadratic variation of the transverse shear strains and transverse shear stresses through each layer. This avoids the need for shear correction coefficients used in the first-order theory.

There are many papers on third-order theories (see [1–32]) and their applications [33–52]. Although many of them seem to differ from each other on the surface, the displacement fields of these theories are related (see Reddy [49]). Here we present the original third-order shear deformation laminate theory of Reddy [25,26] that contains other lower-order laminate theories, including the classical laminate theory and first-order shear deformation laminate theory as special cases. Analytical as well as finite element results of this third-order theory are developed and numerical results are compared with those of the classical and first-order theories.

11.2 A Third-Order Plate Theory

11.2–1 Displacement Field

The third-order plate theory to be developed is based on the same assumptions as the classical and first-order plate theories, except that we relax the assumption on the straightness and normality of a transverse normal after deformation by expanding the displacements (u, v, w) as cubic functions of the thickness coordinate. Figure 11.2–1 shows the kinematics of deformation of a transverse normal on edge $y = 0$.

Consider the displacement field

$$
\begin{aligned}
u &= u_0 + z\left(\varphi_x - c_0\frac{\partial w_0}{\partial x}\right) - z^3 c_1 \varphi_x \\
v &= v_0 + z\left(\varphi_y - c_0\frac{\partial w_0}{\partial y}\right) - z^3 c_1 \varphi_y \\
w &= w_0
\end{aligned}
\tag{11.2 - 1}
$$

where c_1 and c_0 are parameters, called *tracers*, which are introduced to include the displacement fields of CLPT and FSDT as special cases of (1). Clearly, by setting $c_0 = 1$ and $\varphi_x = \varphi_y = 0$ we recover the displacement field of CLPT. The displacement field of FSDT is recovered by setting $c_0 = c_1 = 0$.

Let us introduce the variables ϕ_x and ϕ_y as

$$
\phi_x = \varphi_x - c_0\frac{\partial w_0}{\partial x}\ , \quad \phi_y = \varphi_y - c_0\frac{\partial w_0}{\partial y}
\tag{11.2 - 2}
$$

It is clear that (u_0, v_0, w_0) and (ϕ_x, ϕ_y) have the same physical meaning as in the first-order theory; they denote the displacements and rotations of transverse normals on the plane $z = 0$, respectively:

$$
u_0 = u(x, y, 0, t),\ v_0 = v(x, y, 0, t),\ w_0 = w(x, y, 0, t)
$$

$$
\phi_x = \frac{\partial u}{\partial z}(x, y, 0, t),\ \phi_y = \frac{\partial v}{\partial z}(x, y, 0, t)
\tag{11.2 - 3}
$$

The displacement field (1) now can be expressed in terms of ϕ_x and ϕ_y using the definitions (2):

$$
\begin{aligned}
u(x, y, z, t) &= u_0(x, y, t) + z\phi_x(x, y, t) - c_1 z^3\left(\phi_x + c_0\frac{\partial w_0}{\partial x}\right) \\
v(x, y, z, t) &= v_0(x, y, t) + z\phi_y(x, y, t) - c_1 z^3\left(\phi_y + c_0\frac{\partial w_0}{\partial y}\right) \\
w(x, y, z, t) &= w_0(x, y, t)
\end{aligned}
\tag{11.2 - 4}
$$

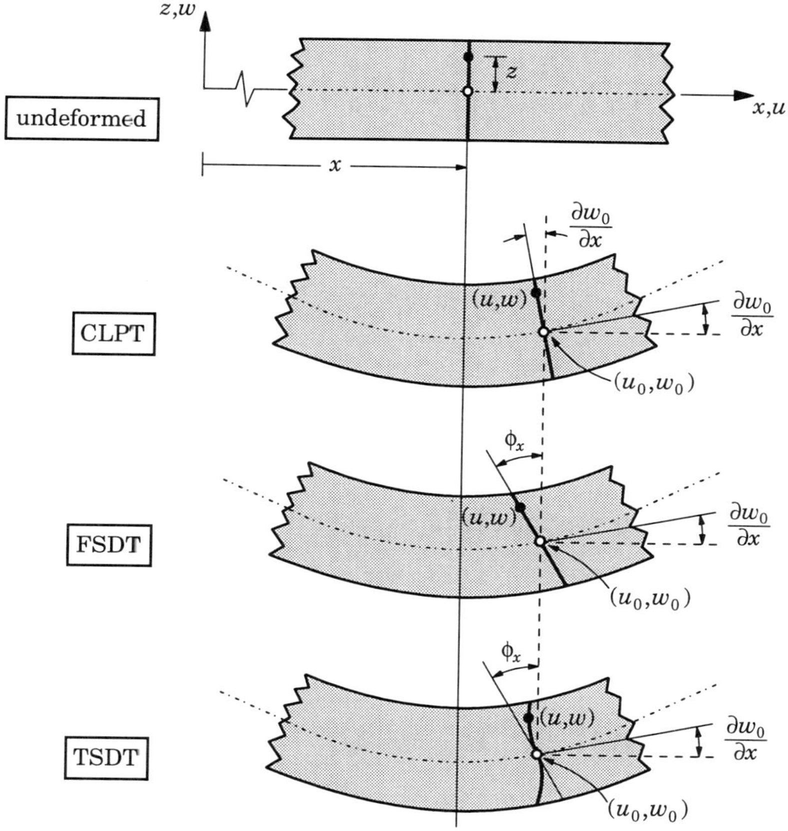

Figure 11.2–1: Deformation of a transverse normal according the classical, first-order, and third-order plate theories.

11.2–2 Strains and Stresses

Substitution of the displacements (1) into the nonlinear strain–displacement relations in Eq. (5.3–7) yields the strains

$$\left\{\begin{array}{c}\varepsilon_{xx}\\ \varepsilon_{yy}\\ \gamma_{xy}\end{array}\right\}=\left\{\begin{array}{c}\varepsilon_{xx}^{(0)}\\ \varepsilon_{yy}^{(0)}\\ \gamma_{xy}^{(0)}\end{array}\right\}+z\left\{\begin{array}{c}\varepsilon_{xx}^{(1)}\\ \varepsilon_{yy}^{(1)}\\ \gamma_{xy}^{(1)}\end{array}\right\}+z^3\left\{\begin{array}{c}\varepsilon_{xx}^{(3)}\\ \varepsilon_{yy}^{(3)}\\ \gamma_{xy}^{(3)}\end{array}\right\} \qquad (11.2-5)$$

$$\left\{\begin{array}{c}\gamma_{yz}\\ \gamma_{xz}\end{array}\right\}=\left\{\begin{array}{c}\gamma_{yz}^{(0)}\\ \gamma_{xz}^{(0)}\end{array}\right\}+z^2\left\{\begin{array}{c}\gamma_{yz}^{(2)}\\ \gamma_{xz}^{(2)}\end{array}\right\} \qquad (11.2-6)$$

where

$$\left\{\begin{array}{c}\varepsilon_{xx}^{(0)}\\ \varepsilon_{yy}^{(0)}\\ \gamma_{xy}^{(0)}\end{array}\right\}=\left\{\begin{array}{c}\frac{\partial u_0}{\partial x}+\frac{1}{2}\left(\frac{\partial w_0}{\partial x}\right)^2\\ \frac{\partial v_0}{\partial y}+\frac{1}{2}\left(\frac{\partial w_0}{\partial x}\right)^2\\ \frac{\partial u_0}{\partial y}+\frac{\partial v_0}{\partial x}+\frac{\partial w_0}{\partial x}\frac{\partial w_0}{\partial y}\end{array}\right\},\quad \left\{\begin{array}{c}\varepsilon_{xx}^{(1)}\\ \varepsilon_{yy}^{(1)}\\ \gamma_{xy}^{(1)}\end{array}\right\}=\left\{\begin{array}{c}\frac{\partial \phi_x}{\partial x}\\ \frac{\partial \phi_y}{\partial y}\\ \frac{\partial \phi_x}{\partial y}+\frac{\partial \phi_y}{\partial x}\end{array}\right\} \qquad (11.2-7a)$$

$$\left\{\begin{array}{c}\varepsilon_{xx}^{(3)}\\ \varepsilon_{yy}^{(3)}\\ \gamma_{xy}^{(3)}\end{array}\right\}=-c_1\left\{\begin{array}{c}\left(\frac{\partial \phi_x}{\partial x}+\frac{\partial^2 w_0}{\partial x^2}\right)\\ \left(\frac{\partial \phi_y}{\partial y}+\frac{\partial^2 w_0}{\partial y^2}\right)\\ \left(\frac{\partial \phi_x}{\partial y}+\frac{\partial \phi_y}{\partial x}+2\frac{\partial^2 w_0}{\partial x \partial y}\right)\end{array}\right\} \qquad (11.2-7b)$$

$$\left\{\begin{array}{c}\gamma_{yz}^{(0)}\\ \gamma_{xz}^{(0)}\end{array}\right\}=\left\{\begin{array}{c}\phi_y+\frac{\partial w_0}{\partial y}\\ \phi_x+\frac{\partial w_0}{\partial x}\end{array}\right\},\quad \left\{\begin{array}{c}\gamma_{yz}^{(2)}\\ \gamma_{xz}^{(2)}\end{array}\right\}=-c_2\left\{\begin{array}{c}\left(\phi_y+\frac{\partial w_0}{\partial y}\right)\\ \left(\phi_x+\frac{\partial w_0}{\partial x}\right)\end{array}\right\} \qquad (11.2-7c)$$

where $c_2 = 3c_1$, and it is assumed that $c_0 = 1$.

The third-order laminated plate theory of Reddy [25,26,49] satisfies the boundary conditions that the transverse shear stresses on the bounding planes of the laminate be zero. The transverse shear stresses in a typical lamina are given by

$$\sigma_{yz}^{(k)} = \bar{Q}_{44}^{(k)}\left(\gamma_{yz}^{(0)}+z^2\gamma_{yz}^{(2)}\right)+\bar{Q}_{45}^{(k)}\left(\gamma_{xz}^{(0)}+z^2\gamma_{xz}^{(2)}\right) \qquad (11.2-8a)$$

$$\sigma_{xz}^{(k)} = \bar{Q}_{45}^{(k)}\left(\gamma_{yz}^{(0)}+z^2\gamma_{yz}^{(2)}\right)+\bar{Q}_{44}^{(k)}\left(\gamma_{xz}^{(0)}+z^2\gamma_{xz}^{(2)}\right) \qquad (11.2-8b)$$

If the transverse stresses σ_{yz} and σ_{xz} are to vanish at the bottom $(z=-h/2)$ and top $(z=h/2)$ of the laminate, we have

$$\bar{Q}_{44}^{(1)}\left(\gamma_{yz}^{(0)}+\frac{h^2}{4}\gamma_{yz}^{(2)}\right)+\bar{Q}_{45}^{(1)}\left(\gamma_{xz}^{(0)}+\frac{h^2}{4}\gamma_{xz}^{(2)}\right)=0 \qquad (11.2-9a)$$

$$\bar{Q}_{44}^{(N)}\left(\gamma_{yz}^{(0)}+\frac{h^2}{4}\gamma_{yz}^{(2)}\right)+\bar{Q}_{45}^{(N)}\left(\gamma_{xz}^{(0)}+\frac{h^2}{4}\gamma_{xz}^{(2)}\right)=0 \qquad (11.2-9b)$$

For an arbitrary lamination scheme, material properties, and fiber orientation, the above equations imply that

$$\gamma_{xz}^{(0)}+\frac{h^2}{4}\gamma_{xz}^{(2)}=0,\quad \gamma_{yz}^{(0)}+\frac{h^2}{4}\gamma_{yz}^{(2)}=0 \qquad (11.2-10)$$

In view of the fact that $\gamma_{xz}^{(2)} = -c_2\gamma_{xz}^{(0)}$ and $\gamma_{yz}^{(2)} = -c_2\gamma_{yz}^{(0)}$, it follows that $c_2 = 4/h^2$. This condition is also sufficient to make the transverse shear stresses σ_{xz} and σ_{yz} zero at the top and bottom of the laminate.

11.2–3 Equations of Motion

The equations of motion of the third-order theory will be derived using the dynamic version of the principle of virtual displacements. The virtual strain energy δU, virtual work done by applied forces δV, and the virtual kinetic energy δK are given by

$$
\begin{aligned}
\delta U =& \int_{\Omega_0} \left\{ \int_{-\frac{h}{2}}^{\frac{h}{2}} \left[\sigma_{xx} \left(\delta\varepsilon_{xx}^{(0)} + z\delta\varepsilon_{xx}^{(1)} - c_1 z^3 \delta\varepsilon_{xx}^{(3)} \right) \right. \right. \\
&+ \sigma_{yy} \left(\delta\varepsilon_{yy}^{(0)} + z\delta\varepsilon_{yy}^{(1)} - c_1 z^3 \delta\varepsilon_{yy}^{(3)} \right) + \sigma_{xy} \left(\delta\gamma_{xy}^{(0)} + z\delta\gamma_{xy}^{(1)} - c_1 z^3 \delta\gamma_{xy}^{(3)} \right) \\
&\left. \left. + \sigma_{xz} \left(\delta\gamma_{xz}^{(0)} + z^2\delta\gamma_{xz}^{(2)} \right) + \sigma_{yz} \left(\delta\gamma_{yz}^{(0)} + z^2\delta\gamma_{yz}^{(2)} \right) \right] dz \right\} dxdy \\
=& \int_{\Omega_0} \left(N_{xx}\delta\varepsilon_{xx}^{(0)} + M_{xx}\delta\varepsilon_{xx}^{(1)} - c_1 P_{xx}\delta\varepsilon_{xx}^{(3)} + N_{yy}\delta\varepsilon_{yy}^{(0)} + M_{yy}\delta\varepsilon_{yy}^{(1)} - c_1 P_{yy}\delta\varepsilon_{yy}^{(3)} \right. \\
&+ N_{xy}\delta\gamma_{xy}^{(0)} + M_{xy}\delta\gamma_{xy}^{(1)} - c_1 P_{xy}\delta\gamma_{xy}^{(3)} \\
&\left. + Q_x\delta\gamma_{xz}^{(0)} - c_2 R_x \delta\gamma_{xz}^{(2)} + Q_y\delta\gamma_{yz}^{(0)} - c_2 R_y\delta\gamma_{yz}^{(2)} \right) dxdy \qquad (11.2-11)
\end{aligned}
$$

$$
\begin{aligned}
\delta V =& -\int_{\Omega_0} \left[q_b(x,y)\delta w(x,y,-\frac{h}{2}) + q_t(x,y)\delta w(x,y,\frac{h}{2}) \right] dxdy \\
&- \int_{\Gamma}\int_{-\frac{h}{2}}^{\frac{h}{2}} \left[\hat{\sigma}_{nn} \left(\delta u_n + z\delta\phi_n - c_1 z^3 \delta\varphi_n \right) \right. \\
&\left. + \hat{\sigma}_{ns} \left(\delta u_s + z\delta\phi_s - c_1 z^3 \delta\varphi_{ns} \right) + \hat{\sigma}_{nr}\delta w_0 \right] dzd\Gamma \\
=& -\int_{\Omega_0} q\delta w_0 dxdy - \int_{\Gamma} (\hat{N}_{nn}\delta u_n + \hat{M}_{nn}\delta\phi_n - c_1\hat{P}_{nn}\delta\varphi_n \\
&+ \hat{N}_{ns}\delta u_s + \hat{M}_{ns}\delta\phi_s - c_1\hat{P}_{ns}\delta\varphi_{ns} + \hat{Q}_n\delta w_0) \, d\Gamma \qquad (11.2-12)
\end{aligned}
$$

$$
\begin{aligned}
\delta K =& \int_{\Omega_0}\int_{-\frac{h}{2}}^{\frac{h}{2}} \rho_0 \left[\left(\dot{u}_0 + z\dot{\phi}_x - c_1 z^3 \dot{\varphi}_x \right) \left(\delta\dot{u}_0 + z\delta\dot{\phi}_x - c_1 z^3 \delta\dot{\varphi}_x \right) \right. \\
&\left. + \left(\dot{v}_0 + z\dot{\phi}_y - c_1 z^3 \dot{\varphi}_y \right) \left(\delta\dot{v}_0 + z\delta\dot{\phi}_y + z^3 \delta\dot{\varphi}_y \right) + \dot{w}_0\delta\dot{w}_0 \right] dv \\
=& \int_{\Omega_0} \left[\left(I_0\dot{u}_0 + I_1\dot{\phi}_x - c_1 I_3\dot{\varphi}_x \right)\delta\dot{u}_0 + \left(I_1\dot{u}_0 + I_2\dot{\phi}_x - c_1 I_4\dot{\varphi}_x \right)\delta\dot{\phi}_x \right. \\
&- c_1\left(I_3\dot{u}_0 + I_4\dot{\phi}_x - c_1 I_6\dot{\varphi}_x \right)\delta\dot{\varphi}_x + \left(I_0\dot{v}_0 + I_1\dot{\phi}_y - c_1 I_3\dot{\varphi}_y \right)\delta\dot{v}_0 \\
&\left. + \left(I_1\dot{v}_0 + I_2\dot{\phi}_y - c_1 I_4\dot{\varphi}_y \right)\delta\dot{\phi}_y - c_1\left(I_3\dot{u}_0 + I_4\dot{\phi}_y - c_1 I_6\dot{\varphi}_y \right)\delta\dot{\varphi}_y \right] dxdy
\end{aligned}
$$

$$(11.2-13)$$

where Ω_0 denotes the midplane of the laminate, and

$$\left\{\begin{matrix} N_{\alpha\beta} \\ M_{\alpha\beta} \\ P_{\alpha\beta} \end{matrix}\right\} = \int_{-\frac{h}{2}}^{\frac{h}{2}} \sigma_{\alpha\beta} \left\{\begin{matrix} 1 \\ z \\ z^3 \end{matrix}\right\} dz, \quad \left\{\begin{matrix} Q_\alpha \\ R_\alpha \end{matrix}\right\} = \int_{-\frac{h}{2}}^{\frac{h}{2}} \sigma_{\alpha z} \left\{\begin{matrix} 1 \\ z^2 \end{matrix}\right\} dz \tag{11.2-14}$$

$$I_i = \int_{-\frac{h}{2}}^{\frac{h}{2}} \rho_0 (z)^i \, dz \quad (i = 0, 1, 2, \cdots, 6) \tag{11.2-15}$$

In Eq. (14), α and β take the symbols x and y. The same definitions hold for the stress resultants with a hat, which are specified.

Substituting for $\delta U, \delta V$, and δK from Eqs. (11)–(13) into the virtual work statement in Eq. (5.4–5), noting that the virtual strains can be written in terms of the generalized displacements using Eqs. (7a,b,c), integrating by parts to relieve the virtual generalized displacements, $\delta u_0, \delta v_0, \delta w_0, \delta\phi_x$, and $\delta\phi_y$ in Ω_0 of any differentiation, and using the fundamental lemma of calculus of variations, we obtain the following Euler–Lagrange equations:

$$\frac{\partial N_{xx}}{\partial x} + \frac{\partial N_{xy}}{\partial y} = I_0 \ddot{u}_0 + J_1 \ddot{\phi}_x - c_1 I_3 \frac{\partial \ddot{w}_0}{\partial x} \tag{11.2-16}$$

$$\frac{\partial N_{xy}}{\partial x} + \frac{\partial N_{yy}}{\partial y} = I_0 \ddot{v}_0 + J_1 \ddot{\phi}_y - c_1 I_3 \frac{\partial \ddot{w}_0}{\partial y} \tag{11.2-17}$$

$$\begin{aligned} &\frac{\partial \bar{Q}_x}{\partial x} + \frac{\partial \bar{Q}_y}{\partial y} + \frac{\partial}{\partial x}\left(N_{xx}\frac{\partial w_0}{\partial x} + N_{xy}\frac{\partial w_0}{\partial y}\right) + \frac{\partial}{\partial y}\left(N_{xy}\frac{\partial w_0}{\partial x} + N_{yy}\frac{\partial w_0}{\partial y}\right) \\ &+ c_1\left(\frac{\partial^2 P_{xx}}{\partial x^2} + 2\frac{\partial^2 P_{xy}}{\partial x \partial y} + \frac{\partial^2 P_{yy}}{\partial y^2}\right) + q = I_0 \ddot{w}_0 - c_1^2 I_6 \left(\frac{\partial^2 \ddot{w}_0}{\partial x^2} + \frac{\partial^2 \ddot{w}_0}{\partial y^2}\right) \\ &+ c_1\left[I_3\left(\frac{\partial \ddot{u}_0}{\partial x} + \frac{\partial \ddot{v}_0}{\partial y}\right) + J_4\left(\frac{\partial \ddot{\phi}_x}{\partial x} + \frac{\partial \ddot{\phi}_y}{\partial y}\right)\right] \end{aligned} \tag{11.2-18}$$

$$\frac{\partial \bar{M}_{xx}}{\partial x} + \frac{\partial \bar{M}_{xy}}{\partial y} - \bar{Q}_x = J_1 \ddot{u}_0 + K_2 \ddot{\phi}_x - c_1 J_4 \frac{\partial \ddot{w}_0}{\partial x} \tag{11.2-19}$$

$$\frac{\partial \bar{M}_{xy}}{\partial x} + \frac{\partial \bar{M}_{yy}}{\partial y} - \bar{Q}_y = J_1 \ddot{v}_0 + K_2 \ddot{\phi}_y - c_1 J_4 \frac{\partial \ddot{w}_0}{\partial y} \tag{11.2-20}$$

where

$$\bar{M}_{\alpha\beta} = M_{\alpha\beta} - c_1 P_{\alpha\beta}, \quad \bar{Q}_\alpha = Q_\alpha - c_2 R_\alpha \tag{11.2-21}$$

$$I_i = \sum_{k=1}^{N} \int_{z_k}^{z_{k+1}} \rho^{(k)} (z)^i \, dz \quad (i = 0, 1, 2, \cdots, 6) \tag{11.2-22a}$$

$$J_i = I_i - c_1 I_{i+2}, \; K_2 = I_2 - 2c_1 I_4 + c_1^2 I_6, \; c_1 = \frac{4}{3h^2}, \; c_2 = 3c_1 \tag{11.2-22b}$$

and (P_{xx}, P_{yy}, P_{xy}) and (R_x, R_y) denote the higher-order stress resultants

$$\left\{\begin{matrix} P_{xx} \\ P_{yy} \\ P_{xy} \end{matrix}\right\} = \int_{-h/2}^{h/2} \left\{\begin{matrix} \sigma_{xx} \\ \sigma_{yy} \\ \sigma_{xy} \end{matrix}\right\} z^3 dz, \quad \left\{\begin{matrix} R_x \\ R_y \end{matrix}\right\} = \int_{-h/2}^{h/2} \left\{\begin{matrix} \sigma_{yz} \\ \sigma_{xz} \end{matrix}\right\} z^2 dz \tag{11.2-23}$$

The primary and secondary variables of the theory are

$$\text{Primary Variables}: \quad u_n,\ u_s,\ w_0,\ \frac{\partial w_0}{\partial n},\ \phi_n,\ \phi_s \tag{11.2-24}$$

$$\text{Secondary Variables}: \quad N_{nn},\ N_{ns},\ \bar{V}_n,\ P_{nn},\ \bar{M}_{nn},\ \bar{M}_{ns} \tag{11.2-25}$$

where

$$\begin{aligned}\bar{V}_n \equiv\ & c_1\left[\left(\frac{\partial P_{xx}}{\partial x}+\frac{\partial P_{xy}}{\partial y}\right)n_x+\left(\frac{\partial P_{xy}}{\partial x}+\frac{\partial P_{yy}}{\partial y}\right)n_y\right]\\ & - c_1\left[\left(I_3\ddot{u}_0+J_4\ddot{\phi}_x-c_1I_6\frac{\partial \ddot{w}_0}{\partial x}\right)n_x+\left(I_3\ddot{v}_0+J_4\ddot{\phi}_y-c_1I_6\frac{\partial \ddot{w}_0}{\partial y}\right)n_y\right]\\ & + (\bar{Q}_xn_x+\bar{Q}_yn_y)+\mathcal{P}(w_0)+c_1\frac{\partial P_{ns}}{\partial s}\end{aligned} \tag{11.2-26}$$

$$\mathcal{P}(w_0)=\left(N_{xx}\frac{\partial w_0}{\partial x}+N_{xy}\frac{\partial w_0}{\partial y}\right)n_x+\left(N_{xy}\frac{\partial w_0}{\partial x}+N_{yy}\frac{\partial w_0}{\partial y}\right)n_y \tag{11.2-27}$$

The stress resultants are related to the strains by the relations

$$\begin{Bmatrix}\{N\}\\ \{M\}\\ \{P\}\end{Bmatrix}=\begin{bmatrix}[A] & [B] & [E]\\ [B] & [D] & [F]\\ [E] & [F] & [H]\end{bmatrix}\begin{Bmatrix}\{\varepsilon^{(0)}\}\\ \{\varepsilon^{(1)}\}\\ \{\varepsilon^{(3)}\}\end{Bmatrix} \tag{11.2-28}$$

$$\begin{Bmatrix}\{Q\}\\ \{R\}\end{Bmatrix}=\begin{bmatrix}[A] & [D]\\ [D] & [F]\end{bmatrix}\begin{Bmatrix}\{\gamma^{(0)}\}\\ \{\gamma^{(2)}\}\end{Bmatrix} \tag{11.2-29}$$

$$(A_{ij},B_{ij},D_{ij},E_{ij},F_{ij},H_{ij})=\sum_{k=1}^{N}\int_{z_k}^{z_{k+1}}\bar{Q}_{ij}^{(k)}\ (1,z,z^2,z^3,z^4,z^6)\ dz \tag{11.2-30a}$$

$$(A_{ij},D_{ij},F_{ij})=\sum_{k=1}^{N}\int_{z_k}^{z_{k+1}}\bar{Q}_{ij}^{(k)}\ (1,z^2,z^6)\ dz \tag{11.2-30b}$$

The stiffnesses in Eq. (30a) are defined for $i,j=1,2,6$ and those in Eq. (30b) are defined for $i,j=4,5$. Note that the matrices in (30a) are of the order 3×3 and those in (30b) are of order 2×2. The coefficients A_{ij},B_{ij}, and D_{ij} were given in terms of the layer stiffnesses $\bar{Q}_{ij}^{(k)}$ and layer coordinates z_{k+1} and z_k in Eqs. (5.3–38b) and (5.4-19). Additional stiffness coefficients are defined by

$$\begin{aligned}E_{ij}&=\frac{1}{4}\sum_{k=1}^{N}\bar{Q}_{ij}^{(k)}\left[(z_{k+1})^4-(z_k)^4\right]\\ F_{ij}&=\frac{1}{5}\sum_{k=1}^{N}\bar{Q}_{ij}^{(k)}\left[(z_{k+1})^5-(z_k)^5\right]\\ H_{ij}&=\frac{1}{7}\sum_{k=1}^{N}\bar{Q}_{ij}^{(k)}\left[(z_{k+1})^7-(z_k)^7\right]\end{aligned} \tag{11.2-31}$$

Note that the stiffnesses E_{ij}, F_{ij} and so on of the third-order theory involve fourth or higher powers of the thickness, and, therefore, they are expected to contribute little to thin laminate solutions. Even for moderately thick laminates the contribution can be small.

This completes the development of the Reddy third-order laminate theory. Note that the equations of motion of the first-order theory are obtained from the present third-order theory by setting $c_1 = 0$. However, the classical plate theory can be obtained from this theory only by replacing ϕ_α with $\varphi_\alpha + \partial w_0/\partial x_\alpha$, which is a differential, not an algebraic relationship. The displacement field in Eq. (4) contains, as special cases, the displacement fields used by other researchers to derive a third-order plate theory, as shown in Table 11.2–1. Therefore, the third-order plate theories reported in the literature, despite their different looks, are equivalent. Many of these theories were developed for only isotropic plates.

Table 11.2–1: Relationship of the displacements of other third-order theories to the one in Eq. (4).

Reference	Displacement Field†	Correspondence with Eq. (4) $\phi_z = \psi_z = \theta_z = 0$
Schmidt [16]	$u_\alpha = u_\alpha^0 - z u_{3,\alpha}^0 + \frac{3}{2} f(z)\varepsilon_\alpha$	$\varphi_\alpha = \frac{3}{2}\varepsilon_\alpha$
Krishna Murty [20]	$u_\alpha = u_\alpha^0 - z u_{3,\alpha}^0 - c_3 f(z)\theta_\alpha$	$\varphi_\alpha = -c_3\theta_\alpha$
Vlasov [5], Jemielita [14], Levinson [21], Reddy [25,26]	$u_\alpha = u_\alpha^0 + f(z)\psi_\alpha - c_1 z^3 u_{3,\alpha}^0$	$\varphi_\alpha = \psi_\alpha + u_{3,\alpha}^0$
Murthy [22]	$u_\alpha = u_\alpha^0 + \frac{z}{4}(5\beta_\alpha + u_{3,\alpha}^0)$ $-c_4 z^3(\beta_\alpha + u_{3,\alpha}^0)$	$\varphi_\alpha = \frac{5}{4}(\beta_\alpha + u_{3,\alpha}^0)$
Reddy [32]	$u_\alpha = u_\alpha^0 + f(z)\hat{\phi}_\alpha - z u_{3,\alpha}^b$ $-c_1 h z^3 u_{3,\alpha}^0$	$\varphi_\alpha = \hat{\phi}_\alpha + u_{3,\alpha}^s$ $w_0 = u_3^s + u_3^b$
Bhimaraddi [27], Reddy [49]	$u_\alpha = u_\alpha^0 - z u_{3,\alpha}^0 + f(z)\phi_\alpha$	$\varphi_\alpha = \phi_\alpha$

† $c_1 = \frac{4}{3h^2}$, $c_3 = \frac{3}{4h}$, $c_4 = \frac{5}{3h^2}$, $f(z) = z(1 - c_1 z^2)$.

11.3 Higher-Order Laminate Stiffness Characteristics

Since a detailed discussion of the laminate stiffnesses was presented in Section 5.5, a brief discussion is presented here for additional laminate stiffnesses introduced in the present third-order theory. We recall that the plane–stress–reduced stiffnesses Q_{ij} in the material coordinate system are given in terms of the engineering constants as

$$Q_{11} = \frac{E_1}{1-\nu_{12}\nu_{21}}\,, \quad Q_{12} = \frac{\nu_{12}E_2}{1-\nu_{12}\nu_{21}}\,, \quad Q_{22} = \frac{E_2}{1-\nu_{12}\nu_{21}}$$

$$Q_{66} = G_{12},\ Q_{44} = G_{23},\ Q_{55} = G_{13} \qquad (11.3-1)$$

where the subscript 1 refers to the fiber direction and 2 to the direction transverse to the fiber.

11.3–1 Single-Layer Plates

Single Isotropic Layer

For a single isotropic layer of material constants E and ν $[G = \frac{E}{2(1+\nu)}]$ and thickness h, the nonzero stiffnesses of Eqs. (11.2–30a,b) become

$$E_{ij} = \bar{B}_{ij} = \bar{E}_{ij} = 0, \quad F_{ij} = \frac{h^5}{80}Q_{ij}, \quad H_{ij} = \frac{h^7}{448}Q_{ij} \qquad (11.3-2a)$$

for $i,j = 1,2,6$, and

$$D_{ij} = \frac{h^3}{12}Q_{ij}, \quad F_{ij} = \frac{h^5}{80}Q_{ij} \quad (i,j = 4,5) \qquad (11.3-2b)$$

Hence, we have

$$F_{11} = \frac{Eh^5}{80(1-\nu^2)}\,, \quad F_{12} = \nu F_{11},\ F_{22} = F_{11},\ F_{66} = \frac{1-\nu}{2}F_{11} \qquad (11.3-3a)$$

$$H_{11} = \frac{Eh^7}{448(1-\nu^2)}\,, \quad H_{12} = \nu H_{11},\ H_{22} = H_{11},\ H_{66} = \frac{1-\nu}{2}H_{11}$$

$$D_{44} = D_{55} = \frac{Gh^3}{12} = \frac{1-\nu}{2}D_{11},\ F_{44} = F_{55} = \frac{Gh^5}{80} = \frac{1-\nu}{2}F_{11} \qquad (11.3-3b)$$

The plate constitutive equations (11.2–28) for the third-order theory become

$$\begin{Bmatrix} N_{xx} \\ N_{yy} \\ N_{xy} \end{Bmatrix} = \begin{bmatrix} A_{11} & \nu A_{11} & 0 \\ \nu A_{11} & A_{11} & 0 \\ 0 & 0 & \frac{1-\nu}{2}A_{11} \end{bmatrix} \begin{Bmatrix} \varepsilon_{xx}^{(0)} \\ \varepsilon_{yy}^{(0)} \\ \gamma_{xy}^{(0)} \end{Bmatrix} \qquad (11.3-4a)$$

$$\begin{Bmatrix} M_{xx} \\ M_{yy} \\ M_{xy} \end{Bmatrix} = \begin{bmatrix} D_{11} & \nu D_{11} & 0 \\ \nu D_{11} & D_{11} & 0 \\ 0 & 0 & \frac{1-\nu}{2} D_{11} \end{bmatrix} \begin{Bmatrix} \varepsilon_{xx}^{(1)} \\ \varepsilon_{yy}^{(1)} \\ \gamma_{xy}^{(1)} \end{Bmatrix} + \begin{bmatrix} F_{11} & \nu F_{11} & 0 \\ \nu F_{11} & F_{11} & 0 \\ 0 & 0 & \frac{1-\nu}{2} F_{11} \end{bmatrix} \begin{Bmatrix} \varepsilon_{xx}^{(3)} \\ \varepsilon_{yy}^{(3)} \\ \gamma_{xy}^{(3)} \end{Bmatrix} \tag{11.3-4b}$$

$$\begin{Bmatrix} P_{xx} \\ P_{yy} \\ P_{xy} \end{Bmatrix} = \begin{bmatrix} F_{11} & \nu F_{11} & 0 \\ \nu F_{11} & F_{11} & 0 \\ 0 & 0 & \frac{1-\nu}{2} F_{11} \end{bmatrix} \begin{Bmatrix} \varepsilon_{xx}^{(1)} \\ \varepsilon_{yy}^{(1)} \\ \gamma_{xy}^{(1)} \end{Bmatrix} + \begin{bmatrix} H_{11} & \nu H_{11} & 0 \\ \nu H_{11} & H_{11} & 0 \\ 0 & 0 & \frac{1-\nu}{2} H_{11} \end{bmatrix} \begin{Bmatrix} \varepsilon_{xx}^{(3)} \\ \varepsilon_{yy}^{(3)} \\ \gamma_{xy}^{(3)} \end{Bmatrix} \tag{11.3-5}$$

$$\begin{Bmatrix} Q_y \\ Q_x \end{Bmatrix} = \frac{1-\nu}{2} \begin{bmatrix} A_{11} & 0 \\ 0 & A_{11} \end{bmatrix} \begin{Bmatrix} \gamma_{yz}^{(0)} \\ \gamma_{xz}^{(0)} \end{Bmatrix} + \frac{1-\nu}{2} \begin{bmatrix} D_{11} & 0 \\ 0 & D_{11} \end{bmatrix} \begin{Bmatrix} \gamma_{yz}^{(2)} \\ \gamma_{xz}^{(2)} \end{Bmatrix} \tag{11.3-6a}$$

$$\begin{Bmatrix} R_y \\ R_x \end{Bmatrix} = \frac{1-\nu}{2} \begin{bmatrix} D_{11} & 0 \\ 0 & D_{11} \end{bmatrix} \begin{Bmatrix} \gamma_{yz}^{(0)} \\ \gamma_{xz}^{(0)} \end{Bmatrix} + \frac{1-\nu}{2} \begin{bmatrix} F_{11} & 0 \\ 0 & F_{11} \end{bmatrix} \begin{Bmatrix} \gamma_{yz}^{(2)} \\ \gamma_{xz}^{(2)} \end{Bmatrix} \tag{11.3-6b}$$

Single Specially Orthotropic Layer

For a single specially orthotropic layer, the stiffnesses can be expressed in terms of the Q_{ij} and thickness h. The nonzero stiffnesses of Eqs. (11.2–30a,b) become ($E_{ij} = G_{ij} = 0; \bar{B}_{ij} = \bar{E}_{ij} = 0$)

$$D_{11} = \frac{Q_{11}h^3}{12}\,,\quad D_{12} = \frac{Q_{12}h^3}{12}\,,\; D_{22} = \frac{Q_{22}h^3}{12}\,,\; D_{66} = \frac{Q_{66}h^3}{12} \tag{11.3-7a}$$

$$F_{11} = \frac{Q_{11}h^5}{80}\,,\quad F_{12} = \frac{Q_{12}h^5}{80}\,,\; F_{22} = \frac{Q_{22}h^5}{80}\,,\; F_{66} = \frac{Q_{66}h^5}{80} \tag{11.3-7b}$$

$$H_{11} = \frac{Q_{11}h^7}{448}\,,\quad H_{12} = \frac{Q_{12}h^7}{448}\,,\; H_{22} = \frac{Q_{22}h^7}{448}\,,\; H_{66} = \frac{Q_{66}h^7}{448} \tag{11.3-7c}$$

The plate constitutive equations for the higher-order stress resultants become (and similar equations hold for N's and M's)

$$\begin{Bmatrix} P_{xx} \\ P_{yy} \\ P_{xy} \end{Bmatrix} = \frac{h^5}{80} \begin{bmatrix} Q_{11} & Q_{12} & 0 \\ Q_{12} & Q_{22} & 0 \\ 0 & 0 & Q_{66} \end{bmatrix} \begin{Bmatrix} \varepsilon_{xx}^{(1)} \\ \varepsilon_{yy}^{(1)} \\ \gamma_{xy}^{(1)} \end{Bmatrix} + \frac{h^7}{448} \begin{bmatrix} Q_{11} & Q_{12} & 0 \\ Q_{12} & Q_{22} & 0 \\ 0 & 0 & Q_{66} \end{bmatrix} \begin{Bmatrix} \varepsilon_{xx}^{(3)} \\ \varepsilon_{yy}^{(3)} \\ \gamma_{xy}^{(3)} \end{Bmatrix} \tag{11.3-8}$$

$$\begin{Bmatrix} R_y \\ R_x \end{Bmatrix} = \frac{h^3}{12} \begin{bmatrix} Q_{44} & 0 \\ 0 & Q_{55} \end{bmatrix} \begin{Bmatrix} \gamma_{yz}^{(0)} \\ \gamma_{xz}^{(0)} \end{Bmatrix} + \frac{h^5}{80} \begin{bmatrix} Q_{44} & 0 \\ 0 & Q_{55} \end{bmatrix} \begin{Bmatrix} \gamma_{yz}^{(2)} \\ \gamma_{xz}^{(2)} \end{Bmatrix} \tag{11.3-9}$$

Single Generally Orthotropic Layer

For a single generally orthotropic layer (*i.e.*, the principal material coordinates do not coincide with those of the plate), the stiffnesses are expressed in terms of the transformed coefficients $\bar{Q}_{ij}$ and thickness h. The plate constitutive equations are

$$\begin{Bmatrix} P_{xx} \\ P_{yy} \\ P_{xy} \end{Bmatrix} = \begin{bmatrix} F_{11} & F_{12} & F_{16} \\ F_{12} & F_{22} & F_{26} \\ F_{16} & F_{26} & F_{66} \end{bmatrix} \begin{Bmatrix} \varepsilon_{xx}^{(1)} \\ \varepsilon_{yy}^{(1)} \\ \gamma_{xy}^{(1)} \end{Bmatrix} + \begin{bmatrix} H_{11} & H_{12} & H_{16} \\ H_{12} & H_{22} & H_{26} \\ H_{16} & H_{26} & H_{66} \end{bmatrix} \begin{Bmatrix} \varepsilon_{xx}^{(3)} \\ \varepsilon_{yy}^{(3)} \\ \gamma_{xy}^{(3)} \end{Bmatrix} \tag{11.3 - 10}$$

$$\begin{Bmatrix} R_{y} \\ R_{x} \end{Bmatrix} = \begin{bmatrix} D_{44} & D_{45} \\ D_{45} & D_{55} \end{bmatrix} \begin{Bmatrix} \gamma_{yz}^{(0)} \\ \gamma_{xz}^{(0)} \end{Bmatrix} + \begin{bmatrix} F_{44} & F_{45} \\ F_{45} & F_{55} \end{bmatrix} \begin{Bmatrix} \gamma_{yz}^{(2)} \\ \gamma_{xz}^{(2)} \end{Bmatrix} \tag{11.3 - 11}$$

The higher-order thermal stress resultants for this case are given by

$$\begin{Bmatrix} P_{xx}^{T} \\ P_{yy}^{T} \\ P_{xy}^{T} \end{Bmatrix} = \sum_{k=1}^{L} \int_{z_k}^{z_{k+1}} \begin{bmatrix} \bar{Q}_{11}^{k} & \bar{Q}_{12}^{k} & \bar{Q}_{16}^{k} \\ \bar{Q}_{12}^{k} & \bar{Q}_{22}^{k} & \bar{Q}_{26}^{k} \\ \bar{Q}_{16}^{k} & \bar{Q}_{26}^{k} & \bar{Q}_{66}^{k} \end{bmatrix}^{(k)} \begin{Bmatrix} \alpha_{xx} \\ \alpha_{yy} \\ 2\alpha_{xy} \end{Bmatrix}^{(k)} \Delta T z^{3} dz \tag{11.3 - 12}$$

Single Anisotropic Layer

For a single anisotropic layer, the stiffnesses are expressed in terms of the coefficients C_{ij} and thickness h. The nonzero higher-order stiffnesses are ($B_{ij} = 0$)

$$F_{ij} = \frac{C_{ij}h^{5}}{80}\,, \quad H_{ij} = \frac{C_{ij}h^{7}}{448} \tag{11.3 - 13}$$

for $i, j = 1, 2, 3, 4, 5$, and 6 [see Eq. (2.4–3a)]. The plate constitutive equations are the same as in Eqs. (10) and (11), except that the plate stiffnesses are given by Eq. (13).

11.3–2 Symmetric Laminates

The force and moment resultants for a symmetric laminate, in general, have the same form as the generally orthotropic single-layer plates [see Eqs. (10) and (11)]. For certain special cases of symmetric laminates, the relations between strains and resultants can be further simplified, as explained next.

Symmetric Laminates with Multiple Isotropic Layers

When isotropic layers of possibly different material properties and thicknesses are arranged symmetrically from both a geometric and a material property standpoint, the resulting laminate will have the following laminate constitutive equations for the third-order theories:

$$\left\{\begin{array}{c} P_{xx} \\ P_{yy} \\ P_{xy} \end{array}\right\} = \begin{bmatrix} F_{11} & F_{12} & 0 \\ F_{12} & F_{11} & 0 \\ 0 & 0 & F_{66} \end{bmatrix} \left\{\begin{array}{c} \varepsilon_{xx}^{(1)} \\ \varepsilon_{yy}^{(1)} \\ \gamma_{xy}^{(1)} \end{array}\right\} + \begin{bmatrix} H_{11} & H_{12} & 0 \\ H_{12} & H_{11} & 0 \\ 0 & 0 & H_{66} \end{bmatrix} \left\{\begin{array}{c} \varepsilon_{xx}^{(3)} \\ \varepsilon_{yy}^{(3)} \\ \gamma_{xy}^{(3)} \end{array}\right\} \tag{11.3 – 14}$$

$$\left\{\begin{array}{c} R_y \\ R_x \end{array}\right\} = \begin{bmatrix} D_{44} & 0 \\ 0 & D_{55} \end{bmatrix} \left\{\begin{array}{c} \gamma_{yz}^{(0)} \\ \gamma_{xz}^{(0)} \end{array}\right\} + \begin{bmatrix} F_{44} & 0 \\ 0 & F_{55} \end{bmatrix} \left\{\begin{array}{c} \gamma_{yz}^{(2)} \\ \gamma_{xz}^{(2)} \end{array}\right\} \tag{11.3 – 15}$$

where the laminate stiffnesses F_{ij} and H_{ij} are defined by Eqs. (5.5–24) with

$$\bar{Q}_{11}^{(k)} = \bar{Q}_{22}^{(k)} = \frac{E^k}{1-\nu_k^2}\,, \quad \bar{Q}_{16}^{(k)} = \bar{Q}_{26}^{(k)} = 0 \tag{11.3 – 16a}$$

$$\bar{Q}_{12}^{(k)} = \frac{\nu_k E^k}{1-\nu_k^2}\,, \quad \bar{Q}_{44}^{(k)} = \bar{Q}_{55}^{(k)} = \bar{Q}_{66}^{(k)} = \frac{E^k}{2(1+\nu_k)} \tag{11.3 – 16b}$$

The thermal stress resultants for this case are given by

$$\left\{\begin{array}{c} P_{xx}^T \\ P_{yy}^T \end{array}\right\} = \sum_{k=1}^{L} \int_{z_k}^{z_{k+1}} \begin{bmatrix} \bar{Q}_{11} & \bar{Q}_{12} \\ \bar{Q}_{12} & \bar{Q}_{22} \end{bmatrix}^{(k)} \left\{\begin{array}{c} \alpha_{xx} \\ \alpha_{yy} \end{array}\right\}^{(k)} \Delta T z^3 \, dz \tag{11.3 – 17}$$

Symmetric Laminates with Multiple Specially Orthotropic Layers

A laminate of multiple specially orthotropic layers that are symmetrically disposed, both from a material and geometric properties standpoint, about the midplane of the laminate do not exhibit coupling between bending and extension. The laminate constitutive equations are again given by Eqs. (13)–(16), where the laminate stiffnesses F_{ij} and H_{ij} are defined by Eqs. (11.2–24) with

$$\bar{Q}_{11}^{(k)} = \frac{E_1^k}{1-\nu_{12}^k \nu_{21}^k}\,, \quad \bar{Q}_{12}^{(k)} = \frac{\nu_{21}^k E_1^k}{1-\nu_{12}^k \nu_{21}^k}\,, \quad \bar{Q}_{22}^{(k)} = \frac{E_2^k}{1-\nu_{12}^k \nu_{21}^k}$$

$$\bar{Q}_{16}^{(k)} = 0,\ \bar{Q}_{26}^{(k)} = 0,\ \bar{Q}_{66}^{(k)} = G_{12}^k,\ \bar{Q}_{44}^{(k)} = G_{23}^k,\ \bar{Q}_{55}^{(k)} = G_{13}^k \tag{11.3 – 18}$$

The thermal stress resultants have the same form as those given in Eq. (17).

11.3–3 Antisymmetric Laminates

Due to the antisymmetry of the lamination scheme but symmetry of the thicknesses of each pair of layers, this class of antisymmetric laminates have the feature that $F_{16} = F_{26} = H_{16} = H_{26} = 0$. The coupling stiffnesses E_{ij} and G_{ij} are not zero. The relations between the stress resultants and the strains are

$$\left\{\begin{matrix} P_{xx} \\ P_{yy} \\ P_{xy} \end{matrix}\right\} = \begin{bmatrix} E_{11} & E_{12} & E_{16} \\ E_{12} & E_{22} & E_{26} \\ E_{16} & E_{26} & E_{66} \end{bmatrix} \left\{\begin{matrix} \varepsilon_{xx}^{(0)} \\ \varepsilon_{yy}^{(0)} \\ \varepsilon_{xy}^{(0)} \end{matrix}\right\} + \begin{bmatrix} F_{11} & F_{12} & 0 \\ F_{12} & F_{22} & 0 \\ 0 & 0 & F_{66} \end{bmatrix} \left\{\begin{matrix} \varepsilon_{xx}^{(1)} \\ \varepsilon_{yy}^{(1)} \\ \varepsilon_{xy}^{(1)} \end{matrix}\right\} + \begin{bmatrix} H_{11} & H_{12} & H_{16} \\ H_{12} & H_{22} & H_{26} \\ H_{16} & H_{26} & H_{66} \end{bmatrix} \left\{\begin{matrix} \varepsilon_{xx}^{(3)} \\ \varepsilon_{yy}^{(3)} \\ \varepsilon_{xy}^{(3)} \end{matrix}\right\} \tag{11.3 – 19}$$

$$\left\{\begin{matrix} R_y \\ R_x \end{matrix}\right\} = \begin{bmatrix} D_{44} & 0 \\ 0 & D_{55} \end{bmatrix} \left\{\begin{matrix} \varepsilon_{yz}^{(0)} \\ \varepsilon_{xz}^{(0)} \end{matrix}\right\} + \begin{bmatrix} F_{44} & 0 \\ 0 & F_{55} \end{bmatrix} \left\{\begin{matrix} \varepsilon_{yz}^{(2)} \\ \varepsilon_{xz}^{(2)} \end{matrix}\right\} \tag{11.3 – 20}$$

Antisymmetric Cross–Ply Laminates

For antisymmetric cross–ply laminates the coupling stiffnesses have the following properties:

$$E_{22} = -E_{11}, \quad \text{and all other} \quad E_{ij} = D_{45} = F_{45} = 0 \tag{11.3 – 21}$$

For regular antisymmetric cross–ply laminates, the coupling coefficients E_{11} can be shown to approach zero as the number of layers increases.

Antisymmetric Angle–Ply Laminates

For antisymmetric angle–ply laminates the stiffnesses can be simplified as

$$D_{45} = F_{45} = F_{16} = F_{26} = H_{16} = H_{26} = E_{11} = E_{22} = E_{12} = E_{66} = 0 \tag{11.3 – 22}$$

For a fixed laminate thickness, the stiffnesses E_{16} and E_{26} go to zero as the number of layers in the laminate increases.

This completes the development of the third-order theory of Reddy. In the next section, we develop the Navier solutions of antisymmetric angle-ply and cross-ply laminates. The Lévy solutions are presented in Section 11.5, and finite element models are discussed in Section 11.6.

11.4 The Navier Solutions

11.4–1 Preliminary Comments

The equations of motion of the third-order theory of Reddy presented in Eqs. (11.2–16)–(11.2–20) are very similar in form to the first-order shear deformation theory. In fact, it is possible to develop the Navier solutions of simply supported antisymmetric cross-ply and angle-ply laminates using the third-order theory (see References 25, 26, and 29). For antisymmetric cross-ply laminates the following stiffnesses are zero:

$$A_{16} = A_{26} = A_{45} = B_{16} = B_{26} = D_{16} = D_{26} = I_1 = 0$$
$$E_{16} = E_{26} = F_{16} = F_{26} = H_{16} = H_{26} = D_{45} = F_{45} = I_3 = I_5 = I_7 = 0 \quad (11.4-1)$$

For antisymmetric angle-ply laminates the following stiffnesses are zero:

$$A_{16} = A_{26} = A_{45} = B_{11} = B_{12} = B_{22} = B_{66} = D_{16} = D_{26} = I_1 = 0$$
$$B_{11} = B_{12} = B_{22} = B_{66} = F_{16} = F_{26} = H_{16} = H_{26} = D_{45} = F_{45} = I_3 = I_5 = I_7 = 0 \quad (11.4-2)$$

The SS–1 boundary conditions for the third-order shear deformation plate theory are (see Figure 11.4–1):

$$u_0(x,0,t) = 0, \quad \phi_x(x,0,t) = 0, \quad u_0(x,b,t) = 0, \quad \phi_x(x,b,t) = 0$$
$$v_0(0,y,t) = 0, \quad \phi_y(0,y,t) = 0, \quad v_0(a,y,t) = 0, \quad \phi_y(a,y,t) = 0$$
$$w_0(x,0,t) = 0, \quad w_0(x,b,t) = 0, \quad w_0(0,y,t) = 0, \quad w_0(a,y,t) = 0 \quad (11.4-3a)$$
$$N_{xx}(0,y,t) = 0, \quad N_{xx}(a,y,t) = 0, \quad N_{yy}(x,0,t) = 0, \quad N_{yy}(x,b,t) = 0$$
$$\bar{M}_{xx}(0,y,t) = 0, \quad \bar{M}_{xx}(a,y,t) = 0, \quad \bar{M}_{yy}(x,0,t) = 0, \quad \bar{M}_{yy}(x,b,t) = 0 \quad (11.4-3b)$$

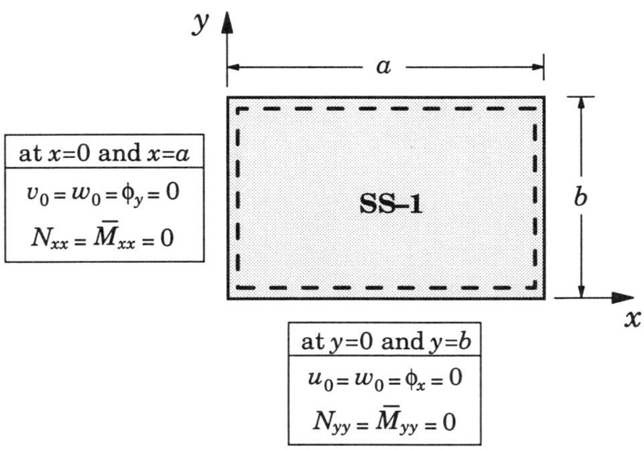

Figure 11.4–1: Simply supported (SS–1) boundary conditions for antisymmetric cross-ply laminates using the third-order shear deformation theory of Reddy.

The SS–2 boundary conditions for the third-order shear deformation plate theory are (see Figure 11.4–2)

$$u_0(0,y,t)=0,\quad u_0(a,y,t)=0,\quad v_0(x,0,t)=0,\quad v_0(x,b,t)=0$$

$$\phi_x(x,0,t)=0,\quad \phi_x(x,b,t)=0,\quad \phi_y(0,y,t)=0,\quad \phi_y(a,y,t)=0$$

$$w_0(x,0,t)=0,\quad w_0(x,b,t)=0,\quad w_0(0,y,t)=0,\quad w_0(a,y,t)=0 \qquad (11.4-4a)$$

$$N_{xy}(0,y,t)=0,\quad N_{xy}(a,y,t)=0,\quad N_{xy}(x,0,t)=0,\quad N_{xy}(x,b,t)=0$$

$$\bar{M}_{xx}(0,y,t)=0,\quad \bar{M}_{xx}(a,y,t)=0,\quad \bar{M}_{yy}(x,0,t)=0,\quad \bar{M}_{yy}(x,b,t)=0 \quad (11.4-4b)$$

In the following sections, we present the Navier solutions of cross-ply laminates for the SS–1 boundary conditions and antisymmetric angle-ply laminates for the SS–2 boundary conditions.

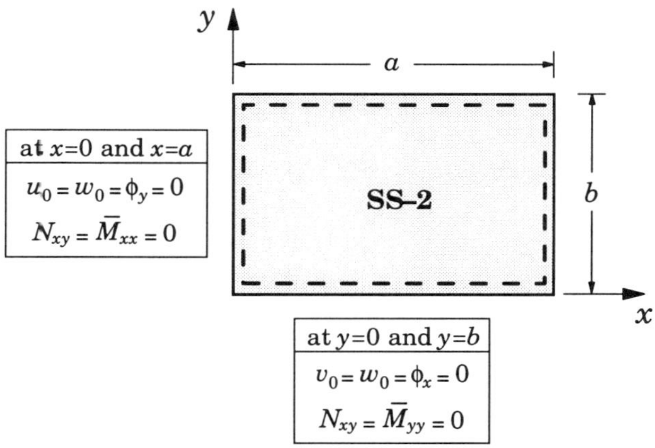

Figure 11.4–2: The simply supported (SS–2) boundary conditions for antisymmetric angle-ply laminates using the third-order shear deformation theory of Reddy.

11.4–2 Antisymmetric Cross-Ply Laminates

The boundary conditions in (3a,b) are satisfied by the following expansions:

$$u_0(x,y,t)=\sum_{n=1}^{\infty}\sum_{m=1}^{\infty} U_{mn}(t)\cos\alpha x\sin\beta y \qquad (11.4-5a)$$

$$v_0(x,y,t)=\sum_{n=1}^{\infty}\sum_{m=1}^{\infty} V_{mn}(t)\sin\alpha x\cos\beta y \qquad (11.4-5b)$$

$$w_0(x,y,t)=\sum_{n=1}^{\infty}\sum_{m=1}^{\infty} W_{mn}(t)\sin\alpha x\sin\beta y \qquad (11.4-5c)$$

$$\phi_x(x,y,t) = \sum_{n=1}^{\infty}\sum_{m=1}^{\infty} X_{mn}(t)\cos\alpha x \sin\beta y \tag{11.4-5d}$$

$$\phi_y(x,y,t) = \sum_{n=1}^{\infty}\sum_{m=1}^{\infty} Y_{mn}(t)\sin\alpha x \cos\beta y \tag{11.4-5e}$$

The transverse load q is also expanded in double Fourier sine series

$$q(x,y,t) = \sum_{n=1}^{\infty}\sum_{m=1}^{\infty} Q_{mn}(t)\sin\alpha x \sin\beta y \tag{11.4-6a}$$

$$Q_{mn}(z,t) = \frac{4}{ab}\int_0^a \int_0^b q(x,y,t)\ \sin\frac{m\pi x}{a}\ \sin\frac{n\pi y}{b}\ dxdy \tag{11.4-6b}$$

Substitution of Eqs. (5) and (6) into Eqs. (11.2–16)–(11.2–20) will show that the Navier solution exists only if the laminate stiffnesses are such that the conditions in Eq. (1) hold. The coefficients $(U_{mn}, V_{mn}, W_{mn}, X_{mn}, Y_{mn})$ of the Navier solution of cross-ply laminates are governed by

$$[\hat{S}]\{\Delta\} + [M]\{\ddot{\Delta}\} = \{F\} \tag{11.4-7a}$$

$$\{\Delta\} = \left\{\begin{matrix} \ddot{U}_{mn} \\ \ddot{V}_{mn} \\ \ddot{W}_{mn} \\ \ddot{X}_{mn} \\ \ddot{Y}_{mn} \end{matrix}\right\}, \quad \{F\} = \left\{\begin{matrix} 0 \\ 0 \\ Q_{mn} \\ 0 \\ 0 \end{matrix}\right\} - \left\{\begin{matrix} \alpha N^1_{mn} \\ \beta N^2_{mn} \\ 0 \\ \alpha \hat{M}^1_{mn} \\ \beta \hat{M}^2_{mn} \end{matrix}\right\} \tag{11.4-7b}$$

where $\hat{s}_{ij}$ and $\hat{m}_{ij}$ are defined by

$$\begin{aligned}
&\hat{s}_{11} = A_{11}\alpha^2 + A_{66}\beta^2, \quad \hat{s}_{12} = (A_{12} + A_{66})\alpha\beta \\
&\hat{s}_{13} = -c_1\left[E_{11}\alpha^2 + (E_{12} + 2E_{66})\beta^2\right]\alpha \\
&\hat{s}_{14} = \hat{B}_{11}\alpha^2 + \hat{B}_{66}\beta^2, \quad \hat{s}_{15} = (\hat{B}_{12} + \hat{B}_{66})\alpha\beta \\
&\hat{s}_{22} = A_{66}\alpha^2 + A_{22}\beta^2, \quad \hat{s}_{24} = \hat{s}_{15} \\
&\hat{s}_{23} = -c_1\left[E_{22}\beta^2 + (E_{12} + 2E_{66})\alpha^2\right]\beta, \quad \hat{s}_{25} = \hat{B}_{66}\alpha^2 + \hat{B}_{22}\beta^2 \\
&\hat{s}_{33} = \bar{A}_{55}\alpha^2 + \bar{A}_{44}\beta^2 + c_1^2\left[H_{11}\alpha^4 + 2(H_{12} + 2H_{66})\alpha^2\beta^2 + H_{22}\beta^4\right] \\
&\hat{s}_{34} = \bar{A}_{55}\alpha - c_1\left[\hat{F}_{11}\alpha^3 + (\hat{F}_{12} + 2\hat{F}_{66})\alpha\beta^2\right] \\
&\hat{s}_{35} = \bar{A}_{44}\beta - c_1\left[\hat{F}_{22}\beta^3 + (\hat{F}_{12} + 2\hat{F}_{66})\alpha^2\beta\right] \\
&\hat{s}_{44} = \bar{A}_{55} + \bar{D}_{11}\alpha^2 + \bar{D}_{66}\beta^2, \quad \hat{s}_{45} = (\bar{D}_{12} + \bar{D}_{66})\alpha\beta \\
&\hat{s}_{55} = \bar{A}_{44} + \bar{D}_{66}\alpha^2 + \bar{D}_{22}\beta^2, \quad \tilde{s}_{33} = \hat{N}_{xx}\alpha^2 + \hat{N}_{yy}\beta^2 \qquad (11.4-8a) \\
&\hat{m}_{11} = I_0, \quad \hat{m}_{22} = I_0, \quad \hat{m}_{33} = I_0 + c_1^2 I_6\left(\alpha^2 + \beta^2\right), \quad \hat{m}_{34} = -c_1 J_4\alpha
\end{aligned}$$

$$\hat{m}_{35} = -c_1 J_4 \beta, \quad \hat{m}_{44} = K_2, \quad \hat{m}_{55} = K_2 \tag{11.4-8b}$$

$$\hat{A}_{ij} = A_{ij} - c_1 D_{ij}, \quad \hat{B}_{ij} = B_{ij} - c_1 E_{ij}, \quad \hat{D}_{ij} = D_{ij} - c_1 F_{ij} \ (i,j = 1,2,6)$$
$$\hat{F}_{ij} = F_{ij} - c_1 H_{ij}, \quad \bar{A}_{ij} = \hat{A}_{ij} - c_1 \hat{D}_{ij} = A_{ij} - 2c_1 D_{ij} + c_1^2 F_{ij} \ (i,j = 1,2,6)$$
$$\bar{D}_{ij} = \hat{D}_{ij} - c_1 \hat{F}_{ij} = D_{ij} - 2c_1 F_{ij} + c_1^2 H_{ij} \ (i,j = 1,2,6)$$
$$\bar{A}_{ij} = \hat{A}_{ij} - c_2 \hat{D}_{ij} = A_{ij} - 2c_2 D_{ij} + c_2^2 F_{ij} \ (i,j = 4,5) \tag{11.4-9}$$

The thermal resultants are defined by [see Eqs. (8.3–11) through (8.3–13)]

$$\begin{Bmatrix} N_{xx}^T \\ N_{yy}^T \\ N_{xy}^T \end{Bmatrix} = \sum_{n=1}^{\infty} \sum_{m=1}^{\infty} \begin{Bmatrix} N_{mn}^1(t) \\ N_{mn}^2(t) \\ N_{mn}^6(t) \end{Bmatrix} \sin \alpha x \ \sin \beta y \tag{11.4-10a}$$

$$\begin{Bmatrix} M_{xx}^T \\ M_{yy}^T \\ M_{xy}^T \end{Bmatrix} = \sum_{n=1}^{\infty} \sum_{m=1}^{\infty} \begin{Bmatrix} M_{mn}^1(t) \\ M_{mn}^2(t) \\ M_{mn}^6(t) \end{Bmatrix} \sin \alpha x \ \sin \beta y \tag{11.4-10b}$$

$$\begin{Bmatrix} P_{xx}^T \\ P_{yy}^T \\ P_{xy}^T \end{Bmatrix} = \sum_{n=1}^{\infty} \sum_{m=1}^{\infty} \begin{Bmatrix} P_{mn}^1(t) \\ P_{mn}^2(t) \\ P_{mn}^6(t) \end{Bmatrix} \sin \alpha x \ \sin \beta y \tag{11.4-10c}$$

where

$$\{N_{mn}(t)\} = \sum_{k=1}^{N} \int_{z_k}^{z_{k+1}} [\bar{Q}]^{(k)} \{\bar{\alpha}\}^{(k)} \ T_{mn}(z,t) \ dz \tag{11.4-11a}$$

$$\{M_{mn}(t)\} = \sum_{k=1}^{N} \int_{z_k}^{z_{k+1}} [\bar{Q}]^{(k)} \{\bar{\alpha}\}^{(k)} \ T_{mn}(z,t) \ z \ dz \tag{11.4-11b}$$

$$\{P_{mn}(t)\} = \sum_{k=1}^{N} \int_{z_k}^{z_{k+1}} [\bar{Q}]^{(k)} \{\bar{\alpha}\}^{(k)} \ T_{mn}(z,t) \ z^3 \ dz \tag{11.4-11c}$$

and

$$\Delta T(x,y,z,t) = \sum_{m=1}^{\infty} \sum_{n=1}^{\infty} \left(T_{mn}^0 + z T_{mn}^1 \right) \sin \alpha x \sin \beta y \tag{11.4-12a}$$

$$\left(T_{mn}^0, T_{mn}^1 \right) = \frac{4}{ab} \int_0^a \int_0^b (T_0, T_1) \ \sin \frac{m\pi x}{a} \ \sin \frac{n\pi y}{b} \ dxdy \tag{11.4-12b}$$

$$\{\hat{M}_{mn}\} = \{M_{mn}\} - c_1 \{P_{mn}\} \tag{11.4-13}$$

The ordinary differential equations (7) in time can be solved for transient response using the Newmark integration procedure described in Chapter 9. Equation (7) can be specialized to static bending analysis, buckling, and natural vibration.

The in-plane stresses in each layer can be computed from the equations [see Eqs. (8.3–29) and (8.3–30)]

$$\begin{Bmatrix} \sigma_{xx} \\ \sigma_{yy} \\ \sigma_{xy} \end{Bmatrix}^{(k)} = \begin{bmatrix} \bar{Q}_{11} & \bar{Q}_{12} & 0 \\ \bar{Q}_{12} & \bar{Q}_{22} & 0 \\ 0 & 0 & \bar{Q}_{66} \end{bmatrix}^{(k)} \left(\begin{Bmatrix} \varepsilon_{xx} \\ \varepsilon_{yy} \\ \gamma_{xy} \end{Bmatrix} - \begin{Bmatrix} \alpha_{xx} \\ \alpha_{yy} \\ 2\alpha_{xy} \end{Bmatrix} \Delta T \right) \tag{11.4 – 14}$$

where

$$\begin{Bmatrix} \varepsilon_{xx} \\ \varepsilon_{yy} \\ \gamma_{xy} \end{Bmatrix} = \begin{Bmatrix} \varepsilon_{xx}^{(0)} \\ \varepsilon_{yy}^{(0)} \\ \gamma_{xy}^{(0)} \end{Bmatrix} + z \begin{Bmatrix} \varepsilon_{xx}^{(1)} \\ \varepsilon_{yy}^{(1)} \\ \gamma_{xy}^{(1)} \end{Bmatrix} + z^3 \begin{Bmatrix} \varepsilon_{xx}^{(1)} \\ \varepsilon_{yy}^{(1)} \\ \gamma_{xy}^{(1)} \end{Bmatrix}$$

$$= \sum_{m=1}^{\infty} \sum_{n=1}^{\infty} \begin{Bmatrix} (R_{mn}^{xx} + zS_{mn}^{xx} + c_1 z^3 T_{mn}^{xx}) \sin \alpha x \sin \beta y \\ (R_{mn}^{yy} + zS_{mn}^{yy} + c_1 z^3 T_{mn}^{yy}) \sin \alpha x \sin \beta y \\ (R_{mn}^{xy} + zS_{mn}^{xy} + c_1 z^3 T_{mn}^{xy}) \cos \alpha x \cos \beta y \end{Bmatrix} \tag{11.4 – 15a}$$

$$\begin{Bmatrix} R_{mn}^{xx} \\ R_{mn}^{yy} \\ R_{mn}^{xy} \end{Bmatrix} = \begin{Bmatrix} -\alpha U_{mn} \\ -\beta V_{mn} \\ \beta U_{mn} + \alpha V_{mn} \end{Bmatrix}, \quad \begin{Bmatrix} S_{mn}^{xx} \\ S_{mn}^{yy} \\ S_{mn}^{xy} \end{Bmatrix} = \begin{Bmatrix} -\alpha X_{mn} \\ -\beta Y_{mn} \\ \beta X_{mn} + \alpha Y_{mn} \end{Bmatrix}$$

$$\begin{Bmatrix} T_{mn}^{xx} \\ T_{mn}^{yy} \\ T_{mn}^{xy} \end{Bmatrix} = \begin{Bmatrix} \alpha X_{mn} + \alpha^2 W_{mn} \\ \beta Y_{mn} + \beta^2 W_{mn} \\ -(\beta X_{mn} + \alpha Y_{mn} + 2\alpha\beta W_{mn}) \end{Bmatrix} \tag{11.4 – 15b}$$

The transverse shear stresses from the constitutive equations are given by

$$\begin{Bmatrix} \sigma_{yz} \\ \sigma_{xz} \end{Bmatrix}^{(k)} = \sum_{m=1}^{\infty} \sum_{n=1}^{\infty} \begin{bmatrix} \bar{Q}_{44} & 0 \\ 0 & \bar{Q}_{55} \end{bmatrix}^{(k)} \left(\begin{Bmatrix} \gamma_{yz}^{(0)} \\ \gamma_{xz}^{(0)} \end{Bmatrix} + z^2 \begin{Bmatrix} \gamma_{yz}^{(2)} \\ \gamma_{xz}^{(2)} \end{Bmatrix} \right)$$

$$= (1 - c_2 z^2) \sum_{m=1}^{\infty} \sum_{n=1}^{\infty} \begin{bmatrix} \bar{Q}_{44} & 0 \\ 0 & \bar{Q}_{55} \end{bmatrix}^{(k)} \begin{Bmatrix} (Y_{mn} + \beta W_{mn}) \sin \alpha x \cos \beta y \\ (X_{mn} + \alpha W_{mn}) \cos \alpha x \sin \beta y \end{Bmatrix} \tag{11.4 – 16}$$

where $c_2 = 4/h^2$. Note that the transverse shear stresses are layerwise quadratic through the thickness.

The transverse shear stresses can also be determined using the equilibrium equations of 3–D elasticity. In the absence of thermal effects they are given by

$$\sigma_{xz}^{(k)}(x, y, z) = \sum_{m=1}^{\infty} \sum_{n=1}^{\infty} \left[(z - z_k)\mathcal{A}_{mn}^{(k)} + \frac{1}{2}(z^2 - z_k^2)\mathcal{B}_{mn}^{(k)} \right.$$
$$\left. + \frac{c_1}{4}(z^4 - z_k^4)\mathcal{E}_{mn}^{(k)} \right] \cos \alpha x \sin \beta y + \sigma_{xz}^{(k-1)}(x, y, z_k) \tag{11.4 – 17a}$$

$$\sigma_{yz}^{(k)}(x,y,z) = \sum_{m=1}^{\infty}\sum_{n=1}^{\infty}\left[(z-z_k)\mathcal{C}_{mn}^{(k)} + \frac{1}{2}(z^2-z_k^2)\mathcal{D}_{mn}^{(k)} + \frac{c_1}{4}(z^4-z_k^4)\mathcal{F}_{mn}^{(k)}\right]\sin\alpha x\cos\beta y + \sigma_{yz}^{(k-1)}(x,y,z_k) \quad (11.4-17b)$$

where

$$\begin{aligned}
\mathcal{A}_{mn}^{(k)} &= \left[\left(\alpha^2\bar{Q}_{11}^{(k)} + \beta^2\bar{Q}_{66}^{(k)}\right)U_{mn} + \alpha\beta\left(\bar{Q}_{12}^{(k)} + \bar{Q}_{66}^{(k)}\right)V_{mn}\right]\\
\mathcal{B}_{mn}^{(k)} &= \left[\left(\alpha^2\bar{Q}_{11}^{(k)} + \beta^2\bar{Q}_{66}^{(k)}\right)X_{mn} + \alpha\beta\left(\bar{Q}_{12}^{(k)} + \bar{Q}_{66}^{(k)}\right)Y_{mn}\right]\\
\mathcal{C}_{mn}^{(k)} &= \left[\alpha\beta\left(\bar{Q}_{12}^{(k)} + \bar{Q}_{66}^{(k)}\right)U_{mn} + \left(\alpha^2\bar{Q}_{66}^{(k)} + \beta^2\bar{Q}_{22}^{(k)}\right)V_{mn}\right]\\
\mathcal{D}_{mn}^{(k)} &= \left[\alpha\beta\left(\bar{Q}_{12}^{(k)} + \bar{Q}_{66}^{(k)}\right)X_{mn} + \left(\alpha^2\bar{Q}_{66}^{(k)} + \beta^2\bar{Q}_{22}^{(k)}\right)Y_{mn}\right]\\
\mathcal{E}_{mn}^{(k)} &= -\left[\alpha^3\bar{Q}_{11}^{(k)} + \alpha\beta^2\left(\bar{Q}_{12}^{(k)} + 2\bar{Q}_{66}^{(k)}\right)\right]W_{mn} - \mathcal{B}_{mn}^{(k)}\\
\mathcal{F}_{mn}^{(k)} &= -\left[\beta^3\bar{Q}_{22}^{(k)} + \alpha^2\beta\left(\bar{Q}_{12}^{(k)} + 2\bar{Q}_{66}^{(k)}\right)\right]W_{mn} - \mathcal{D}_{mn}^{(k)}
\end{aligned} \quad (11.4-18)$$

11.4–3 Antisymmetric Angle-Ply Laminates

The simply supported (SS–2) boundary conditions in (4a,b) are satisfied by

$$u_0(x,y,t) = \sum_{n=1}^{\infty}\sum_{m=1}^{\infty} U_{mn}(t)\cos\alpha x\sin\beta y \quad (11.4-19a)$$

$$v_0(x,y,t) = \sum_{n=1}^{\infty}\sum_{m=1}^{\infty} V_{mn}(t)\sin\alpha x\cos\beta y \quad (11.4-19b)$$

and (w_0, ϕ_x, ϕ_y) have the same expansions as in Eqs. (5c,d,e). Substituting the expansions in Eqs. (19a,b) into Eqs. (11.2–16)–(11.2–20), we obtain equations of the form in (7a)

$$[\hat{S}]\{\Delta\} + [M]\{\ddot{\Delta}\} = \{F\} \quad (11.4-20)$$

with the following coefficients

$$\begin{aligned}
&\hat{s}_{11} = A_{11}\alpha^2 + A_{66}\beta^2, \quad \hat{s}_{12} = (A_{12} + A_{66})\alpha\beta\\
&\hat{s}_{13} = -c_1\left(3E_{16}\alpha^2 + E_{26}\beta^2\right)\beta\\
&\hat{s}_{14} = 2\hat{B}_{16}\alpha\beta, \quad \hat{s}_{15} = \hat{B}_{16}\alpha^2 + \hat{B}_{26}\beta^2\\
&\hat{s}_{22} = A_{66}\alpha^2 + A_{22}\beta^2, \quad \hat{s}_{23} = -c_1\left(E_{16}\alpha^2 + 3E_{26}\beta^2\right)\alpha\\
&\hat{s}_{24} = \hat{s}_{15}, \quad \hat{s}_{25} = 2\hat{B}_{25}\alpha\beta
\end{aligned}$$

$$
\begin{aligned}
\hat{s}_{33} &= \bar{A}_{55}\alpha^2 + \bar{A}_{44}\beta^2 + c_1^2\left[H_{11}\alpha^4 + 2(H_{12}+2H_{66})\alpha^2\beta^2 + H_{22}\beta^4\right]\\
\hat{s}_{34} &= \bar{A}_{55}\alpha - c_1\left[\hat{F}_{11}\alpha^3 + (\hat{F}_{12}+2\hat{F}_{66})\alpha\beta^2\right]\\
\hat{s}_{35} &= \bar{A}_{44}\alpha - c_1\left[\hat{F}_{22}\beta^3 + (\hat{F}_{12}+2\hat{F}_{66})\alpha^2\beta\right]\\
\hat{s}_{44} &= \bar{A}_{55} + \bar{D}_{11}\alpha^2 + \bar{D}_{66}\beta^2, \quad \hat{s}_{45} = (\bar{D}_{12}+\bar{D}_{66})\alpha\beta\\
\hat{s}_{55} &= \bar{A}_{44} + \bar{D}_{66}\alpha^2 + \bar{D}_{22}\beta^2
\end{aligned} \qquad (11.4-21)
$$

The mass and coefficients with hat and overbar are the same as those defined in Eqs. (8) and (9), and thermal effects are not considered.

The in-plane stresses in each layer can be computed from the equations

$$
\begin{Bmatrix}\sigma_{xx}\\ \sigma_{yy}\\ \sigma_{xy}\end{Bmatrix}^{(k)} = \begin{bmatrix}\bar{Q}_{11} & \bar{Q}_{12} & \bar{Q}_{16}\\ \bar{Q}_{12} & \bar{Q}_{22} & \bar{Q}_{26}\\ \bar{Q}_{16} & \bar{Q}_{26} & \bar{Q}_{66}\end{bmatrix}^{(k)} \left(\begin{Bmatrix}\varepsilon_{xx}^{(0)}\\ \varepsilon_{yy}^{(0)}\\ \gamma_{xy}^{(0)}\end{Bmatrix} + z\begin{Bmatrix}\varepsilon_{xx}^{(1)}\\ \varepsilon_{yy}^{(1)}\\ \gamma_{xy}^{(1)}\end{Bmatrix} + z^3\begin{Bmatrix}\varepsilon_{xx}^{(3)}\\ \varepsilon_{yy}^{(3)}\\ \gamma_{xy}^{(3)}\end{Bmatrix}\right) \qquad (11.4-22)
$$

$$
\begin{Bmatrix}\varepsilon_{xx}^{(0)}\\ \varepsilon_{yy}^{(0)}\\ \gamma_{xy}^{(0)}\end{Bmatrix} = \sum_{m=1}^{\infty}\sum_{n=1}^{\infty}\begin{Bmatrix}\alpha U_{mn}\cos\frac{m\pi x}{a}\cos\frac{n\pi y}{b}\\ \beta V_{mn}\cos\frac{m\pi x}{a}\cos\frac{n\pi y}{b}\\ -(\beta U_{mn}+\alpha V_{mn})\sin\frac{m\pi x}{a}\sin\frac{n\pi y}{b}\end{Bmatrix} \qquad (11.4-23a)
$$

$$
\begin{Bmatrix}\varepsilon_{xx}^{(1)}\\ \varepsilon_{yy}^{(1)}\\ \gamma_{xy}^{(1)}\end{Bmatrix} = -\sum_{m=1}^{\infty}\sum_{n=1}^{\infty}\begin{Bmatrix}\alpha X_{mn}\sin\frac{m\pi x}{a}\sin\frac{n\pi y}{b}\\ \beta Y_{mn}\sin\frac{m\pi x}{a}\sin\frac{n\pi y}{b}\\ -(\beta X_{mn}+\alpha Y_{mn})\cos\frac{m\pi x}{a}\cos\frac{n\pi y}{b}\end{Bmatrix} \qquad (11.4-23b)
$$

$$
\begin{Bmatrix}\varepsilon_{xx}^{(3)}\\ \varepsilon_{yy}^{(3)}\\ \gamma_{xy}^{(3)}\end{Bmatrix} = c_1\sum_{m=1}^{\infty}\sum_{n=1}^{\infty}\begin{Bmatrix}(\alpha X_{mn}+\alpha^2 W_{mn})\sin\frac{m\pi x}{a}\sin\frac{n\pi y}{b}\\ (\beta Y_{mn}+\beta^2 W_{mn})\sin\frac{m\pi x}{a}\sin\frac{n\pi y}{b}\\ -(\beta X_{mn}+\alpha Y_{mn}+2\alpha\beta W_{mn})\cos\frac{m\pi x}{a}\cos\frac{n\pi y}{b}\end{Bmatrix} \qquad (11.4-23c)
$$

The transverse stresses determined from the equilibrium equations of 3–D elasticity are

$$
\begin{aligned}
\sigma_{xz}^{(k)}(x,y,z) = \sum_{m=1}^{\infty}\sum_{n=1}^{\infty}&\left[(z-z_k)\bar{\mathcal{A}}_{mn}^{(k)} + \frac{1}{2}(z^2-z_k^2)\bar{\mathcal{B}}_{mn}^{(k)} + \frac{c_1}{4}(z^4-z_k^4)\bar{\mathcal{E}}_{mn}^{(k)}\right]\\
&+ \sigma_{xz}^{(k-1)}(x,y,z_k)
\end{aligned} \qquad (11.4-24a)
$$

$$
\begin{aligned}
\sigma_{yz}^{(k)}(x,y,z) = \sum_{m=1}^{\infty}\sum_{n=1}^{\infty}&\left[(z-z_k)\bar{\mathcal{C}}_{mn}^{(k)} + \frac{1}{2}(z^2-z_k^2)\bar{\mathcal{D}}_{mn}^{(k)} + \frac{c_1}{4}(z^4-z_k^4)\bar{\mathcal{F}}_{mn}^{(k)}\right]\\
&+ \sigma_{yz}^{(k-1)}(x,y,z_k)
\end{aligned} \qquad (11.4-24b)
$$

where

$$
\begin{aligned}
\bar{\mathcal{A}}_{mn}^{(k)} &= \left[\left(\alpha^2\bar{Q}_{11}^{(k)} + \beta^2\bar{Q}_{66}^{(k)}\right)U_{mn} + \alpha\beta\left(\bar{Q}_{12}^{(k)} + \bar{Q}_{66}^{(k)}\right)V_{mn}\right]\sin\alpha x\cos\beta y \\
&\quad + \left[2\alpha\beta\bar{Q}_{16}^{(k)}U_{mn} + \left(\alpha^2\bar{Q}_{16}^{(k)} + \beta^2\bar{Q}_{26}^{(k)}\right)V_{mn}\right]\cos\alpha x\sin\beta y \\
\bar{\mathcal{B}}_{mn}^{(k)} &= \left[\left(\alpha^2\bar{Q}_{11}^{(k)} + \beta^2\bar{Q}_{66}^{(k)}\right)X_{mn} + \alpha\beta\left(\bar{Q}_{12}^{(k)} + \bar{Q}_{66}^{(k)}\right)Y_{mn}\right]\cos\alpha x\sin\beta y \\
&\quad + \left[2\alpha\beta\bar{Q}_{16}^{(k)}X_{mn} + \left(\alpha^2\bar{Q}_{16}^{(k)} + \beta^2\bar{Q}_{26}^{(k)}\right)Y_{mn}\right]\sin\alpha x\cos\beta y \\
\bar{\mathcal{C}}_{mn}^{(k)} &= \left[\left(\alpha^2\bar{Q}_{16}^{(k)} + \beta^2\bar{Q}_{26}^{(k)}\right)U_{mn} + 2\alpha\beta\bar{Q}_{16}^{(k)}V_{mn}\right]\sin\alpha x\cos\beta y \\
&\quad + \left[\alpha\beta\left(\bar{Q}_{12}^{(k)} + \bar{Q}_{66}^{(k)}\right)U_{mn} + \left(\alpha^2\bar{Q}_{66}^{(k)} + \beta^2\bar{Q}_{22}^{(k)}\right)V_{mn}\right]\cos\alpha x\sin\beta y \\
\bar{\mathcal{D}}_{mn}^{(k)} &= \left[\left(\alpha^2\bar{Q}_{16}^{(k)} + \beta^2\bar{Q}_{66}^{(k)}\right)X_{mn} + 2\alpha\beta\bar{Q}_{26}^{(k)}Y_{mn}\right]\cos\alpha x\sin\beta y \\
&\quad + \left[\alpha\beta\left(\bar{Q}_{12}^{(k)} + \bar{Q}_{66}^{(k)}\right)X_{mn} + \left(\alpha^2\bar{Q}_{66}^{(k)} + \beta^2\bar{Q}_{22}^{(k)}\right)Y_{mn}\right]\sin\alpha x\cos\beta y \\
\bar{\mathcal{E}}_{mn}^{(k)} &= -\left[\alpha^3\bar{Q}_{11}^{(k)} + \alpha\beta^2\left(\bar{Q}_{12}^{(k)} + 2\bar{Q}_{66}^{(k)}\right)\right]W_{mn}\cos\alpha x\sin\beta y \\
&\quad - \left(3\bar{Q}_{16}^{(k)}\alpha^2\beta + \bar{Q}_{26}^{(k)}\beta^3\right)W_{mn}\sin\alpha x\cos\beta y - \bar{\mathcal{B}}_{mn}^{(k)} \\
\bar{\mathcal{F}}_{mn}^{(k)} &= -\left(\bar{Q}_{16}^{(k)}\alpha^3 + 3\bar{Q}_{26}^{(k)}\alpha\beta^2\right)W_{mn}\cos\alpha x\sin\beta y \\
&\quad - \left[\alpha^2\beta\left(\bar{Q}_{12}^{(k)} + 2\bar{Q}_{66}^{(k)}\right) + \beta^3\bar{Q}_{22}^{(k)}\right]W_{mn}\sin\alpha x\cos\beta y - \bar{\mathcal{D}}_{mn}^{(k)} \qquad (11.4-25)
\end{aligned}
$$

11.4–4 Numerical Results

Bending Analysis

Tables 11.4–1 and 11.4–2 contain nondimensionalized center deflections and stresses obtained with 3–D elasticity theory (ELS), third-order shear deformation plate theory (TSDT), first-order shear deformation theory (FSDT), and classical laminate plate theory (CLPT) for the following two problems (see Reddy [25]):

1. A three-ply (0/90/0) square ($a/b = 1$) laminate with layers of equal thickness and subjected to sinusoidally distributed transverse load.
2. A four-ply (0/90/90/0) square ($a/b = 1$) laminate with layers of equal thickness and subjected to sinusoidally distributed transverse load.

The material properties of a ply are assumed to be

Material 1:

$$E_1 = 25 \times 10^6 \text{ psi (175 GPa)}, \quad E_2 = 10^6 \text{ psi (7 GPa)}$$

$$G_{12} = G_{13} = 0.5 \times 10^6 \text{ psi } (3.5 \text{ GPa})$$

$$G_{23} = 0.2 \times 10^6 \text{ psi } (1.4 \text{ GPa}), \quad \nu_{12} = \nu_{13} = 0.25 \tag{11.4-26}$$

Table 11.4–1: Nondimensionalized center deflections and stresses in simply supported (SS–1) three-layer (0/90/0) square laminates under sinusoidally distributed transverse load.

a/h	Variable	ELS†	TSDT	FSDT $K=1$	FSDT $K=\frac{5}{6}$	FSDT $K=\frac{3}{4}$	FSDT $K=\frac{1}{2}$	CLPT
	$\bar{w} \times 10^2$	—	1.9218	1.5681	1.7758	1.9122	2.5770	0.4313
4	$\bar{\sigma}_{xx}$	0.755	0.7345	0.4475	0.4370	0.4308	0.4065	0.5387
	$(\bar{\sigma}_{yz})^{\ddagger}$	0.217	0.1832	0.1227	0.1561	0.1793	0.3030	−−
			0.2086	0.1850	0.1968	0.2038	0.2311	0.0823
	$\bar{w} \times 10^2$	—	0.7125	0.6306	0.6693	0.6949	0.8210	0.4313
10	$\bar{\sigma}_{xx}$	0.590	0.5684	0.5172	0.5134	0.5109	0.4993	0.5387
	$\bar{\sigma}_{yz}$	0.123	0.1033	0.0735	0.0915	0.1039	0.1723	−−
			0.1167	0.1065	0.1108	0.1136	0.1267	0.0823
	$\bar{w} \times 10^2$	—	0.4342	0.4333	0.4337	0.4340	0.4353	0.4313
100	$\bar{\sigma}_{xx}$	0.552	0.5390	0.5385	0.5384	0.5384	0.5382	0.5387
	$\bar{\sigma}_{yz}$	0.094	0.0750	0.0586	0.0703	0.0782	0.1174	−−
			0.0827	0.0826	0.0827	0.0827	0.0829	0.0823

† 3–D elasticity solution of Pagano [53].

‡ The second line corresponds to stresses computed from 3-D elasticity equilibrium equations.

The following nondimensionalized quantities are reported in the tables:

$$\begin{aligned}
\bar{w} &= w_0(\frac{a}{2}, \frac{b}{2}) \left(\frac{E_2 h^3}{a^4 q_0}\right), \quad \bar{\sigma}_{xx} = \sigma_{xx}(\frac{a}{2}, \frac{b}{2}, \frac{h}{2}) \left(\frac{h^2}{b^2 q_0}\right) \\
\bar{\sigma}_{yy} &= \sigma_{yy}(\frac{a}{2}, \frac{b}{2}, \frac{h}{4}) \left(\frac{h^2}{b^2 q_0}\right), \quad \bar{\sigma}_{xy} = \sigma_{xy}(0, 0, \frac{h}{2}) \left(\frac{h^2}{b^2 q_0}\right) \\
\bar{\sigma}_{yz} &= \sigma_{yz}(\frac{a}{2}, 0, 0) \left(\frac{h}{b q_0}\right), \quad \bar{\sigma}_{xz} = \sigma_{xz}(0, \frac{b}{2}, 0) \left(\frac{h}{b q_0}\right)
\end{aligned} \tag{11.4-27}$$

The origin of the coordinate system is taken at the lower left corner of the plate.

Table 11.4–2: Nondimensionalized maximum deflections and stresses in simply supported (SS–1) symmetric cross-ply (0/90/90/0) square laminates under sinusoidally distributed transverse load.

a/h	Source	$\bar{w} \times 10^2$	$\bar{\sigma}_{xx}$	$\bar{\sigma}_{yy}$	$\bar{\sigma}_{yz}$	$\bar{\sigma}_{xz}$	$\bar{\sigma}_{xy}$
	ELS†	1.954	0.720	0.663	0.292	0.219	0.0467
4	TSDT	1.894	0.665	0.632	0.239	0.206	0.0440
					0.298	0.231‡	
	FSDT	1.710	0.406	0.576	0.196	0.140	0.0308
					0.280	0.269	
	ELS	0.743	0.559	0.401	0.196	0.301	0.0275
10	TSDT	0.715	0.546	0.389	0.153	0.264	0.0268
					0.192	0.307	
	FDST	0.663	0.4989	0.361	0.130	0.167	0.0241
					0.181	0.318	
	ELS	0.517	0.543	0.308	0.156	0.328	0.0230
20	TSDT	0.506	0.539	0.304	0.123	0.282	0.0228
					0.154	0.330	
	FDST	0.491	0.527	0.296	0.109	0.175	0.0221
					0.150	0.333	
	ELS	0.438	0.539	0.276	0.141	0.337	0.0216
100	TSDT	0.434	0.539	0.271	0.112	0.290	0.0213
					0.139	0.339	
	FDST	0.434	0.538	0.270	0.101	0.178	0.0213
					0.139	0.339	
	CLPT	0.431	0.539	0.269	0.138‡	0.339‡	0.0213

† 3-D elsticity solution of Pagano and Hatfield [54].

‡ Equilibrium-derived stresses.

From the results it is clear that the third-order theory gives more accurate results for deflections and stresses when compared to the first-order shear deformation plate theory with $K = 5/6$. It is known that the shear correction factor K depends on the lamina properties and the stacking sequence. The fact that no shear correction coefficients are needed in the third-order theory makes it more convenient to use. In general, the equilibrium-derived transverse shear stresses compare more favorably with the elasticity solution than those obtained from the constitutive equations for equivalent single-layer theories.

Figure 11.4–3 contains plots of nondimensionalized center deflection, $\bar{w} = w_0(E_2h^3/q_0a^4)$, versus side-to-thickness ratio a/h for Problem 2 (a square, symmetric cross-ply laminate (0/90/90/0) under sinusoidally distributed load; Material 1). Compared to the elasticity solution, the third-order theory underpredicts deflection by 3% while the first-order theory underpredicts by about 12.5% for $a/h = 4$; for $a/h = 10$, the errors are 2.4% in TSDT and 10.8% in FSDT. The errors are much less at lower values of a/h.

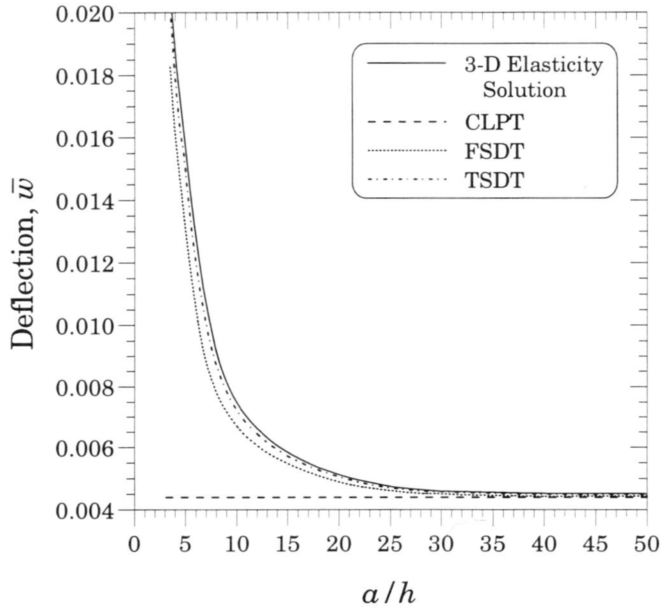

Figure 11.4–3: Plots of nondimensionalized center transverse deflection versus side-to-thickness ratio of a symmetric cross-ply laminate (0/90/90/0) under sinusoidally distributed load (Material 1).

Figures 11.4–4 and 11.4–5 show distributions of nondimensionalized $[\bar{\sigma} = \sigma(h^2/q_0a^2)]$ maximum normal stresses $\bar{\sigma}_{xx}$ and $\bar{\sigma}_{yy}$ predicted by the classical, first-order, and third-order plate theories through the thickness of a square symmetric cross-ply laminate (0/90/90/0) under sinusoidally distributed load (Material 1; $a/h = 4$ and 10). The third-order theory predicts a cubic variation whereas the classical and first-order theories predict linear variation of the stresses. Plots of constitutively derived and equilibrium-derived transverse shear stresses $\bar{\sigma}_{xz} = \sigma_{xz}(h/q_0a)$ and $\bar{\sigma}_{yz} = \sigma_{yz}(h/q_0a)$ are shown as functions of thickness in Figures 11.4–6 and 11.4-7, respectively, for a square, symmetric cross-ply laminate (0/90/90/0) under sinusoidally distributed load (Material 1; $a/h = 10$). The equilibrium-derived stresses are continuous through thickness because they are made to satisfy the interface continuity conditions (in determining the constants of integration), while the stresses computed using constitutive equations are always discontinuous for all equivalent single-layer theories due to the continuity of the transverse shear strains through thickness of the laminate. The third-order theory correctly satisfies vanishing of transverse shear stresses at the top and bottom of the laminate, because the displacement field in TSDT is derived to satisfy these conditions *a priori*.

Figure 11.4–8 shows the effect of material anisotropy on the deflections of antisymmetric cross-ply $(0/90)_n$ $(n = 1,3)$ square laminates under sinusoidal loading (SS–1) for $a/h = 10$, $G_{12} = G_{13} = 0.5E_2$, $G_{23} = 0.2E_2$, $\nu_{12} = 0.25$, and $a/h = 10$. The results predicted by FSDT and TSDT are very close; it is interesting to note that the first-order theory overpredicts deflections for the two-layer case and underpredicts for the six-layer case when compared to TSDT. As noted earlier, bending–stretching coupling is negligible for laminates with six or more layers; hence the deflections of a two-layer laminate without accounting for the coupling are virtually the same as those obtained for the six-layer laminate (see Table 11.4–3).

Table 11.4–3: Maximum deflections, $\bar{w} \times 10^2$, of simply supported (SS–1) antisymmetric cross-ply $(0/90/0/90/\ldots)$ square plates subjected to sinusoidally distributed transverse load.

	$n = 2$		$n = 6$	
$\frac{a}{h}$	FSDT	TSDT	FSDT	TSDT
4	2.1492	1.9985	1.5473	1.5411
10	1.2373	1.2161	0.6354	0.6382
20	1.1070	1.1018	0.5052	0.5060
50	1.0705	1.0697	0.4687	0.4688
100	1.0653	1.0651	0.4635	0.4635
CLPT	1.0636		0.4618	

The deflections of antisymmetric angle-ply $(45/-45)_n$ $(n = 1,3)$ square laminates under sinusoidal loading are presented in Figure 11.4–9 for various ratios of moduli $\frac{E_1}{E_2}$ ($G_{12} = G_{13} = 0.6E_2$, $G_{23} = 0.5E_2$, $\nu_{12} = 0.25$, $a/h = 10$). The effect of coupling between bending and extension is increasingly more pronounced with an increasing degree of material anisotropy. Even at low modulus ratios, the coupling terms cannot be neglected.

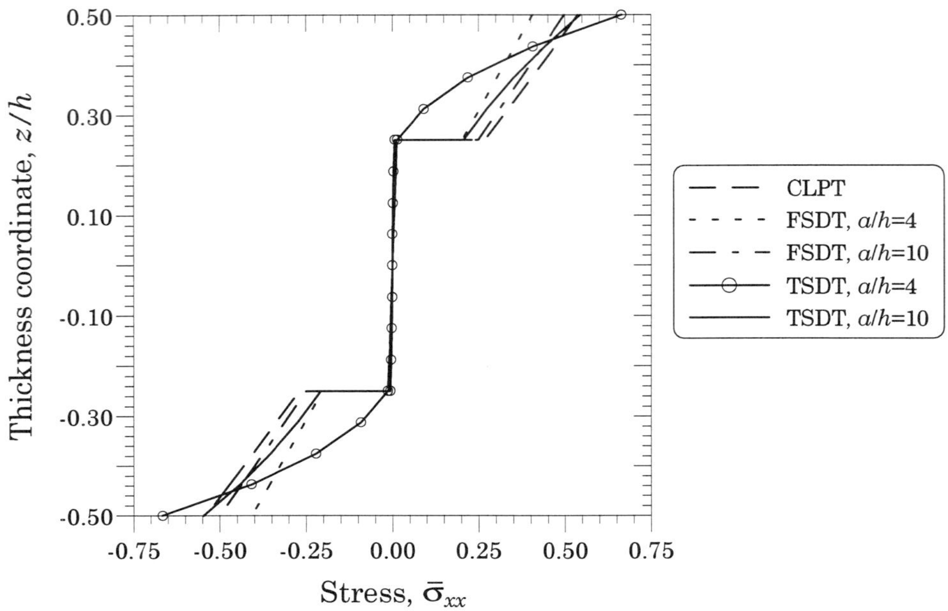

Figure 11.4–4: Comparison of center normal stress $\bar{\sigma}_{xx}$ distributions predicted by the classical, first-order, and third-order plate theories.

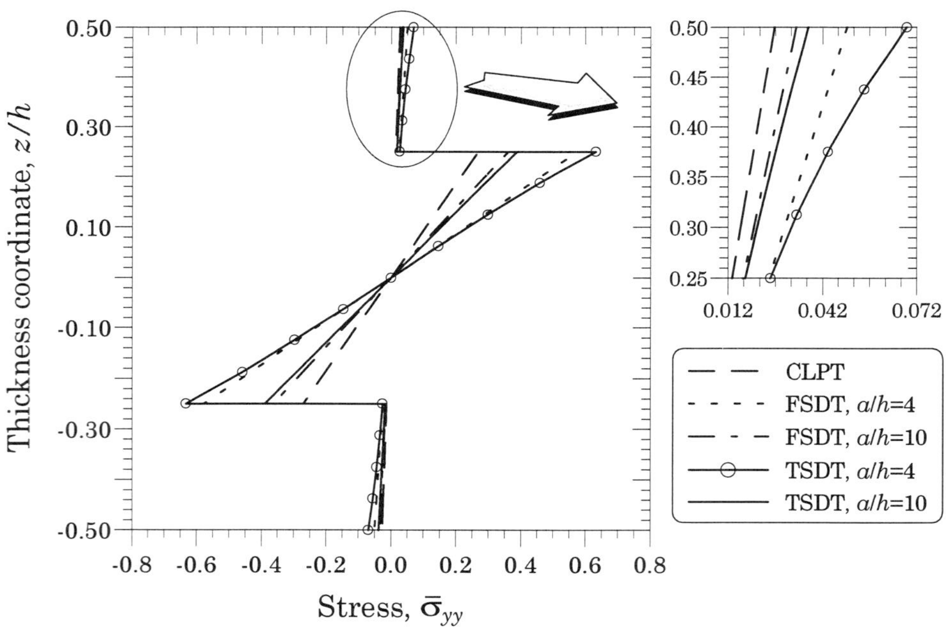

Figure 11.4–5: Comparison of normal stress $\bar{\sigma}_{yy}$ distributions predicted by the classical, first-order, and third-order plate theories.

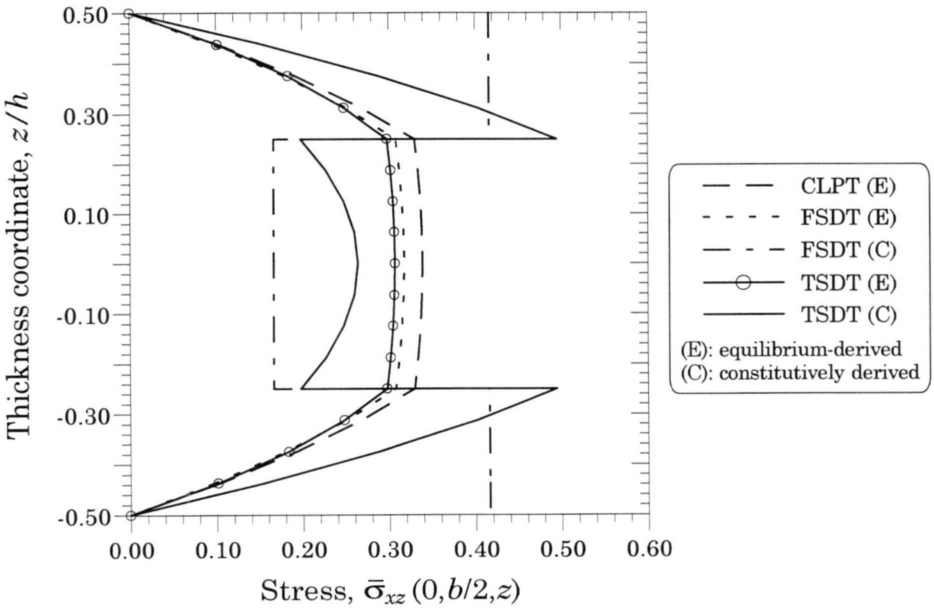

Figure 11.4–6: Plots of constitutively derived (C) and equilibrium-derived (E) transverse shear stresses $\bar{\sigma}_{xz}$ as functions of thickness coordinate.

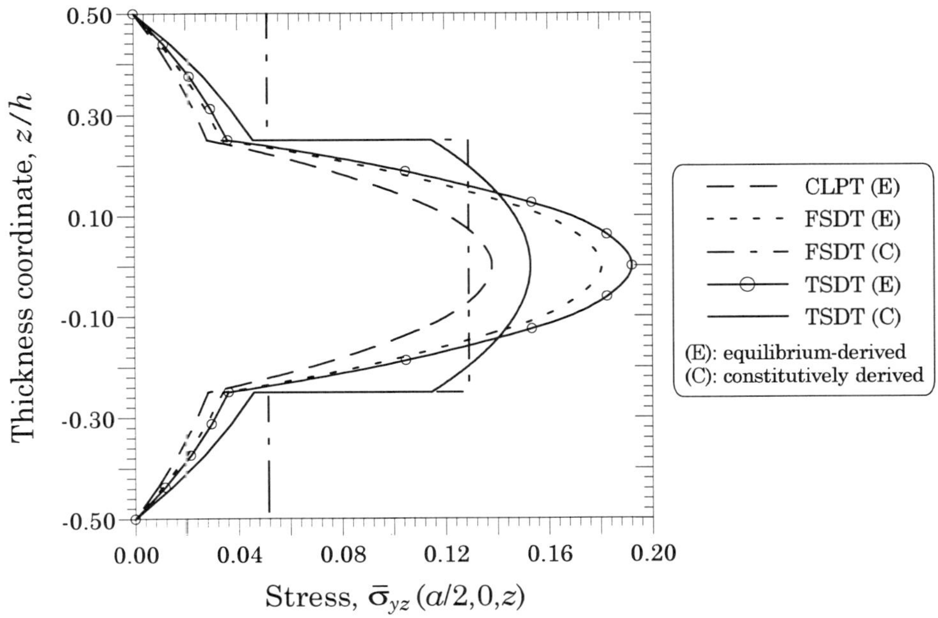

Figure 11.4–7: Plots of constitutively derived (C) and equilibrium-derived (E) transverse shear stresses $\bar{\sigma}_{yz}$ as functions of thickness coordinate.

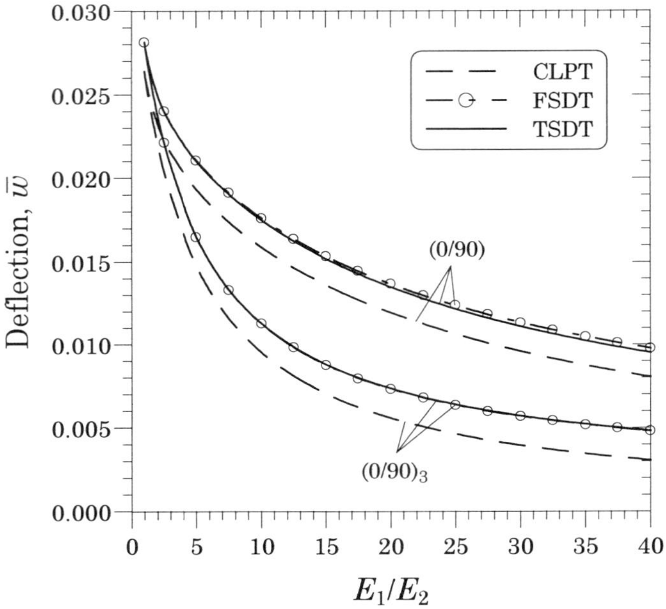

Figure 11.4–8: The effect of material anisotropy on the nondimensionalized deflections $\bar{w}$ of simply supported (SS–1) antisymmetric cross-ply $(0/90)_n$ $(n = 1, 3)$ square laminates.

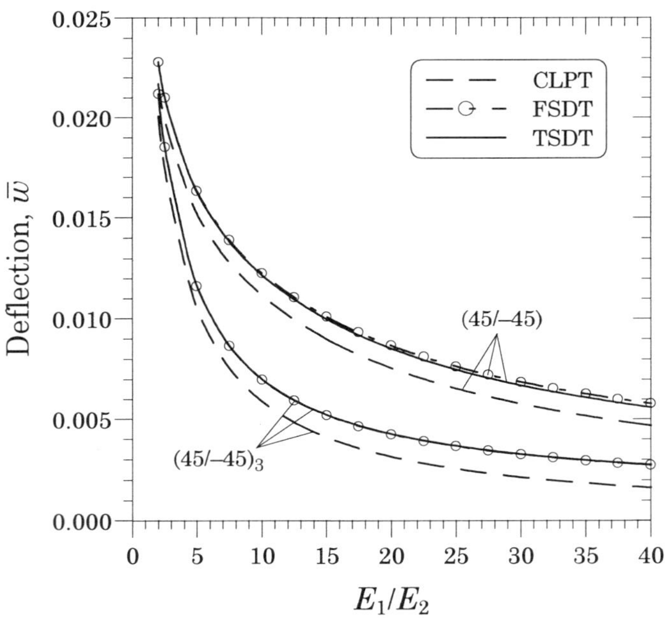

Figure 11.4–9: The effect of material anisotropy on the nondimensionalized deflections $\bar{w}$ of simply supported (SS–2) antisymmetric angle-ply $(45/-45)_n$ $(n = 1, 3)$ square laminates.

Table 11.4–4 contains a comparison of the maximum deflections of two- and six-layer $(\theta/-\theta/\ldots)$ antisymmetric angle-ply laminates under sinusoidal loading with different fiber orientations. The following material properties are assumed

Material 2:

$$\frac{E_1}{E_2} = 40, \quad \frac{G_{12}}{E_2} = 0.6, \quad \frac{G_{23}}{E_2} = 0.5, \quad \nu_{12} = \nu_{13} = 0.25 \qquad (11.4-28)$$

As in the case of antisymmetric cross-ply plates, the coupling causes a significant reduction of the plate stiffness with the most critical case being for $\theta = 45°$. The above observation can be explained by the fact that the magnitudes of the bending-stretching terms $(B_{16}, B_{26}, E_{16}, E_{26}, \ldots)$ are the largest at this particular fiber orientation for a given number of layers.

Table 11.4–4: Maximum deflections, $\bar{w} \times 10^2$, of simply supported (SS–2), antisymmetric angle-ply $(\theta\ /-\theta/\ldots)$ square plates subjected to sinusoidal loading.

		$\theta = 5°$		$\theta = 30°$		$\theta = 45°$	
$\frac{a}{h}$	Source	$n = 2$	$n = 6$	$n = 2$	$n = 6$	$n = 2$	$n = 6$
4	TSDT	1.2625	1.2282	1.0838	0.8851	1.0203	0.8375
	FSDT	1.3165	1.2647	1.2155	0.8994	1.1576	0.8531
10	TSDT	0.4848	0.4485	0.5916	0.3007	0.5581	0.2745
	FSDT	0.4883	0.4491	0.6099	0.2989	0.5773	0.2728
20	TSDT	0.3579	0.3209	0.5180	0.2127	0.4897	0.1905
	FSDT	0.3586	0.3208	0.5224	0.2121	0.4944	0.1899
50	TSDT	0.3215	0.2842	0.4972	0.1878	0.4704	0.1668
	FSDT	0.3216	0.2841	0.4979	0.1877	0.4712	0.1667
100	TSDT	0.3162	0.2789	0.4942	0.1842	0.4676	0.1634
	FSDT	0.3162	0.2789	0.4944	0.1842	0.4678	0.1633
	CLPT	0.3145	0.2771	0.4932	0.1831	0.4667	0.1622

Natural Vibration

Fundamental frequencies, $\bar{\omega} = \omega_{11}(a^2/h)\sqrt{\rho/E_2}$, of simply supported laminates are presented for the following three cases:

1. Four-layer (0/90/90/0) symmetric cross-ply square plate (SS–1).
2. Two- and six-layer (0/90/...) antisymmetric cross-ply square plates (SS–1).
3. Two- and six-layer ($\theta/-\theta/\ldots$) antisymmetric angle-ply square plates (SS–2) with $\theta = 5°, 30°$, and $45°$.

The rotary inertias are included in all cases and theories. Table 11.4–5 contains the nondimensionalized fundamental frequencies $\bar{\omega} = \omega(a^2/h)\sqrt{\rho/E_2}$ of the first laminate as a function of modulus ratio E_1/E_2 ($G_{12} = G_{13} = 0.6E_2$, $G_{23} = 0.5E_2$, $\nu_{12} = 0.25$) for $a/h = 5$ and 10. The fundamental natural frequencies $\bar{\omega}$ of antisymmetric cross-ply laminates (Material 2) as functions of the side-to-thickness are presented in Figure 11.4–10. Table 11.4–6 contains natural frequencies of two-layer and six-layer antisymmetric angle-ply laminates. All these results indicate that there is no significant difference between the predictions of FSDT and TSDT. However, TSDT does not require shear correction factors.

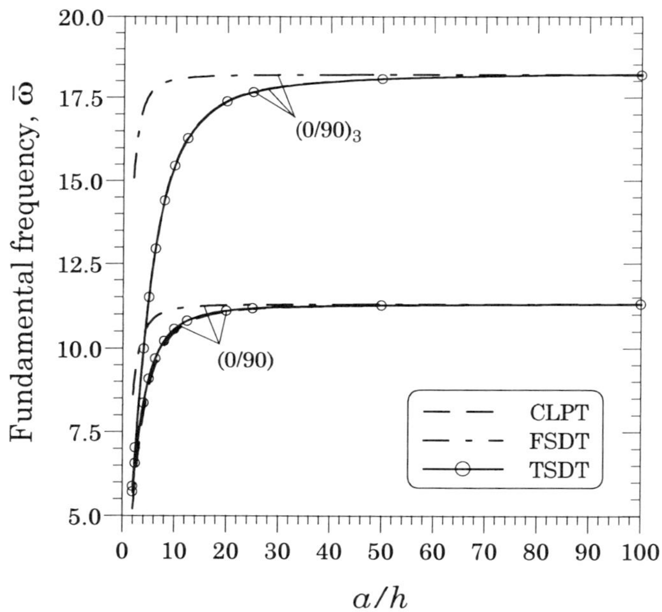

Figure 11.4–10: Plots of nondimensionalized fundamental frequency versus side-to-thickness ratio for cross-ply $(0/90)_n$ ($n = 1, 3$) square laminates.

Table 11.4–5: Nondimensionalized frequencies $\bar{\omega}$ of (0/90/90/0) cross-ply laminates as functions of modulus ratio.

E_1/E_2	a/h	ELS†	TSDT	FSDT	CLPT
3	5	6.618	6.560	6.570	7.299
	10	–	7.247	7.243	7.475
10	5	8.210	8.272	8.298	10.316
	10	–	9.853	9.841	10.563
20	5	9.560	9.526	9.567	13.511
	10	–	12.238	12.218	13.835
30	5	10.272	10.272	10.326	16.084
	10	–	13.892	13.864	16.470
40	5	10.752	10.787	10.854	18.299
	10	–	15.143	15.107	18.738

† Approximate 3-D solution of Noor [55].

Table 11.4–6: Nondimensionalized fundamental frequencies, $\bar{\omega} = \omega_{11}\frac{a^2}{h}\sqrt{\frac{\rho}{E_2}}$, of simply supported (SS–2), antisymmetric angle-ply $(\theta/-\theta/\ldots)$ square plates.

$\frac{a}{h}$	Source	$\theta = 5°$ $n=2$	$\theta = 5°$ $n=6$	$\theta = 30°$ $n=2$	$\theta = 30°$ $n=6$	$\theta = 45°$ $n=2$	$\theta = 45°$ $n=6$
4	TSDT	8.715	8.859	9.446	10.577	9.759	10.895
	FSDT	8.531	8.737	8.917	10.502	9.161	10.805
	CLPT	14.514	16.563	13.012	21.647	13.506	13.766
10	TSDT	14.230	14.848	12.873	18.170	13.263	19.025
	FSDT	14.179	14.840	12.681	18.226	13.044	19.025
	CLPT	17.500	18.819	14.031	23.165	14.439	24.611
20	TSDT	16.656	17.619	13.849	21.648	14.246	22.877
	FSDT	16.641	17.622	13.790	21.679	14.179	22.913
	CLPT	17.754	18.952	14.187	23.320	14.587	24.773
50	TSDT	17.626	18.753	14.174	23.067	14.572	24.480
	FSDT	17.623	18.754	14.164	23.074	14.561	24.488
	CLPT	17.820	18.989	14.231	23.364	14.630	24.819
100	TSDT	17.780	18.935	14.223	23.295	14.621	24.739
	FSDT	17.780	18.935	14.220	23.297	14.618	24.741
	CLPT	17.830	18.995	14.237	23.371	14.636	24.825

Buckling Analysis

The uniaxial critical buckling loads of a four-layer (0/90/90/0) cross-ply plate (Material 2) with various side-to-thickness ratios are compared in Table 11.4–7. The buckling loads of the same laminate as a function of modulus ratio E_1/E_2 are presented in Table 11.4–8. The results are also compared with approximate 3–D elasticity results obtained by Noor [57]. In Figure 11.4–11, the buckling loads of two-layer and six-layer $(0/90/0/\cdots)$ antisymmetric cross-ply laminates are shown as a function of the side-to-thickness ratio. Table 11.4–9 contains critical buckling loads for the two-layer and six-layer antisymmetric angle-ply laminates. Both TSDT and FSDT give very good results and the difference between them is not very significant.

Table 11.4–7: Nondimensionalized uniaxial buckling loads, $\bar{N} = N_{cr}\frac{a^2}{E_2h^3}$, of simply supported (SS–1) symmetric cross-ply (0/90/90/0) square plates.

$\frac{a}{h}$	CLPT	FSDT	TSDT
5	36.160	11.575	11.997
10	36.160	23.453	23.340
20	36.160	31.707	31.660
50	36.160	35.356	35.347
100	36.160	35.955	35.953

Table 11.4–8: Effect of material anisotropy on the uniaxial buckling loads, $\bar{N} = N_{cr}\frac{a^2}{E_2h^3}$, of symmetric cross-ply (0/90/90/0) square plates ($G_{12} = G_{13} = 0.6E_2$, $G_{23} = 0.5E_2$, $\nu_{12} = 0.25$, $a/h = 10$).

E_1/E_2	CLPT	FSDT	TSDT	ELS [57]
3	5.754	5.399	5.114	5.304
10	11.492	9.965	9.777	9.762
20	19.712	15.351	15.298	15.019
30	27.936	19.756	19.957	19.304
40	36.160	23.453	23.340	22.881

Table 11.4–9: Nondimensionalized uniaxial buckling loads, $\bar{N} = N_{cr}\frac{a^2}{E_2h^3}$, of simply supported (SS–2) antisymmetric angle-ply $(\theta/-\theta/\ldots)$ square plates.

$\frac{a}{h}$	Source	$\theta = 5°$ $n=2$	$\theta = 5°$ $n=6$	$\theta = 30°$ $n=2$	$\theta = 30°$ $n=6$	$\theta = 45°$ $n=2$	$\theta = 45°$ $n=6$
5	TSDT	10.674	11.082	11.547	13.546	10.881	12.169
	FSDT	10.384	10.899	10.586	11.986	9.385	11.297
10	TSDT	20.989	22.592	17.127	33.701	18.154	32.405
	FSDT	20.752	22.562	16.613	33.903	17.552	32.525
20	TSDT	28.308	31.577	19.561	47.643	20.691	53.198
	FSDT	28.259	31.587	19.394	47.779	20.495	53.365
50	TSDT	31.519	35.657	20.379	53.951	21.539	60.760
	FSDT	31.511	35.660	20.350	53.981	21.505	60.798
100	TSDT	32.042	36.332	20.502	54.993	21.666	62.022
	FSDT	32.040	36.333	20.494	55.001	21.658	62.032
	CLPT	32.220	36.563	20.543	55.350	21.709	62.455

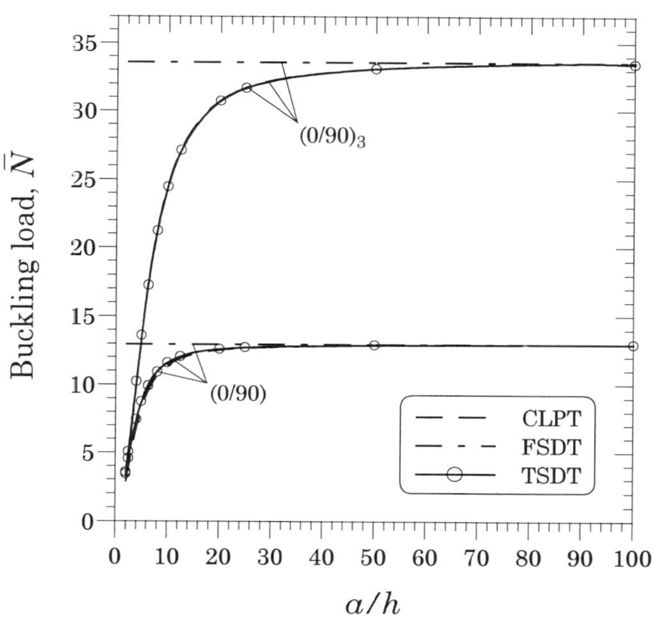

Figure 11.4–11: Nondimensionalized critical buckling load $[\bar{N} = N_{cr}(a^2/E_2h^3)]$ versus side-to-thickness ratio for antisymmetric cross-ply $(0/90)_n$ $(n = 1, 3)$ square laminates subjected to uniaxial compressive load (Material 2).

11.5 Lévy Solutions of Cross-Ply Laminates

11.5–1 Preliminary Comments

The Lévy type solutions for bending, natural vibration, and buckling of rectangular laminates of cross-ply constructions have been developed for the third-order theory of Reddy (see [34,41,42,44,48,50-52]). Here we present the solutions for static bending of cross-ply laminates (see Khdeir and Reddy [50,51]). For additional results, the reader may consult references at the end of the chapter.

For the static case, the governing equations appropriate for the antisymmetric cross-ply laminate construction are given by

$$[L]\{\Delta\} = \{F\} \tag{11.5-1}$$

where

$$\{\Delta\}^T = \{u_0, v_0, w_0, \phi_x, \phi_y\}, \quad \{F\}^T = \{0, 0, q, 0, 0\} \tag{11.5-2a}$$

$$\begin{aligned}
L_{11} &= A_{11}d_1^2 + A_{66}d_2^2, \quad L_{12} = (A_{12} + A_{66})d_1 d_2 \\
L_{13} &= -c_1[E_{11}d_1^3 + (E_{12} + 2E_{66})d_1 d_2^2] \\
L_{14} &= \hat{B}_{11}d_1^2 + \hat{B}_{66}d_2^2, \quad L_{15} = \left(\hat{B}_{12} + \hat{B}_{66}\right) d_1 d_2 \\
L_{22} &= A_{66}d_1^2 + A_{22}d_2^2, \quad L_{23} = -c_1[E_{22}d_2^3 + (E_{12} + 2E_{66})d_1^2 d_2] \\
L_{24} &= L_{15}, \quad L_{25} = \hat{B}_{22}d_2^2 + \hat{B}_{66}d_1^2 \\
L_{33} &= -\bar{A}_{55}d_1^2 - \bar{A}_{44}d_2^2 + c_1^2\left[H_{11}d_1^4 + (2H_{12} + 4H_{66})d_1^2 d_2^2 + H_{22}d_2^4\right] \\
L_{34} &= \left(c_2\hat{D}_{55} - \hat{A}_{55}\right) d_1 - c_1\left[\hat{F}_{11}d_1^3 + \left(\hat{F}_{12} + 2\hat{F}_{66}\right) d_1 d_2^2\right] \\
L_{35} &= \left(c_2\hat{D}_{44} - \hat{A}_{44}\right) d_2 - c_1\left[\hat{F}_{22}d_2^3 + \left(\hat{F}_{12} + 2\hat{F}_{66}\right) d_1^2 d_2\right] \\
L_{44} &= -\bar{A}_{55} + \bar{D}_{11}d_1^2 + \bar{D}_{66}d_2^2, \quad L_{45} = \left(\bar{D}_{12} + \bar{D}_{66}\right) d_1 d_2 \\
L_{55} &= -\bar{A}_{44} + \bar{D}_{22}d_2^2 + \bar{D}_{66}d_1^2
\end{aligned} \tag{11.5-2b}$$

and

$$d_\alpha^i = \frac{\partial^i}{\partial x_\alpha^i}, \quad x_1 = x \text{ and } x_2 = y \tag{11.5-3}$$

11.5–2 Solution Procedure

A generalized Lévy type solution, in conjunction with the state-space concept can be used to determine bending solutions of cross-ply laminated rectangular plates with two parallel edges simply supported and other two having arbitrary combination of boundary conditions. Suppose that the edges $y = 0, b$ are simply supported, while the remaining ones ($x = \pm a/2$) may have arbitrary combinations of free, clamped, and simply supported edge conditions (see Figure 11.5–1).

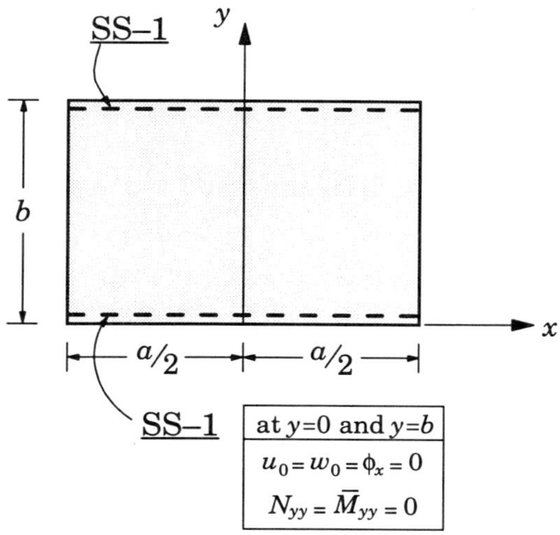

Figure 11.5–1: The coordinate system ($-a/2 \leq x \leq a/2,\ 0 \leq y \leq b$) and boundary conditions used on the simply supported edges for the Lévy solutions of rectangular cross-ply laminates using the third-order shear deformation theory.

The generalized displacements may be expressed as products of undetermined functions and known trigonometric functions so as to identically satisfy the simply supported boundary conditions at $y = 0, b$:

$$u_0 = w_0 = \phi_x = N_{yy} = M_{yy} = P_{yy} = 0 \tag{11.5 – 4}$$

A sinusoidal distribution of the transverse load is considered

$$q(x,y) = q_0 \cos\frac{\pi x}{a} \sin\frac{\pi y}{b} \tag{11.5 – 5}$$

The displacement quantities are represented as

$$\left\{\begin{array}{c} u_0(x,y) \\ v_0(x,y) \\ w_0(x,y) \\ \phi_x(x,y) \\ \phi_y(x,y) \end{array}\right\} = \left\{\begin{array}{c} U_m(x)\sin\beta y \\ V_m(x)\cos\beta y \\ W_m(x)\sin\beta y \\ X_m(x)\sin\beta y \\ Y_m(x)\cos\beta y \end{array}\right\} \tag{11.5 – 6}$$

where $\beta = m\pi/b$.

Substitution of (6) into Eqs. (1) results in the following five differential equations:

$$
\begin{aligned}
U''_m &= C_1U_m + C_2V'_m + C_3W'_m + C_4W'''_m + C_5X_m + C_6Y'_m + f_1^T \sin\alpha x \\
V''_m &= C_7U'_m + C_8V_m + C_9W_m + C_{10}W''_m + C_{11}X'_m + C_{12}Y_m + f_2^T \cos\alpha x \\
W''''_m &= C_{13}U'_m + C_{14}V_m + C_{15}W_m + C_{16}W''_m + C_{17}X'_m + C_{18}Y_m + f_3^T \cos\alpha x \\
X''_m &= C_{19}U_m + C_{20}V'_m + C_{21}W'_m + C_{22}W'''_m + C_{23}X_m + C_{24}Y'_m + f_4^T \sin\alpha x \\
Y''_m &= C_{25}U'_m + C_{26}V_m + C_{27}W_m + C_{28}W''_m + C_{29}X'_m + C_{30}Y_m + f_5^T \cos\alpha x
\end{aligned}
\tag{11.5 – 7}
$$

where primes denote the derivative with respect to x, and the coefficients C_i are defined by

$$
\begin{aligned}
&C_1 = (e_7e_{30} - e_3e_{34})/e_0, \quad C_2 = (e_2e_{30} - e_3e_{29})/e_0 \\
&C_3 = (e_6e_{30} - e_3e_{33})/e_0, \quad C_4 = (e_5e_{30} - e_3e_{32})/e_0 \\
&C_5 = (e_8e_{30} - e_3e_{35})/e_0, \quad C_6 = (e_4e_{30} - e_3e_{31})/e_0 \\
&C_7 = (e_9e_{39} - e_{12}e_{36})/C_0, \quad C_8 = (e_{14}e_{39} - e_{12}e_{41})/C_0 \\
&C_9 = (e_{16}e_{39} - e_{12}e_{43})/C_0, \quad C_{10} = (e_{13}e_{39} - e_{12}e_{40})/C_0 \\
&C_{11} = (e_{11}e_{39} - e_{12}e_{38})/C_0 \quad C_{12} = (e_{15}e_{39} - e_{12}e_{42})/C_0 \\
&C_{19} = (e_1e_{34} - e_7e_{28})/e_0, \quad C_{20} = (e_1e_{29} - e_2e_{28})/e_0 \\
&C_{21} = (e_1e_{33} - e_6e_{28})/e_0, \quad C_{22} = (e_1e_{32} - e_5e_{28})/e_0 \\
&C_{23} = (e_1e_{35} - e_8e_{28})/e_0, \quad C_{24} = (e_1e_{31} - e_4e_{28})/e_0 \\
&C_{25} = (e_{10}e_{36} - e_9e_{37})/C_0, \quad C_{26} = (e_{10}e_{41} - e_{14}e_{37})/C_0 \\
&C_{27} = (e_{10}e_{43} - e_{16}e_{37})/C_0, \quad C_{28} = (e_{10}e_{40} - e_{13}e_{37})/C_0 \\
&C_{29} = (e_{10}e_{38} - e_{11}e_{37})/C_0, \quad C_{30} = (e_{10}e_{42} - e_{15}e_{37})/C_0 \\
&C_{13} = a_0(C_1e_{21} + C_7a_1 + C_{25}a_2 + C_{19}e_{23} + e_{26}) \\
&C_{14} = a_0(C_8a_1 + C_{26}a_2 + e_{27}), \quad C_{15} = a_0(C_9a_1 + C_{27}a_2 + e_{20}) \\
&C_{16} = a_0(e_{18} + C_3e_{21} + C_{21}e_{23} + C_{10}a_1 + C_{28}a_2) \\
&C_{17} = a_0(e_{17} + C_5e_{21} + C_{23}e_{23} + C_{11}a_1 + C_{29}a_2) \\
&C_{18} = a_0(e_{19} + C_{12}a_1 + C_{30}a_2), \quad e_0 = e_3e_{28} - e_1e_{30} \\
&C_0 = e_{12}e_{37} - e_{10}e_{39}, \quad a_0 = -1/\left(C_4e_{21} + C_{22}e_{23} + e_{25}\right) \\
&a_1 = C_2e_{21} + C_{20}e_{23} + e_{22}, \quad a_2 = C_6e_{21} + C_{24}e_{23} + e_{24}
\end{aligned}
\tag{11.5 – 8}
$$

where

$$
\begin{aligned}
&e_1 = A_{11}, \quad e_2 = -\beta(A_{12} + A_{66}), \quad e_3 = \hat{B}_{11} \\
&e_4 = -\beta\left(\hat{B}_{12} + \hat{B}_{66}\right), \quad e_5 = -c_1E_{11} \\
&e_6 = \beta^2c_1(E_{12} + 2E_{66}), \quad e_7 = -\beta^2A_{66} \\
&e_8 = -\beta^2\hat{B}_{66}, \quad e_9 = -e_2, \quad e_{10} = A_{66}, \quad e_{11} = -e_4
\end{aligned}
$$

$$
\begin{aligned}
&e_{12} = \hat{B}_{66}, \quad e_{13} = -c_1\beta(E_{12} + 2E_{66}), \quad e_{14} = -\beta^2 A_{22}\\
&e_{15} = -\beta^2 \hat{B}_{22}, \quad e_{16} = c_1\beta^3 E_{22}\\
&e_{17} = \bar{A}_{55} - c_1\beta^2\left(\hat{F}_{12} + 2\hat{F}_{66}\right)\\
&e_{18} = \bar{A}_{55} + 2c_1^2\beta^2\left(H_{12} + 2H_{66}\right)\\
&e_{19} = -\beta\bar{A}_{44} + c_1\beta^3\hat{F}_{22}, \quad e_{20} = -\beta^2\bar{A}_{44} - c_1^2\beta^4 H_{22}\\
&e_{21} = -e_5, \quad e_{22} = e_{13}, \quad e_{23} = c_1\hat{F}_{11}\\
&e_{24} = -c_1\beta\left(\hat{F}_{12} + 2\hat{F}_{66}\right), \quad e_{25} = -c_1^2 H_{11}\\
&e_{26} = -e_6, \quad e_{27} = e_{16}, \quad e_{28} = e_3, \quad e_{29} = e_4\\
&e_{30} = \bar{D}_{11}, \quad e_{31} = -\beta\left(\bar{D}_{12} + \bar{D}_{66}\right)\\
&e_{32} = -e_{23}, \quad e_{33} = -e_{17}, \quad e_{34} = e_8, \quad e_{35} = -\bar{A}_{55} - \beta^2\bar{D}_{66}\\
&e_{36} = -e_4, \quad e_{37} = e_{12}, \quad e_{38} = -e_{31}, \quad e_{39} = \bar{D}_{66}\\
&e_{40} = e_{24}, \quad e_{41} = e_{15}, \quad e_{42} = -\bar{A}_{44} - \beta^2\bar{D}_{22}, \quad e_{43} = e_{19}
\end{aligned}
\tag{11.5-9}
$$

The stiffnesses with bars and hats are defined in Eq. (11.4–9). The coefficients f_i^T are associated with thermal effects

$$
\begin{aligned}
f_1^T &= \left(\frac{e_{30}}{e_0}g_1 - \frac{e_3}{e_0}g_4\right), \quad f_2^T = \left(\frac{e_{12}}{c_0}g_5 - \frac{e_{39}}{c_0}g_2\right)\\
f_4^T &= \left(\frac{e_1}{e_0}g_4 - \frac{e_{28}}{e_0}g_1\right), \quad f_5^T = \left(\frac{e_{37}}{c_0}g_2 - \frac{e_{10}}{c_0}g_5\right)\\
f_3^T &= a_0\left[a_1 f_2^T + a_2 f_5^T + \alpha\left(e_{21}f_1^T + e_{23}f_4^T\right) - g_3 + q_0\right]
\end{aligned}
\tag{11.5-10a}
$$

$$
\begin{aligned}
g_1 &= \alpha\left(N_{mn}^1\bar{T}_0 + M_{mn}^1\bar{T}_1\right), \quad g_2 = \beta\left(N_{mn}^2\bar{T}_0 + M_{mn}^2\bar{T}_1\right)\\
g_3 &= -c_1\left(\alpha^2 P_{mn}^1 + \beta^2 P_{mn}^2\right)\bar{T}_0, \quad g_4 = \alpha\left(M_{mn}^1 - c_1 P_{mn}^1\right)\bar{T}_0\\
g_5 &= \beta\left(M_{mn}^2 - c_1 P_{mn}^2\right)\bar{T}_0
\end{aligned}
\tag{11.5-10b}
$$

In order to reduce the system of equations (7) to a state–space form, the components of the state vector $\mathbf{Z}(x)$ are defined as follows:

$$
Z_1 = U_m, \quad Z_2 = U_m', \quad Z_3 = V_m, \quad Z_4 = V_m', \quad Z_5 = W_m, \quad Z_6 = W_m', \quad Z_7 = W_m''
$$

$$
Z_8 = W_m''', \quad Z_9 = X_m, \quad Z_{10} = X_m', \quad Z_{11} = Y_m, \quad Z_{12} = Y_m' \tag{11.5-11}
$$

Using the definitions (11), the systems of equations (7) may be converted to the first-order differential operator form

$$
\mathbf{Z}' = A\mathbf{Z} + \mathbf{r} \tag{11.5-12a}
$$

where the matrix $[A]$ is a 12×12 matrix

$$[A] = \begin{bmatrix} 0 & 1 & 0 & 0 & 0 & 0 & 0 & 0 & 0 & 0 & 0 & 0 \\ C_1 & 0 & 0 & C_2 & 0 & C_3 & 0 & C_4 & C_5 & 0 & 0 & C_6 \\ 0 & 0 & 0 & 1 & 0 & 0 & 0 & 0 & 0 & 0 & 0 & 0 \\ 0 & C_7 & C_8 & 0 & C_9 & 0 & C_{10} & 0 & 0 & C_{11} & C_{12} & 0 \\ 0 & 0 & 0 & 0 & 0 & 1 & 0 & 0 & 0 & 0 & 0 & 0 \\ 0 & 0 & 0 & 0 & 0 & 0 & 1 & 0 & 0 & 0 & 0 & 0 \\ 0 & 0 & 0 & 0 & 0 & 0 & 0 & 1 & 0 & 0 & 0 & 0 \\ 0 & C_{13} & C_{14} & 0 & C_{15} & 0 & C_{16} & 0 & 0 & C_{17} & C_{18} & 0 \\ 0 & 0 & 0 & 0 & 0 & 0 & 0 & 0 & 0 & 1 & 0 & 0 \\ C_{19} & 0 & 0 & C_{20} & 0 & C_{21} & 0 & C_{22} & C_{23} & 0 & 0 & C_{24} \\ 0 & 0 & 0 & 0 & 0 & 0 & 0 & 0 & 0 & 0 & 0 & 1 \\ 0 & C_{25} & C_{26} & 0 & C_{27} & 0 & C_{28} & 0 & 0 & C_{29} & C_{30} & 0 \end{bmatrix} \tag{11.5 – 12b}$$

and the load vector $\mathbf{r}$ is

$$\{r\} = \{0, f_1^T \sin \alpha x, 0, f_2^T \cos \alpha x, 0, 0, 0, f_3^T \cos \alpha x, 0, f_4^T \sin \alpha x, 0, f_5^T \cos \alpha x\}^T \tag{11.5 – 13}$$

The solution to Eq. (12a) is given by

$$\mathbf{Z} = e^{Ax}\{\mathbf{K} + \int^x e^{-A\zeta}\mathbf{r}(\zeta)\, d\zeta\} \tag{11.5 – 14}$$

Here $\mathbf{K}$ is a column vector of constants to be determined from the edge conditions, and

$$e^{Ax} = [S]\begin{bmatrix} e^{\lambda_1 x} & \mathbf{0} \\ \mathbf{0} & e^{\lambda_n x} \end{bmatrix}[S]^{-1} \tag{11.5 – 15}$$

where $n = 12$, λ_i denote the distinct eigenvalues of $[A]$, and $[S]$ denotes the matrix of eigenvectors of $[A]$.

The boundary conditions for simply supported (S), clamped (C), and free (F) at the edges $x = \pm a/2$ are

$$\text{S}: \quad v_0 = w_0 = \phi_y = N_{xx} = M_{xx} = P_{xx} = 0 \tag{11.5 – 16}$$

$$\text{C}: \quad u_0 = v_0 = w_0 = \frac{\partial w_0}{\partial x} = \phi_x = \phi_y = 0 \tag{11.5 – 17}$$

$$\begin{aligned} \text{F}: \quad & N_{xx} = N_{xy} = M_{xx} = P_{xx} = M_{xy} - c_1 P_{xy} = 0 \\ & Q_x - c_2 R_x + c_1\left(\frac{\partial P_{xx}}{\partial x} + \frac{\partial P_{xy}}{\partial y}\right) = 0 \end{aligned} \tag{11.5 – 18}$$

where the stress resultants $P_{\alpha\beta}$ and R_α of the third-order theory are defined in Eq. (11.2–14).

11.5–3 Numerical Results

The nondimensionalized center transverse deflections and stresses of two-layer and ten-layer cross-ply laminates under sinusoidally distributed transverse load are presented in Table 11.5–1 for various boundary conditions and side-to-thickness ratio of $a/h = 5$. The nondimensionalized variables used are

$$\bar{w} = w_0(0, b/2)\frac{h^3 E_2}{b^4 q_0} \times 10^2, \quad \bar{\sigma}_{xx} = -\sigma_{xx}(0, \frac{b}{2}, -\frac{h}{2})\frac{h^2}{b^2 q_0}10$$

$$\bar{\sigma}_{yy} = -\sigma_{yy}(0, \frac{b}{2}, \frac{h}{2})\frac{h^2}{b^2 q_0}10, \quad \bar{\sigma}_{yz} = -\sigma_{yz}(0,0,0)\frac{h}{bq_0}10 \qquad (11.5-19)$$

Table 11.5–1: Nondimensionalized deflections and stresses of antisymmetric cross-ply square plates for various boundary conditions ($a/h = 5, E_1/E_2 = 25,\ G_{12} = G_{13} = 0.5E_2,\ G_{23} = 0.2E_2,\ \nu_{12} = 0.25$).

No. of Layers	Theory	Variable	SS	SC	CC	FF	FS	FC
2	TSDT	$\bar{w}$	1.667	1.333	1.088	2.624	2.211	1.733
		$\bar{\sigma}_{xx}$	8.385	6.816	5.679	3.171	5.349	3.727
		$\bar{\sigma}_{yy}$	8.385	6.725	5.505	13.551	11.324	8.919
		$\bar{\sigma}_{yz}$	3.155	2.543	2.095	4.457	3.893	3.048
	FSDT	$\bar{w}$	1.758	1.477	1.257	2.777	2.335	1.897
		$\bar{\sigma}_{xx}$	7.157	5.338	3.911	2.469	4.430	2.434
		$\bar{\sigma}_{yy}$	7.157	6.034	5.153	11.907	9.848	8.047
		$\bar{\sigma}_{yz}$	2.729	2.297	1.958	3.901	3.390	2.748
	CLPT	$\bar{w}$	1.064	0.664	0.429	1.777	1.471	0.980
		$\bar{\sigma}_{xx}$	7.157	5.660	4.800	2.403	4.442	3.042
		$\bar{\sigma}_{yy}$	7.157	4.483	2.914	11.849	9.837	6.560
10	TSDT	$\bar{w}$	1.129	1.001	0.879	1.651	1.450	1.214
		$\bar{\sigma}_{xx}$	6.340	5.196	4.025	2.482	3.946	2.608
		$\bar{\sigma}_{yy}$	6.340	5.635	4.963	9.454	8.252	6.934
		$\bar{\sigma}_{yz}$	3.362	2.974	2.601	4.784	4.234	3.535
	FSDT	$\bar{w}$	1.137	1.045	0.945	1.663	1.460	1.258
		$\bar{\sigma}_{xx}$	5.009	3.707	2.275	1.712	2.957	1.343
		$\bar{\sigma}_{yy}$	5.009	4.628	4.212	7.583	6.590	5.706
		$\bar{\sigma}_{yz}$	2.729	2.498	2.248	3.883	3.437	2.951
	CLPT	$\bar{w}$	0.442	0.266	0.167	0.665	0.579	0.380
		$\bar{\sigma}_{xx}$	5.009	3.829	3.167	1.725	2.986	1.865
		$\bar{\sigma}_{yy}$	5.009	3.025	1.911	7.480	6.531	4.284

For the thick plates considered in this case, there is a significant difference between the results predicted by TSDT and FSDT; FSDT slightly overpredicts deflections and underpredicts stresses.

Table 11.5–2: Nondimensionalized center deflections $\bar{w}$ of cross-ply square plates subjected to sinusoidally distributed temperature distribution.

Laminate	b/h	Theory	SS	SC	CC	FF	FS
0	5	TSDT	1.0711	0.7175	0.3663	2.2812	1.5831
		FSDT	1.0721	0.7613	0.3915	2.2894	1.5859
		CLPT	1.0312	0.4543	0.2443	2.2935	1.6067
	10	TSDT	1.0439	0.5587	0.2871	2.2854	1.5931
		FSDT	1.0440	0.5677	0.2912	2.2928	1.5952
		CLPT	1.0312	0.4543	0.2443	2.2935	1.6067
0/90	5	TSDT	1.1430	0.8190	0.5814	1.2652	1.2068
		FSDT	1.1504	0.8547	0.6231	1.2784	1.2170
		CLPT	1.1504	0.7183	0.4681	1.2639	1.2152
	10	TSDT	1.1485	0.7586	0.5164	1.2693	1.2145
		FSDT	1.1504	0.7703	0.5307	1.2736	1.2176
		CLPT	1.1504	0.7183	0.4681	1.2639	1.2152
0/90/0	5	TSDT	1.0874	0.8032	0.4556	1.6687	1.3805
		FSDT	1.0763	0.8155	0.4578	1.6597	1.3698
		CLPT	1.0312	0.4635	0.2512	1.6645	1.3800
	10	TSDT	1.0499	0.6142	0.3275	1.6632	1.3757
		FSDT	1.0460	0.6037	0.3211	1.6640	1.3737
		CLPT	1.0312	0.4635	0.2512	1.6645	1.3800
0/90/... 10 layers	5	TSDT	1.0336	0.7993	0.5623	1.0733	1.0549
		FSDT	1.0331	0.8191	0.5847	1.0736	1.0549
		CLPT	1.0331	0.6222	0.3914	1.0681	1.0546
	10	TSDT	1.0333	0.7101	0.4880	1.0718	1.0556
		FSDT	1.0331	0.7157	0.4949	1.0722	1.0558
		CLPT	1.0331	0.6222	0.3914	1.0681	1.0546

Tables 11.5–2 and 11.5–3 contain deflections and stresses in cross-ply laminates subjected to sinusoidally distributed temperature field (see [50])

$$T(x, y, z) = z\bar{T}_1 \cos \alpha x \sin \beta y \qquad (11.5-20)$$

The following nondimensionalized variables are used:

$$\bar{w} = w_0(0, b/2)\frac{10}{\alpha_1 \bar{T}_1 b^2}, \quad \bar{\sigma}_{xx} = \sigma_{xx}(0, b/2, -h/2)\frac{10}{\alpha_1 \bar{T}_1 E_2 b} \qquad (11.5-21)$$

$$\bar{\sigma}_{yy} = -\sigma_{yy}(0, b/2, h/2)\frac{10}{\alpha_1 \bar{T}_1 E_2 b}, \quad \bar{\sigma}_{yz} = \sigma_{yz}(0, b, 0)\frac{10}{\alpha_1 \bar{T}_1 E_2 b} \qquad (11.5-22)$$

The difference between the results obtained with TSDT and FSDT is insignificant for side-to-thickness ratios greater than 5.

Table 11.5–3: Nondimensionalized thermal stresses of cross-ply square plates subjected to sinusoidally distributed temperature.

Stress	Laminate	b/h	Theory	SS	SC	CC	FF	FS	FC
$\bar{\sigma}_{xx}$	0/90/0	5	TSDT	0.1154	5.9126	14.6976	1.1490	0.6671	3.8065
			FSDT	0.4072	6.8460	15.6783	1.0220	0.7501	4.5545
			CLPT	0.0526	11.1264	15.2675	1.4489	0.8217	6.4107
		10	TSDT	0.0372	4.4944	7.5416	0.6291	0.3648	2.6993
			FSDT	0.0847	4.7320	7.7020	0.6103	0.3782	2.8579
			CLPT	0.0263	5.5632	7.6338	0.7245	0.4109	3.2052
$\bar{\sigma}_{yy}$	0/90	5	TSDT	-0.5846	2.6319	5.0886	-1.8754	-1.2598	2.6457
			FSDT	-0.6148	1.7502	3.6323	-1.7733	-1.2190	1.7814
			CLPT	-0.6148	5.1916	8.8393	-2.1091	-1.4684	4.9149
		10	TSDT	-0.3036	2.0911	3.6908	-1.0526	-0.7131	2.0023
			FSDT	-0.3074	1.8991	3.3902	-1.0399	-0.7075	1.8168
			CLPT	-0.3074	2.5958	4.4197	-1.0545	-0.7342	2.4575
$\bar{\sigma}_{yz}$	0	5	TSDT	0.1109	0.1981	0.2909	-0.0332	0.0499	0.2332
			FSDT	0.0740	0.1279	0.1921	-0.0229	0.0331	0.1515
		10	TSDT	0.0347	0.0702	0.0903	-0.0085	0.0157	0.0784
			FSDT	0.0231	0.0466	0.0601	-0.0058	0.0104	0.0520

Additional numerical results of the Lévy solutions for TSDT are presented in Section 11.6–3. For additional results the reader may consult the references at the end of the chapter. In the next section we develop the displacement finite element model of the Reddy third-order theory.

11.6 Displacement Finite Element Model

11.6–1 Introduction

Recall from Eq. (11.2–24) that the primary variables of the third-order theory developed in Section 11.2 are $(u_n, u_s, w_0, w_{0,n} = \partial w_0/\partial n, \phi_n, \phi_s)$, where (u_n, u_s) denote in-plane normal and tangential displacements, and (ϕ_n, ϕ_s) are the rotations of a transverse line about the in-plane normal and tangent. A displacement finite element model based on Eqs. (11.2–16) through (11.2–20), with $(u_0, v_0, w_0, w_{0,n}, \phi_x, \phi_y)$ as the primary variables, is called a displacement finite element model (see Phan and Reddy [30]), and it requires the Lagrange interpolation of $(u_0, v_0, \phi_x, \phi_y)$ and Hermite interpolation of w_0. A conforming element will have eight degrees of freedom $(u_0, v_0, w_0, w_{0,x}, w_{0,y}, w_{0,xy}, \phi_x, \phi_y)$ whereas a nonconforming element will have $(u_0, v_0, w_0, w_{0,x}, w_{0,y}, \phi_x, \phi_y)$ seven degrees of freedom per node.

Alternatively, one can develop mixed finite element models of Eqs. (11.2–16)–(11.2–20), in which both generalized displacements and stress resultants become the nodal variables and require only Lagrange interpolation of all variables (see Reddy and his colleagues [58–61]). The number of primary degrees of freedom per node vary from eight to eleven depending on the formulation. The C^0 displacement finite element models of third-order theories can be found in the work of Pandya and Kant [61].

Here we present a displacement finite element model of the third-order theory in Eqs. (11.2–16)–(11.2–20). In view of the detailed discussion of finite element models of the classical and first-order theories presented in Chapter 10, only the salient features of the model are discussed here.

11.6–2 Finite Element Model

The Hamilton's principle or the dynamic version of the principle of virtual displacements [or a weak form of Eqs. (11.2–16)–(11.2–20)] of a typical plate finite element is given by

$$
\begin{aligned}
0 = \int_{\Omega^e} \Big\{ & N_{xx}\delta u_{0,x} + N_{xy}\left(\delta u_{0,y} + \delta v_{0,x}\right) + N_{yy}\delta v_{0,y} + \left(Q_t - c_2 R_x\right)\delta w_{0,x} \\
& + \left(Q_y - c_2 R_y\right)\delta w_{0,y} - c_1\left(P_x \delta w_{0,xx} + P_y \delta w_{0,yy} + 2P_{xy}\delta w_{0,xy}\right) - q\delta w_0 \\
& + \left(Q_x - c_2 R_x\right)\delta\phi_x + \left(M_x - c_1 P_x\right)\delta\phi_{x,x} + \left(M_{xy} - c_1 P_{xy}\right)\delta\phi_{x,y} \\
& + \left(Q_y - c_2 R_y\right)\delta\phi_y + \left(M_{xy} - c_1 P_{xy}\right)\delta\phi_{y,x} + \left(M_{yy} - c_1 P_y\right)\delta\phi_{y,y} \\
& + \left(I_0\ddot{u}_0 + J_1\ddot{\phi}_x - c_1 I_3 \ddot{w}_{0,x}\right)\delta u_0 + \left(I_0\ddot{v}_0 + J_1\ddot{\phi}_y - c_1 I_3\ddot{w}_{0,y}\right)\delta v_0 \\
& + \left(J_1\ddot{u}_0 + K_2\ddot{\phi}_x - c_1 J_4\ddot{w}_{0,x}\right)\delta\phi_x + \left(J_1\ddot{v}_0 + K_2\ddot{\phi}_y - c_1 J_4 \ddot{w}_{0,y}\right)\delta\phi_y + I_0\ddot{w}_0\delta w_0 \\
& + c_1\left(-I_3\ddot{u}_0 - J_4\ddot{\phi}_x + c_1 I_6\ddot{w}_{0,x}\right)\delta w_{0,x} + c_1\left(-I_3\ddot{v}_0 - J_4\ddot{\phi}_y + c_1 I_6 \ddot{w}_{0,y}\right)\delta w_{0,y} \\
& + \left(\hat{N}_{xx} w_{0,x} + \hat{N}_{xy} w_{0,y}\right)\delta w_{0,x} + \left(\hat{N}_{xy} w_{0,x} + \hat{N}_{yy} w_{0,y}\right)\delta w_{0,y} \Big\} dxdy
\end{aligned}
$$

$$- \oint_{\Gamma^e} (N_n \delta u_n + N_{nx} \delta u_{ns})\, ds - \oint_{\Gamma^e} (Q_n \delta w_0 + V_n \delta w_{0,n})\, ds$$
$$- \oint_{\Gamma^e} (M_n \delta \phi_n + M_{ns} \delta \phi_{ns})\, ds \tag{11.6 – 1}$$

where a comma followed by x or y denotes differentiation with respect to the coordinates, the superposed dot denotes differentiation with respect to time, and

$$I_i = \sum_{k=1}^{N} \int_{z_k}^{z_{k+1}} \rho^{(k)} (z)^i \, dz \qquad (i = 0, 1, 2, \cdots, 6) \tag{11.6 – 2}$$

$$c_1 = \frac{4}{3h^2}\,, \quad c_2 = \frac{4}{h^2}\,, \quad J_i = I_i - c_1 I_{i+2}, \quad K_2 = I_2 - 2c_1 I_4 + c_1^2 I_6 \tag{11.6 – 3}$$

The resultants $N_{\alpha\beta}, M_{\alpha\beta}$, and so on are defined in Eq. (11.2–14), and they are known in terms of the generalized displacements through relations (11.2-28), (11.2-29), and (11.2-7).

The generalized displacements are approximated over an element Ω^e by the expressions

$$u_0(x, y, t) = \sum_{i=1}^{m} u_i^e(t) \psi_i^e(x, y)$$
$$v_0(x, y, t) = \sum_{i=1}^{m} v_i^e(t) \psi_i^e(x, y)$$
$$w_0(x, y, t) = \sum_{i=1}^{m} \bar{\Delta}_i^e(t) \varphi_i^e(x, y)$$
$$\phi_x(x, y, t) = \sum_{i=1}^{m} X_i^e(t) \psi_i^e(x, y)$$
$$\phi_y(x, y, t) = \sum_{i=1}^{m} Y_i^e(t) \psi_i^e(x, y) \tag{11.6 – 4}$$

where ψ_i^e denote the Lagrange interpolation functions and φ_i^e are the Hermite interpolation functions. Here we chose the same approximation for the in-plane displacements (u_0, v_0) and rotations (ϕ_x, ϕ_y), although one could use different approximations for these two pairs. In the case of the conforming element, the four nodal values associated with w_0 are

$$\bar{\Delta}_1 = w_0, \quad \bar{\Delta}_2 = \frac{\partial w_0}{\partial x}\,, \quad \bar{\Delta}_3 = \frac{\partial w_0}{\partial y}\,, \quad \bar{\Delta}_4 = \frac{\partial^2 w_0}{\partial x \partial y} \tag{11.6 – 5}$$

For the nonconforming element, the cross derivative is omitted. The conforming rectangular element with linear interpolation of the in-plane displacements and rotations has eight degrees of freedom per node. The corresponding nonconforming element has seven degrees of freedom per node.

Substitution of Eq. (4) into the weak form (1) yields the finite element model

$$\begin{bmatrix} [K^{11}] & [K^{12}] & [K^{13}] & [K^{14}] & [K^{15}] \\ [K^{12}]^T & [K^{22}] & [K^{23}] & [K^{24}] & [K^{25}] \\ [K^{13}]^T & [K^{23}]^T & [K^{33}] & [K^{34}] & [K^{35}] \\ [K^{14}]^T & [K^{24}]^T & [K^{34}]^T & [K^{44}] & [K^{45}] \\ [K^{15}]^T & [K^{25}]^T & [K^{35}]^T & [K^{45}]^T & [K^{55}] \end{bmatrix} \begin{Bmatrix} \{u^e\} \\ \{v^e\} \\ \{\Delta^e\} \\ \{X^e\} \\ \{Y^e\} \end{Bmatrix}$$
$$+ \begin{bmatrix} [M^{11}] & [0] & [M^{13}] & [M^{14}] & [0] \\ [0] & [M^{22}] & [M^{23}] & [0] & [M^{25}] \\ [M^{13}]^T & [M^{23}]^T & [M^{33}] & [M^{34}] & [M^{35}] \\ [M^{14}]^T & [0] & [M^{34}]^T & [M^{44}] & [0] \\ [0] & [M^{25}]^T & [M^{35}]^T & [0] & [M^{55}] \end{bmatrix} \begin{Bmatrix} \{\ddot{u}^e\} \\ \{\ddot{v}^e\} \\ \{\ddot{\Delta}^e\} \\ \{\ddot{X}^e\} \\ \{\ddot{Y}^e\} \end{Bmatrix}$$
$$+ \begin{bmatrix} [0] & [0] & [0] & [0] & [0] \\ [0] & [0] & [0] & [0] & [0] \\ [0] & [0] & [G] & [0] & [0] \\ [0] & [0] & [0] & [0] & [0] \\ [0] & [0] & [0] & [0] & [0] \end{bmatrix} \begin{Bmatrix} \{u^e\} \\ \{v^e\} \\ \{\Delta^e\} \\ \{X^e\} \\ \{Y^e\} \end{Bmatrix} = \begin{Bmatrix} \{F^1\} \\ \{F^2\} \\ \{F^3\} \\ \{F^4\} \\ \{F^5\} \end{Bmatrix} \tag{11.6-6a}$$

or, in compact form, we can write

$$\sum_{\beta=1}^{5} \sum_{j=1}^{n_\beta} \left(K_{ij}^{\alpha\beta} \Delta_j^\beta + M_{ij}^{\alpha\beta} \ddot{\Delta}_j^\beta + S_{ij}^{\alpha\beta} \Delta_j^\beta \right) - F_i^\alpha = 0, \quad i = 1, 2, \ldots, n_\alpha \tag{11.6-6b}$$

where $\alpha = 1, 2, 3, 4, 5$; $n_1 = n_2 = n_4 = n_5 = 4$ and $n_3 = 16$ for the conforming element. The nodal values Δ_j^β, stiffness coefficients $K_{ij}^{\alpha\beta}$, mass coefficients $M_{ij}^{\alpha\beta}$, and geometric stiffness coefficients $S_{ij}^{\alpha\beta}$ are defined by

$$\Delta_j^1 = u_j, \quad \Delta_j^2 = v_j, \quad \Delta_j^3 = \bar{\Delta}_j, \quad \Delta_j^4 = X_j^1, \quad \Delta_j^5 = Y_j^2 \tag{11.6-7}$$

$$\begin{aligned} K_{ij}^{1\alpha} &= \int_{\Omega^e} \left(\frac{\partial \psi_i}{\partial x} N_{1j}^\alpha + \frac{\partial \psi_i}{\partial y} N_{6j}^\alpha \right) dxdy \\ K_{ij}^{2\alpha} &= \int_{\Omega^e} \left(\frac{\partial \psi_i}{\partial x} N_{6j}^\alpha + \frac{\partial \psi_i}{\partial y} N_{2j}^\alpha \right) dxdy \\ K_{ij}^{3\alpha} &= \int_{\Omega^e} \left[\frac{\partial \varphi_i}{\partial x} \hat{Q}_{1j}^\alpha + \frac{\partial \varphi_i}{\partial y} \hat{Q}_{2j}^\alpha - c_1 \left(\frac{\partial^2 \varphi_i}{\partial x^2} P_{1j}^\alpha + 2 \frac{\partial^2 \varphi_i}{\partial x \partial y} P_{6j}^\alpha \right.\right. \\ &\qquad \left.\left. + \frac{\partial^2 \varphi_i}{\partial y^2} P_{2j}^\alpha \right) \right] dxdy \\ K_{ij}^{4\alpha} &= \int_{\Omega^e} \left(\frac{\partial \psi_i}{\partial x} \hat{M}_{1j}^\alpha + \frac{\partial \psi_i}{\partial y} \hat{M}_{6j}^\alpha + \psi_i \hat{Q}_{1j}^\alpha \right) dxdy \\ K_{ij}^{5\alpha} &= \int_{\Omega^e} \left(\frac{\partial \psi_i}{\partial x} \hat{M}_{6j}^\alpha + \frac{\partial \psi_i}{\partial y} \hat{M}_{2j}^\alpha + \psi_i \hat{Q}_{2j}^\alpha \right) dxdy \end{aligned} \tag{11.6-8}$$

$$
\begin{aligned}
M_{ij}^{11} &= I_0 S_{ij}^0, \quad M_{ij}^{12} = 0, \quad M_{ij}^{13} = -c_1 I_3 S_{ij}^{ox}, \quad M_{ij}^{14} = J_1 S_{ij}^0, \quad M_{ij}^{15} = 0 \\
M_{ij}^{22} &= I_0 S_{ij}^0, \quad M_{ij}^{23} = -c_1 I_3 S_{ij}^{oy}, \quad M_{ij}^{24} = 0, \quad M_{ij}^{25} = J_1 S_{ij}^0 \\
M_{ij}^{33} &= I_0 S_{ij}^1 + c_1^2 I_6 \left(S_{ij}^{xx} + S_{ij}^{yy} \right), \quad M_{ij}^{34} = J_4 S_{ji}^{ox}, \; M_{ij}^{35} = J_4 S_{ji}^{oy} \\
M_{ij}^{44} &= K_2 S_{ij}^0, \quad M_{ij}^{45} = 0, \quad M_{ij}^{55} = K_2 S_{ij}^0
\end{aligned} \tag{11.6-9}
$$

$$
\begin{aligned}
S_{ij}^{33} &= G_{ij} = \hat{N}_{xx} S_{ij}^{xx} + \hat{N}_{xy} \left(S_{ij}^{xy} + S_{ij}^{yx} \right) + \hat{N}_{yy} S_{ij}^{yy} \\
F_i^1 &= \oint_{\Gamma^e} \left(N_{xx} n_x + N_{xy} n_y \right) \psi_i \, ds, \quad F_i^2 = \oint_{\Gamma^e} \left(N_{xy} n_x + N_{yy} n_y \right) \psi_i \, ds \\
F_i^4 &= \oint_{\Gamma^e} \left(M_{xx} n_x + M_{xy} n_y \right) \psi_i \, ds, \quad F_i^5 = \oint_{\Gamma^e} \left(M_{xy} n_x + M_{yy} n_y \right) \psi_i \, ds \\
F_i^3 &= \int_{\Omega^e} q \varphi_i \, dxdy + \oint_{\Gamma^e} \left(Q_n \varphi_i + V_n \frac{\partial w_0}{\partial n} \right) \psi_i \, ds
\end{aligned} \tag{11.6-10}
$$

The following notation is used in defining the above coefficients:

$$
\begin{aligned}
S_{ij}^0 &= \int_{\Omega^e} \psi_i \psi_j \, dxdy, \quad S_{ij}^{0x} = \int_{\Omega^e} \psi_i \frac{\partial \varphi_j}{\partial x} \, dxdy \\
S_{ij}^{0y} &= \int_{\Omega^e} \psi_i \frac{\partial \varphi_i}{\partial y} \, dxdy, \quad S_{ij}^1 = \int_{\Omega^e} \varphi_i \varphi_j dxdy \\
S_{ij}^{xx} &= \int_{\Omega^e} \frac{\partial \varphi_i}{\partial x} \frac{\partial \varphi_j}{\partial x} \, dxdy, \quad S_{ij}^{yy} = \int_{\Omega^e} \frac{\partial \varphi_i}{\partial y} \frac{\partial \varphi_j}{\partial y} \, dxdy
\end{aligned} \tag{11.6-11}
$$

$$
\begin{aligned}
N_{1j}^1 &= A_{11} \frac{\partial \psi_j}{\partial x} + A_{16} \frac{\partial \psi_j}{\partial y}, \quad N_{1j}^2 = A_{16} \frac{\partial \psi_j}{\partial x} + A_{12} \frac{\partial \psi_j}{\partial y} \\
N_{1j}^3 &= -c_1 \left(E_{11} \frac{\partial^2 \varphi_j}{\partial x^2} + 2E_{16} \frac{\partial^2 \varphi_j}{\partial x \partial y} + E_{12} \frac{\partial^2 \varphi_j}{\partial^2} \right) \\
N_{1j}^4 &= \hat{B}_{11} \frac{\partial \psi_j}{\partial x} + \hat{B}_{16} \frac{\partial \psi_j}{\partial y}, \quad N_{1j}^5 = \hat{B}_{12} \frac{\partial \psi_j}{\partial x} + \hat{B}_{16} \frac{\partial \psi_j}{\partial y} \\
N_{6j}^1 &= A_{16} \frac{\partial \psi_j}{\partial x} + A_{66} \frac{\partial \psi_j}{\partial y}, \quad N_{6j}^2 = A_{66} \frac{\partial \psi_j}{\partial x} + A_{26} \frac{\partial \psi_j}{\partial y} \\
N_{6j}^3 &= -c_1 \left(E_{16} \frac{\partial^2 \varphi_j}{\partial x^2} + 2E_{66} \frac{\partial^2 \varphi_j}{\partial x \partial y} + E_{26} \frac{\partial^2 \varphi_j}{\partial^2} \right) \\
N_{6j}^4 &= \hat{B}_{16} \frac{\partial \psi_j}{\partial x} + \hat{B}_{66} \frac{\partial \psi_j}{\partial y}, \quad N_{6j}^5 = \hat{B}_{26} \frac{\partial \psi_j}{\partial x} + \hat{B}_{66} \frac{\partial \psi_j}{\partial y} \\
N_{2j}^1 &= A_{12} \frac{\partial \psi_j}{\partial x} + A_{26} \frac{\partial \psi_j}{\partial y}, \quad N_{2j}^2 = A_{26} \frac{\partial \psi_j}{\partial x} + A_{22} \frac{\partial \psi_j}{\partial y} \\
N_{2j}^3 &= -c_1 \left(E_{12} \frac{\partial^2 \varphi_j}{\partial x^2} + 2E_{26} \frac{\partial^2 \varphi_j}{\partial x \partial y} + E_{22} \frac{\partial^2 \varphi_j}{\partial^2} \right) \\
N_{2j}^4 &= \hat{B}_{12} \frac{\partial \psi_j}{\partial x} + \hat{B}_{26} \frac{\partial \psi_j}{\partial y}, \quad N_{2j}^5 = \hat{B}_{22} \frac{\partial \psi_j}{\partial x} + \hat{B}_{26} \frac{\partial \psi_j}{\partial y}
\end{aligned}
$$

$$
\begin{aligned}
\hat{M}^1_{1j} =&\hat{B}_{11}\frac{\partial\psi_j}{\partial x}+\hat{B}_{16}\frac{\partial\psi_j}{\partial y}\ ,\quad \hat{M}^2_{2j}=\hat{B}_{16}\frac{\partial\psi_j}{\partial x}+\hat{B}_{12}\frac{\partial\psi_j}{\partial y}\\
\hat{M}^3_{1j} =&-c_1\left(\hat{F}_{11}\frac{\partial^2\varphi_j}{\partial x^2}+2\hat{F}_{16}\frac{\partial^2\varphi_j}{\partial x\partial y}+\hat{F}_{12}\frac{\partial^2\varphi_j}{\partial^2}\right)\\
\hat{M}^4_{1j} =&\bar{D}_{11}\frac{\partial\psi_j}{\partial x}+\bar{D}_{16}\frac{\partial\psi_j}{\partial y}\ ,\quad \hat{M}^5_{1j}=\hat{\bar{D}}_{12}\frac{\partial\psi_j}{\partial x}+\bar{D}_{16}\frac{\partial\psi_j}{\partial y}\\
\hat{M}^1_{6j} =&\hat{B}_{16}\frac{\partial\psi_j}{\partial x}+\hat{B}_{66}\frac{\partial\psi_j}{\partial y}\ ,\quad \hat{M}^2_{6j}=\hat{B}_{66}\frac{\partial\psi_j}{\partial x}+\hat{B}_{26}\frac{\partial\psi_j}{\partial y}\\
\hat{M}^3_{6j} =&-c_1\left(\hat{F}_{16}\frac{\partial^2\varphi_j}{\partial x^2}+2\hat{F}_{66}\frac{\partial^2\varphi_j}{\partial x\partial y}+\hat{F}_{26}\frac{\partial^2\varphi_j}{\partial^2}\right)\\
\hat{M}^4_{6j} =&\bar{D}_{16}\frac{\partial\psi_j}{\partial x}+\bar{D}_{66}\frac{\partial\psi_j}{\partial y}\ ,\quad \hat{M}^5_{6j}=\bar{D}_{26}\frac{\partial\psi_j}{\partial x}+\bar{D}_{66}\frac{\partial\psi_j}{\partial y}\\
\hat{M}^1_{2j} =&\hat{B}_{12}\frac{\partial\psi_j}{\partial x}+\hat{B}_{26}\frac{\partial\psi_j}{\partial y}\ ,\quad \hat{M}^2_{2j}=\hat{B}_{26}\frac{\partial\psi_j}{\partial x}+\hat{B}_{22}\frac{\partial\psi_j}{\partial y}\\
\hat{M}^3_{2j} =&-c_1\left(\hat{F}_{12}\frac{\partial^2\varphi_j}{\partial x^2}+2F_{26}\frac{\partial^2\varphi_j}{\partial x\partial y}+F_{22}\frac{\partial^2\varphi_j}{\partial^2}\right)\\
\hat{M}^4_{2j} =&\bar{D}_{12}\frac{\partial\psi_j}{\partial x}+\bar{D}_{26}\frac{\partial\psi_j}{\partial y}\ ,\quad \hat{M}^5_{2j}=\bar{D}_{22}\frac{\partial\psi_j}{\partial x}+\bar{D}_{26}\frac{\partial\psi_j}{\partial y}
\end{aligned}
\qquad (11.6-12)
$$

$$
\begin{aligned}
P^1_{1j} =&E_{11}\frac{\partial\psi_j}{\partial x}+E_{16}\frac{\partial\psi_j}{\partial y}\ ,\quad P^2_{1j}=E_{12}\frac{\partial\psi_j}{\partial x}+E_{16}\frac{\partial\psi_j}{\partial y}\\
P^3_{1j} =&-c_1\left(H_{11}\frac{\partial^2\varphi_j}{\partial x^2}+2H_{16}\frac{\partial^2\varphi_j}{\partial x\partial y}+H_{12}\frac{\partial^2\varphi_j}{\partial^2}\right)\\
P^4_{1j} =&\hat{F}_{11}\frac{\partial\psi_j}{\partial x}+\hat{F}_{16}\frac{\partial\psi_j}{\partial y}\ ,\quad P^5_{1j}=\hat{F}_{12}\frac{\partial\psi_j}{\partial x}+\hat{F}_{16}\frac{\partial\psi_j}{\partial y}\\
P^1_{2j} =&E_{12}\frac{\partial\psi_j}{\partial x}+E_{26}\frac{\partial\psi_j}{\partial y}\ ,\quad P^2_{2j}=E_{22}\frac{\partial\psi_j}{\partial x}+E_{26}\frac{\partial\psi_j}{\partial y}\\
P^3_{2j} =&-c_1\left(H_{21}\frac{\partial^2\varphi_j}{\partial x^2}+2H_{26}\frac{\partial^2\varphi_j}{\partial x\partial y}+H_{22}\frac{\partial^2\varphi_j}{\partial^2}\right)\\
P^4_{2j} =&\hat{F}_{12}\frac{\partial\psi_j}{\partial x}+\hat{F}_{26}\frac{\partial\psi_j}{\partial y}\ ,\quad P^5_{2j}=\hat{F}_{22}\frac{\partial\psi_j}{\partial x}+\hat{F}_{26}\frac{\partial\psi_j}{\partial y}\\
P^1_{6j} =&E_{16}\frac{\partial\psi_j}{\partial x}+E_{66}\frac{\partial\psi_j}{\partial y}\ ,\quad P^2_{6j}=E_{26}\frac{\partial\psi_j}{\partial x}+E_{66}\frac{\partial\psi_j}{\partial y}\\
P^3_{6j} =&-c_1\left(H_{16}\frac{\partial^2\varphi_j}{\partial x^2}+2H_{66}\frac{\partial^2\varphi_j}{\partial x\partial y}+H_{26}\frac{\partial^2\varphi_j}{\partial^2}\right)\\
P^4_{6j} =&\hat{F}_{16}\frac{\partial\psi_j}{\partial x}+\hat{F}_{66}\frac{\partial\psi_j}{\partial y}\ ,\quad P^5_{6j}=\hat{F}_{26}\frac{\partial\psi_j}{\partial x}+\hat{F}_{66}\frac{\partial\psi_j}{\partial y}\\
\hat{Q}^3_{1j} =&\bar{A}_{45}\frac{\partial\varphi_j}{\partial x}+\bar{A}_{55}\frac{\partial\varphi_j}{\partial y}\ ,\quad \hat{Q}^3_{2j}=\bar{A}_{44}\frac{\partial\varphi_j}{\partial x}+\bar{A}_{45}\frac{\partial\varphi_j}{\partial y}\\
\hat{Q}^4_{1j} =&\hat{A}_{55}\psi_j,\quad \hat{Q}^5_{1j}=\hat{Q}^4_{2j}=\hat{A}_{45}\psi_j,\quad \hat{Q}^5_{2j}=\hat{A}_{44}\psi_j
\end{aligned}
\qquad (11.6-13)
$$

$$F_i^1 = \oint_{\Gamma^e} \psi_i N_n \, ds, \quad F_i^2 = \oint_{\Gamma^e} \psi_i N_{ns} \, ds$$
$$F_i^4 = \oint_{\Gamma^e} \psi_i M_n \, ds, \quad F_i^5 = \oint_{\Gamma^e} \psi_i M_{ns} \, ds$$
$$F_i^3 = \int_{\Omega^e} q\varphi_i \, dxdy + \oint_{\Gamma^e} \varphi_i Q_n \, ds \qquad (11.6-14)$$

Note that the stiffness matrix evaluation requires the computation of the second derivatives of the interpolation functions used for the transverse deflection. In the present study we use the Lagrange linear interpolation of the geometry (*i.e.*, the same as that used for variables u, v, ϕ_x and ϕ_y):

$$x = \sum_{j=1}^{n} x_j \psi_j, \quad y = \sum_{j=1}^{n} y_j \psi_j \qquad (11.6-15)$$

Thus isoparametric elements are used for $(u_0, v_0, \phi_x, \phi_y)$ and subparametric formulation is used for w_0. We must develop the transformation equations to numerically evaluate the stiffness coefficients. These relations can be found in the textbook by Reddy [62], pp. 435–436. The comments made in connection with the displacement finite element model of the first-order theory also apply for the present element in evaluating the shear stiffness coefficients.

11.6–3 Numerical Results

For the purpose of comparison, the following lamina properties, typical of graphite-epoxy material, are used in all numerical examples presented here:

Material 1 : $E_1 = 25E_2, \; G_{12} = G_{13} = 0.5E_2, \; G_{23} = 0.2E_2, \; \nu_{12} = 0.25$ (11.6 – 16)

Material 2 : $E_1 = 40E_2, \; G_{12} = G_{13} = 0.6E_2, \; G_{23} = 0.5E_2, \; \nu_{12} = 0.25$ (11.6 – 17)

The shear correction coefficients for the first-order theory are taken to be 5/6. The loading, in all cases, is assumed to be sinusoidal: (see Figure 11.5–1 for the geometry and coordinate system.)

$$q(x, y) = q_0 \cos \frac{\pi x}{a} \sin \frac{\pi y}{b} \qquad (11.6-18)$$

The notation SC, for example, refers to the boundary conditions used on the edges $x = \pm a/2$, while the other two edges ($i.e., y = 0, b$), in all cases, are simply supported.

Bending Results

The results for deflections and stresses are presented in tables using the following nondimensional form (see [51]):

$$\begin{aligned}\bar{w} =& w_0(0, b/2)\frac{h^3E_2}{b^4q_0} \times 10^2, \quad \bar{\sigma}_{xx} = -\sigma_{xx}(0, \frac{b}{2}, -\frac{h}{2})\frac{h^2}{b^2q_0}10 \\ \bar{\sigma}_{yy} =& -\sigma_{yy}(0, \frac{b}{2}, \frac{h}{2})\frac{h^2}{b^2q_0}10, \quad \bar{\sigma}_{yz} = -\sigma_{yz}(0, 0, 0)\frac{h}{bq_0}10 \qquad (11.6-19)\end{aligned}$$

where h is the total thickness of the laminate.

Tables 11.6–1 through 11.6–4 contain numerical values of deflections and stresses obtained by the Lévy or the Navier method, and the finite element model. The reduced integration rule is used to evaluate the shear stiffness coefficients. Quarter-plate models with 2×2 mesh are used for SS, CC, and FF boundary conditions, and half-plate models with 4×2 mesh are used for all other boundary conditions. The finite element results are in good agreement with the analytical solutions.

Table 11.6–1: Nondimensionalized center deflection ($\bar{w}$) of antisymmetric cross-ply square plates with various boundary conditions.

No. of Layers	b/h	Theory	Solution	SS	SC	CC	FF	FS	FC
		TSDT	Exact	1.667	1.333	1.088	2.624	2.211	1.733
			FEM	1.667	1.317	1.068	2.647	2.221	1.728
	5	FSDT	Exact	1.758	1.477	1.257	2.777	2.335	1.897
			FEM	1.759	1.478	1.257	2.776	2.334	1.897
		CLPT	Exact	1.064	0.664	0.429	1.777	1.471	0.980
			FEM	1.043	0.648	0.417	1.786	1.465	0.977
2									
		TSDT	Exact	1.216	0.848	0.617	1.992	1.658	1.184
			FEM	1.214	0.838	0.605	2.002	1.662	1.180
	10	FSDT	Exact	1.237	0.883	0.656	2.028	1.687	1.223
			FEM	1.238	0.883	0.657	2.027	1.687	1.223
		CLPT	Exact	1.064	0.664	0.429	1.777	1.471	0.980
			FEM	1.043	0.648	0.417	1.786	1.465	0.977

Table 11.6–1: continued.

No. of Layers	b/h	Theory	Solution	SS	SC	CC	FF	FS	FC
		TSDT	Exact	1.129	1.001	0.879	1.651	1.450	1.214
			FEM	1.135	0.995	0.869	1.670	1.461	1.214
	5	FSDT	Exact	1.137	1.045	0.945	1.663	1.460	1.258
			FEM	1.137	1.045	0.945	1.662	1.460	1.258
		CLPT	Exact	0.442	0.266	0.167	0.665	0.579	0.380
			FEM	0.444	0.266	0.169	0.686	0.593	0.391
10									
		TSDT	Exact	0.616	0.473	0.375	0.916	0.801	0.607
			FEM	0.619	0.471	0.372	0.926	0.808	0.609
		FSDT	Exact	0.615	0.480	0.385	0.915	0.800	0.612
	10		FEM	0.616	0.480	0.386	0.914	0.800	0.612
		CLPT	Exact	0.442	0.266	0.167	0.665	0.579	0.380
			FEM	0.444	0.266	0.169	0.686	0.593	0.391

Table 11.6–2: Nondimensionalized axial stress ($\bar{\sigma}_{xx}$) of antisymmetric cross-ply square plates with various boundary conditions.

No. of Layers	b/h	Theory	Solution	SS	SC	CC	FF	FS	FC
		TSDT	Exact	8.385	6.816	5.679	3.171	5.349	3.727
			FEM	7.669	6.732	5.060	2.722	5.231	3.544
	5	FSDT	Exact	7.157	5.338	3.911	2.469	4.430	2.434
			FEM	6.948	5.465	3.707	2.359	4.479	2.542
		CLPT	Exact	7.157	5.660	4.800	2.403	4.442	3.042
			FEM	6.659	5.782	4.348	2.034	4.288	2.991
2									
		TSDT	Exact	7.468	5.910	4.952	2.624	4.669	3.158
			FEM	6.829	5.914	4.346	2.212	4.537	3.054
	10	FSDT	Exact	7.157	5.494	4.450	2.442	4.435	2.790
			FEM	6.948	5.668	4.222	2.331	4.491	2.895
		CLPT	Exact	7.157	5.660	4.800	2.403	4.442	3.042
			FEM	6.659	5.782	4.348	2.034	4.288	2.991

Table 11.6–2: continued.

No. of Layers	b/h	Theory	Solution	SS	SC	CC	FF	FS	FC
		TSDT	Exact	6.340	5.196	4.025	2.482	3.946	2.608
			FEM	5.762	5.047	3.584	2.124	3.855	2.420
	5	FSDT	Exact	5.009	3.707	2.275	1.712	2.957	1.343
			FEM	4.864	3.755	2.154	1.639	2.993	1.427
		CLPT	Exact	5.009	3.829	3.167	1.725	2.986	1.865
			FEM	4.611	3.911	2.798	1.324	2.806	1.700
10									
		TSDT	Exact	5.346	4.066	3.193	1.924	3.221	1.954
			FEM	4.842	4.030	2.770	1.597	3.102	1.839
		FSDT	Exact	5.009	3.642	2.692	1.723	2.968	1.594
	10		FEM	4.863	3.754	2.550	1.648	3.010	1.674
		CPT	Exact	5.009	3.829	3.167	1.725	2.986	1.865
			FEM	4.611	3.911	2.798	1.324	2.806	1.700

Table 11.6–3: Nondimensionalized axial stress ($\bar{\sigma}_{yy}$) of antisymmetric cross-ply square plates with various boundary conditions.

No. of Layers	b/h	Theory	Solution	SS	SC	CC	FF	FS	FC
		TSDT	Exact	8.385	6.725	5.505	13.551	11.324	8.919
			FEM	7.669	6.285	4.886	13.142	10.182	9.215
	5	FSDT	Exact	7.157	6.034	5.153	11.907	9.848	8.047
			FEM	6.948	5.914	4.990	11.675	9.140	8.367
		CLPT	Exact	7.157	4.483	2.914	11.849	9.837	6.560
			FEM	6.659	4.393	2.615	11.614	8.878	7.181
2		TSDT	Exact	7.468	5.219	3.803	12.295	10.218	7.314
			FEM	6.829	4.932	3.345	11.890	9.138	7.725
	10	FSDT	Exact	7.157	5.109	3.799	11.884	9.847	7.150
			FEM	6.948	5.082	3.661	11.654	9.1201	7.610
		CLPT	Exact	7.157	4.483	2.914	11.849	9.837	6.560
			FEM	6.659	4.393	2.615	11.614	8.878	7.181

Table 11.6–3: continued.

No. of Layers	b/h	Theory	Solution	SS	SC	CC	FF	FS	FC
		TSDT	Exact	6.340	5.635	4.963	9.454	8.252	6.934
			FEM	5.762	5.209	4.410	9.113	7.424	6.994
	5	FSDT	Exact	5.009	4.628	4.212	7.583	6.590	5.706
			FEM	4.864	4.511	4.086	7.429	6.141	5.844
		CLPT	Exact	5.009	3.025	1.911	7.480	6.531	4.284
			FEM	4.611	2.942	1.694	7.395	5.935	4.782
10		TSDT	Exact	5.346	4.110	3.260	8.005	6.987	5.299
			FEM	4.842	3.817	2.871	7.708	6.246	5.483
	10	FSDT	Exact	5.009	3.904	3.135	7.533	6.566	5.029
			FEM	4.863	3.852	3.031	7.384	6.097	5.279
		CLPT	Exact	5.009	3.025	1.911	7.480	6.531	4.284
			FEM	4.611	2.942	1.694	7.395	5.935	4.782

Table 11.6–4: Nondimensionalized transverse shear ($\bar{\sigma}_{yz}$) of antisymmetric cross-ply square plates with various boundary conditions.

No. of Layers	b/h	Theory	Solution	SS	SC	CC	FF	FS	FC
	5	TSDT	Exact	3.155	2.543	2.095	4.457	3.893	3.048
		FSDT	Exact	2.729	2.297	1.958	3.901	3.390	2.748
2									
	10	TSDT	Exact	3.190	2.290	1.725	4.489	3.927	2.805
		FSDT	Exact	2.729	1.993	1.523	3.882	3.383	2.449
	5	TSDT	Exact	3.362	2.974	2.601	4.784	4.234	3.535
		FSDT	Exact	2.729	2.498	2.248	3.883	3.437	2.951
10									
	10	TSDT	Exact	3.408	2.622	2.083	4.814	4.275	3.225
		FSDT	Exact	2.729	2.126	1.708	3.853	3.421	2.605

Natural Vibration and Buckling Results

Tables 11.6–5 and 11.6–6 contain nondimensionalized fundamental frequencies and critical buckling loads, respectively, of antisymmetric cross-ply square laminates for various boundary conditions. Both the Lévy and finite element results are presented in the tables. A 2×2 mesh of nine-node quadratic elements is used in FSDT and a 4×4 mesh of conforming elements is used in TSDT. The rotary inertia is accounted for in the vibration analysis. The first-order theory underpredicts fundamental frequencies and critical buckling loads when compared to the third-order theory. Table 11.6–7 contains natural frequencies of a two-layer (0/90) cantilever plate as predicted by various theories. Figures 11.6–1 and 11.6–2 contain a comparison of the finite element (FEM) results with the closed-form solutions (CFS) for antisymmetric angle-ply plates.

Table 11.6–5: Effect of side-to-thickness ratio on the dimensionless frequencies, $\bar{\omega} = \omega(b^2/h)\sqrt{\rho/E_2}$, of antisymmetric cross-ply square plates (Material 2).

No. of Layers	b/h	Theory	Type of Solution	FF	FS	FC	SS	SC	CC
2	5	TSDT	Exact	6.128	6.387	6.836	9.087	10.393	11.890
			FEM	6.172	6.192	6.648	9.103	10.582	12.053
		FSDT	Exact	5.952	6.213	6.638	8.833	9.822	10.897
			FEM	5.955	6.219	6.646	8.837	9.899	10.906
		CLPT	Exact	7.124	7.450	8.041	10.721	13.627	17.741
			FEM	7.150	7.279	7.802	11.192	15.357	18.694
	10	TSDT	Exact	6.943	7.277	7.810	10.568	12.870	15.709
			FEM	6.915	7.134	7.680	10.594	13.180	15.914
		FSDT	Exact	6.881	7.215	7.741	10.473	12.610	15.152
			FEM	6.886	7.222	7.714	10.480	12.791	15.181
		CLPT	Exact	7.267	7.636	8.228	11.154	14.223	18.543
			FEM	7.262	7.345	7.821	11.383	14.828	19.053
10	5	TSDT	Exact	8.155	8.288	8.966	11.673	12.514	13.568
			FEM	7.989	7.998	8.694	11.664	12.633	13.710
		FSDT	Exact	8.139	8.264	8.919	11.644	12.197	12.923
			FEM	8.143	8.270	8.925	11.647	12.239	12.928
		CLPT	Exact	11.459	11.815	13.618	12.167	23.348	30.855
			FEM	12.156	11.260	11.980	18.624	24.118	31.855
	10	TSDT	Exact	10.893	11.074	11.863	15.771	18.175	20.831
			FEM	10.906	11.088	11.788	15.787	18.214	20.493
		FSDT	Exact	10.900	11.079	11.862	15.779	18.044	20.471
			FEM	10.906	11.088	11.788	15.787	18.214	20.493
		CLPT	Exact	12.680	12.906	13.779	18.492	23.971	31.709
			FEM	12.419	11.283	11.983	18.637	23.991	31.912

Table 11.6–6: Effect of side-to-thickness ratio on the dimensionless critical buckling loads, $\bar{N} = N^0_{yy}\frac{b^2}{E_2h^3}$, of antisymmetric cross-ply square plates under uniaxial compression (Material 2).

No. of Layers	b/h	Theory	Type of Solution	FF	FS	FC	SS	SC	CC
2	5	TSDT	Exact	3.905	4.283	4.908	8.769	10.754	11.490
			FEM	3.979	4.375	5.022	8.985	11.241	12.318
		FSDT	Exact	3.682	4.054	4.632	8.277	9.309	9.757
			FEM	3.719	4.094	4.667	8.328	9.650	9.949
		CLPT	Exact	5.425	6.003	6.968	12.957	21.116	31.280
			FEM	5.616	6.292	7.203	14.520	23.869	37.106
	10	TSDT	Exact	4.940	5.442	6.274	11.562	17.133	21.464
			FEM	5.090	5.621	6.487	12.011	18.257	24.262
		FSDT	Exact	4.851	5.351	6.166	11.353	16.437	20.067
			FEM	4.916	5.420	6.234	11.485	18.338	21.916
		CLPT	Exact	5.425	6.003	6.968	12.957	21.116	31.280
			FEM	5.616	6.292	7.203	14.520	23.869	37.106
10	5	TSDT	Exact	6.780	7.050	8.221	12.109	12.607	13.254
			FEM	6.802	7.089	8.278	12.224	12.800	13.659
		FSDT	Exact	6.750	7.020	8.143	11.494	11.495	11.628
			FEM	6.791	7.064	8.174	11.172	11.181	11.216
		CLPT	Exact	16.426	17.023	19.389	35.232	59.288	89.770
			FEM	16.457	17.141	19.422	36.384	60.406	90.833
	10	TSDT	Exact	12.077	12.506	14.351	25.423	32.885	35.376
			FEM	12.248	12.699	14.568	25.828	33.662	36.657
		FSDT	Exact	12.092	12.524	14.358	25.450	32.614	34.837
			FEM	12.226	12.661	14.480	25.647	33.970	36.129
		CLPT	Exact	16.426	17.023	19.389	35.232	59.288	89.770
			FEM	16.457	17.141	19.422	36.384	60.406	90.833

Table 11.6–7: Fundamental frequencies, $\bar{\omega} = \omega(b^2/h)\sqrt{\rho/E_2}$, of a cantilever plate (0/90) as predicted by various theories (Material 2).

	CLPT		FSDT		TSDT	
b/a	$b/h = 100$	$b/h = 10$	$b/h = 100$	$b/h = 10$	$b/h = 100$	$b/h = 10$
1	2.6285	2.6250	2.6103	2.5334	2.6378	2.5610
2	10.5138	10.4588	10.4318	9.3501	10.5385	9.5988
3	23.6548	23.3775	23.4354	18.8491	23.6666	19.8325

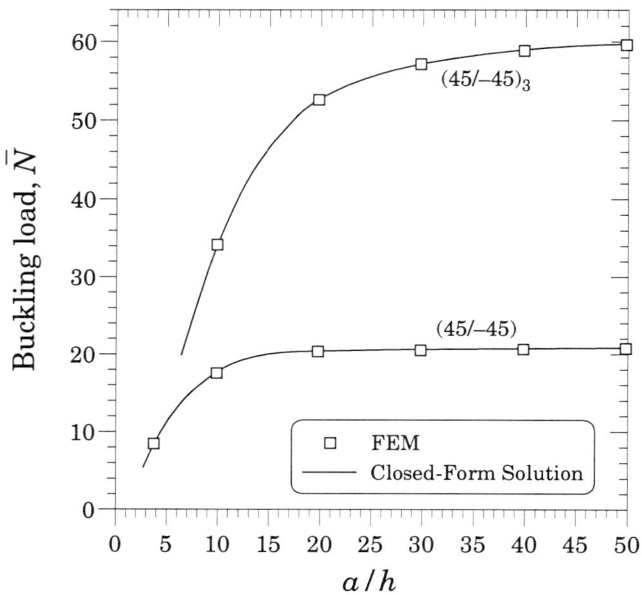

Figure 11.6–1: Nondimensionalized uniaxial critical buckling load versus side-to-thickness ratio for simply supported antisymmetric angle-ply $(45/{-45})_n$ $(n = 1, 3)$ square plates (Material 2).

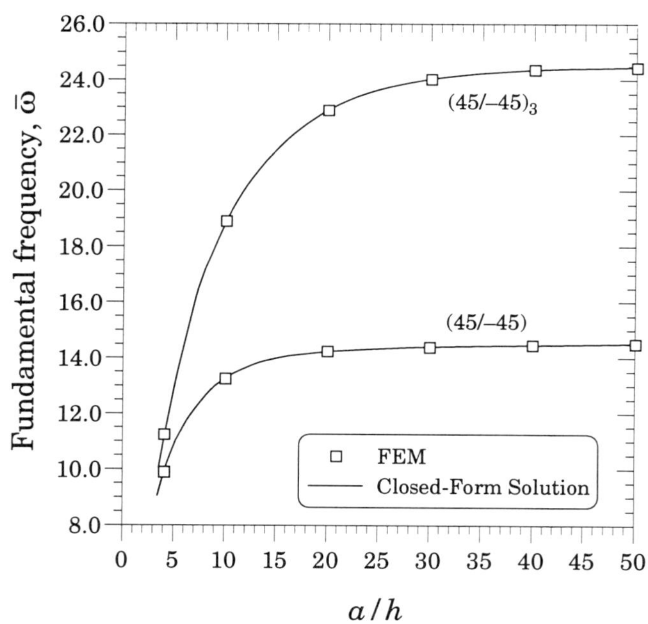

Figure 11.6–2: Nondimensionalized fundamental frequency versus side-to-thickness ratio for simply supported antisymmetric angle-ply $(45/{-45})_n$ $(n = 1, 3)$ square plates (Material 2).

Exercises

11.1 Suppose that the displacements (u, v, w) along the three coordinate axes (x, y, z) in a laminated beam can be expressed as

$$\begin{aligned} u(x,z) &= u_0(x) + z\left[c_0\frac{dw_0}{dx} + c_1\phi(x)\right] + c_2 z^2\psi(x) + c_3\left(\frac{z}{h}\right)^3\left(\phi + \frac{dw_0}{dx}\right) \\ v(x,z) &= 0 \\ w(x,z) &= w_0(x) \end{aligned} \tag{1}$$

where (u_0, w_0) denote the displacements of a point $(x, y, 0)$ along the x and z directions, respectively, ϕ denotes the rotation of a transverse normal about the $y-$axis. Show that the nonzero linear strains are given by

$$\begin{aligned} \varepsilon_{xx} &= \varepsilon_{xx}^{(0)} + z\varepsilon_{xx}^{(1)} + z^2\varepsilon_{xx}^{(2)} + z^3\varepsilon_{xx}^{(3)} \\ \gamma_{xz} &= \gamma_{xz}^{(0)} + z\gamma_{xz}^{(1)} + z^2\gamma_{xz}^{(2)} \end{aligned} \tag{2a}$$

where

$$\varepsilon_{xx}^{(0)} = \frac{du_0}{dx}\ ,\ \varepsilon_{xx}^{(1)} = c_0\frac{d^2w_0}{dx^2} + c_1\frac{d\phi}{dx}\ ,\ \varepsilon_{xx}^{(2)} = c_2\frac{d\psi}{dx}\ ,\ \varepsilon_{xx}^{(3)} = \frac{c_3}{h^3}\left(\frac{d\phi}{dx} + \frac{d^2w_0}{dx^2}\right)$$

$$\gamma_{xz}^{(0)} = (1 + c_0)\frac{dw_0}{dx} + c_1\phi,\ \gamma_{xz}^{(1)} = 2c_2\psi,\ \gamma_{xz}^{(3)} = \frac{3c_3}{h^3}(\phi + \frac{dw_0}{dx}) \tag{2b}$$

11.2 (*Continuation of Exercise 1.*) Use the principle of virtual displacements to derive the equations of equilibrium and the natural and essential boundary conditions associated with the displacement field of Exercise 1. In particular show that

$$\begin{aligned} \delta u_0 :&\quad \frac{dN_x}{dx} + p = 0 \\ \delta\phi :&\quad \frac{d}{dx}(c_1 M_x) + \frac{d}{dx}\left(\frac{c_3}{h^3}P_x\right) - c_1 Q_x - \frac{3c_3}{h^3}S_x = 0 \\ \delta\psi :&\quad \frac{d}{dx}(c_2 L_x) - 2c_2 R_x = 0 \\ \delta w_0 :&\quad \frac{d^2}{dx^2}\left(c_0 M_x + \frac{c_3}{h^3}P_x\right) - (1 + c_0)\frac{dQ_x}{dx} - \frac{3c_3}{h^3}\frac{dS_x}{dx} - f = 0 \end{aligned} \tag{1}$$

and the boundary conditions involve specifying

$$\begin{aligned} N_x &\quad \text{or} \quad u \\ c_1 M_x + \frac{c_3}{h^3}P_x &\quad \text{or} \quad \phi \\ c_2 L_x &\quad \text{or} \quad \psi \\ -\frac{d}{dx}\left(c_0 M_x + \frac{c_3}{h^3}P_x\right) + (1 + c_0)Q_x + \frac{3c_3}{h^3}S_x &\quad \text{or} \quad w \\ c_0 M_x + \frac{c_3}{h^3}P_x &\quad \text{or} \quad \frac{dw}{dx} \end{aligned} \tag{2}$$

where

$$N_x = \int_A \sigma_{xx} dA,\ M_{xx} = \int_A \sigma_{xx} z\, dA,\ L_x = \int_A \sigma_{xx} z^2\, dA,\ P_x = \int_A \sigma_{xx} z^3\, dA \tag{3}$$

$$Q_x = \int_A \sigma_{xz} dA,\ R_x = \int_A \sigma_{xz} z\, dA,\ S_x = \int_A \sigma_{xz} z^2\, dA \tag{4}$$

Note that the displacement field of Exercise 1, hence the equations of equilibrium (3), contain those of the classical (Euler–Bernoulli) beam theory ($c_0 = -1,\ c_1 = 0,\ c_2 = 0,\ c_3 = 0$), the first-order (Timoshenko) beam theory ($c_0 = 0,\ c_1 = 1,\ c_2 = 0,\ c_3 = 0$), and the third-order (Reddy) beam theory ($c_0 = 0,\ c_1 = 1,\ c_2 = 0,\ c_3 = -4h/3$).

11.3 (*Continuation of Exercises 1 and 2*) Assume linear elastic constitutive behavior and show that the laminated beam's constitutive equations are given by

$$\begin{Bmatrix} N_x \\ M_x \\ L_x \\ P_x \end{Bmatrix} = \begin{bmatrix} A_{11} & B_{11} & D_{11} & E_{11} \\ B_{11} & D_{11} & E_{11} & F_{11} \\ D_{11} & E_{11} & F_{11} & G_{11} \\ E_{11} & F_{11} & G_{11} & H_{11} \end{bmatrix} \begin{Bmatrix} \varepsilon_{xx}^{(0)} \\ \varepsilon_{xx}^{(1)} \\ \varepsilon_{xx}^{(2)} \\ \varepsilon_{xx}^{(3)} \end{Bmatrix} \tag{1}$$

$$\begin{Bmatrix} Q_x \\ R_x \\ S_x \end{Bmatrix} = \begin{bmatrix} A_{55} & B_{55} & D_{55} \\ B_{55} & D_{55} & E_{55} \\ D_{55} & E_{55} & F_{55} \end{bmatrix} \begin{Bmatrix} \gamma_{xz}^{(0)} \\ \gamma_{xz}^{(1)} \\ \gamma_{xz}^{(2)} \end{Bmatrix} \tag{2}$$

where

$$(A_{11}, B_{11}, D_{11}, F_{11}, G_{11}, H_{11}) = \int_A E_1(1, z, z^2, z^4, z^5, z^6)\, dA$$
$$(A_{55}, B_{55}, D_{55}, E_{55}, F_{55}) = \int_A G_{13}(1, z, z^2, z^3, z^4)\, dA \tag{3}$$

11.4 Show that for a general laminate composed of multiple isotropic layers, the laminate stiffnesses $F_{16}, F_{26}, E_{16}, E_{26}, F_{16}$, and F_{26} are zero.

11.5 Specialize the equations of Exercise 2 to the case in which $c_0 = 0,\ c_1 = 1,\ c_2 = 0$ and $c_3 = -4h/3$, and express the equations in terms of the displacements.

11.6 Simplify the equations of motion in Eqs. (11.3–38)–(11.3–42) to cylindrical bending of plate strips.

11.7 Specialize the equations of motion and boundary conditions of the third-order theory of Reddy (see Section 11.3–4) to static bending of beams. Discuss the consequence of neglecting the higher-order resultant P_{xx} (but not R_x) on the equations and boundary conditions.

11.8 Develop the Navier solution of the third-order theory of laminated beams derived in Exercise 5.

References for Additional Reading

1. Reddy, J. N., and Chandrasekhara, K., "A Review of the Literature on Finite–Element Modeling of Laminated Composite Plates," *Shock & Vibration Digest,* **17**(4), 3–8 (1985).
2. Reddy, J. N., "A Review of Refined Theories of Laminated Composite Plates," *Shock & Vibration Digest,* **22**(7), 3–17 (1990).
3. Basset, A. B., "On the Extension and Flexure of Cylindrical and Spherical Thin Elastic Shells," *Philosophical Transactions of the Royal Society,* (London) Series A, **181**(6), 433-480 (1890).
4. Hildebrand, F. B., Reissner, E., and Thomas, G. B., "Notes on the Foundations of the Theory of Small Displacements of Orthotropic Shells," NACA TN-1833, Washington, D.C. (1949).
5. Vlasov, B. F., "Ob uravnieniakh izgiba plastinok (On equations of bending of plates)," (in Russian), *Doklady Akademii Nauk Azerbeijanskoi SSR,* **3**, 955–959 (1957).
6. Dong, S. B. and Tso, F. K. W., "On a Laminated Orthotropic Shell Theory Including Transverse Shear Deformation," *Journal of Applied Mechanics,* **39**, 1091–1097 (1972).
7. Librescu, L., "The Elasto-Kinetic Problems in the Theory of Anisotropic Shells and Plates, Part II, Plates Theory," *Revue Roumaine des Sciences Techniques Serie de Mecanique Appliquee,* **7**(3) (1969).
8. Sun, C. T., "Theory of Laminated Plates," *Journal of Applied Mechanics,* **38**, 231–238 (1971).
9. Sun, C. T. and Cheng, N. C., "On the Governing Equations for a Laminated Plate," *Journal of Sound and Vibration,* **21**(3), 307–316 (1972).
10. Sun, C. T. and Whitney, J. M., "On Theories for the Dynamic Response of Laminated Plates," *AIAA Journal,* **11**(2), 178–183 (1973).
11. Whitney, J. M. and Sun, C. T., "A Higher Order Theory for Extensional Motion of Laminated Anisotropic Shells and Plates," *Journal of Sound and Vibration,* **30**, 85–97 (1973).
12. Mau, S. T., "A Refined Laminate Plate Theory," *Journal of Applied Mechanics,* **40**, 606–607 (1973).
13. Nelson, R. B. and Lorch, D. R., "A Refined Theory for Laminated Orthotropic Plates," *Journal of Applied Mechanics*, **41**, 177–183 (1974).
14. Jemielita, G., "Techniczna Teoria Plyt Srednieej Grubbosci," (Technical Theory of Plates with Moderate Thickness), *Rozprawy Inzynierskie (Engineering Transactions), Polska Akademia Nauk,* **23**(3), 483–499 (1975).
15. Librescu, L., *Elastostatics and Kinetics of Anisotropic and Heterogeneous Shell–Type Structures,* Noordhoff, Leyden, Netherlands, 1975.
16. Schmidt, R., "A Refined Nonlinear Theory for Plates with Transverse Shear Deformation," *Journal of the Industrial Mathematics Society,* **27**(1), 23–38 (1977).

17. Lo, K. H., Christensen, R. M., and Wu, E. M., "A High-Order Theory of Plate Deformation: Part 1: Homogeneous Plates," *Journal of Applied Mechanics,* **44**(4), 663–668 (1977).
18. Lo, K. H., Christensen, R. M., and Wu, E. M., "A High-Order Theory of Plate Deformation, Part 2: Laminated Plates," *Journal of Applied Mechanics,* **44**(4), 669–676 (1977).
19. Lo, K. H., Christensen, R. M., and Wu, E. M., "Stress Solution Determination for High Order Plate Theory," *International Journal of Solids and Structures,* **14**, 665–662 (1978).
20. Krishna Murty, A. V., "Higher Order Theory for Vibration of Thick Plates," *AIAA Journal,* **15**(2), 1823–1824 (1977).
21. Levinson, M., "An Accurate, Simple Theory of the Static and Dynamics of Elastic Plates," *Mechanics Research Communications,* **7**(6), 343–350 (1980).
22. Murthy, M. V. V., "An Improved Transverse Shear Deformation Theory for Laminated Anisotropic Plates," NASA Technical Paper, 1903, 1–37 (1981).
23. Levinson, M. and Cooke, D. W., "Thick Rectangular Plates–I. The Generalized Navier Solution," *International Journal of Mechanical Sciences,* **25**(3), 199–205 (1983).
24. Cooke, D. W. and Levinson, M., "Thick Rectangular Plates–II. The Generalized Lévy Solution," *International Journal of Mechanical Sciences,* **25**(3), 207–215 (1983).
25. Reddy, J. N., "A Simple Higher-Order Theory for Laminated Composite Plates," *Journal of Applied Mechanics,* **51**, 745–752 (1984).
26. Reddy, J. N., "A Refined Nonlinear Theory of Plates with Transverse Shear Deformation," *International Journal of Solids Structures,* **20**(9/10), 881–906 (1984).
27. Bhimaraddi, A. and Stevens, L. K., "A Higher Order Theory for Free Vibration of Orthotropic, Homogeneous and Laminated Rectangular Plates," *Journal of Applied Mechanics,* **51**, 195–198 (1984).
28. Di Sciuva, M., "A Refined Transverse Shear Deformation Theory for Multilayered Anisotropic Plates," *Atti della Academia delle Scienze di Torino,* **118**, 269–295 (1984).
29. Reddy, J. N. and Phan, N. D., "Stability and Vibration of Isotropic Orthotropic and Laminated Plates According to a Higher-Order Shear Deformation Theory," *Journal of Sound and Vibration,* **98**, 157–170 (1985).
30. Phan, N. D., and Reddy, J. N., "Analysis of Laminated Composite Plates Using a Higher-Order Shear Deformation Theory," *International Journal for Numerical Methods in Engineering,* **21**, 2201–2219 (1985).
31. Krishna Murty, A. V., "Flexure of Composite Plates," *Composite Structures,* **7**(3), 161–177 (1987).
32. Reddy, J. N., "A Small Strain and Moderate Rotation Theory of Laminated Anisotropic Plates," *Journal of Applied Mechanics,* **54**, 623–626 (1987).

33. Khdeir, A. A., Reddy, J. N., and Librescu, L., "Lévy Type Solutions for Symmetrically Laminated Rectangular Plates Using First-Order Shear Deformation Theory," *Journal of Applied Mechanics,* **54**, 640–642 (1987).
34. Khdeir, A. A., Reddy, J. N., and Librescu, L., "Analytical Solution of a Refined Shear Deformation Theory for Rectangular Composite Plates," *International Journal of Solids and Structures,* **23**(10), 1447–1463 (1987).
35. Librescu, L., Khdeir, A. A., and Reddy, J. N., "A Comprehensive Analysis of the State of Elastic Anisotropic Flat Plates Using Refined Theories," *Acta Mechanica,* **70**, 57–81 (1987).
36. Di Sciuva, M., "An Improved Shear-deformation Theory for Moderately Thick Multilayered Anisotropic Shells and Plates," *Journal of Applied Mechanics,* **54**(3), 589–596 (1987).
37. Bhimaraddi, A., "Dynamic Response Orthotropic, Homogeneous, and Laminated Cylindrical Shells," *AIAA Journal,* **27**(11), 1834–1837 (1985).
38. Khdeir, A. A. and Reddy, J. N., "Dynamic Response of Antisymmetric Angle-Ply Laminated Plates Subjected to Arbitrary Loading," *Journal of Sound and Vibration,* **126**(3), 437–445 (1988).
39. Khdeir, A. A., "Free Vibration of Antisymmetric Angle-Ply Laminated Plates Including Various Boundary Conditions," *Journal of Sound and Vibration,* **122**(2), 377–388 (1988).
40. Khdeir, A. A., "Free Vibration and Buckling of Symmetric Cross-Ply Laminated Plates by an Exact Method," *Journal of Sound and Vibration,* **126**(3), 447–461 (1988).
41. Librescu, L. and Khdeir, A. A., "Analysis of Symmetric Cross-Ply Laminated Elastic Plates Using a Higher-Order Theory, Part I. Stress and Displacement," *Composite Structures,* **9**, 189–213 (1988).
42. Khdeir, A. A. and Librescu, L., "Analysis of Symmetric Cross-Ply Laminated Elastic Plates Using a Higher-Order Theory, Part II. Buckling and Free Vibration," *Composite Structures,* **9**, 259–277 (1988).
43. Librescu, L. and Reddy, J. N., "A Few Remarks Concerning Several Refined Theories of Anisotropic Composite Laminated Plates," *International Journal of Engineering Science,* **27**(5), 515–527 (1989).
44. Khdeir, A. A. and Reddy, J. N., "Exact-Solutions for the Transient Response of Symmetric Cross-Ply Laminates Using a Higher-Order Plate Theory," *Composite Science & Technology,* **34**, 205–224 (1989).
45. Khdeir, A. A. and Reddy, J. N., "On the Forced Motions of Antisymmetric Cross-Ply Laminated Plates," *International Journal of Mechanical Sciences,* **31**(7), 499–510 (1989).
46. Khdeir, A. A., "An Exact Approach to the Elastic State of Stress of Shear Deformable Antisymmetric Angle-Ply Laminated Plates," *Composite Structures,* **11**, 245–258 (1989).
47. Khdeir, A. A., "Stability of Antisymmetric Angle-Ply Laminated Plates," *Journal of Engineering Mechanics,* **115**(5), 952–962 (1989).

48. Khdeir, A. A., "Free Vibration and Buckling of Unsymmetric Cross-Ply Laminated Plates Using a Refined Theory," *Journal of Sound and Vibration,* **128**(3), 377–395 (1989).
49. Reddy, J. N., "A General Non-Linear Third-Order Theory of Plates with Moderate Thickness," *International Journal of Non-Linear Mechanics*, **25**(6), 677–686 (1990).
50. Khdeir, A. A. and Reddy, J. N., "Thermal Stresses and Deflections of Cross-Ply Laminated Plates Using Refined Plate Theories," *Journal of Thermal Stresses*, **14**(4), 419–438 (1991).
51. Khdeir, A. A. and Reddy, J. N., "Analytical Solutions of Refined Plate Theories of Cross-Ply Composite Laminates," *Journal of Pressure Vessel Technology,* **113**(4), 570–578 (1991).
52. Nosier, A. and Reddy, J. N., "On Vibration and Buckling of Symmetric Laminated Plates According to Shear Deformation Theories," *Acta Mechanica*, **94**(3–4), 123–170 (1992).
53. Pagano, N. J., "Exact Solutions for Rectangular Bidirectional Composites and Sandwich Plates," *Journal of Composite Materials*, **4**, 20–34 (1970).
54. Pagano, N. J. and Hatfield, S. J., "Elastic Behavior of Multilayered Bidirectional Composites," *AIAA Journal*, **10**, 931–933 (1972).
55. Noor, A. K., "Free Vibrations of Multilayered Composite Plates," *AIAA Journal,* **11**(7), 1038–1039 (1972).
56. Noor, A. K., "Mixed Finite–Difference Scheme for Analysis of Simply Supported Thick Plates," *Computers and Structures,* **3**, 967–982 (1973).
57. Noor, A. K., "Stability of Multilayered Composite Plates," *Fibre Science and Technology,* **8**, 81–88 (1975).
58. Reddy, J. N., "On Mixed Finite–Element Formulations of a Higher-Order Theory of Composite Laminates," *Finite Element Methods for Plate and Shell Structures*, T. J. R. Hughes and E. Hinton (eds.), Pineridge Press, U.K., 31–57 (1986).
59. Putcha, N. S. and Reddy, J. N., "A Mixed Shear Flexible Finite Element for the Analysis of Laminated Plates," *Computer Methods in Applied Mechanics and Engineering*, **44**, 213–227 (1984).
60. Putcha, N. S. and Reddy, J. N., "A Refined Mixed Shear Flexible Finite Element for the Nonlinear Analysis of Laminated Plates," *Computers & Structures*, **22**(4), 529–538 (1986).
61. Pandya, B. N. and Kant, T., "Flexural Analysis of Laminated Composites Using Refined Higher-Order C^0 Plate Bending Element," *Computer Methods in Applied Mechanics Engineering,* **66**, 173–198 (1988).
62. Reddy, J. N., *An Introduction to the Finite Element Method,* Second Edition, McGraw–Hill, New York, 1993.

Chapter Twelve

Layerwise Theories and Variable Kinematic Models

12.1 Introduction

12.1–1 Motivation

The analysis of fiber-reinforced, laminated composite structures presents the analyst with many challenges. Unlike their homogeneous isotropic counterparts, the heterogeneous anisotropic constitution of laminated composite structures often results in the appearance of many unique phenomena that can occur on vastly different geometric scales, *i.e.*, at the global or laminate level, the ply level, or the fiber/matrix level. For example, the global deformation of laminated composite structures is often characterized by complex coupling between the extension, bending, and shearing modes. Further, due to their characteristically low transverse shear stiffness, composite laminates often exhibit significant transverse shear deformation at lower thickness-to-span ratios than do similar homogeneous isotropic plates and shells. At the ply level, laminated composites often exhibit transverse stress concentrations near material and geometric discontinuities (the so-called free edge effect) that can lead to damage in the form of delamination, matrix cracking, and adhesive joint separation. Once significant damage occurs at the ply level, the kinematic and material description of the problem must be changed before further analysis can proceed. At the fiber/matrix level, stress concentrations can cause fiber/matrix separation, radial matrix cracking, and other forms of cumulative damage that degrade the stiffness of individual laminae, thus causing a complex load redistribution.

When the main emphasis of the analysis is to determine the overall global response of the laminated component, for example, gross deflections, critical buckling loads, fundamental vibration frequencies, and associated mode shapes, such global behavior can often be accurately determined using relatively simple equivalent-single-layer laminate theories (ESL theories), especially for very thin laminates. Two commonly used examples of simple ESL theories are the classical and the first-order shear deformation theories discussed earlier.

As laminated composite materials undergo the transition from secondary structural components to primary critical structural components, the goals of analysis must be broadened to include a highly accurate assessment of localized regions where damage initiation is likely. The simple ESL laminate theories that often prove adequate for modeling secondary structures are of limited value in modeling primary structures for two reasons. First of all, most primary structural components are considerably thicker than secondary components; thus even the determination of the global response may require a refined laminate theory that accounts for thickness effects. Second, the assessment of localized regions of potential damage initiation begins with an accurate determination of the three-dimensional state of stress and strain at the ply level, regardless of whether damage prediction and assessment is desired at the ply level or at the fiber/matrix level. The simple ESL laminate theories are most often incapable of accurately determining the 3-D stress field at the ply level. Thus the analysis of primary composite structural components may require the use of 3-D elasticity theory or a layerwise laminate theory that contains full 3-D kinematics and constitutive relations.

From the equilibrium of interlaminar forces, it follows that the following continuity conditions hold between the stress fields of adjacent layers at their interface (see Figure 12.1–1):

$$\left\{\begin{matrix}\sigma_{xx}\\ \sigma_{yy}\\ \sigma_{xy}\end{matrix}\right\}^{(k)} \neq \left\{\begin{matrix}\sigma_{xx}\\ \sigma_{yy}\\ \sigma_{xy}\end{matrix}\right\}^{(k+1)}, \quad \left\{\begin{matrix}\sigma_{xz}\\ \sigma_{yz}\\ \sigma_{zz}\end{matrix}\right\}^{(k)} = \left\{\begin{matrix}\sigma_{xz}\\ \sigma_{yz}\\ \sigma_{zz}\end{matrix}\right\}^{(k+1)} \qquad (12.1-1)$$

These conditions in turn imply, since $[\bar{Q}^{(k)}] \neq [\bar{Q}^{(k+1)}]$ in general, that the strain fields of adjacent layers satisfy the following conditions:

$$\left\{\begin{matrix}\gamma_{xz}\\ \gamma_{yz}\\ \varepsilon_{zz}\end{matrix}\right\}^{(k)} \neq \left\{\begin{matrix}\gamma_{xz}\\ \gamma_{yz}\\ \varepsilon_{zz}\end{matrix}\right\}^{(k+1)} \qquad (12.1-2)$$

In all equivalent-single layer laminate theories based on assumed displacement fields, it is assumed that the displacements are continuous functions of the thickness coordinate. This in turn results in continuous transverse strains and discontinuous in-plane strains, contrary to the requirement (1). Hence, all stresses in equivalent-single layer theories are discontinuous at layer interfaces. More important, the transverse stresses at the interface of two layers, called *interlaminar stresses*, are discontinuous:

$$\left\{\begin{matrix}\sigma_{xz}\\ \sigma_{yz}\\ \sigma_{zz}\end{matrix}\right\}^{(k)} \neq \left\{\begin{matrix}\sigma_{xz}\\ \sigma_{yz}\\ \sigma_{zz}\end{matrix}\right\}^{(k+1)} \qquad (12.1-3)$$

For thin laminates the error introduced due to discontinuous interlaminar stresses can be negligible. However, for thick laminates, the ESL theories can give erroneous results for all stresses, requiring use of layerwise theories.

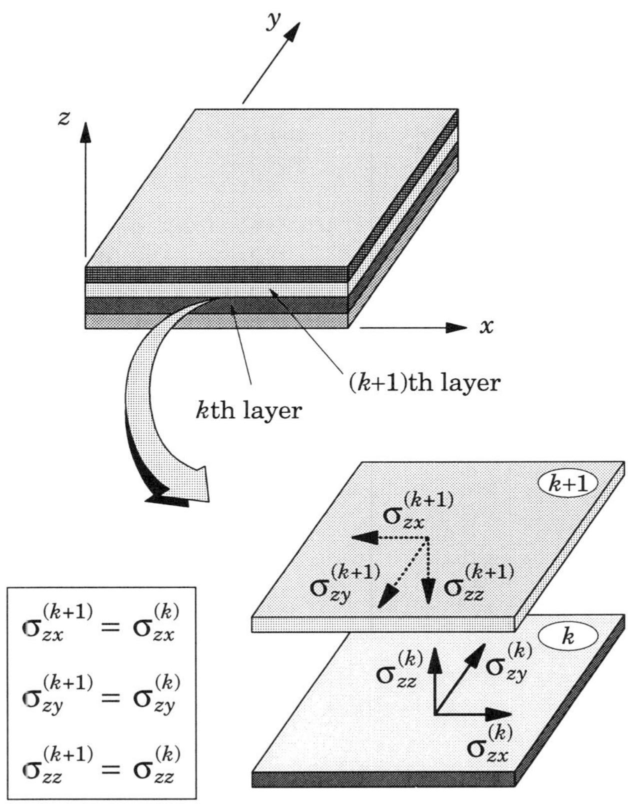

Figure 12.1–1: Equilibrium of interlaminar stresses.

12.1–2 An Overview of Layerwise Theories

In contrast to the ESL theories, the layerwise theories are developed by assuming that the displacement field exhibits only C^0–continuity through the laminate thickness. Thus the displacement components are continuous through the laminate thickness but the derivatives of the displacements with respect to the thickness coordinate may be discontinuous at various points through the thickness, thereby allowing for the possibility of continuous transverse stresses at interfaces separating dissimilar materials. Layerwise displacement fields provide a much more kinematically correct representation of the moderate to severe cross-sectional warping associated with the deformation of thick laminates.

The displacement-based layerwise theories can be subdivided into two classes: (1) the *partial layerwise theories* that use layerwise expansions for the in-plane displacement components but not the transverse displacement component, and (2) the *full layerwise theories* that use layerwise expansions for all three displacement components. Compared to the ESL theories, the partial layerwise theories provide a more realistic description of the kinematics of composite laminates by introducing discrete layer transverse shear effects into the assumed displacement field. The full layerwise theories go one step further by adding both discrete layer transverse shear effects and discrete layer transverse normal effects.

The use of the partial and full layerwise theories for the analysis of thick laminated composite plates is widely accepted. Such theories allow the in-plane displacements to vary in a layerwise manner through the thickness of the laminate. The layerwise theories can represent the zigzag behavior of the in-plane displacements through the thickness. This zigzag behavior is more pronounced for thick laminates where the transverse shear modulus changes abruptly through the thickness and can be seen in the exact 3-D elasticity solutions obtained by Pagano [1,2], Pagano and Hatfield [3], Srinivas and Rao [4,5], Noor [6,7], and Savoia and Reddy [8] for the bending of rectangular laminated plates, and by Varadan and Bhaskar [9] and Ren [10] for the bending of laminated shells.

Whitney [11] used a layerwise quadratic variation of the transverse stresses for improving the gross behavior of laminated plates. This led to a layerwise cubic variation of the in-plane displacements. The use of the necessary continuity conditions resulted in the same number of variables as in the FSDT. However, the equations of equilibrium were taken to be those of the classical lamination theory, and thus they were not variationally consistent. The numerical results obtained for deflections, natural frequencies, and buckling loads were found to be in good agreement with available exact elasticity solutions.

In a series of papers, Swift and Heller [12] studied laminated beams by assuming layerwise constant shear strains and a continuous transverse displacement through the thickness. A similar approach was used by Durocher and Solecki [13] to study transversely isotropic plates with two or three layers. Seide [14], and Choudhuri and Seide [15] extended the work of Swift and Heller to laminated plates (also see [16–20]). The approach involves writing the equilibrium equations for each layer. For example, the equilibrium equations of the kth lamina can be written in terms of the force and moment resultants as

$$\frac{\partial N_{\alpha\beta}^{(k)}}{\partial x_\beta} + \sigma_{\alpha 3}^{(k+1)} - \sigma_{\alpha 3}^{(k)} = 0, \quad \frac{\partial Q_\alpha^{(k)}}{\partial x_\alpha} + \sigma_{33}^{(k+1)} - \sigma_{33}^{(k)} = 0 \qquad (12.1-4a)$$

$$\frac{\partial M_{\alpha\beta}^{(k)}}{\partial x_\beta} + \sigma_{\alpha 3}^{(k+1)} z_{k+1} - \sigma_{\alpha 3}^{(k)} z_k - Q_\alpha^k = 0 \qquad (12.1-4b)$$

for $k = 1, 2, \cdots, N$ and $\alpha, \beta = 1, 2$ ($x_1 = x, x_2 = y, x_3 = z$), where N is the total number of layers, and

$$\left(N_{\alpha\beta}^{(k)}, M_{\alpha\beta}^{(k)}\right) = \int_{z_k}^{z_{k+1}} (1, z)\sigma_{\alpha\beta}^{(k)}\, dz, \quad Q_\alpha^{(k)} = \int_{z_k}^{z_{k+1}} \sigma_{\alpha 3}^{(k)}\, dz \qquad (12.1-5)$$

$$\sigma_{ij}^{(k)} = \sigma_{ij}(x_\beta, z_k) \qquad (12.1-6)$$

Here z_k denotes the $z-$coordinate of the bottom of the kth layer. Then the continuity of the displacements and transverse stresses at layer interfaces and the traction boundary conditions at the top and bottom of the laminate are used to obtain $2(N+1)+1 = 2N+3$ equations in (u_k, v_k, w).

Several other layerwise models for laminated plates have been presented by Mau [21], Chou and Carleone [22], Di Sciuva [23–25], Murakami [26], and Ren [27]. Di Sciuva [23–25] used ideas similar to those of [12–15] to formulate a displacement-based theory, called the *zigzag theory* or *discrete-layer theory*. The displacement field is assumed to be of the form

$$\begin{aligned} u_\alpha(x_\beta, x_3, t) &= u_\alpha^0(x_\beta, t) - x_3 u_{3,\alpha}^0(x_\beta, t) + f_{\alpha\gamma}(x_3)\phi_\gamma(x_\beta, t) \\ u_3(x_\beta, x_3, t) &= u_3^0(x_\beta, t) \end{aligned} \qquad (12.1-7)$$

where α, β, and γ take the values of 1 and 2, and ($x_1 = x,\ x_2 = y,\ x_3 = z$). The functions $f_{\alpha\gamma}$ and ϕ_γ are then determined such that the displacements and transverse stresses are continuous at the layer interfaces. The functions $f_{\alpha\gamma}$ are shown to depend only on x_3 and the layer stiffnesses. The resulting laminate theory contains only five dependent unknowns (also see Zukas and Vinson [28], and Waltz and Vinson [29]), as in the first-order theory or the third-order theory of Reddy. The layerwise cubic model of Ren [27] required two variables more than the FSDT but it produced results which agreed well with those from exact elasticity. These models demonstrated that layerwise functions are necessary for determining the zigzag thickness distribution of the in-plane displacements.

Using an explicit approximation for the transverse shear stresses within each layer, Hsu and Wang [30,31] proposed a layerwise model for laminated cylindrical shells consisting of orthotropic layers. The transverse shear stresses satisfied the traction boundary conditions on the top and bottom surfaces, and the equilibrium conditions at the layer interfaces of the shell. Rath and Das [32] extended this model to symmetrically laminated generally orthotropic shells. However, the number of equations increases with the number of layers.

Mau *et al.* [33] used a layerwise theory in the context of hybrid–stress finite element analysis of thick laminated plates. The theory is based on assumed stresses within each layer, resulting in a large number of variables. Spilker [34,35] used the idea of Mau *et al.* in developing an eight-node hybrid–stress element. In this model, the higher-order distributions of stresses through the thickness were characterized by as many as 67 stress parameters, and different shear strains are assumed within each layer. Using cubic spline functions to approximate the thickness variation of displacements, Hinrichsen and Palazotto [36] proposed a layerwise finite element model for the nonlinear analysis of thick laminated plates. Use of spline functions or Hermite cubic functions, which include continuity of the derivatives, violates the required discontinuity of the interlaminar strains between layers.

A more direct method of achieving a layerwise displacement field was proposed by Reddy [37], who represented the transverse variation of the displacement components in terms of one-dimensional Lagrangian finite elements. The resulting strain field is kinematically correct in that the in-plane strains are continuous through the thickness while the transverse strains are discontinuous through the thickness, thereby allowing for the possibility of continuous transverse stresses as the number of layers is increased. The layerwise field proposed by Reddy is very general in that any desired number of layers, distribution of layers, and order of interpolation can be achieved simply by specifying a particular mesh of one-dimensional finite elements through the thickness. The theory was extended by Barbero, Reddy, and Teply [38] to laminated composite cylindrical shells. Owen and Li [39,40] used the layerwise displacement idea similar to that of Reddy [37] to develop a continuum shell element (also see [41,42]).

Lee *et al.* [43,44] presented a partial layerwise model for laminated plates with a layerwise cubic variation of the in-plane displacements. By imposing the continuity of the interlaminar shear stresses, the number of unknowns is reduced to the same number and type of variables as in FSDT. While the numerical results for the maximum in-plane stresses at the free surfaces showed very good agreement with 3-D elasticity solutions for the cylindrical bending of thick symmetric cross-ply laminates, the displacements and stresses at the interfaces were not accurate enough. The theory predicts even more inaccurate results for unsymmetric laminates. The theory was extended to laminated shells Xavier *et al.* [45,46].

The displacement-based partial layerwise laminate theories, in which the transverse normal strain is neglected, provide a more realistic description of the kinematics of composite laminates when compared to the ESL theories by introducing discrete layer transverse shear effects into the assumed displacement field. However, these models are not capable of accurately determining interlaminar stresses near discontinuities such as holes or cut-outs, traction free edges, and delamination fronts. In modeling these localized effects, inclusion of the transverse normal strain is important for two reasons. First of all, the transverse normal stress is usually a significant, if not dominant, stress in these regions. Secondly, layerwise models that neglect transverse normal strain do not satisfy traction-free boundary conditions for transverse shear stresses at the laminate edge. Examination of the natural boundary conditions associated with the governing differential equations of a partial layerwise theory reveals that the transverse shear stresses satisfy the traction-free boundary conditions at the laminate edge only in the integral sense and not in the local sense (despite the level of refinement through the thickness). In contrast to the partial layerwise theories, full layerwise theories [37] use layerwise expansions for all three displacement components, and thus include both discrete layer transverse shear effects and discrete layer transverse normal effects.

In this chapter we present the displacement-based full layerwise theory of Reddy, develop its finite element model, and present some numerical results to illustrate the accuracy. It should be noted that the full layerwise finite element model is equivalent to the displacement finite element model of 3-D elasticity. Following the layerwise theory, a variable kinematic model that incorporates both equivalent single-layer theories and layerwise theories is described.

12.2 Development of the Theory

12.2–1 Displacement Field

In the layerwise theory of Reddy, the displacement field in the kth layer is written as

$$\begin{aligned} u^k(x,y,z,t) &= \sum_{j=1}^{m} u_j^k(x,y,t)\phi_j^k(z) \\ v^k(x,y,z,t) &= \sum_{j=1}^{m} v_j^k(x,y,t)\phi_j^k(z) \\ w^k(x,y,z,t) &= \sum_{j=1}^{n} w_j^k(x,y,t)\psi_j^k(z) \end{aligned} \qquad (12.2-1)$$

where u^k, v^k, and w^k represent the total displacement components in the x, y and z directions, respectively, of a material point initially located at (x, y, z) in the undeformed laminate, and $\phi_j^k(z)$ and $\psi_j^k(z)$ are continuous functions of the thickness coordinate z. In general, $\psi^k \neq \phi^k$.

The functions $\phi_j^k(z)$ and $\psi_j^k(z)$ are selected to be layerwise continuous functions. For example, they can be chosen to be the one-dimensional Lagrange interpolation functions of the thickness coordinate, in which case,

(u_j^k, v_j^k, w_j^k) denote the values of (u^k, v^k, w^k) at the jth plane (see [37,47–52]). The number of nodes, n, through the layer thickness define the polynomial degree $p = n + 1$ of $\psi_j^k(z)$, which are defined only within the kth numerical layer (see Figure 12.2–1). The functions $u_j^k(x, y, t)$, $v_j^k(x, y, t)$, and $w_j^k(x, y, t)$ represent the displacement components of all points located on the jth plane (defined by $z = z_j$) in the undeformed laminate.

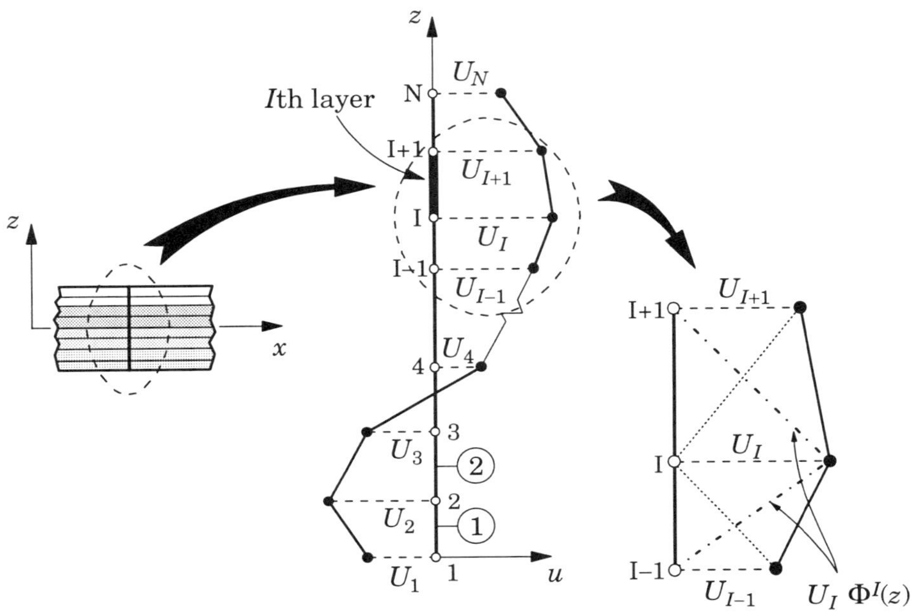

Figure 12.2–1: Displacement representation and the linear approximation functions $\Phi^I(z)$ used in the layerwise theory.

Since the thickness variation of the displacement components is defined in terms of piecewise Lagrangian interpolation functions, the displacement components will be continuous through the laminate thickness, but the transverse strains will be discontinuous across the interface between adjacent thickness subdivisions. This leaves the possibility that the transverse stresses may be continuous across the interface between layers. Note that the use of piecewise Hermite interpolation through the thickness is kinematically incorrect for general laminates since the transverse strains are forced to be continuous through the thickness.

Any desired degree of displacement variation through the thickness is easily obtained by either adding more one-dimensional finite element subdivisions through the thickness (h–refinement) or using higher order Lagrangian interpolation polynomials (p–refinement) through the thickness. The layerwise concept introduced here is very general in that the number of subdivisions through the thickness can be greater than, equal to, or less than the number of material layers through the thickness and each layer can have linear, quadratic, or higher-order polynomial variations of the displacements. Note that the sublaminate concept can be used (*i.e.*, the number of thickness subdivisions is less than the number of material layers); however, each sublaminate will be represented as an equivalent, single, homogeneous layer.

The total displacement field of the laminate can be written as

$$
\begin{aligned}
u(x,y,z,t) &= \sum_{I=1}^{N} U_I(x,y,t)\Phi^I(z)\\
v(x,y,z,t) &= \sum_{I=1}^{N} V_I(x,y,t)\Phi^I(z)\\
w(x,y,z,t) &= \sum_{I=1}^{M} W_I(x,y,t)\Psi^I(z)
\end{aligned}
\qquad (12.2-2)
$$

where (U_I, V_I, W_I) denote the nodal values of (u, v, w), N is the number of nodes and Φ^I are the *global* interpolation functions (see Figure 12.2–1) for the discretization of the in-plane displacements through thickness, and M is the number of nodes and Ψ^I are the global interpolation functions for discretization of the transverse displacement through thickness. For linear and quadratic variation through each numerical layer these functions are given below [53]:

Linear functions $(N = N_e + 1)$:

$$
\begin{aligned}
\Phi^1(z) &= \psi_1^{(1)}(z), \quad z_1 \le z \le z_2\\
\Phi^I(z) &= \begin{cases} \psi_2^{(I-1)}(z), & z_{I-1} \le z \le z_I \\ \psi_1^{(I)}(z), & z_I \le z \le z_{I+1} \end{cases} \quad (I = 2, 3, \cdots, N_e) \qquad (12.2-3a)\\
\Phi^N(z) &= \psi_2^{(N_e)}(z), \quad z_{N-1} \le z \le z_N
\end{aligned}
$$

$$
\psi_1^{(k)} = 1 - \frac{\bar{z}}{h_k}, \quad \psi_2^{(k)} = \frac{\bar{z}}{h_k}, \quad 0 \le \bar{z} \le h_k \qquad (12.2-3b)
$$

Quadratic functions $(N = 2N_e + 1)$:

$$
\begin{aligned}
\Phi^1(z) &= \psi_1^{(1)}(z), \quad z_1 \le z \le z_3\\
\Phi^{2I}(z) &= \psi_2^{(I)}(z), \quad z_{2I-1} \le z \le z_{2I+1} \quad (I = 1, 2, \cdots, N_e)\\
\Phi^{2I+1}(z) &= \begin{cases} \psi_3^{(I)}(z), & z_{2I-1} \le z \le z_{2I+1} \\ \psi_1^{(I+1)}(z), & z_{2I+1} \le z \le z_{2I+3} \end{cases} \quad (I = 1, 2, \cdots, N_e - 1)\\
\Phi^N(z) &= \psi_3^{(N_e)}(z), \quad z_{N-1} \le z \le z_N
\end{aligned}
\qquad (12.2-4a)
$$

$$\psi_1^{(k)} = \left(1 - \frac{z}{h_k}\right)\left(1 - \frac{2z}{h_k}\right), \quad \psi_2^{(k)} = 4\frac{z}{h_k}\left(1 - \frac{z}{h_k}\right), \quad \psi_3^{(k)} = -\frac{z}{h_k}\left(1 - \frac{2z}{h_k}\right)$$

$$z_{2k-1} \leq z \leq z_{2k+1} \qquad (12.2-4b)$$

where h_k is the thickness of the kth layer, $\bar{z} = z - z_t^k$, and z_t^k denotes the $z-$coordinate of top of the kth numerical layer. Independent approximations for the in-plane and transverse displacements are assumed in order to include the possibility of inextensibility of transverse normals. The inextensibility of transverse normals can be included by setting $M = 1$ and $\Psi^1 = 1$ for all z.

12.2–2 Strains and Stresses

The von Kármán nonlinear strains associated with the displacement field (2) are

$$\varepsilon_{xx} = \sum_{I=1}^{N} \frac{\partial U_I}{\partial x}\Phi^I + \frac{1}{2}\left(\sum_{I=1}^{M} \frac{\partial W_I}{\partial x}\Psi^I\right)\left(\sum_{J=1}^{M} \frac{\partial W_J}{\partial x}\Psi^J\right) \qquad (12.2-5a)$$

$$\varepsilon_{yy} = \sum_{I=1}^{N} \frac{\partial V_I}{\partial y}\Phi^I + \frac{1}{2}\left(\sum_{I=1}^{M} \frac{\partial W_I}{\partial y}\Psi^I\right)\left(\sum_{J=1}^{M} \frac{\partial W_J}{\partial y}\Psi^J\right) \qquad (12.2-5b)$$

$$\gamma_{xy} = \sum_{I=1}^{N} \left(\frac{\partial U_I}{\partial y} + \frac{\partial V_I}{\partial x}\right)\Phi^I + \left(\sum_{I=1}^{M} \frac{\partial W_I}{\partial x}\Psi^I\right)\left(\sum_{J=1}^{M} \frac{\partial W_J}{\partial y}\Psi^J\right) \qquad (12.2-5c)$$

$$\varepsilon_{zz} = \sum_{I=1}^{M} W_I \frac{d\Psi^I}{dz} \qquad (12.2-6)$$

$$\gamma_{yz} = \sum_{I=1}^{N} V_I \frac{d\Phi^I}{dz} + \sum_{I=1}^{M} \frac{\partial W_I}{\partial y}\Psi^I \qquad (12.2-7a)$$

$$\gamma_{xz} = \sum_{I=1}^{N} U_I \frac{d\Phi^I}{dz} + \sum_{I=1}^{M} \frac{\partial W_I}{\partial x}\Psi^I \qquad (12.2-7b)$$

Note that the strains are discontinuous at the layer interfaces because of the layerwise definition of the functions Φ^I and Ψ^I.

The stresses in the kth layer may be computed from the 3-D stress–strain equations. For the kth (orthotropic) lamina we have [from Eq. (4.3–18)]

$$\begin{Bmatrix} \sigma_{xx} \\ \sigma_{yy} \\ \sigma_{zz} \\ \sigma_{yz} \\ \sigma_{xz} \\ \sigma_{xy} \end{Bmatrix}^{(k)} = \begin{bmatrix} \bar{C}_{11} & \bar{C}_{12} & \bar{C}_{13} & 0 & 0 & \bar{C}_{16} \\ \bar{C}_{21} & \bar{C}_{22} & \bar{C}_{23} & 0 & 0 & \bar{C}_{26} \\ \bar{C}_{31} & \bar{C}_{32} & \bar{C}_{33} & 0 & 0 & \bar{C}_{36} \\ 0 & 0 & 0 & \bar{C}_{44} & \bar{C}_{45} & 0 \\ 0 & 0 & 0 & \bar{C}_{45} & \bar{C}_{55} & 0 \\ \bar{C}_{16} & \bar{C}_{26} & \bar{C}_{36} & 0 & 0 & \bar{C}_{66} \end{bmatrix}^{(k)} \begin{Bmatrix} \varepsilon_{xx} - \alpha_{xx}\Delta T \\ \varepsilon_{yy} - \alpha_{yy}\Delta T \\ \varepsilon_{zz} - \alpha_{zz}\Delta T \\ \gamma_{yz} \\ \gamma_{xz} \\ \gamma_{xy} - 2\alpha_{xy}\Delta T \end{Bmatrix}^{(k)} \qquad (12.2-8)$$

where the $\bar{C}_{ij}^{(k)}$ are the transformed elastic coefficients in the (x, y, z) system, which are related to the elastic coefficients in the material coordinates, C_{ij} by Eq. (4.3–19). If inextensibility of transverse normals is assumed, one may use the plane stress-reduced stiffness in place of the 3-D stiffnesses. Note that the strains at a layer interface depend on the layer; *i.e.*, $\{\varepsilon\}_P^k \neq \{\varepsilon\}_P^{k+1}$ at a point P on the interface of layers k and $k+1$.

12.2–3 Equations of Motion

The governing equations of motion for the present layerwise theory can be derived using the principle of virtual displacements

$$0 = \int_0^T (\delta U + \delta V - \delta K)\, dt \tag{12.2 – 9}$$

where the virtual strain energy δU, virtual work done by applied forces δV, and the virtual kinetic energy δK are given by

$$\begin{aligned}
\delta U &= \int_{\Omega_0} \left\{ \int_{-\frac{h}{2}}^{\frac{h}{2}} [\, \sigma_{xx}\delta\varepsilon_{xx} + \sigma_{yy}\delta\varepsilon_{yy} + \sigma_{zz}\delta\varepsilon_{zz} + \right.\\
&\qquad \left. \sigma_{xy}\delta\gamma_{xy} + \sigma_{xz}\delta\gamma_{xz} + \sigma_{yz}\delta\gamma_{yz} \,]\, dz \right\} dxdy\\
&= \int_{\Omega_0} \left\{ \sum_{I=1}^{N} \left[N_{xx}^I \frac{\partial \delta U_I}{\partial x} + N_{yy}^I \frac{\partial \delta V_I}{\partial y} + N_{xy}^I \left(\frac{\partial \delta U_I}{\partial y} + \frac{\partial \delta V_I}{\partial x} \right) \right.\right.\\
&\quad \left. + Q_x^I \delta U_I + Q_y^I \delta V_I \right] + \sum_{I=1}^{M} \left[\tilde{Q}_x^I \frac{\partial \delta W_I}{\partial x} + \tilde{Q}_y^I \frac{\partial \delta W_I}{\partial y} \right.\\
&\quad + \tilde{Q}_z^I \delta W_I + \left(\tilde{N}_{xx}^{IJ} \frac{\partial W_I}{\partial x} + \tilde{N}_{xy}^{IJ} \frac{\partial W_I}{\partial x} \right) \frac{\partial \delta W_J}{\partial x}\\
&\quad \left.\left. + \left(\tilde{N}_{xy}^{IJ} \frac{\partial W_I}{\partial x} + \tilde{N}_{yy}^{IJ} \frac{\partial W_I}{\partial y} \right) \frac{\partial \delta W_J}{\partial y} \right] \right\} dxdy
\end{aligned} \tag{12.2 – 10}$$

$$\begin{aligned}
\delta V &= -\int_{\Omega_0} \left[q_b(x,y)\ \delta w(x,y,-\frac{h}{2}) + q_t(x,y)\ \delta w(x,y,\frac{h}{2}) \right] dxdy\\
&\quad - \int_\Gamma \int_{-\frac{h}{2}}^{\frac{h}{2}} [\hat{\sigma}_{nn}\delta u_n + \hat{\sigma}_{ns}\delta u_s + \hat{\sigma}_{nr}\delta w]\, dzds\\
&= -\int_{\Omega_0} (q_b\ \delta W_1 + q_t\ \delta W_M)\, dxdy\\
&\quad - \int_\Gamma \left[\sum_{I=1}^{N} \left(\hat{N}_{nn}^I \delta U_I^n + \hat{N}_{ns}^I \delta U_I^s \right) + \sum_{I=1}^{M} \hat{Q}_n^I \delta W_I \right] ds
\end{aligned} \tag{12.2 – 11}$$

$$\delta K = \int_{\Omega_0} \int_{-\frac{h}{2}}^{\frac{h}{2}} \rho_0 \left(\dot{u}\delta\dot{u} + \dot{v}\delta\dot{v} + \dot{w}\delta\dot{w} \right) dz \; dxdy$$
$$= \int_{\Omega_0} \left[\sum_{I,J=1}^{N} I^{IJ} \left(\dot{U}_I \delta\dot{U}_J + \dot{V}_I \delta\dot{V}_J \right) + \sum_{I,J=1}^{M} \tilde{I}^{IJ} \dot{W}_I \delta\dot{W}_J \right] dxdy \quad (12.2-12)$$

$$\mathcal{N}^I = \sum_{J=1}^{M} \left[\frac{\partial}{\partial x} \left(\tilde{N}_{xx}^{IJ} \frac{\partial W_J}{\partial x} + \tilde{N}_{xy}^{IJ} \frac{\partial W_J}{\partial x} \right) + \frac{\partial}{\partial y} \left(\tilde{N}_{xy}^{IJ} \frac{\partial W_J}{\partial x} + \tilde{N}_{yy}^{IJ} \frac{\partial W_J}{\partial y} \right) \right] \quad (12.2-13)$$

$$\begin{Bmatrix} N_{xx}^I \\ N_{yy}^I \\ N_{xy}^I \end{Bmatrix} = \int_{-\frac{h}{2}}^{\frac{h}{2}} \begin{Bmatrix} \sigma_{xx} \\ \sigma_{yy} \\ \sigma_{xy} \end{Bmatrix} \Phi^I \; dz, \quad \begin{Bmatrix} \tilde{N}_{xx}^{IJ} \\ \tilde{N}_{yy}^{IJ} \\ \tilde{N}_{xy}^{IJ} \end{Bmatrix} = \int_{-\frac{h}{2}}^{\frac{h}{2}} \begin{Bmatrix} \sigma_{xx} \\ \sigma_{yy} \\ \sigma_{xy} \end{Bmatrix} \Psi^I \Psi^J \; dz \quad (12.2-14)$$

$$\begin{Bmatrix} Q_x^I \\ Q_y^I \end{Bmatrix} = \int_{-\frac{h}{2}}^{\frac{h}{2}} \begin{Bmatrix} \sigma_{xz} \\ \sigma_{yz} \end{Bmatrix} \frac{d\Phi^I}{dz} \; dz, \quad \begin{Bmatrix} \tilde{Q}_x^I \\ \tilde{Q}_y^I \\ \tilde{Q}_z^I \end{Bmatrix} = \int_{-\frac{h}{2}}^{\frac{h}{2}} \begin{Bmatrix} \sigma_{xz} \\ \sigma_{yz} \\ \sigma_{zz} \end{Bmatrix} \Psi^I \; dz \quad (12.2-15)$$

$$I^{IJ} = \int_{-\frac{h}{2}}^{\frac{h}{2}} \rho_0 \Phi^I \Phi^J \; dz, \quad \tilde{I}^{IJ} = \int_{-\frac{h}{2}}^{\frac{h}{2}} \rho_0 \Psi^I \Psi^J \; dz \quad (12.2-16)$$

Substituting Eqs. (10)–(12) into Eq. (9), and deriving the Euler–Lagrange equations, we obtain

$$\delta U_I: \quad \frac{\partial N_{xx}^I}{\partial x} + \frac{\partial N_{xy}^I}{\partial y} - Q_x^I = \sum_{J=1}^{N} I^{IJ} \frac{\partial^2 U_J}{\partial t^2} \quad (12.2-17)$$

$$\delta V_I: \quad \frac{\partial N_{xy}^I}{\partial x} + \frac{\partial N_{yy}^I}{\partial y} - Q_y^I = \sum_{J=1}^{N} I^{IJ} \frac{\partial^2 V_J}{\partial t^2} \quad (12.2-18)$$

$$\delta W_I: \quad \frac{\partial \tilde{Q}_x^I}{\partial x} + \frac{\partial \tilde{Q}_y^I}{\partial y} - \tilde{Q}_z^I + \mathcal{N}^I + q_b \delta_{I1} + q_t \delta_{IM} = \sum_{J=1}^{M} \tilde{I}^{IJ} \frac{\partial^2 W_J}{\partial t^2} \quad (12.2-19)$$

The natural (force) boundary conditions of the theory are:

$$N_{nn}^I - \hat{N}_{nn}^I = 0, \quad N_{ns}^I - \hat{N}_{ns}^I = 0, \quad \left(\tilde{Q}_n^I + \tilde{\mathcal{P}}^I \right) - \hat{Q}_n^I = 0 \quad (12.2-20)$$

where

$$N_{nn}^I = N_{xx}^I n_x + N_{xy}^I n_y, \quad N_{ns}^I = N_{xy}^I n_x + N_{yy}^I n_y$$

$$\tilde{Q}_n^I = \tilde{Q}_x^I n_x + \tilde{Q}_y^I n_y, \quad U_I^n = U_I n_x + V_I n_y, \quad U_I^s = -U_I n_x + V_I n_y \quad (12.2-21)$$

$$\tilde{\mathcal{P}}^I = \sum_{J=1}^{M} \left(\tilde{N}_{xx}^{IJ} \frac{\partial W_J}{\partial x} + \tilde{N}_{xy}^{IJ} \frac{\partial W_J}{\partial x} \right) n_x + \left(\tilde{N}_{xy}^{IJ} \frac{\partial W_J}{\partial x} + \tilde{N}_{yy}^{IJ} \frac{\partial W_J}{\partial y} \right) n_y \quad (12.2-22)$$

Thus, the primary variables (displacements) and secondary variables (forces) of the layerwise theory are

$$\begin{aligned} &\text{Primary Variables:} \quad && U_I^n, \ U_I^s; \ W_I \\ &\text{Secondary Variables:} \quad && N_{nn}^I, \ N_{ns}^I; \ \tilde{Q}_n^I + \tilde{\mathcal{P}}^I \end{aligned} \tag{12.2-23}$$

Note that the form of the natural boundary conditions guarantees that the transverse shear stresses will satisfy the traction-free edge conditions on a "local" basis as the number of subdivisions through the thickness is increased (unlike the partial layerwise theories that neglect discrete layer transverse normal strain). The 2-D equations of motion can be expressed in terms of the dependent variables $(U_I(x,y,t), V_I(x,y,t), W_I(x,y,t))$ by replacing the stress components with the 3-D stress–strain relations and strain–displacement relations. The resulting system of $3n$ equations represent the semidiscretized version of Navier's equations of motion.

12.2–4 Laminate Constitutive Equations

The laminate constitutive equations for the layerwise theory can be derived using the lamina constitutive equations (8) and the definition of stress resultants (22). We obtain

$$\begin{aligned}
\begin{Bmatrix} N_{xx}^I \\ N_{yy}^I \\ N_{xy}^I \end{Bmatrix} &= \sum_{k=1}^{N_e} \int_{z_b^k}^{z_t^k} \begin{Bmatrix} \sigma_{xx} \\ \sigma_{yy} \\ \sigma_{xy} \end{Bmatrix}^{(k)} \Phi^I \, dz \\
&= \sum_{k=1}^{N_e} \int_{z_b^k}^{z_t^k} \begin{bmatrix} \bar{C}_{11} & \bar{C}_{12} & \bar{C}_{13} & \bar{C}_{16} \\ \bar{C}_{12} & \bar{C}_{22} & \bar{C}_{23} & \bar{C}_{26} \\ \bar{C}_{16} & \bar{C}_{26} & \bar{C}_{36} & \bar{C}_{66} \end{bmatrix}^{(k)} \begin{Bmatrix} \varepsilon_{xx} - \alpha_{xx}\Delta T \\ \varepsilon_{yy} - \alpha_{yy}\Delta T \\ \varepsilon_{zz} - \alpha_{zz}\Delta T \\ \gamma_{xy} - 2\alpha_{xy}\Delta T \end{Bmatrix}^{(k)} \Phi^I \, dz \\
&= \sum_{J=1}^{N} \begin{bmatrix} A_{11}^{IJ} & A_{12}^{IJ} & \tilde{A}_{13}^{IJ} & A_{16}^{IJ} \\ A_{12}^{IJ} & A_{22}^{IJ} & \tilde{A}_{23}^{IJ} & A_{26}^{IJ} \\ A_{16}^{IJ} & A_{26}^{IJ} & \tilde{A}_{36}^{IJ} & A_{66}^{IJ} \end{bmatrix} \begin{Bmatrix} \frac{\partial U_J}{\partial x} \\ \frac{\partial V_J}{\partial y} \\ W_J \\ \frac{\partial U_J}{\partial y} + \frac{\partial V_J}{\partial x} \end{Bmatrix} - \begin{Bmatrix} N_{xx}^{I(T)} \\ N_{yy}^{I(T)} \\ N_{xy}^{I(T)} \end{Bmatrix} \\
&\quad + \sum_{J,K=1}^{M} \begin{bmatrix} B_{11}^{IJK} & B_{12}^{IJK} & B_{16}^{IJK} \\ B_{12}^{IJK} & B_{22}^{IJK} & B_{26}^{IJK} \\ B_{16}^{IJK} & B_{26}^{IJK} & B_{66}^{IJK} \end{bmatrix} \begin{Bmatrix} \frac{1}{2}\frac{\partial W_J}{\partial x}\frac{\partial W_K}{\partial x} \\ \frac{1}{2}\frac{\partial W_J}{\partial y}\frac{\partial W_K}{\partial y} \\ \frac{\partial W_J}{\partial x}\frac{\partial W_K}{\partial y} \end{Bmatrix}
\end{aligned} \tag{12.2-24}$$

$$\begin{aligned}
\begin{Bmatrix} \tilde{N}_{xx}^{IJ} \\ \tilde{N}_{yy}^{IJ} \\ \tilde{N}_{xy}^{IJ} \end{Bmatrix} &= \sum_{k=1}^{N_e} \int_{z_b^k}^{z_t^k} \begin{Bmatrix} \sigma_{xx} \\ \sigma_{yy} \\ \sigma_{xy} \end{Bmatrix}^{(k)} \Psi^I \Psi^J \, dz \\
&= \sum_{K=1}^{N} \begin{bmatrix} B_{11}^{KIJ} & B_{12}^{KIJ} & \tilde{B}_{13}^{KIJ} & B_{16}^{KIJ} \\ B_{12}^{KIJ} & B_{22}^{KIJ} & \tilde{B}_{23}^{KIJ} & B_{26}^{KIJ} \\ B_{16}^{KIJ} & B_{26}^{KIJ} & \tilde{B}_{36}^{KIJ} & B_{66}^{KIJ} \end{bmatrix} \begin{Bmatrix} \frac{\partial U_K}{\partial x} \\ \frac{\partial V_K}{\partial y} \\ W_K \\ \frac{\partial U_K}{\partial y} + \frac{\partial V_K}{\partial x} \end{Bmatrix} - \begin{Bmatrix} \tilde{N}_{xx}^{IJ(T)} \\ \tilde{N}_{yy}^{IJ(T)} \\ \tilde{N}_{xy}^{IJ(T)} \end{Bmatrix}
\end{aligned}$$

$$+\sum_{K,P=1}^{M}\begin{bmatrix} D_{11}^{IJKP} & D_{12}^{IJKP} & D_{16}^{IJKP} \\ D_{12}^{IJKP} & D_{22}^{IJKP} & D_{26}^{IJKP} \\ D_{13}^{IJKP} & D_{23}^{IJKP} & D_{66}^{IJKP} \end{bmatrix}\begin{Bmatrix} \frac{1}{2}\frac{\partial W_K}{\partial x}\frac{\partial W_P}{\partial x} \\ \frac{1}{2}\frac{\partial W_K}{\partial y}\frac{\partial W_P}{\partial y} \\ \frac{\partial W_K}{\partial x}\frac{\partial W_P}{\partial y} \end{Bmatrix} \qquad (12.2-25)$$

$$\begin{aligned} \begin{Bmatrix} Q_x^I \\ Q_y^I \end{Bmatrix} &= \sum_{k=1}^{N_e}\int_{z_b^k}^{z_t^k}\begin{Bmatrix} \sigma_{xz} \\ \sigma_{yz} \end{Bmatrix}^{(k)}\frac{d\Phi^I}{dz}\,dz \\ &= \sum_{k=1}^{N_e}\int_{z_b^k}^{z_t^k}\begin{bmatrix} \bar{C}_{55} & \bar{C}_{45} \\ C_{45} & \bar{C}_{44} \end{bmatrix}^{(k)}\begin{Bmatrix} \gamma_{xz} \\ \gamma_{yz} \end{Bmatrix}^{(k)}\frac{d\Phi^I}{dz}dz \\ &= \sum_{J=1}^{N}\left(\begin{bmatrix} \bar{A}_{55}^{IJ} & \bar{A}_{45}^{IJ} \\ \bar{A}_{45}^{IJ} & \bar{A}_{44}^{IJ} \end{bmatrix}\begin{Bmatrix} U_J \\ V_J \end{Bmatrix} + \begin{bmatrix} \bar{B}_{55}^{IJ} & \bar{B}_{45}^{IJ} \\ \bar{B}_{45}^{IJ} & \bar{B}_{44}^{IJ} \end{bmatrix}\begin{Bmatrix} \frac{\partial W_J}{\partial x} \\ \frac{\partial W_J}{\partial y} \end{Bmatrix}\right) \qquad (12.2-26) \end{aligned}$$

$$\begin{aligned} \begin{Bmatrix} \tilde{Q}_x^I \\ \tilde{Q}_y^I \end{Bmatrix} &= \sum_{k=1}^{N_e}\int_{z_b^k}^{z_t^k}\begin{Bmatrix} \sigma_{xz} \\ \sigma_{yz} \end{Bmatrix}^{(k)}\Psi^I\,dz \\ &= \sum_{k=1}^{N_e}\int_{z_b^k}^{z_t^k}\begin{bmatrix} \bar{C}_{55} & \bar{C}_{45} \\ C_{45} & \bar{C}_{44} \end{bmatrix}^{(k)}\begin{Bmatrix} \gamma_{xz} \\ \gamma_{yz} \end{Bmatrix}^{(k)}\Psi^I dz \\ &= \sum_{J=1}^{N}\left(\begin{bmatrix} \bar{B}_{55}^{JI} & \bar{B}_{45}^{JI} \\ \bar{B}_{45}^{JI} & \bar{B}_{44}^{JI} \end{bmatrix}\begin{Bmatrix} U_J \\ V_J \end{Bmatrix} + \begin{bmatrix} \bar{D}_{55}^{IJ} & \bar{D}_{45}^{IJ} \\ \bar{D}_{45}^{IJ} & \bar{D}_{44}^{IJ} \end{bmatrix}\begin{Bmatrix} \frac{\partial W_J}{\partial x} \\ \frac{\partial W_J}{\partial y} \end{Bmatrix}\right) \qquad (12.2-27) \end{aligned}$$

$$\begin{aligned} \tilde{Q}_z^I &= \sum_{k=1}^{N_e}\int_{z_b^k}^{z_t^k}\sigma_{zz}^{(k)}\frac{d\Psi^I}{dz}\,dz \\ &= \sum_{k=1}^{N_e}\int_{z_b^k}^{z_t^k}\left(\bar{C}_{13}\varepsilon_{xx} + \bar{C}_{23}\varepsilon_{yy} + \bar{C}_{33}\varepsilon_{zz} + \bar{C}_{36}\varepsilon_{xy}\right)^{(k)}\frac{d\Psi^I}{dz}\,dz \\ &= \sum_{J=1}^{N}\left[\tilde{A}_{13}^{JI}\frac{\partial U_J}{\partial x} + \tilde{A}_{23}^{JI}\frac{\partial V_J}{\partial y} + \hat{A}_{33}^{JI}W_J + \tilde{A}_{36}^{JI}\left(\frac{\partial U_K}{\partial y} + \frac{\partial V_K}{\partial x}\right)\right] - \tilde{Q}_z^{I(T)} \end{aligned} \qquad (12.2-28)$$

where N_e is the number of physical layers in the laminate, z_b^k and z_t^k are coordinates of the bottom and top of the kth layer, and the laminate stiffnesses are defined as follows:

$$A_{ij}^{IJ} = \sum_{k=1}^{N_e}\int_{z_b^k}^{z_t^k}\bar{C}_{ij}^{(k)}\;\Phi^I\Phi^J\;dz,\quad \tilde{A}_{ij}^{IJ} = \sum_{k=1}^{N_e}\int_{z_b^k}^{z_t^k}\bar{C}_{ij}^{(k)}\;\Phi^I\frac{d\Psi^J}{dz}\;dz$$

$$\bar{A}_{ij}^{IJ} = \sum_{k=1}^{N_e}\int_{z_b^k}^{z_t^k}\bar{C}_{ij}^{(k)}\;\frac{d\Phi^I}{dz}\frac{d\Phi^J}{dz}\;dz,\quad \hat{A}_{33}^{IJ} = \sum_{k=1}^{N_e}\int_{z_b^k}^{z_t^k}\bar{C}_{33}^{(k)}\;\frac{d\Psi^I}{dz}\frac{d\Psi^J}{dz}\;dz$$

$$\bar{B}_{ij}^{IJ} = \sum_{k=1}^{N_e}\int_{z_b^k}^{z_t^k}\bar{C}_{ij}^{(k)}\;\frac{d\Phi^I}{dz}\Psi^J\;dz,\quad \bar{D}_{ij}^{IJ} = \sum_{k=1}^{N_e}\int_{z_b^k}^{z_t^k}\bar{C}_{ij}^{(k)}\;\Psi^I\Psi^J\;dz \qquad (12.2-29)$$

$$B_{ij}^{IJK} = \sum_{k=1}^{N_e} \int_{z_b^k}^{z_t^k} \bar{C}_{ij}^{(k)} \; \Phi^I \Psi^J \Psi^K \; dz, \quad \tilde{B}_{ij}^{IJK} = \sum_{k=1}^{N_e} \int_{z_b^k}^{z_t^k} \bar{C}_{ij}^{(k)} \; \Psi^I \Psi^J \frac{d\Psi^K}{dz} \; dz$$

$$D_{ij}^{IJKP} = \sum_{k=1}^{N_e} \int_{z_b^k}^{z_t^k} \bar{C}_{ij}^{(k)} \; \Psi^I \Psi^J \Psi^K \Psi^P \; dz \tag{12.2-30}$$

for $i, j = 1, 2, 6$. Note that when the von Kármán nonlinearity is not included, all laminate stiffnesses with three and four superscripts will not enter the governing equations.

For the nonisothermal case, the thermal stress resultants are defined as

$$\begin{Bmatrix} N_{xx}^{I(T)} \\ N_{yy}^{I(T)} \\ N_{xy}^{I(T)} \end{Bmatrix} = \sum_{k=1}^{N_e} \int_{z_b^k}^{z_t^k} \begin{bmatrix} \bar{C}_{11} & \bar{C}_{12} & \bar{C}_{13} & \bar{C}_{16} \\ \bar{C}_{12} & \bar{C}_{22} & \bar{C}_{23} & \bar{C}_{26} \\ \bar{C}_{16} & \bar{C}_{26} & \bar{C}_{36} & \bar{C}_{66} \end{bmatrix}^{(k)} \begin{Bmatrix} \alpha_{xx} \\ \alpha_{yy} \\ \alpha_{zz} \\ 2\alpha_{xy} \end{Bmatrix}^{(k)} \Delta T \; \Phi^I \; dz$$

$$\begin{Bmatrix} \tilde{N}_{xx}^{IJ(T)} \\ \tilde{N}_{yy}^{IJ(T)} \\ \tilde{N}_{xy}^{IJ(T)} \end{Bmatrix} = \sum_{k=1}^{N_e} \int_{z_b^k}^{z_t^k} \begin{bmatrix} \bar{C}_{11} & \bar{C}_{12} & \bar{C}_{13} & \bar{C}_{16} \\ \bar{C}_{12} & \bar{C}_{22} & \bar{C}_{23} & \bar{C}_{26} \\ \bar{C}_{16} & \bar{C}_{26} & \bar{C}_{36} & \bar{C}_{66} \end{bmatrix}^{(k)} \begin{Bmatrix} \alpha_{xx} \\ \alpha_{yy} \\ \alpha_{zz} \\ 2\alpha_{xy} \end{Bmatrix}^{(k)} \Delta T \; \Psi^I \Psi^J \; dz$$

$$\tilde{Q}_z^{I(T)} = \sum_{k=1}^{N_e} \int_{z_b^k}^{z_t^k} (\bar{C}_{13}\alpha_{xx} + \bar{C}_{23}\alpha_{yy} + \bar{C}_{33}\alpha_{zz} + \bar{2}C_{36}\alpha_{xy})^{(k)} \; \Delta T \; \frac{d\Psi^I}{dz} \; dz \tag{12.2-31}$$

Note that the sum over the number of layers in the definition of the laminate stiffnesses is limited to the numerical layers over which the functions Φ^I and Ψ^I are defined. The functions Φ^I and Ψ^I are defined at the most over two adjacent numerical layers [see Eqs. (3) and (4)]. For example, for the choice of linear interpolation functions, when $\Phi^I = \Psi^I$ $(M = N)$, the following expressions are obtained for a typical layer:

$$[A_{ij}]^{(k)} = \frac{\bar{C}_{ij}^{(k)} h_k}{6} \begin{bmatrix} 2 & 1 \\ 1 & 2 \end{bmatrix}, \quad [\tilde{A}_{ij}]^{(k)} = \frac{\bar{C}_{ij}^{(k)}}{2} \begin{bmatrix} -1 & 1 \\ -1 & 1 \end{bmatrix} \tag{12.2-32a}$$

$$[\bar{A}_{ij}]^{(k)} = \frac{\bar{C}_{ij}^{(k)}}{h_k} \begin{bmatrix} 1 & -1 \\ -1 & 1 \end{bmatrix}, \quad [\hat{A}_{33}]^{(k)} = \frac{\bar{C}_{33}^{(k)}}{h_k} \begin{bmatrix} 1 & -1 \\ -1 & 1 \end{bmatrix} \tag{12.2-32b}$$

$$[\bar{B}_{ij}]^{(k)} = \frac{\bar{C}_{ij}^{(k)}}{2} \begin{bmatrix} 1 & -1 \\ 1 & -1 \end{bmatrix}, \quad [\bar{D}_{ij}]^{(k)} = \frac{\bar{C}_{ij}^{(k)} h_k}{6} \begin{bmatrix} 2 & 1 \\ 1 & 2 \end{bmatrix} \tag{12.2-32c}$$

If a laminate consists of more than one layer, the above matrices must be *assembled* (see Reddy [53]). For example, $[A_{ij}]$ for a two-layer laminate takes the form

$$[A_{ij}] = \frac{1}{6}\begin{bmatrix} 2\bar{C}_{ij}^{(1)}h_1 & \bar{C}_{ij}^{(1)}h_1 & 0 \\ \bar{C}_{ij}^{(1)}h_1 & 2\left(\bar{C}_{ij}^{(1)}h_1 + \bar{C}_{ij}^{(2)}h_2\right) & \bar{C}_{ij}^{(2)}h_2 \\ 0 & \bar{C}_{ij}^{(2)}h_2 & 2\bar{C}_{ij}^{(2)}h_2 \end{bmatrix} \tag{12.2-33}$$

$$[\bar{A}_{ij}] = \begin{bmatrix} \bar{C}_{ij}^{(1)}(h_1)^{-1} & -\bar{C}_{ij}^{(1)}(h_1)^{-1} & 0 \\ -\bar{C}_{ij}^{(1)}(h_1)^{-1} & \bar{C}_{ij}^{(1)}(h_1)^{-1} + \bar{C}_{ij}^{(2)}(h_2)^{-1} & -\bar{C}_{ij}^{(2)}(h_2)^{-1} \\ 0 & -\bar{C}_{ij}^{(2)}(h_2)^{-1} & \bar{C}_{ij}^{(2)}(h_2)^{-1} \end{bmatrix} \tag{12.2-34}$$

$$[\bar{B}_{ij}] = \frac{1}{2}\begin{bmatrix} \bar{C}_{ij}^{(1)} & -\bar{C}_{ij}^{(1)} & 0 \\ \bar{C}_{ij}^{(1)} & 0 & -\bar{C}_{ij}^{(2)} \\ 0 & \bar{C}_{ij}^{(2)} & -\bar{C}_{ij}^{(2)} \end{bmatrix} \tag{12.2-35}$$

Similar expressions hold for the other coefficients in Eq. (30).

12.3 Finite Element Model

12.3–1 Layerwise Model

The displacement finite element model corresponding to the full layerwise theory is developed by substituting an assumed interpolation of the displacement field into the principle of virtual displacements (12.2–9) for a representative finite element of the plate. Suppose that the displacement field is interpolated as

$$\begin{aligned} U_I(x,y,t) &= \sum_{j=1}^{p} U_I^j(t)\psi_j(x,y) \\ V_I(x,y,t) &= \sum_{j=1}^{p} V_I^j(t)\psi_j(x,y) \\ W_I(x,y,t) &= \sum_{j=1}^{q} W_I^j(t)\varphi_j(x,y) \end{aligned} \tag{12.3-1}$$

where p and q are the number of nodes per 2-D element used to approximate the in-plane and transverse deflections, respectively, and $U_I^j(t), V_I^j(t)$, and $W_I^j(t)$ are the values of the displacements U_I, V_I, and W_I, respectively, at the jth node of the 2-D finite element representing the jth plane of the plate element. The functions $\psi_j(x,y)$ $(j = 1, 2, \cdots, p)$ and $\varphi_j(x,y)$ $(j = 1, 2, \cdots, q)$ are the two-dimensional Lagrangian interpolation polynomials associated with the jth node of the two-dimensional finite element.

The semidiscrete finite element equations are obtained by substituting equation (1) into the principle of virtual work in Eq. (12.2–9). Then the fully discretized equations, for the transient case, are obtained using a time approximation scheme, as discussed earlier. Here we limit our discussion to the static problems, and the case in which the same interpolation is used for all three displacements: $\varphi_j = \psi_j$ $(p = q)$ and $\Psi_I = \Phi_I$ $(M = N)$.

The following two–dimensional finite elements $(N = M)$ are used here with isoparametric formulations:

E4: Four-node Lagrange quadrilateral element.

E8: Eight-node serendipity quadrilateral element.

E9: Nine-node Lagrange quadrilateral element.

E12: Twelve-node serendipity quadrilateral element.

E16: Sixteen-node Lagrange quadrilateral element.

Each of these elements may be used in conjunction with one or more linear, quadratic, or cubic (denoted L,Q,C respectively) 1-D Lagrange elements through the thickness to create a wide variety of different layerwise finite elements. We use the notation E12–L6 to denote a two-dimensional E12 element with six linear subdivisions through the thickness; likewise, E9–Q3 denotes a two-dimensional E9 element with three quadratic subdivisions through the thickness.

The full layerwise finite element model is the same as a conventional 3-D displacement finite element model in terms of interpolation capability and problem size for a 3-D body with parallel top and bottom surfaces. A variable thickness plate must be approximated as an elementwise constant-thickness plate in order to use the present element. Virtually in all practical cases, a laminated plate structure, including a structure with dropped plies, is made up of constant-thickness laminae and therefore the present element can be used to model such structures. The layerwise element has some analysis advantages over the conventional 3-D elements. The layerwise format maintains a 2-D type data structure similar to finite element models of 2-D ESL theories. This provides several advantages over conventional 3-D finite element models. First the volume of the input data is reduced. Secondly, the in-plane 2-D mesh and the transverse 1-D mesh can be refined independently of each other without having to reconstruct a 3-D finite element mesh. The 2-D type data structure also allows efficient formulation of the element stiffness matrices as is discussed in the next section.

12.3–2 Full Layerwise Model Versus 3-D Finite Element Model

Since the present layerwise plate model is developed to provide the same modeling capability as a conventional 3-D finite element model of laminated plates, it is informative to investigate the similarities and differences in these two models. First, consider the theories used to develop each model. The conventional 3-D finite element model is based on the 3-D theory of elasticity, and the associated governing equations of motion are Navier's equations. The full layerwise, 2-D laminate theory used to develop the present layerwise model is governed by a set of $3n$ coupled partial differential equations that can be viewed as the semidiscretized version of Navier's equations. Thus the governing equations of motion of the present layerwise theory are necessarily an approximation to the exact 3-D equations of motion; however, the approximation can be made as close as desired by increasing the number of subdivisions through the thickness and/or increasing the order of interpolation through the thickness.

In contrast to the differences that exist between the governing equations of motion for these two theories both finite element models represent fully discretized versions of their respective theories; thus the modeling capabilities of the two finite element models are essentially the same. A comparison of the interpolation capability of the two finite element models reveals a close similarity. Note that both models use actual displacement components (u_1, u_2, u_3) as the primary variables (nodal degrees of freedom) and both require C^0–continuity of these variables across element (or layer) boundaries. Further, if one compares similar element types from the two models, the interpolation of the primary variables is exactly the same. For example, an E9–Q3 layerwise element exhibits the same interpolation as a stack of three, 27 node, Lagrangian hexahedrons; an E12–L6 exhibits the same interpolation as a stack of six, 24 node, hexahedrons with cubic serendipity interpolation in the planar coordinates and linear lagrangian interpolation in the transverse direction. Thus it is not surprising that the resulting global system of equations and subsequent solution produced by these two finite element models are exactly the same as long as the meshes, element types, and integration schemes are comparable.

Obviously the 3-D finite element model is more general than the layerwise finite element model; the latter actually represents a special case of the former. The layerwise model assumes that the displacements, material properties, and element geometry can be approximated by a sum of conveniently separable interpolation functions (*i.e.*, each individual 3-D interpolation function can be written as the product of a 2-D interpolation function and a 1-D interpolation function). This restriction does not imply that the displacement solution itself can be separated into the product of an in-plane function of x and y and an out-of-plane function of z;

however, in computing all element volume integrals, the layerwise model can use separated numerical integration (*i.e.*, the integration with respect to the thickness coordinate is performed independent of the integration with respect to the planar coordinates). The results from a single integration through the thickness can then be used at each Gauss point in the subsequent in-plane integration. This separated integration allows the element stiffness matrix to be computed using only a fraction of the operations required to form the stiffness matrix for a conventional 3-D finite element.

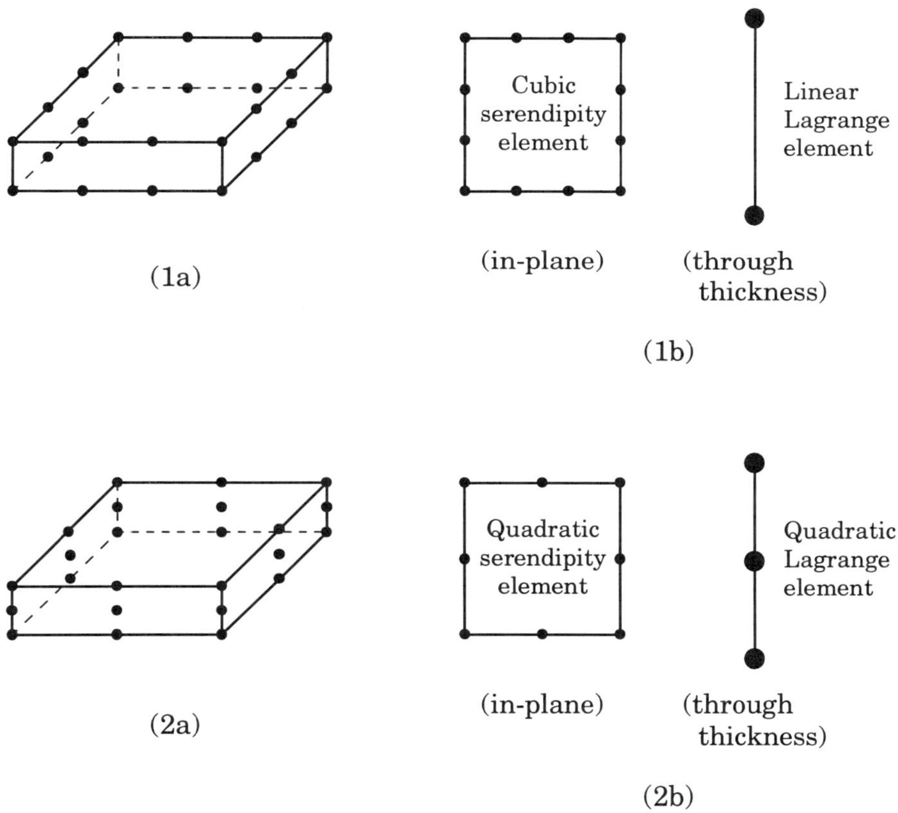

Figure 12.3–1: Various 3-D and layerwise elements used in the numerical comparison.

To illustrate the computational savings of the simplified, separated integration afforded by the layerwise model, the number of operations needed to form the element stiffness matrix for comparable layerwise elements and conventional 3-D elements (see Figure 12.3–1) are tabulated in Table 12.3–1. Full quadrature is used in each case. As illustrated in Table 12.3–1, the layerwise model's separated numerical integration requires significantly fewer operations to form the element stiffness matrix.

Consider elements 2a and 2b in Table 12.3–1. Both have 81 degrees of freedom and exhibit complete quadratic interpolation in all three coordinate directions. To form the element stiffness matrix for element 2a (*i.e.*, 27-node, 3-D isoparametric quadratic hexahedron), a 3-D interpolation function subroutine must be called 27 times since there are 27 Gaussian quadrature points. Each time the 3-D interpolation function subroutine is called, 27 individual interpolation functions must be evaluated. Thus 27^2 or 729 individual interpolation functions must be evaluated. To form the element stiffness matrix for element 2b (*i.e.*, nine-node 2-D layerwise quadrilateral element with a single quadratic layer), a 1-D interpolation function subroutine must be called 3 times and a 2-D interpolation function subroutine must be called 9 times, since there are 3 and 9 Gaussian quadrature points in the 1-D and 2-D elements, respectively. This is obviously less work than required by the 3-D element. The efficiency of the layerwise model increases (relative to the 3-D model) as more elements and/or layers are added, or as the order of the interpolation through the thickness is increased. Note that this efficient, separated integration uses standard Lagrangian interpolation through the thickness; thus certain commonly used 3-D finite elements do not have layerwise counterparts (*e.g.*, the 20-node quadratic serendipity hexahedron element and the 32-node cubic serendipity hexahedron element).

Table 12.3–1 Comparison of the number of operations needed to form the element stiffness matrices for equivalent elements in the conventional 3-D format and the layerwise 2-D format. Full quadrature is used in each case.

Element Type†	Multiplications	Addition	Assignments
1a (3-D)	1,116,000	677,000	511,000
1b (LWPT)	423,000	370,000	106,000
2a (3-D)	1,182,000	819,000	374,000
2b (LWPT)	284,000	270,000	69,000

† Element 1a: 72 degrees of freedom, 24-node 3-D isoparametric hexahedron with cubic in-plane interpolation and linear transverse interpolation.

Element 1b: 72 degrees of freedom, E12–L1 layerwise element.

Element 2a: 81 degrees of freedom, 27-node 3-D isoparametric hexahedron with quadratic interpolation in all three directions.

Element 2b: 81 degrees of freedom, E9–Q1 layerwise element.

12.3–3 Considerations for Modeling Relatively Thin Laminates

First a comment concerning 'thin' and 'thick' plates is in order. In the present study we call a plate domain *thick* if the local span-to-thickness ratio a/h is less than 20. In computing the span-to-thickness ratio, one should use the in-plane dimension of the local domain that is being modeled. Layerwise plate models are primarily intended for thick plate situations where the simple ESL plate theories (*e.g.*, CLPT or FSDT) are known to be inaccurate. In the analysis of thick plates, the added computational expense of a layerwise model is justified by its improved predicting capabilities. Although a layerwise model may be primarily intended for thick plate situations, it is important to determine the limits of the layerwise model's applicability to thin plate situations since finite element models of refined plate theories have proven to be problematic whenever the thickness dimension of the structure is greatly reduced relative to the planar dimensions. In particular, finite elements which possess full three-dimensional modeling capability can exhibit spurious transverse shear stiffness, spurious transverse normal stiffness, and ill-conditioned stiffness matrices as the span-to-thickness ratio of the structure increases. Each of these undesirable phenomena will now be discussed within the context of full layerwise finite elements and selective integration techniques will be suggested to alleviate the spurious transverse stiffnesses.

The spurious shear stiffness phenomenon is caused by an interpolation inconsistency that prevents the finite element from modeling a state of zero transverse shear stress in the presence of general nonzero bending strains. As the plate's span-to-thickness ratio approaches the thin plate limit, the transverse shear deformation must tend toward zero relative to the bending deformation (*i.e.*, the Kirchhoff condition). The degree to which a particular finite element fails in approximating this condition determines the amount of spurious shear stiffness the element will exhibit while deforming in a bending mode. Thus elements which are poor approximators of this condition are known to exhibit shear locking while elements which are better approximators of this condition may simply exhibit a slight overstiffness.

Next we examine the assumed form of the transverse shear strains ε_{xz} and ε_{yz} used in the layerwise finite elements to determine whether or not the Kirchhoff condition can be satisfied in the thin plate limit. Within a specific layer of a typical finite element, the Kirchhoff condition requires that

$$\varepsilon_{yz} = \sum_{I=1}^{M}\sum_{j=1}^{p}\left(V_I^j\frac{d\Phi^I}{dz}(z_k)\psi_j(x_k,y_k) + W_I^j\Phi^I(z_k)\frac{\partial\psi_j}{\partial y}(x_k,y_k)\right) = 0 \quad (12.3-2a)$$

$$\varepsilon_{xz} = \sum_{I=1}^{M}\sum_{j=1}^{p}\left(U_I^j\frac{d\Phi^I}{dz}(z_k)\psi_j(x_k,y_k) + W_I^j\Phi^I(z_k)\frac{\partial\psi_j}{\partial x}(x_k,y_k)\right) = 0 \quad (12.3-2b)$$

where k ranges from 1 to the number of Gaussian integration points in the $xy-$plane of the element; M is the number of nodes through the thickness of the numerical layer; p is the number of nodes in the 2-D element. Note that Eq. (2a) exhibits an interpolation inconsistency with respect to both the y and z directions, while Eq. (2b) exhibits an interpolation inconsistency with respect to both the x and z directions. In general, these interpolation inconsistencies prevent Eqs. (2a,b) from being satisfied at all Gaussian integration points in the element $xy-$plane. Thus, as the span-to-thickness ratio increases, the computed solution tends to suppress the higher-order terms of the interpolation so that the Kirchhoff condition can be satisfied (*i.e.*, spurious constraints are introduced into the model). If a reduced quadrature is used to evaluate terms contributing to the transverse shear energy (*i.e.*, any terms containing a constitutive matrix component $\bar{Q}_{ij}$ where $i, j = 4, 5$), then the Kirchhoff condition will not be enforced as stringently as in the case of full quadrature, since Eqs. (2a,b) will be enforced at fewer points in the domain. Thus the displacements are allowed more freedom to model utilize their full interpolation capability.

In practice, reduced quadrature is required only when integrating the transverse shear terms with respect to x and y (not the thickness coordinate z), because the integrated effect of the transverse interpolation inconsistencies is insignificant compared with the contributions from the in-plane interpolation inconsistencies. Consider expanding Eqs. (2a,b) so that all of the interpolation functions are expressed as summations of simple polynomial terms. As the span-to-thickness ratio is increased, any terms that contain the transverse coordinate (z) will tend toward zero in comparison with terms that do not contain z; therefore, the transverse interpolation inconsistency does not adversely affect the solution and therefore does not need or benefit from reduced quadrature with respect to the z direction.

Another problem that arises when modeling thin plates with finite elements possessing 3-D capability is that the accuracy of the computed transverse normal stress deteriorates rapidly as the span-to-thickness ratio increases. For relatively thick plates, the transverse normal stress is computed quite accurately; however, as the span-to-thickness ratio increases, the computed transverse normal stress actually diverges from the correct value. This phenomenon arises because of an interpolation inconsistency that prevents the finite element from modeling a state of zero transverse normal stress in the presence of general nonzero bending strains.

As the span-to-thickness ratio increases toward the thin plate limit, the strain energy associated with transverse normal strain tends toward zero relative to the bending energy. Thus the transverse normal strain must approach a functional form that is consistent with the combined Poisson effects from the dominating in-plane strains. Setting σ_{zz} equal to zero in the constitutive equations and solving for ε_{zz}, we have

$$\varepsilon_{zz} = -\frac{1}{\bar{Q}_{33}}\left(\bar{Q}_{31}\varepsilon_{xx} + \bar{Q}_{32}\varepsilon_{yy} + 2\bar{Q}_{36}\varepsilon_{xy}\right) \tag{12.3-3a}$$

where $\bar{Q}_{ij}$ are the transformed elastic stiffnesses. On the other hand, from the strain–displacement relations we have

$$\varepsilon_{zz} = \sum_{I=1}^{M} W_I \frac{d\Phi^I}{dz} \tag{12.3-3b}$$

Equations (3a,b) represent a constraint that is imposed on the solution as the thin plate limit is approached.

Expressing the strain component ε_{zz} of Eq. (3b) in terms of the finite element approximations within a typical element layer, we have

$$\begin{aligned}\varepsilon_{zz} &= \sum_{I=1}^{M}\sum_{j=1}^{p} W_I^j \frac{d\Phi^I}{dz}(z_k)\psi_j(x_k, y_k)\\ &= -\frac{1}{\bar{Q}_{33}}\sum_{I=1}^{M}\sum_{j=1}^{p}\Bigg[\bar{Q}_{31}U_I^j\Phi^I(z_k)\frac{\partial\psi_j}{\partial x} + \bar{Q}_{32}V_I^j\Phi^I(z_k)\frac{\partial\psi_j}{\partial y}\\ &\quad + \bar{Q}_{36}\Phi^I(z_k)\left(U_I^j\frac{\partial\psi_j}{\partial y} + V_I^j\frac{\partial\psi_j}{\partial x}\right)\Bigg]_{(x_k,y_k)}\end{aligned} \tag{12.3-4}$$

for $k = 1, 2, \cdots, G$, where G is the number of Gaussian integration points in the element $xy-$plane. Note that ε_{zz} exhibits a higher-order in-plane interpolation and a lower-order transverse interpolation than the in-plane strains. In general, these interpolation inconsistencies prevent Eq. (4) from being satisfied at all Gaussian integration points in the element layer domain. As the span-to-thickness ratio increases toward the thin plate limit, the computed solution tends to suppress the higher-order terms of the interpolation so that Eq. (4) can be satisfied (*i.e.*, spurious constraints are introduced into the model).

If a reduced quadrature is used to evaluate the terms contributing to the energy associated with transverse normal strain (*i.e.*, any terms containing a $\bar{Q}_{3i}$ or $\bar{Q}_{i3}$ constitutive matrix component), then the condition given by Eq. (4) will not be enforced as stringently as in the case of full quadrature, since Eq. (4) will be enforced at fewer points in the domain. Thus the displacements are allowed more freedom to utilize their full interpolation capability.

Although selective integration can efficiently alleviate the problems of spurious transverse shear and normal stresses, such inexact integration can result in element stiffness matrices that have an excess number of zero eigenvalues. Note that the eigenvalues of an element stiffness matrix are representative of the magnitude of the force vector needed to maintain a particular mode of deformation represented by the associated eigenvector. An unconstrained layerwise finite element should have six zero eigenvalues corresponding to the six independent modes of rigid body displacement (three translations and three rotations). Any additional zero eigenvalues are associated with eigenvectors representing spurious deformation modes that produce no strain at the Gaussian integration points used in computing the element stiffness matrix. These spurious deformation modes can be superimposed on the true displacement solution without affecting the system of equations as a whole. The displacements computed in such a system may look nothing like the true solution due to the addition of one or more spurious displacement modes of arbitrary magnitude. The imposition of essential boundary conditions (necessary to prevent rigid body displacement) may or may not prevent these spurious displacement modes from occurring; thus it may be possible for spurious displacement modes to exist unsuppressed in an assembled structure which is constrained against rigid body displacement. Thus elements that possess an excessive number of zero eigenvalues must be used with great care.

An eigenvalue analysis of the layerwise finite elements is performed to determine the effect of various numerical integration schemes on the existence of spurious displacement modes. For this purpose, it is sufficient to consider a layerwise element containing a single layer. The following numerical integration schemes are considered.

F = All terms in the element stiffness matrix are computed using full integration.

S1 = A selective integration scheme in which all terms in the element stiffness matrix which contain the transverse shear stiffnesses $\bar{Q}_{44}$, $\bar{Q}_{45}$, or $\bar{Q}_{55}$ are computed using reduced integration. All remaining terms in the element stiffness matrix are computed using full integration.

S2 = A selective integration scheme in which all terms in the element stiffness matrix which contain the stiffnesses $\bar{Q}_{i3}$, $\bar{Q}_{i4}$, or $\bar{Q}_{i5}$ ($i = 1, 2, .., 6$) are computed using reduced integration. Thus all terms related to transverse shear effects or transverse normal effects are computed using reduced integration. All remaining terms in the element stiffness matrix, including those terms which correspond to strictly in-plane effects, are computed using full integration.

R = All terms in the element stiffness matrix are computed using reduced integration.

Recall from the earlier discussion that integration scheme S1 is used to remove spurious shear stress from the finite element model when the span-to-thickness ratio is large. Integration scheme S2 is used to remove both spurious transverse shear stress and spurious transverse normal stress from the finite element model when the span-to-thickness ratio is large. Integration scheme R, while not necessary, is included for comparison. Also recall that these different integration schemes are only necessary for the inplane integration. Thus all integration through the thickness of the element is performed using full integration.

Table 12.3–2 shows the number of excess zero eigenvalues (*i.e.*, the number of spurious displacement modes) exhibited by various elements under various integration schemes. Note that all three Lagrangian elements (E4, E9, E16) can exhibit spurious displacement modes unless full integration is used. The two serendipity elements do not exhibit any spurious displacement modes except for the special case of a single isolated E8 element under uniform reduced integration.

Table 12.3–2: Number of excess zero eigenvalues that exist for various types of single, unconstrained, layerwise elements under different integration schemes.

Element Type	In-plane Integration Scheme			
	F	S1	S2	R
E4-L1	0	2	3	7
E8-L1	0	0	0	2†
E8-Q1	0	0	0	3†
E9-L1	0	1	2	8
E9-Q2	0	1	3	12
E12-L1	0	0	0	0
E12-Q1	0	0	0	0
E12-C1	0	0	0	0
E16-L1	0	1	2	8
E16-Q1	0	1	3	12
E16-C1	0	1	4	16

† For the E8 elements, these spurious displacement modes can exist only for the case of a single isolated element. If more than one element is present, then no spurious displacement modes can exist.

12.3–4 Bending of a Simply Supported (0/90/0) Laminate

In order to investigate the influence of element type and numerical integration scheme on the accuracy of the layerwise finite element solution, a square, simply supported, symmetric cross-ply (0/90/0) laminated plate subjected to a sinusoidally distributed transverse load on the upper surface is selected. The domain of the plate is $0 < x < a, 0 < y < a, 0 < z < h$. The layers are assumed to be of equal thickness ($h_k = h/3$) and have the following material properties in the principal material coordinate system:

$$E_1 = 25 \times 10^6 \text{ psi}, \quad E_2 = E_3 = 10^6 \text{ psi}$$

$$G_{12} = 0.5{\times}10^6 \text{ psi}, \; G_{13} = G_{23} = 0.2{\times}10^6 \text{ psi}, \; \nu_{12} = \nu_{13} = \nu_{23} = 0.25 \quad (12.3-5)$$

This particular problem has an exact 3-D elasticity solution (see Pagano [1]) which is used to verify the displacements and stresses obtained using the layerwise finite element model. The finite element stresses are computed at the reduced Gauss points within each thickness subdivision of each element, using the computed displacements and the constitutive equations.

Table 12.3–3 Finite element meshes used in the analysis of a simply supported (0/90/0) plates under sinusoidally distributed transverse load.

Mesh No.	Element Type	Mesh Density	Total D.O.F.
1	E4-L6	6×6	1,029
2	E8-Q3	3×3	840
3	E9-Q3	3×3	1,029
4	E12-Q3	2×2	693
5	E16-Q3	2×2	1,029

Due to the symmetry of the problem, only one quadrant ($a/2 < x < a$, $a/2 < y < a$, $0 < z < h$) of the plate is modeled using the uniform, coarse meshes described in Table 12.3–3. The boundary conditions used are

$$\begin{aligned} u(x, a/2, z) =& u(a/2, y, z) = 0 \\ v(a/2, y, z) =& u(x, a/2, z) = 0 \\ w(x, a, z) =& u(a, y, z) = 0 \end{aligned} \quad (12.3-6)$$

Note that all five meshes contain the same nodal density along the edges of the computational domain, while the three meshes of Lagrangian elements contain the exact same nodal distribution throughout the domain. Five different span-to-thickness ratios are considered to test a wide range of plate bending behavior ($a/h = 4, 10, 20, 50, 200$). Two different numerical integration schemes are compared (types F and S2 discussed earlier). Note that while the three Lagrangian elements are capable of exhibiting spurious displacement modes under S2 integration, the simply supported boundary conditions of this particular example problem prevent any of these modes from arising.

A comparison of the computed maximum transverse displacement and exact maximum transverse displacement, which occur at the centroid of the upper surface of the plate are presented in Table 12.3–4. Note that when full quadrature is used, all elements except the E16 element (Mesh 5) exhibit a noticeable amount of artificial shear stiffening as the span-to-thickness ratio increases to 200, with the E4 element (Mesh 1) actually beginning to lock. For large span-to-thickness ratios, the spurious shear stiffening is significantly reduced by using the S2 integration scheme as shown in Table 12.3–4.

Table 12.3–4: Ratio of computed transverse displacement to exact transverse displacement at $(a/2, a/2, h)$ in a square, simply supported (0/90/0) laminate under sinusoidal transverse load.

Mesh	Integ. Scheme	Span-to-Thickness Ratio, a/h 4	10	20	50	200
1	F	0.9815	0.9732	0.9284	0.6975	0.1280
2	F	0.9901	0.9979	0.9981	0.9959	0.9807
3	F	0.9906	0.9985	0.9987	0.9959	0.9845
4	F	0.9880	0.9960	0.9968	0.9953	0.9615
5	F	0.9907	0.9988	1.0006	1.0004	1.0001
1	S2	0.9900	0.9952	0.9945	0.9943	0.9942
2	S2	0.9873	0.9978	0.9992	0.9995	0.9992
3	S2	0.9880	0.9984	0.9998	1.0001	1.0001
4	S2	0.9980	0.9959	0.9966	0.9952	0.9650
5	S2	0.9906	0.9988	0.9997	1.0000	1.0000

Table 12.3–5 contains a comparison of computed in-plane normal stress σ_{xx} with the exact analytical value at the reduced Gauss point nearest to the point $(a/2, a/2, h)$ where σ_{xx} attains a maximum. The computed transverse normal stress σ_{zz} is compared in Table 12.3–6 with the exact analytical value at the reduced Gauss point nearest to the point $(a/2, a/2, h)$ where σ_{zz} attains a maximum. Note that as the span-to-thickness ratio increases, the computed σ_3 diverges for all elements when full quadrature is used. However, the S2 integration scheme allows accurate transverse normal stresses to be computed, even for large span-to-thickness ratios. Table 12.3–7 contains a comparison of computed transverse shear stress σ_{xz} with the exact analytical value at the reduced Gauss point nearest to the point $(a, 0, a/2)$.

Table 12.3–5: Ratio of computed σ_{xx} to exact σ_{xx} in a square, simply supported (0/90/0) laminate under sinusoidally distributed transverse load. The stresses are computed at the reduced Gauss point closest to the centroid of the upper surface of the plate $(a/2, a/2, h)$.

Mesh	Integration Scheme	Span-to-Thickness Ratio, a/h 4	10	20	50	200
1	F	1.0402	0.9727	0.9207	0.6900	0.1265
2	F	1.0377	1.0177	1.0135	0.9922	0.9695
3	F	1.0377	1.0177	1.0135	0.9924	0.9728
4	F	1.0402	1.0055	1.0031	1.0035	0.9828
5	F	0.9960	1.0043	1.0022	1.0042	1.0147
1	S2	1.0481	0.9959	0.9902	0.9888	0.9885
2	S2	1.0380	1.0187	1.0158	0.9997	0.9992
3	S2	1.0380	1.0187	1.0158	0.9997	0.9996
4	S2	1.0401	1.0055	1.0020	1.0000	0.9780
5	S2	1.0038	1.0038	1.0010	1.0004	1.0000

Table 12.3–6: Ratio of computed σ_{zz} to exact σ_{zz} in a square, simply supported (0/90/0) laminate under sinusoidally distributed transverse load. The stress is computed at the reduced Gauss point closest to the centroid of the upper surface of the plate $(a/2, a/2, h)$.

Mesh	Integration Scheme	Span-to-Thickness Ratio, a/h 4	10	20	50	200
1	F	0.9738	0.9823	0.9979	1.0724	1.2549
2	F	0.9887	1.0076	0.9959	0.8184	−4.5795
3	F	0.9887	1.0076	0.9959	0.8181	−4.6023
4	F	0.9897	0.9972	1.0055	1.0525	0.8495
5	F	0.9895	0.9947	0.9958	1.0128	1.2058
1	S2	0.9835	0.9872	0.9879	0.9881	0.9881
2	S2	0.9872	1.0059	1.0058	1.0056	0.9931
3	S2	0.9872	1.0059	1.0058	1.0054	0.9898
4	S2	0.9894	0.9965	1.0031	1.0420	1.6675
5	S2	0.9893	0.9954	0.9985	0.9962	0.9962

Several generalizations can be made from the results presented in Tables 12.3–4 through 12.3–7. For relatively thick plates that undergo both bending and transverse shearing about both in-plane axes, all five element types yield accurate results using full integration and relatively coarse meshes. For relatively thin plates that undergo bending about both in-plane axes, the S2 integration scheme effectively alleviates both the spurious transverse shear stress and spurious transverse normal stress, thus allowing accurate displacements and stresses to be computed. The Lagrangian elements (E4, E9, and E16) can exhibit spurious displacement modes under S2 integration; thus one must be very cautious with these elements since it is difficult to predict whether or not the boundary conditions for a particular problem will prevent these modes from appearing.

Table 12.3–7: Ratio of computed σ_{xz} to exact σ_{xz} in a square, simply supported (0/90/0) laminate under sinusoidal transverse load. The stress is computed at the reduced Gauss point closest to $(a, a/2, h/2)$.

Mesh	Integration Scheme	Span-to-Thickness Ratio, a/h 4	10	20	50	200
1	F	0.9754	0.9714	0.9360	0.7597	0.3267
2	F	0.9860	1.0135	1.0145	1.0021	1.0276
3	F	0.9860	1.0135	1.0145	1.0023	1.0296
4	F	0.9873	0.9999	1.0030	1.0210	1.2883
5	F	0.9867	0.9989	1.0000	1.0014	1.0103
1	S2	0.9857	0.9975	0.9987	0.9990	0.9990
2	S2	0.9861	1.0138	1.0148	0.9988	0.9982
3	S2	0.9861	1.0138	1.0148	0.9987	0.9970
4	S2	0.9873	0.9990	1.0030	1.0205	1.2693
5	S2	0.9868	0.9989	1.0000	1.0000	1.0000

For this simply supported cross-ply (0/90/0) laminate under sinusoidally distributed transverse load, exact solutions exist for the following three theories: (1) 3-D elasticity theory, (2) classical laminate theory (CLPT), and (3) first-order shear deformation theory (FSDT). The exact solutions from these three theories will be compared with the layerwise finite element solution of this problem. One quadrant of the thick plate ($a/h = 4$) is modeled using two finite element meshes that differ only in the refinement through the thickness. Full integration is used for both meshes. Mesh 1 (see Figure 12.3–2) features a 2×2 uniform 2-D mesh of E8–Q3 elements (*i.e.*, one quadratic layer for each distinct material layer) for a total of 441 global degrees of freedom. Mesh 2 (not shown) features a 2×2 uniform 2-D mesh of E8–Q6 elements (*i.e.*, two quadratic layers for each material layer) for a total of 969 global degrees of freedom.

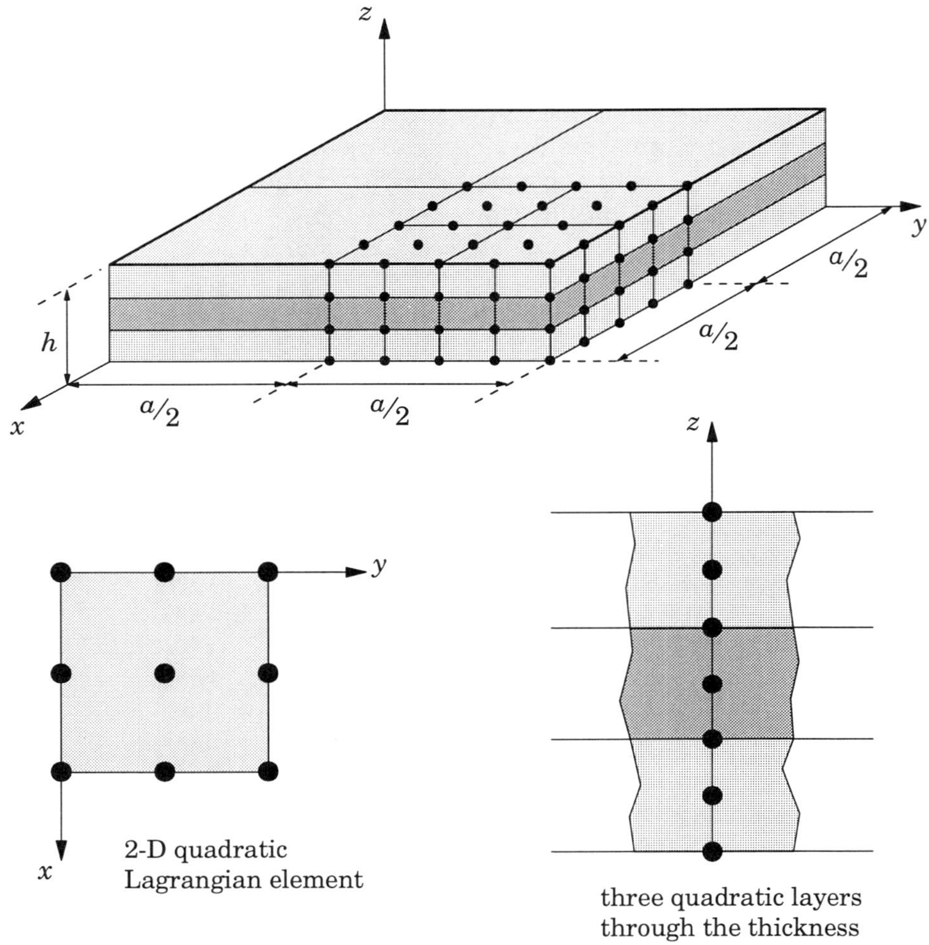

Figure 12.3–2: Mesh 1 of layerwise elements used in the finite element modeling of a quadrant of (0/90/0) laminate.

The layerwise finite element stresses are computed via the constitutive relations at the reduced Gauss points in each finite element. For the E8-Qi elements, the reduced Gauss points correspond to the 2×2 integration points within the domain of each layer of each element. Figures 12.3–3 through 12.3–6 show the distribution of various nondimensionalized stresses σ_{xx}, σ_{zz}, σ_{yz}, and σ_{xz}, respectively, through the thickness of the plate at the reduced Gauss points closest to the position where each stress attains a maximum. Thus for Mesh 1, the stresses are computed at six different points through the thickness, while for Mesh 2, the stresses are computed at 12 different points through the thickness. The nondimensionalized stresses, along with the corresponding reduced Gauss points, are

$$\bar{\sigma}_{xx} = \sigma_{xx}(A, A, z)\frac{h^2}{q_0 a^2}, \quad \bar{\sigma}_{zz} = \sigma_{zz}(A, A, z)\frac{1}{q_0}$$
$$\bar{\sigma}_{yz} = \sigma_{yz}(A, B, z)\frac{h}{q_0 a}, \quad \bar{\sigma}_{xz} = \sigma_{xz}(B, A, z)\frac{h}{q_0 a}$$
$$A = 1.105662\left(\frac{a}{2}\right), \quad B = 1.894338\left(\frac{a}{2}\right) \tag{12.3-7}$$

We note an excellent agreement between the layerwise finite element solution and the exact three-dimensional elasticity solution shown in Figures 12.3–3 through 12.3–6. Although not shown here, the two remaining in-plane stresses, σ_{yy} and σ_{xy}, predicted by the layerwise finite element model also showed very close agreement with the three-dimensional elasticity solution. All of the stress distributions predicted by the single-layer theories (CLPT and FSDT) show considerable error for this thick plate ($a/h = 4$). Note that the transverse shear stresses predicted by CLPT and FSDT are computed via integration of the 3-D equilibrium equations after determining the in-plane stress field. These "equilibrium" shear stresses can also be computed for the layerwise model, thus yielding layerwise smooth shear stresses using relatively coarse refinements through the thickness; however, only the constitutive transverse shear stresses are shown here.

12.3–5 Free Edge Stresses in a $(45/-45)_s$ Laminate

Consider a symmetric angle-ply laminated plate strip $(45/-45)_s$ of length $2a$, width $2b$, and thickness $2h$, and subjected to in-plane displacements along the length at the ends. In the analysis, it is assumed that $a = 10b$ and $b = 4h$. Each of the four material layers is of equal thickness $h_k = h$, and is idealized as a homogeneous, orthotropic material with the following properties (expressed in the principal material coordinate system):

$$E_1 = 20 \times 10^6 \text{ psi}, \; E_2 = E_3 = 2.1 \times 10^6 \text{ psi}$$
$$G_{12} = G_{23} = E_{13} = 0.85 \times 10^6 \text{ psi}, \; \nu_{12} = \nu_{23} = \nu_{13} = 0.21 \tag{12.3-8}$$

The origin of the global coordinate system coincides with the centroid of the 3-D plate. The $x-$coordinate is taken along the length of the plate; the $y-$coordinate is taken along the width, and the $z-$coordinate is taken through the thickness of the plate. The displacement boundary conditions for the laminate are

$$u(a, y, z) = u_0, \quad u(-a, y, z) = 0, \quad v(-a, 0, 0) = v(a, 0, 0) = 0, \quad w(x, y, 0) = 0 \tag{12.3-9}$$

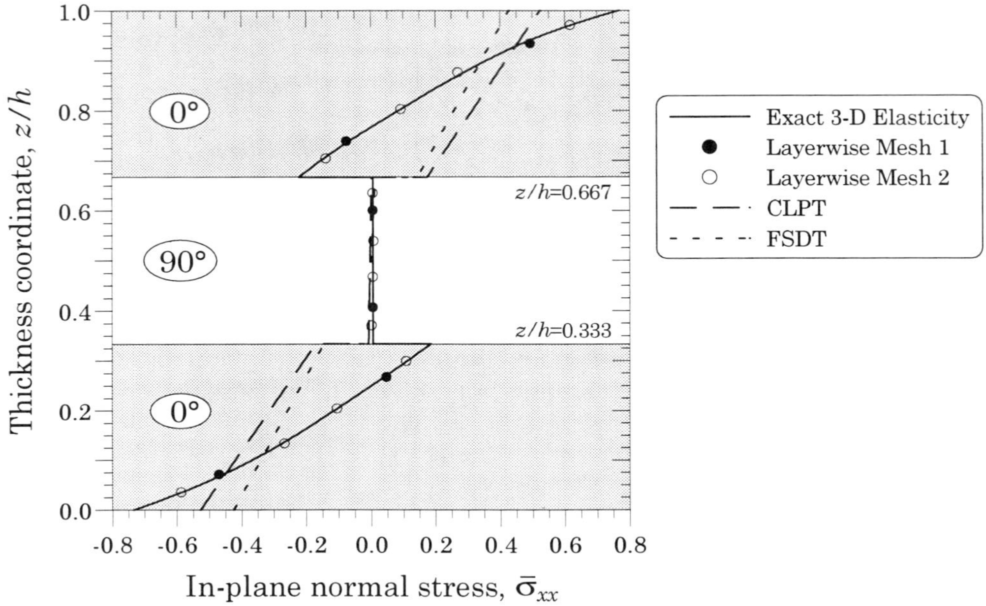

Figure 12.3–3: In-plane normal stress $\bar{\sigma}_{xx}$ distribution through the thickness of a simply supported (0/90/0) laminate subjected to sinusoidally distributed transverse load ($a/h = 4$).

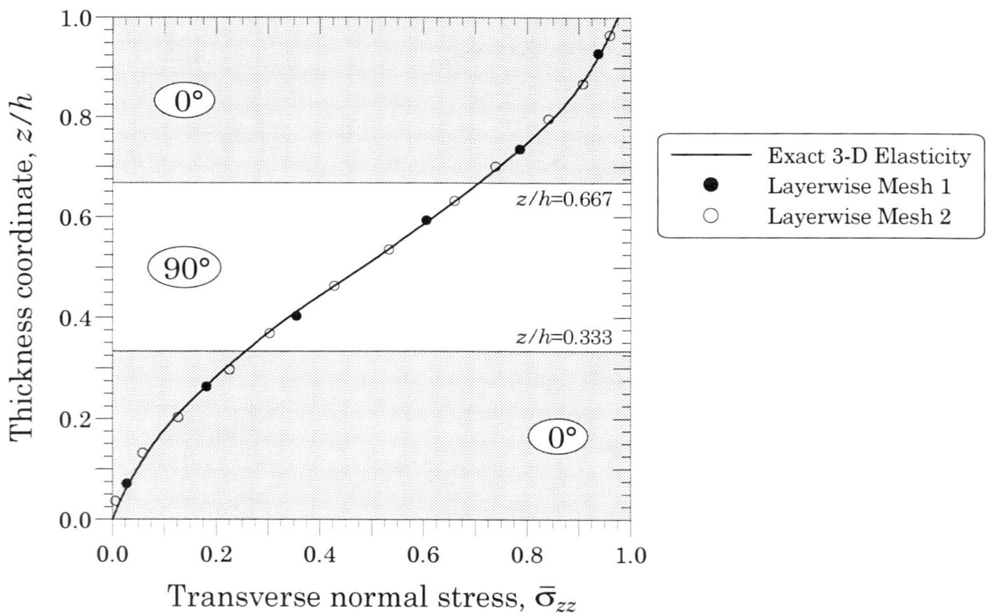

Figure 12.3–4: Transverse normal stress $\bar{\sigma}_{zz}$ distribution through the thickness of a simply supported (0/90/0) laminate subjected to sinusoidally distributed transverse load ($a/h = 4$).

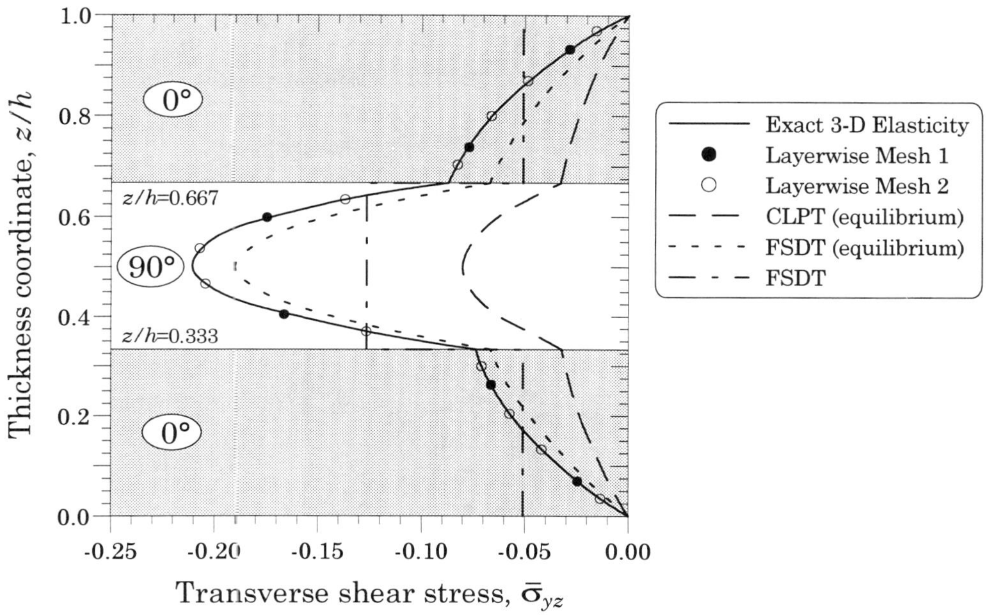

Figure 12.3–5: Transverse shear stress $\bar{\sigma}_{yz}$ distribution through the thickness of a simply supported (0/90/0) laminate subjected to sinusoidally distributed transverse load ($a/h = 4$).

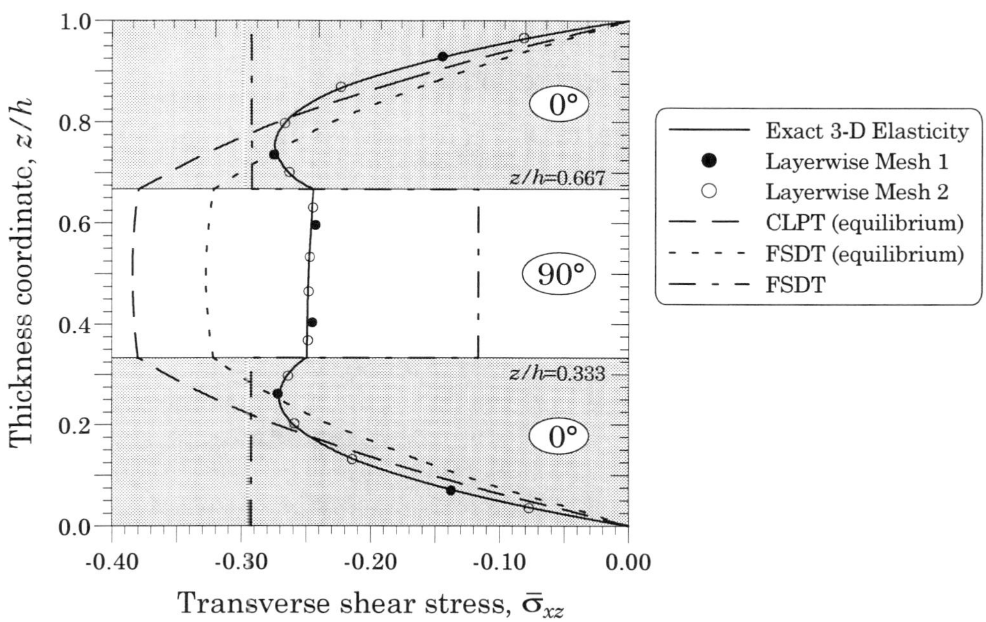

Figure 12.3–6: Transverse shear stress $\bar{\sigma}_{xz}$ distribution through the thickness of a simply supported (0/90/0) laminate subjected to sinusoidally distributed transverse load ($a/h = 4$).

Since the geometry and loading are symmetric about the xy–plane, only the upper half of the strip is modeled. Thus the computational domain is defined by $(-a < x < a,\ -b < y < b,\ 0 < z < h)$. The traction free boundary conditions for the laminate are

$$\sigma_{zz}(x,y,2h) = \frac{\partial \sigma_{zz}}{\partial z}(x,y,0) = 0 \qquad (12.3-10a)$$

$$\sigma_{yz}(x,y,0) = \sigma_{zz}(x,y,h) = 0 \qquad (12.3-10b)$$

$$\sigma_{xz}(x,y,0) = \sigma_{xz}(x,y,h) = 0 \qquad (12.3-10c)$$

$$\sigma_{yz}(x,-b,z) = \sigma_{yz}(x,b,z) = 0 \qquad (12.3-10d)$$

Most of the numerous analytical and numerical studies of the free edge effect have focused on problems that allow the use of quasi–3-D analyses, for example, an infinitely long, symmetric, angle-ply plate strip subjected to an imposed uniform axial strain $[i.e., \varepsilon_1(x,y,z) = \varepsilon_0]$. For the $(45/-45)_s$ laminate, Wang and Choi [57,58] have a quasi–3-D analytical solution and Whitcomb *et al.* [59] have a quasi–3-D finite element solution. In such cases the strains and stresses are assumed to be independent of the axial coordinate x. Thus the analyst only has to be concerned with two independent variables. It should be noted that the purpose of this numerical example is not to compare the efficiency of the layerwise finite element solution with the various quasi–3-D solutions for this particular problem. The purpose of this example is simply to illustrate the ability of the layerwise plate model to accurately describe three-dimensional effects (such as free edge stresses) which are unaccounted for in conventional 2-D plate models.

The layerwise finite element model is used to compute the interlaminar stresses occurring in the boundary region near the traction-free edges in the composite strip. No attempt is made to cast the layerwise finite element model into a quasi–3-D format, thus a full 3-D analysis of the finite length, composite strip is performed. The layerwise finite element solution will obviously be more computationally expensive than a comparable quasi–3-D analysis; however, it is useful to verify the layerwise solution using problems which have received much attention in the literature. The finite element mesh consists of 75, two-dimensional E8–Q6 elements (5 elements along the length, 15 elements across the width, 4 quadratic subdivisions through the thickness of each material layer). All elements have the same length $(2a/5)$; however, the width of the elements decreases as either of the free edges is approached. The widest elements (those elements centered on the x axis) have a width of $0.75b$ (or $12h_k$). The narrowest elements (the last two rows of elements adjacent to either of the free edges) have a width of $0.00782b$ (or $h_k/8$). The results of this analysis are presented in Figures 12.3–7 through 12.3–12, and they are discussed next.

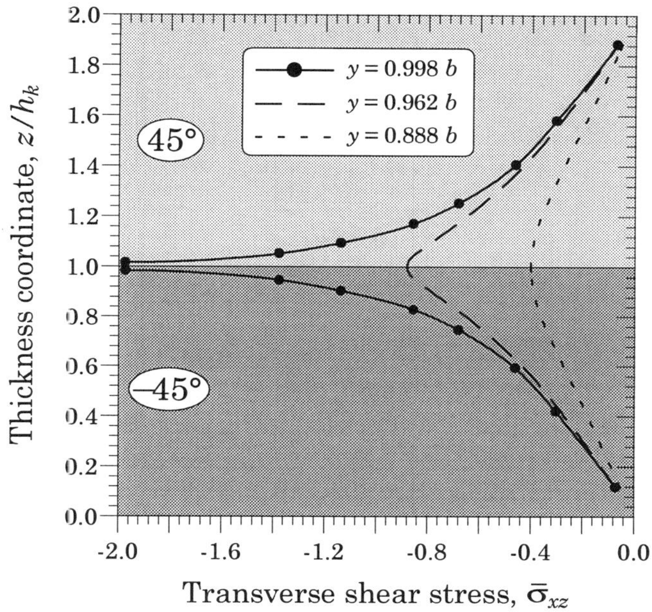

Figure 12.3–7: Distribution of transverse shear stress $\bar{\sigma}_{xz}$ through the thickness of a $(45/–45)_s$ laminate near the free edge ($x = -0.115a$).

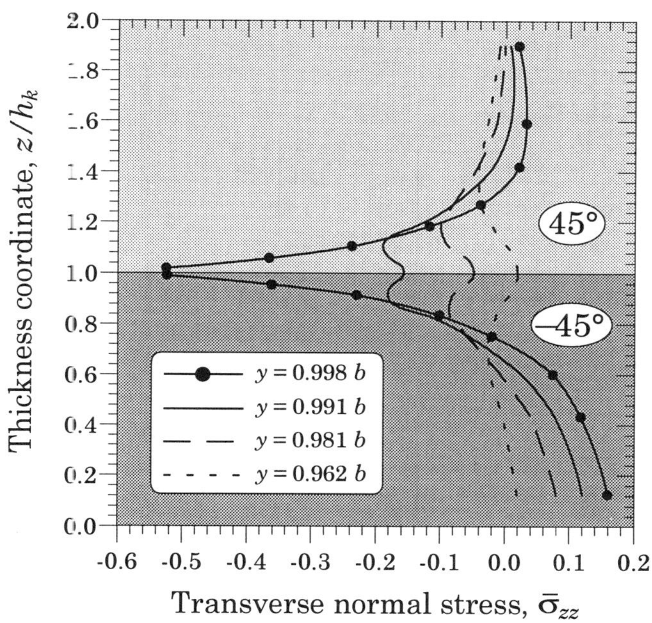

Figure 12.3–8: Distribution of transverse normal stress $\bar{\sigma}_{zz}$ through the thickness of a $(45/–45)_s$ laminate near the free edge ($x = -0.115a$).

Figures 12.3–7 through 12.3–12 show transverse stresses $(\sigma_{zz}, \sigma_{yz}, \sigma_{xz})$ computed using the layerwise finite element model. As in the previous example, the stresses are computed via the constitutive relations at the reduced Gauss points within each thickness subdivision of each element. All stresses have been nondimensionalized according to

$$\bar{\sigma}_{ij} = \frac{20\sigma_{ij}}{\varepsilon_0 E_1}$$

where ε_0 denotes the nominal axial strain induced in the strip by the applied axial displacement of the ends (*i.e.*, $\varepsilon_0 = u_0/2a$). It should be noted that the stresses presented in Figures 12.3–7 through 12.3–12 were taken only from the center row of fifteen elements across the width of the strip (*i.e.*, elements centered on the line $x = a$ in the 2–D mesh. Within these elements, the axial strain (ε_{xx}) experienced a maximum variation of only 0.27% thus the condition of uniform axial strain (as used in the quasi–3-D analyses) is approximately met within this row of elements.

Figures 12.3–7 through 12.3–12 show the variation of the transverse stresses through the thickness of the strip as the free edge is approached. These stresses are computed at the reduced Gauss points which lie closest to the global yz–plane. Note that all three interlaminar stresses exhibit apparent singular behavior near the intersection of the +45/–45 interface and the free edge. In this particular problem, the transverse shear stress σ_{xz} exhibits the largest magnitude of the interlaminar stresses, followed by the transverse normal stress σ_{zz}. The stress distributions in Figures 12.3–7 through 12.3–9 show good qualitative agreement with results reported by Wang and Choi [58], who presented an exact elasticity solution for the associated quasi–3-D problem, and Whitcomb *et al.* [59], who presented a displacement-based finite element solution of the associated quasi–3-D problem (also see [60,61]); however, their results are not reproduced here.

An examination of Figures 12.3–7 through 12.3–9 reveals that traction-free boundary conditions equations (10a) through (10c) are apparently satisfied (within the context of the relatively coarse mesh used). Figure 12.3–9 shows that Eq. (10d) is satisfied over most of the thickness of the free edge, except in the immediate vicinity of the singular point. Note that even though Eq. (10d) is violated near the singular point, this boundary condition is still satisfied in the integral sense. Whitcomb *et al.* [59] suggest that this behavior is caused by the fact that the stress tensor does not have to be symmetric at a singular point (where the stress derivatives are unbounded) while most numerical methods are developed assuming a symmetric stress tensor. Whitcomb *et al.* [59] further reported that the region of boundary condition violation is restricted to the two elements nearest to the singular point and that this region can be made as small as desired through mesh refinement.

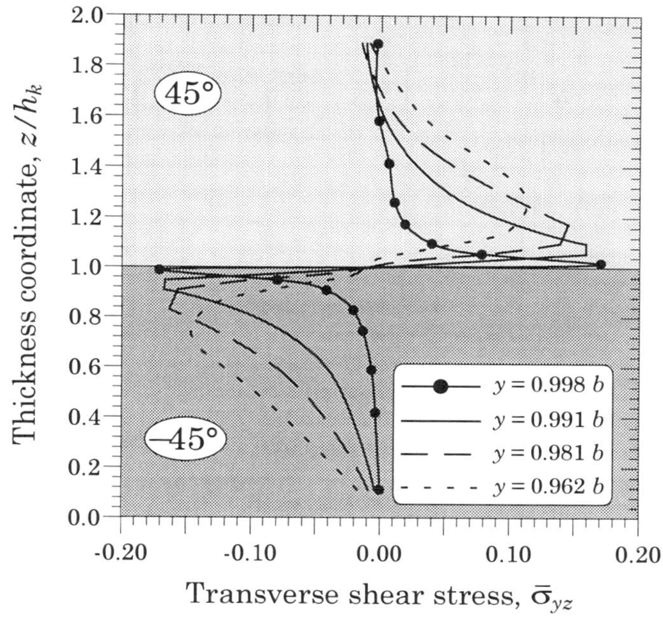

Figure 12.3–9: Distribution of transverse shear stress $\bar{\sigma}_{yz}$ through the thickness of a $(45/\!-\!45)_s$ laminate near the free edge ($x = -0.115a$).

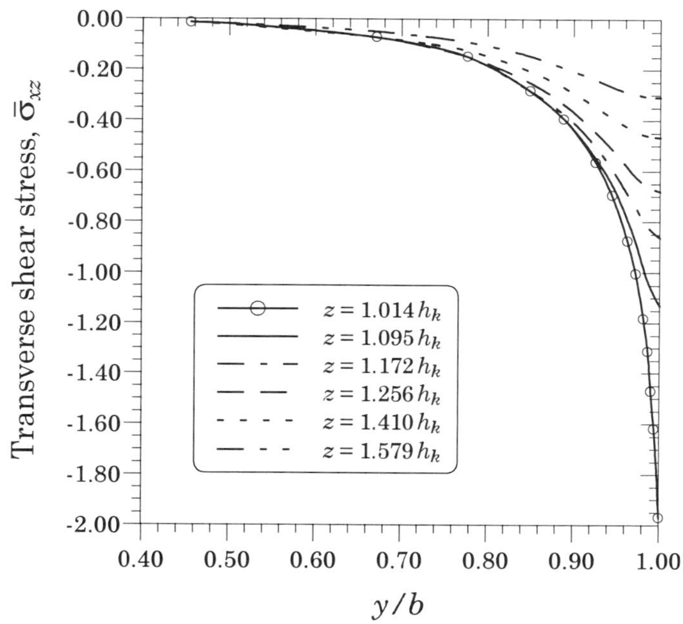

Figure 12.3–10: Distribution of transverse shear stress $\bar{\sigma}_{xz}$ across the width of a $(45/\!-\!45)_s$ laminate near the free edge ($x = -0.115a$).

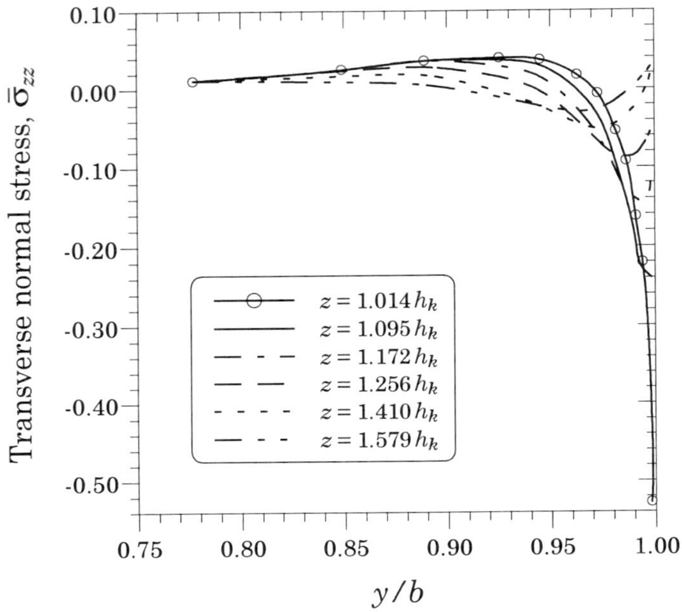

Figure 12.3–11: Distribution of transverse normal stress $\bar{\sigma}_{zz}$ across the width of a $(45/–45)_s$ laminate near the free edge ($x = -0.115a$).

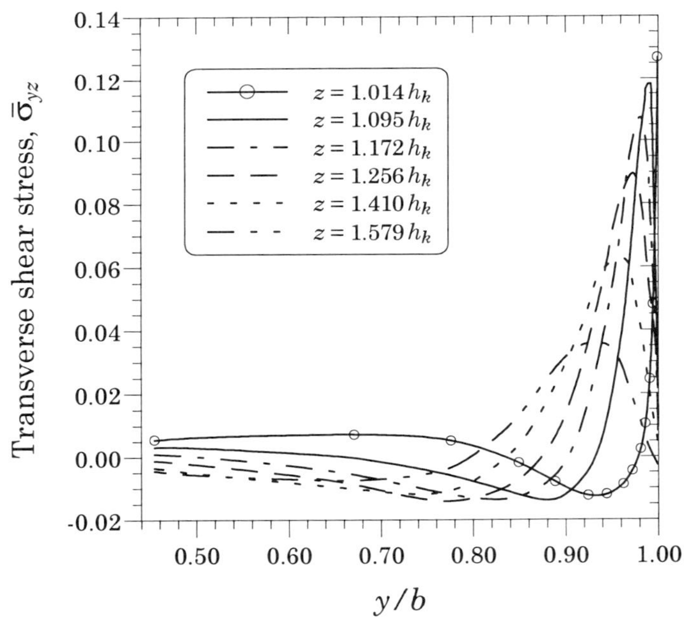

Figure 12.3–12: Distribution of transverse shear stress $\bar{\sigma}_{yz}$ across the width of a $(45/–45)_s$ laminate near the free edge ($x = -0.115a$).

Figures 12.3–10 through 12.3–12 show the variation of the interlaminar stresses along the width coordinate (y) of the composite plate strip at various values of the thickness coordinate (z). There are six lines corresponding to six different z values, all of which occur in the uppermost material layer (+45). Note that all of these transverse stresses decay to zero as the distance from the free edge increases. Also note that only the stresses closest to the 45/–45 interface appear unbounded as the free edge is approached. The transverse shear stress σ_{yz} shown in Figure 12.3–12 appears to satisfy Eq. (10d) except at $z = 1.014h_k$ (again the reduced Gauss point nearest to the singular point). The transverse stress distributions shown in Figures 12.3–10 through 12.3–12 show good qualitative agreement with the quasi–3-D elasticity solution presented by Wang and Choi [58], although their results are not reproduced here.

For this problem, the layerwise finite element stresses agree with the classical laminate theory solution for points sufficiently far from the free edge. For example, the nondimensionalized in-plane stresses at the reduced Gauss points nearest to the centroid of the strip (*i.e.*, $x = 0.1155a$, $y = 0.2165b$) are (depending on the value of z)

Layerwise FEM	CLPT
$2.956 < \sigma_{xx} < 2.975$	$\sigma_{xx} = 2.96$
$1.143 < \sigma_{xy} < 1.164$	$\sigma_{xy} = 1.15$
$0.0 < \sigma_{yy} < 0.012$	$\sigma_{yy} = 0.0$

While highly accurate, the laywerwise models are computationally expensive, thus preventing their general use in modeling entire laminates. Fortunately, unless a laminate is extremely thick, significant three-dimensional states of stress are usually restricted to localized regions near geometric and material discontinuities such as free edges, cut-outs, delamination fronts, and matrix crack fronts, or in localized regions of intense loading. However, the localization of these 3-D stress fields does not diminish their importance, for it is in these very areas that damage initiation and propagation are most likely to occur. Therefore, we must develop procedures where a suitable theory can be used for local as well as global regions.

12.4 Variable Kinematic Formulations

12.4–1 Introduction

The ESL models, partial layerwise models, and full layerwise models each have their own advantages and disadvantages in terms of solution accuracy, solution economy, and ease of implementation. Used alone, none of these three types of models is suitable for general laminate analysis; each is restricted to a limited portion of the composite laminate modeling spectrum. However, by combining all three model types in a multiple model analysis or global–local analysis, a very wide variety of laminate problems can be solved with maximum accuracy and minimal cost. The term "multiple model analysis" is used here to denote any analysis method that uses different mathematical models and/or distinctly different levels of discretization for different subregions of the computational domain. The phrase "global–local analysis" refers to a special case of the more general multiple model analysis; the former term is typically used when there exists a particular subregion of interest that occupies a small portion of the computational domain.

All multiple model methods represent an attempt to distribute limited computational resources in an optimal manner to achieve maximum solution accuracy with minimal solution cost, subject to certain problem-specific constraints. This task often requires the joining of incompatible finite element meshes and/or incompatible mathematical models. Note that for the case of joining incompatible mathematical models, the numerical methods used to implement each of the mathematical models may be the same or different; often the finite element method is used to implement each of the models. The traditional difficulty with multiple model analyses is the maintenance of displacement continuity and force equilibrium along boundaries separating incompatible subregions.

A wide variety of multiple model methods have been reported in the literature. The analysis of composite laminates has provided the incentive for the development of many of the reported multiple model methods [62–72], due mainly to the heterogeneous nature of composite materials and the wide range of scales of interest (*i.e.*, micromechanics level, lamina level, laminate level, and structural component level). In general, the broad spectrum of multiple model methods can be divided into two categories: (1) the sequential or multistep methods, and (2) the simultaneous methods.

Most of the sequential multiple model methods reported to date are developed for global–local analysis. Typically the global region (*i.e.*, the entire computational domain) is analyzed with an economical, yet adequate model (often an ESL laminate model) to determine the displacement or force boundary conditions for a subsequent analysis of the local region (*i.e.*, a small subregion of particular interest). The local region might be modeled with a highly refined mesh of the same ESL laminate elements or it might be modeled with 3-D finite elements. Two-dimensional to three-dimensional sequential global–local methods for laminated composite plates were employed, for example, by Thompson and Griffin [62], who modeled the global region using the first order shear deformation finite elements, while the local region was modeled using 3-D finite elements. The displacement field from the FSDT finite element solution of the global region was used to impose displacement boundary conditions on the local 3-D finite element model. To simplify the analysis, the in-plane finite element discretizations of the global model and the local model were required to be compatible along the global–local boundary. One of the main criticisms of these non-iterating sequential methods [63] is that the influence of the local region on the global region is not accounted for. Specifically, displacement continuity is maintained across the global–local boundary, while the equilibrium of forces along the global–local boundary is not maintained.

This lack of force equilibrium along the global–local boundary has prompted the development of iterative, sequential, multiple model methods by Mao and Sun [64] and Whitcomb and Woo [65,66]. These sequential multiple methods attempt to iteratively establish force equilibrium along the global–local boundary, in addition to imposing displacement continuity. Each of the iterative sequential methods use the same type of mathematical model for both the global region and the local region. The method proposed by Whitcomb and Woo [65,66] requires a compatible finite element discretization along the global–local boundary while the method proposed by Mao and Sun [64] does not.

A number of simultaneous multiple model methods have been reported in the literature [67–72]. These methods are characterized by a simultaneous analysis of the entire computational domain where different subregions are modeled using different mathematical models and/or distinctly different levels of domain discretization. The simultaneous methods explicitly account for the full interaction of the different subregions and are thus directly extendible to nonlinear analysis. One simple type of simultaneous global–local method prompted by composite laminate analysis is the concept of selective ply grouping or sublaminates [67–71]. In this technique, the local region of interest is identified as a specific group of adjacent material plies, within which accurate stresses are desired. The local region spans the entire planar dimensions of the laminate. The global region is identified as that part of the computational domain lying outside the local region. Each of the material plies within the local region is individually modeled with 3-D finite elements, while the remaining plies in the global region are grouped into one or more sublaminates and modeled with 3-D finite elements. This technique amounts to modeling the sublaminates in the global region with an ESL finite element model that assumes transverse shear and transverse normal strains that are C^1–continuous with respect to the thickness coordinate, while the individual plies in the local region are modeled using 3-D finite elements. This ply grouping concept has the disadvantage of requiring the use of 3-D finite elements over the entire planar dimensions of the laminate.

Another means for developing simultaneous multiple model methods is the use of multipoint constraint equations or Lagrange multipliers. In this technique, the variational statement is supplemented with additional integral(s) that serve to enforce compatibility between adjacent subregions. Consequently additional variables (Lagrange multipliers) are introduced into the system of algebraic equations. Within the context of ESL plate and shell models, Aminpour *et al.* [72] used an assumed 1–D interface function in conjunction with a hybrid variational formulation to couple subregions with incompatible mesh discretizations. In this study, only 2-D problems were addressed, and both subregions used the same type of 2-D mathematical model. The subregions differed only in their levels of discretization. While the Lagrange multiplier approach can be used to couple subregions that use different mathematical models (*e.g.*, 2-D/3-D modeling of plates or shells) the general implementation becomes much more cumbersome than a strict 2-D modeling; consequently this method is not often used for connecting different mathematical models.

Currently, most simultaneous 2-D to 3-D modeling of plates and shells is carried out using special transition elements [73–76]. The special transition elements are more convenient than multipoint constraints for joining subregions that use different types of mathematical models; however, they have two disadvantages. First, a different type of transition element is needed for each pair of different mathematical models that might need to be connected in a simultaneous analysis. Further, if the subregion interface has corners, then two different transition elements will be needed to complete the interface, *i.e.*, one type for straight or curved sides and another type for corners. Second, in modeling composite laminates, a significant amount of transitioning (with respect to the transverse coordinate) is required to achieve a discrete layer 3-D representation in localized subregions of interest. The need for transitioning can be avoided by developing a transition element to connect a first-order shear deformation element (both with and without a quadratic thickness stretch) with a stack of 3-D finite elements. Davila [76] developed such a transition element and incorporated the functional interface method [72] to permit abrupt changes in the level of discretization between standard first-order shear deformation elements and the first-order shear deformation edges of the 2-D to 3-D transition elements.

Several types of simultaneous multiple model methods have been based on the hierarchical use of multiple assumed displacement fields. The earliest example of employing multiple assumed displacement fields is due to Mote [77] who combined an assumed global Rayleigh–Ritz field with a local finite element field in the solution of beam and plate problems. Dong [78] generalized the idea of combining classical Rayleigh–Ritz fields and finite element fields and surveyed applications of this technique.

While both full layerwise finite elements and conventional 3-D finite elements permit an accurate determination of 3-D ply level stress fields, they are computationally expensive to use; thus it is most often impractical to discretize an entire laminate with these types of elements. Fortunately, for most laminate applications, significant 3-D stress states are usually present only in localized regions of complex loading or geometric and material discontinuity. To accurately capture these localized 3-D stress fields in a tractable manner, it is usually necessary to resort to a simultaneous multiple model approach in which different subregions of the laminate are described with different types of mathematical models. The objective of such a simultaneous multiple model analysis is to match the most appropriate mathematical model with each subregion based on the physical characteristics, applied loading, expected behavior, and level of solution accuracy desired of each subregion. Thus solution economy can be maximized without sacrificing solution accuracy. In the previously mentioned works that use some form of hierarchical, multiple assumed displacement fields, both displacement fields are based on the same mathematical model, hence the subregions differ only in the level of refinement of the interpolated solution. Reddy and Robbins [79–81] were the first ones to employ hierarchical multiple assumed displacement fields to model different subregions with different mathematical models (*e.g.*, FSDT and LWPT). A discussion of this work is presented next.

12.4–2 Multiple Assumed Displacement Fields

Although simultaneous multiple model methods are simple in concept, the actual implementation of such techniques is complicated and cumbersome due mainly to the need for maintaining displacement continuity across subregion boundaries separating incompatible subdomains. To avoid such difficulties, a new hierarchical, variable kinematic finite element that provides the framework for a very general, robust, simultaneous multiple model methodology for laminated composite plates, is developed. The variable kinematic finite elements possess the following attributes:

1. The kinematics and constitutive relations of the element can be conveniently changed, thus allowing the element to represent a variety of different mathematical models from the very simple to the very complex.

2. Different types of elements can be conveniently connected together in the same computational domain, thus permitting different subregions to be described by different mathematical models. One might also think of the variable kinematic finite element as a very sophisticated, adaptable, transition element that circumvents the need for more than one type of transition element.

The hierarchical, variable kinematic finite element is developed using a multiple assumed displacement field approach, *i.e.*, by superimposing two or more different types of assumed displacement fields in the same finite element domain. In general, the multiple assumed displacement field can be expressed as

$$u_i(x,y,z) = u_i^{ESL}(x,y,z) + u_i^{LWT}(x,y,z) \qquad (12.4-1)$$

where $i = 1,2,3$, and $u_1 = u$, $u_2 = v$, and $u_3 = w$ are the displacement components in the x, y, and z directions, respectively. The reference plane of the plate coincides with the $xy-$plane. The underlying foundation of the displacement field is provided by u_i^{ESL}, which represents the assumed displacement field for any desired equivalent single-layer (ESL) theory. The second term u_i^{LWT} represents the assumed displacement field for any desired full layerwise theory. The layerwise displacement field is included as an optional, incremental enhancement to the basic ESL displacement field, so that the element can have full 3-D modeling capability when needed. Depending on the desired level of accuracy, the element may use none, part, or all of the layerwise field to create a series of different elements having a wide range of kinematic complexity. For example, discrete layer transverse shear effects can be added to the element by including u_1^{LWT} and u_2^{LWT}. Discrete layer transverse normal effects can be added to the element by including u_3^{LWT}. Displacement continuity is maintained between these different types of elements by simply enforcing homogeneous essential boundary conditions on the incremental layerwise variables, thus eliminating the need for multipoint constraints, penalty function methods, or special transition elements. It should be noted that a conventional 3-D finite element displacement field could be used instead of the full layerwise field in equation (1); however, the 2-D data structure of the full layerwise finite elements permits much easier coupling with the 2-D ESL field.

To illustrate the usefulness of a finite element based on the assumed displacement field of Eq. (1), consider a specific case where the individual displacement fields are selected as follows:

u_i^{ESL}: First-Order Shear Deformation Field

$$\begin{aligned} u_1^{ESL}(x,y,z) &= u_0(x,y) + z\phi_x(x,y) \\ u_2^{ESL}(x,y,z) &= v_0(x,y) + z\phi_y(x,y) \\ u_3^{ESL}(x,y,z) &= w_0(x,y) \end{aligned} \qquad (12.4-2)$$

u_i^{LWT}: Layerwise Field of Reddy [37]

$$u_1^{LWT}(x,y,z) = \sum_{I=1}^{N} U_I(x,y)\Phi^I(z)$$
$$u_2^{LWT}(x,y,z) = \sum_{I=1}^{N} V_I(x,y)\Phi^I(z)$$
$$u_3^{LWT}(x,y,z) = \sum_{I=1}^{M} W_I(x,y)\Psi^I(z) \quad (12.4-3)$$

where (U_I, V_I, W_I) denote the nodal values of $(u_1, u_2, u_3)=(u, v, w)$, N is the number of nodes (or $N-1$ is the number of subdivisions) through thickness, and Φ^I are the 1-D (global) interpolation functions for the discretization of the in-plane displacements through thickness; and M and Ψ^I have similar meaning for the discretization of the transverse displacement through thickness. A detailed discussion of displacement–based finite element models of the theory based solely on the displacement field in Eq. (3) was presented in the previous section. It should be noted that the layerwise field given by Eq. (3) is sufficiently general to model any of the deformation modes that can be modeled by the ESL field given in Eq. (2); thus for elements that use all the variables shown in Eqs. (2) and (3), there will be five redundant variables that must be set to zero (or ignored) to permit a unique solution for the remaining variables. The ESL variables are essential for connecting different types of elements. Therefore, the following five of the layerwise variables should be set to zero (see Figure 12.4–1):

$$U_1 = U_N = 0, \quad V_1 = V_N = 0, \quad W_1 = 0 \quad (12.4-4)$$

Parts (a) and (b) of Figure 12.4–1 illustrate a possible in-plane deformation (component $u = u_1$ only) of a transverse normal material line obtained by adding a piecewise linear layerwise displacement field to a first-order shear deformation displacement field. In this particular case, the reference plane is arbitrarily chosen to coincide with the bottom surface of the plate, and there are five nodes (*i.e.*, planes) distributed through the thickness to define the layerwise portion of the composite displacement field. Note that in both parts (a) and (b) of Figure 12.4–1 the layerwise displacements U_I provide an incremental enhancement to the displacement u_1^{ELS} predicted by the first-order shear deformation theory as a result of setting U_1 and U_N to zero. The particular pair of U_I that are zeroed is arbitrary. The same final deformation is achieved in both parts (a) and (b); however, different pairs of U_I are zeroed, thus changing the numerical values of the remaining five nonzero variables. The location of the reference plane ($z = 0$) is arbitrary although its location affects the numerical values of the in-plane first-order shear deformation variables. If the displacements of the first-order shear deformation theory had been set to zero, then the layerwise displacements would be interpreted as total displacements.

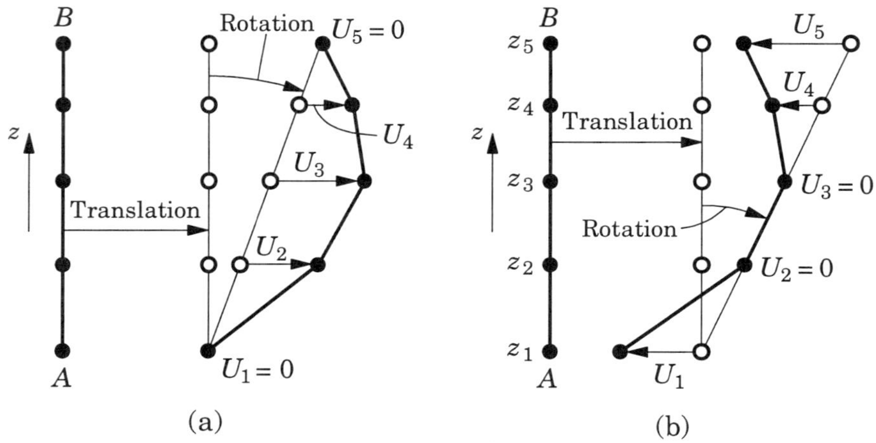

Figure 12.4–1: Superposition of a first-order shear deformation displacement field and a linear layerwise displacement field. In-plane deformation of transverse normal AB. (a) $U_1 = U_5 = 0$. (b) $U_2 = U_3 = 0$.

12.4–3 Incorporation of Delamination Kinematics

A commonly occurring phenomenon in composite laminates is delamination. A delamination is simply a debonding or separation that occurs between individual material plies of a laminate (*i.e.*, an interlaminar crack). Delaminations can occur as a result of manufacturing defects or from interlaminar normal and shear stresses brought about by local anomalies or transverse impacts. Often there may be multiple delaminations distributed through the thickness of a laminate, especially in the region surrounding an impact site. The subject of fracture mechanics has proven to be particularly useful in characterizing the severity of delaminations. For a general review of composite delamination research, the reader is referred to O'Brien [83]. The total energy release rate and its individual components have been successfully used to predict delamination onset and growth by O'Brien [83–86].

Since delamination is a common occurrence in laminates, the kinematics of single and multiple delamination should be incorporated into any general laminate model. The present multiple assumed displacement field of Eq. (1) also permits the modeling of delamination kinematics by using two or more layerwise expansions through the thickness instead of one. In this case, a separate layerwise expansion would be used for each sublaminate created by the delamination(s). Alternatively, one can model the kinematics of delamination by supplementing the composite displacement field with a simple, piecewise constant, discontinuous displacement field that uses unit step functions of the thickness coordinate (see Barbero and Reddy [50]). Both of these methods introduce the same number of additional dependent variables. The second method is chosen here because the additional dependent variables are physically meaningful in that they represent the jump discontinuity in the displacement components across the delamination. Further, the second method allows easier modeling of multiple delaminations and easier implementation of various no-penetration contact algorithms for the delaminated surfaces (see Robbins and Reddy [52]). The supplemented composite displacement field thus becomes

$$u_i(x,y,z) = u_i^{ESL}(x,y,z) + u_i^{LWT}(x,y,z) + u_i^D(x,y,z) \tag{12.4 - 5a}$$

$$\begin{aligned} u_1^D(x,y,z) &= \sum_{I=1}^{D} \mathcal{U}_I(x,y) H^I(z) \\ u_2^D(x,y,z) &= \sum_{I=1}^{D} \mathcal{V}_I(x,y) H^I(z) \\ u_3^D(x,y,z) &= \sum_{I=1}^{D} \mathcal{W}_I(x,y) H^I(z) \end{aligned} \tag{12.4 - 5b}$$

and $H^I(z)$ are the Heaviside step functions, $H^I(z) = 1$ for $z \geq z^I$ and $H^I(z) = 0$ for $z < z^I$, and D is the number of delaminations distributed through the laminate thickness. Equation (5b) represents displacement components that are piecewise constant through the laminate thickness. The delaminations are located at coordinates $z = z^I$ $(I = 1, 2, .., D)$. Note that the I in z^I serves as a superscript and not an exponent, thus distinguishing the locations of the delaminations $(z^I,\ I = 1, 2, ..., D)$ from the locations of the nodes $(z^J,\ J = 1, 2, ..., N)$ in the layerwise expansion. Three dependent variables $(\mathcal{U}_I, \mathcal{V}_I, \mathcal{W}_I)$ are introduced for each delamination. The dependent variables $(\mathcal{U}_I, \mathcal{V}_I, \mathcal{W}_I)$ are interpreted as the jump discontinuities in the displacement components (u_1, u_2, u_3) at $z = z^I$ $(I = 1, 2, .., D)$. The variable $\mathcal{W}_I(x, y)$ is the delamination opening displacement, thus the condition $\mathcal{W}_I \geq 0$ constitutes a no–penetration boundary condition for delaminated surfaces of the I–th delamination. The delamination front for the Ith delamination is defined as a curved or straight line in the xy–plane along which the essential boundary conditions $\mathcal{U}_I = \mathcal{V}_I = \mathcal{W}_I = 0$ are enforced.

The effect of introducing the delamination field of Eq. (5) into the composite displacement field is illustrated in Figure 12.4–2, which shows the x–component of the deformation of a transverse normal material line

AB. The overall deformation of line AB is similar to that of Figure 12.4–1 with the exception of a single delamination located at $z = z^1$, where the jump discontinuity $\mathcal{U}_1$ is introduced. These jump discontinuities can be introduced as many times as desired for multiple delaminations.

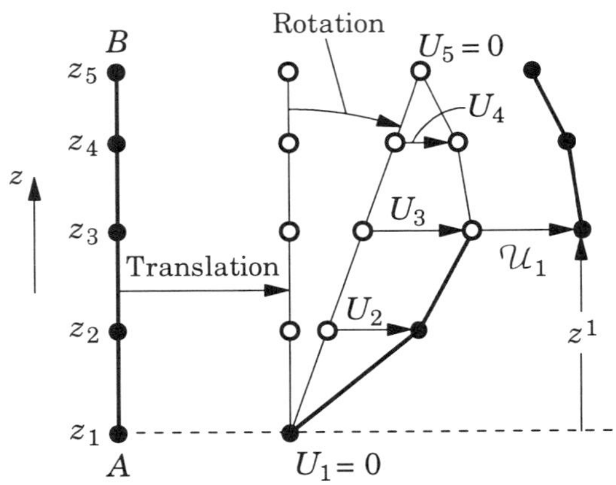

Figure 12.4–2: Superposition of a first-order shear deformation displacement field, a linear layerwise displacement field, and a piecewise constant delamination displacement field. In-plane deformation of transverse normal AB. Delamination occurs at $z = z^1$.

12.4–4 Finite Element Model

A hierarchy of three distinct types of plate elements can be obtained from the composite displacement field of Eq. (4), where the individual displacement expansions are defined by Eqs. (2), (3), and (5). The first and simplest type of element is the first-order shear deformation element (or FSDT element). This element is formed using Eq. (2) while ignoring Eqs. (3) and (5). The second type of element is the Type I layerwise element (or LWT1 element). The LWT1 element is formed using Eqs. (2), (3a), and (3b), while ignoring (3c); thus, it is a partial layerwise element. Four of the layerwise variables should be set to zero (or simply ignored) to remove the redundancy from the composite displacement field (*e.g.*, $U_1 = U_N = V_1 = V_N = 0$). Like the FSDT element, the LWT1 element assumes a state of zero transverse normal stress and thus does not explicitly account for transverse normal strain. This is implicitly achieved by using a reduced constitutive matrix similar to the FSDT element. The inclusion of Eqs. (3a) and (3b) in the LWT1 element provides discrete layer transverse shear effects, unlike the simple gross transverse shear effect included in the FSDT element. Thus the LWT1 element is applicable to thick laminates and often yields results comparable to 3-D finite elements while using approximately two thirds the number of degrees of freedom. The third and most complex element is the Type II layerwise element (or LWT2 element). This element is formed using both Eqs. (2) and (3), thus it is a full layerwise element. The composite displacement field contains five redundant variables, thus five layerwise variables are chosen (*e.g.*, U_1, U_N, V_1, V_N, W_1) and set to zero or simply ignored. The LWT2 element explicitly accounts for all six strain components in a kinematically correct manner, *i.e.*, the in-plane strains are C^1–continuous through the laminate thickness while the transverse strains are C^0–continuous through the laminate thickness. The inclusion of the full layerwise field provides the LWT2 element with both discrete layer transverse shear effects and discrete layer transverse normal effects. The LWT2 element uses a full constitutive matrix ($\bar{Q}_{ij} = \bar{C}_{ij}$), and it is equivalent in accuracy and cost to a stack of conventional 3-D finite elements.

If delaminations are present, then both the LWT1 and LWT2 finite elements can make use of the delamination expansion of Eq. (5). The FSDT element cannot be used to model delamination since the deformation above and below the delamination cannot be separately prescribed due to the C^1–continuity of the FSDT displacement expansion through the thickness.

Note that the various element types are created by hierarchically adding variables to the basic first-order shear deformation field. The matrix form of the finite element equations that result from the hierarchical use of Eqs. (2), (3), and (5b) within a single element domain is given by

$$\begin{bmatrix} [K^{EE}] & [K^{EL}] & [K^{ED}] \\ [K^{LE}] & [K^{LL}] & [K^{LD}] \\ [K^{DE}] & [K^{DL}] & [K^{DD}] \end{bmatrix} \begin{Bmatrix} \{U^E\} \\ \{U^L\} \\ \{U^D\} \end{Bmatrix} = \begin{Bmatrix} \{F^E\} \\ \{F^L\} \\ \{F^D\} \end{Bmatrix} \qquad (12.4-6)$$

where $[K^{EE}]$ represents the element stiffness matrix for an equivalent single-layer FSDT element, $[K^{LL}]$ represents the element stiffness matrix for a full layerwise element, and $[K^{DD}]$ represents the element stiffness matrix for an element based solely on the delamination field of Eq. (5b). The remaining submatrices represent coupling stiffnesses between the three different displacement fields. Based on the particular type of element desired, the appropriate terms in the composite stiffness matrix are identified and computed. Since all three element types possess the first-order shear deformation variables of Eq. (2), these different types of elements can easily be simultaneously connected in the same computational domain by simply setting certain layerwise variables of Eq. (3) to zero along the incompatible boundary.

Figure 12.4–3 illustrates a hypothetical 2-D finite element mesh of variable kinematic finite elements where all three element types (FSDT, LWT1, and LWT2) are simultaneously present. The hierarchical nature of the variable kinematic elements allows interelement compatibility to be achieved by simply enforcing homogeneous boundary conditions on some or all of the incremental layerwise variables along the boundary separating two incompatible subregions. Subregion compatibility can be enforced in a strict sense or a relaxed sense by specifying the essential boundary conditions as defined below (also see Figure 12.4–3).

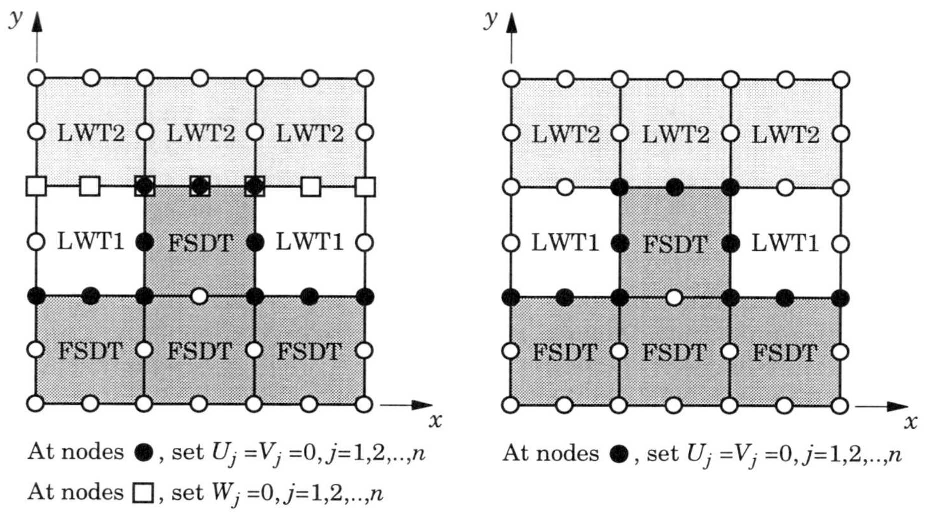

(a) Enforcing **strict** subregion compatibility

(b) Enforcing **relaxed** subregion compatibility

Figure 12.4–3: A simple 2-D mesh of variable kinematic finite elements. All three element types (FSDT, LWT1, and LWT2) are simultaneously present in the mesh.

Strict subregion compatibility (SSC)

At nodes on FSDT/LWT1 boundary, set $U_I = 0,\ V_I = 0$

At nodes on FSDT/LWT2 boundary, set $U_I = 0,\ V_I = 0,\ W_I = 0$

At nodes on LWT1/LWT2 boundary, set $U_I = 0,\ V_I = 0,\ W_I = 0$

for $I = 1, 2, ..., N$.

Relaxed subregion compatibility (RSC)

At nodes on FSDT/LWT1 boundary, set $U_I = 0,\ V_I = 0$

At nodes on FSDT/LWT2 boundary, set $U_I = 0,\ V_I = 0$

At nodes on LWT1/LWT2 boundary, set $U_I = 0,\ V_I = 0$

for $I = 1, 2, ..., N$. When maintaining strict subregion compatibility (SSC), all three displacement components are continuous across all types of subregion boundaries (FSDT/LWT1, LWT1/LWT2, and LWT2/FSDT). In contrast, relaxed subregion compatibility (RSC) maintains total continuity of the in-plane displacement components across all types of subregion boundaries, but does not maintain total continuity of the transverse displacement component across FSDT/LWT2 or LWT1/LWT2 boundaries. This relaxation is often useful for obtaining accurate transverse normal stresses within LWT2 subregions, near FSDT/LWT2 or LWT1/LWT2 boundaries, since it eliminates the transverse pinching or stretching of the laminate near these boundaries, and effectively allows the transverse normal strain to react to the local dominant in-plane strains. Within those portions of LWT2 subregions sufficiently removed from FSDT/LWT2 or LWT1/LWT2 boundaries, both strict and relaxed conditions yield the same stress distributions.

The enforcing of strict or relaxed subregion compatibility via application of the appropriate homogeneous essential boundary conditions is easily automated in a finite element program and can thus be removed from the concern of the user. A significant advantage afforded by the variable kinematic elements is that once the in-plane mesh is defined, the user can then assign any of the three element types (FSDT, LWT1, and LWT2) to any of the elements in the 2-D mesh. Subsequent changes in the type of any element or group of elements can be performed with minimal effort.

12.4–5 Illustrative Examples

The problem of determining the free edge stress fields in laminates subjected to in-plane extension or bending is used to illustrate the variable kinematic finite element (VKFE) model methodology. The free edge problem is ideally suited for global–local analysis, because the 3-D stress field exists only in a boundary region (*i.e.*, free edge) of the laminate and elsewhere only a 2-D stress state exists. Thus, the LWT2 elements can be used in the free edge (local) region and ESL elements can be used everywhere else (global region) to capture the stress fields accurately.

Free Edge Stresses in Laminates in Extension

To demonstrate the accuracy and economy afforded by the variable kinematic finite elements, a global–local analysis is performed to determine the nature of the free edge stress field in three different laminates subjected to axial extension: $(45/-45)_s$, $(45/0/-45/90)_s$, and $(45/0/-45/90/90/-45/0/45)_s$. The three laminates have length $2a$, width $2b$, and thickness $2h$. Each of the three laminates has a length-to-width ratio of 10 ($i.e., a/b = 10$). The material plies in each laminate are of equal thickness h_k. The following geometric differences exist among the three laminates:

$$
\begin{aligned}
(45/{-45})_s \text{ laminate:} \quad & \frac{b}{h} = 4, \; h = 2h_k, \; \frac{b}{h_k} = 8 \\
(45/0/{-45}/90)_s \text{ laminate:} \quad & \frac{b}{h} = 15, \; h = 4h_k, \; \frac{b}{h_k} = 60 \\
(45/0/{-45}/90)_{2s} \text{ laminate:} \quad & \frac{b}{h} = 15, \; h = 8h_k, \; \frac{b}{h_k} = 120 \qquad (12.4-7)
\end{aligned}
$$

Each of the material plies in the three laminates is idealized as a homogeneous, orthotropic material; the material properties (expressed in the principal material coordinate system) are defined below.

Material plies in the four-layer laminate:

$$E_1 = 20 \times 10^6 \text{ psi}, \; E_2 = E_3 = 2.1 \times 10^6 \text{ psi}$$

$$G_{12} = G_{23} = G_{13} = 0.85 \times 10^6 \text{ psi}, \; \nu_{12} = \nu_{23} = \nu_{13} = 0.21 \qquad (12.4-8a)$$

Material plies in the eight- and sixteen-layer laminates:

$$E_1 = 19.5 \times 10^6 \text{ psi}, \; E_2 = E_3 = 1.48 \times 10^6 \text{ psi}$$

$$G_{12} = G_{23} = G_{13} = 0.8 \times 10^6 \text{ psi}, \; \nu_{12} = \nu_{23} = \nu_{13} = 0.3 \qquad (12.4-8b)$$

The origin of the global coordinate system coincides with the centroid of each of the 3-D composite laminates. The $x-$coordinate is taken along the length, the $y-$coordinate is taken across the width, and the $z-$coordinate is taken through the thickness of the laminate. Since the laminate is symmetric about the $xy-$plane, only the upper half of each laminate is modeled. Thus the computational domain is defined by $(-a < x < a, -b < y < b, 0 < z < h)$. The displacement boundary conditions for all three laminates are

$$u_1(a,y,z) = u_0, \quad u_1(-a,y,z) = u_2(-a,0,0) = u_2(a,0,0) = u_3(x,y,0) = 0 \qquad (12.4-9)$$

The slight differences in geometry and material properties among the three laminates allow comparison with solutions published in the literature. For the $(45/{-45})_s$ laminate, Wang and Choi [57,58] have developed a quasi–3-D elasticity solution, while Whitcomb *et al.* [59] produced a solution from a highly refined, quasi–3-D finite element model. For the $(45/0/{-45}/90)_s$ and $(45/0/{-45}/90/90/{-45}/0/45)_s$ laminates, Whitcomb and Raju [61] obtained quasi–3-D finite element solutions using a highly refined mesh.

The variable kinematic finite elements are used in a simultaneous multiple model analysis (global–local analysis) of these three laminates in order to accurately yet efficiently determine the free edge stresses near the middle of one of the two free edges. The global region is modeled using first-order shear deformable elements (FSDT); the local region, where accurate 3-D stresses are desired, is modeled with LWT2 elements.

First the $(45/-45)_s$ laminate will be used to assess the effects of subregion compatibility type (SSC or RSC) and size of the local LWT2 subregion on the accuracy of the computed transverse stresses near the free edge. For this purpose, five different finite element meshes are created. The 2-D, in-plane discretization for all five meshes is exactly the same, consisting of a 5×11 mesh of eight-node, quadratic, 2-D, quadrilateral finite elements (see Figure 12.4–4). All elements have the same length $(2a/5)$; however, the width of the elements decreases as the free edge at (x, b, z) is approached. The widths of the eleven rows of elements, as one moves away from the refined free edge, are $h_k/16, h_k/16, h_k/8, h_k/4, h_k/2, h_k, h_k, 2h_k, 3h_k, 3h_k$, and $5h_k$ (h_k=ply thickness). The five meshes differ only in the width of the local region where LWT2 elements are used. The LWT2 elements used in the local region employ eight quadratic layers through the laminate thickness (four per material layer) as shown in Figure 12.4–5. The thickness of the numerical layers decreases as the +45/−45 interface is approached. From bottom to top, the layer thicknesses are $0.533h_k$, $0.267h_k$, $0.133h_k$, $0.083h_k$, $0.083h_k$, $0.133h_k$, $0.267h_k$, and $0.533h_k$.

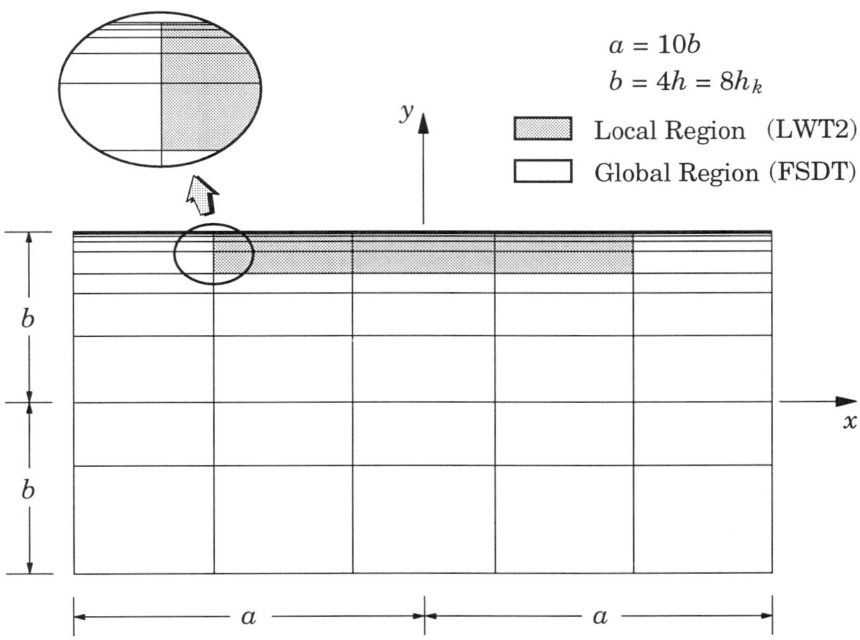

Figure 12.4–4: The 2-D mesh of variable kinematic finite elements used to model a $(45/-45)_s$ laminate under axial extension. All elements are eight-node quadrilaterals.

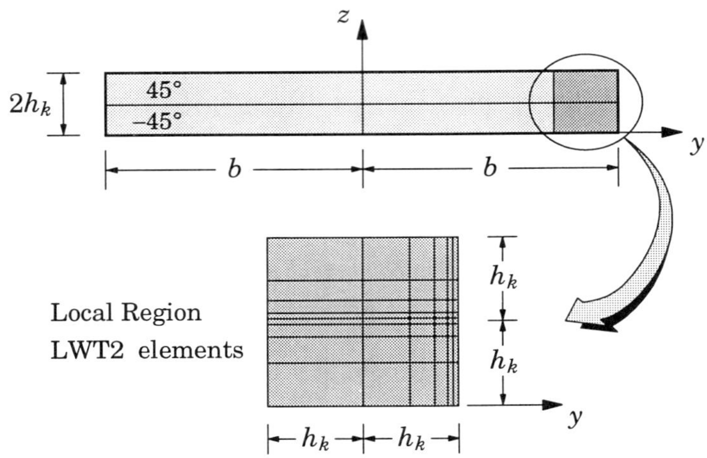

Figure 12.4–5: Discretization within the local LWT2 region, on the yz–plane. The eight-node LWT2 elements in the local region use quadratic layers.

Table 12.4–1: Description of global–local meshes for the $(45/–45)_s$ laminate under axial extension.

Remarks	Mesh 1	Mesh 2	Mesh 3	Mesh 4	Mesh 5
Number of Elements in Local LWT2 Region	3×4	3×5	3×6	3×7	5×11
Width of Local Region	$\frac{1}{2}h_k$	h_k	$2h_k$	$3h_k$	$16h_k$
Length of Local Region	$\frac{6}{5}a$	$\frac{6}{5}a$	$\frac{6}{5}a$	$\frac{6}{5}a$	$2a$
Total Number of Active D.O.F. in VKFE Mesh (Strict Compatibility)	1,986	2,400	2,814	3,228	9,116
Total Number of Active D.O.F. in VKFE Mesh (Relaxed Compatibility)	2,354	2,800	3,246	3,690	9,116

h_k = thickness of a single material ply. All five VKFE meshes have the exact same in-plane discretization (5×11).

Table 12.4–1 summarizes the five meshes used for the $(45/-45)_s$ laminate. Note that mesh 5 is not a global–local mesh. Mesh 5 uses LWT2 elements throughout the entire computational domain, thus serving as a control mesh for judging the accuracy of the four global–local meshes. In meshes 1 through 4, the local region (LWT2 elements) is adjacent to the free edge (x, b, z) and is centered about the plane $(0, y, z)$. In meshes 1 through 4, the length of the local region spans three fifths of the total length of the laminate; however, the width of the local region differs in each mesh ranging from $h_k/2$ to $3h_k$. Two runs are made with each of the four global–local meshes, the first using strict subregion compatibility along the FSDT–LWT2 boundary, and the second using relaxed subregion compatibility along the FSDT–LWT2 boundary.

The stresses are computed via the constitutive relations at the reduced Gauss points within the individual layers of each LWT2 element. All stresses are nondimensionalized as follows:

$$\bar{\sigma}_{ij} = \frac{20\sigma_{ij}}{\varepsilon_0 E_1} \tag{12.4 - 10}$$

where ε_0 is the nominal applied axial strain of $u_0/(2a)$. The stress distributions shown in Figures 12.4–6 through 12.4–8 are generated by computing the nondimensionalized stresses at a series of adjacent reduced Gauss points, and then connecting these points with straight lines.

Figures 12.4–7 and 12.4–8 show the distribution of the interlaminar stresses σ_{zz} and σ_{xz} near the free edge. The results in these two figures were obtained using relaxed subregion compatibility conditions. The stresses presented in Figure 12.4–6 are computed at the reduced Gauss points near the middle of the refined free edge, *i.e.*, along the line $(-0.115a, 0.998b, z)$. This is also the reduced Gauss point located farthest from the FSDT/LWT2 boundary. The stresses presented in Figure 12.4–7 are computed at the reduced Gauss points closest to the line $(0, y, h_k)$, *i.e.*, along the line $(-0.115a, y, 1.014h_k)$. All four of the global local meshes are successful in identifying the spikes in σ_{zz} and σ_{xz} that occur at the 45/–45 interface. The results of meshes 3 and 4 are graphically indistinguishable from the results of the control mesh, mesh 5. While meshes 1 and 2 exhibit some error, they do capture the qualitative nature of the transverse stress distributions near the free edge. In meshes 1 and 2, the transverse shear stresses are predicted more accurately than the transverse normal stresses. The results also indicate that the boundary layer thickness (or the width of the local region) should be at least $2h_k$ to capture both interlaminar stresses accurately.

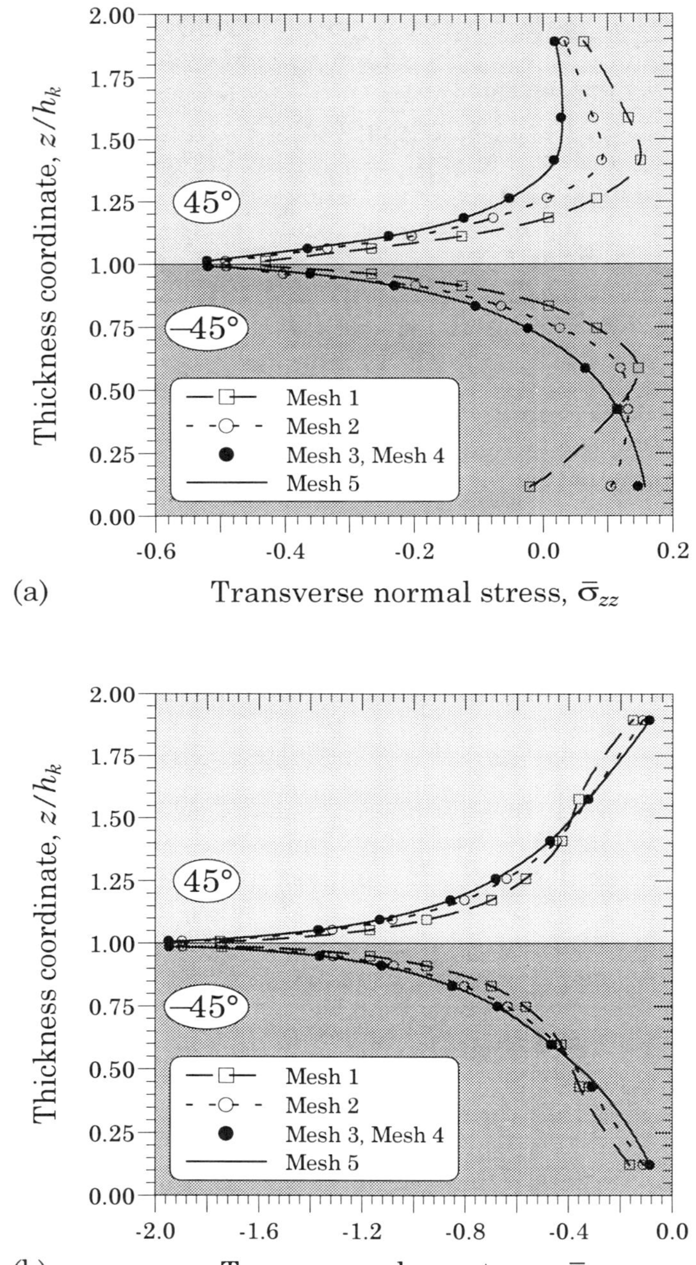

Figure 12.4–6: Interlaminar stress distribution through thickness of the $(45/-45)_s$ laminate near free edge. Results computed for meshes 1 through 5 with relaxed subregion compatibility.

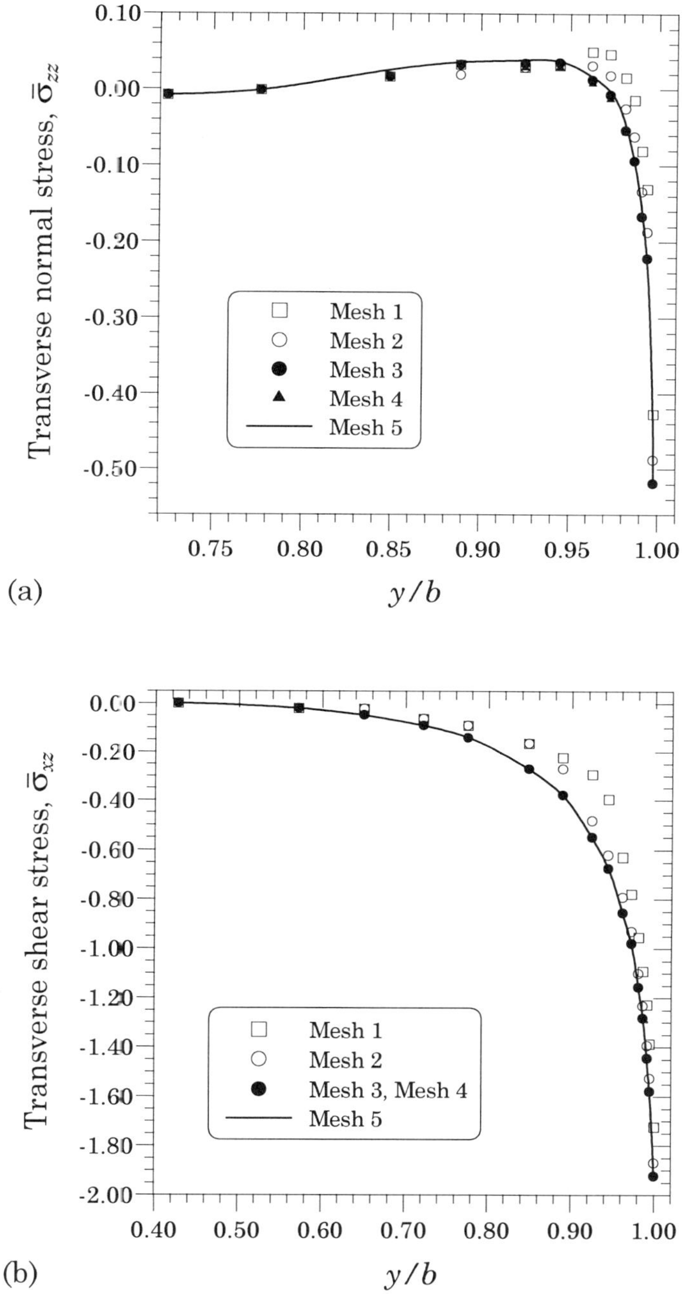

Figure 12.4–7: Interlaminar stress distribution across the width of the (45/–45)$_s$ laminate near the upper 45/–45 interface ($z = 1.014h_k$). Results computed for meshes 1 through 5 with relaxed subregion compatibility.

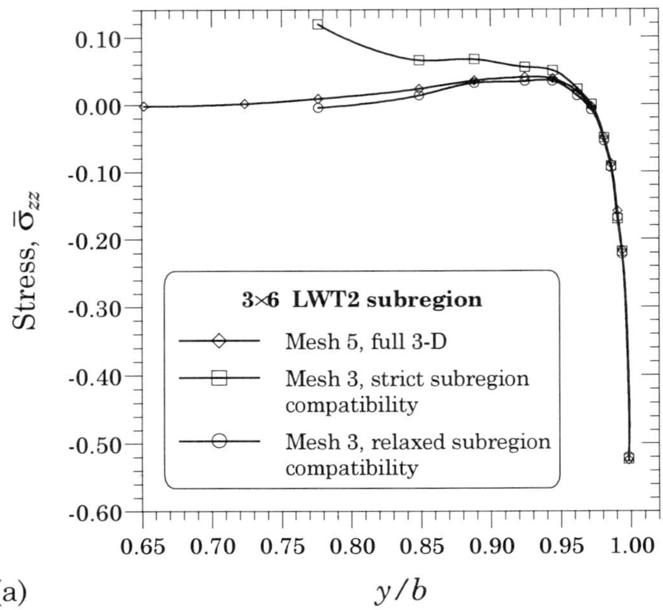

(a)

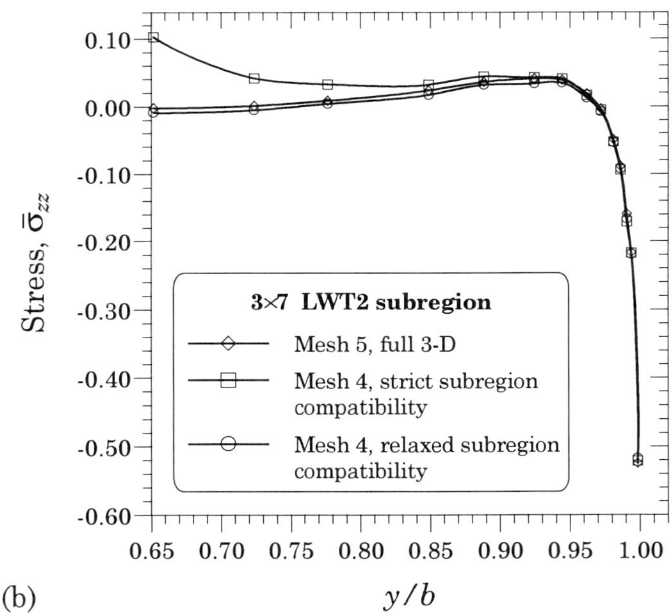

(b)

Figure 12.4–8: Stress distributions across width of the $(45/-45)_s$ laminate near the upper 45/–45 interface ($z = 1.014h_k$). Results computed for (a) mesh 3 and (b) mesh 4 using both strict subregion compatibility and relaxed subregion compatibility.

Figure 12.4–8 shows the effect of subregion compatibility type (strict or relaxed) on the accuracy of the transverse normal stress within the local LWT2 subregion. Only the transverse normal stress distributions are shown since similar transverse shear stress distributions were computed for both strict and relaxed subregion compatibility. Figures 12.4–8 shows the distribution of σ_{zz} across the width of the laminate, near the +45/–45 interface, as the free edge is approached. The stresses are computed along the same line of adjacent reduced Gauss points as in Figure 12.4–7. The use of strict subregion compatibility causes significant error in the transverse normal stress near the FSDT/LWT2 boundary. This error is caused by the enforced transverse inextensibility of the laminate along the FSDT/LWT2 boundary. The FSDT elements enforce a condition of $\varepsilon_{zz} = 0$ on the edges of the LWT2 elements that form the FSDT/LWT2 boundary, thus artificially pinching or stretching the laminate thickness along the FSDT/LWT2 boundary. In contrast, the use of relaxed subregion compatibility allows the edges of the LWT2 elements that lie on the FSDT/LWT2 boundary to expand or contract in the thickness direction in response to the compatible in-plane displacement field. Thus the use of relaxed subregion compatibility permits accurate transverse normal stresses to be computed across the entire width of the LWT2 region, even near the FSDT/LWT2 boundary. The use of relaxed subregion compatibility results in a slight increase in the number of active degrees of freedom since the W_I ($I = 1, 2, ..., N$) are not zeroed along the FSDT/LWT2 boundary (see Table 12.4–1). Thus the analyst may wish to use strict subregion compatibility provided that the LWT2 subregion is sufficiently large and provided that accurate transverse normal stresses are not needed near the FSDT/LWT2 boundary.

To illustrate the accuracy of the variable kinematic elements in determining the free edge stress field for more complex laminates, a simultaneous multiple model analysis (global–local analysis) is performed on an eight-ply $(45/0/{-45}/90)_s$ laminate, and a sixteen-ply $(45/0/{-45}/90/90/{-45}/0/45)_s$ laminate. Both of these laminates are subjected to axial extension similar to the previously examined $(45/{-45})_s$ laminate. The in-plane discretization consists of a 5×11 2-D mesh of eight-node quadrilateral elements as shown previously in Figure 12.4–4. The local region is discretized with a 3×6 mesh of LWT2 elements. For the $(45/0/{-45}/90)_s$ laminate, the LWT2 elements contain 12 quadratic layers (three per material ply). Within each material ply the three quadratic layers have thicknesses of $0.25h_k, 0.5h_k$, and $0.25h_k$ from bottom to top. For the $(45/0/{-45}/90/90/{-45}/0/45)_s$ laminate, the LWT2 elements contain 16 quadratic layers (two per material ply). Within each material ply both of the quadratic layers have thicknesses of $0.5h_k$. The $(45/0/{-45}/90)_s$ model contains 4,382 active degrees of freedom while the $(45/0/{-45}/90/90/{-45}/0/45)_s$ model contains 5,638 active degrees of freedom.

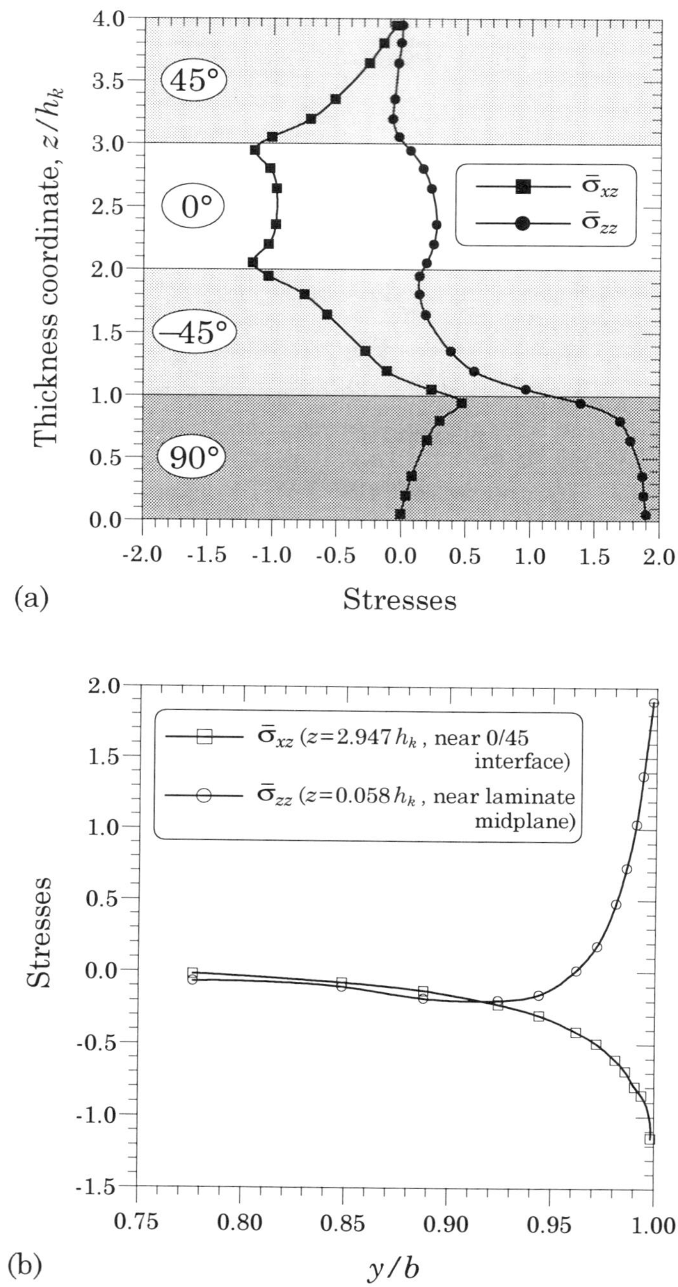

Figure 12.4–9: Interlaminar stress distributions in $(45/0/-45/90)_s$ laminate under axial extension. (a) Through the thickness near the free edge ($x = -0.115a, y = 0.998b$). (b) Across the width ($x = -0.115a$).

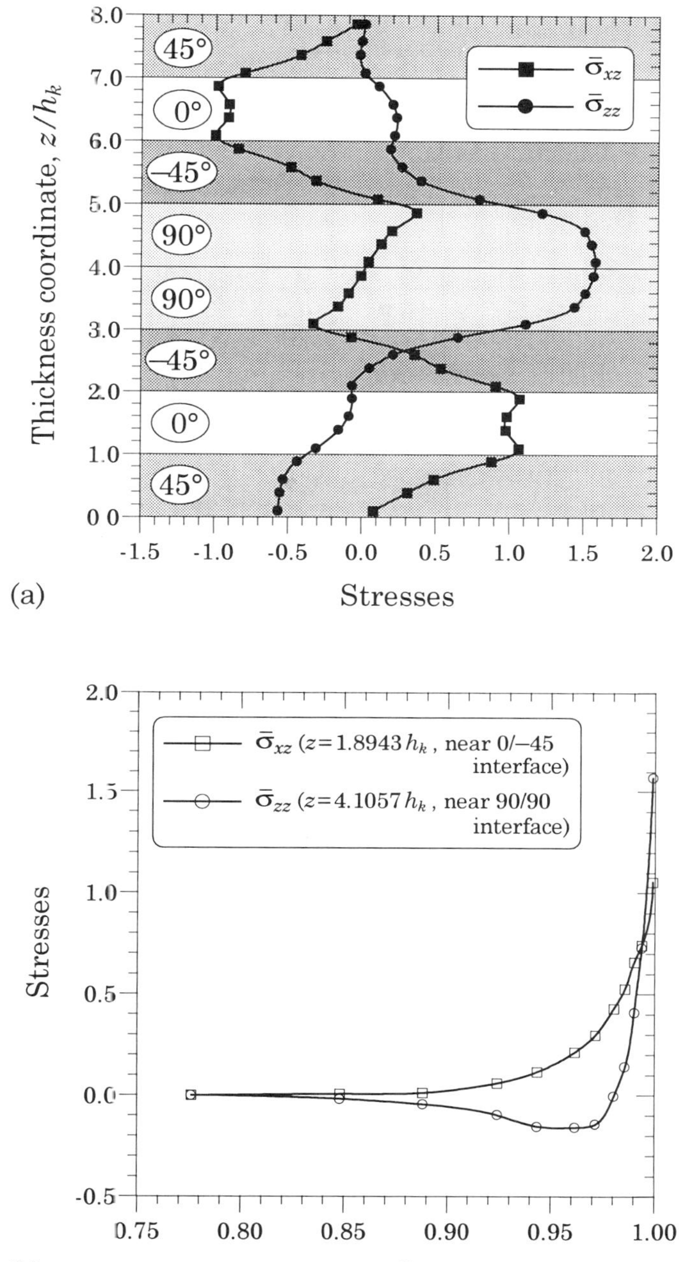

Figure 12.4–10: Interlaminar stress distributions in $(45/0/{-45}/90/90/{-45}/0/45)_s$ laminate under axial extension. (a) Through the thickness near the free edge ($x = -0.115a, y = 0.998b$). (b) Across the width ($x = -0.115a$).

The computed transverse shear stress and transverse normal stress distributions for these two laminates are shown in Figures 12.4–9 and 12.4–10. The present results show excellent agreement with the quasi–3-D finite element solutions of Whitcomb and Raju [61] (not included in the figure). For both laminates the maximum transverse normal stress occurs at the intersection of the 90/90 interface and the free edge, while the maximum transverse shear stress occurs at the intersection of the 45/0 and 0/–45 interfaces with the free edge.

Both of these laminates have enough distinct material plies to make a full 3-D analysis prohibitively expensive, thus a sequential or simultaneous multiple model analysis is the only reasonable alternative. Many laminates have a very large number of distinct material plies, thus even with a multiple model analysis, the investigator may have to resort to using the sublaminate approach (*i.e.*, ply grouping) within the local LWT2 region. In this case the investigator would identify a target group of adjacent material plies that would receive one or more numerical layers each, while the remaining plies are grouped into one or more numerical layers and effectively homogenized. By performing several of these analyses, the investigator can gradually piece together a picture of the 3-D stress state through the laminate thickness within the local LWT2 region.

Free Edge Stresses in a $(45/-45)_s$ Laminate in Bending

All of the previous examples involve the determination of free edge stress fields in laminates subjected to axial tension. To demonstrate the effectiveness of the variable kinematic finite elements for determining free edge stresses in laminates subjected to bending, consider a simply supported $(45/-45)_s$ laminate subjected to a uniform transverse load. The physical dimensions and material properties of the $(45/-45)_s$ laminate will be the same as in the axial extension example, with the exception that the origin of the coordinate system will be placed at the bottom center of one of the ends of the laminated strip. Thus the laminate occupies the domain $(0 < x < 2a, -b < y < b, 0 < z < 2h)$ where $a = 10b = 40h = 80h_k$ and h_k is the thickness of a material ply. The displacement boundary conditions are

$$u_1(a, -b, 0) = u_2(0, y, z) = u_2(2a, y, z) = u_3(0, y, 0) = u_3(2a, y, 0) = 0$$

The uniform transverse load q_0 is applied to the upper surface of the laminate and acts in the negative z direction. Note that there are no planes of symmetry in this problem, thus the computational domain consists of the entire laminate.

To accurately yet efficiently capture the free edge stresses, a global–local analysis is performed using a 2-D mesh of variable kinematic finite elements, where FSDT elements make up the majority of the computational domain and a small patch of LWT2 elements is used to resolve the free edge stress field within a localized region of interest, which is one of the free edges. The objective of this example is to determine the effect of the width of the LWT2 region on the accuracy of the computed free edge stresses.

To investigate the effect of reducing the length of the LWT2 subregion, four different global–local meshes are created. Each mesh has a total in-plane discretization of 9×11 elements (9 elements along the laminate length, 11 elements across the laminate width). As in the previous analyses of the $(45/-45)_s$ laminate under axial extension, the in-plane mesh is highly refined over one of the free edges and is coarse over the other free edge. Note that the collective length of the central three rows of equal length elements is denoted as a_0. The remaining six rows of elements are of equal length $(2a - a_0)/6$. The in-plane discretization of the five meshes differ in the value of a_0; specifically $a_0 = 2h, 4h, 8h$, and $16h$ for meshes 1 through 4, respectively. Each of the four in-plane meshes is used with both a 3×6 LWT2 subregion and 5×6 LWT2 subregion (*i.e.*, three or five LWT2 elements in the x direction and six LWT2 elements in the y direction). The width of the LWT2 subregion is h in each case. Each of the LWT2 elements employs eleven quadratic layers through the laminate thickness: three layers in the bottom $+45°$ ply, five layers in the two collective middle $-45/-45$ plies, and three layers in the top $+45°$ ply. From bottom to top, the layer thicknesses are $0.6h_k, 0.3h_k, 0.1h_k, 0.1h_k, 0.3h_k, 1.2h_k,\ 0.3h_k, 0.1h_k, 0.1h_k, 0.3h_k$, and $0.6h_k$. By comparing the results of the 3×6 and 5×6 LWT2 subregions for each of the four meshes, the effect of LWT2 subregion length on solution accuracy can be established. Each of the global–local meshes uses relaxed subregion compatibility.

The transverse stress distributions obtained with mesh 1 ($a_0 = 2h$) are shown in Figure 12.4–11. Mesh 1 ($a_0 = 2h$) consists of both 3×6 and 5×6 LWT2 subregions. Note that while the response predicted by the 3×6 and 5×6 LWT2 subregions are qualitatively the same, there is a small quantitative difference between the two responses. This quantitative difference indicates that the shorter 3×6 LWT2 subregion (length $= 2h$) is not quite adequate to accurately capture the local 3-D stress field. In particular, Figure 12.4–11b shows that the transverse shear stress distribution predicted with the 3×6 LWT2 subregion is not smooth across the FSDT/LWT2 interface which occurs at $y/b = 0.75$. Figure 12.4–11b also shows that the LWT2 subregion is not quite wide enough (width $= h$) to clearly show the point where σ_{zz} decays to zero. Figures 12.4–12a,b show the same results for mesh 2 where $a_0 = 4h$, with the exception that the width of the LWT2 subregion has been increased to $1.5h$ by using 3×7 and 5×7 LWT2 subregions as opposed to 3×6 and 5×6 LWT2 subregions. Note that the quantitative difference between the 3×7 and 5×7 responses is considerably smaller than for mesh 1. In Figure 12.4–12b, the disruption in the transverse shear stress distribution at the FSDT/LWT2 interface is barely detectable. Further, the increased widths of the LWT2 subregions are adequate to show the complete decay of σ_{zz}.

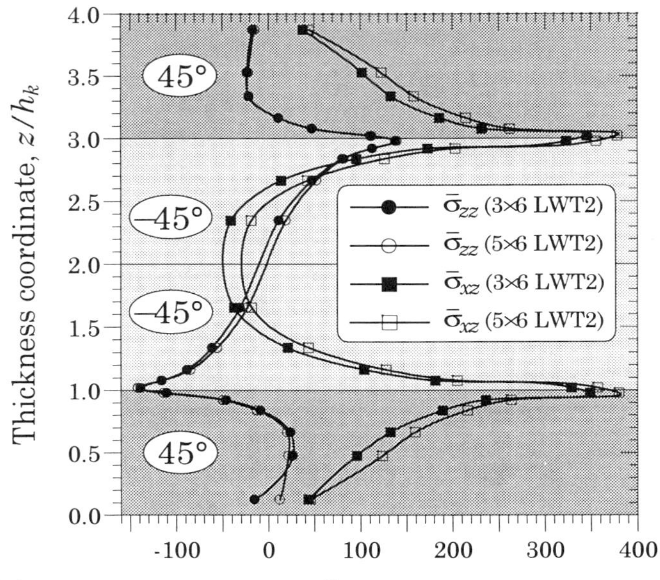

(a)

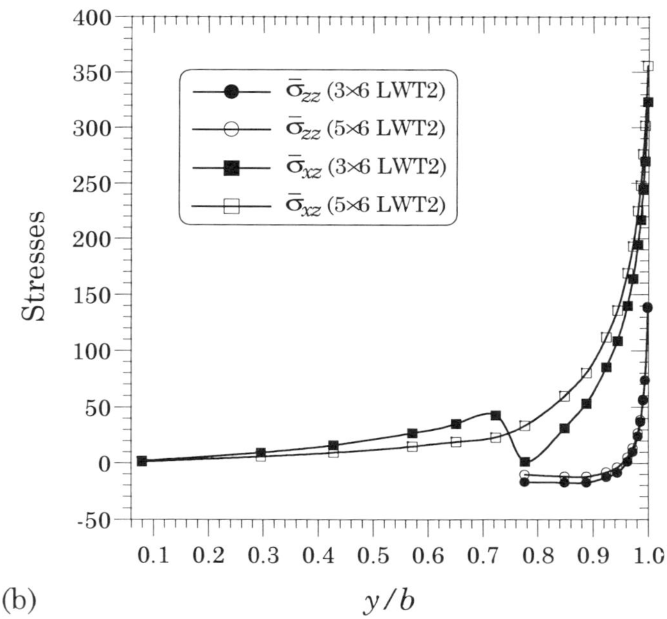

(b)

Figure 12.4–11: Interlaminar stress distributions ($\bar{\sigma} = \sigma/q_0$) in $(45/-45)_s$ laminate in bending (Mesh 1). (a) Through the thickness near the free edge at $x = a$ and $y = b$. (b) Across the width near the upper 45/–45 interface.

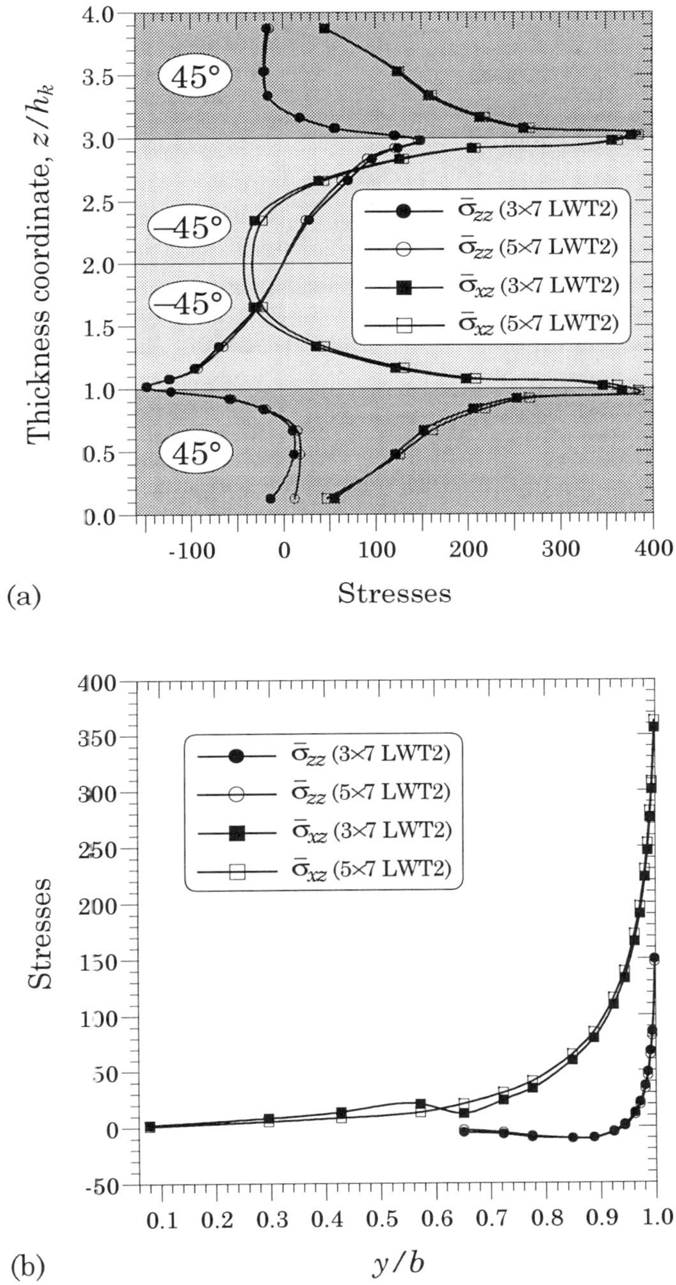

Figure 12.4–12: Interlaminar stress distributions ($\bar{\sigma} = \sigma/q_0$) in $(45/{-45})_s$ laminate in bending (Mesh 2). (a) Through the thickness near the free edge at $x = a$ and $y = b$. (b) Across the width of laminate, near the upper 45/–45 interface.

12.5 Closure

In this chapter a generalized layerwise theory proposed and advanced by the author is described and analytical and finite element solutions of the theory are presented. The layerwise theory of Reddy [37] is based on assumed displacement field, in which the thickness variation is represented using one-dimensional finite elements and thereby reducing the 3-D continuum to a 2-D problem. The procedure has the advantage of using independent approximation of thickness variations from in-plane discretizations. Otherwise, the layerwise theory is indeed the same as the traditional 3-D displacement finite element model.

A hierarchical, displacement-based, global–local finite element model that permits an accurate, efficient, and convenient analysis of localized three-dimensional effects in laminated composite plates is also presented. By superimposing a hierarchy of assumed displacement fields in the same finite element domain, a variable kinematic finite element model is developed. All displacement fields in the hierarchy share the same assumed in-plane variation, but differ in their assumed transverse variation.

The underlying foundation of the variable kinematic element's composite displacement field is provided by a two-dimensional 'equivalent single-layer' plate theory (*e.g.*, the first-order shear deformation theory). The layerwise displacement field of Reddy [37] is included as an optional, incremental enhancement to the displacement field of the two-dimensional plate theory, so that the element can have full three-dimensional modeling capability when needed. Depending on the desired level of accuracy, an element can use none, part, or all of the layerwise field to create a hierarchy of different elements having a wide range of kinematic complexity and representing a number of different mathematical models. Discrete layer transverse shear effects and discrete layer transverse normal effects can be independently added to the element by including appropriate terms from the layerwise field. The delamination kinematics can also be included as described by Barbero and Reddy [50] and Robbins and Reddy [52].

In a 2-D mesh of variable kinematic finite elements, each one of the elements is capable of simulating any of the element types in the hierarchy. Due to the hierarchical nature of the multiple assumed displacement fields, displacement continuity can be maintained between different types of elements in the hierarchy (*i.e.*, elements based on different mathematical models) by simply enforcing homogeneous essential boundary conditions on certain terms in the composite displacement field along the incompatible boundary. This simple process can easily be automated and subsequently removed form the concern of the user/analyst. Thus, in a single 2-D mesh of variable kinematic finite elements, it is possible to designate several different subregions that are described by elements that are based on different mathematical models. The variable kinematic elements circumvent the inconvenience and problems associated with the traditional methods of maintaining displacement continuity across incompatible subdomains (*e.g.*, multipoint constraints, special transition elements, and penalty methods).

References for Additional Reading

1. Pagano, N. J., "Exact Solutions for Composite Laminates in Cylindrical Bending," *Journal of Composite Materials,* **3**, 398–411 (1969).
2. Pagano, N. J., "Exact Solutions for Rectangular Bidirectional Composites and Sandwich Plates," *Journal of Composite Materials,* **4**, 20–34 (1970).
3. Pagano, N. J. and Hatfield, S. J., "Elastic Behavior of Multilayered Bidirectional Composites, *AIAA Journal,* **10**, 931–933 (1972).
4. Srinivas, S., Joga Rao, C. V., and Rao, A. K., "An Exact Analysis for Vibration of Simply Supported Homogeneous and Laminated Thick Rectangular Plates," *Journal of Sound and Vibration,* **12**, 187–199 (1970).
5. Srinivas, S. and Rao, A. K., "Bending, Vibration and Buckling of Simply Supported Thick Orthotropic Rectangular Plates and Laminates," *International Journal of Solids and Structures,* **6**, 1463–1481 (1970).
6. Noor, A. K., "Mixed Finite–Difference Scheme for Analysis of Simply Supported Thick Plates," *Computers and Structures,* **3**, 967–982 (1973).
7. Noor, A. K., "Free Vibrations of Multilayered Composite Plates," *AIAA Journal,* **11**, 1038–1039 (1973).
8. Savoia, M. and Reddy, J. N., "A Variational Approach to Three-Dimensional Elasticity Solutions of Laminated Composite Plates," *Journal of Applied Mechanics,* **59**, S166–S175 (1992).
9. Varadan, T. K. and Bhaskar, K., "Bending of Laminated Orthotropic Cylindrical Shells – An Elasticity Approach," *Composite Structures,* **17**, 141–156 (1991).
10. Ren, J. G., "Exact Solutions for Laminated Cylindrical Shells in Cylindrical Bending," *Composite Science and Technology,* **29**, 169–187 (1987).
11. Whitney, J. M., "The Effect of Transverse Shear Deformation in the Bending of Laminated Plates," *Journal of Composite Materials*, **3**, 534–547 (1969).
12. Swift, G. W. and Heller, R. A., "Layered Beam Analysis," *Journal of the Engineering Mechanics Division, ASCE,* **100**, 267–282 (1974).
13. Durocher, L. L. and Solecki, R., "Bending and Vibration of Isotropic Two-Layer Plates," *AIAA Journal,* **13**, 1522–1523 (1975).
14. Seide, P., "An Improved Approximate Theory for the Bending of Laminated Plates," *Mechanics Today,* **5**, 451–466 (1980).
15. Chaudhuri, R. A. and Seide, P., "Triangular Finite Element for Analysis of Thick Laminated Plates," *International Journal for Numerical Methods in Engineering,* **24**, 1203–1224 (1987).

16. Yu, Y. Y., "A New Theory of Elastic Sandwich Plates—One Dimensional Case," *Journal of Applied Mechanics,* **26,** 415–421 (1959).
17. Boal, J. L. and Reissner, E., "Three-Dimensional Theory of Elastic Plates with Transverse Inextensibility," *Journal of Mathematical Physics,* **39**, 161–181 (1961).
18. Srinivas, S., "A Refined Analysis of Laminated Composites," *Journal of Sound and Vibration,* **30**, 495–507 (1973).
19. Green, A. E. and Naghdi, P. M., "A Theory of Laminated Composite Plates," *IMA Journal of Applied Mathematics,* **29**, 1–23 (1982).
20. Rehfield, L. W. and Valisetty, R. R., "A Comprehensive Theory for Planar Bending of Composite Laminates," *Computers & Structures,* **15**, 441–447 (1983).
21. Mau, S. T., "A Refined Laminate Plate Theory," *Journal of Applied Mechanics,* **40**, 606–607 (1973).
22. Chou, P. C. and Carleone, J., "Transverse Shear in Laminated Plate Theories," *AIAA Journal,* **11**, 1333–1336 (1973) .
23. Di Sciuva, M., "A Refined Transverse Shear Deformation Theory for Multilayered Anisotropic Plates," *Atti della Academia delle Scienze di Torino,* **118**, 269–295 (1984).
24. Di Sciuva, M., "Bending, Vibration, and Buckling of Simply Supported Thick Multilayered Orthotropic Plates: An Evaluation of a New Displacement Model," *Journal of Sound and Vibration,* **105**, 425–442 (1986).
25. Di Sciuva, M., "An Improved Shear-Deformation Theory for Moderately Thick Multilayered Anisotropic Shells and Plates," *Journal of Applied Mechanics,* **54**(3), 589–596 (1987).
26. Murakami, H., "Laminated Composite Plate Theory with Improved In–Plane Responses," *Journal of Applied Mechanics,* **53**, 661–666 (1986).
27. Ren, J. G., "A New Theory of Laminated Plate," *Composite Science and Technology,* **26**, 225–239 (1986).
28. Zukas, J. A. and Vinson, J. R., "Laminated Transversely Isotropic Cylindrical Shells," *Journal of Applied Mechanics,* **38**, 400–407 (1971).
29. Waltz, T. L. and Vinson, J. R., "Interlaminar Stress in Laminated Cylindrical Shells of Composite Materials," *AIAA Journal,* **14**, 1213–1218 (1976).
30. Hsu, T. M. and Wang, J. T. S., "A Theory of Laminated Cylindrical Shells Consisting of Layers of Orthotropic Laminae," *AIAA Journal,* **8**, 2141–2146 (1970).
31. Hsu, T. M. and Wang, J. T. S., "Rotationally Symmetric Vibrations of Orthotropic Layered Cylindrical Shells," *Journal of Sound and Vibration,* **16**(4), 473–487 (1971).
32. Rath, B. K. and Das, Y. C., "Vibration of Laminated Shells," *Journal of Sound and Vibration,* **28**(4), 737–757 (1973).

33. Mau, S. T., Tong, P., and Pian, T. H. H., "Finite Element Solutions for Laminated Plates," *Journal of Composite Materials,* **6**, 304–311 (1972).

34. Spilker, R. L., "A Hybrid–Stress Finite-Element Formulation for Thick Multilayer Laminates," *Computers and Structures,* **11**, 507–514 (1980).

35. Spilker, R. L., "An Invariant Hybrid–Stress Eight-Node Element for Thin and Thick Multilayer Laminated Plates," *International Journal for Numerical Methods in Engineering,* **20**, 573–587 (1984).

36. Hinrichsen, R. L. and Palazotto, A. N., "Nonlinear Finite Element Analysis of Thick Composite Plates Using Cubic Spline Functions," *AIAA Journal,* **24**(11), 1836–1842 (1986).

37. Reddy, J. N., "A Generalization of Two-Dimensional Theories of Laminated Composite Plates," *Communications in Applied Numerical Methods,* **3**, 173–180 (1987).

38. Barbero, E. J., Reddy, J. N., and Teply, J. L., "General Two-Dimensional Theory of Laminated Cylindrical Shells," *AIAA Journal,* **28**(3), 544–553 (1990).

39. Owen, D. R. J. and Li, Z. H., "A Refined Analysis of Laminated Plates by Finite Element Displacement Methods – I. Fundamentals and Static Analysis," *Computers and Structures,* **26**(6), 907–914 (1987).

40. Owen, D. R. J. and Li, Z. H., "A Refined Analysis of Laminated Plates by Finite Element Displacement Methods – II. Vibration and Stability," *Computers and Structures,* **26**(6), 915–923 (1987).

41. Epstein, M. and Glockner, P. G., "Nonlinear Analysis of Multilayered Shells," *International Journal of Solids and Structures,* **13**, 1081–1089 (1977).

42. Epstein, M. and Huttelmaier, H. P., "A Finite Element Formulation for Multilayered and Thick Plates," *Computers and Structures,* **16**, 645–650 (1983).

43. Lee, K. H., Senthilnathan, N. R., Lim, S. P., and Chow, S. T., "An Improved Zig–Zag Model for the Bending of Laminated Composite Plates," *Composite Structures,* **15**, 137–148 (1990).

44. Lee, K. H., Kin, W. Z., and Chow, S. T., "Bi-Directional Bending of Laminated Composite Plates Using an Improved Zig–Zag Model," *Composite Structures,* **28**, 283–294 (1994).

45. Xavier, P. B., Lee, K. H., and Chew, C. H., "An Improved Zigzag Model for the Bending of Laminated Composite Shells," *Composite Structures,* **26**, 123–138 (1993).

46. Xavier, P. B., Chew, C. H., and Lee, K. H., "Buckling and Vibration of Multilayer Orthotropic Composite Shells Using a Simple Higher-Order Layerwise Theory," *International Journal of Solids and Structures,* **32**, 3479–3497 (1995).

47. Reddy, J. N., "On the Generalization of Displacement-Based Laminate Theories," *Applied Mechanics Reviews,* **42**(11), Part 2, S213–S222 (1989).

48. Reddy, J. N., Barbero, E. J., and Teply, J. L., "A Plate Bending Element Based on a Generalized Laminate Theory," *International Journal of Numerical Methods in Engineering,* **28**, 2275–2292 (1989).

49. Barbero, E. J., Reddy, J. N., and Teply, J., "An Accurate Determination of Stresses in Thick Laminates Using a Generalized Plate Theory," *International Journal of Numerical Methods in Engineering,* **29**, 1–14 (1990).

50. Barbero, E. J. and Reddy, J. N., "Modeling of Delamination in Composite Laminates Using a Layer-Wise Plate Theory," *International Journal of Solids and Structures*, **28**(3), 373–388 (1991).

51. Robbins, D. H., Jr. and Reddy, J. N., "Modeling of Thick Composites Using a Layerwise Laminate Theory," *International Journal for Numerical Methods In Engineering,* **36**, 655–677 (1993).

52. Robbins, D. H., Jr. and Reddy, J. N., "The Effects of Kinematic Assumptions on Computed Strain Energy Release Rates for Delaminated Composite Plates," *Mathematical Modelling and Scientific Computing,* **1**(1,2), 50–66 (1993).

53. Reddy, J. N., *An Introduction to the Finite Element Method,* Second Edition, McGraw–Hill, New York, 1992.

54. Goldbery, J. L. and Schwartz, A. J., *Systems of Ordinary Differential Equations, An Introduction,* Harper and Row, New York, 1972.

55. Nosier, A., Kapania, R. K., and Reddy, J. N., "Free Vibration Analysis of Laminated Plates Using a Layerwise Theory," *AIAA Journal,* **31**(12), 2335–2346 (1993).

56. Nosier, A., Kapania, R. K., and Reddy, J. N., "Low–Velocity Impact of Laminated Composites Using a Layerwise Theory," *Computational Mechanics,* **13** (5), 360–379 (1994).

57. Wang, S. S. and Choi, I., "Boundary–Layer Effects in Composite Laminates, Part I: Free Edge Stress Singularities," *Journal of Applied Mechanics,* **49**, 541–548 (1982).

58. Wang, S. S. and Choi, I., "Boundary–Layer Effects in Composite Laminates, Part II: Free Edge Stress Solutions and Basic Characteristics," *Journal of Applied Mechanics,* **49**, 549–560 (1982).

59. Whitcomb, J. D., Raju, I. S., and Goree, J. G., "Reliability of the Finite Element Method for Calculating Free Edge Stresses in Composite Laminates," *Computers and Structures,* **15**(1), 23–37 (1982).

60. Whitcomb, J. D., "Three–Dimensional Analysis of a Post–Buckled Embedded Delamination," *Journal of Composite Materials,* **23**, 470–487 (1989).

61. Whitcomb, J. D. and Raju, I. S., "Analysis of Interlaminar Stresses in Thick Composite Laminates With and Without Edge Delamination," *Delamination and Debonding of Materials,* ASTM STP 876, W. S. Johnson (ed.), American Society for Testing and Materials, Philadelphia, pp. 69–94, 1985.

62. Thompson, D. M. and Griffin, O. H., Jr., "2-D to 3-D Global/Local Finite Element Analysis of Cross-Ply Composite Laminates," *Journal of Reinforced Plastics and Composites,* **9**, 492–502 (1990).
63. Knight, N. F. *et al.*, "Global/Local Methods Research Using a Common Structural Analysis Framework," *Finite Elements in Analysis and Design,* **9**, 91–112 (1991).
64. Mao, K. M. and Sun, C. T., "A Refined Global–Local Finite Element Analysis Method," *International Journal for Numerical Methods in Engineering,* **32**, 29–43 (1991).
65. Whitcomb, J. D. and Woo, K., "Application of Iterative Global/Local Finite Element Analysis. Part 1: Linear Analysis,' *Communications in Numerical Methods in Engineering,* **9**(9), 745–756 (1993).
66. Whitcomb, J. D. and Woo, K., "Application of Iterative Global/Local Finite Element Analysis. Part 2: Geometrically Non-Linear Analysis", *Communications in Numerical Methods in Engineering,* **9**(9), 757–766 (1993).
67. Wang, A. S. D. and Crossman, F. W., "Calculation of Edge Stresses in Multi–Layer Laminates by Sub-Structuring," *Journal of Composite Materials,* **12**, 76–83 (1978).
68. Pagano, N. J. and Soni, S. R., "Global–Local Laminate Variational Model," *International Journal of Solids and Structures,* **19**(3), 207–228 (1983).
69. Jones, R., Callinan, R., Teh, K. K., and Brown, K. C., "Analysis of Multi-Layer Laminates Using Three-Dimensional Super Elements," *International Journal for Numerical Methods in Engineering,* **20**(3), 583–587 (1984).
70. Chang, F. K., Perez, J. L., and Chang, K. Y., "Analysis of Thick Laminated Composites," *Journal of Composite Materials,* **24**, 801–821 (1990).
71. Sun, C. T. and Liao, W. C., "Analysis of Thick Section Composite Laminates Using Effective Moduli," *Journal of Composite Materials,* **24**, 977-993 (1990).
72. Aminpour, A. A., McCleary, S.L., Ransom, J. B., and Housner, J. M., "A Global/Local Analysis Method for Treating Details in Structural Design," *Adaptive, Multilevel, and Hierarchical Computational Strategies,* ASME, AMD Vol. 157, A. K. Noor (ed.), pp. 119-137, 1992.
73. Surana, K. S., "Transition Finite Elements for Three-Dimensional Stress Analysis," *International Journal for Numerical Methods in Engineering,* **15**, 991–1020 (1980).
74. Surana, K. S., "Geometrically Nonlinear Formulation for the Transition Finite Elements for the Three-Dimensional Solid-Shell Transition Finite Elements," *Computers and Structures,* **15**, 549–566 (1982).
75. Liao, C. L., Reddy, J. N., and Engelstad, S. P. "A Solid-Shell Transition Element for Geometrically Nonlinear Analysis of Composite Structures," *International Journal for Numerical Methods in Engineering,* **26**, 1843–1854 (1988).

76. Davila, C., "Solid–to–Shell Transition Elements for the Computation of Interlaminar Stresses," *Second U.S. National Congress on Computational Mechanics,* Washington, D.C., August 16–18, 1993.

77. Mote, C. D., "Global–Local Finite Element," *International Journal for Numerical Methods in Engineering,* **3**, 565–574 (1971).

78. Dong, S. B., "Global-Local Finite Element Methods," *State–of–the–Art Surveys on Finite Element Technology,* A. K. Noor and W. D. Pilkey (eds.), ASME, pp. 451–474, 1983.

79. Reddy, J. N. and Robbins, D. H., Jr., "Analysis of Composite Laminates Using Variable Kinematic Finite Elements," *RBCM–Journal of the Brazilian Society of Mechanical Sciences,* **14**(4), 299–326 (1992).

80. Reddy, J. N. and Robbins, D. H., Jr., "A Simultaneous Multiple Model Approach for the Analysis of Composite Laminates," *Journal of the Aeronautical Society of India,* **45**, 157–177 (1993).

81. Robbins, D. H., Jr. and Reddy, J. N., "Variable Kinematic Modeling of Laminated Composite Plates," *International Journal for Numerical Methods in Engineering,* **39**, 2283–2317 (1996).

82. Teply, J. L., Barbero, E. J., and Reddy, J. N., "Bending, Vibration and Stability of ARALL Laminates Using a Generalized Laminate Plate Theory," *International Journal of Solids and Structures,* **27**(5), 585–599 (1990).

83. O'Brien, T. K., "Characterization of Delamination Onset and Growth in a Composite Laminate," *Damage in Composite Materials,* K. L. Reifsnider (ed.), STP 775, American Society for Testing Materials, Philadelphia, 140–167, 1982.

84. O'Brien, T. K., "Analysis of Local Delaminations and Their Influence on Composite Laminate Behavior," *Delamination and Debonding of Materials,* STP 877, W. S. Johnson (ed.), American Society for Testing Materials, Philadelphia, 282–297, 1985.

85. O'Brien, T.K., "Mixed–Mode Strain–Energy Release Rate Effects on Edge Delamination of Composites," *Effects of Defects in Composite Materials,* ASTM STP 836, American Society for Testing and Materials, Philadelphia, 125–142, 1984.

86. O'Brien, T. K. *et al.*, "Comparisons of Various Configurations of the Edge Delamination Test for Interlaminar Fracture Toughness," *Toughened Composites,* ASTM STP 937, N. J. Johnston (ed.), American Society for Testing and Materials, Philadelphia, 199–221, 1987.

Chapter Thirteen

Nonlinear Analysis of Composite Laminates

13.1 Introduction

The nonlinear partial differential equations governing composite laminates of arbitrary geometries and boundary conditions cannot be solved exactly. Approximate analytical solutions to the large–deflection theory (in von Kármán's sense) of laminated composite plates were obtained by many (see, for example, References 1–12). In most of these studies the effects of shear deformation and rotatory inertia were neglected, and only rectangular or cylindrical geometries were considered. The latter restriction is a direct result of the methods of analysis used, *i.e.*, Rayleigh–Ritz method, Galerkin method, perturbation method, and the double series method cannot be applied to plates of complicated geometries. For example, the classical variational methods (*e.g.*, Rayleigh–Ritz and Galerkin methods) are limited to simple geometries because of the difficulty in constructing the approximation functions for complicated geometries. The use of numerical methods facilitates the solution of these equations for problems of practical importance. Among the numerical methods available for the solution of nonlinear differential equations defined over arbitrary domains, the finite element method is the most practical and robust computational technique ever devised for the analysis of structural problems.

Historically, two distinct approaches have been followed in developing nonlinear finite element models of structural problems. Each of the approaches can be used to develop the displacement type, force type, or mixed/hybrid type finite element models.

The first approach is based on a laminate theory, in which the 3–D elasticity equations are reduced to 2–D equations through certain kinematic assumptions, as described in Chapters 5 and 11. In the nonlinear formulation based on a laminate theory, the geometry of the structure is assumed to remain unchanged during the loading, and the geometric nonlinearity in the form of the von Kármán strains is included. For small strains and moderately large deflections and rotations, these assumptions yield good results. We shall term the elements based on such assumptions *laminated elements* (see References 13–20).

The second approach is based on the 3–D continuum formulation, where any kinematic assumptions are directly introduced through the spatial finite element approximations. Full nonlinear strains or only the von Kármán nonlinear strains are included as desired, and the equations are derived in an incremental form directly. The formulation accounts for geometric changes that occurred during the previous increment of loading. Thus, the geometry is updated between load increments. Finite elements based on this formulation are called *continuum elements* (see References 21, 23, and 26–32).

There are two incremental continuum formulations that are used to determine the deformation and stress states in continuum problems (see Bathe, *et al.* [26] and Crisfield [32]): (1) the *total Lagrangian formulation* and (2) the *updated Lagrangian formulation*. In these formulations, the configuration (*i.e.*, geometry) of the structure for the current load increment is determined from a previously known configuration. In the total Lagrangian formulation, all of the quantities are referred to a fixed, often the undeformed, configuration, and changes in the displacement and stress fields are determined with respect to the reference configuration. The strain and stress measures used in this approach are the Green–Lagrange strain tensor and 2nd Piola–Kirchhoff stress tensor. In the updated Lagrangian formulation, the geometry of the structure from the previous increment is updated using the deformation computed in the current increment, and the updated configuration is used as the reference configuration for the next increment. A direct consequence of this is that differentiations and integrations are performed with respect to this reference configuration. The stress and strain measures used in this approach are the Cauchy stresses and the infinitesimal (Almansi) strains.

The major objective of this chapter is to study the geometrically nonlinear behavior of laminated plates. Towards this objective we develop the displacement finite element models of the classical laminated plate theory (CLPT) and the first–order shear deformation plate theory (FSDT) when the von Kármán nonlinear strains are accounted for, *i.e.*, develop nonlinear laminated plate elements. Alternative finite element models to the displacement model, *i.e.*, mixed and hybrid finite element models, can be found in References 33–42. (For a discussion of the continuum finite elements, one may consult References 21, 23, and 26–32.) The shear deformable nonlinear finite element models presented herein are used to study nonlinear bending, transient behavior, and postbuckling of laminated structures. In view of the detailed development presented in Chapter 10, the discussion on nonlinear finite element models is limited to the nonlinear stiffnesses and solution techniques used to solve nonlinear algebraic equations. For additional details and numerical examples, the reader may consult References 17 and 19–25.

13.2 Nonlinear Stiffness Coefficients

If the von Kármán nonlinear strains [see Eqs. (5.3–8a,b)] are to be included in the finite element model, the procedure discussed in Chapter 10 does not change, except that the resulting finite element equations (10.5–24) and (10.6–6) will be nonlinear. The nonlinear finite element model has the form

$$[K^e]\{\bar{\Delta}^e\} + [M^e]\{\ddot{\bar{\Delta}}^e\} = \{F^e\} \tag{13.2 – 1}$$

The elements of the stiffness matrix $[K^e]$ and mass matrix $[M^e]$ are given in Eq. (10.5-18a,b) for the classical plate theory and in Eq. (10.6–7) for the first–order shear deformation plate theory. The *additional* stiffness coefficients $[\bar{K}^{\alpha 3}]$ $(\alpha = 1, 2, 3)$ for the nonlinear CLPT element are

$$\begin{aligned}
\bar{K}_{ij}^{13} &= \frac{1}{2}\int_{\Omega^e}\left(\frac{\partial\psi_i}{\partial x}\bar{N}_{1j} + \frac{\partial\psi_i}{\partial y}\bar{N}_{6j}\right)\, dxdy\\
\bar{K}_{ij}^{23} &= \frac{1}{2}\int_{\Omega^e}\left(\frac{\partial\psi_i}{\partial x}\bar{N}_{6j} + \frac{\partial\psi_i}{\partial y}\bar{N}_{2j}\right)\, dxdy\\
\bar{K}_{ij}^{31} &= \int_{\Omega^e}\left(\bar{N}_{1i}\frac{\partial\psi_j}{\partial x} + \bar{N}_{6i}\frac{\partial\psi_j}{\partial y}\right)\, dxdy = 2\bar{K}_{ji}^{13}\\
\bar{K}_{ij}^{32} &= \int_{\Omega^e}\left(\bar{N}_{6i}\frac{\partial\psi_j}{\partial x} + \bar{N}_{2i}\frac{\partial\psi_j}{\partial y}\right)\, dxdy = 2\bar{K}_{ji}^{23}
\end{aligned}$$

$$\bar{K}_{ij}^{33} = \int_{\Omega^e}\left[\frac{\partial\varphi_i}{\partial x}\left(\bar{N}_1\frac{\partial\varphi_j}{\partial x} + \bar{N}_6\frac{\partial\varphi_j}{\partial y}\right) + \frac{\partial\varphi_i}{\partial y}\left(\bar{N}_6\frac{\partial\varphi_j}{\partial x} + \bar{N}_2\frac{\partial\varphi_j}{\partial y}\right)\right]\, dxdy \tag{13.2 – 2a}$$

$$\begin{aligned}
\bar{N}_{\alpha j} &= A_{\alpha 1}\frac{\partial w_0}{\partial x}\frac{\partial\varphi_j}{\partial x} + A_{\alpha 2}\frac{\partial w_0}{\partial y}\frac{\partial\varphi_j}{\partial y} + A_{\alpha 6}\left(\frac{\partial w_0}{\partial x}\frac{\partial\varphi_j}{\partial y} + \frac{\partial w_0}{\partial y}\frac{\partial\varphi_j}{\partial x}\right)\\
\bar{N}_{\alpha} &= \frac{1}{2}\left[A_{1\alpha}\left(\frac{\partial w_0}{\partial x}\right)^2 + A_{2\alpha}\left(\frac{\partial w_0}{\partial y}\right)^2 + 2A_{6\alpha}\frac{\partial w_0}{\partial x}\frac{\partial w_0}{\partial y}\right]
\end{aligned} \tag{13.2 – 2b}$$

for $\alpha = 1, 2, 6$. The same expressions hold for the first–order plate theory with φ_i replaced by ψ_i, because w_0 is also approximated by the Lagrange functions ψ_i in FSDT. We note that the nonlinearity of the coefficients $[K^{13}]$, $[K^{23}]$, $[K^{31}]$, $[K^{32}]$, and $[K^{33}]$ is solely due to the transverse deflection, w_0, as can be seen from Eqs. (2a,b). In addition, the stiffness matrix $[K^e]$ is not symmetric for the nonlinear case because

$$\bar{K}_{ij}^{3\alpha} = 2\bar{K}_{ji}^{\alpha 3}, \qquad \alpha = 1, 2 \tag{13.2 – 3}$$

The fully discretized equations are given by [see Eqs. (10.5–26b)]

$$[\hat{K}^e(\bar{\Delta}^e)]_{s+1}\{\bar{\Delta}^e\}_{s+1} = \{\hat{F}^e\} \tag{13.2 – 4}$$

13.3 Solution Methods for Nonlinear Algebraic Equations

13.3–1 Preliminary Comments

The nonlinear algebraic equations (4) must be solved by an iterative method. In iterative methods, the nonlinear equations are linearized by evaluating the nonlinear terms with a known solution. Here we discuss two iterative methods that are commonly used: (1) the Picard method, and (2) the Newton–Raphson method, which is based on Taylor's series (see [43–47]). While the Picard method is simple both in concept and computer implementation, the method uses the stiffness matrix $[\hat{K}^e]$, which is unsymmetric. The Newton–Raphson method is based on the tangent stiffness matrix, which is symmetric for all structural problems. Also, the method has faster convergence for most applications than the Picard method.

13.3–2 Direct Iteration (or Picard) Method

In the direct iteration method, also known as the *Picard method*, the solution vector from the previous iteration is used to evaluate the stiffness matrix, and the solution at the subsequent iteration is determined by solving the assembled equations after the imposition of boundary and initial conditions:

$$[\hat{K}(\{\Delta\}^r)]\{\Delta\}^{r+1} = \{\hat{F}\} \qquad (13.3-1)$$

where $\{\Delta\}^r$ denotes the global solution vector at the rth iteration. Thus, in the direct iteration method, the coefficients $\hat{K}_{ij}$ are obtained by evaluating the element stiffness coefficients $K_{ij}^{\alpha\beta}$ using the solution $\{\Delta\}^r$, or more exactly, w_0 from the previous iteration. The solution at $(r+1)$th iteration is obtained by solving Eq. (1):

$$\{\Delta\}^{r+1} = [\hat{K}(\{\Delta\}^r)]^{-1}\{\hat{F}\} \qquad (13.3-2)$$

At the beginning of the iteration (*i.e.*, $r = 0$), we assume that $\{\Delta\}^0 = \{0\}$ so that the solution at the first iteration is the linear solution, because the nonlinear stiffness matrix reduces to the linear one. The iteration process is continued [*i.e.*, Eq. (2) is solved in each iteration] until the difference between $\{\Delta\}^r$ and $\{\Delta\}^{r+1}$ reduces to a preselected error tolerance. The *error criterion* is of the form

$$\sqrt{\frac{\sum_{I=1}^{N}|\Delta_I^{r+1} - \Delta_I^r|^2}{\sum_{I=1}^{N}|\Delta_I^{r+1}|^2}} < \epsilon \ \ (\ say\ 10^{-3}) \qquad (13.3-3)$$

where N is the total number of nodal generalized displacements in the finite element mesh, and ϵ is the error tolerance.

13.3–3 Newton–Raphson Iteration Method

The Newton–Raphson iterative method is based on Taylor's series expansion of the nonlinear algebraic equation (13.2–4) about the known solution. Suppose that the solution at the rth iteration, $\{\Delta\}^r$, is known. Let

$$\{R\} \equiv [\hat{K}]\{\Delta\} - \{\hat{F}\} = 0 \qquad (13.3-4)$$

where $\{R\}$ is called the *residual*, which is a nonlinear function of the unknown solution $\{\Delta\}$. Expanding $\{R\}$ in Taylor's series about $\{\Delta\}^r$, we obtain

$$\begin{aligned}\{0\} = \{R\} = \{R\}^r &+ \left[\frac{\partial\{R\}}{\partial\{\Delta\}}\right]^r (\{\Delta\}^{r+1} - \{\Delta\}^r) \\ &+ \frac{1}{2!}\left[\frac{\partial^2\{R\}}{\partial\{\Delta\}\partial\{\Delta\}}\right]^r (\{\Delta\}^{r+1} - \{\Delta\}^r)^2 + \ldots \end{aligned} \qquad (13.3-5a)$$

or

$$0 = \{R\}^r + \left([\hat{K}]_r\right)^{tan}\{\delta\Delta\} + O(\{\delta\Delta\}^2) \qquad (13.3-5b)$$

where $O(\cdot)$ denotes the higher-order terms in $\{\delta\Delta\}$, and $[\hat{K}]^{tan}$ is the *tangent stiffness matrix* (or geometric stiffness matrix)

$$\left([\hat{K}]_r\right)^{tan} \equiv \left[\frac{\partial\{R\}}{\partial\{\Delta\}}\right] \quad \text{evaluated at} \quad \{\Delta\} = \{\Delta\}^r \qquad (13.3-6)$$

and the increment of the solution vector is defined by

$$\{\delta\Delta\} = \{\Delta\}^{r+1} - \{\Delta\}^r \qquad (13.3-7)$$

The tangent stiffness matrix $[\hat{K}]^{tan}$ for structural problems is symmetric even though $[\hat{K}]$ (or $[K]$) is not symmetric, as will be shown shortly. Equation (5b) is approximated by neglecting terms of order 2 and higher in the solution increment $\{\delta\Delta\}$. We obtain

$$\left([\hat{K}(\{\Delta\}^r)]\right)^{tan}\{\delta\Delta\} = -\{R\}^r \qquad (13.3-8)$$

Equations (5)–(8) are also applicable to an element. In other words, Eq. (8) can be assembled after the element tangent stiffness matrix and force residual vector are computed. The assembled equations are then solved after imposing the boundary and initial conditions of the problem. We have

$$\begin{aligned}\{\delta\Delta\} &= -\left([\hat{K}(\{\Delta\}^r)]^{tan}\right)^{-1}\{R\}^r\\ &= \left([\hat{K}(\{\Delta\}^r)]^{tan}\right)^{-1}\left(\{\hat{F}\} - [\hat{K}(\{\Delta\}^r)]\{\Delta\}^r\right)\end{aligned} \qquad (13.3-9)$$

and the total solution at the $(r+1)$th iteration is given by

$$\{\Delta\}^{r+1} = \{\Delta\}^r + \{\delta\Delta\} \qquad (13.3-10)$$

The iteration process is continued by solving Eq. (9) until the convergence criteria in Eq. (3) is satisfied or the residual $\{R\}$, measured in the same way as the solution error in Eq. (3), satisfies the error criterion. Note that at the beginning of each iteration the tangent stiffness matrix and residual vector must be updated using the latest available solution $\{\Delta\}$.

If the tangent stiffness matrix is kept constant for a preselected number of iterations but the residual vector is updated during the iteration, there can be a computational saving. Such an approach is called the *modified Newton–Raphson method.* For additional details on iterative methods and the *Riks method* used to accelerate convergence and even compute the load increment for each iteration, the reader may consult References 43–47.

13.3–4 Computation of Tangent Stiffness Coefficients

The coefficients of the tangent stiffness matrix can be computed using the definition in Eq. (6). The procedure to compute the coefficients of the tangent stiffness matrix of the first–order laminated plate theory is illustrated. The nonlinear finite element model of FSDT can be written, for the static case, in the form [see Eq. (10.6–6a)]

$$\begin{bmatrix} [K^{11}] & [K^{12}] & [K^{13}] & [K^{14}] & [K^{15}] \\ [K^{21}] & [K^{22}] & [K^{23}] & [K^{24}] & [K^{25}] \\ [K^{31}] & [K^{32}] & [K^{33}] & [K^{34}] & [K^{35}] \\ [K^{41}] & [K^{42}] & [K^{43}] & [K^{44}] & [K^{45}] \\ [K^{51}] & [K^{52}] & [K^{53}] & [K^{54}] & [K^{55}] \end{bmatrix}^e \begin{Bmatrix} \{u\} \\ \{v\} \\ \{w\} \\ \{S^1\} \\ \{S^2\} \end{Bmatrix}^e = \begin{Bmatrix} \{F^1\} \\ \{F^2\} \\ \{F^3\} \\ \{F^4\} \\ \{F^5\} \end{Bmatrix}^e \qquad (13.3-11)$$

where the stiffness coefficients $K_{ij}^{\alpha\beta}$ and force components F_i^α are defined in Eq. (10.6–7) for the linear case. For the von Kármán nonlinear case, the only coefficients that are affected are $[K^{13}]^e, [K^{23}]^e, [K^{31}]^e$, $[K^{32}]^e$, and $[K^{33}]^e$, which were defined in Eqs. (13.2–2a). Note that the coefficients in Eq. (13.2–2a) are to be added to the corresponding linear coefficients defined in Eq. (10.6–7).

The element tangent stiffness matrix has the same form as the element direct stiffness matrix in Eq. (11). Hence, Eq. (4) for a typical element can be expressed in component form as

$$R_i^\alpha = \sum_{\gamma=1}^{5}\sum_{\ell=1}^{n_\gamma} K_{i\ell}^{\alpha\gamma}\Delta_\ell^\gamma - F_i^\alpha \tag{13.3 – 12}$$

where $\alpha = 1, 2, \cdots, 5$. Here n_γ denotes the number of nodes per element in the Lagrange interpolation of the γth variable [see Eqs. (10.6–3) through (10.6–5)]. In general, different degrees of interpolation are used for (u_0, v_0), w_0, and (ϕ_x, ϕ_y). The components $\{\Delta^\gamma\}$ of the vector of nodal degrees of freedom $\{\Delta\}$ have the meaning

$$\Delta_j^1 = u_j\ ,\quad \Delta_j^2 = v_j \quad \text{for}\ \ j = 1, 2, \cdots, m(= n_1 = n_2)$$

$$\Delta_j^3 = w_j \quad \text{for}\ \ j = 1, 2, \cdots, n(= n_3) \tag{13.3 – 13}$$

$$\Delta_j^4 = S_j^1\ ,\quad \Delta_j^5 = S_j^2 \quad \text{for}\ \ j = 1, 2, \cdots, p(= n_4 = n_5)$$

In the further development we assume that equal interpolation is used for all variables, $m = n = p$. Note that only $[K^{13}]$, $[K^{23}]$, and $[K^{33}]$ are functions of $\Delta_j^3 = w_j$, and all other matrices are independent of Δ_j^α.

The coefficients of the submatrices $[K^{\alpha\beta}]^{tan}$ of the element tangent stiffness matrix can be defined as [see Eq. (6)]

$$\begin{aligned}\left(K_{ij}^{\alpha\beta}\right)^{tan} &\equiv \frac{\partial R_i^\alpha}{\partial \Delta_j^\beta}\\ &= \frac{\partial}{\partial \Delta_j^\beta}\left(\sum_{\gamma=1}^{5}\sum_{\ell=1}^{n} K_{i\ell}^{\alpha\gamma}\Delta_\ell^\gamma - F_i^\alpha\right)\\ &= K_{ij}^{\alpha\beta} + \sum_{\gamma=1}^{5}\sum_{\ell=1}^{n}\frac{\partial K_{i\ell}^{\alpha\gamma}}{\partial \Delta_j^\beta}\Delta_\ell^\gamma - 0\\ &\equiv K_{ij}^{\alpha\beta} + \bar{K}_{ij}^{\alpha\beta}\end{aligned} \tag{13.3 – 14a}$$

where

$$\bar{K}_{ij}^{\alpha\beta} = \sum_{\gamma=1}^{5}\sum_{\ell=1}^{n}\frac{\partial K_{i\ell}^{\alpha\gamma}}{\partial \Delta_j^\beta}\Delta_\ell^\gamma \tag{13.3 – 14b}$$

Thus the tangent stiffness coefficients are the sum of the direct stiffness coefficients $K_{ij}^{\alpha\beta}$ and their derivatives with respect to the nodal values, $\bar{K}_{ij}^{\alpha\beta}$.

Carrying out the operations indicated in the definition (6) of the tangent stiffness coefficients, we obtain

$$\left(K_{ij}^{11}\right)^{tan} = \frac{\partial R_i^1}{\partial \Delta_j^1} = \frac{\partial R_i^1}{\partial u_j} = K_{ij}^{11} + 0 \tag{13.3-15}$$

$$\left(K_{ij}^{12}\right)^{tan} = \frac{\partial R_i^1}{\partial \Delta_j^2} = \frac{\partial R_i^1}{\partial v_j} = K_{ij}^{12} + 0 \tag{13.3-16}$$

$$\begin{aligned}
\left(K_{ij}^{13}\right)^{tan} &= \frac{\partial R_i^1}{\partial \Delta_j^3} = \frac{\partial R_i^1}{\partial w_j} = K_{ij}^{13} + \sum_{\ell=1}^{n} \left(\frac{\partial K_{i\ell}^{13}}{\partial w_j}\right) w_\ell \\
&= K_{ij}^{13} + \frac{1}{2}\sum_{\ell=1}^{n} w_\ell \int_{\Omega^e} \left[\frac{\partial \psi_i}{\partial x}\left(A_{11}\frac{\partial^2 w_0}{\partial w_j \partial x}\frac{\partial \psi_\ell}{\partial x} + A_{12}\frac{\partial^2 w_0}{\partial w_j \partial y}\frac{\partial \psi_\ell}{\partial y}\right.\right. \\
&\quad \left. + A_{16}\frac{\partial^2 w_0}{\partial w_j \partial x}\frac{\partial \psi_\ell}{\partial y} + A_{16}\frac{\partial^2 w_0}{\partial w_j \partial y}\frac{\partial \psi_\ell}{\partial x}\right) \\
&\quad + \frac{\partial \psi_i}{\partial y}\left(A_{16}\frac{\partial^2 w_0}{\partial w_j \partial x}\frac{\partial \psi_\ell}{\partial x} + A_{26}\frac{\partial^2 w_0}{\partial w_j \partial y}\frac{\partial \psi_\ell}{\partial y}\right. \\
&\quad \left.\left. + A_{66}\frac{\partial^2 w_0}{\partial w_j \partial x}\frac{\partial \psi_\ell}{\partial y} + A_{66}\frac{\partial^2 w_0}{\partial w_j \partial y}\frac{\partial \psi_\ell}{\partial x}\right)\right] dxdy \\
&= K_{ij}^{13} + \frac{1}{2}\int_{\Omega^e} \left[\frac{\partial \psi_i}{\partial x}\left(A_{11}\frac{\partial \psi_j}{\partial x}\frac{\partial w_0}{\partial x} + A_{12}\frac{\partial \psi_j}{\partial y}\frac{\partial w_0}{\partial y}\right.\right. \\
&\quad \left. + A_{16}\frac{\partial \psi_j}{\partial x}\frac{\partial w_0}{\partial y} + A_{16}\frac{\partial \psi_j}{\partial y}\frac{\partial w_0}{\partial x}\right) \\
&\quad + \frac{\partial \psi_i}{\partial y}\left(A_{16}\frac{\partial \psi_j}{\partial x}\frac{\partial w_0}{\partial x} + A_{26}\frac{\partial \psi_j}{\partial y}\frac{\partial w_0}{\partial y}\right. \\
&\quad \left.\left. + A_{66}\frac{\partial \psi_j}{\partial x}\frac{\partial w_0}{\partial y} + A_{66}\frac{\partial \psi_j}{\partial y}\frac{\partial w_0}{\partial x}\right)\right] dxdy \\
&= K_{ij}^{13} + K_{ij}^{13} = 2K_{ij}^{13}
\end{aligned} \tag{13.3-17}$$

where we used the identities

$$\frac{\partial w_0}{\partial w_j} = \frac{\partial}{\partial w_j}\left(\sum_{k=1}^{n} w_k \psi_k\right) = \psi_j \tag{13.3-18a}$$

and

$$\sum_{\ell=1}^{n} \frac{\partial \psi_\ell}{\partial x} w_\ell = \frac{\partial w_0}{\partial x}\ , \quad \sum_{\ell=1}^{n} \frac{\partial \psi_\ell}{\partial y} w_\ell = \frac{\partial w_0}{\partial y} \tag{13.3-18b}$$

Similarly, we obtain

$$\left(K_{ij}^{14}\right)^{tan} = \frac{\partial R_i^1}{\partial \Delta_j^4} = K_{ij}^{14}\ , \quad \left(K_{ij}^{15}\right)^{tan} = \frac{\partial R_i^1}{\partial \Delta_j^5} = K_{ij}^{15}$$

$$\left(K_{ij}^{21}\right)^{tan} = \frac{\partial R_i^2}{\partial \Delta_j^1} = K_{ij}^{21}\ , \quad \left(K_{ij}^{22}\right)^{tan} = \frac{\partial R_i^2}{\partial \Delta_j^2} = K_{ij}^{22}\ , \quad \left(K_{ij}^{23}\right)^{tan} = \frac{\partial R_i^2}{\partial \Delta_j^3} = 2K_{ij}^{23}$$

$$\left(K_{ij}^{24}\right)^{tan} = \frac{\partial R_i^2}{\partial \Delta_j^4} = K_{ij}^{24}\ , \quad \left(K_{ij}^{25}\right)^{tan} = \frac{\partial R_i^2}{\partial \Delta_j^5} = K_{ij}^{25}$$

$$\left(K_{ij}^{31}\right)^{tan} = \frac{\partial R_i^3}{\partial \Delta_j^1} = K_{ij}^{31}\ , \quad \left(K_{ij}^{32}\right)^{tan} = \frac{\partial R_i^2}{\partial \Delta_j^2} = K_{ij}^{32}$$

$$\begin{aligned}
\left(K_{ij}^{33}\right)^{tan} =& \frac{\partial R_i^3}{\partial \Delta_j^3} = K_{ij}^{33} + \sum_{\ell=1}^{n}\left[\left(\frac{\partial K_{i\ell}^{31}}{\partial w_j}\right)u_\ell + \left(\frac{\partial K_{i\ell}^{32}}{\partial w_j}\right)v_\ell + \left(\frac{\partial K_{i\ell}^{33}}{\partial w_j}\right)w_\ell\right]\\
=& K_{ij}^{33} + \sum_{\ell=1}^{n} u_\ell \int_{\Omega^e}\left(\frac{\partial \bar{N}_{1i}}{\partial w_j}\frac{\partial \psi_\ell}{\partial x} + \frac{\partial \bar{N}_{6i}}{\partial w_j}\frac{\partial \psi_\ell}{\partial y}\right) dxdy\\
&+ \sum_{\ell=1}^{n} v_\ell \int_{\Omega^e}\left(\frac{\partial \bar{N}_{6i}}{\partial w_j}\frac{\partial \psi_\ell}{\partial x} + \frac{\partial \bar{N}_{2i}}{\partial w_j}\frac{\partial \psi_\ell}{\partial y}\right) dxdy\\
&+ \sum_{\ell=1}^{n} w_\ell \int_{\Omega^e}\left[\frac{\partial \psi_i}{\partial x}\left(\partial \bar{N}_1 \partial w_j \frac{\partial \psi_\ell}{\partial x} + \frac{\partial \bar{N}_6}{\partial w_j}\frac{\partial \psi_\ell}{\partial y}\right)\right.\\
&\qquad\left. + \frac{\partial \psi_i}{\partial y}\left(\frac{\partial \bar{N}_6}{\partial w_j}\frac{\partial \psi_\ell}{\partial x} + \frac{\partial \bar{N}_2}{\partial w_j}\frac{\partial \psi_\ell}{\partial y}\right)\right] dxdy\\
=& \int_{\Omega^e}\left[\frac{\partial \psi_i}{\partial x}\left(N_{xx}\frac{\partial \psi_j}{\partial x} + N_{xy}\frac{\partial \psi_j}{\partial y}\right) + \frac{\partial \psi_i}{\partial y}\left(N_{xy}\frac{\partial \psi_j}{\partial x} + N_{yy}\frac{\partial \psi_j}{\partial y}\right)\right] dxdy\\
&+ \int_{\Omega^e}\left\{\left[A_{11}\left(\frac{\partial w_0}{\partial x}\right)^2 + A_{66}\left(\frac{\partial w_0}{\partial y}\right)^2 + 2A_{16}\frac{\partial w_0}{\partial x}\frac{\partial w_0}{\partial y}\right]\frac{\partial \psi_i}{\partial x}\frac{\partial \psi_j}{\partial x}\right.\\
&+ \left[A_{16}\left(\frac{\partial w_0}{\partial x}\right)^2 + A_{26}\left(\frac{\partial w_0}{\partial y}\right)^2 + (A_{16} + A_{66})\frac{\partial w_0}{\partial x}\frac{\partial w_0}{\partial y}\right]\\
&\left(\frac{\partial \psi_i}{\partial x}\frac{\partial \psi_j}{\partial y} + \frac{\partial \psi_j}{\partial x}\frac{\partial \psi_i}{\partial y}\right)\\
&+ \left.\left[A_{66}\left(\frac{\partial w_0}{\partial x}\right)^2 + A_{22}\left(\frac{\partial w_0}{\partial y}\right)^2 + 2A_{26}\frac{\partial w_0}{\partial x}\frac{\partial w_0}{\partial y}\right]\frac{\partial \psi_i}{\partial y}\frac{\partial \psi_j}{\partial y}\right\} dxdy
\end{aligned} \tag{13.3-19}$$

where [see Eq. (13.2–2b); replace φ with ψ for FSDT]

$$\frac{\partial \bar{N}_{\alpha\ell}}{\partial w_j} = A_{\alpha 1}\frac{\partial \psi_j}{\partial x}\frac{\partial \psi_\ell}{\delta x} + A_{\alpha 2}\frac{\partial \psi_j}{\partial y}\frac{\partial \psi_\ell}{\partial y} + A_{\alpha 6}\left(\frac{\partial \psi_j}{\partial x}\frac{\partial \psi_\ell}{\partial y} + \frac{\partial \psi_j}{\partial y}\frac{\partial \psi_\ell}{\partial x}\right) \tag{13.3-20}$$

$$\frac{\partial \bar{N}_{\alpha}}{\partial w_j} = A_{1\alpha}\frac{\partial w_0}{\partial x}\frac{\partial \psi_j}{\partial x} + A_{2\alpha}\frac{\partial w_0}{\partial y}\frac{\partial \psi_j}{\partial y} + A_{6\alpha}\left(\frac{\partial w_0}{\partial x}\frac{\partial \psi_j}{\partial y} + \frac{\partial w_0}{\partial y}\frac{\partial \psi_j}{\partial x}\right) \tag{13.3-21}$$

From Eqs. (15)–(19), it is clear that the element tangent stiffness matrix is symmetric.

13.4 Computational Aspects and Numerical Examples

13.4–1 Membrane Locking

Linear (four–node) and quadratic (eight–node and nine–node) Lagrange finite elements of the first–order shear deformation plate theory, with isoparametric formulation, are employed in the examples discussed. The element has, in general, five degrees of freedom associated with the generalized displacements $(u_0, v_0, w_0, \phi_x, \phi_y)$ per node. As discussed in Chapter 10, selective reduced integration must be used in the numerical evaluation of stiffness coefficients associated with the transverse shear coefficients, *i.e.*, those involving transverse shear stiffnesses, A_{44}, A_{45}, and A_{55}, to eliminate shear locking. A 1×1 Gauss rule for the linear element and 2×2 Gauss rule for the quadratic elements are used in the evaluation of the transverse shear stiffness coefficients.

For nonlinear problems, the FSDT element also experiences *membrane locking*. The membrane locking can be explained by considering, for simplicity, the Timoshenko beam finite element (see Exercise 13.12). When the element is used to analyze pure bending deformation, it should experience no axial (or membrane) strain:

$$\varepsilon_{xx}^0 \equiv \frac{du_0}{dx} + \frac{1}{2}\left(\frac{dw_0}{dx}\right)^2 = 0 \quad \text{(for pure bending)} \qquad (13.4-1)$$

In order that the above constraint be satisfied for independent approximations of u_0 and w_0, the term should cancel the second term in Eq. (1). This in turn requires that

$$\text{degree of polynomial of } \frac{du_0}{dx} \sim \text{degree of polynomial of } \left(\frac{dw_0}{dx}\right)^2 \qquad (13.4-2)$$

If both variables are approximated with sufficiently higher-order polynomials, the coefficients in the polynomials get adjusted to satisfy the constraint (1). Also, when both u_0 and w_0 are approximated using linear polynoimals, the correspondence (2), hence constraint (1), is automatically satisfied; however, when quadratic interpolation of both u_0 and w_0 is used, then $\frac{du_0}{dx}$ is linear and $(\frac{dw_0}{dx})^2$ is quadratic and there is no possibility of canceling the coefficient in quadratic term. If u_0 is interpolated with cubic polynomials and w_0 is interpolated with quadratic polynomials, we have

$$\frac{du_0}{dx} \text{ (quadratic)} \quad \sim \quad \left(\frac{dw_0}{dx}\right)^2 \text{ (quadratic)} \qquad (13.4-3)$$

Thus the constraint (1) is again satisfied. In summary, the element does not experience membrane locking for the following two cases:

(1) u_0 is linear and w is linear

(2) u_0 is cubic and w_0 is quadratic

and it experiences locking when both u_0 and w_0 are interpolated using quadratic polynomials.

Since quadratic approximation of u_0 and w_0 is common in practice, it is necessary to find a way to avoid membrane locking of the element. It is found that, for this case, the membrane locking can be avoided by using selective integrations of the terms of the form in ε_{xx}. For example, consider the coefficient $(K_{ij}^{22})^{tan}$ (see Exercise 13.14):

$$\left(K_{ij}^{22}\right)^{tan} = \int_{x_A}^{x_B} E_{xx}^b A \left[\frac{du_0}{dx} + \frac{3}{2}\left(\frac{dw_0}{dx}\right)^2\right] \frac{d\psi_i}{dx}\frac{d\psi_j}{dx} dx \tag{13.4-4}$$

For quadratic interpolation of u_0 and w_0, the first term is a cubic ($p = 3$) polynomial and the second term is a fourth-order ($p = 4$) polynomial. Thus the exact evaluation of the first term requires $NGP = (p+1)/2 = 2$ and the second term requires $NGP = [(p+1)/2] = 3$, where NGP denotes the number of Gauss points. Thus, for constant $E_{xx}^b A$, three–point Gauss quadrature yields exact values of both integrals. However, the two–point Gauss rule would yield an exact value for the first term

$$\int_{x_A}^{x_B} \frac{du_0}{dx}\frac{d\psi_i}{dx}\frac{d\psi_j}{dx}\, dx$$

and at the same time the second term

$$\int_{x_A}^{x_B} \left(\frac{dw_0}{dx}\right)^2 \frac{d\psi_i}{dx}\frac{d\psi_j}{dx}\, dx$$

is approximated as the same degree polynomial as the first term. This amounts to using an interpolation for w_0 that satisfies the constraint $\varepsilon_{xx}^0 = 0$.

The discussion presented above for the Timoshenko beam element also applies to the Mindlin plate element (*i.e.*, the FSDT element). Of course, if the mesh of quadratic elements is sufficiently refined, the membrane locking disappears. Here we present some numerical examples of laminated plates and shells using the nonlinear shear deformable laminated plate finite element presented in Section 13.2. A shear correction coefficient $K = 5/6$ is used here.

The problems presented here illustrate certain features characteristic to composite laminates. These include:

(1) the effect of geometric nonlinearity on static deflections,

(2) the use of biaxial symmetry boundary conditions in quarter plate models of rectangular laminates,

(3) postbuckling response of laminates under in-plane compression,

(4) nonlinear transient response of composite laminates, and

(5) postbuckling and progressive failure analysis of composite panels subjected to in-plane compression.

All of the problems are selected from the author's publications, and additional examples can be found in the references cited at the end of the chapter.

13.4–2 Nonlinear Bending of Laminated Rectangular Plates

First we consider the nonlinear bending of an isotropic ($\nu = 0.3$) square plate under uniformly distributed transverse load. The following geometric boundary conditions, a type of simple support (SS–3), are used:

$$u_0 = v_0 = w_0 = 0 \quad \text{on all four edges} \tag{13.4-5}$$

In the finite element method, when a generalized displacement (primary variable) is not specified at a boundary point, the corresponding generalized force (secondary variable) is implicitly included as zero at that point. Thus, since ϕ_x and ϕ_y are not specified in SS–3, it follows that

$$\begin{aligned} &\text{on } y = 0, b: \qquad M_{xy} = M_{yy} = 0 \\ &\text{on } x = 0, a: \qquad M_{xx} = M_{xy} = 0 \end{aligned} \tag{13.4-6}$$

Figures 13.4–1 and 13.4–2 contain plots of the nondimensionalized center deflection, $\bar{w} = w_0/h$, and nondimensionalized center stress, $\bar{\sigma}_{xx} = \sigma_{xx}(a^2/Eh^2)$, as functions of the load parameter, $\bar{P} = q_0a^4/Eh^4$ for a simply supported (SS–3) isotropic square plate under uniformly distributed load of intensity q_0. The double Fourier-series solution of Lévy [48], the finite–difference solution of Wang [49], and the Galerkin solution of Yamaki [50] are compared with the finite element solutions obtained with 2×2 meshes of linear (L) and nine-node quadratic (Q) elements. Both three (w_0, ϕ_x, ϕ_y) and five $(u_0, v_0, w_0, \phi_x, \phi_y)$ degrees of freedom models were studied. Omission of the in-plane displacements (u_0, v_0) amounts to constraining the in-plane extension everywhere in the plate domain, and hence it will behave as a stiffer plate than one in which the in-plane displacement degrees of freedom are included.

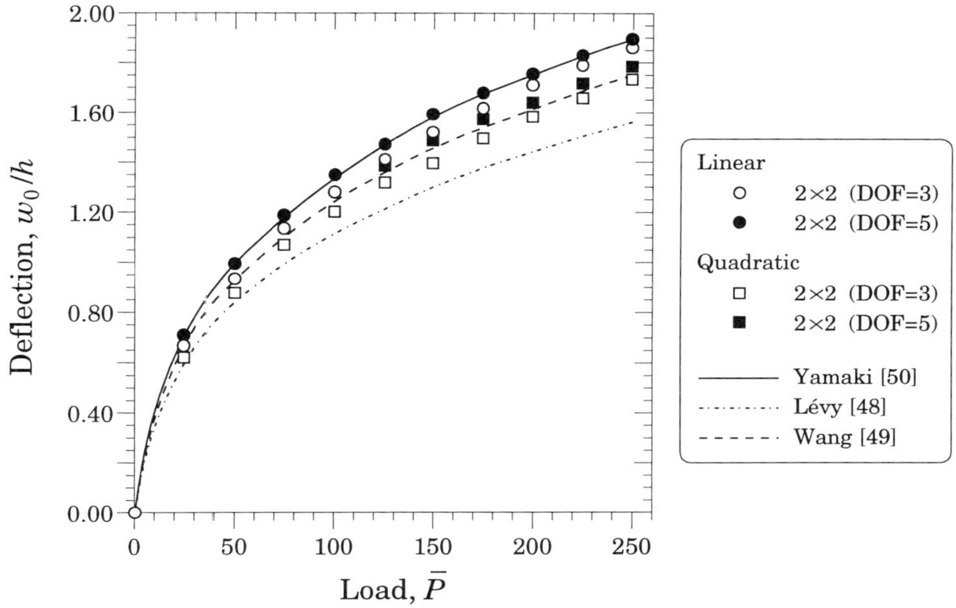

Figure 13.4–1: Load–deflection curves ($\bar{P}$ vs. w_0/h) for simply supported (SS–3) isotropic square plates under uniformly distributed transverse load q_0.

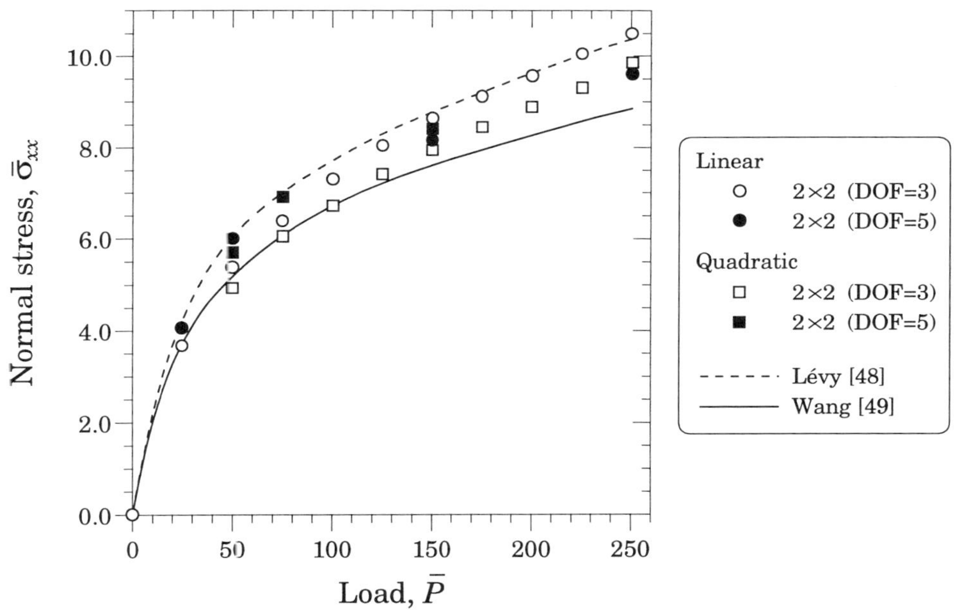

Figure 13.4–2: Load versus normal stress ($\bar{P}$ vs. $\bar{\sigma}_{xx}$) for simply supported (SS–3) isotropic square plates under uniformly distributed transverse load q_0.

Figures 13.4–3 and 13.4–4 contain the load-deflection plots for a simply supported orthotropic square plate and a clamped cross-ply (0/90/90/0) square laminate, respectively. In both cases, a uniformly distributed transverse load is used. The geometry and material properties used are given below [8].

Orthotropic plate:

$$\begin{aligned} a = b &= 12 \text{ in.}, \quad h = 0.138 \text{ in.} \\ E_1 &= 3 \times 10^6 \text{ psi}, \quad E_2 = 1.28 \times 10^6 \text{ psi} \\ G_{12} &= G_{23} = G_{13} = 0.37 \times 10^6 \text{ psi} \\ \nu_{12} &= \nu_{23} = \nu_{13} = 0.32 \end{aligned} \tag{13.4-7}$$

Cross-ply laminate:

$$\begin{aligned} a = b &= 12 \text{ in.}, \quad h = 0.096 \text{ in.} \\ E_1 &= 1.8282 \times 10^6 \text{ psi}, \quad E_2 = 1.8315 \times 10^6 \text{ psi} \\ G_{12} &= G_{13} = G_{23} = 0.3125 \times 10^6 \text{ psi} \\ \nu_{12} &= \nu_{23} = \nu_{13} = 0.2395 \end{aligned} \tag{13.4-8}$$

For the simply supported orthotropic plate, the shear deformation element yields results that are in excellent agreement with the experimental data of Zaghloul and Kennedy [8], while the classical plate bending element underpredicts deflections even for this thin plate (a/h is about 120). For the clamped cross ply laminate, the finite element results are not in as good agreement with the experimental data. This is believed to be due to possible differences between the actual material properties and boundary conditions and those used in the finite element model. It is much more difficult to simulate fully clamped boundary conditions ($u_0 = v_0 = w_0 = \phi_x = \phi_y = 0$ on all edges) in an experiment.

Next, results of large-deflection bending of two-layer and six-layer cross-ply and angle-ply rectangular plates ($a/h = 40$) are presented. The plates are assumed to be clamped ($u_0 = v_0 = w_0 = \phi_x = \phi_y = 0$) and subjected to uniformly distributed transverse load. The following material properties of a lamina are assumed:

$$E_1 = 40E_2, \quad G_{12} = G_{13} = 0.6E_2, \quad G_{23} = 0.5E_2, \quad \nu_{12} = 0.25 \tag{13.4-9}$$

A three degrees of freedom model is used. Figures 13.4–5 and 13.4–6 contain plots of nondimensionalized center deflection versus the load parameter $\bar{P} = 10^{-4} q_0 (2b/h)^4 / E_2$ for cross-ply and angle-ply laminates, respectively. The degree of nonlinearity exhibited by the six-layer plates is very small. Since the bending–stretching coupling is significant in two-layer plates and the in-plane displacements are omitted in the model, the results for two-layer plates should not be considered as typical. They are included only for comparison.

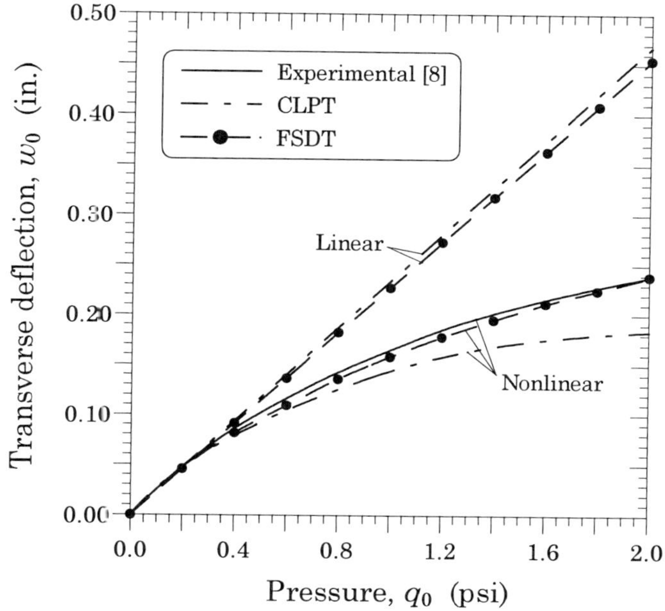

Figure 13.4–3: Load-deflection curves (q_0 vs. w_0) for simply supported (SS–1) orthotropic square plates under uniformly distributed transverse load q_0.

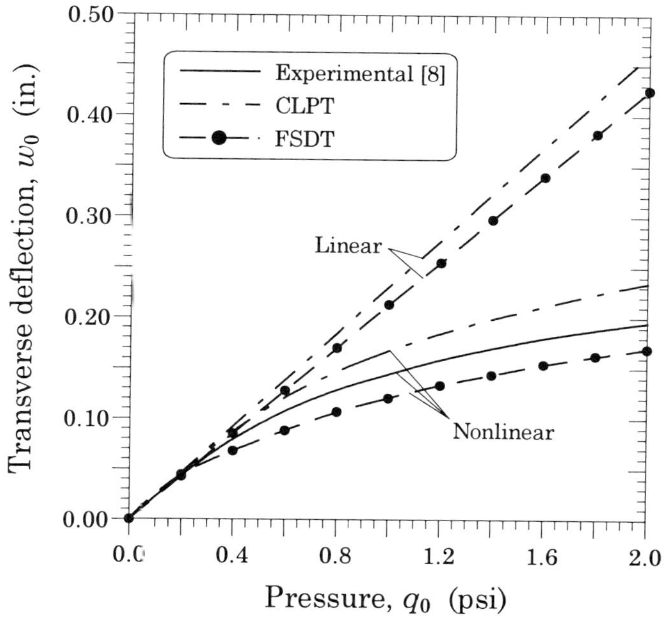

Figure 13.4–4: Load-deflection curves (q_0 vs. w_0) for clamped cross-ply (0/90/90/0) square plates under uniformly distributed transverse load q_0.

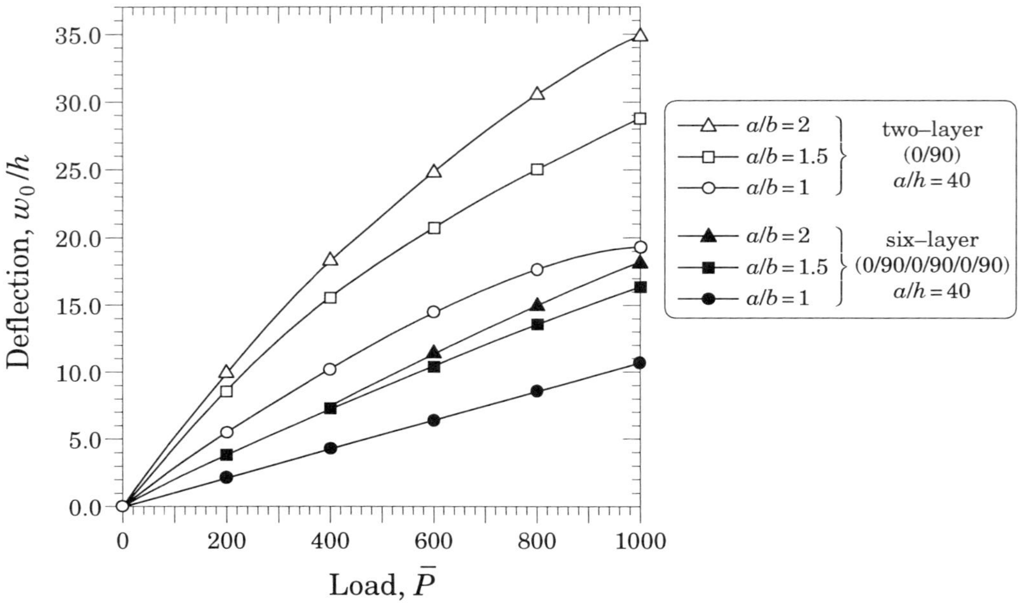

Figure 13.4–5: Load-deflection curves ($\bar{P}$ vs. w_0/h) for clamped two- and six-layer antisymmetric cross-ply plates ($a/h = 40$) under uniformly distributed transverse load q_0.

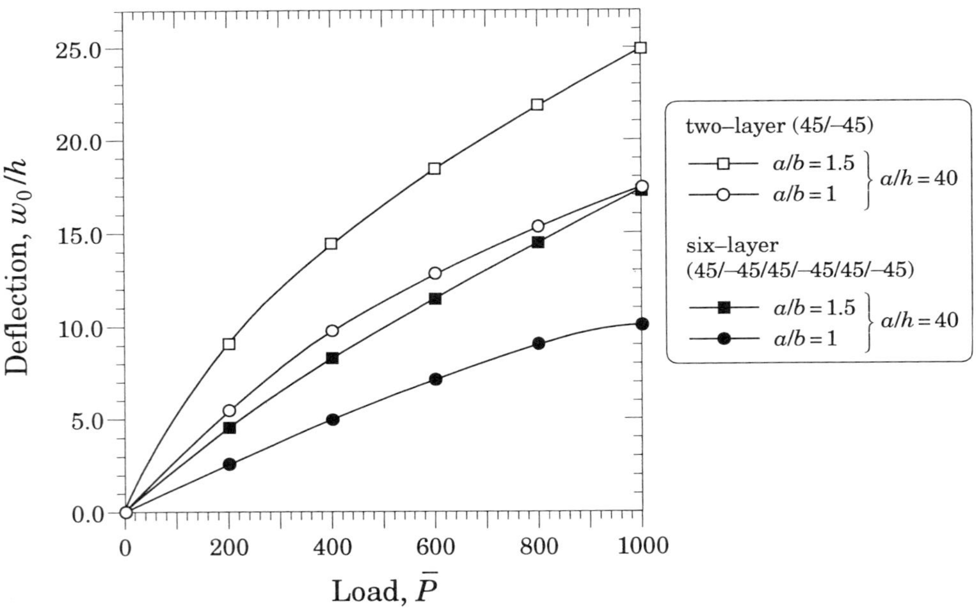

Figure 13.4–6: Load-deflection curves ($\bar{P}$ vs. w_0/h) for clamped two- and six-layer antisymmetric angle-ply ($45/–45/\cdots$) plates ($a/h = 40$) under uniformly distributed transverse load q_0.

13.4–3 Effect of Symmetry Boundary Conditions on Nonlinear Response

As discussed in Chapters 8, 9, and 10, the Navier solutions of the linear theories can be developed for antisymmetric cross-ply plates with SS–1 boundary conditions and antisymmetric angle-ply laminates with the SS–2 boundary conditions. The Navier solutions can be used to determine the conditions on deflections and forces along the lines of biaxial symmetry, *i.e.*, along the lines $x = a/2$ and $y = b/2$ of a rectangular plate of dimension $a \times b$ and with the origin of the (x, y) coordinate system being at the lower left corner of the plate. If the symmetry conditions implied by the Navier solutions are used in the linear finite element analysis of a quarter plate, as was done in Chapter 10, one obtains correct full plate solutions. When quarter-plate models with the geometric boundary conditions implied by the Navier solutions on the lines of symmetry are used in the nonlinear finite element analysis, results do not agree, in general, with those of the corresponding full-plate models (see Reddy [51]).

To illustrate this point a two-layer (45/–45) square laminate (a = 1000 mm, h = 2 mm) under uniform transverse load and the following layer properties is considered:

$$E_1 = 250 \text{ GPa}, \quad E_2 = 20 \text{ GPa}, \quad G_{12} = G_{13} = 10 \text{ GPa}$$

$$G_{23} = 4 \text{ GPa}, \quad \nu_{12} = \nu_{13} = 0.25 \qquad (13.4-10)$$

The load-deflection curves obtained from the quarter-plate and full-plate analyses are shown in Figure 13.4–7. Meshes of 2×2 and 4×4 nine-node quadratic elements based on the first-order shear deformation plate theory are used to model the quarter and full plates, respectively. The following boundary conditions along the lines of symmetry were used:

$$\begin{aligned} &\text{at } x = a/2: \quad && v_0 = \phi_x = 0\ ; \quad N_{xx} = M_{xy} = 0 \\ &\text{at } y = b/2: \quad && u_0 = \phi_y = 0\ ; \quad N_{yy} = M_{xy} = 0 \end{aligned} \qquad (13.4-11)$$

where the coordinate system is fixed at the lower left corner of the laminate.

Note that the force boundary conditions are included in the finite element model in an integral sense. It is clear from the results that the use of a quarter-plate model with the symmetry conditions (11) yields larger deflections than those obtained from the full-plate model. The discrepancy increases with the intensity of the transverse load. This discrepancy can be explained in the light of the symmetry conditions (11) used to model the quarter plate. As noted earlier, the symmetry conditions are derived from the Navier solution for the linear theory. For the angle-ply case, the assumed solution is of the form [see Eqs. (9.3–2)]

$$u_0 = \sum_{m,n=1}^{\infty} U_{mn} \sin\frac{m\pi x}{a}\cos\frac{n\pi y}{b}$$
$$v_0 = \sum_{m,n=1}^{\infty} V_{mn} \cos\frac{m\pi x}{a}\sin\frac{n\pi y}{b}$$
$$w_0 = \sum_{m,n=1}^{\infty} W_{mn} \sin\frac{m\pi x}{a}\sin\frac{n\pi y}{b}$$
$$\phi_x = \sum_{m,n=1}^{\infty} S^1_{mn} \cos\frac{m\pi x}{a}\sin\frac{n\pi y}{b}$$
$$\phi_y = \sum_{m,n=1}^{\infty} S^2_{mn} \sin\frac{m\pi x}{a}\cos\frac{n\pi y}{b} \qquad (13.4-12)$$

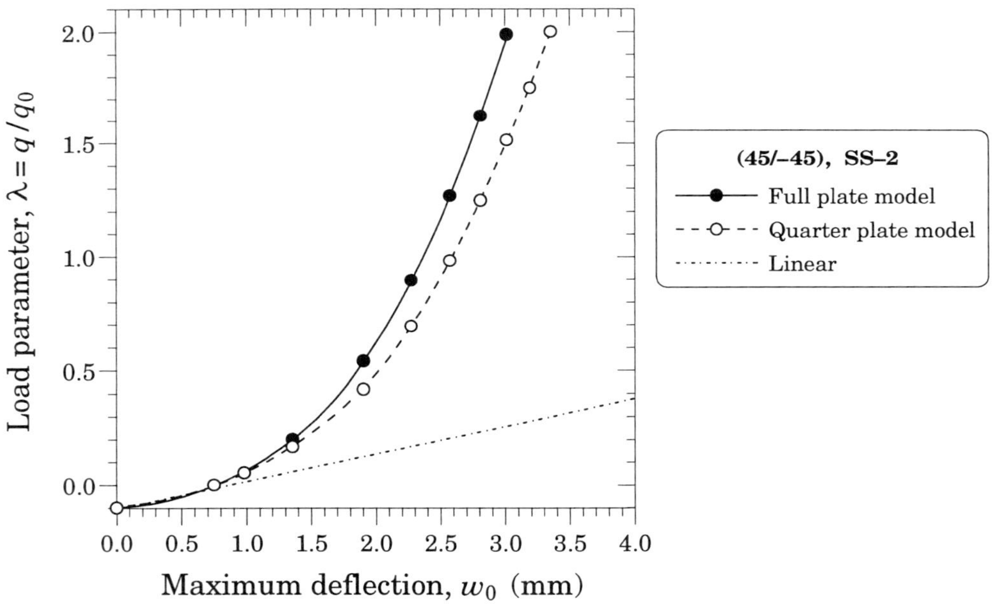

Figure 13.4–7: Load-deflection curves (λ vs. w_0) for full-plate and quarter-plate models of simply supported (SS–2) antisymmetric angle-ply (45/–45) laminates.

The resultant forces N_{xx} and N_{yy} and moment M_{xy} are given by (note that $A_{16} = A_{26} = A_{45} = 0$, $B_{11} = B_{12} = B_{22} = B_{66} = 0$, and $D_{16} = D_{26} = 0$)

$$N_{xx} = A_{11}\left[\frac{\partial u_0}{\partial x} + \frac{1}{2}\left(\frac{\partial w_0}{\partial x}\right)^2\right] + A_{12}\left[\frac{\partial v_0}{\partial y} + \left(\frac{\partial w_0}{\partial y}\right)^2\right] + B_{16}\left(\frac{\partial \phi_x}{\partial y} + \frac{\partial \phi_y}{\partial x}\right) \quad (13.4-13a)$$

$$N_{yy} = A_{12}\left[\frac{\partial u_0}{\partial x} + \frac{1}{2}\left(\frac{\partial w_0}{\partial x}\right)^2\right] + A_{22}\left[\frac{\partial v_0}{\partial y} + \left(\frac{\partial w_0}{\partial y}\right)^2\right] + B_{26}\left(\frac{\partial \phi_x}{\partial y} + \frac{\partial \phi_y}{\partial x}\right) \quad (13.4-13b)$$

$$M_{xy} = B_{16}\left[\frac{\partial u_0}{\partial x} + \frac{1}{2}\left(\frac{\partial w_0}{\partial x}\right)^2\right] + B_{26}\left[\frac{\partial v_0}{\partial y} + \left(\frac{\partial w_0}{\partial y}\right)^2\right] + D_{66}\left(\frac{\partial \phi_x}{\partial y} + \frac{\partial \phi_y}{\partial x}\right) \quad (13.4-13c)$$

The expressions in Eqs. (13a–c), for the nonlinear case, indicate that the force boundary conditions in (11) are satisfied. For example, N_{xx} and N_{yy} have the form

$$N_{xx}(a/2, y) \approx A_{12}\frac{n^2\pi^2 W_{mn}^2}{2b^2}\cos^2\frac{n\pi y}{b} \neq 0$$
$$N_{yy}(x, b/2) \approx A_{12}\frac{m^2\pi^2 W_{mn}^2}{2a^2}\cos^2\frac{m\pi x}{a} \neq 0 \quad (13.4-14)$$

When a quarter plate model is used without specifying u_0 on line $x = a/2$ and v_0 on line $y = b/2$, in the finite element analysis it is automatically taken that the natural boundary conditions $N_{xx} = 0$ on $x = a/2$ and $N_{yy} = 0$ on $y = b/2$ are specified. The quarter-plate model with zero in-plane forces $N_{xx}(a/2, y)$ and $N_{yy}(x, b/2)$ simulates the plate as more flexible than the full-plate model, in which the in-plane forces are not taken to be zero on the lines of symmetry.

For antisymmetric cross-ply laminates, a quarter-plate model with the symmetry conditions implied by the Navier solution gives the same solution as the full-plate model for both linear and nonlinear theories. This is due to the fact that the zero force boundary conditions are satisfied in an integral sense for cross-ply laminates.

13.4–4 Nonlinear Response Under In-plane Compressive Loads

Another interesting characteristic of composite laminates is their behavior under compressive loads. Most often the critical buckling loads are determined through an eigenvalue analysis. The critical buckling loads can also be determined from geometric nonlinear analysis, where the critical buckling load is taken to be the so-called limit load.

First we consider an angle-ply (45/–45) laminate with the following geometric parameters and material properties:

$$a = b = 1,000 \text{ mm}, \quad h = 2 \text{ mm}$$

$$E_1 = 40E_2, \quad E_2 = 6.25 \text{ GPa}, \quad G_{12} = G_{13} = 0.82E_2, \quad G_{23} = 0.52E_2$$

$$\nu_{12} = \nu_{13} = 0.24, \quad \nu_{23} = 0.49 \qquad (13.4-15)$$

For SS–2 type simply supported boundary conditions, the uniaxial buckling load can be determined analytically (see Chapter 9). Here we use 4×4 mesh of nine-node elements in the full plate to determine the critical buckling load, and the same mesh is used to determine the nonlinear response under applied in-plane compressive load $N_{yy} = \lambda_0 N_{yy}^0$, where λ_0 is the critical buckling load determined from the eigenvalue analysis. Figure 13.4–8 contains a plot of the maximum out-of-plane deflection w_0 (mm) versus load parameter λ ($N_{yy}^0 = 10.85$) N/m). It is clear the load-deflection curve exhibits a limit point, which is the same as the critical buckling load determined from the eigenvalue analysis.

Next we consider antisymmetric cross-ply laminates. The geometry and materials properties used are the same as those used for the angle-ply laminate. The SS–1 type simply supported boundary conditions and 2×2 mesh of nine-node elements in a quarter plate are used to determine the critical buckling loads N_0 by eigenvalue analysis and load-deflection curves in the nonlinear analysis under in-plane load $N_{yy} = \lambda_0 N_{yy}^0$ ($N_{yy}^0 = 6.25$) N/m). Figure 13.4–9 contains load-deflection curves for two-, four-, six-, and eight-layer laminates. The critical buckling loads are indicated on the load-deflection curves for comparison. Unlike the angle-ply laminates, the cross-ply laminates do not exhibit clear limit-load behavior.

13.4–5 Nonlinear Response of Antisymmetric Cross-Ply Laminated Plate Strips

Unlike isotropic metallic plates, composite plates exhibit quite different nonlinear behavior. For example, the geometric nonlinear effects could be very significant even at small loads and deflections, depending on the lamination scheme and boundary conditions (see [24,25,52]). To illustrate the point we analyze an antisymmetric cross-ply square laminate (90/0) with two opposite edges pinned ($u_0 = 0$) or hinged ($u_0 \neq 0$) and the other two edges free, and subjected to uniformly distributed transverse load.

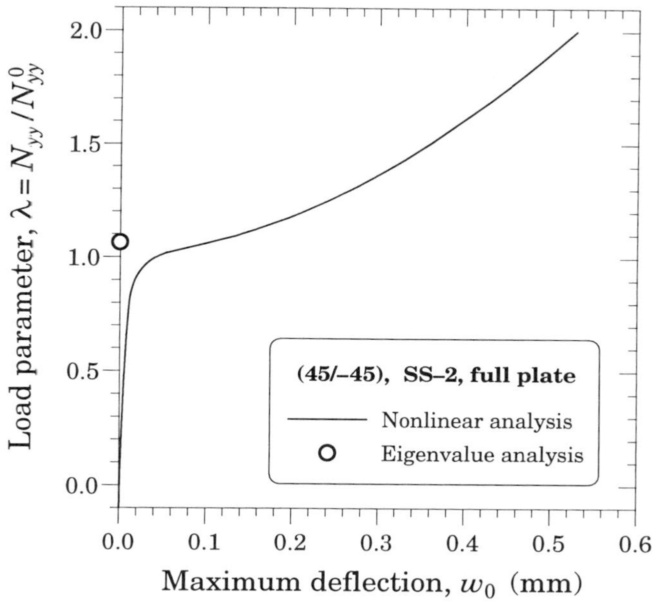

Figure 13.4–8: The load-deflection curve (λ vs. w_0) of a simply supported (SS–2) two-layer antisymmetric angle-ply (45/–45) laminate under uniformly distributed in-plane compressive edge load.

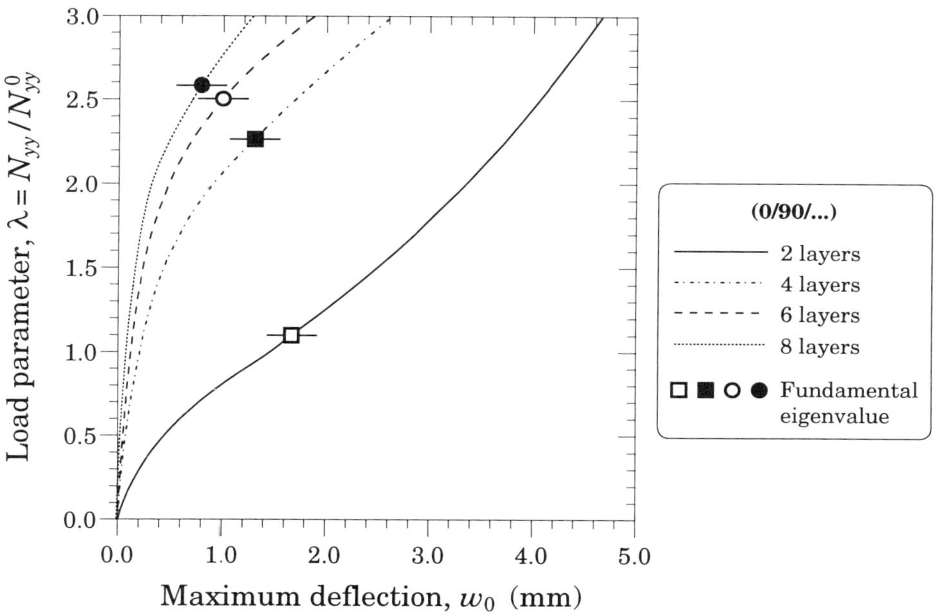

Figure 13.4–9: The load-deflection curve (λ vs. w_0) of a simply supported (SS–1) antisymmetric cross-ply ($0/90/\cdots$) laminate under uniformly distributed in-plane compressive edge load.

The geometry, finite element mesh, and boundary conditions for the pinned-pinned and hinged-hinged cases are shown in Figure 13.4–10. The material properties and geometric parameters used are

$$E_1 = 20 \text{ msi}, \; E_2 = 1.4 \text{ msi}, \; \nu_{12} = 0.3, \; G_{12} = G_{13} = G_{23} = 0.7 \text{ msi}$$

$$a = 9 \text{ in.}, \; b = 1.5 \text{ in.}, \; h = 0.4 \text{ in.} \qquad (13.4-16)$$

The results of the nonlinear analysis are presented in Table 13.4–1. For hinged–hinged boundary conditions, the plate strip is essentially in pure bending and hence the axial force $N_{xx} = 0$. Therefore, the solution is independent of the sign of the applied load. For a pinned plate strip, the axial force N_{xx} is not zero; it is

$$N_{xx} = A_{11}\left[\frac{du_0}{dx} + \frac{1}{2}\left(\frac{dw_0}{dx}\right)^2\right] + B_{11}\frac{d\phi_x}{dx}\,, \quad B_{11} < 0 \qquad (13.4-17)$$

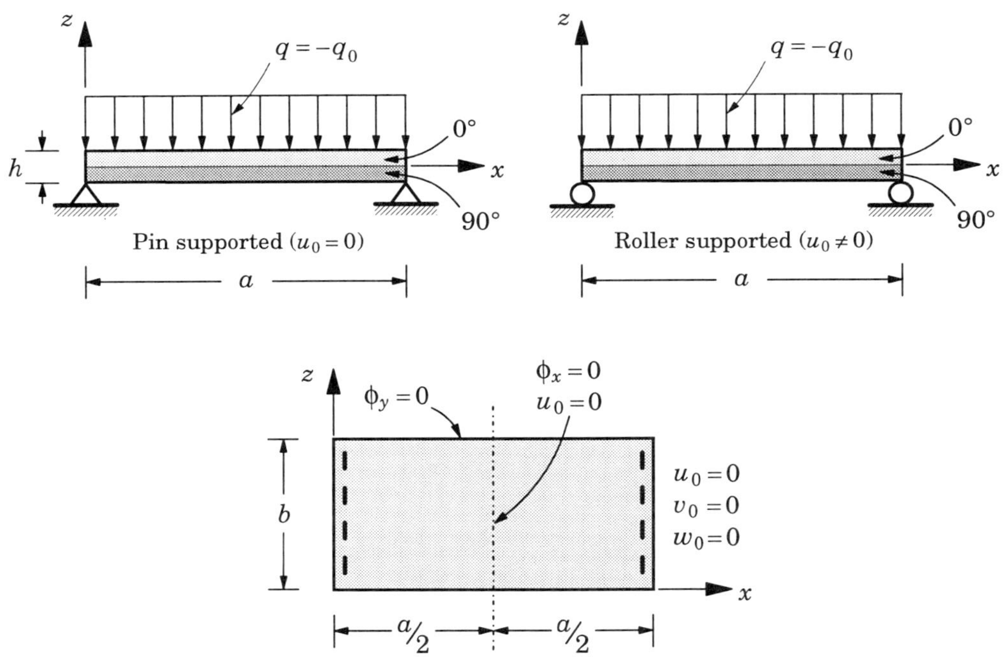

Figure 13.4–10: Geometry, loading, and boundary conditions used for cylindrical bending of a cross-ply plate strip.

Table 13.4–1: Transverse deflections, w_0/h, of cylindrical bending of a (90/0) laminate under uniformly distributed transverse load.

Load		Pinned		Hinged	
P_o	Linear*	Nonlinear*	Nonlinear†	Case 1*	Case 2†
0.005	-0.235	-0.159	0.475	-0.429	0.429
0.01	-0.470	-0.255	0.673	-0.858	0.858
0.02	-0.940	-0.386	0.847	-1.710	1,710
0.03	-1.41	-0.480	0.954	-2.550	2.55
0.04	-1.88	-0.555	1.034	-3.370	3.37
0.05	-2.35	-0.618	1.100	-4.190	4.19
0.10	-4.70	-0.845	1.327	-7.920	7.92
0.25	-11.75	-1.233	1.705	-16.16	16.17
0.50	-23.50	-1.609	2.075	-24.82	24.82
0.75	-35.25	-1.870	2.332	-30.87	30.87
1.0	-47.00	-2.078	2.532	-35.69	35.69
2.0	-94.00	-2.665	3.117	-49.56	49.56
3.0	-141.00	-3.075	3.525	-59.65	59.65
4.0	-188.00	-3.402	3.850	-68.00	68.00
5.0	-235.00	-3.675	4.125	-75.33	75.33

*For negative load values. † For positive load values.

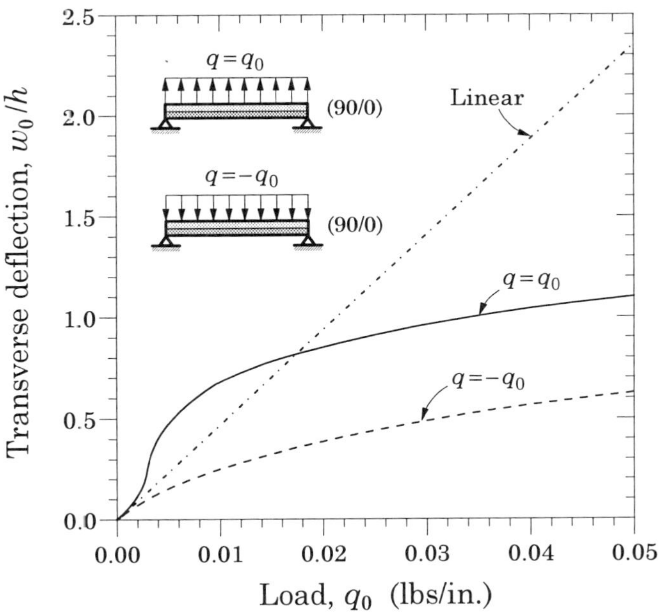

Figure 13.4–11: Load-deflection curves (q_0 vs. w_0) for cylindrical bending of a cross-ply (90/0) plate strip with pinned-pinned edges.

For small values of the positive load, the expression containing the A_{11} coefficient is small compared to the expression containing the B_{11} coefficient, which is negative for $0 < x < 4.5$. Hence, N_{xx} is compressive and increases the transverse deflection — analogous to the transverse deflection of a plate strip under an axial compressive load and a transverse load. Thus, the nonlinear solution is larger than the linear solution for small values of the load. As the load is increased, the A_{11} expression becomes larger than the B_{11} expression, and N_{xx} becomes positive. This stiffens the structure and the nonlinear solution becomes smaller than the linear solution. The load deflection curves for the first few load steps are shown in Figure 13.4–11. For a negative load, $B_{11}\frac{d\phi_x}{dx}$ is positive, and the two terms in N_{xx} add up; this yields a larger axial force and therefore a stiffer structure than for the positive load case. Therefore the deflection is lower than that for the case of positive load.

13.4–6 Nonlinear Transient Analysis of Composite Plates

Nonlinear transient response of laminated composite plates was reported by Reddy [20]. Here we present a few examples from this paper. For additional examples, the reader may consult Reference 20. In all examples discussed here, suddenly applied transverse step loads are considered, and the initial conditions are taken to be zero. In the nonlinear transient analysis, there are three loops. The iterative loop for convergence of the solution is the innermost, followed by loops on time increments and the load increments.

The first example involves a simply supported isotropic square plate subjected to suddenly applied uniformly distributed transverse load. The following geometric, material, and load parameters were used:

$$a = b = 243.8 \text{ cm}, \quad h = 0.635 \text{ cm}, \quad \rho = 2.547 \times 10^{-6} \text{ N-s}^2/\text{cm}^4$$
$$E_1 = E_2 = 7.031 \times 10^5 \text{ N/cm}^2, \quad \nu_{12} = 0.25$$
$$q_0 = 4.882 \times 10^{-4} \text{ N/cm}^2, \quad \Delta t = 0.005 \text{ s} \qquad (13.4-18)$$

Figure 13.4–12 shows the center deflection w_0 as a function of time for four different values of the load and two different time steps. The amplitude increases while the period of response decreases with an increase in the intensity of load. Load versus the maximum deflection is also shown in the figure. This problem may serve as a reference for verification of the geometric nonlinear option of a finite element program.

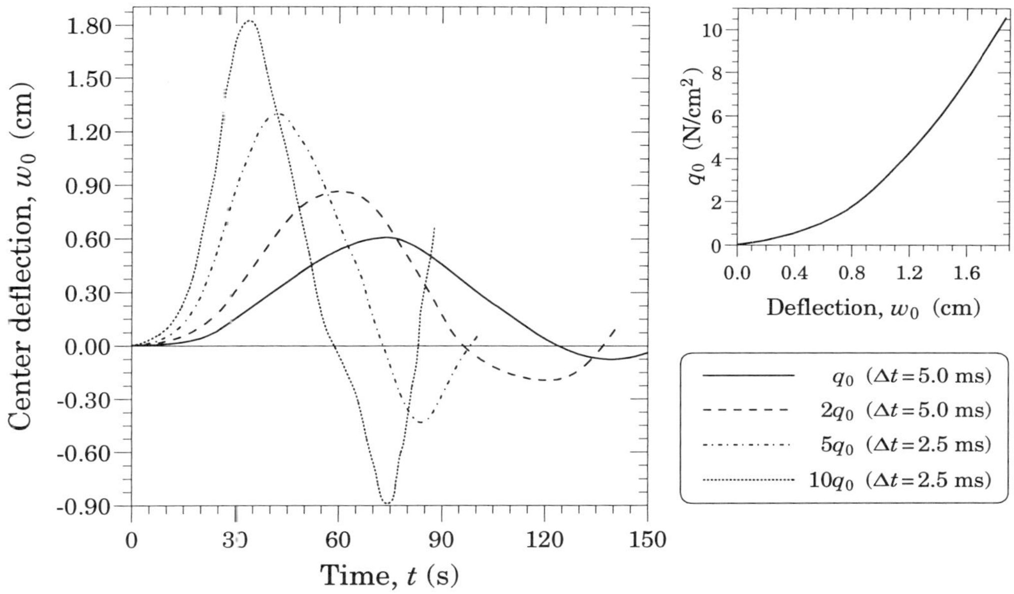

Figure 13.4–12: Center deflection versus time for nonlinear transient analysis of an isotropic, simply supported, square plate.

Next we consider nonlinear transient analysis of simply supported cross-ply (0/90) and angle-ply (45/-45) plates. Figure 13.4–13 contains plots of center deflections ($\bar{w} = w_0 \times 10^3$ and $\hat{w} = w_0 \times 10^2$) of simply supported (SS–1) cross-ply laminated rectangular plate ($a = \sqrt{2}, b = 1, h = 0.2$, $\rho = 1.0$, in English units of inches and pounds) under uniformly distributed transverse patch loading of intensity $q_0 = 10^{-2}$ over the area $0 \leq (x, y) \leq 0.2$. The material properties were assumed to be

$$E_1 = 25E_2, \quad G_{12} = G_{13} = 0.5E_2, \quad G_{23} = 0.2E_2, \quad \nu_{12} = 0.25 \qquad (13.4-19)$$

A time step of $\Delta t = 0.1$ was used. A nonuniform 4×4 mesh of nine-node quadratic elements in quarter plate was used. Figure 13.4–13 contains results of both five and three degrees of freedom models. The three degrees of freedom element models the plate stiffer than the five degrees of freedom element.

Figure 13.4–14 contains plots of center deflection versus time for simply supported (SS–2), square ($a = b = 243.8$ cm, $h = 0.635$ cm, $\rho = 2.547 \times 10^{-6}\text{Ns}^2/\text{cm}^4$), angle-ply (45/–45) plates under uniformly distributed pressure loading ($q_0 = 50 \times 10^{-4}\text{N/cm}^2$). The material properties in Eq. (19) were used. The effect of geometric nonlinearity is obvious from the figure.

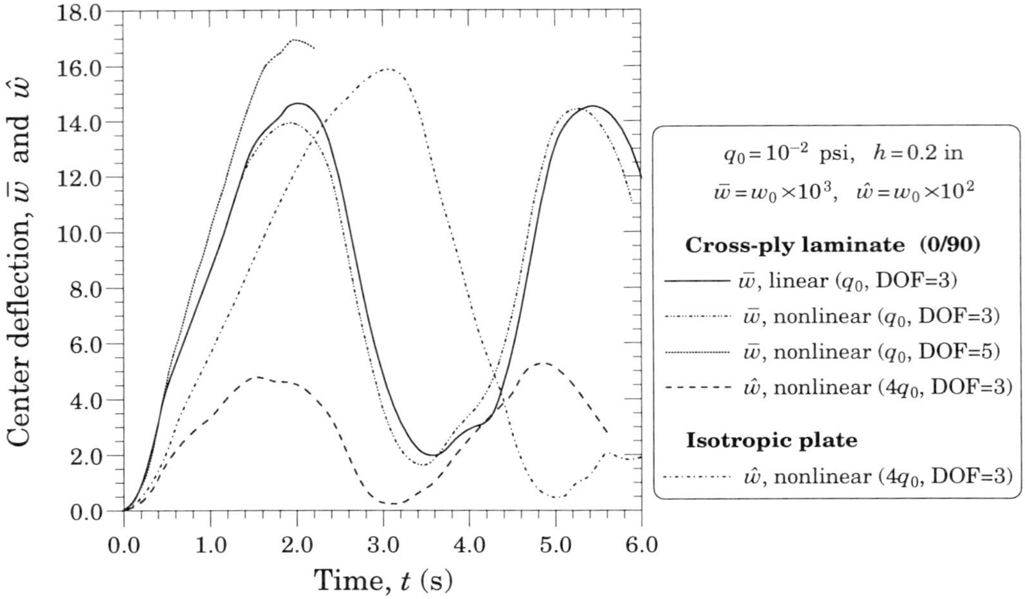

Figure 13.4–13: Center deflection versus time for nonlinear transient analysis of a simply supported (SS–1) cross-ply (0/90) plate.

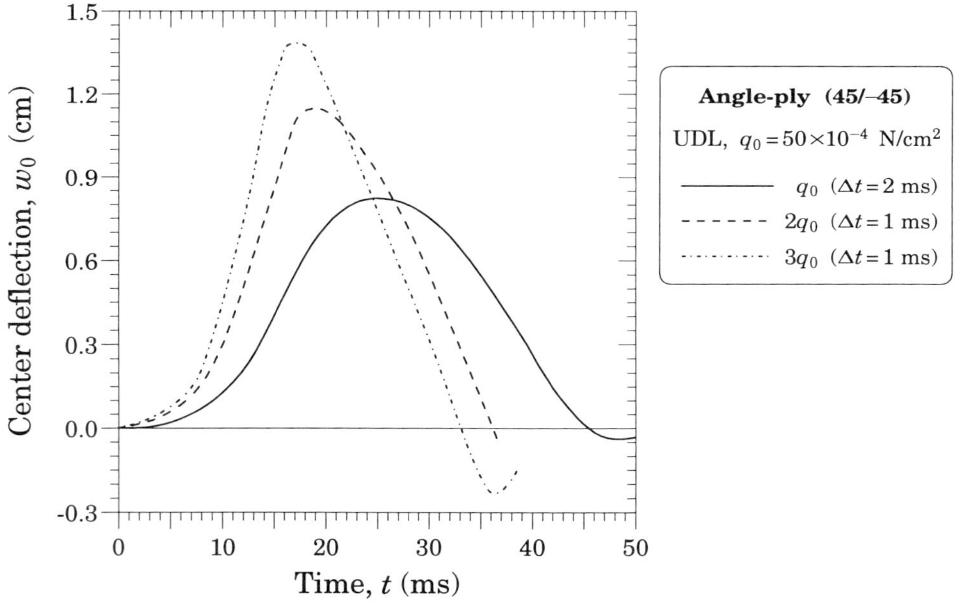

Figure 13.4–14: Center deflection versus time for nonlinear transient analysis of a simply supported (SS–2) angle-ply (45/–45) square plate under suddenly applied uniformly distributed transverse load.

13.4–7 Postbuckling Response and Progressive Failure of Composite Panels in Compression

Preliminary Comments

The classical lamination theory, in which the transverse shear effects are neglected, is often used to analyze laminated composite structures. Because of low moduli and strengths in transverse directions compared to that of in-plane directions, composite laminates may fail due to transverse stresses. Indeed, it is found that (see [54,55]) composite laminates loaded in compression fail due to high interlaminar stresses. Therefore, shear deformable plate and shell elements are needed to provide information regarding the through-thickness strength of composite structures. Insight gained by using these elements may aid in the characterization of failure modes of composite panels.

Here we discuss a study by Engelstad, Reddy, and Knight [55] of the postbuckling response of several graphite–epoxy panels loaded in axial compression. The study makes comparisons between the experimentally obtained and analytically determined postbuckling response of composite panels.

Experimental Study

The postbuckling and failure characteristics of flat, rectangular graphite–epoxy panels, with and without holes, and loaded in axial compression have been examined in an experimental study by Starnes and Rouse [54]. The panels were fabricated from commercially available unidirectional Thornel 300 graphite–fiber tapes preimpregnated with 450°K cure Narmco 5208 thermosetting epoxy resin. Typical lamina properties for this graphite–epoxy system are

$$E_1 = 131.0 \text{ GPa } (19,000 \text{ ksi}), \quad E_2 = 13.0 \text{ GPa } (1,890 \text{ ksi})$$

$$G_{12} = 6.4 \text{ GPa } (930 \text{ ksi}), \quad \nu_{12} = 0.38, \ h_k = 0.14 \text{ mm } (0.0055 \text{ in.}) \qquad (13.4-20)$$

where h_k denotes ply thickness. Each panel was loaded in axial compression using a 1.33 MN (300 kips) capacity hydraulic testing machine. The loaded ends of the panels were clamped by fixtures during testing and the unloaded edges were simply supported by knife-edge restraints to prevent the panels from buckling as wide columns. Most panels exhibited postbuckling strength and failed along a nodal line of the buckling mode in a transverse shear failure mode (see [54]). However, a different failure mode was observed for some of the 24-ply panels with holes. These panels failed along a transverse line passing through the hole, and failed soon after buckling.

Here we analyze two panels, denoted C4 and H4 in Reference 54 (see Figure 13.4–15). The finite element results are compared with the experimental results of Reference 54. Panel C4 is 50.8 cm by 17.8 cm (20.0 in. long and 7.0 in. wide), 24-ply laminate, $[\pm 45/0_2/\pm 45/0_2/\pm 45/0/90]_s$ (orthotropic). Panel C4 was observed in the test to buckle into two longitudinal half-waves and one transverse half-wave. The second panel, Panel H4, is a 50.8 cm by 14.0 cm (20.0 in. long by 5.5 in. wide) 24-ply laminate $[\pm 45/0/90]_{3s}$ (quasi-isotropic). A 1.91 cm diameter (0.75 in. diameter) hole is located 19.1 cm (7.5 in.) from one of the loaded edges and along the panel centerline. Panel H4 was observed in the test to buckle into four longitudinal half-waves and one transverse half-wave with the hole located near the buckle crest of the second longitudinal half-wave.

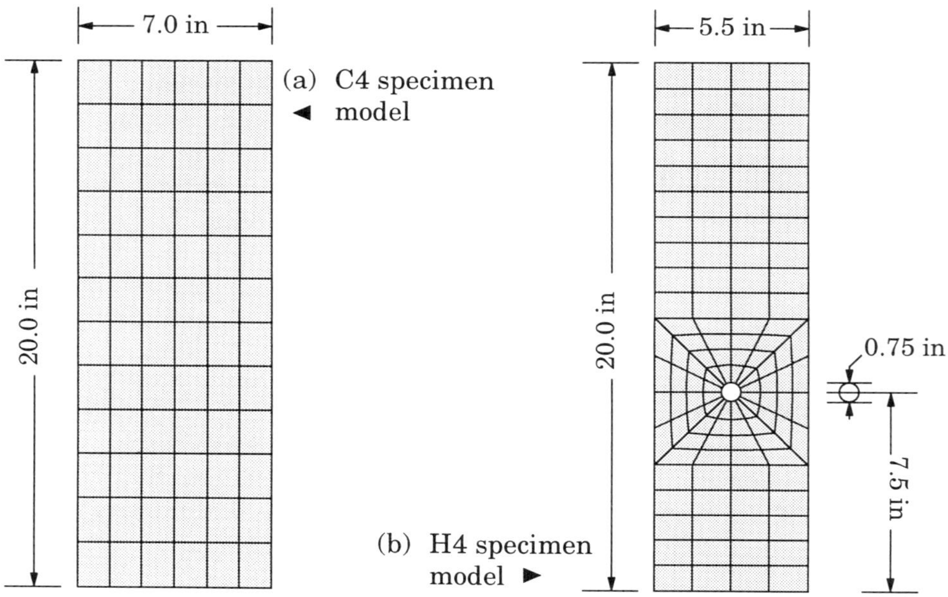

Figure 13.4–15: Geometry and finite elements meshes of the C4 and H4 composite panels used in the postbuckling study.

Finite Element Models

Finite element models of these panels were developed in Reference 55 using nine-node quadratic shell elements of Chao and Reddy [21], which are based on continuum formulation of a laminated 3-D shell element, denoted 9CR. The incremental equations of a continuous medium are formulated using the principle of virtual displacements and the total Lagrangian description. The continuum shell element is obtained from the three-dimensional solid element by imposing the same two constraints as in FSDT:

(1) straight lines normal to the midsurface before deformation remain straight, but not normal, after deformation; (2) the transverse normal components of strain, and hence stress, are ignored in the development. These assumptions reduce a 3-D theory into a 2-D theory, and the resulting shell element is called a *3-D degenerated shell element.* The nonlinear formulation admits large displacements and rotations of the shell element and small strains, since the thickness does not change and the normal does not distort. The final incremental equations of equilibrium for an element are of the form

$$([K_L] + [K_{NL}])\{\delta\Delta\} = \{R\} - \{F\} \tag{13.4 – 21a}$$

where $\{\delta\Delta\}$ is the vector of incremental nodal displacements, $([K_L], [K_{NL}])$ are the linear and nonlinear parts of the stiffness matrix, and $\{F\}$ is the force vector:

$$\begin{aligned} [K_L] &= \int_{\Omega^e} [B_L]^T [C][B_L]\, dv \\ [K_{NL}] &= \int_{\Omega^e} [B_{NL}][S][B_{NL}]\, dv \\ \{F\} &= \int_{\Omega^e} [B_L]\{\hat{S}\}\, dv \end{aligned} \tag{13.4 – 21b}$$

In these equations, $[B_L]$ and $[B_{NL}]$ are linear and nonlinear strain–displacement transformation matrices, $[C]$ is the constitutive elasticity matrix, $[S]$ and $\{\hat{S}\}$ are the matrix and vector of second Piola–Kirchhoff stresses, and $\{R\}$ is the external load vector. All matrix elements refer to the deformed state and are measured with respect to the original undeformed configuration.

To evaluate the integrals in Eq. (21) (*e.g.*, $[K_L]$, $[K_{NL}]$, and $\{F\}$), we use Gauss quadrature in the surface directions of the shell, but explicit integration in the thickness direction. Thus the thickness direction integration for matrices $[K_L]$ and $[K_{NL}]$ gives the following laminate stiffnesses:

$$\begin{aligned} [A] &= \sum_{k=1}^{P} [C']_k (\zeta_{k+1} - \zeta_k) \\ [B] &= \frac{1}{2} \sum_{k=1}^{P} [C']_k (\zeta_{k+1}^2 - \zeta_k^2) \\ [D] &= \frac{1}{3} \sum_{k=1}^{P} [C']_k (\zeta_{k+1}^3 - \zeta_k^3) \end{aligned} \tag{13.4 – 22}$$

Here ζ_k is the thickness coordinate of the bottom of the kth lamina, P is the number of laminae, $[C']_k$ is the constitutive matrix for the kth lamina in the principal material coordinates, which has the form

$$[C'] = f([\bar{Q}], \theta) \qquad (13.4-23)$$

where $\bar{Q}_{ij}$ are the plane stress-reduced elastic coefficients in the material coordinates and θ is the fiber orientation angle. For an orthotropic lamina, $\bar{Q}_{ij}$ can be expressed as

$$\bar{Q}_{11} = \frac{E_1}{1-\nu_{12}\nu_{21}}, \quad \bar{Q}_{12} = \frac{\nu_{12}E_2}{1-\nu_{12}\nu_{21}}, \quad \bar{Q}_{22} = \frac{E_2}{1-\nu_{12}\nu_{21}}$$

$$\bar{Q}_{44} = G_{13}K, \quad \bar{Q}_{55} = G_{23}K, \quad \bar{Q}_{66} = G_{12} \qquad (13.4-24)$$

where E_1, ν_{12}, ν_{21}, E_2, G_{13}, G_{23}, and G_{12} are the lamina material properties and K is the shear correction factor, which is taken to be 5/6.

The finite element model used in Reference 55 consisted of six elements per buckle half-wave in each direction. Hence, the finite element model of Panel C4 consists of 12 nine-node quadrilateral elements along the panel length. Figure 13.4–15a shows the model used for the C4 specimen. The finite element model of Panel H4 is different, due to the presence of the hole. This model has four "rings" of elements around the hole with each ring subdivided into 16 elements, as shown in Figure 13.4–15b. The total numbers of nine–node quadrilateral elements in the finite element models of Panels C4 and H4 are 72 and 124, respectively.

In order to proceed beyond the critical buckling point in the analysis of each panel, an initial geometric imperfection, typically the same shape as the first linear buckling mode, was assumed in the finite element analysis. The amplitude of each mode was selected to be 1-5% of the total laminate thickness. This allows efficient progress past the critical buckling point, but does not affect the results in the postbuckling range.

Failure Analysis

The maximum stress and Tsai–Wu failure criteria are used (see [13,14,56–59]). In the *maximum stress criterion*, failure is assumed to occur if any one of the following conditions are satisfied:

$$\sigma_1 \geq X_T, \quad \sigma_4 \geq R, \quad \sigma_2 \geq Y_T, \quad \sigma_5 \geq S, \quad \sigma_3 \geq Z_T, \quad \sigma_6 \geq T \qquad (13.4-25)$$

where (σ_1, σ_2, σ_3) are the normal stress components, (σ_4, σ_5, σ_6) are shear stress components, (X_T, Y_T, Z_T) are the lamina normal strengths in tension (T) along the (1, 2, 3) directions, and (R, S, T) are the shear strengths in the (23, 13, 12) planes, respectively. When (σ_1, σ_2, σ_3) are compressive, they should be compared with (X_C, Y_C, Z_C), which are normal strengths in compression (C) along the (1, 2, 3) principal material directions, respectively.

The *Tsai–Wu criterion* is given by

$$\mathcal{F} \equiv \sum_{i=1}^{6} F_i\sigma_i + \sum_{i=1}^{6}\sum_{j=1}^{6} F_{ij}\sigma_i\sigma_j \geq 1 \tag{13.4-26}$$

$$\begin{aligned}
F_1 &= \frac{1}{X_T} - \frac{1}{X_C}\,, \quad F_2 = \frac{1}{Y_T} - \frac{1}{Y_C}\,, \quad F_3 = \frac{1}{Z_T} - \frac{1}{Z_C} \\
F_{11} &= \frac{1}{X_T X_C}\,, \quad F_{22} = \frac{1}{Y_T Y_C}\,, \quad F_{33} = \frac{1}{Z_T Z_C} \\
F_{44} &= \frac{1}{R^2}\,, \quad F_{55} = \frac{1}{S^2}\,, \quad F_{66} = \frac{1}{T^2}\,, \quad F_{12} = -\frac{1}{2}\frac{1}{\sqrt{X_T X_C Y_T Y_C}} \\
F_{13} &= -\frac{1}{2}\frac{1}{\sqrt{X_T X_C Z_T Z_C}}, \quad F_{23} = -\frac{1}{2}\frac{1}{\sqrt{Y_T Y_C Z_T Z_C}}
\end{aligned} \tag{13.4-27}$$

where σ_i denote the stress components referred to the principal material coordinates.

In reality, laminate failure occurs due to propagation of damage as the load is increased. To model this effect, a progressive failure approach is used in the nonlinear finite element analysis. At each load step, Gauss point stresses are used in the selected failure criterion. If failure occurred at a Gauss point, a modification of the lamina properties was made at that Gauss point, which results in reduced stiffnesses $[A], [B]$, and $[D]$ of the laminate [see Eq. (22)]. For example, for the maximum stress criterion, if the σ_1 stress exceeds the longitudinal tensile strength X_T, then the longitudinal modulus E_1 at that point is reduced to zero. For the Tsai–Wu criterion, if failure occurs, then the following expressions are used to determine the failure mode:

$$\begin{aligned}
H_1 &= F_1\sigma_1 + F_{11}\sigma_1^2, \quad H_2 = F_2\sigma_2 + F_{22}\sigma_2^2 \\
H_4 &= F_{44}\sigma_4^2, \quad H_5 = F_{55}\sigma_5^2, \quad H_6 = F_{66}\sigma_6^2
\end{aligned} \tag{13.4-28}$$

The largest H_i term is selected as the dominant failure mode and the corresponding modulus is reduced to zero. Thus H_1 corresponds to the modulus E_1, H_2 to E_2, H_4 to G_{23}, H_5 to G_{13}, and H_6 to G_{23}. As a consequence of this reduction, engineering material properties are updated as failure progresses. An outline of the steps used in the analysis is given below.

1. After nonlinear iterative displacement convergence is achieved, calculate stresses in the global (x, y, z) coordinates at the middle of each layer at each Gauss point.

2. Transform the stresses to the principal material coordinates.

3. Compute the failure index, $\mathcal{F}$.

4. If failure occurs (*i.e.*, $\mathcal{F} \geq 1$), (a) identify the maximum H_i, (b) reduce the appropriate lamina moduli at that Gauss point, and (c) recompute laminate stiffnesses and restart the nonlinear analysis at the same load step (*i.e.*, return to Step 1).

5. If no failure occurs, proceed to the next load step.

The end shortening of the panel is monitored as in a compression test. The failure load is defined to be that load for which the panel undergoes large end shortening for small increments of load.

Results for Panel C4

Comparison between test results from Reference 54 and finite element results from Reference 55 for Panel C4 are shown in Figure 13.4–16. The figure shows end shortening u_0, normalized by the analytical end shortening u_{cr} at buckling (Figure 13.4–16a); out-of-plane deflection w_0 near a point of maximum deflection, normalized by the panel thickness h (Figure 13.4–16b); and the longitudinal surface strains e near a point of maximum out-of-plane deflection, normalized by the analytical buckling strain e_{cr}. These are all shown as functions of the applied load P, normalized by the theoretical buckling load P_{cr}. The circles in the figure represent test data, and the curves denote the results of the nonlinear finite element analyses. These experimental and finite element results agree well up to failure of the panel. The postbuckling response exhibits large out-of-plane deflections (nearly three times the panel thickness; see Figure 13.4–16b) and high longitudinal strains from front and back surfaces (nearly three times the analytical buckling strain; see Figure 13.4–16c).

Figure 13.4–17a contains a contour plot of the out-of-plane deflections generated from the finite element analysis at an applied load of $2.1P_{cr}$. Figure 13.4–17b contains a photograph of the Moiré fringe pattern from Reference 54 corresponding to the out-of-plane deflections observed during the testing of Panel C4 at the same load. These results indicate that the out-of-plane deflections from both test and analysis have the same pattern over the entire panel. Both patterns indicate two longitudinal half-waves with a buckling-mode nodal line at panel midlength. Stress distributions in each layer of the laminate were calculated using the nonlinear finite element results in order to determine the failure loads. The stresses were determined using the constitutive relations for both the in-plane and transverse components. In addition, the transverse shear stress distributions were also obtained by integrating the equilibrium equations, wherein the in-plane stresses were computed using the constitutive relations.

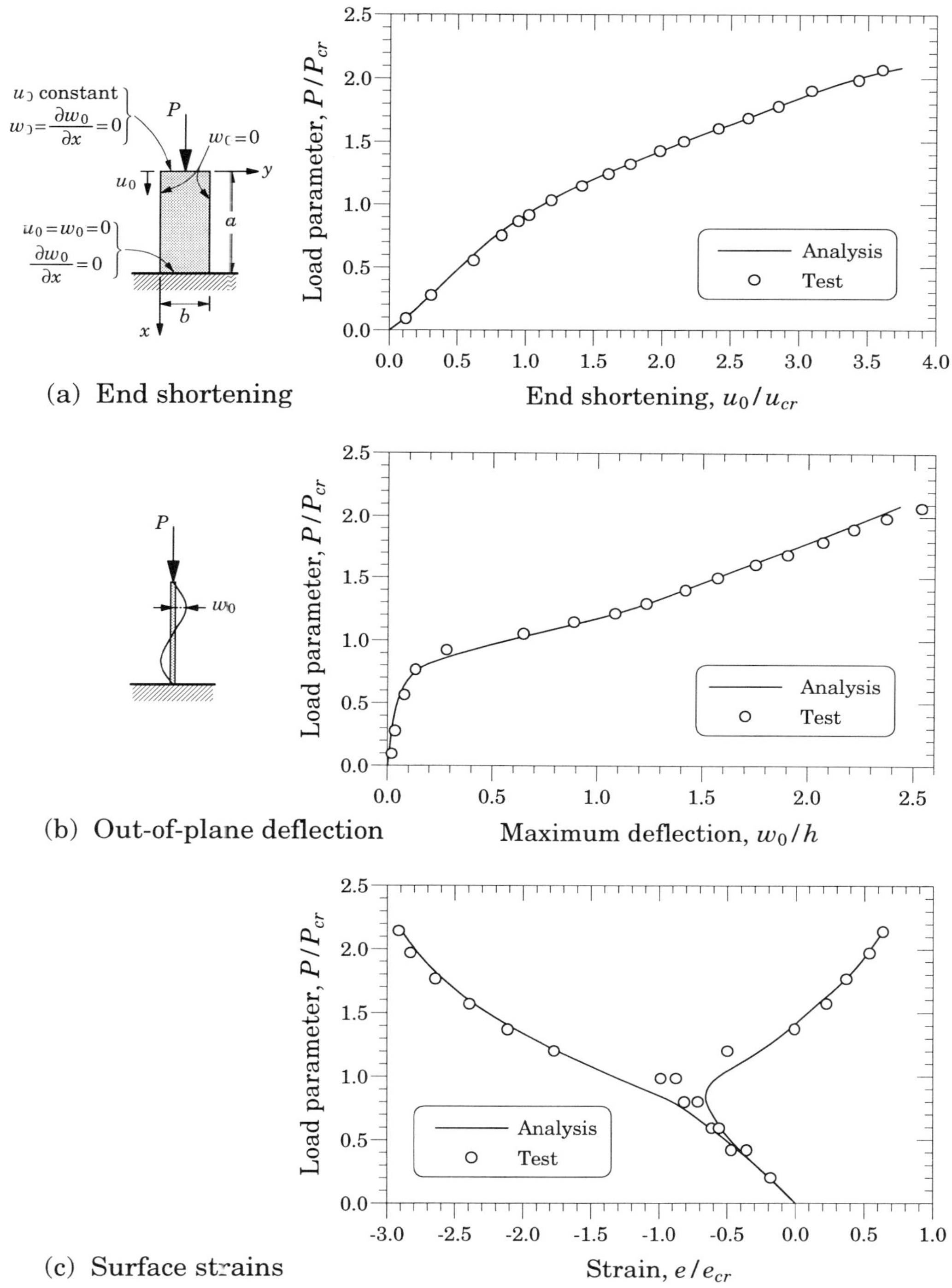

Figure 13.4–16: Postbuckling response characteristics of panel C4.

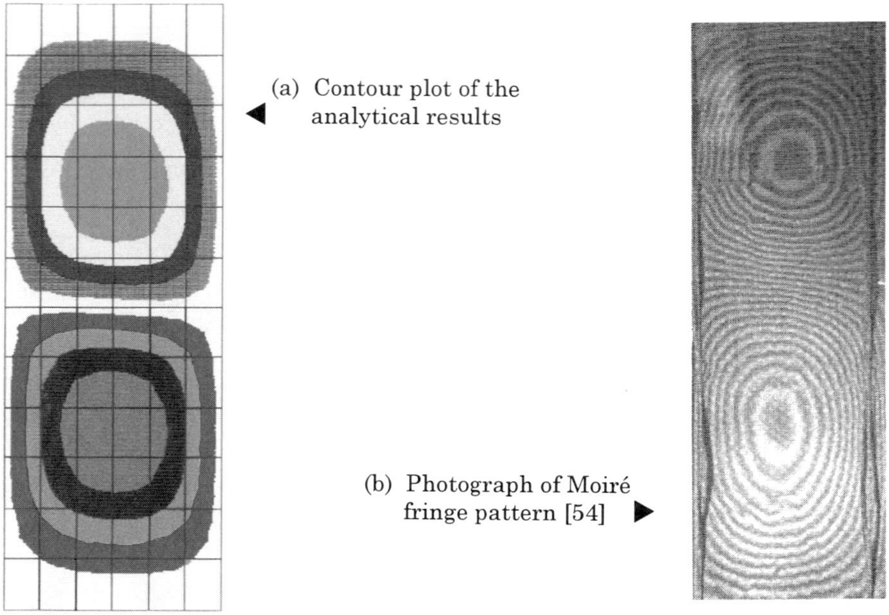

Figure 13.4–17: Comparison of experimental (Moiré) and analytical out-of-plane deflection patterns for panel C4.

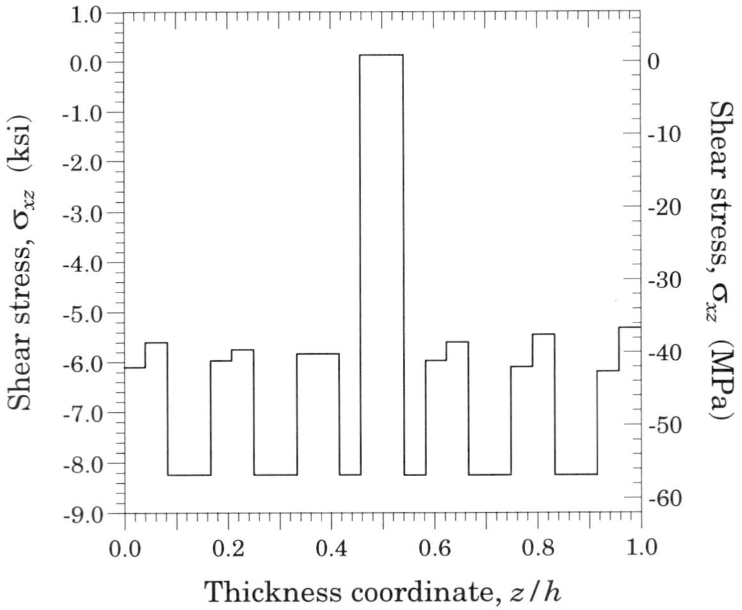

Figure 13.4–18: Transverse shear stress, σ_{xz}, distribution through the thickness of panel C4.

Figure 13.4–18 shows the distribution of the peak σ_{xz} stress through the thickness direction z, normalized by the laminate thickness h for $P = 2.1\ P_{cr}$. It is clear that the 0° layers carry the highest transverse shear load. Figure 13.4–19 contains the distribution of the axial stress σ_{xx} in the third layer of the laminate (a 0° ply) at panel midlength for three values of the applied load. At the buckling load, the axial stress is nearly uniform across the panel. Although the axial stress is large, the values are well below the material allowable values in tension $X_T = 1400$ MPa (203 ksi) and compression $X_C = 1138$ MPa (165 ksi). The contour plot of axial stress σ_{xx} over the entire panel in this 0° ply for an applied load of $2.1P_{cr}$ indicates (not shown here; see References 13 and 55) that high compressive axial stresses occur along the longitudinal edges of the panel.

Figure 13.4–20 shows the distribution of the transverse shear stress in the third layer of the laminate (a 0° ply) at panel midlength for three values of the applied load. The solid curves represent the transverse shear stress distributions obtained using the constitutive relations, and the dashed curves denote the transverse shearing stress distributions obtained from the equilibrium equations. Both methods give very similar results. At the buckling load, the peak transverse shear stress occurs near the center of the panel. After buckling, the transverse shear stresses σ_{xz} redistribute towards the edges of the panel. The peak values of the transverse shear stress σ_{xz} approach the material allowable value of $S = T = 62$ MPa (9 ksi) for $P = 2.1P_{cr}$, indicating the panel failure due to transverse shear stress. A contour plot (not shown here) of the distribution of the transverse shear stress σ_{xz} over the entire panel in this 0° ply for an applied load of $P = 2.1P_{cr}$ indicates that high transverse shear stresses occur along the buckling-mode nodal line.

This failure mode can be further explained through a close examination of the Green–Lagrange strain component

$$\varepsilon_{xz} = \frac{1}{2}\left(\frac{\partial u_1}{\partial z} + \frac{\partial u_3}{\partial x} + \frac{\partial u_1}{\partial x}\frac{\partial u_1}{\partial z} + \frac{\partial u_2}{\partial x}\frac{\partial u_2}{\partial z} + \frac{\partial u_3}{\partial x}\frac{\partial u_3}{\partial z}\right) \qquad (13.4-29)$$

in conjunction with the displacement field of the first-order shear deformation theory

$$\begin{aligned} u_1(x,y,z) &= u_0(x,y) + z\phi_x \\ u_2(x,y,z) &= v_0(x,y) + z\phi_y \\ u_3(x,y,z) &= w_0(x,y) \end{aligned} \qquad (13.4-30)$$

Substituting of the displacements from Eq. (30) into the strain in Eq. (29) and noting that ϕ_y is zero along a buckling-mode nodal line, we obtain

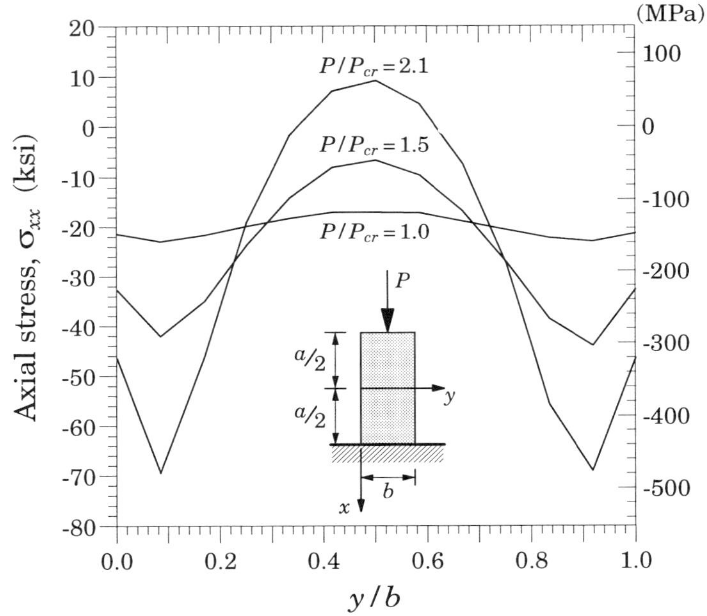

Figure 13.4–19: Axial stress, σ_{xx}, distributions in the third layer from the surface (0° ply) of panel C4.

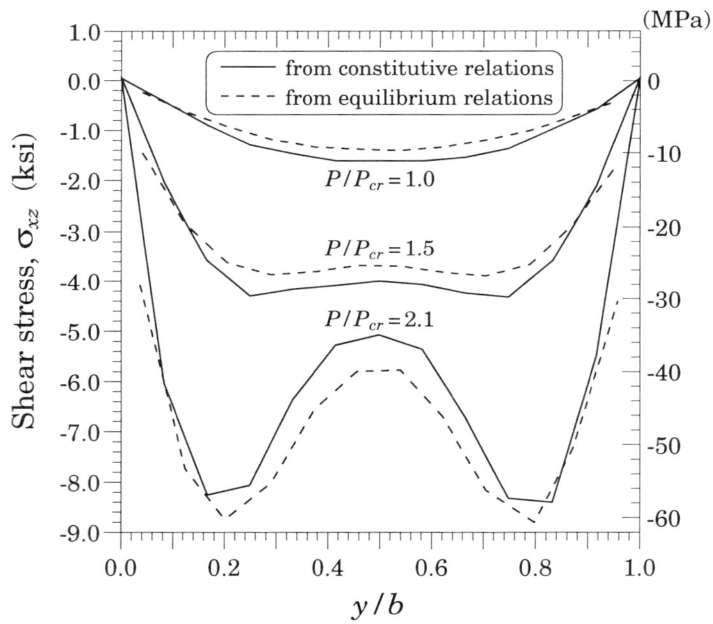

Figure 13.4–20: Transverse shear stress, σ_{xz}, distributions in the third layer from the surface (0° ply) of panel C4.

$$\varepsilon_{xz} = \frac{1}{2}\left(\phi_x + \frac{\partial w_0}{\partial x} + \frac{\partial u_0}{\partial x}\phi_x\right) \qquad (13.4-31)$$

The quantity $\frac{\partial w_0}{\partial x}$ (out-of-plane deflection gradient) is largest along a buckling-mode nodal line and the quantity $\frac{\partial u_0}{\partial x}$ (related to the membrane strain) is largest along the panel edges. A similar examination of the other transverse shearing strain ε_{yz} leads to the conclusion that the transverse shearing strain ε_{xz} is the dominant one.

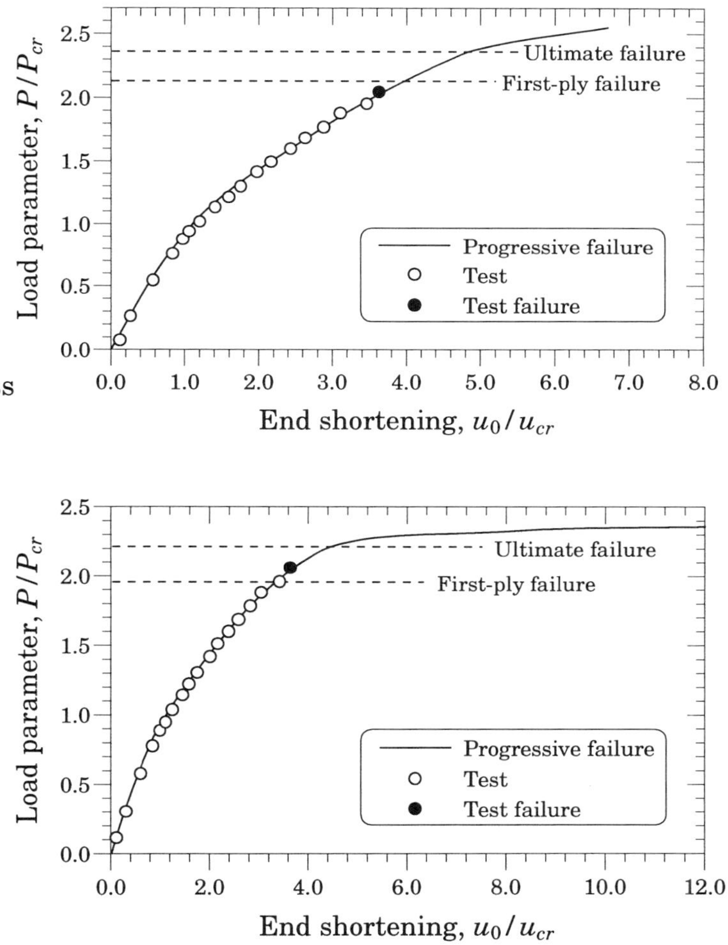

(a) Maximum stress failure criterion

(b) Tsai–Wu failure criterion

Figure 13.4–21: Progressive failure results of panel C4.

Figures 13.4–21a and 13.4–21b present the progressive failure results for Panel C4, using the maximum stress and Tsai–Wu failure criteria, respectively. In addition to the strengths already mentioned, the other allowables used are transverse tensile strength, $X_T = 80.9$ MPa (11.7 ksi); transverse compressive strength, $X_C = 189.0$ MPa (27.4 ksi); and in-plane shear strength, $T = 69.0$ MPa (10.0 ksi). At some point in the analysis a dramatic change in slope indicates an inability of the panel to support additional load. This location is identified as the failure load. Figures 13.4–21a and 13.4–21b show that the Tsai–Wu criterion estimates the experimental failure more closely than the maximum stress criterion. This is attributed to the presence of stress interaction terms in the Tsai–Wu criterion failure index.

Results for Panel H4

Panel H4 was analyzed to investigate deformation and failure of a panel with a hole. An imperfection of 1% of panel thickness times mode 1 was used to proceed into the postbuckling range. Figure 13.4–22a contains a comparison of end shortening obtained numerically and experimentally. Figures 13.4–22b and 13.4–22c show the longitudinal surface strains e (both top and bottom surfaces) across the panel at the hole for a load of $0.90P_{cr}$ and $1.39P_{cr}$, respectively. These results are in good agreement with experimental results from Reference 54.

It should be noted that if uniformly reduced or selective reduced integration were used in the analysis of this panel, it would predict spurious modes. These occur because of zero energy modes, and lack of restraint of the model around the hole. Element distortion around the hole could be another contributing factor. It is necessary to use full integration to alleviate this problem, and the mesh should be sufficiently refined so that element locking effects are negligible.

For panel H4, once again, a transverse shear mechanism develops along nodal lines away from the hole. However, the peak stress only approaches 48.3 MPa (7.0 ksi) at the experimental failure load. At this load the in-plane shear stress approaches its allowable around the hole. Simultaneous first-ply failure occurs due to σ_{xy} and σ_{xx} components around the hole edge. Thus the failure mode is not a dominant transverse shear mode as for Panels C4 and C10, but a more complex interacting mode with a dominant in-plane shear component. Progressive failure results are shown in Figures 13.4–23a and 13.4–23b. The Tsai–Wu criterion is, once again, in better agreement with experimental results.

This completes the discussion of the postbuckling behavior of laminated panels. References 13 and 55 contain additional results of the study. Results of nonlinear finite element analysis of laminated plates and shells can also be found in References 15, 17–25, and 42.

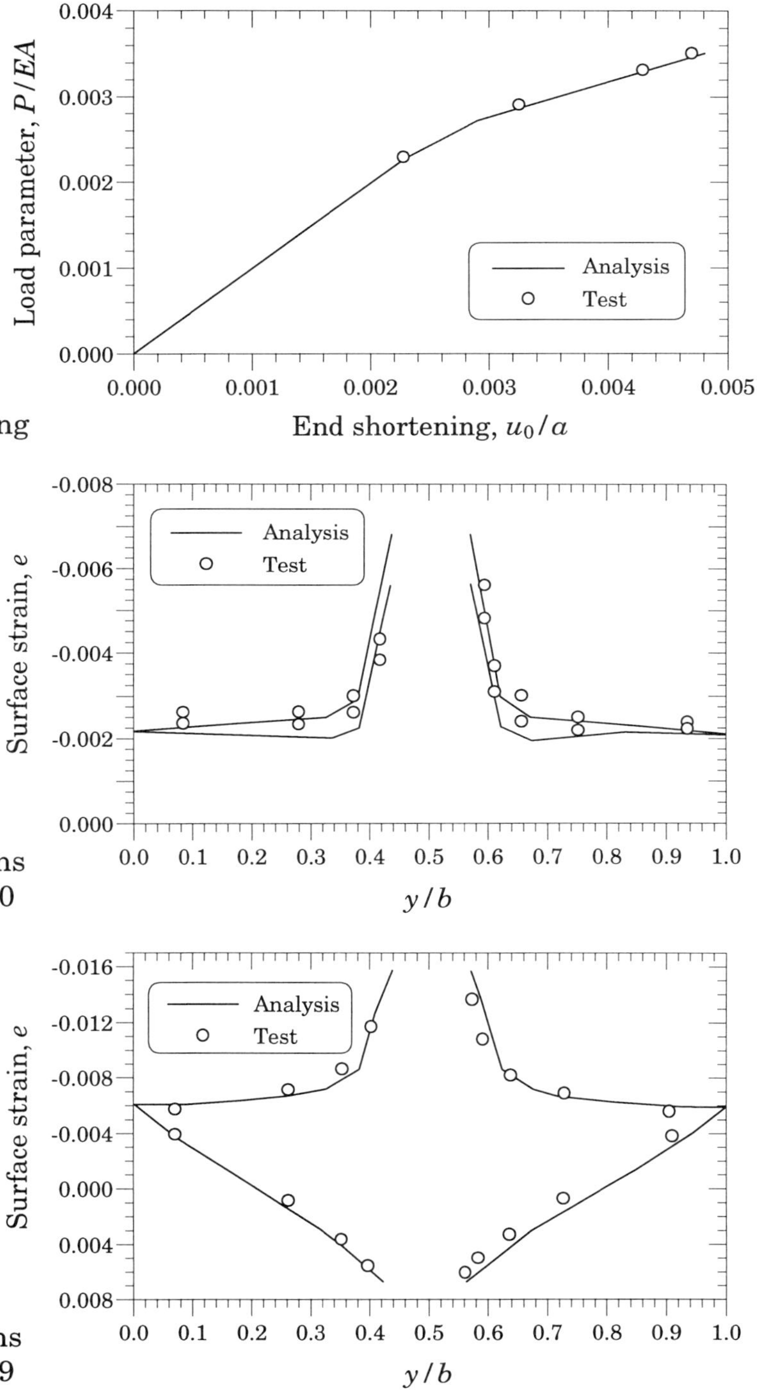

(a) End shortening

(b) Surface strains at $P/P_{cr} = 0.90$

(c) Surface strains at $P/P_{cr} = 1.39$

Figure 13.4–22: Postbuckling response characteristics of panel H4.

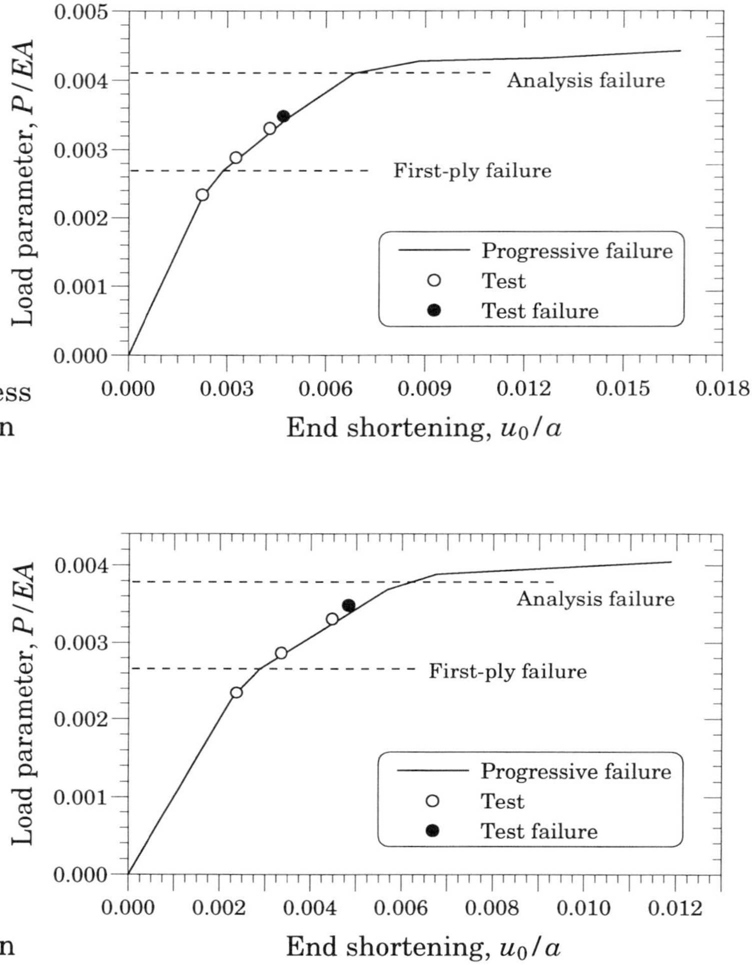

(a) Maximum stress failure criterion

(b) Tsai–Wu failure criterion

Figure 13.4–23: Progressive failure results of panel H4.

13.5 Closure

The objective of the chapter was to introduce the concept of geometric nonlinearity, develop finite element models of the von Kármán nonlinear plate theories, and study the influence of nonlinearities on bending, transient and buckling response and ultimate failure of laminated plates. The von Kármán nonlinear formulations of laminated plates using the classical and first-order shear deformation theories are presented. The Picard and Newton–Raphson iterative methods of solution are discussed, and the tangent stiffness matrix coefficients of the FSDT element are derived. Numerical results of the nonlinear analysis using the FSDT elements are presented to illustrate the influence of symmetry boundary conditions on the nonlinear response of angle-ply laminates, effect of the geometric nonlinearity on the response of antisymmetric cross-ply plate strips, nonlinear transient response of laminated plates, and postbuckling response and progressive failure analysis of laminated panels under compressive load. The last example illustrates how both geometric and material nonlinearities influence the response of composite panels (see References 55 and 60).

Exercises

13.1 Consider the nonlinear differential equation

$$-\frac{d}{dx}\left(u\frac{du}{dx}\right) = f(x), \quad 0 < x < 1 \tag{1a}$$

$$\frac{du}{dx}(0) = 0, \quad u(1) = \sqrt{2} \tag{1b}$$

Show that the finite element model is given by

$$[K^e]\{u^e\} = \{F^e\} \tag{2a}$$

$$K_{ij}^e = \int_{x_A}^{x_B} u\frac{d\psi_i}{dx}\frac{d\psi_j}{dx}\, dx, \quad F_i^e = P_i^e + \int_{x_e}^{x_B} f\psi_i\, dx \tag{2b}$$

13.2 Compute the tangent coefficient matrix for the finite element model of Exercise 13.1. Note that in an iterative solution of the problem, the initial guess for u should be *nonzero*. Explain why.

13.3 Consider the displacement field of the Euler–Bernoulli beam theory (see Chapter 6):

$$u_1(x,z) = u_0(x) - z\frac{dw_0}{dx}, \quad u_2 = 0, \quad u_3(x,z) = w_0(x) \tag{1}$$

Show that the von Kármán nonlinear strains are given by

$$\varepsilon_{xx} = \frac{du_0}{dx} + \frac{1}{2}\left(\frac{dw_0}{dx}\right)^2 - z\frac{d^2w_0}{dx^2} \tag{2}$$

and that the total potential energy associated with a laminated beam is

$$\Pi(u_0, w_0) = \int_0^L \left\{ \frac{E_{xx}^b A}{2}\left[\frac{du_0}{dx} + \frac{1}{2}\left(\frac{dw_0}{dx}\right)^2\right]^2 + \frac{E_{xx}^b I_{yy}}{2}\left(\frac{d^2w_0}{dx^2}\right)^2 \right\} dx$$
$$- \int_0^L \hat{q} w_0\, dx \tag{3}$$

where $\hat{q} = qb$ is the distributed transverse load, b is the width, and $A = bh$ is the area of cross section of the beam.

13.4 Show that the Euler–Lagrange equations associated with the displacement field of Exercise 13.3 are

$$-\frac{dN_{xx}}{dx} = 0, \quad -\frac{dM_{xx}}{dx^2} - \frac{d}{dx}\left(N_{xx}\frac{dw_0}{dx}\right) - q = 0 \tag{1}$$

where N_{xx} and M_{xx} are the force and moment resultants

$$N_{xx} = \int_A \sigma_{xx}\, dx, \quad M_{xx} = \int_A \sigma_{xx} z\, dx \tag{2}$$

Show that the force and moment resultants for symmetrically laminated beams can be expressed in terms of the displacements as

$$N_{xx} = E_{xx}^b A\left[\frac{du_0}{dx} + \frac{1}{2}\left(\frac{dw_0}{dx}\right)^2\right], \quad M_{xx} = -E_{xx}^b I_{yy}\frac{d^2w_0}{dx^2} \tag{3}$$

13.5 Use the equations of equilibrium of Exercise 13.4 to derive the the following weak forms for a beam finite element, $x_A < x < x_B$:

$$\begin{aligned} 0 = &\int_{x_A}^{x_B} E_{xx}^b A\frac{d\delta u_0}{dx}\left[\frac{du_0}{dx} + \frac{1}{2}\left(\frac{dw_0}{dx}\right)^2\right] dx \\ &- P_1\delta u_0(x_A) - P_2\delta u_0(x_B) \end{aligned} \tag{1}$$

$$\begin{aligned} 0 = &\int_{x_A}^{x_B}\left\{E_{xx}^b I_{yy}\frac{d\delta^2 w_0}{dx^2}\frac{d^2w_0}{dx^2} + E_{xx}^b A\frac{d\delta w_0}{dx}\frac{dw_0}{dx}\left[\frac{du_0}{dx} + \frac{1}{2}\left(\frac{dw_0}{dx}\right)^2\right]\right\} dx \\ &- \int_{x_A}^{x_B} q\delta w_0 dx - \bar{Q}_1 w_0 x_A - \bar{Q}_3 w_0(x_B) - \bar{Q}_2\theta(x_A) - \bar{Q}_3\theta(x_B) \end{aligned} \tag{2}$$

Define the secondary variables P_i and $\bar{Q}_i$ in terms of the displacements.

13.6 Develop the nonlinear finite element model of a laminated beam using the weak forms given in Exercise 13.5. Assume finite element interpolation of u_0 and w_0 in the form

$$u_0(x) = \sum_{j=1}^{2} u_j\psi_j(x), \quad w_0(x) = \sum_{j=1}^{4}\bar{\Delta}_j\varphi_j(x) \tag{1a}$$

$$\bar{\Delta}_1 = w_1, \quad \bar{\Delta}_2 = \theta_1, \quad \bar{\Delta}_3 = w_2, \quad \bar{\Delta}_4 = \theta_2 \tag{1b}$$

and ψ_j are the linear Lagrange interpolation functions, φ_j are the Hermite cubic interpolation functions, and $\theta = -\frac{dw_0}{dx}$. In particular, show that

$$0 = \sum_{j=1}^{2} K_{ij}^{11} u_j + \sum_{j=1}^{4} K_{ij}^{12}\bar{\Delta}_j - F_i^1 \quad (i = 1, 2) \tag{3a}$$

$$0 = \sum_{j=1}^{2} K_{ij}^{21} u_j + \sum_{j=1}^{4} K_{ij}^{22} \bar{\Delta}_j - F_i^2 \quad (i = 1, 2, 3, 4) \tag{3b}$$

$$\begin{aligned}
K_{ij}^{11} &= \int_{x_A}^{x_B} E_{xx}^b A \frac{d\psi_i}{dx}\frac{d\psi_j}{dx}\, dx \\
K_{ij}^{12} &= \int_{x_A}^{x_B} \frac{1}{2} E_{xx}^b A \frac{dw_0}{dx}\frac{d\psi_i}{dx}\frac{d\varphi_j}{dx}\, dx \\
K_{ij}^{21} &= \int_{x_A}^{x_B} E_{xx}^b A \frac{dw_0}{dx}\frac{d\varphi_i}{dx}\frac{d\psi_j}{dx}\, dx \\
K_{ij}^{22} &= \int_{x_A}^{x_B} E_{xx}^b A \frac{1}{2}\left(\frac{dw_0}{dx}\right)^2 \frac{d\varphi_i}{dx}\frac{d\varphi_j}{dx}\, dx \\
&+ \int_{x_A}^{x_B} E_{xx}^b A \frac{d^2\varphi_i}{dx^2}\frac{d^2\varphi_j}{dx^2}\, dx \\
F_i^1 &= P_i\ , \quad F_i^2 = \int_{x_A}^{x_B} q\psi_i\, dx + \bar{Q}_i
\end{aligned} \tag{4}$$

Note: $[K^{12}]^T \neq [K^{21}]$; hence, the element stiffness matrix is unsymmetric. Equations (3a,b) can be written in matrix form as

$$\begin{bmatrix} [K^{11}] & [K^{12}] \\ [K^{21}] & [K^{22}] \end{bmatrix} \left\{ \begin{matrix} \{u\} \\ \{\bar{\Delta}\} \end{matrix} \right\} = \left\{ \begin{matrix} \{F^1\} \\ \{F^2\} \end{matrix} \right\} \tag{5}$$

13.7 Use matrix notation

$$\begin{aligned}
\left\{ \begin{matrix} u_0 \\ w_0 \end{matrix} \right\} &= \left\{ \begin{matrix} \sum_{j=1}^{2} u_j \psi_j \\ \sum_{j=1}^{4} \bar{\Delta}_j \varphi_j \end{matrix} \right\} = \begin{bmatrix} \psi_1 & 0 & 0 & \psi_2 & 0 & 0 \\ 0 & \varphi_1 & \varphi_2 & 0 & \varphi_3 & \varphi_4 \end{bmatrix} \left\{ \begin{matrix} u_1 \\ \bar{\Delta}_1 \\ \bar{\Delta}_2 \\ u_2 \\ \bar{\Delta}_3 \\ \bar{\Delta}_4 \end{matrix} \right\} \\
&\equiv [\Psi]\{\Delta\}
\end{aligned} \tag{1}$$

$$\left\{ \begin{matrix} \frac{du_0}{dx} + \frac{1}{2}\left(\frac{dw_0}{dx}\right)^2 \\ \frac{d^2 w_0}{dx^2} \end{matrix} \right\} = \left\{ \begin{matrix} \frac{du_0}{dx} \\ \frac{d^2 w_0}{dx^2} \end{matrix} \right\} + \left\{ \begin{matrix} \frac{1}{2}\left(\frac{dw_0}{dx}\right)^2 \\ 0 \end{matrix} \right\} \tag{2a}$$

$$\left\{ \begin{matrix} \frac{du_0}{dx} \\ \frac{d^2 w_0}{dx^2} \end{matrix} \right\} = \begin{bmatrix} \psi_1' & 0 & 0 & \psi_2' & 0 & 0 \\ 0 & \varphi_1'' & \varphi_2'' & 0 & \varphi_3'' & \varphi_4'' \end{bmatrix} \{\Delta\} \equiv [B_L]\{\Delta\} \tag{2b}$$

$$\begin{aligned}
\left\{ \begin{matrix} \frac{1}{2}\left(\frac{dw_0}{dx}\right)^2 \\ 0 \end{matrix} \right\} &= \frac{1}{2}\frac{dw_0}{dx} \begin{bmatrix} 0 & \varphi_1' & \varphi_2' & 0 & \varphi_3' & \varphi_4' \\ 0 & 0 & 0 & 0 & 0 & 0 \end{bmatrix} \{\Delta\} \\
&= \frac{1}{2}[B_N]\{\Delta\}
\end{aligned} \tag{2c}$$

$$\varphi_i' \equiv \frac{d\varphi_i}{dx}, \qquad \varphi_i'' \equiv \frac{d^2\varphi_i}{dx^2}$$

and express the total potential energy functional for an element as

$$\Pi^e(\{\Delta\}) = \int_{x_A}^{x_B} \frac{1}{2}\left(([B_L] + \frac{1}{2}[B_N])\{\Delta\}\right)^T [D]([B_L] + \frac{1}{2}[B_N])\{\Delta\}\, dx$$
$$- \int_{x_A}^{x_B} ([\Psi]\{\Delta\})^T \{ {0 \atop q} \} dx - \{\Delta\}^T\{Q\} \tag{3}$$

$$\Pi^e(\{\Delta\}) = \frac{1}{2}\{\Delta\}^T \Big\{ \left(\int_{x_A}^{x_B} [B_L]^T[D][B_L]dx \right) + \frac{1}{2}\left(\int_{x_A}^{x_B} [B_L]^T[D][B_N]dx \right)$$
$$-\{\Delta\}^T \left(\int_{x_A}^{x_B} [\Psi]^T \{ {0 \atop q} \} dx + \{Q\} \right) \tag{4}$$

$$[D] = \begin{bmatrix} E_{xx}^b A & 0 \\ 0 & E_{xx}^b I_{yy} \end{bmatrix}, \quad \{Q\} = \begin{Bmatrix} P_1 \\ \bar{Q}_1 \\ \bar{Q}_2 \\ P_2 \\ \bar{Q}_3 \\ \bar{Q}_4 \end{Bmatrix} = \begin{Bmatrix} Q_1 \\ Q_2 \\ Q_3 \\ Q_4 \\ Q_5 \\ Q_6 \end{Bmatrix} \tag{4}$$

13.8 Use the principle of the minimum total potential energy and show that

$$\{0\} = ([K_L] + [K_{NL}] + [K_N])\{\Delta\} - \{F\} \tag{5a}$$

where

$$[K_L] = \int_{x_A}^{x_B} [B_L]^T[D][B_L]\, dx$$
$$[K_{NL}] = \frac{1}{2}\int_{x_A}^{x_B} [B_L]^T[D][B_N]\, dx + \int_{x_A}^{x_B} [B_N]^T[D][B_L]\, dx$$
$$[K_N] = \frac{1}{2}\int_{x_A}^{x_B} [B_N]^T[D][B_N]\, dx$$
$$\{F\} = \int_{x_A}^{x_B} [\Psi]^T \begin{Bmatrix} 0 \\ q \end{Bmatrix} dx + \{Q\} \tag{5b}$$

13.9 Show that the tangent stiffness matrix of the Euler–Bernoulli beam element is

$$[K]^{tan} = \frac{\partial^2}{\partial\{\Delta\}^2}(\Pi(\{\Delta\})) \tag{1a}$$

where

$$[K]^{tan} = [K] + \frac{\partial}{\partial\{\Delta\}}[K]$$
$$= [K] + \frac{1}{2}\int_{x_A}^{x_B} [B_L]^T[D][B_N]\, dx + \int_{x_A}^{x_B} [B_N]^T[D][B_N]\, dx$$
$$= [K_L] + [\hat{K}_{NL}] + \frac{3}{2}[K_N] \tag{1b}$$

and

$$\hat{K}_{NL} = K_{NL} + \frac{1}{2}\int_{x_A}^{x_B} [B_L]^T[D][B_N]dx \tag{1c}$$

13.10 Beginning with the displacement field

$$u_1(x,z) = u_0(x) + z\phi_x(x), \quad u_2(x,z) = 0, \quad u_3(x,z) = w_0(x) \tag{1}$$

show that the equations governing the Timoshenko beam theory with the von Kármán nonlinearity are given by

$$-\frac{dN_{xx}}{dx} = 0, \quad -\frac{dM_{xx}}{dx} + Q_x = 0$$

$$-\frac{d}{dx}\left(N_{xx}\frac{dw_0}{dx}\right) - \frac{dQ_x}{dx} - q = 0 \tag{2}$$

where N_{xx}, Q_x and M_{xx} are the force and moment resultants

$$N_{xx} = \int_A \sigma_{xx}\, dx, \quad Q_x = \int_A \sigma_{xz}\, dx, \quad M_{xx} = \int_A \sigma_{xx} z\, dx \tag{3}$$

Show that the force and moment resultants can be expressed in terms of the displacements as

$$\begin{aligned} N_{xx} &= E_{xx}^b A\left[\frac{du_0}{dx} + \frac{1}{2}\left(\frac{dw_0}{dx}\right)^2\right] \\ M_{xx} &= E_{xx}^b I_{yy}\frac{d\phi_x}{dx} \\ Q_x &= KG_{xz}^b A\left(\phi_x + \frac{dw_0}{dx}\right) \end{aligned} \tag{4}$$

13.11 Show that the weak forms of the Timoshenko beam theory with the von Kármán nonlinearity are given by

$$\begin{aligned} 0 &= \int_{x_A}^{x_B} E_{xx}^b A\frac{d\delta u_0}{dx}\left[\frac{du_0}{dx} + \frac{1}{2}\left(\frac{dw_0}{dx}\right)^2\right] dx \\ &\quad - P_1\delta u_0(x_A) - P_2\delta u_0(x_B) \end{aligned} \tag{1}$$

$$\begin{aligned} 0 &= \int_{x_A}^{x_B} \left(\frac{d\delta w_0}{dx}N_{xx}\frac{dw_0}{dx} + \frac{d\delta w_0}{dx}Q_x - \delta w_0 q\right) dx \\ &\quad - Q_1\delta w_0(x_A) - Q_3\delta w_0(x_B) \end{aligned} \tag{2}$$

$$\begin{aligned} 0 &= \int_{x_A}^{x_B} \left[E_{xx}^b I_{yy}\frac{d\delta\phi_x}{dx}\frac{d\phi_x}{dx} + G_{xz}^b AK\delta\phi_x\left(\phi_x + \frac{dw_0}{dx}\right)\right] dx \\ &\quad - Q_2\delta\phi_x(x_A) - Q_4\delta\phi_x(x_B) \end{aligned} \tag{3}$$

13.12 Assume that the generalized displacements (u_0, w_0, ϕ_x) of the Timoshenko beam theory are approximated by

$$u_0 = \sum_{i=1}^{m} u_i\psi_i^1, \quad w_0 = \sum_{i=1}^{n} w_i\psi_i^2, \quad \phi_x = \sum_{i=1}^{p} S_i\psi_i^3 \tag{1}$$

and show that the finite element model is of the form

$$\begin{bmatrix} [K^{11}] & [K^{12}] & [K^{13}] \\ [K^{21}] & [K^{22}] & [K^{23}] \\ [K^{31}] & [K^{32}] & [K^{33}] \end{bmatrix} \begin{Bmatrix} \{u\} \\ \{w\} \\ \{S\} \end{Bmatrix} = \begin{Bmatrix} \{F^1\} \\ \{F^2\} \\ \{F^3\} \end{Bmatrix} \tag{2}$$

$$\begin{aligned}
K_{ij}^{11} &= \int_{x_A}^{x_B} E_{xx}^b A \frac{d\psi_i^1}{dx}\frac{d\psi_j^1}{dx}\,dx, \quad K_{ij}^{12} = \frac{1}{2}\int_{x_A}^{x_B} E_{xx}^b A\left(\frac{dw_0}{dx}\right)\frac{d\psi_i^1}{dx}\frac{d\psi_j^2}{dx}\,dx \\
K_{ij}^{22} &= \frac{1}{2}\int_{x_A}^{x_B} E_{xx}^b A\left(\frac{dw_0}{dx}\right)^2 \frac{d\psi_i^1}{dx}\frac{d\psi_j^2}{dx}\,dx + \int_{x_A}^{x_B} G_{xz}^b AK\frac{d\psi_i^2}{dx}\frac{d\psi_j^2}{dx}\,dx \\
K_{ij}^{23} &= \int_{x_A}^{x_B} G_{xz}^b AK\frac{d\psi_i}{dx}\psi_i^3\psi_j^3\,dx \\
K_{ij}^{33} &= \int_{x_A}^{x_B} E_{xx}^b I_{yy}\frac{d\psi_i^3}{dx}\frac{d\psi_j^3}{dx}\,dx + \int_{x_A}^{x_B} G_{xz}^b AK\psi_i^3\psi_j^3\,dx \\
K_{ij}^{13} &= 0, \quad K_{ij}^{21} = 2K_{ji}^{12}, \quad K_{ij}^{32} = K_{ji}^{23}
\end{aligned} \tag{3}$$

13.13 Evaluate the direct stiffness coefficients $[K^{\alpha\beta}]$ $(\alpha, \beta = 1, 2, 3)$ of the nonlinear Timoshenko beam finite element assuming linear but equal interpolation of u_0, w_0, and ϕ_x.

13.14 Compute the tangent stiffness matrix coefficients associated with the nonlinear Timoshenko beam finite element. *Ans:* The tangent stiffness coefficients are

$$\begin{aligned}
\left(K_{ij}^{11}\right)^{tan} &= \int_{x_A}^{x_B} E_{xx}^b A\frac{d\psi_i}{dx}\frac{d\psi_j}{dx}dx \\
\left(K_{ij}^{12}\right)^{tan} &= \int_{x_A}^{x_B} E_{xx}^b A\left(\frac{dw_0}{dx}\right)\frac{d\psi_i}{dx}\frac{d\psi_j}{dx}dx = 2K_{ij}^{12} \\
\left(K_{ij}^{13}\right)^{tan} &= K_{ij}^{13} = 0 \\
\left(K_{ij}^{21}\right)^{tan} &= \int_{x_A}^{x_B} E_{xx}^b A\frac{dw_0}{dx}\frac{d\psi_i}{dx}\frac{d\psi_j}{dx}\,dx = K_{ij}^{21} \\
\left(K_{ij}^{22}\right)^{tan} &= \int_{x_A}^{x_B} \left\{E_{xx}^b A\left[\frac{du_0}{dx} + \frac{3}{2}\left(\frac{dw_0}{dx}\right)^2\right]\frac{d\psi_i}{dx}\frac{d\psi_j}{dx}\right. \\
&\qquad \left. + KG_{xz}^b A\frac{d\psi_i}{dx}\frac{d\psi_j}{dx}\right\}dx \\
\left(K_{ij}^{23}\right)^{tan} &= K_{ij}^{23}, \quad \left(K_{ij}^{31}\right)^{tan} = K_{ij}^{31} \\
\left(K_{ij}^{32}\right)^{tan} &= K_{ij}^{32}, \quad \left(K_{ij}^{33}\right)^{tan} = K_{ij}^{33}
\end{aligned}$$

References for Additional Reading

1. Khot, N. S., "Buckling and Postbuckling Behavior of Composite Cylindrical Shells Under Axial Compression," *AIAA Journal*, **8**, 229–235 (1970).
2. Khot, N. S., Venkayya, V. B., and Berke, L., "Buckling and Postbuckling of Initially Imperfect Orthotropic Cylindrical Shells Under Axial Compression and Internal Pressure," in *Instability of Continuous Systems*, IUTAM Symposium, Leipholz, H. (ed.), Springer–Verlag, Berlin, pp. 392-398, 1971.
3. Bennett, J. A., "Nonlinear Vibration of Simply Supported Angle-Ply Laminated Plates," *AIAA Journal*, **9**, 1997–2003 (1971).
4. Bert, C. W., "Nonlinear Vibration of a Rectangular Plate Arbitrarily Laminated of Anisotropic Material," *Journal of Applied Mechanics*, **40**, 452–458 (1973).
5. Chandra, R. and Raju, B. B., "Large Amplitude Flexural Vibration of Cross Ply Laminated Composite Plates," *Fibre Science and Technology*, **8**, 243–263 (1975).
6. Chandra, R., "Large Deflection Vibration of Cross–Ply Laminated Plates with Certain Edge Conditions," *Journal of Sound and Vibration*, **47**(4), 509–514 (1976).
7. Zaghloul, S. A. and Kennedy, J. B., "Nonlinear Analysis of Unsymmetrically Laminated Plates," *Journal of Engineering Mechanics Division, ASCE*, **101**(EM3), 169–185 (1975).
8. Zaghloul, S. A. and Kennedy, J. B., "Nonlinear Behavior of Symmetrically Laminated Plates," *Journal of Applied Mechanics*, **42**, 234–236 (1975).
9. Chia, C. Y. and Prabhakara, M. K., "Large Deflection of Unsymmetric Cross–Ply and Angle–Ply Plates," *Journal of Mechanical Engineering Science*, **18**(4), 179–183 (1976).
10. Iu, V. P. and Chia, C. Y., "Effect of Transverse Shear on Nonlinear Vibration and Post–Buckling of Antisymmetric Cross–Ply Imperfect Cylindrical Shells," *International Journal of Solids and Structures*, **30**, 705–718 (1988).
11. Reddy, J. N. and Savoia, M., "The Layerwise Shell Theory for Post-Buckling of Circular Cylindrical Shells," *AIAA Journal*, **30**, 2148–2154 (1992).
12. Savoia, M. and Reddy, J. N., "Post-Buckling Behavior of Stiffened Cross-Ply Cylindrical Shells," *Journal of Applied Mechanics*, **61**, pp. 998–1000 (1994).
13. Ochoa, O. O. and Reddy, J. N., *Finite Element Analysis of Composite Laminates*, Kluwer, Netherlands, 1992.
14. Reddy, J. N. and Miravete, A., *Practical Analysis of Composite Laminates*, CRC Press, Boca Raton, Florida, 1995.
15. Noor, A. K. and Hartley, S. J., "Effect of Shear Deformation and Anisotropy on the Non-Linear Response of Composite Plates," *Developments in Composite Materials - 1*, G. Holister (ed), Applied Science Publishers, Barking, pp. 55–56, 1977.

16. Reddy, J. N., "Simple Finite Elements with Relaxed Continuity for Nonlinear Analysis of Plates," *Proceedings of the Third International Conference on Finite Element Methods*, Australia, pp. 265–281, 1979.

17. Reddy, J. N. and Chao, W. C., "Large Deflection and Large Amplitude Free Vibrations of Laminated Composite–Material Plates," *Symposium on Computational Methods in Nonlinear Structural and Solid Mechanics,* Washington, D.C., October 6–8, 1980.

18. Chang, T. Y. and Sawamiphakdi, K., "Large Deformation Analysis of Laminated Shells by Finite Element Method," *Computers and Structures,* **13**, 331–340 (1981).

19. Reddy, J. N., "Analysis of Layered Composite Plates Accounting for Large Deflections and Transverse Shear Strains," in *Recent Advances in Nonlinear Computational Mechanics*, E. Hinton *al.* (eds.), pp. 155–202, 1982.

20. Reddy, J. N., "Geometrically Nonlinear Transient Analysis of Laminated Composite Plates," *AIAA Journal*, **21**(4), 621–629 (1983).

21. Chao, W. C. and Reddy, J. N., "Analysis of Laminated Composite Shells Using a Degenerated 3–D Element," *International Journal for Numerical Methods in Engineering*, **20**, 1991–2007 (1984).

22. Putcha, N. S. and Reddy, J. N., "A Refined Mixed Shear Flexible Finite Element for the Nonlinear Analysis of Laminated Plates," *Computers and Structures*, **22** (2), 529–538 (1986).

23. Liao, C. L. and Reddy, J. N., "Continuum–Based Stiffened Composite Shell Element for Geometrically Nonlinear Analysis," *AIAA Journal*, **27** (1), 95–101 (1989).

24. Reddy, J. N., "On Refined Computational Models of Composite Laminates," *International Journal for Numerical Methods in Engineering*, **27**, 361–382 (1989).

25. Barbero, E. J. and Reddy, J. N., "Nonlinear Analysis of Composite Laminates Using a Generalized Laminate Plate Theory," *AIAA Journal*, **28**(11), 1987–1994 (1990).

26. Bathe, K. J., Ramm, E., and Wilson, E. L., "Finite Element Formulations for Large Deformation Dynamic Analysis," *International Journal for Numerical Methods in Engineering*, **9**, 353–386 (1975).

27. Horrigmoe, G. and Bergan, P. G., "Incremental Variational Principle and Finite Element Models for Nonlinear Problems," *Computer Methods in Applied Mechanics in Engineering,* **7**, 201–217 (1976).

28. Wunderlich, W., "Incremental Formulations for Geometrically Nonlinear Problems," in *Formulations and Algorithms in Finite Element Analysis,* K. J. Bathe *et al.* (eds.), M.I.T. Press, Massachusetts, pp. 193–239, 1977.

29. Ramm, E., "A Plate/Shell Element for Large Deflections and Rotations," in *Formulations and Computational Algorithms in Finite Element Analysis,* K. J. Bathe *al.* (eds.), M.I.T. Press, Massachusetts, 1977.

30. Stricklin, J. A., Von Riesemann, W. A., Tillerson, J. R., and Haisler, W. E., "Static Geometrical and Material Nonlinear Analysis," *Advances in Computer Methods in Structural Mechanics and Design,* J. T. Oden *et al.* (eds.), University of Alabama in Huntsville, Huntsville, Alabama, pp. 301–324, 1972.

31. Bathe, K. J., *Finite Element Procedures*, Prentice–Hall, Englewood Cliffs, New Jersey, 1996.

32. Crisfield, M. A., *Non–Linear Finite Element Analysis of Solids and Structures,* John Wiley, London, 1991.

33. Pian, T. H. H. and Boland, P., "Formulation of Large Deflection Shell Analysis by Assumed Stress Finite Element Method," in *Formulations and Computational Algorithms in Finite Element Analysis*, K. J. Bathe, *et al.* (eds.), M.I.T. Press, Massachusetts, 1977.

34. Lee, S. W. and Pian, T. H. H., "Improvement of Plate and Shell Finite Elements by Mixed Formulations," *AIAA Journal*, **16**, 29–34 (1978).

35. Spilker, R. L. and Munir, N. I., "The Hybrid–Stress Model for Thin Plates," *International Journal for Numerical Methods in Engineering,* **15**, 1239-1260 (1980).

36. Bathe, K. J., Dvorkin, E. N., and Ho, L. W., "On Discrete–Kirchhoff and Isoparametric Shell Elements for Nonlinear Analysis — An Assessment," *Computers and Structures*, **16**, 89–98 (1983).

37. Putcha, N. S. and Reddy, J. N., "A Mixed Finite Element for the Analysis of Laminated Plates," in *Advances in Aerospace Structures, Materials and Dynamics*, U. Yuceoglu *et al.* (eds.), ASME AD-06, pp. 31-39 (1983).

38. Putcha, N. S. and Reddy, J. N., "Stability and Natural Vibration Analysis of Laminated Plates by Using a Mixed Element Based on a Refined Plate Theory," *Journal of Sound and Vibration*, **104** (2), 285–300 (1986).

39. Reddy, J. N., "On Mixed Finite–Element Formulations of a Higher–Order Theory of Composite Laminates," *Finite Element Methods for Plate and Shell Structures*, T. J. R. Hughes and E. Hinton (eds.), Pineridge Press, U.K., pp. 31–57, 1986.

40. Park, K. C. and Stanley, G. M., "A Curved C° Shell Element Based on Assumed Natural-Coordinate Strains," *Journal of Applied Mechanics*, **53**, 278–290 (1986).

41. Reddy, J. N. and Heyliger, P. R., "A Mixed Updated Lagrangian Formulation for Plane Elastic Problems," *Journal of Composites Technology and Research*, **9**(4), 131–140 (1987).

42. Liao, C. L. and Reddy, J. N., "Analysis of Anisotropic, Stiffened Composite Laminates Using a Continuum–Based Shell Element," *Computers and Structures*, **34**(6), 805–815 (1990).

43. Batoz, J. L. and Dhatt, G., "Incremental Displacement Algorithms for Nonlinear Problems," *International Journal for Numerical Methods in Engineering,* **14**, 1262–1266 (1979).

44. Riks, E., "An Incremental Approach to the Solution of Snapping and Buckling Problem," *International Journal of Solids and Structures,* **15**, 529–551 (1979).

45. Ramm, E., "Strategies for Tracing the Nonlinear Response Near Limit Points," *Nonlinear Finite Element Analysis in Structural Mechanics,* W. Wunderlich *et al.* (eds.), Springer–Verlag, Berlin, pp. 63–89, 1981.

46. Crisfield, M. A., "A Fast Incremental/Iterative Solution Procedure That Handles Snap–Through," *Computers and Structures,* **13**, 55–62 (1981).

47. Wempner, G. A., "Discrete Approximations Related to Nonlinear Theories of Solids," *International Journal of Solids and Structures,* **7**, 1581–1599 (1971).

48. Lévy, S., "Bending of Rectangular Plates with Large Deflections," Report No. 737, NACA (1942).

49. Wang, C. T., "Bending of Rectangular Plates with Large Deflections," Report No. 1462, NACA (1948).

50. Yamaki, N., "Influence of Large Amplitudes on Flexural Vibrations of Elastic Plates," *ZAMM,* **41**, 501–510 (1967).

51. Reddy, J. N., "A Note on Symmetry Considerations in the Transient Response of Unsymmetrically Laminated Anisotropic Plates," *International Journal for Numerical Methods in Engineering,* **20**, 175–194 (1984).

52. Sun, C. T. and Chin, H., "Analysis of Asymmetric Composite Laminates," *AIAA Journal,* **26**(6), 714–718 (1988).

53. Bathe, K. J. and Ho, L. W., "A Simple Effective Element for the Analysis of General Shell Structures," *Computers and Structures,* **13**, 673–681 (1981).

54. Starnes, J. H., Jr. and Rouse, M., "Postbuckling and Failure Characteristics of Selected Flat Rectangular Graphite–Epoxy Plates Loaded in Compression," AIAA Paper No. 81-0543 (1981).

55. Engelstad, S. P., Reddy, J. N., and Knight, N. F., Jr., "Postbuckling Response and Failure Prediction of Graphite–Epoxy Plates Loaded in Compression," *AIAA Journal,* **30**(8), 2106–2113 (1992).

56. Tsai, S. W., "A Survey of Macroscopic Failure Criteria for Composite Materials," *Journal of Reinforced Plastics and Composites,* **3**, 40–62 (1984).

57. Hill, R., "A Theory of the Yielding and Plastic Flow of Anisotropic Metals," *Proceedings of the Royal Society,* Series A, **193**, 281–297 (1948).

58. Wu, E. M., "Phenomenological Anisotropic Failure Criterion," *Composite Materials,* **2**, 353–431 (1974).

59. Azzi, V. D. and Tsai, S. W., "Anisotropic Strength of Composites," *Experimental Mechanics,* **5**, 283–288 (1965).

60. Knight, N. F., Jr., "Factors Influencing Nonlinear Static Response Prediction and Test-Analysis Correlation for Composite Panels," *Composite Structures,* **29,** 13–25 (1994).

SUBJECT INDEX